Student Study Guide/ Solutions Manual

to accompany

Organic Chemistry

Fifth Edition

Janice Gorzynski Smith
University of Hawai'i at Mānoa

and

Erin R. Smith

STUDENT STUDY GUIDE/SOLUTIONS MANUAL TO ACCOMPANY
ORGANIC CHEMISTRY, FIFTH EDITION

Published by McGraw-Hill Education, 2 Penn Plaza, New York, NY 10121. Copyright © 2017 by McGraw-Hill Education. All rights reserved. Printed in the United States of America. Previous editions © 2014 and 2011. No part of this publication may be reproduced or distributed in any form or by any means, or stored in a database or retrieval system, without the prior written consent of McGraw-Hill Education, including, but not limited to, in any network or other electronic storage or transmission, or broadcast for distance learning.

Some ancillaries, including electronic and print components, may not be available to customers outside the United States.

This book is printed on acid-free paper.

1 2 3 4 5 6 7 8 9 0 DOH/DOH 1 0 9 8 7 6 5 4 3 2 1

ISBN 978-1-259-63706-3
MHID 1-259-63706-9

All credits appearing on page or at the end of the book are considered to be an extension of the copyright page.

The Internet addresses listed in the text were accurate at the time of publication. The inclusion of a website does not indicate an endorsement by the authors or McGraw-Hill Education, and McGraw-Hill Education does not guarantee the accuracy of the information presented at these sites.

mheducation.com/highered

Contents

Chapter 1	Structure and Bonding	1-1
Chapter 2	Acids and Bases	2-1
Chapter 3	Introduction to Organic Molecules and Functional Groups	3-1
Chapter 4	Alkanes	4-1
Chapter 5	Stereochemistry	5-1
Chapter 6	Understanding Organic Reactions	6-1
Chapter 7	Alkyl Halides and Nucleophilic Substitution	7-1
Chapter 8	Alkyl Halides and Elimination Reactions	8-1
Chapter 9	Alcohols, Ethers, and Related Compounds	9-1
Chapter 10	Alkenes	10-1
Chapter 11	Alkynes	11-1
Chapter 12	Oxidation and Reduction	12-1
Chapter 13	Mass Spectrometry and Infrared Spectroscopy	13-1
Chapter 14	Nuclear Magnetic Resonance Spectroscopy	14-1
Chapter 15	Radical Reactions	15-1
Chapter 16	Conjugation, Resonance, and Dienes	16-1
Chapter 17	Benzene and Aromatic Compounds	17-1
Chapter 18	Reactions of Aromatic Compounds	18-1
Chapter 19	Carboxylic Acids and the Acidity of the O−H Bond	19-1
Chapter 20	Introduction to Carbonyl Chemistry; Organometallic Reagents; Oxidation and Reduction	20-1
Chapter 21	Aldehydes and Ketones—Nucleophilic Addition	21-1
Chapter 22	Carboxylic Acids and Their Derivatives— Nucleophilic Acyl Substitution	22-1
Chapter 23	Substitution Reactions of Carbonyl Compounds at the α Carbon	23-1
Chapter 24	Carbonyl Condensation Reactions	24-1
Chapter 25	Amines	25-1
Chapter 26	Carbon–Carbon Bond-Forming Reactions in Organic Synthesis	26-1
Chapter 27	Pericyclic Reactions	27-1
Chapter 28	Carbohydrates	28-1
Chapter 29	Amino Acids and Proteins	29-1
Chapter 30	Synthetic Polymers	30-1
Chapter 31	Lipids	31-1

本書『スミス有機化学（第5版）問題の解き方［英語版］』は、"Student Study Guide/Solutions Manual to accompany Organic Chemistry Fifth Edition"のリプリント版です。そのため、日本語版の章番号と本書の章番号のあいだに相違がございます。読者の皆様にはお手数をおかけして恐縮ですが、以下のように読み替えてご利用くださいますようお願い申し上げます。

「30章脂質」について ⇒本書の chapter31（p.31-1〜31-17）を参照
　［例］問題 30.1 の答え⇒本書の 31.1 を参照

「31章合成ポリマー」について ⇒本書の chapter30（p.30-1〜30-22）を参照
　［例］問題31.1の答え⇒本書の30.1を参照

以上
化学同人 編集部

Structure and Bonding 1–1

Chapter 1: Structure and Bonding

Chapter Review

Important facts

- **The general rule of bonding:** Atoms strive to attain a complete outer shell of valence electrons (Section 1.2). H "wants" 2 electrons. Second-row elements "want" 8 electrons.

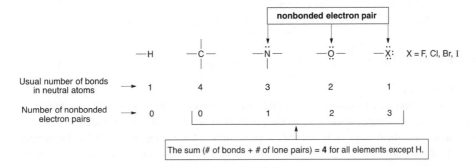

- **Formal charge** (FC) is the difference between the number of valence electrons on an atom and the number of electrons it "owns" (Section 1.3C). See Sample Problem 1.3 for a stepwise example.

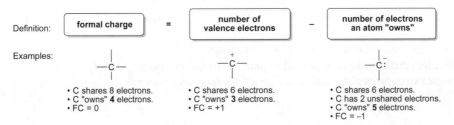

- **Curved arrow notation** shows the movement of an electron pair. The tail of the arrow always begins at an electron pair, either in a bond or a lone pair. The head points to where the electron pair "moves" (Section 1.6).

- **Electrostatic potential plots** are color-coded maps of electron density, indicating electron rich (red) and electron deficient (blue) regions (Section 1.12).

Chapter 1–2

The importance of Lewis structures (Sections 1.3–1.5)

A properly drawn Lewis structure shows the number of bonds and lone pairs present around each atom in a molecule. In a valid Lewis structure, each H has two electrons, and each second-row element has no more than eight. This is the first step needed to determine many properties of a molecule.

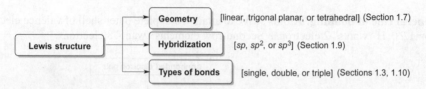

Resonance (Section 1.6)

The basic principles:
- Resonance occurs when a compound cannot be represented by a single Lewis structure.
- Two resonance structures differ *only* in the position of nonbonded electrons and π bonds.
- The resonance hybrid is the only accurate representation for a resonance-stabilized compound. A hybrid is more stable than any single resonance structure because electron density is delocalized.

The difference between resonance structures and isomers:
- Two **isomers** differ in the arrangement of *both* atoms and electrons.
- **Resonance structures** differ *only* in the *arrangement of electrons*.

Geometry and hybridization

The number of groups around an atom determines both its geometry (Section 1.7) and hybridization (Section 1.9).

Number of groups	Geometry	Bond angle (°)	Hybridization	Examples
2	linear	180	sp	BeH$_2$, HC≡CH
3	trigonal planar	120	sp^2	BF$_3$, CH$_2$=CH$_2$
4	tetrahedral	109.5	sp^3	CH$_4$, NH$_3$, H$_2$O

Structure and Bonding 1–3

Drawing organic molecules (Section 1.8)

- Shorthand methods are used to abbreviate the structure of organic molecules.

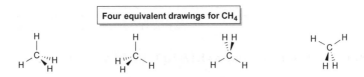

- A carbon bonded to four atoms is tetrahedral in shape. The best way to represent a tetrahedron is to draw two bonds in the plane, one in front, and one behind.

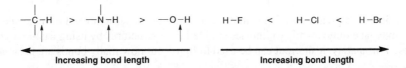

Bond length

- Bond length decreases as you go from left to right across a row and increases down a column of the periodic table (Section 1.7A).

$$-\overset{|}{\underset{|}{C}}-H \quad > \quad -\overset{|}{N}-H \quad > \quad -O-H \qquad H-F \quad < \quad H-Cl \quad < \quad H-Br$$

Increasing bond length   Increasing bond length

- Bond length decreases as the number of electrons between two nuclei increases (Section 1.11A).

$$CH_3-CH_3 \quad < \quad CH_2=CH_2 \quad < \quad H-C\equiv C-H$$

Increasing bond length

- Bond length increases as the percent *s*-character decreases (Section 1.11B).

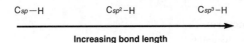

Chapter 1–4

- Bond length and bond strength are inversely related. Shorter bonds are stronger bonds (Section 1.11).

| longest C–C bond
weakest bond | $-\overset{\mid}{\underset{\mid}{C}}-\overset{\mid}{\underset{\mid}{C}}-$ | $\overset{\diagdown}{\diagup}C=C\overset{\diagup}{\diagdown}$ | $-C\equiv C-$ | shortest C–C bond
strongest bond |

Increasing bond strength

- Sigma (σ) bonds are generally stronger than π bonds (Section 1.10).

$-\overset{\mid}{\underset{\mid}{C}}-\overset{\mid}{\underset{\mid}{C}}-$ $\overset{\diagdown}{\diagup}C=C\overset{\diagup}{\diagdown}$ $-C\equiv C-$

1 strong σ bond   1 stronger σ bond   1 stronger σ bond
                       1 weaker π bond   2 weaker π bonds

Electronegativity and polarity (Sections 1.12, 1.13)

- Electronegativity increases from left to right across a row and decreases down a column of the periodic table.
- A polar bond results when two atoms with different electronegativities are bonded together. Whenever C or H is bonded to N, O, or any halogen, the bond is polar.
- A polar molecule has either one polar bond, or two or more bond dipoles that reinforce.

Drawing Lewis structures: A shortcut

Chapter 1 devotes a great deal of time to drawing valid Lewis structures. For molecules with many bonds, it may take quite awhile to find acceptable Lewis structures by using trial-and-error to place electrons. Fortunately, a shortcut can be used to figure out how many bonds are present in a molecule.

Shortcut on drawing Lewis structures—Determining the number of bonds:
    [1] Count up the number of valence electrons.
    [2] Calculate how many electrons are needed if there are no bonds between atoms and every atom has a filled shell of valence electrons; that is, hydrogen gets two electrons, and second-row elements get eight.
    [3] Subtract the number obtained in Step [1] from the sum obtained in Step [2]. **This difference tells how many electrons must be shared** to give every H two electrons and every second-row element eight. Because there are two electrons per bond, dividing this difference by two tells how many bonds are needed.

To draw the Lewis structure:
    [1] Arrange the atoms as usual.
    [2] Count up the number of valence electrons.
    [3] Use the shortcut to determine how many bonds are present.
    [4] Draw in the two-electron bonds to all the H's first. Then, draw the remaining bonds between other atoms, making sure that no second-row element gets more than eight electrons and that you use the total number of bonds determined previously.

Structure and Bonding 1–5

[5] Finally, place unshared electron pairs on all atoms that do not have an octet of electrons, and calculate formal charge. You should have now used all the valence electrons determined in the first step.

Example: Draw all valid Lewis structures for CH₃NCO using the shortcut procedure.

[1] Arrange the atoms.

```
    H
    |
H   C   N   C   O
    |
    H
```

- In this case the arrangement of atoms is implied by the way the structure is drawn.

[2] Count up the number of valence electrons.

3 H's	×	1 electron per H	=	3 electrons
2 C's	×	4 electrons per C	=	8 electrons
1 N	×	5 electrons per N	=	5 electrons
1 O	×	6 electrons per O	=	+ 6 electrons
				22 electrons total

[3] Use the shortcut to figure out how many bonds are needed.

- Number of electrons needed if there were no bonds:

3 H's	×	2 electrons per H	=	6 electrons
4 second-row elements	×	8 electrons per element	=	+ 32 electrons
				38 electrons needed if there were no bonds

- Number of electrons that must be shared:

```
   38 electrons
 − 22 electrons
```
16 electrons must be shared

- Every bond requires two electrons, so 16/2 = **8 bonds are needed.**

[4] Draw all possible Lewis structures.
- Draw the bonds to the H's first (three bonds). Then add five more bonds. Arrange them between the C's, N, and O, making sure that no atom gets more than eight electrons. There are three possible arrangements of bonds; that is, there are three resonance structures.
- Add additional electron pairs to give each atom an octet and check that all 22 electrons are used.

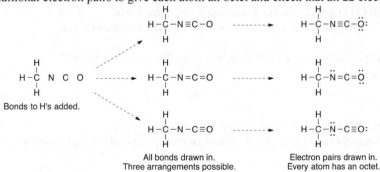

All bonds drawn in.   Electron pairs drawn in.
Three arrangements possible.   Every atom has an octet.

Chapter 1–6

- Calculate the formal charge on each atom.

$$H-\overset{\overset{\displaystyle H}{|}}{\underset{\underset{\displaystyle H}{|}}{C}}-N\equiv C-\ddot{\ddot{O}}: \qquad \longleftrightarrow \qquad H-\overset{\overset{\displaystyle H}{|}}{\underset{\underset{\displaystyle H}{|}}{C}}-\ddot{N}=C=\ddot{O} \qquad \longleftrightarrow \qquad H-\overset{\overset{\displaystyle H}{|}}{\underset{\underset{\displaystyle H}{|}}{C}}-\ddot{N}-C\equiv O:$$

$$+1 \quad -1 \qquad\qquad\qquad\qquad\qquad\qquad\qquad\qquad -1 \quad +1$$

- You can evaluate the Lewis structures you have drawn. The middle structure is the best resonance structure, because it has no charged atoms.

Note: This method works for compounds that contain second-row elements in which every element gets an octet of electrons. It does NOT necessarily work for compounds with an atom that does not have an octet (such as BF₃), or compounds that have elements located in the third row and later in the periodic table.

Practice Test on Chapter Review

1.a. Which compound(s) contain a labeled atom with a +1 formal charge? All lone pairs of electrons have been drawn in.

1. ⬡=N(CH₃)₂ 2. (ring with $\ddot{O}$–CH₃) 3. $CH_3-\ddot{C}-CH_3$

4. Both (1) and (2) have labeled atoms with a +1 charge.
5. Compounds (1), (2), and (3) all contain labeled atoms with a +1 formal charge.

b. Which of the following compounds is a valid resonance structure for **A?**

A: (ring with $\ddot{O}CH_3$, + on ring)

1. (ring, + top, $\ddot{O}CH_3$) 2. (ring, +, $\ddot{O}CH_3$) 3. (ring, +, $\overset{+}{O}CH_3$)

4. Both (1) and (2) are valid resonance structures for **A.**
5. Cations (1), (2), and (3) are all valid resonance structures for **A.**

c. Which species contains a labeled carbon atom that is sp^2 hybridized?

1. ⬡⁺ 2. ⬡=C=CH₂ 3. ⬡⁻

4. Both (1) and (2) contain labeled sp^2 hybridized atoms.
5. Species (1), (2), and (3) all contain labeled sp^2 hybridized carbon atoms.

Structure and Bonding 1–7

d. Which of the following compounds has a net dipole?

1. $CH_3CH_2NHCH_2CH_3$ 4. Compounds (1) and (2) both have net dipoles.
2. $CH_3CH_2CH_2OH$ 5. Compounds (1), (2), and (3) all have net dipoles.
3. $FCH_2CH_2CH_2F$

2. Rank the labeled bonds in order of increasing bond length. Label the shortest bond as **1,** the longest bond as **4,** and the bonds of intermediate length as **2** and **3.**

3. Answer the following questions about compounds **A–D.**

a. What is the hybridization of the labeled atom in **A?**
b. What is the molecular shape around the labeled atom in **B?**
c. In what type of orbital does the lone pair in **C** reside?
d. What orbitals are used to form bond [1] in **D?**
e. Which orbitals are used to form the carbon–oxygen double bond [2] in **D?**

4. Draw an acceptable Lewis structure for CH_3NO_3. Assume that the atoms are arranged as drawn.

```
      H    O
  H   C  O  N  O
      H
```

5. Follow the curved arrows and draw the product with all the needed charges and lone pairs.

Answers to Practice Test

1. a. 4   2. **A** – 1   3. a. sp^3 4. One possibility: 5.
   b. 1   **B** – 4   b. trigonal planar
   c. 1   **C** – 2   c. sp^2
   d. 5   **D** – 3   d. $C sp^3$–$C sp^2$
                       e. $C sp$–$O sp^2$,
                          $C p$–$O p$

Chapter 1–8

Answers to Problems

1.1 Two isotopes differ in the number of neutrons. Two isotopes have the same number of protons and electrons, group number, and number of valence electrons. The **mass number** is the number of protons and neutrons. The **atomic number** is the number of protons and is the same for all isotopes.

	Nitrogen-14	Nitrogen-13
a. number of protons = atomic number for N = 7	7	7
b. number of neutrons = mass number – atomic number	7	6
c. number of electrons = number of protons	7	7
d. group number	5A	5A
e. number of valence electrons	5	5

1.2 **Ionic bonds** form when an element on the far left side of the periodic table transfers an electron to an element on the far right side of the periodic table. **Covalent bonds** result when two atoms *share* electrons.

a. F–F covalent

b. Li⁺ Br⁻ ionic

c. H–C–C–H All C–H and C–C bonds are covalent.

d. Na⁺ :N–H Both N–H bonds are covalent. ionic

e. Na⁺ :O–C–H ionic All other bonds are covalent.

1.3 Atoms with one, two, three, or four valence electrons form one, two, three, or four bonds, respectively. Atoms with five or more valence electrons form [8 – (number of valence electrons)] bonds.

a. O 8 – 6 valence e⁻ = 2 bonds

b. Al 3 valence e⁻ = 3 bonds

c. Br 8 – 7 valence e⁻ = 1 bond

d. Si 4 valence e⁻ = 4 bonds

1.4 [1] Arrange the atoms with the H's on the periphery.
[2] Count the valence electrons.
[3] Arrange the electrons around the atoms. Give the H's 2 electrons first, and then fill the octets of the other atoms.
[4] Assign formal charges (Section 1.3C).

a. [1] H H
        H C C H
        H H

[2] Count valence e⁻.
2 C x 4 e⁻ = 8
6 H x 1 e⁻ = 6
total e⁻ = 14

[3] H H
     H–C–C–H
     H H

All 14 e⁻ used.
All second-row elements have an octet.

b. [1] H
        H C N H
        H H

[2] Count valence e⁻.
1 C x 4 e⁻ = 4
5 H x 1 e⁻ = 5
1 N x 5 e⁻ = 5
total e⁻ = 14

[3] H
     H–C–N–H
     H H

12 e⁻ used.
N needs 2 more electrons for an octet.

→ H
   H–C–N–H
   H H

Structure and Bonding 1–9

c.

[1]
H
H C Cl
H

[2] Count valence e⁻.
$1\,C \times 4\,e^- = 4$
$3\,H \times 1\,e^- = 3$
$1\,Cl \times 7\,e^- = 7$
total e⁻ $= 14$

[3]
H
H–C–Cl
H

8 e⁻ used.
Cl needs 6 more
electrons for an octet.

→

H
H–C–Cl:
H

Complete octet.

1.5 Follow the directions from Answer 1.4.

a. HCN

H C N

Count valence e⁻.
$1\,C \times 4\,e^- = 4$
$1\,H \times 1\,e^- = 1$
$1\,N \times 5\,e^- = 5$
total e⁻ $= 10$

H–C–N → H–C≡N:

4 e⁻ used.   Complete N
               and C octets.

b. H₂CO

H C O
H

Count valence e⁻.
$1\,C \times 4\,e^- = 4$
$2\,H \times 1\,e^- = 2$
$1\,O \times 6\,e^- = 6$
total e⁻ $= 12$

H–C–O → H–C=O:
  | |
  H   H

6 e⁻ used.   Complete O
               and C octets.

c. HOCH₂CO₂H

        H O
H O C C O H
        H

Count valence e⁻.
$2\,C \times 4\,e^- = 8$
$4\,H \times 1\,e^- = 4$
$3\,O \times 6\,e^- = 18$
total e⁻ $= 30$

H O
| ‖
H–O–C–C–O–H
   |
   H

16 e⁻ used.

→

H :O:
| ‖
H–Ö–C–C–Ö–H
   |
   H

Complete octets.

1.6 Formal charge (FC) = number of valence electrons – [number of unshared electrons + (½)(number of shared electrons)]

a.
$$\left[\begin{array}{c} H \\ | \\ H-N-H \\ | \\ H \end{array} \right]^+$$

$5 - [0 + (1/2)(8)] = +1$

b.
$5 - [0 + 1/2(8)] = +1$

$CH_3-N\equiv C:$

$4 - [0 + (1/2)(8)] = 0$ $4 - [2 + (1/2)(6)] = -1$

c.
$6 - [2 + 1/2(6)] = +1$

:Ö=Ö–Ö:

$6 - [4 + (1/2)(4)] = 0$ $6 - [6 + (1/2)(2)] = -1$

1.7

a. CH₃O⁻

        H
[1] H C O
        H

[2] Count valence e⁻.
$1\,C \times 4\,e^- = 4$
$3\,H \times 1\,e^- = 3$
$1\,O \times 6\,e^- = 6$
total e⁻ $= 13$
Add 1 for (–) charge $= 14$

[3]
H
H–C–O
H

8 e⁻ used.

→

H
H–C–Ö:
H

[4]
H
H–C–Ö:⁻
H

Assign charge.

Chapter 1–10

b. HC_2^- [1] H C C   [2] Count valence e⁻.
$2 C \times 4 e^- = 8$
$1 H \times 1 e^- = 1$
total e⁻ = 9
Add 1 for (–) charge = 10

[3] H–C–C ⟶ H–C≡C: [4] H–C≡C:⁻
4 e⁻ used.   Assign charge.

c. $(CH_3NH_3)^+$ [1] H H
H C N H
H H

[2] Count valence e⁻.
$1 C \times 4 e^- = 4$
$6 H \times 1 e^- = 6$
$1 N \times 5 e^- = 5$
total e⁻ = 15
Subtract 1 for (+) charge = 14

[3]
```
  H   H
  |   |
H–C – N–H
  |   |
  H   H
```
14 e⁻ used.

[4]
```
  H   H
  |   |+
H–C – N–H
  |   |
  H   H
```
Assign charge.

d. $(CH_3NH)^-$ [1] H
H C N H
H

[2] Count valence e⁻.
$1 C \times 4 e^- = 4$
$4 H \times 1 e^- = 4$
$1 N \times 5 e^- = 5$
total e⁻ = 13
Add 1 for (–) charge = 14

[3]
```
  H
  |
H–C–N–H
  |
  H
```
10 e⁻ used.

[4]
```
  H    ..
  |   ..
H–C–N–H
  |   ..
  H
```
Complete octet and
assign charge.

1.8

a. ≡O: b. =O̲—   c. =O:

$6 - [2 + (1/2)6] = +1$ $6 - [2 + (1/2)6] = +1$ $6 - [4 + (1/2)4] = 0$

1.9

a. $C_2H_4Cl_2$ (two isomers)

Count valence e⁻.
$2 C \times 4 e^- = 8$
$4 H \times 1 e^- = 4$
$2 Cl \times 7 e^- = 14$
total e⁻ = 26

```
  H  H  ..            H  H  ..
  |  |  ..            |  |  ..
H–C–C–Cl:          H–C–C–Cl:
  |  ..  ..          ..  |
  H  :Cl:          :Cl:  H
      ..            ..
```

b. C_3H_8O (three isomers)

Count valence e⁻.
$3 C \times 4 e^- = 12$
$8 H \times 1 e^- = 8$
$1 O \times 6 e^- = 6$
total e⁻ = 26

```
  H  H  H                   H               H  H     H
  |  |  |                   |               |  |     |
H–C–C–C–O̤–H        H :O̤: H          H–C–C–O̤–C–H
  |  |  |              |  |  |            |  |     |
  H  H  H          H–C–C–C–H          H  H     H
                      |  |  |
                      H  H  H
```

c. C_3H_6 (two isomers)

Count valence e⁻.
$3 C \times 4 e^- = 12$
$6 H \times 1 e^- = 6$
total e⁻ = 18

```
  H     H                 H  H
  |     |                  C
H–C–C=C              H–C–C–H
  |  |  |              |     |
  H  H  H              H     H
```

1.10 Two different definitions:

• **Isomers** have the same molecular formula and a *different* arrangement of atoms.
• **Resonance structures** have the same molecular formula and the *same* arrangement of atoms.

Structure and Bonding 1–11

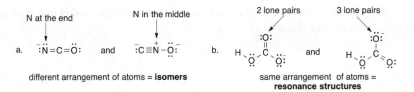

1.11 **Isomers** have the same molecular formula and a *different* arrangement of atoms.
Resonance structures have the same molecular formula and the *same* arrangement of atoms.

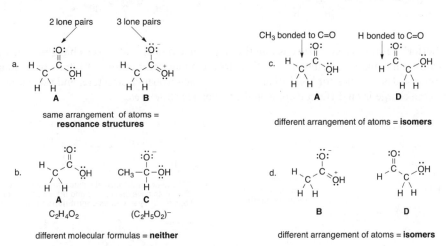

1.12 Curved arrow notation shows the movement of an electron pair. The tail begins at an electron pair (a bond or a lone pair) and the head points to where the electron pair moves.

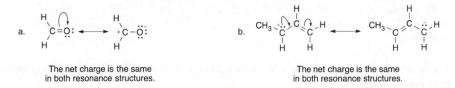

1.13 Compare the resonance structures to see what electrons have "moved." **Use one curved arrow to show the movement of each electron pair.**

Chapter 1–12

1.14 To draw another resonance structure, **move electrons only in multiple bonds and lone pairs** and keep the number of unpaired electrons constant.

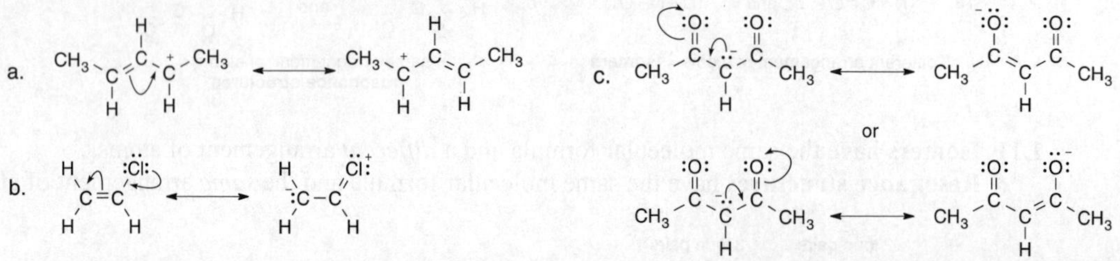

a.

b.

c.

or

1.15 **A "better" resonance structure is one that has more bonds and fewer charges.** The better structure is the major contributor and all others are minor contributors. To draw the resonance hybrid, use dashed lines for bonds that are in only one resonance structure, and use partial charges when the charge is on different atoms in the resonance structures.

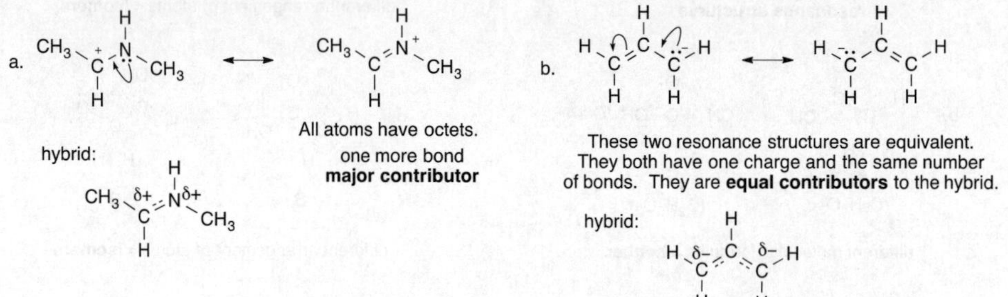

a.

All atoms have octets.
one more bond
major contributor

hybrid:

b.

These two resonance structures are equivalent.
They both have one charge and the same number
of bonds. They are **equal contributors** to the hybrid.

hybrid:

1.16

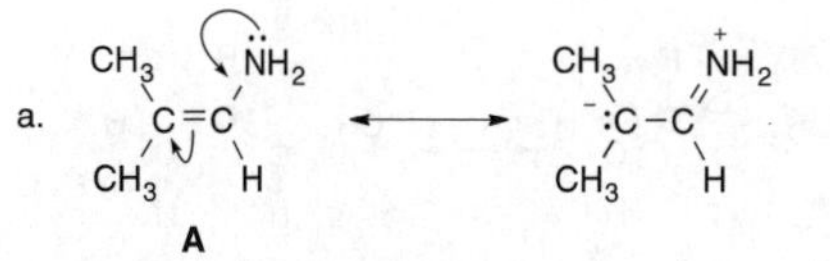

a.

A

b. The N atom in **B** has four atoms and no lone pairs, so there is no way to move the electrons to give the N another bond.

1.17 To predict the geometry around an atom, **count the number of groups (atoms + lone pairs),** making sure to draw in any needed lone pairs or hydrogens: 2 groups = linear, 3 groups = trigonal planar, 4 groups = tetrahedral.

Structure and Bonding 1–13

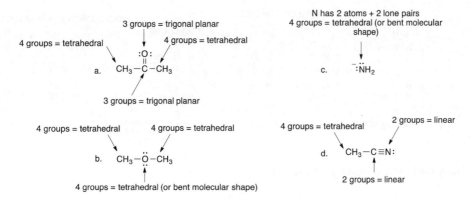

1.18 To predict the bond angle around an atom, **count the number of groups (atoms + lone pairs)**, making sure to draw in any needed lone pairs or hydrogens: 2 groups = 180°, 3 groups = 120°, 4 groups = 109.5°.

1.19 To predict the geometry around an atom, use the rules in Answer 1.17.

1.20 Reading from left-to-right, draw the molecule as a Lewis structure. Always check that carbon has four bonds and all heteroatoms have an octet by adding any needed lone pairs.

Chapter 1–14

$(CH_3)_3CCH(OH)CH_2CH_3$

$(CH_3)_2CHCHO$

$(HOCH_2)_2CH(CH_2)_3C(CH_3)_2CH_2CH_3$

a. [Lewis structure]

$CH_3(CH_2)_4CH(CH_3)_2$

b. [Lewis structure]

c. [Lewis structure]

double bond
needed to give
C an octet

d. [Lewis structure]

1.21 Draw the Lewis structure of lactic acid.

$CH_3CH(OH)CO_2H \longrightarrow$ [Lewis structure]

1.22 In shorthand or skeletal drawings, **all line junctions or ends of lines represent carbon atoms.**
The carbons are all tetravalent.

a. [skeletal structure]
3 H's
1 H
0 H's
1 H
octinoxate
(2-ethylhexyl 4-methoxycinnamate)
$C_{18}H_{26}O_3$

b. [skeletal structure]
1 H
0 H's
0 H's
3 H's
avobenzone
$C_{20}H_{22}O_3$

1.23 In shorthand or skeletal drawings, **all line junctions or ends of lines represent carbon atoms**.
Convert by writing in all carbons, and then adding hydrogen atoms to make the carbons
tetravalent.

a. [structure]   b. [structure]   c. [structure]   d. [structure]

1.24

[structure of quinine]

HO

CH_3O

quinine

$C_{20}H_{24}N_2O_2$
molecular formula

Structure and Bonding 1–15

1.25 A charge on a carbon atom takes the place of one hydrogen atom. **A negatively charged C has one lone pair, and a positively charged C has none.**

a. positive charge
no lone pairs
no H's needed

b. negative charge
one lone pair
one H needed

c. positive charge
no lone pairs
one H needed

d. negative charge
one lone pair
one H needed

1.26 Draw each indicated structure. Recall that in the skeletal drawings, a carbon atom is located at the intersection of any two lines and at the end of any line.

a. $(CH_3)_2C=CH(CH_2)_4CH_3$ =

c. = $(CH_3)_2CH(CH_2)_2CONHCH_3$

b.

d. $HO(CH_2)_2CH=CHCO_2CH(CH_3)_2$

1.27 To determine the orbitals used in bonding, **count the number of groups** (atoms + lone pairs): 4 groups = sp^3, 3 groups = sp^2, 2 groups = sp, H atom = $1s$ (no hybridization). All covalent single bonds are σ, and all double bonds contain one σ and one π bond.

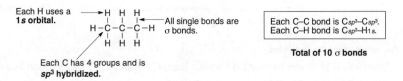

1.28 [1] Draw a valid Lewis structure for each molecule.
[2] **Count the number of groups** around each atom: 4 groups = sp^3, 3 groups = sp^2, 2 groups = sp, H atom = $1s$ (no hybridization).

Note: **Be and B** (Groups 2A and 3A) do not have enough valence e⁻ to form an octet, **and do not form an octet in neutral molecules.**

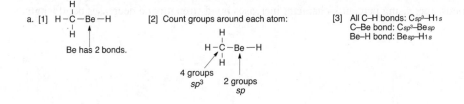

Chapter 1–16

b. [1] $CH_3-\overset{\overset{\displaystyle CH_3}{|}}{B}-CH_3$

B forms 3 bonds.

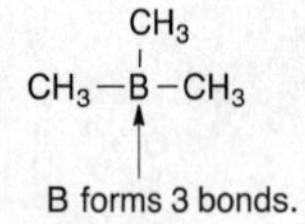

[2] Count groups around each atom:

$CH_3-\overset{\overset{\displaystyle CH_3}{|}}{B}-CH_3$

4 groups   3 groups
sp^3 sp^2

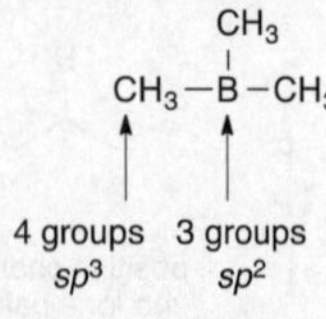

[3] All C–H bonds: $C_{sp^3}-H_{1s}$
C–B bonds: $C_{sp^3}-B_{sp^2}$

c. [1] $H-\overset{\overset{\displaystyle H}{|}}{\underset{\underset{\displaystyle H}{|}}{C}}-\overset{..}{\underset{..}{O}}-\overset{\overset{\displaystyle H}{|}}{\underset{\underset{\displaystyle H}{|}}{C}}-H$

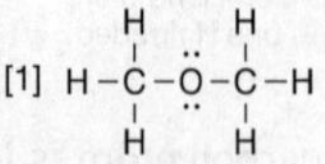

[2] Count groups around each atom:

$H-\overset{\overset{\displaystyle H}{|}}{\underset{\underset{\displaystyle H}{|}}{C}}-\overset{..}{\underset{..}{O}}-\overset{\overset{\displaystyle H}{|}}{\underset{\underset{\displaystyle H}{|}}{C}}-H$

4 groups
sp^3

4 groups
sp^3

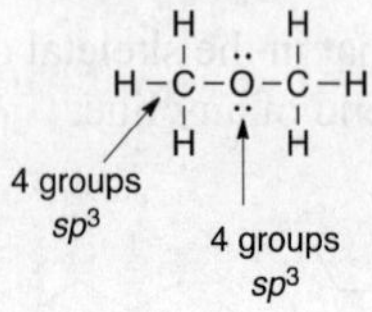

[3] All C–H bonds: $C_{sp^3}-H_{1s}$
C–O bonds: $C_{sp^3}-O_{sp^3}$

1.29 To determine the hybridization, **count the number of groups** around each atom: 4 groups = sp^3, 3 groups = sp^2, 2 groups = sp, H atom = $1s$ (no hybridization).

a. 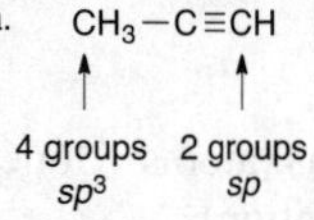 $CH_3-C\equiv CH$

4 groups   2 groups
sp^3 sp

b.

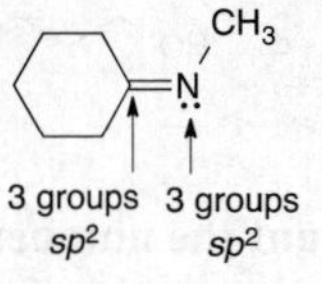

3 groups 3 groups
sp^2 sp^2

c. $CH_2=C=CH_2$

3 groups 2 groups
sp^2 sp

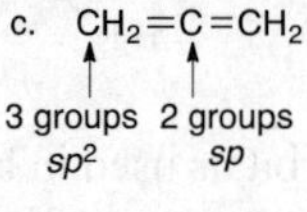

1.30

a.

Five sp^2 hybridized C's are labeled.

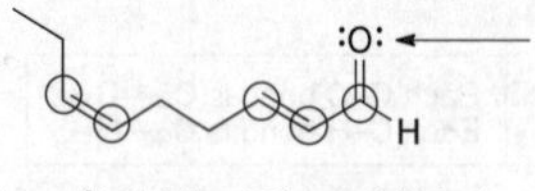

b. O has three groups (one atom + two lone pairs), so it is sp^2 hybridized.
c. $C_{sp^2}-O_{sp^2}$ C_p-O_p

d, e. Draw in all H atoms to count σ bonds. Each C–H and C–C single bond is a σ bond. Each double bond has one σ bond and one π bond.

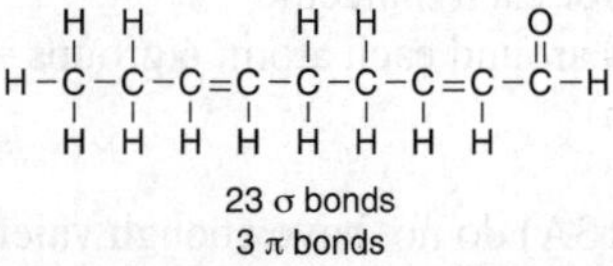

$H-\overset{\overset{\displaystyle H}{|}}{\underset{\underset{\displaystyle H}{|}}{C}}-\overset{\overset{\displaystyle H}{|}}{\underset{\underset{\displaystyle H}{|}}{C}}-\overset{\overset{\displaystyle H}{|}}{C}=\overset{\overset{\displaystyle H}{|}}{C}-\overset{\overset{\displaystyle H}{|}}{\underset{\underset{\displaystyle H}{|}}{C}}-\overset{\overset{\displaystyle H}{|}}{\underset{\underset{\displaystyle H}{|}}{C}}-\overset{\overset{\displaystyle H}{|}}{C}=\overset{\overset{\displaystyle H}{|}}{C}-\overset{\overset{\displaystyle O}{\|}}{C}-H$

23 σ bonds
3 π bonds

1.31 Single bonds are weaker and longer than double bonds, which are weaker and longer than triple bonds. Increasing percent *s*-character increases bond strength and decreases bond length.

Structure and Bonding 1–17

a. [cyclohexyl−C≡C−cyclohexenyl] ←— double bond
   ↑
   triple bond
   The triple bond is shorter than the double bond.

b. single bond
   ↓
   H
   N—(ring)—N= ←— double bond
              |
              —C≡N
   The C=N bond is shorter than the C–N bond.

c.
   :O:
   ‖
   cyclohexyl–C–H   or   cyclohexyl–C(H)(ÖH)
            ↑   ↑
         C_{sp^2}–H$_{1s}$ C_{sp^3}–H$_{1s}$
         33% s-character   25% s-character
         shorter bond

d. cyclohexyl=N–H   or   cyclohexyl–N(H)(H)
             ↑   ↑
         N_{sp^2}–H$_{1s}$ N_{sp^3}–H$_{1s}$
         33% s-character   25% s-character
         shorter bond

1.32 **Electronegativity increases from left-to-right across a row of the periodic table and decreases down a column.** Look at the relative position of the atoms to determine their relative electronegativity.

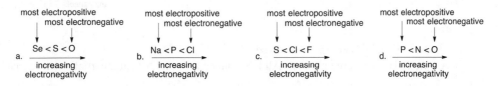

a. Se < S < O   b. Na < P < Cl   c. S < Cl < F   d. P < N < O
   increasing   increasing   increasing   increasing
   electronegativity electronegativity electronegativity electronegativity

1.33 Dipoles result from unequal sharing of electrons in covalent bonds. More electronegative atoms "pull" electron density towards them, making a dipole. **Dipole arrows point towards the atom of higher electron density.**

a. δ+ δ−
   H–F
   ⟶

b. \δ+ |δ−
   δ+ B–C—
      / |
   ⟶

c. δ−| δ+
   —C–Li
     |
   ⟵

d. δ+| δ−
   —C–Cl
     |
   ⟶

1.34 Polar molecules result from a net dipole. To determine polarity, draw the molecule in three dimensions around any polar bonds, draw in the dipoles, and look to see whether the dipoles cancel or reinforce.

Chapter 1–18

a. [Br atom pulls e− density — electronegative atom pulls e− density.]
   δ− Br, δ+C, H, H — net dipole
   All C–H bonds have no dipole.
   one polar bond
   net dipole = **polar molecule**

b. Br–C(H)(H)–Br — re-draw → C with Br, Br, H, H in 3D
   δ+ H H, δ− Br Br — resulting dipole = **polar molecule**

   Note: You must draw the molecule in three dimensions to observe the net dipole. In the Lewis structure, it appears the dipoles would cancel out, when in fact they add to make a polar molecule.

c. δ−F, δ−F, δ−F, δ−F around δ+C — no resulting dipole = **nonpolar molecule**
   Four polar bonds cancel.

d. δ−Cl, δ+C=Cδ+, δ−Cl, H, H — resulting dipole = **polar molecule**

e. δ−Cl, H, δ+C=Cδ+, H, Clδ− — no resulting dipole = **nonpolar molecule**
   Two polar bonds are **equal and opposite** and cancel.

1.35

a. The two circled C's are sp^3 hybridized.
b. All the C–H bonds are nonpolar. All H's bonded to O and N bear a partial positive charge (δ+).

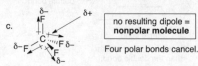

c. (one possibility — resonance structure shown)

1.36

a.
   skeletal structure (HO–, H₂NHN–, with phenyl ring bearing two OH groups, and COOH)

b. Circled carbons are sp^3 hybridized. All others are sp^2 hybridized.
c. Each N is surrounded by three atoms and a lone pair, making it sp^3 hybridized and trigonal pyramidal in molecular shape.
d.
   (structure with H–O, H–N–N, H, H, O, O–H, O–H labeled)   11 polar bonds shown in bold

Structure and Bonding 1–19

1.37

a, b, c.

$C_6H_8O_7$

14 lone pairs, 2 lone pairs on each O

d. Each C that is part of a C=O is sp^2 hybridized, so there are three sp^2 C's.

e. Orbitals:

[1] C=O, Csp^2–Osp^2 and Cp–Op

[2] C–C, Csp^3–Csp^2

[3] O–H, Osp^3–$H1s$

[4] C–O, Csp^3–Osp^3

1.38

a, b, c.

$C_{11}H_{14}O_3$

6 lone pairs, 2 lone pairs on each O

d. The sp^2 hybridized C's (seven) are labeled with circles.

e. Orbitals:

[1] C–C, Csp^3–Csp^2

[2] C–C, Csp^3–Csp^3

[3] C–H, Csp^3–$H1s$

[4] C–H, Csp^2– $H1s$

1.39 Formal charge (FC) = number of valence electrons – [number of unshared electrons + (½)(number of shared electrons)]. C is in group 4A.

a. CH_2=$\ddot{C}H$

$4 - [0 + (1/2)(8)] = 0$ $4 - [2 + (1/2)(6)] = -1$

b. H–$\ddot{C}$–H

$4 - [2 + (1/2)(4)] = 0$ $4 - [1 + (1/2)(6)] = 0$

c. H–$\overset{\cdot}{C}$–H with H

$4 - [0 + (1/2)(8)] = 0$

d. H–C–C with H's

$4 - [0 + (1/2)(6)] = +1$

1.40 Formal charge (FC) = number of valence electrons – [number of unshared electrons + (½)(number of shared electrons)]. N is in group 5A and O is in group 6A.

a. cyclohexyl–N:

$5 - [4 + (1/2)(4)] = -1$

b. $:\ddot{N}$=N=$\ddot{N}:$

$5 - [0 + (1/2)(8)] = +1$

$5 - [4 + (1/2)(4)] = -1$

$5 - [4 + (1/2)(4)] = -1$

c. cyclohexyl=O–H

$6 - [2 + (1/2)(6)] = +1$

$5 - [2 + (1/2)(6)] = 0$

d. cyclopentyl–N=$\ddot{O}:$

$6 - [4 + (1/2)(4)] = 0$

1.41 Follow the steps in Answer 1.4 to draw Lewis structures.

Chapter 1–20

a. CH$_2$N$_2$

valence e$^-$
1 C x 4 e$^-$ = 4
2 H x 1 e$^-$ = 2
2 N x 5 e$^-$ = 10
total e$^-$ = 16

H–C=N=N:̈$^-$ (with + on N)
 |
 H
 or
H–C̈–N≡N: (with – on C, + on N)
 |
 H

c. CH$_3$CNO

valence e$^-$
2 C x 4 e$^-$ = 8
3 H x 1 e$^-$ = 3
1 N x 5 e$^-$ = 5
1 O x 6 e$^-$ = 6
total e$^-$ = 22

    H
    |
H–C–C≡N–Ö:$^-$ or   H–C–C=N=Ö:
    | (+ on N) | (– on C, + on N)
    H   H

or H–C–C–N≡O: (2– on C, + on N, + on O)
     |
     H

b. CH$_3$NO$_2$

valence e$^-$
1 C x 4 e$^-$ = 4
3 H x 1 e$^-$ = 3
1 N x 5 e$^-$ = 5
2 O x 6 e$^-$ = 12
total e$^-$ = 24

    H   H
    | |
H–C–N–Ö:$^-$ or H–C–N=Ö:
    | ‖ | |
    H :O: H :Ö:
(+ on N)   (+ on N)

d. $^-$CH$_2$CN

valence e$^-$
2 C x 4 e$^-$ = 8
2 H x 1 e$^-$ = 2
1 N x 5 e$^-$ = 5
1 for (–) charge = 1
total e$^-$ = 16

H–C=C=N:̈$^-$ or   H–C̈$^-$–C≡N:
 | |
 H   H

1.42 Follow the steps in Answer 1.4 to draw Lewis structures.

a. (CH$_3$CH$_2$)$_2$O

[1]
H H   H H
H C C O C C H
  H H   H H

[2] Count valence e$^-$.
1 O x 6 e$^-$ = 6
10 H x 1 e$^-$ = 10
4 C x 4 e$^-$ = 16
total e$^-$ = 32

[3]
H H   H H
 | | | |
H–C–C–O–C–C–H
 | | | |
H H   H H
28 e$^-$ used.

[4]
H H   H H
 | | | |
H–C–C–Ö–C–C–H
 | | | |
H H   H H
Add lone pairs.

b. CH$_2$CHCN

[1]
H H
C C C N
H

[2] Count valence e$^-$.
1 N x 5 e$^-$ = 5
3 H x 1 e$^-$ = 3
3 C x 4 e$^-$ = 12
total e$^-$ = 20

[3]
H H
 | |
C–C–C–N
 |
H
12 e$^-$ used.

[4]
H H
 | |
C=C–C≡N:̈
 |
H
Add lone pairs and π bonds.

c. (HOCH$_2$)$_2$CO

[1]
   H O H
H O C C C O H
   H   H

[2] Count valence e$^-$.
3 O x 6 e$^-$ = 18
6 H x 1 e$^-$ = 6
3 C x 4 e$^-$ = 12
total e$^-$ = 36

[3]
   H O H
   | ‖ |
H–O–C–C–C–O–H
   | |
   H   H
22 e$^-$ used.

[4]
   H :O: H
   | ‖ |
H–Ö–C–C–C–Ö–H
   | |
   H   H
Add lone pairs and π bonds.

d. (CH$_3$CO)$_2$O

[1]
   H O   O H
H C C O C C H
   H   H

[2] Count valence e$^-$.
3 O x 6 e$^-$ = 18
6 H x 1 e$^-$ = 6
4 C x 4 e$^-$ = 16
total e$^-$ = 40

[3]
H O   O H
 | ‖   ‖ |
H–C–C–O–C–C–H
 | |
 H   H
24 e$^-$ used.

[4]
H :O: :O: H
 | ‖   ‖ |
H–C–C–Ö–C–C–H
 | |
 H   H
Add lone pairs and π bonds.

1.43

a.

b. Two of the possible resonance structures:

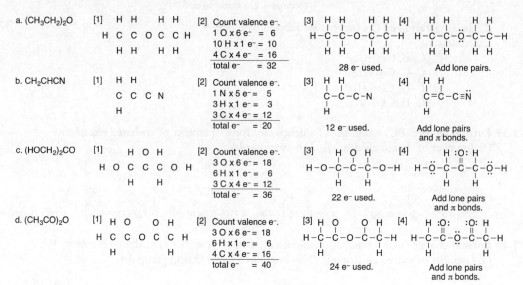

1.44 Isomers must have a different arrangement of atoms. Compounds are drawn as Lewis structures with no implied geometry.

a. Two isomers of molecular formula C₃H₇Cl

```
  :Cl: H  H              H :Cl: H
   |   |  |              |  |  |
 H-C - C -C-H          H-C -C -C-H
   |   |  |              |  |  |
   H   H  H              H  H  H
```

b. Three isomers of molecular formula C₂H₄O

```
  :Ö-H            :O: H            H    :O:
   |              ||  |             \  /   \
 H-C=C-H        H-C - C-H            C - C
   |                |               / ‾‾‾ \
   H                H              H  H    H
```

c. Four isomers of molecular formula C₃H₉N

```
  H H H                 H H   H
  | | |                 | |   |
H-C-C-C-N-H          H-C-C - N-C-H
  | | | |              | |   | |
  H H H H              H H   H H

     H     H            H  H  H
     |     |            |  |  |
   H-C - N-C-H        H-C -C -C-H
     |   | |            |  |  |
     H   | H            H  |  H
         |                 |
       H-C-H             H-N:
         |                 |
         H                 H
```

1.45

Nine isomers of C₃H₆O:

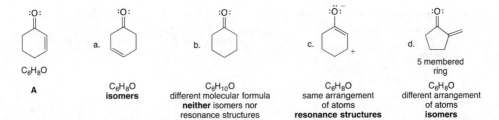

1.46 Use the definition of isomers and resonance structures in Answer 1.10.

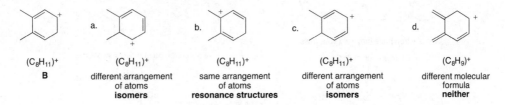

1.47 Use the definitions of isomers and resonance structures in Answer 1.9.

(C₈H₁₁)⁺
B

a. (C₈H₁₁)⁺
different arrangement of atoms
isomers

b. (C₈H₁₁)⁺
same arrangement of atoms
resonance structures

c. (C₈H₁₁)⁺
different arrangement of atoms
isomers

d. (C₈H₉)⁺
different molecular formula
neither

Chapter 1–22

1.48 Use the definitions of isomers and resonance structures in Answer 1.9.

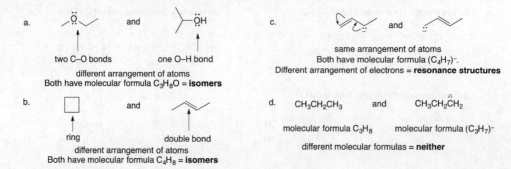

1.49 Compare the resonance structures to see what electrons have "moved." **Use one curved arrow to show the movement of each electron pair.**

1.50 Curved arrow notation shows the movement of an electron pair. The tail begins at an electron pair (a bond or a lone pair) and the head points to where the electron pair moves.

1.51 Use the rules in Answer 1.14.

Structure and Bonding 1–23

c.

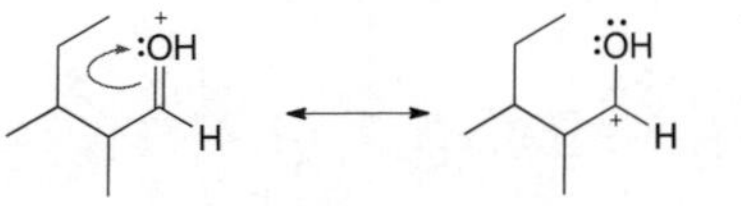

One electron pair moves = one arrow

hybrid

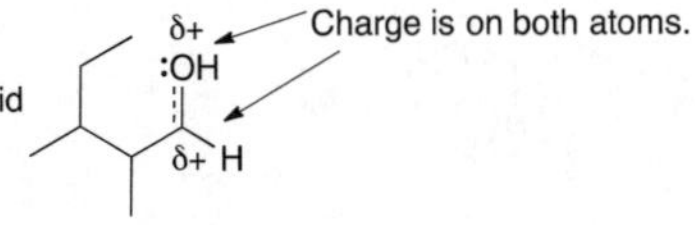

Charge is on both atoms.

$\delta+$:OH

$\delta+$ H

C–O bond has
partial double bond character.

1.52 For the compounds where the arrangement of atoms is not given, first draw a Lewis structure. Then use the rules in Answer 1.14.

a. O_3

Count valence e⁻.
3 O x 6 e⁻ = 18
total e⁻ = 18

b. NO_3^- (a central N atom)

Count valence e⁻.
1 N x 5 e⁻ = 5
3 O x 6 e⁻ = 18
(–) charge = 1
total e⁻ = 24

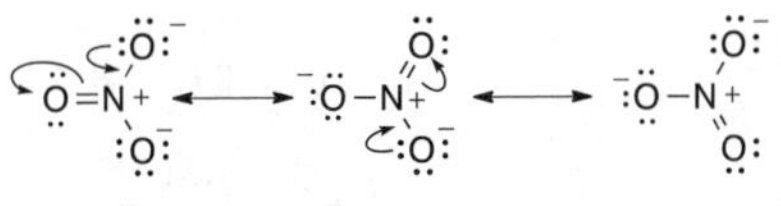

c. N_3^-

Count valence e⁻.
3 N x 5 e⁻ = 15
(–) charge = 1
total e⁻ = 16

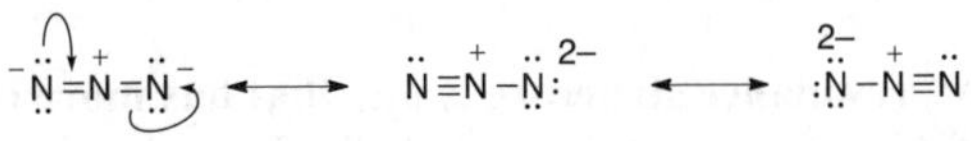

d.

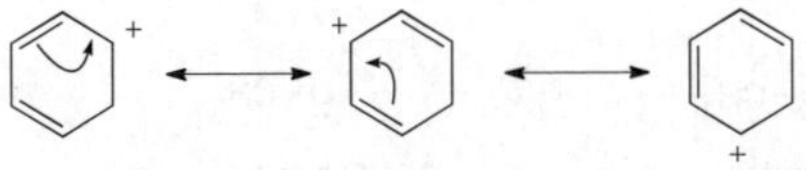

e.

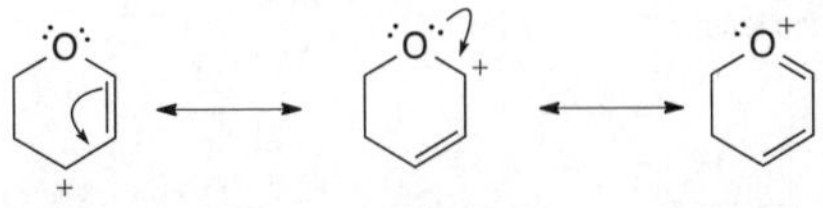

1.53 a. No additional Lewis structures can be drawn for **B.**
b. Additional Lewis structures can be drawn for **A, C,** and **D,** which are all examples of X=Y–Z* resonance.

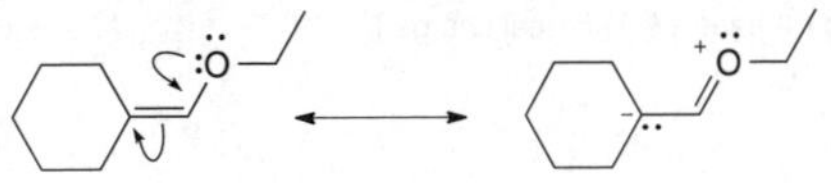

A

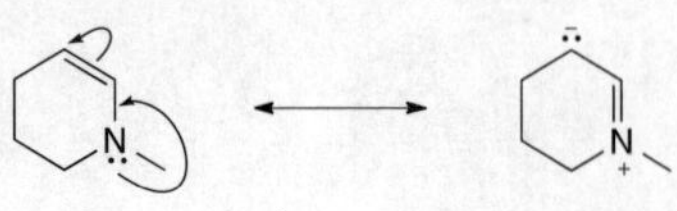

Chapter 1–24

C

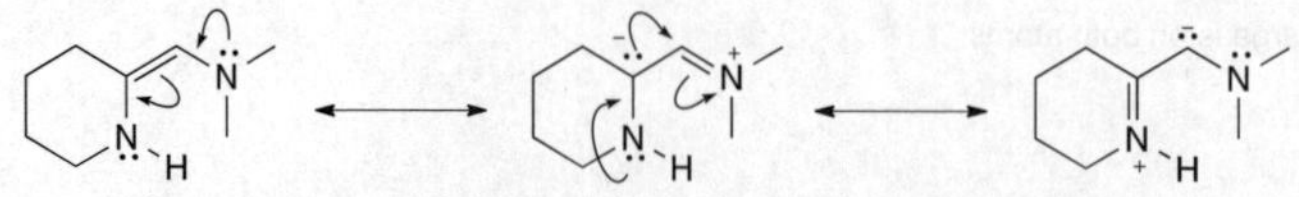

D

1.54 To draw the **resonance hybrid,** use the rules in Answer 1.15.

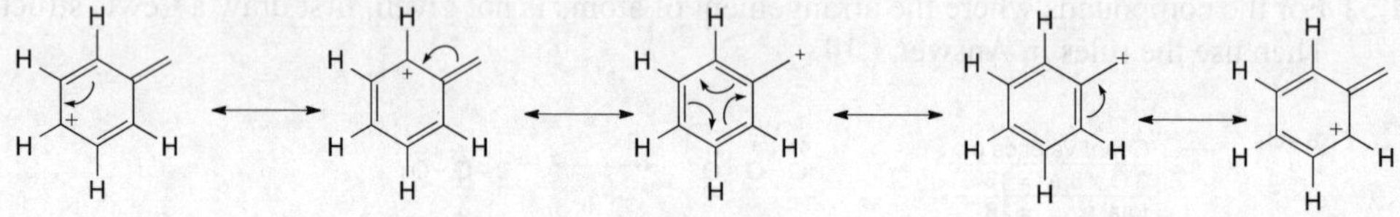

resonance hybrid

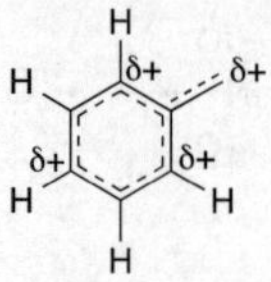

1.55 A "better" resonance structure is one that has more bonds and fewer charges. The better structure is the major contributor and all others are minor contributors.

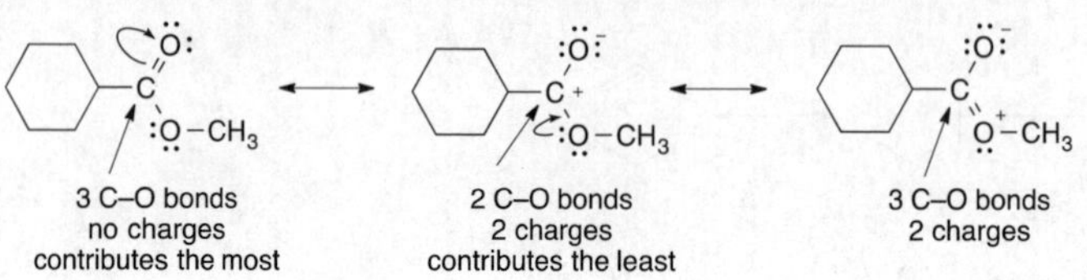

3 C–O bonds
no charges
contributes the most

2 C–O bonds
2 charges
contributes the least

3 C–O bonds
2 charges

1.56

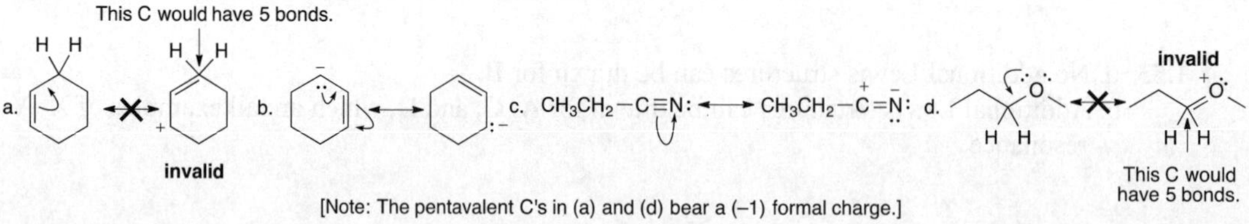

This C would have 5 bonds.

a.

invalid

b.

c. $CH_3CH_2-C\equiv N:$ ⟷ $CH_3CH_2-\overset{+}{C}=\overset{..}{\underset{..}{N}}:$

d.

invalid

This C would
have 5 bonds.

[Note: The pentavalent C's in (a) and (d) bear a (−1) formal charge.]

Structure and Bonding 1–25

1.57 Use the rules in Answer 1.18.

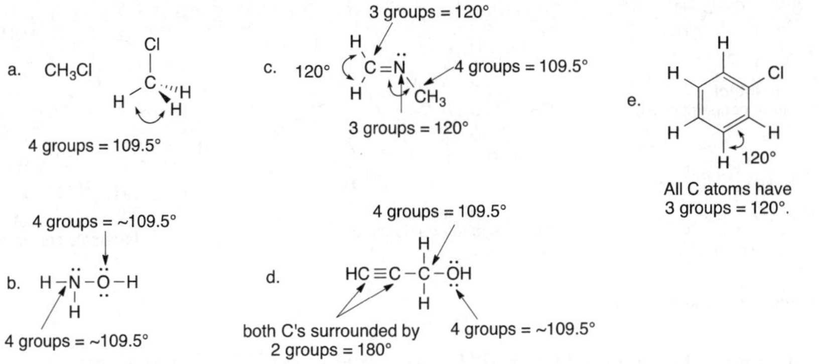

a. CH_3Cl

4 groups = 109.5°

b. $H-\overset{..}{\underset{H}{N}}-\overset{..}{\underset{..}{O}}-H$

4 groups = ~109.5°
4 groups = ~109.5°

c. 120° $\overset{H}{\underset{H}{C}}=\overset{..}{N}$ CH₃

3 groups = 120°
4 groups = 109.5°
3 groups = 120°

d. $HC\equiv C-\overset{H}{\underset{H}{C}}-\overset{..}{O}H$

both C's surrounded by 2 groups = 180°
4 groups = 109.5°
4 groups = ~109.5°

e.

120°
All C atoms have 3 groups = 120°.

1.58 To predict the geometry around an atom, use the rules in Answer 1.17.

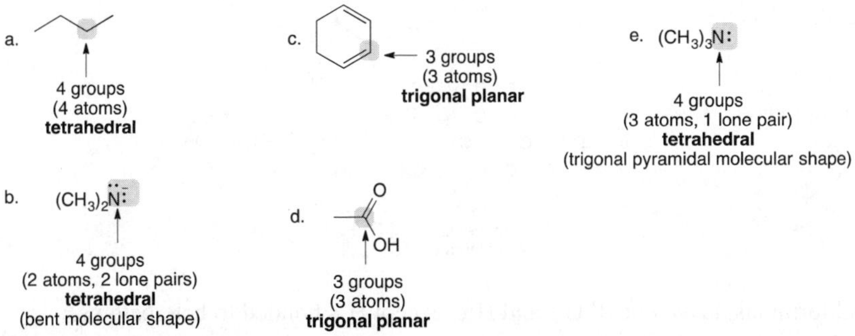

a.
4 groups
(4 atoms)
tetrahedral

b. $(CH_3)_2\overset{..}{N}:$
4 groups
(2 atoms, 2 lone pairs)
tetrahedral
(bent molecular shape)

c.
3 groups
(3 atoms)
trigonal planar

d.
3 groups
(3 atoms)
trigonal planar

e. $(CH_3)_3\overset{..}{N}:$
4 groups
(3 atoms, 1 lone pair)
tetrahedral
(trigonal pyramidal molecular shape)

1.59 In shorthand or skeletal drawings, **all line junctions or ends of lines represent carbon atoms.** The C's are all tetravalent. All H's bonded to C's are drawn in the following structures. C's labeled with (*) have no H's bonded to them.

a.

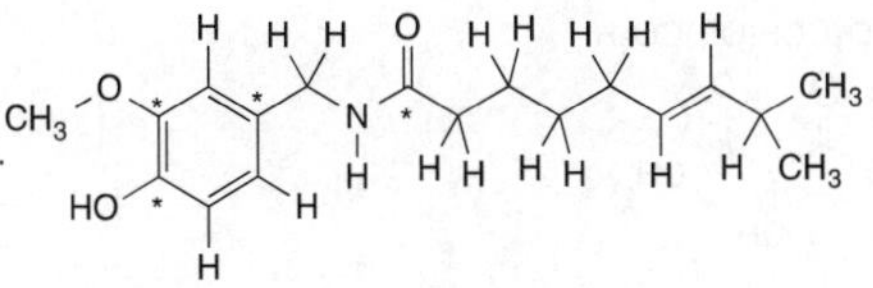

b.

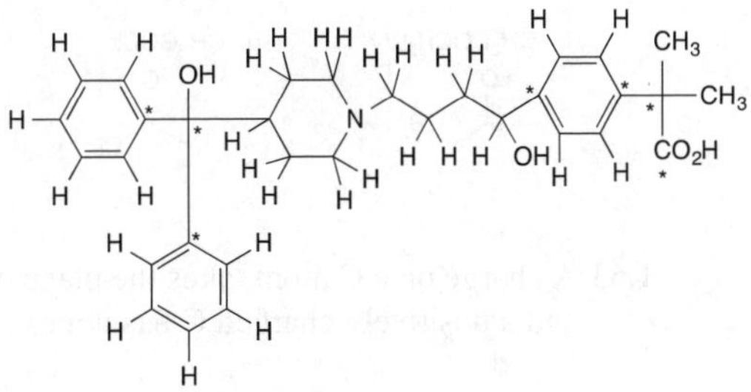

1.60 In shorthand or skeletal drawings, **all line junctions or ends of lines represent carbon atoms.** Convert by writing in all C's, and then adding H's to make the C's tetravalent.

Chapter 1–26

a. menthol (isolated from peppermint oil)

b. myrcene (isolated from bayberry)

c. ethambutol (drug used to treat tuberculosis)

d. estradiol (a female sex hormone)

1.61 In skeletal formulas, leave out all C's and H's, except H's bonded to heteroatoms.

a. $(CH_3)_2CHCH_2CH_2CH(CH_3)_2$

b. $CH_3CH(Cl)CH(OH)CH_3$

c. $CH_3(CH_2)_2C(CH_3)_2CH(CH_3)CH(CH_3)CH(Br)CH_3$

d. limonene (oil of lemon)

1.62 In skeletal formulas, leave out all C's and H's, except H's bonded to heteroatoms.

a. $CH_3CONHCH_3$

b. CH_3COCH_2Br

c. $(CH_3)_3COH$

d. CH_3COCl

e. $CH_3COCH_2CO_2H$

f. $HO_2CCH(OH)CO_2H$

1.63 A charge on a C atom takes the place of one H atom. A negatively charged C has one lone pair, and a positively charged C has none.

a. (benzene anion) b. (acetaldehyde anion) c. (methylcyclohexadienyl cation) d. $CH_3-C\equiv \overset{+}{N}H$ e. (oxocarbenium/ammonium structure)

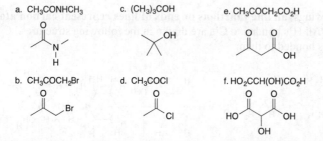

Structure and Bonding 1–27

1.64 To determine the hybridization around the labeled atoms, use the procedure in Answer 1.29.

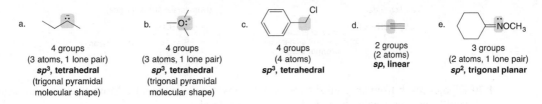

a. 4 groups (3 atoms, 1 lone pair) **sp^3, tetrahedral** (trigonal pyramidal molecular shape)

b. 4 groups (3 atoms, 1 lone pair) **sp^3, tetrahedral** (trigonal pyramidal molecular shape)

c. 4 groups (4 atoms) **sp^3, tetrahedral**

d. 2 groups (2 atoms) **sp, linear**

e. 3 groups (2 atoms, 1 lone pair) **sp^2, trigonal planar**

1.65 To determine what orbitals are involved in bonding, use the procedure in Answer 1.27.

a. C_{sp^2}–H$_{1s}$, σ: C_{sp^2}–C_{sp^2}, π: C_p–C_p, C_{sp^2}–C_{sp^3}

b. C_{sp^2}–C_{sp^3}, :Ö:, σ: C_{sp^2}–O$_{sp^2}$, π: C_p–O_p

c. H–C≡C–C=N–CH$_3$, C_{sp}–H$_{1s}$, σ: C_{sp}–C_{sp^2}, σ: C_{sp}–C_{sp}, π: C_p–C_p, π: C_p–C_p, σ: C_{sp^3}–N$_{sp^2}$

1.66

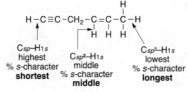

ketene CH$_2$=C=O

σ: C_{sp}–C_{sp^2}, π: C_p–C_p, sp, sp^2, sp^2

σ: C_{sp}–O$_{sp^2}$, π: C_p–O_p

σ bonds: sp^2, 1s, sp, sp^2, sp^2, sp^2

π bonds: π bond, π bond

[For clarity, only the large bonding lobes of the hybrid orbitals are drawn.]

1.67 To determine relative bond length, use the rules in Answer 1.31. All C's and H's are drawn in for emphasis.

H–C≡C–CH$_2$–C=C–C–H with H's

C_{sp}–H$_{1s}$ highest % s-character **shortest**

C_{sp^2}–H$_{1s}$ middle % s-character middle

C_{sp^3}–H$_{1s}$ lowest % s-character **longest**

1.68

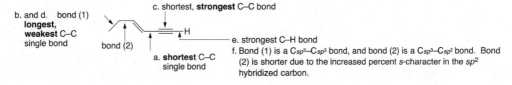

b. and d. bond (1) **longest, weakest** C–C single bond

c. shortest, **strongest** C–C bond

a. **shortest** C–C single bond

bond (2)

e. strongest C–H bond

f. Bond (1) is a C_{sp^3}–C_{sp^3} bond, and bond (2) is a C_{sp^3}–C_{sp^2} bond. Bond (2) is shorter due to the increased percent s-character in the sp^2 hybridized carbon.

Chapter 1–28

1.69 Percent *s*-character determines the strength of a bond. **The higher percent *s*-character of an orbital used to form a bond, the stronger the bond.**

vinyl chloride

$CH_2=CH-Cl$

33% *s*-character
higher percent *s*-character
stronger bond

C_{sp^2}

chloroethane (ethyl chloride)

CH_3-CH_2-Cl

25% *s*-character

C_{sp^3}

1.70 Dipoles result from unequal sharing of electrons in covalent bonds. More electronegative atoms "pull" electron density towards them, making a dipole.

a. $\overset{\delta+}{NH_2}-\overset{\delta-}{OH}$ b. $\delta+$ —$\overset{\delta-}{NH_2}$ c. $\delta-$ —Li $\delta+$

1.71 Use the directions from Answer 1.34.

a. $CHBr_3$ b. $CH_3CH_2OCH_2CH_3$ c. net dipole   d. no net dipole

net dipole

net dipole

1.72

aspirin

a. molecular formula $C_9H_8O_4$
b. eight lone pairs total
c. C labeled with a circle is sp^3 hybridized. All other C's are sp^2 hybridized.
d. three possible resonance structures:

caffeine

a. molecular formula $C_8H_{10}N_4O_2$
b. eight lone pairs total
c. C's labeled with a circle are sp^3 hybridized. All other C's are sp^2 hybridized.
d. three possible resonance structures:

Structure and Bonding 1–29

1.73

```
                          1 σ and 2 π bonds
                             polar bond
                                 ↓
             σ
     (essentially) nonpolar
                ↓
    tetrahedral  →  CH₃─C≡N:
    sp³ hybridized      ↑        ↑
                      linear    linear
                      sp hybridized   sp hybridized
                                The lone pair is in an
                                sp hybrid orbital.
```
All C–H bonds are nonpolar σ bonds.
All H's use a 1s orbital in bonding.

1.74 a. sp^2
b. Each C is trigonal planar; the ring is flat, drawn as a hexagon.
c. [benzene resonance structures]

d. Benzene is stable because of its two resonance structures that contribute equally to the hybrid. (This is only part of the story. We'll learn more about benzene's unusual stability in Chapter 17.)

1.75

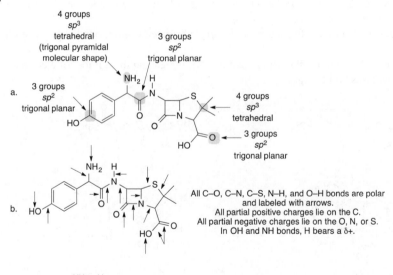

All C–O, C–N, C–S, N–H, and O–H bonds are polar and labeled with arrows.
All partial positive charges lie on the C.
All partial negative charges lie on the O, N, or S.
In OH and NH bonds, H bears a δ+.

6 π bonds

d. **33% s-character = sp^2 hybridized**

These C–H bonds have a C atom with 33% s-character.

Chapter 1–30

1.76

a, b, c.

3 groups
sp^2
trigonal planar
lone pair in sp^2 orbital

4 groups
sp^3
tetrahedral
(trigonal pyramidal molecular shape)
lone pair in sp^3 orbital

d.

constitutional isomer

e.

resonance structure

1.77 a.

longest C–N bond

shortest
C–N bond   longest C–C bond

b. The C–C bonds in the CH$_2$CH$_3$ groups are the longest because they are formed from sp^3 hybridized C's.

c. The shortest C–C bond is labeled with an asterisk (*) because it is formed from orbitals with the highest percent s-character (Csp–Csp^2).

d. The longest C–N bond is formed from the sp^3 hybridized C atom bonded to a N atom [labeled in part (a)].

e. The shortest C–N bond is the triple bond (C≡N); increasing the number of electrons between atoms decreases bond length.

f.

g.

1.78

$\overset{+}{C}H_3$
3 groups
sp^2 trigonal planar
plot **A**

| The blue region is evidence of the electron-poor cation. |

$:\!\overset{-}{C}H_3$
4 groups
sp^3 tetrahedral
(The molecular shape is trigonal pyramidal.)
plot **B**

| The red region is evidence of the electron-rich anion. |

1.79 If the N atom is *sp*² hybridized, the lone pair occupies a *p* orbital, which can overlap with the π bond of the adjacent C=O. This allows electron density to delocalize, which is a stabilizing feature.

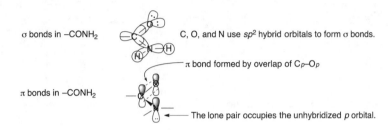

1.80

a.

CH₃—OH [1]
↑
*sp*³
25% *s*-character
The lower percent *s*-character makes this bond longer.

CH₃—C(=O)—OH [2]
↑
*sp*²
33% *s*-character

b.

CH₃—C(=O)—Ö:⁻ [3], [4] ⟷ CH₃—C(—Ö:⁻)=O   CH₃—C(⋯Ö:)⋯Ö: hybrid (δ–, δ–)

two resonance structures

Bonds [3] and [4] are both equivalent in length, because the anion is resonance stabilized, and the C–O bond of the hybrid is a composite of one single bond and one double bond. Both resonance structures contribute equally to the hybrid. Since each C–O bond in the hybrid has partial double bond character, it is shorter than the C–O bond labeled [2].

Chapter 1–32

1.81 Ten additional resonance structures are drawn. (There are more possibilities.)

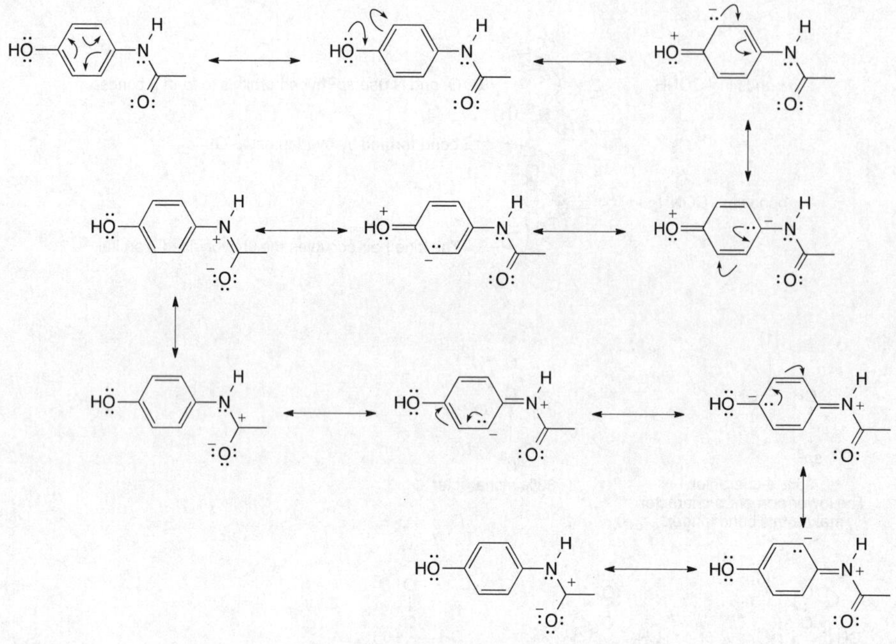

1.82 Polar bonds result from unequal sharing of electrons in covalent bonds. Normally we think of more electronegative atoms "pulling" more of the electron density towards them, making a dipole. In looking at a C_{sp^2}–C_{sp^3} bond, the atom with a higher percent s-character will "pull" more of the electron density towards it, creating a small dipole.

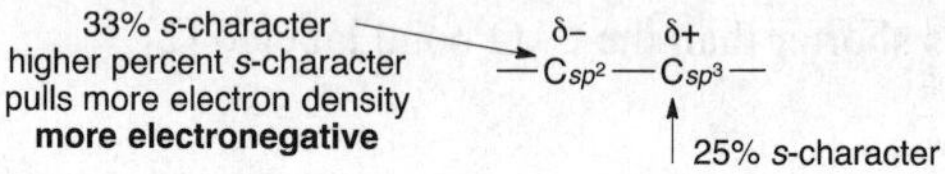

33% s-character
higher percent s-character
pulls more electron density
more electronegative

$\delta-$ $\delta+$
C_{sp^2} — C_{sp^3} —

25% s-character

1.83

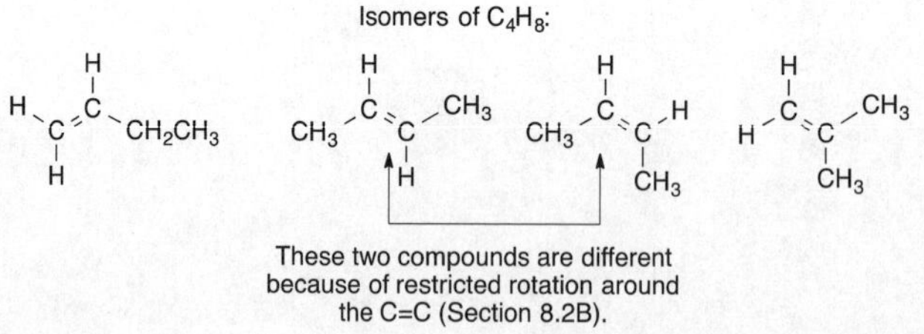

Isomers of C_4H_8:

These two compounds are different because of restricted rotation around the C=C (Section 8.2B).

Structure and Bonding 1–33

1.84 Carbocation **A** is more stable than carbocation **B** because resonance distributes the positive charge over two carbons. Delocalizing electron density is stabilizing. **B** has no possibility of resonance delocalization.

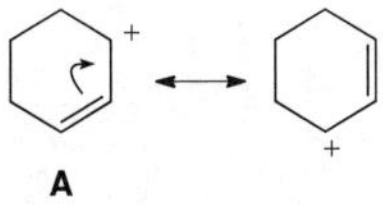

No resonance structures

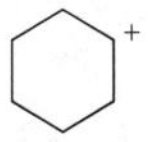

A   **B**

1.85

a.

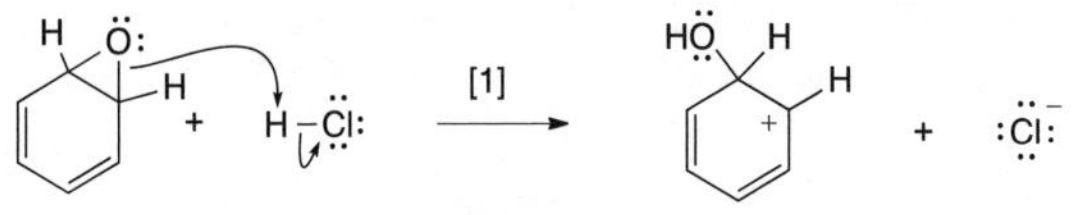

[1]

b.

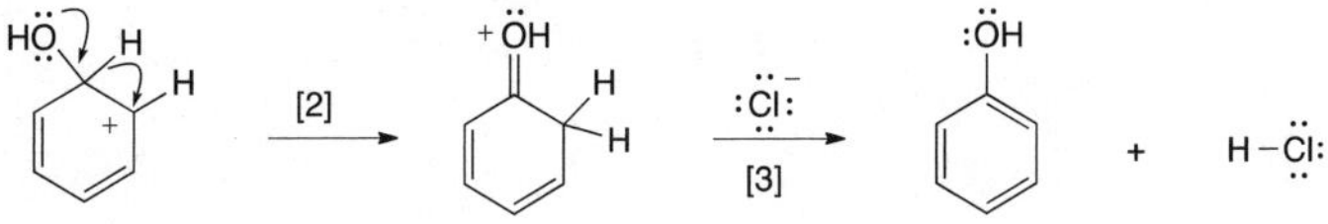

[2]

X   [3]   phenol

Chapter 2: Acids and Bases

Chapter Review

A comparison of Brønsted–Lowry and Lewis acids and bases

Type	Definition	Structural feature	Examples
Brønsted–Lowry acid (2.1)	proton donor	a proton	HCl, H_2SO_4, H_2O, CH_3CO_2H, TsOH
Brønsted–Lowry base (2.1)	proton acceptor	a lone pair or a π bond	$^-$OH, $^-OCH_3$, H$^-$, $^-NH_2$, $CH_2=CH_2$
Lewis acid (2.8)	electron pair acceptor	a proton, or an unfilled valence shell, or a partial (+) charge	BF_3, $AlCl_3$, HCl, CH_3CO_2H, H_2O
Lewis base (2.8)	electron pair donor	a lone pair or a π bond	$^-$OH, $^-OCH_3$, H$^-$, $^-NH_2$, $CH_2=CH_2$

Acid–base reactions

[1] A Brønsted–Lowry acid donates a proton to a Brønsted–Lowry base (2.2).

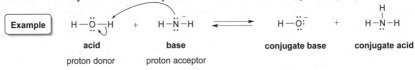

[2] A Lewis base donates an electron pair to a Lewis acid (2.8).

- Electron-rich species react with electron-poor ones.
- Nucleophiles react with electrophiles.

Important facts

- Definition: $pK_a = -\log K_a$. The **lower** the pK_a, the **stronger** the acid (2.3).

NH_3 versus H_2O
$pK_a = 38$ $pK_a = 15.7$
                              lower pK_a = stronger acid

Chapter 2–2

- The stronger the acid, the weaker the conjugate base (2.3).

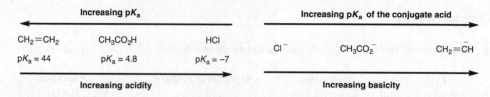

- In proton transfer reactions, equilibrium favors the weaker acid and the weaker base (2.4).

- An acid can be deprotonated by the conjugate base of any acid having a **higher pK_a** (2.4).

Acid	pK_a	Conjugate base	
CH$_3$CO$_2$–H	4.8	CH$_3$CO$_2$$^-$	
CH$_3$CH$_2$O–H	16	CH$_3$CH$_2$O$^-$	These bases
HC≡CH	25	HC≡C$^-$	can deprotonate
H–H	35	H$^-$	CH$_3$CO$_2$–H.

higher pK_a than CH$_3$CO$_2$–H

Factors that determine acidity (2.5)

[1] **Element effects** (2.5A)   The acidity of H–A increases both left-to-right across a row and down a column of the periodic table.

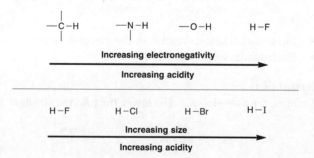

Acids and Bases 2–3

[2] **Inductive effects** (2.5B)	The acidity of H–A increases with the presence of electron-withdrawing groups in A.

CH$_3$CH$_2$OH   ⟶   CH$_3$CH$_2$O$^-$
weaker acid   **No additional electronegative atoms stabilize the conjugate base.**

CF$_3$CH$_2$OH   ⟶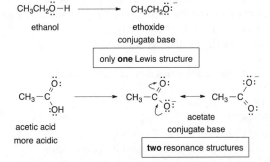
stronger acid

CF$_3$ withdraws electron density, stabilizing the conjugate base.

[3] **Resonance effects** (2.5C)	The acidity of H–A increases when the conjugate base A:$^-$ is resonance stabilized.

CH$_3$CH$_2$Ö–H   ⟶   CH$_3$CH$_2$Ö:$^-$
ethanol   ethoxide
                               conjugate base

only **one** Lewis structure

CH$_3$–C(=Ö)(:ÖH)   ⟶   CH$_3$–C(=Ö)(:Ö:$^-$)   ⟷   CH$_3$–C(:Ö:$^-$)(=Ö:)
acetic acid   acetate
more acidic   conjugate base

two resonance structures

[4] **Hybridization effects** (2.5D)	The acidity of H–A increases as the percent *s*-character of the A:$^-$ increases.

CH$_3$CH$_3$ CH$_2$=CH$_2$ H–C≡C–H
ethane   ethylene   acetylene
pK_a = 50   pK_a = 44   pK_a = 25

⟶ **Increasing acidity**

Chapter 2–4

Practice Test on Chapter Review

1.a. Given the pK_a data, which of the following bases is strong enough to deprotonate C_6H_5OH ($pK_a = 10$) so that the equilibrium lies to the right?

Compound	pK_a	
H_3O^+	−1.7	1. NaOH
NH_4^+	9.4	2. $NaNH_2$
H_2O	15.7	3. NH_3
NH_3	38	4. Compounds (1) and (2) are strong enough to deprotonate C_6H_5OH.
		5. Compounds (1), (2), and (3) are all strong enough to deprotonate C_6H_5OH.

b. Which of the following statements is true about pK_a, acidity, and basicity?
   1. A higher pK_a means the acid is less acidic.
   2. In an acid–base reaction, the equilibrium lies on the side of the acid with the higher pK_a.
   3. A lower pK_a value for the acid means the conjugate base is more basic.
   4. Statements (1) and (2) are both true.
   5. Statements (1), (2), and (3) are all true.

c. Which of the following species can be Lewis acids?
   1. BCl_3
   2. CH_3OH
   3. $(CH_3)_3C^+$
   4. Both (1) and (2) can be Lewis acids.
   5. Species (1), (2), and (3) can all be Lewis acids.

2. Answer the following questions about compounds **A–D.**

a. Which compound is the strongest acid?
b. Which compound forms the strongest conjugate base?
c. The conjugate base of **C** is strong enough to remove a proton on which compound(s), so that the equilibrium favors the products?

3. (a) Which compound is the strongest Brønsted–Lowry acid? (b) Which compound is the weakest Brønsted–Lowry acid?

Acids and Bases 2–5

4. Draw all the products formed in the following reactions.

a. [structure: phenyl-CH2-CH(CH3)-NHCH3] + HBr ⟶

b. [structure: 4-chloro-3-nitrophenol] + Na+ ⁻NH2 ⟶

5. Draw the product(s) formed in the following Lewis acid–base reaction.

(CH3)2ĊH + CH3OH ⟶

Answers to Practice Test

1. a. 4   2. a. **D**   3. a. **C**   4. a.   5.
   b. 4   b. **A**   b. **B**
   c. 5   c. **B, D**   b.

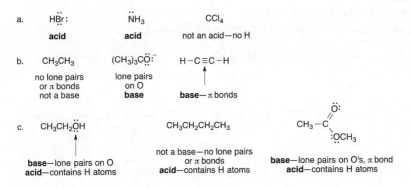

Answers to Problems

2.1 Brønsted–Lowry acids are **proton donors** and must contain a hydrogen atom.
Brønsted–Lowry bases are **proton acceptors** and must have an available electron pair (either a lone pair or a π bond).

a. HB̈r: N̈H3   CCl4
    acid   **acid**   not an acid—no H

b. CH3CH3   (CH3)3CÖ:⁻   H–C≡C–H
    no lone pairs   lone pairs
    or π bonds   on O   ↑
    not a base   **base**   **base—π bonds**

c. CH3CH2ÖH   CH3CH2CH2CH3   CH3–C(=Ö:)–ÖCH3
         ↑   not a base—no lone pairs
    base—lone pairs on O   or π bonds   **base—lone pairs on O's, π bond**
    acid—contains H atoms **acid—contains H atoms** **acid—contains H atoms**

2.2 A Brønsted–Lowry base accepts a proton to form the conjugate acid. A Brønsted–Lowry acid loses a proton to form the conjugate base.

a. NH3 ⟶ NH4+   b. HBr   ⟶ Br⁻
    Cl⁻ ⟶ HCl   HSO4⁻ ⟶ SO4²⁻
    (CH3)2C=O ⟶ (CH3)2C=ÖH⁺   CH3OH ⟶ CH3O⁻

Chapter 2–6

2.3 a. True.

$$CH_2{=}CH_2 + H^+ \longrightarrow CH_3CH_2^+$$
    base   conjugate acid

b. False. $CH_3CH_2^-$ cannot be the conjugate base of $CH_3CH_2^+$ because they both have the same number of H's and a conjugate base must have one fewer H.

c. False. $CH_2{=}CH_2$ and $CH_3CH_2^-$ differ by the presence of H^-, not H^+.

d. True.

$$CH_2{=}CH_2 \xrightarrow{\text{Remove } H^+} CH_2{=}\overset{-}{C}H$$
    acid   conjugate base

e. True.

$$CH_3CH_2^- + H^+ \longrightarrow CH_3CH_3$$
    base   conjugate acid

2.4 The Brønsted–Lowry base accepts a proton to form the conjugate acid. The Brønsted–Lowry acid loses a proton to form the conjugate base. Use curved arrows to show the movement of electrons (***NOT protons***). Re-draw the starting materials if necessary to clarify the electron movement.

a.

    acid   base   conjugate base   conjugate acid

b.

    acid   base   conjugate base   conjugate acid

2.5 To draw the products:
[1] Find the acid and base.
[2] Transfer a proton from the acid to the base.
[3] Check that the charges on each side of the arrows are balanced.

a.

    acid   base $HO{-}CH_3$ (–)1 charge on each side

b.

    acid   base H_2 (–)1 charge on each side

c.

    base   acid   net neutral on each side

d.

    base   acid   net neutral on each side

Acids and Bases 2–7

2.6 Draw the products in each reaction as in Answer 2.4.

a. [structure: pentan-2-ol] + HCl → [structure: protonated OH₂⁺] + Cl⁻

b. [ether structure] + HCl → [protonated ether] + Cl⁻

c. CH₃CH₂–N(CH₃)–CH₃ + HCl → CH₃CH₂–N⁺(H)(CH₃)–CH₃ + Cl⁻

d. pyrrolidine (NH) + HCl → pyrrolidinium (N⁺H₂) + Cl⁻

2.7 The smaller the pK_a, the stronger the acid. The larger the K_a, the stronger the acid.

a. CH₃CH₂CH₃   or   CH₃CH₂OH
    pK_a = 50   pK_a = 16
                                    ↑
                          smaller pK_a
                          stronger acid

b. phenol (OH)   or   toluene (CH₃)
    K_a = 10⁻¹⁰ K_a = 10⁻⁴¹
         ↑
    larger K_a
    stronger acid

2.8 To convert from K_a to pK_a, take (–) the log of the K_a; **pK_a = –log K_a**.
To convert pK_a to K_a, take the antilog of (–) the pK_a.

a. $K_a = 10^{-10}$ $K_a = 10^{-21}$ $K_a = 5.2 \times 10^{-5}$
        ↓   ↓   ↓
    pK_a = 10   pK_a = 21   pK_a = 4.3

b. pK_a = 7   pK_a = 11   pK_a = 3.2
        ↓   ↓   ↓
    $K_a = 10^{-7}$ $K_a = 10^{-11}$ $K_a = 6.3 \times 10^{-4}$

2.9 Since **strong acids form weak conjugate bases,** the basicity of conjugate bases increases with increasing pK_a of their acids. Find the pK_a of each acid from Table 2.1 and then rank the acids in order of increasing pK_a. This will also be the order of increasing basicity of their conjugate bases.

a. ←——— Increasing acidity
          H₂O   NH₃   CH₄
    pK_a = 15.7   38   50
    conjugate bases: ⁻OH ⁻NH₂ ⁻CH₃
          ———→ Increasing basicity

b. ←——— Increasing acidity
          HC≡CH   CH₂=CH₂   CH₄
    pK_a = 25   44   50
    conjugate bases: ⁻C≡CH ⁻CH=CH₂ ⁻CH₃
          ———→ Increasing basicity

2.10 Use the definitions in Answer 2.9 to compare the acids. The smaller the pK_a, the larger the K_a and the stronger the acid. When a stronger acid dissolves in water, the equilibrium lies further to the right.

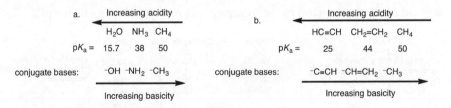

HCO₂H
formic acid
pK_a = 3.8

(CH₃)₃CCO₂H
pivalic acid
pK_a = 5.0

a. smaller pK_a = larger K_a
b. smaller pK_a = stronger acid
c. weaker acid = stronger conjugate base
d. stronger acid = equilibrium further to the right

Chapter 2–8

2.11 To estimate the pK_a of the indicated bond, find a similar bond in the pK_a table (H bonded to the same atom with the same hybridization).

a. cyclohexyl-NH-H (arrow on N–H)
For NH$_3$, pK_a is 38.
estimated pK_a = 38

b. cyclohexyl-O-H (arrow on O–H)
For CH$_3$CH$_2$OH, pK_a is 16.
estimated pK_a = 16

c. Br-CH$_2$-C(=O)-O-H (arrow on O–H)
For CH$_3$COOH, pK_a is 4.8.
estimated pK_a = 5

2.12 Label the acid and the base and then transfer a proton from the acid to the base. To determine if the reaction will proceed as written, compare the pK_a of the acid on the left with the conjugate acid on the right. **The equilibrium always favors the formation of the weaker acid and the weaker base.**

a. CH$_2$=CH$_2$ + H:$^-$ ⇌ CH$_2$=$\ddot{\text{C}}$H$^-$ + H$_2$
   acid   base   conjugate base   conjugate acid
   pK_a = 44   pK_a = 35
   weaker acid
Equilibrium favors the **starting materials**.

b. CH$_4$ + :ÖH$^-$ ⇌ :CH$_3^-$ + H$_2$Ö:
   acid   base   conjugate base   conjugate acid
   pK_a = 50   pK_a = 15.7
   weaker acid
Equilibrium favors the **starting materials**.

c. CH$_3$CO$_2$H + CH$_3$CH$_2$Ö:$^-$ ⇌ CH$_3$CO$_2^-$ + CH$_3$CH$_2$ÖH
   acid   base   conjugate base   conjugate acid
   pK_a = 4.8   pK_a = 16
                                                      weaker acid
Equilibrium favors the **products**.

d. :Cl:$^-$ + CH$_3$CH$_2$ÖH ⇌ HCl: + CH$_3$CH$_2$Ö:$^-$
   base   acid   conjugate acid   conjugate base
          pK_a = 16   pK_a = –7
          weaker acid
Equilibrium favors the **starting materials**.

2.13 An acid can be deprotonated by the conjugate base of any acid with a higher pK_a.

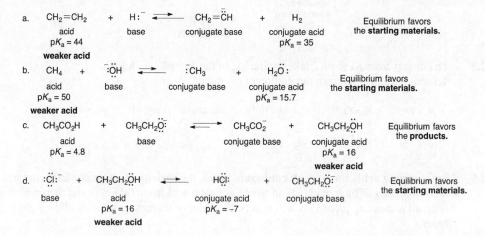

CH$_3$CN
pK_a = 25
Any base having a conjugate acid with a pK_a higher than 25 can deprotonate this acid.

Base	Conjugate acid	pK_a
NaH	H$_2$	35
Na$_2$CO$_3$	HCO$_3^-$	10.2
NaOH	H$_2$O	15.7
NaNH$_2$	NH$_3$	38
NaHCO$_3$	H$_2$CO$_3$	6.4

Only NaH and NaNH$_2$ are strong enough to deprotonate acetonitrile.

2.14 The acidity of H–Z **increases left-to-right across a row and down a column** of the periodic table.

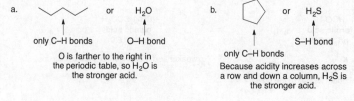

a. CH$_3$CH$_2$CH$_2$CH$_3$ or H$_2$O
   only C–H bonds   O–H bond
O is farther to the right in the periodic table, so H$_2$O is the stronger acid.

b. cyclopentane or H$_2$S
   only C–H bonds   S–H bond
Because acidity increases across a row and down a column, H$_2$S is the stronger acid.

2.15

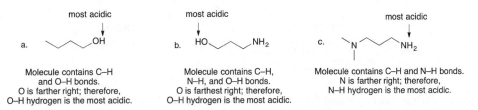

Because acidity increases across a row and down a column, the order of acidity is N–H < O–H < S–H.

2.16 Look at the element bonded to the acidic H and decide its acidity based on the periodic trends. **Farther to the right and down the periodic table is more acidic.**

a. Molecule contains C–H and O–H bonds. O is farther right; therefore, O–H hydrogen is the most acidic.

b. Molecule contains C–H, N–H, and O–H bonds. O is farthest right; therefore, O–H hydrogen is the most acidic.

c. Molecule contains C–H and N–H bonds. N is farther right; therefore, N–H hydrogen is the most acidic.

2.17 The acidity of HA increases left-to-right across the periodic table. Pseudoephedrine contains C–H, N–H, and O–H bonds. The O–H bond is most acidic.

pseudoephedrine skeletal structure

2.18 Compare the most acidic protons in each compound to determine the stronger acid.

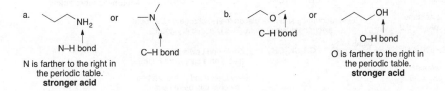

a. N–H bond — N is farther to the right in the periodic table. **stronger acid** / C–H bond

b. C–H bond / O–H bond — O is farther to the right in the periodic table. **stronger acid**

2.19 **More electronegative atoms stabilize the conjugate base, making the acid stronger.**
Compare the electron-withdrawing groups on the acids below to decide which is a stronger acid **(more electronegative groups = more acidic).**

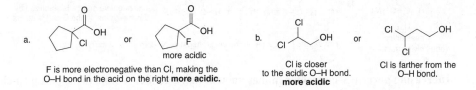

a. F is more electronegative than Cl, making the O–H bond in the acid on the right **more acidic.**

b. Cl is closer to the acidic O–H bond. **more acidic** / Cl is farther from the O–H bond.

Chapter 2–10

c.   or

more acidic
NO$_2$ is electron withdrawing, making the
O–H bond in the acid on the
right **more acidic.**

2.20   **More electronegative groups stabilize the conjugate base, making the acid stronger.**

HOCH$_2$CO$_2$H   CH$_3$CO$_2$H

an α-hydroxy acid   acetic acid
The extra OH group contains an electronegative O, which
stabilizes the conjugate base.
stronger acid

2.21   HBr is a stronger acid than HCl because Br is farther down a column of the periodic table, and
the larger Br$^-$ anion is more stable than the smaller Cl$^-$ anion. In these acids the H is bonded
directly to the halogen. In HOCl and HOBr, the H is bonded to O, and the halogens Cl and Br
exert an inductive effect. In this case, the more electronegative Cl stabilizes $^-$OCl more than the
less electronegative Br stabilizes $^-$OBr. Thus, HOCl forms the more stable conjugate base,
making it the stronger acid.

2.22   The acidity of an acid increases when the conjugate base is resonance stabilized. Compare the
conjugate bases of acetone and propane to explain why acetone is more acidic.

2 resonance structures
more stable conjugate base
Acetone is more acidic.

acetone
pK_a = 19.2

One resonance structure places the (–) charge on the more
electronegative O atom. This is especially good.

CH$_3$CH$_2$CH$_3$ $\xrightarrow{\text{base}}$ CH$_3$CH$_2$ĊH$_2$ only one Lewis structure
less stable conjugate base

propane
pK_a = 50

(Any C–H bond in the starting
material can be removed.)

2.23   The acidity of an acid increases when the conjugate base is resonance stabilized. Acetonitrile
has a resonance-stabilized conjugate base, which accounts for its acidity.

The negative charge is stabilized by
delocalization on the C and N atoms.

acetonitrile
(one Lewis structure)

Having the (–) charge on the electronegative N atom adds stability.

2.24   **Increasing percent s-character makes an acid more acidic.** Compare the percent s-character
of the carbon atoms in each of the C–H bonds in question. A stronger acid has a weaker
conjugate base.

Acids and Bases 2–11

a.

sp hybridized C
50% s-character
more acidic

base

sp³ hybridized C
25% s-character

base

or

**stronger
conjugate base**

b.

sp² hybridized C
33% s-character
more acidic

base

or

sp³ hybridized C
25% s-character

base

or

**stronger
conjugate base**

2.25 To compare the acids, first **look for element effects.** Then identify electron-withdrawing groups, resonance, or hybridization differences.

a.

C is farthest left in
the periodic table.
**CH bond is
least acidic.**

**intermediate
acidity**

O is farthest right in
the periodic table.
**OH bond is
most acidic.**

c.

C is farthest left in
the periodic table.
**CH bond is
least acidic.**

**intermediate
acidity**

O is farthest right in
the periodic table.
**OH bond is
most acidic.**

b.

OH group
least acidic

**intermediate
acidity**

Br is electron withdrawing
and the conjugate base
is resonance stabilized.
most acidic

2.26 Look at the element bonded to the acidic H and decide its acidity based on the periodic trends. **Farther to the right and down the periodic table is more acidic.**

a.

THC
tetrahydrocannabinol
The molecule contains C–H
and O–H bonds.
O is farther right; therefore,
O–H hydrogen is the most acidic.

b.

ketoprofen
The molecule contains C–H
and O–H bonds.
O is farther right; therefore,
O–H hydrogen is the most acidic.

Chapter 2–12

2.27

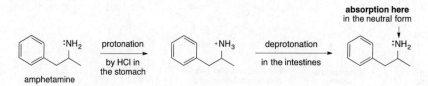

2.28 Draw the products of proton transfer from the acid to the base.

a. $(CH_3)_2CH\ddot{O}-H$ + Na^+ $H:^-$ ⇌ $(CH_3)_2CH\ddot{O}:^-$ Na^+ + H_2
        acid   base   conjugate base   conjugate acid

b. $(CH_3)_2CH\ddot{O}-H$ + $H-OSO_3H$ ⇌ $(CH_3)_2CH\overset{+}{O}H_2$ + HSO_4^-
        base   acid   conjugate acid   conjugate base

c. $(CH_3)_2CH\ddot{O}-H$ + Li^+ $^-\ddot{N}[CH(CH_3)_2]_2$ ⇌ $(CH_3)_2CH\ddot{O}:^-$ Li^+ + $H\ddot{N}[CH(CH_3)_2]_2$
        acid   base   conjugate base   conjugate acid

d. $(CH_3)_2CH\ddot{O}-H$ + $H-OCOCH_3$ ⇌ $(CH_3)_2CH\overset{+}{O}H_2$ + $^-OCOCH_3$
        base   acid   conjugate acid   conjugate base

2.29 To cross a cell membrane, amphetamine must be in its neutral (not ionic) form.

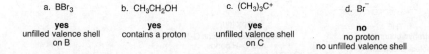

2.30 Lewis bases are electron pair donors: they contain a lone pair or a π bond.

a. $\ddot{N}H_3$ b. $CH_3CH_2CH_3$ c. $H:^-$ d. $H-C\equiv C-H$

yes—has lone pair   **no**—no lone pair or π bond   **yes**—has lone pair   **yes**—has 2 π bonds

2.31 Lewis acids are electron pair acceptors. Most Lewis acids contain a proton or an unfilled valence shell of electrons.

a. BBr_3 b. CH_3CH_2OH c. $(CH_3)_3C^+$ d. Br^-

yes unfilled valence shell on B   **yes** contains a proton   **yes** unfilled valence shell on C   **no** no proton no unfilled valence shell

Acids and Bases 2–13

2.32 Label the Lewis acid and Lewis base and then draw the curved arrows.

a.

Lewis acid
unfilled valence shell
on B

Lewis base
lone pairs
on O

new bond

b.

Lewis acid
unfilled valence
shell on C

Lewis base
lone pairs
on O

2.33 A Lewis acid is also called an **electrophile.** When a Lewis base reacts with an electrophile other than a proton, it is called a **nucleophile.** Label the electrophile and nucleophile in the starting materials and then draw the products.

a.

$+$ BBr_3 $\longrightarrow$

Lewis base
nucleophile
lone pairs
on O

Lewis acid
electrophile
unfilled valence shell
on B

b.

$+$ $AlCl_3$ $\longrightarrow$

Lewis base
nucleophile
lone pairs
on O

Lewis acid
electrophile
unfilled valence shell
on Al

2.34 Draw the product of each reaction by using an electron pair of the Lewis base to form a new bond to the Lewis acid.

a. $CH_3CH_2-\overset{..}{N}-CH_2CH_3$ $+$ $B(CH_3)_3$ $\longrightarrow$
        |
        CH_2CH_3

Lewis base
nucleophile
lone pair
on N

Lewis acid
electrophile
unfilled valence shell
on B

b. $CH_3CH_2-\overset{..}{N}-CH_2CH_3$ $+$ $^+C(CH_3)_3$ $\longrightarrow$
        |
        CH_2CH_3

Lewis base
nucleophile
lone pair
on N

Lewis acid
electrophile
unfilled valence shell
on C

c. $CH_3CH_2-\overset{..}{N}-CH_2CH_3$ $+$ $AlCl_3$ $\longrightarrow$
        |
        CH_2CH_3

Lewis base
nucleophile
lone pair
on N

Lewis acid
electrophile
unfilled valence shell
on Al

Chapter 2–14

2.35 Curved arrows begin at the Lewis base and point towards the Lewis acid.

Lewis base
contains a lone pair

Lewis acid
contains a proton

2.36 a, b. Since acidity increases from left-to-right across a row of the periodic table and propranolol has C–H, N–H, and O–H bonds, the O–H bond is most acidic. NaH is a base that removes the most acidic OH proton.

most acidic

propranolol
skeletal structure

c, d. Of the atoms with lone pairs (N and O), N is to the left in the periodic table, making it the most basic site. HCl is an acid, which protonates the most basic site.

2.37 a, b. Using periodic trends, the N–H bond of amphetamine is most acidic. NaH is a base that removes an acidic proton on N, forming H_2 in the process.

amphetamine
skeletal structure   **most acidic**

c. HCl protonates the lone pair on N (the most basic site).

2.38 To draw the conjugate acid of a Brønsted–Lowry base, **add a proton to the base.**

a. HCO_3^- $\xrightarrow{H^+}$ H_2CO_3

b.

c.

d.

Acids and Bases 2–15

2.39 To draw the conjugate base of a Brønsted–Lowry acid, **remove a proton from the acid.**

a. HCO_3^- $\xrightarrow{-H^+}$ CO_3^{2-}

c. [structure: butanoic acid] $\xrightarrow{-H^+}$ [carboxylate anion]

b. [structure: pentan-2-amine cation with $\overset{+}{N}H_3$] $\xrightarrow{-H^+}$ [pentan-2-amine with NH_2]

d. [cyclohexyl acetylene] $\xrightarrow{-H^+}$ [cyclohexyl acetylide anion]

2.40 To draw the products of an acid–base reaction, transfer a proton from the acid (H_2SO_4 in this case) to the base.

a. [cyclopentanol] $-\overset{..}{\underset{..}{O}}H$ + $H-OSO_3H$ ⟶ [cyclopentyl oxonium $\overset{+}{O}$ with two H] + HSO_4^-

b. [cyclopentyl] $-\overset{..}{N}H_2$ + $H-OSO_3H$ ⟶ [cyclopentyl $-\overset{+}{N}H_3$] + HSO_4^-

c. [cyclopentyl] $-\overset{..}{\underset{..}{O}}CH_3$ + $H-OSO_3H$ ⟶ [cyclopentyl $\overset{+}{O}$ with CH_3 and H] + HSO_4^-

d. [pyrrolidine $\overset{..}{N}-CH_3$] + $H-OSO_3H$ ⟶ [pyrrolidinium $\overset{+}{N}$ with H and CH_3] + HSO_4^-

2.41 To draw the products of an acid–base reaction, transfer a proton from the acid to the base (^-OH in this case).

a. [cyclohexyl] $-\overset{..}{\underset{..}{O}}-H$ + $K^+ \, ^{:}\overset{..}{O}H$ ⟶ [cyclohexyl] $-\overset{..}{\underset{..}{O}}{:}^- \, K^+$ + H_2O

b. [cyclohexyl carboxylic acid with $\overset{..}{O}{:}$ and $\overset{..}{O}-H$] + $K^+ \, ^{:}\overset{..}{O}H$ ⟶ [carboxylate with $\overset{..}{O}{:}$ and $^:\overset{..}{O}{:}^- \, K^+$] + H_2O

c. [cyclohexyl] $-C\equiv C-H$ + $K^+ \, ^{:}\overset{..}{O}H$ ⟶ [cyclohexyl] $-C\equiv C{:}^- \, K^+$ + H_2O

d. CH_3-[benzene ring]$-\overset{..}{\underset{..}{O}}-H$ + $K^+ \, ^{:}\overset{..}{O}H$ ⟶ CH_3-[benzene ring]$-\overset{..}{\underset{..}{O}}{:}^- \, K^+$ + H_2O

2.42 Label the Brønsted–Lowry acid and Brønsted–Lowry base in the starting materials and **transfer a proton from the acid to the base** for the products.

a. [carboxylic acid with $\overset{..}{O}{:}$ and $\overset{..}{O}-H$]
acid
+ $CH_3\overset{..}{\underset{..}{O}}{:}^-$
base
⇌ [carboxylate with $\overset{..}{O}{:}$ and $\overset{..}{O}{:}^-$]
conjugate base
+ $CH_3\overset{..}{O}H$
conjugate acid

Chapter 2–16

b.

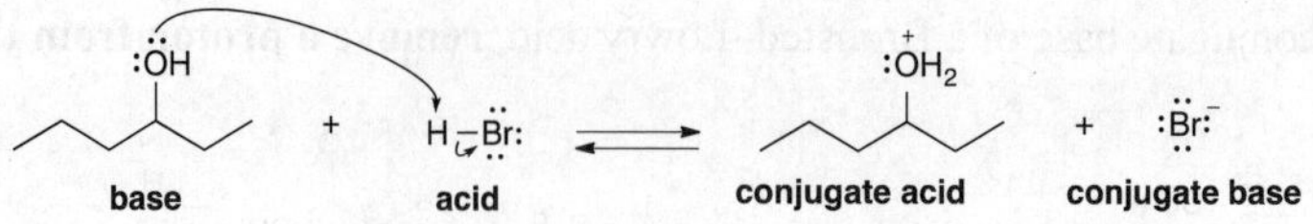

base   acid   conjugate acid   conjugate base

c.

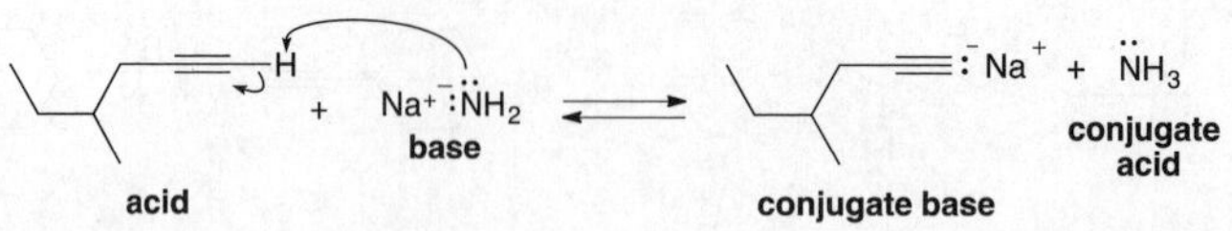

acid   base   conjugate base   conjugate acid

d.

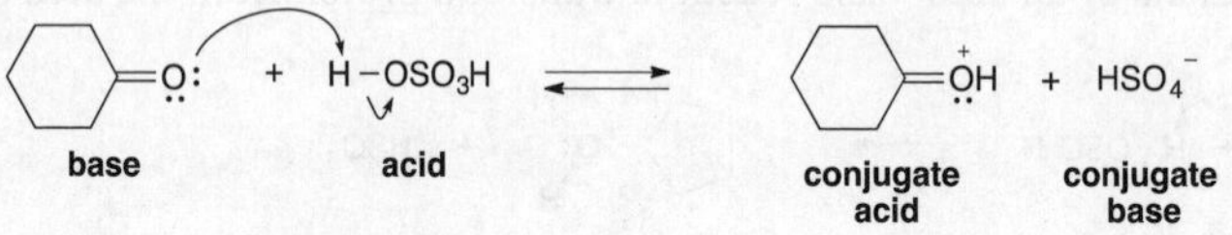

base   acid   conjugate acid   conjugate base

2.43 Draw the products of proton transfer from acid to base.

a.

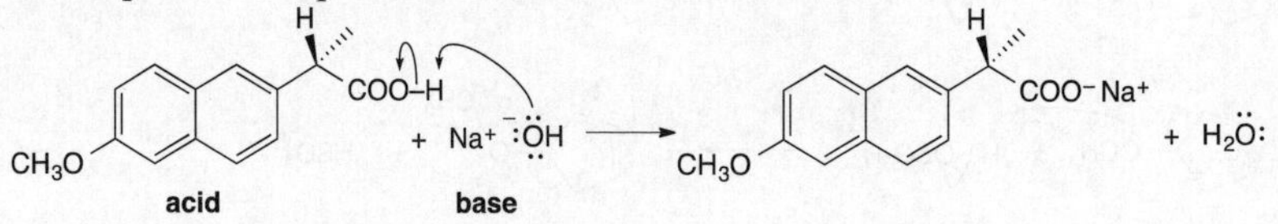

acid   base

b.

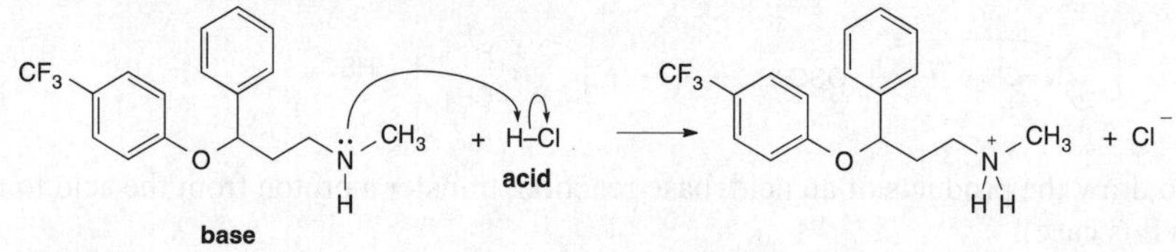

base

2.44 Draw the products of proton transfer from acid to base.

base   acid

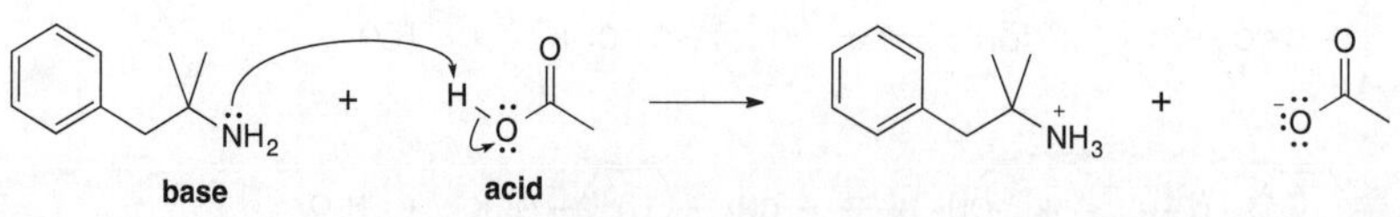

base   acid

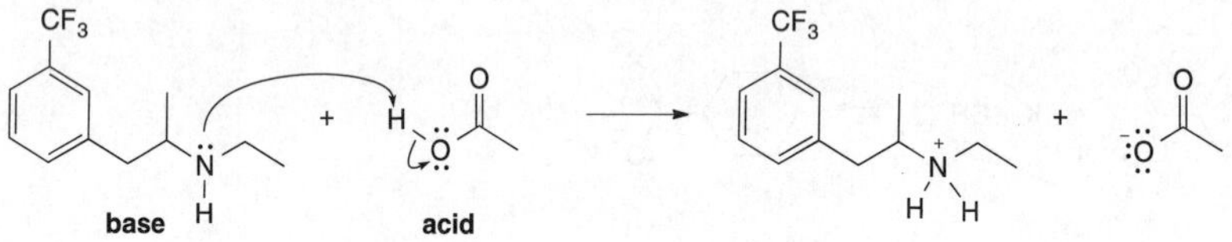

2.45 To convert pK_a to K_a, take the antilog of (−) the pK_a.

a. H_2S
$pK_a = 7.0$
$K_a = 10^{-7}$

b. $ClCH_2COOH$
$pK_a = 2.8$
$K_a = 1.6 \times 10^{-3}$

c. HCN
$pK_a = 9.1$
$K_a = 7.9 \times 10^{-10}$

Acids and Bases 2–17

2.46 To convert from K_a to pK_a, take (–) the log of the K_a; **pK_a = –log K_a.**

a. ⬡—$CH_2\overset{+}{N}H_3$

$K_a = 4.7 \times 10^{-10}$
p$K_a = 9.3$

b. ⬡—$\overset{+}{N}H_3$

$K_a = 2.3 \times 10^{-5}$
p$K_a = 4.6$

c. CF_3COOH

$K_a = 5.9 \times 10^{-1}$
p$K_a = 0.23$

2.47 An acid can be deprotonated by the conjugate base of any acid with a higher pK_a.

$CH_3CH_2CH_2C{\equiv}CH$
pK_a = 25
Any base having a conjugate
acid with a pK_a higher than
25 can deprotonate this acid.

Base	Conjugate acid	pK_a
H_2O	H_3O^+	–1.7
NaOH	H_2O	15.7
$NaNH_2$	NH_3	38
NH_3	NH_4^+	9.4
NaH	H_2	35
CH_3Li	CH_4	50

Only $NaNH_2$, NaH, and CH_3Li
are strong enough to
deprotonate the acid.

2.48 $^-$OH can deprotonate any acid with a pK_a < 15.7.

a. HCOOH

p$K_a = 3.8$
stronger acid
deprotonated

b. H_2S

p$K_a = 7.0$
stronger acid
deprotonated

c. ⬡—CH_3

p$K_a = 41$
weaker acid

d. CH_3NH_2

p$K_a = 40$
weaker acid

These acids are too weak to be
deprotonated by $^-$OH.

2.49 Draw the products and then compare the pK_a of the acid on the left and the conjugate acid on the right. **The equilibrium lies towards the side having the acid with a higher pK_a (weaker acid).**

a. $CH_3\ddot{N}H_2$ + H–OSO_3H ⇌ $CH_3\overset{+}{N}H_3$ + HSO_4^- **products favored**

p$K_a = -9$ p$K_a = 10.7$

b. [acyl with Ö: and Ö–H] + Na^+ :$\ddot{C}l$:$^-$ ⇌ [acyl with Ö: and Ö:$^-$ Na^+] + H$\ddot{C}l$:   **starting material favored**

p$K_a = {\sim}5$ p$K_a = -7$

c. ⬡—$\overset{..}{O}$–H + Na^+ HCO_3^- ⇌ ⬡—$\ddot{O}$:$^-$ Na^+ + H_2CO_3 **starting material favored**

p$K_a = 10$ p$K_a = 6.4$

d. H–C≡C–H + $CH_3CH_2^-Li^+$ ⇌ H–C≡C:$^-$ Li^+ + CH_3CH_3 **products favored**

p$K_a = 25$ p$K_a = 50$

Chapter 2–18

2.50 Compare element effects first and then resonance, hybridization, and electron-withdrawing groups to determine the relative strengths of the acids.

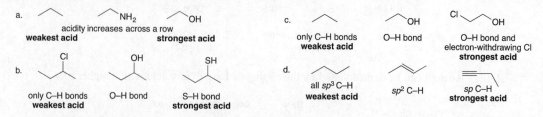

2.51 The strongest acid has the weakest conjugate base.

a. Draw the conjugate acid.
Increasing acidity of conjugate acids:
$CH_3CH_3 < CH_3NH_2 < CH_3OH$

increasing basicity: $CH_3O^- < CH_3\overline{N}H < CH_3\overline{C}H_2$

b. Draw the conjugate acid.
Increasing acidity of conjugate acids:
$CH_4 < H_2O < HBr$

increasing basicity: $Br^- < HO^- < CH_3^-$

c. Draw the conjugate acid.
Increasing acidity of conjugate acids:

[structures: CH₃CH₂OH < CH₃COOH < ClCH₂COOH]

increasing basicity:

[structures: ClCH₂COO⁻ < CH₃COO⁻ < CH₃CH₂O⁻]

d. Draw the conjugate acid.
Increasing acidity of conjugate acids:

[cyclohexyl-CH₂CH₃ < cyclohexyl-CH=CH₂ < cyclohexyl-C≡CH]

increasing basicity:

[cyclohexyl-C≡C:⁻ < cyclohexyl-CH=CH⁻ < cyclohexyl-CH₂CH₂⁻ (with lone pair)]

2.52 More electronegative atoms stabilize the conjugate base by an electron-withdrawing inductive effect, making the acid stronger. Thus, an O atom increases the acidity of an acid.

[piperidinium structure]
$pK_a = 11.1$

[morpholinium structure]
The O atom makes this cation the stronger acid.
$pK_a = 8.33$

2.53

[pentan-2-one structure with H_a and H_b labeled]
pentan-2-one

H_a is more acidic than H_b because loss of H_a forms a resonance-stabilized conjugate base.

[mechanism showing deprotonation by :B giving resonance-stabilized conjugate base with two resonance structures]
resonance-stabilized conjugate base

B: [deprotonation of H_b]
→ [enolate with only one Lewis structure]
only one Lewis structure

Acids and Bases 2–19

2.54

pK_a = 50

pK_a = 43

pK_a = 19.2

The negative charge on O is good. This makes this resonance structure especially good.

conjugate base:

one Lewis structure
weakest acid

two resonance structures
negative charge delocalized
on two carbons

two resonance structures
negative charge delocalized
on one O and one C
strongest acid

2.55 To draw the conjugate acid, look for the most basic site and protonate it. To draw the conjugate base, look for the most acidic site and remove a proton.

conjugate acid

most basic site

A

most acidic proton

conjugate base

2.56 Estimate the pK_a of **B** as 16. A difference of 10^5 in acidity is a difference of 5 pK_a units.

a.

most acidic H
pK_a ~ 5

b.

most acidic
pK_a ~ 25

c.

most acidic
pK_a ~ 16

2.57 Remove the most acidic proton to form the conjugate base. Protonate the most basic electron pair to form the conjugate acid.

a.

only O–H bond
most acidic proton

ibuprofen

conjugate base:

b.

most basic electron pair

Increasing basicity:

conjugate acid:

cocaine

Chapter 2–20

2.58 Compare the isomers.

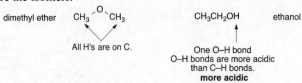

2.59 Compare the Lewis structures of the conjugate bases when each H is removed. The more stable base makes the proton more acidic.

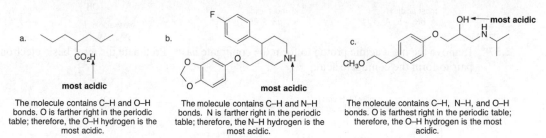

2.60 Look at the element bonded to the acidic H and decide its acidity based on the periodic trends. **Farther to the right across a row and down a column of the periodic table is more acidic.**

a. most acidic (CO₂H)
The molecule contains C–H and O–H bonds. O is farther right in the periodic table; therefore, the O–H hydrogen is the most acidic.

b. most acidic (NH)
The molecule contains C–H and N–H bonds. N is farther right in the periodic table; therefore, the N–H hydrogen is the most acidic.

c. OH ← most acidic
The molecule contains C–H, N–H, and O–H bonds. O is farthest right in the periodic table; therefore, the O–H hydrogen is the most acidic.

2.61 Use element effects, inductive effects, and resonance to determine which protons are the most acidic. The H's of the CH₃ group are least acidic, because they are bonded to an sp^3 hybridized C and the conjugate base formed by their removal is not resonance stabilized.

Acids and Bases 2–21

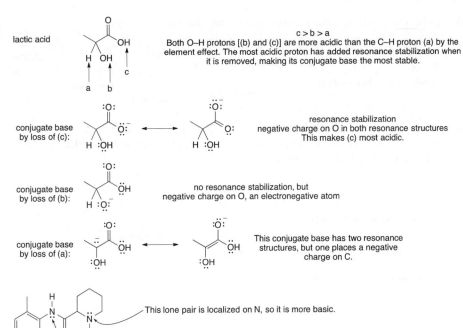

2.62

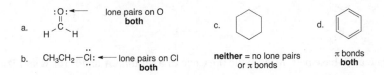

2.63 *Lewis bases* are electron pair donors: they contain a lone pair or a π bond. *Brønsted–Lowry bases* are proton acceptors: to accept a proton they need a lone pair or a π bond. This means Lewis bases are also Brønsted–Lowry bases.

a. H−C(=O:)−H ← lone pairs on O
both

b. CH₃CH₂−Cl: ← lone pairs on Cl
both

c. cyclohexane
neither = no lone pairs or π bonds

d. benzene
π bonds
both

2.64 A *Lewis acid* is an electron pair acceptor and usually contains a proton or an unfilled valence shell of electrons. A *Brønsted–Lowry acid* is a proton donor and must contain a hydrogen atom. All Brønsted–Lowry acids are Lewis acids, though the reverse may not be true.

a. H₃O⁺	b. Cl₃C⁺	c. BCl₃	d. BF₄⁻
both contains a H	**Lewis acid** unfilled valence shell on C	**Lewis acid** unfilled valence shell on B	**neither** no H or unfilled valence shell

Chapter 2–22

2.65 Label the Lewis acid and Lewis base and then draw the products.

a. :Cl⁻ + BCl₃ ⟶ Cl—B⁻—Cl (Cl, Cl)

    Lewis base   Lewis acid

    new bond

b. (structure) + :OH ⟶ (product) new bond

    Cl: / Lewis base

    Lewis acid

2.66 A Lewis acid is also called an **electrophile.** When a Lewis base reacts with an electrophile other than a proton, it is called a **nucleophile.** Label the electrophile and nucleophile in the starting materials and then draw the products.

a. (S structure) + AlCl₃ ⟶ (⁻AlCl₃ / ⁺S structure)

    nucleophile   electrophile

b. (=O structure) + BF₃ ⟶ (⁺=O—BF₃⁻ structure)

    nucleophile   electrophile

c. (cyclohexyl cation) + H₂O ⟶ (cyclohexyl-OH₂⁺)

    electrophile   nucleophile

2.67 Draw the product of each reaction.

a. (carbocation) + CH₃OH ⟶ (product with ⁺O—CH₃, H)

b. (carbocation) + (CH₃)₂O ⟶ (product with ⁺O—CH₃, CH₃)

c. (carbocation) + (CH₃)₂NH ⟶ (product with N⁺—CH₃, H, CH₃)

2.68

A   B   C   D

a. A conjugate acid–base pair differ in the presence of a proton. **A** and **B** (or **B** and **D**) represent a conjugate acid–base pair.

b. **A** and **D** are resonance structures because the atom placement is the same, but the electron pairs are placed differently.

c. **A** and **C** (or **C** and **D**) are constitutional isomers because they have the same molecular formula (C_5H_9NO), but the bonds are different. **A** and **D** have an N–H bond, and **C** has an O–H bond.

2.69

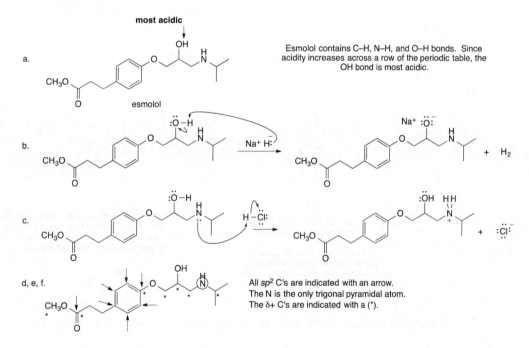

Chapter 2–24

2.72

[1] → no additional stabilization

or

[2] more basic N → resonance-stabilized cation

Path [2] is favored because a resonance-stabilized conjugate acid is formed. The N that is part of the C=N is therefore more basic.

2.73 Draw the product of protonation of either O or N and compare the conjugate acids. When acetamide reacts with an acid, the O atom is protonated because it results in a resonance-stabilized conjugate acid.

protonate O ⟷ resonance stabilization of the + charge O is more readily protonated because the product is resonance stabilized.

acetamide

protonate N → no other resonance structure

2.74

$pK_a = 2.86$

$pK_a = 5.70$

This group destabilizes the second negative charge.

δ+ stabilizes the (–) charge of the conjugate base.

The nearby COOH group serves as an electron-**withdrawing** group to stabilize the negative charge. This makes the first proton **more** acidic than CH_3COOH.

COO^- now acts as an electron-**donor** group which destabilizes the conjugate base, making removal of the second proton more difficult and thus it is **less** acidic than CH_3COOH.

2.75 The COOH group of glycine gives up a proton to the basic NH_2 group to form the zwitterion.

a. acts as a base →

glycine   acts as an acid

proton transfer →

zwitterion form

Acids and Bases 2–25

b. [reaction showing H₃N⁺–CH₂–COO⁻ (most basic site) + H–Cl → H₃N⁺–CH₂–COOH + Cl⁻]

c. [reaction showing zwitterion (most acidic site) + Na⁺ ⁻:ÖH → H₂N–CH₂–COO⁻ + Na⁺ + H₂Ö:]

2.76 Use curved arrows to show how the reaction occurs.

[1] [cyclohexenone with α-H being removed by ⁻:ÖH → enolate resonance structures]

[2] [enolate + H–Ö–H → cyclohexenone + ⁻:ÖH]

Protonate the negative charge on this carbon to form the product.

2.77 Compare the OH bonds in vitamin C and decide which one is the most acidic.

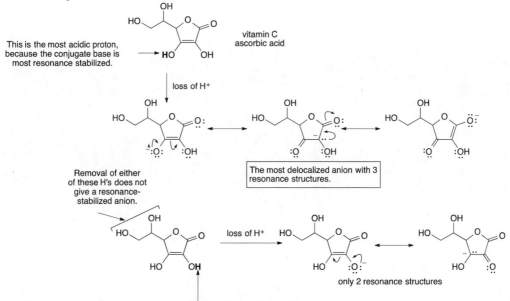

Chapter 2–26

2.78

M

N

only a minor contributor to the
hybrid because of charges

less resonance stabilized than **N**

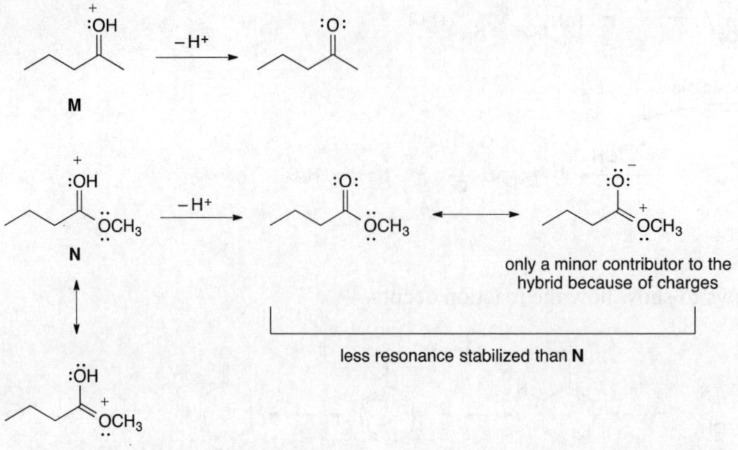

N has two resonance structures with the same number of bonds and charges, so both contribute
approximately equally to the hybrid. This makes **N** more resonance stabilized than its conjugate
base, and less willing to give up a proton than **M,** which has no similar resonance stabilization. Thus
M is a stronger acid than **N.** (Resonance structures that break the C=O bond are not drawn in this
solution, because they are possible for both compounds.)

Introduction to Organic Molecules and Functional Groups 3–1

Chapter 3: Introduction to Organic Molecules and Functional Groups

Chapter Review

Classifying carbon atoms, hydrogen atoms, alcohols, alkyl halides, amines, and amides (3.2)

- Carbon atoms are classified by the number of carbons bonded to them; a 1° carbon is bonded to one other carbon, and so forth.
- Hydrogen atoms are classified by the type of carbon atom to which they are bonded; a 1° hydrogen is bonded to a 1° carbon, and so forth.
- Alkyl halides and alcohols are classified by the type of carbon to which the OH or X group is bonded; a 1° alcohol has an OH group bonded to a 1° carbon, and so forth.
- Amines and amides are classified by the number of carbons bonded to the nitrogen atom; a 1° amine has one carbon–nitrogen bond, and so forth.

Types of intermolecular forces (3.3)

Type of force	Cause	Examples
van der Waals (VDW)	Due to the interaction of temporary dipoles <ul><li>Larger surface area, stronger forces</li><li>Larger, more polarizable atoms, stronger forces</li></ul>	All organic compounds
dipole–dipole (DD)	Due to the interaction of permanent dipoles	$(CH_3)_2C{=}O$, H_2O
hydrogen bonding (HB or H-bonding)	Due to the electrostatic interaction of a H atom in an O–H, N–H, or H–F bond with another N, O, or F atom.	H_2O
ion–ion	Due to the interaction of two ions	NaCl, LiF

Increasing strength (downward)

Physical properties

Property	Observation
Boiling point (3.4A)	• For compounds of comparable molecular weight, the stronger the forces the higher the bp. VDW — MW = 72 — bp = 36 °C VDW, DD — MW = 72 — bp = 76 °C VDW, DD, HB — MW = 74 — bp = 118 °C Increasing strength of intermolecular forces Increasing boiling point

Chapter 3–2

- For compounds with similar functional groups, the larger the surface area, the higher the bp.

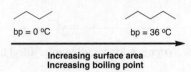

Increasing surface area
Increasing boiling point

- For compounds with similar functional groups, the more polarizable the atoms, the higher the bp.

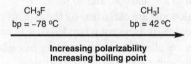

Increasing polarizability
Increasing boiling point

Melting point (3.4B)

- For compounds of comparable molecular weight, the stronger the forces the higher the mp.

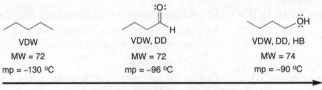

Increasing strength of intermolecular forces
Increasing melting point

- For compounds with similar functional groups, the more symmetrical the compound, the higher the mp.

mp = –160 °C   mp = –17 °C

Increasing symmetry
Increasing melting point

Introduction to Organic Molecules and Functional Groups 3–3

Solubility (3.4C)	Types of water-soluble compounds: • Ionic compounds • Organic compounds having ≤ 5 C's, and an O or N atom for hydrogen bonding (for a compound with one functional group) Types of compounds soluble in organic solvents: • Organic compounds regardless of size or functional group • Examples:
	Key: VDW = van der Waals, DD = dipole–dipole, HB = hydrogen bonding MW = molecular weight

butane — CCl₄ → soluble; H₂O → insoluble

acetone (:O:) — CCl₄ → soluble; H₂O → soluble

Reactivity (3.8)

- **Nucleophiles react with electrophiles.**
- Electronegative heteroatoms create electrophilic carbon atoms that react with nucleophiles.
- Lone pairs and π bonds are nucleophilic sites that react with electrophiles.

CH_3CH_2-Cl ($\delta+$) ⬡$-OH$ ($\delta+$) — **electrophilic site**

$CH_3-\overset{..}{\underset{..}{O}}-CH_3$ $CH_3-\overset{CH_3}{\underset{|}{N}}-CH_3$ — **basic and nucleophilic site**

Practice Test on Chapter Review

1.a. Which of the following compounds exhibits dipole–dipole interactions?

1. CH_2Cl_2
2. $(CH_3)_2O$
3. $CH_3CH_2CH_2-OH$

4. Compounds (1) and (2) both exhibit dipole–dipole interactions.

5. Compounds (1), (2), and (3) all exhibit dipole–dipole interactions.

b. Which of the following compounds can hydrogen bond both to another molecule of itself and to water?

1. (dioxolane with $-OCH_3$ group)

2. (ring with NH)

3. $CH_3C\equiv N$

4. Compounds (1) and (2) can each hydrogen bond to another molecule like itself and water.

5. Each of the three compounds, (1), (2), and (3), can hydrogen bond to another molecule like itself and to water.

Chapter 3–4

c. Which of the following compounds has the highest boiling point?

1. CH₃CH₂CH₂OCH₂CH₃
2. CH₃(CH₂)₄CH₃
3. CH₃(CH₂)₄NH₂
4. (CH₃)₂CHCH(CH₃)NH₂
5. (CH₃)₂CHCH(CH₃)₂

d. Which statement(s) is (are) true about the following compounds?

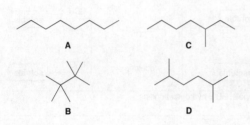

1. The boiling point of **A** is higher than the boiling point of **B**.
2. The melting point of **B** is higher than the melting point of **C**.
3. The boiling point of **A** is higher than the boiling point of **D**.
4. Statements (1) and (2) are both true.
5. Statements (1), (2), and (3) are all true.

2. Consider the anti-hypertensive agent atenolol drawn below.

a. What is the hybridization of the N atom labeled with **A**?
b. What is the shape around the C atom labeled with **F**?
c. What orbitals are used to form the bond labeled with **E**?
d. Which bond (**B**, **C**, or **D**) is the longest bond?
e. Would you predict atenolol to be soluble in water?
f. Which of the labeled H atoms (H_a, H_b, or H_c) is most acidic?

Answers to Practice Test

1. a. 5
   b. 2
   c. 3
   d. 5

2. a. sp^3
   b. trigonal planar
   c. C_{sp2}–C_{sp3}
   d. **D**
   e. yes
   f. H_b

Introduction to Organic Molecules and Functional Groups 3–5

Answers to Problems

3.1

$$CH_3CH_2-\overset{\cdot\cdot}{\underset{\cdot\cdot}{O}}H \xrightarrow{\text{H}-OSO_3H} CH_3CH_2-\overset{+}{\underset{\cdot\cdot}{O}}H_2 \quad + \quad HSO_4^- \qquad CH_3CH_3 \xrightarrow{H_2SO_4} \text{no reaction}$$

$$CH_3CH_2-\overset{\cdot\cdot}{\underset{\cdot\cdot}{O}}H \xrightarrow{Na^+ H:^-} CH_3CH_2-\overset{\cdot\cdot}{\underset{\cdot\cdot}{O}}:^- Na^+ \quad + \quad H_2 \qquad CH_3CH_3 \xrightarrow{NaH} \text{no reaction}$$

3.2 To classify a carbon atom as 1°, 2°, 3°, or 4°, **determine how many carbon atoms it is bonded to (1° C = bonded to one other C, 2° C = bonded to two other C's, 3° C = bonded to three other C's, 4° C = bonded to four other C's).** Re-draw if necessary to see each carbon clearly.

To classify a hydrogen atom as 1°, 2°, or 3°, **determine if it is bonded to a 1°, 2°, or 3° C (a 1° H is bonded to a 1° C; a 2° H is bonded to a 2° C; a 3° H is bonded to a 3° C).** Re-draw if necessary.

a. [1] [2] [3] [4]

b. [1] $CH_3CH_2CH_2CH_3$ [2] [3] $CH_3-\overset{CH_3}{\underset{CH_3}{\overset{|}{C}}}-\overset{CH_3}{\underset{CH_3}{\overset{|}{C}}}-CH_3$ [4]

3.3 Use the definition of 1°, 2°, 3°, or 4° carbon atoms from Answer 3.2.

Chapter 3–6

3.4 A 1° ROH or RX has the functional group bonded to a 1° C; a 2° ROH or RX has the functional group bonded to a 2° C; a 3° ROH or RX has the functional group bonded to a 3° C.

a. [structure with OH, HH, labeled 1°]   b. [cyclohexane with F, labeled 3°]   c. [structure with Br, H, labeled 2°]   d. [cyclopentane with H, OH, labeled 2°]

3.5 Use the definitions from Answer 3.4.

[structure of dexamethasone with labels: 2°, OH ← 1°, OH ← 3°, HO, F, H, 3°, O]

dexamethasone

3.6 Amines are classified as 1°, 2°, or 3° by the number of alkyl groups bonded to the *nitrogen* atom.

a. [structure with labels: 2° amine, 1° amine, 2° amine, NH₂, H₂N, 1° amine]   b. [structure with labels: C₆H₅, N, 3° amine, O]

3.7

a. [structure: HO, NH₂, labeled 3° alcohol, 1° amine]   b. [structure with labels: 3° amine, N, OH, 1° alcohol]

3.8 A 1° amide has one carbon–nitrogen bond; a 2° amide has two carbon–nitrogen bonds; a 3° amide has three carbon–nitrogen bonds.

[large peptide-like structure with labels: 2°, 3°, 2°, 3°, 3°, with rings including pyrrolidine, phenyl, and thiazole]

Introduction to Organic Molecules and Functional Groups 3–7

3.9 Identify the functional groups based on Tables 3.1, 3.2, and 3.3.

shikimic acid: alcohols (hydroxy groups), alkene (double bond), carboxylic acid

oseltamivir: ether, alkene (double bond), ester, amide, amine

3.10 One possible structure for each functional group:

a. aldehyde = R–C(=O)–H → CH₃CH₂CH₂CHO

b. ketone = R–C(=O)–R → CH₃CH₂C(=O)CH₃

c. carboxylic acid = R–C(=O)–OH → CH₃CH₂CH₂CO₂H

d. ester = R–C(=O)–O–R → CH₃C(=O)OCH₃

3.11 One possible structure for each description:

a. C₅H₁₀O — aldehyde, ketone

b. C₆H₁₀O — ketone with alkene (two structures shown)

3.12 Summary of forces:
- **All compounds exhibit van der Waals forces (VDW).**
- **Polar molecules have dipole–dipole forces (DD).**
- **Hydrogen bonding (H-bonding)** can occur only when a **H is bonded to an O, N, or F.**

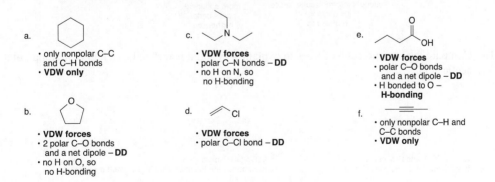

a. cyclohexane
- only nonpolar C–C and C–H bonds
- **VDW only**

b. tetrahydrofuran
- **VDW forces**
- 2 polar C–O bonds and a net dipole – **DD**
- no H on O, so no H-bonding

c. triethylamine
- **VDW forces**
- polar C–N bonds – **DD**
- no H on N, so no H-bonding

d. vinyl chloride
- **VDW forces**
- polar C–Cl bond – **DD**

e. carboxylic acid
- **VDW forces**
- polar C–O bonds and a net dipole – **DD**
- H bonded to O – **H-bonding**

f. alkyne
- only nonpolar C–H and C–C bonds
- **VDW only**

3.13 One principle governs boiling point:
- **Stronger intermolecular forces = higher bp.**
  Increasing intermolecular forces: van der Waals < dipole–dipole < hydrogen bonding

Two factors affect the strength of van der Waals forces, and thus affect bp:
- **Increasing surface area = increasing bp.**
  Longer molecules have a larger surface area. Any branching decreases the surface area of a molecule.
- **Increasing polarizability = increasing bp.**

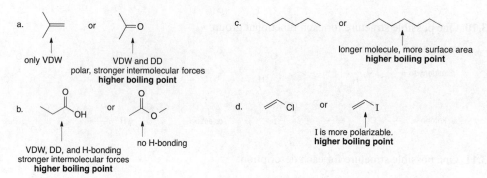

3.14 Increasing intermolecular forces: van der Waals < dipole–dipole < hydrogen bonding

CH₃CH₂–C(=O)–NH₂
N–H bonds allow for hydrogen bonding.
stronger intermolecular forces
higher boiling point

H–C(=O)–N(CH₃)(CH₃)
no hydrogen bonding
weaker intermolecular forces

3.15

a. (pentane) or (propyl-NH₂)
more polar
stronger intermolecular forces
(H-bonding)
higher mp

b. (neopentane-like) or (2-methylhexane)
more spherical
packs better
higher mp

3.16 Compare the intermolecular forces to explain why sodium acetate has a higher melting point than acetic acid.

CH₃–C(=O)–OH
acetic acid

CH₃–C(=O)–O⁻ Na⁺
sodium acetate

a. VDW, DD, and H-bonding
b. not ionic, lower melting point

a. VDW, DD, ionic bonds
b. Ionic bonds are the strongest: **higher melting point.**

Introduction to Organic Molecules and Functional Groups 3–9

3.17 A compound is water soluble if it is ionic or if it has an O or N atom and ≤ 5 C's.

a. an O atom that can H-bond with water
≤ 5 C's
water soluble

b. nonpolar
not water soluble

c. an N atom that can H-bond to H₂O, but > 5 C's
not water soluble

3.18 Hydrophobic portions will primarily be hydrocarbon chains. **Hydrophilic** portions will be polar. Circled regions are **hydrophilic** because they are polar.
All other regions are **hydrophobic** because they have only C and H.

C norethindrone — can H-bond to itself (OH)

D arachidonic acid — can H-bond to H₂O, can H-bond to H₂O

3.19 Like dissolves like.
- To be **soluble in water**, a molecule must be ionic, or have a polar functional group capable of H-bonding for every 5 C's.
- Organic compounds are generally **soluble in organic solvents** regardless of size or functional group.

a. vitamin B₃ (niacin) — polar, polar
soluble in water due to two polar functional groups and only 6 C's in the molecule

b. vitamin K₁ (phylloquinone) — nonpolar, nonpolar long hydrocarbon chain
soluble in organic solvents two polar C–O bonds but the compound has > 10 C's
water insoluble

3.20 a.

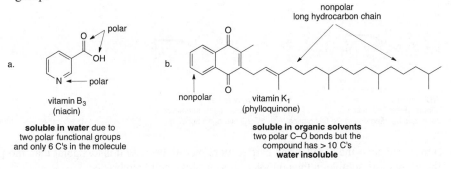

alcohol — OH, HO, amide, carboxylic acid

b. The amide, carboxylic acid, and both alcohols can all hydrogen bond with water.
c. Since pantothenic acid has only nine carbons with four functional groups that can hydrogen bond, pantothenic acid is a water-soluble vitamin.

Chapter 3–10

3.21 A soap contains both a long hydrocarbon chain and a carboxylic acid salt.

a.

short chain
carboxylic acid salt

c.

no salt

ionic salt

b.

long chain
This is a soap because it contains both a
long chain and a carboxylic acid salt.

3.22 Detergents have a polar head consisting of oppositely charged ions, and a nonpolar tail consisting of C–C and C–H bonds, just like soaps do. Detergents clean by having the **hydrophobic ends of molecules surround grease**, while the **hydrophilic portion of the molecule interacts with the polar solvent** (usually water).

a detergent

SO_3^- Na$^+$

nonpolar tail
hydrophobic
This end interacts with the
grease to dissolve it.

polar head
ionic – hydrophilic
This end interacts with the water solvent
to maintain the micelle's solubility in water.

3.23

a.

morphine
VDW forces
dipole–dipole interactions
hydrogen bonding due to the OH groups

heroin
VDW forces
dipole–dipole interactions
NO hydrogen bonding

b. Heroin can cross the blood–brain barrier more readily because it is less polar than morphine, and therefore more soluble in the nonpolar interior of the cell membrane.

3.24 The noble gas xenon consists of uncharged atoms that exhibit only van der Waals interactions, so it is very soluble in the nonpolar interior of the cell membrane. It can thus cross the blood–brain barrier and act as an anesthetic.

3.25 Because the interior of a cell membrane is nonpolar, aspirin crosses a cell membrane as a neutral carboxylic acid, by the general rule that "like dissolves like."

Introduction to Organic Molecules and Functional Groups 3–11

3.26 Electronegative heteroatoms like N, O, or X make a carbon atom an *electrophile*.
A lone pair on a heteroatom makes it basic and nucleophilic.
Pi (π) bonds create *nucleophilic* sites and are more easily broken than σ bonds.

a. C bonded to Br **electrophilic**

b. **nucleophilic** C's bonded to S are **electrophilic.**

c. **nucleophilic**
All lone pairs—**nucleophilic** sites.
All atoms labeled with δ+ are **electrophilic** sites.

3.27 Electrophiles and nucleophiles react with each other.

a. electrophile + nucleophile ⟶ **YES**

b. nucleophile + nucleophile ⟶ **NO**

c. electrophile + nucleophile ⟶ **YES**

d. nucleophile + electrophile ⟶ **YES**

3.28

a. amine, amide, aromatic ring, ester, carboxylic acid

b, c. H's that are boxed in can hydrogen bond to O of H_2O.
Atoms labeled with (*) can hydrogen bond to H of H_2O.

3.29

a, c. alkene, aromatic ring, three ethers
Electrophilic carbons are labeled with (*).

b. Three OH's allow for H-bonding.
Stronger intermolecular forces mean a higher bp and mp.

Chapter 3–12

3.30

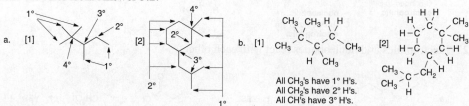

a, c. labels: alkene, aldehyde
The most electrophilic C is labeled with *.

b. This isomer has an OH, giving more opportunities for H-bonding (through both O and H atoms), and probably making it more H_2O soluble.

3.31 Use the rules from Answer 3.2.

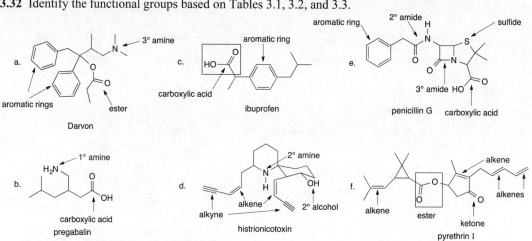

b. [1] All CH_3's have 1° H's. All CH_2's have 2° H's. All CH's have 3° H's.

3.32 Identify the functional groups based on Tables 3.1, 3.2, and 3.3.

a. Darvon — aromatic rings, 3° amine, ester
b. pregabalin — 1° amine, carboxylic acid
c. ibuprofen — carboxylic acid, aromatic ring
d. histrionicotoxin — 2° amine, alkene, alkyne, 2° alcohol
e. penicillin G — aromatic ring, 2° amide, sulfide, 3° amide, carboxylic acid
f. pyrethrin I — alkene, alkenes, ester, ketone, alkene

3.33 A cyclic ester is called a lactone. A cyclic amide is called a lactam.

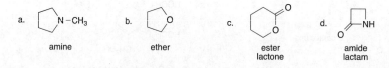

a. amine
b. ether
c. ester / lactone
d. amide / lactam

Introduction to Organic Molecules and Functional Groups 3–13

3.34

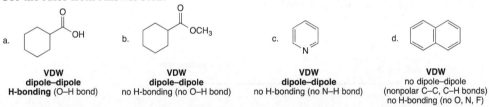

3.35 Draw the constitutional isomers and identify the functional groups.

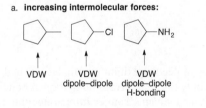

3.36 Use the rules from Answer 3.12.

a.
**VDW
dipole–dipole
H-bonding (O–H bond)**

b.
**VDW
dipole–dipole
no H-bonding (no O–H bond)**

c.
**VDW
dipole–dipole
no H-bonding (no N–H bond)**

d.
**VDW
no dipole–dipole
(nonpolar C–C, C–H bonds)
no H-bonding (no O, N, F)**

3.37 Increasing intermolecular forces: van der Waals < dipole–dipole < H-bonding

a. **increasing intermolecular forces:**

VDW   VDW   VDW
         dipole–dipole dipole–dipole
                        H-bonding

b. **increasing intermolecular forces:**

VDW   VDW   VDW
         dipole–dipole dipole–dipole
                        H-bonding

3.38

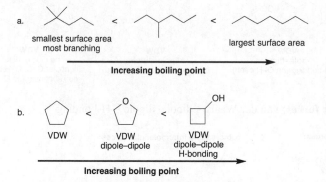

indinavir

Indinavir can hydrogen bond to another molecule of itself at boxed-in sites. Indinavir can hydrogen bond to water at both the boxed-in sites and the sites labeled with an arrow.

3.39 **A** = VDW forces; **B** = H-bonding; **C** = ion–ion interactions; **D** = H-bonding; **E** = H-bonding; **F** = VDW forces.

3.40

a.

aldehyde   ketone   ether   alcohol

b. The alcohol is the highest boiling, because it is the only isomer that can hydrogen bond with another molecule of itself.

3.41 Use the principles from Answer 3.13.

a.

smallest surface area
most branching   largest surface area

Increasing boiling point

b.

```
VDW    <    VDW          <    VDW
              dipole–dipole      dipole–dipole
                                 H-bonding
```

Increasing boiling point

3.42 In CH₃CH₂NHCH₃, there is a N–H bond, so the molecules exhibit intermolecular hydrogen bonding, whereas in (CH₃)₃N the N is bonded only to C, so there is no hydrogen bonding. The hydrogen bonding in CH₃CH₂NHCH₃ makes it have much **stronger intermolecular forces** than (CH₃)₃N. As intermolecular forces increase, the boiling point of a molecule of the same molecular weight increases.

Introduction to Organic Molecules and Functional Groups 3–15

3.43   Stronger forces, higher mp.

menthone
VDW
dipole–dipole
lower melting point

menthol
VDW
dipole–dipole
H-bonding
stronger forces
higher melting point

3.44   Stronger forces, higher mp.

VDW

VDW
DD

VDW
DD
H-bonding

Increasing intermolecular forces

Increasing melting point

3.45   Boiling point is determined solely by the strength of the intermolecular forces. Because benzene has a smaller size, it has less surface area and weaker VDW interactions and therefore a lower boiling point than toluene. The increased melting point for benzene can be explained by symmetry: benzene is much more symmetrical than toluene. More symmetrical molecules can pack more tightly together, increasing their melting point. Symmetry has no effect on boiling point.

benzene
bp = 80 °C
mp = 5 °C

and

toluene
bp = 111 °C
mp = –93 °C

very symmetrical
closer packing in solid form
higher mp

less symmetrical
lower mp

3.46 Increasing polarity = increasing water solubility.

polar
no H-bonding

polar
H-bonding to H_2O,
not itself

polar and
H-bonding
More opportunities
for H-bonding with its
O atom and its H on O.

Chapter 3–16

3.47 Look for two things:
- To H-bond to another molecule of itself, the molecule must contain a **H bonded to O, N, or F.**
- To H-bond with water, a molecule needs **only to contain an O, N, or F.**

Only (c) can H-bond to another molecule like itself.
c. $CH_3CH_2CONH_2$

These molecules can H-bond with water. All of these molecules have an O or N atom.
b. $(CH_3CH_2)_3N$
c. $CH_3CH_2CONH_2$
d. $CH_3CH_2COOCH_3$

3.48 Draw the molecules in question and look at the intermolecular forces involved.

no H bonded to O

diethyl ether

H bonded to O: hydrogen bonding

butan-1-ol

VDW forces
dipole–dipole forces

VDW forces
dipole–dipole forces
H-bonding

- Both have $\leq$ 5 C's and an electronegative O atom, so they can H-bond to water, making them soluble in water.
- Only butan-1-ol can H-bond to another molecule of itself, and this increases its boiling point.

3.49 Use the solubility rule from Answer 3.19.

a.

caffeine
many polar bonds with N and O atoms
many opportunities for H-bonding
water soluble

c.

sucrose
many polar bonds with O
11 O's and 12 C's
many opportunities for H-bonding with H_2O
water soluble

b.

mestranol
CH_3O
2 polar functional groups
but > 10 C's
not water soluble

d.

carotatoxin
1 polar functional group
but > 10 C's
not water soluble

Introduction to Organic Molecules and Functional Groups 3–17

3.50 Water solubility is determined by polarity. Polar molecules are soluble in water, while nonpolar molecules are soluble in organic solvents.

Arrows indicate polar functional groups.

a.

vitamin E
only 2 polar functional groups
many nonpolar C–C and C–H bonds (29 C's)
soluble in organic solvents
insoluble in H$_2$O

b.

pyridoxine
vitamin B$_6$
many polar bonds and few nonpolar bonds
soluble in H$_2$O
It is also **soluble in organic solvents** because it
is organic, but is probably more soluble in H$_2$O.

3.51 Compare the functional groups in the two components of sunscreen. Dioxybenzone will most likely be washed off in water because it contains two hydroxy groups and is more water soluble.

avobenzone
two ketones
one ether

dioxybenzone
two hydroxy groups
one ketone
one ether
more water soluble

3.52 Because of the O atoms, PEG is capable of hydrogen bonding with water, which makes PEG water soluble and suitable for a product like shampoo. PVC cannot hydrogen bond to water, so PVC is water insoluble, even though it has many polar bonds. Because PVC is water insoluble, it can be used to transport and hold water.

H-bond →

poly(ethylene glycol)
PEG
water soluble

no H-bonding

poly(vinyl chloride)
PVC
water insoluble

3.53 Molecules that dissolve in water are readily excreted from the body in urine, whereas less polar molecules that dissolve in organic solvents are soluble in fatty tissue and are retained for longer periods. Compare the solubility properties of THC and ethanol to determine why drug screenings can detect THC and not ethanol weeks after introduction to the body.

Chapter 3–18

tetrahydrocannabinol
THC

THC has relatively few polar
bonds compared to the number
of nonpolar bonds, making it
soluble in organic solvents
and therefore **soluble in fatty tissue.**

ethanol

Ethanol has 1 O atom and
only 2 C's, making it
soluble in water.

Due to their solubilities, **THC is retained much longer in the fatty tissue of the body,** being slowly excreted over many weeks, while ethanol is excreted rapidly in urine after ingestion.

3.54 Compare the intermolecular forces of crack and cocaine hydrochloride. Stronger intermolecular forces increase both the boiling point and the water solubility.

ionic bond

cocaine (crack)
neutral organic molecule

cocaine hydrochloride
a salt

The molecules are identical except for the ionic bond in cocaine hydrochloride. Ionic forces are extremely strong forces, and therefore the cocaine hydrochloride salt has a much **higher boiling point and is more water soluble.** Because the salt is highly water soluble, it can be injected directly into the bloodstream, where it dissolves. Crack is smoked because it can dissolve in the organic tissues of the nasal passages and lungs.

3.55

a.

HCl

b.

five functional groups that have
many opportunities for H-bonding
water soluble

ionic salt
more water soluble

c. Because the hydrochloride salt is ionic and therefore more water soluble, it is more readily transported in the bloodstream.

Introduction to Organic Molecules and Functional Groups 3–19

3.56 Use the rules from Answer 3.26.

a.

nucleophilic

$\delta+$ I : $\delta-$

electrophilic

b.

nucleophilic

$\delta+$ | $\delta+$
O

electrophilic

c.

All the C=C's are
nucleophilic.

d.

:O: ← nucleophilic
‖
CH$_3$ — C — Cl:
$\delta+$

electrophilic

(All lone pairs on O and
Cl are nucleophilic.)

3.57

a.

+ Br⁻ ⟶ **NO**

nucleophilic

nucleophilic

c.

+ ⁻OH ⟶ **NO**

nucleophilic

nucleophilic

b.

Cl + ⁻CN ⟶ **YES**

nucleophilic

electrophilic

d.

+ H$_3$O⁺ ⟶ **YES**

nucleophilic electrophilic

3.58 More rigid cell membranes have phospholipids with *fewer* C=C's. Each C=C introduces a bend in the molecule, making the phospholipids pack less tightly. Phospholipids without C=C's can pack very tightly, making the membrane less fluid and more rigid.

The double bonds introduce kinks in the chain,
making packing of the hydrocarbon chains less
efficient. This makes the cell membrane formed
from them more fluid.

3.59 **B** is ionic, so it does not "dissolve" in the nonpolar interior of a cell membrane and it cannot cross the blood–brain barrier. **A** is a neutral organic molecule with polar bonds, so it can enter the nonpolar interior of the cell membrane and cross the blood–brain barrier.

3.60

a, b.

2° amine

ester

carboxylic acid

3° amide

aromatic ring

aromatic ring

quinapril

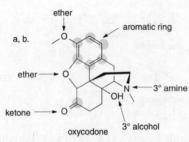

c. Quinapril can hydrogen bond to water at all N and O atoms (boxed in).
d. Quinapril can hydrogen bond to acetone at its O–H and N–H bonds (shown with arrows).
e. The most acidic H is the lone H bonded to O, which is part of a carboxylic acid.
f. The most basic site is the N atom of the amine (circled).

3.61

c. The lone OH proton is the most acidic site.
d. The lone N atom is the most basic site.
e. The N atom is surrounded by three atoms and a lone pair (four groups), so it is sp^3 hybridized.
f. Oxycodone contains seven sp^2 hybridized C's labeled with gray circles.

3.62 Because the O atom in tetrahydrofuran is in a ring, the C atoms bonded to it are kept away from the lone pairs on O. This allows the O atom to more readily hydrogen bond with water, thus increasing its solubility in water.

Introduction to Organic Molecules and Functional Groups 3–21

3.63

a.

amide   amine

aromatic ring

aromatic ring

e. An isomer that can hydrogen bond should
have a higher boiling point.

hydrogen bonding possible

b, c, f, g.

most basic atom

The N atom of the amide is less basic because the lone
pair is part of resonance with the C=O.

Atoms that can hydrogen bond to H_2O are boxed in.
Electrophilic carbons are labeled with (*).

most acidic H, $pK_a \sim 25$
All H's are bonded to C's. Removal of the H on the C adjacent to the C=O results in a resonance-
stabilized anion.

d. Fentanyl exhibits van der Waals and dipole–dipole interactions but no
hydrogen bonding, because there is no H bonded to O or N.

3.64

A

B

The OH and CHO groups are close enough that they can
intramolecularly H-bond to each other. Because the two
polar functional groups are involved in intramolecular H-
bonding, they are less available for H-bonding to H_2O.
This makes **A** less H_2O soluble than **B**, whose two
functional groups are both available for H-bonding to the
H_2O solvent.

The OH and the CHO are too far
apart to intramolecularly H-bond
to each other, leaving more
opportunity to H-bond with
solvent.

3.65

a. melting point

fumaric acid

Fumaric acid has its two larger COOH groups on opposite
ends of the molecule, and in this way it can pack better
in a lattice than maleic acid, giving it a **higher mp.**

b. solubility

maleic acid

Maleic acid is more polar, giving it
greater **H_2O solubility.** The bond dipoles in
fumaric acid cancel.

Chapter 3–22

c. removal of the first proton (pK_{a1})

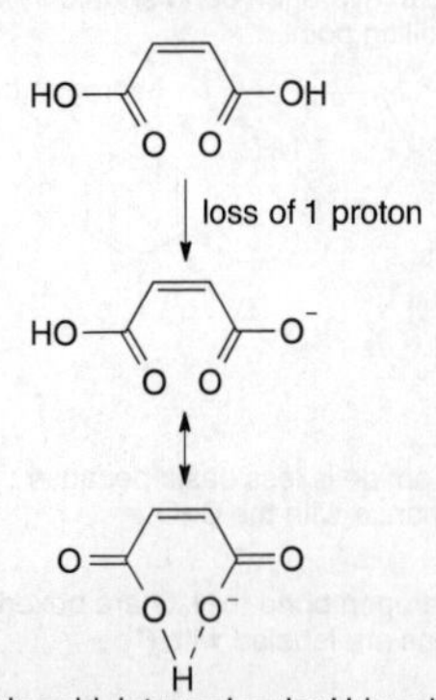

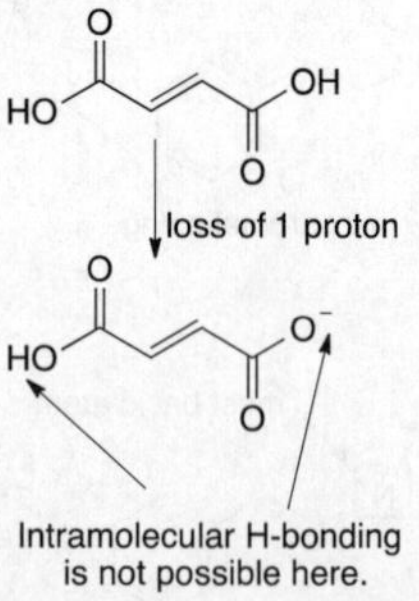

loss of 1 proton

loss of 1 proton

Intramolecular H-bonding
is not possible here.

In maleic acid, intramolecular H-bonding
stabilizes the conjugate base after one H is
removed, making maleic acid more acidic
than fumaric acid.

d. removal of the second proton (pK_{a2})

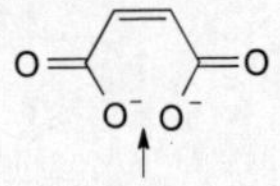

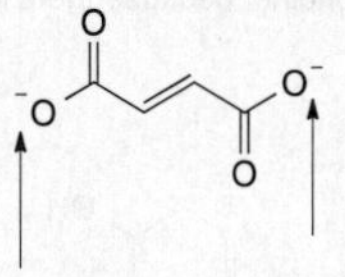

Now the dianion is held in close proximity
in maleic acid, and this destabilizes the conjugate
base. Thus, removing the second H in maleic
acid is harder, making it a weaker acid than
fumaric acid for removal of the second proton.

The two negative charges are much farther
apart. This makes the dianion from fumaric
acid more stable and thus pK_{a2} is lower for
fumaric acid than maleic acid.

Alkanes 4–1

Chapter 4: Alkanes

Chapter Review

General facts about alkanes (4.1–4.3)

- Alkanes are composed of **tetrahedral**, sp^3 hybridized C's.
- There are two types of alkanes: acyclic alkanes having molecular formula C_nH_{2n+2}, and cycloalkanes having molecular formula C_nH_{2n}.
- Alkanes have only **nonpolar C–C and C–H bonds** and no functional group, so they undergo few reactions.
- Alkanes are named with the suffix *-ane*.

Names of alkyl groups (4.4A)

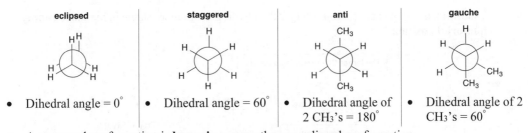

Conformations in acyclic alkanes (4.9, 4.10)

- Alkane conformations can be classified as **staggered, eclipsed, anti,** or **gauche** depending on the relative orientation of the groups on adjacent carbons.

eclipsed	staggered	anti	gauche
Dihedral angle = 0°	Dihedral angle = 60°	Dihedral angle of 2 CH₃'s = 180°	Dihedral angle of 2 CH₃'s = 60°

- A staggered conformation is **lower in energy** than an eclipsed conformation.
- An anti conformation is **lower in energy** than a gauche conformation.

Chapter 4–2

Types of strain

- **Torsional strain**—an increase in energy due to eclipsing interactions (4.9).
- **Steric strain**—an increase in energy when atoms are forced too close to each other (4.10).
- **Angle strain**—an increase in energy when tetrahedral bond angles deviate from 109.5° (4.11).

Two types of isomers

[1] **Constitutional isomers**—isomers that differ in the way the atoms are connected to each other (4.1A).
[2] **Stereoisomers**—isomers that differ only in the way atoms are oriented in space (4.13B).

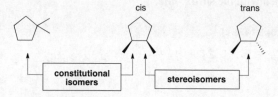

Conformations in cyclohexane (4.12, 4.13)

- Cyclohexane exists as **two chair conformations** in rapid equilibrium at room temperature.
- Each carbon atom on a cyclohexane ring has **one axial** and **one equatorial hydrogen**. Ring-flipping converts axial H's to equatorial H's, and vice versa.

- In substituted cyclohexanes, groups larger than hydrogen are more stable in the **more roomy equatorial position**.

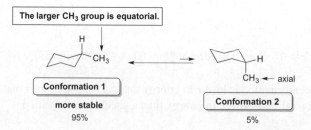

Alkanes 4–3

- Disubstituted cyclohexanes with substituents on different atoms exist as two possible stereoisomers.
  - The **cis** isomer has two groups on the **same side** of the ring, either both up or both down.
  - The **trans** isomer has two groups on **opposite sides** of the ring, one up and one down.

Oxidation–reduction reactions (4.14)

- **Oxidation** results in an **increase in the number of C–Z bonds** or a **decrease in the number of C–H bonds.**

  CH₃CH₂—OH → CH₃COOH
  ethanol   acetic acid

  Increase in C–O bonds = **oxidation**

- **Reduction** results in a **decrease in the number of C–Z bonds** or an **increase in the number of C–H bonds.**

  ethylene → ethane

  Increase in C–H bonds = **reduction**

Practice Test on Chapter Review

1.a. Which statement is true about compounds **A–D** below?

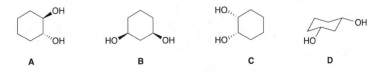

1. **A** and **C** are stereoisomers.
2. **B** and **D** are identical.
3. **A** and **B** are stereoisomers.
4. Statements (1) and (2) are both true.
5. Statements (1), (2), and (3) are all true.

b. Which of the following statements is true about *cis*-1-isopropyl-2-methylcyclohexane?
  1. The more stable conformation has the isopropyl group in the equatorial position and the methyl group in the axial position.
  2. The more stable conformation has both the methyl and isopropyl groups in the equatorial position.
  3. *cis*-1-Isopropyl-2-methylcyclohexane is a stereoisomer of *trans*-1-isopopyl-3-methyl-cyclohexane.
  4. Statements (1) and (2) are true.
  5. Statements (1), (2), and (3) are all true.

Chapter 4–4

c. Rank the following conformations in order of *increasing energy.*

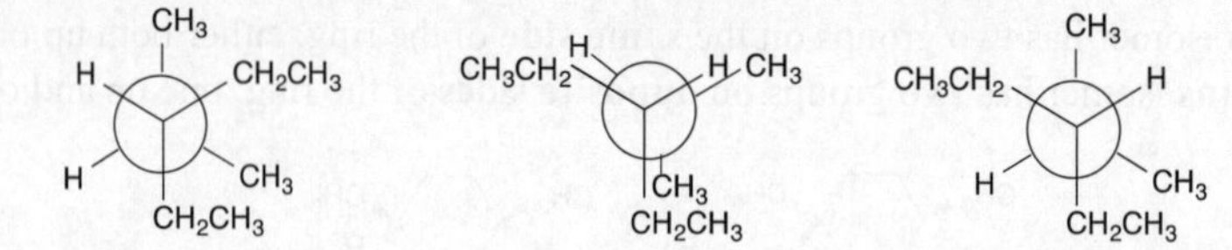

A

B

C

1. C < B < A
2. C < A < B

3. A < C < B
4. B < A < C

5. A < B < C

2. Give the IUPAC name for each of the following compounds.

a.

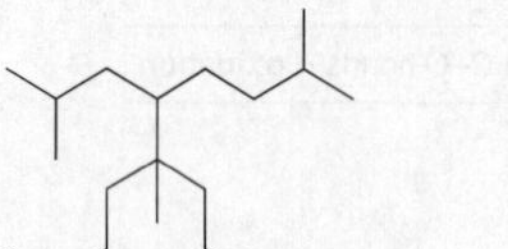

b.

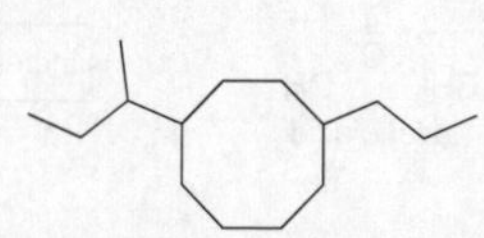

3. How are the molecules in each pair related? Are they constitutional isomers, stereoisomers, identical, or not isomers?

a. 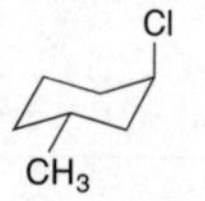   and

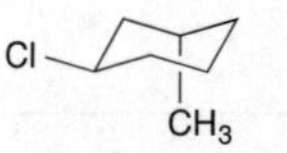

b. 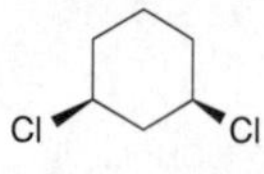   and

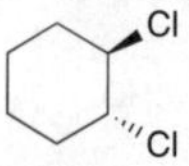

c. 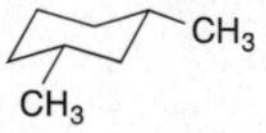   and

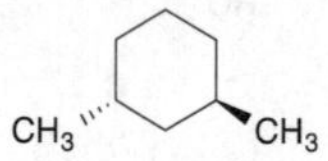

4. Rank the following conformations in order of increasing energy. Label the conformation of lowest energy as **1,** the highest energy as **4,** and the conformations of intermediate energy as **2** and **3.**

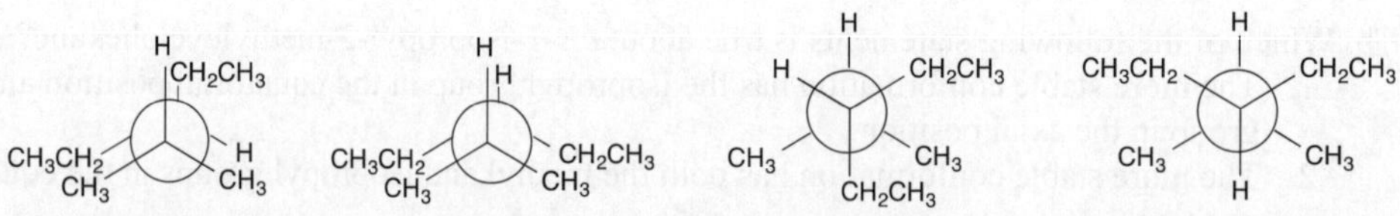

A   **B**   **C**   **D**

Alkanes 4–5

5. Consider the following disubstituted cyclohexane drawn below:

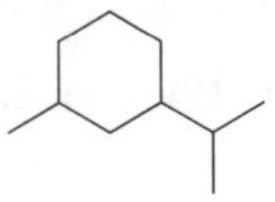

a. Draw the more stable chair conformation for the cis isomer.
b. Draw the more stable chair conformation for the trans isomer.

Answers to Practice Test

1. a. 4

   b. 1

   c. 2

2. a. 5-isobutyl-2,6-dimethyl-6-propyldecane

   b. 1-*sec*-butyl-4-propylcyclooctane

3. a. identical

   b. constitutional isomers

   c. stereoisomers

4. A–3
   B–4
   C–2
   D–1

5. a.

   b.

Answers to Problems

4.1 The general molecular formula for an acyclic alkane is C_nH_{2n+2}.

Number of C atoms $= n$	$2n + 2$	Number of H atoms
23	$2(23) + 2 =$	48
25	$2(25) + 2 =$	52
27	$2(27) + 2 =$	56

4.2 2-Methylbutane has 4 C's in a row with a 1 C branch.

a. 2-methylbutane

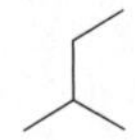

b. 2-methylbutane

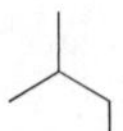

c. re-draw → 2-methylbutane

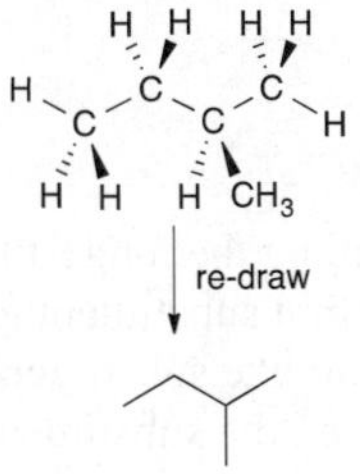

d. 5 C's in a row
   pentane

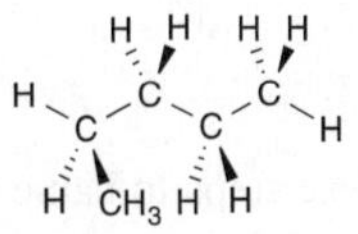

Chapter 4–6

4.3 Constitutional isomers differ in the way the atoms are connected to each other. To draw all the constitutional isomers:
  [1] Draw all of the C's in a long chain.
  [2] Take off one C and use it as a substituent. (Don't add it to the end carbon: this re-makes the long chain.)
  [3] Take off two C's and use these as substituents, etc.

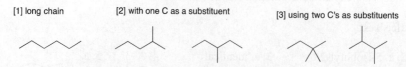

4.4 Draw each alkane to satisfy the requirements.

4.5

A — 6 C chain, CH₃ group on C3
B — identical
C — identical
D — isomer, CH₃ bonded to C2
E — identical
F — isomer, 7 C chain

4.6 Use the steps from Answer 4.3 to draw the constitutional isomers.

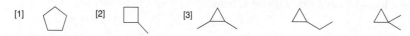

4.7 Follow these steps to name an alkane:
  [1] **Name the parent chain** by finding the longest C chain.
  [2] **Number the chain** so that the first substituent gets the lower number. Then **name and number all substituents,** giving like substituents a prefix (di, tri, etc.).
  [3] **Combine all parts,** alphabetizing the substituents, ignoring all prefixes except *iso*.

Alkanes 4–7

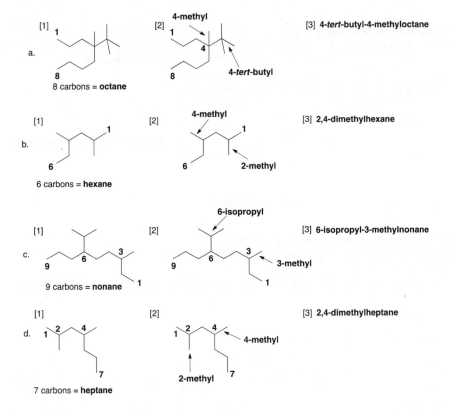

4.8 Use the steps in Answer 4.7 to name each alkane.

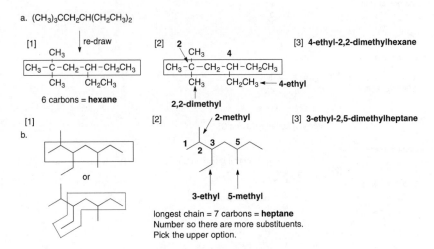

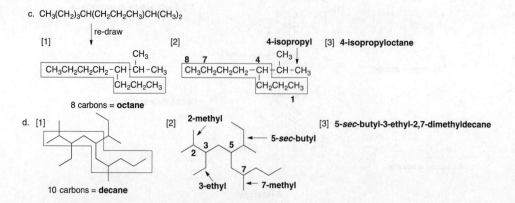

4.9 To work backwards from a name to a structure:

[1] Find the parent name and draw that number of C's. Use the suffix to identify the functional group (-ane = alkane).

[2] Arbitrarily number the C's in the chain or ring. Add the substituents to the appropriate C's.

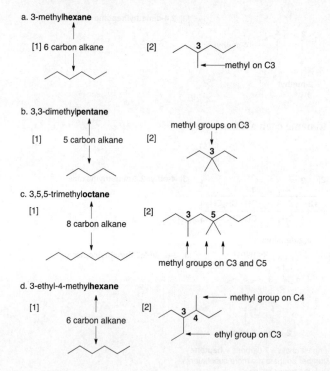

e. 3-ethyl-5-isobutyl**nonane**

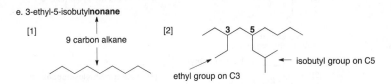

4.10 Use the steps in Answer 4.7 to name each alkane.

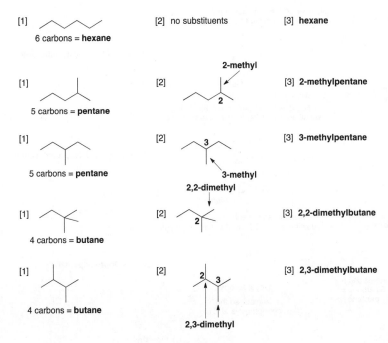

4.11 Follow these steps to name a cycloalkane:
[1] **Name the parent cycloalkane** by counting the C's in the ring and adding cyclo-.
[2] **Numbering:**
    a. **Number around the ring** beginning at a substituent and giving the second substituent the lower number.
    b. **Number to assign the lower number to the substituents alphabetically.**
    c. **Name and number all substituents,** giving like substituents a prefix (di, tri, etc.).
[3] **Combine all parts,** alphabetizing the substituents, ignoring all prefixes except *iso*.
(Remember: If a carbon chain has more C's than the ring, the chain is the parent, and the ring is a substituent.)

Chapter 4–10

a. [1]   [2] **1,1-dimethyl**   [3] **1,1-dimethylcyclohexane**

6 carbons in ring =
cyclohexane

Number so the
substituents are at C1.

b. [1]   [2] **1,2,3-trimethyl**   [3] **1,2,3-trimethylcyclopentane**

5 carbons in ring =
cyclopentane

Number so the first substituent
is at C1, second at C2.

c. [1]   [2]   [3] **1-butyl-4-methylcyclohexane**

6 carbons in ring =
cyclohexane

1-butyl

4-methyl

Number so the earlier alphabetical
substituent is at C1, **b**utyl before **m**ethyl.

d. [1]   [2] **1-*sec*-butyl**   [3] **1-*sec*-butyl-2-isopropylcyclohexane**

6 carbons in ring =
cyclohexane

2-isopropyl

Number so the earlier alphabetical
substituent is at C1, **b**utyl before **i**sopropyl.

e. [1]   [2]   [3] **1-cyclopropylpentane**

longest chain =
5 carbons =
pentane

1-cyclopropyl

Number so the
cyclopropyl is at C1.

f. [1]   [2] **1,1-dimethyl**   [3] **3-butyl-1,1-dimethylcyclohexane**

6 carbons in ring =
cyclohexane

3-butyl

Number so the two
methyls are at C1.

4.12 To draw the structures, use the steps in Answer 4.9.

a. 1,2-dimethyl**cyclobutane**

[1] 4 carbon cycloalkane [2] methyl groups on C1 and C2 [3]

b. 1,1,2-trimethyl**cyclopropane**

[1] 3 carbon cycloalkane [2] 3 CH₃'s [3]

c. 4-ethyl-1,2-dimethyl**cyclohexane**

[1] 6 carbon cycloalkane [2] ethyl on C4, 2 CH₃'s [3]

d. 1-*sec*-butyl-3-isopropyl**cyclopentane**

[1] 5 carbon cycloalkane [2] isopropyl, *sec*-butyl [3]

e. 1,1,2,3,4-pentamethyl**cycloheptane**

[1] 7 carbon cycloalkane [2] 5 CH₃'s [3]

4.13 **Compare the number of C's and surface area to determine relative boiling points.** Rules:
[1] Increasing number of C's = increasing boiling point.
[2] Increasing surface area = increasing boiling point (branching decreases surface area).

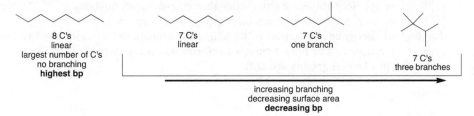

8 C's
linear
largest number of C's
no branching
highest bp

7 C's
linear

7 C's
one branch

7 C's
three branches

increasing branching
decreasing surface area
decreasing bp

Increasing boiling point: (CH₃)₃CCH(CH₃)₂ < CH₃CH₂CH₂CH₂CH(CH₃)₂ < CH₃(CH₂)₅CH₃ < CH₃(CH₂)₆CH₃

4.14 To draw a Newman projection, visualize the carbons as one in front and one in back of each other. The C–C bond is not drawn. There is only one staggered and one eclipsed conformation.

4.15 Re-draw each Newman projection as a skeletal structure, remembering that a Newman projection shows the substituents bonded to two C's but not the C's themselves.

4.16

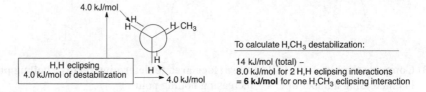

To calculate H,CH$_3$ destabilization:

14 kJ/mol (total) −
8.0 kJ/mol for 2 H,H eclipsing interactions
= **6 kJ/mol** for one H,CH$_3$ eclipsing interaction

4.17 To determine the energy of conformations, keep two things in mind:
[1] Staggered conformations are more stable than eclipsed conformations.
[2] Minimize steric interactions: keep large groups away from each other.
The highest energy conformation is the eclipsed conformation in which the two largest groups are eclipsed. The lowest energy conformation is the staggered conformation in which the two largest groups are anti.

Alkanes 4–13

rotation here

1
staggered
most stable

60°

2
eclipsed

60°

3
staggered
most stable

60°

6
eclipsed
least stable

60°

5
staggered

60°

4
eclipsed
least stable

60°

4.18 Use the criteria in problem 4.17 to determine the relative energy.

A   **B**   **C**   **D**

Staggered conformations **A** and **D** are lower in energy than eclipsed conformations **B** and **C**. **D** is lower in energy than **A** because it has two gauche interactions and **A** has three. **C** is higher in energy than **B** because all the larger groups (CH_3 and CH_2CH_3) are eclipsed.
Ranking: **D < A < B < C**

4.19 To determine the most and least stable conformations, use the rules from Answer 4.17.

a. **1,2-dichloroethane**
   $ClCH_2—CH_2Cl$

   rotation here

1
staggered, anti

60°

2
eclipsed

60°

3
staggered, gauche

60°

6
eclipsed

60°

5
staggered, gauche

60°

4
eclipsed

60°

Chapter 4–14

b.

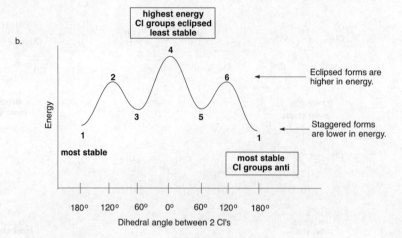

4.20 Add the energy increase for each eclipsing interaction to determine the destabilization.

a.

1 H,H interaction = 4.0 kJ/mol
2 H,CH₃ interactions
(2 × 6.0 kJ/mol) = 12.0 kJ/mol
────────────────────────────────
Total destabilization = **16 kJ/mol**

b.

3 H,CH₃ interactions
(3 × 6.0 kJ/mol) = **18 kJ/mol**
Total destabilization

4.21 Two points:
- Axial bonds point up or down, while equatorial bonds point out.
- An *up* carbon has an axial *up* bond, and a *down* carbon has an axial *down* bond.

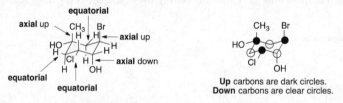

Up carbons are dark circles.
Down carbons are clear circles.

4.22

a. axial CH₃ ... OH equatorial

b. equatorial CH₃ ... OH axial

c. HO ... OH three equatorial OH's

Alkanes 4–15

4.23 Draw the second chair conformation by flipping the ring.
- The *up* carbons become *down* carbons, and the axial bonds become equatorial bonds.
- Axial bonds become equatorial, but *up* bonds stay *up*; that is, an axial *up* bond becomes an equatorial *up* bond.
- The conformation with **larger groups equatorial is the more stable** conformation and is present in higher concentration at equilibrium.

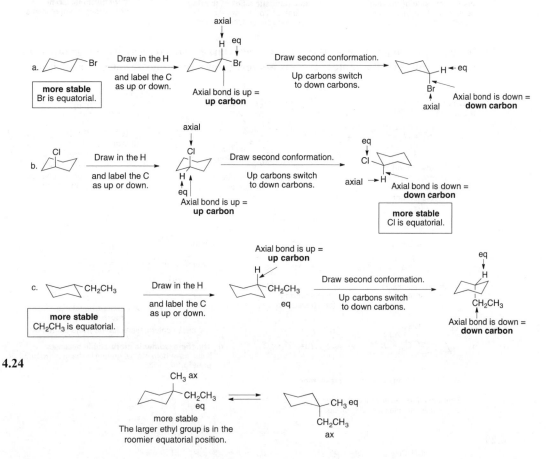

4.24

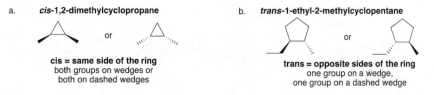

4.25 Wedges represent "up" groups in front of the page, whereas dashed wedges are "down" groups in back of the page. Cis groups are on the same side of the ring, whereas trans groups are on opposite sides of the ring.

a. ***cis*-1,2-dimethylcyclopropane**

cis = same side of the ring
both groups on wedges or
both on dashed wedges

b. ***trans*-1-ethyl-2-methylcyclopentane**

trans = opposite sides of the ring
one group on a wedge,
one group on a dashed wedge

4.26 Cis and trans isomers are stereoisomers.

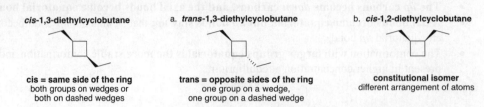

cis-1,3-diethylcyclobutane
cis = same side of the ring
both groups on wedges or
both on dashed wedges

a. trans-1,3-diethylcyclobutane
trans = opposite sides of the ring
one group on a wedge,
one group on a dashed wedge

b. cis-1,2-diethylcyclobutane
constitutional isomer
different arrangement of atoms

4.27 To classify a compound as a cis or trans isomer, **classify each non-hydrogen group as up or down. Groups on the same side = cis isomer; groups on opposite sides = trans isomer.**

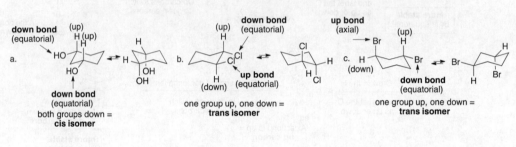

a. down bond (equatorial) / H (up) / HO / HO / down bond (equatorial)
both groups down =
cis isomer

b. (up) / down bond (equatorial) / Cl / Cl / up bond (equatorial) (down)
one group up, one down =
trans isomer

c. up bond (axial) / Br / H (down) / down bond (equatorial) / Br
one group up, one down =
trans isomer

4.28

a. groups on same side — **cis isomer**   groups on opposite sides — **trans isomer** (one possibility)

b. cis: CH₃ / H / CH₃ / H ⇌ CH₃ / CH₃ / H / H
two chair conformations for the **cis isomer**
Same stability because they both have one equatorial, one axial CH₃ group.

c. trans: CH₃ / H / H / CH₃ ⇌ H / CH₃ / CH₃ / H
both groups equatorial
more stable
two chair conformations for the **trans isomer**

d. The **trans isomer is more stable** because it can have both methyl groups in the more roomy **equatorial** position.

4.29

a. CH₂CH₃ ← axial CH₂CH₃ / CH₃
1,1-disubstituted

b. CH₃ ← up (axial) / H / CH₂CH₃ ← up / H
cis-1,2-disubstituted

c. up → CH₃CH₂ / H / H / CH₃ ← down (equatorial)
trans-1,3-disubstituted

d. up (equatorial) → CH₃CH₂ / H / H / CH₃ ← down
trans-1,4-disubstituted

Alkanes 4–17

4.30 *Oxidation* results in an *increase* in the number of C–Z bonds, or a *decrease* in the number of C–H bonds.
Reduction results in a *decrease* in the number of C–Z bonds, or an *increase* in the number of C–H bonds.

a. Decrease in the number of C–H bonds.
Increase in the number of C–O bonds.
Oxidation

b. Decrease in the number of C–O bonds.
Increase in the number of C–H bonds.
Reduction

c. No change in the number of C–O or C–H bonds. **Neither**

d. Decrease in the number of C–O bonds.
Increase in the number of C–H bonds.
Reduction

4.31 The products of a combustion reaction of a hydrocarbon are always the same: **CO₂ and H₂O.**

a. CH₃CH₂CH₃ + 5 O₂ —flame→ 3 CO₂ + 4 H₂O + heat

b. ⬡ + 9 O₂ —flame→ 6 CO₂ + 6 H₂O + heat

4.32 "Like dissolves like." Beeswax is a lipid, so it will be more soluble in nonpolar solvents. H₂O is very polar, ethanol is slightly less polar, and chloroform is least polar. Beeswax is most soluble in the least polar solvent.

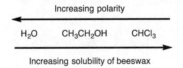

4.33 Re-draw the model as a skeletal structure. The longest chain has 15 C's, making it a derivative of pentadecane. Numbering from either direction gives the same numbers.

15 C's ----→ pentadecane
4 methyl groups at C2, C6, C10, and C14
2,6,10,14-tetramethylpentadecane

4.34 Re-draw each alkane as a skeletal structure, and use the steps in Answer 4.7 to name each compound. Classify C's as in Section 3.2.

Chapter 4–18

a.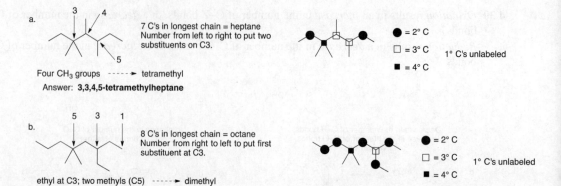

Four CH₃ groups ----→ tetramethyl
7 C's in longest chain = heptane
Number from left to right to put two substituents on C3.

● = 2° C
□ = 3° C
■ = 4° C
1° C's unlabeled

Answer: **3,3,4,5-tetramethylheptane**

b.

8 C's in longest chain = octane
Number from right to left to put first substituent at C3.

ethyl at C3; two methyls (C5) ----→ dimethyl

● = 2° C
□ = 3° C
■ = 4° C
1° C's unlabeled

Answer: **3-ethyl-5,5-dimethyloctane**

4.35 Re-draw the ball-and-stick model as a chair form.

a.

b. CH₃ on C1 and Br on C2 are both down, making them cis.
c. Br on C2 and CH(CH₃)₂ on C4 are both down, making them cis.
d. Second chair form:

This chair form is less stable because two groups [Br and CH(CH₃)₂] are in the more crowded axial position.

4.36

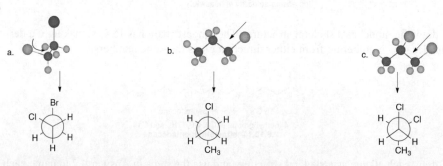

4.37
  a. Five constitutional isomers of molecular formula C₄H₈:

b. Nine constitutional isomers of molecular formula C_7H_{16}:

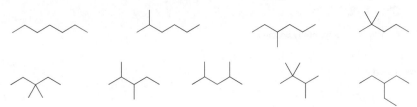

c. Twelve constitutional isomers of molecular formula C_6H_{12} containing one ring:

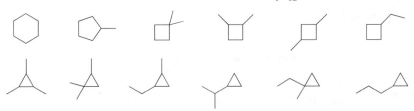

4.38 Use the steps in Answers 4.7 and 4.11 to name the alkanes.

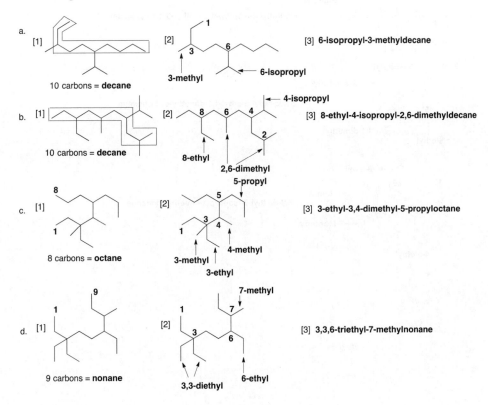

Chapter 4–20

e. 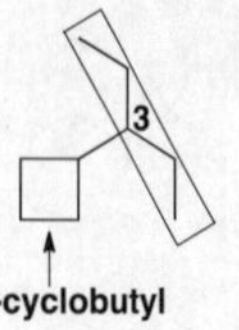 **3-cyclobutylpentane**

3-cyclobutyl

f. 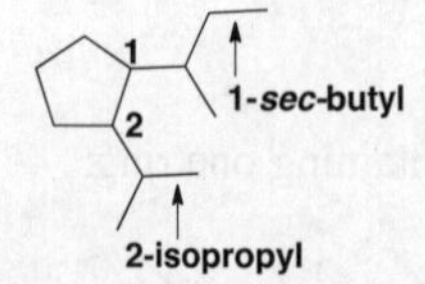 1-*sec*-butyl

2-isopropyl

1-*sec*-butyl-2-isopropylcyclopentane

g. 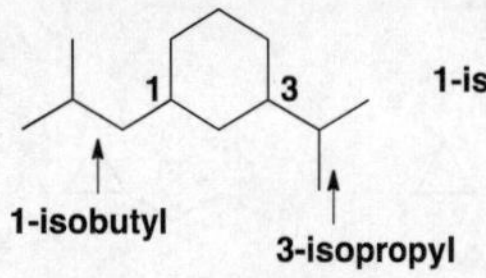 **1-isobutyl-3-isopropylcyclohexane**

1-isobutyl

3-isopropyl

h. 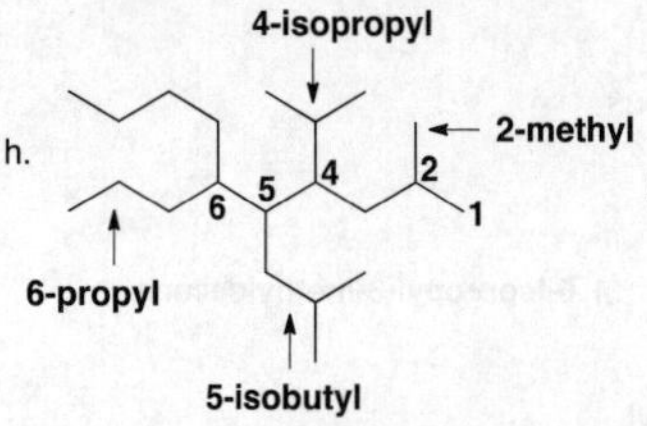 4-isopropyl

← 2-methyl

5-isobutyl-4-isopropyl-2-methyl-6-propyldecane

6-propyl

5-isobutyl

i. 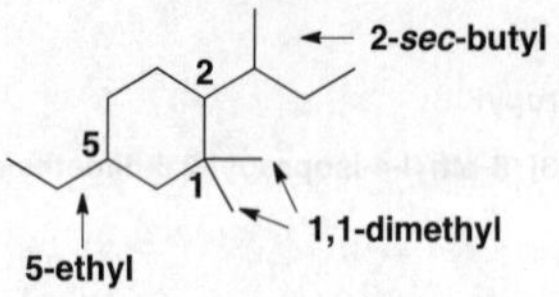 ← 2-*sec*-butyl

2-*sec*-butyl-5-ethyl-1,1-dimethylcyclohexane

← 1,1-dimethyl

5-ethyl

j. 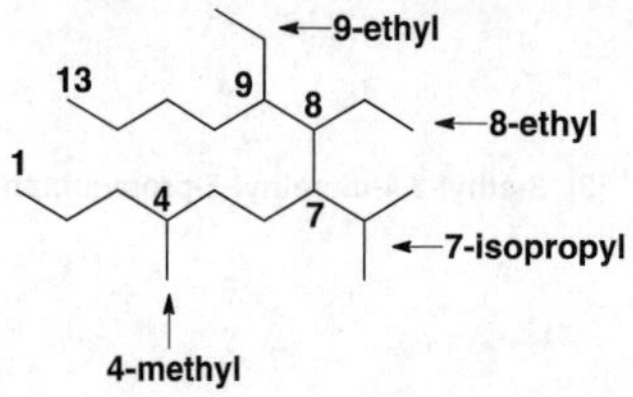 ←9-ethyl

← 8-ethyl

8,9-diethyl-7-isopropyl-4-methyltridecane

←7-isopropyl

4-methyl

4.39

2,2-dimethylheptane — 2,2-dimethyl

3,3-dimethylheptane — 3,3-dimethyl

4,4-dimethylheptane — 4,4-dimethyl

3,4-dimethylheptane

2,4-dimethylheptane

2,5-dimethylheptane

2,3-dimethylheptane

3,5-dimethylheptane

2,6-dimethylheptane

4.40 Use the steps in Answer 4.9 to draw the structures.

a. 3-ethyl-2-methyl**hexane**

[1] 6 C chain

[2] methyl on C2, ethyl on C3

b. *sec*-butyl**cyclopentane**

[1] 5 C ring

[2]

c. 4-isopropyl-2,4,5-trimethyl**undecane**

[1] 11 C chain

[2] isopropyl on C4, methyls on C2, C4, and C5

d. cyclobutylcycloheptane

[1] 7 C cycloalkane

[2]

e. 3-ethyl-1,1-dimethyl**cyclohexane**

[1] 6 C cycloalkane

[2] ethyl on C3, 2 methyl groups on C1

f. 4-butyl-1,1-diethyl**cyclooctane**

[1] 8 C cycloalkane

[2] 2 ethyl groups

Chapter 4–22

g. 6-isopropyl-2,3-dimethyl**dodecane**
[1] 12 C alkane
[2] methyl on C2, isopropyl on C6, methyl on C3

h. 2,2,6,6,7-pentamethyl**octane**
[1] 8 C alkane
[2] 5 methyl groups

i. *cis*-1-ethyl-3-methyl**cyclopentane**
[1] 5 C ring
[2] ethyl on C1, methyl on C3 or

j. *trans*-1-*tert*-butyl-4-ethyl**cyclohexane**
[1] 6 C ring
[2]

4.41 Draw the compounds.

a. 2,2-dimethyl-4-ethylheptane
alphabetized incorrectly
ethyl before **methyl**
4-ethyl-2,2-dimethylheptane

b. 5-ethyl-2-methylhexane
Longest chain was not chosen = **heptane**
2,5-dimethylheptane

c. 2-methyl-2-isopropylheptane
Longest chain was not chosen = **octane**
2,3,3-trimethyloctane

d. 1,5-dimethylcyclohexane
Numbered incorrectly.
Re-number so methyls are at C1 and C3.

1,3-dimethylcyclohexane

e. 1-ethyl-2,6-dimethylcycloheptane
Numbered incorrectly.
Re-number so methyls are at C1 and C4.
2-ethyl-1,4-dimethylcycloheptane

f. 5,5,6-trimethyloctane
Numbered incorrectly.
Re-number so methyls are at C3 and C4.
3,4,4-trimethyloctane

g. 3-butyl-2,2-dimethylhexane
Longest chain not chosen = **octane**
4-*tert*-butyloctane

h. 1,3-dimethylbutane
Longest chain not chosen = **pentane**
2-methylpentane

4.42

a. 4-isopropylheptane (re-draw)

b. 3-ethyl-3-methylpentane (re-draw)

c. 4,4-diethyl-5-methyloctane (re-draw)

4.43 Use the rules from Answer 4.13.

most branching
lowest boiling point

least branching
highest boiling point

4.44 a.

CH₃(CH₂)₆CH₃
no branching = higher surface area
higher boiling point

(CH₃)₃CC(CH₃)₃
branching = lower surface area
lower boiling point
more spherical, better packing =
higher melting point

b. There is a 159 °C difference in the melting points, but only a 20 °C difference in the boiling points because the symmetry in (CH₃)₃CC(CH₃)₃ allows it to pack more tightly in the solid, thus requiring more energy to melt. In contrast, once the compounds are in the liquid state, symmetry is no longer a factor, the compounds are isomeric alkanes, and the boiling points are closer together.

4.45

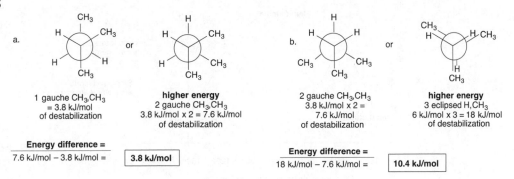

a. 1 gauche CH₃,CH₃
= 3.8 kJ/mol
of destabilization

or

higher energy
2 gauche CH₃,CH₃
3.8 kJ/mol x 2 = 7.6 kJ/mol
of destabilization

Energy difference =
7.6 kJ/mol – 3.8 kJ/mol = **3.8 kJ/mol**

b. 2 gauche CH₃,CH₃
3.8 kJ/mol x 2 =
7.6 kJ/mol
of destabilization

or

higher energy
3 eclipsed H,CH₃
6 kJ/mol x 3 = 18 kJ/mol
of destabilization

Energy difference =
18 kJ/mol – 7.6 kJ/mol = **10.4 kJ/mol**

4.46 Use the rules from Answer 4.17 to determine the most and least stable conformations.

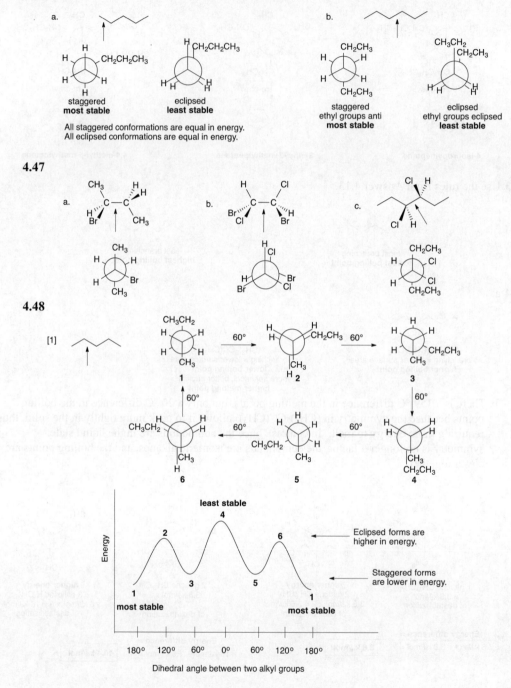

4.49 Two types of strain:

- *Torsional strain* is due to eclipsed groups on adjacent carbon atoms.
- *Steric strain* is due to overlapping electron clouds of large groups (e.g. gauche interactions).

a.
two sites
three bulky methyl groups close =
steric strain

b.
eclipsed conformation =
torsional strain

c.
two bulky ethyl groups close =
steric strain
eclipsed conformation =
torsional strain

4.50 The barrier to rotation is equal to the difference in energy between the highest energy eclipsed and lowest energy staggered conformations of the molecule.

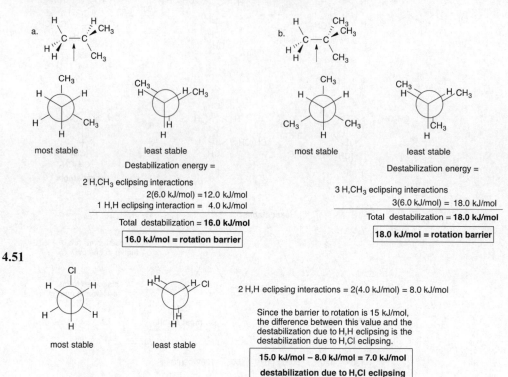

4.52 The gauche conformation can intramolecularly hydrogen bond, making it the more stable conformation.

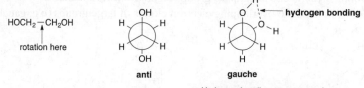

Alkanes 4–27

[2]

a. **axial** Br — H **eq** / CH$_3$ **eq** / H **axial**

b. **up** Br — H / CH$_3$ **up** / H — both up = **cis**

c. Br / CH$_3$

d. **ax** Br — H **eq** / CH$_3$ **eq** / H **ax** ⇌ **ax** CH$_3$ **eq** Br — H **eq** / H **ax**

[3]

a. **axial** H / HO— **eq** —OH **eq** / H **axial**

b. H / HO— **up** —OH **down** / H — one up, one down = **trans**

c. HO / OH

d. **ax** H / **eq** HO— —OH **eq** / H **ax** ⇌ **eq** H — OH **ax** / H **eq** / OH **ax**

4.54

ax H / **ax** H — CH$_3$ **eq** / CH$_3$ **eq** — both groups equatorial **more stable** ⇌ **eq** H — H **eq** / CH$_3$ **ax** CH$_3$ **ax**

4.55 A **cis isomer** has two groups on the **same side** of the ring. The two groups can be drawn both up or both down. Only one possibility is drawn. A **trans isomer** has one group on one side of the ring and one group on the other side. Either group can be drawn on either side. Only one possibility is drawn.

[1]

a. **cis**   **trans**

[2]

a. **cis**   **trans**

[3]

a. **cis**   **trans**

Chapter 4–28

b. cis isomer

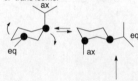

both groups equatorial
more stable

b. cis isomer

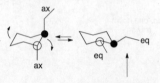

larger group equatorial
more stable

b. cis isomer

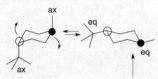

larger group equatorial
more stable

c. trans isomer

larger group equatorial
more stable

c. trans isomer

both groups equatorial
more stable

c. trans isomer

both groups equatorial
more stable

d.

The cis isomer is more
stable than the trans
because one conformation has
both groups equatorial.

d.

The trans isomer is more
stable than the cis
because one conformation has
both groups equatorial.

d.

The trans isomer is more
stable than the cis
because one conformation has
both groups equatorial.

4.56

Re-draw to see
axial and equatorial.

all equatorial
menthol

Alkanes 4–29

4.57 a.

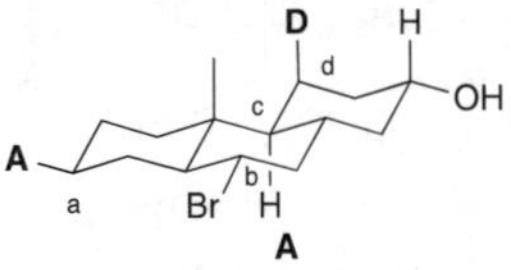

CH_3 is axial and up. OH is equatorial and down, so the two groups are trans.

b. A substituent (**A**) on C_a that is cis to the existing CH_3 must be up, so it is equatorial.
c. An equatorial Br on C_b is down, so it is cis to the OH group.
d. A H on C_c must be axial and down, making it cis to the OH group.
e. A substituent (**D**) on C_d that is trans to the OH must be up and axial.

4.58

a.

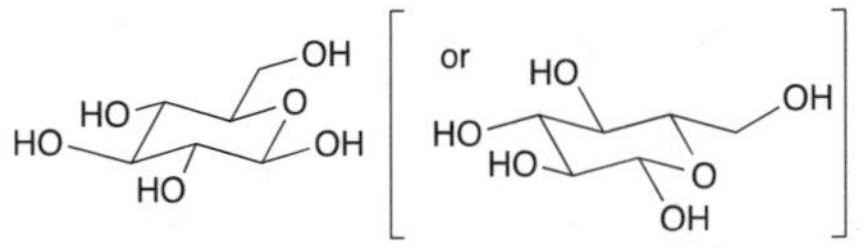

most stable
All groups are equatorial.

b.

c.

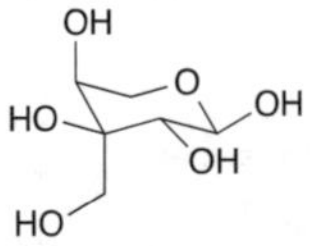

constitutional isomer

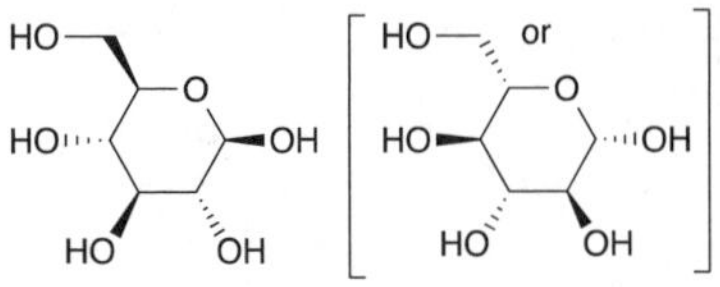

d.

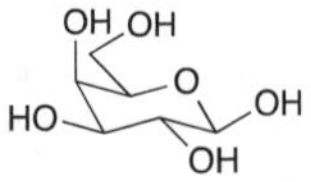

4.59

a.

and

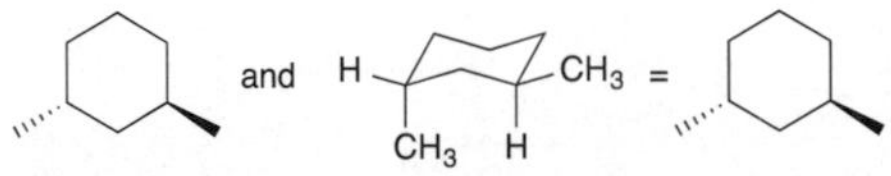

1 down, 1 up =
trans

1 down, 1 up =
trans

same arrangement in three dimensions
identical

b.

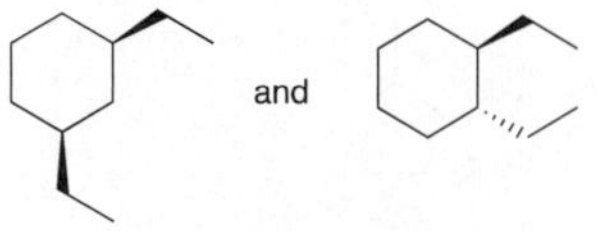

and

same molecular formula $C_{10}H_{20}$
different connectivity
constitutional isomers

c.

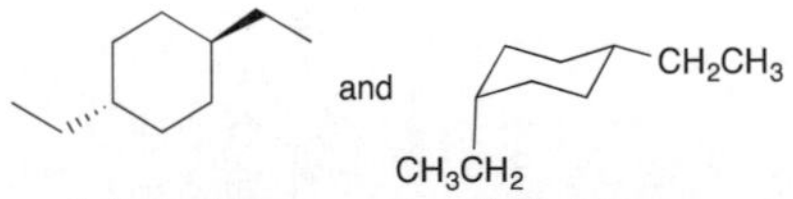

and

1 down, 1 up =
trans

both down =
cis

different arrangement in three dimensions
stereoisomers

d.

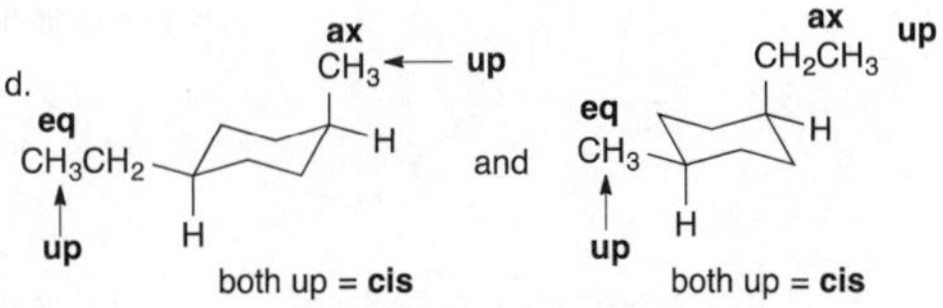

and

both up = **cis**

both up = **cis**

same arrangement in three dimensions
identical

Chapter 4–30

4.60

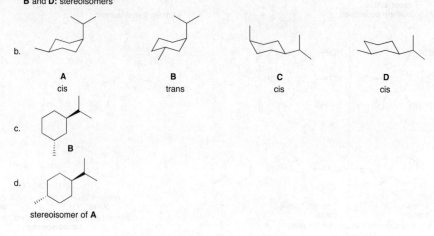

4.61

a. **A** and **B**: constitutional isomers
A and **C**: identical
B and **D**: stereoisomers

4.62

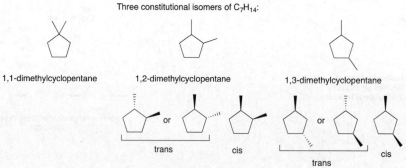

Alkanes 4–31

4.63 Use the definitions from Answer 4.30 to classify the reactions.

a. benzene → benzyl bromide (PhCH₂Br)
Increase in the number of C–Z bonds. **Oxidation**

b. cyclohexanol → cyclohexene
Loss of one C–O bond *and* one C–H bond. **Neither**

4.64 Use the rule from Answer 4.31.

a. 2-methylhexane $\xrightarrow[11\,O_2]{\text{flame}}$ 7 CO₂ + 8 H₂O + heat

b. pentane $\xrightarrow[(13/2)\,O_2]{\text{flame}}$ 4 CO₂ + 5 H₂O + heat

4.65

a. benzene $\xrightarrow{[1]}$ an arene oxide (2 C–O bonds) $\xrightarrow{[2]}$ phenol (1 C–O bond, 1 C–H bond)

[1] increase in C–O bonds, **oxidation reaction**
[2] loss of 1 C–O bond, loss of 1 C–H bond
neither

b. Phenol is more water soluble than benzene because it is **polar (contains an O–H group) and can hydrogen bond with water,** whereas benzene is nonpolar and cannot hydrogen bond.

4.66

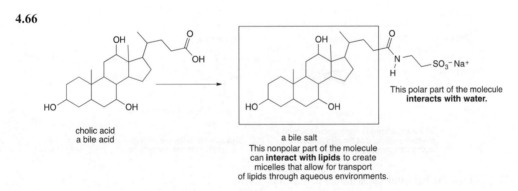

cholic acid
a bile acid

a bile salt
This nonpolar part of the molecule can **interact with lipids** to create micelles that allow for transport of lipids through aqueous environments.

This polar part of the molecule **interacts with water.**

4.67 The mineral oil can prevent the body's absorption of important fat-soluble vitamins. The vitamins dissolve in the mineral oil, and are thus not absorbed. Instead, they are expelled with the mineral oil.

4.68 Cyclopropane has larger angle strain than cyclobutane because the internal angles in the three-membered ring (60°) are smaller than they are in cyclobutane. Although cyclobutane is not flat, as shown in Figure 4.11, there are more C–H bonds than there are in cyclopropane, so there are more sites of torsional strain. Thus cyclopropane has more angle strain but less torsional strain. The result is that both cyclopropane and cyclobutane have roughly similar strain energies.

4.69 The amide in the four-membered ring has 90° bond angles giving it angle strain, which makes it more reactive.

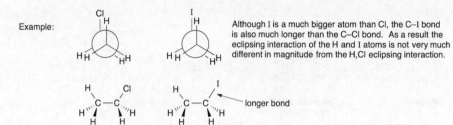

4.70

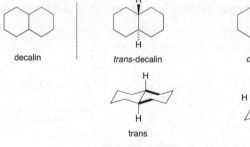

Although I is a much bigger atom than Cl, the C–I bond is also much longer than the C–Cl bond. As a result the eclipsing interaction of the H and I atoms is not very much different in magnitude from the H,Cl eclipsing interaction.

4.71

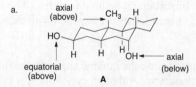

The trans isomer is more stable because the carbon groups at the ring junction are both in the favorable equatorial position.

This bond is axial, creating unfavorable 1,3-diaxial interactions.

4.72 Re-draw the ball-and-stick model using chair forms.

a.

b. All bonds above the ring are on wedges and all bonds below the ring are on dashed wedges.

4.73

a, b.

CH₃ ← axial
Cl ← axial
Cl ← equatorial

B

c. The circled H's at one ring fusion are cis. The boxed in CH₃ and H at the second ring fusion are trans.

4.74

pentylcyclopentane

(1,1-dimethylpropyl)cyclopentane

(2-methylbutyl)cyclopentane

(2,2-dimethylpropyl)cyclopentane

(1-methylbutyl)cyclopentane

(1-ethylpropyl)cyclopentane

(1,2-dimethylpropyl)cyclopentane

(3-methylbutyl)cyclopentane

4.75

a.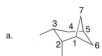
2,3-dimethylbicyclo[3.1.1]heptane

c.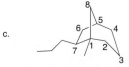
1-methyl-7-propylbicyclo[3.2.1]octane

b.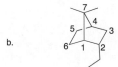
2-ethyl-7,7-dimethylbicyclo[2.2.1]heptane

d.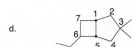
6-ethyl-3,3-dimethylbicyclo[3.2.0]heptane

Stereochemistry 5–1

Chapter 5: Stereochemistry

Chapter Review

Isomers are different compounds with the same molecular formula (5.2, 5.11).

[1] **Constitutional isomers**—isomers that differ in the way the atoms are connected to each other. They have:
  - different IUPAC names
  - the same or different functional groups
  - different physical and chemical properties.

[2] **Stereoisomers**—isomers that differ only in the way atoms are oriented in space. They have the same functional group and the same IUPAC name except for prefixes such as cis, trans, *R*, and *S*.
  - **Enantiomers**—stereoisomers that are nonsuperimposable mirror images of each other (5.4).
  - **Diastereomers**—stereoisomers that are not mirror images of each other (5.7).

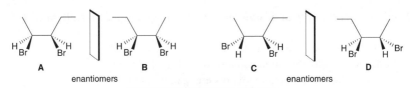

Assigning priority (5.6)

- Assign priorities (1, 2, 3, or 4) to the atoms bonded directly to the stereogenic center in order of decreasing atomic number. The atom of *highest* atomic number gets the *highest* priority (1).
- If two atoms on a stereogenic center are the *same,* assign priority based on the atomic number of the atoms bonded to these atoms. *One* atom of higher atomic number determines a higher priority.
- If two isotopes are bonded to the stereogenic center, assign priorities in order of decreasing *mass* number.
- To assign a priority to an atom that is part of a multiple bond, treat a multiply bonded atom as an equivalent number of singly bonded atoms.

- The stereogenic center is bonded to Br, Cl, C, and H.
- The stereogenic center is *not* bonded directly to I.

- CH(CH$_3$)$_2$ gets the highest priority because the C is bonded to two other C's.

- OH gets the highest priority because O has the highest atomic number.
- CO$_2$H (three bonds to O) gets higher priority than CH$_2$OH (one bond to O).

Chapter 5–2

Some basic principles

- When a compound and its mirror image are **superimposable,** they are **identical achiral compounds.** A plane of symmetry in one conformation makes a compound achiral (5.3).
- When a compound and its mirror image are **not superimposable,** they are **different chiral compounds** called **enantiomers.** A chiral compound has no plane of symmetry in any conformation (5.3).
- A **tetrahedral stereogenic center** is a carbon atom bonded to four different groups (5.4, 5.5).
- For *n* **stereogenic centers,** the maximum number of stereoisomers is 2^n (5.7).

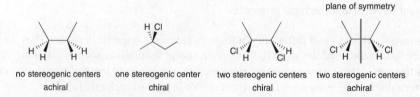

Optical activity is the ability of a compound to rotate plane-polarized light (5.12).

- An optically active solution contains a chiral compound.
- An optically inactive solution contains one of the following:
  - an achiral compound with no stereogenic centers.
  - a meso compound—an achiral compound with two or more stereogenic centers.
  - a racemic mixture—an equal amount of two enantiomers.

The prefixes *R* and *S* compared with *d* and *l*

The prefixes *R* and *S* are labels used in nomenclature. Rules on assigning *R,S* are found in Section 5.6.
- An enantiomer has every stereogenic center opposite in configuration.
- A diastereomer of this same compound has one stereogenic center with the same configuration and one that is opposite.

The prefixes *d* (or +) and *l* (or −) tell the direction a compound rotates plane-polarized light (5.12).
- *d* (or +) stands for dextrorotatory, rotating polarized light clockwise.
- *l* (or −) stands for levorotatory, rotating polarized light counterclockwise.

Stereochemistry 5–3

The physical properties of isomers compared (5.12)

Type of isomer	Physical properties
Constitutional isomers	Different
Enantiomers	Identical except the direction of rotation of polarized light
Diastereomers	Different
Racemic mixture	Possibly different from either enantiomer

Equations

- Specific rotation (5.12C):

  $$\text{specific rotation} = [\alpha] = \frac{\alpha}{l \times c}$$

  α = observed rotation (°)
  l = length of sample tube (dm)
  c = concentration (g/mL)

  [dm = decimeter
  1 dm = 10 cm]

- Enantiomeric excess (5.12D):

  $$ee = \% \text{ of one enantiomer} - \% \text{ of other enantiomer}$$
  $$= \frac{[\alpha] \text{ mixture}}{[\alpha] \text{ pure enantiomer}} \times 100\%$$

Practice Test on Chapter Review

1.a. Which of the following statements is true for compounds **A–D** below?

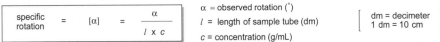

1. **A** and **B** are separable by physical methods such as distillation.
2. **A** and **C** are separable by physical methods such as distillation.
3. **A** and **D** are separable by physical methods such as distillation.
4. Statements (1) and (2) are both true.
5. Statements (1), (2), and (3) are all true.

b. Which of the following statements is true about compounds **A–C** below?

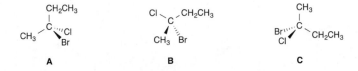

Chapter 5–4

1. **A** and **B** are enantiomers.
2. **A** and **C** are enantiomers.
3. An equal mixture of **B** and **C** is optically active.
4. Statements (1) and (2) are true.
5. Statements (1), (2), and (3) are all true.

c. Which compound is a diastereomer of **A**?

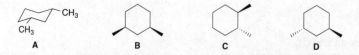

1. **B** only
2. **C** only
3. **D** only
4. Both **B** and **C**
5. Compounds **B**, **C**, and **D**

2. Rank the following four groups around a stereogenic center in order of decreasing priority. Rank the highest priority group as **1**, the lowest priority group as **4**, and the two groups of intermediate priority as **2** and **3**.

$$\text{—CH}_2\text{Cl} \qquad \text{—CH}_2\text{CH}_2\text{Br} \qquad \text{—COOH} \qquad \text{—CH}_2\text{OH}$$

  A   B   C   D

3. Label each stereogenic center in the following compound as *R* or *S*.

4. State how the compounds in each pair are related to each other. Choose from constitutional isomers, enantiomers, diastereomers, or identical compounds.

a.

b.

c.

5. The enantiomeric excess of a mixture of **A** and **B** is 62% with **A** in excess. How much of **A** and **B** are present in the mixture?

Answers to Practice Test

1. a. 1	2. A–1	3. a. *S*	4. a. diastereomers	5. 81% **A**
b. 2	B–4	b. *S*	b. enantiomers	19% **B**
c. 3	C–2		c. identical	
	D–3			

Answers to Problems

5.1 Cellulose consists of long chains held together by intermolecular hydrogen bonds forming sheets that stack in extensive three-dimensional arrays. Most of the OH groups in cellulose are in the interior of this three-dimensional network, unavailable for hydrogen bonding to water. Thus, even though cellulose has many OH groups, its three-dimensional structure prevents many of the OH groups from hydrogen bonding with the solvent and this makes it water insoluble.

5.2 Constitutional isomers have atoms bonded to different atoms.
Stereoisomers differ only in the three-dimensional arrangement of atoms.

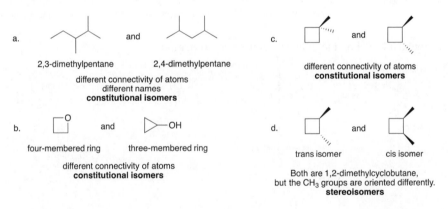

5.3 Draw the mirror image of each molecule by drawing a mirror plane and then drawing the molecule's reflection. **A chiral molecule is one that is not superimposable on its mirror image.** A molecule with one stereogenic center is always chiral. A molecule with zero stereogenic centers is not chiral (in general).

Chapter 5–6

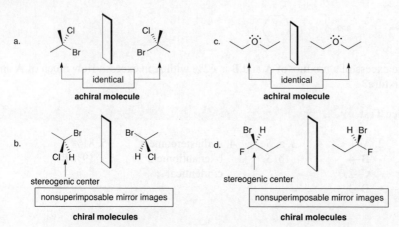

5.4 A plane of symmetry cuts the molecule into **two identical halves.**

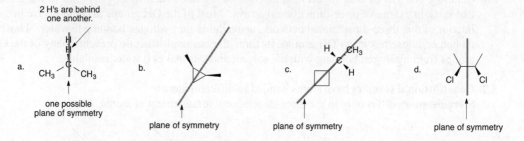

5.5 Rotate around a C–C bond to draw an eclipsed conformation.

Stereochemistry 5–7

5.6 To locate a stereogenic center, omit all C's with two or more H's, all *sp* and *sp*² hybridized atoms, and all heteroatoms. [In Chapter 25, we will learn that the N atoms of ammonium salts (R₄N⁺ X⁻) can sometimes be stereogenic centers.] Then evaluate any remaining atoms. A tetrahedral stereogenic center has a carbon bonded to **four different groups.**

a.

c.

e.

b.

d.

f.

* = stereogenic center

5.7 Use the directions from Answer 5.6 to locate the stereogenic centers.

aliskiren

4 C's bonded to 4 different groups
4 stereogenic centers

5.8 Find the C bonded to four different groups in each molecule. At the stereogenic center, draw two bonds in the plane of the page, one in front (on a wedge), and one behind (on a dashed wedge). Then draw the mirror image (enantiomer).

a.

stereogenic center

mirror images
nonsuperimposable
enantiomers

b.

stereogenic center

mirror images
nonsuperimposable
enantiomers

c.

stereogenic center

mirror images
nonsuperimposable
enantiomers

5.9 Use the directions from Answer 5.6 to locate the stereogenic centers.

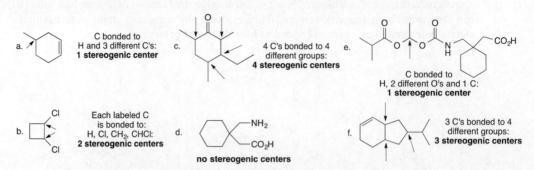

5.10

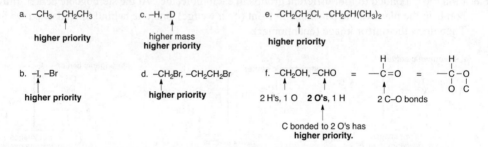

All stereogenic C's are circled. Each C is *sp³* hybridized and bonded to 4 different groups.

5.11 Assign priority based on atomic number: atoms with a higher atomic number get a higher priority. If two atoms are the same, look at what they are bonded to and assign priority based on the atomic number of these atoms.

a. –CH₃, –CH₂CH₃
           ↑
   higher priority

b. –I, –Br
     ↑
   higher priority

c. –H, –D
       ↑
   higher mass
   higher priority

d. –CH₂Br, –CH₂CH₂Br
     ↑
   higher priority

e. –CH₂CH₂Cl, –CH₂CH(CH₃)₂
        ↑
   higher priority

f. –CH₂OH, –CHO = H–C=O = H–C–O
                                                                                        O C
   2 H's, 1 O   2 O's, 1 H   2 C–O bonds
                 ↑
   C bonded to 2 O's has
   higher priority.

5.12 Rank by decreasing priority. Lower atomic number = lower priority.

Highest priority = 1, Lowest priority = 4

		priority			priority
a. –COOH	C = second lowest atomic number	3	b. –H	H = lowest atomic number	4
–H	H = lowest atomic number	4	–CH₃	C bonded to 3 H's	3
–NH₂	N = second highest atomic number	2	–Cl	Cl = highest atomic number	1
–OH	O = highest atomic number	1	–CH₂Cl	C bonded to 2 H's + 1 Cl	2

decreasing priority: –OH, –NH₂, –COOH, –H   decreasing priority: –Cl, –CH₂Cl, –CH₃, –H

Stereochemistry 5–9

		priority				priority
c. $-CH_2CH_3$	C bonded to 2 H's + **1 C**	2	d. $-CH=CH_2$	C bonded to 1 H + **2 C's**	2
$-CH_3$	C bonded to 3 H's	3	$-CH_3$	C bonded to 3 H's	3
$-H$	H = lowest atomic number	4	$-C\equiv CH$	C bonded to **3 C's**	1
$-CH(CH_3)_2$	C bonded to 1 H + **2 C's**	1	$-H$	H = lowest atomic number	4

decreasing priority: $-CH(CH_3)_2$, $-CH_2CH_3$, $-CH_3$, $-H$ **decreasing priority: $-C\equiv CH$, $-CH=CH_2$, $-CH_3$, $-H$**

5.13 To assign *R* or *S* to the molecule, first rank the groups. The lowest priority group must be oriented behind the page. If tracing a circle from (1) → (2) → (3) proceeds in the clockwise direction, then the stereogenic center is labeled *R;* if the circle is counterclockwise, then it is labeled *S*.

a.

b.
lowest priority forward
clockwise
It looks like *R*, but reverse answer
R → S

c.

d.
lowest priority forward
clockwise
It looks like *R*, but reverse answer
R → S

5.14

clopidogrel

clockwise
R isomer

counterclockwise
S isomer
Plavix

5.15 a, b. Re-draw lisinopril as a skeletal structure, locate the stereogenic centers, and assign *R,S*.

three stereogenic centers
All have the *S* configuration.

5.16 The maximum number of stereoisomers $= 2^n$ where $n =$ the number of stereogenic centers.

     a. 3 stereogenic centers   b. 8 stereogenic centers
        $2^3 = 8$ stereoisomers $2^8 = 256$ stereoisomers

Chapter 5–10

5.17

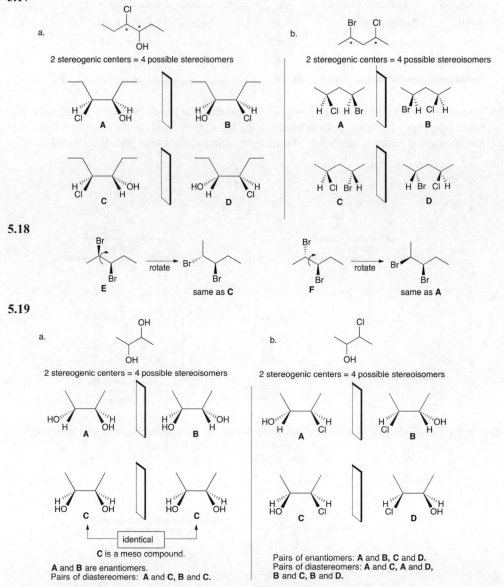

5.20 **A meso compound must have at least two stereogenic centers. Usually a meso compound has a plane of symmetry.** You may have to rotate around a C–C bond to see the plane of symmetry clearly.

Stereochemistry 5–11

a.

2 stereogenic centers
plane of symmetry
meso compound

b.

2 stereogenic centers
no plane of symmetry
not a meso compound

c.

rotate

2 stereogenic centers
plane of symmetry
meso compound

5.21 Use the definition in Answer 5.20 to draw the meso compounds.

a.

plane of symmetry

b.

plane of symmetry

c.

plane of symmetry

5.22 The enantiomer has the exact opposite R,S designations. Diastereomers with two stereogenic centers have one center the same and one different.

If a compound is **R,S:**

Its enantiomer is: **S,R** ◄──────── Exact opposite: R and S interchanged.

Its diastereomers are: **R,R and S,S** ◄──────── One designation remains the same, the other changes.

5.23 The enantiomer has the exact opposite R,S designations. For diastereomers, at least one of the R,S designations is the same, but not all of them.

a. (2R,3S)-hexane-2,3-diol and (2R,3R)-hexane-2,3-diol
　　One changes; one remains the same:
　　　diastereomers
b. (2R,3R)-hexane-2,3-diol and (2S,3S)-hexane-2,3-diol
　　Both R's change to S's:
　　　enantiomers
c. (2R,3S,4R)-hexane-2,3,4-triol and (2S,3R,4R)-hexane-2,3,4-triol
　　Two change; one remains the same:
　　　diastereomers

5.24 The enantiomer must have the exact opposite R,S designations. For diastereomers, at least one of the R,S designations is the same, but not all of them.

a.

sorbitol

b.

A
One changes; three remain the same.
diastereomer

c.

B
All stereogenic centers change.
enantiomer

Chapter 5–12

5.25 Meso compounds generally have a plane of symmetry. They cannot have just one stereogenic center.

a. no plane of symmetry
not a meso compound

b. plane of symmetry
meso compound

c. no plane of symmetry
not a meso compound

5.26

a.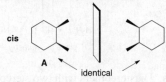
2 stereogenic centers =
4 stereoisomers maximum

Draw the cis and trans isomers:

cis
A   identical   (mirror)

trans
B   C

Pair of enantiomers: **B** and **C**.
Pairs of diastereomers: **A** and **B**, **A** and **C**.

Only 3 stereoisomers exist.

c.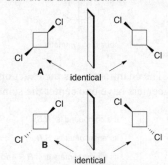

Draw the cis and trans isomers:

A   identical

B   identical

Pair of diastereomers: **A** and **B**.

Only 2 stereoisomers exist.

b.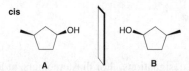
2 stereogenic centers =
4 stereoisomers maximum

Draw the cis and trans isomers:

cis
A   B

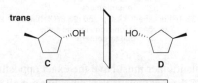

trans
C   D

Pairs of enantiomers: **A** and **B**, **C** and **D**.
Pairs of diastereomers: **A** and **C**, **A** and **D**, **B** and **C**, **B** and **D**.

All 4 stereoisomers exist.

Stereochemistry 5–13

5.27 Four facts:
- **Enantiomers** are mirror image isomers.
- **Diastereomers** are stereoisomers that are not mirror images.
- **Constitutional isomers** have the same molecular formula but the atoms are bonded to different atoms.
- **Cis and trans isomers** are always diastereomers.

a.

same molecular formula
same *R,S* designation:
identical

c.

1,4- isomer   1,3-isomer
constitutional isomers

b.

same molecular formula,
opposite configuration at one
stereogenic center
enantiomers

d.

trans   **cis**

Both 1,3 isomers,
cis and trans:
diastereomers

5.28

(*S*)-alanine

$[\alpha]$ = +8.5

mp = 297 °C

a. Mp = same as the *S* isomer.
b. The mp of a racemic mixture is often different from the melting point of the enantiomers.
c. −8.5, same as *S* but opposite sign
d. Zero. A racemic mixture is optically inactive.
e. Solution of pure (*S*)-alanine: **optically active**
   Equal mixture of (*R*)- and (*S*)-alanine: **optically inactive**
   75% (*S*)- and 25% (*R*)-alanine: **optically active**

5.29

$$[\alpha] = \frac{\alpha}{l \times c}$$

α = observed rotation
l = length of tube (dm)
c = concentration (g/mL)

$$[\alpha] = \frac{10°}{1 \text{ dm} \times (1 \text{ g/10 mL})} = +100 = \text{specific rotation}$$

5.30 Enantiomeric excess = *ee* = % of one enantiomer − % of other enantiomer.

a. 95% − 5% = **90% *ee***   b. 85% − 15% = **70% *ee***

5.31
a. 90% *ee* means 90% excess of **A** and 10% racemic mixture of **A** and **B** (5% each); therefore, **95% A and 5% B.**
b. 99% *ee* means 99% excess of **A** and 1% racemic mixture of **A** and **B** (0.5% each); therefore, **99.5% A and 0.5% B.**
c. 60% *ee* means 60% excess of **A** and 40% racemic mixture of **A** and **B** (20% each); therefore, **80% A and 20% B.**

Chapter 5–14

5.32

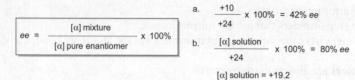

a. $\dfrac{+10}{+24} \times 100\% = 42\%$ *ee*

b. $\dfrac{[\alpha]\text{ solution}}{+24} \times 100\% = 80\%$ *ee*

$[\alpha]$ solution = +19.2

5.33

a. $\dfrac{[\alpha]\text{ mixture}}{+3.8} \times 100\% = 60\%$ *ee*

$[\alpha]$ mixture = +2.3

b. % one enantiomer – % other enantiomer = *ee*
80% – 20% = 60% *ee*

80% dextrorotatory (+) enantiomer
20% levorotatory (–) enantiomer

5.34 • **Enantiomers have the same physical properties** (mp, bp, solubility), and rotate the plane of polarized light to an equal extent, but in opposite directions.
• **Diastereomers have different physical properties.**
• **A racemic mixture is optically inactive.**

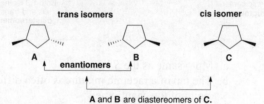

A and **B** are diastereomers of **C**.

a. The bp's of **A** and **B** are the same. The bp's of **A** and **C** are different.
b. Pure **A**: optically active
Pure **B**: optically active
Pure **C**: optically inactive
Equal mixture of **A** and **B**: optically inactive
Equal mixture of **A** and **C**: optically active
c. There would be two fractions: one containing **A** and **B** (optically inactive), and one containing **C** (optically inactive).

5.35

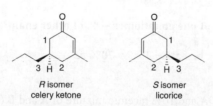

R isomer
celery ketone

S isomer
licorice

5.36

a, b.

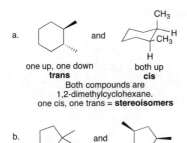

three stereogenic centers

5.37

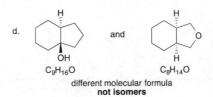

a. **A and B,** same *R,S* assignment, identical
b. **A and C,** opposite *R,S* assignment, enantiomers
c. **A and D,** one stereogenic center different, diastereomers
d. **C and D,** one stereogenic center different, diastereomers

5.38 Use the definitions from Answer 5.2.

a.
one up, one down
trans
— and —
both up
cis
Both compounds are 1,2-dimethylcyclohexane.
one cis, one trans = **stereoisomers**

b.
same molecular formula C_7H_{14}
different connectivity
constitutional isomers

c.
same molecular formula $C_{10}H_{16}O$
different connectivity
constitutional isomers

d.
$C_9H_{16}O$ and $C_8H_{14}O$
different molecular formula
not isomers

Chapter 5–16

5.39 Use the definitions from Answer 5.3.

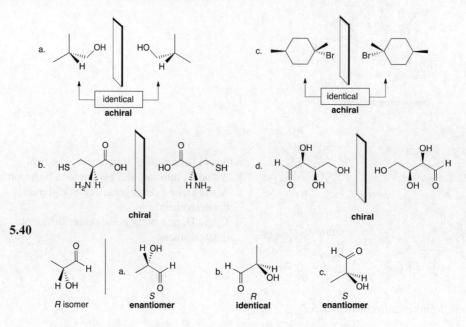

5.40

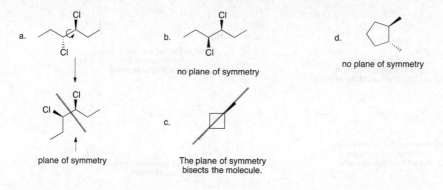

5.41 A plane of symmetry cuts the molecule into **two identical halves**.

Stereochemistry 5–17

5.42 Use the directions from Answer 5.6 to locate the stereogenic centers.

5.43 Stereogenic centers are circled.

amoxicillin

norethindrone

heroin

5.44

a. amphetamine

b. ketoprofen

5.45 Assign priority based on the rules in Answer 5.11.

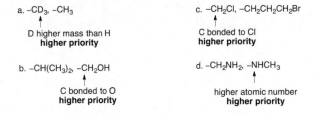

a. –CD₃, –CH₃
  ↑
  D higher mass than H
  higher priority

b. –CH(CH₃)₂, –CH₂OH
  ↑
  C bonded to O
  higher priority

c. –CH₂Cl, –CH₂CH₂CH₂Br
  ↑
  C bonded to Cl
  higher priority

d. –CH₂NH₂, –NHCH₃
  ↑
  higher atomic number
  higher priority

Chapter 5–18

5.46 Assign priority based on the rules in Answer 5.11.

a. –F > –OH > –NH$_2$ > –CH$_3$

b. –(CH$_2$)$_3$CH$_3$ > –CH$_2$CH$_2$CH$_3$ > –CH$_2$CH$_3$ > –CH$_3$

c. –NH$_2$ > –CH$_2$NHCH$_3$ > –CH$_2$NH$_2$ > –CH$_3$

d. –COOH > –CHO > –CH$_2$OH > –H

e. –Cl > –SH > –OH > –CH$_3$

f. –C≡CH > –CH=CH$_2$ > –CH(CH$_3$)$_2$ > –CH$_2$CH$_3$

5.47 Use the rules in Answer 5.13 to assign *R* or *S* to each stereogenic center.

a. 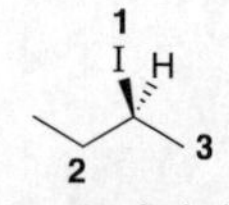
counterclockwise
S isomer

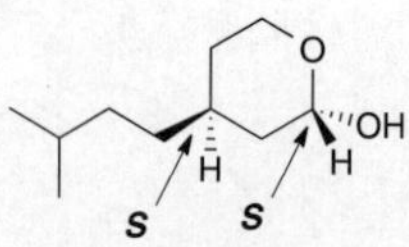

b.
clockwise, but H in front
S isomer

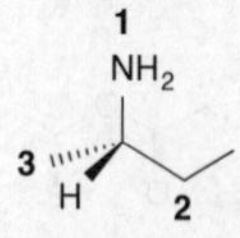

c. 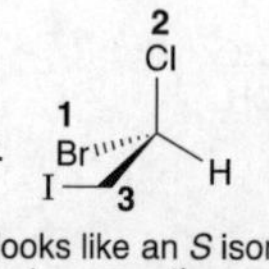
It looks like an *S* isomer, but we must reverse the answer, *S* to *R*.
R isomer

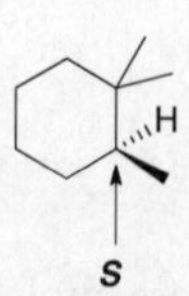

d. 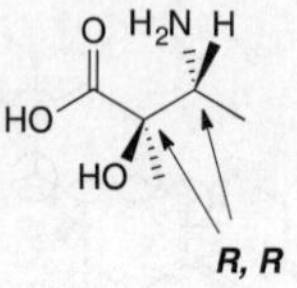
R, R

e.
S

f. 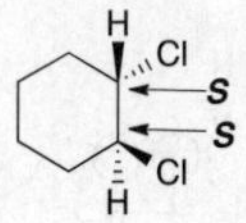
S
S

g.
S   *S*

h. 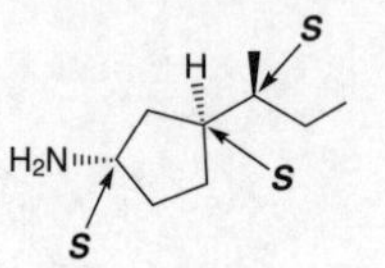
S
S
S

5.48

a. 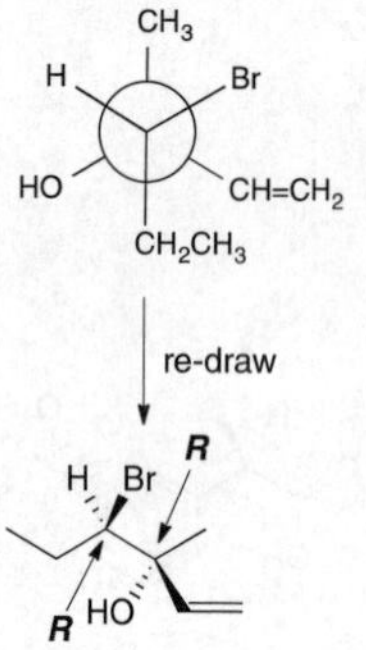
re-draw
R
R

b. 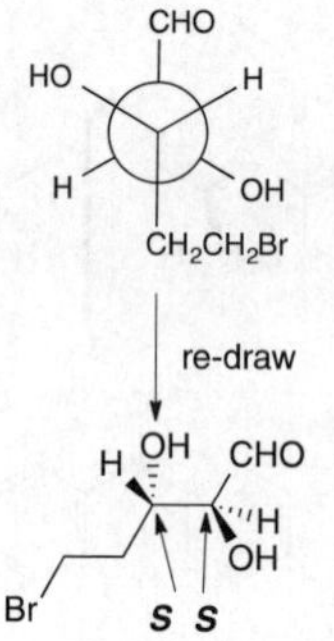
re-draw
S *S*

5.49

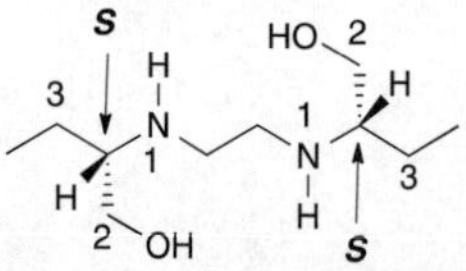

5.50

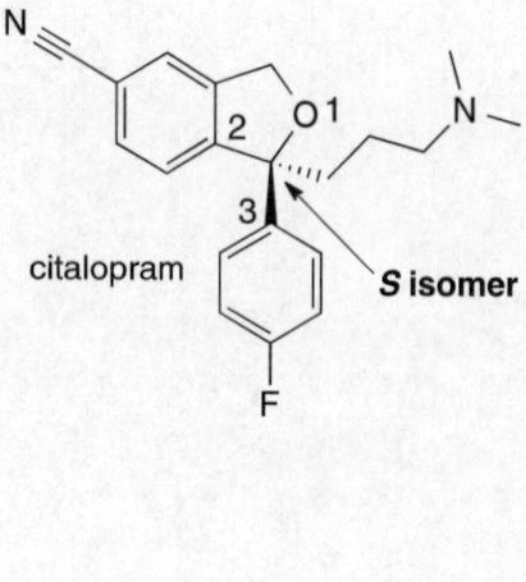

citalopram
S isomer

Stereochemistry 5–19

5.51

a. (R)-3-methylhexane

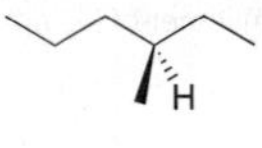

c. (3R,5S,6R)-5-ethyl-3,6-dimethylnonane

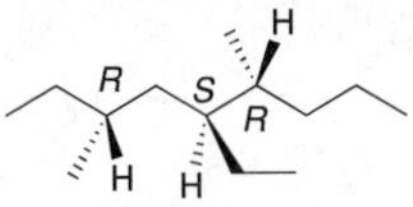

b. (4R,5S)-4,5-diethyloctane

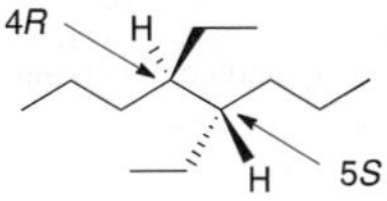

d. (3S,6S)-6-isopropyl-3-methyldecane

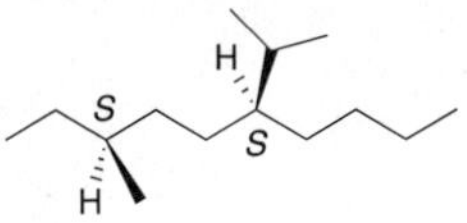

5.52

a.

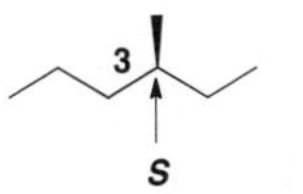

(S)-3-methylhexane

b.

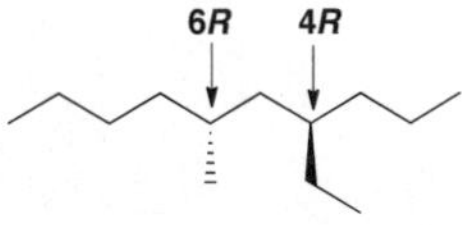

(4R,6R)-4-ethyl-6-methyldecane

c.

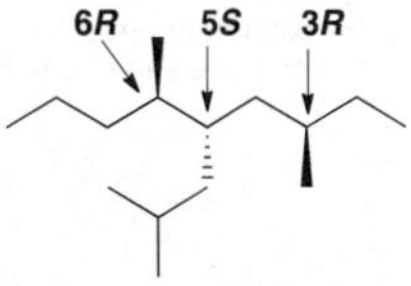

(3R,5S,6R)-5-isobutyl-3,6-dimethylnonane

5.53

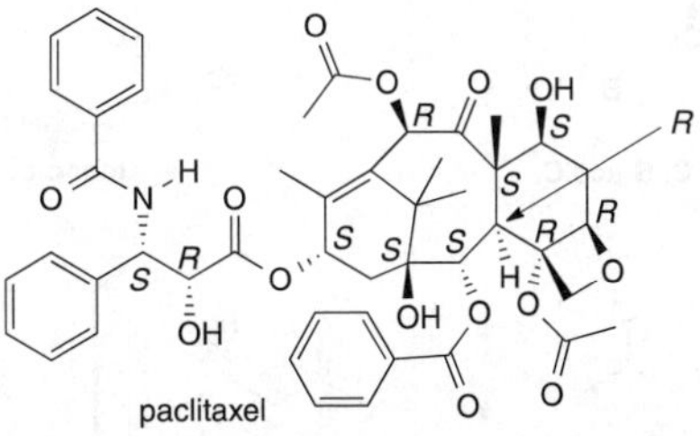

paclitaxel

5.54

a.

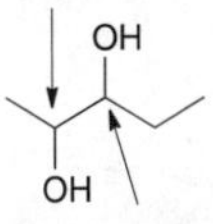

2 stereogenic centers
$2^2 = 4$ possible stereoisomers

b.

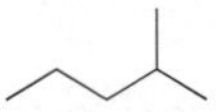

0 stereogenic centers

c.

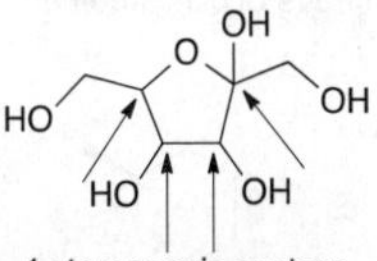

4 stereogenic centers
$2^4 = 16$ possible stereoisomers

5.55

a. 1R,2S — ephedrine

b. 1S,2S — pseudoephedrine

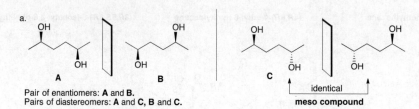

c. Ephedrine and pseudoephedrine are diastereomers. One stereogenic center is the same; one is different.

d. ← e. enantiomer of (–)-ephedrine

← diastereomer of (–)-ephedrine

5.56

a.
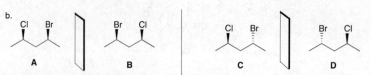

Pair of enantiomers: **A** and **B**.
Pairs of diastereomers: **A** and **C**, **B** and **C**.

identical
meso compound

b.
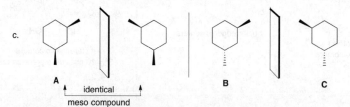

Pairs of enantiomers: **A** and **B**, **C** and **D**.
Pairs of diastereomers: **A** and **C**, **A** and **D**, **B** and **C**, **B** and **D**.

c.

identical
meso compound

Pair of enantiomers: **B** and **C**.
Pairs of diastereomers: **A** and **B**, **A** and **C**.

Stereochemistry 5–21

d.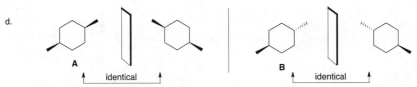

Pair of diastereomers: **A** and **B**.
Meso compounds: **A** and **B**.

5.57

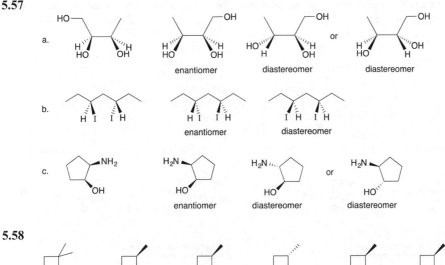

5.58

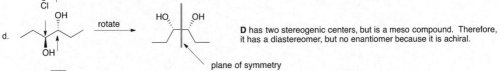

5.59 Explain each statement.

a. All molecules have a mirror image, but only chiral molecules have enantiomers. **A** is not chiral, and therefore, does not have an enantiomer.

b. **B** has one stereogenic center, and therefore, has an enantiomer. Only compounds with two or more stereogenic centers have diastereomers.

c. **C** is chiral and has two stereogenic centers, and therefore, has both an enantiomer and a diastereomer.

d. **D** has two stereogenic centers, but is a meso compound. Therefore, it has a diastereomer, but no enantiomer because it is achiral.

e. **E** has two stereogenic centers, but is a meso compound. Therefore, it has a diastereomer, but no enantiomer because it is achiral.

Chapter 5–22

5.60

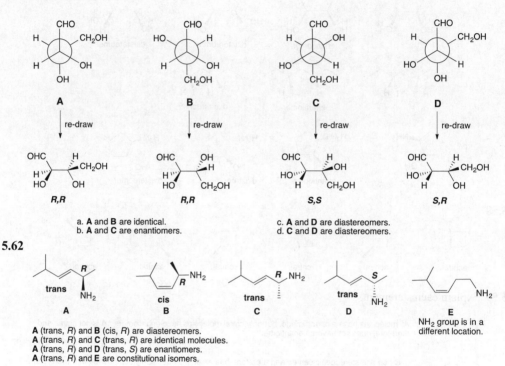

5.61 Re-draw each Newman projection and determine the *R,S* configuration. Then determine how the molecules are related.

a. **A** and **B** are identical.
b. **A** and **C** are enantiomers.
c. **A** and **D** are diastereomers.
d. **C** and **D** are diastereomers.

5.62

A (trans, *R*) and **B** (cis, *R*) are diastereomers.
A (trans, *R*) and **C** (trans, *R*) are identical molecules.
A (trans, *R*) and **D** (trans, *S*) are enantiomers.
A (trans, *R*) and **E** are constitutional isomers.

5.63

a. enantiomers

b. one different configuration, diastereomers (3S, 2R and 2R,3R)

c. mirror images not superimposable, enantiomers

d. enantiomers

e. 2S,3S and 2S,3S — identical

f. 1,4-trans and 1,4-cis — diastereomers

5.64

a. **A** and **B** are constitutional isomers.
   A and **C** are constitutional isomers.
   B and **C** are diastereomers (cis and trans).
   C and **D** are enantiomers.

b.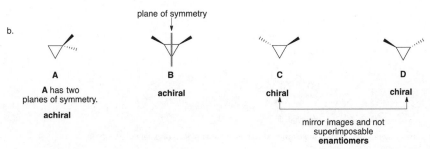

   A — A has two planes of symmetry. achiral
   B — achiral (plane of symmetry)
   C — chiral
   D — chiral
   C and D: mirror images and not superimposable, enantiomers

c. Alone, **C** and **D** would be optically active.
d. **A** and **B** have a plane of symmetry.
e. **A** and **B** have different boiling points.
   B and **C** have different boiling points.
   C and **D** have the same boiling point.
f. **B** is a meso compound.
g. An equal mixture of **C** and **D** is optically inactive because it is a racemic mixture.
   An equal mixture of **B** and **C** would be optically active.

Chapter 5–24

5.65

quinine

$$ee = \frac{[\alpha] \text{ mixture}}{[\alpha] \text{ pure enantiomer}} \times 100\%$$

quinine = **A**
quinine's enantiomer = **B**

a.

$$\frac{-50}{-165} \times 100\% = 30\% \; ee$$

$$\frac{-83}{-165} \times 100\% = 50\% \; ee$$

$$\frac{-120}{-165} \times 100\% = 73\% \; ee$$

b. 30% *ee* = 30% excess one compound (**A**)
remaining 70% = mixture of 2 compounds (35% each **A** and **B**)
Amount of **A** = 30 + 35 = **65%**
Amount of **B** = **35%**

50% *ee* = 50% excess one compound (**A**)
remaining 50% = mixture of 2 compounds (25% each **A** and **B**)
Amount of **A** = 50 + 25 = **75%**
Amount of **B** = **25%**

73% *ee* = 73% excess of one compound (**A**)
remaining 27% = mixture of 2 compounds (13.5% each **A** and **B**)
Amount of **A** = 73 + 13.5 = **86.5%**
Amount of **B** = **13.5%**

c. $[\alpha] = +165$
d. 80% – 20% = 60% *ee*

e. $60\% = \dfrac{[\alpha] \text{ mixture}}{-165} \times 100\%$

$[\alpha]$ mixture = –99

5.66

amygdalin

a. The 11 stereogenic centers are circled. Maximum number of stereoisomers = 2^{11} = 2048
b. Enantiomers of mandelic acid:

c. 60% – 40% = 20% *ee*
20% = $[\alpha]$ mixture/–154 x 100%
$[\alpha]$ mixture = –31

d. $ee = \dfrac{+50}{+154} \times 100\% = 32\% \; ee$

$[\alpha]$ for (*S*)-mandelic acid = +154

32% excess of the *S* enantiomer
68% of racemic *R* and *S* = 34% *S* and 34% *R*

S enantiomer: 32% + 34% = 66%
R enantiomer = 34%

Stereochemistry 5–25

5.67

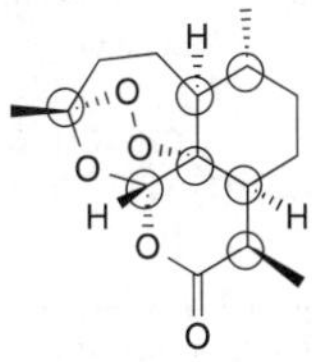

artemisinin

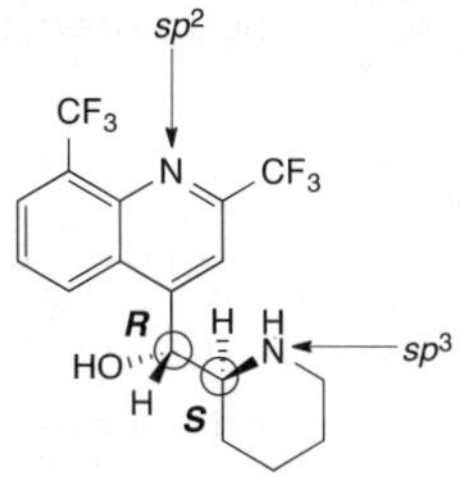

mefloquine

a. Each stereogenic center is circled.
b. The stereogenic centers in mefloquine are labeled.
c. Artemisinin has seven stereogenic centers.
   $2^n = 2^7 = 128$ possible stereoisomers
d. One N atom in mefloquine is sp^2 and one is sp^3.
e. Two molecules of artemisinin cannot intermolecularly H-bond because there are no O–H or N–H bonds.

f.

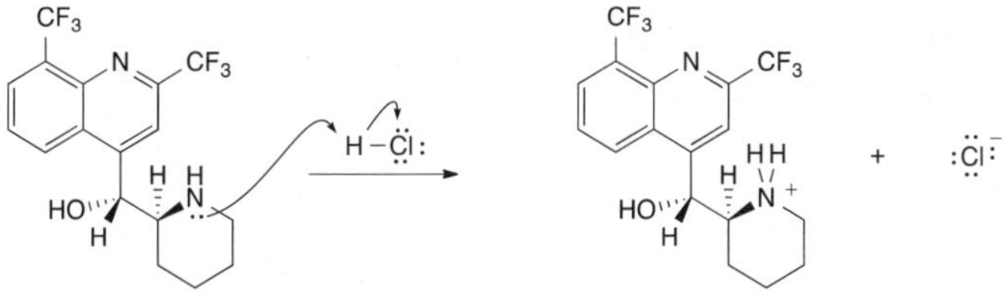

5.68

a. Each stereogenic center is circled.

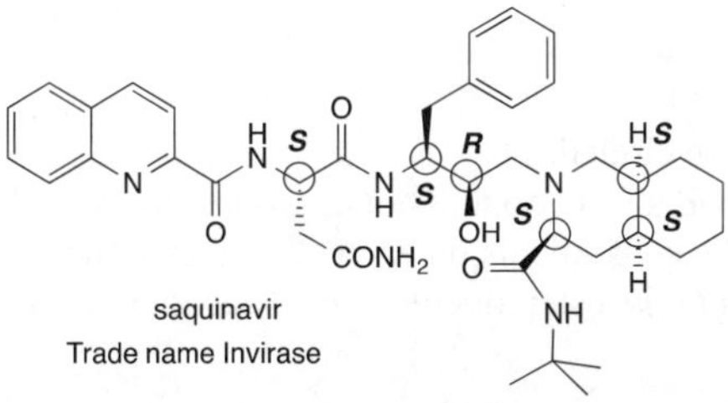

saquinavir
Trade name Invirase

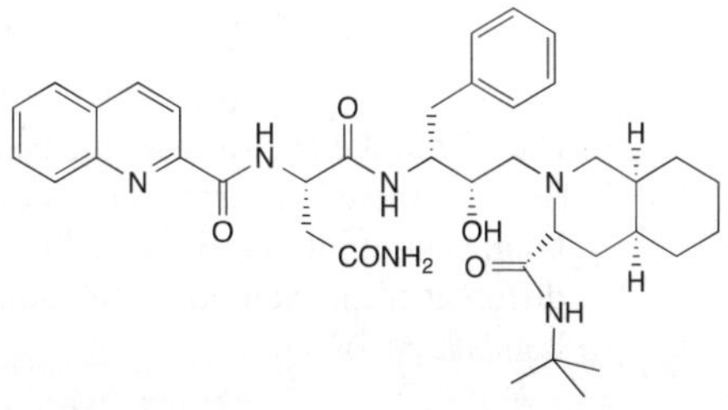

b. enantiomer

c. diastereomer

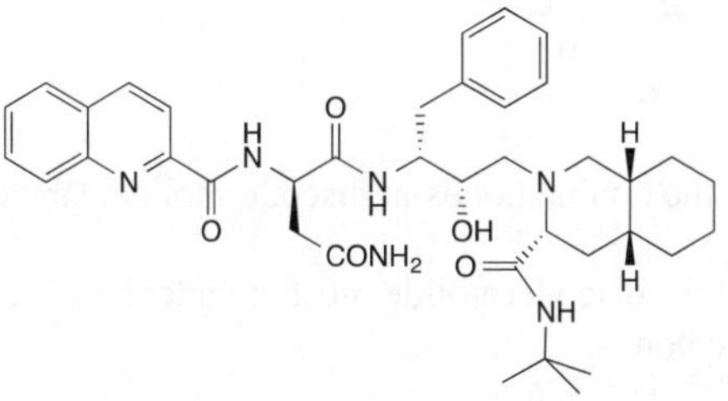

d. constitutional isomer

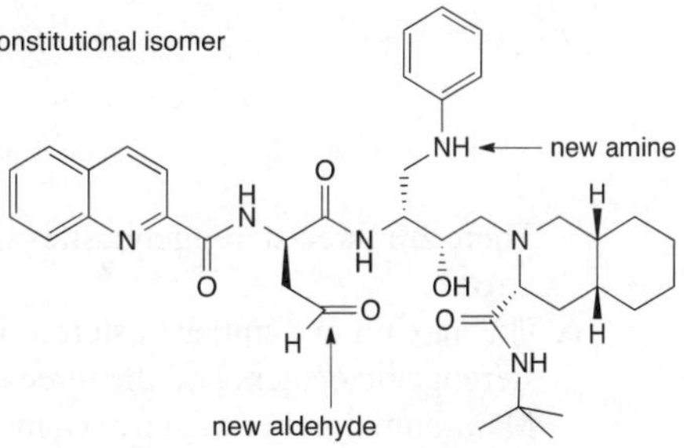

5.69 Allenes contain an *sp* hybridized carbon atom doubly bonded to two other carbons. This makes the double bonds of an allene perpendicular to each other. When each end of the allene has two like substituents, the allene contains two planes of symmetry and it is achiral. When each end of the allene has two different groups, the allene has no plane of symmetry and it becomes chiral.

Chapter 5–26

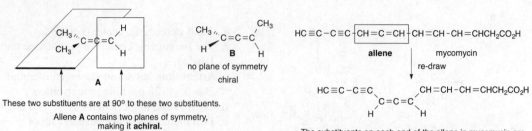

5.70

[structure with labels: H₂N, S, S, O, R, NH₂, HO, S, S, R, N, N, HO, OH — six stereogenic centers]

5.71

[structure of discodermolide]

a. The 13 tetrahedral stereogenic centers are circled.
b. Because there is restricted rotation around a C–C double bond, groups on the end of the double bond cannot interconvert. Whenever the substituents on each end of the double bond are different from each other, the double bond is a stereogenic site. Thus, the following two double bonds are isomers:

[two C=C structures with R/H substituents]
These compounds are isomers.

There are three stereogenic sites due to the double bonds in discodermolide, labeled with arrows.
c. The maximum number of stereoisomers for discodermolide must include the 13 tetrahedral stereogenic centers and the three double bonds.
Maximum number of stereoisomers = 2^{16} = 65,536.

5.72 When the spiro compound has a plane of symmetry, it is achiral.

a. achiral   b. chiral   c. achiral   d. chiral

5.73

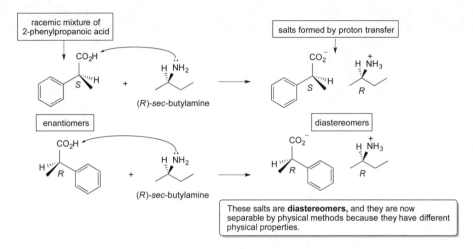

Understanding Organic Reactions 6–1

Chapter 6: Understanding Organic Reactions

Chapter Review

Writing organic reactions (6.1)

- Use curved arrows to show the movement of electrons. Full-headed arrows are used for electron pairs and half-headed arrows are used for single electrons.

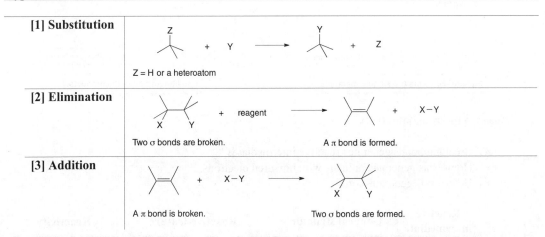

- Reagents can be drawn either on the left side of an equation or over an arrow. Catalysts are drawn over or under an arrow.

Types of reactions (6.2)

[1] Substitution	![substitution reaction: R3C-Z + Y → R3C-Y + Z, where Z = H or a heteroatom]
[2] Elimination	![elimination: R2C(X)-C(Y)R2 + reagent → R2C=CR2 + X-Y. Two σ bonds are broken. A π bond is formed.]
[3] Addition	![addition: R2C=CR2 + X-Y → R2C(X)-C(Y)R2. A π bond is broken. Two σ bonds are formed.]

Important trends

Values compared	Trend
Bond dissociation energy and **bond strength**	The *higher* the bond dissociation energy, the *stronger* the bond (6.4). Increasing size of the halogen → CH$_3$—F CH$_3$—Cl CH$_3$—Br CH$_3$—I $\Delta H°$ = 456 kJ/mol 351 kJ/mol 293 kJ/mol 234 kJ/mol ← Increasing bond strength

E_a and **reaction rate**	The *larger* the energy of activation, the *slower* the reaction (6.9A). 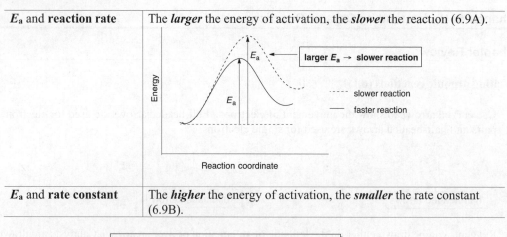
E_a and **rate constant**	The *higher* the energy of activation, the *smaller* the rate constant (6.9B).

Equilibrium always favors the species *lower* in energy.

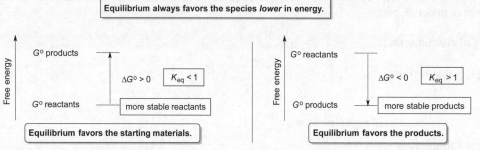

Reactive intermediates (6.3)

- Breaking bonds generates reactive intermediates.
- Homolysis generates radicals with unpaired electrons.
- Heterolysis generates ions.

Reactive intermediate	General structure	Reactive feature	Reactivity
radical		unpaired electron	electrophilic
carbocation		positive charge; only six electrons around C	electrophilic
carbanion		net negative charge; lone electron pair on C	nucleophilic

Understanding Organic Reactions 6–3

Energy diagrams (6.7, 6.8)

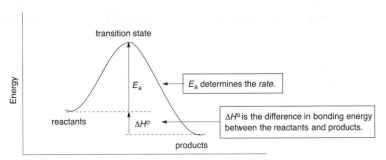

Conditions favoring product formation (6.5, 6.6)

Variable	Value	Meaning
K_{eq}	$K_{eq} > 1$	More product than starting material is present at equilibrium.
$\Delta G°$	$\Delta G° < 0$	The energy of the products is **lower** than the energy of the reactants.
$\Delta H°$	$\Delta H° < 0$	The bonds in the products are **stronger** than the bonds in the reactants.
$\Delta S°$	$\Delta S° > 0$	The product is **more disordered** than the reactant.

Equations (6.5, 6.6)

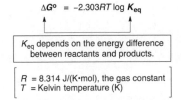
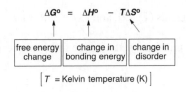

Factors affecting reaction rate (6.9)

Factor	Effect
energy of activation	higher E_a → slower reaction
concentration	higher concentration → faster reaction
temperature	higher temperature → faster reaction

Chapter 6–4

Practice Test on Chapter Review

1. Label each statement as TRUE (T) or FALSE (F) for a reaction with $K_{eq} = 0.5$ and $E_a = 18$ kJ/mol. Ignore entropy considerations.

   a. The reaction is faster than a reaction with $K_{eq} = 8$ and $E_a = 18$ kJ/mol.
   b. The reaction is faster than a reaction with $K_{eq} = 0.5$ and $E_a = 12$ kJ/mol.
   c. $\Delta G°$ for the reaction is a positive value.
   d. The starting materials are lower in energy than the products of the reaction.
   e. The reaction is exothermic.

2.a. Which of the following statements is true about an endothermic reaction, ignoring entropy considerations?

   1. The bonds in the products are stronger than the bonds in the starting materials.
   2. $K_{eq} < 1$.
   3. A catalyst speeds up the rate of the reaction and gives a larger amount of product.
   4. Statements (1) and (2) are both true.
   5. Statements (1), (2), and (3) are all true.

   b. Which of the following statements is true about a reaction with $K_{eq} = 10^3$ and $E_a = 2.5$ kJ/mol? Ignore entropy considerations.

   1. The reaction is faster than a reaction with $E_a = 4$ kJ/mol.
   2. The starting materials are higher in energy than the products of the reaction.
   3. $\Delta G°$ is positive.
   4. Statements (1) and (2) are both true.
   5. Statements (1), (2), and (3) are all true.

3.a. Draw the transition state for the following reaction.

   b. Draw the transition state for the following one-step elimination reaction.

Understanding Organic Reactions 6–5

Answers to Practice Test

1. a. F
   b. F
   c. T
   d. T
   e. F

2. a. 2
   b. 4

3. a. $\left[\begin{array}{c} \overset{\delta+}{\diagup\diagdown} \cdots \overset{H}{\underset{H}{\diagup}} \\ H\text{--}\overset{..}{O}\text{:}\delta+ \end{array} \right]^{\ddagger}$ b. $\left[\begin{array}{c} \diagdown\diagup\diagdown \overset{\cdots\ddot{Br}:\delta-}{=} \\ H\cdots\overset{|}{O}CH_3 \\ \delta- \end{array} \right]^{\ddagger}$

Answers to Problems

6.1 [1] In a **substitution reaction,** one group replaces another.
[2] In an **elimination reaction,** elements of the starting material are lost and a π bond is formed.
[3] In an **addition reaction,** elements are added to the starting material.

a. cyclohexanol → cyclohexyl bromide
   Br replaces OH = substitution reaction

b. cyclohexanone → cyclohexanol
   addition of 2 H's
   addition reaction

c. acetone → chloroacetone
   Cl replaces H = substitution reaction

d. 2-butanol → 2-butene
   elements lost (H + OH)
   π bond formed
   elimination reaction

6.2 The elements of Cl and OH are added to a π bond, so the reaction is an addition.

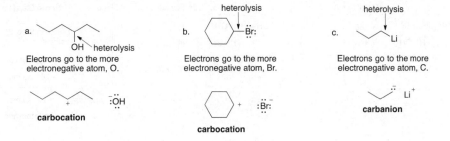

6.3 **Heterolysis** means one atom gets both of the electrons when a bond is broken. A carbocation is a C with a positive charge, and a carbanion is a C with a negative charge.

a. heterolysis
   Electrons go to the more electronegative atom, O.
   + carbocation   :ÖH

b. heterolysis
   Electrons go to the more electronegative atom, Br.
   + carbocation   :Br:⁻

c. heterolysis
   Electrons go to the more electronegative atom, C.
   carbanion   Li⁺

Chapter 6–6

6.4 Use **full-headed arrows** to show the movement of electron pairs, and **half-headed arrows** to show the movement of single electrons.

a.

b.

6.5

6.6 Increasing number of electrons between atoms = increasing bond strength = increasing bond dissociation energy = decreasing bond length.
Increasing size of an atom = increasing bond length = decreasing bond strength.

a.

    —OH   or   —SH

higher bond dissociation energy

S is larger than O.
longer, weaker bond

b.

higher bond dissociation energy

single bond fewer electrons

6.7 **To determine $\Delta H°$ for a reaction:**
[1] Add the bond dissociation energies for all bonds *broken* in the equation (+ values).
[2] Add the bond dissociation energies for all of the bonds *formed* in the equation (– values).
[3] *Add the energies together* to get the $\Delta H°$ for the reaction.
A positive $\Delta H°$ means the reaction is *endothermic*. A negative $\Delta H°$ means the reaction is *exothermic*.

a. $CH_3CH_2-Br + H_2O \longrightarrow CH_3CH_2-OH + HBr$

[1] Bonds broken

	$\Delta H°$ (kJ/mol)
CH_3CH_2-Br	+ 285
$H-OH$	+ 498
Total	+ 783 kJ/mol

[2] Bonds formed

	$\Delta H°$ (kJ/mol)
CH_3CH_2-OH	– 393
$H-Br$	– 368
Total	– 761 kJ/mol

[3] Overall $\Delta H° =$

sum in Step [1]
+
sum in Step [2]

+ 783 kJ/mol
– 761 kJ/mol

ANSWER: + 22 kJ/mol

endothermic

Understanding Organic Reactions 6–7

b. $CH_4 + Cl_2 \longrightarrow CH_3Cl + HCl$

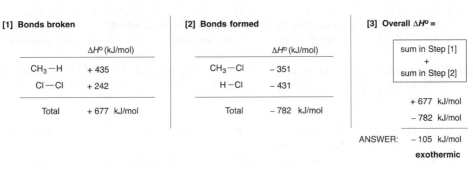

6.8 Use the directions from Answer 6.7. In determining the number of bonds broken or formed, you must take into account the coefficients needed to balance an equation.

a. $CH_4 + 2 O_2 \longrightarrow CO_2 + 2 H_2O$

b. $2 CH_3CH_3 + 7 O_2 \longrightarrow 4 CO_2 + 6 H_2O$

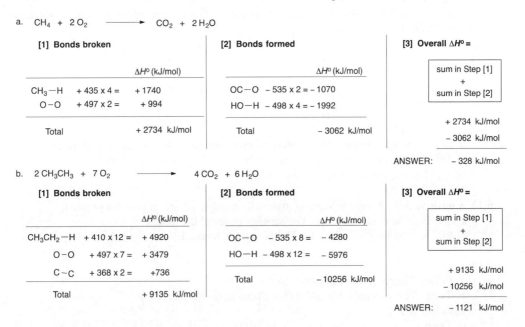

6.9 Use the following relationships to answer the questions:
If $K_{eq} = 1$, then $\Delta G° = 0$; if $K_{eq} > 1$, then $\Delta G° < 0$; if $K_{eq} < 1$, then $\Delta G° > 0$.

a. A negative value of $\Delta G°$ means the equilibrium favors the product and $K_{eq} > 1$. Therefore, $K_{eq} = 1000$ is the answer.
b. A lower value of $\Delta G°$ means a larger value of K_{eq}, and the products are more favored. $K_{eq} = 10^{-2}$ is larger than $K_{eq} = 10^{-5}$, so $\Delta G°$ is lower.

Chapter 6–8

6.10 Use the relationships from Answer 6.9.

    a. $K_{eq} = 5.5$. $K_{eq} > 1$ means that the equilibrium favors the **product.**
    b. $\Delta G° = 40$ kJ/mol. A positive $\Delta G°$ means the equilibrium favors the **starting material.**

6.11 When the *product* is lower in energy than the *starting material,* the equilibrium favors the *product.* When the *starting material* is lower in energy than the *product,* the equilibrium favors the *starting material.*

    a. $\Delta G°$ **is positive,** so the equilibrium favors the starting material. Therefore the *starting material is lower in energy than the product.*
    b. $K_{eq} > 1$, so the equilibrium favors the product. Therefore the *product is lower in energy than the starting material.*
    c. $\Delta G°$ **is negative,** so the equilibrium favors the product. Therefore the *product is lower in energy than the starting material.*
    d. $K_{eq} < 1$, so the equilibrium favors the starting material. Therefore *the starting material is lower in energy than the product.*

6.12

    a. $K_{eq} > 1$, so the **product** (the conformation on the right) is favored at equilibrium.
    b. $\Delta G°$ for this process must be **negative,** because the product is favored.
    c. $\Delta G°$ is somewhere between 0 and −6 kJ/mol.

6.13 A positive $\Delta H°$ favors the starting material. A negative $\Delta H°$ favors the product.
    a. $\Delta H°$ is positive (80 kJ/mol). The starting material is favored.
    b. $\Delta H°$ is negative (−40 kJ/mol). The product is favored.

6.14
    a. **False.** The reaction is endothermic.
    b. **True.** This assumes that $\Delta G°$ is approximately equal to $\Delta H°$.
    c. **False.** $K_{eq} < 1$.
    d. **True.**
    e. **False.** The starting material is favored at equilibrium.

6.15
    a. **True.**
    b. **False.** $\Delta G°$ for the reaction is negative.
    c. **True.**
    d. **False.** The bonds in the product are stronger than the bonds in the starting material.
    e. **True.**

6.16

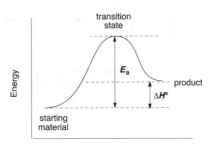

6.17 A transition state is drawn with dashed lines to indicate the partially broken and partially formed bonds. Any atom that gains or loses a charge contains a partial charge in the transition state.

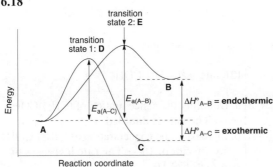

6.18

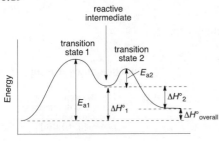

a. Reaction **A–C** is exothermic. Reaction **A–B** is endothermic.
b. Reaction **A–C** is faster.
c. Reaction **A–C** generates a lower-energy product.
d. See labels.
e. See labels.
f. See labels.

6.19

a. Two steps, because there are two energy barriers.
b. See labels.
c. See labels.
d. One reactive intermediate is formed (see label).
e. The first step is rate-determining, because its transition state is at higher energy.
f. The overall reaction is endothermic, because the energy of the products is higher than the energy of the reactants.

Chapter 6–10

6.20

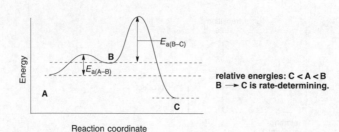

relative energies: C < A < B
B → C is rate-determining.

6.21 E_a, **concentration, and temperature affect reaction rate.** $\Delta H°$, $\Delta G°$, and K_{eq} do not affect reaction rate.

a. E_a = **4 kJ/mol** corresponds to a faster reaction rate.
b. A temperature of **25 °C** will have a faster reaction rate, because a higher temperature corresponds to a faster reaction.
c. **No change:** K_{eq} does not affect reaction rate.
d. **No change:** $\Delta H°$ does not affect reaction rate.

6.22
a. **False.** The reaction occurs at the same rate as a reaction with K_{eq} = 8 and E_a = 80 kJ/mol.
b. **False.** The reaction is slower than a reaction with K_{eq} = 0.8 and E_a = 40 kJ/mol.
c. **True.**
d. **True.**
e. **False.** The reaction is endothermic.

6.23 All reactants in the rate equation determine the rate of the reaction.

[1] rate = $k[CH_3CH_2Br][^-OH]$

a. Tripling the concentration of CH_3CH_2Br only → **The rate is tripled.**
b. Tripling the concentration of ^-OH only → **The rate is tripled.**
c. Tripling the concentration of both $CH_3CH_2CH_2Br$ and ^-OH → **The rate increases by a factor of 9 (3 × 3 = 9).**

[2] rate = $k[(CH_3)_3COH]$

a. Doubling the concentration of $(CH_3)_3COH$ → **The rate is doubled.**
b. Increasing the concentration of $(CH_3)_3COH$ by a factor of 10 → **The rate increases by a factor of 10.**

6.24 The rate equation is determined by the rate-determining step.

a. [structure with Br] + ^-OH → [alkene] + H_2O + Br^- one step
   rate = $k[CH_3CH_2CH_2Br][^-OH]$

b. [tert-butyl Br] slow → [carbocation] + Br^- ^-OH fast → [alkene] + H_2O two steps
   The slow step determines the rate equation.
   rate = $k[(CH_3)_3CBr]$

Understanding Organic Reactions 6–11

6.25 A catalyst is not used up or changed in the reaction. It only speeds up the reaction rate.

OH and H are added to
the starting material.

a. $CH_2=CH_2$ $\xrightarrow[H_2SO_4]{H_2O}$ CH_3CH_2OH

H_2SO_4 is not used up = **catalyst.**

I^- not used up = **catalyst.**

b. CH_3Cl $\xrightarrow[^-OH]{I^-}$ CH_3OH

^-OH substitutes for Cl^-.

6.26

a.
radical
$+$ $\cdot H$

b.
$CH_3\overset{+}{C}H_2$ $+$ ^-OH
carbocation

6.27

propane

$CH_3-CH_2CH_3$ $\longrightarrow$ $\cdot CH_3$ $+$ $\cdot CH_2CH_3$

$\Delta H^\circ = 356$ kJ/mol
This bond is formed from two
sp^3 hybridized C's.

propene

$CH_3-CH=CH_2$ $\longrightarrow$ $\cdot CH_3$ $+$ $\cdot CH=CH_2$

$\Delta H^\circ = 385$ kJ/mol
This bond is formed from one sp^2 and one sp^3
hybridized C. The higher percent s-character in one
C makes a stronger bond; thus, the bond
dissociation energy is higher.

6.28 Use the directions from Answer 6.1.

a.
HO   OH
elements lost
(H + OH)

π bond formed

elimination reaction

b.
O

H   OH
addition of 2 H's
addition reaction

6.29 Use the rules in Answer 6.4 to draw the arrows.

a.
$:\overset{..}{\underset{..}{O}}:$ $:\overset{..}{\underset{..}{Cl}}:$

$\longrightarrow$
$+$ $:\overset{..}{\underset{..}{Cl}}:$

b.
$+$ $:\overset{..}{\underset{..}{Br}}-\overset{..}{\underset{..}{Br}}:$ $\longrightarrow$
$:\overset{..}{\underset{..}{Br}}:$
$+$ $:\overset{..}{\underset{..}{Br}}\cdot$

c. $CH_3CH_2-\overset{..}{\underset{..}{Br}}:$ $+$ $^-:\overset{..}{\underset{..}{OH}}$ $\longrightarrow$ $CH_3CH_2\overset{..}{\underset{..}{OH}}$ $+$ $:\overset{..}{\underset{..}{Br}}:^-$

d.
$\overset{+}{}$ H $+$ $^-:\overset{..}{\underset{..}{OH}}$ $\longrightarrow$ $+$ $H_2\overset{..}{\underset{..}{O}}:$

6.30

a. [reaction scheme]

b. [reaction scheme]

6.31

a. [reaction scheme showing A → B → C]

b. The conversion of **A** to **C** is an addition because a π bond is broken and an O atom is added.

c. [resonance structures]

6.32 Draw the curved arrows to identify the product **X**.

[reaction scheme: A + H–Br → [1] → B + Br⁻ → [2] → X]

6.33 Follow the curved arrows to identify the intermediate **Y**.

[reaction scheme: C → [1] → D → [2] → Y]

6.34 Use the rules from Answer 6.6.

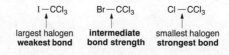

Understanding Organic Reactions 6–13

6.35 Use the directions from Answer 6.7.

a. ·OH + CH$_4$ ⟶ ·CH$_3$ + H$_2$O

[1] Bonds broken		[2] Bonds formed		[3] Overall $\Delta H°$ =
	$\Delta H°$ (kJ/mol)		$\Delta H°$ (kJ/mol)	+ 435 kJ/mol
CH$_3$–H	+ 435 kJ/mol	H–OH	– 498 kJ/mol	– 498 kJ/mol
				ANSWER: – 63 kJ/mol

b. CH$_3$–OH + HBr ⟶ CH$_3$–Br + H$_2$O

[1] Bonds broken		[2] Bonds formed		[3] Overall $\Delta H°$ =
	$\Delta H°$ (kJ/mol)		$\Delta H°$ (kJ/mol)	
CH$_3$–OH	+ 389	CH$_3$–Br	– 293	+ 757 kJ/mol
H–Br	+ 368	H–OH	– 498	– 791 kJ/mol
Total	+ 757 kJ/mol	Total	– 791 kJ/mol	ANSWER: – 34 kJ/mol

6.36

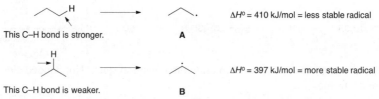

6.37 The more stable radical is formed by a reaction with a smaller $\Delta H°$.

This C–H bond is stronger. → **A** $\Delta H°$ = 410 kJ/mol = less stable radical

This C–H bond is weaker. → **B** $\Delta H°$ = 397 kJ/mol = more stable radical

Because the bond dissociation for cleavage of the C–H bond to form radical **A** is higher, more energy must be added to form it. This makes **A** higher in energy and therefore less stable than **B**.

6.38 Use the rules from Answer 6.11.
a. K_{eq} = 0.5. K_{eq} is less than one, so the **starting material** is favored.
b. $\Delta G°$ = –100 kJ/mol. $\Delta G°$ is less than 0, so the **product** is favored.
c. $\Delta H°$ = 8.0 kJ/mol. $\Delta H°$ is positive, so the **starting material** is favored.
d. K_{eq} = 16. K_{eq} is greater than one, so the **product** is favored.
e. $\Delta G°$ = 2.0 kJ/mol. $\Delta G°$ is greater than zero, so the **starting material** is favored.
f. $\Delta H°$ = 200 kJ/mol. $\Delta H°$ is positive, so the **starting material** is favored.
g. $\Delta S°$ = 8 J/(K·mol). $\Delta S°$ is greater than zero, so the **product** is more disordered and favored.
h. $\Delta S°$ = –8 J/(K·mol). $\Delta S°$ is less than zero, so the **starting material** is more disordered and favored.

Chapter 6–14

6.39

a. A negative ΔG° must have $K_{eq} > 1$. $K_{eq} = 10^2$.
b. $K_{eq} = [\text{products}]/[\text{reactants}] = [1]/[5] = 0.2 = K_{eq}$. ΔG° is positive.
c. A negative ΔG° has $K_{eq} > 1$, and a positive ΔG° has $K_{eq} < 1$. $\Delta G^\circ = -8$ kJ/mol will have a larger K_{eq}.

6.40

R	K_{eq}
$-CH_2CH_3$	23
$-C(CH_3)_3$	4000

a. The equatorial conformation is present in the larger amount at equilibrium, because the K_{eq} is greater than 1.
b. The cyclohexane with the $-C(CH_3)_3$ group will have the greater amount of equatorial conformation at equilibrium, because this group has the higher K_{eq}.
c. The cyclohexane with the $-CH_2CH_3$ group will have the greater amount of axial conformation at equilibrium, because this group has the lower K_{eq}.
d. The cyclohexane with the $-C(CH_3)_3$ group will have the more negative ΔG°, because it has the larger K_{eq}.
e. The larger the R group, the more favored the equatorial conformation.

6.41 Reactions resulting in an increase in entropy are favored. When a single molecule forms two molecules, there is an increase in entropy.

a.

increased number of molecules
ΔS° is positive.
products favored

b. $CH_3CO_2CH_3$ + H_2O ⟶ CH_3CO_2H + CH_3OH

no change in the number of molecules
neither favored

6.42 Use the directions in Answer 6.17 to draw the transition state. Nonbonded electron pairs are drawn in at reacting sites.

6.43

a.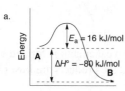
Reaction coordinate
- one step **A→B**
- exothermic because **B** lower than **A**

b.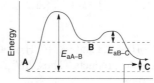
Reaction coordinate
- two steps
- **A** lowest energy
- **B** highest energy
- $E_{a(A-B)}$ **is rate-determining,** because the transition state for Step [1] is higher in energy.

6.44

a. $CH_3\text{—}H + \cdot \ddot{C}\ddot{l}: \longrightarrow \cdot CH_3 + H\ddot{C}\ddot{l}:$

b. $\cdot Cl + CH_4 \longrightarrow \cdot CH_3 + HCl$

[1] Bonds broken	$\Delta H°$ (kJ/mol)
$CH_3\text{—}H$	+ 435 kJ/mol

[2] Bonds formed	$\Delta H°$ (kJ/mol)
H–Cl	– 431 kJ/mol

[3] Overall $\Delta H°$ =
+ 435 kJ/mol
– 431 kJ/mol
ANSWER: + 4 kJ/mol

c.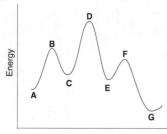
Reaction coordinate

d. E_a for the reverse reaction is the difference in energy between the products and the transition state, 12 kJ/mol.

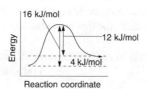
Reaction coordinate

6.45

a. **B, D,** and **F** are transition states.
b. **C** and **E** are reactive intermediates.
c. The overall reaction has **three steps.**
d. **A–C** is endothermic.
    C–E is exothermic.
    E–G is exothermic.
e. The overall reaction is exothermic.

6.46

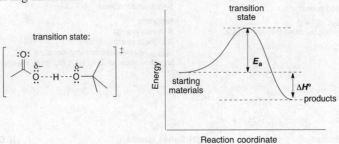

Because pK_a (CH$_3$CO$_2$H) = 4.8 and pK_a [(CH$_3$)$_3$COH] = 18, the weaker acid is formed as product, and equilibrium favors the products. Thus, $\Delta H°$ is negative, and the products are lower in energy than the starting materials.

6.47

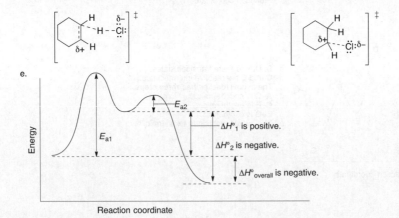

a. Step [1] breaks one π bond and the H–Cl bond, and one C–H bond is formed. $\Delta H°$ for this step should be positive, because more bonds are broken than formed.
b. Step [2] forms one bond. $\Delta H°$ for this step should be negative, because one bond is formed and none is broken.
c. Step [1] is rate-determining, because it is more difficult.
d. Transition state for Step [1]:   Transition state for Step [2]:

e.

Understanding Organic Reactions 6–17

6.48 E_a, concentration, catalysts, rate constant, and temperature affect reaction rate so (c), (d), (e), (g), and (h) affect rate.

6.49
a. **rate = k[CH₃Br][NaCN]**

a. **rate = $k[\text{CH}_3\text{Br}][\text{NaCN}]$**
b. Double [CH₃Br] = **rate doubles.**
c. Halve [NaCN] = **rate halved.**
d. Increase both [CH₃Br] and [NaCN] by factor of 5 = [5][5] = **rate increases by a factor of 25.**

6.50

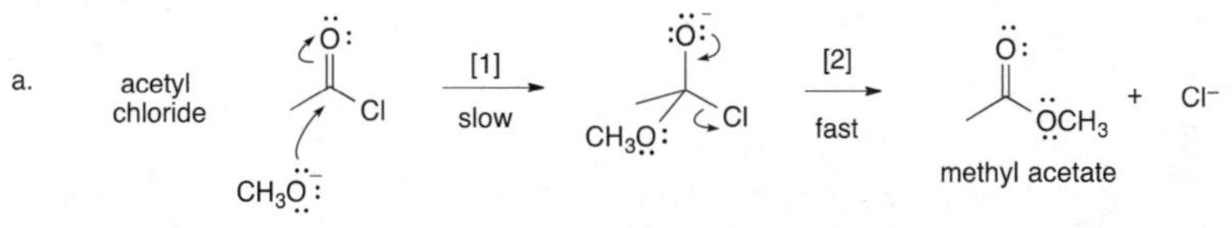

a.

b. Only the slow step is included in the rate equation: **Rate = $k[\text{CH}_3\text{O}^-][\text{CH}_3\text{COCl}]$**
c. CH_3O^- is in the rate equation. Increasing its concentration by 10 times would increase the rate by **10 times.**
d. When both reactant concentrations are increased by 10 times, the rate increases by **100 times (10 × 10 = 100).**
e. This is a **substitution reaction** (OCH₃ substitutes for Cl).

6.51
a. **True:** Increasing temperature increases reaction rate.
b. **True:** If a reaction is fast, then it has a large rate constant.
c. **False: Corrected**—There is no relationship between $\Delta G°$ and reaction rate.
d. **False: Corrected**—When the E_a is large, *the rate constant is small.*
e. **False: Corrected**—There is no relationship between K_{eq} and reaction rate.
f. **False: Corrected**—Increasing the concentration of a reactant increases the rate of a reaction *only if the reactant appears in the rate equation.*

6.52

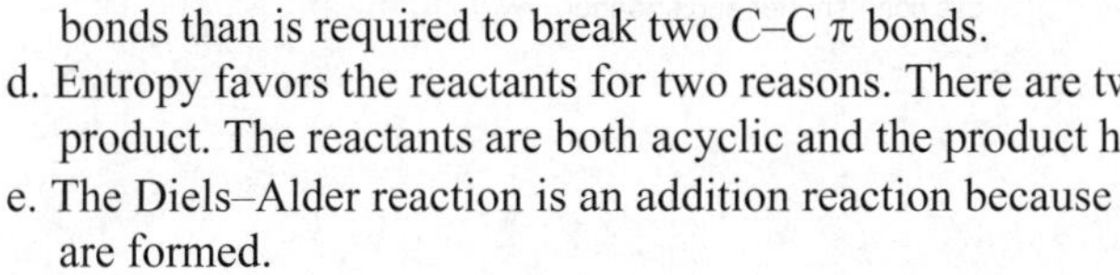

a.

b. Two π bonds in **A** are broken and one π bond in **B** is broken. Two new σ bonds in **C** are formed (in bold), as well as a new π bond.
c. The reaction should be exothermic because more energy is released in forming two new C–C σ bonds than is required to break two C–C π bonds.
d. Entropy favors the reactants for two reasons. There are two molecules of reactant and only one product. The reactants are both acyclic and the product has a ring with fewer degrees of freedom.
e. The Diels–Alder reaction is an addition reaction because π bonds are broken and new σ bonds are formed.

Chapter 6–18

6.53
a. The first mechanism has one step: **Rate = $k[(CH_3)_3CI][^-OH]$**
b. The second mechanism has two steps, but only the first step would be in the rate equation, because it is slow and therefore rate-determining: **Rate = $k[(CH_3)_3CI]$**
c. Possibility [1] is second order; possibility [2] is first order.
d. These rate equations can be used to show which mechanism is plausible by changing the concentration of ^-OH. If this affects the rate, then possibility [1] is reasonable. If it does not affect the rate, then possibility [2] is reasonable.

e.

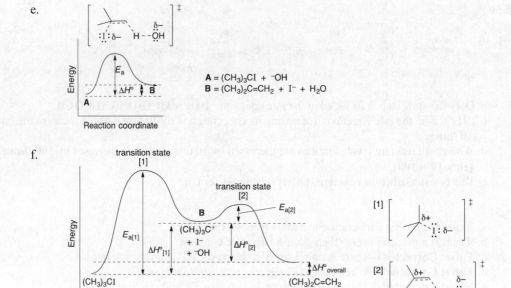

6.54 The difference in both the acidity and the bond dissociation energy of CH_3CH_3 versus $HC\equiv CH$ is due to the same factor: percent s-character. The difference results because one process is based on homolysis and one is based on heterolysis.
Bond dissociation energy:

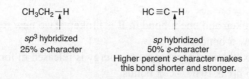

Understanding Organic Reactions 6–19

Acidity: To compare acidity, we must compare the stability of the conjugate bases:

$CH_3\overset{-}{C}H_2$

sp^3 hybridized
25% s-character

$HC\equiv C^{-}$

sp hybridized
50% s-character
Now a higher percent s-character
stabilizes the conjugate base making the
starting acid more acidic.

6.55 a. Re-draw **A** to see more clearly how cyclization occurs.

σ bonds formed

re-draw

A

σ bond formed

B

b.

O

6.56

a.

H_a

H_b

b.

H_a

H_b

H_a

H_a

c. C–H_a is weaker than C–H_b because the carbon radical formed when the C–H_a bond is broken is highly resonance stabilized. This means the bond dissociation energy for C–H_a is lower.

6.57 In Reaction [1], the number of molecules of reactants and products stays the same, so entropy is not a factor. In Reaction [2], a single molecule of starting material forms two molecules of products, so entropy increases. This makes $\Delta G°$ more favorable, thus increasing K_{eq}.

Chapter 6–20

6.58

ethyl acetate

$+ H_2O$ $K_{eq} = 4$

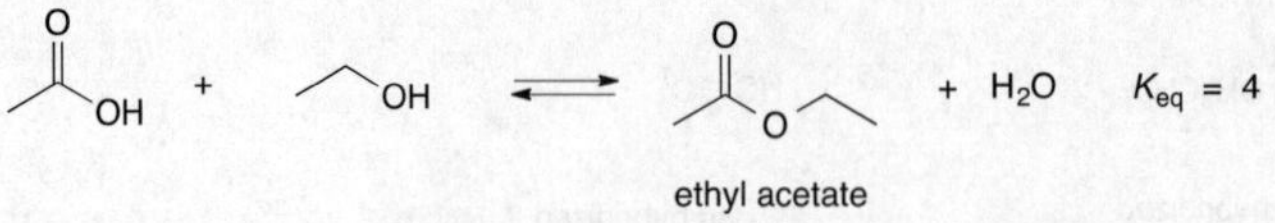

To increase the yield of ethyl acetate, H_2O can be removed from the reaction mixture, or there can be a large excess of one of the starting materials.

6.59

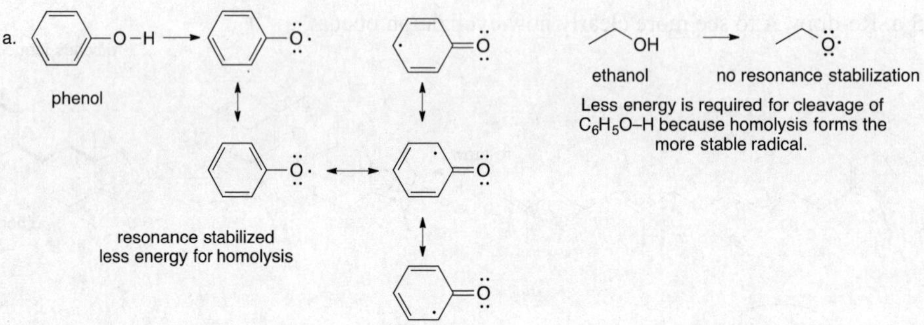

a.

phenol

ethanol   no resonance stabilization

Less energy is required for cleavage of C_6H_5O–H because homolysis forms the more stable radical.

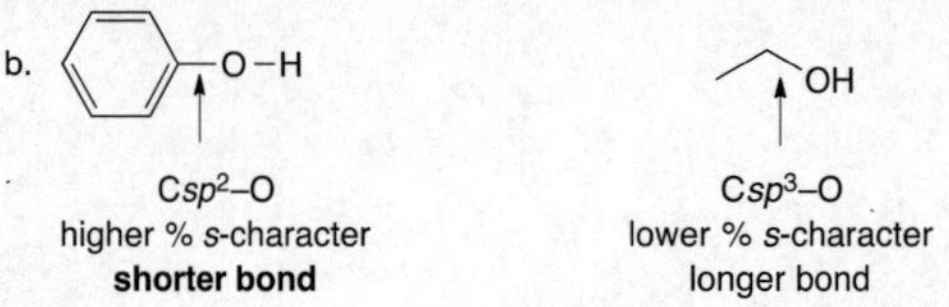

resonance stabilized
less energy for homolysis

b.

Csp^2–O
higher % s-character
shorter bond

Csp^3–O
lower % s-character
longer bond

Chapter 7 Alkyl Halides and Nucleophilic Substitution

Chapter Review

General facts about alkyl halides

- Alkyl halides contain a halogen atom X bonded to an sp^3 hybridized carbon (7.1).
- Alkyl halides are named as halo alkanes, with the halogen as a substituent (7.2).
- Alkyl halides have a polar C–X bond, so they exhibit dipole–dipole interactions but are incapable of intermolecular hydrogen bonding (7.3).
- The polar C–X bond containing an electrophilic carbon makes alkyl halides reactive towards nucleophiles and bases (7.5).

The central theme (7.6)

- Nucleophilic substitution is one of the two main reactions of alkyl halides. A nucleophile replaces a leaving group on an sp^3 hybridized carbon.

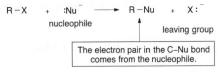

- One σ bond is broken and one σ bond is formed.
- There are two possible mechanisms: S_N1 and S_N2.

S_N1 and S_N2 mechanisms compared

	S_N2 mechanism	S_N1 mechanism
[1] Mechanism	• One step (7.11B)	• Two steps (7.12B)
[2] Alkyl halide	• Order of reactivity: $CH_3X >$ $RCH_2X > R_2CHX > R_3CX$ (7.11D)	• Order of reactivity: $R_3CX >$ $R_2CHX > RCH_2X > CH_3X$ (7.12D)
[3] Rate equation	• rate = $k[RX][:Nu^-]$ • second-order kinetics (7.11A)	• rate = $k[RX]$ • first-order kinetics (7.12A)
[4] Stereochemistry	• backside attack of the nucleophile (7.11C) • inversion of configuration at a stereogenic center	• trigonal planar carbocation intermediate (7.12C) • racemization at a stereogenic center
[5] Nucleophile	• favored by stronger nucleophiles (7.15B)	• favored by weaker nucleophiles (7.15B)
[6] Leaving group	• better leaving group → faster reaction (7.15C)	• better leaving group → faster reaction (7.15C)
[7] Solvent	• favored by polar aprotic solvents (7.15D)	• favored by polar protic solvents (7.15D)

Chapter 7–2

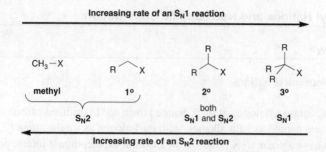

Important trends

- The best leaving group is the weakest base. Leaving group ability increases left-to-right across a row and down a column of the periodic table (7.7).

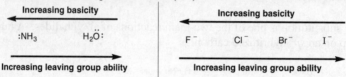

- Nucleophilicity decreases across a row of the periodic table (7.8A).

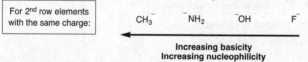

- Nucleophilicity decreases down a column of the periodic table in polar aprotic solvents (7.8C).

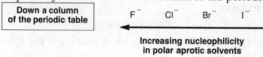

- Nucleophilicity increases down a column of the periodic table in polar protic solvents (7.8C).

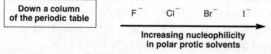

- The stability of a carbocation increases as the number of R groups bonded to the positively charged carbon increases (7.13).

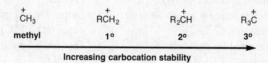

Alkyl Halides and Nucleophilic Substitution 7–3

Important principles

Principle	Example
• Electron-donating groups (such as R groups) stabilize a positive charge (7.13A).	• 3° Carbocations (R$_3$C$^+$) are more stable than 2° carbocations (R$_2$CH$^+$), which are more stable than 1° carbocations (RCH$_2$$^+$).
• Steric hindrance decreases nucleophilicity but not basicity (7.8B).	• (CH$_3$)$_3$CO$^-$ is a stronger base but a weaker nucleophile than CH$_3$CH$_2$O$^-$.
• Hammond postulate: In an endothermic reaction, the more stable product is formed faster. In an exothermic reaction, this fact is not necessarily true (7.14).	• S$_N$1 reactions are faster when more stable (more substituted) carbocations are formed, because the rate-determining step is endothermic.
• Planar, sp^2 hybridized atoms react with reagents from both sides of the plane (7.12C).	• A trigonal planar carbocation reacts with nucleophiles from both sides of the plane.

Practice Test on Chapter Review

1. Give the IUPAC name for the following compound, including the appropriate R,S prefix.

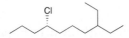

2.a. Which of the following carbocations is the most stable?

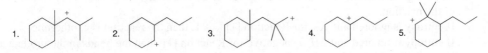

b. Which of the following anions is the best leaving group?

   1. CH$_3$$^-$ 2. $^-$OH   3. H$^-$ 4. $^-$NH$_2$ 5. Cl$^-$

c. Which species is the strongest nucleophile in polar protic solvents?

   1. F$^-$ 2. $^-$OH   3. Cl$^-$ 4. H$_2$O   5. $^-$SH

d. Which of the following statements is true about the given reaction?

Chapter 7–4

1. The reaction follows second-order kinetics.
2. The rate of the reaction increases when the solvent is changed from CH₃CH₂OH to DMSO.
3. The rate of the reaction increases when the leaving group is changes from Br to F.
4. Statements (1) and (2) are both true.
5. Statements (1), (2), and (3) are all true.

3. Rank the following compounds in order of *increasing* reactivity in an S$_N$1 reaction. Rank the *least reactive* compound as **1**, the *most reactive* compound as **4**, and the compounds of intermediate reactivity as **2** and **3**.

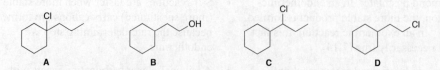

4. Consider the following two nucleophilic substitution reactions, labeled Reaction [1] and Reaction [2]. (Only the starting materials are drawn.) Then answer True (T) or False (F) to each of the following statements.

   Reaction [1]   (CH₃CH₂)₃CBr   +   CH₃OH   ⎯⎯⎯⎯→

   Reaction [2]   CH₃CH₂CH₂Br   +   ⁻OCH₃   ⎯⎯⎯⎯→

   a. The rate equation for Reaction [1] is rate = k[(CH₃CH₂)₃CBr][CH₃OH].
   b. Changing the leaving group from Br⁻ to Cl⁻ decreases the rate of both reactions.
   c. Changing the solvent from CH₃OH to (CH₃)₂S=O increases the rate of Reaction [2].
   d. Doubling the concentration of both CH₃CH₂CH₂Br and ⁻OCH₃ in Reaction [2] doubles the rate of the reaction.
   e. If entropy is ignored and K$_{eq}$ for Reaction [1] is < 1, then the reaction is exothermic.
   f. If entropy is ignored and ΔH° is negative for Reaction [1], then the bonds in the product are stronger than the bonds in the starting materials.
   g. The energy diagram for Reaction [2] exhibits only one energy barrier.

5. Draw the organic products formed in the following reactions. **Use wedges and dashed wedges to show stereochemistry in compounds with stereogenic centers.**

   a. [structure with Br, D, H]   ⁻C≡C–H ⎯⎯→

   b. [structure with Cl]   ⎯⎯CH₃OH⎯⎯→ (Consider substitution only.)

   c. [structure with Br]   ⎯⎯NaOH⎯⎯→

Alkyl Halides and Nucleophilic Substitution 7–5

d.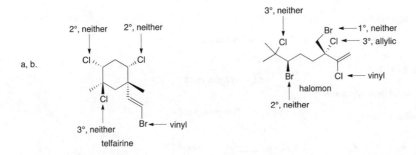

Answers to Practice Test

1. (S)-7-chloro-3-ethyldecane

2. a. 4
   b. 5
   c. 5
   d. 4

3. A–4
   B–1
   C–3
   D–2

4. a. F
   b. T
   c. T
   d. F
   e. F
   f. T
   g. T

5. a., b., c., d.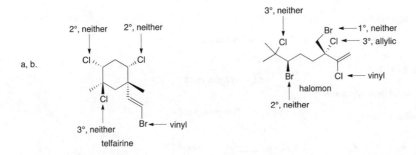

Answers to Problems

7.1 Classify the alkyl halide as 1°, 2°, or 3° **by counting the number of carbons bonded directly to the carbon bonded to the halogen.**

a, b.

(telfairine labels: 2°, neither; 2°, neither; 3°, neither; Br ← vinyl)

halomon labels: 3°, neither; 1°, neither; 3°, allylic; Cl ← vinyl; 2°, neither

7.2 To name a compound with the IUPAC system:
   [1] **Name the parent** chain by finding the longest carbon chain.
   [2] **Number the chain** so the first substituent gets the lower number. Then **name and number all substituents,** giving like substituents a prefix (di, tri, etc.). **To name the halogen substituent, change the -*ine* ending to -*o*.**
   [3] **Combine all parts,** alphabetizing substituents, and ignoring all prefixes except iso.

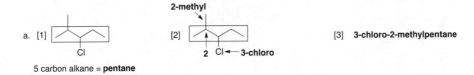

a. [1] 5 carbon alkane = **pentane**   [2] 2-methyl, 2 Cl ← 3-chloro   [3] **3-chloro-2-methylpentane**

Chapter 7–6

b.

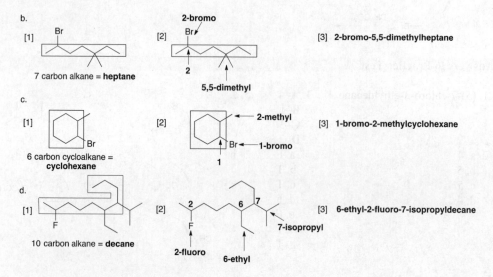

c.

d.

7.3 To work backwards from a name to a structure:
[1] Find the parent name and draw that number of carbons. Use the suffix to identify the functional group (**-ane = alkane**).
[2] Arbitrarily number the carbons in the chain. Add the substituents to the appropriate carbon.

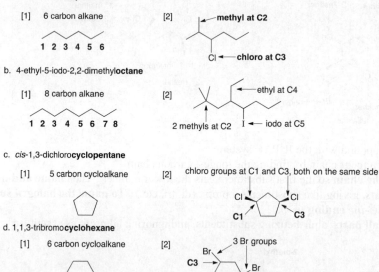

e. **sec-butyl bromide**

[1]   4 carbon alkyl group   [2]

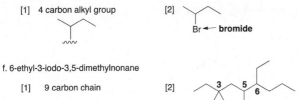

Br ← bromide

f. 6-ethyl-3-iodo-3,5-dimethylnonane

[1]   9 carbon chain   [2]

7.4 a. Because an sp^2 hybridized C has a higher percent s-character than an sp^3 hybridized C, it holds electron density closer to C. This pulls a little more electron density towards C, away from Cl, and thus a C_{sp^2}–Cl bond is less polar than a C_{sp^3}–Cl bond.

b.
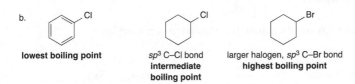

lowest boiling point sp^3 C–Cl bond intermediate boiling point   larger halogen, sp^3 C–Br bond **highest boiling point**

7.5 a. Since chondrocole A has 10 C's and only one functional group capable of hydrogen bonding to water (an ether), it is insoluble in H₂O. Because it is organic, it is soluble in CH₂Cl₂.

b.
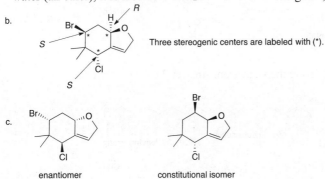

Three stereogenic centers are labeled with (*).

c.

enantiomer   constitutional isomer

7.6 To draw the products of a nucleophilic substitution reaction:
[1] **Find the sp^3 hybridized electrophilic carbon** with a leaving group.
[2] **Find the nucleophile** with lone pairs or electrons in π bonds.
[3] **Substitute the nucleophile for the leaving group** on the electrophilic carbon.

a.
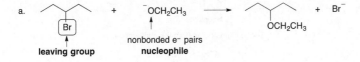

leaving group   nonbonded e⁻ pairs nucleophile

b.
Cl
Na+ -OH
OH
+ Na+Cl–
leaving group
nonbonded e– pairs
nucleophile

c.
I
+
N3–
N3
+ I–
leaving group
nonbonded e– pairs
nucleophile

d.
Br
+
Na+ -CN
CN
+ Na+ Br–
leaving group
nonbonded e– pairs
nucleophile

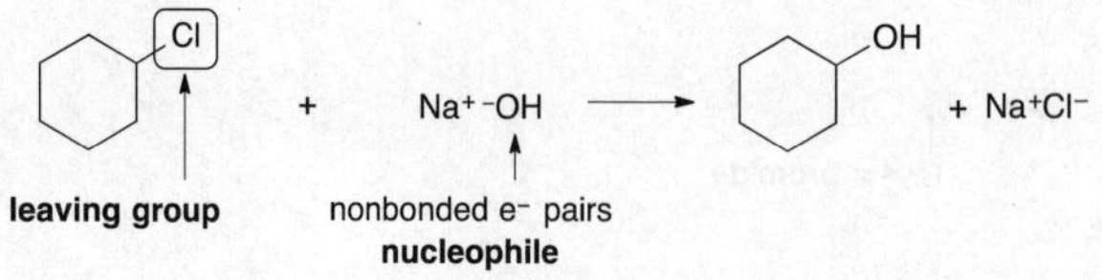

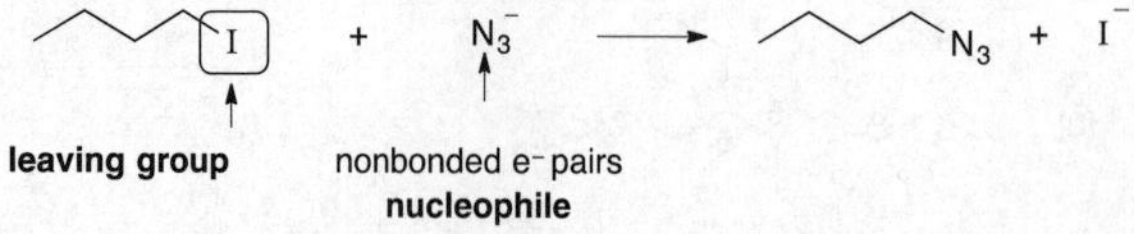

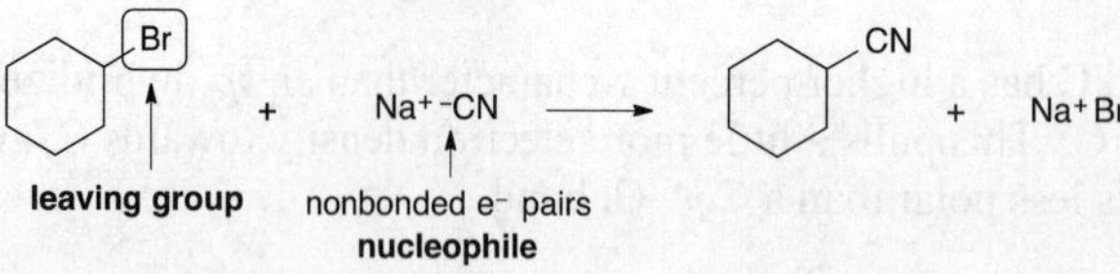

7.7 Use the steps from Answer 7.6 and then draw the proton transfer reaction.

a.
Br
+ : N(CH2CH3)3
substitution
+N(CH2CH3)3 + Br–
nucleophile
leaving group

b.
Cl
+
H2Ö:
substitution
nucleophile
leaving group
H
O+
H
:Cl:–
proton
transfer
ÖH
+ HCl

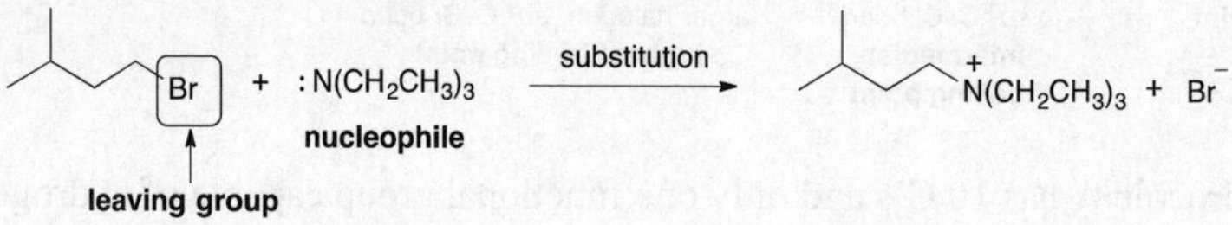

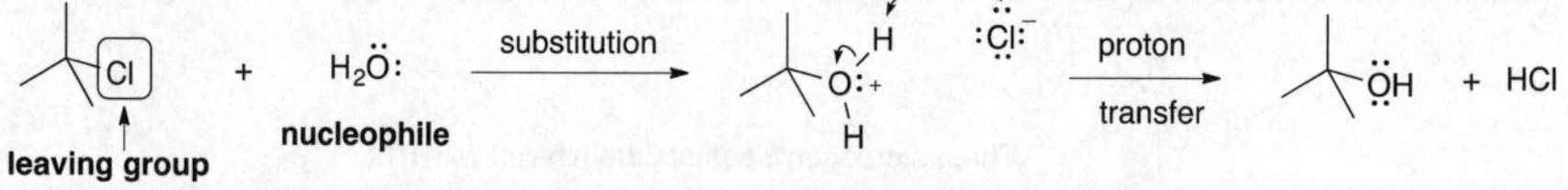

7.8 Draw the structure of CPC using the steps from Answer 7.6.

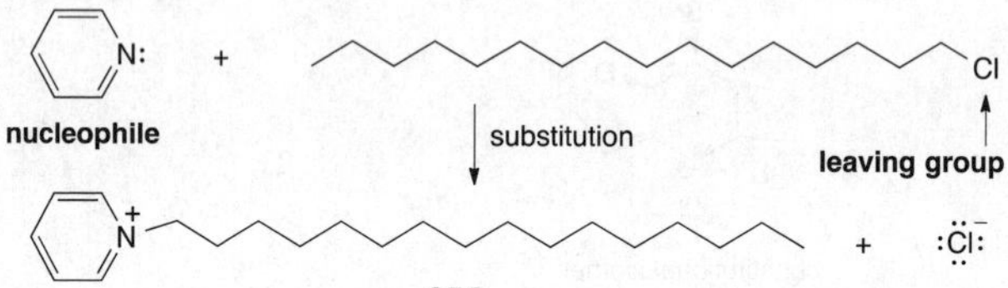

N:
+
Cl
nucleophile
substitution
leaving group
N+
+
:Cl:–

CPC

7.9

Cl
Cl
A
Cl
N
S
ticlopidine
These atoms come from the nucleophile.
NH
S

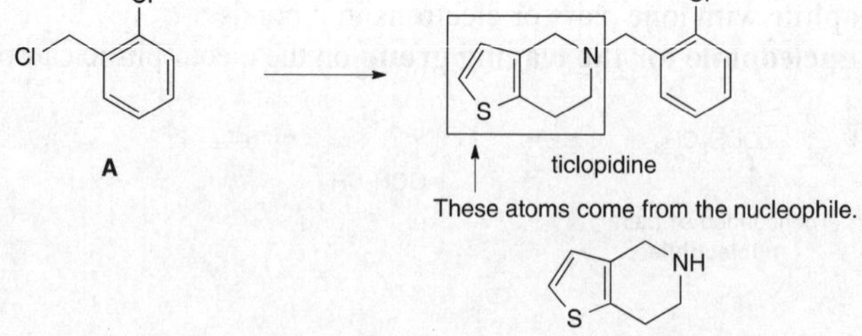

7.10 Compare the leaving groups based on these trends:
- Better leaving groups are weaker bases.
- A neutral leaving group is always better than its conjugate base.

7.11 Good leaving groups include Cl⁻, Br⁻, I⁻, and H₂O.

7.12 To decide whether the equilibrium favors the starting material or the products, **compare the nucleophile and the leaving group.** The reaction proceeds towards the weaker base.

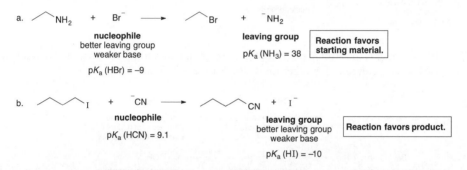

7.13 Use these three rules to find the stronger nucleophile in each pair:
  [1] Comparing two nucleophiles having the *same attacking atom*, the stronger base is a stronger nucleophile.
  [2] **Negatively charged nucleophiles** are always **stronger than their conjugate acids.**
  [3] **Nucleophilicity decreases left to right across a row of the periodic table,** when comparing species of similar charge.

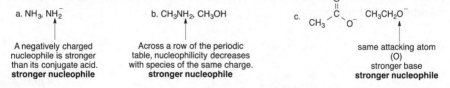

7.14 *Polar protic solvents* are capable of hydrogen bonding, and therefore must contain a **H bonded to an electronegative O or N.** *Polar aprotic solvents* are incapable of hydrogen bonding, and therefore do not contain any O–H or N–H bonds.

Chapter 7–10

a. HO—CH2CH2—OH
contains 2 O–H bonds
polar protic

b. CH3CH2—O—CH2CH3
no O–H bonds
polar aprotic

c. CH3C(=O)OCH2CH3
no O–H bonds
polar aprotic

7.15 • In *polar protic solvents,* **the trend in nucleophilicity is opposite to the trend in basicity** down a column of the periodic table, so nucleophilicity increases.
  • In *polar aprotic solvents,* **the trend is identical to basicity,** so nucleophilicity decreases down a column.

a. Br⁻ and Cl⁻ in polar protic solvent

farther down the column
**more nucleophilic
in protic solvent**

c. HS⁻ and F⁻ in polar protic solvent

farther down the column
and left in the row
**more nucleophilic
in protic solvent**

In polar protic solvents:
nucleophilicity increases

O   F
S   nucleophilicity increases

b. ⁻OH and Cl⁻ in polar aprotic solvent

farther up the column
and to the left in the row
more basic
more nucleophilic

In polar aprotic solvents:
nucleophilicity increases

O   F
    Cl   nucleophilicity increases

7.16 The stronger base is the stronger nucleophile, except in polar protic solvents, where nucleophilicity increases down a column. For other rules, see Answers 7.13 and 7.15.

a. H2O   ⁻OH   ⁻NH2
    no charge   negatively charged   negatively charged
    **weakest   **intermediate   farther left in periodic table
    nucleophile**   nucleophile**   **strongest nucleophile**

b. Br⁻   F⁻   ⁻OH
    Basicity decreases down a Basicity decreases   **strongest nucleophile**
    column in polar aprotic solvents. across a row.
    weakest nucleophile   **intermediate nucleophile**

c. H2O   CH3COO⁻   ⁻OH
    weakest nucleophile weaker base than ⁻OH   **strongest nucleophile**
                            intermediate nucleophile

7.17 To determine what nucleophile is needed to carry out each reaction, look at the product to see what has replaced the leaving group.

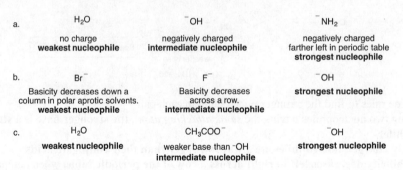

a. R-Br → R-SH
SH replaces Br.
HS⁻ is needed.

b. R-Br → R-OCH2CH3
OCH2CH3 replaces Br.
CH3CH2O⁻ is needed.

c. R-Br → R-OC(=O)CH3
OCOCH3 replaces Br.
CH3COO⁻ is needed.

d. R-Br → R-C≡CH
C≡CH replaces Br.
HC≡C⁻ is needed.

7.18 The general rate equation for an S$_N$2 reaction is rate = k[RX][:Nu$^-$].

   a. [RX] is tripled, and [:Nu$^-$] stays the same: **rate triples.**
   b. Both [RX] and [:Nu$^-$] are tripled: **rate increases by a factor of 9 (3 × 3 = 9).**
   c. [RX] is halved, and [:Nu$^-$] stays the same: **rate halved.**
   d. [RX] is halved, and [:Nu$^-$] is doubled: **rate stays the same (1/2 × 2 = 1).**

7.19 All S$_N$2 reactions have one step. The transition state in an S$_N$2 reaction has **dashed bonds to both the leaving group and the nucleophile,** and must contain partial charges.

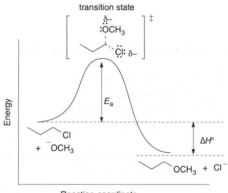

7.20 To draw the products of S$_N$2 reactions, **replace the leaving group by the nucleophile, and then draw the stereochemistry with *inversion* at the stereogenic center.**

7.21 *Increasing* the number of R groups *increases* crowding of the transition state and *decreases* the rate of an S$_N$2 reaction.

a. 2° alkyl halide   or   1° alkyl halide **faster reaction**

b. 2° alkyl halide **faster reaction**   or   3° alkyl halide

Chapter 7–12

7.22

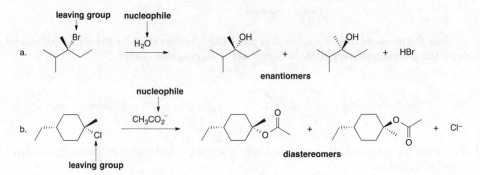

7.23 In a first-order reaction, **the rate changes with any change in [RX].** The rate is independent of any change in [:Nu⁻].
  a. [RX] is tripled, and [:Nu⁻] stays the same: **rate triples.**
  b. Both [RX] and [:Nu⁻] are tripled: **rate triples.**
  c. [RX] is halved, and [:Nu⁻] stays the same: **rate is halved.**
  d. [RX] is halved, and [:Nu⁻] is doubled: **rate is halved.**

7.24 In S$_N$1 reactions, racemization always occurs at a stereogenic center. Draw two products, with the two possible configurations at the stereogenic center.

7.25 Carbocations are classified by the number of R groups bonded to the carbon: 0 R groups = methyl, 1 R group = 1°, 2 R groups = 2°, and 3 R groups = 3°.

  a. 2 R groups
     2° carbocation
  b. 1 R group
     1° carbocation
  c. 3 R groups
     3° carbocation
  d. 2 R groups
     2° carbocation

7.26 For carbocations: Increasing number of R groups = Increasing stability.

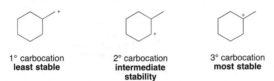

1° carbocation
least stable

2° carbocation
intermediate stability

3° carbocation
most stable

7.27 For carbocations: Increasing number of R groups = Increasing stability.

1° carbocation
least stable

2° carbocation
intermediate stability

3° carbocation
most stable

7.28 The rate of an S$_N$1 reaction increases with increasing alkyl substitution.

a. 3° alkyl halide
    faster S$_N$1 reaction   or   1° alkyl halide
    slower S$_N$1 reaction

b. 3° alkyl halide
    faster S$_N$1 reaction   or   2° alkyl halide
    slower S$_N$1 reaction

7.29 • For **methyl and 1° alkyl halides,** only S$_N$2 will occur.
• For **2° alkyl halides,** S$_N$1 and S$_N$2 will occur.
• For **3° alkyl halides,** only S$_N$1 will occur.

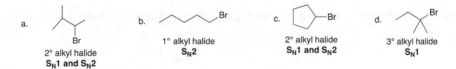

a. 2° alkyl halide
S$_N$1 and S$_N$2

b. 1° alkyl halide
S$_N$2

c. 2° alkyl halide
S$_N$1 and S$_N$2

d. 3° alkyl halide
S$_N$1

7.30 • Draw the product of nucleophilic substitution for each reaction.
• For **methyl and 1° alkyl halides,** only S$_N$2 will occur.
• For **2° alkyl halides,** S$_N$1 and S$_N$2 will occur and other factors determine which mechanism operates.
• For **3° alkyl halides,** only S$_N$1 will occur.

Chapter 7–14

a.

3° alkyl halide
only S_N1

CH_3OH

OCH_3 + HCl

c.

2° alkyl halide
**Both S_N1 and S_N2
are possible.**

Strong nucleophile
favors S_N2.

$CH_3CH_2O^-$

OCH_2CH_3

+ I^-

b.

1° alkyl halide
only S_N2

^-SH

SH + Br^-

d.

2° alkyl halide
**Both S_N1 and S_N2
are possible.**

Weak nucleophile
favors S_N1.

CH_3OH

OCH_3 + HBr

7.31 First decide whether the reaction will proceed via an S_N1 or S_N2 mechanism. Then draw the products with stereochemistry.

a.

H   Br

2° alkyl halide
S_N1 and S_N2

+ H_2O

Weak nucleophile
favors S_N1.

H   OH   HO   H

enantiomers

+ HBr

**S_N1 = racemization at the
stereogenic C**

b.

Cl

H   D

1° alkyl halide
S_N2 only

+ $^-:C\equiv C-H$

$HC\equiv C$

D   H

+ Cl^-

S_N2 = inversion at the stereogenic C

7.32 Compounds with better leaving groups react faster. Weaker bases are better leaving groups.

a.

Cl   or   I

**weaker base
better leaving group**

c.

OH   or $\overset{+}{O}H_2$

**weaker base
better leaving group**

b.

Br   or   I

**weaker base
better leaving group**

d.

OH   or

**weaker base
better leaving group**

7.33   • **Polar protic solvents** favor the S_N1 mechanism by solvating the intermediate carbocation and halide.
   • **Polar aprotic solvents** favor the S_N2 mechanism by making the nucleophile stronger.

a. CH_3CH_2OH
polar protic solvent
contains an O–H bond
favors S_N1

b. CH_3CN
polar aprotic solvent
no O–H or N–H bond
favors S_N2

c. CH_3COOH
polar protic solvent
contains an O–H bond
favors S_N1

d. $CH_3CH_2OCH_2CH_3$
polar aprotic solvent
no O–H or N–H bond
favors S_N2

Alkyl Halides and Nucleophilic Substitution 7–15

7.34 Compare the solvents in the reactions below. **For the solvent to increase the reaction rate of an S_N1 reaction, the solvent must be *polar protic*. For the solvent to increase the reaction rate of an S_N2 reaction, the solvent must be *polar aprotic*.**

a.

$3^{\circ} RX - S_N1$ reaction

CH$_3$OH
Polar protic solvent
increases the rate of an
S_N1 reaction.

b.

$1^{\circ} RX - S_N2$ reaction

DMF [HCON(CH$_3$)$_2$]
Polar aprotic solvent
increases the rate of an
S_N2 reaction.

c.

$2^{\circ} RX$ strong nucleophile
S_N2 reaction

HMPA [(CH$_3$)$_2$N]$_3$P=O
Polar aprotic solvent
increases the rate of an
S_N2 reaction.

7.35 To predict whether the reaction follows an S_N1 or S_N2 mechanism:
   [1] **Classify RX as a methyl, 1°, 2°, or 3° halide.** (Methyl, 1° = S_N2; 3° = S_N1; 2° = either.)
   [2] **Classify the nucleophile as strong or weak.** (Strong favors S_N2; weak favors S_N1.)
   [3] **Classify the solvent as polar protic or polar aprotic.** (Polar protic favors S_N1; polar aprotic favors S_N2.)

a.

1° alkyl halide
S_N2

S_N2 reaction

b.

2° alkyl halide   Strong nucleophile
S_N1 or S_N2   **favors S_N2.**

S_N2 reaction = *inversion* at the stereogenic center
The leaving group was "up."
The nucleophile attacks from below.

c.

3° alkyl halide   Weak nucleophile
S_N1   **favors S_N1.**

S_N1 reaction

d.

3° alkyl halide   Weak nucleophile
S_N1   **favors S_N1.**

S_N1 reaction
forms **two enantiomers.**

7.36

Chapter 7–16

7.37 Vinyl carbocations are even less stable than 1° carbocations.

vinyl carbocation
least stable

1° carbocation
intermediate stability

2° carbocation
most stable

7.38 Convert each ball-and-stick model to a skeletal or condensed structure and draw the reactants.

a. carbon framework — CN (nucleophile)

~~~Cl →(Na+ -CN)→ ~~~CN

b. carbon framework — SH (nucleophile)

>~Cl →(Na+ -SH)→ >~SH

c. carbon framework — OH (nucleophile)

cyclohexyl-Cl →(Na+ -OH)→ cyclohexyl-OH

d. carbon framework — C≡CH (nucleophile)

~Cl →(Na+ -C≡CH)→ ~C≡CH

7.39

CH₃O:⁻ + Cl—CH₂CH₃ ⟶ CH₃ÖCH₂CH₃   CH₃CH₂Ö:⁻ + Cl—CH₃ ⟶ CH₃ÖCH₂CH₃

7.40 Use the directions from Answer 7.2 to name the compounds.

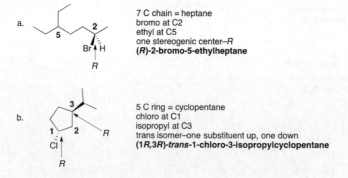

a. 7 C chain = heptane
bromo at C2
ethyl at C5
one stereogenic center–R
(R)-2-bromo-5-ethylheptane

b. 5 C ring = cyclopentane
chloro at C1
isopropyl at C3
trans isomer–one substituent up, one down
(1R,3R)-trans-1-chloro-3-isopropylcyclopentane

7.41

a.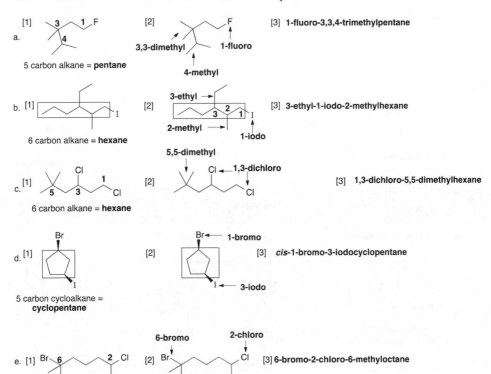

7.42 Use the directions from Answer 7.2 to name the compounds.

Chapter 7–18

f. [1]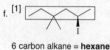
6 carbon alkane = **hexane**
(Indicate the R,S designation also)

[2]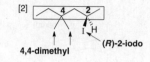
4,4-dimethyl   (R)-2-iodo

[3] **(R)-2-iodo-4,4-dimethylhexane**
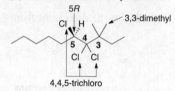
Clockwise
R

7.43 To work backwards to a structure, use the directions in Answer 7.3.

a. 3-bromo-4-ethyl**heptane**

[structure: 3-bromo, 4-ethyl on heptane chain with carbons 3,4 labeled; Br at C3]

b. 1,1-dichloro-2-methyl**cyclohexane**

1,1-dichloro
2-methyl

c. 1-bromo-4-ethyl-3-fluoro**octane**

[structure: 4-ethyl, 1-bromo, 3-fluoro on octane chain with carbons 1,3,4 labeled]

d. (S)-3-iodo-2-methyl**nonane**

[structure: 2-methyl, 3-iodo on nonane chain, 3S]

e. (1R,2R)-*trans*-1-bromo-2-chloro**cyclohexane**

[cyclohexane with 1R Br and 2R Cl, trans]

f. (R)-4,4,5-trichloro-3,3-dimethyl**decane**

[structure: 5R, 3,3-dimethyl, 4,4,5-trichloro on decane]

7.44

1-chloro 1°
1-chloropentane

2° 3-chloro
3-chloropentane

3-methyl
1-chloro 1°
1-chloro-3-methylbutane

1-chloro 1°
1-chloro-2,2-dimethylpropane

2-chloro 3°
2-methyl
2-chloro-2-methylbutane

Two stereoisomers

2-chloro 2°

2-chloropentane
[* denotes stereogenic center]

→ Clockwise "4" in back = **R**   Clockwise "4" in front = **S**

Two stereoisomers

3-methyl
2° 2-chloro
2-chloro-3-methylbutane
[* denotes stereogenic center]

→ Counterclockwise "4" in front = **R**   Counterclockwise "4" in back = **S**

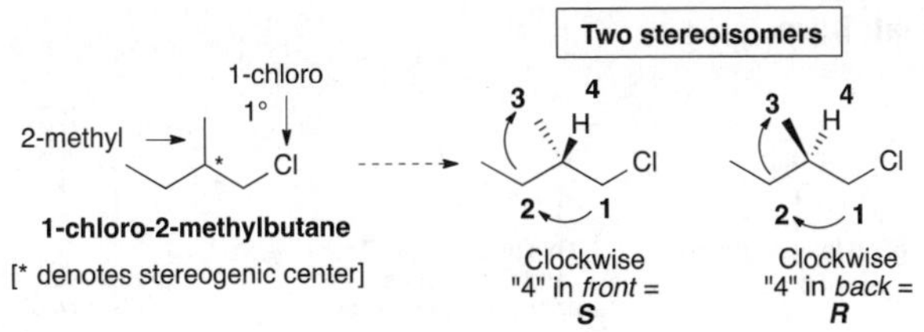

Alkyl Halides and Nucleophilic Substitution 7–19

Two stereoisomers

1-chloro
1°

2-methyl →

1-chloro-2-methylbutane

[* denotes stereogenic center]

Clockwise
"4" in *front* =
S

Clockwise
"4" in *back* =
R

7.45

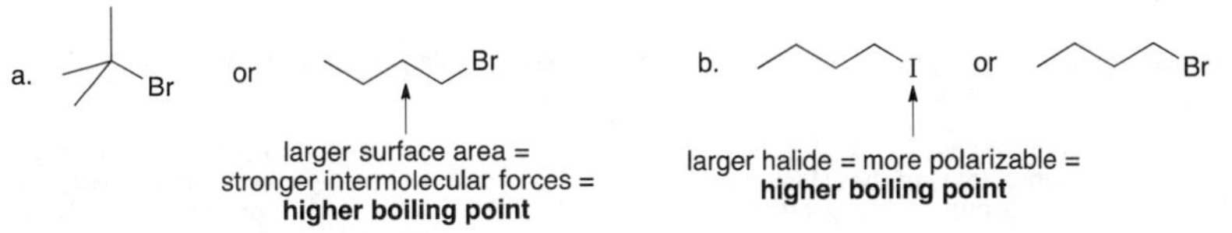

a. or

larger surface area =
stronger intermolecular forces =
higher boiling point

b. or

larger halide = more polarizable =
higher boiling point

7.46 Use the steps from Answer 7.6 and then draw the proton transfer reaction, when necessary.

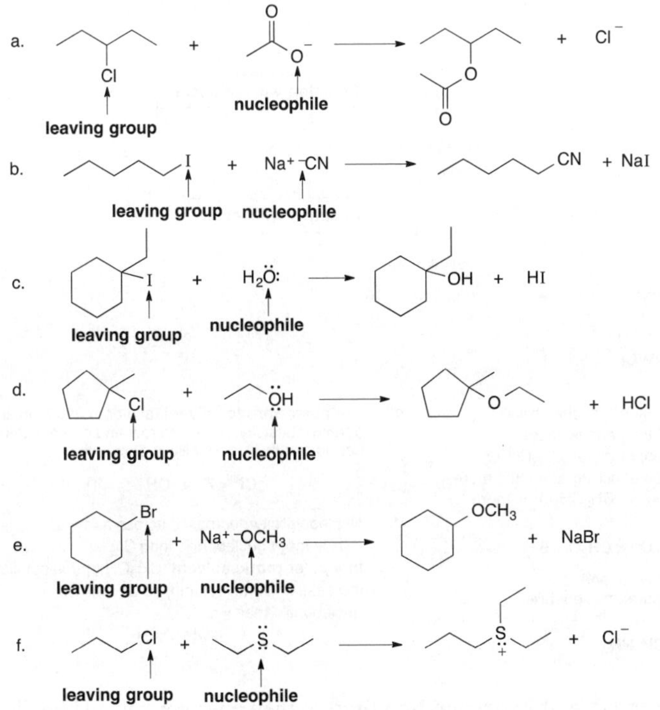

a.

Cl
leaving group

nucleophile

+ Cl⁻

b.

leaving group   nucleophile

Na⁺ ⁻CN

CN + NaI

c.

leaving group   nucleophile

H₂Ö:

OH + HI

d.

leaving group   nucleophile

ÖH

+ HCl

e.

leaving group   nucleophile

Na⁺ ⁻OCH₃

OCH₃

+ NaBr

f.

leaving group   nucleophile

Cl +

Ṣ

+ Cl⁻

7.47 A good leaving group is a weak base.

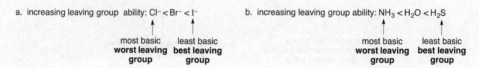

a. bad leaving group, ⁻OH is a strong base.
b. Cl⁻ good leaving group, weak base
c. This has only C–C and C–H bonds. no good leaving group
d. good leaving group, H₂O is a weak base.

7.48 Use the rules from Answer 7.10.

a. increasing leaving group ability: Cl⁻ < Br⁻ < I⁻
   - Cl⁻: most basic, worst leaving group
   - I⁻: least basic, best leaving group

b. increasing leaving group ability: NH₃ < H₂O < H₂S
   - NH₃: most basic, worst leaving group
   - H₂S: least basic, best leaving group

7.49 Compare the nucleophile and the leaving group in each reaction. The reaction will occur if it proceeds towards the weaker base. Remember that the stronger the acid (lower pK_a), the weaker the conjugate base.

a. cyclohexyl-NH₂ + I⁻ ⟶ (X) cyclohexyl-I + ⁻NH₂   **Reaction will not occur.**
   - I⁻: weaker base, pK_a (HI) = –10
   - ⁻NH₂: stronger base, pK_a (NH₃) = 38

b. R–I + ⁻O–CH₃ ⟶ R–O–CH₃ + I⁻   **Reaction will occur.**
   - ⁻O–CH₃: stronger base, pK_a (CH₃OH) = 15.5
   - I⁻: weaker base, pK_a (HI) = –10

7.50 Use the directions in Answer 7.13.

a. • In a **polar protic solvent** (CH₃OH), nucleophilicity increases down a column of the periodic table, so: CH₃CH₂S⁻ is more nucleophilic than CH₃CH₂O⁻.
   • For two species with the same attacking atom, the more basic is the more nucleophilic, so CH₃CH₂O⁻ is more nucleophilic than CH₃COO⁻.

   CH₃COO⁻ < CH₃CH₂O⁻ < CH₃CH₂S⁻

b. Compare the nucleophilicity of N, S, and O. In a polar aprotic solvent (acetone) nucleophilicity parallels basicity.

   CH₃SH < CH₃OH < CH₃NH₂

c. In a **polar aprotic solvent** (acetone), nucleophilicity parallels basicity. Across a row and down a column of the periodic table nucleophilicity decreases.

   Cl⁻ < F⁻ < ⁻OH

d. Nucleophilicity decreases across a row so ⁻SH is more nucleophilic than Cl⁻. In a **polar protic solvent** (CH₃OH), nucleophilicity increases down a column, so Cl⁻ is more nucleophilic than F⁻.

   F⁻ < Cl⁻ < ⁻SH

7.51 *Polar protic solvents* are capable of hydrogen bonding, so they must contain a H bonded to an electronegative O or N. *Polar aprotic solvents* are incapable of hydrogen bonding, so they do not contain any O–H or N–H bonds.

a. (CH₃)₂CHOH

contains O–H bond
protic

c. CH₂Cl₂

no O–H or N–H bond
aprotic

e. N(CH₃)₃

no O–H or N–H bond
aprotic

b. CH₃NO₂

no O–H or N–H bond
aprotic

d. NH₃

contains N–H bond
protic

f. HCONH₂

contains an N–H bond
protic

7.52

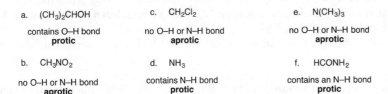

The amine N is more nucleophilic because the electron pair is localized on the N.

The amide N is less nucleophilic because the electron pair is delocalized by resonance.

7.53

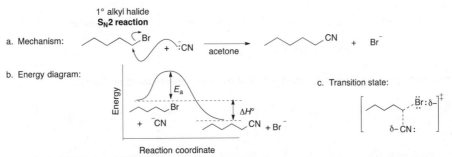

a. Mechanism:

b. Energy diagram:

c. Transition state:

d. Rate equation: one-step reaction with both nucleophile and alkyl halide in the only step:
   rate = k[R–Br][⁻CN]

e. [1] The leaving group is changed from Br⁻ to I⁻:
   Leaving group becomes less basic → a better leaving group → faster reaction.

   [2] The solvent is changed from acetone to CH₃CH₂OH:
   Solvent changed to polar protic → decreases reaction rate.

   [3] The alkyl halide is changed from CH₃(CH₂)₄Br to CH₃CH₂CH₂CH(Br)CH₃:
   Changed from 1° to 2° alkyl halide → the alkyl halide gets more crowded and the reaction rate decreases.

   [4] The concentration of ⁻CN is increased by a factor of 5.
   Reaction rate will increase by a factor of 5.

   [5] The concentration of both the alkyl halide and ⁻CN are increased by a factor of 5:
   Reaction rate will increase by a factor of 25 (5 × 5 = 25).

Chapter 7–22

7.54 a. CH₃CH₂Br reacts faster than CH₃CH₂Cl because Br⁻ is a better leaving group than Cl⁻.
b. The S_N2 reaction with NaOH is faster than the S_N2 reaction with NaOCOCH₃ because ⁻OH is a stronger nucleophile than ⁻OCOCH₃.
c. The S_N2 reaction is faster in the polar aprotic solvent DMSO because the nucleophile ⁻OCH₃ is stronger.

7.55 All S_N2 reactions proceed with backside attack of the nucleophile. When nucleophilic attack occurs at a stereogenic center, inversion of configuration occurs.

a.

b.

No bond to the stereogenic center is broken, because the leaving group is not bonded to the stereogenic center.

c.

[* denotes a stereogenic center]

7.56 For carbocations: **Increasing number of R groups = Increasing stability.**

a. 1° carbocation — least stable; 2° carbocation — intermediate stablity; 3° carbocation — most stable

b. 1° carbocation — least stable; 2° carbocation — intermediate stablity; 3° carbocation — most stable

7.57 Both **A** and **B** are resonance stabilized, but the N atom in **B** is more basic and therefore more willing to donate its electron pair.

A

B

more basic N atom

This resonance form stabilizes the carbocation more than the equivalent resonance structure for **A**. Thus, **B** is more stable than **A**.

7.58

a. Mechanism: S_N1 only

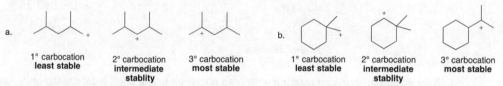

b. Energy diagram:

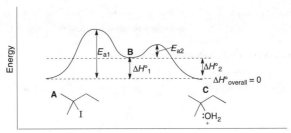

c. Transition states:

$$\left[\begin{array}{c} \delta+ \\ \vdots \ddot{I}\colon \delta- \end{array} \right]^{\ddagger} \quad \left[\begin{array}{c} \delta+ \\ :OH_2 \\ \delta+ \end{array} \right]^{\ddagger}$$

d. Rate equation: **rate = $k[(CH_3)_2CICH_2CH_3]$**

e. [1] Leaving group changed from I^- to Cl^-: **rate decreases** because I^- is a better leaving group.
   [2] Solvent changed from H_2O (polar protic) to DMF (polar aprotic):
       rate decreases because polar protic solvent favors S_N1.
   [3] Alkyl halide changed from 3° to 2°: **rate decreases** because 2° carbocations are less stable.
   [4] $[H_2O]$ increased by factor of five: **no change in rate** because H_2O is not in rate equation.
   [5] [R–X] and $[H_2O]$ increased by factor of five: **rate increases** by a factor of five. (Only the concentration of R–X affects the rate.)

7.59 a. S_N1 reaction of $(CH_3)_3CI$ is faster than S_N1 reaction with $(CH_3)_3CCl$ because I^- is a better leaving group than Cl^-.
      b. S_N1 reaction is faster with the 3° alkyl halide $(CH_3)_3CBr$ than with the 1° alkyl halide $(CH_3)_2CHCH_2Br$.
      c. S_N1 reaction is faster with the polar protic solvent H_2O rather than the aprotic solvent DMSO.

7.60

a. [structure with Br] + CH_3CH_2OH → [structure with OCH_2CH_3] + [structure with CH_3CH_2O] + HBr

b. [cyclohexane with Br] + H_2O → [cyclohexane with OH] + [cyclohexane with OH] + HBr

7.61 The 1° alkyl halide is also allylic, so it forms a resonance-stabilized carbocation. Increasing the stability of the carbocation by resonance increases the rate of the S_N1 reaction.

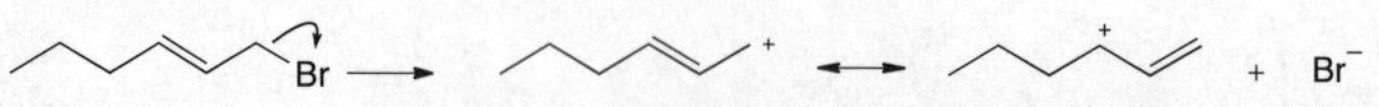

Chapter 7–24

resonance-stabilized carbocation

Use each resonance structure individually to continue the mechanism:

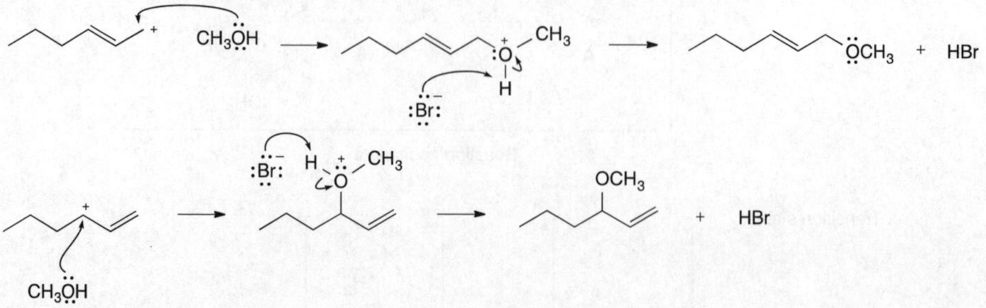

7.62

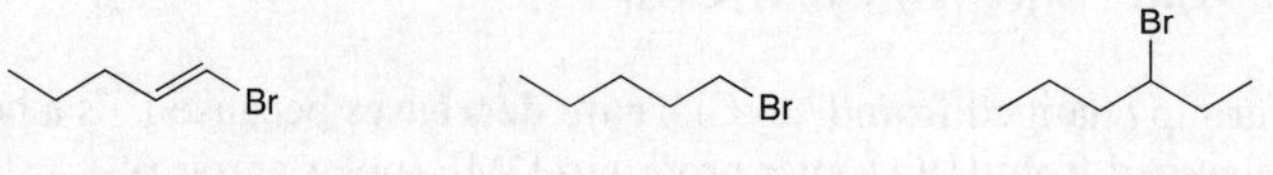

A   B   C

Vinyl halides like **A** do not react by either an S_N1 or S_N2 mechanism. With S_N2, the rate is $3° < 2° < 1°$, and with S_N1, the rate is $1° < 2° < 3°$.

a. S_N2 reactivity: **A < C < B**
b. S_N1 reactivity: **A < B < C**

7.63

a.

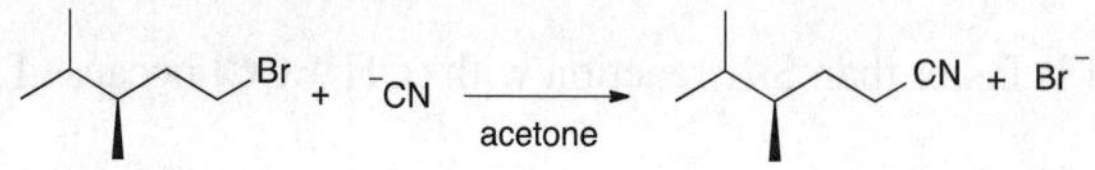

      1° alkyl halide
      S_N2 only

b.

      2° alkyl halide   strong nucleophile   reaction at a stereogenic center
      S_N1 and S_N2   polar aprotic solvent   **inversion of configuration**
                     Both favor S_N2.

c.

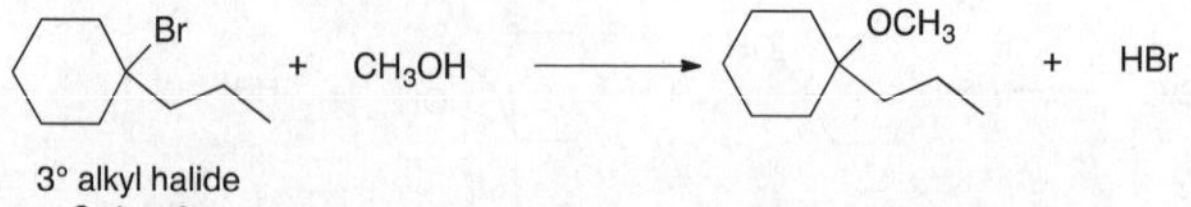

      3° alkyl halide
      S_N1 only

d.

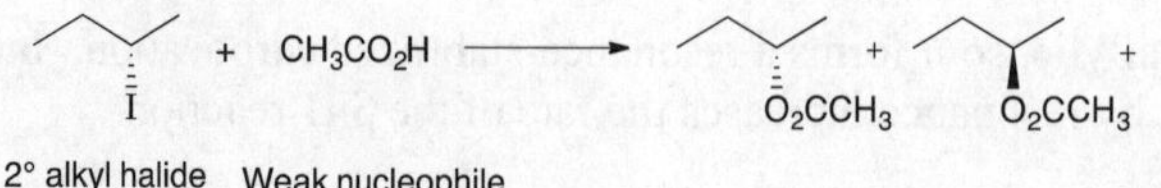

      2° alkyl halide   Weak nucleophile   reaction at a stereogenic center
      S_N1 and S_N2   **favors S_N1.**   **racemization of product**

Alkyl Halides and Nucleophilic Substitution 7–25

e.

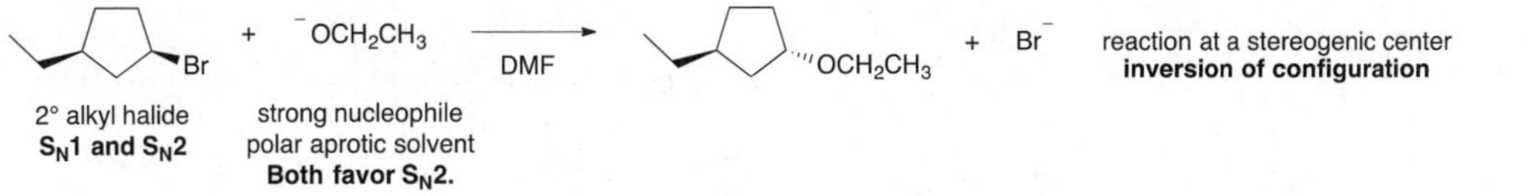

2° alkyl halide
S_N1 and S_N2

strong nucleophile
polar aprotic solvent
Both favor S_N2.

DMF

+ Br⁻   reaction at a stereogenic center
inversion of configuration

f.

2° alkyl halide
S_N1 and S_N2

Weak nucleophile
favors S_N1.

+ CH₃CH₂OH

+ HCl   two products – **diastereomers**
Nucleophile attacks
from above and below.

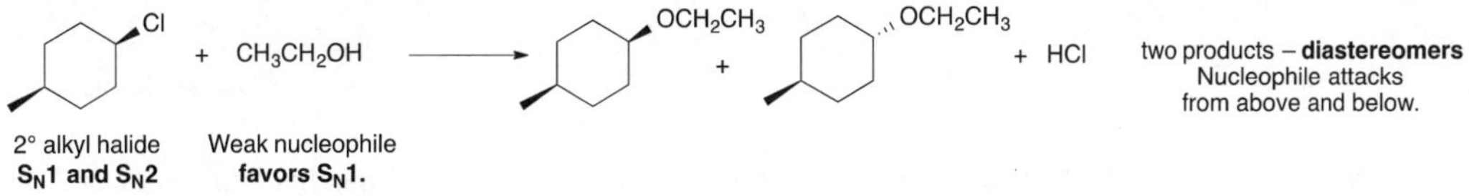

7.64 The reaction follows an S_N2 mechanism.

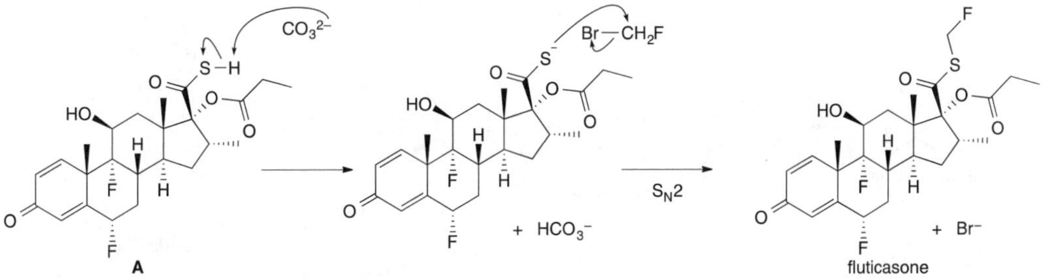

CO_3^{2-}

Br—CH₂F

A

S_N2

+ HCO_3^-

+ Br⁻

fluticasone

7.65 An S_N1 mechanism means the reaction occurs in a stepwise fashion by way of a carbocation.

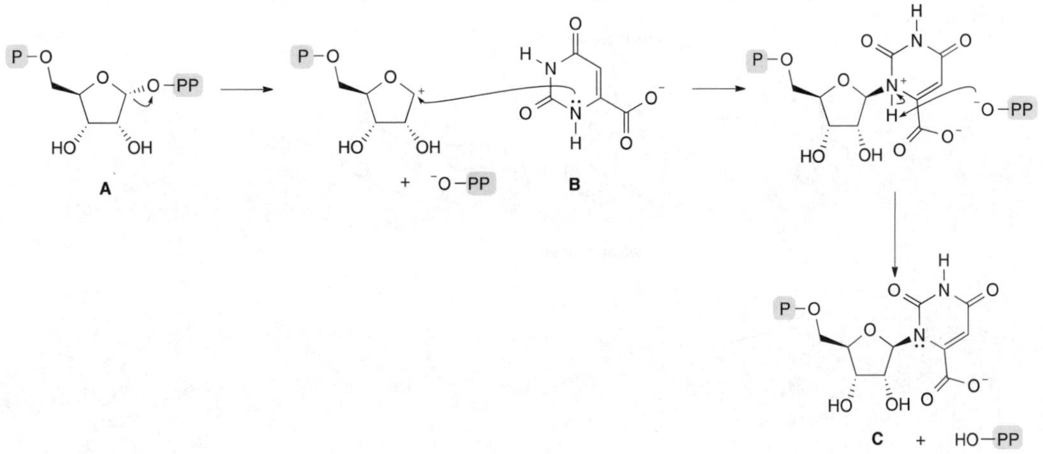

A

+ ⁻O—PP   B

C   +   HO—PP

7.66

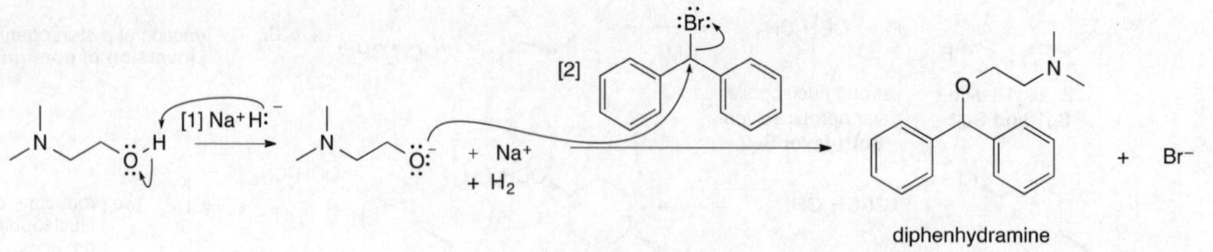
[1] Na+ H:−
[2]
:Br:
+ Na+
+ H_2
+ Br−
diphenhydramine

7.67 First decide whether the reaction will proceed via an SN1 or SN2 mechanism (Answer 7.38), and then draw the mechanism.

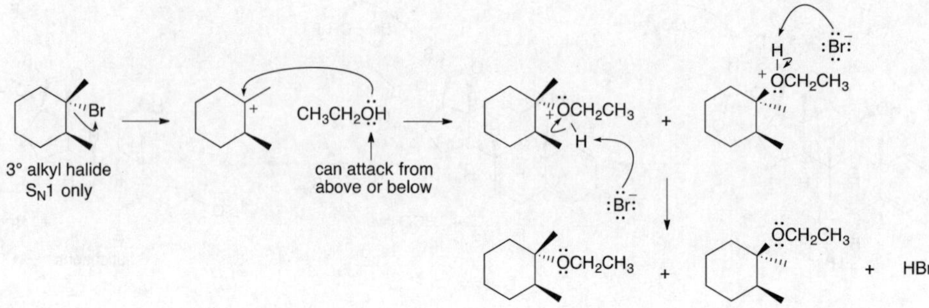
3° alkyl halide
SN1 only
CH3CH2OH
can attack from
above or below
OCH2CH3
:Br:−
+ HBr

7.68

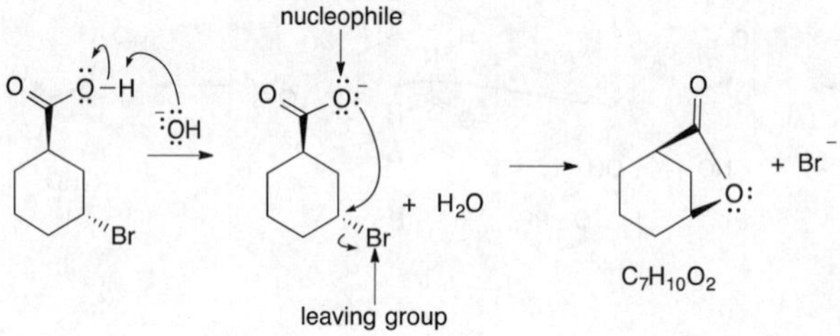
nucleophile
:OH
+ H2O
leaving group
+ Br−
C7H10O2

7.69

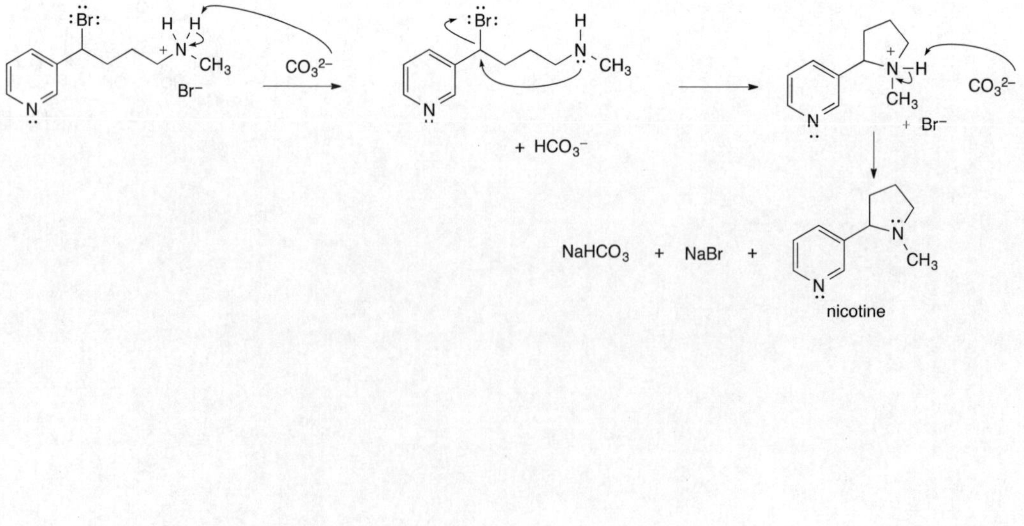
:Br:
Br−
CO3 2−
CH3
+ HCO3−
+ Br−
NaHCO3 + NaBr +
nicotine

Alkyl Halides and Nucleophilic Substitution 7–27

7.70

a. Two diastereomers (**C** and **D**) are formed as products from the two enantiomers of **A**.

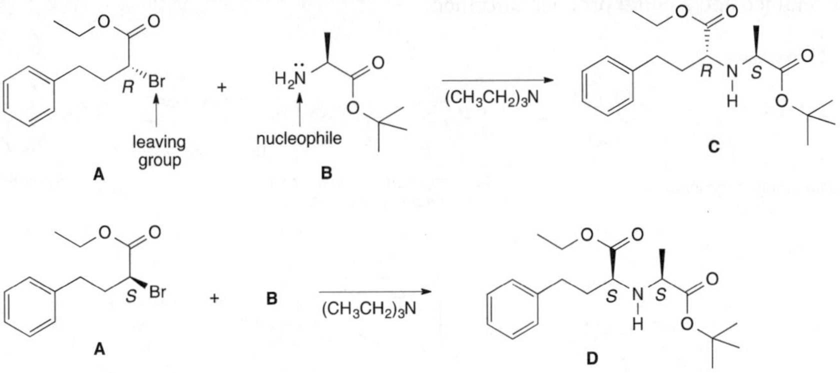

b.

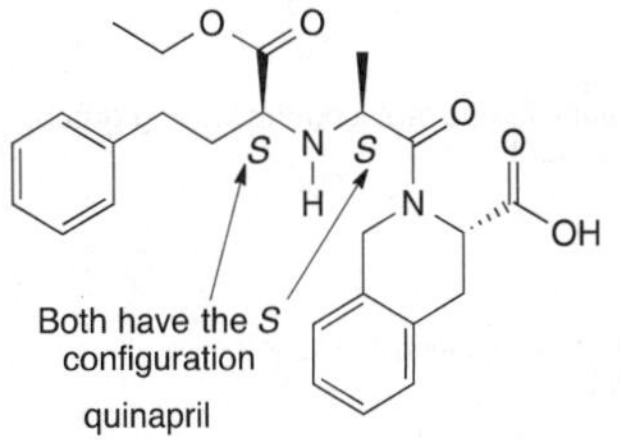

Since both stereogenic centers in **D** have the S configuration and the corresponding stereogenic centers in quinapril are also S, **D** is needed to synthesize the drug.

Both have the S configuration

quinapril

7.71

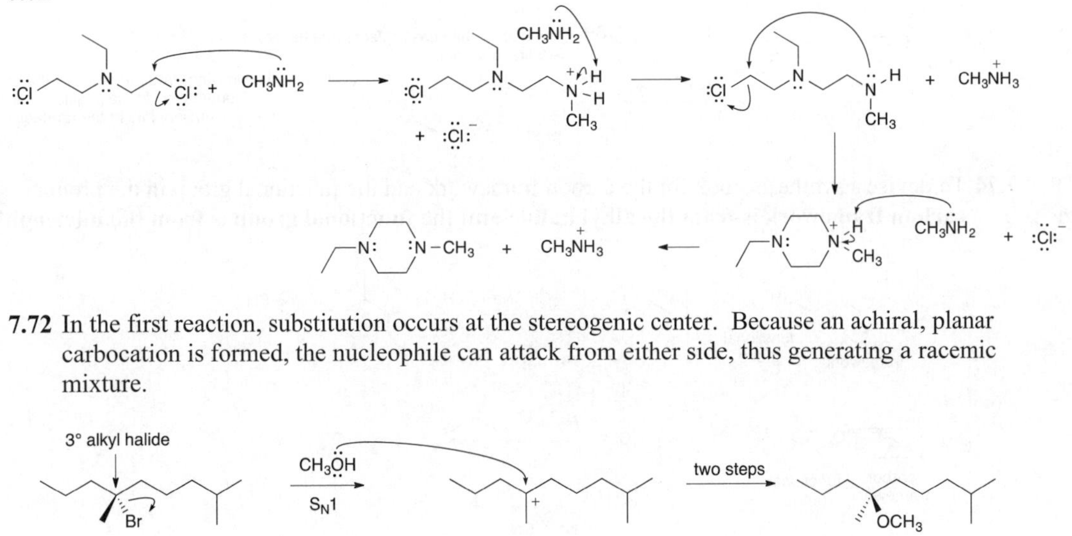

7.72 In the first reaction, substitution occurs at the stereogenic center. Because an achiral, planar carbocation is formed, the nucleophile can attack from either side, thus generating a racemic mixture.

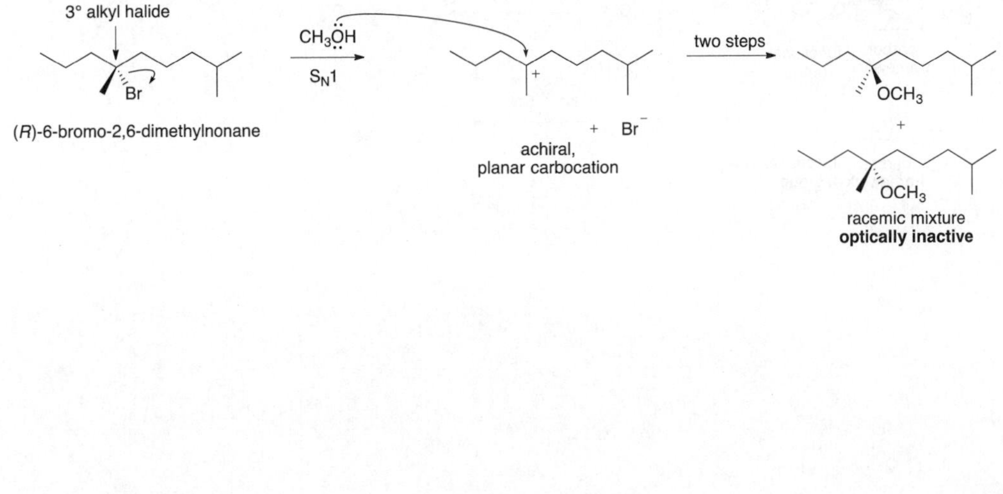

3° alkyl halide

(R)-6-bromo-2,6-dimethylnonane

two steps

achiral, planar carbocation

racemic mixture
optically inactive

Chapter 7–28

In the second reaction, the starting material contains a stereogenic center, but the nucleophile does not attack at that carbon. Because a bond to the stereogenic center is not broken, the configuration is retained and a chiral product is formed.

3° alkyl halide

CH$_3$ÖH

S$_N$1

two steps

(R)-2-bromo-2,5-dimethylnonane

Reaction does not occur at the stereogenic center.

+ Br⁻

OCH$_3$ — configuration retained

optically active

7.73

a. The nucleophile has replaced the leaving group. Missing reagent:

⁻O

b. The nucleophile has replaced the leaving group. Missing reagent:

⁻C≡CH

c. N$_3^-$ The nucleophile has replaced the halide. Starting material:

Cl

d. ⁻SH The nucleophile has replaced the halide. Starting material:

Cl — The leaving group must have the opposite orientation to the position of the nucleophile in the product.

7.74 To devise a synthesis, look for the carbon framework and the functional group in the product. **The carbon framework is from the alkyl halide and the functional group is from the nucleophile.**

a.
carbon framework   functional group

Cl Na⁺ ⁻SH → SH

b.
carbon framework   functional group

Cl Na⁺ ⁻O →

c.
carbon framework   functional group

Cl Na⁺ ⁻CN → CN

Alkyl Halides and Nucleophilic Substitution 7–29

d.

carbon framework / functional group

2° halide

$Na^+ \, ^-O$

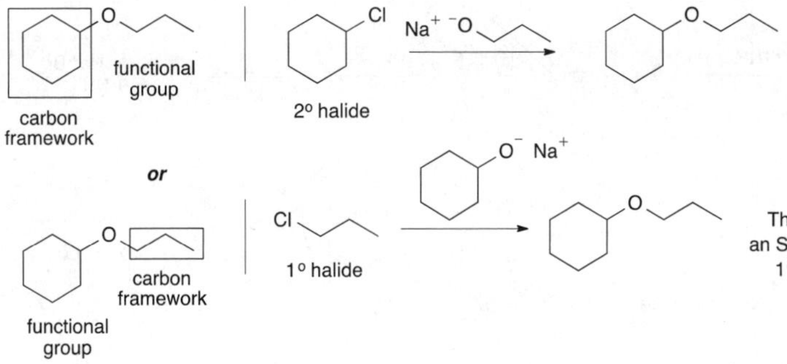

This path is preferred.
The strong nucleophile favors an S_N2 reaction so an unhindered 1° alkyl halide reacts faster.

or

functional group / carbon framework

1° halide

e.

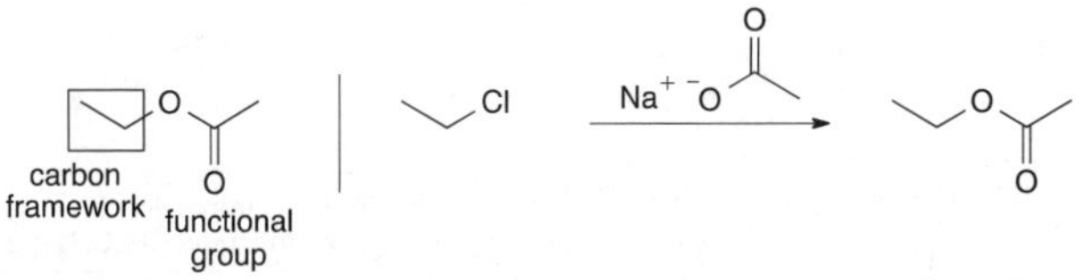

carbon framework / functional group

7.75

B
very crowded 3° halide

$Na^+ : \ddot{O}CH_3$

C

E

D

A
unhindered methyl halide

E

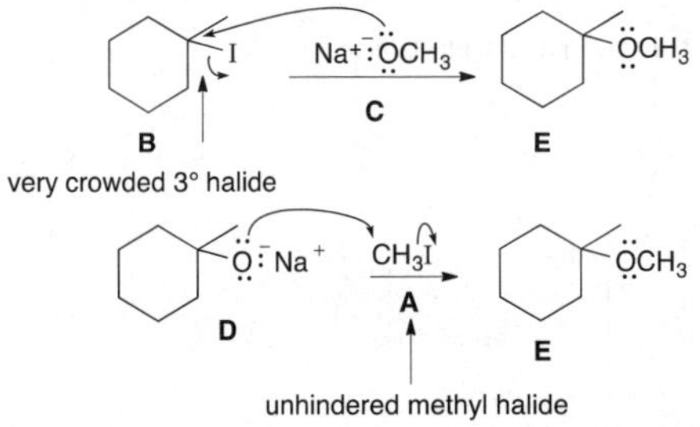

preferred method
The strong nucleophile favors S_N2 reaction, so the alkyl halide should be unhindered for a faster reaction.

7.76

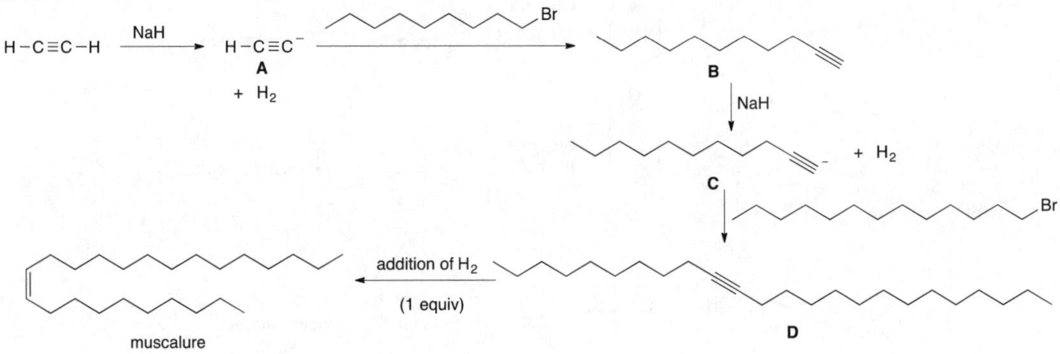

$H-C{\equiv}C-H$ —NaH→ $H-C{\equiv}C^-$
A
$+ \, H_2$

B

NaH

C $+ \, H_2$

D

addition of H_2
(1 equiv)

muscalure

7.77

a.

(Chapter 9)

[1] $Na^+H:^-$

[2] $CH_3-\ddot{B}r:$

$+ \, Na^+ \, Br^-$
$+ \, H_2$

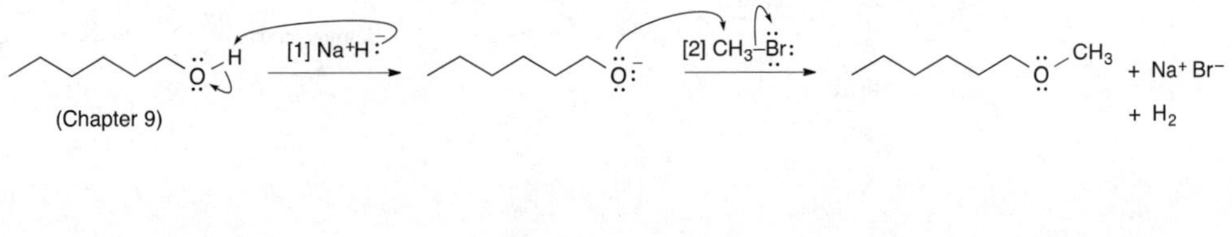

Chapter 7–30

b. (Chapter 11)

c. (Chapter 23)

7.78

quinuclidine — The three alkyl groups are "tied back" in a ring, making the electron pair more available.

triethylamine — This electron pair is more hindered by the three CH$_2$CH$_3$ groups. These bulky groups around the N cause steric hindrance and this decreases nucleophilicity.

This electron pair on quinuclidine is much more available than the one on triethylamine.

less steric hindrance
more nucleophilic

7.79

minor product

major product

+ NaBr

7.80

a. base → intramolecular S$_N$2 →

b. base → intramolecular S$_N$2 →

Alkyl Halides and Nucleophilic Substitution 7–31

c.

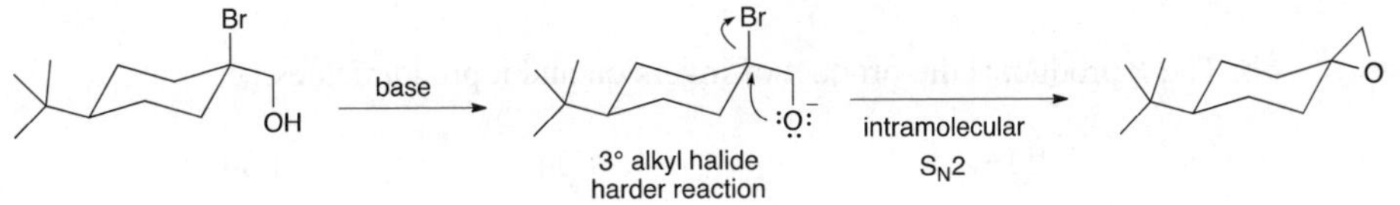

base

3° alkyl halide
harder reaction

intramolecular
S_N2

d.

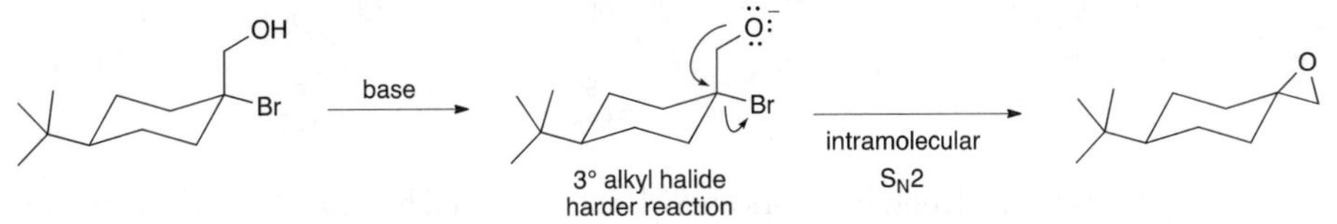

base

3° alkyl halide
harder reaction

intramolecular
S_N2

7.81

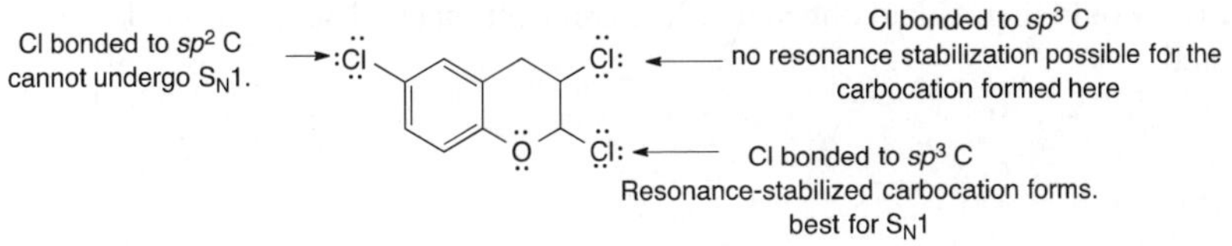

Cl bonded to sp^2 C
cannot undergo S_N1.

Cl bonded to sp^3 C
no resonance stabilization possible for the
carbocation formed here

Cl bonded to sp^3 C
Resonance-stabilized carbocation forms.
best for S_N1

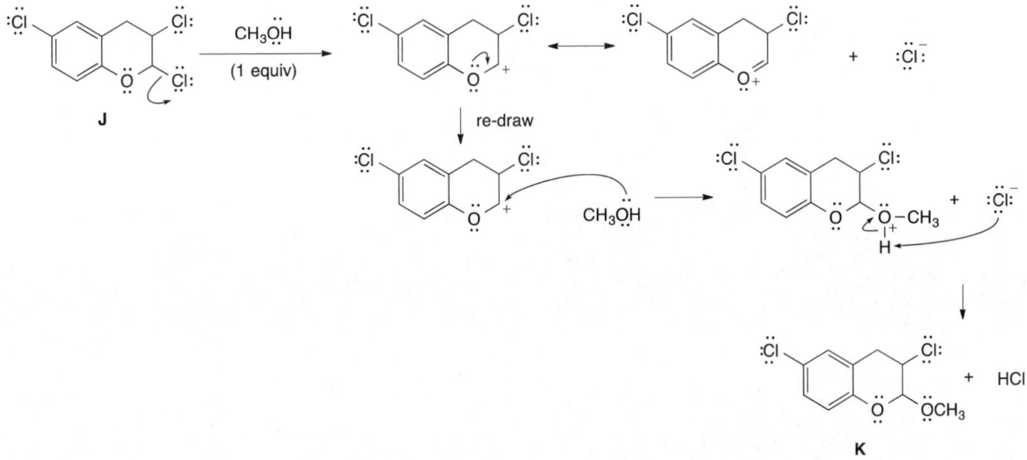

J

CH_3OH
(1 equiv)

re-draw

CH_3OH

+ HCl

K

7.82

a.

(S)-1-phenylpropan-1-ol
$[\alpha] = -48$

(R)-1-phenylpropan-1-ol
$[\alpha] = +48$

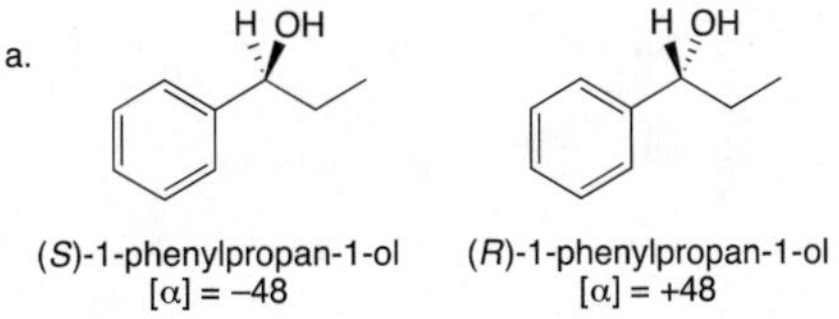

$$ee = \frac{[\alpha]\ \text{mixture}}{[\alpha]\ \text{pure enantiomer}} \times 100\%$$

$$= \frac{+5.0}{+48} \times 100\% = 10.\%\ \text{excess of } R \text{ isomer}$$

90% racemic mixture = 45% R and 45% S
Total R isomer = 45 + 10 = 55% R isomer

Chapter 7–32

b. The *R* product is the product of inversion and it predominates.

c. The weak nucleophile favors an S$_N$1 reaction, which occurs by way of an intermediate carbocation. Perhaps there is more inversion than retention because H$_2$O attacks the intermediate carbocation while the Br⁻ leaving group is still in the vicinity of the carbocation. The Br⁻ would then shield one side of the carbocation and backside attack would be slightly favored.

Alkyl Halides and Elimination Reactions 8–1

Chapter 8 Alkyl Halides and Elimination Reactions

Chapter Review

A comparison between nucleophilic substitution and β-elimination

Nucleophilic substitution—A nucleophile attacks a carbon atom (7.6).

β-Elimination—A base attacks a proton (8.1).

Similarities	Differences
• In both reactions RX acts as an electrophile, reacting with an electron-rich reagent. • Both reactions require a **good leaving group X:** willing to accept the electron density in the C–X bond.	• In substitution, a nucleophile attacks a single carbon atom. • In elimination, a Brønsted–Lowry base removes a proton to form a π bond, and two carbons are involved in the reaction.

The importance of the base in E2 and E1 reactions (8.9)

The strength of the base determines the mechanism of elimination.
• Strong bases favor E2 reactions.
• Weak bases favor E1 reactions.

Chapter 8–2

E1 and E2 mechanisms compared

	E2 mechanism	E1 mechanism
[1] Mechanism	• one step (8.4B)	• two steps (8.6B)
[2] Alkyl halide	• rate: $R_3CX > R_2CHX >$ RCH_2X (8.4C)	• rate: $R_3CX > R_2CHX >$ RCH_2X (8.6C)
[3] Rate equation	• rate = k[RX][B:] • second-order kinetics (8.4A)	• rate = k[RX] • first-order kinetics (8.6A)
[4] Stereochemistry	• anti periplanar arrangement of H and X (8.8)	• trigonal planar carbocation intermediate (8.6B)
[5] Base	• favored by strong bases (8.4B)	• favored by weak bases (8.6C)
[6] Leaving group	• better leaving group → faster reaction (8.4B)	• better leaving group → faster reaction (Table 8.4)
[7] Solvent	• favored by polar aprotic solvents (8.4B)	• favored by polar protic solvents (Table 8.4)
[8] Product	• more substituted alkene favored (Zaitsev rule, 8.5)	• more substituted alkene favored (Zaitsev rule, 8.6C)

Summary chart on the four mechanisms: S_N1, S_N2, E1, and E2 (8.11)

Alkyl halide type	Conditions	Mechanism
1° RCH_2X	strong nucleophile	S_N2
	strong bulky base	E2
2° R_2CHX	strong base and nucleophile	S_N2 + E2
	strong bulky base	E2
	weak base and nucleophile	S_N1 + E1
3° R_3CX	weak base and nucleophile	S_N1 + E1
	strong base	E2

Zaitsev rule

• β-Elimination affords the more stable product having the more substituted double bond.
• Zaitsev products predominate in E2 reactions except when a cyclohexane ring prevents trans diaxial arrangement.

Alkyl Halides and Elimination Reactions 8–3

Practice Test on Chapter Review

1. Which of the following is true about an E1 reaction?
   1. The reaction is faster with better leaving groups.
   2. The reaction is fastest with 3° alkyl halides.
   3. The reaction is faster with stronger bases.
   4. Statements (1) and (2) are true.
   5. Statements (1), (2), and (3) are all true.

2. Consider the S_N2 and E1 reaction mechanisms. What effect on the rate of the reaction is observed when each of the following changes is made? Fill in each box of the table with one of the following phrases: **increases, decreases,** or **remains the same.**

Change	S_N2 mechanism	E1 mechanism
a. The alkyl halide is changed from $(CH_3)_3CBr$ to $CH_3CH_2CH_2CH_2Br$.		
b. The solvent is changed from $(CH_3)_2CO$ to CH_3CH_2OH.		
c. The nucleophile/base is changed from ^-OH to H_2O.		
d. The alkyl halide is changed from CH_3CH_2Cl to CH_3CH_2I.		
e. The concentration of the base/nucleophile is increased by a factor of five.		

3. Rank the following compounds in order of *increasing* **reactivity in an E2 elimination** reaction. Rank the *most reactive* compound as **3,** the *least reactive* compound as **1,** and the compound of intermediate reactivity as **2.**

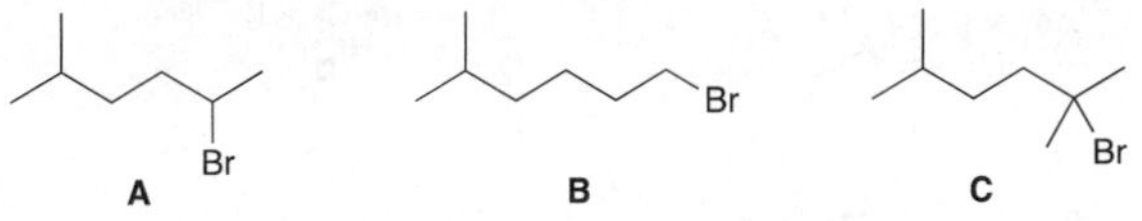

A   B   C

4. Draw the organic products formed in the following reactions.

a. 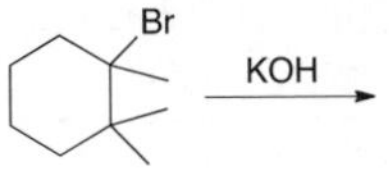   KOH

c. 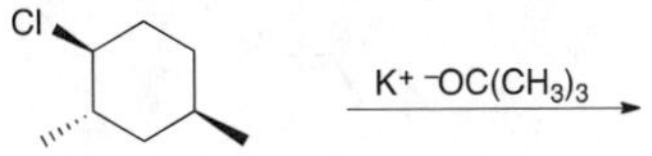   K+ ⁻OC(CH₃)₃

b. 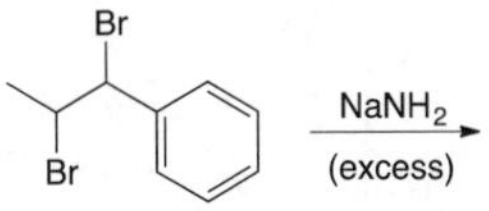   NaNH₂ (excess)

d. 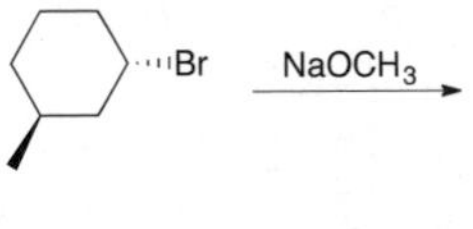   NaOCH₃

Chapter 8–4

5.a. Fill in the appropriate alkyl halide needed to synthesize the following compound as a single product using the given reagents.

C $\xrightarrow{\text{K}^+ \; ^-\text{OC(CH}_3)_3}$

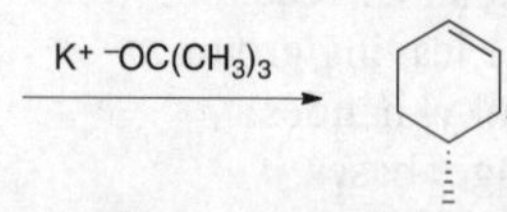

b. What starting material is needed for the following reaction? The starting material must yield product cleanly, in one step without any other organic side products.

D $\xrightarrow{\text{K}^+ \; ^-\text{OC(CH}_3)_3}$ $CH_3CH_2CH{=}CHCH_3$
(cis and trans mixture)

6. Draw all products formed in the following reaction.

$\xrightarrow{\text{CH}_3\text{OH}}$

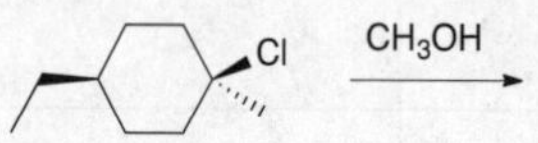

Answers to Practice Test

1. 4

2.   S$_N$2   E1
   a. increases   decreases
   b. decreases   increases
   c. decreases   same
   d. increases   increases
   e. increases   same

3. A–2
   B–1
   C–3

4.

a.

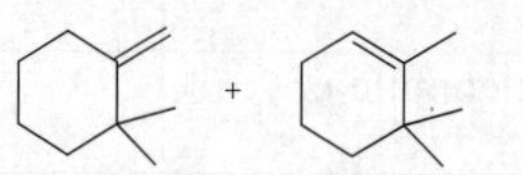

b.

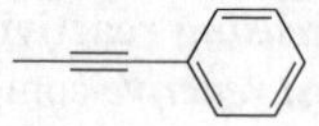

c.

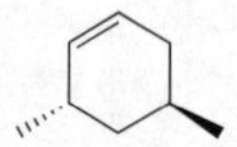

d.

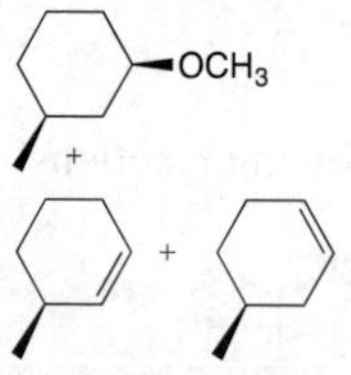

5.

a.

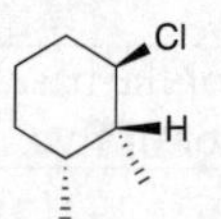

b. 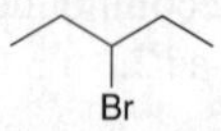
(or Cl or I)
D

6.

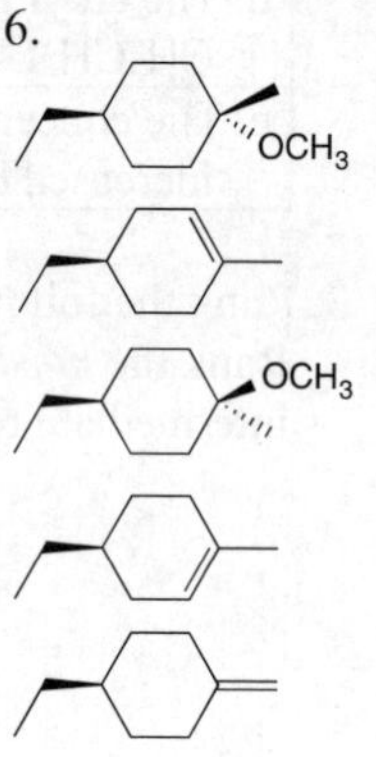

Alkyl Halides and Elimination Reactions 8–5

Answers to Problems

8.1 • The carbon bonded to the leaving group is the **α carbon.** Any carbon bonded to it is a **β carbon.**
• **To draw the products of an elimination reaction:** Remove the leaving group from the α carbon and a H from the β carbon and form a π bond.

a.

$K^+ \, ^-OC(CH_3)_3$

b.

$K^+ \, ^-OC(CH_3)_3$

$+ \quad CH_3CH = C(CH_3)CH_2CH_3$

c.

$K^+ \, ^-OC(CH_3)_3$

$+$

8.2 **Alkenes are classified by the number of carbon atoms bonded to the double bond.** A monosubstituted alkene has one carbon atom bonded to the double bond, a disubstituted alkene has two carbon atoms bonded to the double bond, etc.

a.

4 C's bonded to C=C
tetrasubstituted

2 C's bonded to each C=C
disubstituted

3 C's bonded to each C=C
trisubstituted

vitamin A

b.

vitamin D$_3$

3 C's bonded to each C=C
trisubstituted

2 C's bonded to the C=C
disubstituted

8.3 To have stereoisomers at a C=C, the two groups on each end of the double bond must be different from each other.

a.

two CH$_3$ groups
no stereoisomers
possible

b.

two CH$_3$ groups
no stereoisomers
possible

two different groups on both ends
stereoisomers possible

c.

two different groups
on both ends
stereoisomers
possible

two identical groups on one end
no stereoisomers
possible

Chapter 8–6

8.4

a.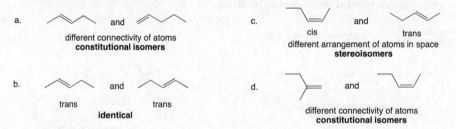
2 CH₃'s on one end → ← 2 H's on one end
Only this C=C exhibits stereoisomerism.

b. A diastereomer has a different 3-D arrangement of groups but the carbon skeleton and the double bonds must stay in the original positions.

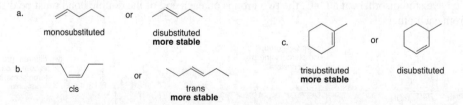

diastereomer — different arrangement of groups around this double bond

8.5 Two definitions:
- **Constitutional isomers** differ in the connectivity of the atoms.
- **Stereoisomers** differ only in the 3-D arrangement of the atoms in space.

a. and
different connectivity of atoms
constitutional isomers

b. and
trans   trans
identical

c. and
cis   trans
different arrangement of atoms in space
stereoisomers

d. and
different connectivity of atoms
constitutional isomers

8.6 Two rules to predict the relative stability of alkenes:
[1] Trans alkenes are generally more stable than cis alkenes.
[2] The stability of an alkene increases as the number of R groups on the C=C increases.

a. or
monosubstituted   disubstituted
more stable

b. or
cis   trans
more stable

c. or
trisubstituted   disubstituted
more stable

8.7

Alkene **A** is more stable than alkene **B** because the double bond in **A** is in a six-membered ring. The double bond in **B** is in a four-membered ring, which has considerable angle strain due to the small ring size.

8.8 In an E2 mechanism, four bonds are involved in the single step. Use curved arrows to show these simultaneous actions:
[1] The base attacks a hydrogen on a β carbon.
[2] A π bond forms.
[3] The leaving group comes off.

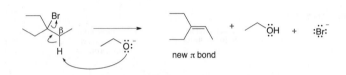

8.9 In both cases, the rate of elimination decreases.

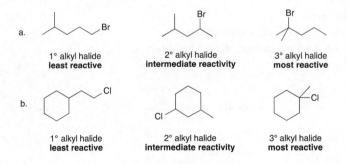

8.10 As the number of R groups on the carbon with the leaving group increases, the rate of an E2 reaction increases.

8.11 Use the following characteristics of an E2 reaction to answer the questions:
[1] E2 reactions are second order and one step.
[2] More substituted halides react faster.
[3] Reactions with strong bases or better leaving groups are faster.
[4] Reactions with polar aprotic solvents are faster.

Chapter 8–8

Rate equation: rate = k[RX][Base]
   a. tripling the concentration of the alkyl halide = **rate triples**
   b. halving the concentration of the base = **rate is halved**
   c. changing the solvent from CH_3OH to DMSO = **rate increases** (Polar aprotic solvent is better for E2.)
   d. changing the leaving group from I^- to Br^- = **rate decreases** (I^- is a better leaving group.)
   e. changing the base from ^-OH to H_2O = **rate decreases** (weaker base)
   f. changing the alkyl halide from CH_3CH_2Br to $(CH_3)_2CHBr$ = **rate increases** (More substituted halide reacts faster.)

8.12 According to the Zaitsev rule, the major product in a β-elimination reaction has the *more* substituted double bond.

a.
loss of H and Br

trisubstituted
major product

+

(+ stereoisomer)
disubstituted
minor product

b.
loss of H and Br

trisubstituted
minor product

+

tetrasubstituted
major product

+

disubstituted
minor product

c.
loss of H and Cl

monosubstituted
minor product

+

(+ stereoisomer)
disubstituted
major product

d.
loss of H and Cl

trisubstituted
ONLY product

8.13 An E1 mechanism has two steps:
   [1] The leaving group comes off, creating a carbocation.
   [2] A base pulls off a proton from a β carbon, and a π bond forms.

+ $CH_3\ddot{O}H$ [1]   + $CH_3\ddot{O}H$ + Cl^- [2]   + $CH_3\overset{+}{O}H_2$ + Cl^-

transition state [1]:

$$\left[\begin{array}{c} \delta+ \\ :\ddot{C}l:\ \delta- \end{array} \right]^{\ddagger}$$

transition state [2]:

$$\left[\begin{array}{c} \delta+ \\ H \\ H\ \overset{..}{O}\ CH_3 \\ \delta+ \end{array} \right]^{\ddagger}$$

Alkyl Halides and Elimination Reactions 8–9

8.14 According to the Zaitsev rule, the major product in a β-elimination reaction has the *more* substituted double bond.

a.

(+ stereoisomer)
trisubstituted
major product

+

disubstituted

b.

tetrasubstituted
major product

+

disubstituted

+

trisubstituted

8.15 Use the following characteristics of an **E1 reaction** to answer the questions:
[1] E1 reactions are first order and two steps.
[2] More substituted halides react faster.
[3] Weaker bases are preferred.
[4] Reactions with better leaving groups are faster.
[5] Reactions in polar protic solvents are faster.

Rate equation: rate = k[RX]. The base doesn't affect rate.
  a. doubling the concentration of the alkyl halide = **rate doubles**
  b. doubling the concentration of the base = **no change** (The base is not in the rate equation.)
  c. changing the alkyl halide from $(CH_3)_3CBr$ to $CH_3CH_2CH_2Br$ = **rate decreases** (More substituted halides react faster.)
  d. changing the leaving group from Cl^- to Br^- = **rate increases** (better leaving group)
  e. changing the solvent from DMSO to CH_3OH = **rate increases** (Polar protic solvent favors E1.)

8.16 Both S$_N$1 and E1 reactions occur by forming a carbocation. To draw the products:
[1] **For the S$_N$1 reaction,** substitute the nucleophile for the leaving group.
[2] **For the E1 reaction,** remove a proton from a β carbon and create a new π bond.

a.

leaving
group

nucleophile
and base

$+$ H_2O

S$_N$1 product

E1 products

b.

leaving
group

nucleophile
and base

S$_N$1 product

E1 products

Chapter 8–10

8.17 The E2 elimination reactions will occur in the anti periplanar orientation as drawn. To draw the product of elimination, maintain the orientation of the remaining groups around the C=C.

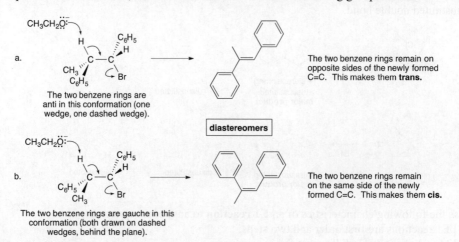

8.18 Note: The Zaitsev products predominate in E2 elimination *except* when substituents on a cyclohexane ring prevent a **trans diaxial** arrangement of H and X.

Alkyl Halides and Elimination Reactions 8–11

b.

two
conformations

A

Use this conformation.
It has Cl axial and
one axial H.

B

only one axial H
on a β carbon

−OH

[loss of H(β₁) + Cl]

disubstituted
only product

8.19 Draw the chair conformations of *cis*-1-chloro-2-methylcyclohexane and its trans isomer. For E2 elimination reactions to occur, **there must be a H and X trans diaxial to each other.**

Two conformations of the cis isomer:

A
reacting conformation (axial Cl)

This reacting conformation has only one group axial,
making it more stable and present in a higher
concentration than **B**. This makes a **faster
elimination reaction with the cis isomer.**

Two conformations of the trans isomer:

B
reacting conformation (axial Cl)

This conformation is less stable than **A**,
because both CH₃ and Cl are axial.
**This slows the rate of elimination from
the trans isomer.**

8.20 **E2 reactions are favored by strong negatively charged bases** and occur with 1°, 2°, and 3° halides, with 3° being the most reactive.
E1 reactions are favored by weaker neutral bases and do not occur with 1° halides because they would have to form highly unstable carbocations.

a.

+ ⁻OCH₃ ⟶

strong negatively
charged base
E2

b.

+ H₂O ⟶

weak neutral
base
E1

c.

+ CH₃OH ⟶

weak neutral
base
E1

d.

+ ⁻OC(CH₃)₃ ⟶

strong negatively
charged base
E2

8.21 Draw the alkynes that result from removal of two equivalents of HX.

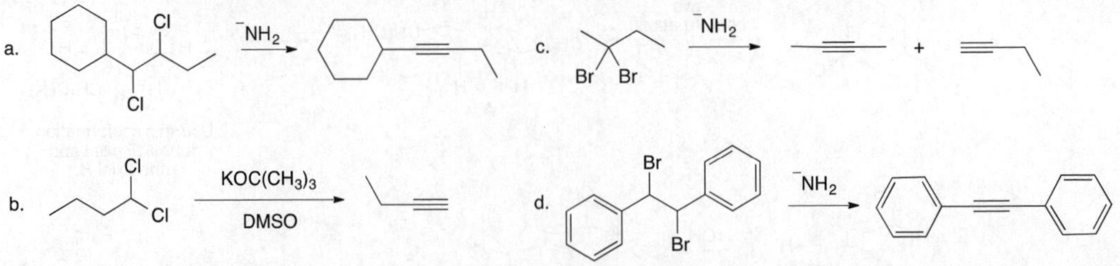
a.
Cl
Cl
$^-NH_2$
b.
Cl
Cl
$KOC(CH_3)_3$
DMSO
c.
Br Br
$^-NH_2$
+
d.
Br
Br
$^-NH_2$

8.22

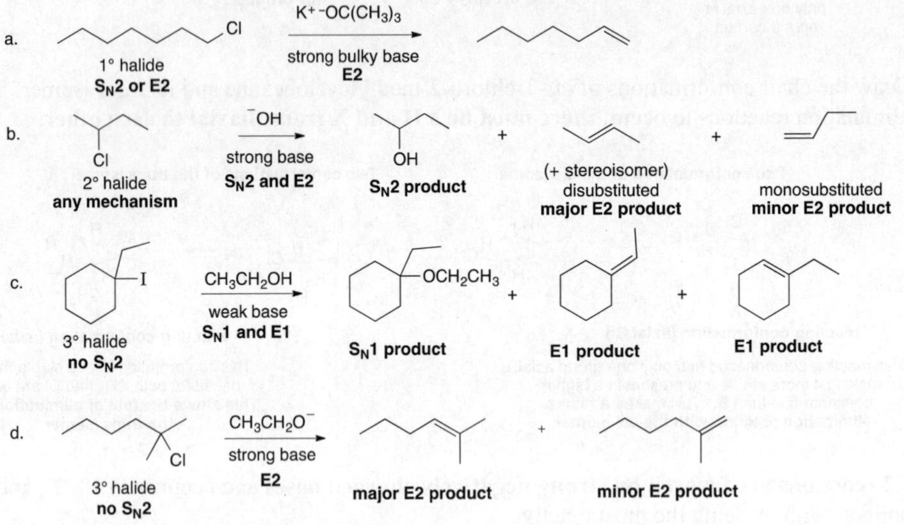
a.
Cl
1° halide
S_N2 or E2
$K^+ {}^-OC(CH_3)_3$
strong bulky base
E2
b.
Cl
2° halide
any mechanism
^-OH
strong base
S_N2 and E2
OH
S_N2 product
+
(+ stereoisomer)
disubstituted
major E2 product
+
monosubstituted
minor E2 product
c.
I
3° halide
no S_N2
CH_3CH_2OH
weak base
S_N1 and E1
OCH_2CH_3
S_N1 product
+
E1 product
+
E1 product
d.
Cl
3° halide
no S_N2
$CH_3CH_2O^-$
strong base
E2
major E2 product
+
minor E2 product

8.23

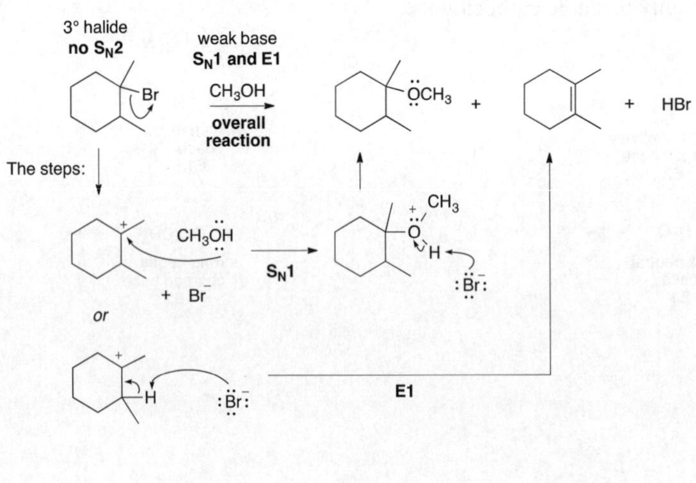
3° halide
no S_N2
weak base
S_N1 and E1
Br
CH_3OH
overall
reaction
OCH_3
+
+ HBr
The steps:
CH_3OH
+ Br^-
S_N1
CH_3
H
:Br:
or
H
:Br:
E1

Alkyl Halides and Elimination Reactions 8–13

8.24 More substituted alkenes are more stable. Trans alkenes are generally more stable than cis alkenes. Order of stability:

| **B** | **C** | **A** |
| least stable | | most stable |

8.25 The trans isomer (**E**) reacts faster. During elimination, Br must be axial to give trans diaxial elimination. In the trans isomer, the more stable conformation has the bulky *tert*-butyl group in the more roomy equatorial position and Br in the axial position. In the cis isomer (**D**), elimination can occur only when both the *tert*-butyl and Br groups are axial, a conformation that is not energetically favorable.

cis
D
cis-1-bromo-3-*tert*-butylcyclohexane

This conformation must react, but it contains two axial groups.

trans
E
trans-1-bromo-3-*tert*-butylcyclohexane

preferred conformation

8.26 Translate each model to a structure and arrange H and Br to be anti periplanar.

a.

rotate

H and Br 180°
away from each other

E2

b.

rotate

H and Br 180°
away from each other

E2

Chapter 8–14

8.27

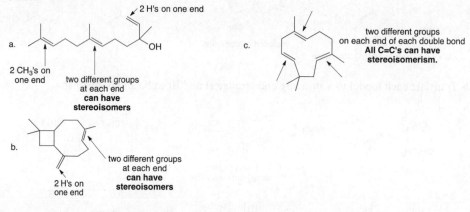

8.28 To give only one product in an elimination reaction, **the starting alkyl halide must have only one type of β carbon with H's.**

8.30 Use the definitions in Answer 8.5.

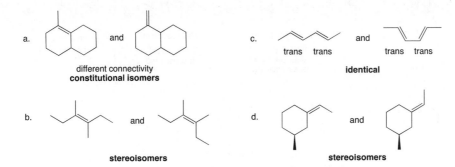

8.31 Use the rules from Answer 8.6 to rank the alkenes.

B	A	C
monosubstituted	disubstituted	trisubstituted
least stable	intermediate stability	most stable

8.32 A larger negative value for Δ*H*° means the reaction is more exothermic. Because both but-1-ene and *cis*-but-2-ene form the same product (butane), these data show that but-1-ene was higher in energy to begin with, **because more energy is released in the hydrogenation reaction.**

8.33

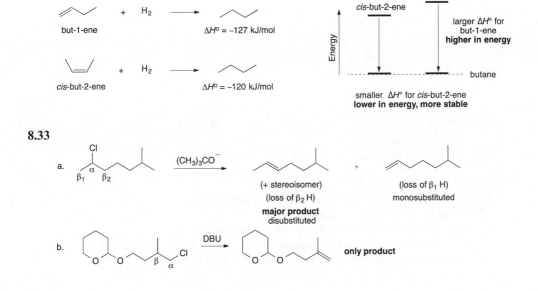

Chapter 8–16

c.

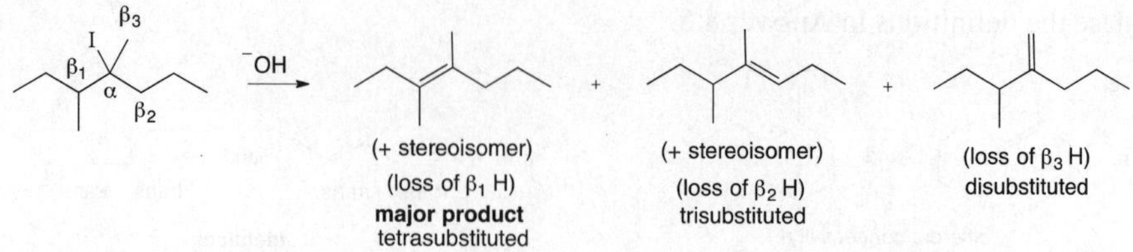

(+ stereoisomer)
(loss of β_1 H)
major product
tetrasubstituted

(+ stereoisomer)
(loss of β_2 H)
trisubstituted

(loss of β_3 H)
disubstituted

d.

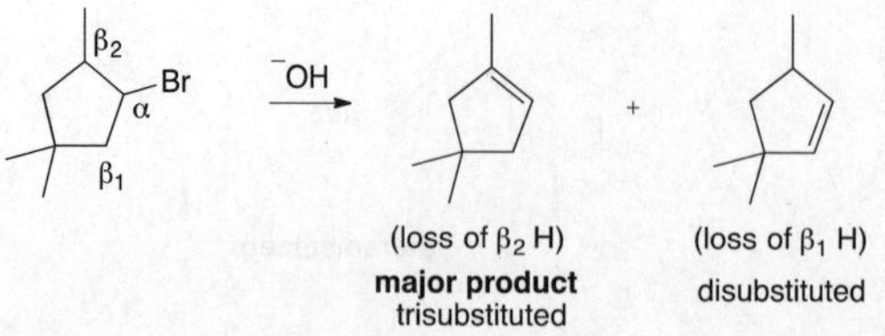

(loss of β_2 H)
major product
trisubstituted

(loss of β_1 H)
disubstituted

8.34 To give only one alkene as the product of elimination, the alkyl halide must have either:
- only one β carbon with a hydrogen atom, or
- all identical β carbons, so the resulting elimination products are identical

a.

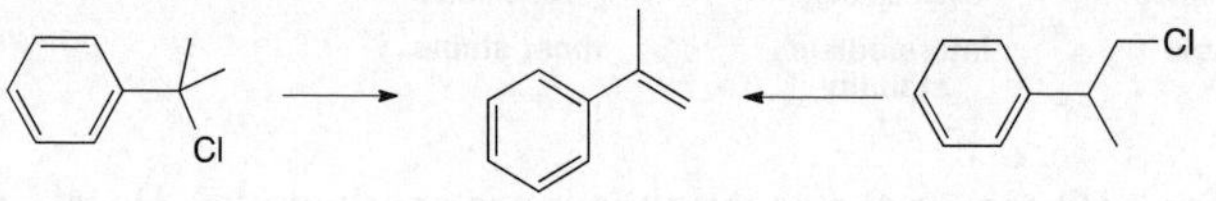

b.

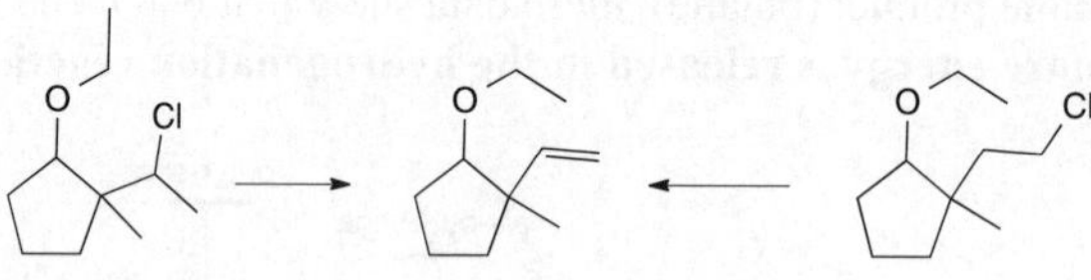

c.

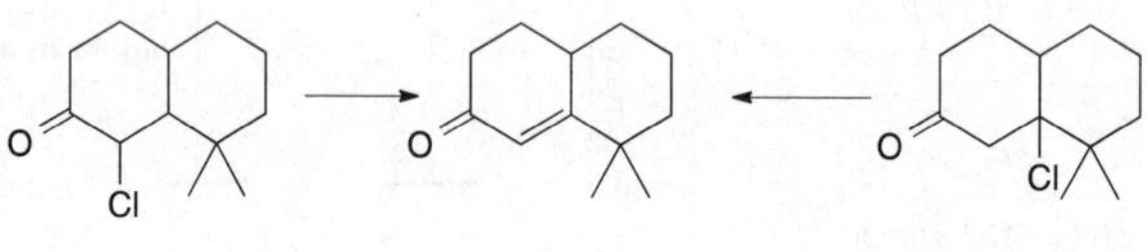

8.35

a. Mechanism:

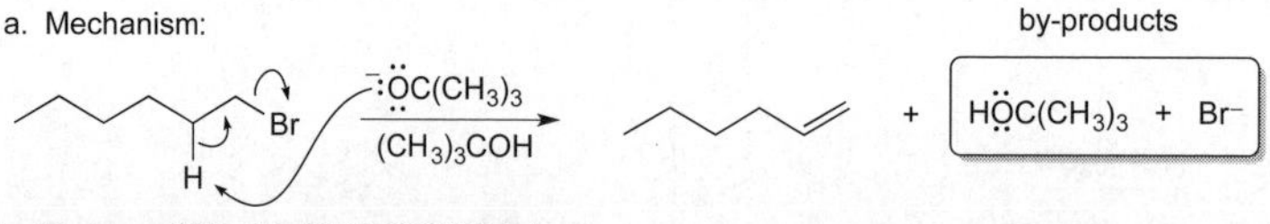

by-products

b. Rate = k[R–Br][$^-$OC(CH$_3$)$_3$]
   [1] Solvent changed to DMF (polar aprotic) = **rate increases**
   [2] [$^-$OC(CH$_3$)$_3$] decreased = **rate decreases**
   [3] Base changed to $^-$OH = **rate decreases** (weaker base)
   [4] Halide changed to 2° = **rate increases** (More substituted RX reacts faster.)
   [5] Leaving group changed to I$^-$ = **rate increases** (better leaving group)

Alkyl Halides and Elimination Reactions 8–17

8.36

1-chloro-1-methyl-
cyclopropane

The dehydrohalogenation of an alkyl halide usually forms the more stable alkene. In this case, **A**
is more stable than **B** even though **A** contains a disubstituted C=C whereas **B** contains a
trisubstituted C=C. The double bond in **B** is part of a three-membered ring, and is less stable than
A because of severe angle strain around both C's of the double bond.

8.37

a.

**trans isomer more stable
major product**

b.

**trans isomer more stable
major product**

8.38

a.

tetrasubstituted
major product   trisubstituted   disubstituted

b.

trisubstituted
This isomer is more stable—
large groups farther away.
major product   trisubstituted   disubstituted

c.

disubstituted   trisubstituted
major product

8.39 Use the rules from Answer 8.20.

a.

Br

2° halide

$^-$OCH$_3$

strong base

E2

(+ cis isomer)

b.

Br

2° halide

CH$_3$OH

weak base

E1

(+ cis isomer)

c.

1° halide

$^-$OC(CH$_3$)$_3$

strong base

E2

Chapter 8–18

d.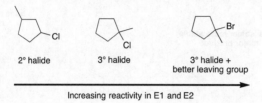

e. [figure: 2° halide with ⁻OH / strong base / E2 giving cyclohexene product]

f. [figure: 2° halide with ⁻OH / strong base / E2 giving two cyclohexene products]

8.40 The order of reactivity is the same for both E2 and E1: 1° < 2° < 3°.

[figure: 2° halide, 3° halide, 3° halide + better leaving group — Increasing reactivity in E1 and E2]

8.41

a. 1-chloro-1-methylcyclohexane — 3° RX reacts faster in E2 (strong base)
   1-chloro-3-methylcyclohexane

b. 2° RX   3° RX reacts faster in E1 (weak base)

c. The mechanism is E2 because ⁻OH is a strong base, so the reaction is faster in a polar aprotic solvent (DMSO).

8.42

bromocyclodecane → cis-cyclodecene

In a ten-membered ring, the cis isomer is more stable and, therefore, the preferred elimination product. The trans isomer is less stable because strain is introduced when two ends of the double bond are connected in a trans arrangement in this medium-sized ring.

Alkyl Halides and Elimination Reactions 8–19

8.43

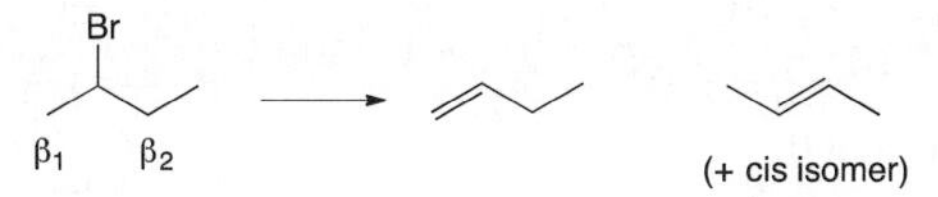

	but-1-ene from loss of β₁ H	but-2-ene from loss of β₂ H
Na⁺ ⁻OCH₂CH₃	19%	81%
K⁺ ⁻OC(CH₃)₃	33%	67%

The H's on the CH_2 group of the β_2 carbon are more sterically hindered than the H's on the CH_3 group of the β_1 carbon. Because $K^+{}^-OC(CH_3)_3$ is a much bulkier base than $Na^+{}^-OCH_2CH_3$, it is easier to remove the more accessible H on β_1, giving $K^+{}^-OC(CH_3)_3$ a higher percentage of but-1-ene than $Na^+{}^-OCH_2CH_3$.

8.44 H and Br must be anti during the E2 elimination. Rotate if necessary to make them anti; then eliminate.

a.

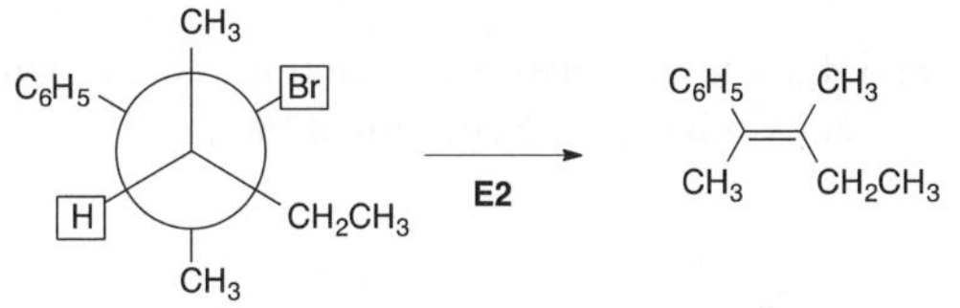

b.

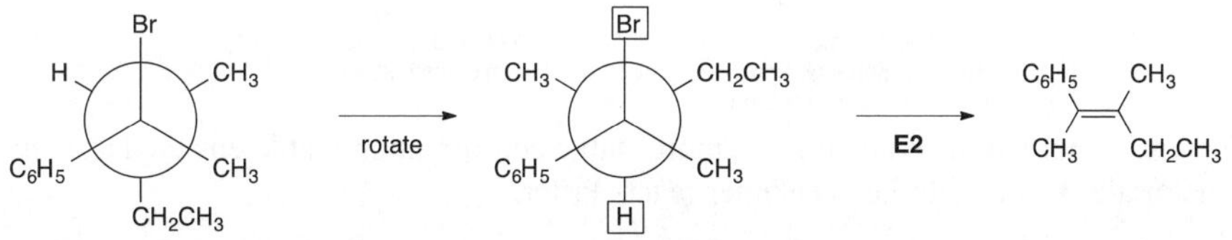

c.

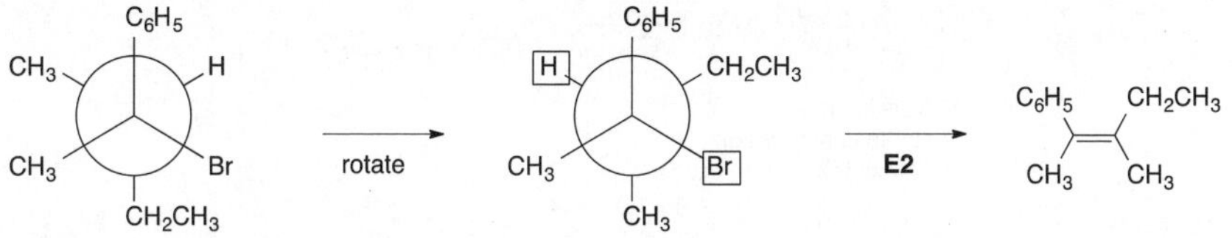

8.45

a.

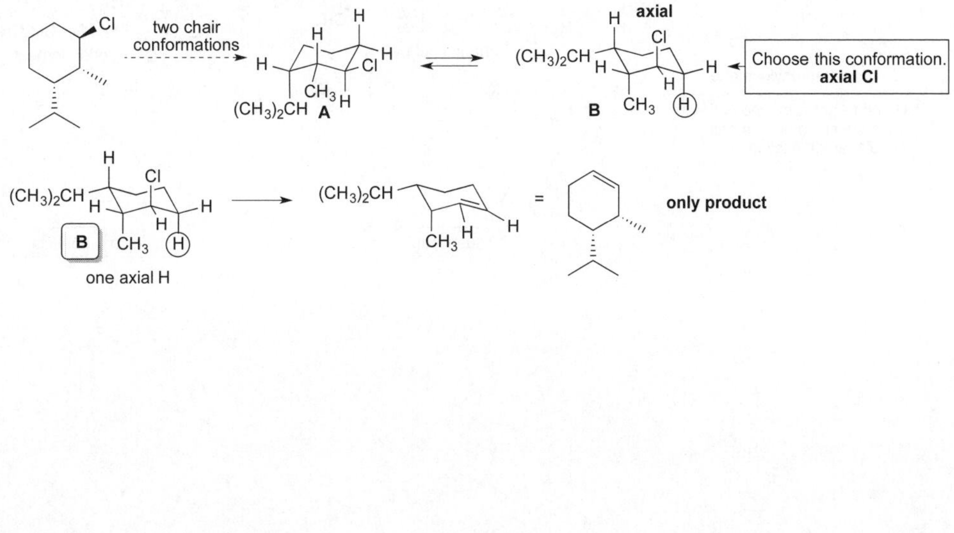

Chapter 8–20

b.

two chair conformations

A

H axial
Cl

Choose this conformation.
axial Cl

B

two axial H's

B

(loss of β_1 H)
major product
trisubstituted

+ (loss of β_2 H)

↓ re-draw

↓ re-draw

+

8.46 To react by E2, the Br must be axial and this can only happen in the trans isomer when the large *tert*-butyl group is also axial, an energetically unfavorable conformation.

trans isomer
two equatorial groups
more stable conformation

Br
two axial groups
highly destabilized

The cis isomer has an axial Br in its more stable conformation that keeps the large *tert*-butyl group equatorial. As a result, the cis isomer reacts faster.

cis isomer
more stable conformation
axial Br

8.47

a.

C2   C3
2-chloro-3-methylpentane

H and Cl are arranged anti in each stereoisomer, for anti periplanar elimination.

A   enantiomers   B

C   enantiomers   D

–HCl   –HCl   –HCl   –HCl

identical

identical

Alkyl Halides and Elimination Reactions 8–21

    b. Two different alkenes are formed as products.
    c. The products are diastereomers: Two enantiomers (**A** and **B**) give identical products. **A** and **B** are diastereomers of **C** and **D**. Each pair of enantiomers gives a single alkene. Thus, diastereomers give diastereomeric products.

8.48 To undergo an E2 reaction the Cl must be axial and the conformation with an axial Cl is highly unstable. Thus, no E2 reaction can occur because the needed conformation is too destabilized.

A

more stable conformation
All substituents are equatorial.

highly destabilized
All substituents are axial.

8.49 The alkyl chloride must have a Cl and H anti periplanar with all the substituents having the same arrangement that they have in the alkene product.

or

8.50

a.

$$\xrightarrow[\text{excess}]{\text{NaNH}_2}$$

b.

$$\xrightarrow[\text{excess}]{\text{NaNH}_2}$$

c.

$$\xrightarrow[\text{(excess)}]{\text{NaNH}_2}$$

d.

$$\xrightarrow[\text{excess}]{\text{NaNH}_2}$$

Chapter 8–22

8.51

a.

b.

c.

8.52

2,3-dibromobutane

$sp\ sp$
A

$CH_3CH=C=CH_2$
sp
B

C

8.53 Use the "Summary chart on the four mechanisms: S$_N$1, S$_N$2, E1, or E2" on p. 8–2 to answer the questions.

a. Both S$_N$1 and E1 involve carbocation intermediates.
b. Both S$_N$1 and E1 have two steps.
c. S$_N$1, S$_N$2, E1, and E2 all have increased reaction rates with better leaving groups.
d. Both S$_N$2 and E2 have increased rates when changing from CH$_3$OH (a protic solvent) to (CH$_3$)$_2$SO (DMSO—an aprotic solvent).
e. In S$_N$1 and E1 reactions, the rate depends on only the alkyl halide concentration.
f. Both S$_N$2 and E2 are concerted reactions.
g. CH$_3$CH$_2$Br and NaOH react by an S$_N$2 mechanism.
h. Racemization occurs in S$_N$1 reactions.
i. In S$_N$1, E1, and E2 mechanisms, 3° alkyl halides react faster than 1° or 2° halides.
j. E2 and S$_N$2 reactions follow second-order rate equations.

8.54

a.

1° halide
S$_N$2 or E2

$^-OC(CH_3)_3$
sterically
hindered base
E2

b.

1° halide
S$_N$2 or E2

$^-OCH_2CH_3$
strong
nucleophile
S$_N$2

Alkyl Halides and Elimination Reactions 8–23

c.

dihalide

$\xrightarrow[\substack{\text{(2 equiv)} \\ \text{strong base}}]{^-NH_2}$

d.

1° halide
S$_N$2 or E2

$\xrightarrow[\substack{\text{sterically} \\ \text{hindered} \\ \text{base} \\ \textbf{E2}}]{\text{DBU}}$

e.

2° halide
S$_N$1, S$_N$2, E1, E2

$\xrightarrow[\substack{\text{sterically} \\ \text{hindered} \\ \text{base} \\ \textbf{E2}}]{^-OC(CH_3)_3}$

major product   +

f.

3° halide
no S$_N$2

$\xrightarrow[\text{weak base}]{CH_3CH_2OH}$

S$_N$1 product   +   +   E1 products

g.

diahlide

$\xrightarrow{\text{2 NaNH}_2}$

h.

3° halide
no S$_N$2

$\xrightarrow[\text{weak base}]{H_2O}$

S$_N$1 product   +   (+ stereoisomer)   +   E1 product
                          E1 product

8.55 [1] NaOCOCH$_3$ is a good nucleophile and weak base, and substitution is favored. [3] KOC(CH$_3$)$_3$ is a strong, bulky base that reacts by E2 elimination when there is a β hydrogen in the alkyl halide.

a. $\xrightarrow{\text{[1] NaOCOCH}_3}$ OCOCH$_3$ b. $\xrightarrow{\text{[1] NaOCOCH}_3}$ OCOCH$_3$

$\xrightarrow{\text{[2] NaOCH}_3}$ OCH$_3$ $\xrightarrow{\text{[2] NaOCH}_3}$ OCH$_3$ +

S$_N$2   E2

$\xrightarrow{\text{[3] KOC(CH}_3)_3}$ $\xrightarrow{\text{[3] KOC(CH}_3)_3}$

Chapter 8–24

c.

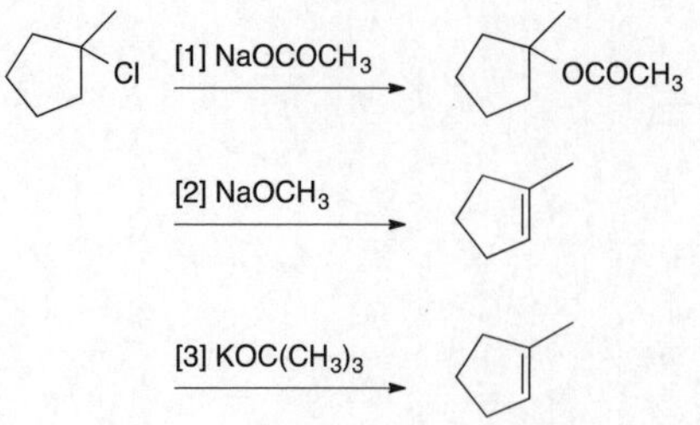

[1] NaOCOCH$_3$

[2] NaOCH$_3$

[3] KOC(CH$_3$)$_3$

8.56

a.

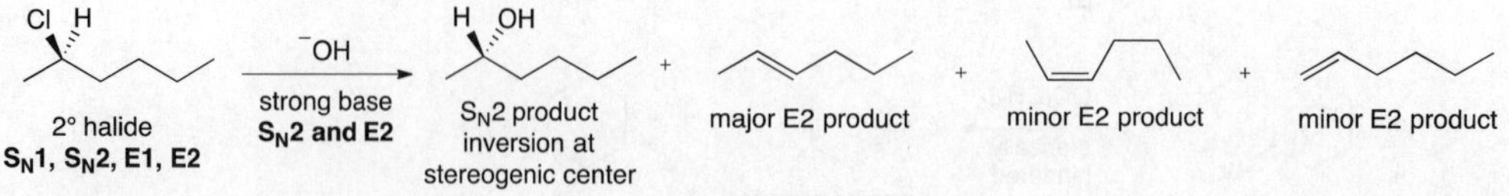

2° halide
S$_N$1, S$_N$2, E1, E2

$^-$OH

strong base
S$_N$2 and E2

S$_N$2 product
inversion at
stereogenic center

+ major E2 product + minor E2 product + minor E2 product

b.

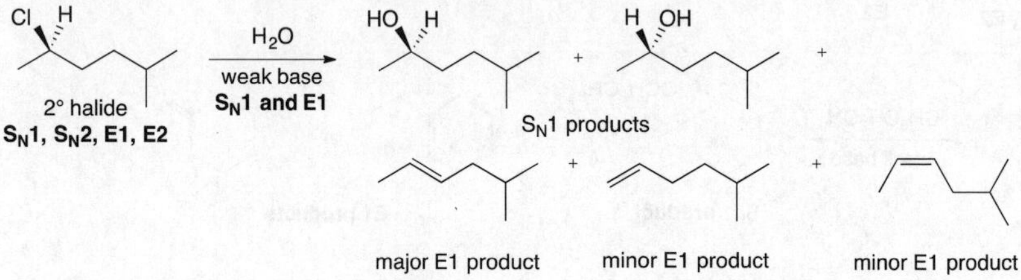

2° halide
S$_N$1, S$_N$2, E1, E2

H$_2$O

weak base
S$_N$1 and E1

+

S$_N$1 products

major E1 product + minor E1 product + minor E1 product

c.

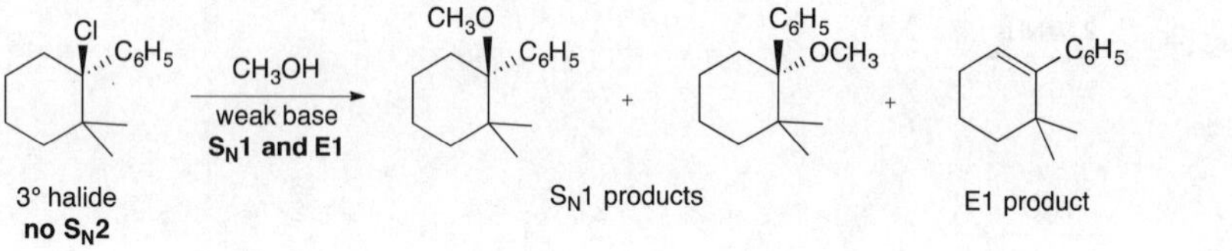

3° halide
no S$_N$2

CH$_3$OH

weak base
S$_N$1 and E1

+

S$_N$1 products

E1 product

d.

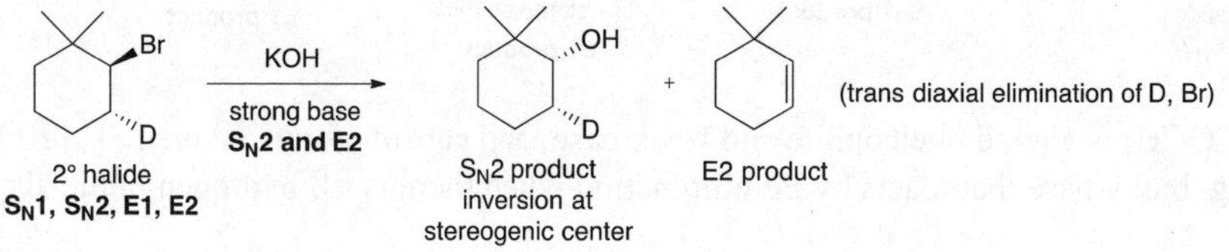

2° halide
S$_N$1, S$_N$2, E1, E2

KOH

strong base
S$_N$2 and E2

S$_N$2 product
inversion at
stereogenic center

+

E2 product

(trans diaxial elimination of D, Br)

8.57

a.

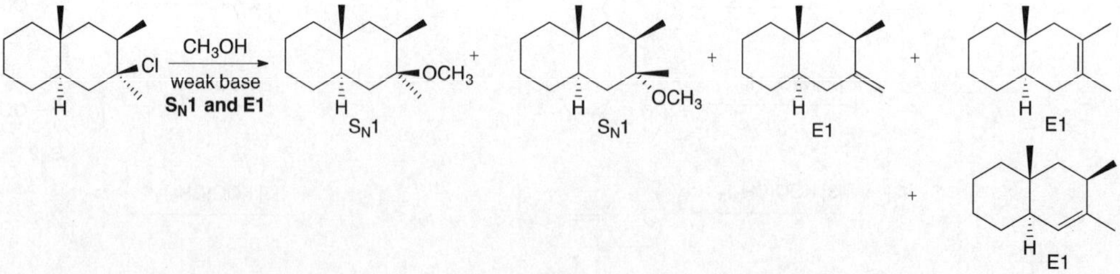

CH$_3$OH

weak base
S$_N$1 and E1

OCH$_3$

S$_N$1

+

OCH$_3$

S$_N$1

+

E1

+

E1

+

E1

b.

KOH

strong base
E2

+

+

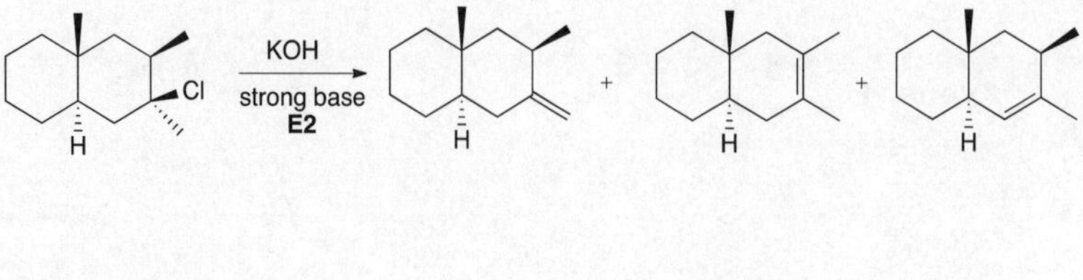

Alkyl Halides and Elimination Reactions 8–25

8.58

a.

3° halide   strong bulky base **E2**

major product
more substituted alkene

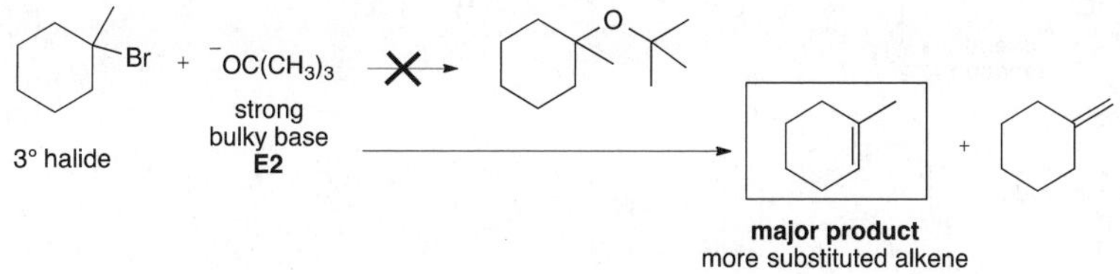

No substitution occurs with a strong bulky base and a 3° RX. The C with the leaving group is too crowded for an S_N2 substitution to occur. Elimination occurs instead by an E2 mechanism.

b.

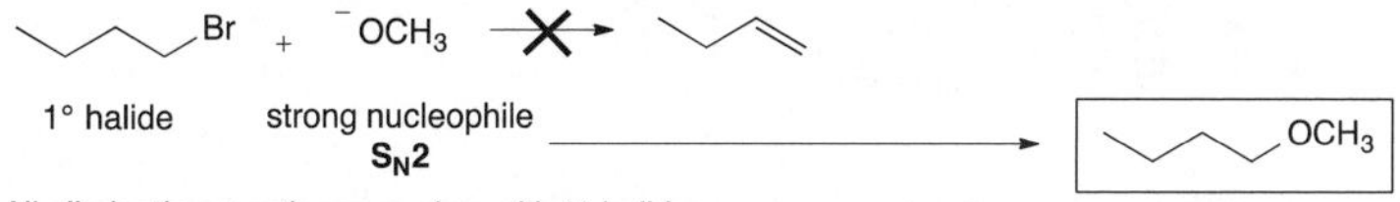

1° halide   strong nucleophile **S_N2**

All elimination reactions are slow with 1° halides.
The strong nucleophile reacts by an S_N2 mechanism instead.

c.

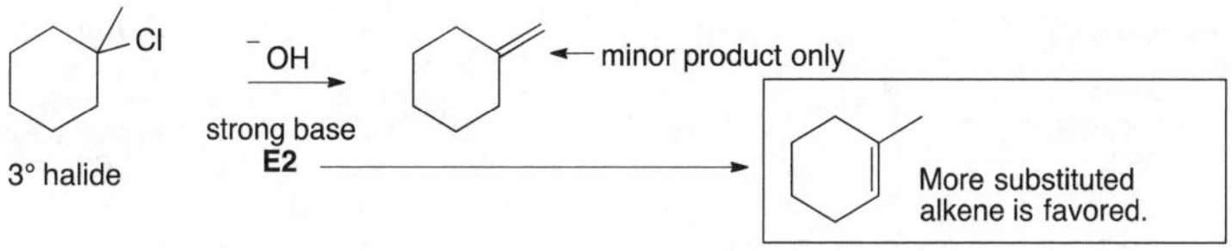

3° halide   strong base **E2**

← minor product only

More substituted alkene is favored.

d.

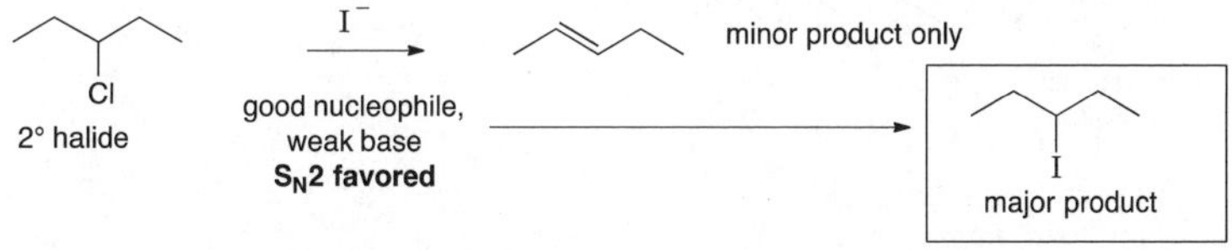

2° halide   good nucleophile, weak base **S_N2 favored**

minor product only

major product

The 2° halide can react by an E2 or S_N2 reaction with a negatively charged nucleophile or base. Since I⁻ is a weak base, substitution by an S_N2 mechanism is favored.

8.59

3° halide, weak base:
S_N1 and E1

a.

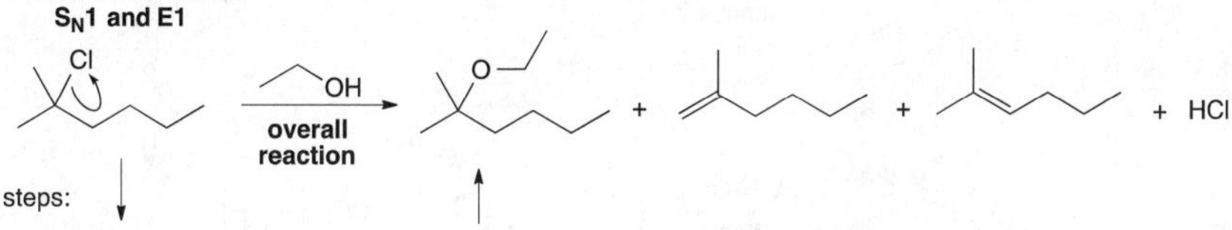

overall reaction

+   +   + HCl

The steps:

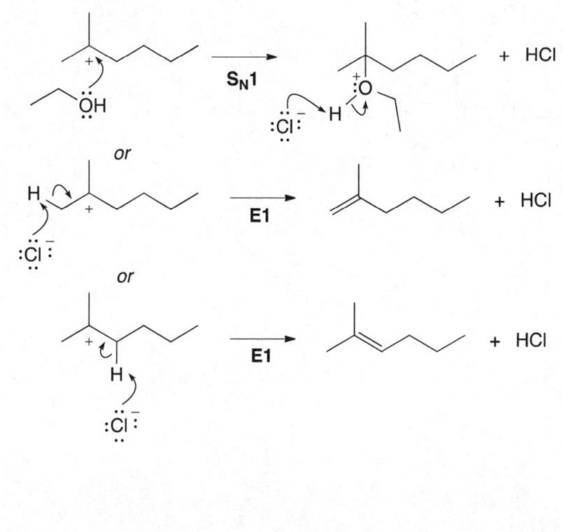

S_N1   + HCl   Any base (such as CH_3CH_2OH or Cl⁻) can be used to remove a proton to form an alkene. If Cl⁻ is used, HCl is formed as a reaction by-product. If CH_3CH_2OH is used, $(CH_3CH_2OH_2)^+$ is formed instead.

or

E1   + HCl

or

E1   + HCl

Chapter 8–26

b.

3° halide
strong base
E2

Each product:

or

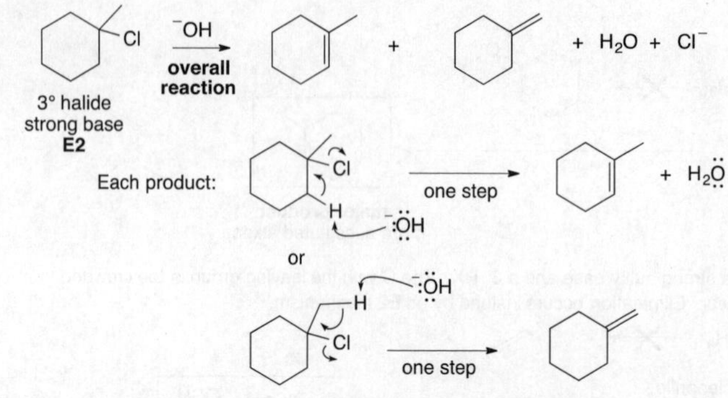

8.60 Draw the products of each reaction with the 1° alkyl halide.

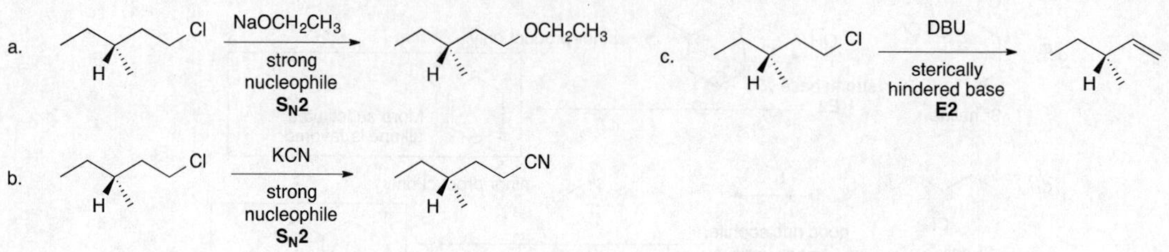

a.

strong
nucleophile
S_N2

c.

sterically
hindered base
E2

b.

strong
nucleophile
S_N2

8.61

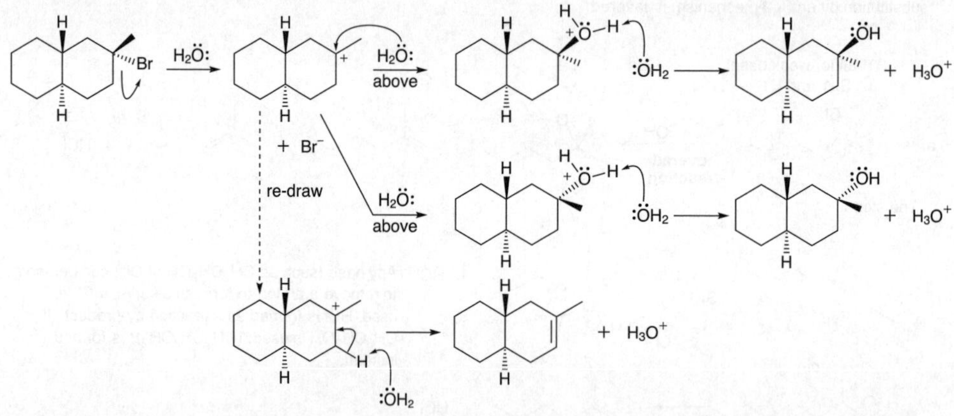

Alkyl Halides and Elimination Reactions 8–27

8.62

good nucleophile

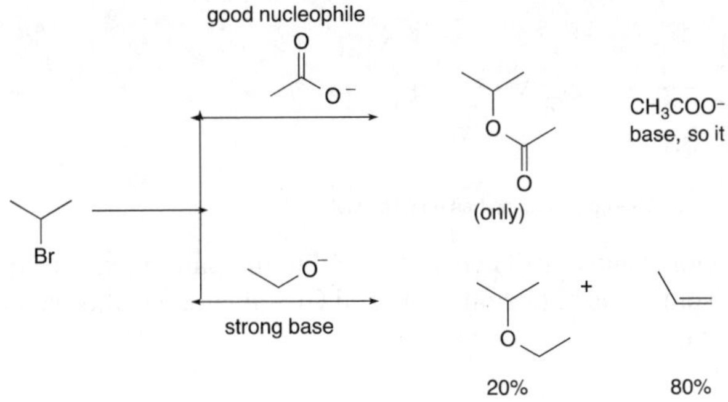

CH_3COO^- is a good nucleophile and a weak base, so it favors substitution by S_N2.

(only)

strong base

20%   80%

The strong base gives both S_N2 and E2 products, but because the 2° RX is somewhat hindered to substitution, the E2 product is favored.

8.63

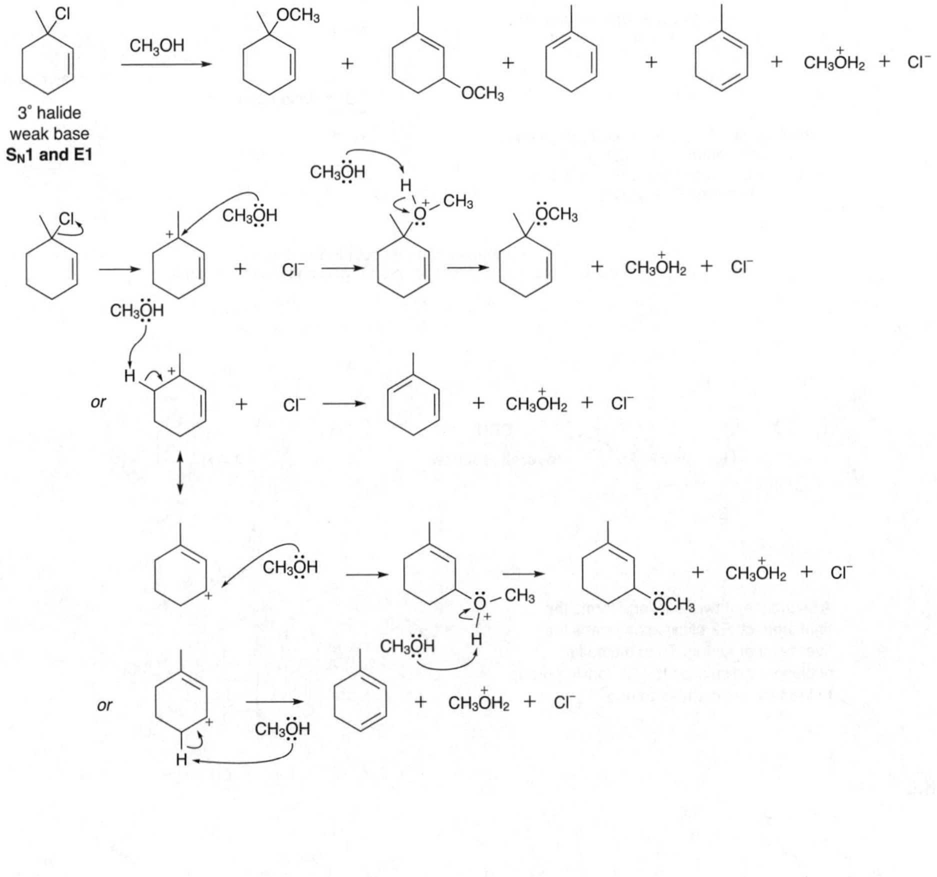

CH_3OH

+ $CH_3\overset{+}{O}H_2$ + Cl^-

3° halide
weak base
S_N1 and E1

$CH_3\overset{..}{O}H$

+ Cl^- → → + $CH_3\overset{+}{O}H_2$ + Cl^-

$CH_3\overset{..}{O}H$

or → + Cl^- → + $CH_3\overset{+}{O}H_2$ + Cl^-

$CH_3\overset{..}{O}H$ → → + $CH_3\overset{+}{O}H_2$ + Cl^-

$CH_3\overset{..}{O}H$

or → + $CH_3\overset{+}{O}H_2$ + Cl^-

$CH_3\overset{..}{O}H$

Chapter 8–28

8.64 E2 elimination needs a leaving group and a hydrogen in the **trans diaxial** position.

Two different
conformations:

This conformation has Cl's
axial, but no H's axial.

This conformation has no Cl's axial.

For elimination to occur, a cyclohexane must have a H and Cl in the trans diaxial arrangement.
Neither conformation of this isomer has both atoms—H and Cl—axial; thus, this isomer only
slowly loses HCl by elimination.

8.65

H and Br are *anti periplanar.*
Elimination can occur.

CH_3O^-
$-HBr$

major product

Elimination can occur here.

H (in the ring) and Br are **NOT** *anti periplanar.*
Elimination can**not** occur using this H.
Instead elimination must occur with the
H on the CH_3 group.

CH_3O^-
$-HBr$

Elimination cannot occur in the ring
because the required anti periplanar geometry is not present.

8.66

leaving group

B:

DBN
overall reaction

E2

S_N2

A sequence of two reactions forms the
final product: E2 elimination opens the
five-membered ring. Then the sulfur
nucleophile displaces the Cl^- leaving group
to form the six-membered ring.

8.67

a.

Δ

$SeOC_6H_5$ and H are on the same
side of the ring.
syn elimination

b.

rotate

Zn

Both Br atoms are on the opposite sides
of the C–C bond.
anti elimination

8.68 One equivalent of $NaNH_2$ removes one mole of HBr in an anti periplanar fashion from each dibromide. Two modes of elimination are possible for each compound.

a.

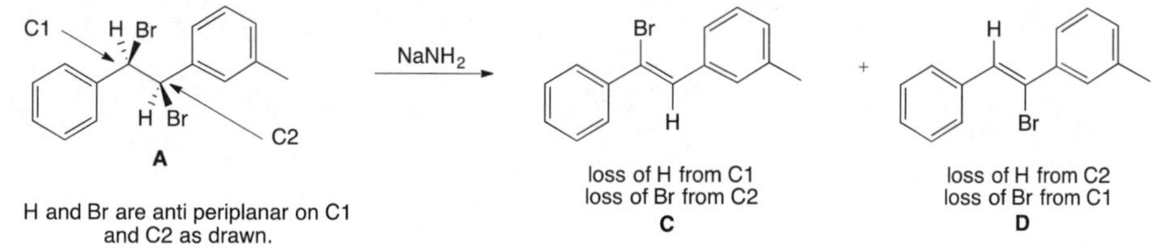
C1
H Br
H Br
C2
A
H and Br are anti periplanar on C1 and C2 as drawn.
NaNH2
Br
H
loss of H from C1
loss of Br from C2
C
+
H
Br
loss of H from C2
loss of Br from C1
D

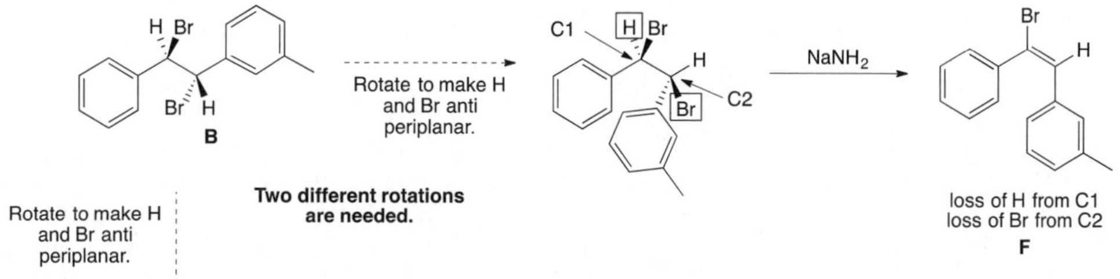
H Br
Br H
B
Rotate to make H and Br anti periplanar.
C1
H Br
H
Br
C2
NaNH2
Br
H
loss of H from C1
loss of Br from C2
F
Two different rotations are needed.
Rotate to make H and Br anti periplanar.

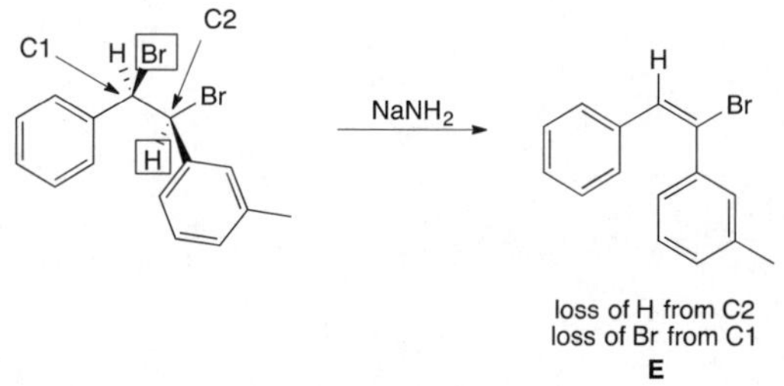
C2
C1
H Br
Br
H
NaNH2
H
Br
loss of H from C2
loss of Br from C1
E

b. **C** and **F** are diastereomers.
D and **E** are diastereomers.
C and **D** are constitutional isomers.
E and **F** are constitutional isomers.

Alcohols, Ethers, and Related Compounds 9–1

Chapter 9 Alcohols, Ethers, and Related Compounds

Chapter Review

General facts about ROH, ROR, and epoxides

- All three compounds contain an O atom that is sp^3 hybridized and tetrahedral (9.2).

$$CH_3 \overset{\ddot{O}}{\frown} H$$
109°
an alcohol

$$CH_3 \overset{\ddot{O}}{\frown} CH_3$$
111°
an ether

60° $\overset{\ddot{O}}{\triangle}$
H H H H
an epoxide

- All three compounds have polar C–O bonds, but only alcohols have an O–H bond for intermolecular hydrogen bonding (9.4).

hydrogen bond

- Alcohols and ethers do not contain a good leaving group. Nucleophilic substitution can occur only after the OH (or OR) group is converted to a better leaving group (9.7A).

$$R-\ddot{O}H \; + \; H-Cl \; \rightleftharpoons \; R-\overset{+}{O}H_2 \; + \; Cl^-$$

strong acid

weak base
good leaving group

- Epoxides have a leaving group located in a strained three-membered ring, making them reactive to strong nucleophiles and acids HZ that contain a nucleophilic atom Z (9.16).

leaving group

With strong nucleophiles,
:Nu⁻

[1]
:Nu⁻

H–OH

[2]

+ ⁻OH

A new reaction of carbocations (9.9)

- Less stable carbocations rearrange to more stable carbocations by shift of a hydrogen atom or an alkyl group. Besides rearrangement, carbocations also react with nucleophiles (7.12) and bases (8.6).

1,2-shift

R
(or H)

R
(or H)

Chapter 9–2

Preparation of alcohols, ethers, and epoxides (9.6)

[1] Preparation of alcohols

$R-X$ + ^-OH ⟶ $R\!-\!OH$ + X^-

- The mechanism is S_N2.
- The reaction works best for CH_3X and 1° RX.

---

[2] Preparation of alkoxides (a Brønsted–Lowry acid–base reaction)

$R-O-H$ + Na^+H^- ⟶ $R\!-\!O^-$ Na^+ + H_2

alkoxide

---

[3] Preparation of ethers (Williamson ether synthesis)

$R-X$ + $^-OR'$ ⟶ $R\!-\!OR'$ + X^-

- The mechanism is S_N2.
- The reaction works best for CH_3X and 1° RX.

---

[4] Preparation of epoxides (intramolecular S_N2 reaction)

- A two-step reaction sequence:
  [1] Removal of a proton with base forms an alkoxide.
  [2] Intramolecular S_N2 reaction forms the epoxide.

Reactions of alcohols

[1] Dehydration to form alkenes

   a. Using strong acid (9.8, 9.9)

$\underset{H\ \ OH}{-C-C-}$ $\xrightarrow[\text{TsOH}]{\text{H}_2\text{SO}_4 \text{ or}}$ $C\!=\!C$ + H_2O

- Order of reactivity: $R_3COH > R_2CHOH > RCH_2OH$.
- The mechanism for 2° and 3° ROH is E1; carbocations are intermediates and rearrangements occur.
- The mechanism for 1° ROH is E2.
- The Zaitsev rule is followed.

---

   b. Using POCl₃ and pyridine (9.10)

$\underset{H\ \ OH}{-C-C-}$ $\xrightarrow[\text{pyridine}]{\text{POCl}_3}$ $C\!=\!C$ + H_2O

- The mechanism is E2.
- No carbocation rearrangements occur.

Alcohols, Ethers, and Related Compounds 9–3

[2] Reaction with HX to form RX (9.11)

$$R-OH \ + \ H-X \ \longrightarrow \ \boxed{R-X} \ + \ H_2O$$

- Order of reactivity: $R_3COH > R_2CHOH > RCH_2OH$.
- The mechanism for 2° and 3° ROH is S_N1; carbocations are intermediates and rearrangements occur.
- The mechanism for CH_3OH and 1° ROH is S_N2.

[3] Reaction with other reagents to form RX (9.12)

$$R-OH \ + \ SOCl_2 \ \xrightarrow{\text{pyridine}} \ \boxed{R-Cl}$$

$$R-OH \ + \ PBr_3 \ \longrightarrow \ \boxed{R-Br}$$

- Reactions occur with CH_3OH and 1° and 2° ROH.
- The reactions follow an S_N2 mechanism.

[4] Reaction with tosyl chloride to form alkyl tosylates (9.13A)

R—OH + Cl—S(=O)(=O)—⟨C₆H₄⟩—CH₃ $\xrightarrow{\text{pyridine}}$ R—O—S(=O)(=O)—⟨C₆H₄⟩—CH₃

$$\boxed{R-OTs}$$

- The C–O bond is not broken, so the configuration at a stereogenic center is retained.

Reactions of alkyl tosylates

Alkyl tosylates undergo either substitution or elimination depending on the reagent (9.13B).

- Substitution is carried out with strong :Nu⁻, so the mechanism is S_N2.
- Elimination is carried out with strong bases, so the mechanism is E2.

Reactions of ethers

Only one reaction is useful: Cleavage with strong acids (9.14)

$$R-O-R' \ + \ \underset{\substack{\text{(2 equiv)} \\ (X = \text{Br or I})}}{H-X} \ \longrightarrow \ \boxed{R-X} \ + \ \boxed{R'-X} \ + \ H_2O$$

- With 2° and 3° R groups, the mechanism is S_N1.
- With CH_3 and 1° R groups, the mechanism is S_N2.

Chapter 9–4

Reactions involving thiols and sulfides (9.15)

[1] Preparation of thiols

$$R-X \ + \ {}^-SH \ \longrightarrow \ \boxed{R-SH} \ + \ X^-$$

- The mechanism is S_N2.
- The reaction works best for CH_3X and 1° RX.

[2] Oxidation and reduction involving thiols

a. Oxidation of thiols to disulfides

$$R-SH \ \xrightarrow{\ Br_2 \ or \ I_2\ } \ \boxed{RS-SR}$$

b. Reduction of disulfides to thiols

$$RS-SR \ \xrightarrow[\text{HCl}]{\text{Zn}} \ \boxed{R-SH}$$

[3] Preparation of sulfides

$$R-X \ + \ {}^-SR' \ \longrightarrow \ \boxed{R-SR'} \ + \ X^-$$

- The mechanism is S_N2.
- The reaction works best for CH_3X and 1° RX.

[4] Reaction of sulfides to form sulfonium ions

$$R'_2S \ + \ R-X \ \longrightarrow \ \boxed{R'_2\overset{+}{S}-R} \ + \ X^-$$

- The mechanism is S_N2.
- The reaction works best for CH_3X and 1° RX.

Reactions of epoxides

Epoxide rings are opened with nucleophiles :Nu$^-$ and acids HZ (9.16).

- The reaction occurs with backside attack, resulting in trans or anti products.
- With :Nu$^-$, the mechanism is S_N2, and nucleophilic attack occurs at the *less* substituted C.
- With HZ, the mechanism is between S_N1 and S_N2, and attack of Z$^-$ occurs at the *more* substituted C.

Alcohols, Ethers, and Related Compounds 9–5

Practice Test on Chapter Review

1. Give the IUPAC name for each of the following compounds.

a.

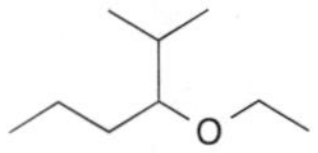

b.

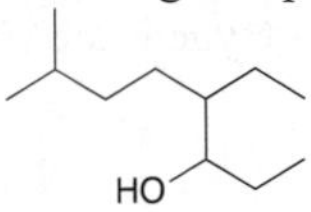

2. Draw the organic products formed in each reaction. Draw all stereogenic centers using wedges and dashed wedges.

a.

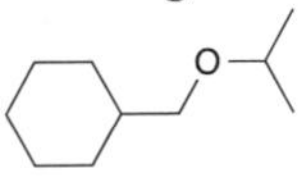

$$\xrightarrow[\text{(2 equiv)}]{\text{HI}}$$

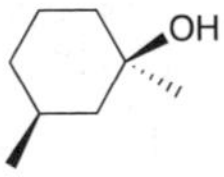

b.

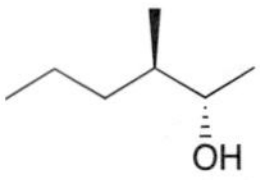

$$\xrightarrow[\text{[2] } CH_3CH_2Br]{\text{[1] NaH}}$$

c.

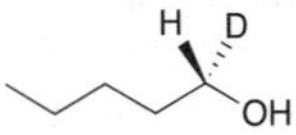

$$\xrightarrow[\text{[2] NaOCH}_3]{\text{[1] TsCl, pyr}}$$

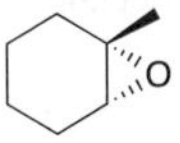

d.

$$\xrightarrow{\text{HBr}}$$

e.

$$\xrightarrow[\text{[2] } H_2O]{\text{[1] NaCN}}$$

f.

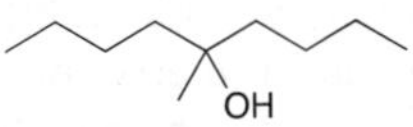

$$\xrightarrow{H_2SO_4}$$

3. What starting material is needed for the following reaction?

$$\xrightarrow[\text{[2] NaOCH}_3]{\text{[1] TsCl, pyridine}}$$ ᐧᐧᐧOCH$_3$

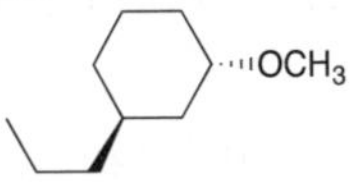

4. What alkoxide and alkyl halide are needed to make the following ether?

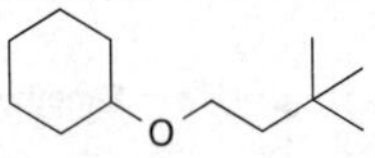

Answers to Practice Test

1.a. 3-ethoxy-2-methylhexane

b. 4-ethyl-7-methyl-octan-3-ol

2.

a. 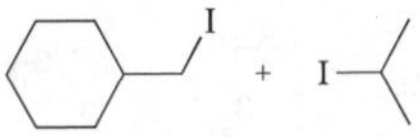   I   +   I

b. 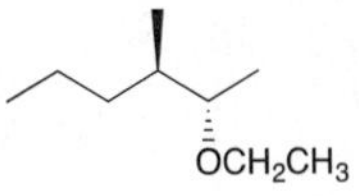   ŌCH$_2$CH$_3$

c.   D H   OCH$_3$

d.   Br   +   Br

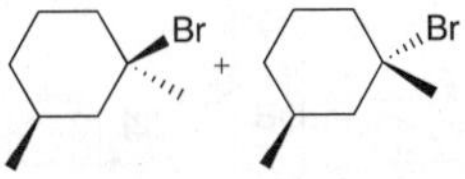

e. 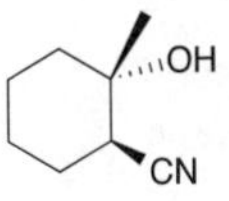   ᐧᐧᐧOH   CN

f. 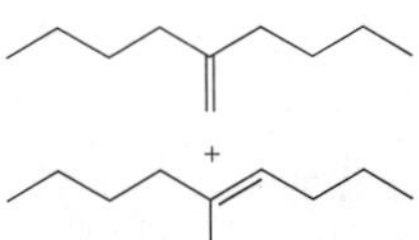   (+ stereoisomer)

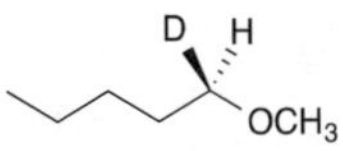

3. 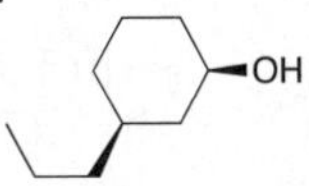   ᐧᐧᐧOH

4. 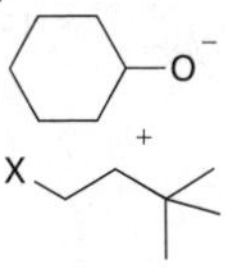   O$^-$ +   X

Chapter 9–6

Answers to Problems

9.1 **Alcohols** are classified as 1°, 2°, or 3°, depending on the number of carbon atoms bonded to the carbon with the OH group. Five ether oxygens are circled.

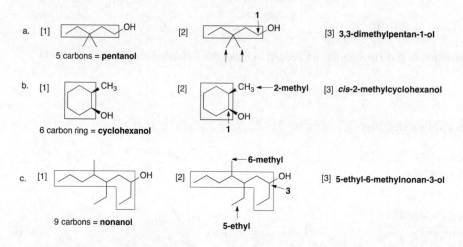

9.2 To name an alcohol:
 [1] **Find the longest chain that has the OH group as a substituent.** Name the molecule as a derivative of that number of carbons by changing the *-e* ending of the alkane to the suffix *-ol*.
 [2] **Number the carbon chain to give the OH group the lower number.** When the OH group is bonded to a ring, the ring is numbered beginning with the OH group, and the "1" is usually omitted.
 [3] Apply the other rules of nomenclature to complete the name.

a. [1] 5 carbons = **pentanol**   [2]   [3] **3,3-dimethylpentan-1-ol**

b. [1] 6 carbon ring = **cyclohexanol**   [2] 2-methyl   [3] *cis*-2-methylcyclohexanol

c. [1] 9 carbons = **nonanol**   [2] 6-methyl, 5-ethyl   [3] **5-ethyl-6-methylnonan-3-ol**

Alcohols, Ethers, and Related Compounds 9–7

9.3 To work backwards from a name to a structure:
  [1] Find the parent name and draw its structure.
  [2] Add the substituents to the long chain.

  a. 7,7-dimethyl**octan-4-ol**

  c. 2-*tert*-butyl-3-methyl**cyclohexanol**

  b. 5-methyl-4-propyl**heptan-3-ol**

  d. *trans*-**cyclohexane-1,2-diol**

9.4 To name simple ethers:
  [1] Name both alkyl groups bonded to the oxygen.
  [2] Arrange these names alphabetically and add the word ***ether***. For symmetrical ethers, name the alkyl group and add the prefix ***di***.

  To name ethers using the IUPAC system:
  [1] Find the two alkyl groups bonded to the ether oxygen. The smaller chain becomes the substituent, named as an alkoxy group.
  [2] Number the chain to give the lower number to the first substituent.

  a. **common name:**
  ↑ methyl ↑ butyl
  butyl methyl ether

  IUPAC name:
  ← 4 C's, butane
  substituent: methoxy
  1-methoxybutane

  b. **common name:**
  ↑ cyclohexyl ↑ methyl
  cyclohexyl methyl ether

  IUPAC name:
  ← methoxy substituent
  6 C's, cyclohexane
  methoxycyclohexane

  c. **common name:**
  ↑ propyl ↑ propyl
  dipropyl ether

  IUPAC name:
  ↑ propoxy ↑ propane
  1-propoxypropane

9.5 Three ways to name epoxides:
[1] Epoxides are named as derivatives of oxirane, the simplest epoxide.
[2] Epoxides can be named by considering the oxygen as a substituent called an **epoxy** group, bonded to a hydrocarbon chain or ring. Use two numbers to designate which two atoms the oxygen is bonded to.
[3] Epoxides can be named as **alkene oxides** by mentally replacing the epoxide oxygen by a double bond. Name the alkene (Chapter 10) and add the word *oxide.*

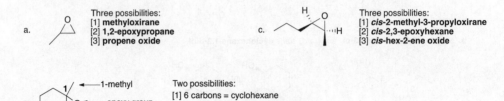

a. Three possibilities:
[1] **methyloxirane**
[2] **1,2-epoxypropane**
[3] **propene oxide**

c. Three possibilities:
[1] *cis*-**2-methyl-3-propyloxirane**
[2] *cis*-**2,3-epoxyhexane**
[3] *cis*-**hex-2-ene oxide**

b. 1-methyl ← epoxy group

Two possibilities:
[1] 6 carbons = cyclohexane
1,2-epoxy-1-methylcyclohexane
[2] **1-methylcyclohexene oxide**

9.6 Two rules for boiling point:
[1] **The stronger the intermolecular forces the higher the bp.**
[2] **Bp increases as the extent of the hydrogen bonding increases.** For alcohols with the same number of carbon atoms: 3° ROH < 2° ROH < 1° ROH.

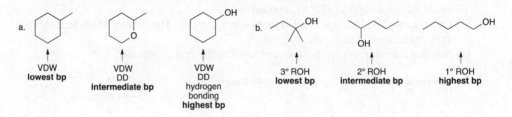

a.
VDW
lowest bp

VDW
DD
intermediate bp

VDW
DD
hydrogen bonding
highest bp

b.
3° ROH
lowest bp

2° ROH
intermediate bp

1° ROH
highest bp

9.7
Strong nucleophiles (like ⁻CN) favor S_N2 reactions. The use of crown ethers in nonpolar solvents increases the nucleophilicity of the anion, and this increases the rate of the S_N2 reaction. The nucleophile does not appear in the rate equation for the S_N1 reaction. Nonpolar solvents cannot solvate carbocations, so this disfavors S_N1 reactions as well.

9.8
Draw the products of substitution in the following reactions by substituting OH or OR for X in the starting material.

a. ~~~Br + ⁻OH ⟶ ~~~OH + Br⁻   **alcohol**

b. ~~~Cl + ⁻OCH₃ ⟶ ~~~OCH₃ + Cl⁻   **unsymmetrical ether**

c. (cyclohexyl-CH₂-I) + ⁻O-C(CH₃)₂ ⟶ (cyclohexyl-CH₂-O-C(CH₃)₂) + I⁻   **unsymmetrical ether**

d. ~~~Br + ⁻OCH₂CH₃ ⟶ ~~~OCH₂CH₃ + Br⁻   **unsymmetrical ether**

Alcohols, Ethers, and Related Compounds 9–9

9.9 Two possible routes to **X** are shown. Path [2] with a 1° alkyl halide is preferred. Path [1] cannot occur because the leaving group would be bonded to an sp^2 hybridized C, making it an unreactive aryl halide.

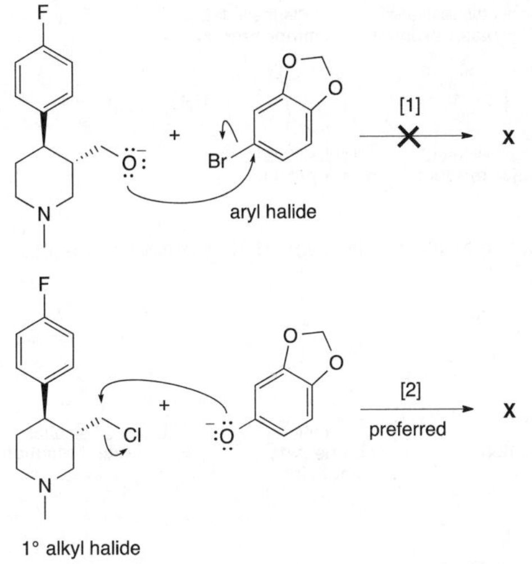

aryl halide

1° alkyl halide

9.10 NaH and NaNH₂ are strong bases that will remove a proton from an alcohol, creating a nucleophile.

a.

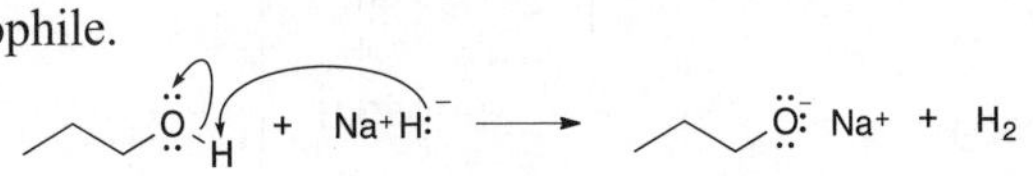

b.

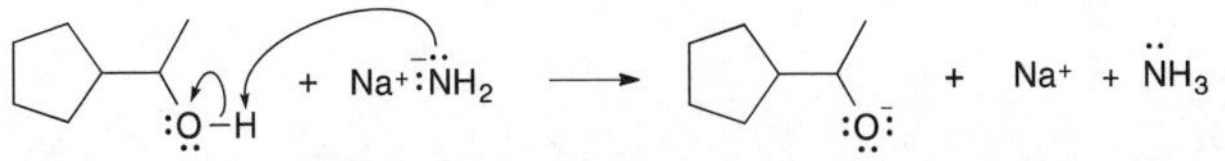

c.

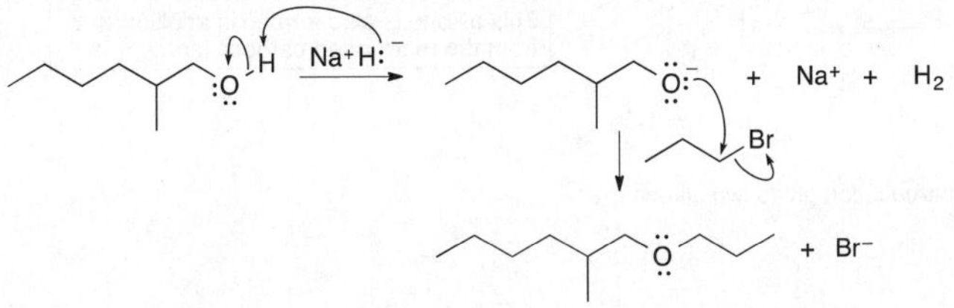

d.

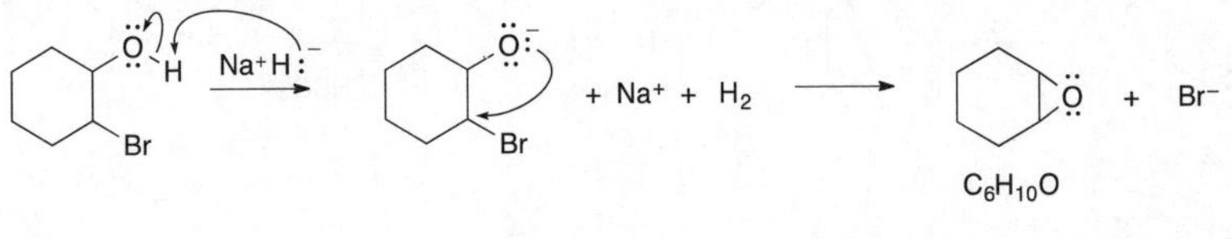

$C_6H_{10}O$

9.11 Dehydration follows the Zaitsev rule, so the more stable, more substituted alkene is the major product.

a.

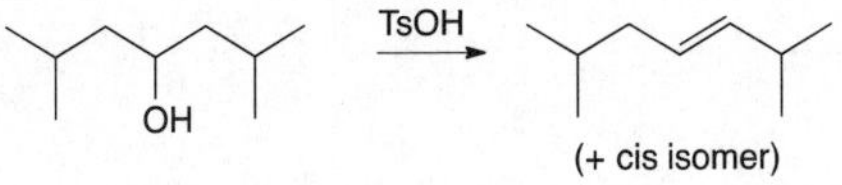

TsOH

(+ cis isomer)

Chapter 9–10

b.

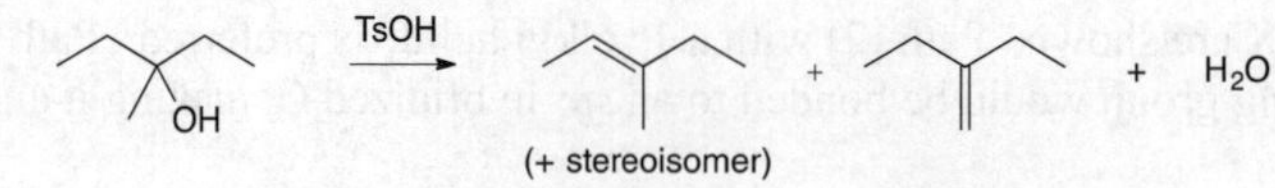

(+ stereoisomer)
trisubstituted
major product

disubstituted
minor product

+ H₂O

c.

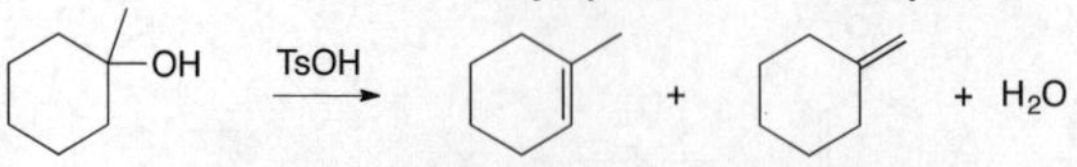

trisubstituted
major product

disubstituted
minor product

+ H₂O

9.12 The rate of dehydration increases as the number of R groups increases.

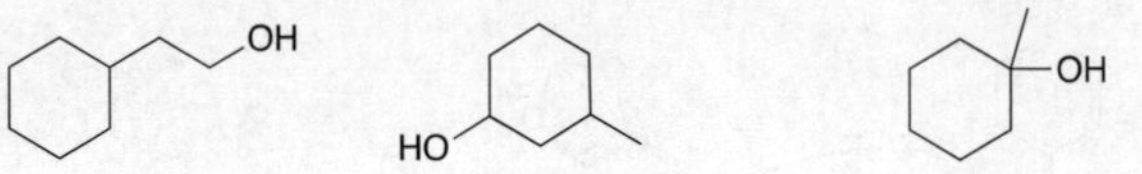

1° alcohol
slowest reaction

2° alcohol
intermediate reactivity

3° alcohol
fastest reaction

9.13

transition state [1]:

transition state [2]:

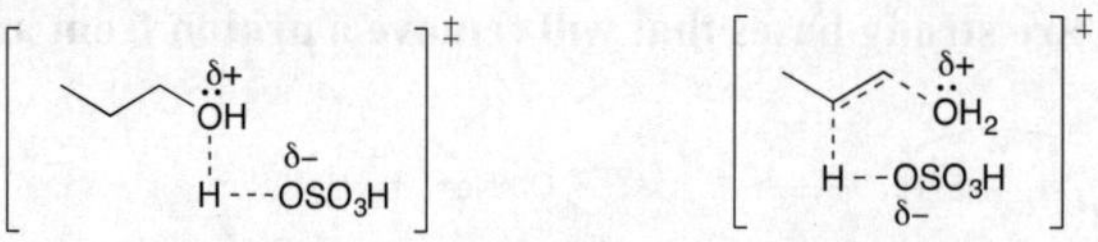

9.14

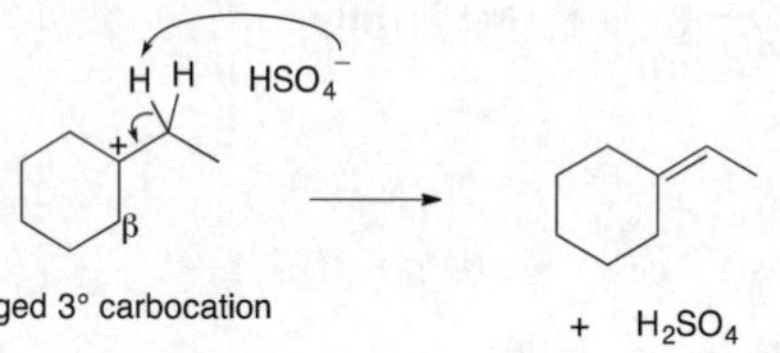

rearranged 3° carbocation

+ H₂SO₄

This alkene is also formed in addition to **Y** from the rearranged carbocation.

The initially formed 2° carbocation gives two alkenes:

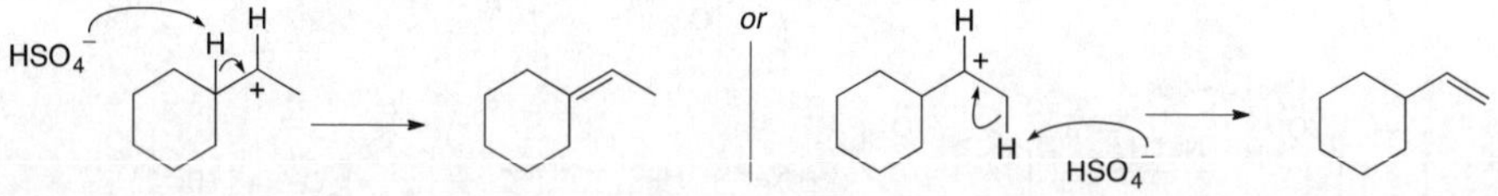

Alcohols, Ethers, and Related Compounds 9–11

9.15

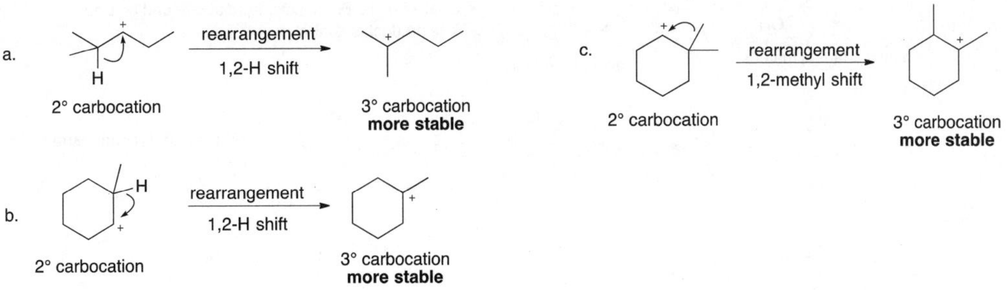

a.

2° carbocation → rearrangement 1,2-H shift → 3° carbocation **more stable**

c.

2° carbocation → rearrangement 1,2-methyl shift → 3° carbocation **more stable**

b.

2° carbocation → rearrangement 1,2-H shift → 3° carbocation **more stable**

9.16

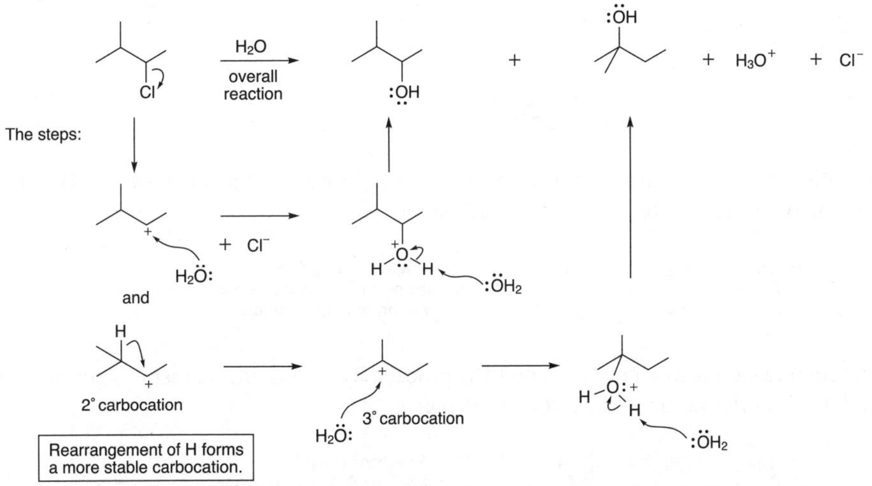

H_2O overall reaction

:ÖH   +   :ÖH   + H_3O^+ +   Cl$^-$

The steps:

+ Cl$^-$ $H_2\ddot{O}$:

and

H

2° carbocation

| Rearrangement of H forms a more stable carbocation. |

:ÖH$_2$

H—Ö:—H   :ÖH$_2$

$H_2\ddot{O}$:   3° carbocation

H—Ö:+—H   :ÖH$_2$

9.17

a. OH → HCl → Cl + H_2O

c. OH → HBr → Br + H_2O

b. OH → HI → I + H_2O

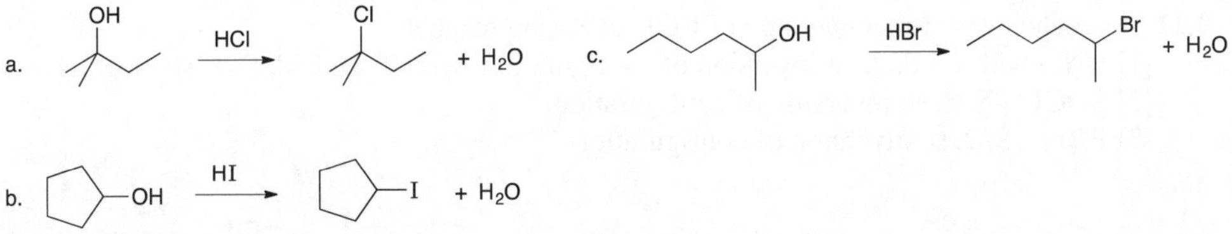

9.18 • **CH$_3$OH and 1° alcohols** follow an S$_N$2 mechanism, which results in inversion of configuration.
• **Secondary (2°) and 3° alcohols** follow an S$_N$1 mechanism, which results in racemization at a stereogenic center.

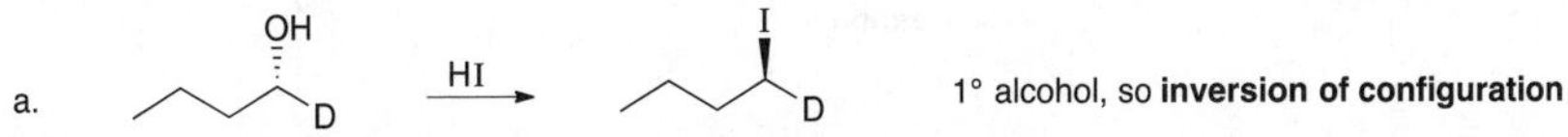

a. OH⁄⁄D → HI → I⁄D   1° alcohol, so **inversion of configuration**

Chapter 9–12

b.

3° alcohol, so Br⁻ attacks from above and below.
The product is achiral.

achiral starting material   achiral product

c.

3° alcohol = **racemization**

9.19

a.

HCl

c.

HCl

(product formed
after a 1,2-H shift)

b.

HCl

(product formed after
a 1,2-CH$_3$ shift)

9.20 Substitution reactions of alcohols using SOCl$_2$ proceed by an S$_N$2 mechanism. Therefore, there is **inversion of configuration** at a stereogenic center.

SOCl$_2$ / pyridine

Reactions using SOCl$_2$
proceed by an S$_N$2 mechanism =
inversion of configuration.

9.21 Substitution reactions of alcohols using PBr$_3$ proceed by an S$_N$2 mechanism. Therefore, there is inversion of configuration at a stereogenic center.

PBr$_3$

Reactions using PBr$_3$
proceed by an S$_N$2 mechanism =
inversion of configuration.

9.22 Stereochemistry for conversion of ROH to RX by reagent:
[1] **HX**—with 1°, S$_N$2, so inversion of configuration; with 2° and 3°, S$_N$1, so racemization.
[2] **SOCl$_2$**—S$_N$2, so inversion of configuration.
[3] **PBr$_3$**—S$_N$2, so inversion of configuration.

a.

SOCl$_2$ / pyridine

c.

PBr$_3$

S$_N$2 =
inversion

b.

HI

3° alcohol, S$_N$1 =
racemization

Alcohols, Ethers, and Related Compounds 9–13

9.23 To do a two-step synthesis with this starting material:
[1] Convert the OH group into a good leaving group (by using either PBr₃ or SOCl₂).
[2] Add the nucleophile for the SN2 reaction.

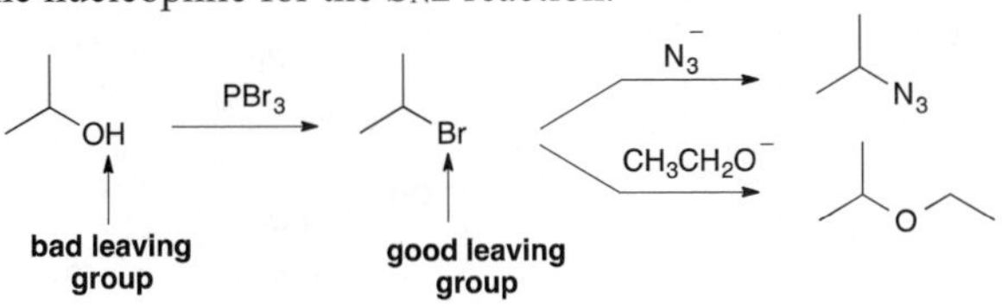

9.24

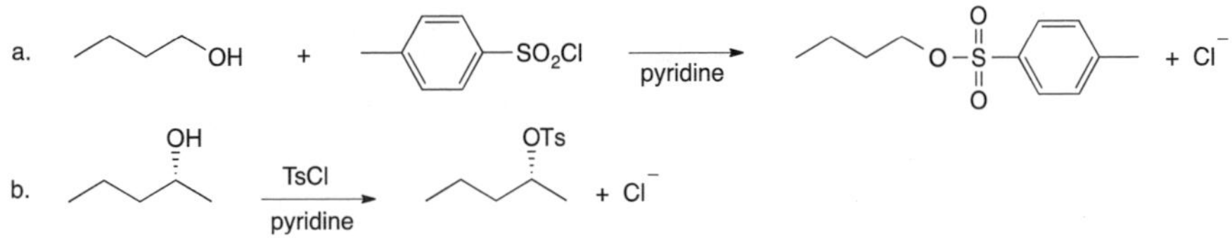

a.

b.

9.25

a.

b.

c.

(Substitution is favored over elimination.)

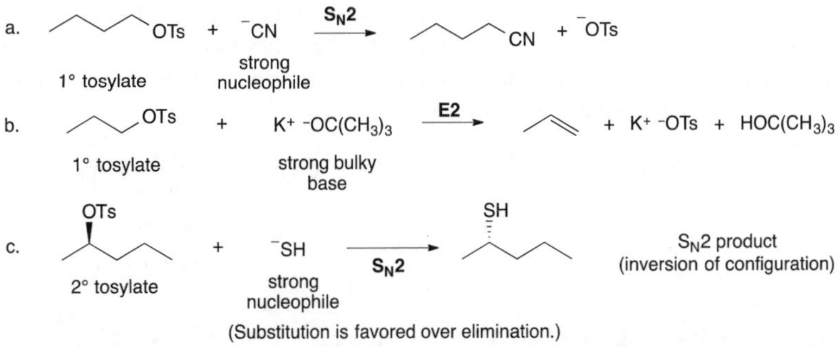

9.26

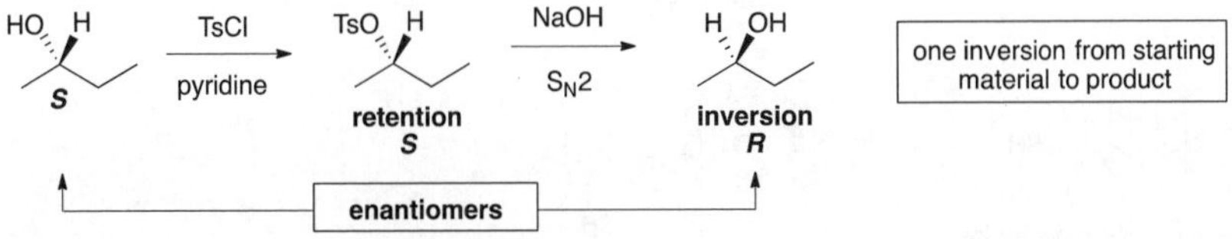

9.27 These reagents can be classified as follows:
[1] SOCl₂, PBr₃, HCl, and HBr replace OH with X by a substitution reaction.
[2] Tosyl chloride (TsCl) makes OH a better leaving group by converting it to OTs.
[3] Strong acids (H₂SO₄) and POCl₃ (pyridine) result in elimination by dehydration.

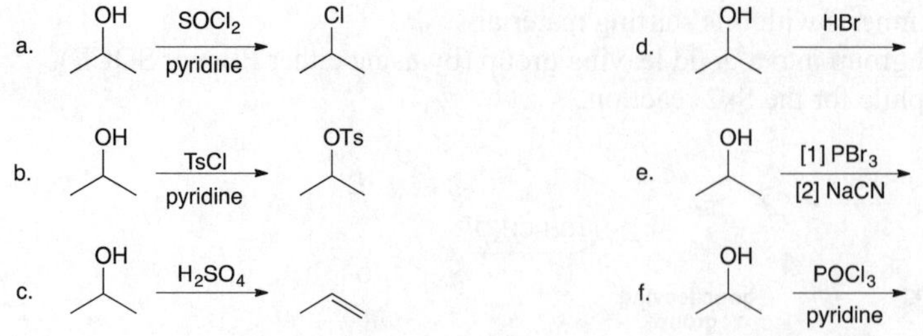

Chapter 9–14

a.

d.

b.

e.

c.

f.

9.28

a.

c.

b.

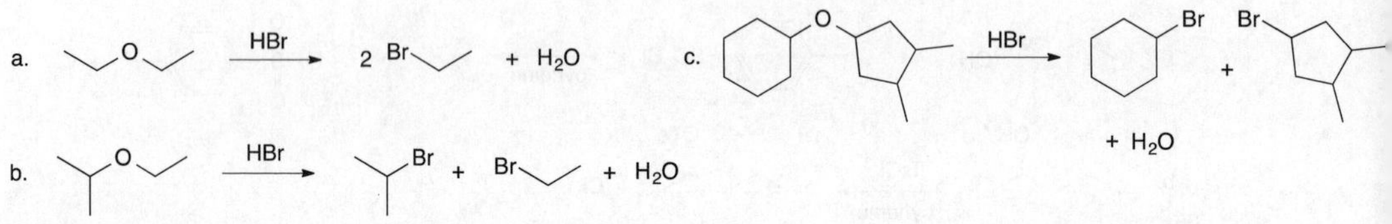

9.29 Ether cleavage can occur by either an S_N1 or S_N2 mechanism, but neither mechanism can occur when the ether O atom is bonded to an aromatic ring. An S_N1 reaction would require formation of a highly unstable carbocation on a benzene ring, a process that does not occur. An S_N2 reaction would require backside attack through the plane of the aromatic ring, which is also not possible. Thus, cleavage of the Ph–OCH$_3$ bond does not occur.

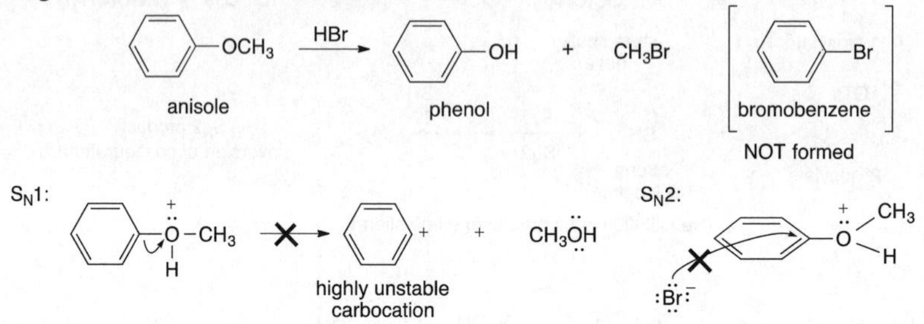

anisole   phenol   bromobenzene

NOT formed

S_N1:

highly unstable carbocation

S_N2:

9.30

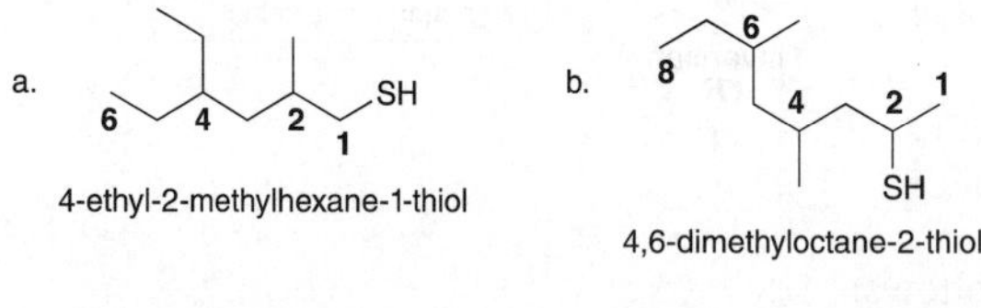

a.

4-ethyl-2-methylhexane-1-thiol

b.

4,6-dimethyloctane-2-thiol

9.31

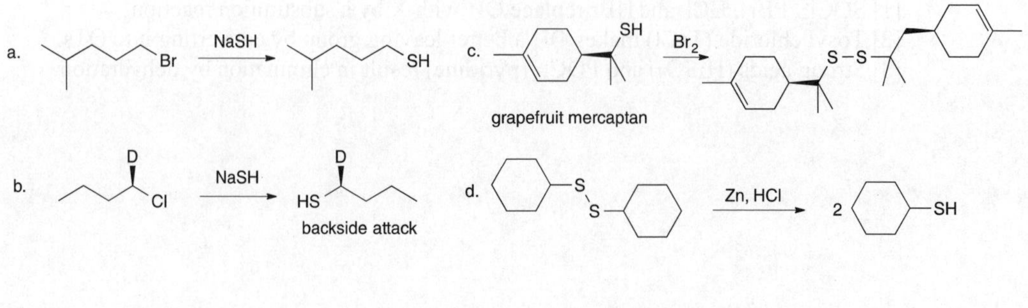

a.

c.

grapefruit mercaptan

b.

backside attack

d.

9.32

a. 1-ethylthiobutane or butyl ethyl sulfide

b. 2-methyl-1-methylthiocyclopentane

9.33

a. Cyclohexyl-SH [1] NaH / [2] CH₃Br → cyclohexyl-S-CH₃

b. ethyl sec-butyl sulfide + isobutyl chloride → sulfonium salt + Cl⁻

9.34 Two rules for the reaction of an epoxide:
[1] Nucleophiles attack from the **back side** of the epoxide.
[2] Negatively charged nucleophiles attack at the **less substituted carbon.**

a. epoxide [1] CH₃CH₂O⁻ / [2] H₂O → trans-1-ethoxy-2-hydroxycyclopentane

Attack here: less substituted C, backside attack

b. epoxide [1] H–C≡C⁻ / [2] H₂O → product

Attack here: less substituted C, backside attack

9.35 In both isomers, ⁻OH attacks from the back side at either C–O bond.

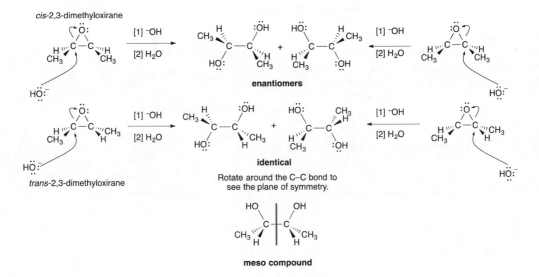

enantiomers (from cis-2,3-dimethyloxirane)

identical (from trans-2,3-dimethyloxirane)

Rotate around the C–C bond to see the plane of symmetry.

meso compound

Chapter 9–16

9.36 Remember the difference between negatively charged nucleophiles and neutral nucleophiles:
- **Negatively charged nucleophiles attack first,** followed by protonation, and the nucleophile attacks at the *less* substituted carbon.
- **Neutral nucleophiles have protonation first,** followed by nucleophilic attack at the *more* substituted carbon.

Trans or anti products are always formed, regardless of the nucleophile.

a.
$$\xrightarrow{\text{HBr}}$$
neutral nucleophile:
attack at **more**
substituted C

c.
$$\xrightarrow[\text{H}_2\text{SO}_4]{\text{CH}_3\text{CH}_2\text{OH}}$$
neutral nucleophile:
attack at **more**
substituted C

b.
$$\xrightarrow[\text{[2] H}_2\text{O}]{\text{[1] }^-\text{CN}}$$
negatively charged
nucleophile:
attack at **less**
substituted C

d.
$$\xrightarrow[\text{[2] CH}_3\text{OH}]{\text{[1] CH}_3\text{O}^-}$$
negatively charged
nucleophile:
attack at **less**
substituted C

9.37

a.
HO   **3**
1
6 C ring ⟶ cyclohexanol
2 CH₃'s at C3
3,3-dimethylcyclohexanol

b.
3 2 1
ethyl isobutyl ether
or
1-ethoxy-2-methylpropane

c.
1
2
1,2-epoxy-1-ethylcyclopentane
or
1-ethylcyclopentene oxide

9.38

a. (1*R*,2*R*)-2-isobutylcyclopentanol

b. 2° alcohol

HO
A

c. stereoisomer

HO
(1*R*,2*S*)-2-isobutylcyclopentanol

d. constitutional isomer

OH
(1*S*,3*S*)-3-isobutylcyclopentanol

e. constitutional isomer with an ether

O
butoxycyclopentane

Alcohols, Ethers, and Related Compounds 9–17

f.

[1] A + NaH →

[2] + H₂SO₄ → +

[3] + POCl₃ / pyridine → +

[4] + HCl →

[5] + SOCl₂ / pyridine →

[6] + TsCl / pyridine →

9.39

a. HO,H compound + HBr → Br,H (S) + H,Br (R) 2° alcohol S_N1 = racemization

b. HO,H compound + PBr₃ → H,Br (R) PBr₃ follows S_N2 = inversion.

c. HO,H compound + HCl → Cl,H (S) + H,Cl (R) 2° alcohol S_N1 = racemization

d. HO,H compound + SOCl₂/pyridine → H,Cl (R) SOCl₂ follows S_N2 = inversion.

9.40 Use the directions from Answer 9.2.

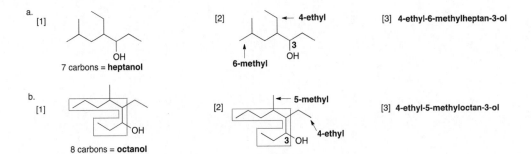

a. [1] 7 carbons = **heptanol** [2] ← 4-ethyl, ↑ 6-methyl, 3 OH [3] **4-ethyl-6-methylheptan-3-ol**

b. [1] 8 carbons = **octanol** [2] ← 5-methyl, 4-ethyl, 3 OH [3] **4-ethyl-5-methyloctan-3-ol**

Chapter 9–18

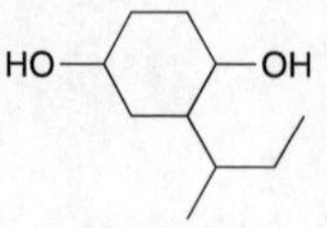

c.

[1] HO—⬡—OH
cyclohexanediol

[2] HO—⬡(4)(1)—OH
(3)(2)

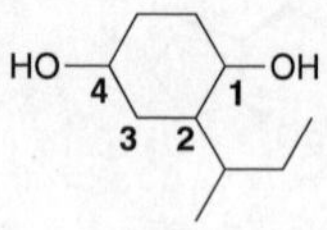

[3] **2-*sec*-butylcyclohexane-1,4-diol**

d.

[1] OH
OH   OH
7 carbons = **heptanetriol**

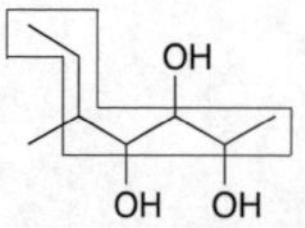

[2] OH
(4)(3)(2)
OH   OH
5-methyl

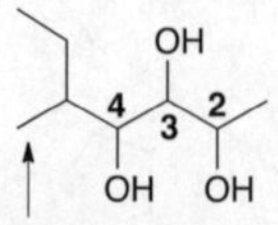

[3] **5-methylheptane-2,3,4-triol**

e.

[1] HO''⬠
5 carbons = **cyclopentanol**

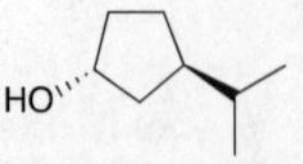

[2] HO''⬠(1)
3-isopropyl

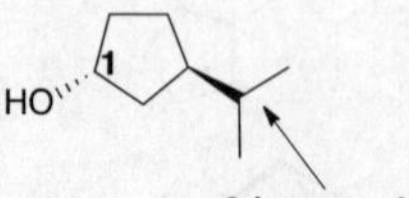

[3] ***trans*-3-isopropylcyclopentanol**

9.41 Use the rules from Answers 9.4 and 9.5.

a.

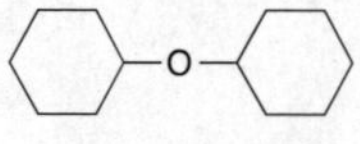

⬡—O—⬡
dicyclohexyl ether

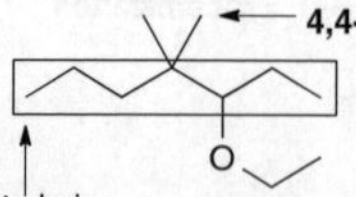

c.

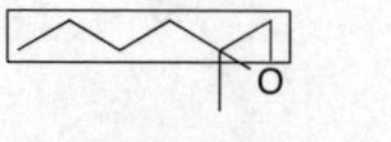

◁▷—◁▷
O
1,2-epoxy-2-methylhexane
or **2-butyl-2-methyloxirane**
or **2-methylhexene oxide**

e.

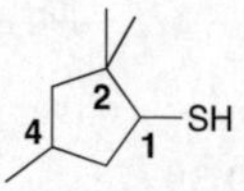

(2)
(4)(1)—SH
2,2,4-trimethylcyclopentanethiol

b.

◀—— **4,4-dimethyl**
O
longest chain =
heptane
substituent =
3-ethoxy
3-ethoxy-4,4-dimethylheptane

d.

S
(8)(4)
(3)(1)
4-ethylthio-3-methyloctane

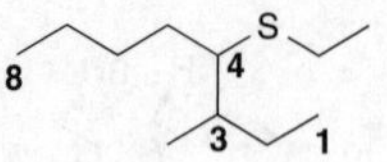

Alcohols, Ethers, and Related Compounds 9–19

9.42 Use the directions from Answer 9.3.

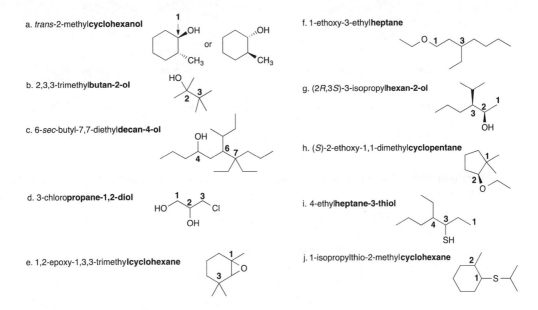

a. *trans*-2-methyl**cyclohexanol**

b. 2,3,3-trimethyl**butan-2-ol**

c. 6-*sec*-butyl-7,7-diethyl**decan-4-ol**

d. 3-chloro**propane-1,2-diol**

e. 1,2-epoxy-1,3,3-trimethyl**cyclohexane**

f. 1-ethoxy-3-ethyl**heptane**

g. (2*R*,3*S*)-3-isopropyl**hexan-2-ol**

h. (*S*)-2-ethoxy-1,1-dimethyl**cyclopentane**

i. 4-ethyl**heptane-3-thiol**

j. 1-isopropylthio-2-methyl**cyclohexane**

9.43 Melting points depend on intermolecular forces and symmetry. (CH₃)₂CHCH₂OH has a lower melting point than CH₃CH₂CH₂CH₂OH because branching decreases surface area and makes (CH₃)₂CHCH₂OH less symmetrical, so it packs less well. Although (CH₃)₃COH has the most branching and least surface area, it is the most symmetrical, so it packs best in a crystalline lattice, giving it the highest melting point.

–108 °C	–90 °C	26 °C
lowest melting point	intermediate melting point	highest melting point

9.44 Stronger intermolecular forces increase boiling point. All of the compounds can hydrogen bond, but both diols have more opportunity for hydrogen bonding because they have two OH groups, making their bp's higher than the bp of butan-1-ol. Propane-1,2-diol can also intramolecularly hydrogen bond. Intramolecular hydrogen bonding decreases the amount of intermolecular hydrogen bonding, so the bp of propane-1,2-diol is somewhat lower.

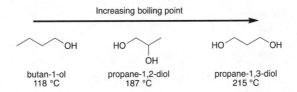

Increasing boiling point

butan-1-ol 118 °C   propane-1,2-diol 187 °C   propane-1,3-diol 215 °C

Chapter 9–20

9.45

a. (structure) OH → H₂SO₄ → (alkene)   f. (structure) OH → PBr₃ → (structure) Br

b. (structure) OH → NaH → (structure) O⁻ Na⁺   g. (structure) OH → TsCl / pyridine → (structure) OTs

c. (structure) OH → HCl / ZnCl₂ → (structure) Cl   h. (structure) OH → [1] NaH → (structure) O⁻ Na⁺ [2] (structure) Br → (structure) O

d. (structure) OH → HBr → (structure) Br   i. (structure) OH → [1] TsCl → (structure) OTs → [2] NaSH → (structure) SH

e. (structure) OH → SOCl₂ / pyridine → (structure) Cl   j. (structure) OH → POCl₃ / pyridine → (alkene)

9.46 Dehydration follows the Zaitsev rule, so the more stable, more substituted alkene is the major product.

a. (structure) OH → TsOH → (structure) + (structure)

   tetrasubstituted **major product**   disubstituted

b. (structure) OH → TsOH → (structure) + (structure)

c. (structure) OH → TsOH → (structure) + (structure)

   trisubstituted **major product**   disubstituted

d. (structure) OH → TsOH → (structure) + (structure)

   tetrasubstituted **major product**   disubstituted

   ┌─────────────────────────┐
   │ two products formed   │
   │ by carbocation rearrangement │
   └─────────────────────────┘

9.47 OTs is a good leaving group and will easily be replaced by a nucleophile. Draw the products by substituting the nucleophile in the reagent for OTs in the starting material.

a. (structure) OTs → CH₃SH / S_N2 → (structure) S + HOTs

b. (structure) OTs → NaOCH₂CH₃ / S_N2 → (structure) O + Na⁺ ⁻OTs

c. (structure) OTs → NaOH / S_N2 → (structure) OH + Na⁺ ⁻OTs

d. (structure) OTs → K⁺ ⁻OC(CH₃)₃ / E2 → (structure) + (CH₃)₃COH + K⁺ ⁻OTs

Alcohols, Ethers, and Related Compounds 9–21

9.48

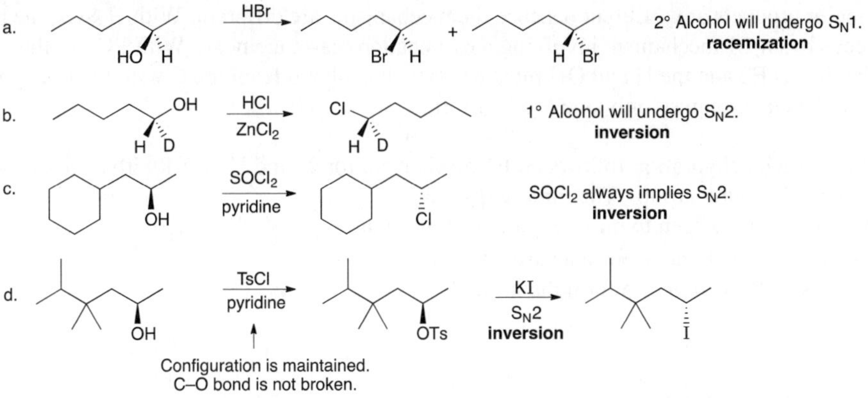

a. HBr → 2° Alcohol will undergo S$_N$1. **racemization**

b. HCl / ZnCl$_2$ → 1° Alcohol will undergo S$_N$2. **inversion**

c. SOCl$_2$ / pyridine → SOCl$_2$ always implies S$_N$2. **inversion**

d. TsCl / pyridine → OTs — KI / S$_N$2 / **inversion** → I

Configuration is maintained.
C–O bond is not broken.

9.49

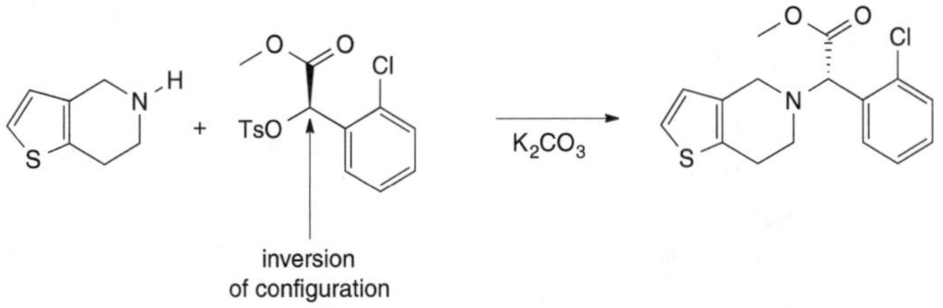

+ TsO (inversion of configuration) — K$_2$CO$_3$ →

inversion of configuration

9.50

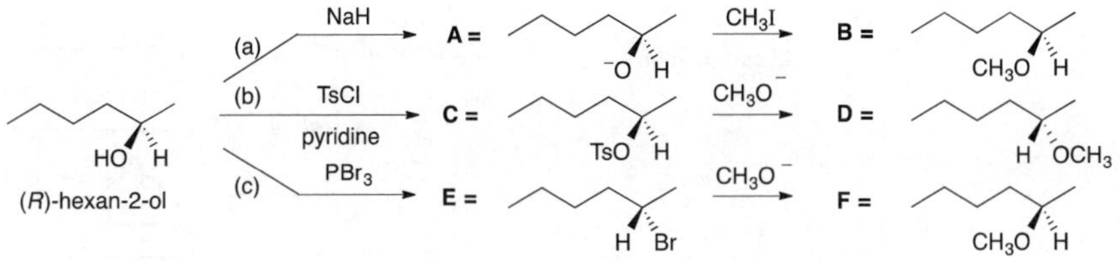

(a) NaH → A = ⎯ CH$_3$I → B =

(b) TsCl / pyridine → C = ⎯ CH$_3$O$^-$ → D =

(c) PBr$_3$ → E = ⎯ CH$_3$O$^-$ → F =

(R)-hexan-2-ol

Routes (a) and (c) give identical products, labeled **B** and **F**.

9.51

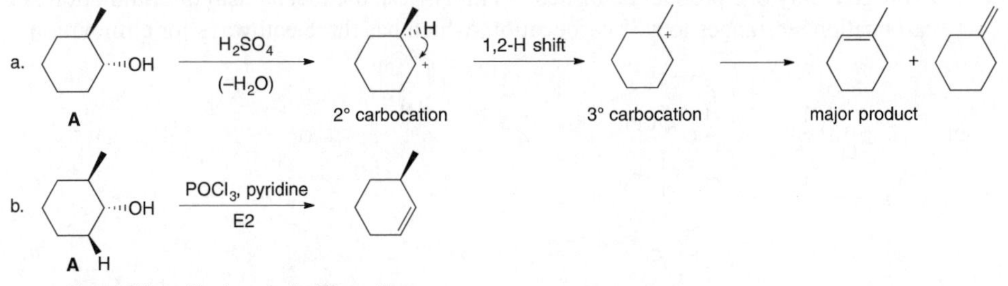

a. H$_2$SO$_4$ (–H$_2$O) → 2° carbocation — 1,2-H shift → 3° carbocation → major product

A

b. POCl$_3$, pyridine / E2 →

A H
H and OH must be trans in the E2 reaction.

Chapter 9–22

The major products are different because the mechanisms are different. With H_2SO_4, the reaction proceeds by an E1 mechanism involving a carbocation rearrangement. With $POCl_3$, the mechanism is E2 and the H and OH must be trans diaxial. No H on the C with the CH_3 group is trans to the OH group, so only one product forms.

9.52 Acid-catalyzed dehydration follows an E1 mechanism for 2° and 3° ROH with an added step to make a good leaving group. The three steps are:
[1] Protonate the oxygen to make a good leaving group.
[2] Break the C–O bond to form a carbocation.
[3] Remove a β hydrogen to form the π bond.

9.53 With $POCl_3$ (pyridine), elimination occurs by an E2 mechanism. Because only one carbon has a β hydrogen, only one product is formed. With H_2SO_4, the mechanism of elimination is E1. A 2° carbocation rearranges to a 3° carbocation, which has three pathways for elimination.

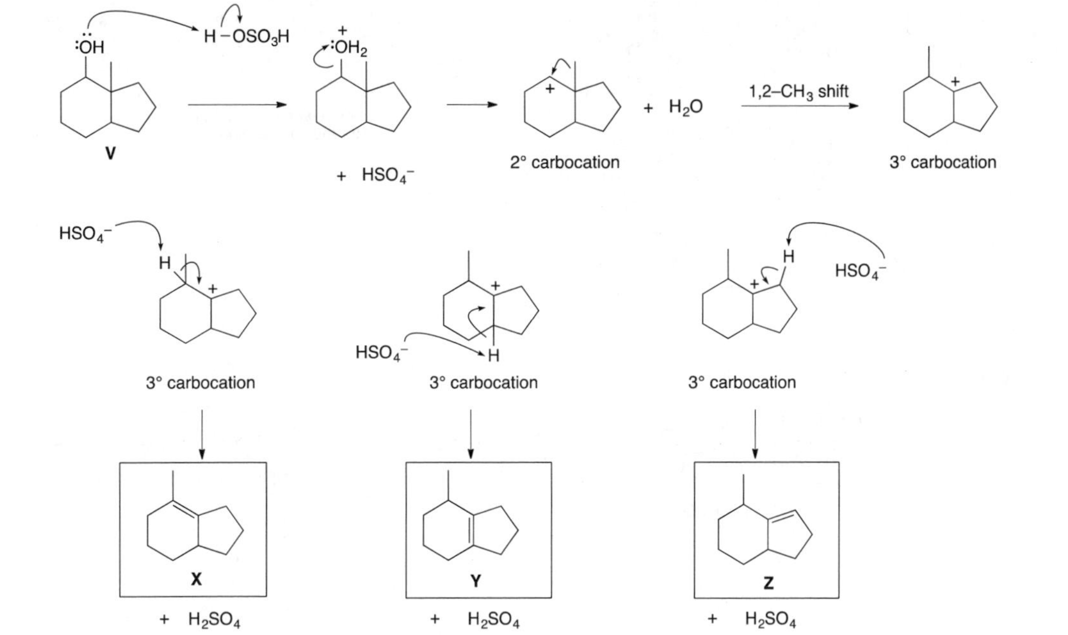

Alcohols, Ethers, and Related Compounds 9–23

9.54 To draw the mechanism:

[1] Protonate the oxygen to make a good leaving group.

[2] Break the C–O bond to form a carbocation.

[3] Look for possible rearrangements to make a more stable carbocation.

[4] Remove a β hydrogen to form the π bond.

Dark and light circles are meant to show where the carbons in the starting material appear in the product.

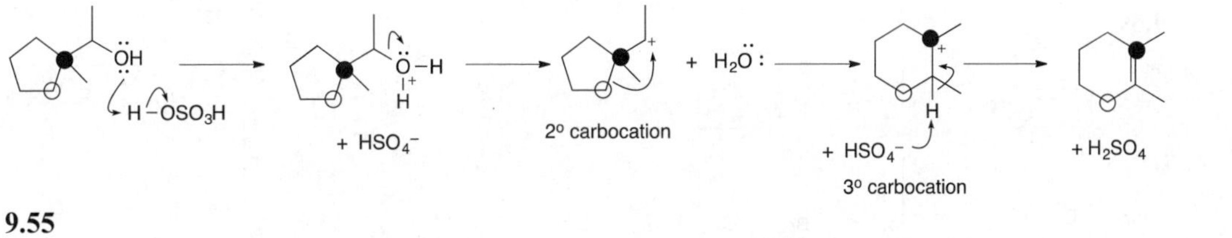

9.55

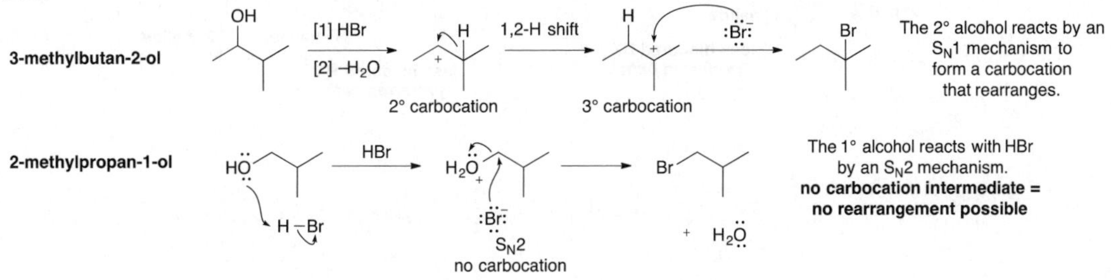

3-methylbutan-2-ol

The 2° alcohol reacts by an S_N1 mechanism to form a carbocation that rearranges.

2-methylpropan-1-ol

The 1° alcohol reacts with HBr by an S_N2 mechanism. **no carbocation intermediate = no rearrangement possible**

Chapter 9–24

9.56

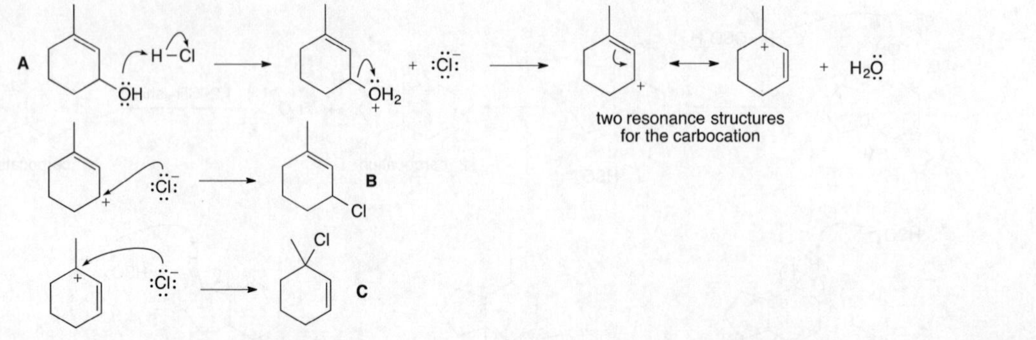

two resonance structures
for the carbocation

9.57

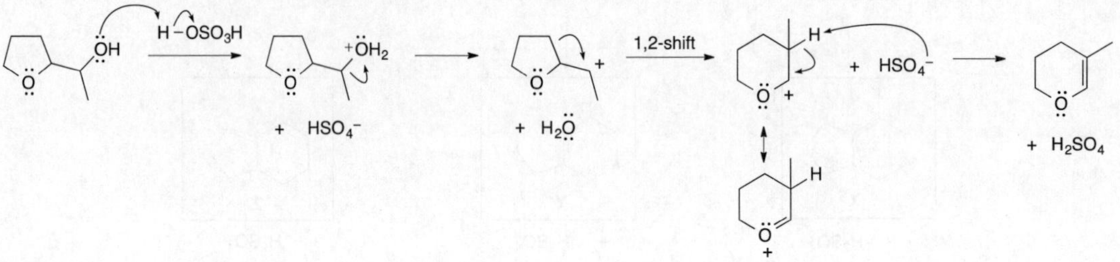

1,2-shift

9.58

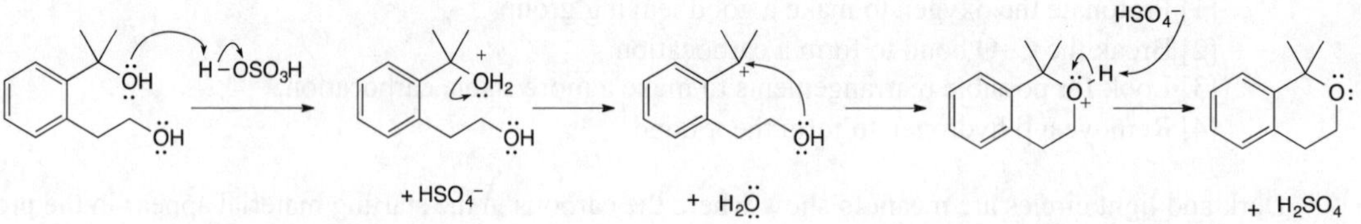

9.59

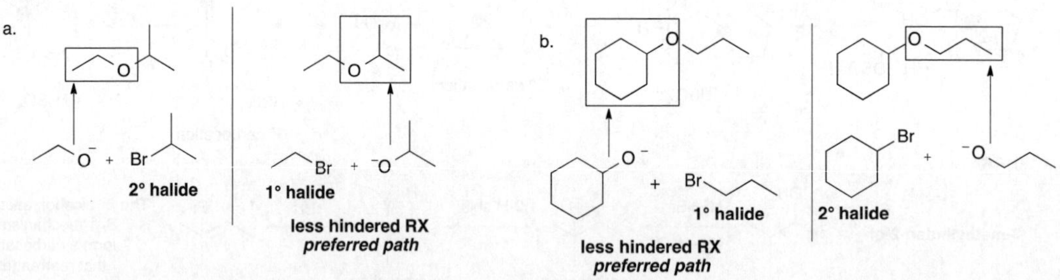

a.

2° halide   1° halide

less hindered RX
preferred path

b.

1° halide   2° halide

less hindered RX
preferred path

Alcohols, Ethers, and Related Compounds 9–25

c.

1° halide | 1° halide

Neither path preferred.

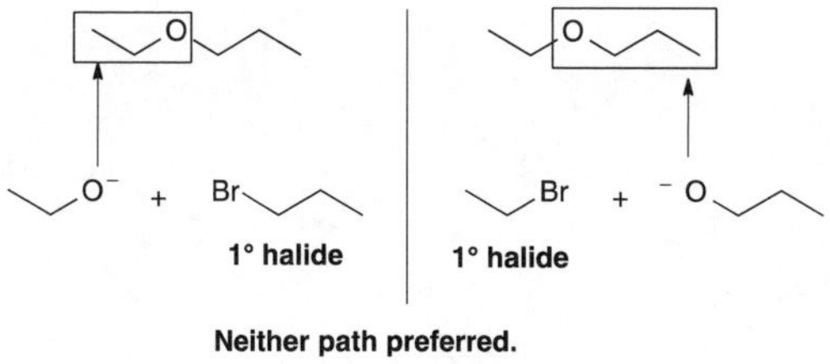

9.60 A tertiary halide is too hindered and an aryl halide is too unreactive to undergo a Williamson ether synthesis.

Two possible sets of starting materials:

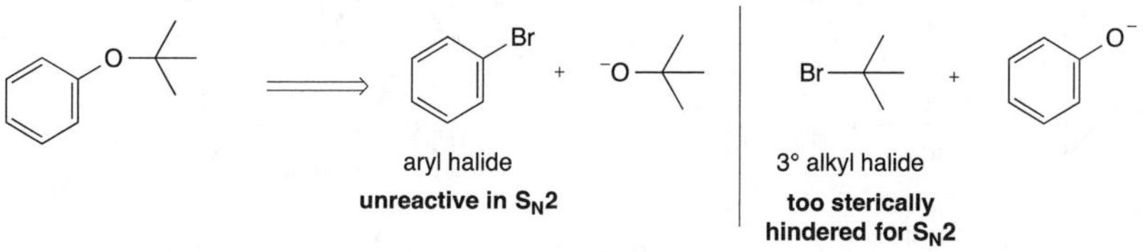

aryl halide
unreactive in S_N2

3° alkyl halide
**too sterically
hindered for S_N2**

9.61

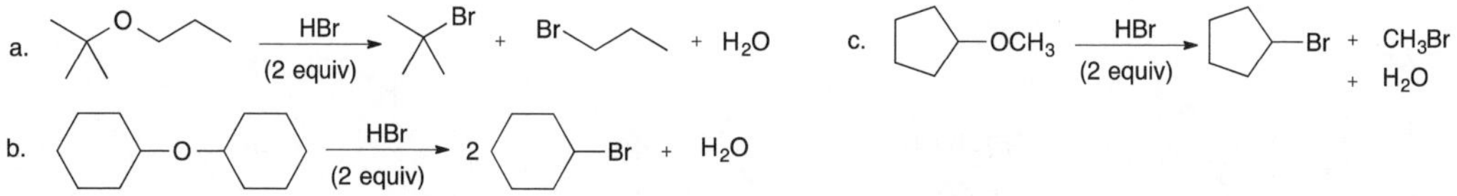

a. $\xrightarrow{\text{HBr} \atop \text{(2 equiv)}}$ $+$ $+$ H_2O

c. $\xrightarrow{\text{HBr} \atop \text{(2 equiv)}}$ Br $+$ CH_3Br $+$ H_2O

b. $\xrightarrow{\text{HBr} \atop \text{(2 equiv)}}$ 2 Br $+$ H_2O

9.62

a.

overall reaction

$+$ $H_2\ddot{O}:$

The steps:

$+$ $\ddot{I}:^-$

$+$ $\ddot{I}:^-$

b. Cl $\xrightarrow{\text{Na}^+\text{H}:^-}$ Cl $+$ H_2 $+$ Na^+ $\longrightarrow$ $+$ H_2 $+$ $NaCl$

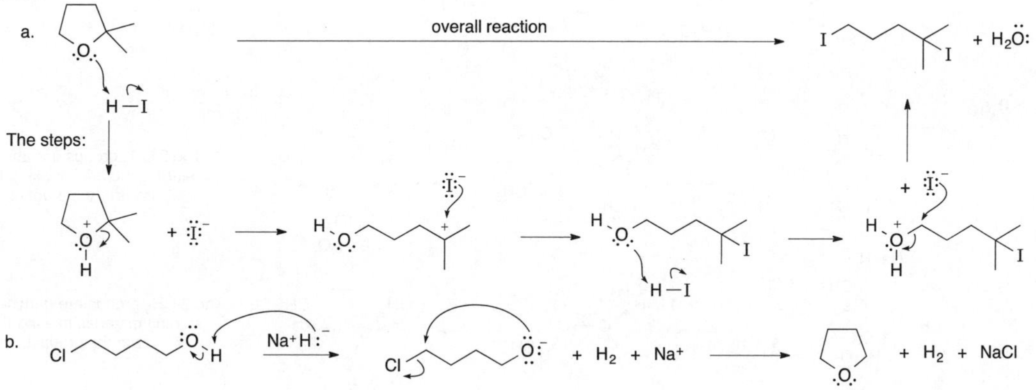

Chapter 9–26

9.63

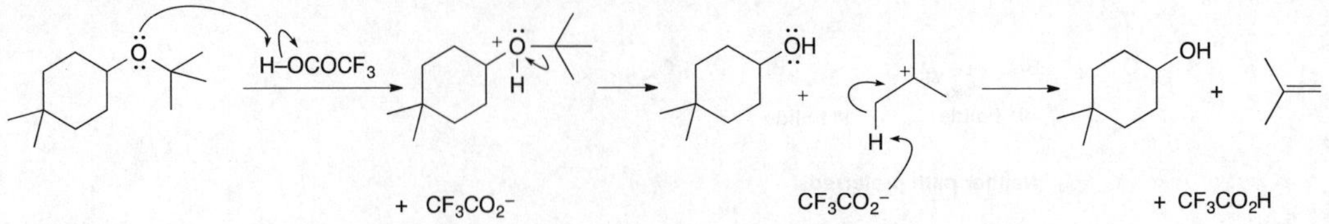

+ $CF_3CO_2^-$

$CF_3CO_2^-$

+ CF_3CO_2H

9.64

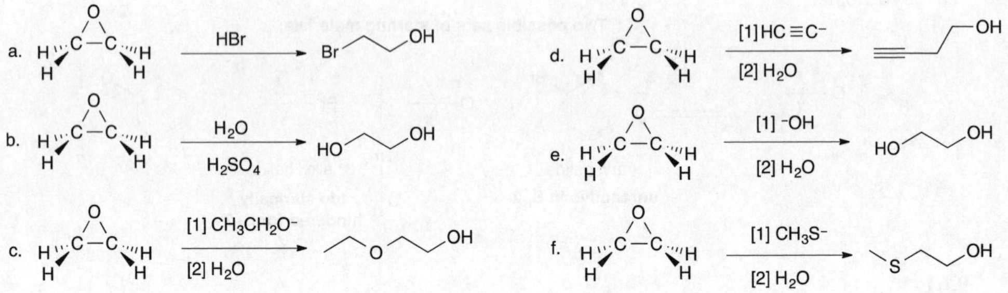

a. $\xrightarrow{HBr}$ Br⌒OH

b. $\xrightarrow[H_2SO_4]{H_2O}$ HO⌒OH

c. $\xrightarrow[{[2] H_2O}]{[1] CH_3CH_2O^-}$ ⌒O⌒OH

d. $\xrightarrow[{[2] H_2O}]{[1]HC\equiv C^-}$ ≡⌒OH

e. $\xrightarrow[{[2] H_2O}]{[1]\ ^-OH}$ HO⌒OH

f. $\xrightarrow[{[2] H_2O}]{[1] CH_3S^-}$ ⌒S⌒OH

9.65

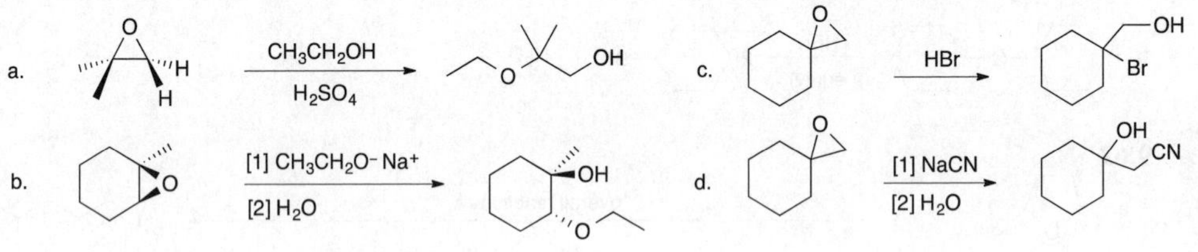

a. $\xrightarrow[H_2SO_4]{CH_3CH_2OH}$ ⌒O⌒OH

b. $\xrightarrow[{[2] H_2O}]{[1] CH_3CH_2O^-Na^+}$

c. $\xrightarrow{HBr}$

d. $\xrightarrow[{[2] H_2O}]{[1] NaCN}$

9.66

a.

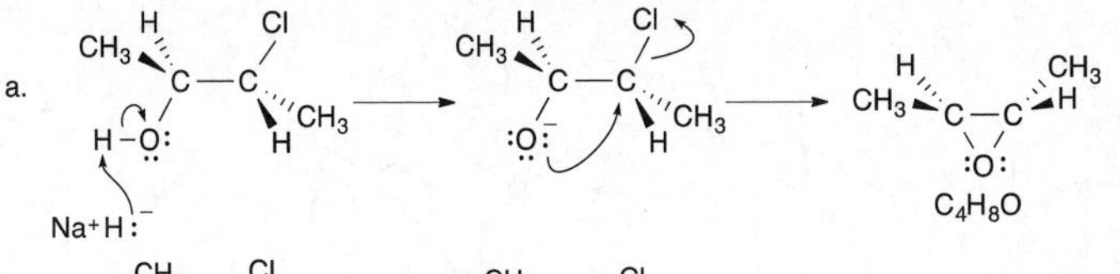

The 2 CH$_3$ groups are anti in the starting material, making them trans in the product.

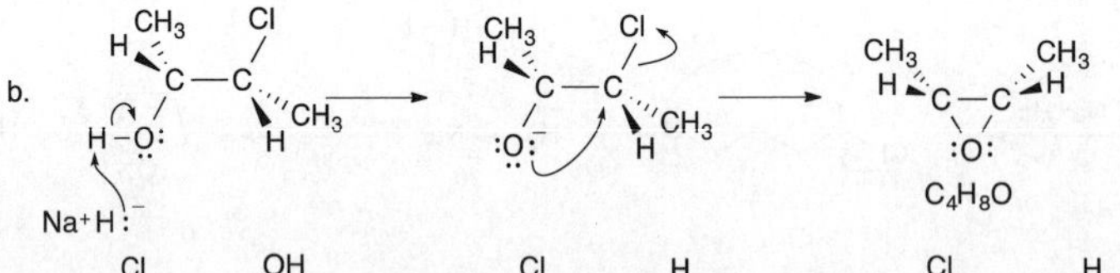

b.

The 2 CH$_3$ groups are gauche in the starting material, making them cis in the product.

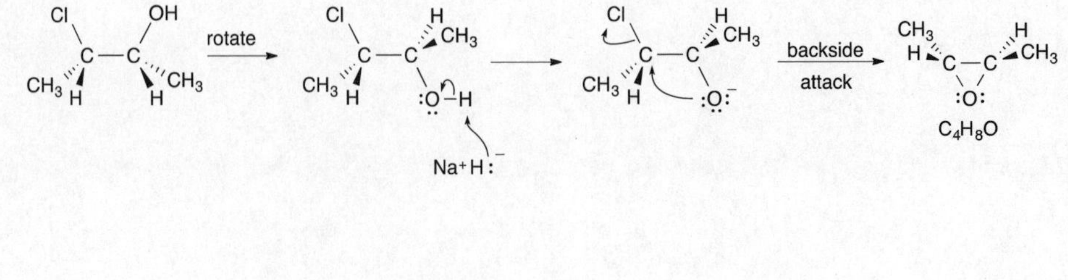

c.

Alcohols, Ethers, and Related Compounds 9–27

9.67 First, use the names to draw the structures of the starting material and both products. Because the product has two OH groups, one OH must come from the epoxide oxygen, and one must come from the nucleophile, either ⁻OH or H_2O.

With ⁻OH, the nucleophile attacks at the less substituted end of the epoxide to form the *R* isomer of the product.

(*R*)-2-ethyl-2-methyloxirane

[1] Na⁺ ⁻OH
[2] H_2O

from the nucleophile

(*R*)-2-methylbutane-1,2-diol

With H_2O and H_2SO_4, H_2O attacks at the more substituted end of the epoxide, from the back side, and the *S* isomer is formed.

(*R*)-2-ethyl-2-methyloxirane

H_2O
H_2SO_4

← from the nucleophile

(*S*)-2-methylbutane-1,2-diol

9.68

+ HB⁺

9.69

a. KOC(CH₃)₃
Bulky base favors E2.

b. HBr
Keep the stereochemistry at the stereogenic center [*] the same here because no bond to it is broken .

c. Br₂

d. KSH

e. PBr₃
S_N2 inversion

f. TsCl
pyridine
$CH_3CO_2^-$

Chapter 9–28

g.

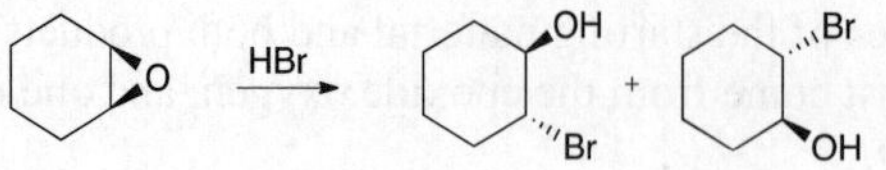

h.

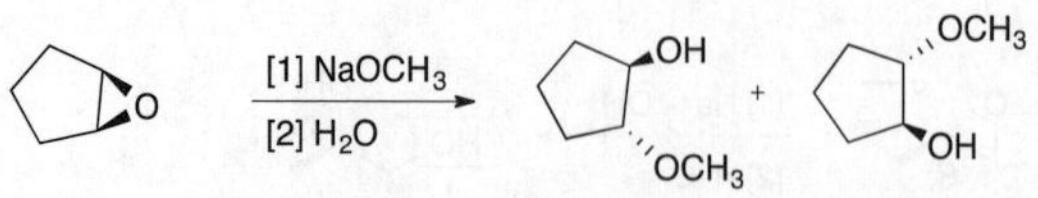

i.

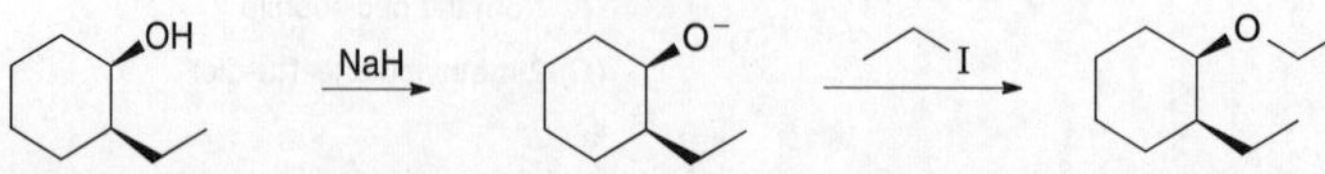

j.

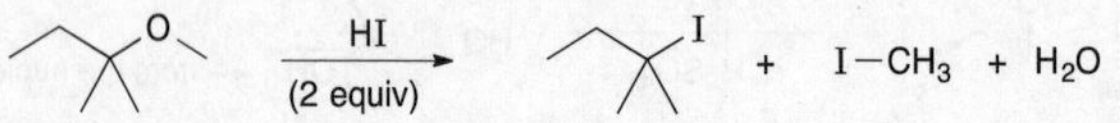

k.

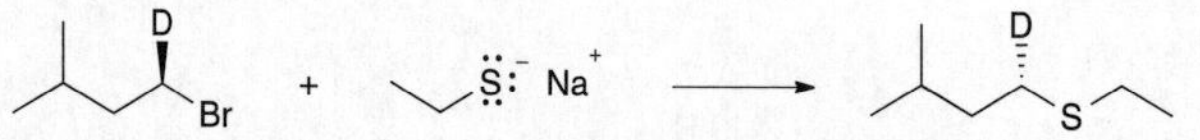

l.

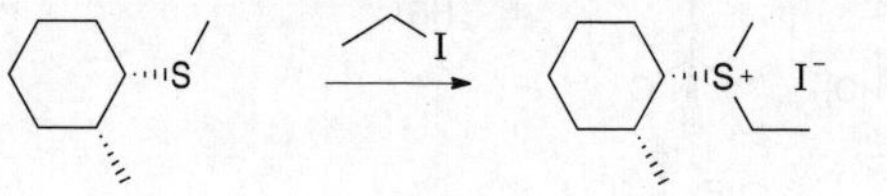

9.70

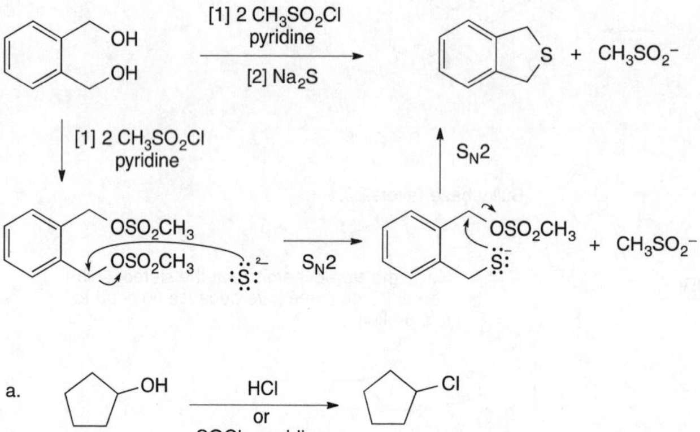

9.71

a.

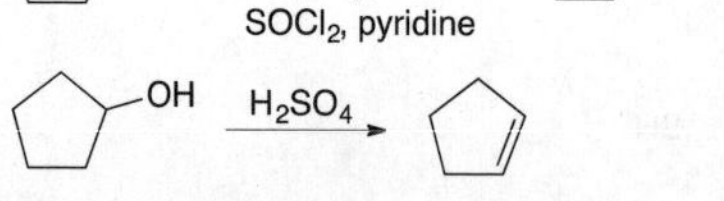

b.

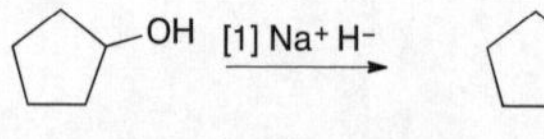

c.

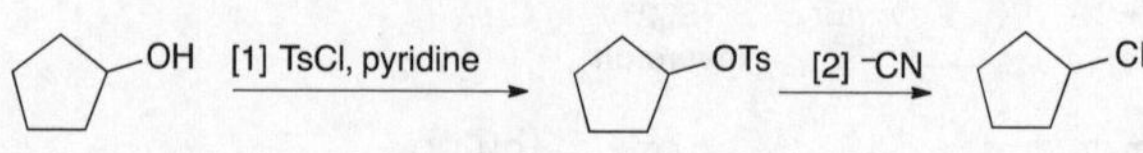

d.

Make OH a good leaving group (use TsCl); then add ⁻CN.

Alcohols, Ethers, and Related Compounds 9–29

9.72

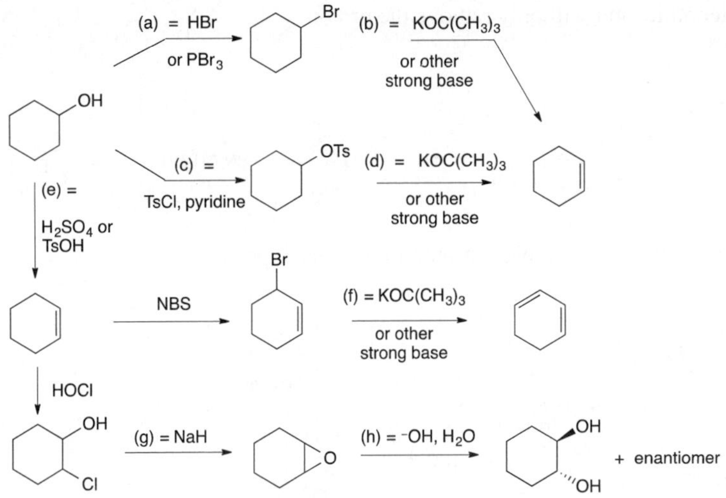

(a) = HBr or PBr$_3$

(b) = KOC(CH$_3$)$_3$ or other strong base

(c) = TsCl, pyridine

(d) = KOC(CH$_3$)$_3$ or other strong base

(e) = H$_2$SO$_4$ or TsOH

NBS

(f) = KOC(CH$_3$)$_3$ or other strong base

HOCl

(g) = NaH

(h) = ⁻OH, H$_2$O

+ enantiomer

9.73

NaH

proton transfer

propranolol

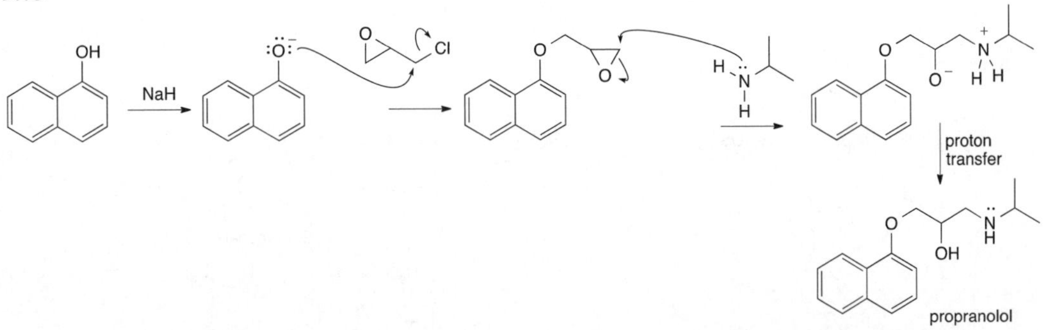

9.74 With the cis isomer, ⁻OH acts as a nucleophile to displace Br⁻ from the back side, forming a trans diol (**A**). With the trans isomer, the two functional groups are arranged in a manner that allows an intramolecular S$_N$2. ⁻OH removes a proton to form an alkoxide, which can then displace Br⁻ by intramolecular backside attack to afford an ether (**B**). Such a reaction is not possible with the cis isomer because the nucleophile and leaving group are on the same side.

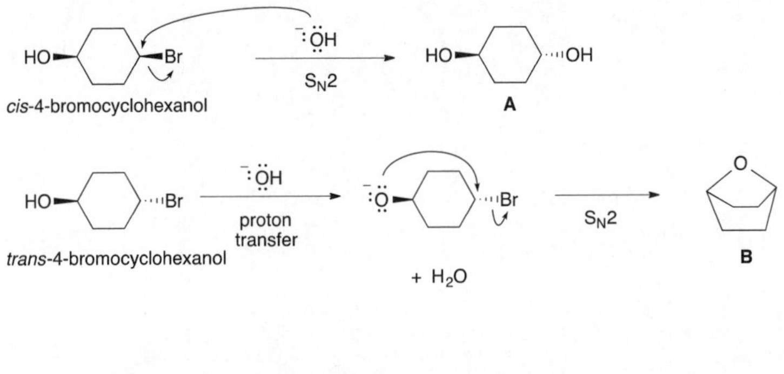

cis-4-bromocyclohexanol

S$_N$2

A

trans-4-bromocyclohexanol

proton transfer

+ H$_2$O

S$_N$2

B

Chapter 9–30

9.75 If the base is not bulky, it can react as a nucleophile and open the epoxide ring. The bulky base cannot act as a nucleophile, and will only remove the proton.

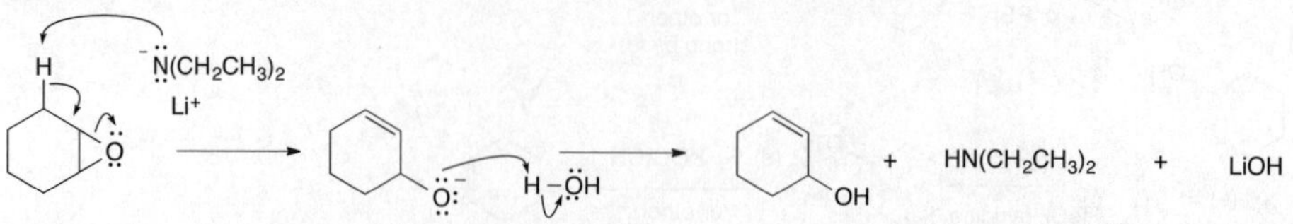

9.76 First form the 2° carbocation. Then lose a proton to form each product.

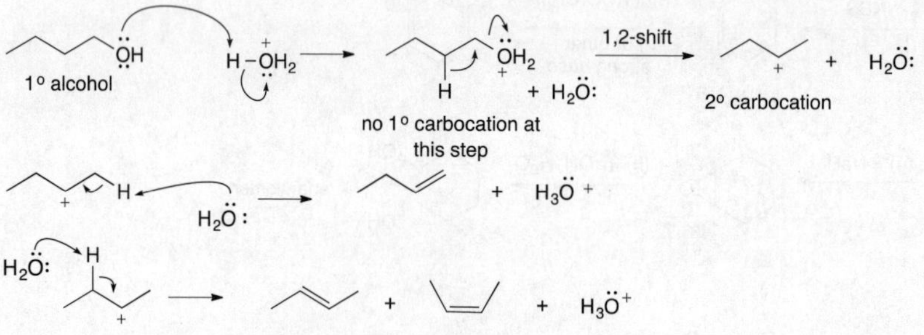

1° alcohol

no 1° carbocation at this step

1,2-shift

2° carbocation

9.77

+ HSO$_4$$^-$

2° carbocation

1,2-shift

3° and resonance-stabilized allylic carbocation

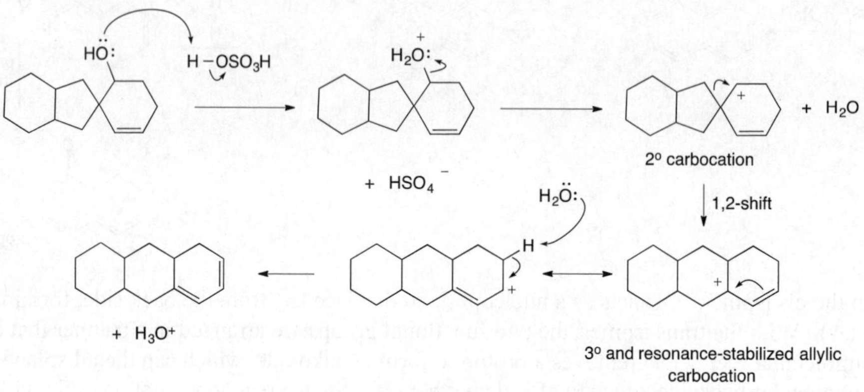

Alcohols, Ethers, and Related Compounds 9–31

9.78

a.

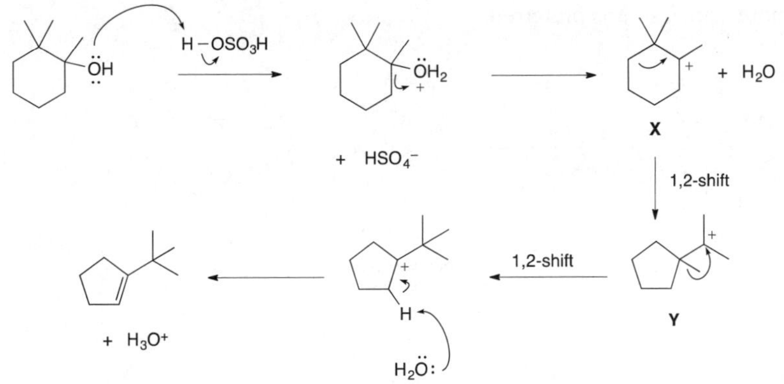

b. Other elimination products can form from carbocations **X** and **Y**.

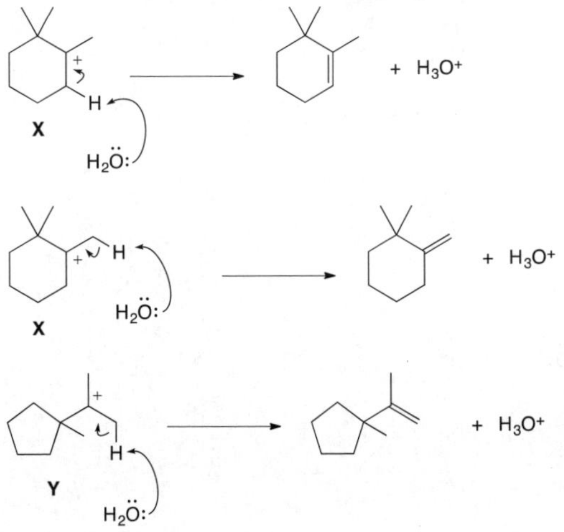

9.79

a.

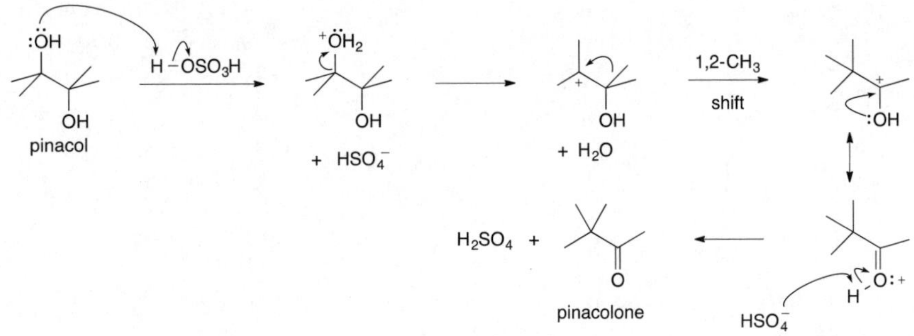

Chapter 9–32

b. Two different carbocations can form. The carbocation with the (+) charge adjacent to the benzene rings (**A**) is more stable, so it is preferred.

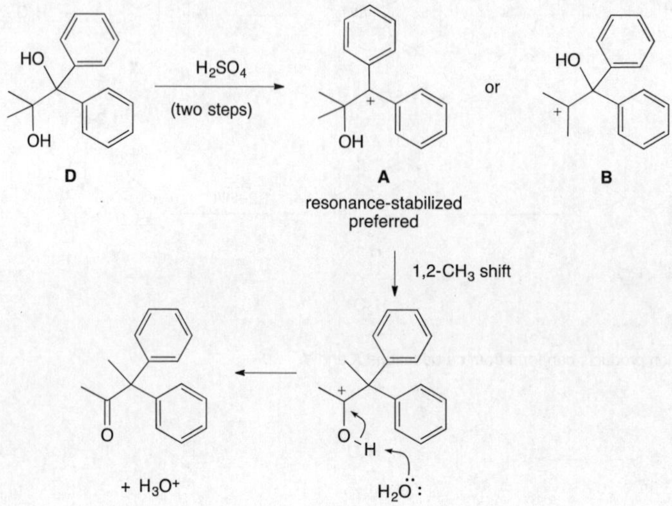

D

H_2SO_4

(two steps)

A
resonance-stabilized
preferred

or

B

1,2-CH$_3$ shift

+ H$_3$O$^+$

H$_2$O:

9.80

Na$^+$ $^-$:OH

+ I$^-$

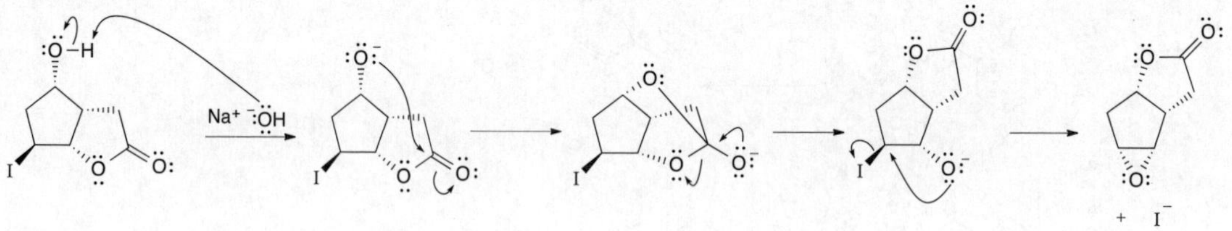

Alcohols, Ethers, and Related Compounds 9–33

9.81 The conversion of **X** to **Y** requires two operations. **X** contains both a nucleophile (NH$_2$) and a leaving group (OSO$_2$CH$_3$), so an intramolecular S$_N$2 reaction forms an aziridine. Because the aziridine is strained, the amine nucleophile (CH$_2$=CHCH$_2$NH$_2$) opens the ring by backside attack, resulting in the trans stereochemistry of the two N's on the six-membered ring.

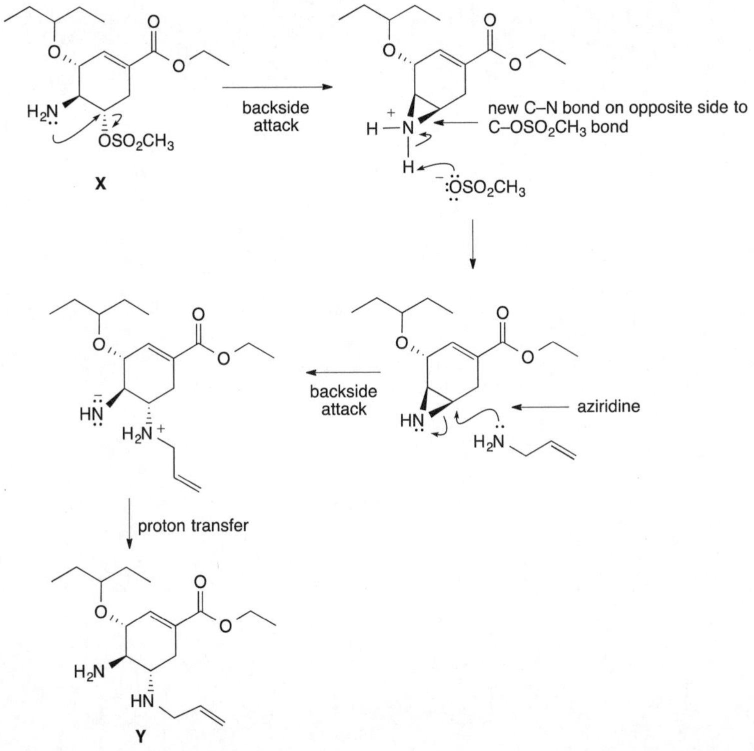

backside attack

new C–N bond on opposite side to C–OSO$_2$CH$_3$ bond

X

:ÖSO$_2$CH$_3$

backside attack

aziridine

proton transfer

Y

Alkenes 10–1

Chapter 10 Alkenes

Chapter Review

General facts about alkenes

- Alkenes contain a carbon–carbon double bond consisting of a stronger σ bond and a weaker π bond. Each carbon is sp^2 hybridized and trigonal planar (10.1).
- Alkenes are named using the suffix *-ene* (10.3).
- Alkenes with different groups on each end of the double bond exist as a pair of diastereomers, identified by the prefixes *E* and *Z* (10.3B).

- Alkenes have weak intermolecular forces, giving them low mp's and bp's, and making them water insoluble. A cis alkene is more polar than a trans alkene, giving it a slightly higher boiling point (10.4).

```
                cis-but-2-ene                    trans-but-2-ene
                 CH₃    CH₃                      CH₃     H
                   C=C                              C=C
more polar isomer  H     H                       H     CH₃    ← less polar isomer
                 a small net dipole              no net dipole
                    higher bp                       lower bp
```

- A π bond is electron rich and much weaker than a σ bond, so alkenes undergo addition reactions with electrophiles (10.8).

Stereochemistry of alkene addition reactions (10.8)

A reagent XY adds to a double bond in one of three different ways:

- **Syn addition**—X and Y add from the same side.

  C=C →(H–BH₂)→ C–C with H and BH₂ on same side
  - Syn addition occurs in **hydroboration.**

- **Anti addition**—X and Y add from opposite sides.

  C=C →(X₂ or X₂, H₂O)→ C–C with X and X(OH) on opposite sides
  - Anti addition occurs in **halogenation** and **halohydrin formation.**

Chapter 10–2

- **Both syn and anti addition** occur when carbocations are intermediates.

  "C=C" + H–X or H₂O, H⁺ → "C–C" (H, X(OH)) and "C–C" (H, X(OH))

  - Syn and anti addition occur in **hydrohalogenation** and **hydration.**

Addition reactions of alkenes

[1] Hydrohalogenation—Addition of HX (X = Cl, Br, I) (10.9–10.11)

R–CH=CH₂ + H–X → alkyl halide (X on R-substituted C)

- The mechanism has two steps.
- Carbocations are formed as intermediates.
- Carbocation rearrangements are possible.
- Markovnikov's rule is followed. H bonds to the less substituted C to form the more stable carbocation.
- Syn and anti addition occur.

[2] Hydration and related reactions—Addition of H₂O or ROH (10.12)

R–CH=CH₂ + H–OH →(H₂SO₄) alcohol (OH on R-substituted C)

R–CH=CH₂ + H–OR →(H₂SO₄) ether (OR on R-substituted C)

For both reactions:
- The mechanism has three steps.
- Carbocations are formed as intermediates.
- Carbocation rearrangements are possible.
- Markovnikov's rule is followed. H bonds to the less substituted C to form the more stable carbocation.
- Syn and anti addition occur.

[3] Halogenation—Addition of X₂ (X = Cl or Br) (10.13–10.14)

R–CH=CH₂ + X–X → vicinal dihalide

- The mechanism has two steps.
- Bridged halonium ions are formed as intermediates.
- No rearrangements occur.
- Anti addition occurs.

[4] Halohydrin formation—Addition of OH and X (X = Cl, Br) (10.15)

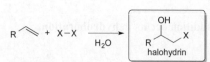

- The mechanism has three steps.
- Bridged halonium ions are formed as intermediates.
- No rearrangements occur.
- X bonds to the less substituted C.
- Anti addition occurs.
- NBS in DMSO and H₂O adds Br and OH in the same fashion.

Alkenes 10–3

[5] **Hydroboration–oxidation**—Addition of H_2O (10.16)

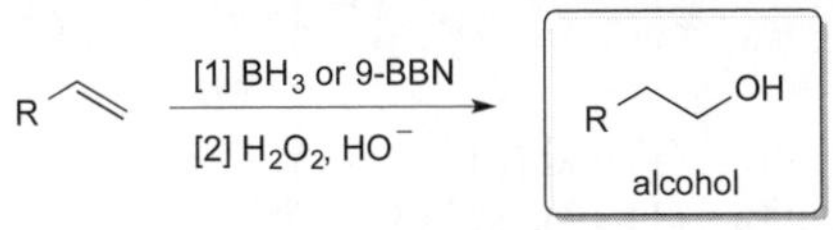

R ⟍ $\xrightarrow[\text{[2] } H_2O_2,\ HO^-]{\text{[1] } BH_3 \text{ or 9-BBN}}$ R ⟍⟋OH

alcohol

- Hydroboration has a one-step mechanism.
- No rearrangements occur.
- OH bonds to the less substituted C.
- Syn addition of H_2O results.

Practice Test on Chapter Review

1. Give the IUPAC name for the following compounds.

a.

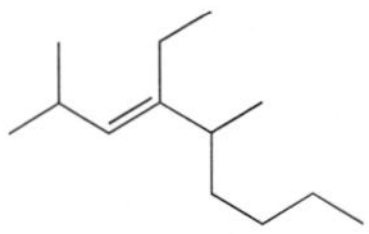

b.

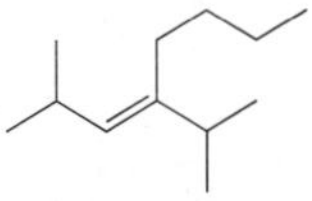

2. Draw the organic products formed in the following reactions. Draw all stereogenic centers using wedges and dashes.

a.  $\xrightarrow{\text{HCl}}$

b. 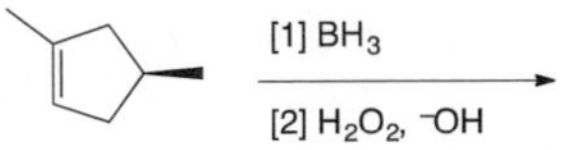 $\xrightarrow[\text{[2] } H_2O_2,\ ^-OH]{\text{[1] } BH_3}$

c. 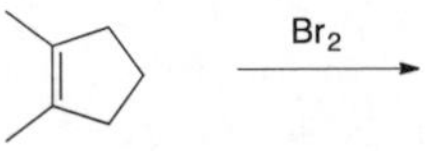 $\xrightarrow{\text{Br}_2}$

d. 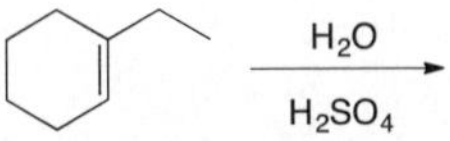 $\xrightarrow[\text{H}_2SO_4]{\text{H}_2O}$

3. Fill in the table with the stereochemistry observed in the reaction of an alkene with each reagent. Choose from syn, anti, or both syn and anti.

Reagent	Stereochemistry
a. [1] 9-BBN; [2] H_2O_2, ^-OH	
b. H_2O, H_2SO_4	
c. Cl_2, H_2O	
d. HI	
e. Br_2	

Chapter 10–4

4. a. In which of the following reactions are carbocation rearrangements observed?
   1. hydrohalogenation
   2. halohydrin formation
   3. hydroboration–oxidation
   4. Carbocation rearrangements are observed in reactions (1) and (2).
   5. Carbocation rearrangements are observed in reactions (1), (2), and (3).

   b. Which of the following products are formed when HCl is added to 3-methylpent-1-ene?
   1. 2-chloro-2-methylpentane
   2. 3-chloro-3-methylpentane
   3. 1-chloro-3-methylpentane
   4. Both (1) and (2) are formed.
   5. Products (1), (2), and (3) are all formed.

Answers to Practice Test

1. a. (*E*)-4-ethyl-2,5-
      dimethylnon-3-ene
   b. (*E*)-4-isopropyl-2-
      methyloct-3-ene

2.

a.

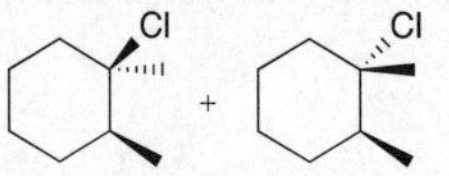

b.

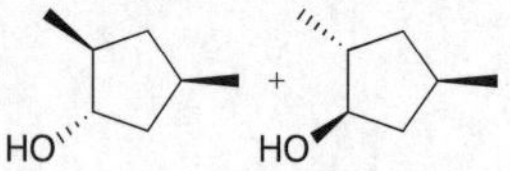

c.

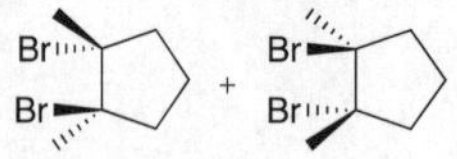

d.

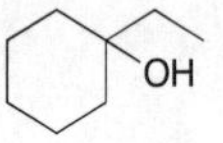

3. a. syn
   b. both
   c. anti
   d. both
   e. anti

4. a. 1
   b. 2

Answers to Problems

10.1

Six alkenes of molecular formula C_5H_{10}:

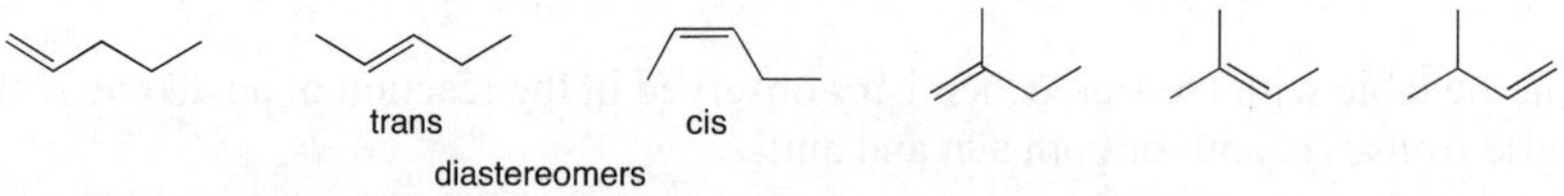

trans   cis

diastereomers

10.2 To determine the number of degrees of unsaturation:
   [1] Calculate the maximum number of H's $(2n + 2)$.
   [2] Subtract the actual number of H's from the maximum number.
   [3] Divide by two.

Alkenes 10–5

a. C_6H_6
[1] maximum number of H's = $2n + 2 = 2(6) + 2 = 14$
[2] subtract actual from maximum = $14 - 6 = 8$
[3] divide by two = $8/2 =$ **4 degrees of unsaturation**

b. C_8H_{18}
[1] maximum number of H's = $2n + 2 = 2(8) + 2 = 18$
[2] subtract actual from maximum = $18 - 18 = 0$
[3] divide by two = $0/2 =$ **0 degrees of unsaturation**

c. C_7H_8O
Ignore the O.
[1] maximum number of H's = $2n + 2 = 2(7) + 2 = 16$
[2] subtract actual from maximum = $16 - 8 = 8$
[3] divide by two = $8/2 =$ **4 degrees of unsaturation**

d. $C_7H_{11}Br$
Because of Br, add one more H ($11 + 1$ H = 12 H's).
[1] maximum number of H's = $2n + 2 = 2(7) + 2 = 16$
[2] subtract actual from maximum = $16 - 12 = 4$
[3] divide by two = $4/2 =$ **2 degrees of unsaturation**

e. C_5H_9N
Because of N, subtract one H ($9 - 1$ H = 8 H's).
[1] maximum number of H's = $2n + 2 = 2(5) + 2 = 12$
[2] subtract actual from maximum = $12 - 8 = 4$
[3] divide by two = $4/2 =$ **2 degrees of unsaturation**

10.3 Use the directions from Answer 10.2. Ignore O in calculating degrees of unsaturation. Add one more H for each halogen. Subtract one H for each N.

a. $C_{19}H_{21}N_3O$ is equivalent to $C_{19}H_{18}$ when calculating degrees of unsaturation.
For 19 C's, the maximum number of H's = $2n + 2 = 2(19) + 2 = 40$ H's
Subtract actual from maximum = $40 - 18 = 22$ H's fewer
Divide by two = $22/2 = 11$ degrees of unsaturation

b. $C_{17}H_{16}F_6N_2O$ is equivalent to $C_{17}H_{20}$ when calculating degrees of unsaturation.
For 17 C's, the maximum number of H's = $2n + 2 = 2(17) + 2 = 36$ H's
Subtract actual from maximum = $36 - 20 = 16$ H's fewer
Divide by two = $16/2 = 8$ degrees of unsaturation

10.4 To name an alkene:
[1] Find the longest chain that contains the double bond. Change the ending from -*ane* to -*ene.*
[2] Number the chain to give the double bond the lower number. The alkene is named by the first number.
[3] Apply all other rules of nomenclature.

To name a cycloalkene:
[1] When a double bond is located in a ring, it is always located between C1 and C2. Omit the "1" in the name. Change the ending from -*ane* to -*ene.*
[2] Number the ring clockwise or counterclockwise to give the first substituent the lower number.
[3] Apply all other rules of nomenclature.

a. [1] 5 C chain with double bond
pentene

[2] **1** ← 3-methyl
pent-1-ene

[3] **3-methylpent-1-ene**

b. [1] 7 C chain with double bond
heptene

[2] **3**
hept-3-ene
3-ethyl

[3] **3-ethylhept-3-ene**

Chapter 10–6

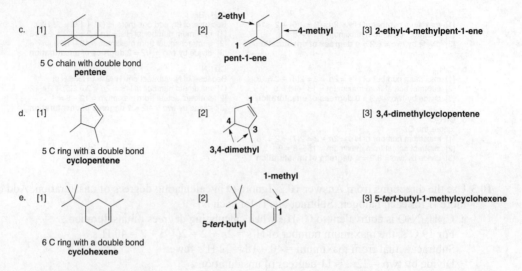

10.5 Use the rules from Answer 10.4 to name the compounds. Enols are named to give the OH the lower number. Compounds with two C=C's are named with the suffix **-adiene**.

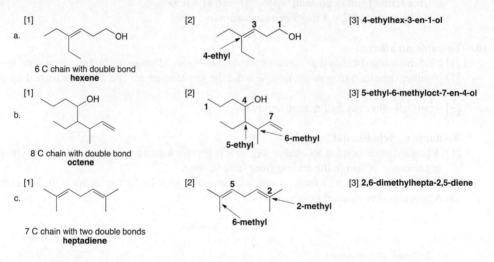

10.6 To label an alkene as *E* or *Z*:
 [1] **Assign priorities** to the two substituents *on each end* using the rules for *R,S* nomenclature.
 [2] **Assign *E* or *Z*** depending on the location of the two higher priority groups.
  • The *E* prefix is used when the two higher priority groups are on **opposite sides** of the double bond.
  • The *Z* prefix is used when the two higher priority groups are on the **same side** of the double bond.

Alkenes 10–7

a. higher priority → [C=C with Cl, Br] ← higher priority
Two higher priority groups are on opposite sides: **E isomer.**

b. higher priority → [alkene] ← higher priority
Two higher priority groups are on the same side: **Z isomer.**

c. higher priority → OCH₃ ; higher priority → C₆H₅ ; higher priority ← (on ring)
kavain
In both double bonds, the two higher priority groups are on opposite sides: **E isomers.**

10.7

[structure of 11-cis-retinal with labels: E, E, Z, E, Z on double bonds, CHO group]

skeletal structure of 11-*cis*-retinal

10.8 To work backwards from a name to a structure:
[1] Find the parent name and functional group and draw, remembering that the double bond is between C1 and C2 for cycloalkenes.
[2] Add the substituents to the appropriate carbons.

a. (*Z*)-4-ethylhept-3-ene — 7 carbons; The double bond is between C3 and C4. 4-ethyl. The higher priority groups are on the same side = **Z**.

b. (*E*)-3,5,6-trimethyloct-2-ene — 8 carbons; The double bond is between C2 and C3. 3,5,6-trimethyl. The higher priority groups are on opposite sides = **E**.

c. (*Z*)-2-bromo-1-iodohex-1-ene — 6 carbons; The double bond is between C1 and C2. The higher priority groups are on the same side = **Z**.

10.9

(2*Z*,6*E*)-3-ethyl-7-methyldeca-2,6-dien-1-ol

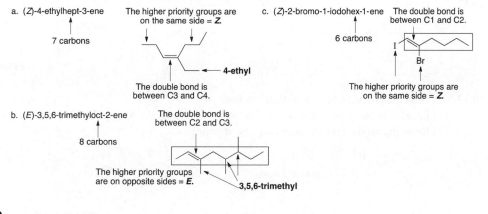

10 C chain with OH at C1
2 C=C's beginning at C2, C6

Chapter 10–8

10.10 To rank the isomers by increasing boiling point:
Look for polarity differences: *small net dipoles* make an alkene more polar, giving it a higher boiling point than an alkene with *no net dipole*. Cis isomers have a higher boiling point than their trans isomers.

All dipoles cancel.
smallest surface area
no net dipole
lowest bp

Two dipoles cancel.
no net dipole
trans isomer
intermediate bp

Two dipoles reinforce.
net dipole
cis isomer
highest bp

10.11 Increasing number of double bonds = decreasing melting point.

stearidonic acid
4 double bonds
lowest melting point

stearic acid
no double bonds
highest melting point

linolenic acid
3 double bonds
intermediate melting point

10.12

a.

H_2SO_4 →   +

b.

$NaOCH_2CH_3$ →   +

(+ cis isomer)

10.13 To draw the products of an addition reaction:
[1] Locate the two bonds that will be broken in the reaction. Always break the π bond.
[2] Draw the product by forming two new σ bonds.

a.

HCl →   two new σ bonds

c.

HCl →   two new σ bonds

b.

HCl →   two new σ bonds

10.14 Addition reactions of HX occur in two steps:
[1] The double bond attacks the H atom of HX to form a carbocation.
[2] X^- attacks the carbocation to form a C–X bond.

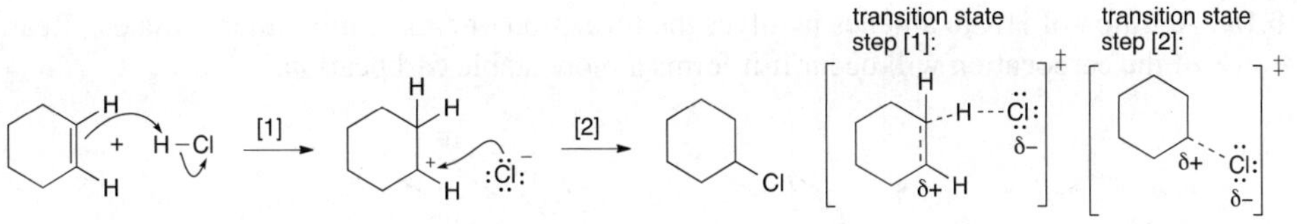

Alkenes 10–9

transition state step [1]:

transition state step [2]:

10.15 Addition to alkenes follows Markovnikov's rule: When HX adds to an unsymmetrical alkene, the H bonds to the C that has more H's to begin with.

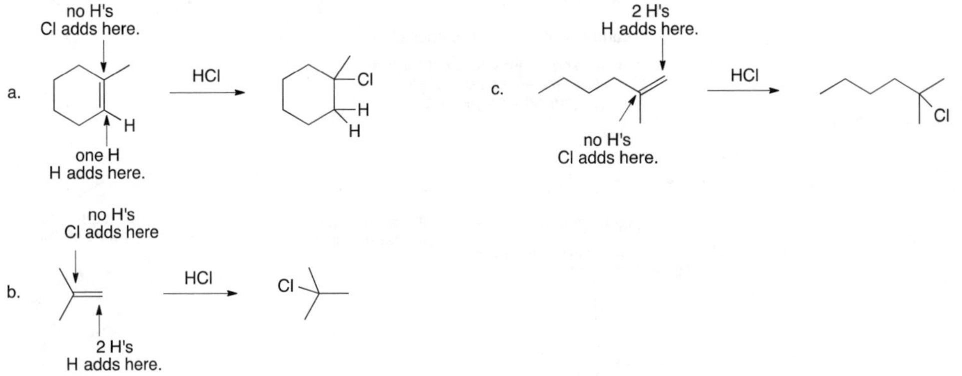

a.

no H's
Cl adds here.

HCl →

one H
H adds here.

b.

no H's
Cl adds here

HCl →

2 H's
H adds here.

c.

2 H's
H adds here.

HCl →

no H's
Cl adds here.

10.16 To determine which alkene will react faster, draw the carbocation that forms in the rate-determining step. The more stable, more substituted the carbocation, the lower the E_a to form it and the faster the reaction.

3° carbocation
faster reaction

2° carbocation
slower reaction

10.17 Look for rearrangements of a carbocation intermediate to explain these results.

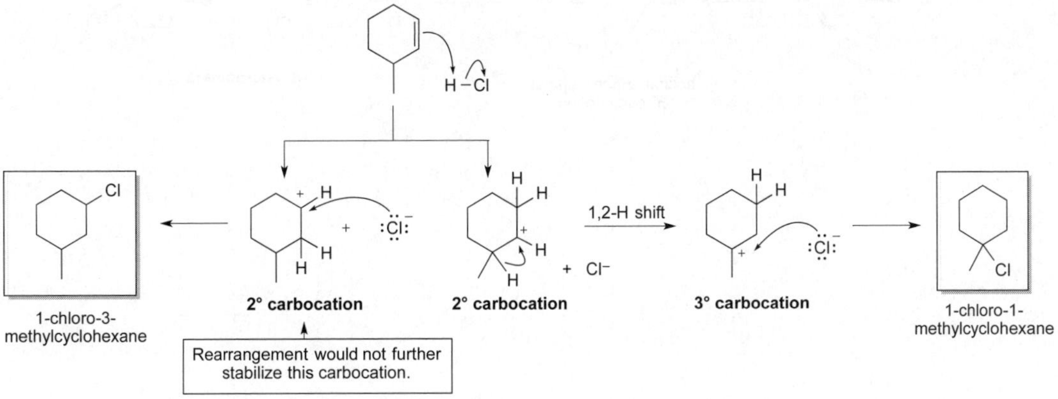

1-chloro-3-methylcyclohexane

2° carbocation

2° carbocation

1,2-H shift

3° carbocation

1-chloro-1-methylcyclohexane

+ Cl⁻

Rearrangement would not further stabilize this carbocation.

Chapter 10–10

10.18 Addition of HX to alkenes involves the formation of carbocation intermediates. Rearrangement of the carbocation will occur if it forms a more stable carbocation.

a.

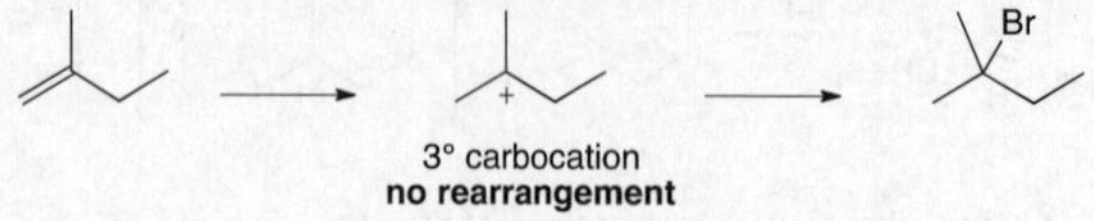

3° carbocation
no rearrangement

b.

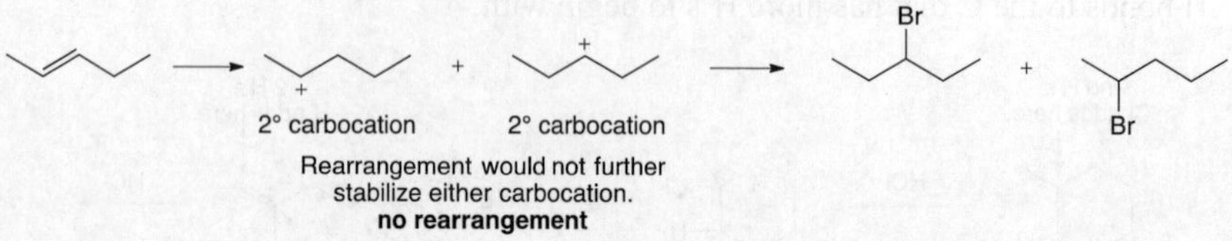

2° carbocation   2° carbocation
Rearrangement would not further
stabilize either carbocation.
no rearrangement

c.

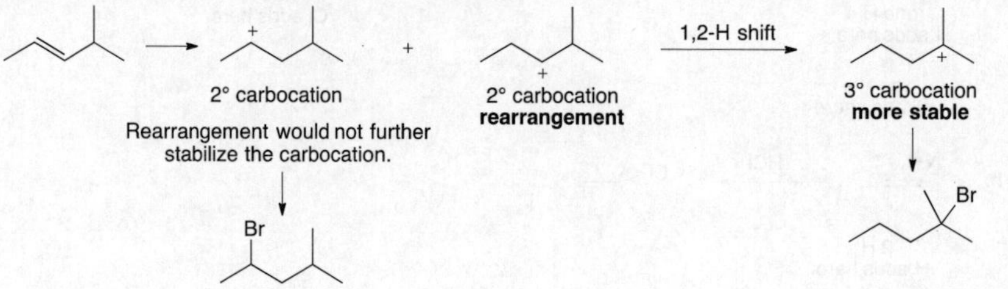

2° carbocation   2° carbocation   1,2-H shift   3° carbocation
rearrangement   **more stable**
Rearrangement would not further
stabilize the carbocation.

10.19 To draw the products, remember that addition of HX proceeds via a carbocation intermediate.

a.

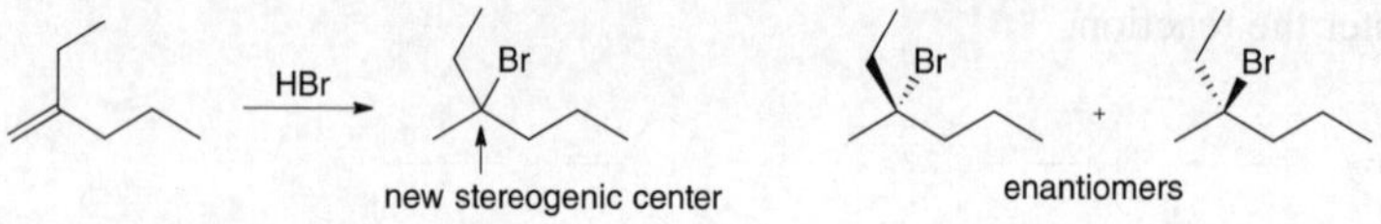

HBr

Br
new stereogenic center   Br   Br
enantiomers

b.

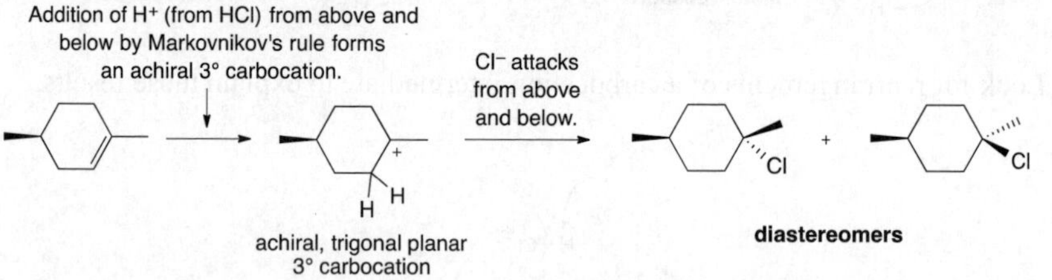

Addition of H+ (from HCl) from above and
below by Markovnikov's rule forms
an achiral 3° carbocation.

Cl⁻ attacks
from above
and below.

H   H
achiral, trigonal planar
3° carbocation

Cl   Cl
diastereomers

10.20 The product of syn addition will have H and Cl both up or down (both on wedges or both on dashed wedges), while the product of anti addition will have one up and one down (one wedge, one dashed wedge).

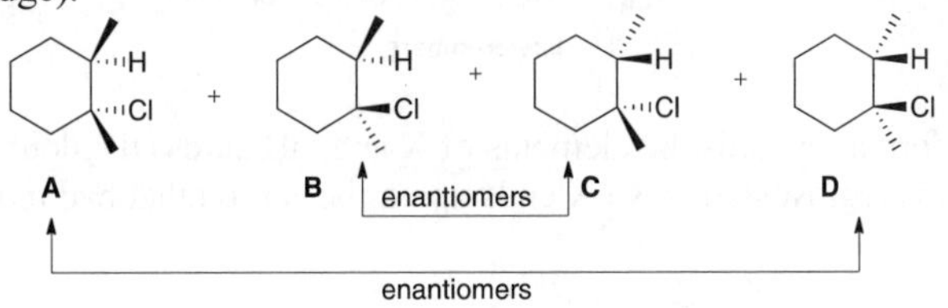
H
Cl
+
A
B
C
D
enantiomers
enantiomers

10.21

a.
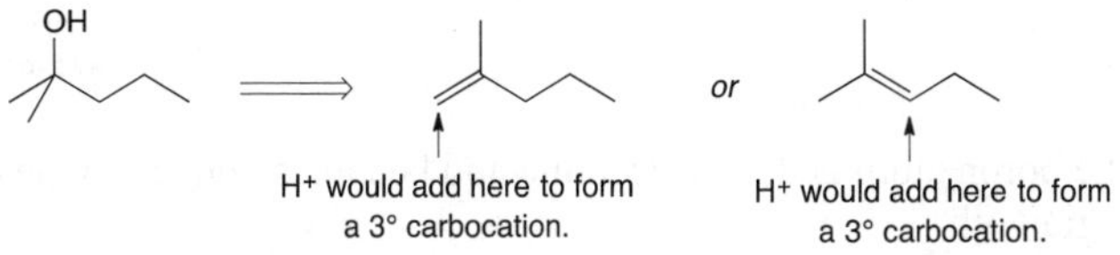
OH
or
H+ would add here to form
a 3° carbocation.
H+ would add here to form
a 3° carbocation.

b.
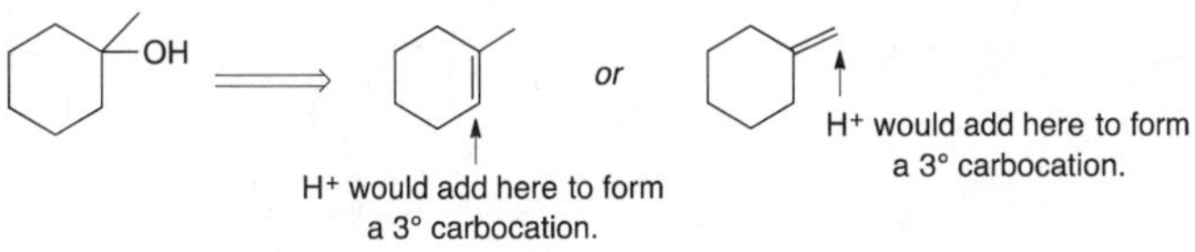
OH
or
H+ would add here to form
a 3° carbocation.
H+ would add here to form
a 3° carbocation.

c.
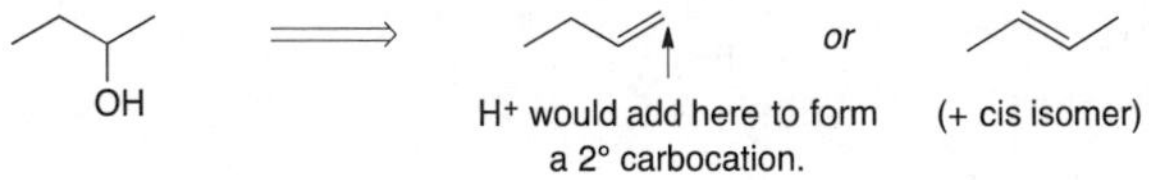
OH
or
H+ would add here to form
a 2° carbocation.
(+ cis isomer)

10.22

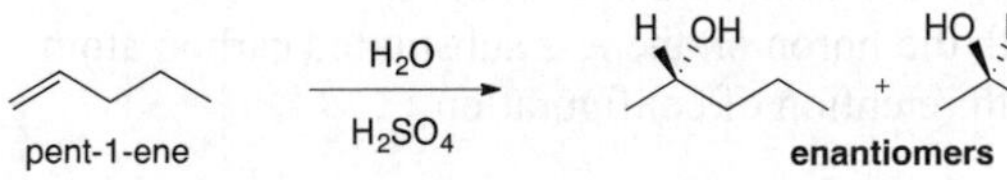
pent-1-ene
H2O
H2SO4
H OH
HO H
+
enantiomers

10.23 Halogenation of an alkene adds two elements of X in an anti fashion.

a.
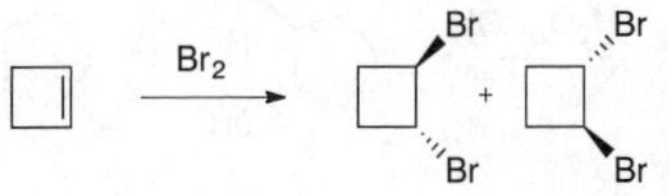
Br2
Br
Br
+
Br
Br

b.
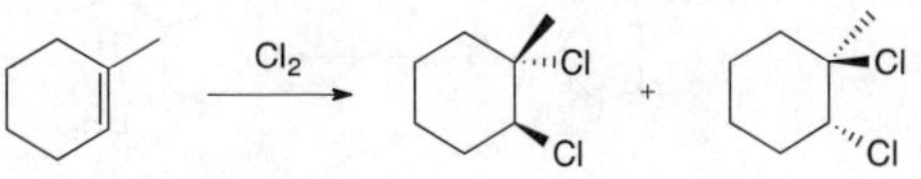
Cl2
Cl
Cl
+
Cl
Cl

10.24 To draw the products of halogenation of an alkene, remember that the halogen adds to both ends of the double bond but only anti addition occurs.

a.
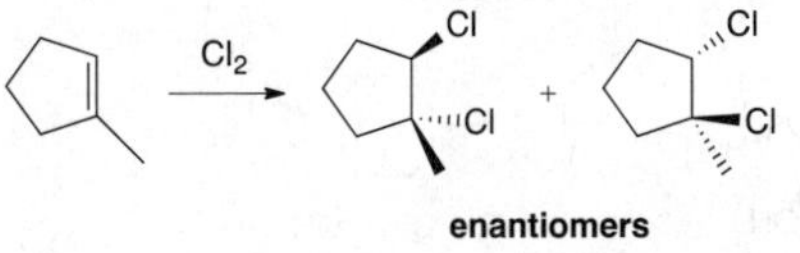
Cl2
Cl
Cl
+
Cl
Cl
enantiomers

b.
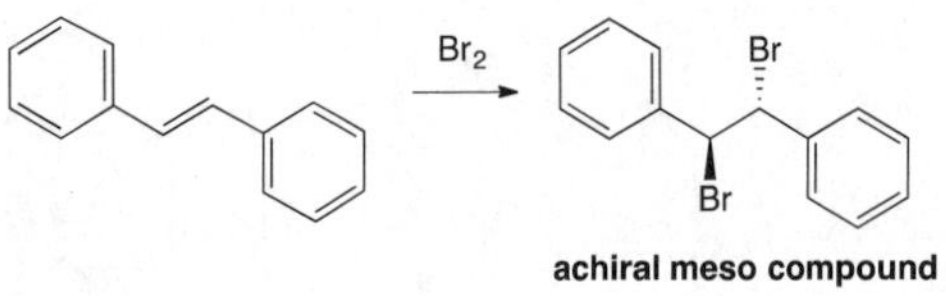
Br2
Br
Br
achiral meso compound

Chapter 10–12

c.

diastereomers

10.25 Halohydrin formation adds the elements of X and OH across the double bond in an anti fashion. The reaction is regioselective, so X ends up on the carbon that had more H's to begin with.

a.

$\xrightarrow[\text{DMSO, H}_2\text{O}]{\text{NBS}}$

b.

$\xrightarrow[\text{H}_2\text{O}]{\text{Cl}_2}$

Cl bonds to the carbon with more H's to begin with.

10.26 In hydroboration the boron atom is the electrophile and becomes bonded to the carbon atom that had more H's to begin with.

a.

$\xrightarrow{\text{BH}_3}$

C with more H's. B will add here.

c.

$\xrightarrow{\text{BH}_3}$

C with more H's. B will add here.

b.

$\xrightarrow{\text{BH}_3}$

C with more H's. B will add here.

10.27 The hydroboration–oxidation reaction occurs in two steps:
[1] Syn addition of BH₃, with the boron on the less substituted carbon atom
[2] OH replaces the BH₂ with retention of configuration.

a.

$\xrightarrow{\text{BH}_3}$ BH₂ $\xrightarrow{\text{H}_2\text{O}_2, \ ^-\text{OH}}$ OH

b.

$\xrightarrow{\text{H}_2\text{O}_2, \ ^-\text{OH}}$

c.

$\xrightarrow{\text{BH}_3}$

$\downarrow \text{H}_2\text{O}_2, \ ^-\text{OH}$

10.28 Remember that hydroboration results in addition of OH on the less substituted C.

a. [structure: 2-methylcyclopentanol] ← [methylenecyclopentane]

c. [structure: 1-cyclohexyl-ethanol with OH] ← [ethylidenecyclohexane]

b. [structure: 4-methylheptan-3-ol with OH] ← [3-methylhept-3-ene] (*E* or *Z* isomer can be used.)

10.29

a. [methylenecyclopentane] —H₂O, H₂SO₄→ [1-methylcyclopentanol, OH on more substituted C] — Hydration places the OH on the more substituted carbon.

[methylenecyclopentane] —[1] BH₃ [2] H₂O₂, HO⁻→ [(cyclopentyl)methanol] — Hydroboration–oxidation places the OH on the less substituted carbon.

b. [isopropenylcyclohexane] —H₂O, H₂SO₄→ [2-cyclohexylpropan-2-ol] — Hydration places the OH on the more substituted carbon.

[isopropenylcyclohexane] —[1] BH₃ [2] H₂O₂, HO⁻→ [2-cyclohexylpropan-1-ol] — Hydroboration–oxidation places the OH on the less substituted carbon.

c. [4-methylpent-1-ene type: 2,4-dimethyl-1-butene] —H₂O, H₂SO₄→ [2,4-dimethylpentan-2-ol] — Hydration places the OH on the more substituted carbon.

[same alkene] —[1] BH₃ [2] H₂O₂, HO⁻→ [2,4-dimethylpentan-1-ol, HO on less substituted C] — Hydroboration–oxidation places the OH on the less substituted carbon.

10.30 There are always two steps in this kind of question:
  [1] **Identify the functional group and decide what types of reactions it undergoes** (e.g., substitution, elimination, or addition).
  [2] **Look at the reagent and determine if it is an electrophile, nucleophile, acid, or base.**

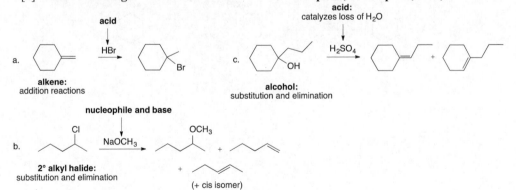

a. [methylenecyclohexane] —acid, HBr→ [1-bromo-1-methylcyclohexane]
   alkene: addition reactions

b. [2-chlorobutane-type, 2° alkyl halide] —nucleophile and base, NaOCH₃→ [2-methoxy derivative + alkene (+ cis isomer)]
   2° alkyl halide: substitution and elimination

c. [tertiary alcohol] —acid: catalyzes loss of H₂O, H₂SO₄→ [alkene] + [alkene]
   alcohol: substitution and elimination

Chapter 10–14

10.31 To devise a synthesis:
[1] Look at the starting material and decide what reactions it can undergo.
[2] Look at the product and decide what reactions could make it.

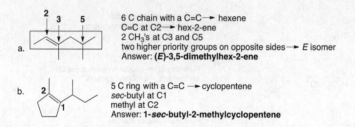

10.32 Convert each ball-and-stick model to a skeletal structure and then name the molecule.

a.
6 C chain with a C=C → hexene
C=C at C2 → hex-2-ene
2 CH₃'s at C3 and C5
two higher priority groups on opposite sides → *E* isomer
Answer: (***E***)-3,5-dimethylhex-2-ene

b.
5 C ring with a C=C → cyclopentene
sec-butyl at C1
methyl at C2
Answer: **1-*sec*-butyl-2-methylcyclopentene**

10.33

a. The two higher priority groups are on the same side of the C=C, making it a **Z** alkene.

b. Add H₂O in a Markovnikov fashion to form two products.

Alkenes 10–15

10.34

8-methylnon-1-ene

a. Br_2 →

b. Br_2 / H_2O →

c. Br_2 / CH_3OH →

10.35 Use the directions from Answer 10.2 to calculate degrees of unsaturation.

a. C_6H_8
[1] maximum number of H's = $2n + 2 = 2(6) + 2 = 14$
[2] subtract actual from maximum = $14 - 8 = 6$
[3] divide by 2 = $6/2$ = **3 degrees of unsaturation**

b. $C_{40}H_{56}$
[1] maximum number of H's = $2n + 2 = 2(40) + 2 = 82$
[2] subtract actual from maximum = $82 - 56 = 26$
[3] divide by 2 = $26/2$ = **13 degrees of unsaturation**

c. $C_{10}H_{16}O_2$
Ignore both O's.
[1] maximum number of H's = $2n + 2 = 2(10) + 2 = 22$
[2] subtract actual from maximum = $22 - 16 = 6$
[3] divide by 2 = $6/2$ = **3 degrees of unsaturation**

d. C_8H_9Br
Because of Br, add one H $(9 + 1 = 10$ H's$)$.
[1] maximum number of H's = $2n + 2 = 2(8) + 2 = 18$
[2] subtract actual from maximum = $18 - 10 = 8$
[3] divide by 2 = $8/2$ = **4 degrees of unsaturation**

e. C_8H_9ClO
Ignore the O; count Cl as one more H $(9 + 1 = 10$ H's$)$.
[1] maximum number of H's = $2n + 2 = 2(8) + 2 = 18$
[2] subtract actual from maximum = $18 - 10 = 8$
[3] divide by 2 = $8/2$ = **4 degrees of unsaturation**

f. $C_7H_{11}N$
Because of N, subtract one H $(11 - 1 = 10$ H's$)$.
[1] maximum number of H's = $2n + 2 = 2(7) + 2 = 16$
[2] subtract actual from maximum = $16 - 10 = 6$
[3] divide by 2 = $6/2$ = **3 degrees of unsaturation**

g. C_4H_8BrN
Because of Br, add one H, but subtract one for N
$(8 + 1 - 1 = 8$ H's$)$.
[1] maximum number of H's = $2n + 2 = 2(4) + 2 = 10$
[2] subtract actual from maximum = $10 - 8 = 2$
[3] divide by 2 = $2/2$ = **1 degree of unsaturation**

h. $C_{10}H_{18}ClNO$
Add one H because of Cl and subtract 1 H for N
$(18 + 1 - 1 = 18)$.
[1] maximum number of H's = $2n + 2 = 2(10) + 2 = 22$
[2] subtract actual from maximum = $22 - 18 = 4$
[3] divide by 2 = $4/2$ = **2 degrees of unsaturation**

10.36 First determine the number of degrees of unsaturation in the compound. Then decide which combinations of rings and π bonds could exist.

$C_{10}H_{14}$
[1] maximum number of H's = $2n + 2 = 2(10) + 2 = 22$
[2] subtract actual from maximum = $22 - 14 = 8$
[3] divide by two = $8/2$ = **4 degrees of unsaturation**

possibilities:
4 π bonds
3 π bonds + 1 ring
2 π bonds + 2 rings
1 π bond + 3 rings
4 rings

Chapter 10–16

10.37

a.

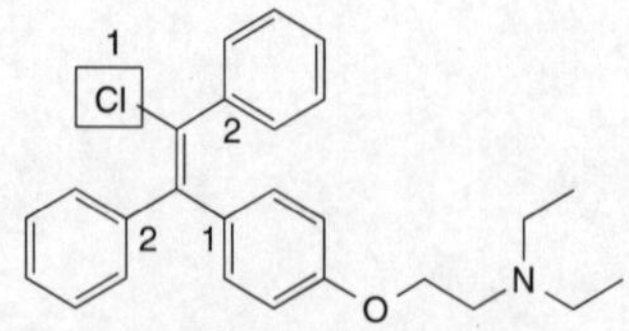

enclomiphene
E

b.

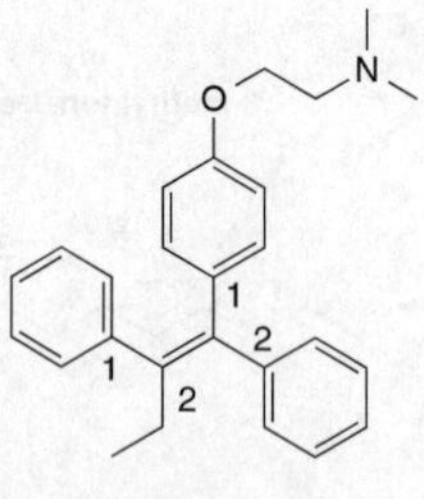

tamoxifen
Z

c.

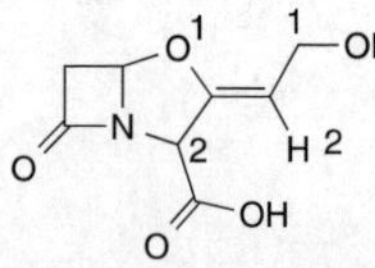

clavulanic acid
Z

10.38 Name the alkenes using the rules in Answers 10.4 and 10.6.

a.

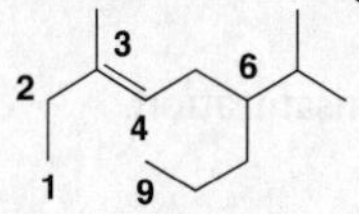

(*E*)-6-isopropyl-3-methylnon-3-ene

9 C chain with a double bond =
nonene

b.

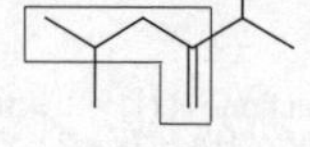

2-isopropyl

pent-1-ene

4-methyl

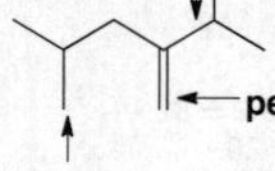

2-isopropyl-4-methylpent-1-ene

5 C chain with a double bond =
pentene

c.

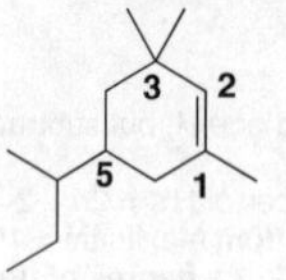

5-*sec*-butyl-1,3,3-trimethylcyclohexene

6 C ring with a double bond =
hexene

d.

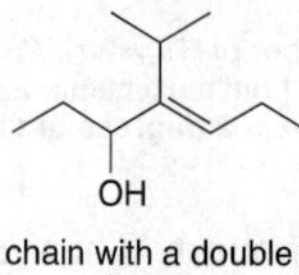

4-isopropyl →

E **double bond** (higher priority groups on opposite
sides with bold bonds)

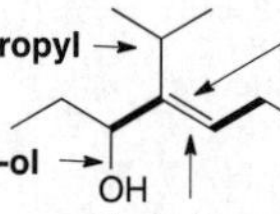

(*E*)-4-isopropylhept-4-en-3-ol

3-ol →

hept-4-ene

7 C chain with a double bond =
heptene

e.

cyclohex-2-ene

1-ol

5-*sec*-butyl

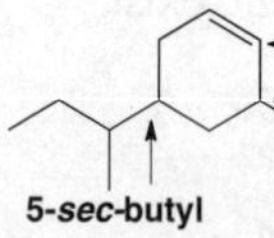

5-*sec*-butylcyclohex-2-enol

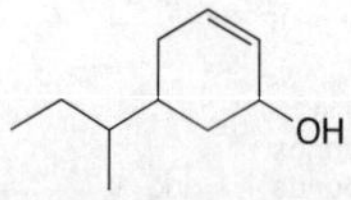

6 C ring with a double bond =
cyclohexene

f.

(*E*)-5-ethyl-3,4-dimethylnon-2-ene

10.39 Use the directions from Answer 10.8.

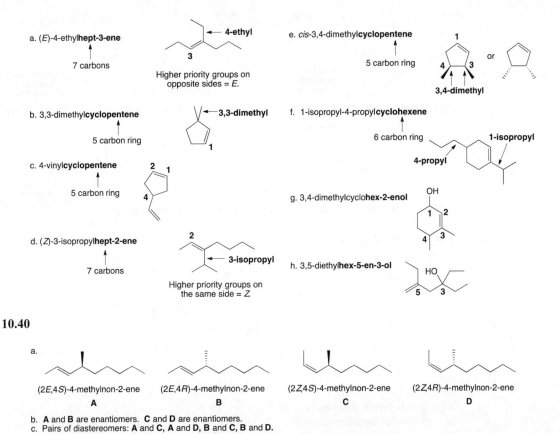

10.40

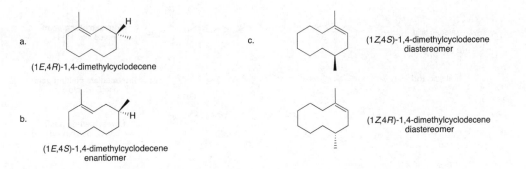

b. **A** and **B** are enantiomers. **C** and **D** are enantiomers.
c. Pairs of diastereomers: **A** and **C**, **A** and **D**, **B** and **C**, **B** and **D**.

10.41

a. (1E,4R)-1,4-dimethylcyclodecene

b. (1E,4S)-1,4-dimethylcyclodecene
enantiomer

c. (1Z,4S)-1,4-dimethylcyclodecene
diastereomer

(1Z,4R)-1,4-dimethylcyclodecene
diastereomer

Chapter 10–18

10.42 Name the alkene from which the epoxide can be derived and add the word *oxide*.

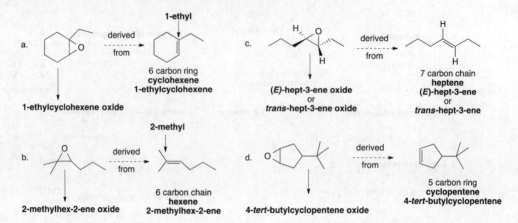

10.43 a, b. *E,Z* and *R,S* designations are shown.

c. Nine double bonds that can be *E* or *Z*
Six tetrahedral stereogenic centers
Maximum possible number of stereoisomers = 2^{15}

10.44

stearic acid — **highest melting point** no double bonds

elaidic acid — **intermediate melting point** one *E* double bond

oleic acid — **lowest melting point** one *Z* double bond

10.45

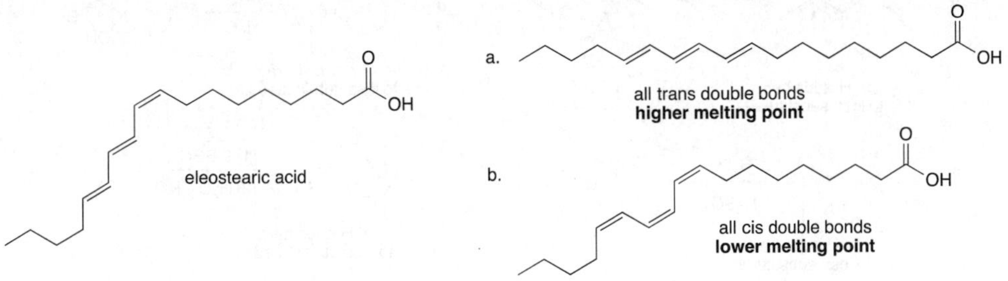
eleostearic acid
a.
all trans double bonds
higher melting point
b.
all cis double bonds
lower melting point
O
OH

10.46

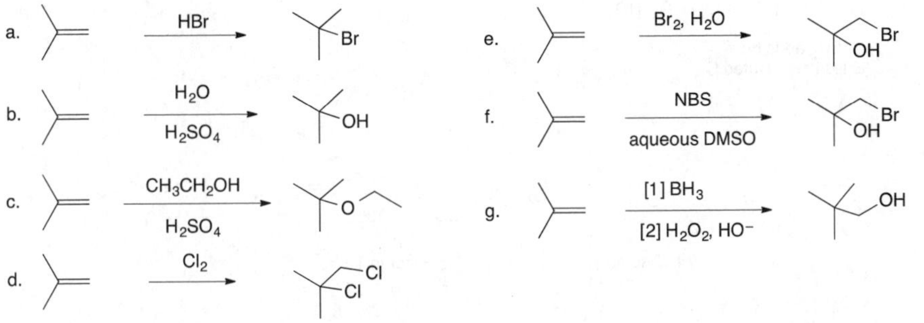
a. HBr Br
b. H_2O H_2SO_4 OH
c. CH_3CH_2OH H_2SO_4 O
d. Cl_2 Cl Cl
e. Br_2, H_2O Br OH
f. NBS aqueous DMSO Br OH
g. [1] BH_3 [2] H_2O_2, HO^- OH

10.47

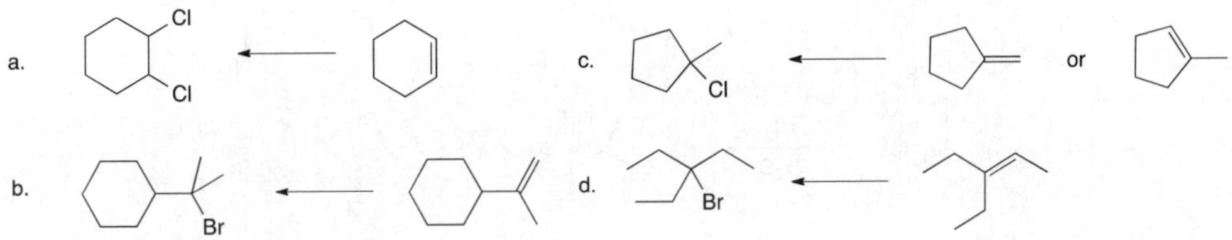
a. Cl Cl
b. Br
c. Cl or
d. Br

10.48 Hydroboration–oxidation results in addition of an OH group on the *less* substituted carbon, whereas acid-catalyzed addition of H_2O results in the addition of an OH group on the *more* substituted carbon.

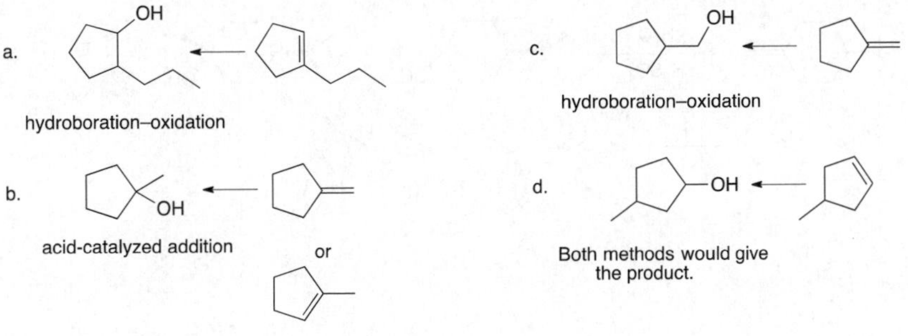
a. OH
hydroboration–oxidation
b. OH
acid-catalyzed addition
or
c. OH
hydroboration–oxidation
d. OH
Both methods would give the product.

10.49

a. 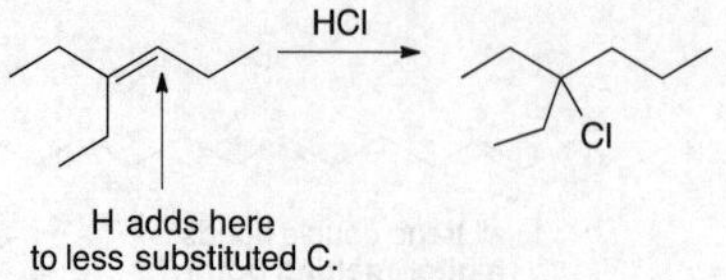
HCl
Cl
H adds here
to less substituted C.

d. 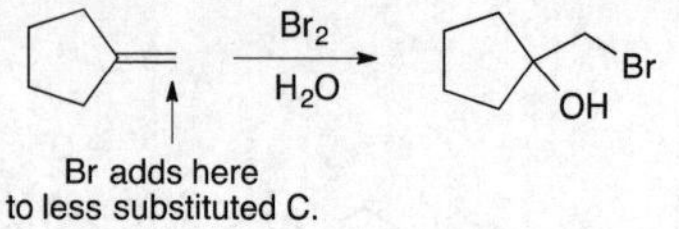
Br2
H2O
Br
OH
Br adds here
to less substituted C.

b. 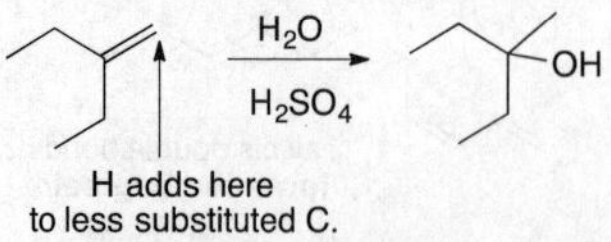
H2O
H2SO4
OH
H adds here
to less substituted C.

e. 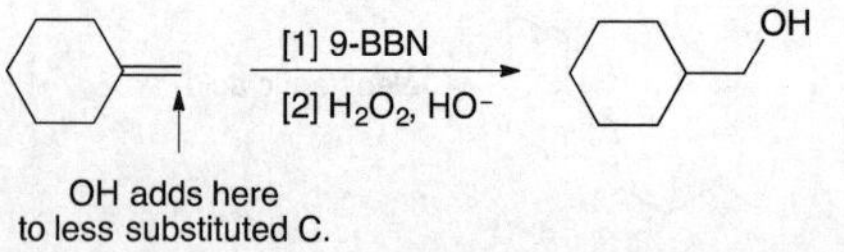
[1] 9-BBN
[2] H2O2, HO–
OH
OH adds here
to less substituted C.

c. 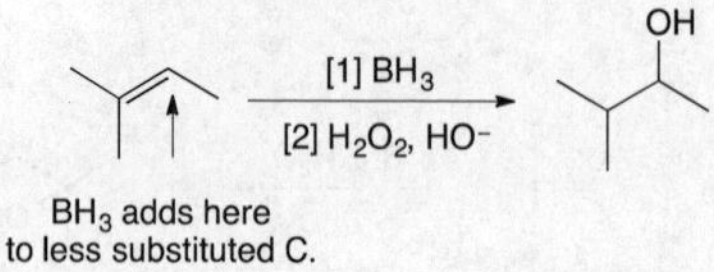
[1] BH3
[2] H2O2, HO–
OH
BH3 adds here
to less substituted C.

f. 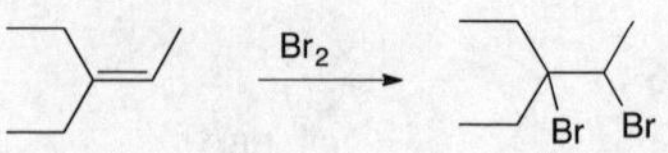
Br2
Br
Br

10.50

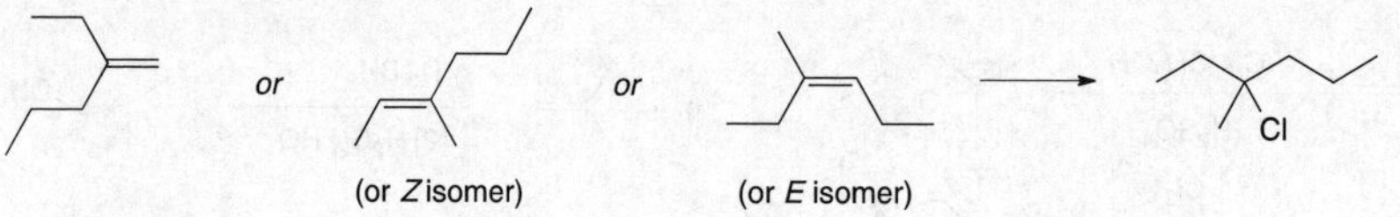
or
or
(or Z isomer)
(or E isomer)
Cl

10.51

a. 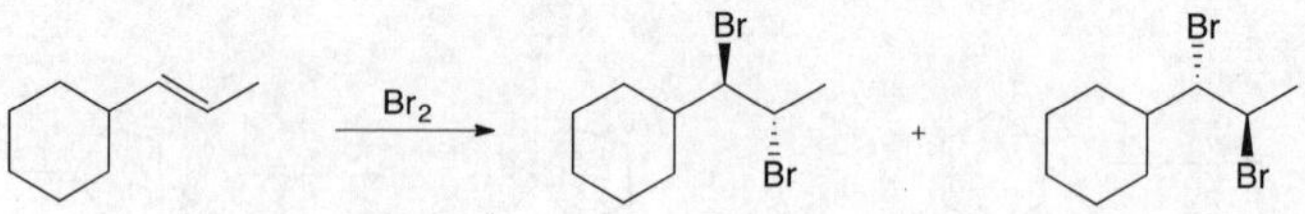
Br2
Br
Br
+
Br
Br

b. 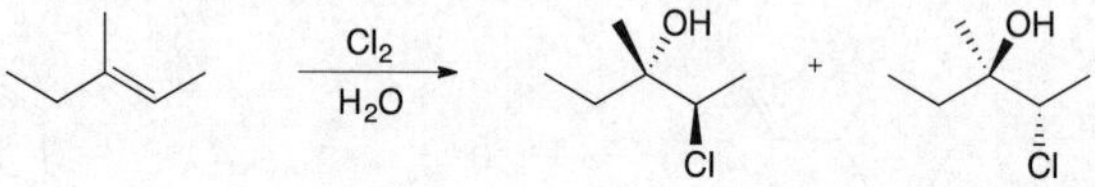
Cl2
H2O
OH
Cl
+
OH
Cl

c. 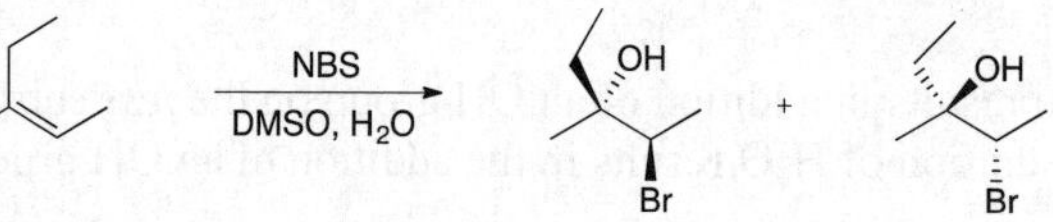
NBS
DMSO, H2O
OH
Br
+
OH
Br

10.52

a. 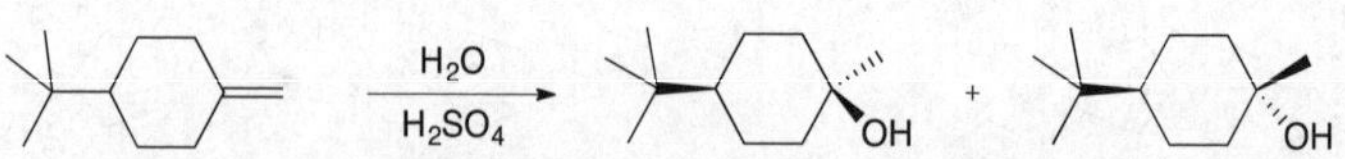
H2O
H2SO4
OH
+
OH

b. 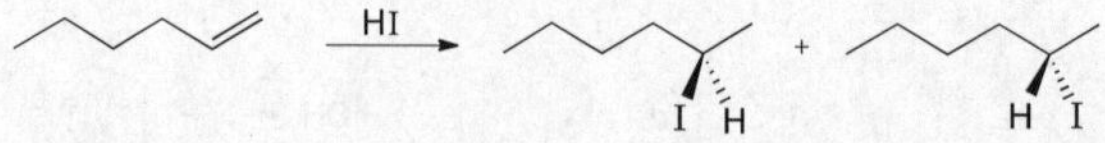
HI
I H
+
H I

Alkenes 10–21

c. 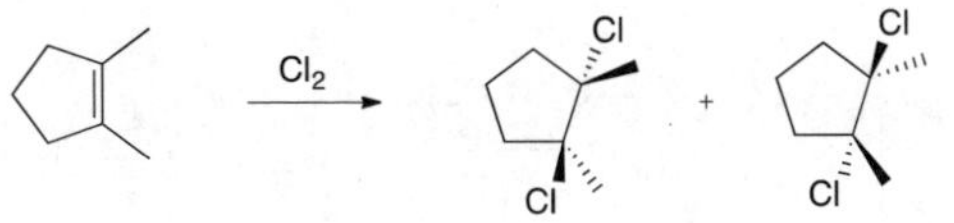 only anti addition

d. 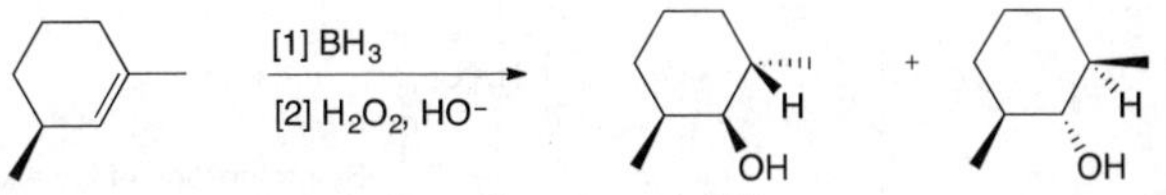 [1] BH₃ → [2] H₂O₂, HO⁻ only syn addition

e. 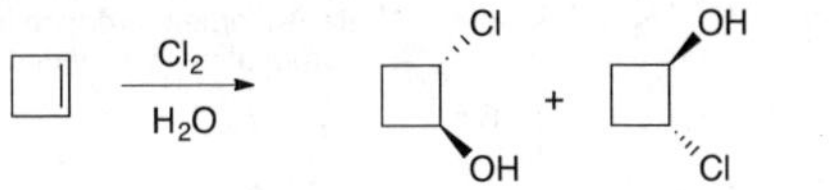 Cl₂ / H₂O only anti addition

f. 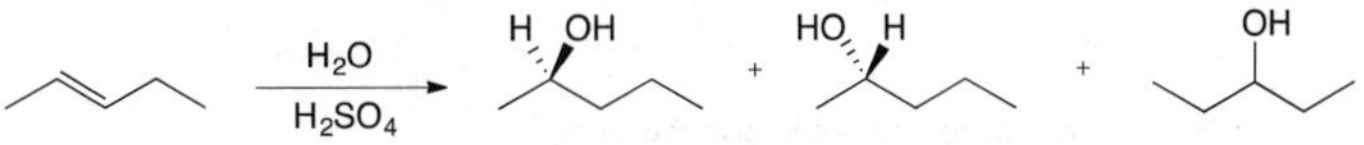 H₂O / H₂SO₄

10.53 The alkene that forms the more stable carbocation when H⁺ is added reacts faster. Both alkenes form 2° carbocations, but the carbocation from **B** is also resonance stabilized by the benzene ring. As a result, **B** reacts faster.

A → 2° carbocation

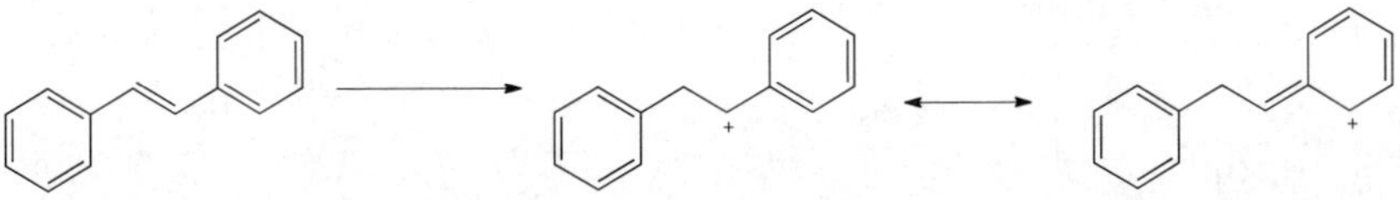

B → + 3 more resonance structures

10.54 Draw each reaction. (a) The cis isomer of oct-4-ene gives two enantiomers on addition of Br₂. (b) The trans isomer gives a meso compound.

a. 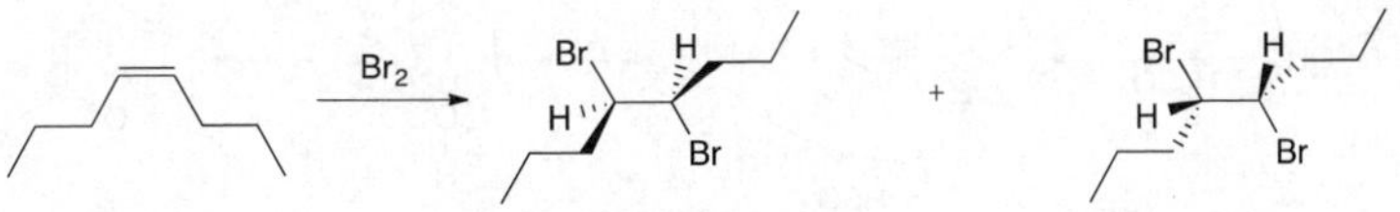 *cis*-oct-4-ene → Br₂ → (4R,5R)-4,5-dibromooctane + (4S,5S)-4,5-dibromooctane

enantiomers

b. 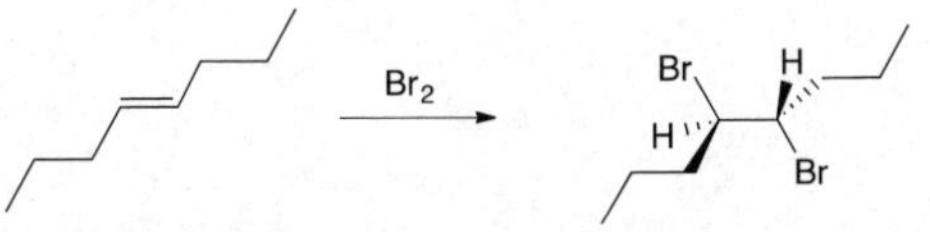 *trans*-oct-4-ene → Br₂ → (4R,5S)-4,5-dibromooctane

meso compound

10.55

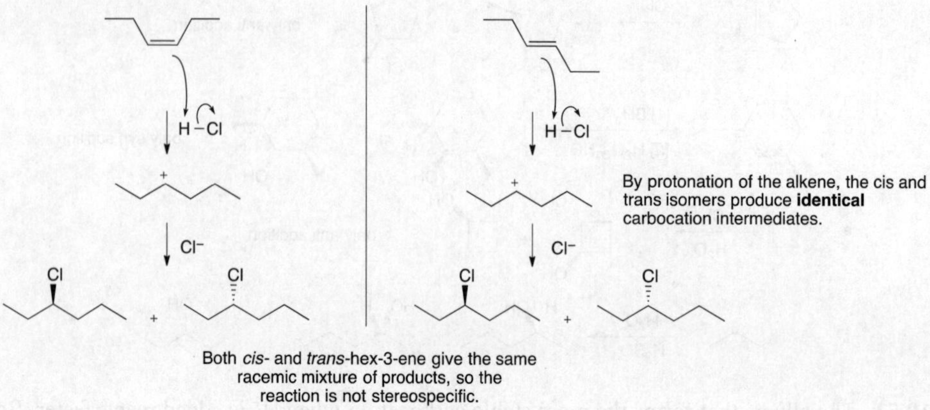
H–Cl
H–Cl
Cl⁻
Cl⁻
Cl
Cl
Cl
Cl
By protonation of the alkene, the cis and trans isomers produce **identical** carbocation intermediates.
Both *cis*- and *trans*-hex-3-ene give the same racemic mixture of products, so the reaction is not stereospecific.

10.56

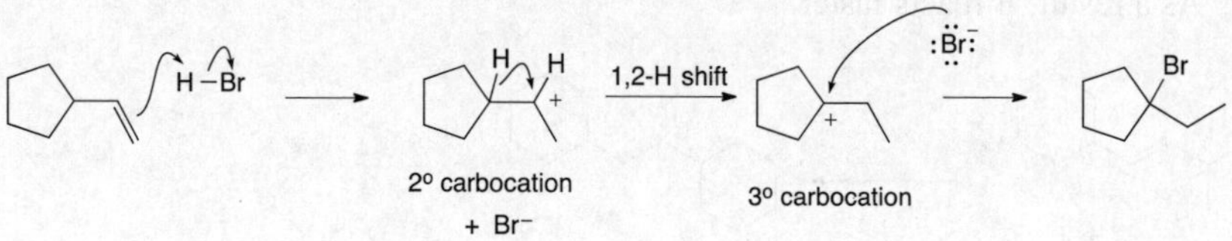
H–Br
1,2-H shift
:Br:⁻
Br
2° carbocation
+ Br⁻
3° carbocation

10.57

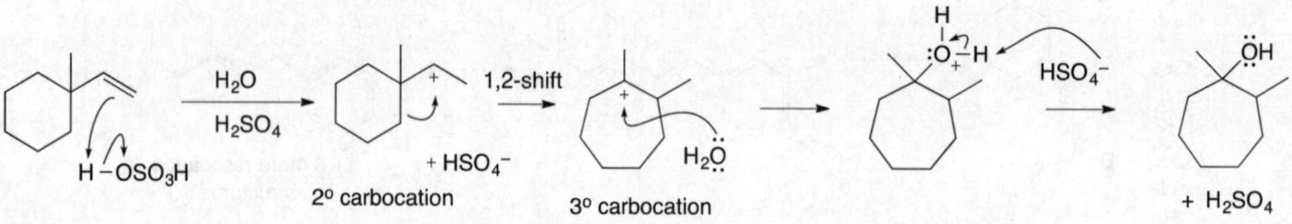
H₂O
H₂SO₄
H–OSO₃H
+ HSO₄⁻
2° carbocation
1,2-shift
3° carbocation
H₂O
HSO₄⁻
OH
+ H₂SO₄

10.58

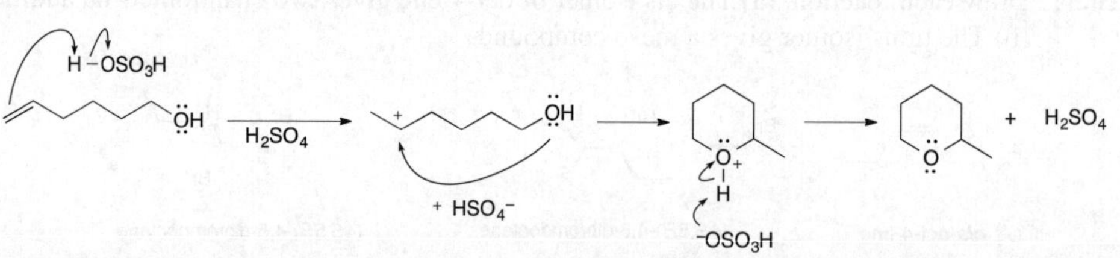
H–OSO₃H
OH
H₂SO₄
OH
+ HSO₄⁻
⁻OSO₃H
+ H₂SO₄

10.59

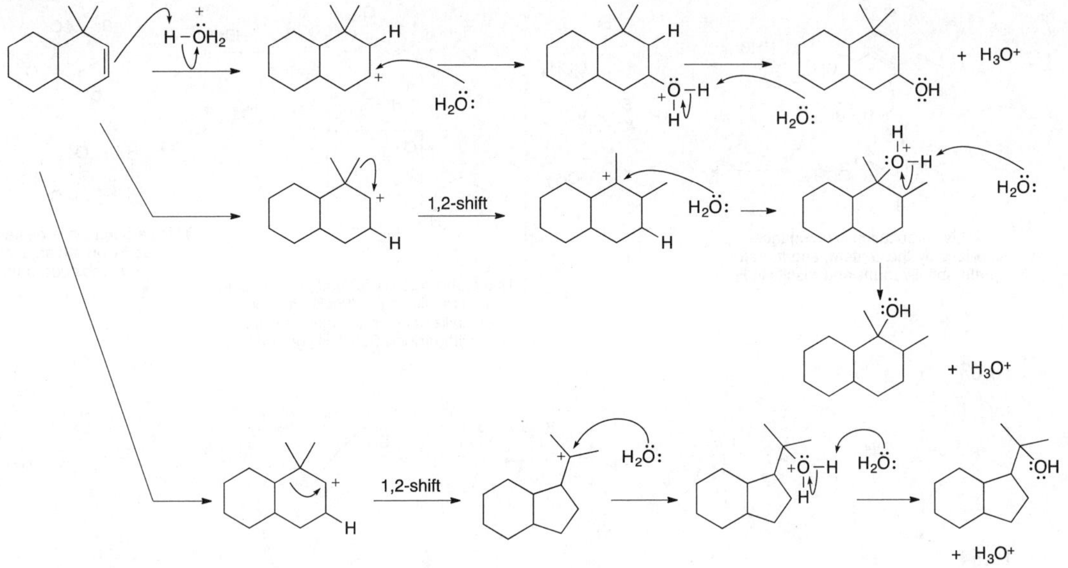
H–ÖH2
H2Ö:
H
1,2-shift
+ H3O+
:ÖH
+ H3O+
+ H3O+

10.60 The isomerization reaction occurs by protonation and deprotonation.

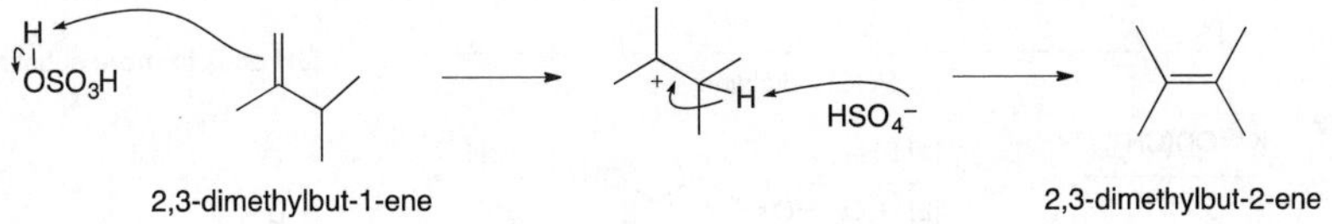
H
OSO3H
HSO4−
2,3-dimethylbut-1-ene
2,3-dimethylbut-2-ene

10.61

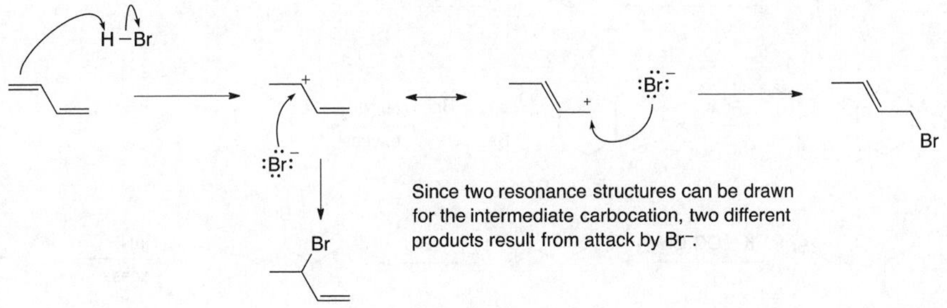
H–Br
:Br:−
Br
Since two resonance structures can be drawn for the intermediate carbocation, two different products result from attack by Br−.

Chapter 10–24

10.62

A

HBr →

B

This carbocation is resonance stabilized by the O atom, and therefore preferentially forms and results in **B**.

C

HBr →

D

This carbocation is formed preferentially and results in product **D**. It is not destabilized by an adjacent electron-withdrawing COOCH$_3$ group.

This carbocation is destabilized by the δ+ on the adjacent C, so it does not form.

10.63

+ :Br$^-$ + :Br$^-$ + HBr

10.64

a.   OH $\xrightarrow{PBr_3}$ Br $\xrightarrow{K^+ \ ^-OC(CH_3)_3}$ $\xrightarrow{H_2O \ + \ H_2SO_4}$ OH

POCl$_3$, pyridine

OH adds to **more** substituted C.

b. Br $\xrightarrow{K^+ \ ^-OC(CH_3)_3}$ $\xrightarrow[{[2] \ H_2O_2, \ HO^-}]{[1] \ BH_3}$ OH $\xrightarrow{NaH}$ O$^-$ + H$_2$ $\xrightarrow{CH_3I}$ OCH$_3$

c. I $\xrightarrow{K^+ \ ^-OC(CH_3)_3}$ $\xrightarrow{HCl}$ Cl

d. $\xrightarrow{Br_2}$ Br, Br $\xrightarrow[\text{excess}]{NaNH_2}$

e. Br $\xrightarrow{K^+ \ ^-OC(CH_3)_3}$ $\xrightarrow{Br_2}$ Br, Br $\xrightarrow[\text{excess}]{NaNH_2}$

f. Cl $\xrightarrow{K^+ \ ^-OC(CH_3)_3}$ $\xrightarrow[H_2O]{Cl_2}$ OH, Cl $\xrightarrow{NaH}$ O

10.65

a. $\xrightarrow[H_2O]{Cl_2}$ Cl, OH $\xrightarrow{Na^+ \ H^-}$ Cl, O$^-$ + H$_2$ → O

Alkenes 10–25

b. HBr → (Br) NaCN → (CN)

c. (Br) NaOH → (OH) NaH → (O⁻) I → (O)

(from b.)

d. (O) [1] ⁻SH / [2] H₂O → (OH, ''SH) + enantiomer

(from a.)

10.66 Having two rings joined together as in **A** and **B** creates a very rigid ring system and constrains bond angles. Evidently the bond angles around the C=C in **A** are close enough to the trigonal planar bond angle of 120°, so **A** is stable. With **B**, however, the C=C is located at a carbon shared by both rings and the bond angles around the C=C deviate greatly from the desired angle, so **B** is *not* a stable compound.

10.67 a. Br$_2$ adds in an anti fashion to form a meso dibromide. Rotate around the C–C bond to place H and Br anti periplanar in the second step. HBr can be eliminated in two ways, but both give the same product.

meso compound

rotate

H and Br must be anti.

KOH (–HBr)

Two higher priority groups are on opposite sides = **E.**

b. Br$_2$ addition forms two enantiomers. Anti periplanar elimination of H and Br gives the same alkene from both compounds.

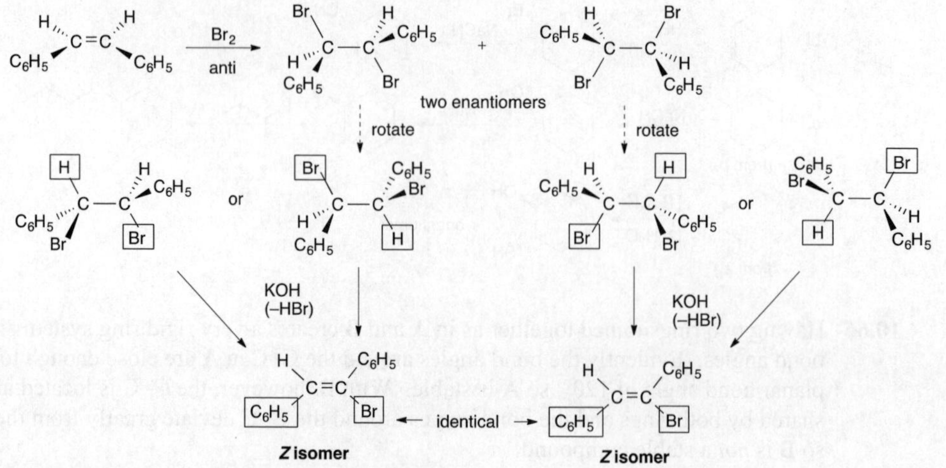

Chapter 10–26

two enantiomers

rotate   rotate

or   or

KOH (–HBr)   KOH (–HBr)

Z isomer ← identical → **Z isomer**

c. The products in (a) and (b) are diastereomers.

10.68

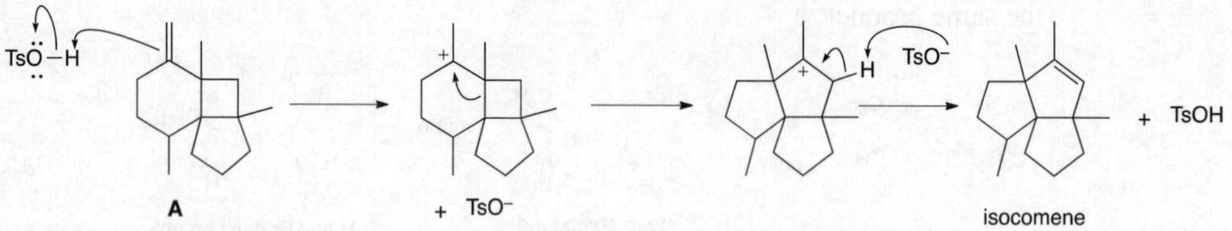

A   + TsO⁻   isocomene   + TsOH

10.69

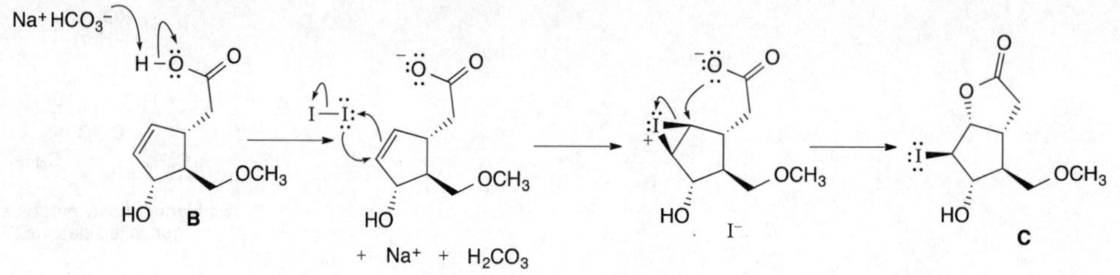

B   + Na⁺ + H₂CO₃   I⁻   **C**

10.70

2° carbocation

+ HSO₄⁻   + H₂O

1,2-H shift

3° carbocation

+ H₂SO₄

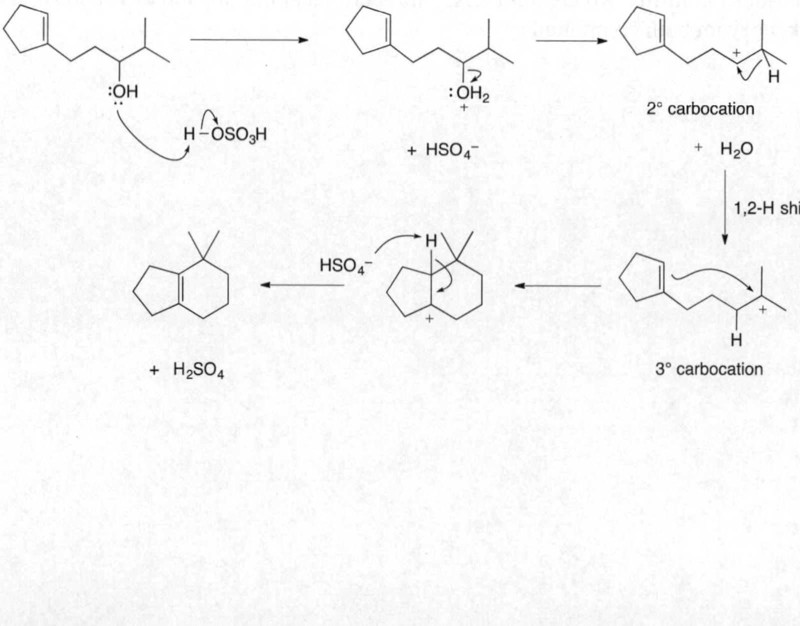

10.71

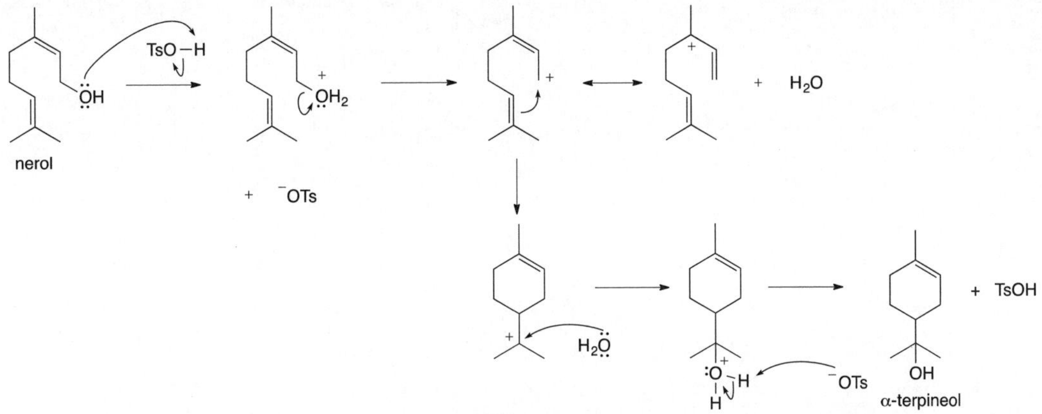

TsO—H

nerol

+ ⁻OTs

OH_2 ⁺

+ H_2O

H_2O

⁻OTs

OH

α-terpineol

+ TsOH

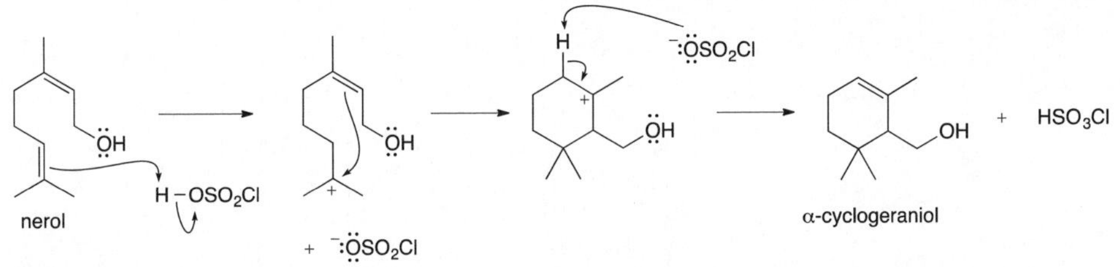

nerol

H—OSO_2Cl

:OSO_2Cl

:OSO_2Cl

H

OH

OH

α-cyclogeraniol

+ HSO_3Cl

10.72

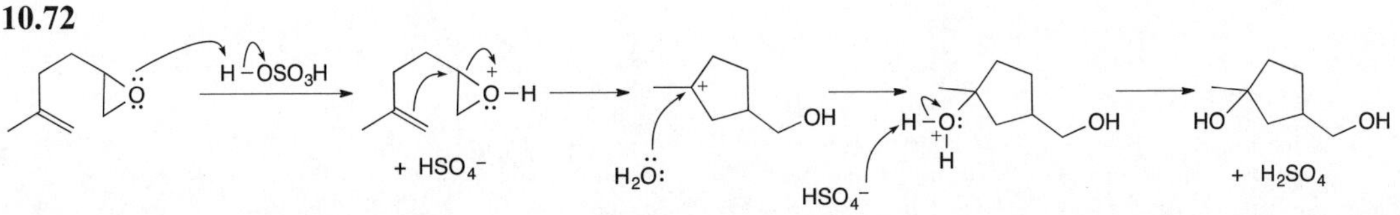

H—OSO_3H

+ HSO_4^-

H_2O:

O—H

OH

HSO_4^-

H—O: ⁺

H

OH

HO

OH

+ H_2SO_4

Alkynes 11–1

Chapter 11 Alkynes

Chapter Review

General facts about alkynes

- Alkynes contain a carbon–carbon triple bond consisting of a strong σ bond and two weak π bonds. Each carbon is *sp* hybridized and linear (11.1).

$$H-C\equiv C-H$$
acetylene

180°

sp hybridized

- Alkynes are named using the suffix *-yne* (11.2).
- Alkynes have weak intermolecular forces, giving them low mp's and low bp's, and making them water insoluble (11.3).
- Since its weaker π bonds make an alkyne electron rich, alkynes undergo addition reactions with electrophiles (11.6).

Addition reactions of alkynes

[1] Hydrohalogenation—Addition of HX (X = Cl, Br, or I) (11.7)

$$R-\!\!\!\equiv\ \xrightarrow[\text{(2 equiv)}]{H-X}$$

X X

R

geminal dihalide

- Markovnikov's rule is followed. H bonds to the *less* substituted C in order to form the more stable carbocation.

[2] Halogenation—Addition of X_2 (X = Cl or Br) (11.8)

$$R-\!\!\!\equiv\ \xrightarrow[\text{(2 equiv)}]{X-X}$$

X X

R X

X

tetrahalide

- Bridged halonium ions are formed as intermediates.
- Anti addition of X_2 occurs.

[3] Hydration—Addition of H_2O (11.9)

$$R-\!\!\!\equiv\ \xrightarrow[\substack{H_2SO_4 \\ HgSO_4}]{H_2O}$$

OH

R

enol

O

R

ketone

- Markovnikov's rule is followed. H bonds to the *less* substituted C in order to form the more stable carbocation.
- The unstable enol that is first formed rearranges to a carbonyl group.

Chapter 11–2

[4] Hydroboration–oxidation—Addition of H_2O (11.10)

$$R-\!\!\!\equiv\!\!\!- \xrightarrow[\text{[2] } H_2O_2,\ HO^-]{\text{[1] } R_2BH} \left[\begin{array}{c} R \diagdown\!\!\diagup^{H} \\ OH \\ \text{enol} \end{array} \rightleftharpoons \begin{array}{c} R \diagdown\!\!\diagup^{H} \\ O \\ \text{aldehyde} \end{array} \right]$$

- The unstable enol, first formed after oxidation, rearranges to a carbonyl group.

Reactions involving acetylide anions

[1] Formation of acetylide anions from terminal alkynes (11.6B)

$$R-C\equiv C-H \ + \ :B \ \rightleftharpoons \ R-C\equiv C:^- \ + \ HB^+$$

- Typical bases used for the reaction are $NaNH_2$ and NaH.

[2] Reaction of acetylide anions with alkyl halides (11.11A)

$$H-C\equiv C:^- \ + \ R-X \ \longrightarrow \ H-C\equiv C-R \ + \ X^-$$

- The reaction follows an S_N2 mechanism.
- The reaction works best with CH_3X and RCH_2X.

[3] Reaction of acetylide anions with epoxides (11.11B)

$$H-C\equiv C:^- \ \xrightarrow[\text{[2] } H_2O]{\text{[1] } \triangle\!\!O} \ H-C\equiv C-CH_2CH_2OH$$

- The reaction follows an S_N2 mechanism.
- Ring opening occurs from the back side at the *less* substituted end of the epoxide.

Alkynes 11–3

Practice Test on Chapter Review

1. Draw the structure for the compound with the following IUPAC name: 5-*tert*-butyl-6,6-dimethyl-non-3-yne.

2.a. Which of the following compounds is an enol tautomer of compound **A?**

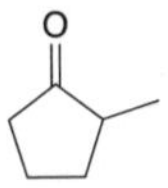

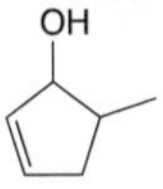

   4. **A** can be a tautomer of both compounds (1) and (2).
   5. **A** can be a tautomer of compounds (1), (2), and (3).

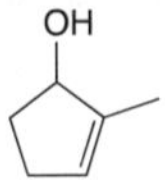

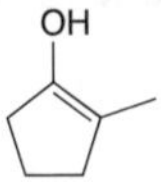

   b. Which of the following bases is strong enough to deprotonate $CH_3C{\equiv}CH$ (propyne, $pK_a = 25$)? The pK_a's of the conjugate acids of the bases are given in parentheses.

   1. CH_3Li ($pK_a = 50$)
   2. $NaOCH_3$ ($pK_a = 15.5$)
   3. $NaOCOCH_3$ ($pK_a = 4.8$)
   4. The bases in (1) and (2) are both strong enough.
   5. The bases in (1), (2), and (3) are all strong enough.

3. Draw the organic products formed in the following reactions.

a. 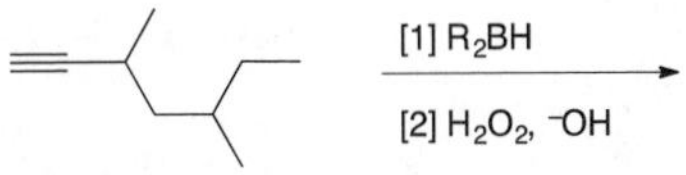
   [1] R_2BH
   [2] H_2O_2, ⁻OH

b. 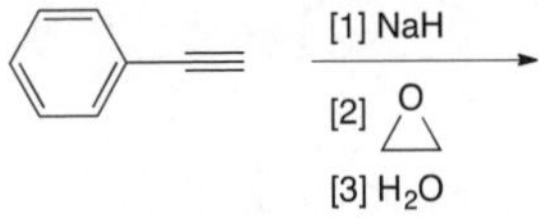
   [1] NaH
   [2] (epoxide)
   [3] H_2O

c. 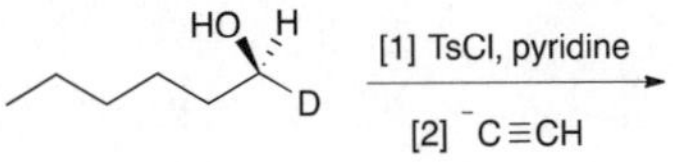
   HO H
   D 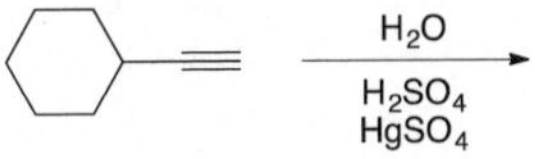
   [1] TsCl, pyridine
   [2] ⁻$C{\equiv}CH$

d.
   H_2O
   H_2SO_4
   $HgSO_4$

e. 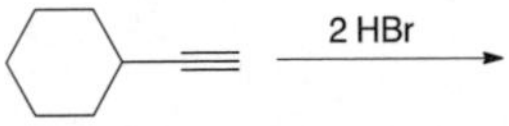
   2 HBr

Chapter 11–4

4. Draw two different enol tautomers for the following compound.

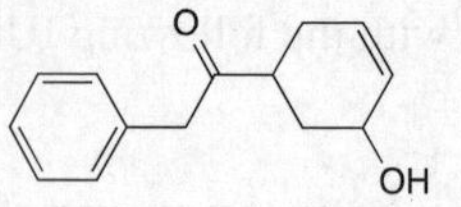

5. What acetylide anion and alkyl halide are needed to make the following alkyne?

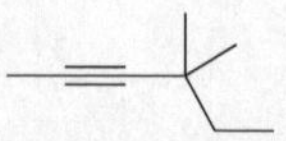

Answers to Practice Test

1.

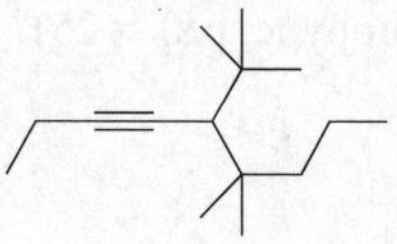

2.a. 2
b. 1

3.

a.

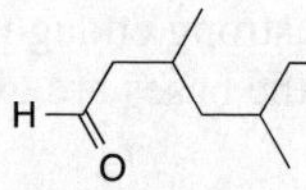

b.

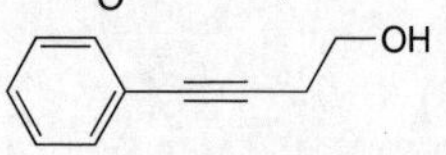

c.   HC≡C D
          H

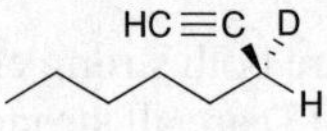

d.

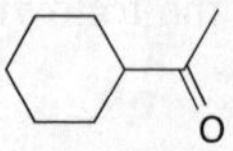

e.   Br Br

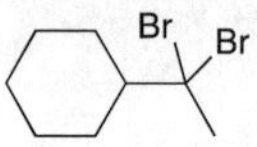

4.

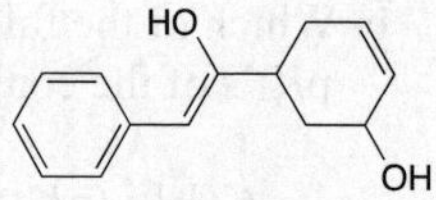

HO
(E + Z)

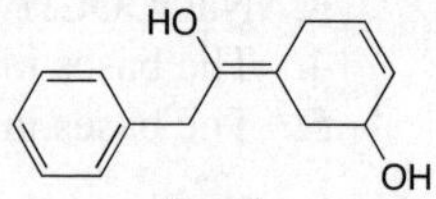

HO
(E + Z)

5.

CH₃X   +

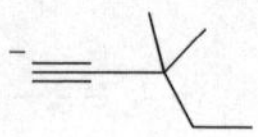

Answers to Problems

11.1

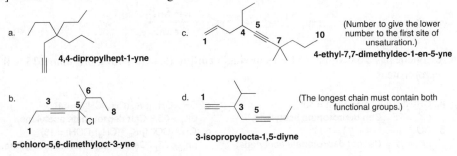

nepheliosyne B

a. The most acidic H is part of the COOH group of the carboxylic acid.
b. The shortest C–C σ bond is formed from C atoms with a higher percent *s*-character. The labeled C–C bond is formed from an *sp* hybridized orbital and an *sp²* hybridized orbital.
c. Nepheliosyne B has four triple bonds and five double bonds = 13 degrees of unsaturation.
d. Six bonds (labeled with x) are formed from C*sp*–C*sp³*.
e. All triple bonds are internal except the C≡C at the end of the chain, which is boxed in.

11.2 To name an alkyne:
[1] Find the longest chain that contains both atoms of the triple bond, change the *-ane* ending of the parent name to *-yne*, and number the chain to give the first carbon of the triple bond the lower number.
[2] Name all substituents following the other rules of nomenclature.

a. **4,4-dipropylhept-1-yne**

c. **4-ethyl-7,7-dimethyldec-1-en-5-yne** (Number to give the lower number to the first site of unsaturation.)

b. **5-chloro-5,6-dimethyloct-3-yne**

d. **3-isopropylocta-1,5-diyne** (The longest chain must contain both functional groups.)

11.3 To work backwards from a name to a structure:
[1] Find the parent name and the functional group.
[2] Add the substituents to the appropriate carbon.

a. *trans*-2-ethynyl**cyclopentanol**
5 C ring with OH at C1

OH ← OH on C1
ethynyl at C2

or

Chapter 11–6

b. 4-*tert*-butyl**dec-5-yne**

10 C chain with
a triple bond

c. 3,3,5-trimethyl**cyclononyne**

9 C ring with a
triple bond at C1

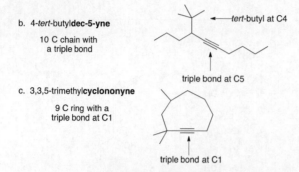

11.4 Two factors cause the boiling point increase. The linear *sp* hybridized C's of the alkyne allow for more van der Waals attraction between alkyne molecules. Also, a triple bond is more polarizable than a double bond, which increases the van der Waals forces between two alkyne molecules as well.

11.5 To convert an alkene to an alkyne:
   [1] Make a vicinal dihalide from the alkene by addition of X_2.
   [2] Add base to remove two equivalents of HX and form the alkyne.

11.6 Acetylene has a pK_a of 25, so **bases having a conjugate acid with a pK_a *above* 25** will be able to deprotonate it.

   a. CH_3NH^- [pK_a (CH_3NH_2) = 40]
   pK_a > 25 = **Can deprotonate acetylene.**
   b. CO_3^{2-} [pK_a (HCO_3^-) = 10.2]
   pK_a < 25 = **Cannot deprotonate acetylene.**
   c. $CH_2=CH^-$ [pK_a ($CH_2=CH_2$) = 44]
   pK_a > 25 = **Can deprotonate acetylene.**
   d. $(CH_3)_3CO^-$ {pK_a [$(CH_3)_3COH$] = 18}
   pK_a < 25 = **Cannot deprotonate acetylene.**

11.7 To draw the products of reactions with HX:
   • Add two moles of HX to the triple bond, following Markovnikov's rule.
   • Both X's end up on the more substituted C.

Alkynes 11–7

a.

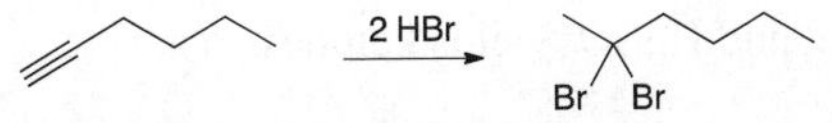

b.

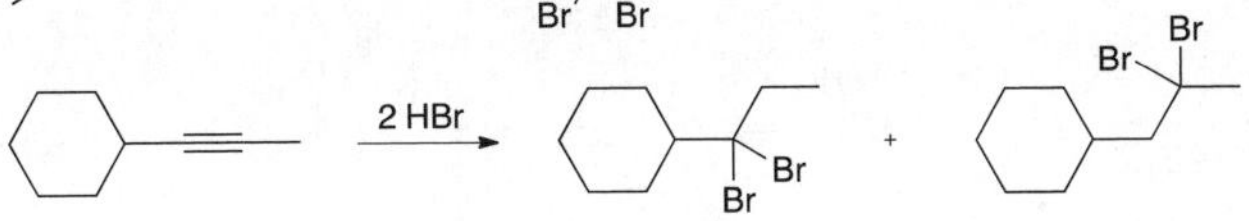

c.

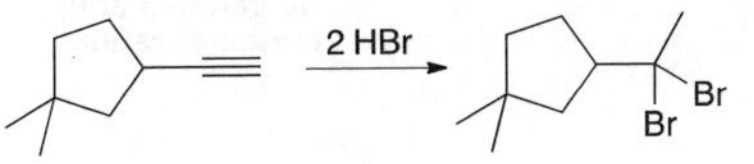

11.8

a.    c.

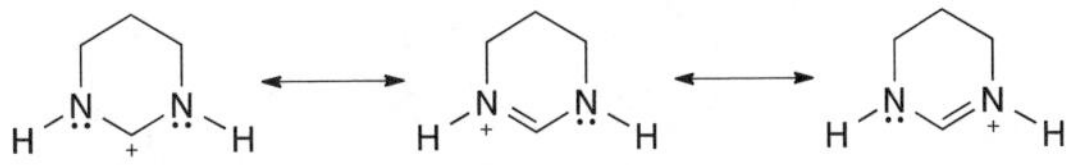

b.

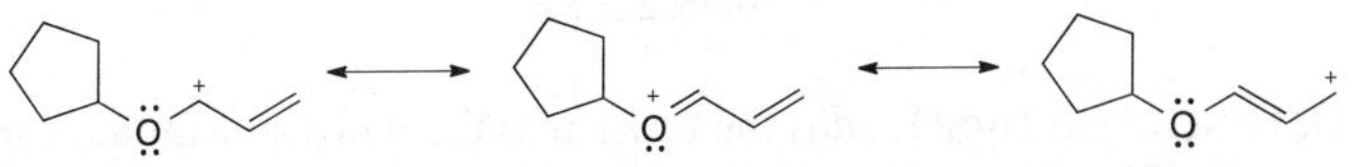

11.9 Addition of one equivalent of X_2 to alkynes forms trans dihalides. Addition of two equivalents of X_2 to alkynes forms tetrahalides.

a.

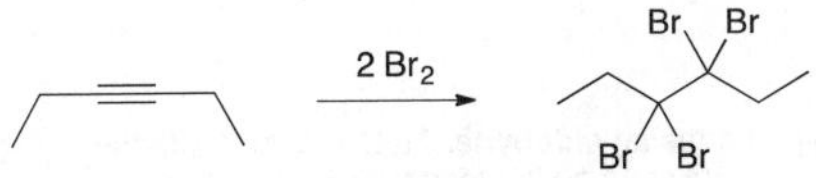

b. 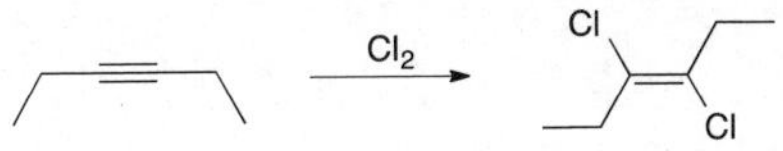   **trans dihalide**

11.10

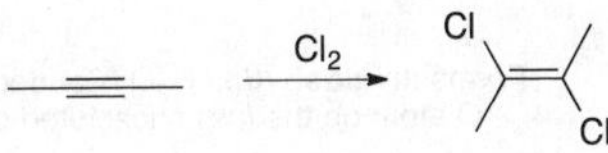

The two Cl atoms are electron withdrawing, making the π bond less electron rich and therefore less reactive with an electrophile.

11.11 **To draw the keto form of each enol:**
   [1] Change the C–OH to a C=O at one end of the double bond.
   [2] Add a proton at the other end of the double bond.

a. 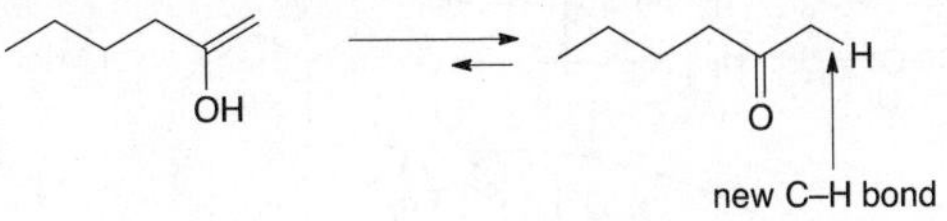   new C–H bond   c.   new C–H bond

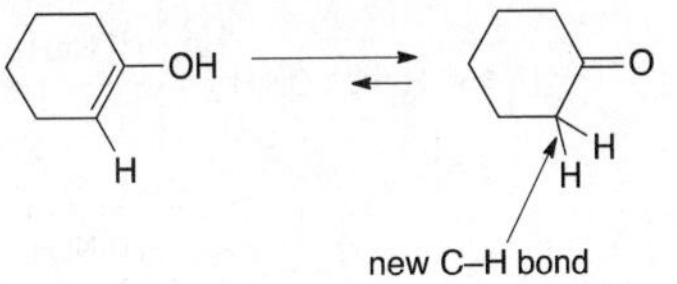

b. 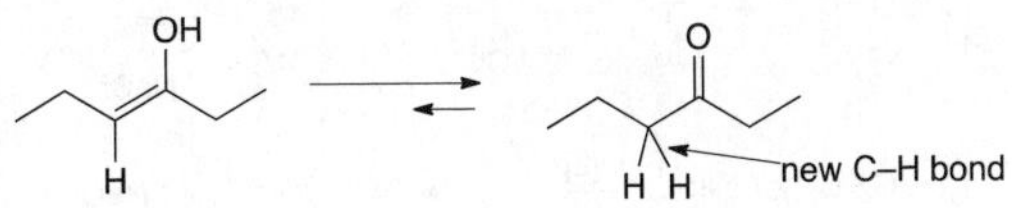   new C–H bond

Chapter 11–8

11.12 The treatment of alkynes with H_2O, H_2SO_4, and $HgSO_4$ yields ketones.

$$\xrightarrow[\text{H}_2\text{SO}_4,\ \text{HgSO}_4]{\text{H}_2\text{O}}$$

Two enols form.
(*E* and *Z* isomers)

two ketones after
tautomerization

11.13

a.

enol tautomers

b.

constitutional isomers,
but not tautomers

11.14 Reaction with H_2O, H_2SO_4, and $HgSO_4$ adds the oxygen to the *more* substituted carbon. Reaction with [1] R_2BH, [2] H_2O_2, ^-OH adds the oxygen to the *less* substituted carbon.

a.

$$\xrightarrow[\text{H}_2\text{SO}_4,\ \text{HgSO}_4]{\text{H}_2\text{O}}$$

Forms a **ketone.** H_2O is added with the O atom on the *more* substituted carbon.

$$\xrightarrow[\text{[2] H}_2\text{O}_2,\ \text{HO}^-]{\text{[1] R}_2\text{BH}}$$

Forms an **aldehyde.** H_2O is added with the O atom on the *less* substituted carbon.

b.

$$\xrightarrow[\text{H}_2\text{SO}_4,\ \text{HgSO}_4]{\text{H}_2\text{O}}$$

Forms a **ketone.** H_2O is added with the O atom on the *more* substituted carbon.

$$\xrightarrow[\text{[2] H}_2\text{O}_2,\ \text{HO}^-]{\text{[1] R}_2\text{BH}}$$

Forms an **aldehyde.** H_2O is added with the O atom on the *less* substituted carbon.

11.15

a. $H-C\equiv C-H$ $\xrightarrow{\text{[1] NaH}}$ $H-C\equiv C:^-$ + H_2 $\xrightarrow{\text{[2]}}$ + NaCl

b.

$\xrightarrow{\text{[1] NaH}}$ $-C\equiv C:^-$ + H_2 $\xrightarrow{\text{[2]}}$ + NaBr

1° alkyl halide
substitution product

$\xrightarrow{\text{[1] NaNH}_2}$ $-C\equiv C:^-$ + NH_3 $\xrightarrow{\text{[2]}}$ + NaCl

3° alkyl halide
elimination product

Alkynes 11–9

11.16

a. [retrosynthesis] → isobutyl-Cl + ⁻C≡C−H   **terminal alkyne — only one possibility**
   1° RX

b. [internal alkyne, positions 1 and 2]
   [1] CH₃Cl + ⁻≡−propyl
   [2] ≡⁻ + Cl-butyl   1° RX
   internal alkyne — two possibilities

c. [tBu-C≡C-CH₂-] → tBu-C≡C⁻ + Cl-CH₂CH₃   1° RX   **internal alkyne — only one possibility**

   ⟹✗ tBu-Cl + ⁻≡−Et   **The 3° alkyl halide would undergo elimination.**
   3° RX
   too crowded for S$_N$2 reaction

11.17

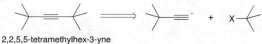

HC≡C−H —Na⁺H:⁻→ HC≡C:⁻   —Br-R→ H−≡−R   —Na⁺H:⁻→ :⁻≡−R   + H₂
           + H₂   + Na⁺Br⁻
           + Na⁺

[followed by attack on iBu-Br giving 2,2,5,5-... style internal alkyne + Br⁻]

11.18

2,2,5,5-tetramethylhex-3-yne ⟹ tBu−C≡C⁻ + X−tBu

The 3° alkyl halide is too crowded for nucleophilic substitution. Instead, it would undergo elimination with the acetylide anion.

11.19

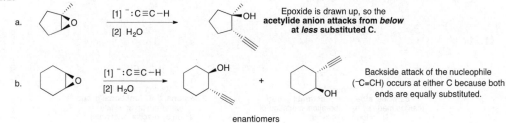

a. cyclopentene oxide [1] ⁻:C≡C−H / [2] H₂O → trans-2-ethynylcyclopentanol
   Epoxide is drawn up, so the **acetylide anion attacks from below** at *less* substituted C.

b. cyclohexene oxide [1] ⁻:C≡C−H / [2] H₂O → two trans-2-ethynylcyclohexanol enantiomers

   Backside attack of the nucleophile (−C≡CH) occurs at either C because both ends are equally substituted.

Chapter 11–10

11.20

a.

b.

c.

d.

e.

f.

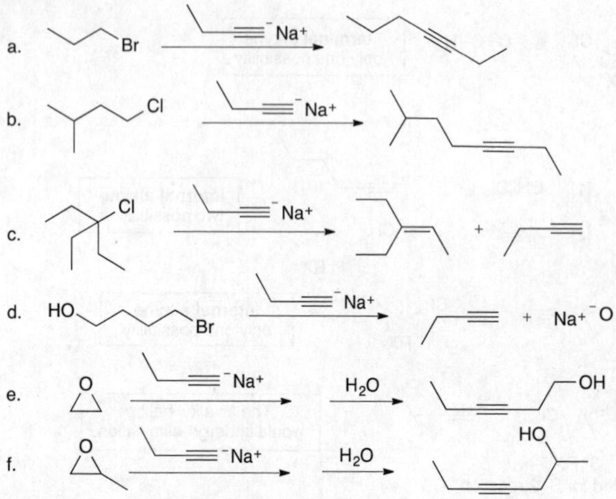

11.21 To use a retrosynthetic analysis:

[1] **Count the number of carbon atoms** in the starting material and product.

[2] **Look at the functional groups** in the starting material and product.

- Determine what types of reactions can form the product.
- Determine what types of reactions the starting material can undergo.

[3] **Work backwards** from the product to make the starting material.

[4] Write out the synthesis in the synthetic direction.

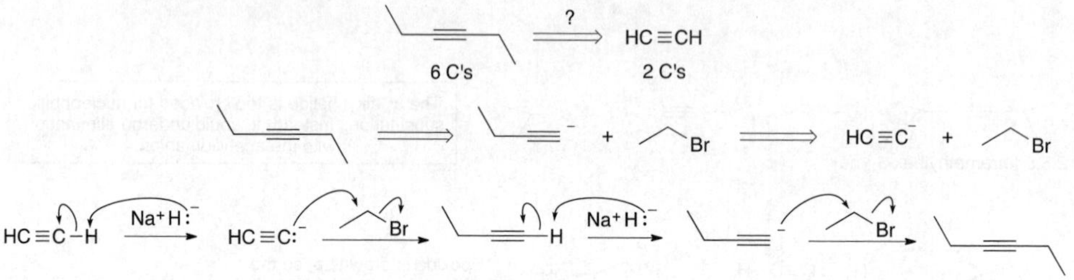

11.22

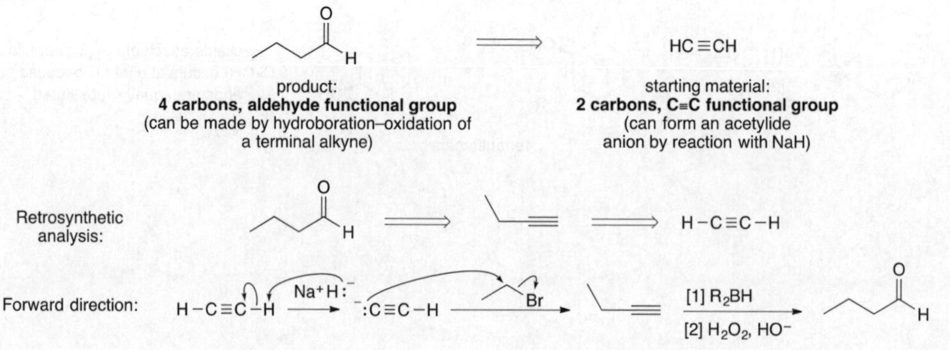

product:
4 carbons, aldehyde functional group
(can be made by hydroboration–oxidation of
a terminal alkyne)

starting material:
2 carbons, C≡C functional group
(can form an acetylide
anion by reaction with NaH)

Retrosynthetic
analysis:

Forward direction:

11.23

a.

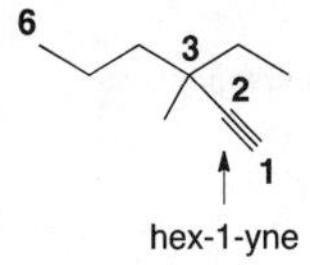
6
3
2
1
hex-1-yne

6 C alkyne → hexyne
C≡C at C1 → hex-1-yne
CH_3 and CH_2CH_3 at C3
3-ethyl-3-methylhex-1-yne

b.

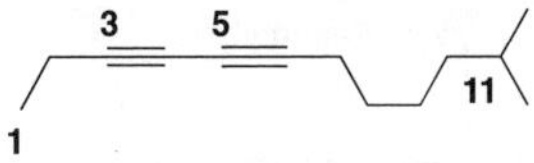
3
5
11
1

12 C chain with 2 C≡C's → dodecadiyne
C≡C's at C3 and C5 → dodeca-3,5-diyne
CH_3 at C11
11-methyldodeca-3,5-diyne

11.24

a.

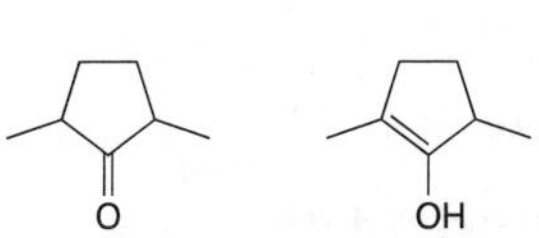
O
OH

keto form enol form

b.

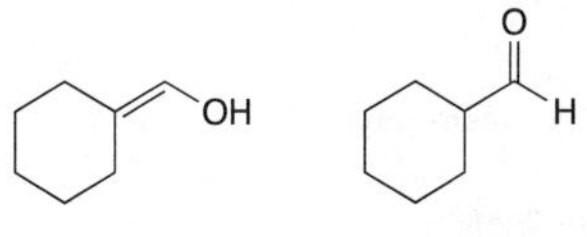
OH
O
H

enol form keto form

11.25

a, b.

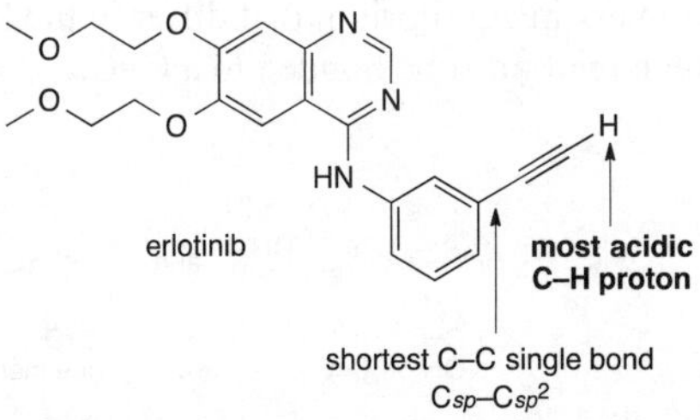
O
O
O
O
N
N
HN
H
erlotinib
most acidic
C–H proton
shortest C–C single bond
Csp–Csp2

c.
d. * = *sp* hybridized C
e. 3 < 1 < 2

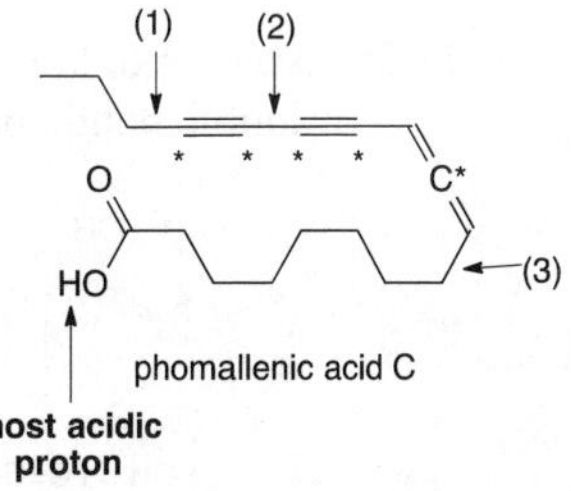
(1)
(2)
O
C*
HO
(3)
phomallenic acid C
most acidic
proton

11.26 Use the rules from Answer 11.2 to name the alkynes.

a.

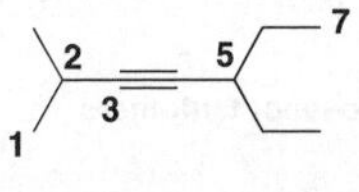
7
2
5
3
1

5-ethyl-2-methylhept-3-yne

b.

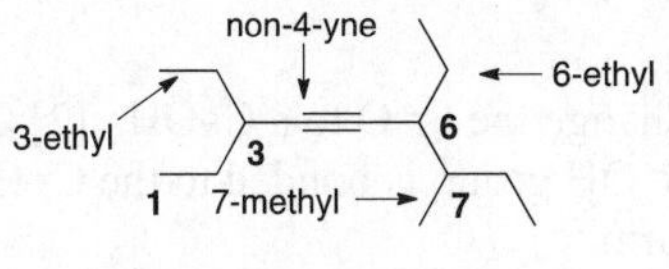
non-4-yne
6-ethyl
3-ethyl
3
6
1
7-methyl
7

3,6-diethyl-7-methylnon-4-yne

c.

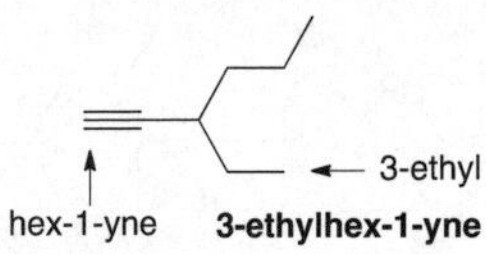
3-ethyl
hex-1-yne

3-ethylhex-1-yne

d.

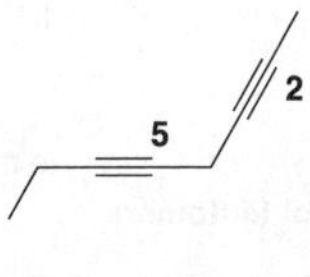
2
5

octa-2,5-diyne

e.

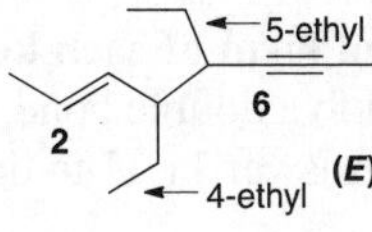
5-ethyl
6
2
4-ethyl

(*E*)-4,5-diethyldec-2-en-6-yne

f.

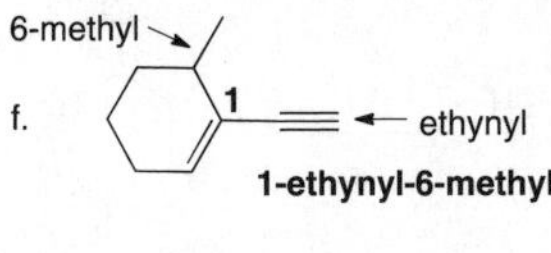
6-methyl
1
ethynyl

1-ethynyl-6-methylcyclohexene

Chapter 11–12

11.27 Use the directions from Answer 11.3 to draw each structure.

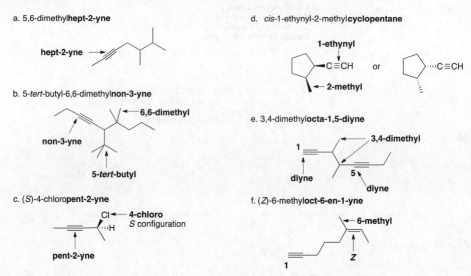

11.28 Keto–enol tautomers are constitutional isomers in equilibrium that differ in the location of a double bond and a hydrogen. The OH in an enol must be bonded to a C=C.

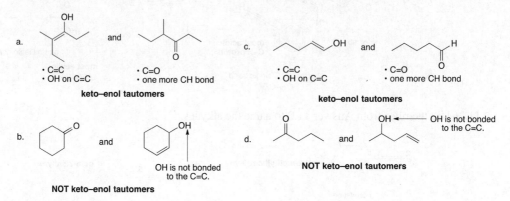

11.29 To draw the enol form of each keto form: [1] Change the C=O to a C–OH. [2] Change one single C–C bond to a double bond, making sure the OH group is bonded to the C=C. Use the directions from Answer 11.11 to draw each keto form.

Alkynes 11–13

11.30 Tautomers are constitutional isomers that are in equilibrium and differ in the location of a double bond and a hydrogen atom.

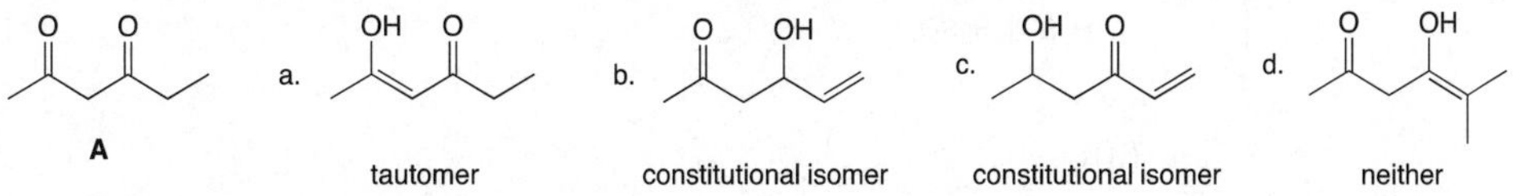

A

a. tautomer

b. constitutional isomer

c. constitutional isomer

d. neither

11.31

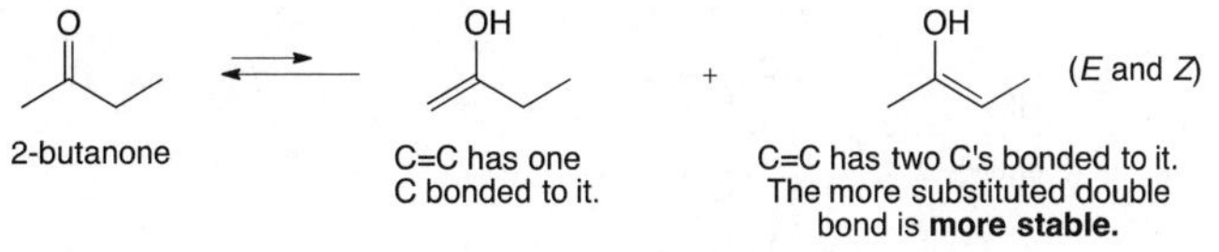

2-butanone

C=C has one C bonded to it.

C=C has two C's bonded to it. The more substituted double bond is **more stable.**

(*E* and *Z*)

11.32

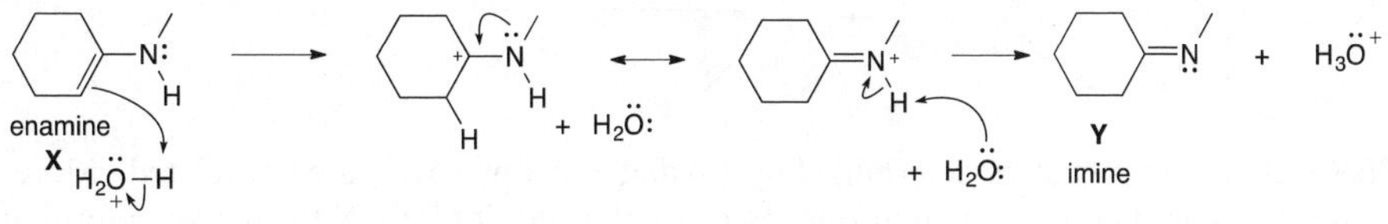

enamine
X

$H_2\ddot{O}$

+ $H_2\ddot{O}$:

+ $H_2\ddot{O}$:

Y
imine

+ $H_3\ddot{O}^+$

11.33

a. HCl (2 equiv)

b. HBr (2 equiv)

c. Cl_2 (2 equiv)

d. H_2O, H_2SO_4, $HgSO_4$

e. [1] R_2BH [2] H_2O_2, HO^-

f. NaH

g. [1] $^-NH_2$ [2] CH_3CH_2Br

h. [1] $^-NH_2$ [2] (epoxide) [3] H_2O

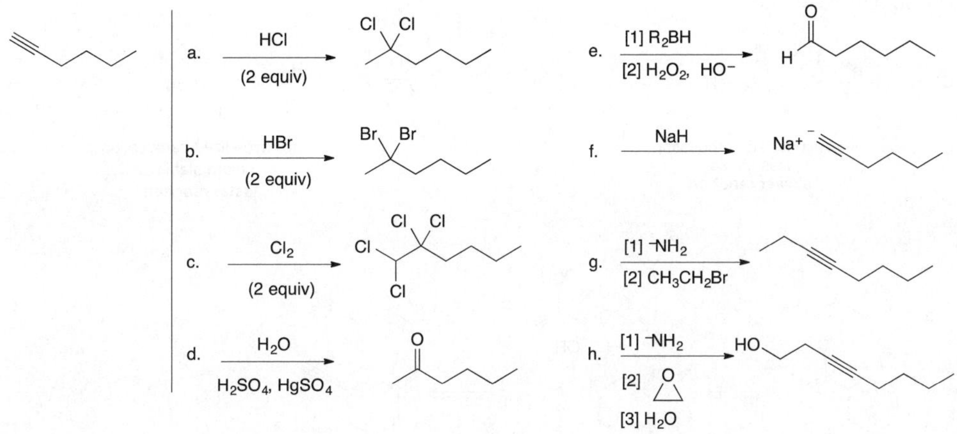

Chapter 11–14

11.34

a.

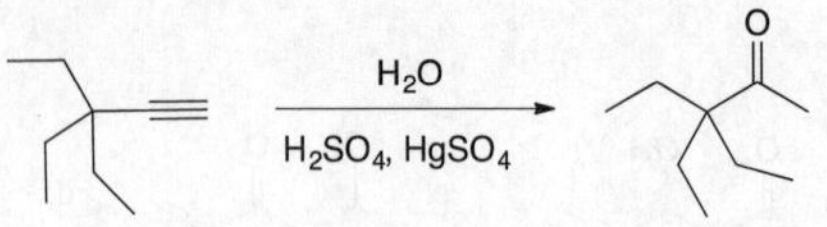

$$\xrightarrow[\text{H}_2\text{SO}_4,\ \text{HgSO}_4]{\text{H}_2\text{O}}$$

b.

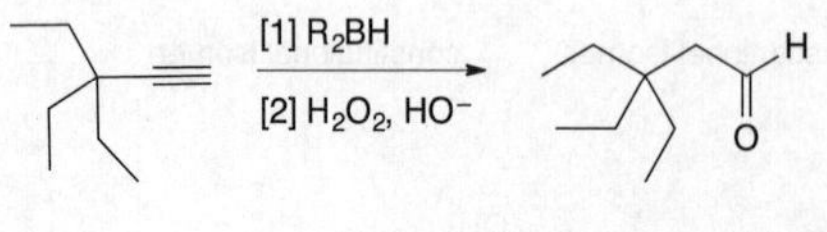

$$\xrightarrow[\text{[2]}\ \text{H}_2\text{O}_2,\ \text{HO}^-]{\text{[1]}\ \text{R}_2\text{BH}}$$

c.

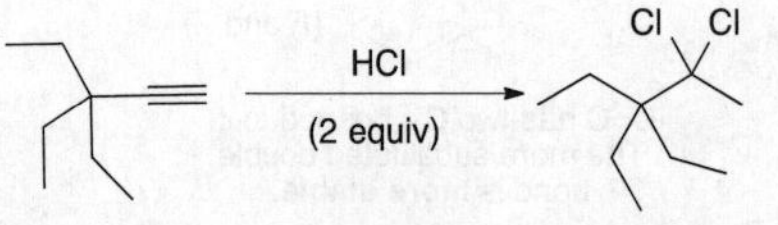

$$\xrightarrow[\text{(2 equiv)}]{\text{HCl}}$$

d.

$$\xrightarrow[\text{[2]}\ \text{CH}_3\text{CH}_2\text{Br}]{\text{[1]}\ \text{NaH}}$$

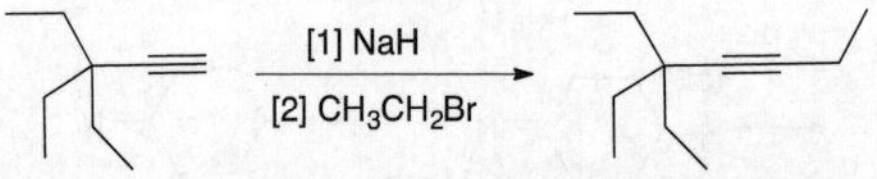

11.35 Reaction rate (which is determined by E_a) and enthalpy ($\Delta H°$) are not related. More exothermic reactions are not necessarily faster. Because the addition of HX to an alkene forms a more stable carbocation in an endothermic, rate-determining step, this carbocation is formed faster by the Hammond postulate.

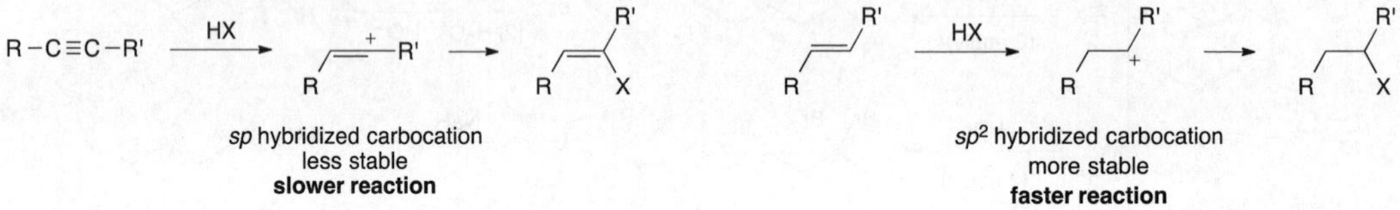

$$R-C\equiv C-R' \xrightarrow{\text{HX}} \qquad \longrightarrow \qquad$$

sp hybridized carbocation
less stable
slower reaction

$$\xrightarrow{\text{HX}} \qquad \longrightarrow \qquad$$

sp² hybridized carbocation
more stable
faster reaction

11.36

a.

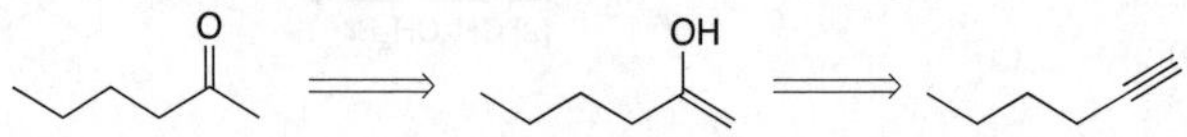

b.

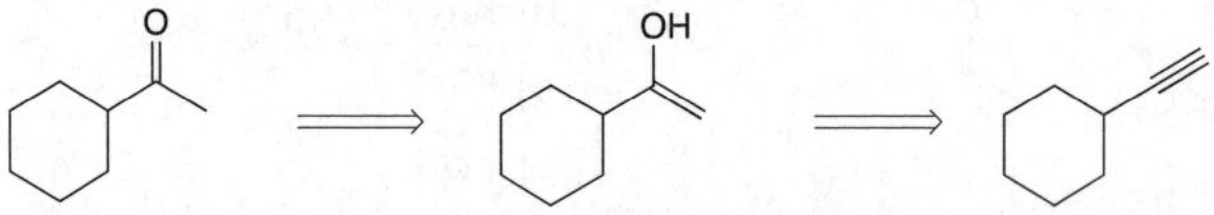

c.

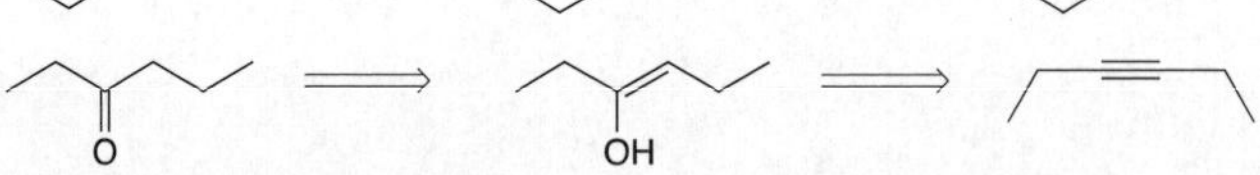

d.

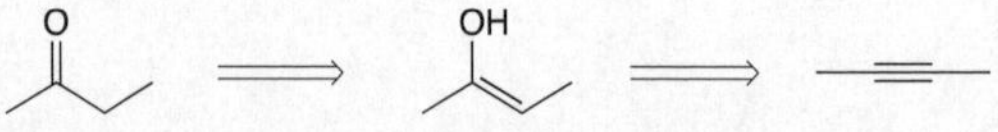

Alkynes 11–15

11.37

a.

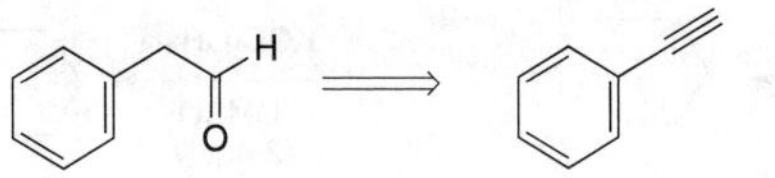

b.

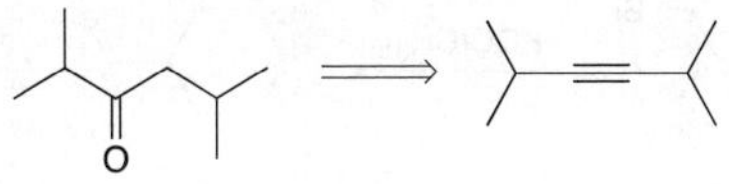

11.38

a. 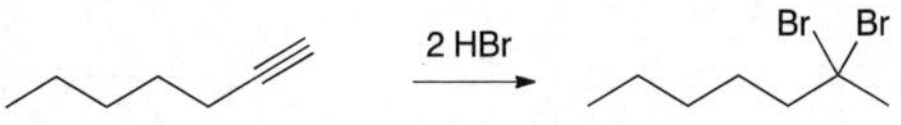
2 HBr

b. 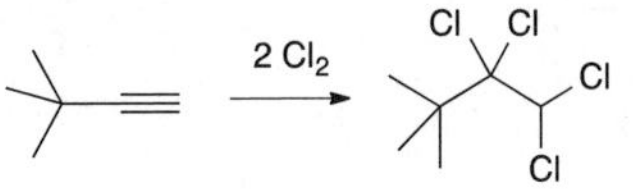
2 Cl₂

c. 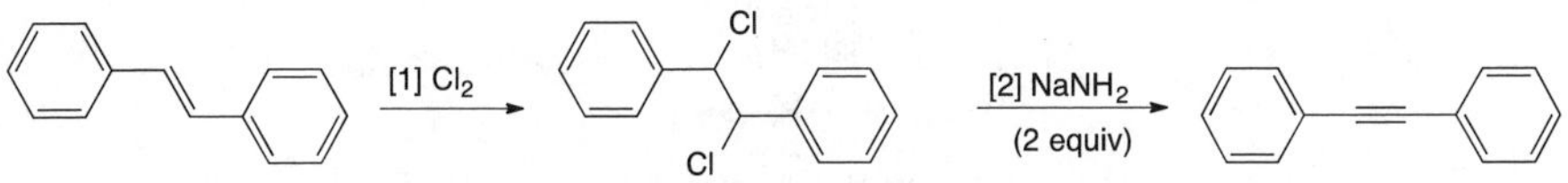
[1] Cl₂   [2] NaNH₂ (2 equiv)

d. 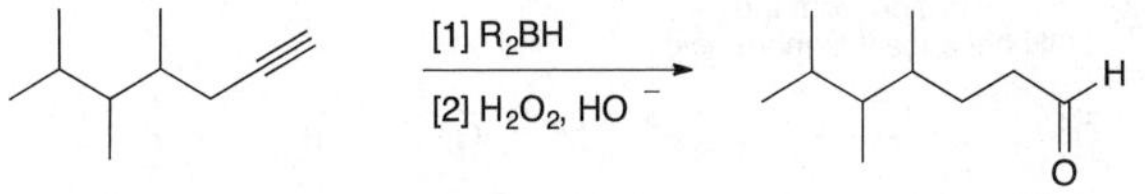
[1] R₂BH
[2] H₂O₂, HO⁻

e.   HC≡C⁻   +   D₂O ⟶ HC≡CD   +   DO⁻

f. 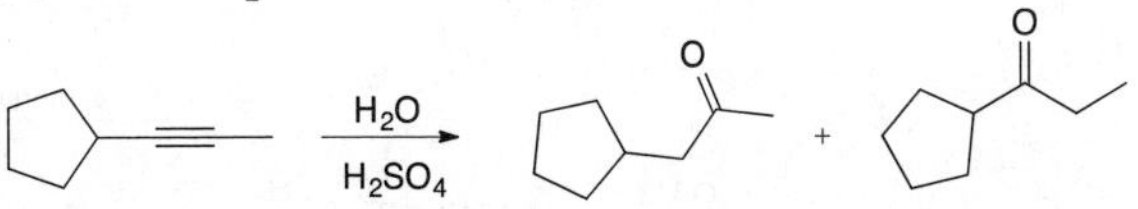
H₂O
H₂SO₄

g. 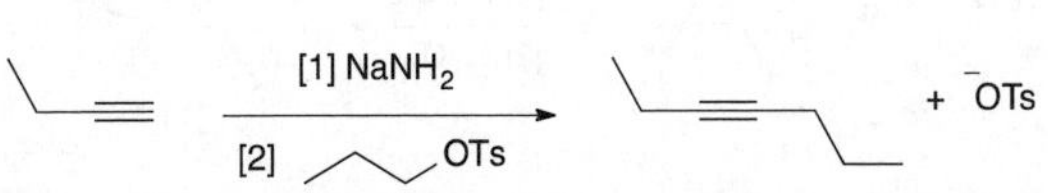
[1] NaNH₂
[2]   OTs
+ ⁻OTs

h.
[1] HC≡C⁻   [2] HO–H

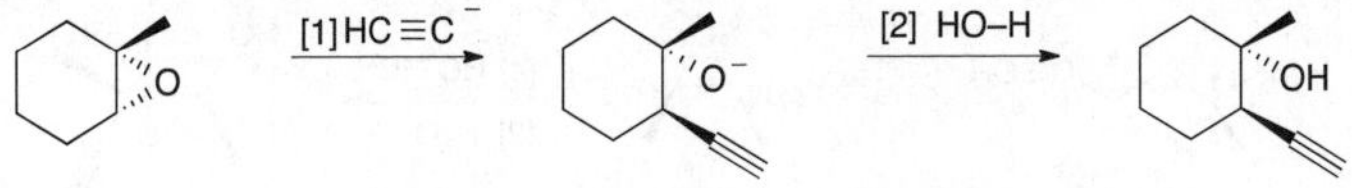

i. 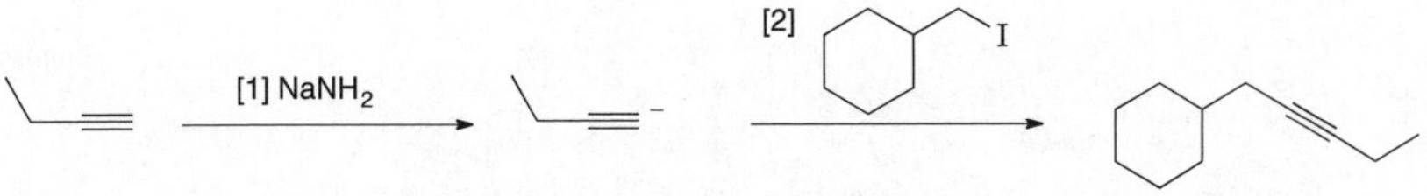
[1] NaNH₂   [2] cyclohexyl-CH₂I

j. 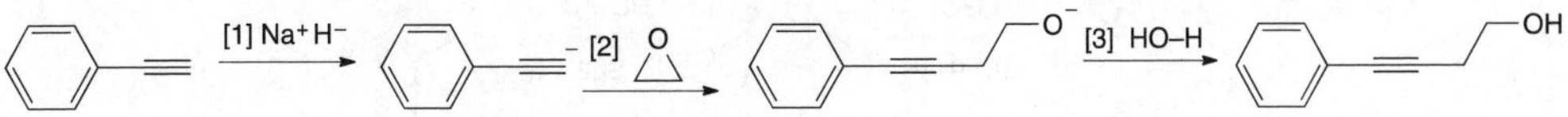
[1] Na⁺H⁻   [2] epoxide   [3] HO–H

11.39

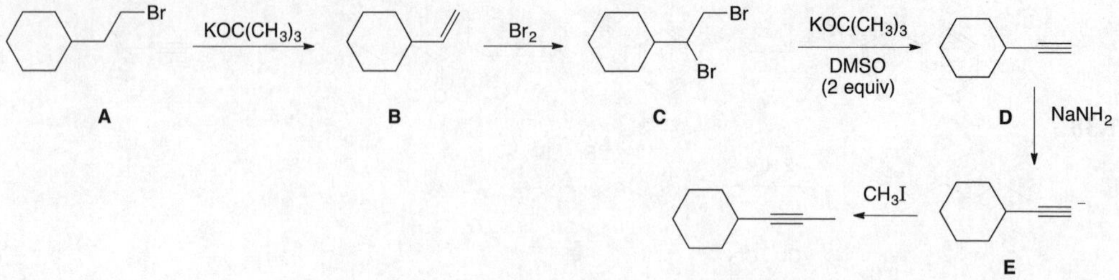
Br
KOC(CH3)3
Br2
Br
Br
KOC(CH3)3
DMSO
(2 equiv)
NaNH2
CH3I
A
B
C
D
E

11.40

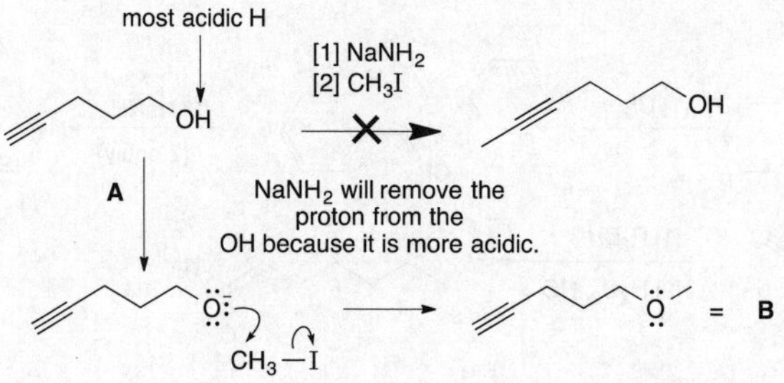
most acidic H
OH
[1] NaNH2
[2] CH3I
OH
A
NaNH2 will remove the
proton from the
OH because it is more acidic.
CH3—I
= B

11.41

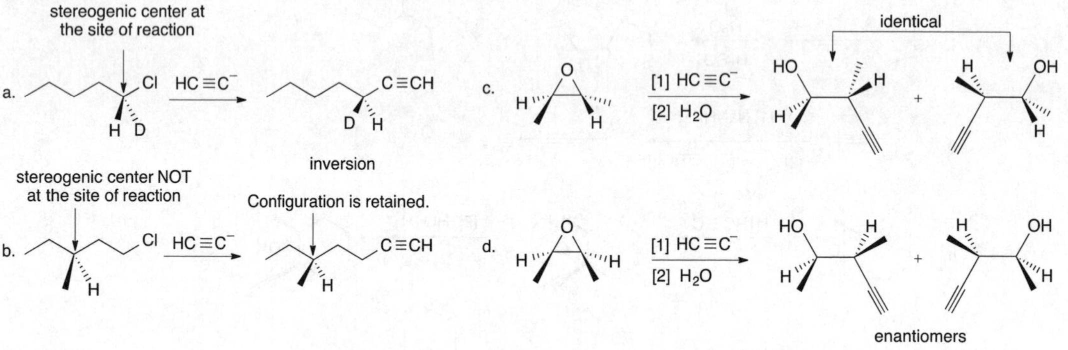
stereogenic center at
the site of reaction
a.
Cl
HC≡C⁻
C≡CH
H D
D H
inversion
stereogenic center NOT
at the site of reaction
b.
Cl
HC≡C⁻
Configuration is retained.
C≡CH
H
H
c.
[1] HC≡C⁻
[2] H2O
identical
HO
OH
d.
[1] HC≡C⁻
[2] H2O
HO
OH
enantiomers

11.42

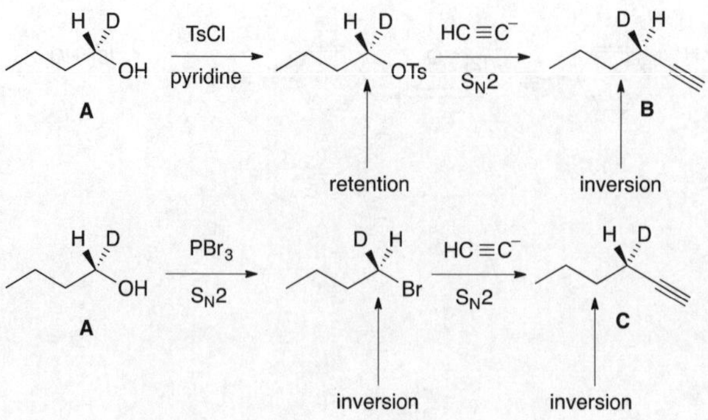
H D
OH
A
TsCl
pyridine
H D
OTs
HC≡C⁻
SN2
D H
B
retention
inversion
H D
OH
A
PBr3
SN2
D H
Br
HC≡C⁻
SN2
H D
C
inversion
inversion

11.43

11.44

11.45
A carbanion is more stable when its lone pair is in an orbital with a higher percentage of the smaller *s* orbital. A carbocation is more stable when its positive charge is due to a vacant orbital with a lower percentage of the smaller *s* orbital. In HC≡C⁺, the positively charged C uses two *p* orbitals to form two π bonds. If the σ bond is formed using an *sp* hybrid orbital, the second hybrid orbital would have to remain vacant, a highly unstable situation.

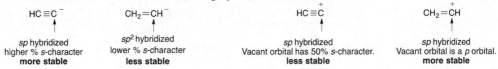

Chapter 11–18

11.46

a.

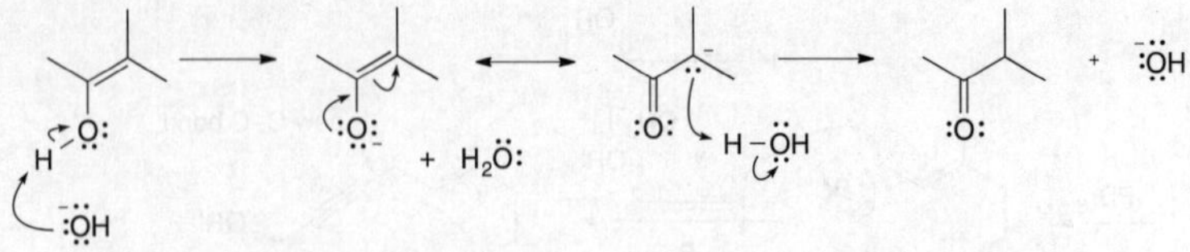

b.

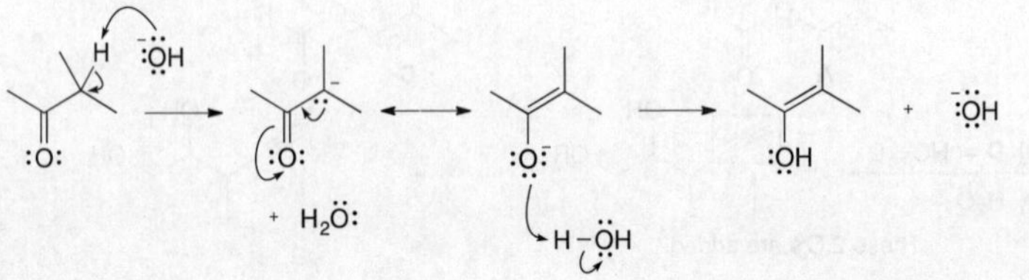

11.47

a.

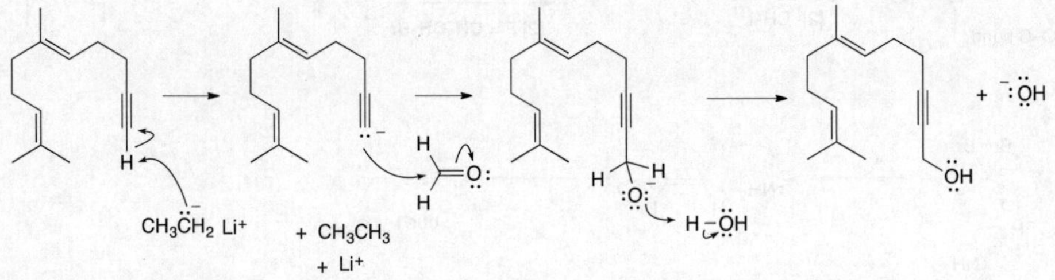

$CH_3\overset{-}{C}H_2$ Li^+ + CH_3CH_3
   + Li^+

b.

   + HSO_4^- resonance structures

   + H_2SO_4

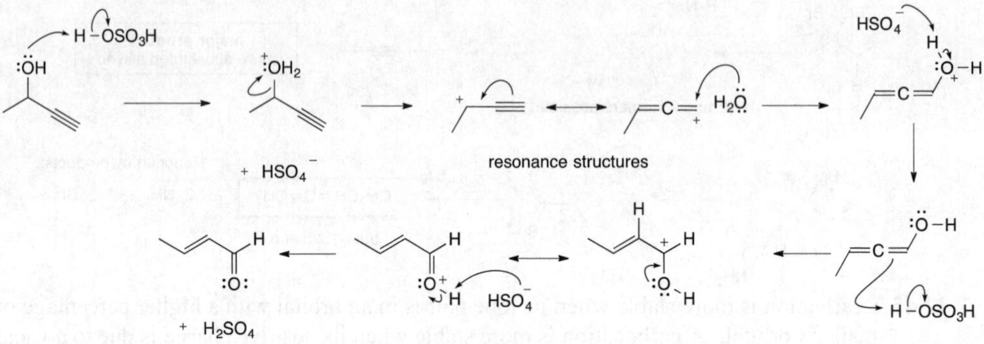

Alkynes 11–19

11.48

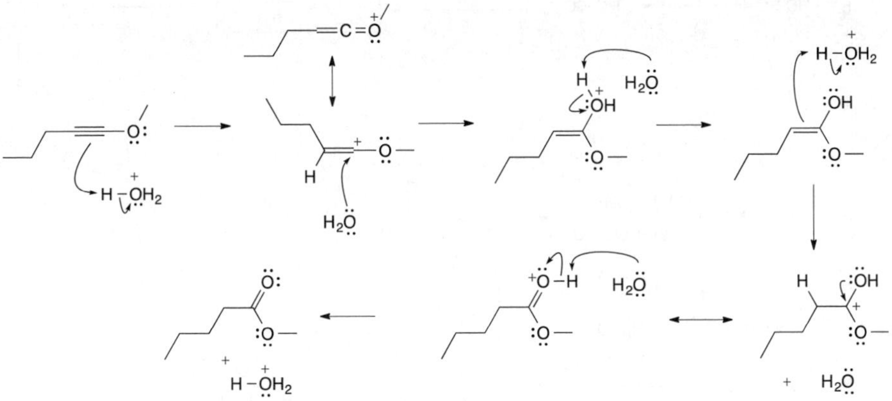

11.49

a.

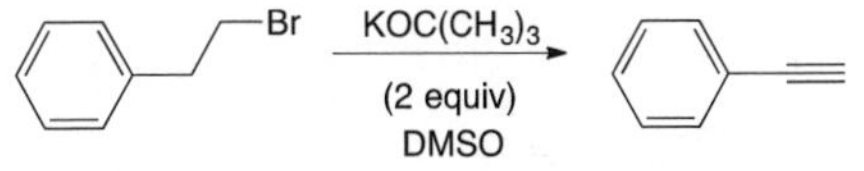

b.

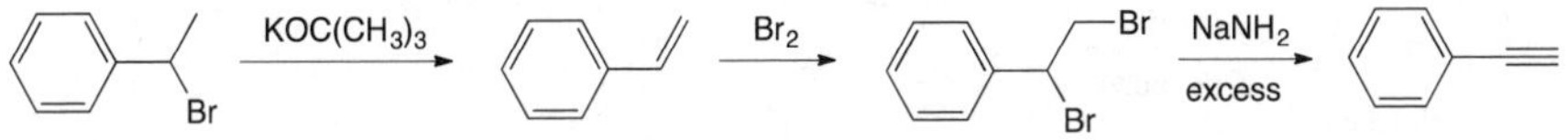

c.

11.50 The alkyl halides must be methyl or 1°.

a.

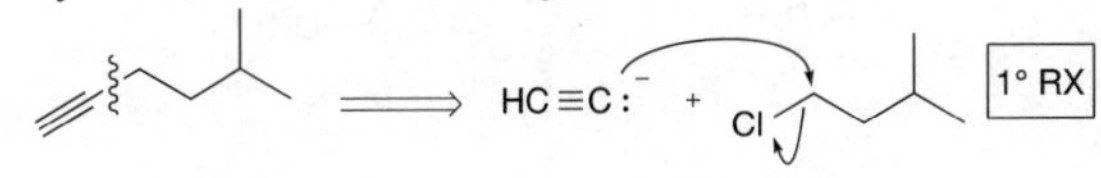

b.

c.

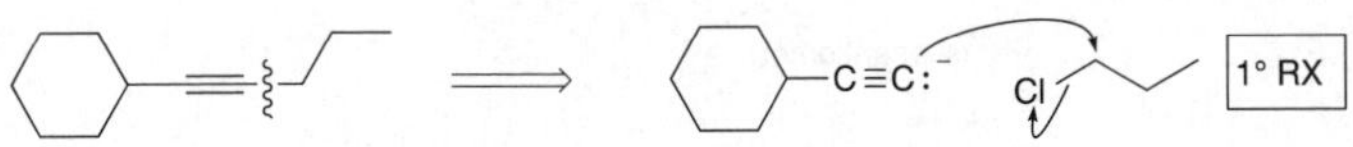

11.51

a.

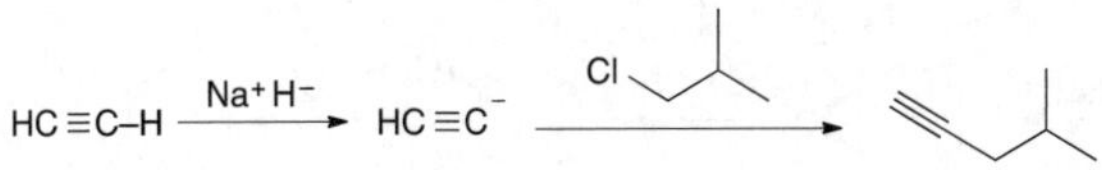

Chapter 11–20

b. $HC \equiv C–H$ $\xrightarrow{Na^+H^-}$ $HC \equiv C^-$ [reaction scheme with propyl chloride, alkyne intermediate, NaH, and isobutyl chloride]

c. [alkyne from b.] $\xrightarrow[\text{[2] }H_2O_2,\ HO^-]{\text{[1] }R_2BH}$ [aldehyde product]
(from b.)

d. [alkyne from b.] $\xrightarrow[\substack{H_2SO_4 \\ HgSO_4}]{H_2O}$ [ketone product]
(from b.)

e. [alkynide from b.] $\xrightarrow{\text{propyl Cl}}$ [internal alkyne] $\xrightarrow[H_2SO_4,\ HgSO_4]{H_2O}$ [ketone product]
(from b.)

11.52

a. [alkene] $\xrightarrow{Cl_2}$ [dichloride] $\xrightarrow{2\ ^-NH_2}$ [alkyne]

b. [alkyne] $\xrightarrow[\text{(2 equiv)}]{HBr}$ [dibromide]
(from a.)

c. [diene] $\xrightarrow{Br_2}$ [dibromide]

d. [alkyne] $\xrightarrow{NaH}$ [alkynide] $\xrightarrow[\text{[2] }H_2O]{\text{[1] epoxide}}$ [alkynol with OH]
(from a.)

e. [alkynide] $\xrightarrow[\text{[2] }H_2O]{\text{[1] epoxide}}$ [cyclohexanol product] (+ enantiomer)
(from d.)

11.53

[cyclohexene] $\xrightarrow[H_2O]{Br_2}$ [bromohydrin] $\xrightarrow{NaH}$ [epoxide + $:C\equiv CH$ from $HC\equiv CH$ / NaH] $\longrightarrow$ [alkynyl cyclohexanolate] $\xrightarrow{H_2O}$ [alkynyl cyclohexanol]

+ enantiomer   + enantiomer

Alkynes 11–21

11.54

a. $HC \equiv CH$ — NaH → $HC \equiv C^-$ — (hexyl bromide) → (alkyne) — NaH → (alkyne anion $^-$) — $\downarrow$ (ethyl Br) → (product)

b. (from a.) — [1] epoxide / [2] H_2O → (alkynol, OH) — [1] NaH / [2] ethyl Br → (ether product)

11.55

(ethyl Br) — $K^+ {}^-OC(CH_3)_3$ → $CH_2 = CH_2$ — Br_2 → (BrCH$_2$CH$_2$Br) — $NaNH_2$ (2 equiv) → $HC \equiv CH$ — NaH → $HC \equiv C^-$ — $\downarrow$ (ethyl Br)

(ketone product) ← H_2O / H_2SO_4, $HgSO_4$ — (alkyne) ← (ethyl Br) — (alkyne anion $^-$) ← NaH — (alkyne)

11.56

a. (propanol, OH) — H_2SO_4 → (propene) — Br_2 → (dibromide, Br, Br) — $2\ {}^-NH_2$ → (alkyne) — NaH → (alkyne anion $^-$) → (product)

(propanol, OH) — $SOCl_2$ → (propyl Cl)

b. (propene, from a.) — Cl_2 / H_2O → (chlorohydrin, OH, Cl) — NaH → (epoxide, O) — [1] (alkyne anion $^-$, from a.) / [2] H_2O → (product, HO)

11.57

(ethanol, OH) — H_2SO_4 → $CH_2 = CH_2$ — Br_2 → (BrCH$_2$CH$_2$Br) — $NaNH_2$ (2 equiv) → $HC \equiv CH$ — NaH → $HC \equiv C^-$ — $\downarrow$

(ethyl Br) ← PBr_3 — (ethanol, OH)

[1] NaH / [2] epoxide O / [3] H_2O — $HC \equiv C$ — → (product, HO)

(ethanol, OH) — H_2SO_4 → $CH_2 = CH_2$ — Br_2 / H_2O → (bromohydrin, Br, OH) — NaH → (epoxide, O) →

Chapter 11–22

11.58 Two resonance structures can be drawn for an enol.

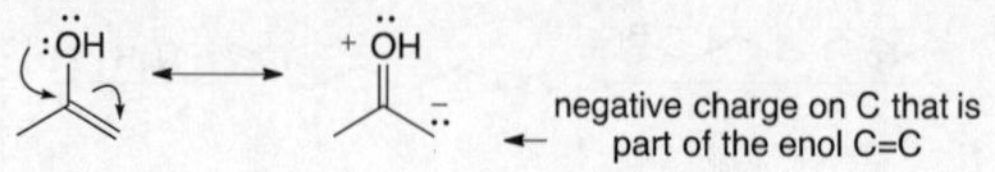

negative charge on C that is
part of the enol C=C

Because the second resonance structure places an electron pair (and therefore a negative charge) on an enol carbon, the C=C is more nucleophilic than the C=C of an alkene for which no additional resonance forms can be drawn. Thus, the OH group donates electron density to the C=C by a resonance effect.

11.59

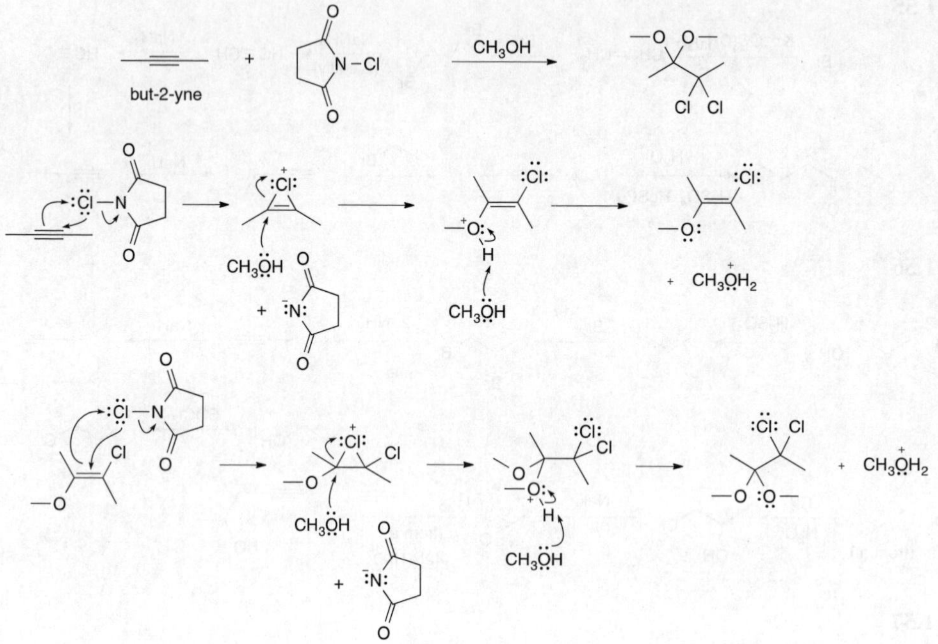

but-2-yne

CH₃OH

11.60

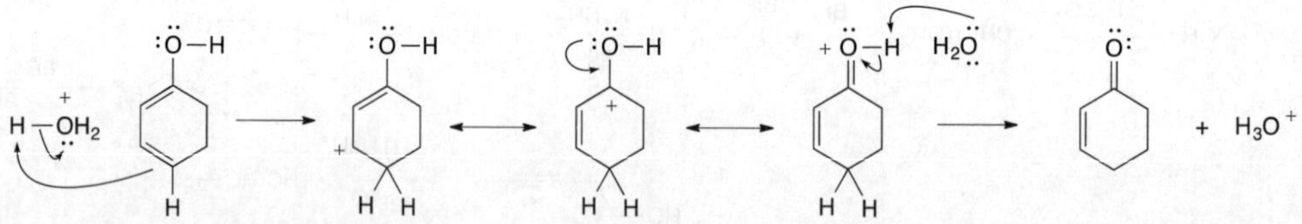

Alkynes 11–23

11.61

$H_2\ddot{O}$ + Br⁻

+ $H_2\ddot{O}$

+ $H_3\ddot{O}^+$ + $H_2\ddot{O}$:

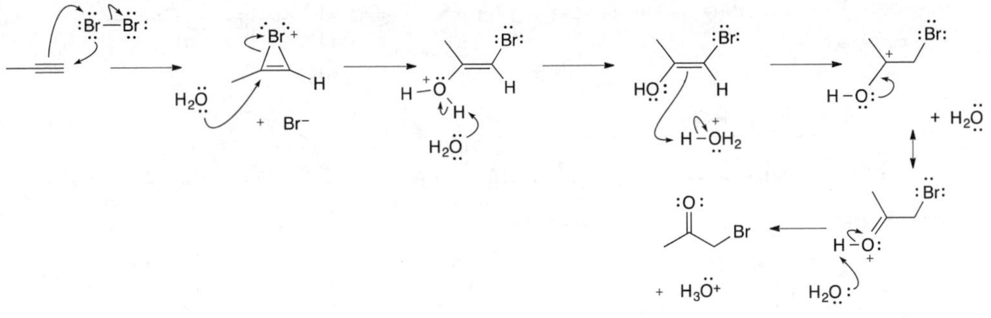

11.62

Only this carbocation forms because it is resonance stabilized. The positive charge is delocalized on oxygen.

TsOH

Ts$\ddot{O}$⁻H

+ TsO⁻

(not formed)   not resonance stabilized

re-draw

$CH_3\ddot{O}H$

$CH_3\ddot{O}H$

X   NOT   Y

+ $CH_3\overset{+}{\underset{}{O}}H_2$

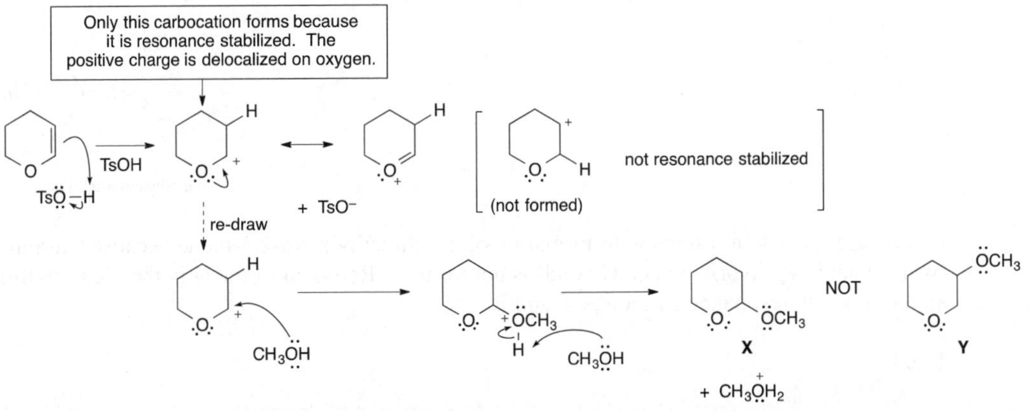

11.63 A more stable internal alkyne can be isomerized to a less stable terminal alkyne under these reaction conditions because when $CH_3CH_2C{\equiv}CH$ is first formed, it contains an *sp* hybridized C–H bond, which is more acidic than any proton in $CH_3–C{\equiv}C–CH_3$. Under the reaction conditions, this proton is removed with base. Formation of the resulting acetylide anion drives the equilibrium to favor its formation. Protonation of this acetylide anion gives the less stable terminal alkyne.

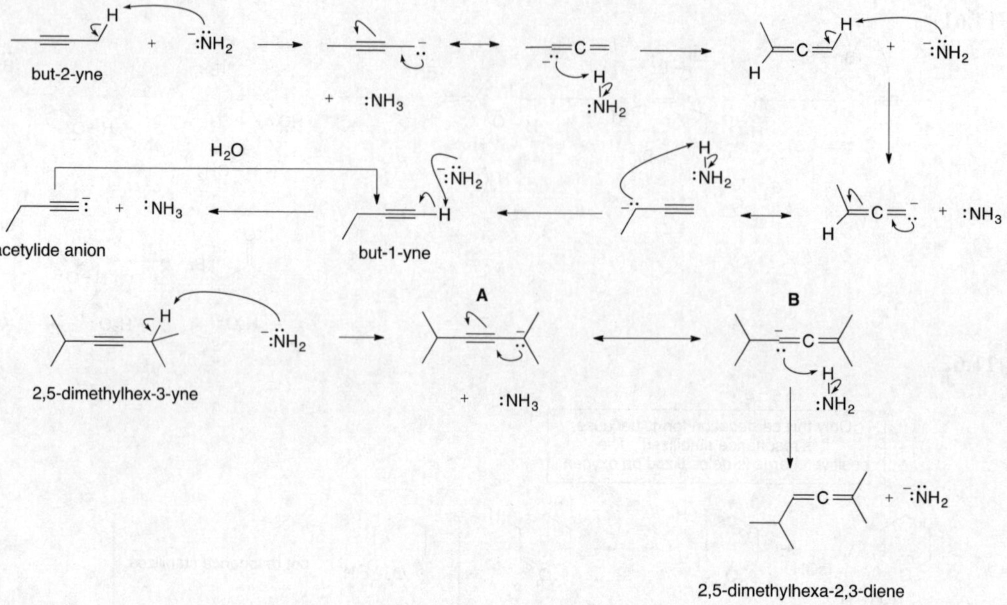

Chapter 11–24

but-2-yne

+ :NH$_2^-$ → ⟷ → + :NH$_2^-$

+ :NH$_3$

:NH$_2$

H$_2$O

acetylide anion + :NH$_3$ ← ⟷ ← ⟷ + :NH$_3$

but-1-yne

:NH$_2$

:NH$_2$

2,5-dimethylhex-3-yne

:NH$_2^-$ → ⟷

A

B

+ :NH$_3$

:NH$_2$

+ :NH$_2^-$

2,5-dimethylhexa-2,3-diene

In this case the reaction stops with formation of 2,5-dimethylhexa-2,3-diene because a terminal alkyne (with an acidic *sp* hybridized C–H bond) is not formed. Removal of the H in the diene re-forms the anion shown in resonance structures **A** and **B**.

11.64

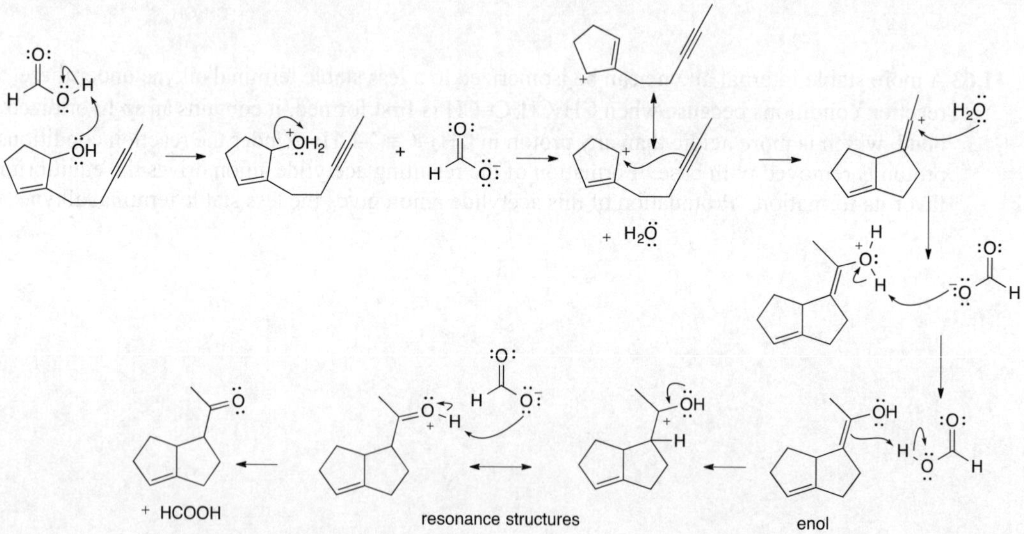

+ H$_2$Ö

+ H$_2$Ö

+ HCOOH

resonance structures

enol

Alkynes 11–25

11.65 In the presence of acid, (R)-α-methylbutyrophenone enolizes to form an achiral enol.

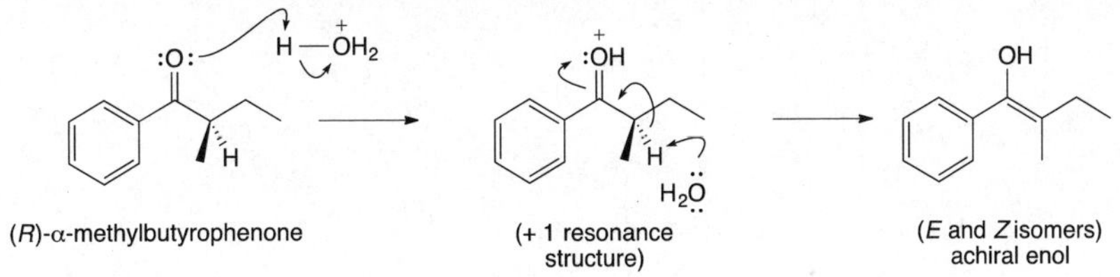

(*R*)-α-methylbutyrophenone   (+ 1 resonance
                                      structure)

(*E* and *Z* isomers)
achiral enol

The achiral enol can then be protonated from above or below the plane to form a racemic mixture that is optically inactive.

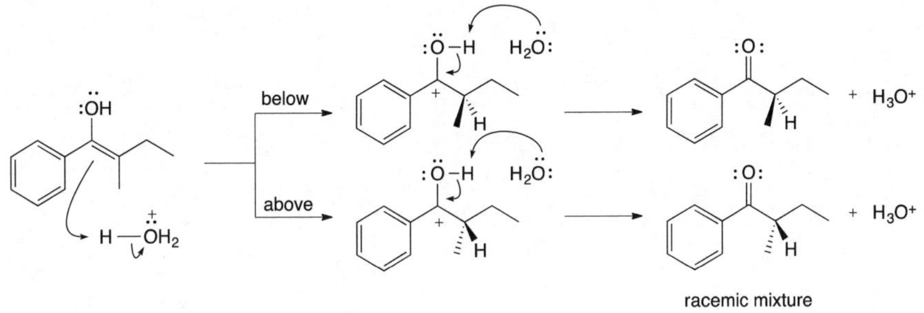

racemic mixture

Oxidation and Reduction 12–1

Chapter 12 Oxidation and Reduction

Chapter Review

Summary: Terms that describe reaction selectivity

- A **regioselective reaction** forms predominately or exclusively one constitutional isomer (Section 8.5).

major product
trisubstituted alkene

minor product
disubstituted alkene

- A **stereoselective reaction** forms predominately or exclusively one stereoisomer (Section 8.5).

Na⁺ ⁻OCH₂CH₃

trans alkene
major product

cis alkene
minor product

- An **enantioselective reaction** forms predominately or exclusively one enantiomer (Section 12.15).

allylic alcohol

Sharpless
reagent

or

One enantiomer is favored.

Definitions of oxidation and reduction

Oxidation reactions result in:
- an increase in the number of C–Z bonds, *or*
- a decrease in the number of C–H bonds.

Reduction reactions result in:
- a decrease in the number of C–Z bonds, *or*
- an increase in the number of C–H bonds.

[Z = an element more electronegative than C]

Reduction reactions

[1] Reduction of alkenes—Catalytic hydrogenation (12.3)

H_2

Pd, Pt, or Ni

alkane

- **Syn addition** of H_2 occurs.
- Increasing alkyl substitution on the C=C decreases the rate of reaction.

Chapter 12–2

[2] Reduction of alkynes

a. R−≡−R →(2 H₂, Pd-C) alkane
- Two equivalents of H₂ are added and four new C–H bonds are formed (12.5A).

b. R−≡−R →(H₂, Lindlar catalyst) cis alkene
- **Syn addition** of H₂ occurs, forming a **cis** alkene (12.5B).
- The Lindlar catalyst is deactivated so that reaction stops after one equivalent of H₂ has been added.

c. R−≡−R →(Na, NH₃) trans alkene
- **Anti addition** of H₂ occurs, forming a **trans** alkene (12.5C).

[3] Reduction of alkyl halides (12.6)

R−X →([1] LiAlH₄, [2] H₂O) R−H alkane
- The reaction follows an **S_N2** mechanism.
- CH₃X and RCH₂X react faster than more substituted RX.

[4] Reduction of epoxides (12.6)

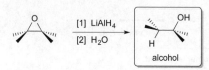

- The reaction follows an **S_N2** mechanism.
- In unsymmetrical epoxides, H⁻ (from LiAlH₄) attacks at the less substituted carbon.

Oxidation reactions

[1] Oxidation of alkenes

a. Epoxidation (12.8)

>=< + RCO₃H → epoxide
- The mechanism has **one step**.
- **Syn addition** of an O atom occurs.
- The reaction is stereospecific.

b. Anti dihydroxylation (12.9A)

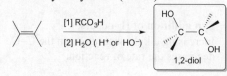

- Ring opening of an epoxide intermediate with ⁻OH or H₂O forms a 1,2-diol with two OH groups added in an **anti** fashion.

Oxidation and Reduction 12–3

c. Syn dihydroxylation (12.9B)

[1] OsO$_4$; [2] NaHSO$_3$, H$_2$O

or

[1] OsO$_4$, NMO; [2] NaHSO$_3$, H$_2$O

or

KMnO$_4$, H$_2$O, HO$^-$

1,2-diol

- Each reagent adds two new C–O bonds to the C=C in a **syn** fashion.

d. Oxidative cleavage (12.10)

[1] O$_3$

[2] Zn, H$_2$O or CH$_3$SCH$_3$

ketone   aldehyde

- Both the σ and π bonds of the alkene are cleaved to form two carbonyl groups.

[2] Oxidative cleavage of alkynes (12.11)

a.

R—≡—R'
internal alkyne

[1] O$_3$

[2] H$_2$O

carboxylic acids

- The σ bond and both π bonds of the alkyne are cleaved.

b.

R—≡—H
terminal alkyne

[1] O$_3$

[2] H$_2$O

+ CO$_2$

[3] Oxidation of alcohols (12.12, 12.13)

a.

R   OH
1° alcohol

PCC

or

HCrO$_4$–
Amberlyst A-26
resin

aldehyde

- Oxidation of a 1° alcohol with PCC or HCrO$_4^-$ (Amberlyst A-26 resin) stops at the aldehyde stage. Only one C–H bond is replaced by a C–O bond.

b.

R   OH
1° alcohol

CrO$_3$

H$_2$SO$_4$, H$_2$O

carboxylic acid

- Oxidation of a 1° alcohol under harsher reaction conditions—CrO$_3$ (or Na$_2$Cr$_2$O$_7$ or K$_2$Cr$_2$O$_7$) + H$_2$O + H$_2$SO$_4$—affords a RCO$_2$H. Two C–H bonds are replaced by two C–O bonds.

c.

OH
R   R
2° alcohol

PCC or CrO$_3$

or

HCrO$_4$–
Amberlyst A-26
resin

ketone

- A 2° alcohol has only one C–H bond on the carbon bearing the OH group, so all Cr^{6+} reagents—PCC, CrO$_3$, Na$_2$Cr$_2$O$_7$, K$_2$Cr$_2$O$_7$, or HCrO$_4^-$ (Amberlyst A-26 resin)— oxidize a 2° alcohol to a ketone.

[4] Asymmetric epoxidation of allylic alcohols (12.15)

OH

(CH$_3$)$_3$C−OOH

Ti[OCH(CH$_3$)$_2$]$_4$

with (−)-DET   or   with (+)-DET

Chapter 12–4

Practice Test on Chapter Review

1.a. Compound **X** has a molecular formula of C_9H_{12} and contains no triple bonds. **X** is hydrogenated to a compound of molecular formula C_9H_{14} with excess H_2 and a palladium catalyst. What can be said about **X**?

     1. **X** has four rings.   4. **X** has one ring and three double bonds.
     2. **X** has three rings and one double bond.   5. **X** has four double bonds.
     3. **X** has two rings and two double bonds.

  b. Syn addition to an alkene occurs exclusively with which reagents?

     1. OsO_4
     2. $KMnO_4$, H_2O, ^-OH
     3. mCPBA, then H_2O, ^-OH
     4. Both reagents (1) and (2) give syn addition exclusively.
     5. Reagents (1), (2), and (3) give syn addition exclusively.

  c. Which of the following reagents adds to an alkene exclusively in an anti fashion?

     1. Br_2 4. Reagents (1) and (2) both add in an anti fashion.
     2. H_2, Pd-C   5. Reagents (1), (2), and (3) all add in an anti fashion.
     3. BH_3, then H_2O_2, ^-OH

2. Label each statement as True (T) or False (F).

    a. PCC oxidizes 1° alcohols to aldehydes.
    b. CrO_3 oxidizes 2° alcohols to ketones.
    c. Treatment of hex-2-yne with Na in NH_3 forms *cis*-hex-2-ene.
    d. Reduction of propene oxide with $LiAlH_4$ forms propan-1-ol.
    e. Ozonolysis of 2-methyloct-2-ene forms one ketone and one aldehyde.
    f. mCPBA is an oxidizing agent that converts alkenes to trans diols.
    g. Oct-1-en-5-yne reacts with H_2 and Pd-C, but does not react with H_2 and Lindlar catalyst.
    h. Treatment of cyclohexene with OsO_4 affords an optically inactive product mixture that contains two enantiomers.

3. Label each reagent as an oxidizing agent, reducing agent, or neither.

    a. O_3
    b. $LiAlH_4$
    c. mCPBA
    d. H_2O, H_2SO_4
    e. PCC
    f. Na, NH_3

Oxidation and Reduction 12–5

4. Draw the organic products formed in each reaction and indicate stereochemistry when necessary.

a. 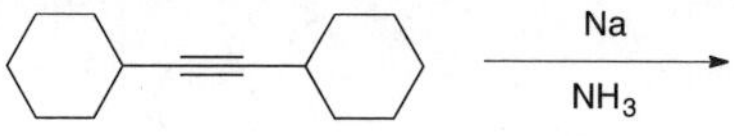
$$\xrightarrow[\text{NH}_3]{\text{Na}}$$

b. 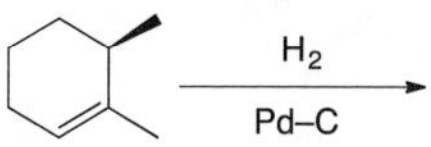
$$\xrightarrow[\text{Pd–C}]{\text{H}_2}$$

c.
$$\xrightarrow{\text{PCC}}$$

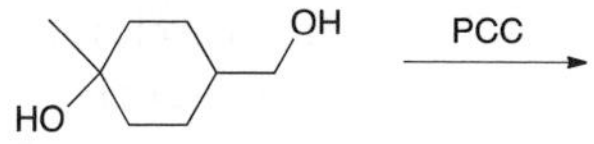

d. 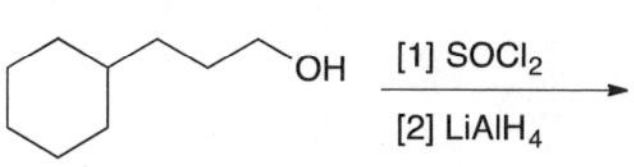
$$\xrightarrow[\text{[2] LiAlH}_4]{\text{[1] SOCl}_2}$$

5 a. Fill in the appropriate starting material (including any needed stereochemistry) in the following reaction.

$$\xrightarrow[\text{[2] H}_2\text{O, NaHSO}_3]{\text{[1] OsO}_4}$$ + enantiomer

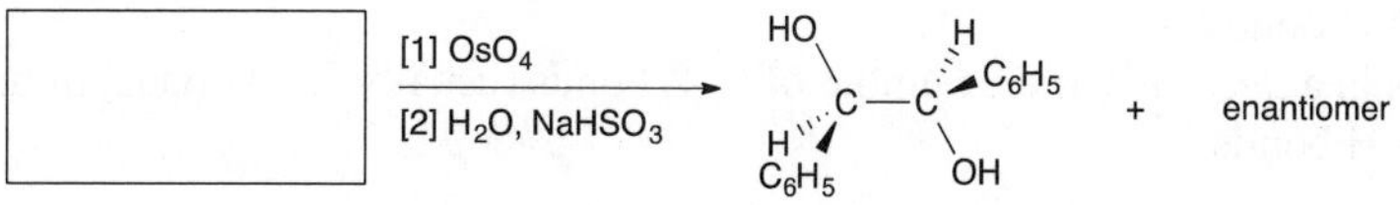

b. Fill in the appropriate reagent in the following reaction.

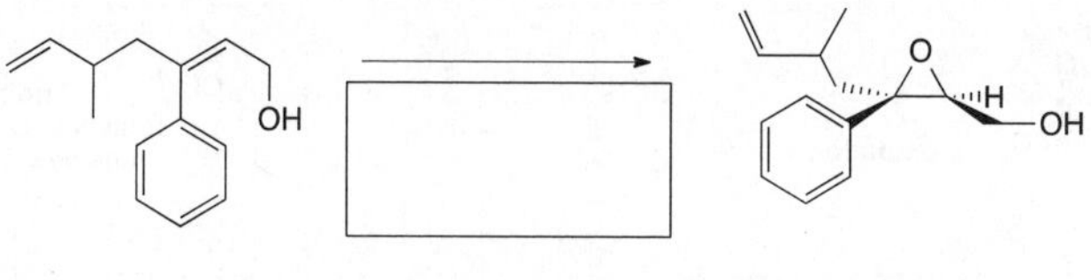

c. What starting material is needed for the following reaction?

$$\xrightarrow[\text{[2] (CH}_3)_2\text{S}]{\text{[1] O}_3}$$

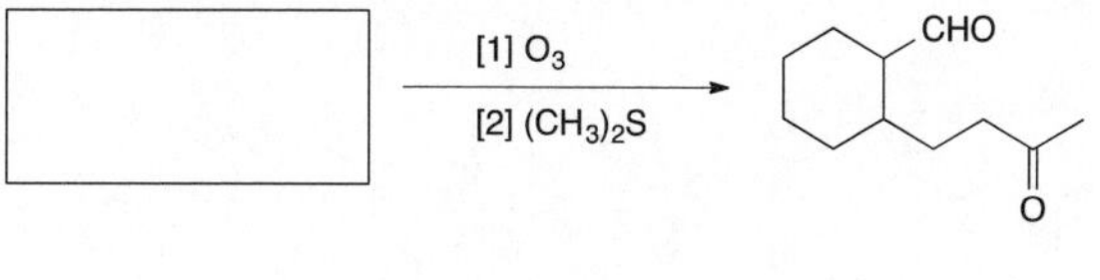

Chapter 12–6

Answers to Practice Test

1.a. 2
  b. 4
  c. 1

2.a. T
  b. T
  c. F
  d. F
  e. T
  f. F
  g. F
  h. F

3.a. oxidizing
  b. reducing
  c. oxidizing
  d. neither
  e. oxidizing
  f. reducing

4.
  a.
  b.
  c.
  d.

5.
  a.
  b. Sharpless reagent
     (+)-DET
  c.

Chapter 12: Answers to Problems

12.1 *Oxidation* results in an *increase* in the number of C–Z bonds (usually C–O bonds) *or* a *decrease* in the number of C–H bonds.
Reduction results in a *decrease* in the number of C–Z bonds (usually C–O bonds) *or* an *increase* in the number of C–H bonds.

a. **reduction**

b. **oxidation**

c. **oxidation**
    (two new C–H bonds, three new C–O bonds)

d. **neither**
    (one new C–H bond and one new C–Cl bond)

12.2 **Hydrogenation** is the addition of hydrogen. When alkenes are hydrogenated, they are *reduced* by the addition of H_2 to the π bond. To draw the alkane product, add a H to each C of the double bond.

a.

b.

c.

12.3 Draw the alkenes that form each alkane when hydrogenated.

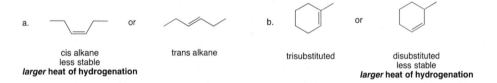

12.4 Cis alkenes are less stable than trans alkenes, so they have larger heats of hydrogenation. Increasing alkyl substitution increases the stability of a C=C, thus decreasing the heat of hydrogenation.

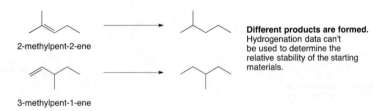

12.5 Hydrogenation products must be identical to use hydrogenation data to evaluate the relative stability of the starting materials.

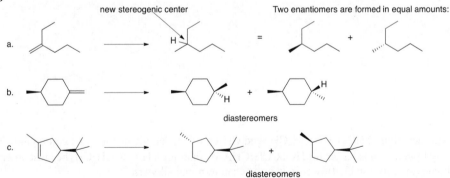

12.6

12.7

Compound	Molecular formula before hydrogenation	Molecular formula after hydrogenation	Number of rings	Number of π bonds
A	C$_{10}$H$_{12}$	C$_{10}$H$_{16}$	3	2
B	C$_4$H$_8$	C$_4$H$_{10}$	0	1
C	C$_6$H$_8$	C$_6$H$_{12}$	1	2

12.8

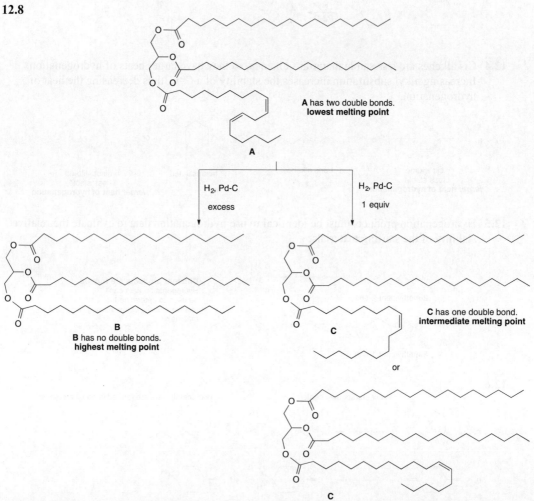

12.9
Hydrogenation of HC≡CCH$_2$CH$_2$CH$_3$ and CH$_3$C≡CCH$_2$CH$_3$ yields the same compound. The heat of hydrogenation is larger for HC≡CCH$_2$CH$_2$CH$_3$ than for CH$_3$C≡CCH$_2$CH$_3$ because internal alkynes are more stable (lower in energy) than terminal alkynes.

Oxidation and Reduction 12–9

12.10 In the presence of the Lindlar catalyst, alkynes react with H_2 to form cis alkenes. Alkenes do not react with the Lindlar catalyst.

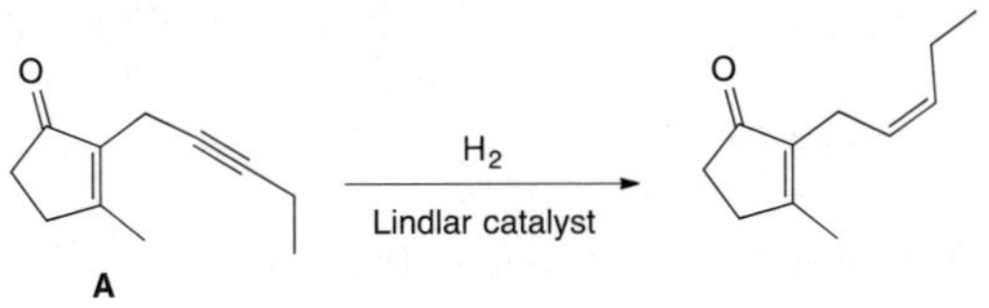

cis-jasmone

(perfume component isolated from jasmine flowers)

12.11 In the presence of Pd-C, H_2 adds to alkenes and alkynes to form alkanes. In the presence of the Lindlar catalyst, alkynes react with H_2 to form cis alkenes. Alkenes do not react with the Lindlar catalyst.

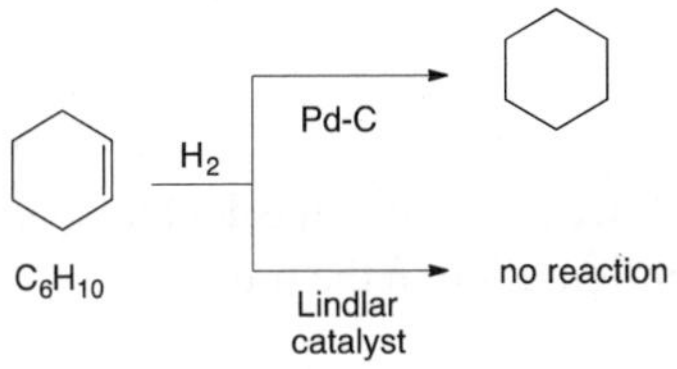

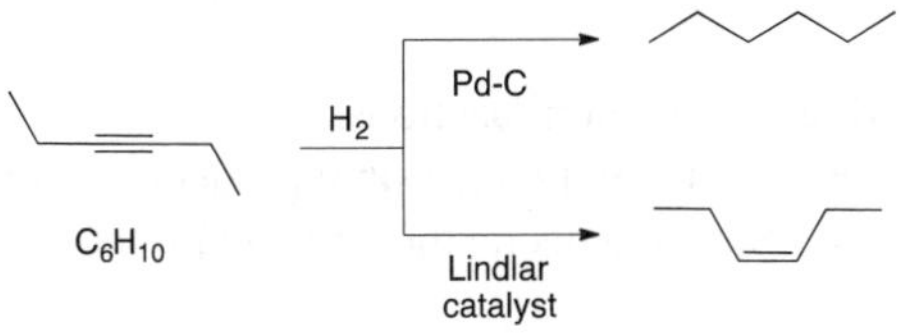

12.12 Use the directions from Answer 12.11.

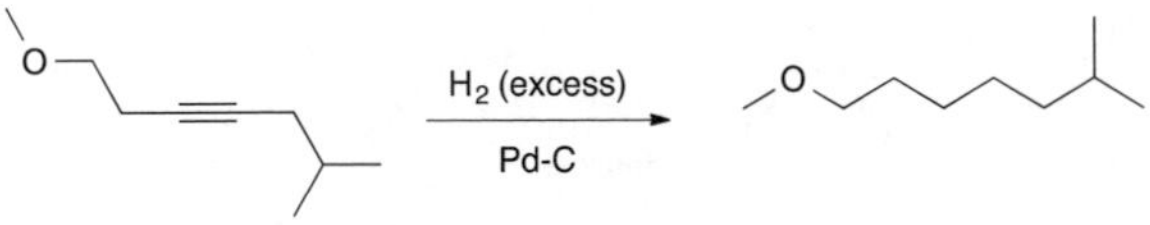

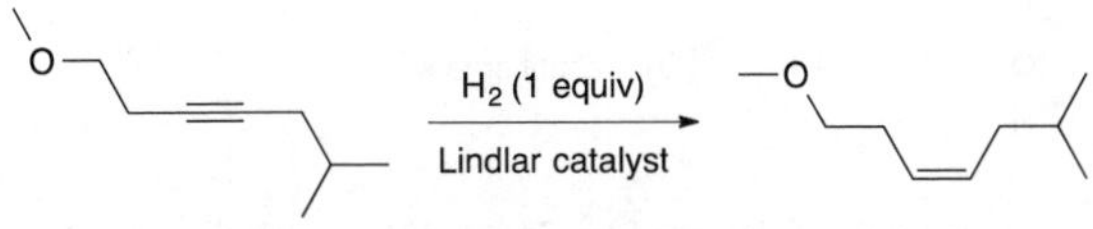

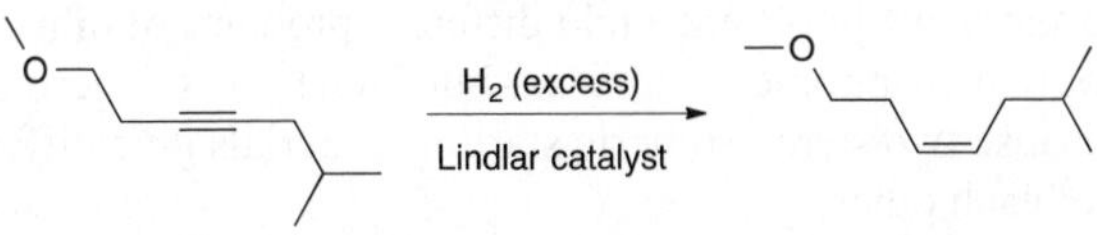

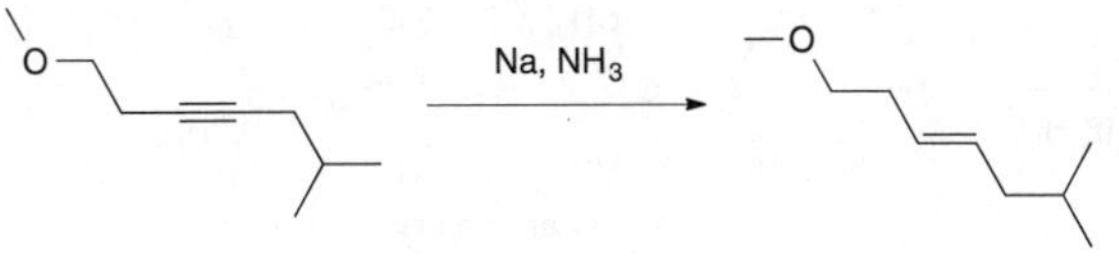

12.13

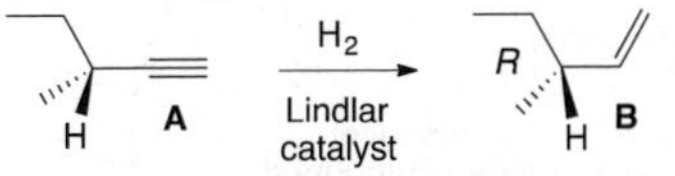

Chapter 12–10

12.14 LiAlH₄ reduces alkyl halides to alkanes and epoxides to alcohols.

a. [structure with Cl] → [1] LiAlH₄ / [2] H₂O → [structure with H]

H replaces Cl.

b. [bicyclic epoxide] → [1] LiAlH₄ / [2] H₂O → [cyclopentane with OH and H]

12.15 To draw the product, add an O atom across the π bond of the C=C.

a. [structure] → mCPBA → [epoxide]

b. [methylenecyclohexane] → mCPBA → [spiro epoxide]

c. [octahydronaphthalene] → mCPBA → [epoxide]

12.16 For epoxidation reactions:
- There are two possible products: O adds from above and below the double bond.
- Substituents on the C=C retain their original configuration in the products.

a. [propene] → mCPBA → [epoxide] + [epoxide] enantiomers

b. [cis-alkene] → mCPBA → [epoxide] + [epoxide] identical

c. [methylcyclohexene] → mCPBA → [epoxide] + [epoxide] enantiomers

12.17 Treatment of an alkene with a peroxyacid followed by H₂O, HO⁻ adds two hydroxy groups in an **anti** fashion. *cis*-But-2-ene and *trans*-but-2-ene yield different products of dihydroxylation. *cis*-But-2-ene gives a mixture of two enantiomers and *trans*-but-2-ene gives a meso compound. The reaction is stereospecific because two stereoisomeric starting materials give different products that are also stereoisomers of each other.

cis-but-2-ene → [1] RCO₃H / [2] H₂O, HO⁻ → [product] + [product] enantiomers

trans-but-2-ene → [1] RCO₃H / [2] H₂O, HO⁻ → [product] + [product] identical meso compound

Oxidation and Reduction 12–11

12.18 Treatment of an alkene with OsO₄ adds two hydroxy groups in a **syn** fashion. *cis*-But-2-ene and *trans*-but-2-ene yield different stereoisomers in this dihydroxylation, so the reaction is stereospecific.

cis-but-2-ene [1] OsO₄ / [2] NaHSO₃, H₂O → identical meso compound

trans-but-2-ene [1] OsO₄ / [2] NaHSO₃, H₂O → enantiomers

12.19 To draw the oxidative cleavage products:
- **Locate all the π bonds** in the molecule.
- **Replace all C=C's with *two* C=O's**.

Replace this π bond with two C=O's.

a. [1] O₃ / [2] Zn, H₂O → ketone + aldehyde   One **ketone** and one **aldehyde** are formed.

b. [1] O₃ / [2] Zn, H₂O → + Two **aldehydes** are formed.

c. [1] O₃ / [2] Zn, H₂O → A **dicarbonyl** compound is formed.

12.20 To find the alkene that yields the oxidative cleavage products:
- **Find the two carbonyl groups** in the products.
- **Join the two carbonyl carbons** together with a double bond. This is the double bond that was broken during ozonolysis.

a. Join these two C's.

b. Join these two C's.

c. only   With only one product, the alkene must be symmetrical around the double bond. Join this C to the same C in another identical molecule.

Chapter 12–12

12.21

a. [diagram: benzene → [1] O_3 / [2] CH_3SCH_3 → products]

b. [diagram: cyclohexadiene → [1] O_3 / [2] CH_3SCH_3 → products]

c. [diagram: cyclohexene → [1] O_3 / [2] CH_3SCH_3 → products]

12.22 To draw the products of oxidative cleavage of alkynes:
- **Locate the triple bond.**
- For internal alkynes, **convert the *sp* hybridized C to COOH.**
- For terminal alkynes, the *sp* hybridized C–H becomes CO$_2$.

a. [diagram: internal alkyne → [1] O_3 / [2] H_2O → products]

  internal alkyne

b. [diagram: internal alkyne (diphenylacetylene) → [1] O_3 / [2] H_2O → products]

  internal alkyne

  identical compounds

c. terminal alkyne   internal alkyne

[diagram → [1] O_3 / [2] H_2O → CO$_2$ + products]

12.23

a. CO$_2$ + [diagram of carboxylic acid] ⟹ [diagram of terminal alkyne]

b. [diagram] ⟹ [diagram]

c. [diagram of products] ⟹ [diagram of diyne]

d. [diagram of diacid] ⟹ [diagram of cyclic alkyne]

12.24 For the **oxidation of alcohols,** remember:
- **1° Alcohols** are oxidized to aldehydes with PCC.
- **1° Alcohols** are oxidized to carboxylic acids with oxidizing agents like CrO$_3$ or Na$_2$Cr$_2$O$_7$.
- **2° Alcohols** are oxidized to ketones with all Cr^{6+} reagents.

Oxidation and Reduction 12–13

a. ⟶ PCC

c. ⟶ CrO₃ / H₂SO₄, H₂O

b. ⟶ PCC

d. ⟶ CrO₃ / H₂SO₄, H₂O

12.25 Upon treatment with $HCrO_4^-$–Amberlyst A-26 resin:
- **1° Alcohols** are oxidized to aldehydes.
- **2° Alcohols** are oxidized to ketones.

a. ⟶ $HCrO_4^-$ / Amberlyst A-26 resin

b. HO— —OH ⟶ $HCrO_4^-$ / Amberlyst A-26 resin ⟶ O= =O

c. ⟶ $HCrO_4^-$ / Amberlyst A-26 resin

12.26

a. ⟶ NaOCl ⟶ + NaCl + H_2O

The by-products of the reaction with sodium hypochlorite are water and table salt (NaCl), as opposed to the by-products with $HCrO_4^-$– Amberlyst A-26 resin, which contain carcinogenic Cr^{3+} metal.

b. Oxidation with NaOCl has at least two advantages over oxidation with CrO_3, H_2SO_4 and H_2O. Because no Cr^{6+} is used as oxidant, there are no Cr by-products that must be disposed of. Also, CrO_3 oxidation is carried out in corrosive inorganic acids (H_2SO_4) and oxidation with NaOCl avoids this.

12.27 To draw the products of a **Sharpless epoxidation:**
- With the C=C horizontal, draw the allylic alcohol with the OH on the **top right** of the alkene.
- Add the new oxygen **above** the plane if (−)-DET is used and **below** the plane if (+)-DET is used.

a. ⟶ $(CH_3)_3C-OOH$ / $Ti[OCH(CH_3)_2]_4$ / (+)-DET ⟶

(+)-DET adds O **below** the plane.

b. ⟶ re-draw ⟶ $(CH_3)_3C-OOH$ / $Ti[OCH(CH_3)_2]_4$ / (−)-DET ⟶

(−)-DET adds O **above** the plane.

Chapter 12–14

12.28 Sharpless epoxidation needs an ***allylic alcohol*** as the starting material. Alkenes with no allylic OH group will not undergo reaction with the Sharpless reagent.

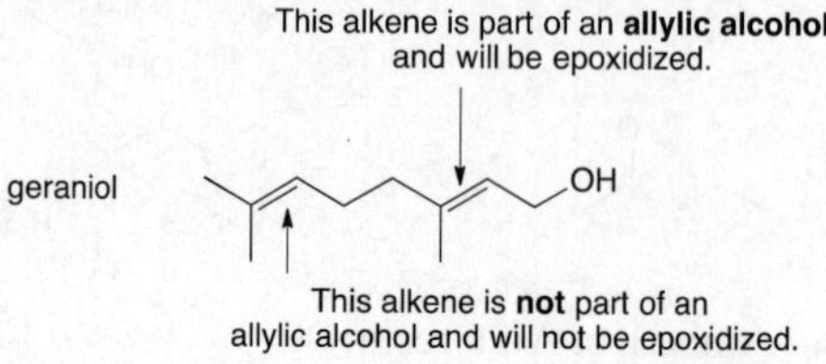

This alkene is part of an **allylic alcohol** and will be epoxidized.

geraniol

This alkene is **not** part of an allylic alcohol and will not be epoxidized.

12.29

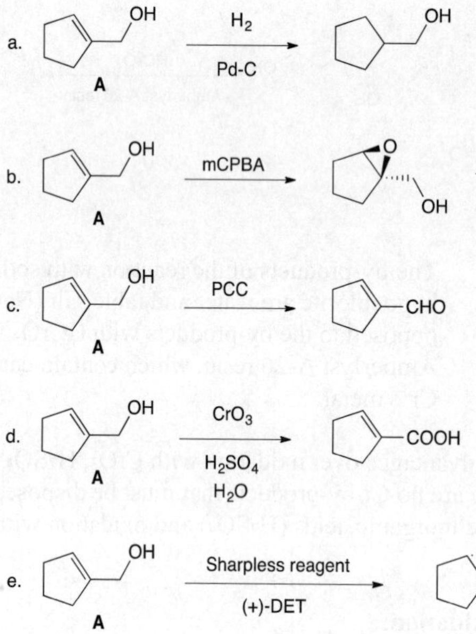

a. A $\xrightarrow[\text{Pd-C}]{\text{H}_2}$

b. A $\xrightarrow{\text{mCPBA}}$ +

c. A $\xrightarrow{\text{PCC}}$ —CHO

d. A $\xrightarrow[\substack{\text{H}_2\text{SO}_4 \\ \text{H}_2\text{O}}]{\text{CrO}_3}$ —COOH

e. A $\xrightarrow[\text{(+)-DET}]{\text{Sharpless reagent}}$

12.30

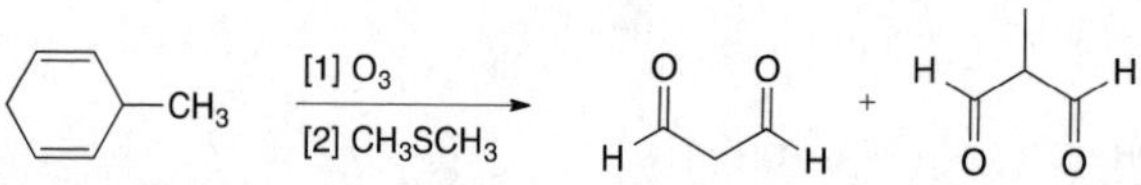

—CH$_3$ $\xrightarrow[\text{[2] CH}_3\text{SCH}_3]{\text{[1] O}_3}$ +

12.31

HC≡CH $\xrightarrow[\text{[2] CH}_3\text{CH}_2\text{Br}]{\text{[1] NaH}}$ $\xrightarrow[\substack{\text{[2]} \triangle \\ \text{[3] H}_2\text{O}}]{\text{[1] NaH}}$ —OH $\xrightarrow[\text{NH}_3]{\text{Na}}$ —OH

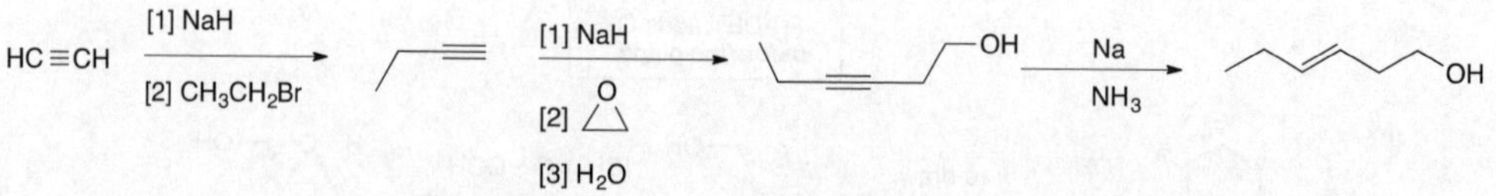

Oxidation and Reduction 12–15

12.32 Use the rules from Answer 12.1.

a. [epoxycyclohexane] → [cyclohexanol] **reduction**

b. HO—⟨⟩—OH → O=⟨⟩=O **oxidation**

c. [CH(OCH₃)₂ compound] → [α-Br aldehyde] **oxidation** (one new C–Br bond)

d. [cyclopentane with two OCH₃] → [cyclopentene with OCH₃] **neither** (one fewer C–O bond and one fewer C–H bond)

12.33 Use the principles from Answer 12.2 and draw the products of syn addition of H₂ from above and below the C=C.

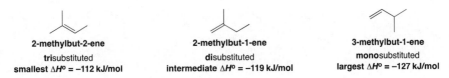

12.34 Increasing alkyl substitution increases alkene stability, thus decreasing the heat of hydrogenation.

2-methylbut-2-ene	2-methylbut-1-ene	3-methylbut-1-ene
trisubstituted	**di**substituted	**mono**substituted
smallest $\Delta H° = -112$ kJ/mol	intermediate $\Delta H° = -119$ kJ/mol	largest $\Delta H° = -127$ kJ/mol

12.35

A possible structure:

a. Compound **A**: molecular formula C_5H_8: hydrogenated to C_5H_{10}.
2 degrees of unsaturation, 1 is hydrogenated.
1 ring and 1 π bond

b. Compound **B**: molecular formula $C_{10}H_{16}$: hydrogenated to $C_{10}H_{18}$.
3 degrees of unsaturation, 1 is hydrogenated.
2 rings and 1 π bond

c. Compound **C**: molecular formula C_8H_8: hydrogenated to C_8H_{16}.
5 degrees of unsaturation, 4 are hydrogenated.
1 ring and 4 π bonds

Chapter 12–16

12.36

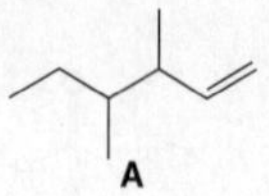

A

a. **mono**substituted
   largest heat of hydrogenation
b. **fastest** reaction rate

c. [1] O₃ / [2] Zn, H₂O →

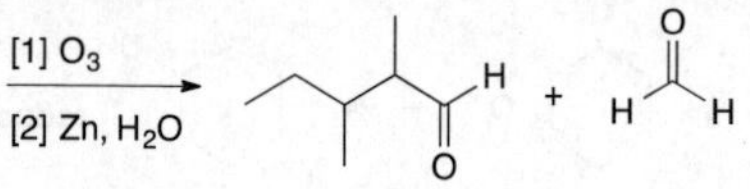

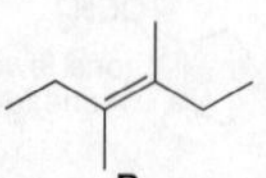

B

a. **tetra**substituted
   smallest heat of hydrogenation
b. **slowest** reaction rate

c. [1] O₃ / [2] Zn, H₂O → identical

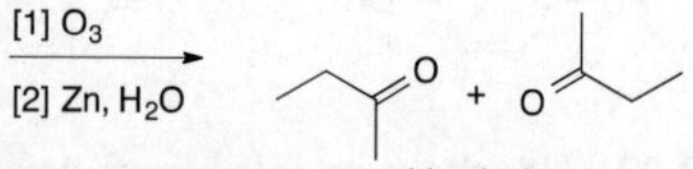

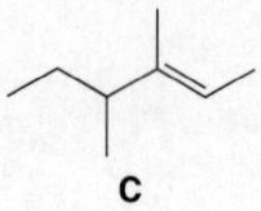

C

a. **tri**substituted
   intermediate heat of hydrogenation
b. **intermediate** reaction rate

c. [1] O₃ / [2] Zn, H₂O →

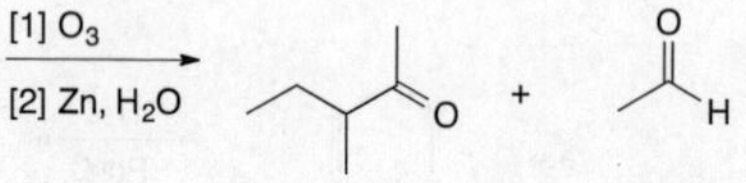

12.37

a.

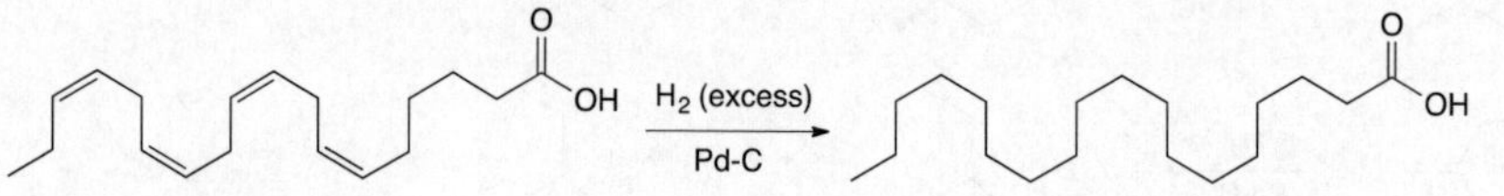

stearidonic acid   H₂ (excess) / Pd-C →   stearic acid

b.

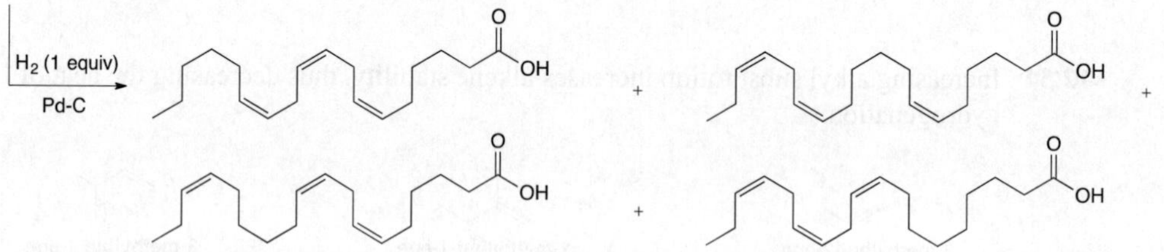

H₂ (1 equiv) / Pd-C →

c.

trans

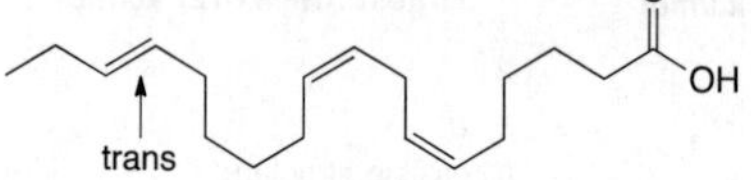

one possibility

d.

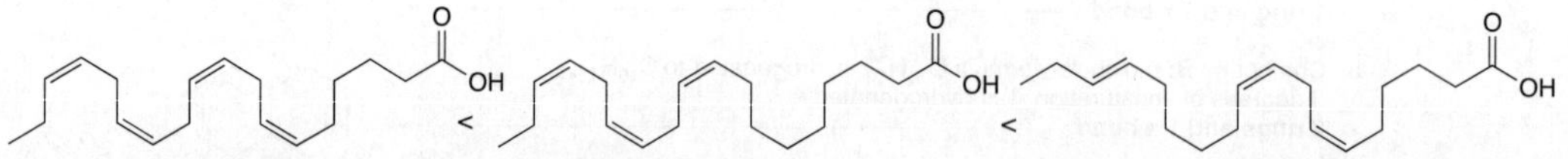

stearidonic acid
4 cis C=C's

<   product in (b)
3 cis C=C's

<   product in (c)
2 cis and one trans C=C's

Oxidation and Reduction 12–17

12.38

a. (cyclopentene) $\xrightarrow[\text{Pd-C}]{\text{H}_2}$ (cyclopentane)

b. (cyclopentane) $\xrightarrow[\text{Lindlar catalyst}]{\text{H}_2}$ no reaction

c. (cyclopentene) $\xrightarrow[\text{NH}_3]{\text{Na}}$ no reaction

d. (cyclopentene) $\xrightarrow{\text{CH}_3\text{CO}_3\text{H}}$ (epoxide)

e. (cyclopentene) $\xrightarrow[\text{[2] H}_2\text{O, HO}^-]{\text{[1] CH}_3\text{CO}_3\text{H}}$ (trans diol) + (trans diol) anti addition

f. (cyclopentene) $\xrightarrow[\text{[2] NaHSO}_3, \text{H}_2\text{O}]{\text{[1] OsO}_4 + \text{NMO}}$ (cis diol) syn addition

g. (cyclopentene) $\xrightarrow[\text{H}_2\text{O, HO}^-]{\text{KMnO}_4}$ (cis diol) syn addition

h. (cyclopentene) $\xrightarrow[\text{[2] H}_2\text{O}]{\text{[1] LiAlH}_4}$ no reaction

i. (cyclopentene) $\xrightarrow[\text{[2] CH}_3\text{SCH}_3]{\text{[1] O}_3}$ (dialdehyde)

j. (cyclopentene) $\xrightarrow[\substack{\text{Ti[OCH(CH}_3)_2]_4 \\ (-)\text{-DET}}]{(\text{CH}_3)_3\text{COOH}}$ no reaction

k. (cyclopentene) $\xrightarrow{\text{mCPBA}}$ (epoxide)

l. (epoxide) $\xrightarrow[\text{[2] H}_2\text{O}]{\text{[1] LiAlH}_4}$ (cyclopentanol)

12.39

a. $\xrightarrow[\text{Pd-C}]{\text{H}_2}$ +

b. $\xrightarrow{\text{mCPBA}}$ +

c. $\xrightarrow{\text{PCC}}$

d. $\xrightarrow[\text{H}_2\text{SO}_4, \text{H}_2\text{O}]{\text{CrO}_3}$

e. $\xrightarrow[\substack{\text{Ti[OCH(CH}_3)_2]_4 \\ (+)\text{-DET}}]{(\text{CH}_3)_3\text{COOH}}$

f. $\xrightarrow[\substack{\text{Ti[OCH(CH}_3)_2]_4 \\ (-)\text{-DET}}]{(\text{CH}_3)_3\text{COOH}}$

g. $\xrightarrow{\text{[1] PBr}_3}$ Br $\xrightarrow[\text{[3] H}_2\text{O}]{\text{[2] LiAlH}_4}$

h. $\xrightarrow[\text{Amberlyst A-26}]{\text{HCrO}_4^-}$

Chapter 12–18

12.40

a.

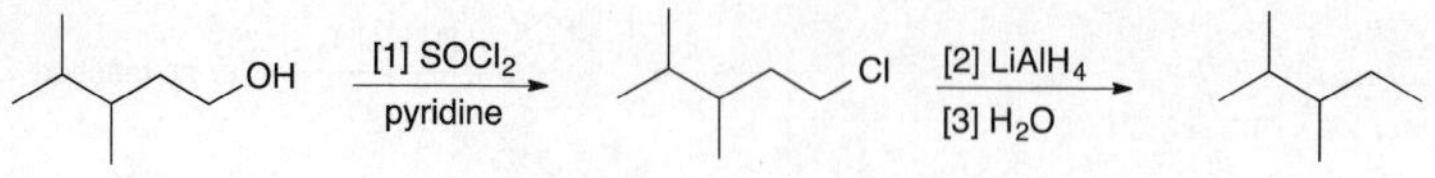

b.

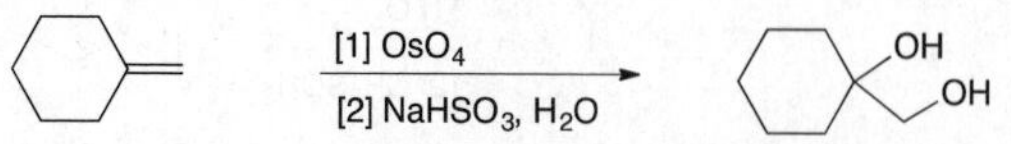

c.

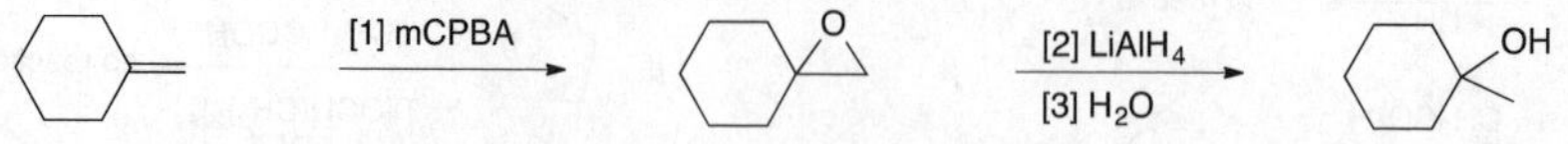

d.

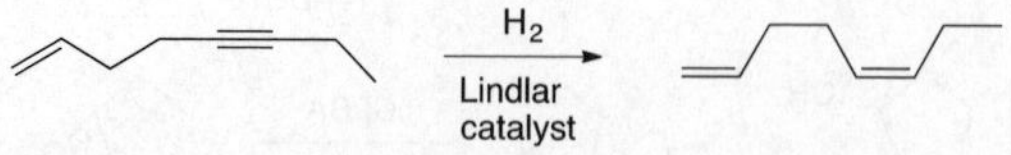

12.41

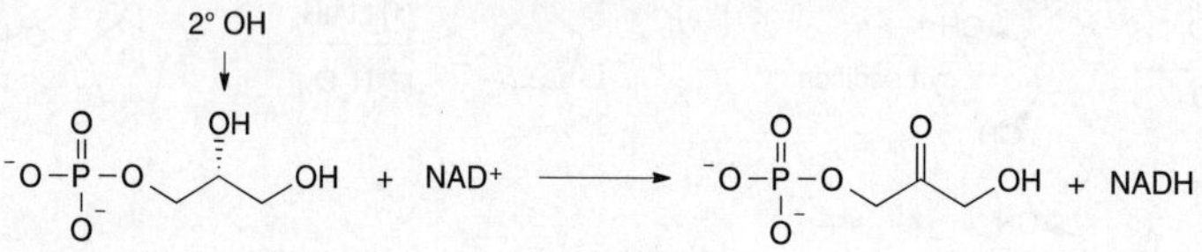

With NAD$^+$ the 2° alcohol is oxidized to a ketone, and NAD$^+$ is reduced to NADH.

(R)-glycerol phosphate

12.42 LiAlH$_4$ attacks at the less substituted end of an unsymmetrical epoxide to form an alcohol with an OH on the more substituted carbon.

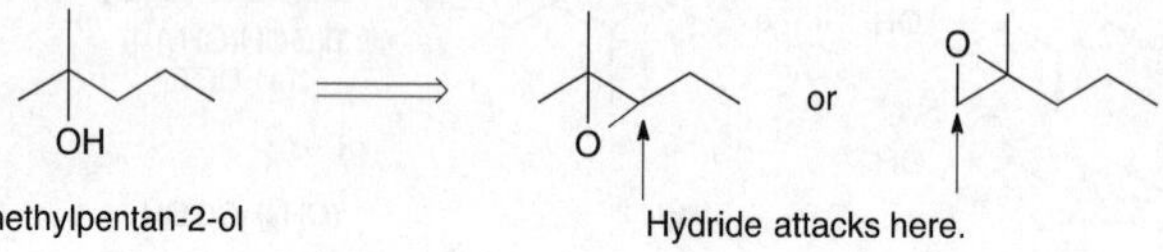

2-methylpentan-2-ol

Hydride attacks here.

12.43 The two sides of the C=C of **A** are different. D$_2$ adds only from above, so this side must be less sterically hindered. Other reagents will also add from the same side.

a.

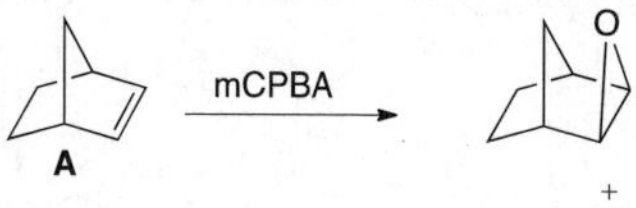

b.

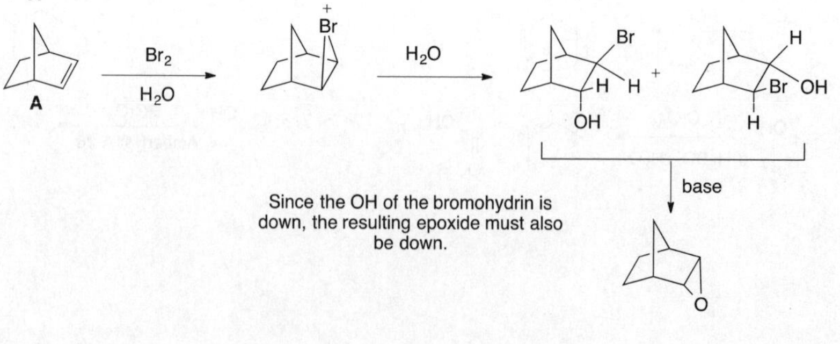

Since the OH of the bromohydrin is down, the resulting epoxide must also be down.

Oxidation and Reduction 12–19

12.44 Alkenes treated with [1] OsO$_4$ followed by NaHSO$_3$ in H$_2$O will undergo **syn** addition, whereas alkenes treated with [2] CH$_3$CO$_3$H followed by ⁻OH in H$_2$O will undergo **anti** addition.

a. [1]

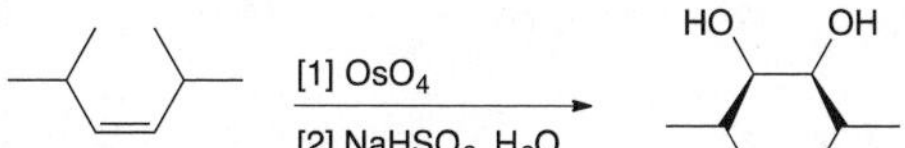

[1] OsO$_4$
[2] NaHSO$_3$, H$_2$O

[2]

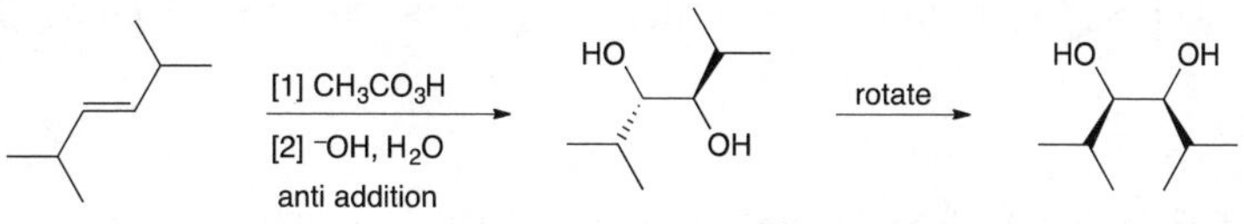

[1] CH$_3$CO$_3$H
[2] ⁻OH, H$_2$O
anti addition

rotate

b. [1]

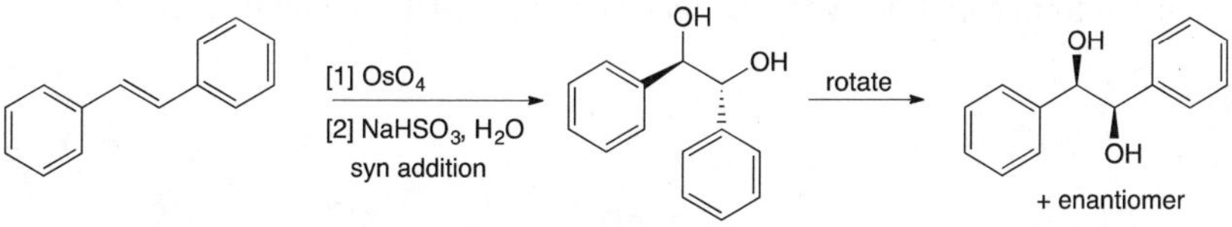

[1] OsO$_4$
[2] NaHSO$_3$, H$_2$O
syn addition

rotate

+ enantiomer

[2]

[1] CH$_3$CO$_3$H
[2] ⁻OH, H$_2$O

+ enantiomer

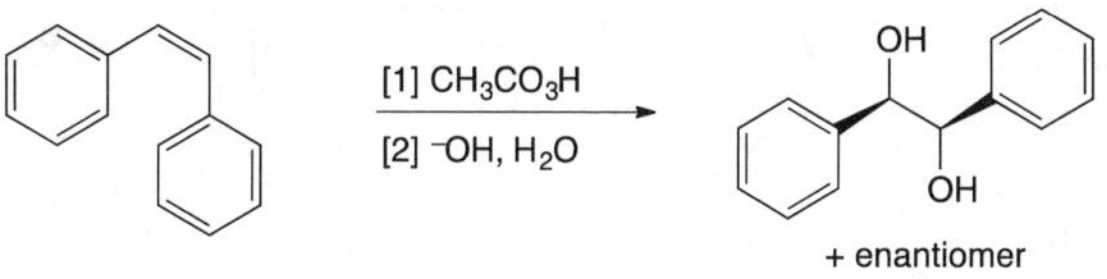

c. [1]

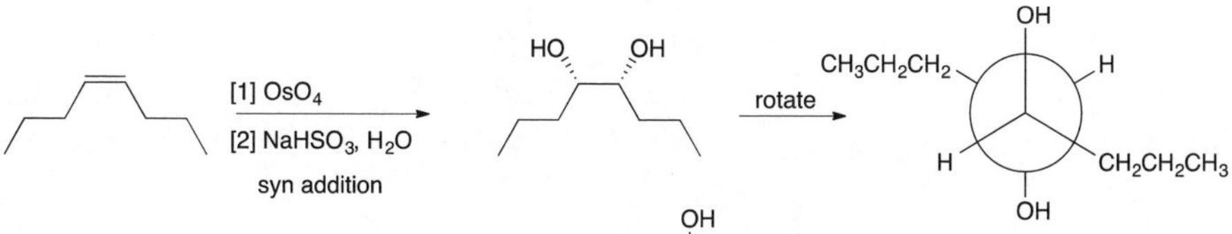

[1] OsO$_4$
[2] NaHSO$_3$, H$_2$O
syn addition

rotate

CH$_3$CH$_2$CH$_2$ · · · CH$_2$CH$_2$CH$_3$

[2]

[1] CH$_3$CO$_3$H
[2] ⁻OH, H$_2$O

CH$_3$CH$_2$CH$_2$ · · · CH$_2$CH$_2$CH$_3$

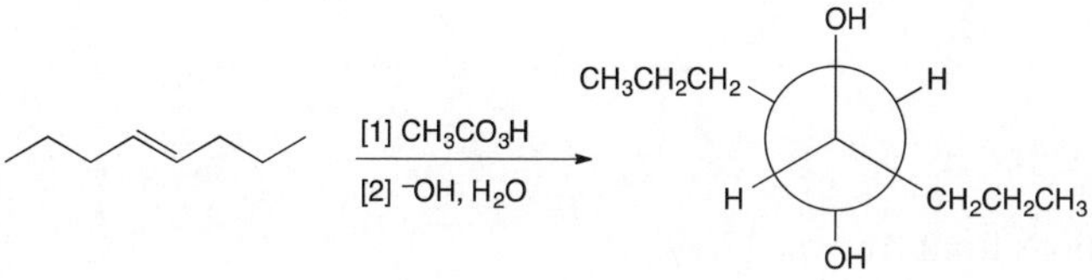

12.45

A

H$_3$Al—H
Li⁺

+ AlH$_3$ + Li⁺

H—OH

+ :ÖH

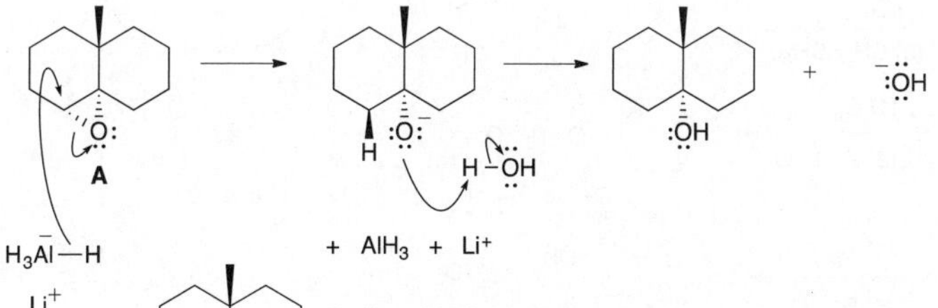

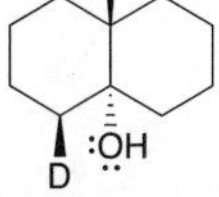

D⁻ (from LiAlD$_4$) opens the epoxide ring from the back side, so it is oriented on a wedge in the final product.

Chapter 12–20

12.46

a. A →(CH₃SO₂Cl)

b. B

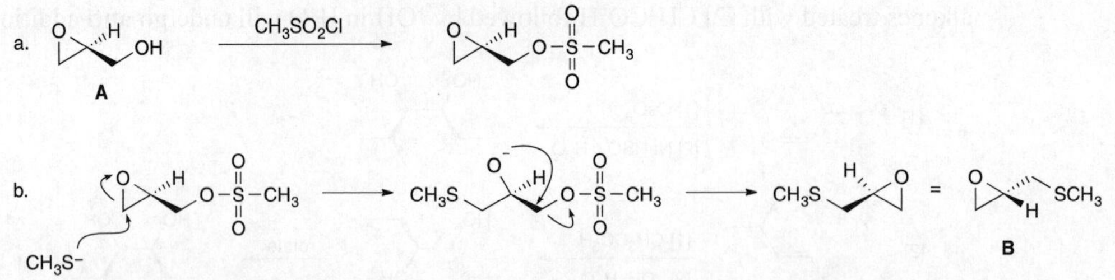

c. To form chiral **A**, Sharpless reagent with (+)-DET could be used.

12.47

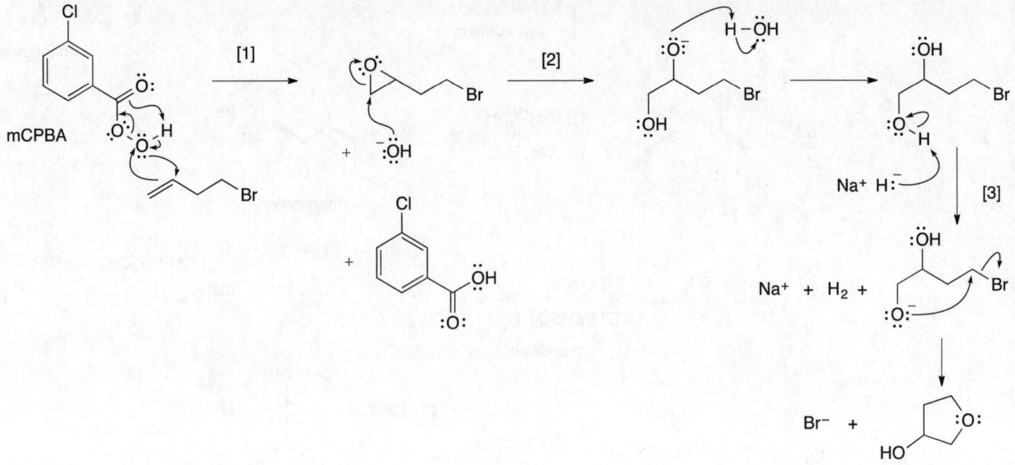

mCPBA [1] → + :ÖH +

[2] → → Na⁺ H:⁻ [3]

Na⁺ + H₂ + → Br⁻ +

12.48 Use the directions from Answer 12.19.

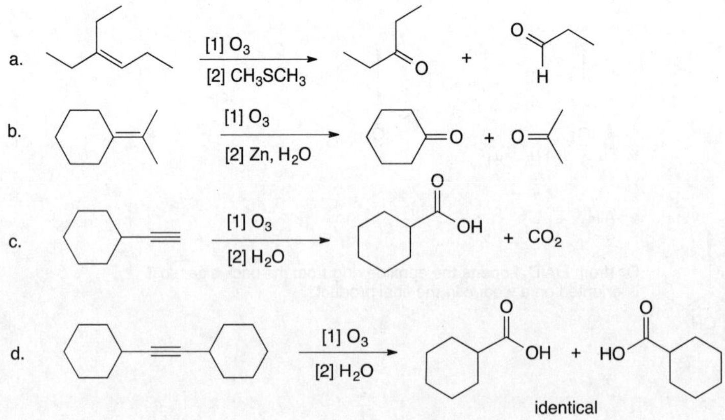

a. $\dfrac{[1]\ O_3}{[2]\ CH_3SCH_3}$

b. $\dfrac{[1]\ O_3}{[2]\ Zn,\ H_2O}$

c. $\dfrac{[1]\ O_3}{[2]\ H_2O}$ + CO₂

d. $\dfrac{[1]\ O_3}{[2]\ H_2O}$ identical

Oxidation and Reduction 12–21

12.49

a. [structure] =O and [structure] =O ⟹ [structure]

c. Join these two C's. [structure] + CO_2 ⟹ [structure]

b. [structure] and 2 equivalents of CH_2=O ⟹ [structure]

Join both of these C's to a C from formaldehyde.

formaldehyde C

d. Join these two C's. [structure] + [structure] ⟹ [structure]

12.50 Use the directions from Answer 12.20.

a. $C_{10}H_{18}$ $\xrightarrow{\text{[1] } O_3 \text{ [2] } CH_3SCH_3}$ [structure] ⟹ [structure]

2 degrees of unsaturation

Join these two C's.

one ring + one π bond

b. $C_{10}H_{16}$ $\xrightarrow{\text{[1] } O_3 \text{ [2] } CH_3SCH_3}$ [structure] ⟹ [structure]

3 degrees of unsaturation

two rings + one π bond

12.51

a. [structure] **squalene**

A   B   B   C   B   B   A

$\xrightarrow{\text{[1] } O_3 \text{ [2] Zn, } H_2O}$

[structure] + [structure] + [structure]

2 equiv (from portion **A**)   4 equiv (from portion **B**)   1 equiv (from portion **C**)

b. [structure] COOH

linolenic acid

$\xrightarrow{\text{[1] } O_3 \text{ [2] Zn, } H_2O}$

[structure] + [structure] COOH

2 equiv

c. [structure] $\xrightarrow{\text{[1] } O_3 \text{ [2] Zn, } H_2O}$ [structure] + [structure] + [structure]

zingiberene

Chapter 12–22

12.52

a. [structure: cyclooctadiene] **A** C_8H_{12} → [1] O_3 / [2] CH_3SCH_3 → [structure: succinaldehyde, OHC–CH₂CH₂–CHO]

b. [structure] C_6H_{10} **B** → H_2 (excess) / Pd-C → [structure] [structure] → [1] $NaNH_2$ / [2] CH_3I → [structure] C_7H_{12} **C**

12.53

$C_{10}H_{16}$ → H_2 / Pd-C → [structure] 2,6-dimethyloctane

3 degrees of unsaturation

The hydrogenation reaction tells you that both oximene and myrcene have 3 π bonds (and no rings). Use this carbon backbone and add in the double bonds based on the oxidative cleavage products.

Oximene: $(CH_3)_2C{=}O$ $CH_2{=}O$ $CH_2(CHO)_2$ $CH_3\text{-}C(O)\text{-}CHO$ ⟹ [structure]

Myrcene: $(CH_3)_2C{=}O$ $CH_2{=}O$ [structure: OHC–CH₂–C(O)–CH₂–CHO] ⟹ [structure]
   (2 equiv)

12.54

[structure] $C_{10}H_{18}O$ **A** → H_2SO_4 → [structure] $C_{10}H_{16}$ **B** + [structure] $C_{10}H_{16}$ **C** → H_2 / Pd-C → [structure] decalin

ozonolysis

[structure] **D** [structure] $C_{10}H_{16}O_2$ **E**

12.55 Hydrogenation of DHA forms $CH_3(CH_2)_{20}COOH$, so DHA is a 22-carbon fatty acid. The ozonolysis products show where the double bonds are located.

[structure] DHA
A B B B B B C

[1] O_3 / [2] Zn, H_2O

[structure] (from portion **A**) + [structure] 5 equiv (from portion **B**) + [structure] (from portion **C**)

Oxidation and Reduction 12–23

12.56 The stereogenic center (labeled with *) in both structures can be *R* or *S*.

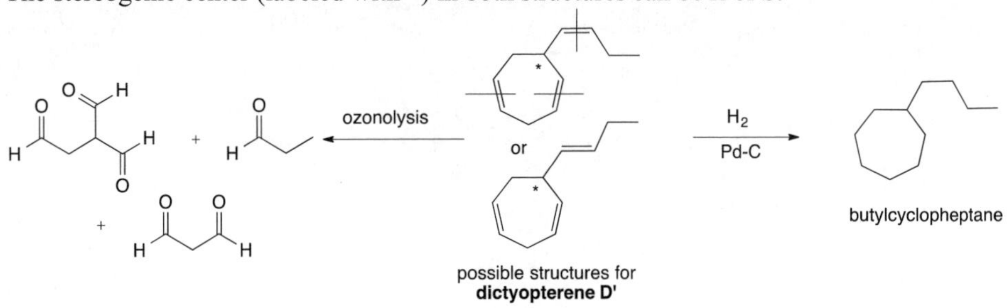

ozonolysis

or

possible structures for
dictyopterene D'

H₂
Pd-C

butylcycloheptane

12.57

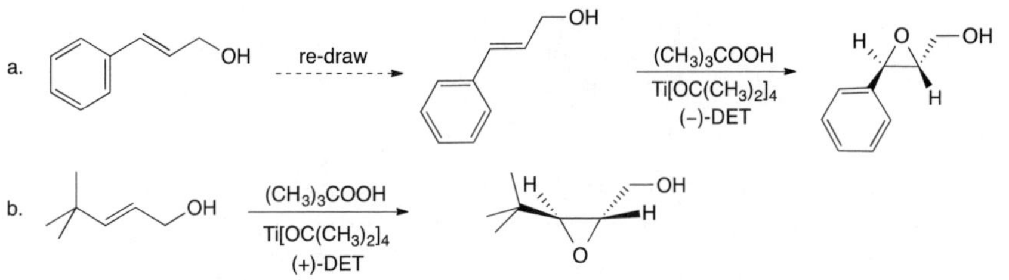

a. re-draw

(CH₃)₃COOH
Ti[OC(CH₃)₂]₄
(–)-DET

b. (CH₃)₃COOH
Ti[OC(CH₃)₂]₄
(+)-DET

12.58

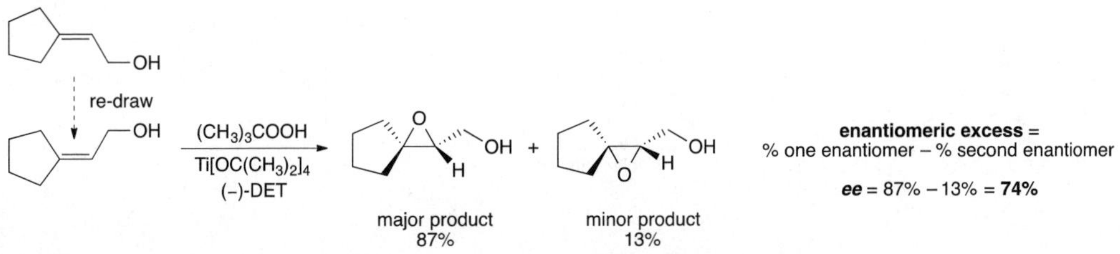

re-draw

(CH₃)₃COOH
Ti[OC(CH₃)₂]₄
(–)-DET

major product
87%

+

minor product
13%

enantiomeric excess =
% one enantiomer – % second enantiomer

ee = 87% – 13% = **74%**

12.59

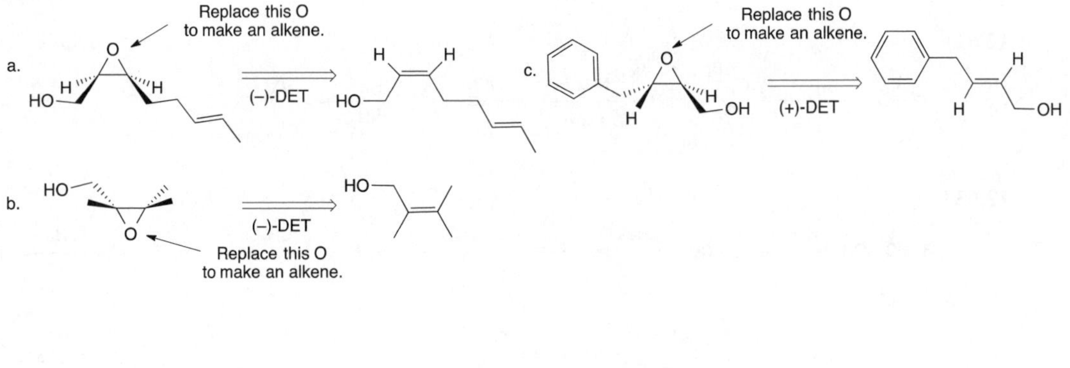

a. Replace this O
to make an alkene.

(–)-DET

c. Replace this O
to make an alkene.

(+)-DET

b. (–)-DET

Replace this O
to make an alkene.

Chapter 12–24

12.60

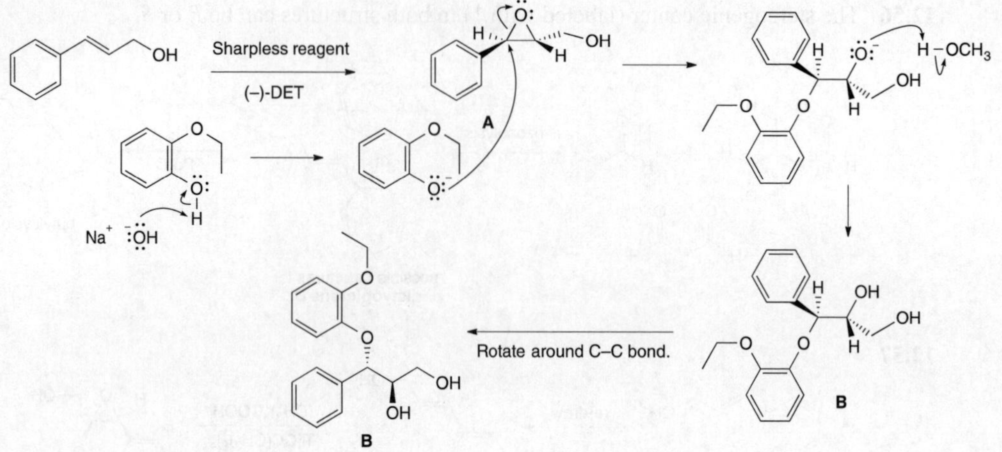

12.61

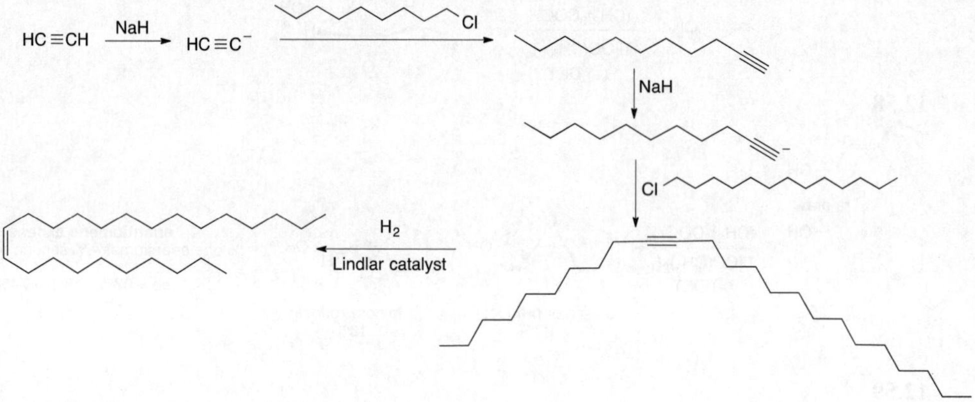

12.62

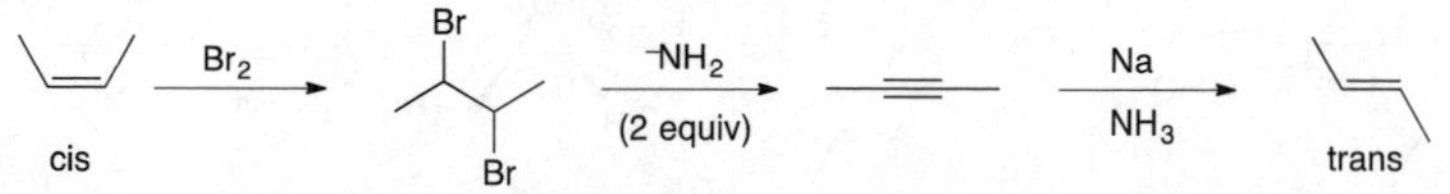

12.63

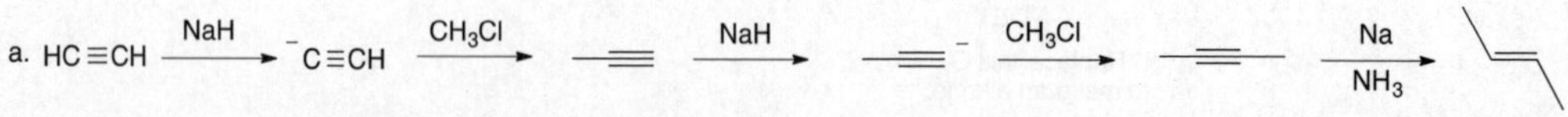

a.

Oxidation and Reduction 12–25

b. 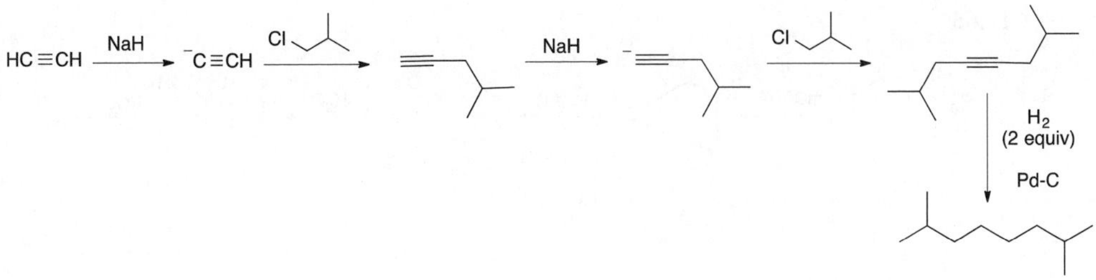 $HC \equiv CH \xrightarrow{NaH} {}^{-}C \equiv CH$

c. $HC \equiv CH \xrightarrow{NaH} HC \equiv C^{-}$

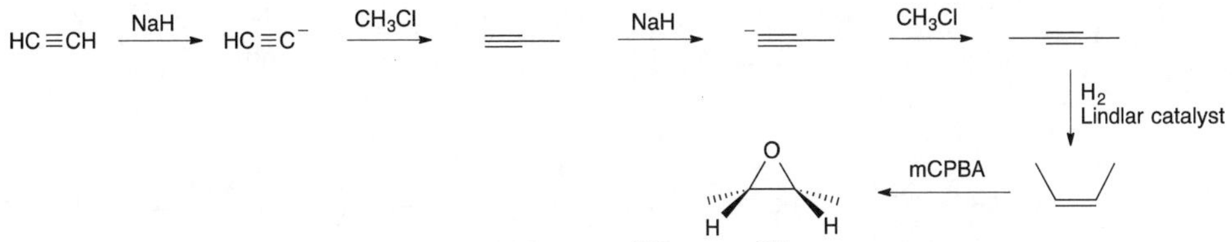

d. 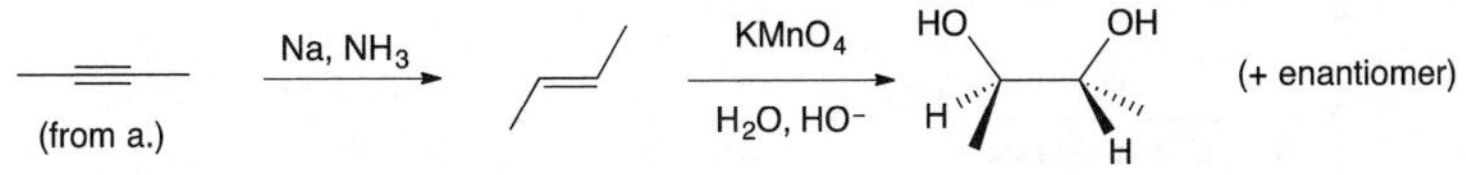 (from a.) $\xrightarrow{Na, NH_3}$ $\xrightarrow[H_2O, HO^-]{KMnO_4}$ (+ enantiomer)

12.64

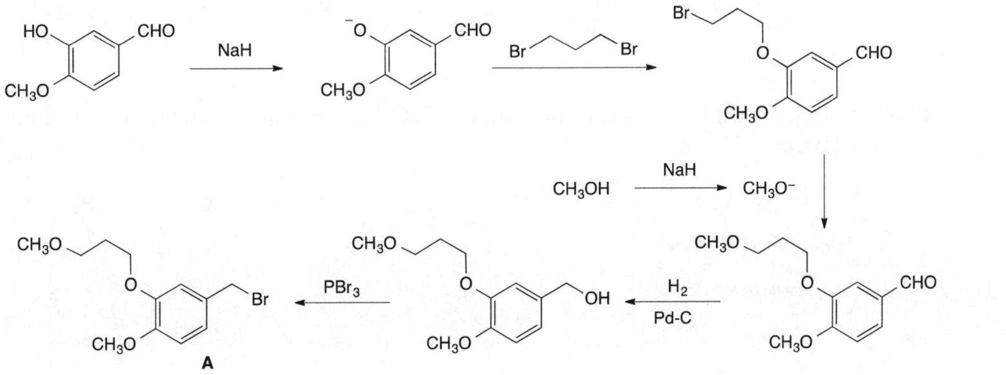

12.65

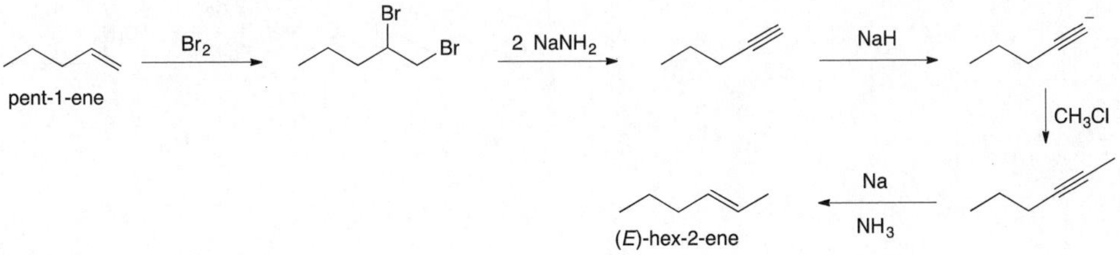

Chapter 12–26

12.66

a.

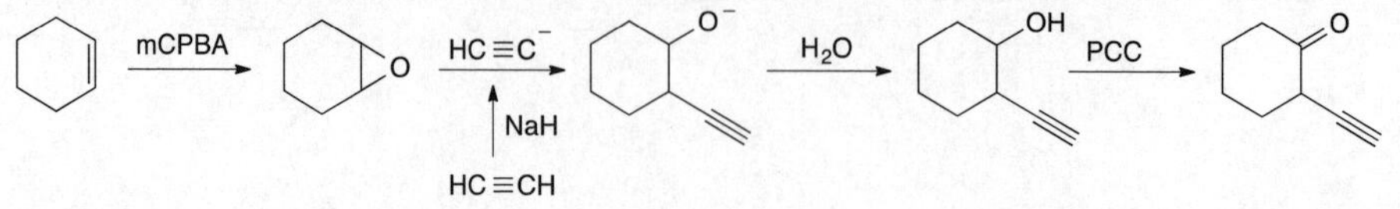

b.

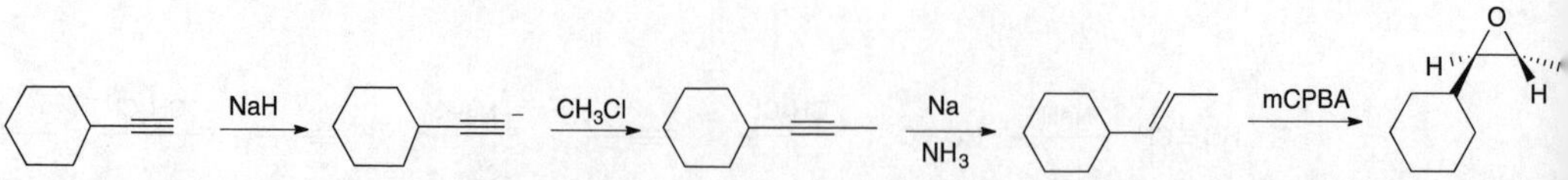

(+ enantiomer)

c.

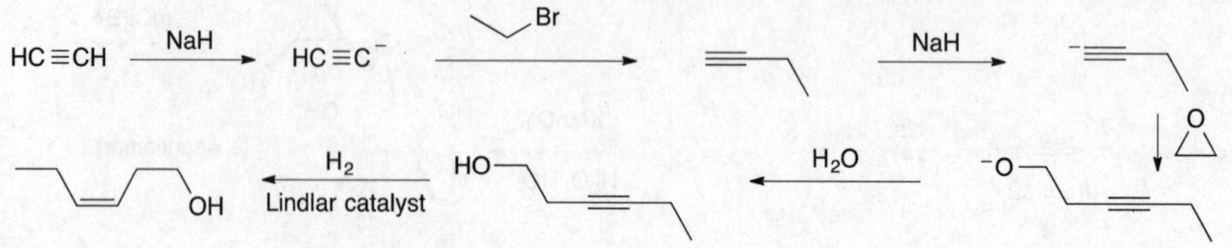

d. 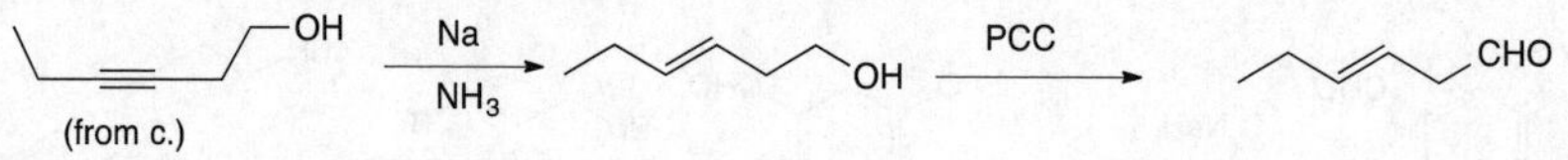 (from c.)

12.67 Determine which 2 C's can come from HC≡CH and two alkyl halides that are unhindered (1° or CH$_3$X).

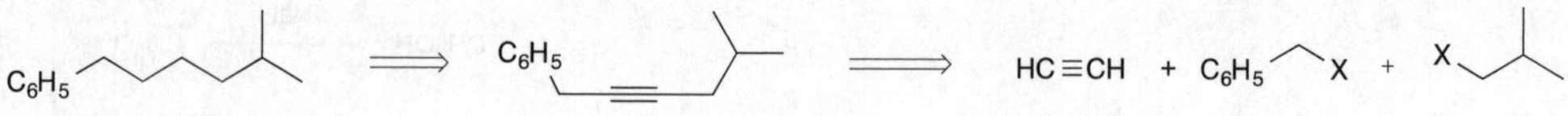

C$_6$H$_5$
1-phenyl-5-methylhexane

In synthetic direction:

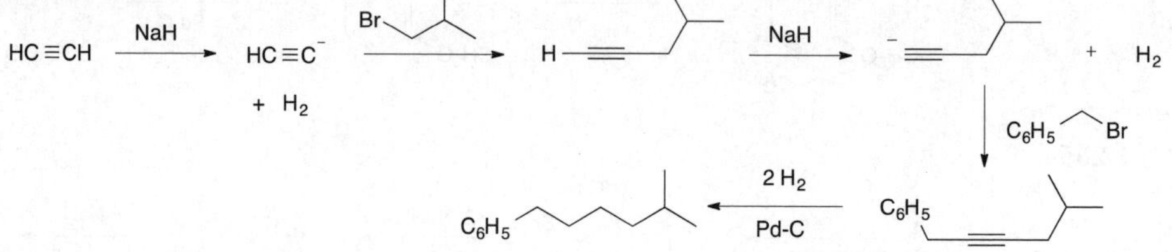

$$HC \equiv CH \xrightarrow{NaH} HC \equiv C^-$$
$$+ H_2$$

Oxidation and Reduction 12–27

12.68

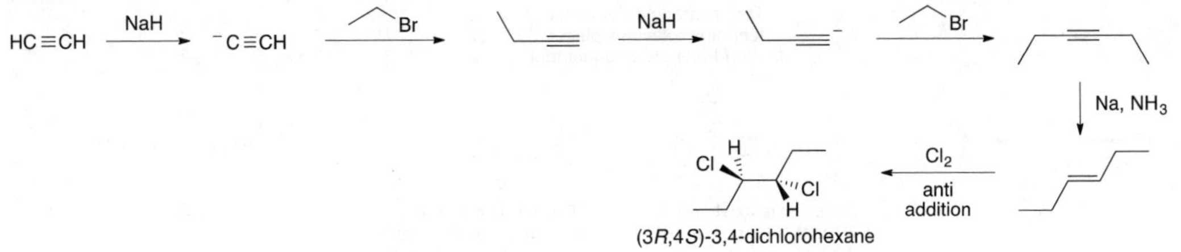

(3R,4S)-3,4-dichlorohexane

12.69

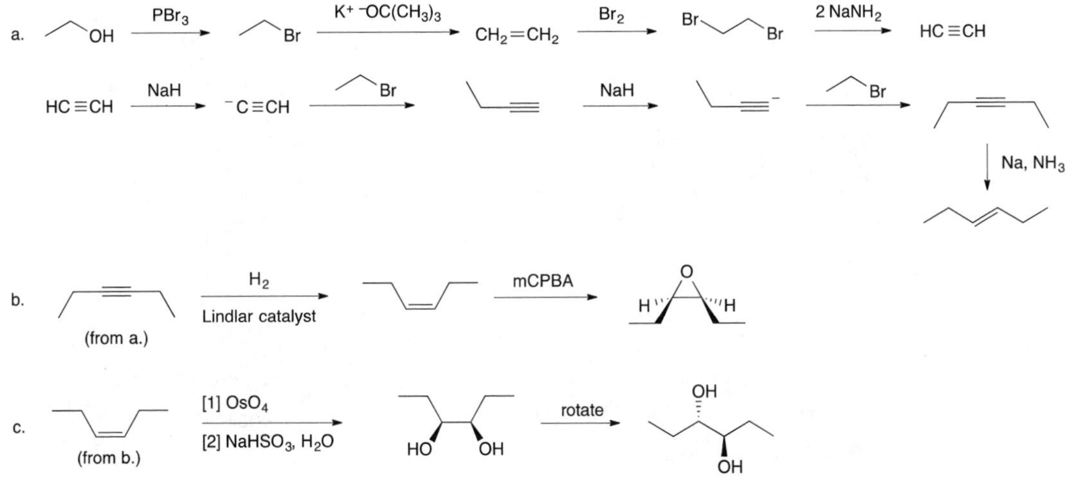

a.

b.

(from a.)

c.

(from b.)

12.70

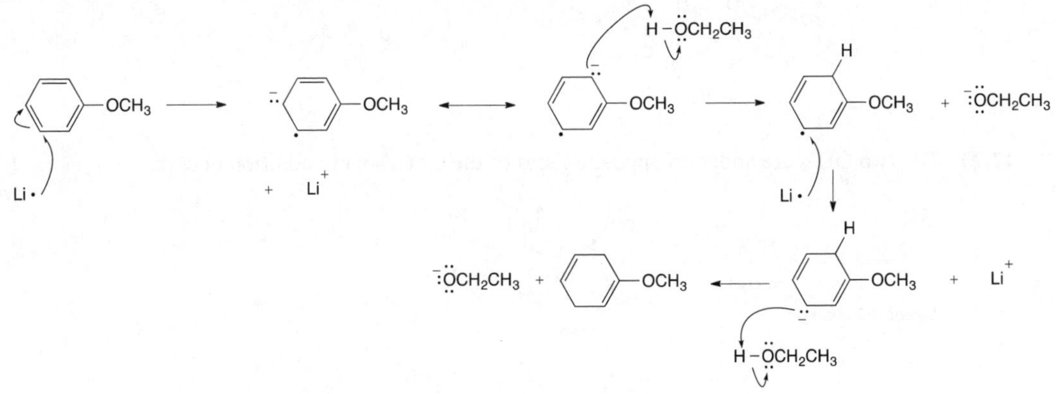

12.71

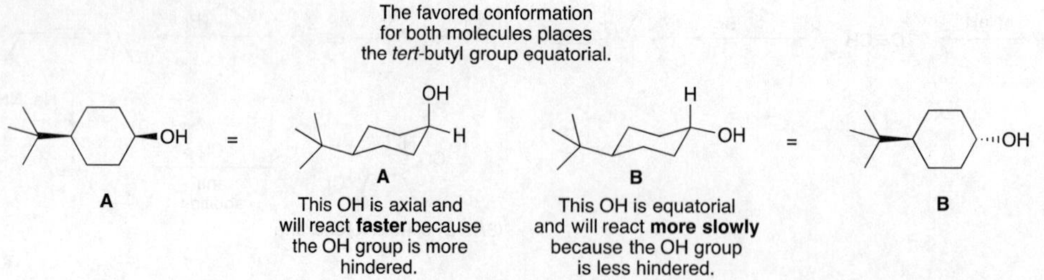
The favored conformation
for both molecules places
the *tert*-butyl group equatorial.
OH
H
OH
=
A
A
This OH is axial and
will react **faster** because
the OH group is more
hindered.
B
This OH is equatorial
and will react **more slowly**
because the OH group
is less hindered.
B

12.72

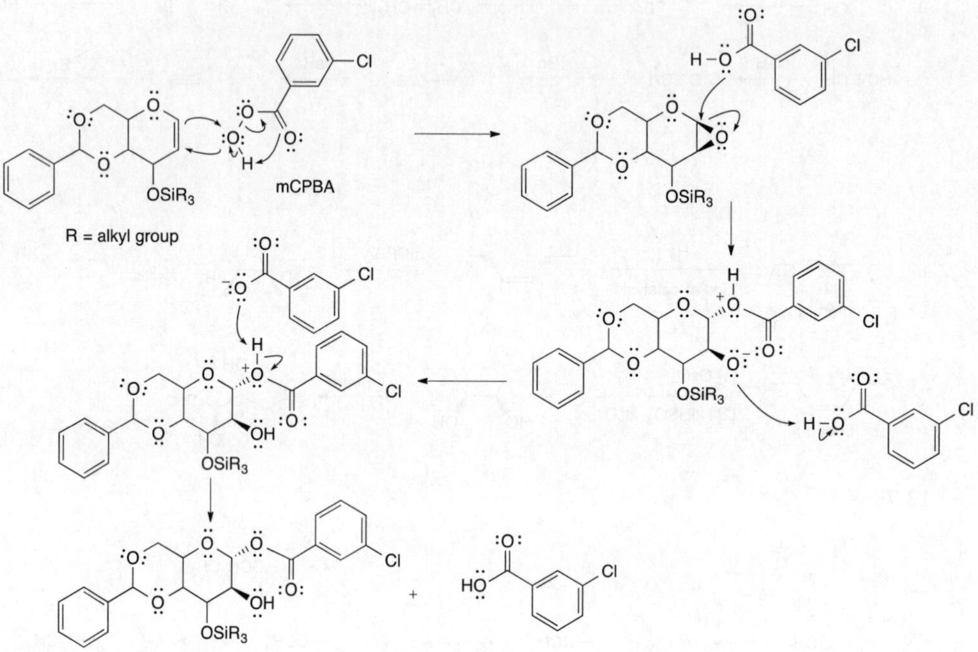
Cl
mCPBA
OSiR3
R = alkyl group
Cl
OSiR3
H
Cl
OSiR3
OH
OSiR3
Cl
+

12.73 The two OH's are added to opposite faces of the C=C, so anti addition occurs.

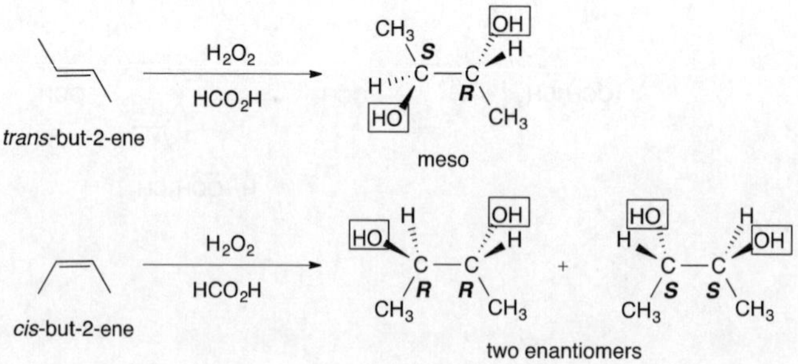
H2O2
HCO2H
trans-but-2-ene
CH3
OH
H
CH3
HO
meso
H2O2
HCO2H
cis-but-2-ene
two enantiomers

Oxidation and Reduction 12–29

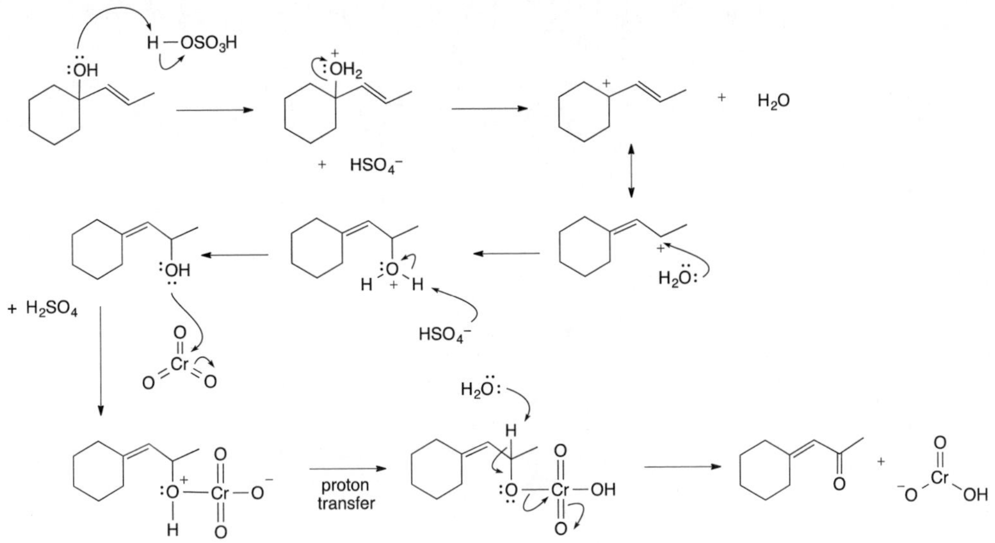

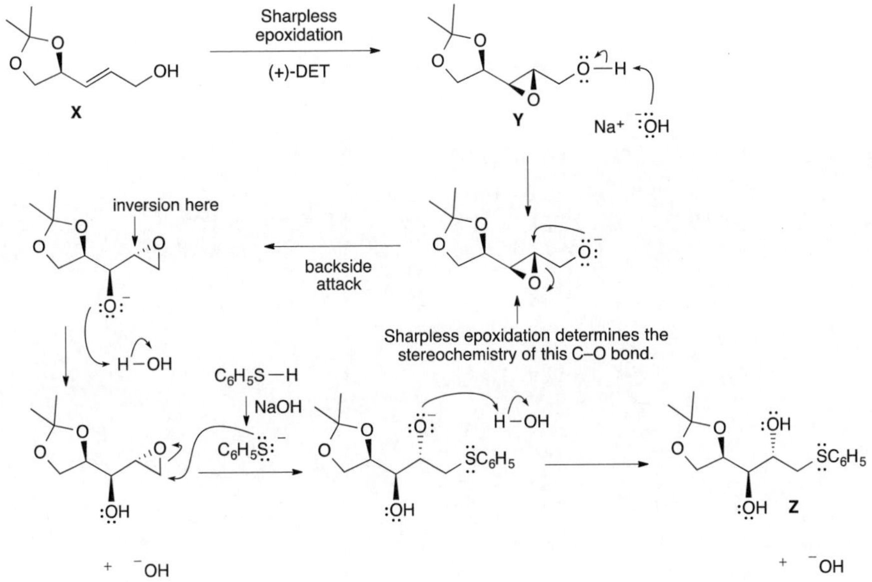

12.74

12.75

Chapter 13 Mass Spectrometry and Infrared Spectroscopy

Chapter Review

Mass spectrometry (MS)
- Mass spectrometry measures the molecular weight of a compound (13.1A).
- The mass of the molecular ion (**M**) = the molecular weight of a compound. Except for isotope peaks at M + 1 and M + 2, the molecular ion has the highest mass in a mass spectrum (13.1A).
- The base peak is the tallest peak in a mass spectrum (13.1A).
- A compound with an odd number of N atoms gives an odd molecular ion. A compound with an even number of N atoms (including zero) gives an even molecular ion (13.1B).
- Organic chlorides show two peaks for the molecular ion (M and M + 2) in a 3:1 ratio (13.2).
- Organic bromides show two peaks for the molecular ion (M and M + 2) in a 1:1 ratio (13.2).
- The fragmentation of radical cations formed in a mass spectrometer gives lower molecular weight fragments, often characteristic of a functional group (13.3).
- High-resolution mass spectrometry gives the molecular formula of a compound (13.4A).

Electromagnetic radiation
- The wavelength (λ) and frequency (ν) of electromagnetic radiation are *inversely* related by the following equations, where c is the speed of light: $\lambda = c/\nu$ or $\nu = c/\lambda$ (13.5).
- The energy (E) of a photon is proportional to its frequency; the higher the frequency, the higher the energy (h = Planck's constant): $E = h\nu$ (13.5).

Infrared spectroscopy (IR, 13.6 and 13.7)
- Infrared spectroscopy identifies functional groups.
- IR absorptions are reported in wavenumbers:

$$\text{wavenumber} = \tilde{\nu} = 1/\lambda$$

- The functional group region from **4000–1500 cm^{-1}** is the most useful region of an IR spectrum.
- C–H, O–H, and N–H bonds absorb at high frequency, ≥ 2500 cm^{-1}.
- As bond strength increases, the wavenumber of an absorption increases; thus triple bonds absorb at higher wavenumber than double bonds.

$$
\begin{array}{cc}
\text{C=C} & \text{C}\equiv\text{C} \\
\sim 1650 \text{ cm}^{-1} & \sim 2250 \text{ cm}^{-1}
\end{array}
$$

Increasing bond strength
Increasing $\tilde{\nu}$

- The higher the percent *s*-character, the stronger the bond, and the higher the wavenumber of an IR absorption.

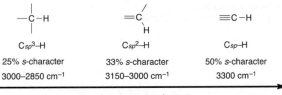

Increasing percent *s*-character
Increasing $\tilde{\nu}$

Chapter 13–2

Practice Test on Chapter Review

1.a. Which compound has a molecular ion at 112 and a peak at 1720 cm^{-1} in its IR spectrum?

1.   2.   3.

    4. Compounds (1) and (2) both fit these criteria.
    5. Compounds (1), (2), and (3) all fit these criteria.

b. Which of the following compounds has peaks at 3300, 3000, and 2250 cm^{-1} in its IR spectrum?

1.   2.   3.

    4. Compounds (1) and (2) have these peaks in their IR spectra.
    5. Compounds (1), (2), and (3) all contain these peaks in their IR spectra.

c. What is the base peak in a mass spectrum?
    1.   the peak due to the radical cation formed when a molecule loses an electron
    2.   the tallest peak in the mass spectrum
    3.   the peak due to the fragment with the largest m/z ratio
    4.   Both (1) and (2) describe the base peak.
    5.   Statements (1), (2), and (3) all describe the base peak.

d. Which compounds are possible structures for a molecule that has a molecular ion at 150 in its mass spectrum?

1.   2.   3.

    4. Both (1) and (2) are possible structures.
    5. Compounds (1), (2), and (3) are all possible structures.

e. Which compounds exhibit prominent M + 2 peaks in their mass spectra?

1.   2.   3.

    4. Both (1) and (2) show M + 2 peaks.
    5. Compounds (1), (2), and (3) all show M + 2 peaks.

Mass Spectrometry and Infrared Spectroscopy 13–3

2. Answer each question with the number that corresponds to one of the following regions of an IR spectrum.

   1. 4000–2500 cm^{-1}
   2. 2500–2000 cm^{-1}
   3. 2000–1500 cm^{-1}
   4. < 1500 cm^{-1}

   a. This region is called the fingerprint region of an IR spectrum.
   b. The OH group of propan-1-ol absorbs in this region.
   c. The C=N of $C_6H_5CH_2CH=NCH_3$ absorbs in this region.
   d. An unsymmetrical C≡C absorbs in this region.
   e. An *sp* hybridized C–H bond absorbs in this region.
   f. Ethyl benzoate ($C_6H_5CO_2CH_2CH_3$) absorbs in all regions of the IR except this one.

3. Answer True (T) or False (F).
   a. IR spectroscopy is useful for determining the molecular weight of a compound.
   b. A C–H bond that absorbs at 3140 cm^{-1} is stronger than a C–H bond that absorbs at 2950 cm^{-1}.
   c. A compound with a molecular ion at 109 contains a N atom.
   d. A compound with a base peak at 57 must contain a N atom.
   e. But-2-yne shows an IR absorption at 2250 cm^{-1}.
   f. Propan-1-ol shows an IR absorption at 3200–3600 cm^{-1}.
   g. An ether shows no IR absorptions at 3200–3600 or 1700 cm^{-1}.
   h. In its mass spectrum, a compound that has a molecular ion with two peaks of approximately equal intensity at 124 and 126 contains chlorine.

Answers to Practice Test

1.a. 4	2.a. 4	3.a. F	e. F
b. 1	b. 1	b. T	f. T
c. 2	c. 3	c. T	g. T
d. 4	d. 2	d. F	h. F
e. 4	e. 1		
	f. 2		

Chapter 13–4

Answers to Problems

13.1   The molecular ion formed from each compound is equal to its molecular weight.

a. C_3H_6O
molecular weight = **58**
molecular ion (*m/z*) = **58**

b. $C_{10}H_{20}$
molecular weight = **140**
molecular ion (*m/z*) = **140**

c. $C_8H_8O_2$
molecular weight = **136**
molecular ion (*m/z*) = **136**

d. $C_{10}H_{15}N$
molecular weight = **149**
molecular ion (*m/z*) = **149**

13.2   Some possible formulas for each molecular ion:
a.   Molecular ion at 72: C_5H_{12}, C_4H_8O, $C_3H_4O_2$
b.   Molecular ion at 100: C_8H_4, C_7H_{16}, $C_6H_{12}O$, $C_5H_8O_2$
c.   Molecular ion at 73: $C_4H_{11}N$, $C_2H_7N_3$

13.3   Use the molecular ion to propose molecular formulas that contain C, H, and O. Then determine which formula has five degrees of unsaturation.

M at 218: $C_{18}H_2$ (not enough H's)

$C_{17}H_{14} \xrightarrow[+O]{-CH_4} C_{16}H_{10}O$ (not enough H's)

$C_{16}H_{26} \xrightarrow[+O]{-CH_4} C_{15}H_{22}O$ (five degrees of unsaturation)
Answer

13.4   To calculate the molecular ions you would expect for compounds with Cl, calculate the molecular weight using each of the two most common isotopes of Cl (^{35}Cl and ^{37}Cl).

a. $C_4H_9{}^{35}Cl = $ **92**
$C_4H_9{}^{37}Cl = $ **94**
Two peaks in 3:1 ratio at *m/z* 92 and 94

c. $C_4H_{11}N = $ **73**
One peak at *m/z* 73

b. $C_3H_7F = $ **62**
One peak at *m/z* 62

d. $C_4H_4N_2 = $ **80**
One peak at *m/z* 80

13.5   Convert the ball-and-stick model to a skeletal structure and determine the molecular formula. Calculate the molecular weight using each of the two common isotopes for Br (^{79}Br and ^{81}Br).

$C_6H_{11}{}^{79}Br = $ **162**
$C_6H_{11}{}^{81}Br = $ **164**
Two peaks in a 1:1 ratio at *m/z* 162 and 164

$C_6H_{11}Br$

13.6   After calculating the mass of the molecular ion, draw the structure and determine which C–C bond is broken to form fragments of the appropriate mass-to-charge ratio.

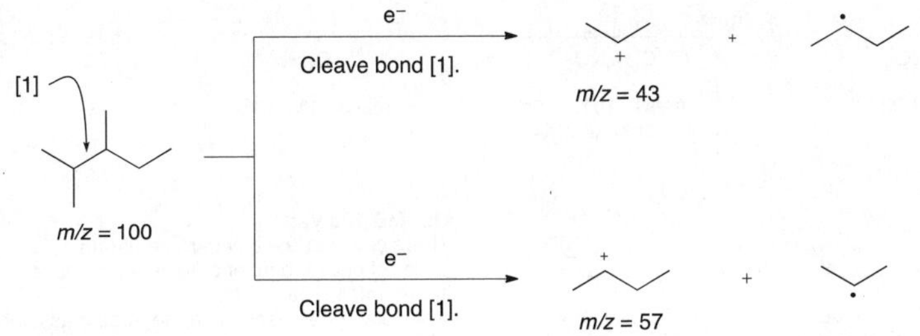

Mass Spectrometry and Infrared Spectroscopy 13–5

[1]

$m/z = 100$

Cleave bond [1].

$m/z = 43$

Cleave bond [1].

$m/z = 57$

13.7

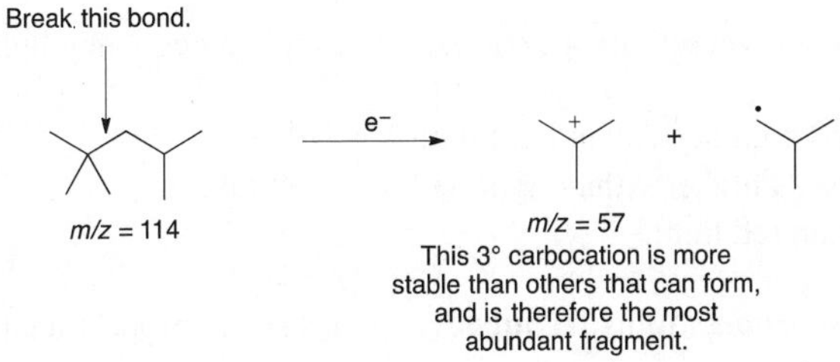

Break this bond.

$m/z = 114$

$m/z = 57$
This 3° carbocation is more
stable than others that can form,
and is therefore the most
abundant fragment.

13.8

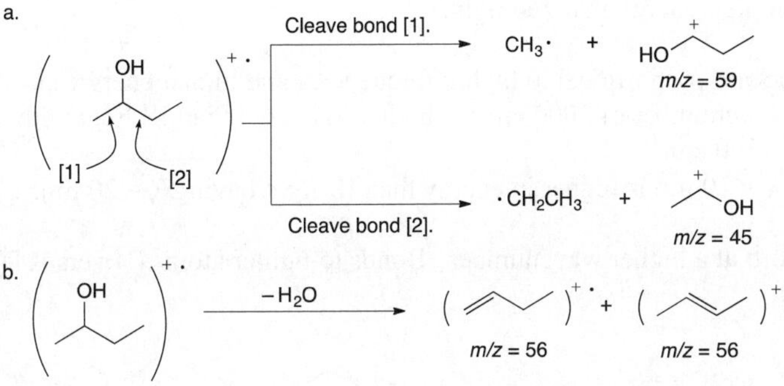

a.

Cleave bond [1].

$CH_3 \cdot$ +

$m/z = 59$

[1]   [2]

$\cdot CH_2CH_3$ +

$m/z = 45$

Cleave bond [2].

b.

$-H_2O$

$m/z = 56$ $m/z = 56$

13.9

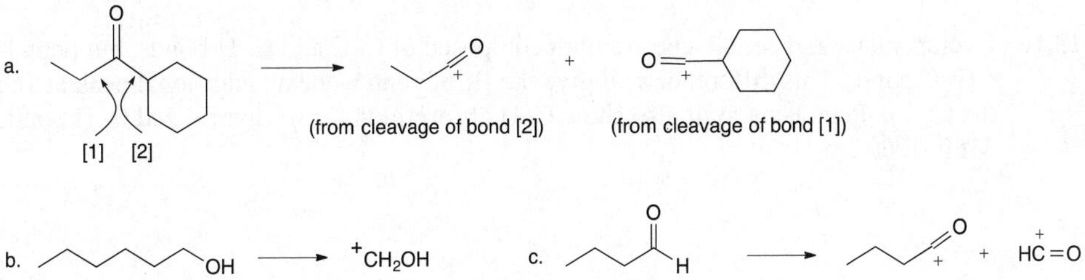

a.

[1]   [2]

(from cleavage of bond [2])   (from cleavage of bond [1])

b.

$^+CH_2OH$

c.

+ $HC=O$

Chapter 13–6

13.10 Use the exact mass values given in Table 13.1 to calculate the exact mass of each compound.

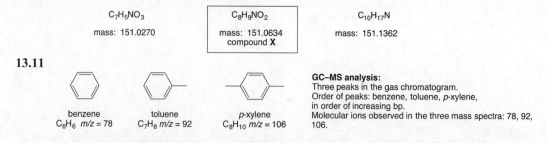

13.11

benzene
C_6H_6 $m/z = 78$

toluene
C_7H_8 $m/z = 92$

p-xylene
C_8H_{10} $m/z = 106$

GC–MS analysis:
Three peaks in the gas chromatogram.
Order of peaks: benzene, toluene, *p*-xylene, in order of increasing bp.
Molecular ions observed in the three mass spectra: 78, 92, 106.

13.12 Wavelength and frequency are inversely proportional. The higher frequency light will have a shorter wavelength.
  a. Light having $\lambda = 10^2$ nm has a higher ν than light with $\lambda = 10^4$ nm.
  b. Light having $\lambda = 100$ nm has a higher ν than light with $\lambda = 100$ μm.
  c. Blue light has a higher ν than red light.

13.13 The **energy of a photon** is *proportional* to its **frequency,** and inversely proportional to its wavelength.
  a. Light having $\nu = 10^8$ Hz is of higher energy than light having $\nu = 10^4$ Hz.
  b. Light having $\lambda = 10$ nm is of higher energy than light having $\lambda = 1000$ nm.
  c. Blue light is of higher energy than red light.

13.14 Higher wavenumbers are proportional to higher frequencies and higher energies.
  a. IR light with a wavenumber of 3000 cm^{-1} is higher in energy than IR light with a wavenumber of 1500 cm^{-1}.
  b. IR light having $\lambda = 10$ μm is higher in energy than IR light having $\lambda = 20$ μm.

13.15 Stronger bonds absorb at a higher wavenumber. Bonds to lighter atoms (H versus D) absorb at higher wavenumber.

13.16 Cyclopentane and pent-1-ene are both composed of C–C and C–H bonds, but pent-1-ene also has a C=C bond. This difference will give the IR of pent-1-ene an additional peak at 1650 cm^{-1} (for the C=C). Pent-1-ene will also show C–H absorptions for sp^2 hybridized C–H bonds at 3150–3000 cm^{-1}.

13.17 Look at the functional groups in each compound below to explain how each IR is different.

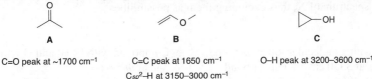

A
C=O peak at ~1700 cm^{-1}

B
C=C peak at 1650 cm^{-1}
C$_{sp^2}$–H at 3150–3000 cm^{-1}

C
O–H peak at 3200–3600 cm^{-1}

13.18 a. Compound **A** has peaks at ~3150 (sp^2 hybridized C–H), 3000–2850 (sp^3 hybridized C–H), and 1650 (C=C) cm^{-1}.
b. Compound **B** has a peak at 3000–2850 (sp^3 hybridized C–H) cm^{-1}.

13.19 All compounds show an absorption at 3000–2850 cm^{-1} due to the sp^3 hybridized C–H bonds. Additional peaks in the functional group region for each compound are shown.

a.
no additional peaks

b.
O–H bond at 3600–3200 cm^{-1}

c.
C$_{sp^2}$–H at 3150–3000 cm^{-1}
C=C bond at 1650 cm^{-1}

d.
C=O bond at ~1700 cm^{-1}

e.
O–H at 3600–3200 cm^{-1}
N–H at 3500–3200 cm^{-1}
C$_{sp^2}$–H at 3150–3000 cm^{-1}
C=O at ~1700 cm^{-1}
C=C at 1650 cm^{-1}
aromatic ring at 1600, 1500 cm^{-1}

13.20

C$_{sp^2}$–H at 3150–3000 cm^{-1}
C$_{sp^3}$–H at 3000–2850 cm^{-1}
C=O at ~1700 cm^{-1}
C=C at 1650 cm^{-1}
O–H above 3000 cm^{-1}
[The OH of a COOH is much broader than the OH of an alcohol and occurs at 3500–2500 cm^{-1} (see Chapter 19).]

Chapter 13–8

13.21 Possible structures are (a) CH₃COOCH₂CH₃ and (c) CH₃CH₂COOCH₃. Compounds (b) and (d) also have an OH group that would give a strong absorption at ~3600–3200 cm⁻¹, which is absent in the IR spectrum of **X**, thus excluding them as possibilities.

13.22
a. Hydrocarbon with a molecular ion at m/z = 68
   IR absorptions at 3310 cm⁻¹ = Csp–H bond
   3000–2850 cm⁻¹ = Csp^3–H bonds
   2120 cm⁻¹ = C≡C bond
   Molecular formula: C₅H₈

b. Compound with C, H, and O with a molecular ion at m/z = 60
   IR absorptions at 3600–3200 cm⁻¹ = O–H bond
   3000–2850 cm⁻¹ = Csp^3–H bonds
   Molecular formula: C₃H₈O

13.23

a.
   Csp^2–H at 3000–3150 cm⁻¹
   Csp^3–H at 2850–3000 cm⁻¹
   C=C at 1650 cm⁻¹

b. O–H at 3200–3600 cm⁻¹
   Csp^2–H at 3000–3150 cm⁻¹
   Csp^3–H at 2850–3000 cm⁻¹
   C=C at 1650 cm⁻¹

13.24

Cleave bond [1]. → ·CH₃ + [fragment] m/z = 127
Cleave bond [2]. → [fragment] + [fragment] m/z = 85
Cleave bond [3]. → ·CH₂CH₃ + [fragment] m/z = 113

m/z = 142

13.25

a. molecular formula: C₆H₆
   molecular ion (m/z): **78**

b. molecular formula: C₁₀H₁₆
   molecular ion (m/z): **136**

c. molecular formula: C₅H₁₀O
   molecular ion (m/z): **86**

d. molecular formula: C₅H₁₁Cl
   molecular ions (m/z): **106, 108**

e. molecular formula: C₈H₁₇Br
   molecular ions (m/z): **192, 194**

13.26

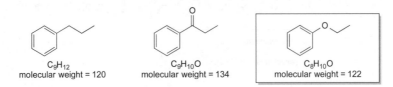

13.27 Examples are given for each molecular ion.
  a. molecular ion 102: C_8H_6, $C_6H_{14}O$, $C_5H_{10}O_2$, $C_5H_{14}N_2$
  b. molecular ion 98: C_8H_2, C_7H_{14}, $C_6H_{10}O$, $C_5H_6O_2$
  c. molecular ion 119: C_8H_9N, $C_6H_5N_3$
  d. molecular ion 74: C_6H_2, $C_4H_{10}O$, $C_3H_6O_2$

13.28 Likely molecular formula, C_8H_{16} (one degree of unsaturation—one ring or one π bond).

Four structures with m/z = 112

13.29 Use the molecular ion to propose molecular formulas for a compound with only C and H. Then determine which formula has four degrees of unsaturation.

M at 204: C_{17} (no H's)

$C_{16}H_{12}$ (not enough H's)

$C_{15}H_{24}$ (four degress of unsaturation)
Answer

13.30 Use the directions from Answer 13.3 and propose a formula with two degrees of unsaturation.

M at 154: $C_{12}H_{10}$ (not enough H's)

$C_{11}H_{22}$ $\xrightarrow[+\text{O}]{-\text{CH}_4}$ $C_{10}H_{18}O$ (two degrees of unsaturation)
Answer

13.31

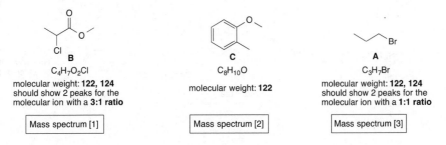

13.32

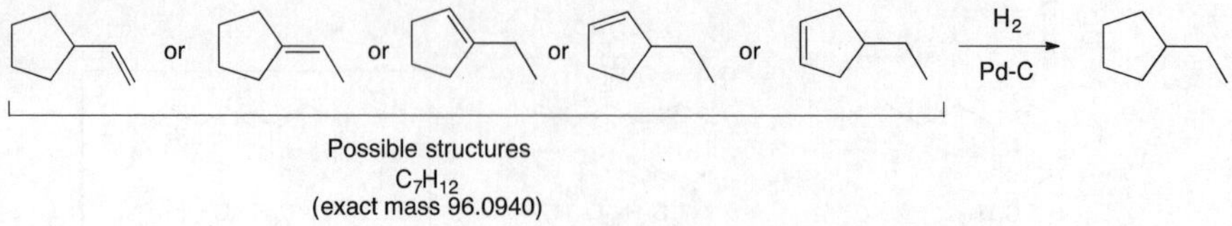
or
or
or
or
H_2
Pd-C
Possible structures
C_7H_{12}
(exact mass 96.0940)

13.33

a.

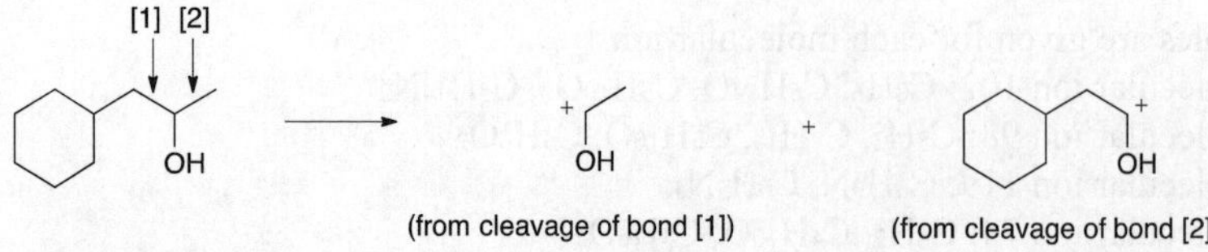
[1] [2]
OH
OH
OH
(from cleavage of bond [1])
(from cleavage of bond [2])

b.

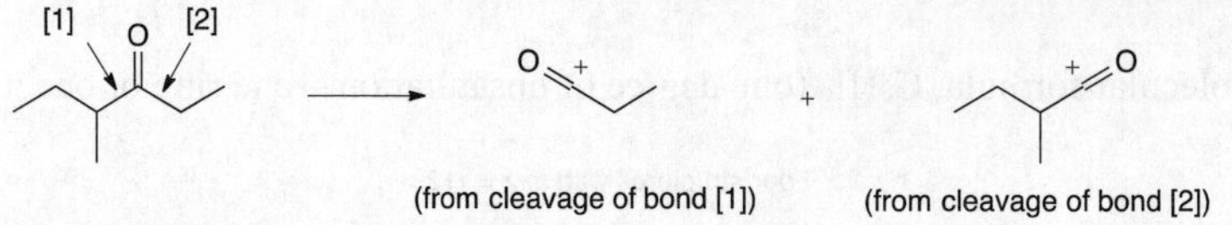
[1]
O
[2]
O
O
(from cleavage of bond [1])
(from cleavage of bond [2])

c.

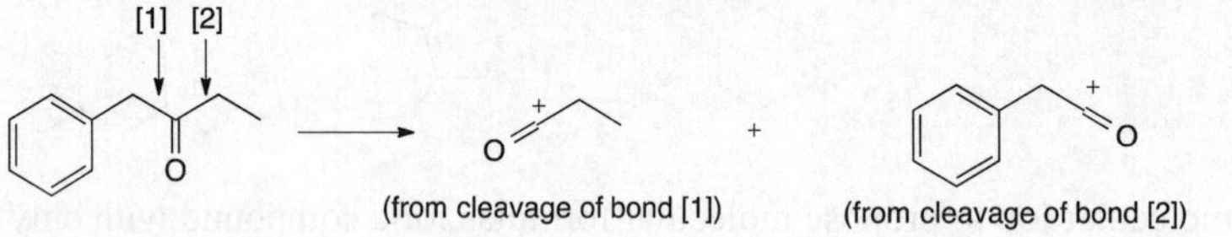
[1] [2]
O
O
O
(from cleavage of bond [1])
(from cleavage of bond [2])

13.34 2,2-Dimethylbutane shows a peak at m/z = 57 in it s mass spectrum due to a 3° carbocation. 2,3-Dimethylbutane shows a peak at m/z = 43 due to a 2° carbocation.

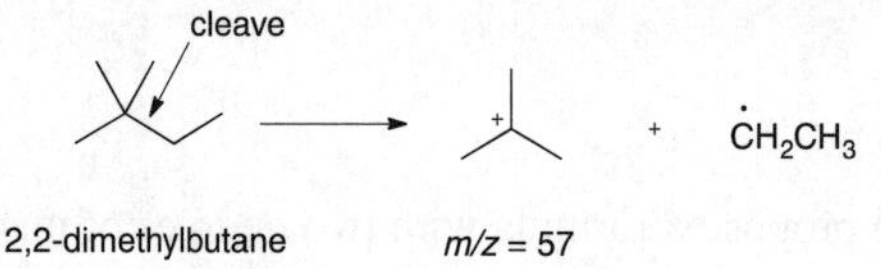
cleave
$\dot{C}H_2CH_3$
2,2-dimethylbutane
m/z = 57

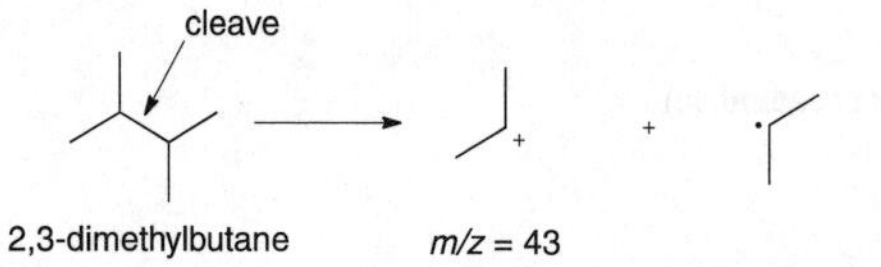
cleave
2,3-dimethylbutane
m/z = 43

13.35

a.

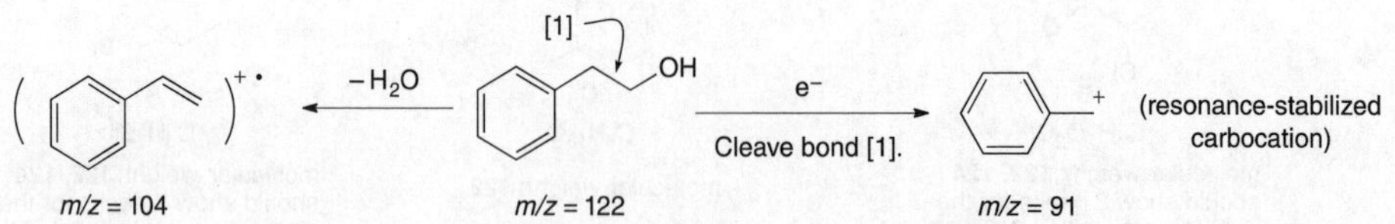
[1]
OH
– H_2O
e⁻
Cleave bond [1].
(resonance-stabilized
carbocation)
m/z = 104
m/z = 122
m/z = 91

Mass Spectrometry and Infrared Spectroscopy 13–11

b.

Cleave bond [1].

m/z = 71

Cleave bond [2].

m/z = 41

Cleave bond [3].

$^+CH_2-OH$

m/z = 31

m/z = 68   − H$_2$O   [1] [3]   [2]   m/z = 86

13.36

Ketone **A**

m/z = 128

α cleavage

·CH$_2$CH$_3$ +

m/z = 99

This is ketone **A** because α cleavage gives a fragment with m/z of 99.

Ketone **B**

m/z = 128

α cleavage

·CH$_3$ +

m/z = 113

This is ketone **B** because α cleavage gives a fragment with m/z of 113.

13.37 One possible structure is drawn for each set of data:

a. a compound that contains a benzene ring and has a molecular ion at m/z = 107

C$_7$H$_9$N

b. a hydrocarbon that contains only sp^3 hybridized carbons and a molecular ion at m/z = 84

C$_6$H$_{12}$

c. a compound that contains a carbonyl group and gives a molecular ion at m/z = 114

C$_7$H$_{14}$O

d. a compound that contains C, H, N, and O and has an exact mass for the molecular ion at 101.0841

C$_5$H$_{11}$NO

13.38 Use the values given in Table 13.1 to calculate the exact mass of each compound. C$_8$H$_{11}$NO$_2$ (exact mass 153.0790) is the correct molecular formula.

13.39 Alpha cleavage of a 1° alcohol (RCH$_2$OH) forms an alkyl radical (R•) and a resonance-stabilized carbocation with m/z = 31.

$+$ CH$_2$OH ⟷ CH$_2$=$\overset{+}{O}$H   m/z = 31

resonance-stabilized carbocation

Chapter 13–12

13.40 An ether fragments by α cleavage because the resulting carbocation is resonance stabilized.

resonance-stabilized carbocations

13.41

a.   stronger bond
     higher ṽ absorption

b.   stronger bond
     higher ṽ absorption

c.   stronger bond
     higher ṽ absorption

13.42 Locate the functional groups in each compound. Use Table 13.2 to determine what IR absorptions each would have.

a.
C_{sp}–H at 3300 cm^{-1}
C_{sp^3}–H at 2850–3000 cm^{-1}
C–C triple bond at 2250 cm^{-1}

c.
C_{sp^3}–H at 2850–3000 cm^{-1}
C=O at 1700 cm^{-1}

b.
O–H at 3200–3600 cm^{-1}
C_{sp^3}–H at 2850–3000 cm^{-1}

d.
O–H at > 3000 cm^{-1}
C_{sp^2}–H at 3000–3150 cm^{-1}
C=O at ~1700 cm^{-1}
phenyl group at 1600, 1500 cm^{-1}

The OH of the RCOOH is even broader than the OH of an alcohol (3500–2500 cm^{-1}), as we will learn in Chapter 19.

13.43

a.   and
C=C bond   C≡C bond
1650 cm^{-1} 2250 cm^{-1}

C_{sp^2}–H at 3150–3000 cm^{-1} C_{sp}–H at 3300 cm^{-1}

c.   and
no C=O bond   C=O bond
                             ~1700 cm^{-1}

b.   and
O–H bond   no O–H bond
> 3000 cm^{-1}
[See note on OH in Answer 13.42.]

d.   and
C_{sp}–H bond
3300 cm^{-1}

13.44 The IR absorptions above 1500 cm^{-1} are different for each of the narcotics.

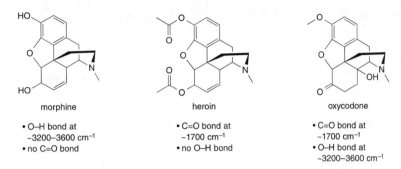

morphine
- O–H bond at ~3200–3600 cm^{-1}
- no C=O bond

heroin
- C=O bond at ~1700 cm^{-1}
- no O–H bond

oxycodone
- C=O bond at ~1700 cm^{-1}
- O–H bond at ~3200–3600 cm^{-1}

13.45 The three compounds show differences in their IR spectra.

cyclohexanone
C=O at ~1700 cm^{-1}

cyclohex-2-enol
C=C at ~1650 cm^{-1}
O–H at ~3200–3600 cm^{-1}
C$_{sp^2}$–H at ~3150–3000 cm^{-1}

cyclohexanol
O–H at ~3200–3600 cm^{-1}

13.46 Look for a **change in functional groups** from starting material to product to see how IR could be used to determine when the reaction is complete.

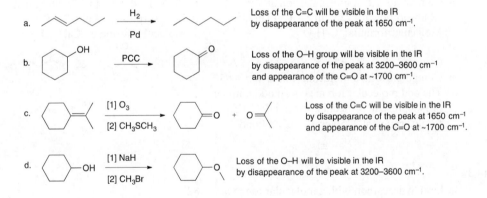

a. Loss of the C=C will be visible in the IR by disappearance of the peak at 1650 cm^{-1}.

b. Loss of the O–H group will be visible in the IR by disappearance of the peak at 3200–3600 cm^{-1} and appearance of the C=O at ~1700 cm^{-1}.

c. Loss of the C=C will be visible in the IR by disappearance of the peak at 1650 cm^{-1} and appearance of the C=O at ~1700 cm^{-1}.

d. Loss of the O–H will be visible in the IR by disappearance of the peak at 3200–3600 cm^{-1}.

Chapter 13–14

13.47 In addition to C_{sp^3}–H at ~3000–2850 cm^{-1}:

Spectrum [1]:

(B)
C=C peak at 1650 cm^{-1}
C_{sp^2}–H at ~3150 cm^{-1}

Spectrum [2]:

(F)
OH at 3600–3200 cm^{-1}

Spectrum [3]:

(D)
No other peaks above 1500 cm^{-1}

Spectrum [4]:

(C)
C_{sp^2}–H at ~3150 cm^{-1}
Phenyl peaks at 1600 and 1500 cm^{-1}

Spectrum [5]:

(A)
OH at ~3500–2500 cm^{-1}
C=O at ~1700 cm^{-1}

Spectrum [6]:

(E)
C=O at ~1700 cm^{-1}

13.48

a. Compound with a molecular ion at $m/z = 72$
IR absorption at 1725 cm^{-1} = C=O bond
Molecular formula: C_4H_8O

b. Compound with a molecular ion at $m/z = 55$
The odd molecular ion means an odd number
of N's present. Molecular formula: C_3H_5N
IR absorption at 2250 cm^{-1} = C≡N bond

c. Compound with a molecular ion at $m/z = 74$
IR absorption at 3600–3200 cm^{-1} = O–H bond
Molecular formula: $C_4H_{10}O$

13.49

Chiral hydrocarbon with a molecular ion at $m/z = 82$
Molecular formula: C_6H_{10}
IR absorptions at 3300 cm^{-1} = C_{sp}–H bond
3000–2850 cm^{-1} = C_{sp^3}–H bonds
2250 cm^{-1} = C≡C bond

stereogenic center

Two possible enantiomers:

or

Mass Spectrometry and Infrared Spectroscopy 13–15

13.50 The chiral compound **Y** has a strong absorption at 2970–2840 cm^{-1} in its IR spectrum due to sp^3 hybridized C–H bonds. The two peaks of equal intensity at 136 and 138 indicate the presence of a Br atom. The molecular formula is C$_4$H$_9$Br. Only one constitutional isomer of this molecular formula has a stereogenic center:

Y = [structures] and [structure]

two possible enantiomers

13.51 The molecular ion of 192 suggests C$_{12}$H$_{16}$O$_2$ as a possible molecular formula. IR absorption at 1721 cm^{-1} is due to a C=O, and the absorptions around 3000 cm^{-1} are due to C$_{sp^2}$–H and C$_{sp^3}$–H. The compound is an ester, formed in the following manner.

[reaction scheme]

H
C$_{12}$H$_{16}$O$_2$

13.52

[reaction scheme]

Z

m/z = 92; molecular formula C$_7$H$_8$
IR absorptions at:
3150–2950 cm^{-1} = C$_{sp^3}$–H and C$_{sp^2}$–H bonds
1605 cm^{-1} and 1496 cm^{-1} due to phenyl group

13.53

[reaction mechanism scheme]

B

13.54 The molecular ion of 144 suggests C$_8$H$_{16}$O$_2$ as a possible molecular formula for **X.** The IR absorption at 1739 cm^{-1} is due to a C=O, and the absorptions at less than 3000 cm^{-1} are due to C$_{sp^3}$–H.

[reaction scheme]

2-methylpropan-1-ol

C$_8$H$_{16}$O$_2$
X
possible structure

Chapter 13–16

13.55

J

$C_6H_{12}O$
$m/z = 100$
IR absorption at 2962 cm^{-1} = C_{sp^3}–H bonds
1718 cm^{-1} = C=O bond

fragments:

α cleavage product
$m/z = 43$

α cleavage product
$m/z = 85$

The fragment at $m/z = 57$ could be due to $(C_4H_9)^+$ or $(C_3H_5O)^+$.

13.56

K

C_7H_9N
$m/z = 107$
IR absorptions at 3373 and 3290 cm^{-1} = N–H

3062 cm^{-1} = C_{sp^2}–H bonds
2920 cm^{-1} = C_{sp^3}–H bonds
1600 cm^{-1} = benzene ring

The odd molecular ion indicates the presence of a N atom.

L
C_7H_6O
$m/z = 106$

$m/z = 105$ +

$m/z = 77$ +

IR absorption at 3068 cm^{-1} = C_{sp^2}–H bonds on ring
2820 cm^{-1} and 2736 cm^{-1} = C–H of RCHO (Appendix E)
1703 cm^{-1} = C=O bond
1600 cm^{-1} = aromatic ring

13.57

Possible structures of **P**:

C_7H_7ClO

$m/z = 142, 144$
IR absorption at 3096–2837 cm^{-1} = C_{sp^3}–H bonds and C_{sp^2}–H bonds
1582 cm^{-1} and 1494 cm^{-1} = benzene ring
The peak at M + 2 shows the presence of Cl or Br. Since Cl$_2$ is a reactant, the compound presumably contains Cl.

13.58 The mass spectrum has a molecular ion at 71. The odd mass suggests the presence of an odd number of N atoms; likely formula, C$_4$H$_9$N. The IR absorption at ~3300 cm^{-1} is due to N–H and the 3000–2850 cm^{-1} is due to sp^3 hybridized C–H bonds.

W

13.59 Because the carbonyl absorption of an amide is at lower wavenumber than the carbonyl absorption of an ester, the C=O of the amide must be weaker and have more single bond character. This can be explained by resonance. Although both an ester and amide are resonance stabilized, the N atom of the amide is more basic, making it more willing to donate its electron pair.

![amide and ester resonance structures]

Because the amide carbonyl has more single bond character, the bond is weaker and it absorbs at lower wavenumber.

13.60 The α,β-unsaturated carbonyl compound has three resonance structures, two of which place a single bond between the C and O atoms. This means that the C–O bond has partial single bond character, making it weaker than a regular C=O bond, and moving the absorption to lower wavenumber.

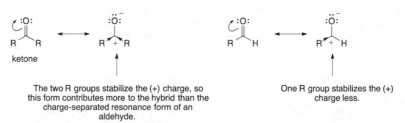

three resonance structures for cyclohex-2-enone

13.61 If a ketone carbonyl absorbs at lower wavenumber than an aldehyde carbonyl, the ketone carbonyl is weaker and has more single bond character. This can be explained by the fact that R groups are electron donating and stabilize an adjacent (+) charge.

![ketone and aldehyde resonance structures]

As a result, the charge-separated resonance form of a ketone, which contains a C–O single bond, contributes more to the hybrid of a ketone, making the C=O weaker and shifting the absorption to lower wavenumber.

13.62

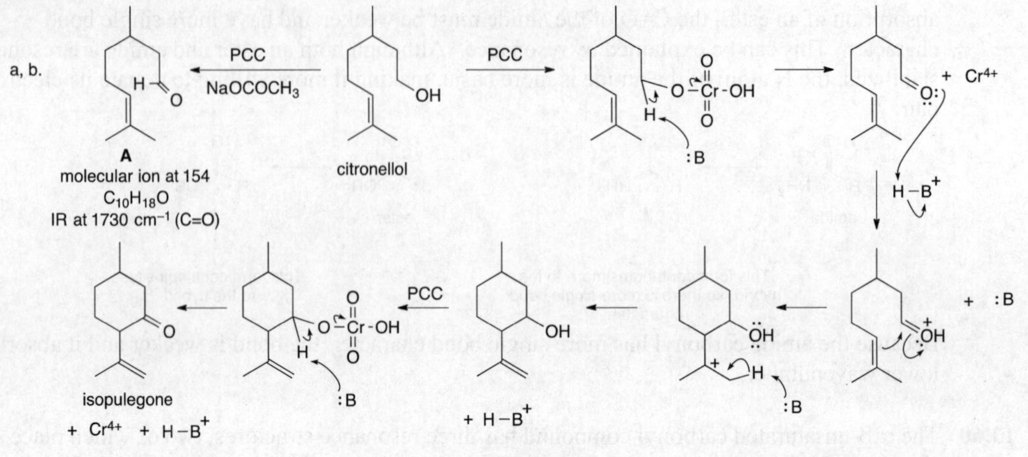
a, b.
PCC
NaOCOCH3
PCC
citronellol
A
molecular ion at 154
C10H18O
IR at 1730 cm−1 (C=O)
Cr4+
isopulegone
+ Cr4+ + H–B+
+ H–B+
PCC

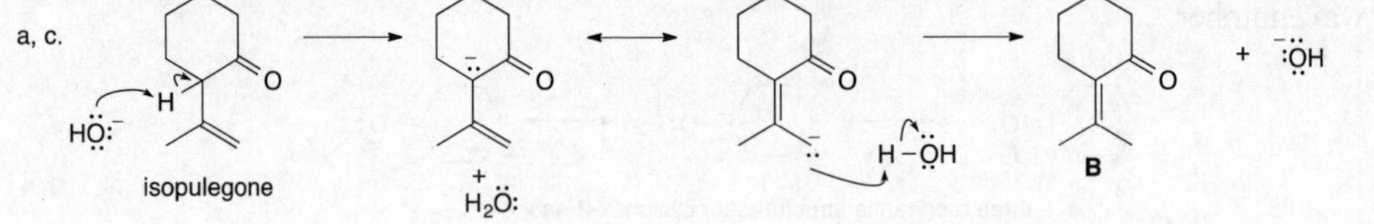
a, c.
isopulegone
B

Chapter 14 Nuclear Magnetic Resonance Spectroscopy

Chapter Review

¹H NMR spectroscopy
[1] The **number of signals** equals the number of different types of protons (14.2).

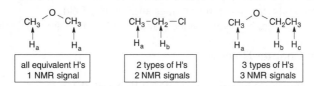

[2] The **position of a signal** (its chemical shift) is determined by shielding and deshielding effects.
- Shielding shifts an absorption upfield; deshielding shifts an absorption downfield.
- Electronegative atoms withdraw electron density, deshield a nucleus, and shift an absorption downfield (14.3).

- Loosely held π electrons can either shield or deshield a nucleus. Protons on benzene rings and double bonds are deshielded and absorb downfield, whereas protons on triple bonds are shielded and absorb upfield (14.4).

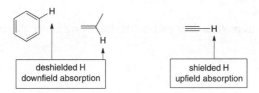

[3] The **area under an NMR signal** is proportional to the number of absorbing protons (14.5).

[4] **Spin–spin splitting** tells about nearby nonequivalent protons (14.6–14.8).
- Equivalent protons do not split each other's signals.
- A set of n nonequivalent protons on the same carbon or adjacent carbons split an NMR signal into $n + 1$ peaks.
- OH and NH protons do not cause splitting (14.9).
- When an absorbing proton has two sets of nearby nonequivalent protons that are equivalent to each other, use the $n + 1$ rule to determine splitting.
- When an absorbing proton has two sets of nearby nonequivalent protons that are not equivalent to each other, the number of peaks in the NMR signal = $(n + 1)(m + 1)$. In flexible alkyl chains, peak overlap often occurs, resulting in $n + m + 1$ peaks in an NMR signal.

Chapter 14–2

^{13}C NMR spectroscopy (14.11)

[1] The number of signals equals the number of different types of carbon atoms. All signals are single
lines.

[2] The relative position of ^{13}C signals is determined by shielding and deshielding effects.
- Carbons that are sp^3 hybridized are shielded and absorb upfield.
- Electronegative elements (N, O, and X) shift absorptions downfield.
- The carbons of alkenes and benzene rings absorb downfield.
- Carbonyl carbons are highly deshielded, and absorb farther downfield than other carbon types.

Practice Test on Chapter Review

1.a. Which of the following statements is true about ^1H NMR absorptions?
   1.   A signal that occurs at 1800 Hz on a 300 MHz NMR spectrometer occurs at 3000 Hz on a
        500 MHz NMR spectrometer.
   2.   A signal that occurs at 3.3 ppm on a 60 MHz NMR absorbs at 198 Hz upfield from TMS.
   3.   A signal that occurs at 600 Hz is downfield from a signal that occurs at 800 Hz.
   4.   Statements (1) and (2) are both true.
   5.   Statements (1), (2), and (3) are all true.

  b. Which of the following statements is true about ^1H NMR spectroscopy?
   1.   Electronegative elements shield a nucleus so an absorption shifts downfield.
   2.   A triplet is due to a proton that has four adjacent nonequivalent protons.
   3.   Circulating π electrons create a magnetic field that reinforces the applied field in the vicinity of
        the protons in benzene.
   4.   Statements (1) and (2) are both true.
   5.   Statements (1), (2), and (3) are all true.

2. How many different types of protons does each of the following molecules contain?

3. Into how many peaks will each of the circled protons be split in a proton NMR spectrum?

Nuclear Magnetic Resonance Spectroscopy 14–3

4. How many lines are presents in the ^{13}C NMR spectrum of each compound?

a. CH_3O ⟍⟍⟍⟍

c. ⟍⟍⟍⟍⟍ Cl

e. ⟍⟍⟍⟍⟍

b. ⟍⟍⟍⟍⟍⟍

d. ⟍⟍⟍O⟍⟍

f. ⟍⟍⟍

5. With reference to the ^1H NMR absorptions in the following compound, (a) which proton absorbs farthest upfield; (b) which proton absorbs farthest downfield?

H_a —H_b H_d
O
O H_c

6. With reference to the ^{13}C NMR absorptions in the following compound, (a) which carbon absorbs farthest downfield; (b) which carbon absorbs farthest upfield?

C_b O
C_a → F C_c
C_d

Answers to Practice Test

1. a. 1	2.a. 5	3.a. 8	4.a. 6	5.a. H_b	6.a. C_c
b. 3	b. 5	b. 3	b. 4	b. H_c	b. C_a
	c. 4	c. 7	c. 6		
	d. 5	d. 4	d. 4		
	e. 4	e. 4	e. 4		
	f. 5	f. 3	f. 5		
	g. 3	g. 8			
	h. 9				

Answers to Problems

14.1 Use the formula δ = [observed chemical shift (Hz)/ν of the NMR (MHz)] to calculate the chemical shifts.

a. **CH_3 protons:**
δ = [1715 Hz] / [500 MHz]
= 3.43 ppm

OH proton:
δ = [1830 Hz] / [500 MHz]
= 3.66 ppm

b. The positive direction of the δ scale is downfield from TMS. The CH_3 protons absorb upfield from the OH proton.

14.2 Calculate the chemical shifts as in Answer 14.1.

a. one signal:
δ = [1017 Hz] / [300 MHz]
= 3.39 ppm

second signal:
δ = [1065 Hz] / [300 MHz]
= 3.55 ppm

b. one signal:
3.39 = [x Hz] / [500 MHz]
x = 1695 Hz

second signal:
3.55 = [x Hz] / [500 MHz]
x = 1775 Hz

Chapter 14–4

14.3 To determine if two H's are equivalent replace each by an atom X. If this yields the same compound or mirror images, the two H's are equivalent. Each kind of H will give one NMR signal.

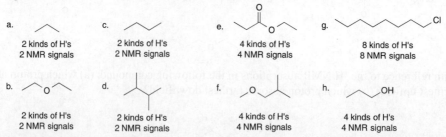

14.4 Draw in all of the H's and compare them. If two H's are cis and trans to the same group, they are equivalent.

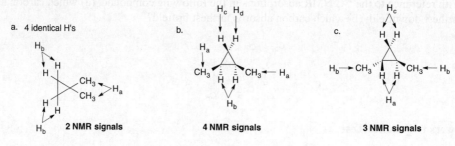

14.5

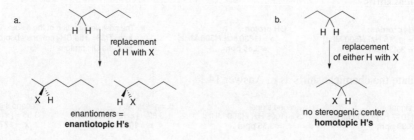

14.6 If replacement of H by X yields the same compound, the protons are **homotopic**.
If replacement of H with X yields enantiomers, the protons are **enantiotopic**.
If replacement of H with X yields diastereomers, the protons are **diastereotopic**. In general, if the compound has **one stereogenic center**, the protons in a CH$_2$ group are **diastereotopic**.

Nuclear Magnetic Resonance Spectroscopy 14–5

c.

Pick one configuration at the existing stereogenic center.

diastereomers =
diastereotopic H's

14.7 The two protons of a CH_2 group are different from each other if the compound has one stereogenic center. Replace one proton with X and compare the products.

a. The stereogenic center makes the H's in the CH_2 group diastereotopic and therefore different from each other.

5 NMR signals

b.

5 NMR signals

c.

7 NMR signals

14.8 Decreased electron density deshields a nucleus and the absorption goes downfield. Absorption also shifts downfield with increasing alkyl substitution.

a. $FCH_2CH_2CH_2Cl$
F is more electronegative than Cl. The CH_2 group adjacent to the F is more deshielded and the H's will absorb farther downfield.

b. $CH_3CH_2CH_2CH_2OCH_3$
The CH_2 group adjacent to the O will absorb farther downfield because it is closer to the electronegative O atom.

c. $CH_3OC(CH_3)_3$
The CH_3 group bonded to the O atom will absorb farther downfield.

14.9

a.

3 types of protons:
$H_b < H_c < H_a$

b.

3 types of protons:
$H_c < H_a < H_b$

c.

3 types of protons:
$H_c < H_a < H_b$

14.10 a. False. When a nucleus is strongly shielded, the effective field is smaller than the applied field and the absorption shifts upfield.

b. True.

c. False. A nucleus that is strongly deshielded requires a higher field strength for resonance. Alternatively, a nucleus that is strongly *shielded* requires a lower field strength for resonance.

d. False. A nucleus that is strongly shielded absorbs at a smaller δ value. Alternatively, a nucleus that is strongly *deshielded* absorbs at a larger δ value.

Chapter 14–6

14.11

a. H–C≡C–**H**ₐ   CH₂=CH–C**H**ᵦ   CH₃–CH₂–C**H**ᵧ

Hᵧ protons are shielded because they are bonded to an sp^3 C.
Hₐ is shielded because it is bonded to an sp C.
Hᵦ protons are deshielded because they are bonded to an sp^2 C.

$H_c < H_a < H_b$

b. CH₃–C(=O)–OCH₂CH₃
   Hₐ   **H**ᵦ **H**ᵧ

Hᵧ protons are shielded because they are bonded to an sp^3 C.
Hₐ protons are deshielded slightly because the CH₃ group is bonded to a C=O.
Hᵦ protons are deshielded because the CH₂ group is bonded to an O atom.

$H_c < H_a < H_b$

14.12
An integration ratio of 2:3 means that there are two types of hydrogens in the compound, and that the ratio of one type to another type is 2:3.

a. CH₃CH₂Cl
2 types of H's
3:2 - YES

b. (CH₃)₂CHCH₃
2 types of H's
6:2 or 3:1 - no

c. CH₃CH₂OCH₂CH₃
2 types of H's
6:4 or 3:2 - YES

d. CH₃OCH₂CH₂OCH₃
2 types of H's
6:4 or 3:2 - YES

14.13

downfield absorption
closer to O
↓
CH₃O₂CCH₂CH₂CO₂CH₃
A
ratio of absorbing signals 2:3

Signal [1] = 4 H = 2.64
Signal [2] = 6 H = 3.69 ← 6 H's with downfield absorption

downfield absorption
closer to O
↓
CH₃CO₂CH₂CH₂O₂CCH₃
B
ratio of absorbing signals 3:2

Signal [1] = 6 H = 2.09
Signal [2] = 4 H = 4.27 ← 4 H's with downfield absorption

14.14
To determine the **splitting pattern** for a molecule:
- Determine the number of different kinds of protons.
- Nonequivalent protons on the same C or adjacent C's split each other.
- Apply the $n + 1$ rule.

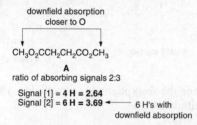

a. CH₃CH₂–C(=O)Cl
   Hₐ Hᵦ
Hₐ: 3 peaks - triplet
Hᵦ: 4 peaks - quartet

c. CH₃–C(=O)–CH₂CH₂Br
   Hₐ   Hᵦ Hᵧ
Hₐ: 1 peak - singlet
Hᵦ: 3 peaks - triplet
Hᵧ: 3 peaks - triplet

e. (CH₃)₂C=CHH
   Hₐ, Hᵦ
Hₐ: 2 peaks - doublet
Hᵦ: 2 peaks - doublet

b. (CH₃)₂CBr₂... CH₃–CHBr–CHBr–H
   Hₐ   Hᵦ
Hₐ: 2 peaks - doublet
Hᵦ: 4 peaks - quartet

d. Hₐ→H   Cl
       C=C
    Br   H←Hᵦ
Hₐ: 2 peaks - doublet
Hᵦ: 2 peaks - doublet

f. ClCH₂–CH(O–)(O–)
   Hₐ   Hᵦ
Hₐ: 2 peaks - doublet
Hᵦ: 3 peaks - triplet

14.15 Use the directions from Answer 14.14.

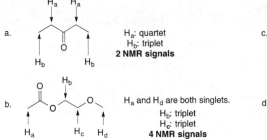

a. H$_a$: quartet
H$_b$: triplet
2 NMR signals

b. H$_a$ and H$_d$ are both singlets.
H$_b$: triplet
H$_c$: triplet
4 NMR signals

c. H$_a$: doublet
H$_b$: quartet
2 NMR signals

d. H$_a$: triplet
H$_b$: doublet
H$_c$: singlet
3 NMR signals

14.16 CH$_3$CH$_2$Cl

chemical shift (ppm)

There are two kinds of protons, and they can split each other. The CH$_3$ signal will be split by the CH$_2$ protons into 2 + 1 = 3 peaks. It will be upfield from the CH$_2$ protons because it is farther from the Cl. The CH$_2$ signal will be split by the CH$_3$ protons into 3 + 1 = 4 peaks. It will be downfield from the CH$_3$ protons because the CH$_2$ protons are closer to the Cl. The ratio of integration units will be 3:2.

14.17

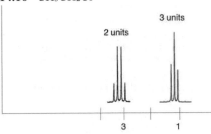

a. split by 6 equivalent H's
6 + 1 = **7 peaks**

b. H$_a$: split by 2 H's
3 peaks
H$_c$: split by 4 equivalent H's
5 peaks
H$_b$: split by 2 sets of H's
(3 + 1)(2 + 1) = **12 peaks (maximum)**
Since this is a flexible alkyl chain, the signal due to H$_b$ will have peak overlap, and 3 + 2 + 1 = **6 peaks** will likely be visible.

c. H$_a$: split by 1 H
2 peaks
H$_b$: split by 2 sets of H's
(1 + 1)(2 + 1) = **6 peaks**

d. H$_a$: split by 2 different H's
(1+1)(1+1) = **4 peaks**
H$_b$: split by 2 different H's
(1+1)(1+1) = **4 peaks**
H$_c$: split by 2 different H's
(1+1)(1+1) = **4 peaks**

14.18

a. H$_a$: singlet at ~3 ppm
H$_b$: quartet at ~3.5 ppm
H$_c$: triplet at ~1 ppm

b. H$_a$: triplet at ~1 ppm
H$_b$: quartet at ~2 ppm
H$_c$: septet at ~3.5 ppm
H$_d$: doublet at ~1 ppm

c. H$_a$: singlet at ~3 ppm
H$_b$: triplet at ~3.5 ppm
H$_c$: quintet at ~1.5 ppm

d. H$_a$: triplet at ~1 ppm
H$_b$: multiplet (8 peaks) at ~2.5 ppm
H$_c$: triplet at ~5 ppm

Chapter 14–8

14.19

trans-1,3-dichloropropene

2 H$_c$ protons

Splitting diagram for H$_b$

1 **trans** H$_a$ proton splits H$_b$ into
1 + 1 = 2 peaks
a doublet

J_{ab} = 13.1 Hz

2 H$_c$ protons split H$_b$ into
2 + 1 = 3 peaks
Now it's a doublet of triplets.

J_{bc} = 7.2 Hz

14.20

C$_3$H$_4$Cl$_2$

A
H$_a$: 1.75 ppm, doublet, 3 H, J = 6.9 Hz
H$_b$: 5.89 ppm, quartet, 1 H, J = 6.9 Hz

B
signal at 4.16 ppm, singlet, 2 H
signal at 5.42 ppm, doublet, 1 H, J = 1.9 Hz
signal at 5.59 ppm, doublet, 1 H, J = 1.9 Hz

singlet
doublet
doublet

14.21 Remember that OH (or NH) protons do not split other signals, and are not split by adjacent protons.

a.
singlet
singlet
singlet
3 NMR signals

b.
triplet   triplet
singlet
12 peaks (maximum)
6 peaks (more likely, resulting from peak overlap)
4 NMR signals

c.
doublet
singlet
7 peaks
3 NMR signals

14.22

H$_d$
5 H's on benzene ring

A

H$_a$: doublet at ~1.4 due to the CH$_3$ group, split into two peaks by one adjacent nonequivalent H (H$_c$).

H$_b$: singlet at ~2.7 due to the OH group. OH protons are not split by nor do they split adjacent protons.

H$_c$: quartet at ~4.7 due to the CH group, split into four peaks by the adjacent CH$_3$ group.

H$_d$: multiplets at ~7.2–7.4 due to five protons on the benzene ring.

14.23

palau'amine

H$_a$: one adjacent nonequivalent H, so two peaks

H$_b$: one adjacent nonequivalent H, so two peaks

H$_c$: H$_c$ is located on a N atom, so there is no splitting and it appears as one peak.

H$_d$: H$_d$ has one nonequivalent H on the same carbon and one on an adjacent carbon, so it is split into (1 + 1)(1 + 1) = 4 peaks (a doublet of doublets).

Nuclear Magnetic Resonance Spectroscopy 14–9

14.24 Use these steps to propose a structure consistent with the molecular formula, IR, and NMR data.
- Calculate the **degrees of unsaturation.**
- Use the IR data to determine what types of **functional groups** are present.
- Determine the number of different **types of protons.**
- Calculate the **number of H's** giving rise to each signal.
- Analyze the **splitting pattern** and put the molecule together.
- Use the **chemical shift** information to check the structure.

- Molecular formula $C_7H_{14}O_2$

  $2n + 2 = 2(7) + 2 = 16$
  $16 - 14 = 2/2 = $ **1 degree of unsaturation**
  1 π bond or 1 ring

- IR peak at 1740 cm^{-1}

  C=O absorption is around 1700 cm^{-1} (causes the degree of unsaturation).
  No signal at 3200–3600 cm^{-1} means there is no O–H bond.

- NMR data:

absorption	ppm	relative area
singlet	1.2	9 ⎯⎯⎯⎯▶ **9 H's**
triplet	1.3	3 ⎯⎯⎯⎯▶ **3 H's** (probably a CH_3 group)
quartet	4.1	2 ⎯⎯⎯⎯▶ **2 H's** (probably a CH_2 group)

  - 3 kinds of H's
  - number of H's per signal

    Because the sum of the relative areas equals the number of
    absorbing H's (9 + 3 + 2 = 14), the relative area shows the
    actual number of absorbing H's: 9 H's, 3 H's and 2 H's.
  - splitting pattern

    The singlet (9 H) is likely from a *tert*-butyl group: $-\overset{\overset{\displaystyle CH_3}{|}}{\underset{\underset{\displaystyle CH_3}{|}}{C}}-CH_3$

    The CH_3 and CH_2 groups split each other: CH_3-CH_2-

- Join the pieces together.

  Pick this structure due to the chemical shift data.
  The CH_2 group is shifted downfield (4 ppm), so it
  is close to the electron-withdrawing O.

Chapter 14–10

14.25

- Molecular formula: C_3H_8O ➢ Calculate degrees of unsaturation
$$2n + 2 = 2(3) + 2 = 8$$
$$8 - 8 = \textbf{0 degrees of unsaturation}$$

- IR peak at 3200–3600 cm^{-1} ➢ Peak at 3200–3600 cm^{-1} is due to an **O–H bond.**
- NMR data:
  - doublet at ~1.2 (6 H)
  - singlet at ~2.2 (1 H)
  - septet at ~4 (1 H)

3 types of H's
septet from 1 H ◄──── split by 6 H's
singlet from 1 H
doublet from 6 H's ◄──── split by 1 H
from the O–H proton

➢ Put information together:

HO⟨structure⟩

14.26 Identify each compound from the ^{1}H NMR data.

a. $CH_2=CHCOCH_3$ ──HCl──▶
A
CH₃ (singlet), triplet at 3.05, H H, H H, Cl, O
triplet at 3.6

b. $(CH_3)_2C=O$ ──base/H_2O──▶ B
singlet at 2.5
singlet at 1.3
singlet at 3.8
singlet at 2.2
OH

14.27 Each different kind of carbon atom will give a different ^{13}C NMR signal.

a. C_b C_a / C_a C_b
2 kinds of C's
2 ^{13}C NMR signals

b. Each C is different.
4 kinds of C's
4 ^{13}C NMR signals

c. C_b C_b / C_a C_c C_c C_a
same groups on both sides of O
3 kinds of C's
3 ^{13}C NMR signals

d. Each C is different.
4 kinds of C's
4 ^{13}C NMR signals

14.28

a. Cl, Cl, H_c, H_d
2 different H's
H_a and H_b
4 ^{1}H NMR signals

H_b, Cl, Cl, H_a H_a
2 ^{1}H NMR signals

H_b Cl, Cl, H_c H_a
3 ^{1}H NMR signals

Cl Cl
all H's identical
1 ^{1}H NMR signal

b. Cl, Cl
Each C is different.
3 kinds of C's
3 ^{13}C NMR signals

C_b, Cl, Cl, C_a C_a
2 kinds of C's
2 ^{13}C NMR signals

Cl, Cl
Each C is different.
3 kinds of C's
3 ^{13}C NMR signals

Cl Cl, C_a C_b C_a
2 kinds of C's
2 ^{13}C NMR signals

c. Although the number of ^{13}C signals cannot be used to distinguish these isomers, each isomer exhibits a different number of signals in its ^1H NMR spectrum. As a result, the isomers are distinguishable by ^1H NMR spectroscopy.

14.29

These 2 C's are different because they are cis and trans to different groups.

Every carbon is different so there are 10 lines for the 10 C atoms.

14.30 Electronegative elements shift absorptions downfield. The carbons of alkenes, benzene rings, and carbonyl groups are also shifted downfield.

a. The C closer to the electronegative O will be farther downfield.

b. The C of the CBr₂ group has two bonds to electronegative Br atoms and will be farther downfield.

c. The carbonyl carbon is highly deshielded and will be farther downfield.

d. The C atom that is part of the double bond will be farther downfield.

14.31
a. In order of lowest to highest chemical shift:

$C_d < C_a < C_c < C_b$

b. In order of lowest to highest chemical shift:

$C_a < C_b < C_c$

14.32

- molecular formula $C_4H_8O_2$
  $2n + 2 = 2(4) + 2 = 10$
  $10 - 8 = 2/2 = $ **1 degree of unsaturation**
- no IR peaks at 3200–3600 or 1700 cm^{-1}
  no O–H or C=O
- ^1H NMR spectrum at 3.69 ppm
  only one kind of proton
- ^{13}C NMR spectrum at 67 ppm
  only one kind of carbon

This structure satisifies all the data. One ring is one degree of unsaturation. All carbons and protons are identical.

14.33

- molecular formula C_4H_8O
  $2n + 2 = 2(4) + 2 = 10$
  $10 - 8 = 2/2 = $ **1 degree of unsaturation**
- ^{13}C NMR signal at > 160 ppm due to C=O

- molecular formula C_4H_8O
  $2n + 2 = 2(4) + 2 = 10$
  $10 - 8 = 2/2 = $ **1 degree of unsaturation**
- all ^{13}C NMR signals at < 160 ppm
  NO C=O

Chapter 14–12

14.34

A
a. 4 ¹H NMR signals
b. 5 ¹³C NMR signals (including the 4° C)

B
a. 6 ¹H NMR signals
b. 7 ¹³C NMR signals (including the carbonyl C)

14.35

C
a. 4 ¹H NMR signals
b. H$_a$: 1 adjacent H, so 2 peaks
H$_b$: 2 adjacent H's, so 3 peaks
H$_c$: 3 adjacent H's, so 4 peaks
H$_d$: 2 adjacent H's, so 3 peaks

D
a. 5 ¹H NMR signals
b. H$_a$: singlet
H$_b$: 2 adjacent H's, so 3 peaks
H$_c$: 2 adjacent H's, so 3 peaks
H$_d$: 1 nonequivalent H on the same C, so 2 peaks
H$_e$: 1 nonequivalent H on the same C, so 2 peaks

14.36 Use the directions from Answer 14.3.

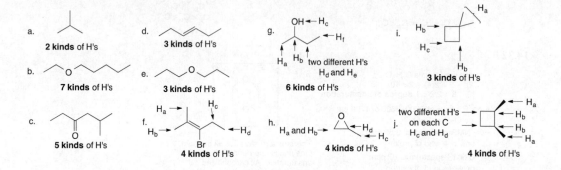

14.37

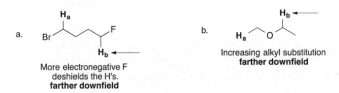

a. caffeine — **4 NMR signals**
b. vanillin — **6 NMR signals**
c. thymol — **7 NMR signals**
d. capsaicin — **15 NMR signals**

14.38

δ (in ppm) = [observed chemical shift (Hz)] / ν of the NMR (MHz)]

a. 2.5 = x Hz/300 MHz
   x = 750 Hz
b. ppm = 1200 Hz/300 MHz
   = 4 ppm
c. 2.0 = x Hz/300 MHz
   x = 600 Hz

14.39

a. The chemical shift in δ is independent of the operating frequency, so there is no change in δ when the ν is increased.

b. When the operating ν increases, the ν of an absorption increases as well, because the two quantities are proportional.

c. Coupling constants are independent of the operating ν, so the J value in Hz remains the same.

14.40 Use the directions from Answer 14.8.

a. More electronegative F deshields the H's. **farther downfield**

b. Increasing alkyl substitution **farther downfield**

14.41

$H_a : H_b$ = 3:2 $H_a : H_b$ = 3:1

different ratio of peak areas

14.42

a. H_a protons split by 1 H = **doublet**
   H_b proton split by 3 H's = **quartet**

b. both CH_2 groups split each other = **triplets**

c. H_a protons split by 1 H = **doublet**
   H_b proton split by 2 H's = **triplet**

Chapter 14–14

d.

H_a protons split by 1 H = **doublet**
H_b proton split by 6 H's = **septet**
H_c protons split by 3 H's = **quartet**
H_d protons split by 2 H's = **triplet**

g.

H_a protons split by 2 H's = **triplet**
H_c protons split by 2 H's = **triplet**
H_b protons split by CH_3 + CH_2 protons = **12 peaks**
(maximum)
Since H_b is located in a flexible alkyl chain, it is
likely that peak overlap occurs, so that only 3 + 2 +
1 = 6 peaks will be observed.

e.

H_a protons split by 2 CH_2 groups =
quintet
H_b protons split by 2 H's = **triplet**

h.

H_a: split by CH_3 group + H_b
= **8 peaks** (maximum)
H_b: split by 2 H's = **triplet**

f.

H_a protons split by 2 H's = **triplet**
H_b protons split by CH_3 + CH_2 protons = **12 peaks**
(maximum)
H_c protons split by 2 different CH_2 groups = **9 peaks**
(maximum)
H_d protons split by 2 H's = **triplet**
Since H_b and H_c are located in a flexible alkyl chain, it
is likely that peak overlap occurs, so that the following
is observed: H_b (3 + 2 + 1 = 6 peaks), H_c (2 + 2 + 1 = 5
peaks).

i.

H_a: split by 1 H = **doublet**
H_b: split by 1 H = **doublet**

j.

H_a: split by H_b + H_c =
doublet of doublets (4 peaks)
H_b: split by H_a + H_c =
doublet of doublets (4 peaks)
H_c: split by CH_3, H_a + H_b = **16 peaks**

14.43

H_a: split by 1 H = **doublet**
H_b: split by 1 H = **doublet**

H_a and H_b are geminal.

H_a: split by 1 H = **doublet**
H_b: split by 1 H = **doublet**

H_a and H_b are trans.

Both compounds exhibit two doublets for the H's on the C=C, but the
coupling constants ($J_{geminal}$ and J_{trans}) are different. $J_{geminal}$ is much
smaller than J_{trans} (0–3 Hz versus 11–18 Hz).

14.44

a.

H_a: 1 adjacent H_b = **doublet**
H_b: 1 adjacent H_a = **doublet**
H_c: no adjacent H's = **singlet**

b.

H_a: no adjacent H's = **singlet**
H_b: 1 adjacent H_c = **doublet**
H_c: 1 adjacent H_b and 2 adjacent H_d's
(1+1)(2 + 1) = **6 peaks**
H_d: 1 adjacent H_c = **doublet**

14.45

J_{ab} = 11.8 Hz
J_{bc} = 0.9 Hz
J_{ac} = 18 Hz

H$_a$: doublet of doublets at 5.7 ppm. Two large J values are seen for the H's cis (J_{ab} = 11.8 Hz) and trans (J_{ac} = 18 Hz) to H$_a$.

H$_b$: doublet of doublets at ~6.2 ppm. One large J value is seen for the cis H (J_{ab} = 11.8 Hz). The geminal coupling (J_{bc} = 0.9 Hz) is hard to see.

H$_c$: doublet of doublets at ~6.6 ppm. One large J value is seen for the trans H (J_{ac} = 18 Hz). The geminal coupling (J_{bc} = 0.9 Hz) is hard to see.

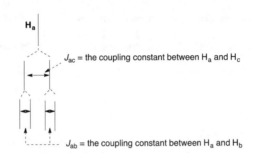

Splitting diagram for H$_a$

1 **trans** H$_c$ proton splits H$_a$ into
1 + 1 = 2 peaks
a doublet

1 **cis** H$_b$ proton splits H$_a$ into
1 + 1 = 2 peaks
Now it's a doublet of doublets.

J_{ac} = the coupling constant between H$_a$ and H$_c$

J_{ab} = the coupling constant between H$_a$ and H$_b$

14.46

Four constitutional isomers of **C$_4$H$_9$Br**:

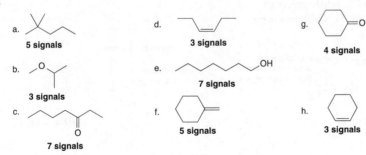

4 different C's   4 different C's   2 different C's   3 different C's

14.47

The O atom of an ester donates electron density, so the carbonyl carbon has less δ+, making it less deshielded than the carbonyl carbon of an aldehyde or ketone. Therefore, the carbonyl carbon of an aldehyde or ketone is more deshielded and absorbs farther downfield.

14.48

a. 5 signals

b. 3 signals

c. 7 signals

d. 3 signals

e. 7 signals

f. 5 signals

g. 4 signals

h. 3 signals

Chapter 14–16

14.49

a.

C_a C_b C_c

$C_a < C_b < C_c$

b.

C_a C_b C_c

$C_b < C_c < C_a$

14.50

19 ppm   62 ppm

a. $CH_3CH_2CH_2CH_2OH$

14 ppm   35 ppm

16 ppm   205 ppm

b. $(CH_3)_2CHCHO$

41 ppm

143 ppm   23 ppm

c. $CH_2=CHCH(OH)CH_3$

113 ppm   69 ppm

14.51

a.

C_a
C_c C_b
C_b
C_c
C_a

3 different C's
3 signals

b.

C_a C_a
C_b C_b
C_c C_c

3 signals

C_b C_a
C_d C_c
C_b
C_a

4 signals

C_d C_a
C_c
C_e C_b C_c
C_a

5 signals

14.52

OH

1-hydroxybutan-2-one

A

HO

4-hydroxybutan-2-one

B

The answers for parts (a)–(d) are the same for both compounds.
a. molecular ion for $C_4H_8O_2 = 88$
b. IR absorptions at 3200–3600 (OH), ~3000 (CH), and ~1700 (C=O) cm^{-1}.
c. Four lines in ^{13}C NMR spectrum
d. Four signals in ^1H NMR spectrum

e.

quartet at ~ 2.1 ppm   singlet at ~3.5 ppm
singlet anywhere in
the 1–5 ppm region
OH

triplet at ~1.0 ppm

A

triplet at ~3.5 ppm

singlet anywhere in
the 1–5 ppm region   singlet at ~2.0 ppm

HO

triplet at ~2.1 ppm

B

14.53 Use the directions from Answer 14.24.

a. **C₄H₈Br₂:** 0 degrees of unsaturation
   IR peak at 3000–2850 cm⁻¹: **Csp^3–H bonds**
   NMR: singlet at 1.87 ppm (6 H) (2 CH₃ groups)
          singlet at 3.86 ppm (2 H) (CH₂ group)

b. **C₃H₆Br₂:** 0 degrees of unsaturation
   IR peak at 3000–2850 cm⁻¹: **Csp^3–H bonds**
   NMR: quintet at 2.4 ppm (split by 2 CH₂ groups)
          triplet at 3.5 ppm (split by 2 H's)

c. **C₅H₁₀O₂:** 1 degree of unsaturation
   IR peak at 1740 cm⁻¹: **C=O**
   NMR: triplet at 1.15 ppm (3 H) (CH₃ split by 2 H's)
          triplet at 1.25 ppm (3 H) (CH₃ split by 2 H's)
          quartet at 2.30 ppm (2 H) (CH₂ split by 3 H's)
          quartet at 4.72 ppm (2 H) (CH₂ split by 3 H's)

d. **C₆H₁₄O:** 0 degrees of unsaturation
   IR peak at 3600–3200 cm⁻¹: **O–H**
   NMR: triplet at 0.8 ppm (6 H) (2 CH₃ groups split by CH₂ groups)
          singlet at 1.0 ppm (3 H) (CH₃)
          quartet at 1.5 ppm (4 H) (2 CH₂ groups split by CH₃ groups)
          singlet at 1.6 ppm (1 H) (O–H proton)

e. **C₆H₁₄O:** 0 degrees of unsaturation
   IR peak at 3000–2850 cm⁻¹: **Csp^3–H bonds**
   NMR: doublet at 1.10 ppm (relative area = 6)
          (from 12 H's)
          septet at 3.60 ppm (relative area = 1)
          (from 2 H's)

f. **C₃H₆O:** 1 degree of unsaturation
   IR peak at 1730 cm⁻¹: **C=O**
   NMR: triplet at 1.11 ppm
          multiplet at 2.46 ppm
          triplet at 9.79 ppm

14.54
Two isomers of C₉H₁₀O: 5 degrees of unsaturation (benzene ring likely)

Compound A:
IR absorption at 1742 cm⁻¹: **C=O**
NMR data:
Absorptions:
singlet at 2.15 (3 H) (CH₃ group)
singlet at 3.70 (2 H) (CH₂ group)
broad singlet at 7.20 (5 H)
    (likely a monosubstituted benzene ring)

Compound B:
IR absorption at 1688 cm⁻¹: **C=O**
NMR data:
Absorptions:
triplet at 1.22 (3 H) (CH₃ group split by 2 H's)
quartet at 2.98 (2 H) (CH₂ group split by 3 H's)
multiplet at 7.28–7.95 (5 H)
    (likely a monosubstituted benzene ring)

14.55 IR absorptions:
    3088–2897 cm⁻¹: sp^2 and sp^3 hybridized C–H
    1740 cm⁻¹: C=O
    1606 cm⁻¹: benzene ring

Chapter 14–18

triplet at 2.91

5 H
multiplet
7.20–7.35

singlet at 2.02

triplet at 4.25

$C_{10}H_{12}O_2$
W

14.56 IR absorption at 1713 cm^{-1} is due to C=O.

doublet at 1.09

triplet at 2.43

CH_3 H H

CH_3

CH_3 ← *triplet at 0.91*

H
O

H H

septet at 2.60

$C_7H_{14}O$
V

multiplet at 1.6

14.57

Compound C:
  molecular ion 146 (molecular formula $C_6H_{10}O_4$)
  IR absorption at 1762 cm^{-1}: **C=O**
  ^{1}H NMR data:
    Absorptions:
    H_a: doublet at 1.47 (3 H) (CH$_3$ group adjacent to CH)
    H_b: singlet at 2.07 (6 H) (2 CH$_3$ groups)
    H_c: quartet at 6.84 (1 H adjacent to CH$_3$)

H_c

O
‖
CH_3—C—O

H
|
C
|
CH_3

O
‖
O—C—CH_3

H_b

H_a

H_b

14.58

[1] LiC≡CH
[2] H$_2$O

H_b
OH
|
CH_3—C—C≡CH
|
CH_3

H_c

H_a

H_a

D

Compound D:
  molecular ion 84 (molecular formula C_5H_8O)
  IR absorptions at 3600–3200 cm^{-1}: OH
                          3303 cm^{-1}: Csp–H
                          2938 cm^{-1}: Csp^3–H
                          2120 cm^{-1}: C≡C
  ^{1}H NMR data:
    Absorptions:
    H_a: singlet at 1.53 (6 H) (2 CH$_3$ groups)
    H_b: singlet at 2.37 (1 H)
    H_c: singlet at 2.43 (1 H)   alkynyl CH and OH

Nuclear Magnetic Resonance Spectroscopy 14–19

14.59

Compound E:
$C_4H_8O_2$:
 1 degree of unsaturation
IR absorption at 1743 cm^{-1}: **C=O**
NMR data:
   H_a: quartet at 4.1 (**2 H**)
   H_b: singlet at 2.0 (**3 H**)
   H_c: triplet at 1.4 (**3 H**)

$$CH_3 - \overset{\overset{\displaystyle O}{\|}}{C} - OCH_2CH_3$$

H_b H_a H_c

Compound F:
$C_4H_8O_2$:
 1 degree of unsaturation
IR absorption at 1730 cm^{-1}: **C=O**
NMR data:
   H_a: singlet at 4.1 (**2 H**)
   H_b: singlet at 3.4 (**3 H**)
   H_c: singlet at 2.1 (**3 H**)

$$CH_3 - \overset{\overset{\displaystyle O}{\|}}{C} - CH_2OCH_3$$

H_c H_a H_b

14.60

Compound H:
$C_8H_{11}N$:
 4 degrees of unsaturation
IR absorptions at 3365 cm^{-1}: N–H
         3284 cm^{-1}: N–H
         3026 cm^{-1}: Csp^2–H
         2932 cm^{-1}: Csp^3–H
         1603 cm^{-1}: due to benzene
         1497 cm^{-1}: due to benzene

NMR data:
   multiplet at 7.2–7.4 ppm, **5 H** on a benzene ring
   H_a: triplet at 2.9 ppm, **2 H**, split by 2 H's
   H_b: triplet at 2.8 ppm, **2 H**, split by 2 H's
   H_c: singlet at 1.1 ppm, **2 H**, no splitting (on NH$_2$)

Compound I:
$C_8H_{11}N$:
 4 degrees of unsaturation
IR absorptions at 3367 cm^{-1}: N–H
         3286 cm^{-1}: N–H
         3027 cm^{-1}: Csp^2–H
         2962 cm^{-1}: Csp^3–H
         1604 cm^{-1}: due to benzene
         1492 cm^{-1}: due to benzene

NMR data:
   multiplet at 7.2–7.4 ppm, **5 H** on a benzene ring
   H_a: quartet at 4.1 ppm, **1 H**, split by 3H's
   H_b: singlet at 1.45 ppm, **2 H**, no splitting (NH$_2$)
   H_c: doublet at 1.4 ppm, **3 H**, split by 1 H

14.61

a. $C_9H_{10}O_2$:
     5 degrees of unsaturation
   IR absorption at 1718 cm^{-1}: **C=O**
   NMR data:
      multiplet at 7.4–8.1 ppm, **5 H** on a benzene ring
      quartet at 4.4 ppm, **2 H,** split by 3 H's
      triplet at 1.3 ppm, **3 H,** split by 2 H's

downfield due to the O atom

b. C_9H_{12}:
     4 degrees of unsaturation
   IR absorption at 2850–3150 cm^{-1}:
      C–H bonds
   NMR data:
      singlet at 7.1–7.4 ppm, **5 H,** benzene
      septet at 2.8 ppm, **1 H,** split by 6 H's
      doublet at 1.3 ppm, **6 H,** split by 1 H

Chapter 14–20

14.62 IR absorption at 1717 cm^{-1} is due to a C=O.

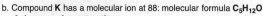

14.63 IR absorption at 1730 cm^{-1} is due to a C=O. Eight lines in the ^{13}C NMR spectrum means there are eight different types of C.

14.64

a. Compound **J** has a molecular ion at 72: molecular formula **C$_4$H$_8$O**
  1 degree of unsaturation
  IR spectrum at 1710 cm^{-1}: **C=O**
  ^{1}H NMR data (ppm):
   1.0 (triplet, 3 H), split by 2 H's
   2.1 (singlet, 3 H)
   2.4 (quartet, 2 H), split by 3 H's

b. Compound **K** has a molecular ion at 88: molecular formula **C$_5$H$_{12}$O**
  0 degrees of unsaturation
  IR spectrum at 3600–3200 cm^{-1}: **O–H bond**
  ^{1}H NMR data (ppm):
   0.9 (triplet, 3 H), split by 2 H's
   1.2 (singlet, 6 H), due to 2 CH$_3$ groups
   1.5 (quartet, 2 H), split by 3 H's
   1.6 (singlet, 1 H), due to the OH proton

14.65

Compound **L** has a molecular ion at 90: molecular formula **C$_4$H$_{10}$O$_2$**
  0 degrees of unsaturation
  IR absorptions at 2992 and 2941 cm^{-1}: Csp^3–H
  ^{1}H NMR data (ppm):
   H$_a$: 1.2 (doublet, 3 H), split by 1 H
   H$_b$: 3.3 (singlet, 6 H), due to 2 CH$_3$ groups
   H$_c$: 4.8 (quartet, 1 H), split by 3 adjacent H's

Nuclear Magnetic Resonance Spectroscopy 14–21

14.66

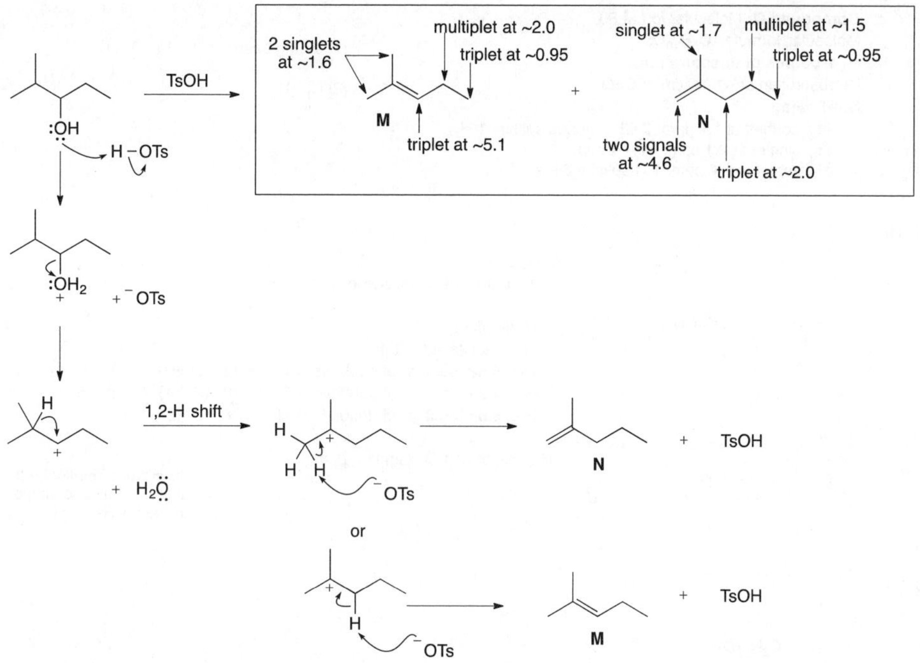

14.67

Compound **O** has a molecular formula $C_{10}H_{12}O$.
    5 degrees of unsaturation
    IR absorption at 1687 cm^{-1}
    ^{1}H NMR data (ppm):
        H_a: 1.0 (triplet, 3 H), due to CH_3 group, split by 2 adjacent H's
        H_b: 1.7 (sextet, 2 H), split by CH_3 and CH_2 groups
        H_c: 2.9 (triplet, 2 H), split by 2 H's
        7.4–8.0 (multiplet, 5 H), benzene ring

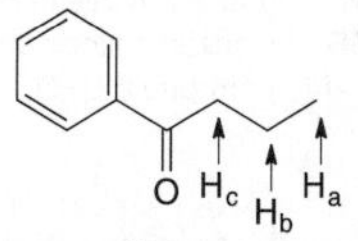

14.68

Compound **P** has a molecular formula $C_5H_9ClO_2$.
    1 degree of unsaturation
    ^{13}C NMR shows 5 different C's, including a C=O.
    ^{1}H NMR data (ppm):
        H_a: 1.3 (triplet, 3 H), split by 2 H's
        H_b: 2.8 (triplet, 2 H), split by 2 H's
        H_c: 3.7 (triplet, 2 H), split by 2 H's
        H_d: 4.2 (quartet, 2 H), split by CH_3 group

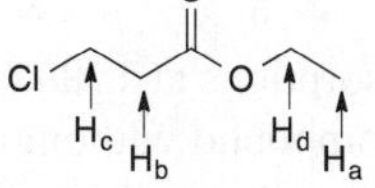

Chapter 14–22

14.69

Compound Q: Molecular ion at 86.
Molecular formula: **C$_5$H$_{10}$O**:
    1 degree of unsaturation
IR absorption at ~1700 cm^{-1}: **C=O**
NMR data:
    H$_a$: doublet at 1.1 ppm, 2 CH$_3$ groups split by 1 H
    H$_b$: singlet at 2.1 ppm, CH$_3$ group
    H$_c$: septet at 2.6 ppm, 1 H split by 6 H's

14.70

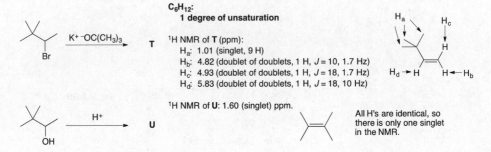

C$_6$H$_{12}$:
    1 degree of unsaturation
^{1}H NMR of **T** (ppm):
    H$_a$: 1.01 (singlet, 9 H)
    H$_b$: 4.82 (doublet of doublets, 1 H, J = 10, 1.7 Hz)
    H$_c$: 4.93 (doublet of doublets, 1 H, J = 18, 1.7 Hz)
    H$_d$: 5.83 (doublet of doublets, 1 H, J = 18, 10 Hz)

^{1}H NMR of **U**: 1.60 (singlet) ppm.

All H's are identical, so there is only one singlet in the NMR.

14.71

a.

C$_6$H$_{12}$O$_2$:
    1 degree of unsaturation
IR peak at 1740 cm^{-1}: **C=O**
^{1}H NMR 2 signals: 2 types of H's
^{13}C NMR: 4 signals: 4 kinds of C's, including one at ~170 ppm due a C=O

b.

C$_6$H$_{10}$:
    2 degrees of unsaturation
IR peak at 3000 cm^{-1}: **Csp^3–H bonds**
    peak at 3300 cm^{-1}: **C$_{sp}$–H bond**
    peak at ~2150 cm^{-1}: **C≡C bond**
^{13}C NMR: 4 signals: 4 kinds of C's

14.72

    a. Because **A** has no absorptions at 1700 cm^{-1} or 3600–3200 cm^{-1}, it has no C=O or OH. An oxygen-containing compound without these functional groups is an ether (or an epoxide). Because **B** is formed from a reaction with HCl, **A** must contain an epoxide, because ethers are unreactive with HCl.

doublet at 3.5 — H, CH$_3$ ← doublet at 1.4
H ← quartet of doublets at 3.0
singlet at 3.8 → CH$_3$O
2 doublets at 6.9 and 7.2
A

Nuclear Magnetic Resonance Spectroscopy 14–23

b. Epoxide **A** is equally substituted by R groups on both C's. With HCl, the epoxide is protonated first and then backside attack by Cl⁻ forms the chlorohydrin. Attack at the C adjacent to the benzene ring is preferred because the δ+ in the transition state at this carbon can be delocalized on the benzene ring.

The benzene ring and CH_3 group must be trans in the epoxide to give the correct configuration at the two stereogenic centers in the product.

14.73 A second resonance structure for *N,N*-dimethylformamide places the two CH_3 groups in different environments. One CH_3 group is cis to the O atom, and one is cis to the H atom. This gives rise to two different absorptions for the CH_3 groups.

N,N-dimethylformamide

14.74

18-Annulene has 18 π electrons that create an **induced magnetic field** similar to the 6 π electrons of benzene. 18-Annulene has 12 protons that are oriented on the outside of the ring (labeled H_o), and 6 protons that are oriented inside the ring (labeled H_i). The induced magnetic field reinforces the external field in the vicinity of the protons on the outside of the ring. These H_o protons are deshielded, so they absorb downfield (8.9 ppm). In contrast, the induced magnetic field is opposite in direction to the applied magnetic field in the vicinity of the protons on the inside of the ring. This shields the H_i protons, so they absorb very far upfield, at −1.8 ppm, which is even higher than TMS.

Chapter 14–24

14.75

C_a
stereogenic center

3-methylbutan-2-ol

Replace a CH$_3$ group
with X.

or

Replace C_a.   Replace C_b.

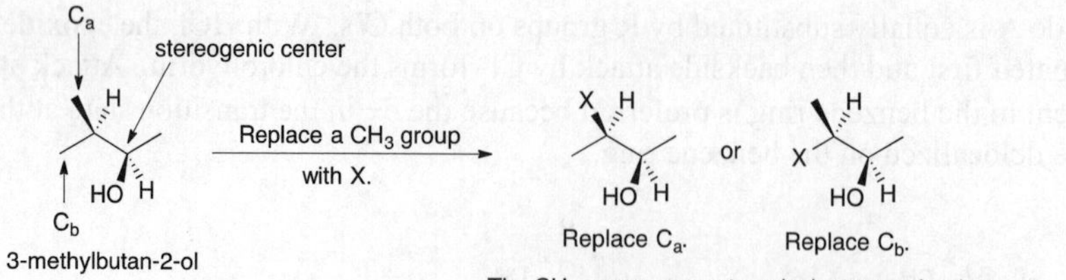

The CH$_3$ groups are not equivalent to each other,
because replacement of each by X forms two diastereomers.

Thus, every C in this compound is different
and there are five ^{13}C signals.

14.76

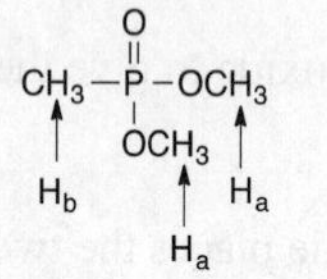

$$CH_3-P-OCH_3$$
O
OCH$_3$
H_b   H_a
H_a

One P atom splits each nearby CH$_3$ into a doublet
by the $n + 1$ rule, making two doublets.

All 6 H$_a$ protons are equivalent.

14.77 a. Splitting pattern:

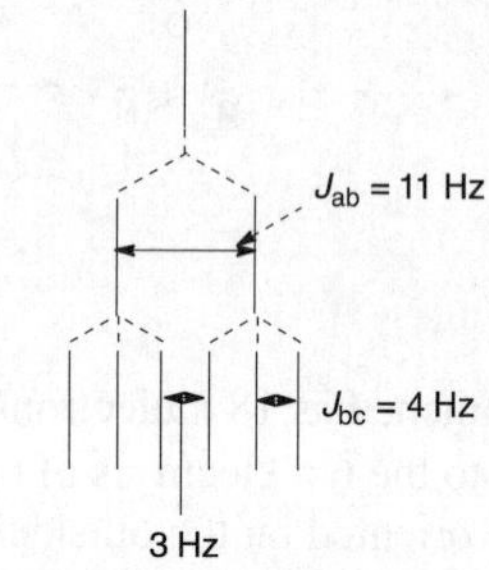

J_{ab} = 11 Hz

J_{bc} = 4 Hz

3 Hz

b. Three resonance structures can be drawn for cyclohex-2-enone.

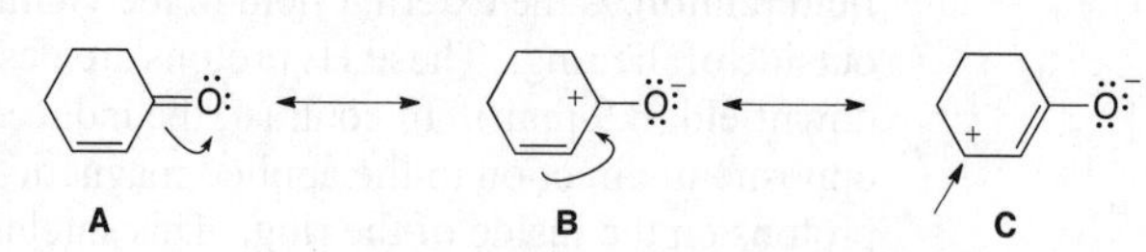

A   B   C

Resonance structure **C** places a (+) charge on one C of the C=C, deshielding the H attached to
it and shifting the absorption downfield.

Chapter 15 Radical Reactions

Chapter Review

General features of radicals
- A radical is a reactive intermediate with an unpaired electron (15.1).
- A carbon radical is sp^2 hybridized and trigonal planar (15.1).
- The stability of a radical increases as the number of C's bonded to the radical carbon increases (15.1).

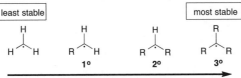

- Allylic radicals are stabilized by resonance, making them more stable than 3° radicals (15.10).

two resonance structures for the allyl radical

Radical reactions

[1] Halogenation of alkanes (15.4)

- The reaction follows a radical chain mechanism.
- The weaker the C–H bond, the more readily the H is replaced by X.
- Chlorination is faster and less selective than bromination (15.6).
- Radical substitution results in racemization at a stereogenic center (15.8).

[2] Allylic halogenation (15.10)

- The reaction follows a radical chain mechanism.

[3] Radical addition of HBr to an alkene (15.13)

- A radical addition mechanism is followed.
- Br bonds to the less substituted carbon atom to form the more substituted, more stable radical.

Chapter 15–2

[4] Radical polymerization of alkenes (15.14)

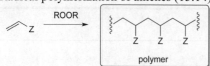

- A radical addition mechanism is followed.

Practice Test on Chapter Review

1.a. Which alkyl halide(s) can be made in good yield by radical halogenation of an alkane?

    4. Both (1) and (2) can be made in good yield.
    5. Compounds (1), (2), and (3) can all be made in good yield.

b. In which of the following reactions will rearrangement *not* occur?
    1. halogenation of an alkane with Cl_2 and heat
    2. addition of Cl_2 to an alkene
    3. addition of HCl to an alkene
    4. Rearrangements do not occur in reactions (1) and (2).
    5. Rearrangements do not occur in reactions (1), (2), and (3).

c. Which labeled H is most easily abstracted in a radical halogenation reaction?

    1. H_a
    2. H_b
    3. H_c
    4. H_d
    5. H_e

d. Which of the labeled C–H bonds in the following compound has the smallest bond dissociation energy?

    1. C–H_a
    2. C–H_b
    3. C–H_c
    4. C–H_d
    5. C–H_e

2. (a) Which radical is the most stable? (b) Which radical is the least stable?

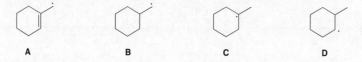

Radical Reactions 15–3

3. Draw all of the organic products formed in each reaction. Indicate stereochemistry in part (c).

a. 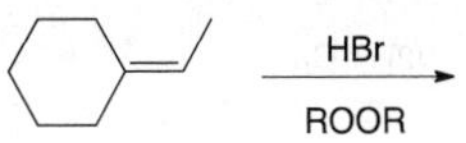 $\xrightarrow[\text{ROOR}]{\text{HBr}}$

b. 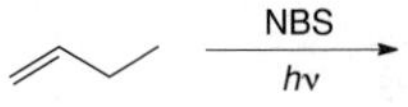 $\xrightarrow[\text{h}\nu]{\text{NBS}}$

c. 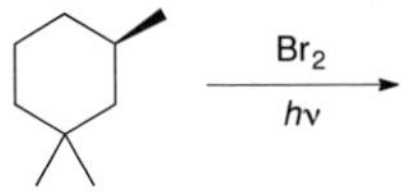 $\xrightarrow[\text{h}\nu]{\text{Br}_2}$

4. What monomer is needed to make the following polymer?

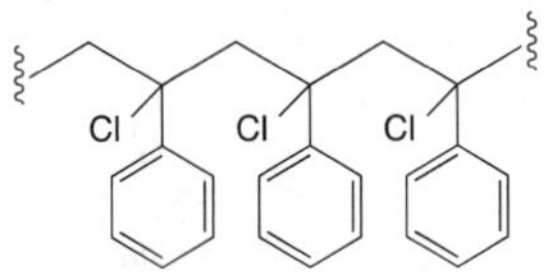

5. In each box, fill in the appropriate reagents needed to carry out the given reaction. This question involves reactions from Chapter 15, as well as previous chapters.

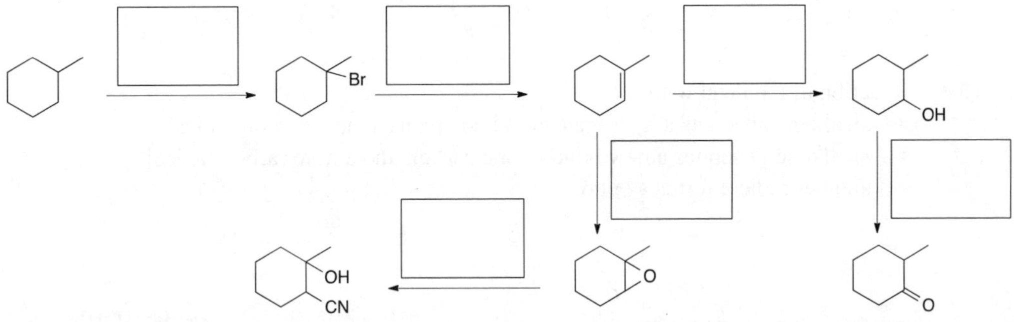

Answers to Practice Test

1.a. 2
  b. 4
  c. 2
  d. 3

2.a. **A**
  b. **B**

3.a.

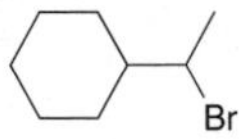

3.b.

(+ cis isomer)

3.c.

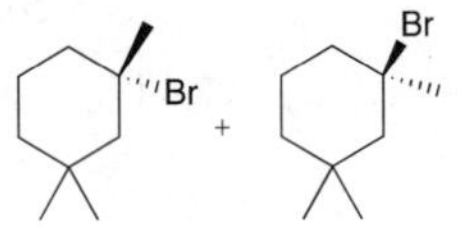

4.

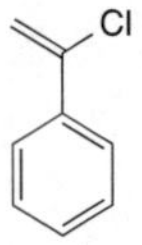

Chapter 15–4

5.

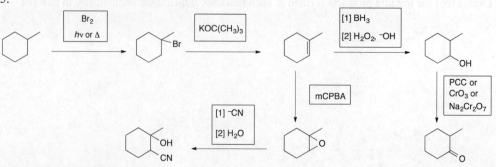

Answers to Problems

15.1 1° Radicals are on C's bonded to one other C; 2° radicals are on C's bonded to two other C's; 3° radicals are on C's bonded to three other C's.

a. 2° radical   b. 3° radical   c. 2° radical   d. 1° radical

15.2 The stability of a radical increases as the number of alkyl groups bonded to the radical carbon increases. Draw the most stable radical.

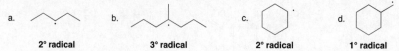

15.3 Reaction of a radical with:
- an alkane abstracts a hydrogen atom and creates a new carbon radical.
- an alkene generates a new bond to one carbon and a new carbon radical.
- another radical forms a bond.

a. ⌬ + :Cl· ⟶ ⌬· + H–Cl: c. :Cl· —:Cl·→ :Cl–Cl:

b. CH₂=CH₂ + :Cl· ⟶ :Cl–CH₂–CH₂·   d. :Cl· + ·O–O· ⟶ :Cl–O–O·

15.4 Monochlorination is a radical substitution reaction in which a Cl replaces a H, thus generating an alkyl halide.

a. ☐ + Cl₂ /Δ → ☐–Cl

b. CH₃CH₂CH₂CH₂CH₃ + Cl₂/Δ → CH₃CH₂CH₂CH₂CH₂Cl + CH₃CH₂CH₂CH(Cl)CH₃ + CH₃CH₂CH(Cl)CH₂CH₃

c. (CH₃)₃CH + Cl₂/Δ → (CH₃)₂CHCH₂Cl + (CH₃)₃CCl

15.5

A: (CH₃)₃C–CH₃ + Cl₂ —Δ→ (CH₃)₃C–CH₂Cl

B: CH₃CH₂CH₂CH₂CH₃ + Cl₂ —Δ→ CH₃CH₂CH₂CH(Cl)CH₃ + CH₃CH₂CH(Cl)CH₂CH₃ + CH₃CH₂CH₂CH₂CH₂Cl

15.6

Initiation: :Br–Br: —hv or Δ→ :Br· + ·Br:

Propagation:
CH₃–H + ·Br: ⟶ ·CH₃ + H–Br:

·CH₃ + :Br–Br: ⟶ CH₃–Br: + ·Br:

Termination:
:Br· + ·Br: ⟶ :Br–Br:

or

·CH₃ + ·CH₃ ⟶ CH₃–CH₃

or

·CH₃ + ·Br: ⟶ CH₃–Br:

15.7 The rate-determining step for halogenation reactions is formation of CH₃· + HX.

CH₃–H + ·I: ⟶ ·CH₃ + H–I: $\Delta H° = +138$ kJ/mol

1 bond broken   1 bond formed   This reaction is more endothermic and
+435 kJ/mol   –297 kJ/mol   has a higher E_a than a similar reaction
                                       with Cl₂ or Br₂.

15.8 The **weakest C–H bond** in each alkane is the **most readily cleaved** during radical halogenation.

a. 3° most reactive   b. 3° most reactive   c. 2° most reactive

15.9 To draw the product of bromination:
- Draw out the starting material and find the most reactive C–H bond (on the most substituted C).
- The major product is formed by **cleavage of the *weakest* C–H bond.**

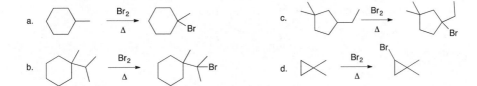

Chapter 15–6

15.10 If 1° C–H and 3° C–H bonds were equally reactive there would be nine times as much
(CH₃)₂CHCH₂Cl as (CH₃)₃CCl because the ratio of 1° to 3° H's is 9:1. The fact that the ratio is
only 63:37 shows that the 1° C–H bond is less reactive than the 3° C–H bond. (CH₃)₂CHCH₂Cl
is still the major product, though, because there are nine 1° C–H bonds and only one 3° C–H
bond.

15.11

a. (structure) →[Br₂, Δ] (structure with Br)   c. (structure) →[H₂O, H₂SO₄] (structure with OH)
(from b.)

b. (structure with Br) →[(CH₃)₃CO⁻K⁺] (alkene structure)
(from a.)

15.12

a. (cyclohexane) →[Br₂, hv] (bromocyclohexane) →[K⁺ ⁻OC(CH₃)₃] (cyclohexene) →[Br₂] (dibromocyclohexane) + enantiomer

b. (cyclohexene) →[mCPBA] (epoxide, O)
(from a.)

15.13 The reaction does not occur at the stereogenic center, so leave it as is.

(R)-2-bromobutane →[Cl₂, hv] (product, S) + (product, R)

15.14

a. (pentane) →[Cl₂, Δ] (product) + (product) + (product) + (product)

b. (methylcyclopropane) →[Cl₂, Δ] (product) + (product) + (product) + (product) + (product)

c. (structure) →[Cl₂, hv] (product) + (product) + (product)

d. (structure, H Cl) →[Cl₂, hv] (product) + (product) + (product)
(Consider attack at C2 and C3 only.)

Radical Reactions 15–7

15.15

Chain propagation:

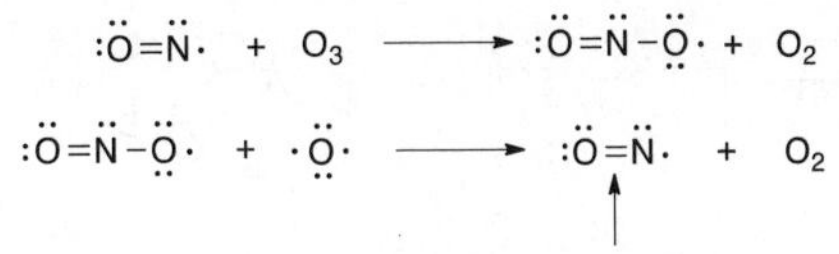

$$:\ddot{O}=\dot{N}\cdot \ + \ O_3 \longrightarrow :\ddot{O}=\ddot{N}-\ddot{O}\cdot \ + \ O_2$$

$$:\ddot{O}=\ddot{N}-\ddot{O}\cdot \ + \ \cdot\ddot{O}\cdot \longrightarrow :\ddot{O}=\dot{N}\cdot \ + \ O_2$$

The radical is re-formed.

15.16 Draw the resonance structure by moving the π bond and the unpaired electron. The hybrid is drawn with dashed lines for bonds that are in one resonance structure but not another. The symbol δ· is used on any atom that has an unpaired electron in any resonance structure.

a.

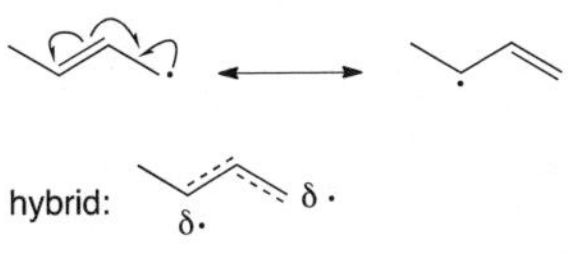

hybrid:

c.

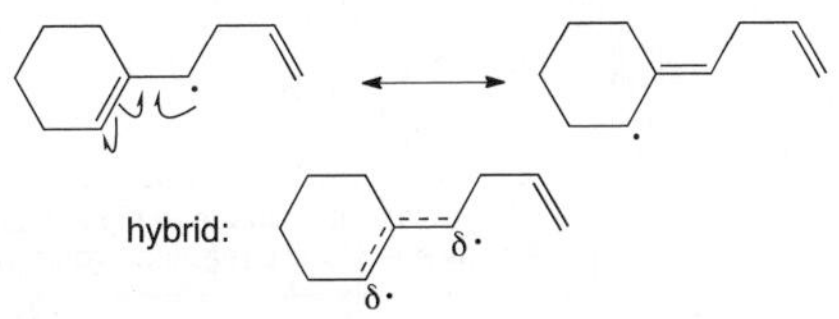

hybrid:

b.

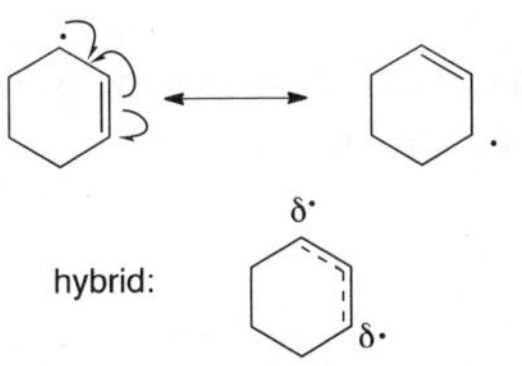

hybrid:

d.

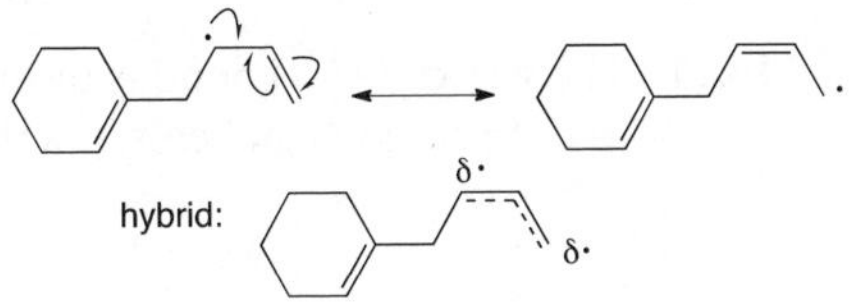

hybrid:

15.17 Reaction of an alkene with NBS or Br₂ + hν yields allylic substitution products.

a. NBS / hν

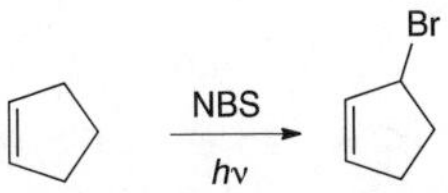

b. NBS / hν

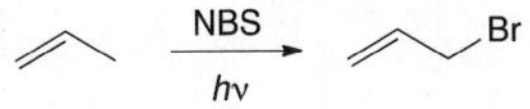

c. Br₂

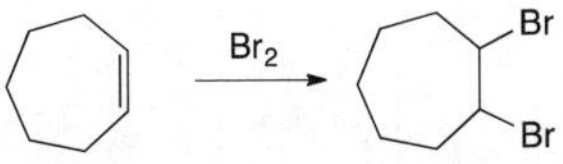

15.18

a. NBS / hν   (+ Z isomer)

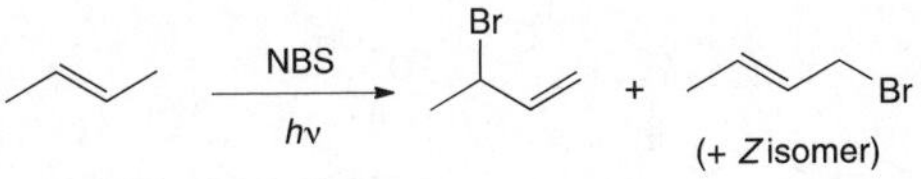

c. NBS / hν   (+ Z isomer)

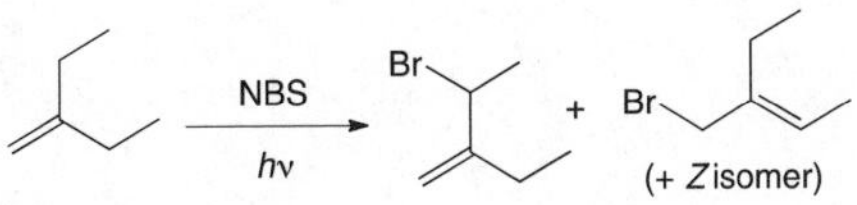

b. NBS / hν

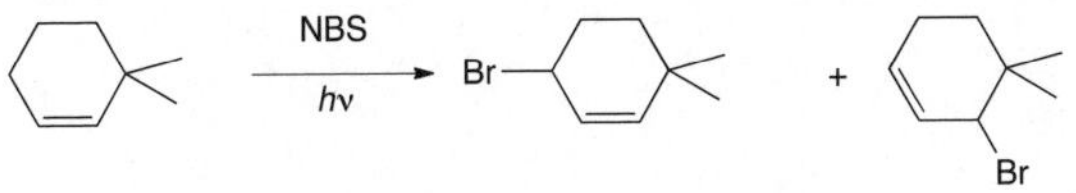

Chapter 15–8

15.19

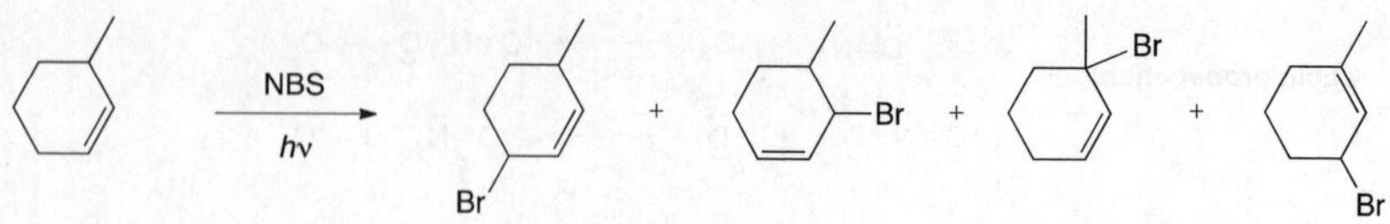

15.20 Reaction of an alkene with NBS + *h*ν yields allylic substitution products.

a.

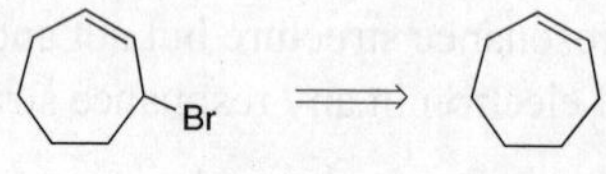

one possible product: **high yield**

b.

c.

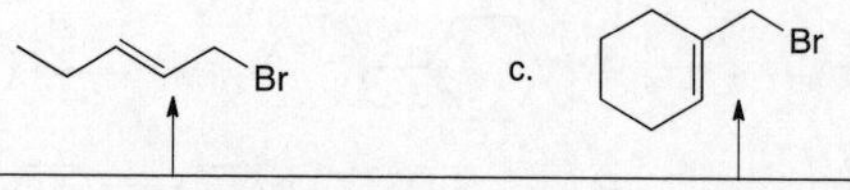

Cannot be made in high yield by allylic halogenation.
Any alkene starting material would yield a mixture of allylic halides.

15.21 The weakest C–H bond is most readily cleaved. To draw the hydroperoxide products, add OOH to each carbon that bears a radical in one of the resonance structures.

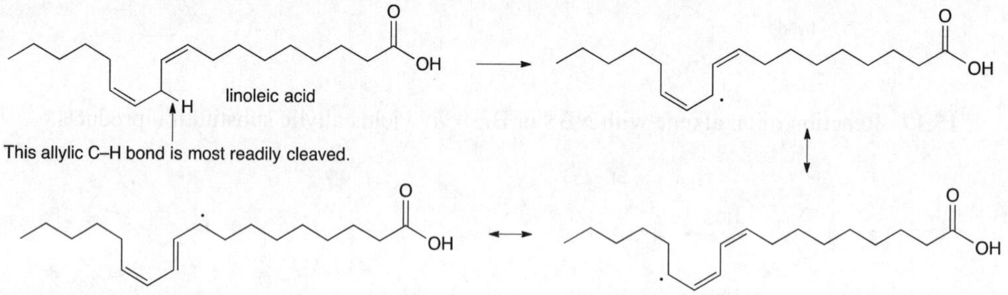

linoleic acid

This allylic C–H bond is most readily cleaved.

hydroperoxide products:

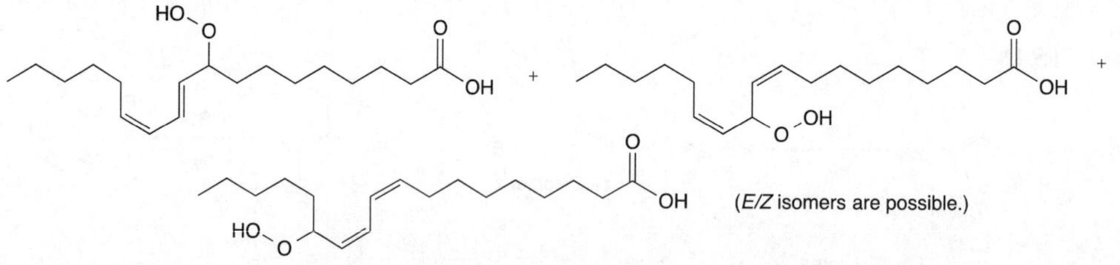

(*E/Z* isomers are possible.)

Radical Reactions 15–9

15.22

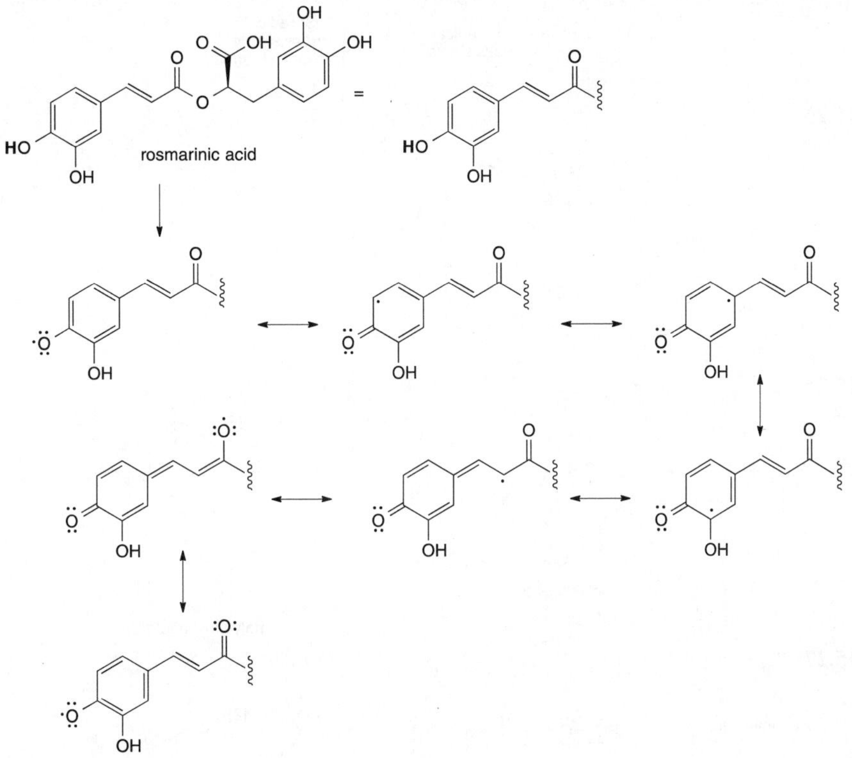

rosmarinic acid

15.23

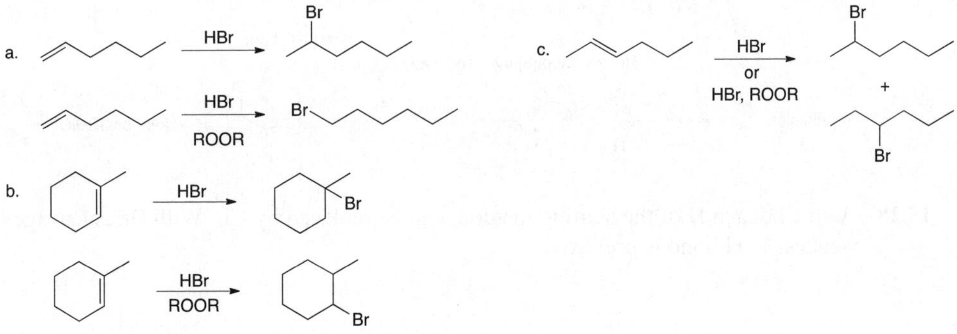

a.

b.

c.

15.24 In addition of HBr under radical conditions:
- Br· adds first to form the more stable radical.
- Then H· is added to the carbon radical.

Chapter 15–10

2 radical possibilities: [1° radical less stable] or [3° radical more stable — This radical forms.] → product

15.25

a. (1-ethylcyclohexene) + HBr / ROOR → (2-bromo-1-ethylcyclohexane)

b. (1-ethylcyclohexene) + HBr → (1-bromo-1-ethylcyclohexane)

c. (1-ethylcyclohexene) + Br₂ → (1,2-dibromo-1-ethylcyclohexane)

15.26

a. styrene → polystyrene

b. poly(vinyl acetate) ⟹ vinyl acetate

15.27

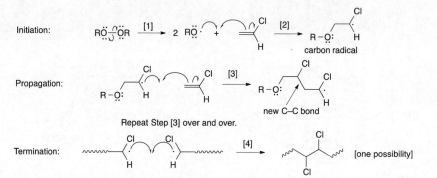

15.28 With Cl₂, each H of the starting material can be replaced by Cl. With Br₂, cleavage of the weakest C–H bond is preferred.

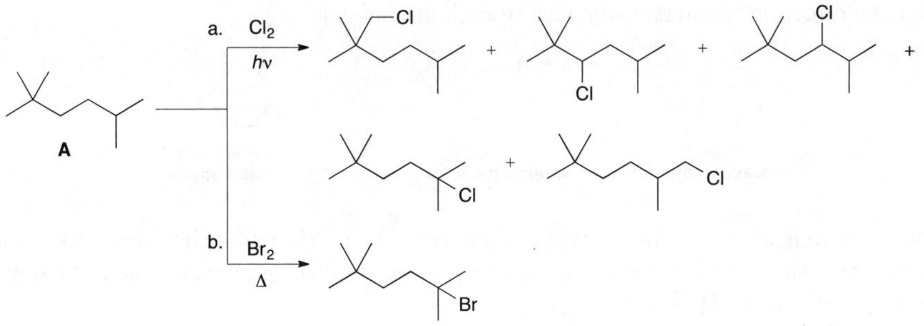

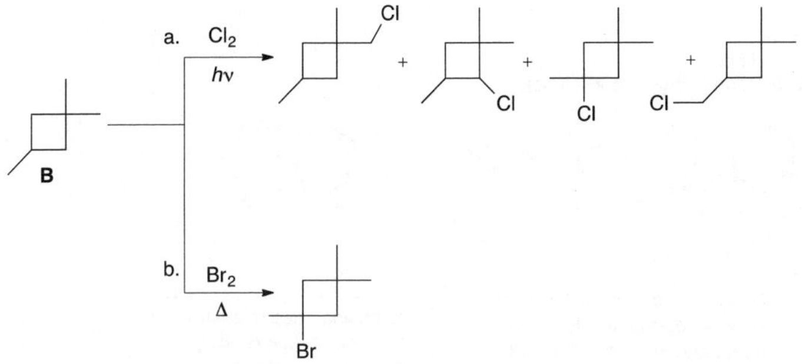

15.29

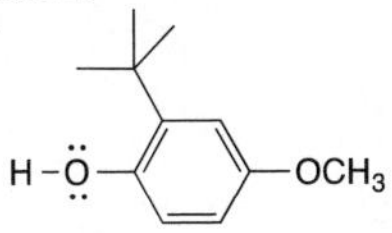

Abstraction of the phenol H
produces a resonance-stabilized radical.

BHA

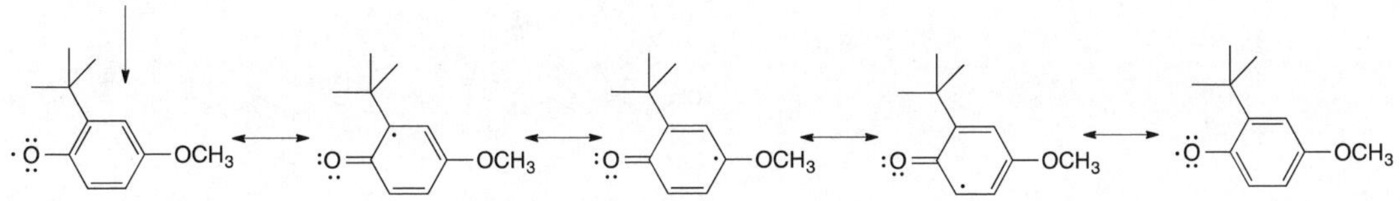

15.30

a. increasing bond strength: 2 < 3 < 1

b. and c.

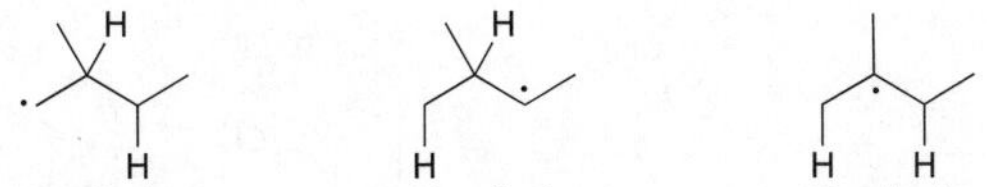

1° radical
least stable

2° radical
intermediate stability

3° radical
most stable

d. increasing ease of H abstraction: 1 < 3 < 2

Chapter 15–12

15.31 Use the directions from Answer 15.2 to rank the radicals.

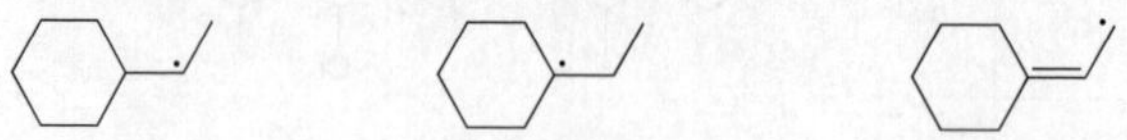

2° radical
least stable

3° radical
intermediate stability

allylic radical
most stable

15.32 Draw the radical formed by cleavage of the benzylic C–H bond. Then draw all of the resonance structures. Having more resonance structures (five in this case) makes the radical more stable, and the benzylic C–H bond weaker.

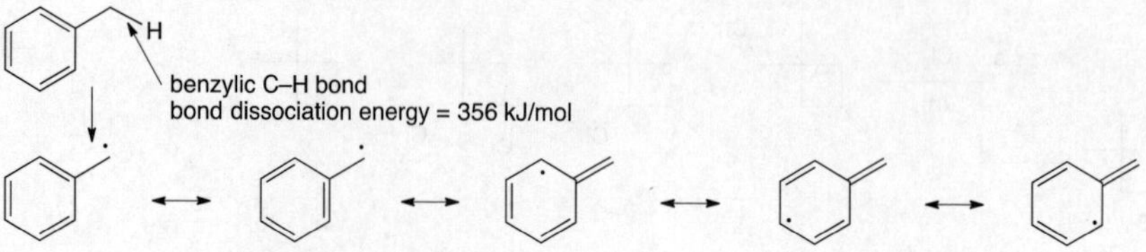

benzylic C–H bond
bond dissociation energy = 356 kJ/mol

15.33

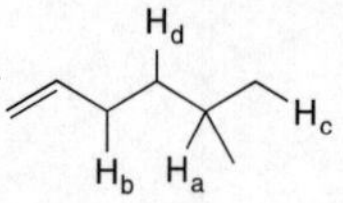

H_a = bonded to an sp^3 3° carbon
H_b = bonded to an allylic carbon
H_c = bonded to an sp^3 1° carbon
H_d = bonded to an sp^3 2° carbon

Increasing ease of abstraction:
$H_c < H_d < H_a < H_b$

15.34

a.

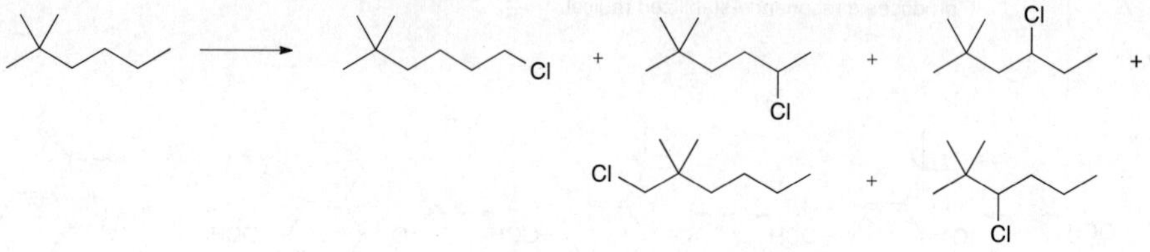

b.

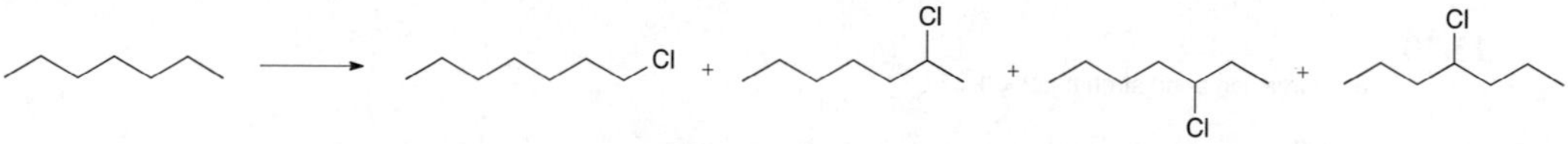

c.

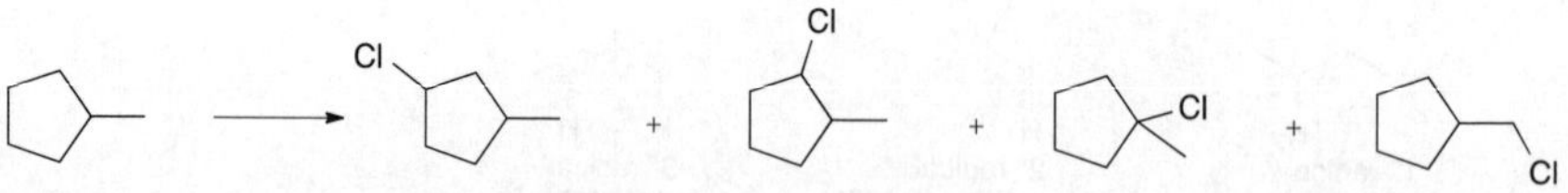

Radical Reactions 15–13

15.35 To draw the product of bromination:
- Draw out the starting material and find the most reactive C–H bond (on the most substituted C).
- The major product is formed by **cleavage of the *weakest* C–H bond.**

a.

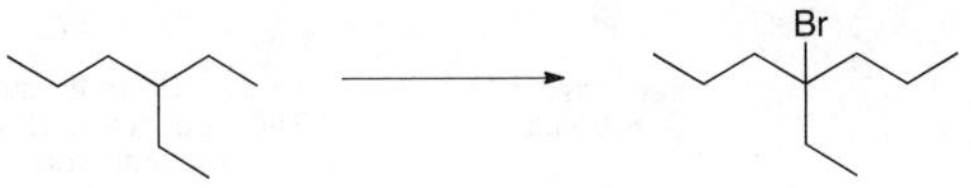

c.

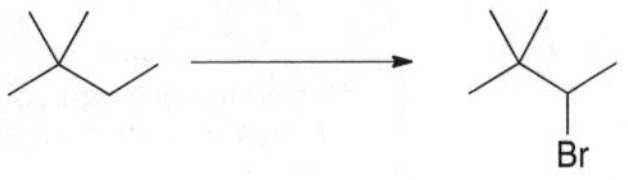

b.

15.36 Draw all of the alkane isomers of C_6H_{14} and their products on chlorination. Then determine which letter corresponds to which alkane.

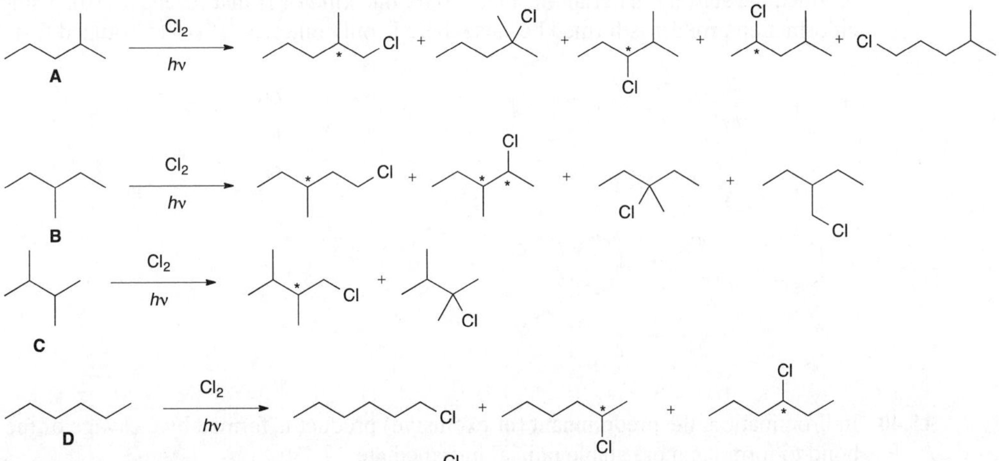

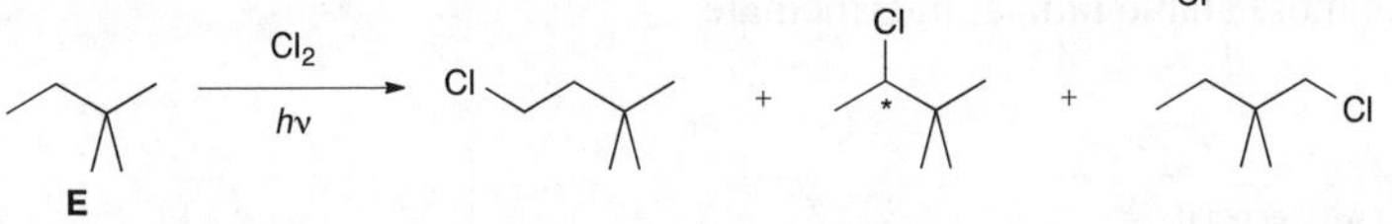

A

Cl_2
$h\nu$

B

Cl_2
$h\nu$

C

Cl_2
$h\nu$

D

Cl_2
$h\nu$

E

Cl_2
$h\nu$

[* = stereogenic center]

15.37 Halogenation replaces a C–H bond with a C–X bond. To find the alkane needed to make each of the alkyl halides, replace the X with a H.

a.

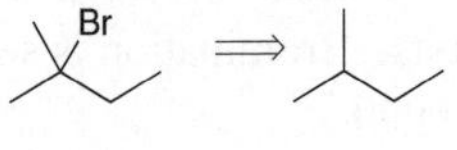

b.

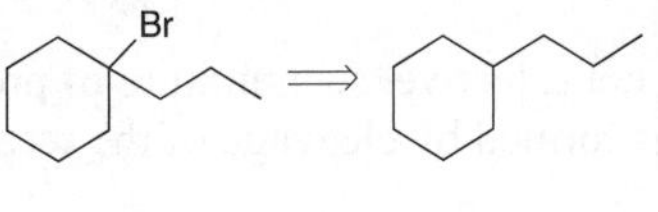

c.

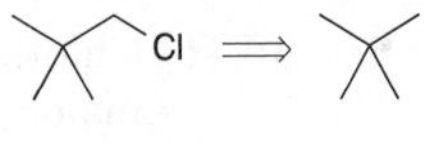

Chapter 15–14

15.38 For an alkane to yield one major product on monohalogenation with Cl$_2$, all of the hydrogens must be identical in the starting material. For an alkane to yield one major product on bromination, it must have a more substituted carbon in the starting material.

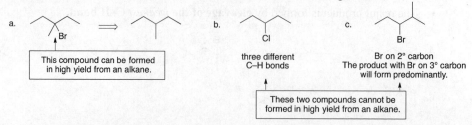

15.39 A single constitutional isomer is formed in both halogenations in (b) and (c). Bromination often forms a single product by cleavage of the weakest C–H bond. For chlorination to form a single product, the starting material must have only one kind of H that reacts. In (b), a single chlorination product is formed because there is only one type of sp^3 hybridized C–H bond.

15.40 In bromination, the predominant (or exclusive) product is formed by cleavage of the weaker C–H bond to form the more stable radical intermediate.

15.41 Chlorination is not selective, so a mixture of products results. Bromination is selective, and the major product is formed by cleavage of the weakest C–H bond.

Radical Reactions 15–15

b. 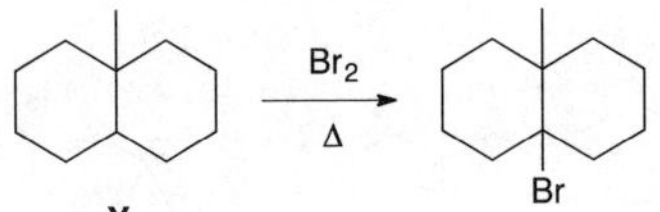 [structure] $\xrightarrow[\Delta]{Br_2}$ [structure with Br]   **Y**

c. 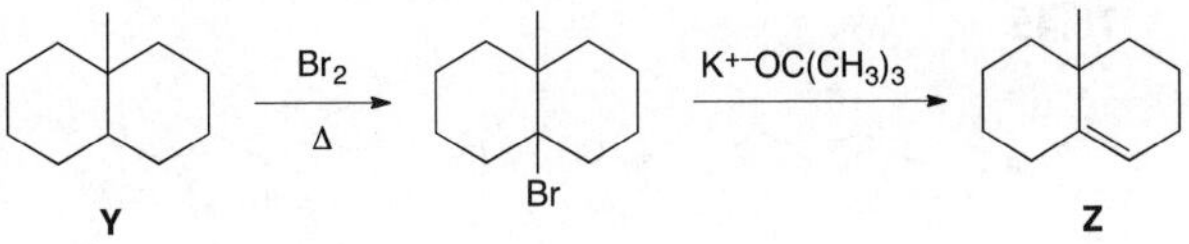 [structure] $\xrightarrow[\Delta]{Br_2}$ [structure with Br] $\xrightarrow{K^+\,{}^-OC(CH_3)_3}$ [structure]   **Z**

Y

15.42 Draw the resonance structures by moving the π bonds and the radical.

a.  [structures]

b. 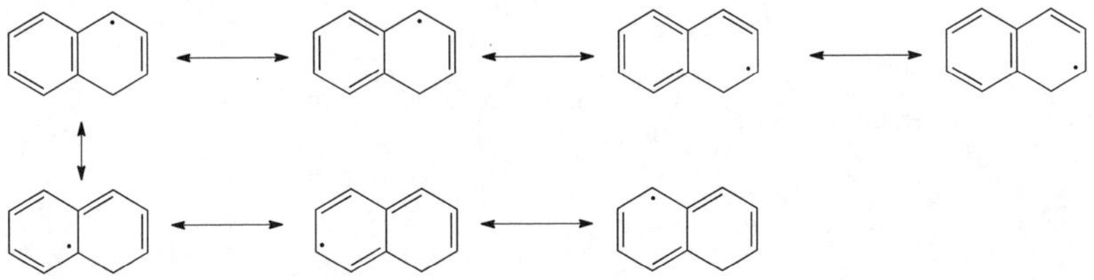 [structures]

c.  [structures]

15.43 Reaction of an alkene with NBS + $h\nu$ yields allylic substitution products.

a. 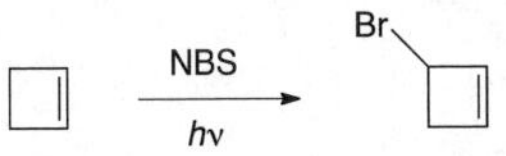 [structure] $\xrightarrow[h\nu]{NBS}$ [structure with Br]

b. 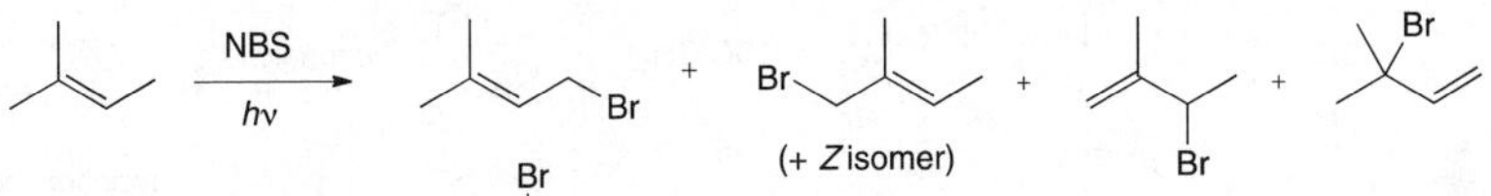 [structure] $\xrightarrow[h\nu]{NBS}$ [structure with Br] + [structure with Br] (+ Z isomer) + [structure with Br] + [structure with Br]

c. 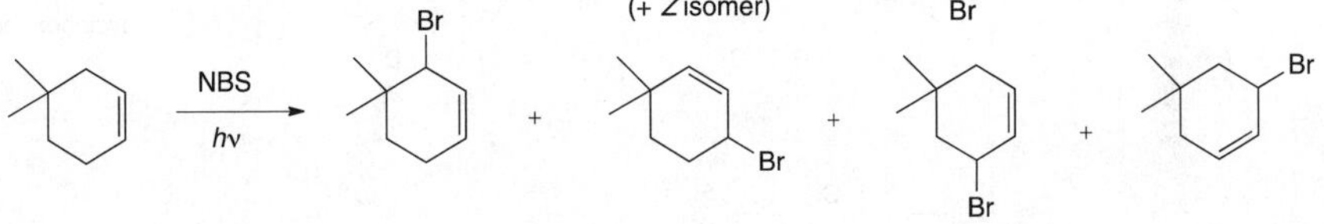 [structure] $\xrightarrow[h\nu]{NBS}$ [structure with Br] + [structure with Br] + [structure with Br] + [structure with Br]

15.44

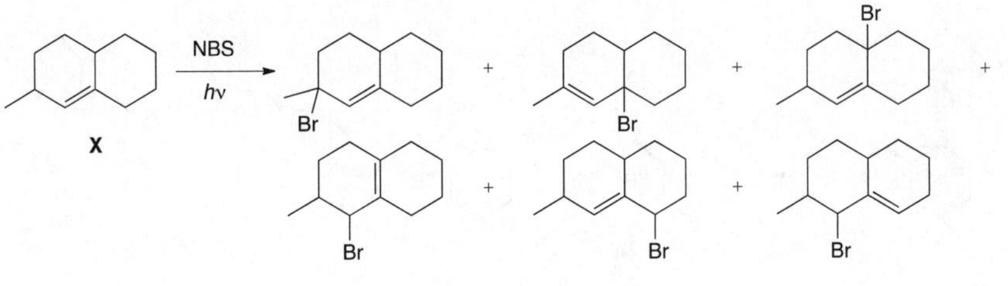

[structure] **X** $\xrightarrow[h\nu]{NBS}$ [structures with Br] +

Chapter 15–16

15.45

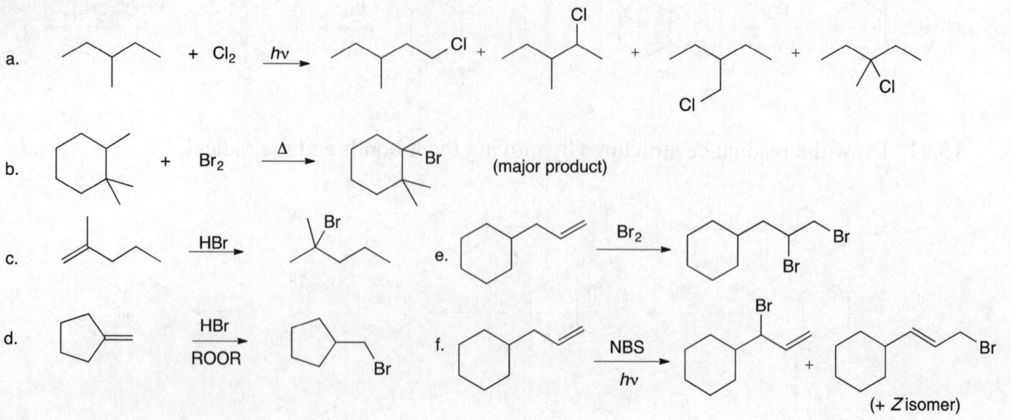

a.

b. (major product)

c.

d.

e.

f. (+ Z isomer)

15.46

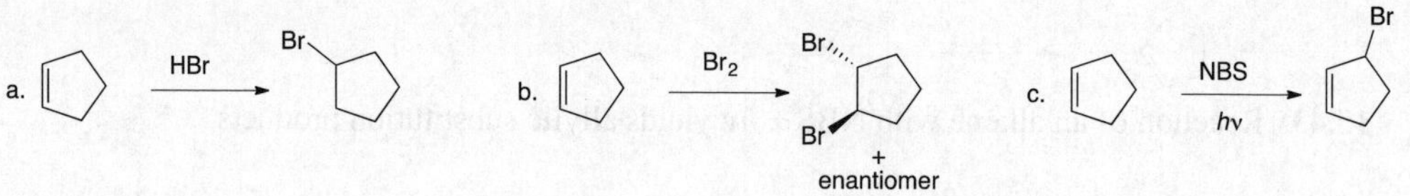

a.

b. + enantiomer

c.

15.47

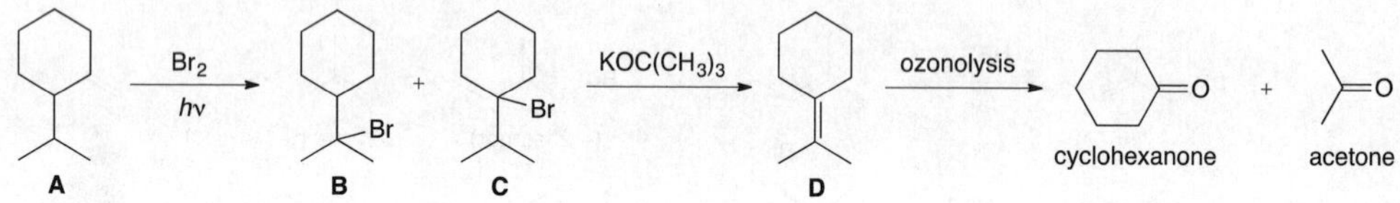

A   B   C   D   cyclohexanone   +   acetone

15.48

a.

b.

c.

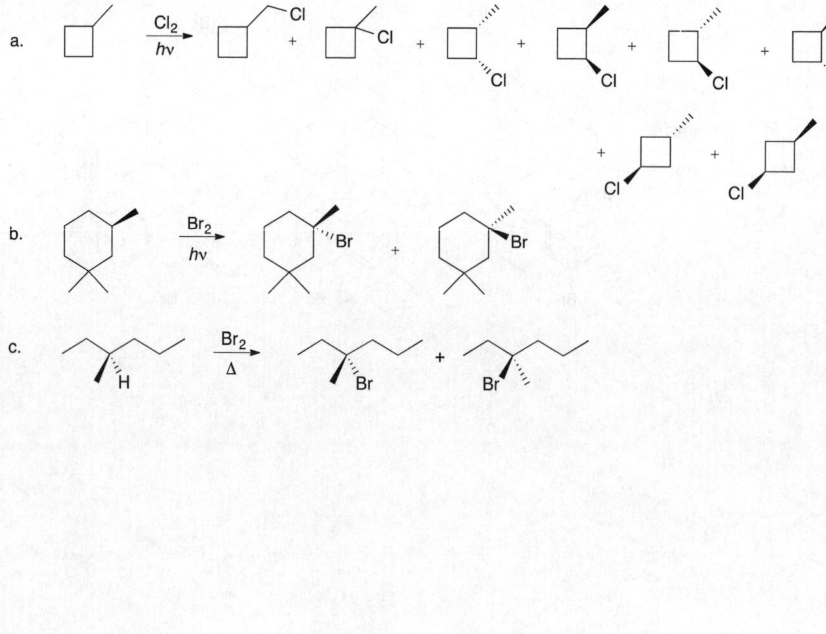

Radical Reactions 15–17

d.

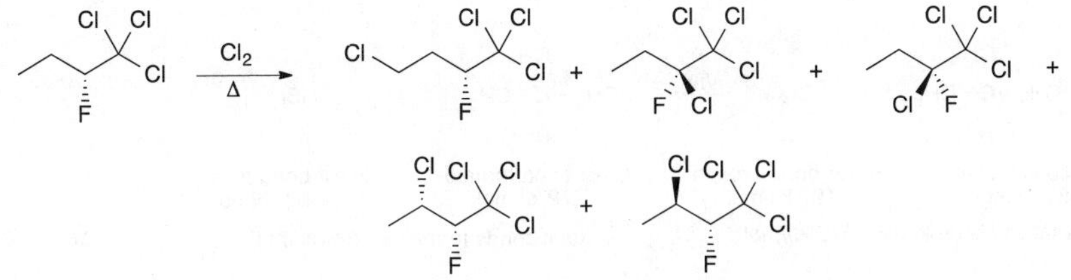

Cl₂
Δ

15.49

a.

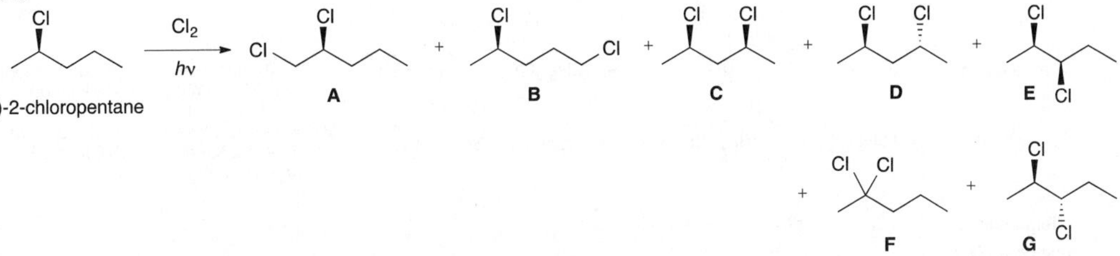

Cl₂
hν

(R)-2-chloropentane

A   B   C   D   E

F   G

b. There would be seven fractions, because each molecule drawn has different physical properties.
c. Fractions **A, B, D, E,** and **G** would show optical activity.

15.50

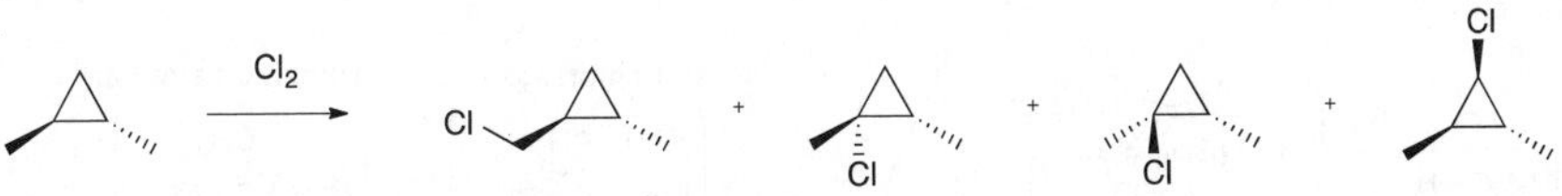

Cl₂

15.51

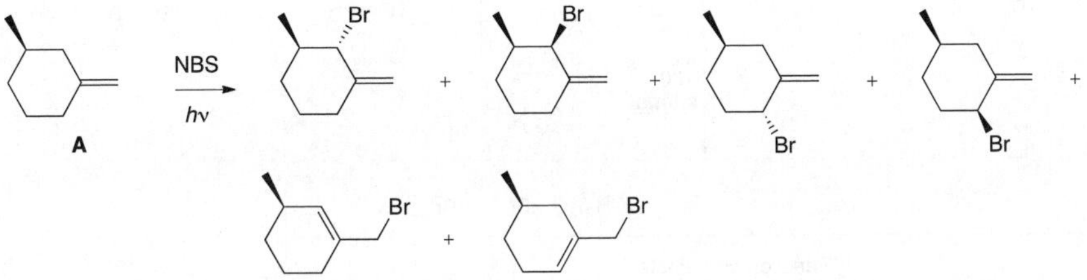

A

NBS
hν

15.52

a.

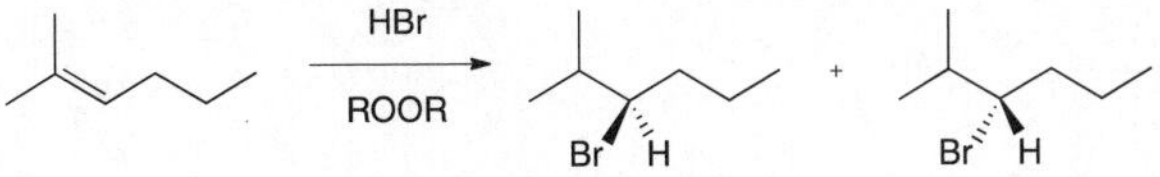

HBr
ROOR

b.

HBr
ROOR

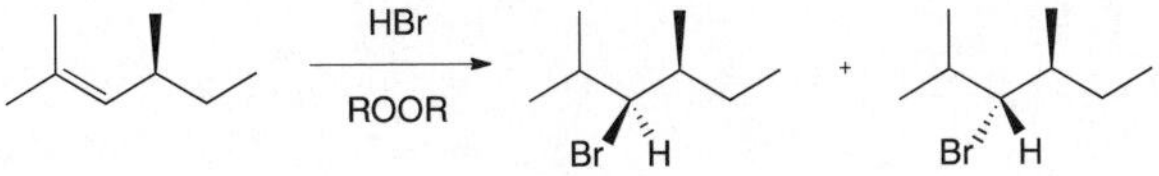

Chapter 15–18

15.53

a.

$$CH_3-\underset{H}{\underset{|}{\overset{CH_3}{\overset{|}{C}}}}-CH_3 + Br_2 \xrightarrow{h\nu} CH_3-\underset{Br}{\underset{|}{\overset{CH_3}{\overset{|}{C}}}}-CH_3 + HBr$$

C–H bond broken +381 kJ/mol
Br–Br bond broken +192 kJ/mol
total bonds broken = +573 kJ/mol

C–Br bond formed −272 kJ/mol
H–Br bond formed −368 kJ/mol
total bonds formed = −640 kJ/mol

$\Delta H° = -67$ kJ/mol

b. Initiation:

$:\ddot{\text{Br}}\frown\ddot{\text{Br}}: \xrightarrow[\text{or }\Delta]{h\nu} :\ddot{\text{Br}}\cdot + \cdot\ddot{\text{Br}}:$

Propagation:

$(CH_3)_3C\frown H + \cdot\ddot{\text{Br}}: \longrightarrow (CH_3)_3C\cdot + H-\ddot{\text{Br}}:$

$(CH_3)_3C\cdot + :\ddot{\text{Br}}\frown\ddot{\text{Br}}: \longrightarrow (CH_3)_3C-\ddot{\text{Br}}: + \cdot\ddot{\text{Br}}:$

c. $\Delta H° =$ (bonds broken) − (bonds formed)
  = (+381 kJ/mol) + (−368 kJ/mol)
  = +13 kJ/mol

$\Delta H° =$ (bonds broken) − (bonds formed)
  = (+192 kJ/mol) + (−272 kJ/mol)
  = −80 kJ/mol

Termination:
(one possibility)

$:\ddot{\text{Br}}\cdot + \cdot\ddot{\text{Br}}: \longrightarrow :\ddot{\text{Br}}-\ddot{\text{Br}}:$

d, e.

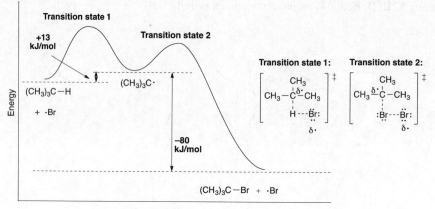

Radical Reactions 15–19

15.54

Initiation:

NBS

Propagation:

(from NBS)

(from NBS)

$+$ H$-$Br:

Br

Br

$+$ $\cdot$Br:

Termination:
(one possibility)

:Br$\cdot$ $+$ $\cdot$Br : $\longrightarrow$:Br$-$Br:

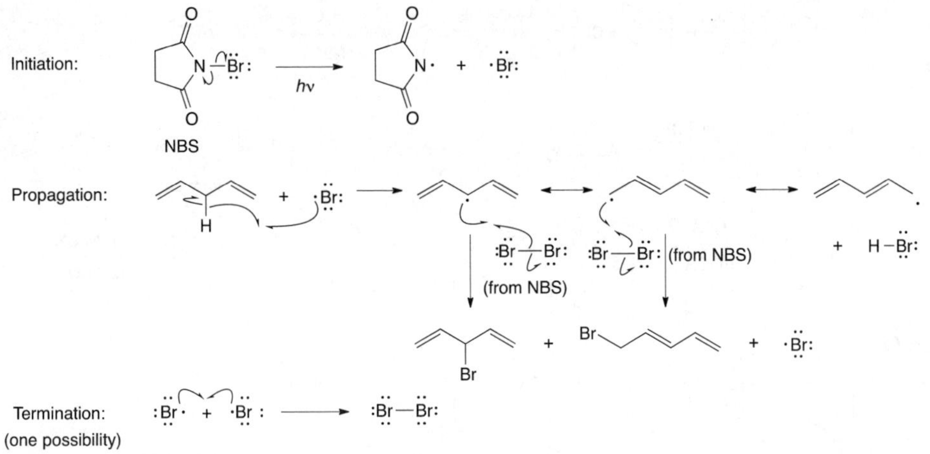

15.55 Calculate the $\Delta H°$ for the propagation steps of the reaction of CH$_4$ with I$_2$ to show why it does not occur at an appreciable rate.

CH$_3$$-$H $+$ $\cdot$İ: $\longrightarrow$ ĊH$_3$ $+$ H$-$İ: $\Delta H° = +138$ kJ/mol

+435 kJ/mol -297 kJ/mol

| This step is highly endothermic, making it difficult for chain propagation to occur over and over again. |

ĊH$_3$ $+$:İ$-$İ: $\longrightarrow$ CH$_3$$-$I $+$ $\cdot$İ: $\Delta H° = -83$ kJ/mol

+151 kJ/mol -234 kJ/mol

15.56

H$-$Br

3,3-dimethylbut-1-ene   2° carbocation   1,2-CH$_3$ shift   3° carbocation   2-bromo-2,3-dimethylbutane

$+$ Br$^-$ $+$ Br$^-$ Br$^-$

3,3-dimethylbut-1-ene   Br$\cdot$ Br   HBr   Br $+$ Br$\cdot$

HBr   The 2° radical does   1-bromo-3,3-dimethylbutane
+   NOT rearrange.
peroxide

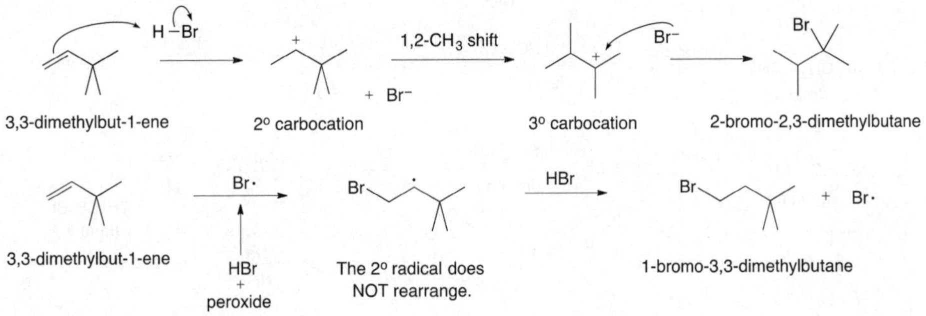

Addition of HBr without added peroxide occurs by an ionic mechanism and forms a 2° carbocation, which rearranges to a more stable 3° carbocation. The addition of H$^+$ occurs first, followed by Br$^-$. Addition of HBr with added peroxide occurs by a radical mechanism and forms a 2° radical that does not rearrange. In the radical mechanism, Br$\cdot$ adds first, followed by H$\cdot$.

Chapter 15–20

15.57

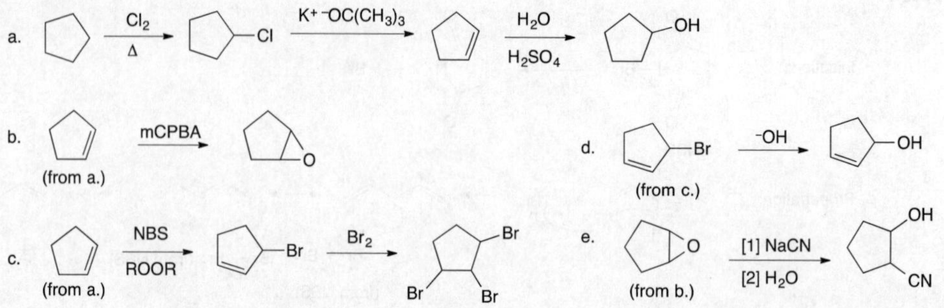

a.

b. (from a.)   d. (from c.)

c. (from a.)   e. (from b.)

15.58

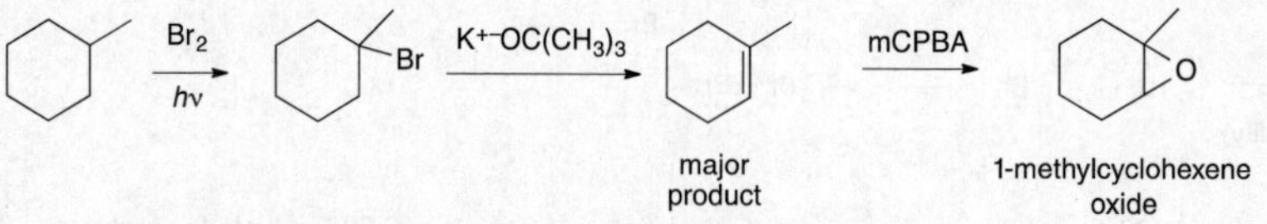

major
product

1-methylcyclohexene
oxide

15.59

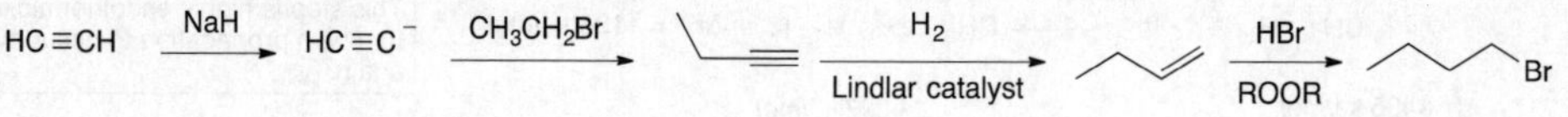

$HC \equiv CH$ $\xrightarrow{NaH}$ $HC \equiv C^-$ $\xrightarrow{CH_3CH_2Br}$ $\xrightarrow[\text{Lindlar catalyst}]{H_2}$ $\xrightarrow[\text{ROOR}]{HBr}$

15.60

a. CH_3-CH_3 $\xrightarrow{Br_2}{hv}$ CH_3CH_2Br $\xrightarrow{K^+ {}^-OC(CH_3)_3}$ $CH_2=CH_2$ $\xrightarrow{Br_2}$ $BrCH_2-CH_2Br$ $\xrightarrow{2\ NaNH_2}$ $HC \equiv CH$

$HC \equiv CH$ (from a.)
$\downarrow$ NaH

b. $CH_2=CH_2$ $\xrightarrow{mCPBA}$ $\xrightarrow[{[2]\ H_2O}]{[1]\ HC \equiv C^-}$
(from a.)

c. $HC \equiv CH$ $\xrightarrow{NaH}$ $HC \equiv C^-$ $\xrightarrow[\text{(from a.)}]{CH_3CH_2Br}$ $\xrightarrow{NaH}$
(from a.)

$\downarrow$ CH_3CH_2Br
(from a.)

$\xleftarrow[NH_3]{Na}$

d. (from c.) $\xrightarrow[\substack{H_2SO_4 \\ HgSO_4}]{H_2O}$

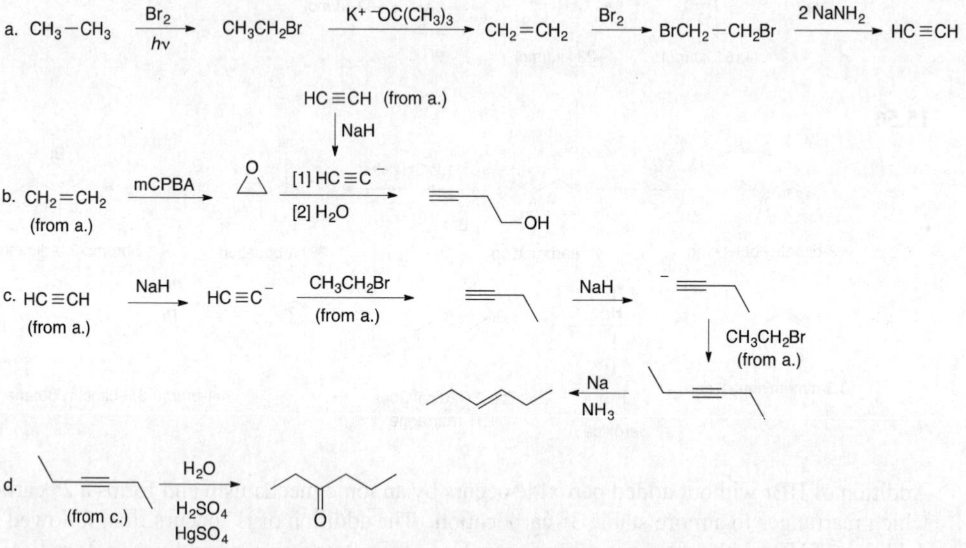

15.61

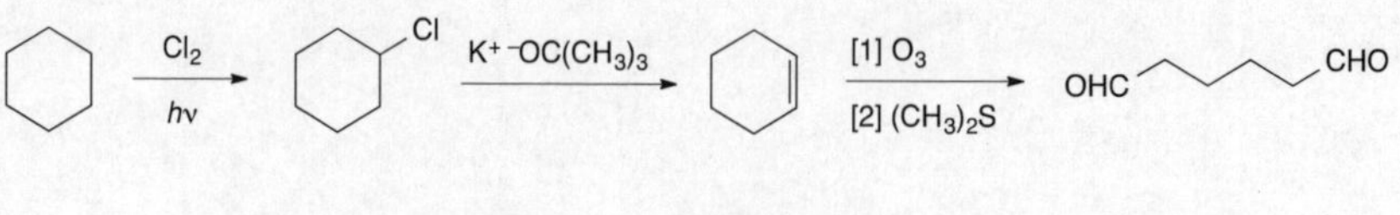

Radical Reactions 15–21

15.62

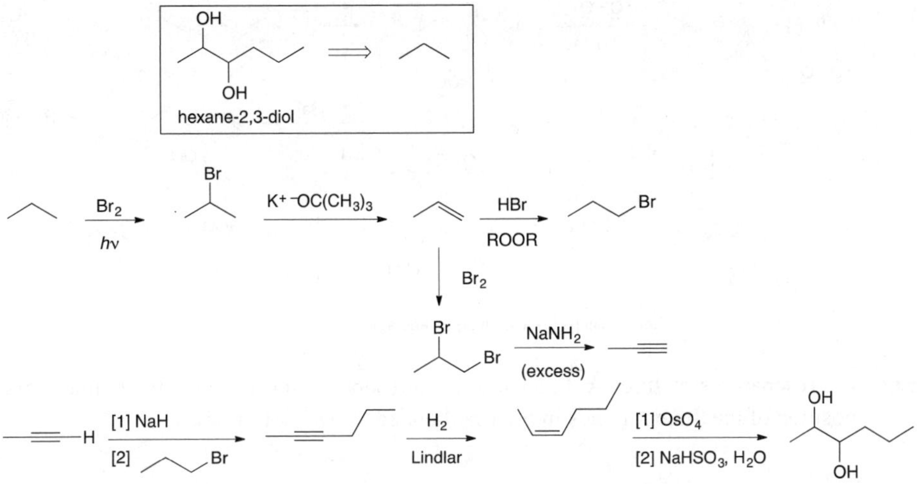

OH

OH
hexane-2,3-diol

Br$_2$
$h\nu$

Br

K$^+$ $^-$OC(CH$_3$)$_3$

HBr
ROOR

Br

Br$_2$

Br
Br

NaNH$_2$
(excess)

—H

[1] NaH
[2] ⌢Br

H$_2$
Lindlar

[1] OsO$_4$
[2] NaHSO$_3$, H$_2$O

OH

OH

15.63

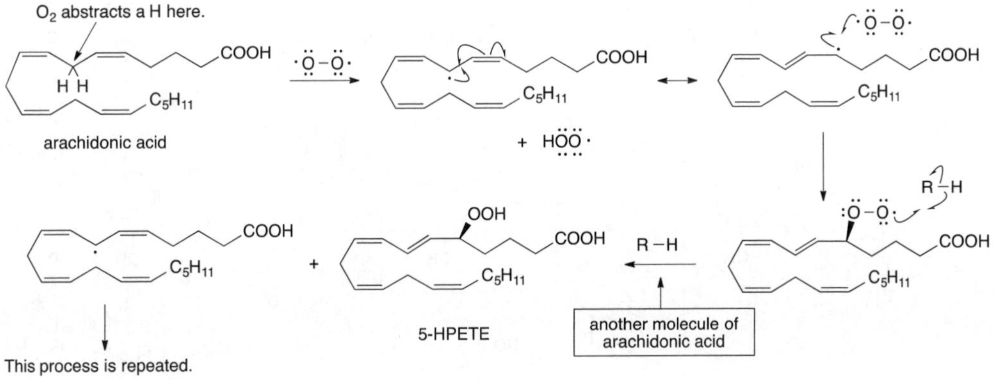

O$_2$ abstracts a H here.

COOH

H H
C$_5$H$_11$

arachidonic acid

$\cdot \ddot{O}-\ddot{O} \cdot$

COOH

C$_5$H$_11$

+ H$\ddot{O}\ddot{O} \cdot$

$\cdot \ddot{O}-\ddot{O} \cdot$

COOH

C$_5$H$_11$

R—H

COOH

C$_5$H$_11$

+

OOH

COOH

C$_5$H$_11$

5-HPETE

R—H

another molecule of
arachidonic acid

$:\ddot{O}-\ddot{O}:$ R—H

COOH

C$_5$H$_11$

This process is repeated.

15.64

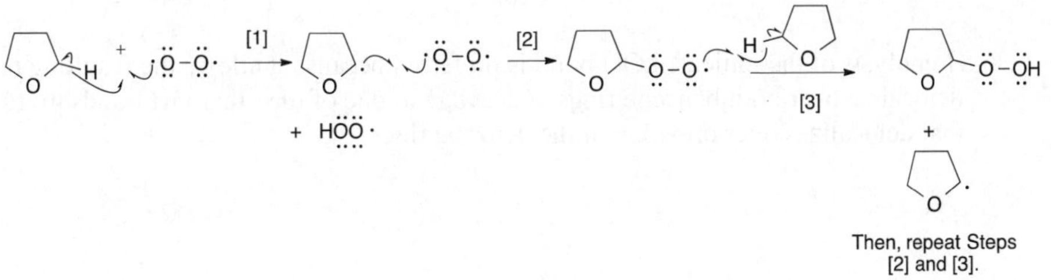

O

H

+ $\cdot \ddot{O}-\ddot{O} \cdot$

[1]

O

$\ddot{O}-\ddot{O} \cdot$

[2]

O

$\ddot{O}-\ddot{O}:$

H

O

[3]

O

$\ddot{O}-\ddot{O}H$

+ H$\ddot{O}\ddot{O} \cdot$

+

O $\cdot$

Then, repeat Steps
[2] and [3].

15.65

a.

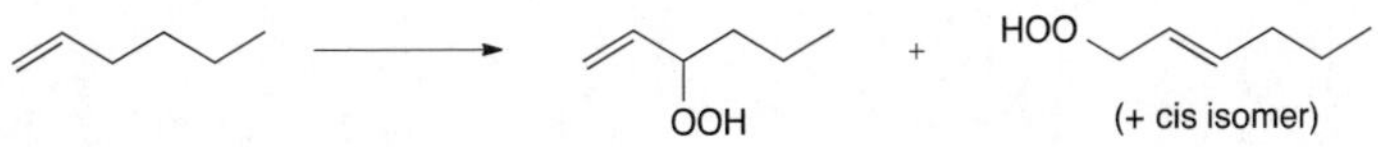

OOH

+

HOO

(+ cis isomer)

Chapter 15–22

b.

O₂ abstracts a H here.

[structures and reaction scheme]

[1]

+ HOO·

[2a]   [3a]   + R·   [RH = hex-1-ene]

OOH

[2b]   [3b]   HOO   + R·

Repeat Steps [2] and [3] again and again.

15.66 For resonance structures **A–F,** an additional resonance form can be drawn that moves the position of the three π bonds in the ring bonded to two OH groups.

a.

HO   HO   HO

OH   **A**   OH   **B**   OH   **C**

HO   HO   HO

OH   **F**   OH   **E**   OH   **D**

HO   HO   HO

OH   OH   OH

b. Homolysis of the indicated OH bond is preferred because it allows the resulting radical to delocalize over both benzene rings. Cleavage of one of the other OH bonds gives a radical that delocalizes over only one of the benzene rings.

Radical Reactions 15–23

15.67 Abstraction of the labeled H forms a highly resonance-stabilized radical. Four of the possible resonance structures are drawn.

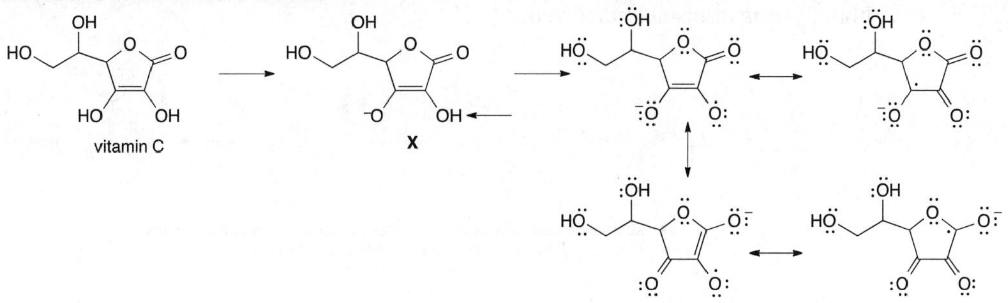

vitamin C → X

15.68 The monomers used in radical polymerization always contain double bonds.

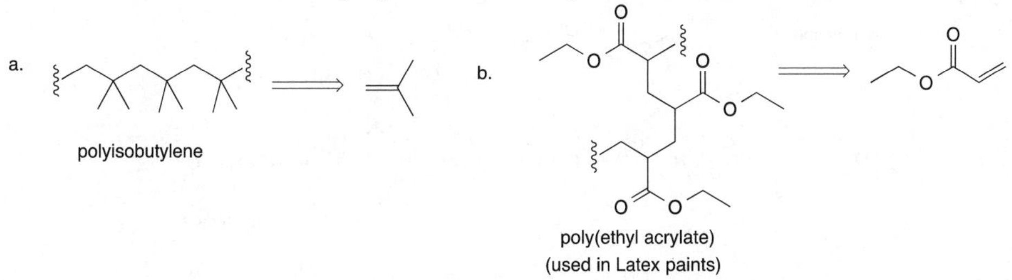

a.

polyisobutylene

b.

poly(ethyl acrylate)
(used in Latex paints)

15.69

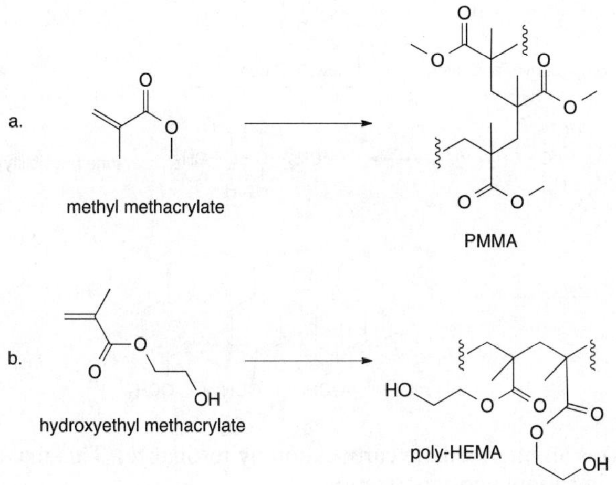

a.

methyl methacrylate → PMMA

b.

hydroxyethyl methacrylate → poly-HEMA

Chapter 15–24

15.70 Polystyrene has H atoms bonded to benzylic carbons—that is, carbons bonded directly to a benzene ring. These C–H bonds are unusually weak because the radical that results from homolysis is resonance stabilized.

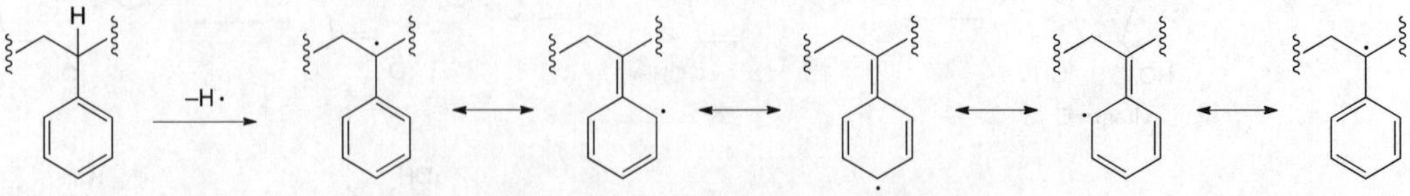

No such resonance stabilization is possible for the radical that results from C–H bond cleavage in polyethylene.

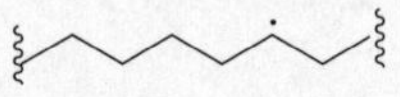

15.71

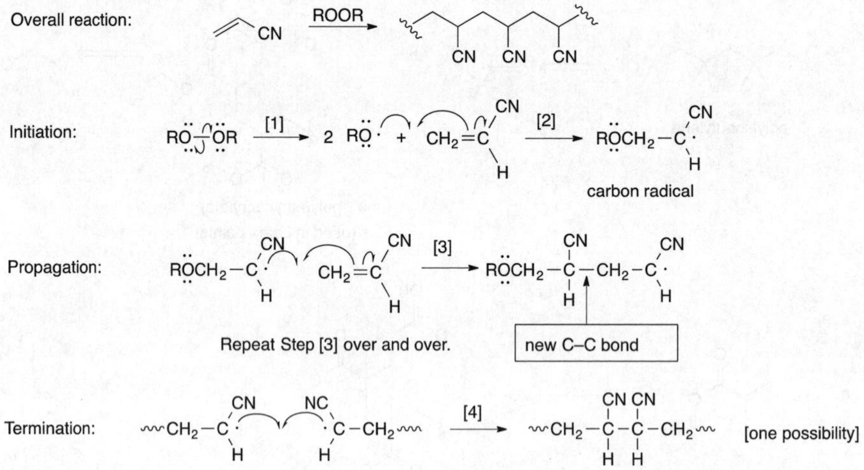

Overall reaction:

Initiation:

carbon radical

Propagation:

Repeat Step [3] over and over.

new C–C bond

Termination:

[one possibility]

15.72

a. CH$_3$O—⬡—CH=CH$_2$

A

OCH$_3$ OCH$_3$ OCH$_3$

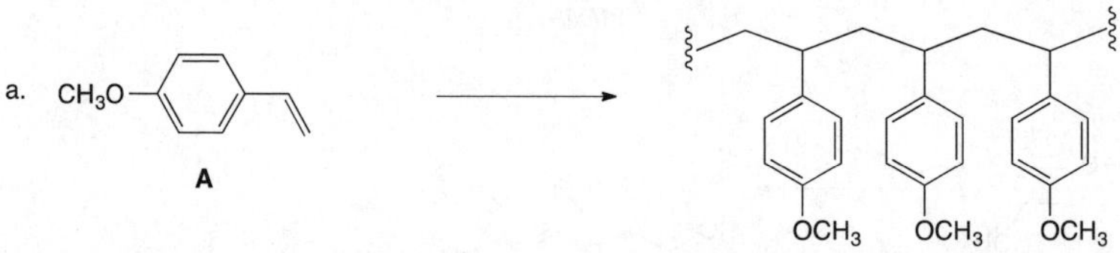

b. The OCH$_3$ group stabilizes an intermediate carbocation by resonance. This makes **A** react faster than styrene in cationic polymerization.

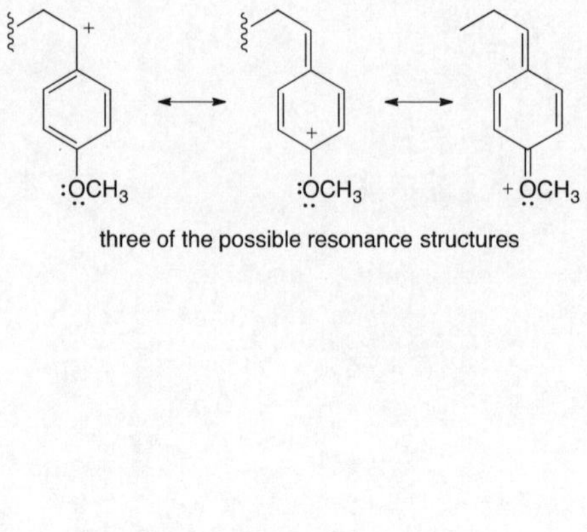

:OCH$_3$:OCH$_3$ $^+$OCH$_3$

three of the possible resonance structures

Radical Reactions 15–25

15.73

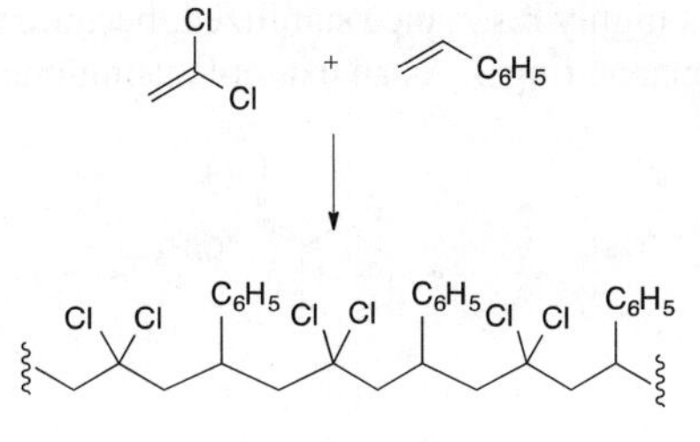

alternating copolymer

15.74

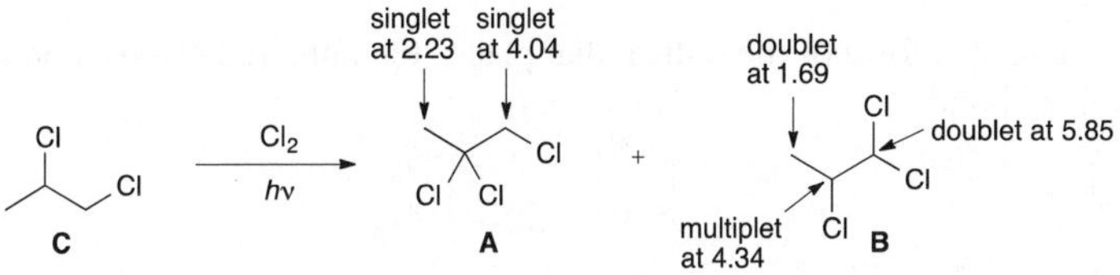

singlet at 2.23 singlet at 4.04

doublet at 1.69

doublet at 5.85

multiplet at 4.34

C   A   B

15.75

triplet

quintet
minor product

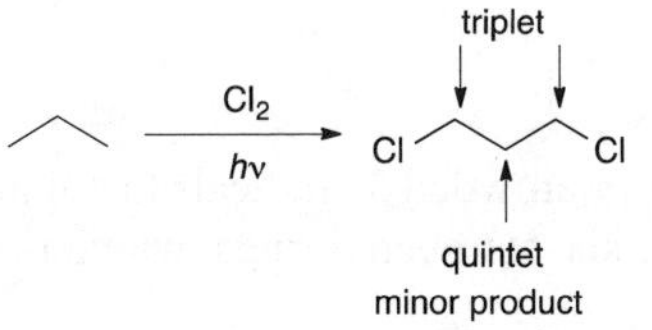

Molecular formula $C_3H_6Cl_2$
Integration: relative area 2:1
Because the compound has 6 H's and the sum of the relative areas is 3,
each absorption is due to twice as many H's.
　One signal is due to 4 H's.
　The second signal is due to 2 H's (2 x 1).
^{1}H NMR data:
　quintet at 2.2 (2 H's) split by 4 H's
　triplet at 3.7 (4 H's) split by 2 H's

15.76

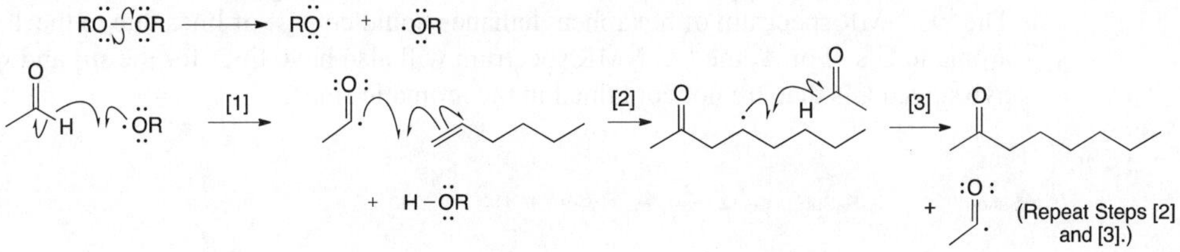

+ H–ÖR

+ :O: ‥ (Repeat Steps [2] and [3].)

Chapter 15–26

15.77

    a. The triphenylmethyl radical is highly resonance stabilized, because the radical can be delocalized on each of the benzene rings. As an example using one ring:

In addition, the radical is very sterically hindered, making it
difficult to undergo reactions.

    b. First, draw the resonance form of the radical that places the unpaired electron on the C that forms the new C–C bond.

    c. Hexaphenylethane formation would require two very crowded 3° radicals to combine. The formation of **A** results from a radical on one of the six-membered rings, which is much more accessible for reaction.

    d. The ^{1}H NMR spectrum of hexaphenylethane should show signals only in the aromatic region (7–8 ppm), whereas the ^{1}H NMR spectrum of **A** will also have signals for the sp^2 hybridized C–H bonds (4.5–6.0 ppm) of the alkenes, as well as the single H on the sp^3 hybridized carbon. The ^{13}C NMR spectrum of hexaphenylethane should consist of lines due to the 4° C's and the aromatic C's. For **A**, the ^{13}C NMR spectrum will also have lines for the sp^3 and sp^2 hybridized C's that are not contained in the aromatic rings.

15.78

Radical Reactions 15–27

15.79

A

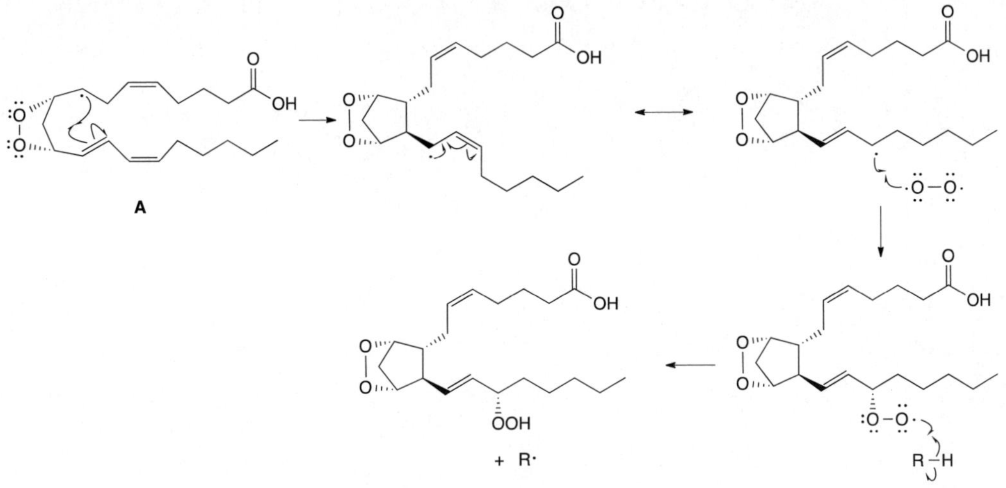

Conjugation, Resonance, and Dienes 16–1

Chapter 16: Conjugation, Resonance, and Dienes

Chapter Review

Conjugation and delocalization of electron density

- The overlap of p orbitals on three or more adjacent atoms allows electron density to delocalize, thus adding stability (16.1).
- An allyl carbocation ($CH_2=CHCH_2^+$) is more stable than a 1° carbocation because of p orbital overlap (16.2).
- In a system X=Y–Z:, Z is generally sp^2 hybridized to allow the lone pair to occupy a p orbital, making the system conjugated (16.5).

Four common examples of resonance (16.3)

[1] The three atom "allyl" system: X=Y–Z $\longleftrightarrow$ X–Y=Z $*$ = +, –, •, or ••

[2] Conjugated double bonds:

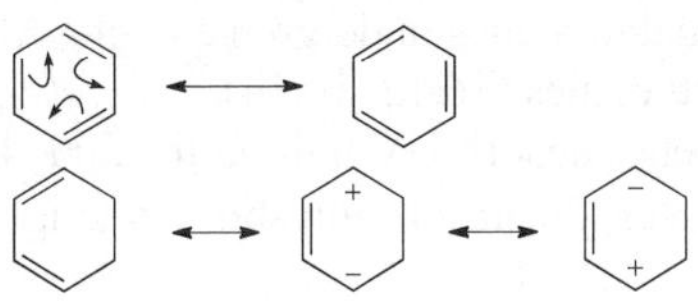

[3] Cations having a positive charge adjacent to a lone pair:

[4] Double bonds having one atom more electronegative than the other: X=Y $\longleftrightarrow$ X–Y: [Electronegativity of Y > X]

Rules on evaluating the relative "stability" of resonance structures (16.4)

[1] Structures with more bonds and fewer charges are better.

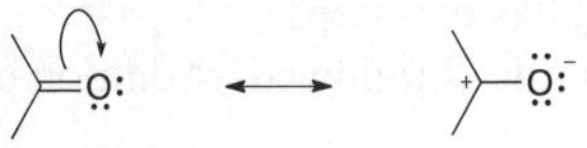

all neutral atoms
one more bond
bettter resonance structure

[2] Structures in which every atom has an octet are better.

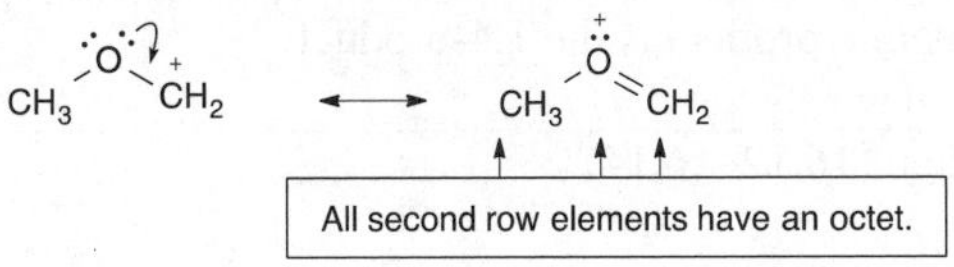

All second row elements have an octet.

Chapter 16–2

[3] Structures that place a negative charge on a more electronegative element are better.

The (–) charge is on the more electronegative O atom.

better resonance structure

The unusual properties of conjugated dienes

[1] The C–C σ bond joining the two double bonds is unusually short (16.8).

[2] Conjugated dienes are more stable than similar isolated dienes. $\Delta H°$ of hydrogenation is smaller for a conjugated diene than for an isolated diene converted to the same product (16.9).

[3] The reactions are unusual:
- Electrophilic addition affords products of 1,2-addition and 1,4-addition (16.10, 16.11).
- Conjugated dienes undergo the Diels–Alder reaction, a reaction that does not occur with isolated dienes (16.12–16.14).

[4] Conjugated dienes absorb UV light in the 200–400 nm region. As the number of conjugated π bonds increases, the absorption shifts to longer wavelength (16.15).

Reactions of conjugated dienes

[1] Electrophilic addition of HX (X = halogen) (16.10, 16.11)

HX
(1 equiv)

1,2-product
kinetic product

+

1,4-product
thermodynamic product

- The mechanism has two steps.
- Markovnikov's rule is followed. Addition of H^+ forms the *more* stable allylic carbocation.
- The 1,2-product is the kinetic product. When H^+ adds to the double bond, X^- adds to the end of the allylic carbocation to which it is closer (C2 not C4). The kinetic product is formed faster at low temperature.
- The thermodynamic product has the more substituted, more stable double bond. The thermodynamic product predominates at equilibrium. With buta-1,3-diene, the thermodynamic product is the 1,4-product.

[2] Diels–Alder reaction (16.12–16.14)

Δ

1,3-diene   dienophile

The three new bonds are labeled in **bold.**

- The reaction forms two σ bonds and one π bond in a six-membered ring.
- The reaction is initiated by heat.
- The mechanism is concerted: all bonds are broken and formed in a single step.
- The diene must react in the *s*-cis conformation (16.13A).
- Electron-withdrawing groups in the dienophile increase the reaction rate (16.13B).
- The stereochemistry of the dienophile is retained in the product (16.13C).
- Endo products are preferred (16.13D).

Practice Test on Chapter Review

1.a. Which of the following statements is true about the Diels–Alder reaction?
  1. The reaction is faster with electron-donating groups in the dienophile.
  2. The reaction is endothermic.
  3. The diene must adopt the *s*-cis conformation.
  4. Statements (1) and (2) are true.
  5. Statements (1), (2), and (3) are all true.

b. Which of the following statements is true about the absorption of ultraviolet light by unsaturated systems?
  1. Penta-1,4-diene requires light having a wavelength < 200 nm for electron promotion.
  2. Cyclohexa-1,3-diene absorbs ultraviolet light with a wavelength > 200 nm.
  3. As the number of conjugated π bonds increases, the energy difference between the excited state and ground state decreases.
  4. Statements (1) and (2) are true.
  5. Statements (1), (2), and (3) are all true.

c. Which of the following compounds contains a labeled carbon atom that is sp^2 hybridized?

  1. **A** only
  2. **B** only
  3. **C** only
  4. **A** and **B**
  5. **A**, **B**, and **C**

d. Which of the following represent valid resonance structures for **A**?

  4. Both (1) and (2) are valid resonance structures.
  5. Structures (1), (2), and (3) are all valid.

Chapter 16–4

2. Name the following compounds and indicate the conformation around the σ bond that joins the two double bonds.

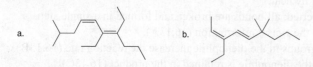

a.   b.

3.a. Consider the four hydrocarbons (**A–D**) drawn below. [1] Which compound absorbs the *shortest* wavelength of ultraviolet light? [2] Which compound absorbs the *longest* wavelength of ultraviolet light?

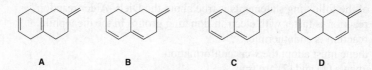

    A   B   C   D

b. Consider the four dienes (**A–D**) drawn below. [1] Which diene is *most* reactive in the Diels–Alder reaction? [2] Which diene is the *least* reactive in the Diels–Alder reaction?

    A   B   C   D

4. Draw the organic products formed in each reaction. In part (b), label the kinetic and thermodynamic products.

a.

b.

c.

Conjugation, Resonance, and Dienes 16–5

5. What diene and dienophile are needed to synthesize the following Diels–Alder adducts?

a.

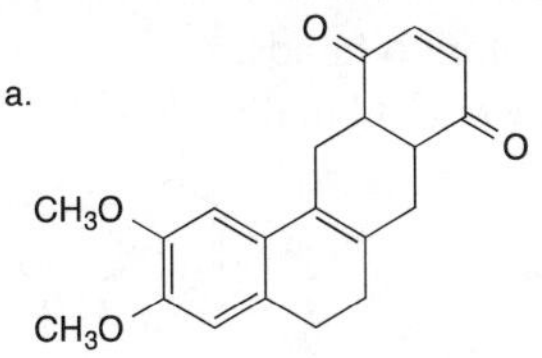

b.

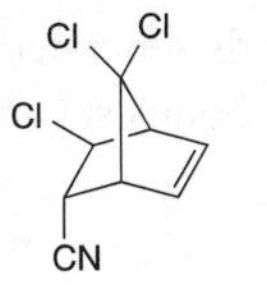

Answers to Practice Test

1.a. 3
   b. 5
   c. 4
   d. 1

2.a. (4Z,6E)-6,7-diethyl-2-
      methyldeca-4,6-diene, s-trans

   b. (2Z,4E)-3-ethyl-6,6-
      dimethylnona-2,4-diene,
      s-trans

3.a. [1] **A**; [2] **C**
   b. [1] **D**; [2] **A**

4.a.

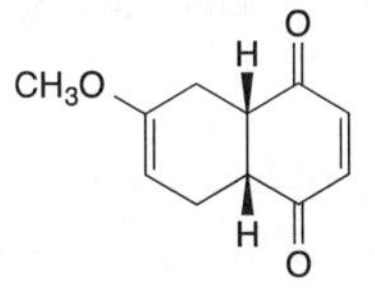

(both H's up or both H's down)

b.

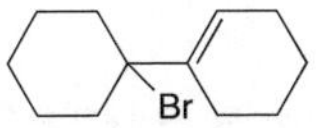

kinetic

+

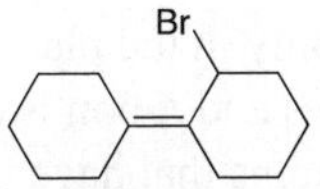

thermodynamic

c.

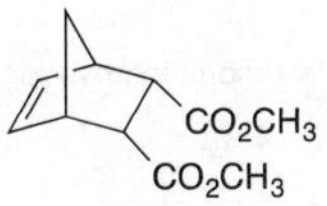

5.a.

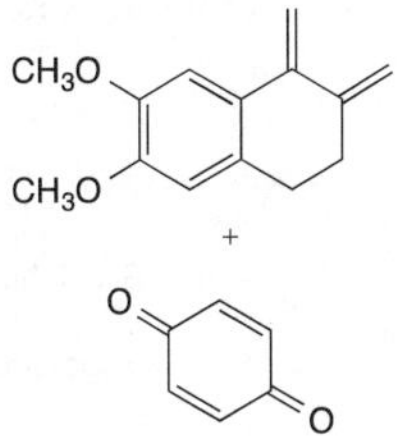

+

b.

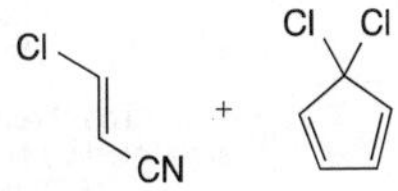

+

Answers to Problems

16.1 **Isolated dienes** have two double bonds separated by two or more σ bonds.
Conjugated dienes have two double bonds separated by only one σ bond.

a. One σ bond separates two double bonds = **conjugated diene.**

b. Two σ bonds separate two double bonds = **isolated diene.**

c. One σ bond separates two double bonds = **conjugated diene.**

d. Four σ bonds separate two double bonds = **isolated diene.**

16.2 **Conjugation** occurs when there are overlapping *p* orbitals on three or more adjacent atoms. Double bonds separated by two σ bonds are not conjugated.

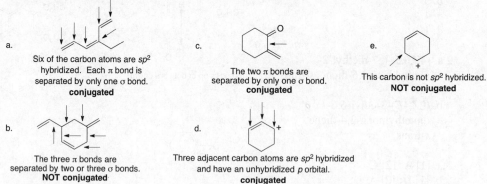

a. Six of the carbon atoms are *sp*² hybridized. Each π bond is separated by only one σ bond. **conjugated**

b. The three π bonds are separated by two or three σ bonds. **NOT conjugated**

c. The two π bonds are separated by only one σ bond. **conjugated**

d. Three adjacent carbon atoms are *sp*² hybridized and have an unhybridized *p* orbital. **conjugated**

e. This carbon is not *sp*² hybridized. **NOT conjugated**

16.3 Two resonance structures differ only in the placement of electrons. All σ bonds stay in the same place. Nonbonded electrons and π bonds can be moved. To draw the hybrid:
- Use a dashed line between atoms that have a π bond in one resonance structure and not the other.
- Use a δ symbol for atoms with a charge or radical in one structure but not the other.

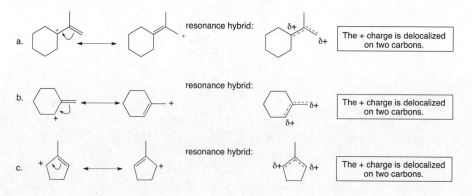

a. resonance hybrid: The + charge is delocalized on two carbons.

b. resonance hybrid: The + charge is delocalized on two carbons.

c. resonance hybrid: The + charge is delocalized on two carbons.

Conjugation, Resonance, and Dienes 16–7

16.4 S$_N$1 reactions proceed via a carbocation intermediate. Draw the carbocation formed on loss of Cl and compare. The more stable the carbocation, the faster the S$_N$1 reaction.

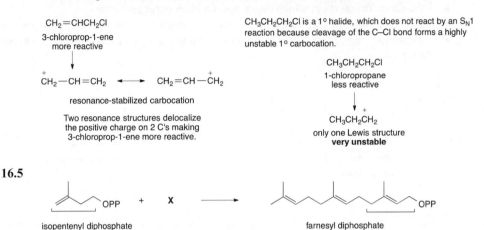

16.5

isopentenyl diphosphate + **X** → farnesyl diphosphate

Five C's of farnesyl diphosphate come from isopentenyl diphosphate, so the remaining 10 C's come from **X**. If the reaction is analogous to the formation of geranyl diphosphate, **X** must contain 10 C's and have an allylic diphosphate.

16.6

a. Move the charge and the double bond.

b. Move the charge and the double bond.

c. Move the lone pair.

d. Move the charge and the double bond.

Chapter 16–8

16.7 To compare the resonance structures remember:
- Resonance structures with **more bonds** are better.
- Resonance structures in which **every atom has an octet** are better.
- Resonance structures with **neutral atoms** are better than those with charge separation.
- Resonance structures that place a **negative charge on a more electronegative atom** are better.

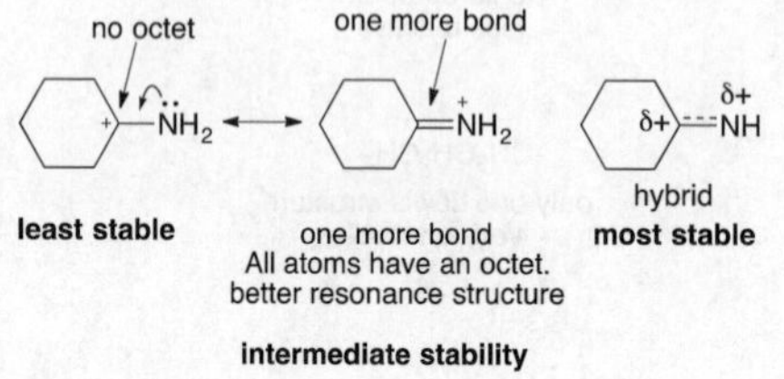

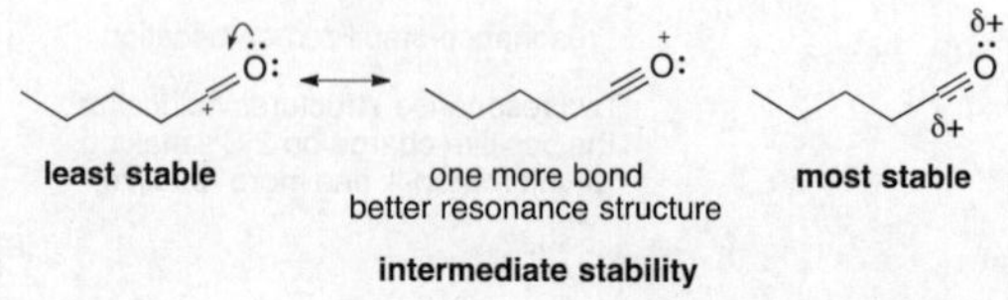

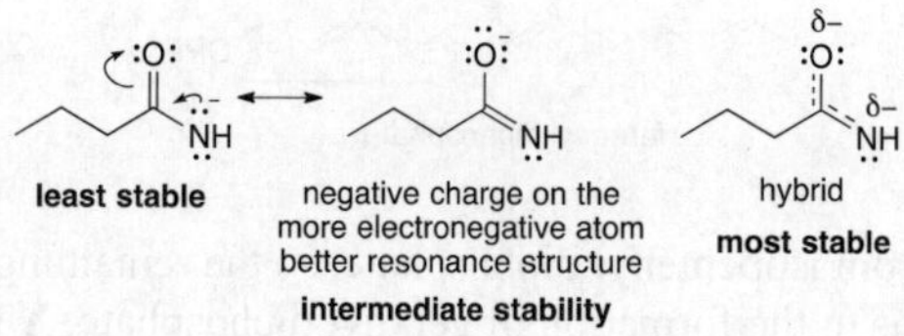

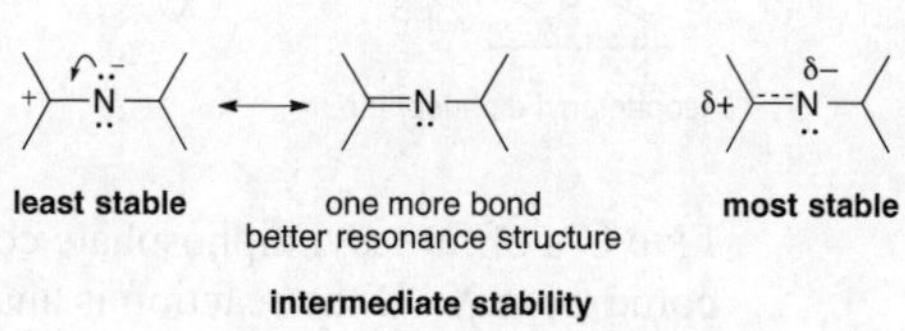

16.8

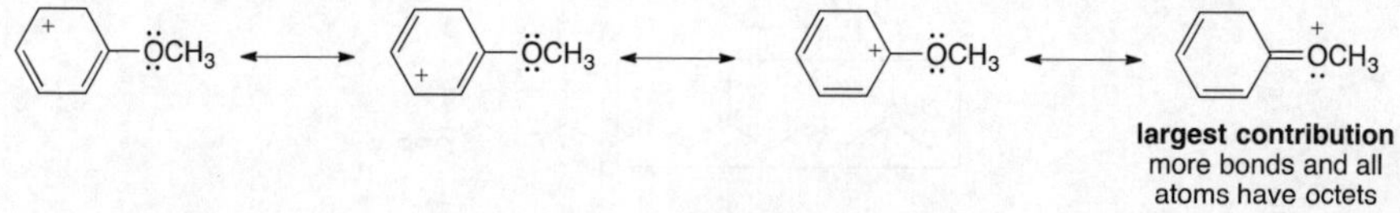

16.9 Remember that in any allyl system, there must be *p* orbitals to delocalize the lone pair.

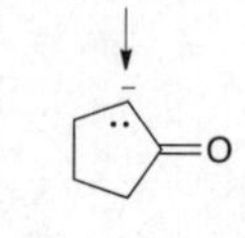

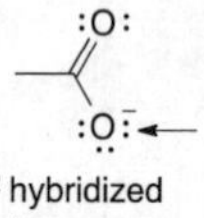

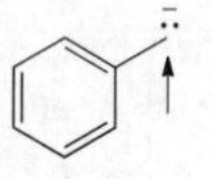

16.10 The ***s*-cis** conformation has two double bonds on the **same side** of the single bond.
The ***s*-trans** conformation has two double bonds on **opposite sides** of the single bond.

Conjugation, Resonance, and Dienes 16–9

a. (2E,4E)-**octa-2,4-diene** in the s-trans conformation

double bonds on opposite sides
s-trans

b. (3E,5Z)- **nona-3,5-diene** in the s-cis conformation

double bonds on the same side
s-cis

c. (3Z,5Z)-4,5-dimethyl**deca-3,5-diene** in both the s-cis and s-trans conformations

s-cis →

s-trans →

16.11

s-cis

conjugated

s-trans

conjugated →

conjugated →

E E

OH

isolated

O isolated

OH

isolated

HO

Z

Z

isolated

isolated

16.12 Bond length depends on hybridization and percent s-character. Bonds with a higher percent s-character have smaller orbitals and are shorter.

$HC \equiv C - C \equiv CH$

sp hybridized carbons
50% s-character
shortest bond

$CH_2 = CH - CH = CH_2$

sp^2 hybridized carbons
33% s-character
intermediate length

$CH_3 - CH_3$

sp^3 hybridized carbons
25% s-character
longest bond

16.13 Two equivalent resonance structures delocalize the π bond and the negative charge.

hybrid:

These bond lengths are equal because they are identical.

16.14 The **less stable** (higher energy) **diene** has the **larger heat of hydrogenation.** Isolated dienes are higher in energy than conjugated dienes, so they will have a larger heat of hydrogenation.

a.   or

Double bonds separated by
one σ bond = **conjugated diene**
smaller heat of hydrogenation

Double bonds separated by
two σ bonds = **isolated diene**
larger heat of hydrogenation

Chapter 16–10

b.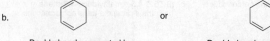

Double bonds separated by one σ bond = **conjugated diene** **smaller** heat of hydrogenation

or

Double bonds separated by two σ bonds = **isolated diene** **larger** heat of hydrogenation

16.15 Isolated dienes are higher in energy than conjugated dienes. Compare the location of the double bonds in the compounds below.

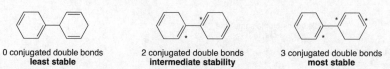

0 conjugated double bonds
least stable

2 conjugated double bonds
intermediate stability

3 conjugated double bonds
most stable

16.16 Conjugated dienes react with HX to form 1,2- and 1,4-products.

a. [reaction of 1,3-butadiene + HCl → 1,2-product (+ Z isomer) + 1,4-product (+ Z isomer)]

b. [cyclohexadiene (isolated diene) + HCl → chlorocyclohexene]

c. [1,3-cyclohexadiene + HCl → 3-chlorocyclohexene]

d. [methylcyclohexadiene + HCl → A + B + C]

This double bond is more reactive, so **C** is probably a minor product because it results from HCl addition to the less reactive double bond.

16.17 The mechanism for addition of DCl has two steps:
 [1] **Addition of D⁺** forms a resonance-stabilized carbocation.
 [2] **Nucleophilic attack of Cl⁻** forms 1,2- and 1,4-products.

[mechanism scheme showing 1,3-cyclohexadiene + D–Cl → resonance-stabilized carbocation intermediates with D, then Cl⁻ attack giving 1,2- and 1,4-products]

Conjugation, Resonance, and Dienes 16–11

16.18 Label the products as 1,2- or 1,4-products. The 1,2-product is the kinetic product, and the 1,4-product, which has the more substituted double bond, is the thermodynamic product.

16.19 To draw the products of a Diels–Alder reaction:
[1] Find the 1,3-diene and the dienophile.
[2] Arrange them so the diene is on the left and the dienophile is on the right.
[3] Cleave three bonds and use arrows to show where the new bonds will be formed.

a.

diene   dienophile

b.

diene
Rotate to
make it *s*-cis.

dienophile

c.

dienophile   diene
Rotate to
make it *s*-cis.

16.20 For a diene to be reactive in a Diels–Alder reaction, **a diene must be able to adopt an *s*-cis conformation.**

s-trans
**cannot rotate
unreactive**

rotate

s-cis
reactive

s-cis
most reactive

The diene is always in the
s-cis conformation.

16.21 Zingiberene reacts much faster than β-sesquiphellandrene as a Diels–Alder diene because its diene is constrained in the *s*-cis conformation. The diene in β-sesquiphellandrene is constrained in the *s*-trans conformation, so it is unreactive in the Diels–Alder.

Chapter 16–12

16.22 Electron-withdrawing substituents in the dienophile increase the reaction rate.

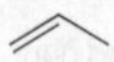

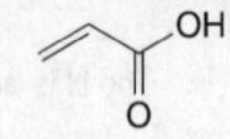

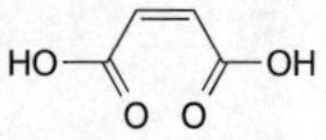

no electron-withdrawing groups
least reactive

one electron-withdrawing group
intermediate reactivity

two electron-withdrawing groups
most reactive

16.23 A cis dienophile forms a cis-substituted cyclohexene.
A trans dienophile forms a trans-substituted cyclohexene.

a.   cis dienophile   cis-substituted products

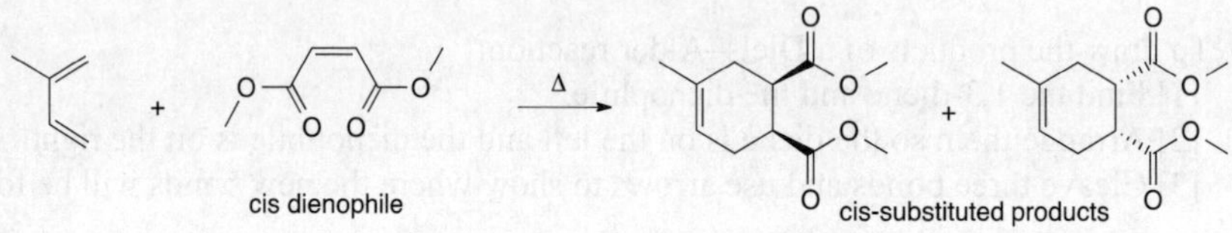

b.   trans dienophile   trans-substituted products

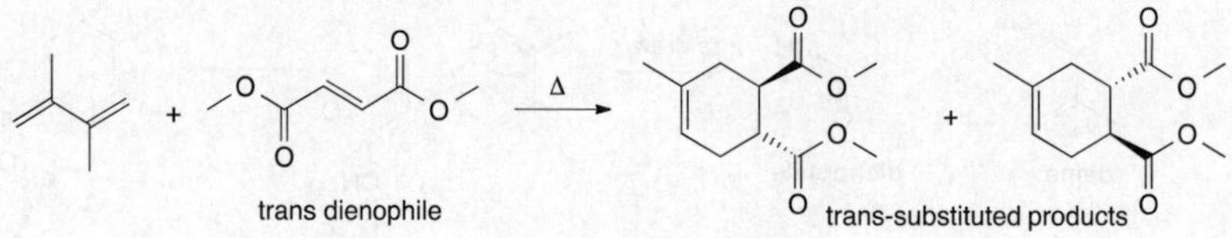

c.   cis dienophile   identical   cis-substituted product

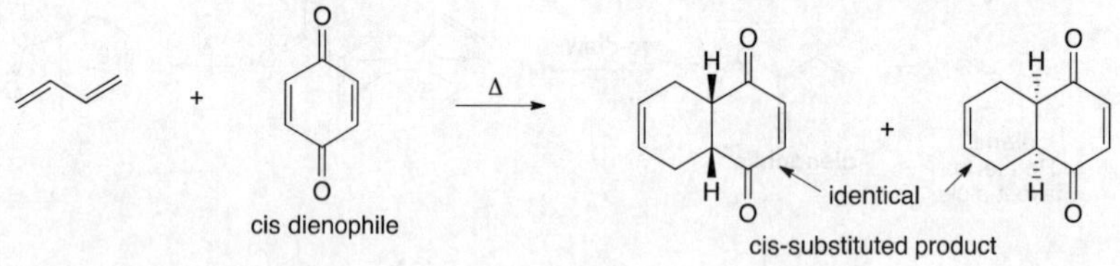

16.24 The **endo product** (with the substituents under the plane of the new six-membered ring) is the preferred product.

a.   **endo** substituent

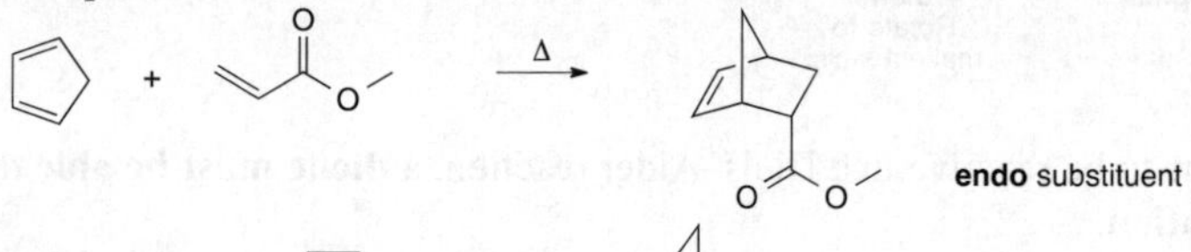

b.   both groups **endo**

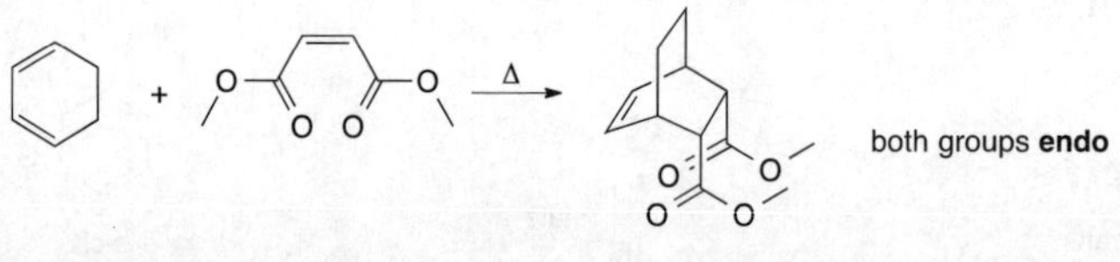

16.25 To find the diene and dienophile needed to make each of the products:
[1] Find the six-membered ring with a C–C double bond.
[2] Draw three arrows to work backwards.
[3] Follow the arrows to show the diene and dienophile.

Conjugation, Resonance, and Dienes 16–13

a.

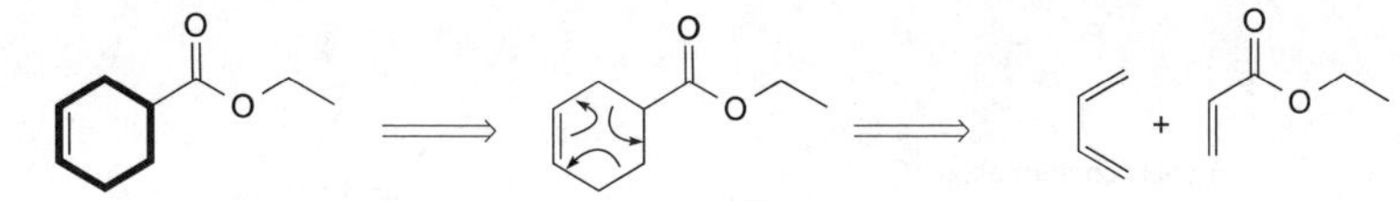

b.

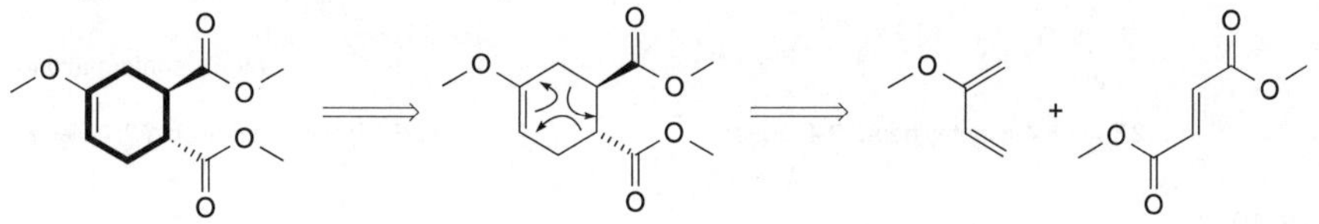

c.

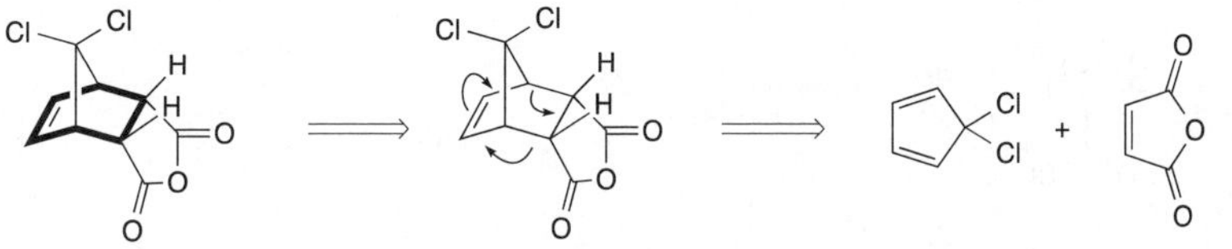

16.26

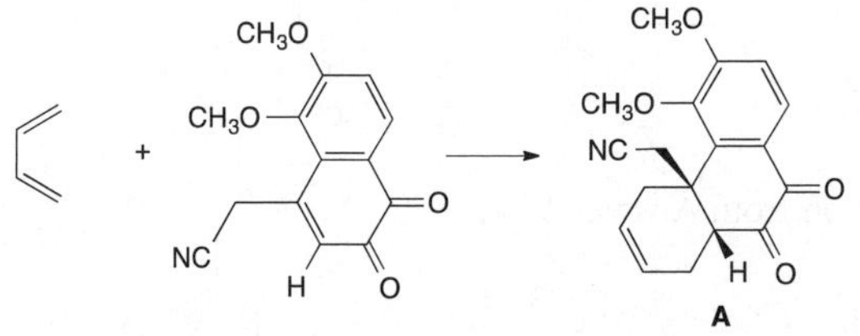

(+ enantiomer)

A

16.27 Conjugated molecules absorb light at a longer wavelength than molecules that are not conjugated.

a. 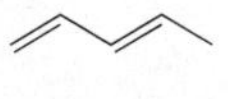   or

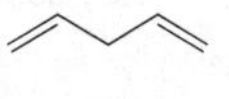

conjugated
longer wavelength

not conjugated

b. 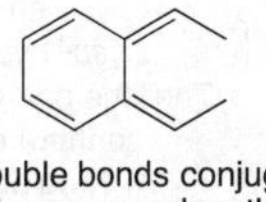   or

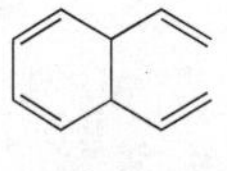

all double bonds conjugated
longer wavelength

one set of
conjugated dienes

16.28 Sunscreens contain conjugated systems to absorb UV radiation from sunlight. Look for conjugated systems in the compounds below.

a.

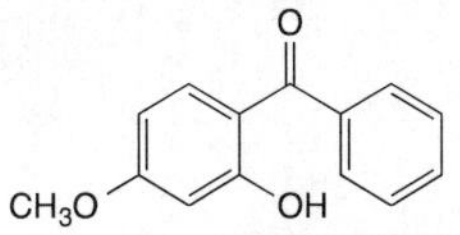

conjugated system
could be a sunscreen

b.

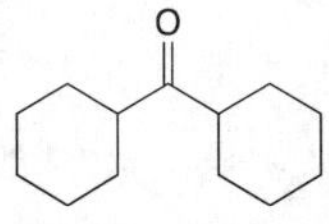

not a conjugated system

c.

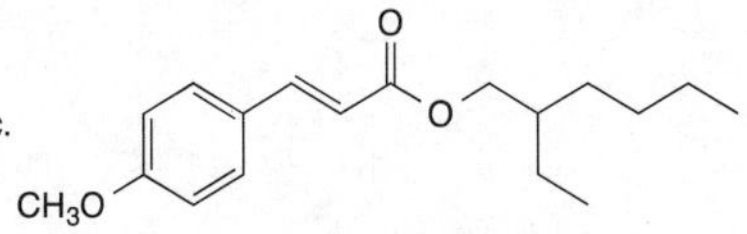

conjugated system
could be a sunscreen

Chapter 16–14

16.29

a. (2Z,4E)-3,4-dimethylhepta-2,4-diene — s-cis conformation

b. (2E,4Z)-3,4-dimethylocta-2,4-diene — s-trans conformation

16.30

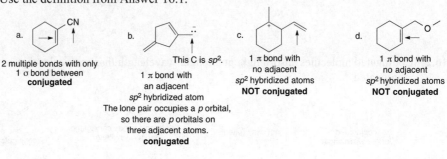

16.31 Use the definition from Answer 16.1.

a. 2 multiple bonds with only 1 σ bond between
conjugated

b. This C is sp². 1 π bond with an adjacent sp² hybridized atom
The lone pair occupies a p orbital, so there are p orbitals on three adjacent atoms.
conjugated

c. 1 π bond with no adjacent sp² hybridized atoms
NOT conjugated

d. 1 π bond with no adjacent sp² hybridized atoms
NOT conjugated

16.32

a. [resonance structures]

b. [resonance structures]

c. [resonance structures]

d. [resonance structures]

Conjugation, Resonance, and Dienes 16–15

e.

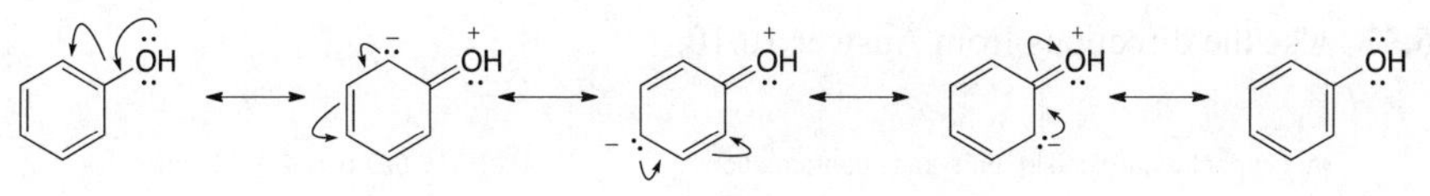

f.

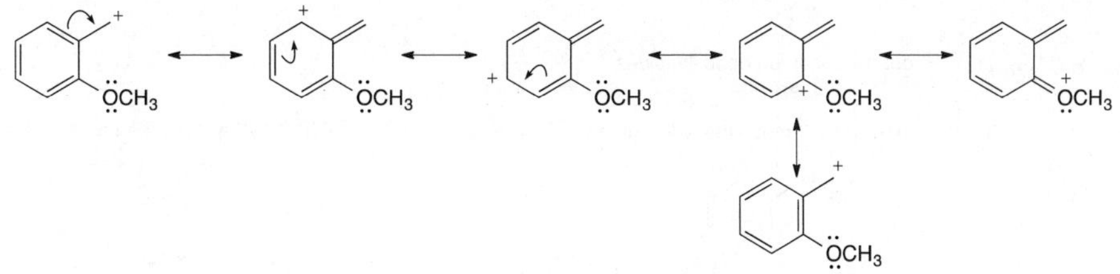

16.33 No additional resonance structures can be drawn for compounds (a) and (d).

b.

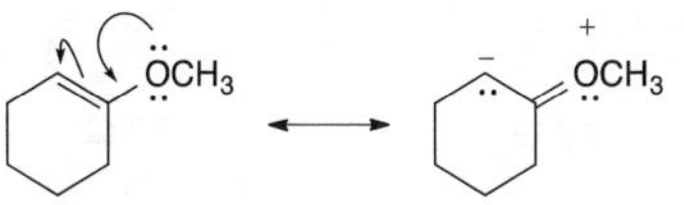

c.

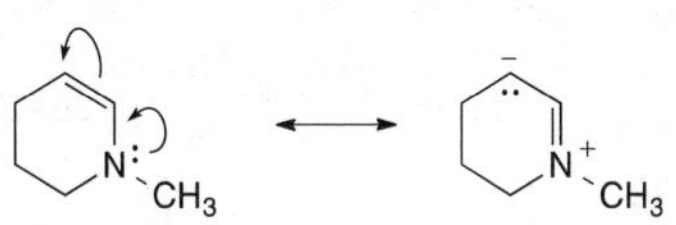

16.34

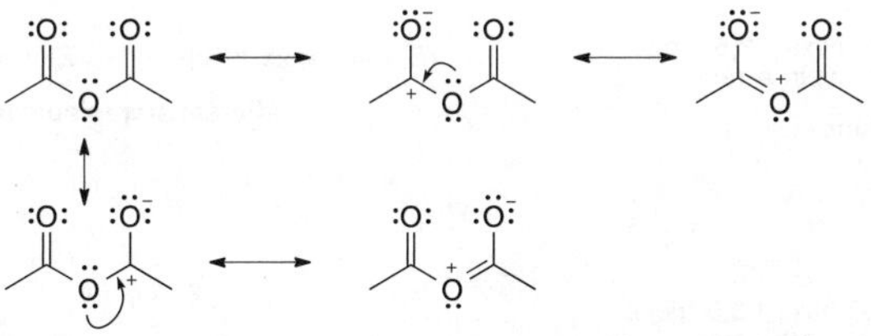

16.35

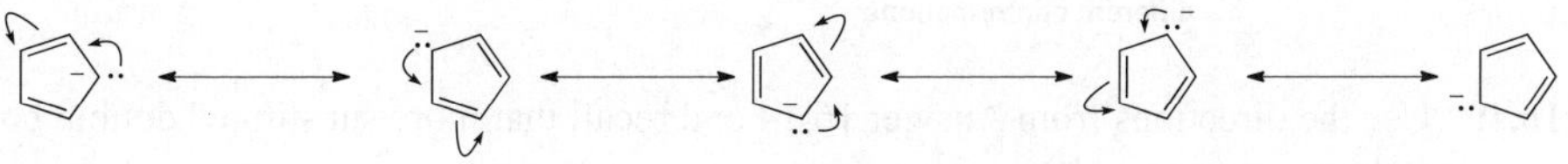

resonance hybrid:

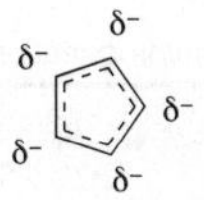

Five resonance structures delocalize the negative charge on five C's, making them all equivalent.

All of the carbons are identical in the anion.

16.36

Draw the products of cleavage of the bond.

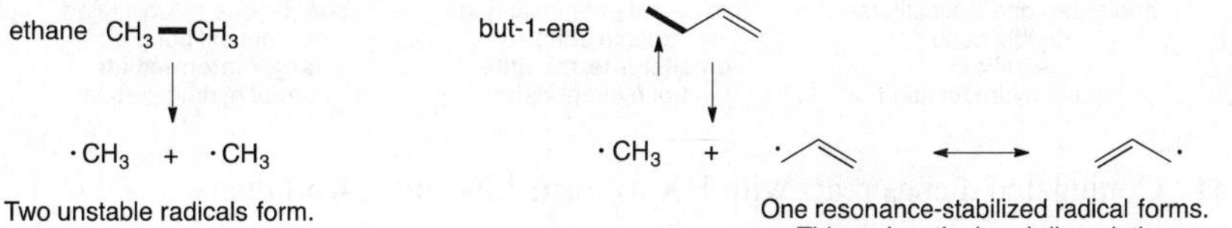

ethane $CH_3 \!-\! CH_3$

but-1-ene

$\cdot CH_3 \;+\; \cdot CH_3$

Two unstable radicals form.

$\cdot CH_3 \;+\;$

One resonance-stabilized radical forms. This makes the bond dissociation energy lower because a more stable radical is formed.

Chapter 16–16

16.37 Use the directions from Answer 16.10.

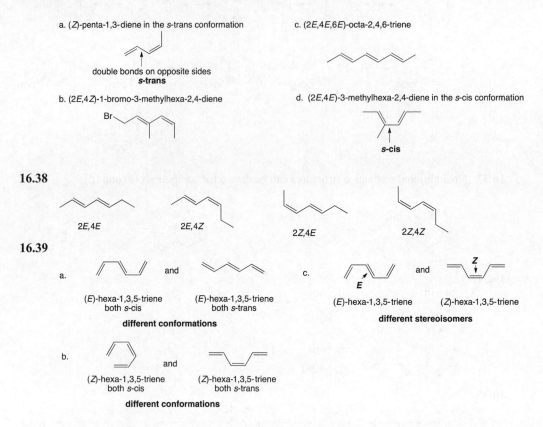

16.38

16.39

16.40 Use the directions from Answer 16.14 and recall that more substituted double bonds are more stable.

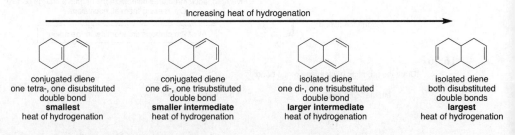

16.41 Conjugated dienes react with HX to form 1,2- and 1,4-products.

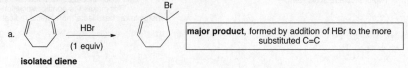

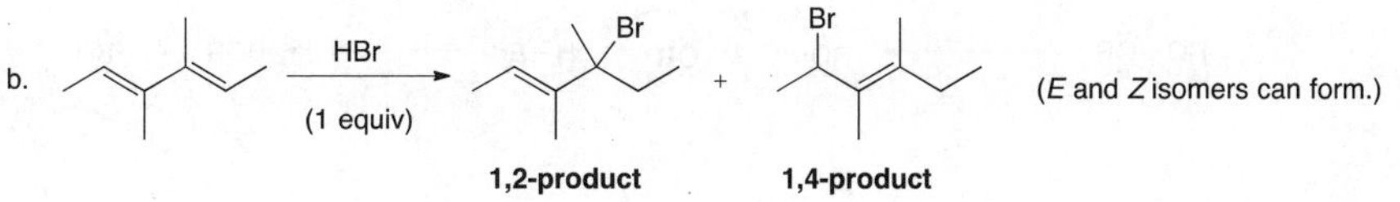
b.
HBr
(1 equiv)
Br
+
Br
(E and Z isomers can form.)
1,2-product
1,4-product

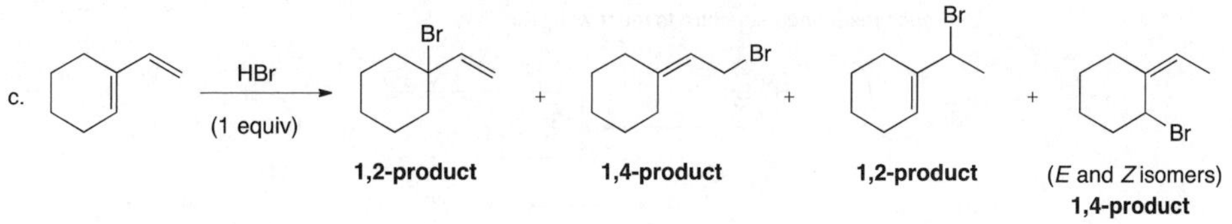
c.
HBr
(1 equiv)
Br
+
Br
+
Br
+
Br
1,2-product
1,4-product
1,2-product
(E and Z isomers)
1,4-product

16.42

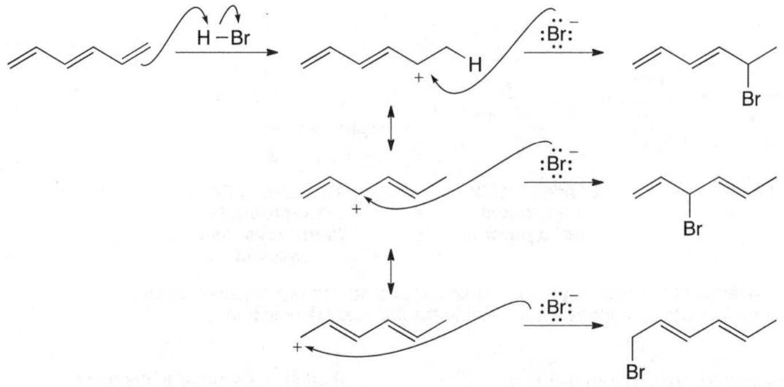
H–Br
H
:Br:⁻
Br
Br
Br

16.43

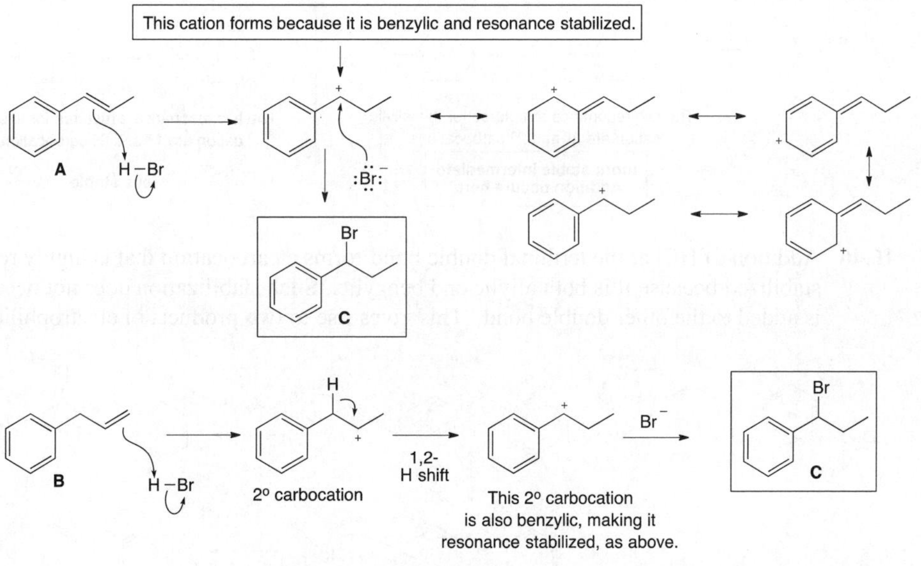
This cation forms because it is benzylic and resonance stabilized.
A
H–Br
:Br:⁻
Br
C
B
H–Br
H
2° carbocation
1,2-
H shift
Br⁻
This 2° carbocation
is also benzylic, making it
resonance stabilized, as above.
Br
C

16.44 To draw the mechanism for the reaction of a diene with HBr and ROOR, recall from Chapter 15 that when an alkene is treated with HBr under these radical conditions, the Br ends up on the carbon with more H's to begin with.

Chapter 16–18

Use each resonance structure to react with HBr.

16.45

a, b.

Y is the kinetic product because of the proximity effect. H and Cl add across two adjacent atoms.
Z is the thermodynamic product because it has a more stable trisubstituted double bond.

Addition occurs at the labeled double bond due to the stability of the carbocation intermediate.

If addition occurred at the other C=C, the following allylic carbocation would form:

c.

The two resonance structures for this allylic cation are 3° and 2° carbocations.

**more stable intermediate
Addition occurs here.**

The two resonance structures for this allylic cation are 1° and 2° carbocations.

less stable

16.46 Addition of HCl at the terminal double bond forms a carbocation that is highly resonance stabilized because it is both allylic and benzylic. Such stabilization does not occur when HCl is added to the other double bond. This gives rise to two products of electrophilic addition.

1,2-product

1,4-product

(+ three more resonance structures that delocalize the positive charge onto the benzene ring)

16.47 There are two possible products:

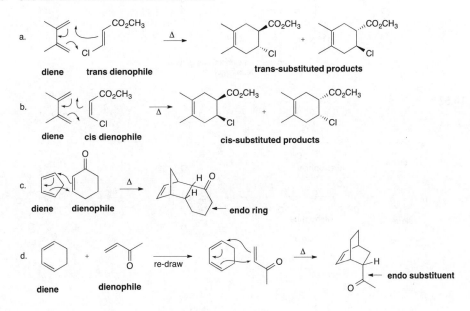

The 1,2-product is always the kinetic product because of the proximity effect. In this case, it is also the thermodynamic (more stable) product because it contains a more highly substituted C=C (trisubstituted) than the 1,4-product (disubstituted). Thus, the 1,2-product is the major product at high and low temperature.

16.48 The electron pairs on O can be donated to the double bond through resonance. This increases the electron density of the double bond, making it less electrophilic and therefore less reactive in a Diels–Alder reaction.

16.49 Use the directions from Answer 16.19.

16.50 Use the directions from Answer 16.25.

b. 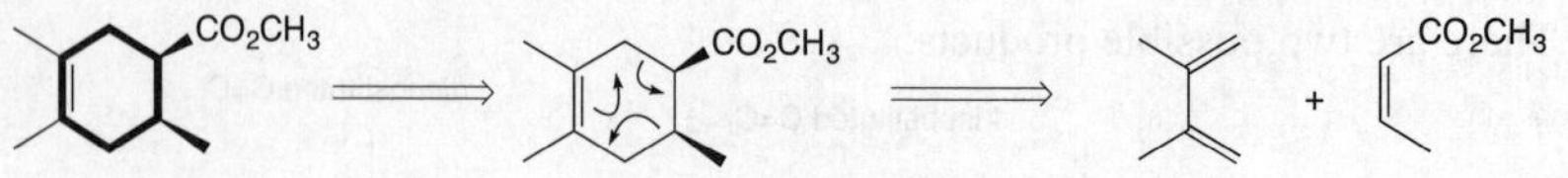
CO2CH3
CO2CH3
CO2CH3
+

c. 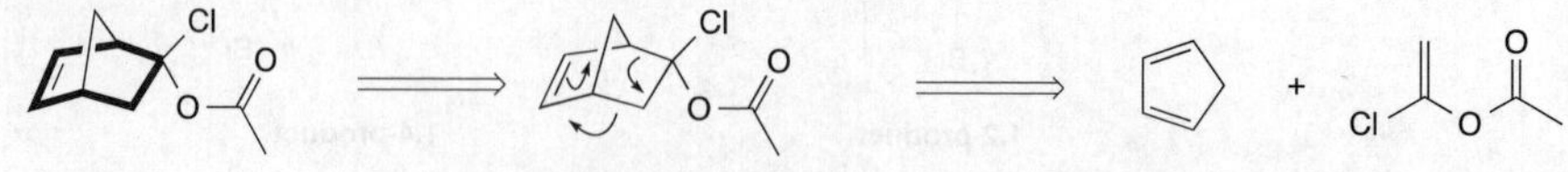
Cl
Cl
O
O
O
+
O
Cl
O

d. 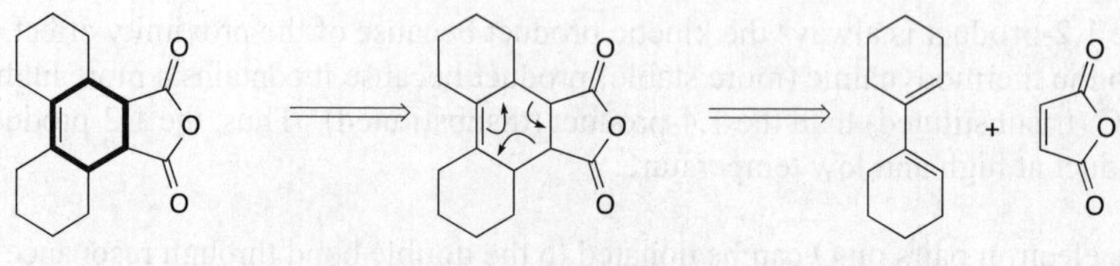
O
O
O
O
O
O
+
O
O
O

16.51

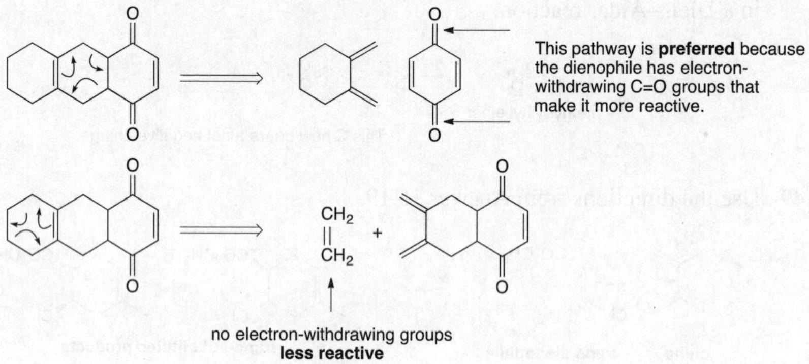
This pathway is **preferred** because the dienophile has electron-withdrawing C=O groups that make it more reactive.
CH2
CH2
+
no electron-withdrawing groups
less reactive

16.52

a. 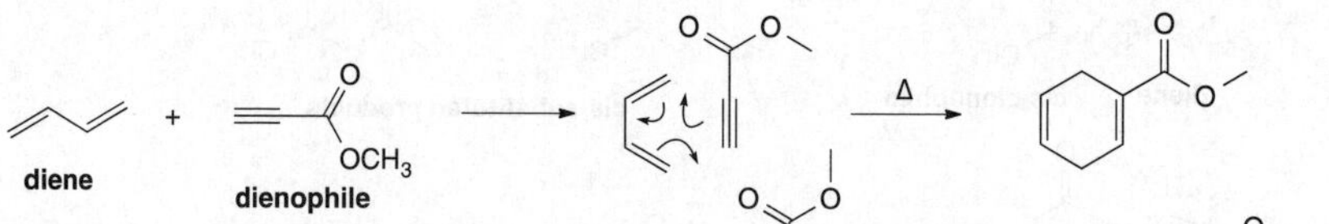
+
OCH3
diene
dienophile
Δ

b. 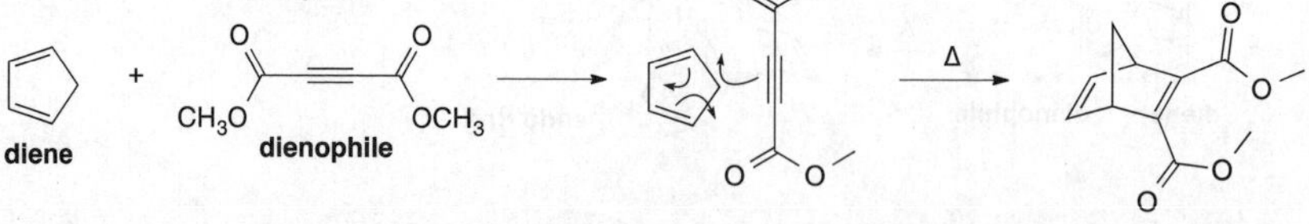
+
CH3O
OCH3
diene
dienophile
Δ

Conjugation, Resonance, and Dienes 16–21

16.53

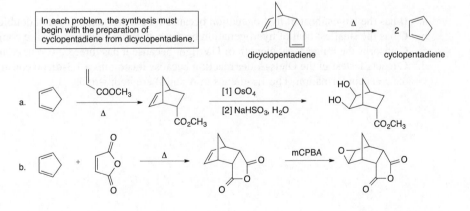

16.54

16.55

Chapter 16–22

c.

16.56

a. diene dienophile

b. diene dienophile

16.57 A transannular Diels–Alder reaction forms a tricyclic product from a monocyclic starting material.

16.58

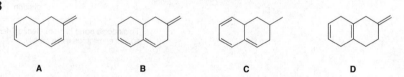

A   B   C   D

a. **D** has the largest heat of hydrogenation because it contains no conjugated double bonds.
b. **C** has the smallest heat of hydrogenation because all three double bonds are conjugated.
c. **C** absorbs the longest wavelength of UV light because it has three C=C's in conjugation.
d. **C** reacts fastest in the Diels-Alder reaction because it contains a 1,3-diene constrained in the *s*-cis conformation. The 1,3-dienes in **A** and **B** are both *s*-trans.

16.59

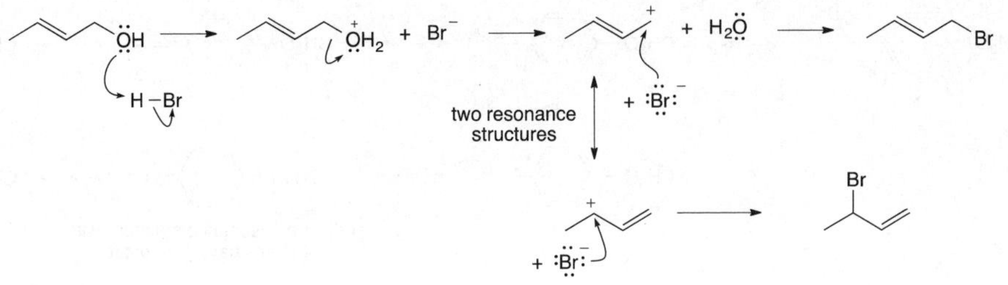
OH
H–Br
OH2
+ Br⁻
+ H2O
Br
two resonance structures
+ :Br:
Br
+ :Br:

16.60

a.

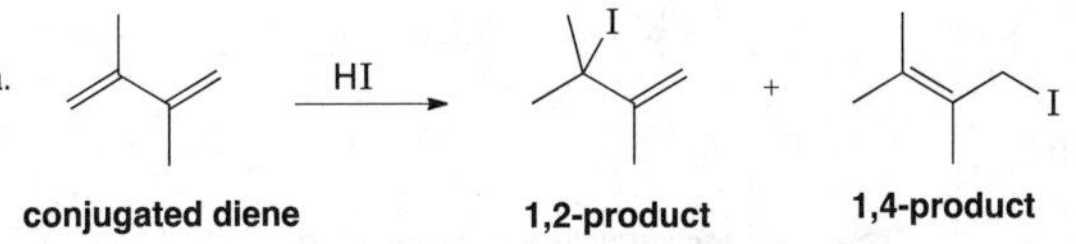
HI
I
+
I
conjugated diene
1,2-product
1,4-product

b.

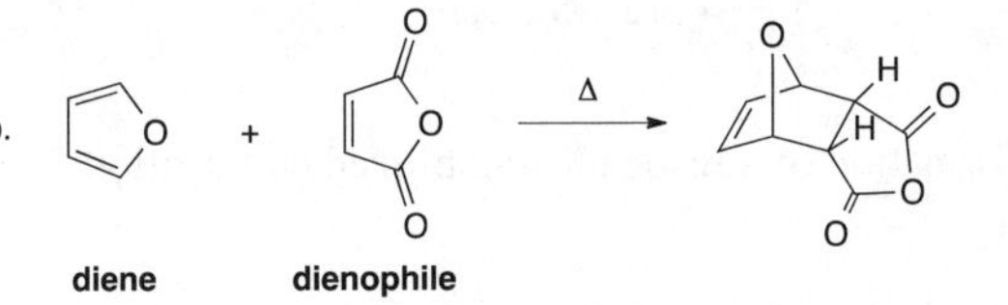
O
+
O
O
O
Δ
O
H
H
O
O
O
diene
dienophile

c.

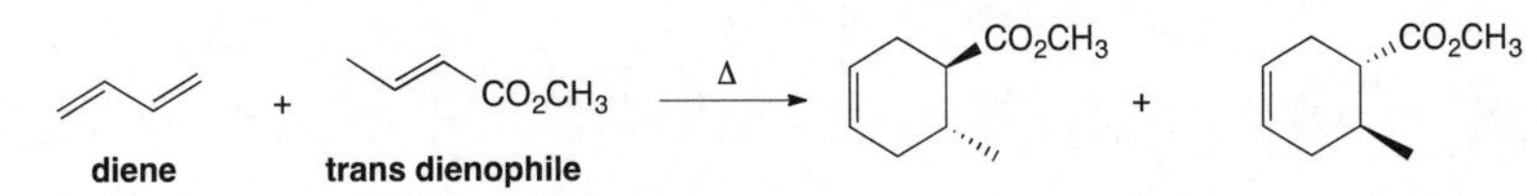
+
CO2CH3
Δ
CO2CH3
+
CO2CH3
diene
trans dienophile

d.

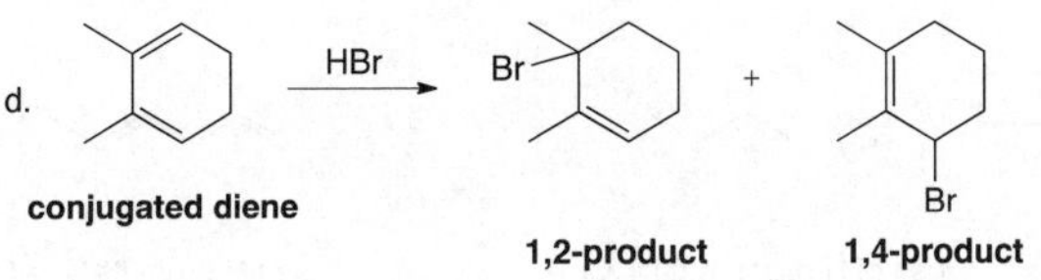
HBr
Br
+
Br
conjugated diene
1,2-product
1,4-product

16.61

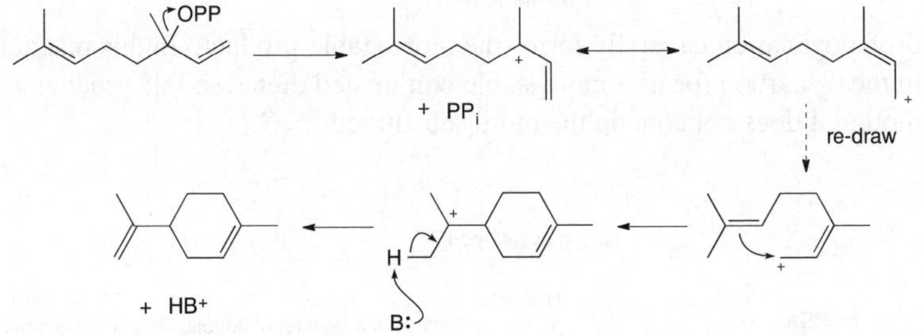
OPP
+ PPi
re-draw
+ HB+
H
B:

16.62 The more stable the carbocation, the faster the S_N1 reaction. The carbocation from **A** is more stable because the lone pairs on the O atom of the OCH_3 afford additional resonance stabilization.

Chapter 16–24

more stable carbocation

[chemical structures showing resonance structures with CH₃O groups]

A

+ Br⁻

additional resonance structure with
all atoms having an octet

less stable carbocation

B

+ Br⁻

This resonance structure is destabilized
because the (+) charge is adjacent to the
δ+ on the C=O atom.

16.63 The mechanism is E1, with formation of a resonance-stabilized carbocation.

[chemical mechanism structures]

H–A   + A:⁻   + H₂Ö   A:⁻   + H–A

16.64

a.

[chemical structures with Cl, β_1, β_2 labels]

loss of H (from β_1 C) + Cl
conjugated
more stable
major product

loss of H (from β_2 C) + Cl
more substituted

b. Dehydrohalogenation generally forms the more stable product. In this reaction, loss of H from the β_1 carbon forms a more stable conjugated diene, so this product is preferred even though it does not contain the more substituted C=C.

16.65

singlet at 1.42 ppm

[chemical structure of epoxide product]

isoprene   mCPBA

Each H is a doublet of doublets in the 5.2–5.4 ppm region.

two doublets at 2.7–2.9 ppm

doublet of doublets at 5.5–5.7 ppm

16.66

[Figure: Reaction of Br-substituted alkene with H₂O giving product B]
- 1 H: doublet of doublets at 6.0 ppm
- Each H is a doublet of doublets in the 4.9–5.2 ppm region.
- 6 H: singlet at 1.3 ppm
- OH peak at 1.5 ppm
- The IR shows an OH absorption at 3200–3600 cm⁻¹.

16.67

[Structures A and B shown]

All double bonds in **A** are conjugated, whereas the double bonds in the circled ring in **B** are not conjugated. As a result, **A** absorbs at a longer wavelength of UV light.

16.68 C and D absorb UV light because they contain conjugated π systems.

[Structures C (conjugated) and D (conjugated) shown]

16.69

The phenol makes ferulic acid an antioxidant. Loss of H forms a highly stabilized phenoxy radical that inhibits radical formation during oxidation.

[Structure of ferulic acid]

The highly conjugated π system makes it a sunscreen.

16.70 There are two possible modes of addition of HBr to allene. When H⁺ adds to the terminal carbon, a 2° vinyl carbocation is formed, which affords 2-bromoprop-1-ene after nucleophilic attack.

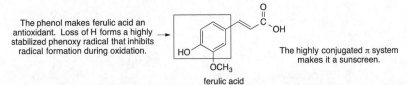

[1] CH₂=C=CH₂ + H–Br → 2° vinyl carbocation + :Br:⁻ → 2-bromoprop-1-ene

Chapter 16–26

When H[+] adds to the middle carbon, an intermediate carbocation with a (+) charge adjacent to the C=C is formed. This carbocation is not resonance stabilized (at least initially), because the two C=C's of allene are oriented 90° to each other, a geometry that does not allow for overlap of the C=C with the empty p orbital of the carbocation.

These C=C's are oriented 90° to each other.

[2] $CH_2=C=CH_2$ ⟶ 1° carbocation ⟶ Br

H—Br

+ :Br:[−]

As a result, path [1] forms a more stable carbocation and is preferred.

16.71

cyclohexanamine

$\ddot{N}H_2$

N is surrounded by 3 atoms and 1 lone pair, so it is sp^3 hybridized.

aniline

$\ddot{N}H_2$ ⟷ $=NH_2^+$ + other resonance structures

N has a lone pair on an atom adjacent to a C=C. N must be sp^2 hybridized to delocalize the lone pair by resonance.

Basicity is a measure of how willing an atom is to donate an electron pair. The lone pair on the N in cyclohexanamine is localized on N, whereas the lone pair on the N in aniline is delocalized onto the benzene ring. As a result, the lone pair in cyclohexanamine is much more available for donation to a proton than the lone pair in aniline. This makes cyclohexanamine much more basic than aniline.

16.72

O_3 cleaves the C=C.

Endo product formed.

[1] O_3
[2] $(CH_3)_2S$

The Diels–Alder reaction establishes the stereochemistry of the four carbons on the six-membered ring. All four carbon atoms bonded to the six-membered ring are on the same side.

16.73

CO_2CH_3

Δ

Diels–Alder reaction

CH_3O_2C

A

Δ

loss of CO_2

CO_2CH_3

B

+ $O=C=O$

Conjugation, Resonance, and Dienes 16–27

16.74

16.75 A retro Diels–Alder reaction forms a conjugated diene. An intramolecular Diels–Alder reaction then forms **N**.

Benzene and Aromatic Compounds 17–1

Chapter 17 Benzene and Aromatic Compounds

Chapter Review

Comparing aromatic, antiaromatic, and nonaromatic compounds (17.7)

- **Aromatic compound**
  - A cyclic, planar, completely conjugated compound that contains $4n + 2$ π electrons ($n = 0, 1, 2, 3$, and so forth).
  - An aromatic compound is more stable than a similar acyclic compound having the same number of π electrons.

- **Antiaromatic compound**
  - A cyclic, planar, completely conjugated compound that contains $4n$ π electrons ($n = 0, 1, 2, 3$, and so forth).
  - An antiaromatic compound is less stable than a similar acyclic compound having the same number of π electrons.

- **Nonaromatic compound**
  - A compound that lacks one (or more) of the requirements to be aromatic or antiaromatic.

Properties of aromatic compounds
- Every carbon in an aromatic ring has a *p* orbital to delocalize electron density (17.2).
- Aromatic compounds are unusually stable. $\Delta H°$ for hydrogenation is much less than expected, given the number of degrees of unsaturation (17.6).
- Aromatic compounds do not undergo the usual addition reactions of alkenes (17.6).
- ^{1}H NMR spectra show highly deshielded protons because of ring currents (17.4).

Examples of aromatic compounds with 6 π electrons (17.8)

benzene   pyridine   pyrrole   cyclopentadienyl anion   tropylium cation

Examples of compounds that are not aromatic (17.8)

not cyclic   not planar   not completely conjugated

Chapter 17–2

Practice Test on Chapter Review

1. Give the IUPAC name for each of the following compounds.

a.

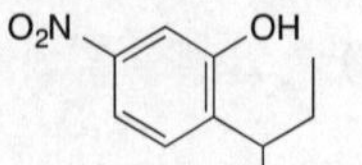

b.

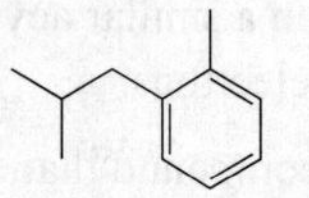

c.

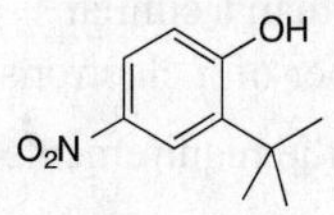

2. Label each compound as aromatic, nonaromatic, or antiaromatic. Choose only **one** possibility. Assume all completely conjugated rings are planar.

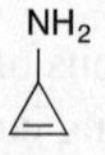

a.

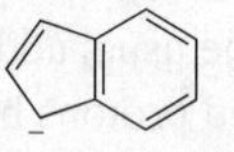

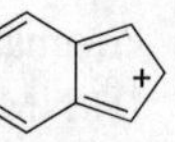

b.

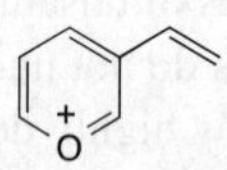

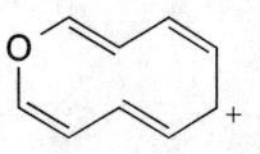

c.

d.

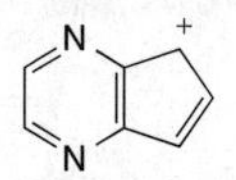

e.

f.

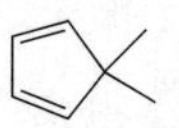

g.

h.

i.

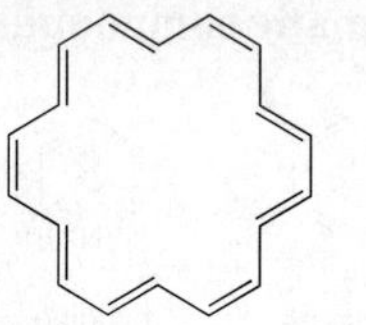

j.

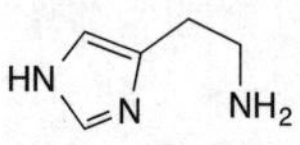

3. Answer the following questions about compounds **A–E** drawn below.

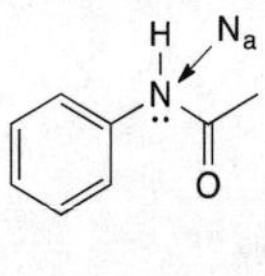

A

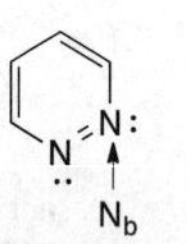

B

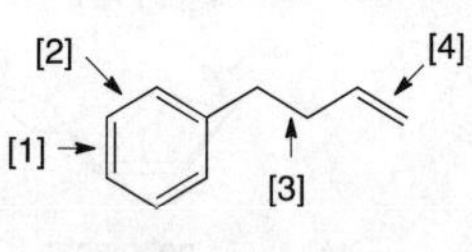

C

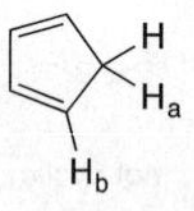

D

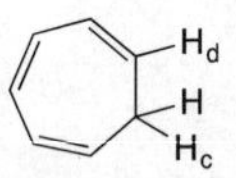

E

Benzene and Aromatic Compounds 17–3

a. How is nitrogen N_a in compound **A** hybridized?
b. In what type of orbital does the lone pair on N_a reside?
c. How is nitrogen N_b in compound **B** hybridized?
d. In what type of orbital does the lone pair on N_b reside?
e. Which of the labeled bonds in compound **C** is the shortest?
f. Which of the labeled bonds in compound **C** is the longest?
g. When considering both compounds **D** and **E**, which of the labeled hydrogen atoms (H_a, H_b, H_c, or H_d) is the most acidic? Give only **one** answer.
h. When considering both compounds **D** and **E**, which of the labeled hydrogen atoms (H_a, H_b, H_c, or H_d) is the least acidic? Give only **one** answer.

Answers to Practice Test

1.a. 2-*sec*-butyl-5-nitrophenol
  b. *o*-isobutyltoluene
  c. 2-*tert*-butyl-4-nitrophenol

2.a. nonaromatic
  b. aromatic
  c. nonaromatic
  d. antiaromatic
  e. aromatic
  f. nonaromatic
  g. aromatic
  h. aromatic
  i. antiaromatic
  j. aromatic

3.a. sp^2
  b. p
  c. sp^2
  d. sp^2
  e. 4
  f. 3
  g. H_a
  h. H_c

Answers to Problems

17.1 Move the electrons in the π bonds to draw all major resonance structures.

diphenhydramine

17.2 Look at the hybridization of the atoms involved in each bond. Carbons in a benzene ring are surrounded by three groups and are sp^2 hybridized.

Chapter 17–4

17.3
- To name a benzene ring with **one substituent,** name the substituent and add the word *benzene*.
- To name a **disubstituted ring,** select the correct prefix (ortho = 1,2; meta = 1,3; para = 1,4) and alphabetize the substituents. Use a common name if it is a derivative of that monosubstituted benzene.
- To name a **polysubstituted ring,** number the ring to give the lowest possible numbers and then follow other rules of nomenclature.

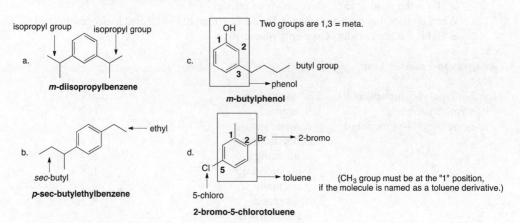

17.4 Work backwards to draw the structures from the names.

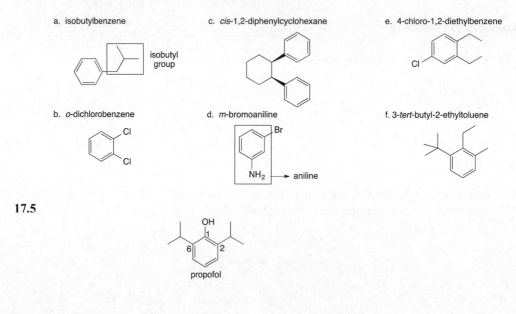

17.5

Benzene and Aromatic Compounds 17–5

17.6

Molecular formula $C_{10}H_{14}O_2$: 4 degrees of unsaturation
IR absorption at 3150–2850 cm^{-1}: sp^2 and sp^3 hybridized C–H bonds
NMR absorptions (ppm):
   1.4 (triplet, 6 H)
   4.0 (quartet, 4 H)
   6.8 (singlet, 4 H)

17.7   Count the different types of carbons to determine the number of ^{13}C NMR signals.

a.

4 types of C's in the benzene ring
6 signals

b.

All C's are different.
7 signals

c.

4 signals

17.8   The less stable compound has a larger heat of hydrogenation.

A

benzene ring, more stable
smaller $\Delta H°$

B

no benzene ring, less stable
larger $\Delta H°$

17.9   The protons on sp^2 hybridized carbons in aromatic hydrocarbons are highly deshielded and absorb at 6.5–8 ppm, whereas hydrocarbons that are not aromatic show an absorption at 4.5–6 ppm, typical of protons bonded to the C=C of an alkene.

a.

aromatic ring
H's ~6.5–8 ppm

b.

not aromatic
alkene H's ~4.5–6 ppm

c.

protons on isolated C=C
~4.5–6 ppm

protons on benzene ring
~6.5–8 ppm

17.10   To be aromatic, a ring must have $4n + 2$ π electrons.

16 π e$^-$
$4n$
$4(4) = 16$
antiaromatic

20 π e$^-$
$4n$
$4(5) = 20$
antiaromatic

22 π e$^-$
$4n + 2$
$4(5) + 2 = 22$
aromatic

17.11

Chapter 17–6

17.12 In determining if a heterocycle is aromatic, count a nonbonded electron pair if it makes the ring aromatic in calculating $4n + 2$. Lone pairs on atoms already part of a multiple bond cannot be delocalized in a ring, and so they are never counted in determining aromaticity.

a. Count one lone pair from O.
$4n + 2 = 4(1) + 2 = 6$
aromatic

b. no lone pair from O
$4n + 2 = 4(1) + 2 = 6$
aromatic

c. With one lone pair from each O there would be 8 π electrons. If O's are sp^3 hybridized, the ring is not completely conjugated.
not aromatic

d. Both N atoms are part of a double bond, so the lone pairs cannot be counted: there are 6 π electrons.
$4n + 2 = 4(1) + 2 = 6$
aromatic

17.13

quinine (antimalarial drug)

N is sp^3 hybridized and the lone pair is in an sp^3 hybrid orbital.

N is sp^2 hybridized and the lone pair is *not* part of the aromatic ring. This means it occupies an sp^2 hybrid orbital.

17.14

a. The five-membered ring is aromatic because it has 6 π electrons, two from each π bond and two from the N atom that is not part of a double bond.

sitagliptin

b, c.

sp^2 hybridized N lone pair in *p* orbital

sp^3 hybridized N lone pair in sp^3 orbital

sp^2 hybridized N lone pair in *p* orbital

sp^2 hybridized N lone pair in sp^2 orbital

sp^2 hybridized N lone pair in sp^2 orbital

17.15 Compare the conjugate base of cyclohepta-1,3,5-triene with the conjugate base of cyclopentadiene. Remember that the compound with the more stable conjugate base will have a lower pK_a.

cyclohepta-1,3,5-triene
pK_a = 39

8 π Electrons make this conjugate base especially unstable (**antiaromatic**).

Because the conjugate base is unstable, the pK_a of cyclohepta-1,3,5-triene is **high**.

cyclopentadiene

6 π electrons aromatic conjugate base **very stable anion**

Because the conjugate base is very stable, the pK_a of cyclopentadiene is much **lower**.

17.16 The compound with the most stable conjugate base is the most acidic.

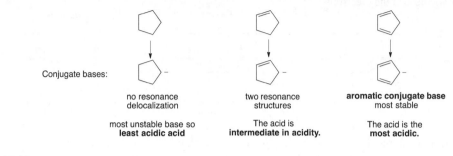

17.17

[resonance structures shown]

17.18 To be aromatic, the ions must have $4n + 2$ π electrons. Ions in (b) and (c) do not have the right number of π electrons to be aromatic.

a. [triangle cation] +
  2 π electrons
  $4(0) + 2 = 2$
  aromatic

d. [cyclic structure]
  10 π electrons
  $4(2) + 2 = 10$
  aromatic

17.19

A = [macrocyclic structure with triple bonds and H]

The NMR indicates that **A** is aromatic. (a) The C's of the triple bond are *sp* hybridized. (b) Each triple bond has one set of electrons in *p* orbitals that overlap with other *p* orbitals on adjacent atoms in the ring. This overlap allows electrons to delocalize. Each C of the triple bonds also has a *p* orbital in the plane of the ring. The electrons in these *p* orbitals are localized between the C's of the triple bond, not delocalized in the ring. (c) Although **A** has 24 π e⁻ total, only 18 e⁻ are delocalized around the ring.

17.20 In using the inscribed polygon method, always draw the vertex pointing down.

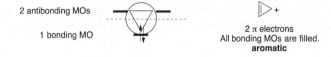

2 antibonding MOs
1 bonding MO

[triangle cation] +
2 π electrons
All bonding MOs are filled.
aromatic

Chapter 17–8

17.21 Draw the inscribed pentagons with the vertex pointing down. Then draw the molecular orbitals (MOs) and add the electrons.

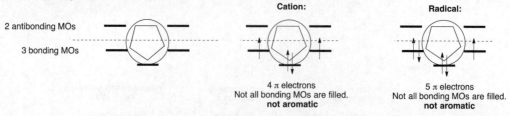

17.22 C_{60} would exhibit only one ^{13}C NMR signal because all the carbons are identical.

17.23

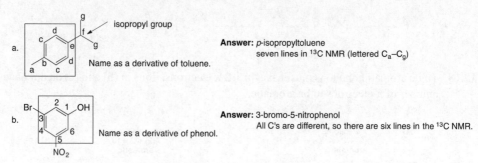

a. **Answer:** *p*-isopropyltoluene
seven lines in ^{13}C NMR (lettered C_a–C_g)

b. **Answer:** 3-bromo-5-nitrophenol
All C's are different, so there are six lines in the ^{13}C NMR.

17.24

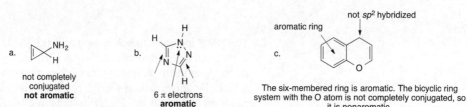

a. not completely conjugated
not aromatic

b. 6 π electrons
aromatic

c. The six-membered ring is aromatic. The bicyclic ring system with the O atom is not completely conjugated, so it is nonaromatic.

17.25

a. If the Kekulé description of benzene was accurate, only one product would form in Reaction [1], but there would be four (not three) dibromobenzenes (**A**–**D**), because adjacent C–C bonds are different—one is single and one is double. Thus, compounds **A** and **B** would *not* be identical. **A** has two Br's bonded to the same double bond, but **B** has two Br's on different double bonds.

b. In the resonance description, only one product would form in Reaction [1], because all C's are identical, but only three dibromobenzenes (ortho, meta, and para isomers) are possible. **A** and **B** are identical because each C–C bond is identical and intermediate in bond length between a C–C single and C–C double bond.

17.26

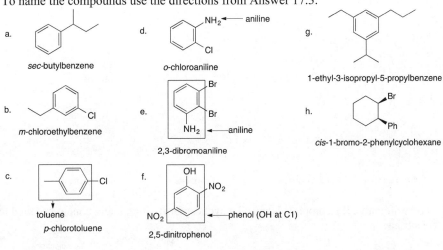

17.27 To name the compounds use the directions from Answer 17.3.

a. sec-butylbenzene

b. m-chloroethylbenzene

c. toluene → p-chlorotoluene

d. o-chloroaniline (NH₂ ← aniline)

e. 2,3-dibromoaniline (NH₂ ← aniline)

f. 2,5-dinitrophenol (phenol (OH at C1))

g. 1-ethyl-3-isopropyl-5-propylbenzene

h. cis-1-bromo-2-phenylcyclohexane

17.28

a. p-dichlorobenzene

b. p-iodoaniline

c. o-bromonitrobenzene

d. 2,6-dimethoxytoluene

e. 2-phenylprop-2-en-1-ol

f. trans-1-benzyl-3-phenylcyclopentane

Chapter 17–10

17.29

a, b. constitutional isomers of molecular formula **C₈H₉Cl** and names of the trisubstituted benzenes

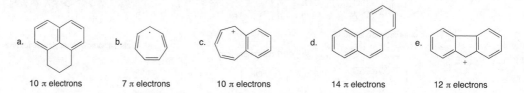

stereoisomers for this isomer

2-chloro-1,3-dimethyl-benzene

1-chloro-2,3-dimethyl-benzene

4-chloro-1,2-dimethyl-benzene

1-chloro-2,4-dimethyl-benzene

1-chloro-3,5-dimethyl-benzene

2-chloro-1,4-dimethyl-benzene

c. stereoisomers

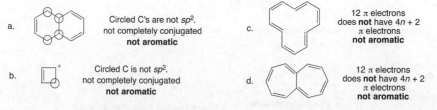

17.30 Count the electrons in the π bonds. Each π bond holds two electrons.

a. 10 π electrons
b. 7 π electrons
c. 10 π electrons
d. 14 π electrons
e. 12 π electrons

17.31 To be aromatic, the compounds must be cyclic, planar, completely conjugated, and have $4n + 2$ π electrons.

a. Circled C's are not sp^2.
not completely conjugated
not aromatic

b. Circled C is not sp^2.
not completely conjugated
not aromatic

c. 12 π electrons
does **not** have $4n + 2$ π electrons
not aromatic

d. 12 π electrons
does **not** have $4n + 2$ π electrons
not aromatic

17.32 In determining if a heterocycle is aromatic, count a nonbonded electron pair if it makes the ring aromatic in calculating $4n + 2$. Lone pairs on atoms already part of a multiple bond cannot be delocalized in a ring, so they are never counted in determining aromaticity.

Benzene and Aromatic Compounds 17–11

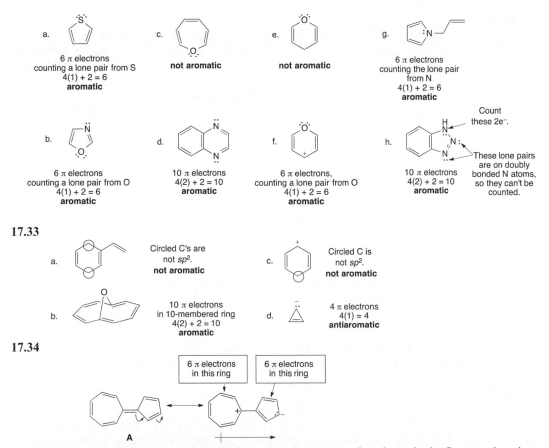

17.34

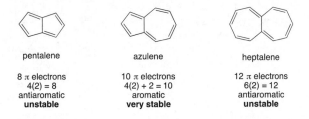

A resonance structure can be drawn for **A** that places a negative charge in the five-membered ring and a positive charge in the seven-membered ring. This resonance structure shows that each ring has six π electrons, making it aromatic. The molecule possesses a dipole such that the seven-membered ring is electron deficient and the five-membered ring is electron rich.

17.35 Each compound is completely conjugated. A compound with $4n + 2$ π electrons is especially stable, whereas a compound with $4n$ π electrons is especially unstable.

Chapter 17–12

17.36

a. Each N atom is *sp*² hybridized.
b. The three unlabeled N atoms are *sp*² hybridized with lone pairs in one of the *sp*² hybrid orbitals. The labeled N has its lone pair in a *p* orbital.
c. 10 π electrons
d. Purine is cyclic, planar, completely conjugated, and has 10 π electrons [4(2) + 2], so it is aromatic.

17.37

a. Every N atom is either part of a double bond or is bonded directly to a double bond, so each N atom is *sp*² hybridized.
b. Electron pairs on boxed-in N atoms are located in *p* orbitals to overlap with adjacent π bonds. Lone pairs on all other N atoms are in *sp*² hybrid orbitals.
c. The bicyclic ring system contains 10 π electrons from the five double bonds, making it aromatic. All lone pairs on N atoms in the rings are not delocalized.

17.38

a. 16 total π electrons
b. 14 π electrons delocalized in the ring. [Note: Two of the electrons in the triple bond are localized between two C's, perpendicular to the π electrons delocalized in the ring.]
c. By having two of the *p* orbitals of the C–C triple bond co-planar with the *p* orbitals of all the C=C's, the total number of π electrons delocalized in the ring is 14. 4(3) + 2 = 14, so the ring is **aromatic.**

17.39 A second resonance structure can be drawn for the six-membered ring that gives it three π bonds, thus making it aromatic with six π electrons.

6 π electrons

17.40 The rate of an S_N1 reaction increases with increasing stability of the intermediate carbocation.

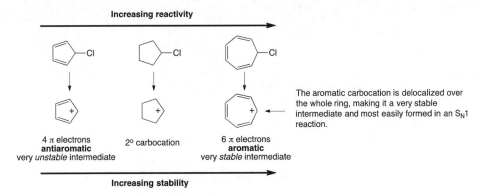

17.41

[figure: mechanism showing Na+H⁻ abstracting D from cyclopentadiene to form cyclopentadienyl anion (with D substituent), shown with resonance structures, reacting with H—OH to give cyclopentadiene-D + Na⁺ ⁻OH. "Two additional resonance structures are not drawn."]

17.42 α-Pyrone reacts like benzene because it is aromatic. A second resonance structure can be drawn showing how the ring has six π electrons. Thus, α-pyrone undergoes reactions characteristic of aromatic compounds—that is, substitution rather than addition.

[figure: α-pyrone resonance structures, labeled "α-pyrone" and "6 π electrons"]

17.43

a. [three resonance structures of cyclopropenyl radical]
cyclopropenyl radical

b. [resonance structures of pyrrole]
pyrrole

Chapter 17–14

c. [resonance structures of phenanthrene]

phenanthrene

17.44

Naphthalene can be drawn as three resonance structures:

[three resonance structures of naphthalene with bonds (a) and (b) labeled]

In two of the resonance structures bond (a) is a double bond, and bond (b) is a single bond. Therefore, bond (b) has more single bond character, making it longer.

17.45

a. [resonance structures of pyrrole]

pyrrole

Pyrrole is less resonance stabilized than benzene because four of the resonance structures have charges, making them less good.

b. [resonance structures of furan]

furan

Furan is less resonance stabilized than pyrrole because its O atom is less basic, so it donates electron density less "willingly." Thus, charge-separated resonance forms are more minor contributors to the hybrid than the charge-separated resonance forms of pyrrole.

17.46 The compound with the more stable conjugate base is the stronger acid. Draw and compare the conjugate bases of each pair of compounds.

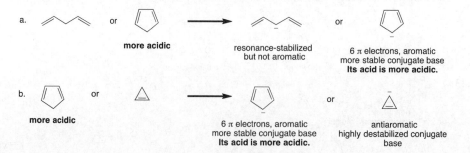

17.47

[Reaction: indene + NaNH₂ → indenyl anion Na⁺ + NH₃]

The conjugate base of indene has 10 π electrons, making it aromatic and very stable. Therefore, indene is more acidic than many hydrocarbons.

17.48

A: H_b is most acidic because its conjugate base is aromatic (6 π electrons).

B: H_c is least acidic because its conjugate base has 8 π electrons, making it antiaromatic.

17.49

pyrrole → conjugate base: Both pyrrole and the conjugate base of pyrrole have 6 π electrons in the ring, making them both aromatic. Thus, deprotonation of pyrrole does not result in a gain of aromaticity because the starting material is aromatic to begin with.

cyclopentadiene (more acidic) → conjugate base: Cyclopentadiene is not aromatic, but the conjugate base has 6 π electrons and is therefore aromatic. This makes the C–H bond in cyclopentadiene more acidic than the N–H bond in pyrrole, because deprotonation of cyclopentadiene forms an aromatic conjugate base.

17.50

a.

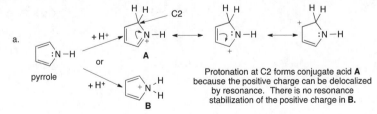

Protonation at C2 forms conjugate acid **A** because the positive charge can be delocalized by resonance. There is no resonance stabilization of the positive charge in **B**.

Chapter 17–16

b. Loss of a proton from **A** (which is not aromatic) gives two electrons to N, and forms pyrrole, which has six π electrons that can then delocalize in the five-membered ring, making it aromatic. This makes deprotonation a highly favorable process, and **A** more acidic.

A
pK_a = 0.4

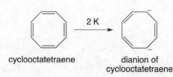 Both **C** and its conjugate base pyridine are aromatic. Since **C** has six π electrons, it is already aromatic to begin with, so there is less to be gained by deprotonation, and **C** is thus less acidic than **A**.

pK_a = 5.2
C

17.51

a.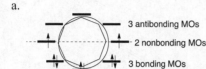

cyclooctatetraene and its 8 π electrons

b. Even if cyclooctatetraene were flat, it has two unpaired electrons in its HOMOs (nonbonding MOs), so it cannot be aromatic.
c. The dianion has 10 π electrons.
d. The two additional electrons fill the nonbonding MOs; that is, all the bonding and nonbonding MOs are filled with electrons in the dianion.
e. The dianion is aromatic because its HOMOs are completely filled, and it has no electrons in antibonding MOs.

17.52

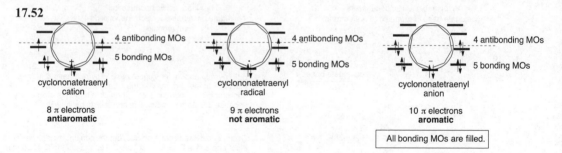

17.53 The number of different types of C's = the number of signals.

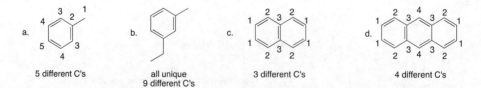

17.54 Draw the three isomers and count the different types of carbons in each. Then match the structures with the data.

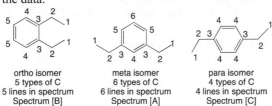

ortho isomer
5 types of C
5 lines in spectrum
Spectrum [B]

meta isomer
6 types of C
6 lines in spectrum
Spectrum [A]

para isomer
4 types of C
4 lines in spectrum
Spectrum [C]

17.55

a. $C_{10}H_{14}$: IR absorptions at 3150–2850 (sp^2 and sp^3 hybridized C–H), 1600, and 1500 (due to a benzene ring) cm^{-1}

^{1}H NMR data:

Absorption	ppm	# of H's	Explanation
doublet	1.2	6	6 H's adjacent to 1 H
singlet	2.3	3	CH$_3$
septet	3.1	1	1 H adjacent to 6 H's
multiplet	7–7.4	4	a disubstituted benzene ring

(CH$_3$)$_2$CH group

You can't tell from these data where the two groups are on the benzene ring. They are not para, because the para arrangement usually gives two sets of distinct peaks (resembling two doublets), so there are two possible structures—ortho and meta isomers.

b. C_9H_{12}: ^{13}C NMR signals at 21, 127, and 138 ppm → means three different types of C's.
^{1}H NMR shows two types of H's: 9 H's probably means 3 CH$_3$ groups; the other 3 H's are very deshielded so they are bonded to a benzene ring.
Only one possible structure fits:

c. C_8H_{10}: IR absorptions at 3108–2875 (sp^2 and sp^3 hybridized C–H), 1606, and 1496 (due to a benzene ring) cm^{-1}

^{1}H NMR data: Structure:

Absorption	ppm	# of H's	Explanation
triplet	1.3	3	3 H's adjacent to 2 H's
quartet	2.7	2	2 H's adjacent to 3 H's
multiplet	7.3	5	a monosubstituted benzene ring

Chapter 17–18

17.56

a. Compound **A**: Molecular formula $C_8H_{10}O$
   IR absorption at 3150–2850 (sp^2 and sp^3 hybridized C–H) cm^{-1}
   ^{1}H NMR data:

Absorption	ppm	# of H's	Explanation	Structure:
triplet	1.4	3	3 H's adjacent to 2 H's	
quartet	3.95	2	2 H's adjacent to 3 H's	
multiplet	6.8–7.3	5	a monosubstituted benzene ring	

b. Compound **B**: Molecular formula $C_9H_{10}O_2$
   IR absorption at 1669 (C=O) cm^{-1}
   ^{1}H NMR data:

Absorption	ppm	# of H's	Explanation	Structure:
singlet	2.5	3	CH_3 group	
singlet	3.8	3	CH_3 group	
doublet	6.9	2	2 H's on a benzene ring	
doublet	7.9	2	2 H's on a benzene ring	

It would be hard to distinguish these two compounds with the given data.

17.57

3 equivalent H's
singlet at ~2.3 ppm

*3 other H's on benzene ring (arrows with *) at ~6.9 ppm

thymol

IR absorptions:
3500–3200 cm^{-1} (O–H)
3150–2850 cm^{-1} (C–H bonds)
1621 and 1585 cm^{-1} (benzene ring)

1 H
septet at ~3.2 ppm

6 equivalent H's
doublet at ~1.2 ppm

basic structure of thymol

Thymol must have this basic structure, given the NMR and IR data because it is a trisubstituted benzene ring with one singlet and two doublets in the NMR at ~6.9 ppm. However, which group [OH, CH_3, or $CH(CH_3)_2$] corresponds to X, Y, and Z is not readily distinguished with the given data. The correct structure for thymol is given.

Benzene and Aromatic Compounds 17–19

17.58

^{13}C NMR has four lines that are located in the aromatic region (~110–155 ppm), corresponding to the four different types of carbons in the aromatic ring of the para isomer. The ortho and meta isomers have six different C's, and so six lines would be expected for each of them.

17.59 Because tetrahydrofuran has a higher boiling point and is more water soluble, it must be more polar and have stronger intermolecular forces than furan. There are two contributing factors. One lone pair on furan's O atom is delocalized on the five-membered ring to make it aromatic. This makes it less available for H-bonding with water and other intermolecular interactions. Also, the C–O bonds in furan are less polar than the C–O bonds in tetrahydrofuran because of hybridization. The sp^2 hybridized C's of furan pull a little more electron density towards them than do the sp^3 hybridized C's of tetrahydrofuran. This counteracts the higher electronegativity of O compared to C to a small extent.

17.60

rizatriptan

a. Rizatriptan contains three aromatic rings.
b. The circled N atom is sp^3 hybridized. All other N atoms are either part of a π bond or bonded to a π bond, making them sp^2 hybridized.
c. The lone pair on the circled N atom occupies an sp^3 hybrid orbital. The lone pairs on the boxed-in N atoms are in a p orbital. The lone pairs on the unlabeled N atoms occupy an sp^2 hybrid orbital.

d.

one additional resonance structure

Chapter 17–20

e.

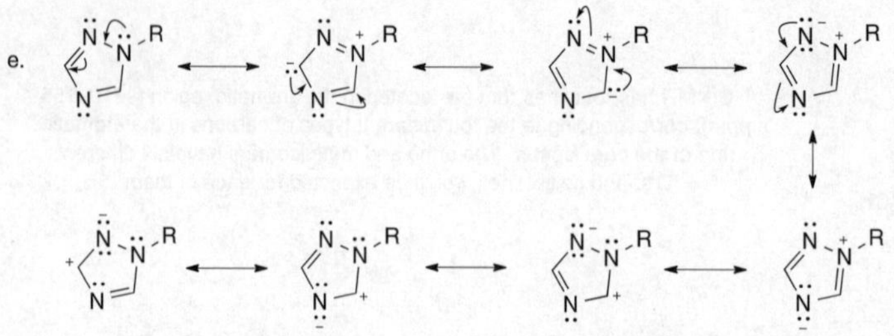

17.61
a, b.

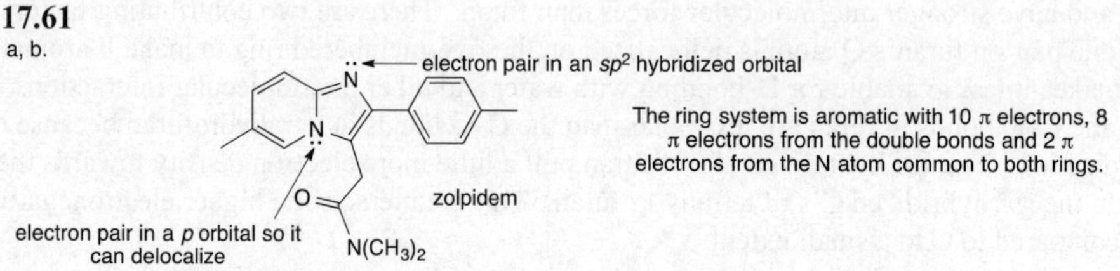

← electron pair in an sp^2 hybridized orbital

The ring system is aromatic with 10 π electrons, 8 π electrons from the double bonds and 2 π electrons from the N atom common to both rings.

electron pair in a p orbital so it can delocalize

zolpidem

$N(CH_3)_2$

c.

$N(CH_3)_2$

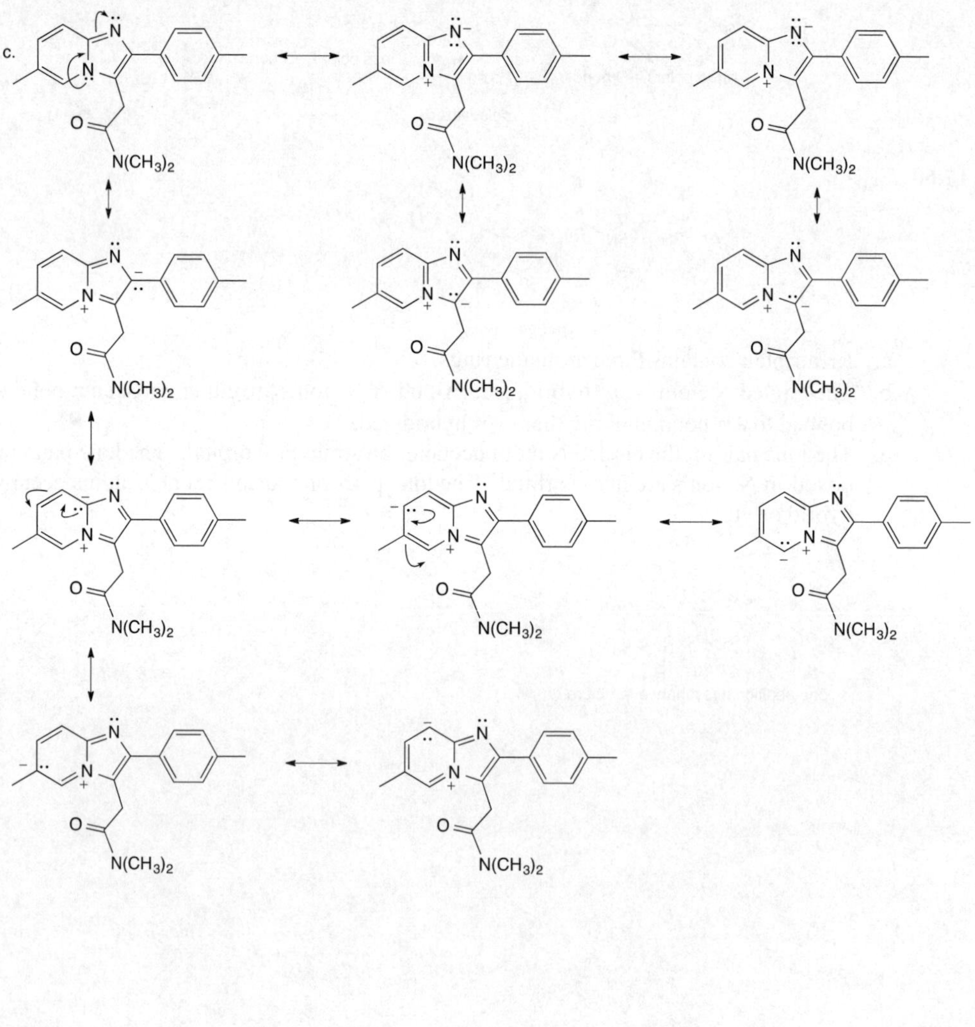

Benzene and Aromatic Compounds 17–21

17.62

a.

curcumin

The enol form is more stable because the enol double bond makes a highly conjugated system. The enol OH can also intramolecularly hydrogen bond to the nearby carbonyl O atom.

sp³
keto form

b.

The enol O–H proton is more acidic than an alcohol O–H proton because the conjugate base is resonance stabilized.

c. Curcumin is colored because it has many conjugated π electrons, which shift absorption of light from the UV to the visible region.

d. Curcumin is an antioxidant because it contains a phenol. Homolytic cleavage affords a resonance-stabilized phenoxy radical, which can inhibit oxidation from occurring, much like vitamin E and BHT in Chapter 15.

(+ other resonance structures)

phenoxy radical
Resonance delocalizes the radical on the ring and C chain of curcumin.

17.63

a. Pyrazole rings are aromatic because they have six π electrons—two from the lone pair on the N atom that is not part of the double bond and four from the double bonds.

b.

sp²
p

Chapter 17–22

c.

d. The N atom in the NH bond in the pyrazole ring is sp^2 hybridized with 33% s-character, increasing the acidity of the N–H bond. The N–H bond of CH_3NH_2 contains an sp^3 hybridized N atom (25% s-character).

17.64 Both **A** and **B** are cyclic, and if the lone pair of electrons on N is in a p orbital, they are completely conjugated with 10 π electrons, a number that satisfies Hückel's rule. To be aromatic, **A** and **B** must be planar, and the internal bond angles of **A** and **B** would be much larger than 120°, the theoretical bond angle of sp^2 hybridized C's. The fact that **A** is aromatic means that the lone pair on N occupies a p orbital, so it can delocalize onto the nine-membered ring. The stabilization gained by being aromatic is greater than any angle strain. With **B**, the lone pair on N is also delocalized onto the C=O, making it less available for donation to the ring, so the ring is not aromatic.

Benzene and Aromatic Compounds 17–23

17.65 With 14 π electrons in the double bonds, the system is aromatic [4(3) + 2 = 14 π electrons]. The ring current generated by the circulating π electrons deshields the protons on the C=C's, so they absorb downfield (8.14–8.67 ppm). The CH₃ groups, however, are very shielded because they lie above and below the plane, so they absorb far upfield (–4.25 ppm). The dianion of **C** now has 16 π electrons, making it antiaromatic, so the position of the absorptions reverses. The C=C protons are now shielded (–3 ppm), and the CH₃ protons are now deshielded (21 ppm).

17.66 A second resonance structure for **A** shows that the ring is completely conjugated and has six π electrons, making it aromatic and especially stable. A similar charge-separated resonance structure for **B** makes the ring completely conjugated, but gives the ring four π electrons, making it antiaromatic and especially unstable.

17.67 The conversion of carvone to carvacrol involves acid-catalyzed isomerization of two double bonds and tautomerization of a ketone to an enol tautomer. In this case the enol form is part of an aromatic phenol. Each isomerization of a C=C involves Markovnikov addition of a proton, followed by deprotonation.

Chapter 17–24

17.68 Resonance structures for triphenylene:

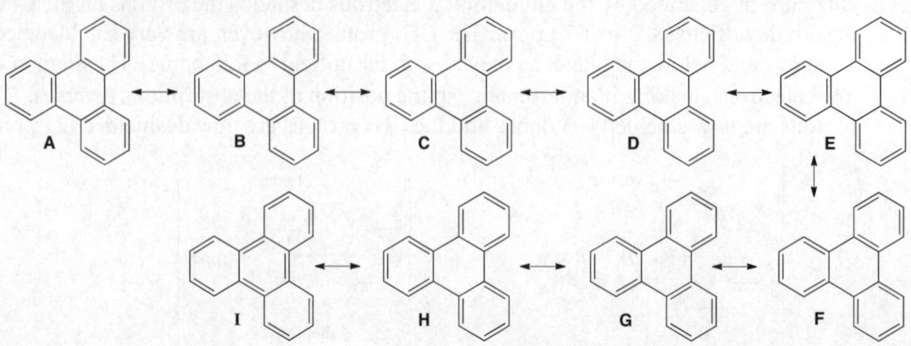

Resonance structures **A–H** all keep three double and three single bonds in the three six-membered rings on the periphery of the molecule. This means that each ring behaves like an isolated benzene ring undergoing substitution rather than addition because the π electron density is delocalized within each six-membered ring. Only resonance structure **I** does not have this form. Each C–C bond of triphenylene has four (or five) resonance structures in which it is a single bond and four (or five) resonance structures in which it is a double bond.

Resonance structures for phenanthrene:

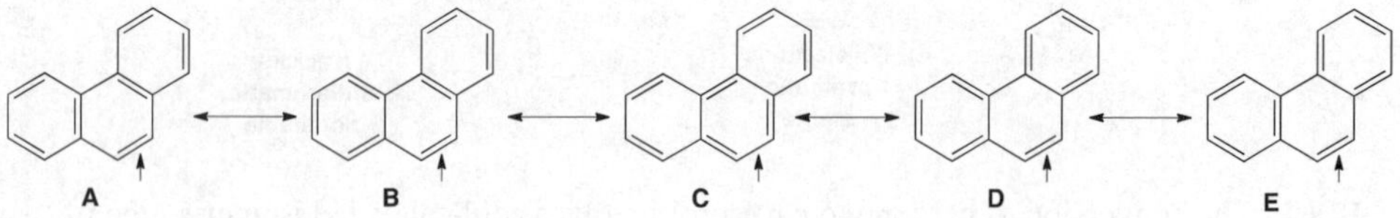

With phenanthrene, however, four of the five resonance structures keep a double bond at the labeled C's. (Only **C** does not.) This means that these two C's have more double bond character than other C–C bonds in phenanthrene, making them more susceptible to addition rather than substitution.

17.69

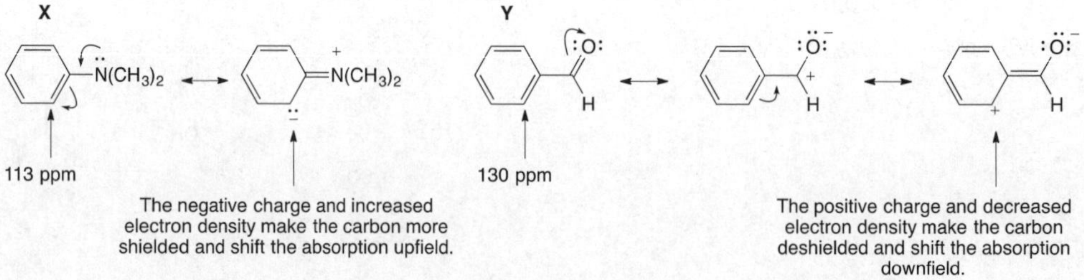

113 ppm

The negative charge and increased electron density make the carbon more shielded and shift the absorption upfield.

130 ppm

The positive charge and decreased electron density make the carbon deshielded and shift the absorption downfield.

Reactions of Aromatic Compounds 18–1

Chapter 18 Reactions of Aromatic Compounds

Chapter Review

Mechanism of electrophilic aromatic substitution (18.2)

- Electrophilic aromatic substitution follows a two-step mechanism. Reaction of the aromatic ring with an electrophile forms a carbocation, and loss of a proton regenerates the aromatic ring.
- The first step is rate-determining.
- The intermediate carbocation is stabilized by resonance; a minimum of three resonance structures can be drawn. The positive charge is always located ortho or para to the new C–E bond.

| (+) ortho to E | (+) para to E | (+) ortho to E |

Three rules describing the reactivity and directing effects of common substituents (18.7–18.9)

[1] All ortho, para directors except the halogens activate the benzene ring.
[2] All meta directors deactivate the benzene ring.
[3] The halogens deactivate the benzene ring.

Summary of substituent effects in electrophilic aromatic substitution (18.6–18.9)

	Substituent	Inductive effect	Resonance effect	Reactivity	Directing effect
[1]	R R = alkyl	donating	none	activating	ortho, para
[2]	Z : Z = N or O	withdrawing	donating	activating	ortho, para
[3]	X : X = halogen	withdrawing	donating	deactivating	ortho, para
[4]	Y (δ+ or +)	withdrawing	withdrawing	deactivating	meta

Chapter 18–2

Five examples of electrophilic aromatic substitution

[1] Halogenation—Replacement of H by Cl or Br (18.3)

$$\underset{\text{[X = Cl, Br]}}{\xrightarrow[\text{FeX}_3]{X_2}}$$

aryl chloride   or   aryl bromide

- Polyhalogenation occurs on benzene rings substituted by OH and NH_2 (and related substituents) (18.10A).

[2] Nitration—Replacement of H by NO_2 (18.4)

$$\xrightarrow[\text{H}_2\text{SO}_4]{\text{HNO}_3}$$

nitro compound

[3] Sulfonation—Replacement of H by SO_3H (18.4)

$$\xrightarrow[\text{H}_2\text{SO}_4]{\text{SO}_3}$$

benzenesulfonic acid

[4] Friedel–Crafts alkylation—Replacement of H by R (18.5)

$$\xrightarrow[\text{AlCl}_3]{\text{RCl}}$$

alkyl benzene (arene)

- Rearrangements can occur.
- Vinyl halides and aryl halides are unreactive.
- The reaction does not occur on benzene rings substituted by meta deactivating groups or NH_2 groups (18.10B).
- Polyalkylation can occur.

Variations:

[1] with alcohols

$$\xrightarrow[\text{H}_2\text{SO}_4]{\text{ROH}}$$

[2] with alkenes

$$\xrightarrow[\text{H}_2\text{SO}_4]{}$$

Reactions of Aromatic Compounds 18–3

[5] Friedel–Crafts acylation—Replacement of H by RCO (18.5)

- The reaction does not occur on benzene rings substituted by meta deactivating groups or NH_2 groups (18.10B).

Nucleophilic aromatic substitution (18.13)
[1] Nucleophilic substitution by an addition–elimination mechanism

X = F, Cl, Br, I
A = electron-withdrawing group

- The mechanism has two steps.
- Strong electron-withdrawing groups at the ortho or para position are required.
- Increasing the number of electron-withdrawing groups increases the rate.
- Increasing the electronegativity of the halogen increases the rate.

[2] Nucleophilic substitution by an elimination–addition mechanism

X = halogen

- Reaction conditions are harsh.
- Benzyne is formed as an intermediate.
- Product mixtures may result.

Other reactions of benzene derivatives
[1] Benzylic halogenation (18.14)

Br_2
$h\nu$ or Δ
or
NBS
$h\nu$ or ROOR

benzylic bromide

[2] Oxidation of alkyl benzenes (18.15A)

$KMnO_4$

benzoic acid

- A benzylic C–H bond is needed for reaction.

Chapter 18–4

[3] Reduction of ketones to alkyl benzenes (18.15B)

$$\text{PhCOR} \xrightarrow[\text{NH}_2\text{NH}_2,\ ^-\text{OH}]{\text{Zn(Hg), HCl} \ \text{or}} \text{alkyl benzene (PhCH}_2\text{R)}$$

[4] Reduction of nitro groups to amino groups (18.15C)

$$\text{PhNO}_2 \xrightarrow[\begin{array}{c}\text{or}\\\text{Fe, HCl}\\\text{or}\\\text{Sn, HCl}\end{array}]{\text{H}_2,\ \text{Pd-C}} \text{aniline (PhNH}_2)$$

Practice Test on Chapter Review

1.a. Which of the following statements is true about an ethoxy substituent ($-OCH_2CH_3$) on a benzene ring?
   1. OCH_2CH_3 increases the rates of both electrophilic substitution and nucleophilic substitution.
   2. OCH_2CH_3 decreases the rates of both electrophilic substitution and nucleophilic substitution.
   3. OCH_2CH_3 increases the rate of electrophilic substitution and decreases the rate of nucleophilic substitution.
   4. OCH_2CH_3 decreases the rate of electrophilic substitution and increases the rate of nucleophilic substitution.
   5. None of these statements is true.

 b. Which of the following statements is true about a $-CO_2CH_3$ group on a benzene ring?
   1. CO_2CH_3 increases the rates of both electrophilic substitution and nucleophilic substitution.
   2. CO_2CH_3 decreases the rates of both electrophilic substitution and nucleophilic substitution.
   3. CO_2CH_3 increases the rate of electrophilic substitution and decreases the rate of nucleophilic substitution.
   4. CO_2CH_3 decreases the rate of electrophilic substitution and increases the rate of nucleophilic substitution.
   5. None of these statements is true.

Reactions of Aromatic Compounds 18–5

c. Which of the following is *not* a valid resonance structure for the carbocation that results from ortho attack of an electrophile on C₆H₅C(CH₃)=CH₂?

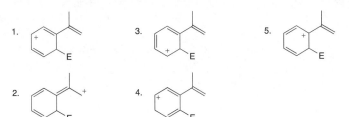

2. Draw the organic products formed in the following reactions.

a. [1] CH₃CH₂CH₂COCl, AlCl₃ [2] Zn(Hg), HCl

b. –CN, HNO₃/H₂SO₄, then Fe/HCl

c. KMnO₄

d. CH₃O–, HNO₃/H₂SO₄, then H₂/Pd-C

e. –OCH₃ (with Br substituent), SO₃/H₂SO₄

f. (benzamide with Br), Br₂ (1 equiv), FeBr₃

g. O₂N– –Cl, NaOCH₂CH₃

h. (isopropyl–Cl aromatic), NaOH, Δ

3. (a) Considering the compound drawn below, which ring is *most* reactive in electrophilic aromatic substitution? (b) Which ring is the *least* reactive in electrophilic aromatic substitution?

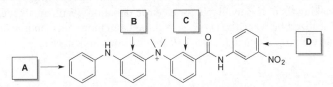

4. Classify each substituent as [1] ortho, para activating, [2] ortho, para deactivating, or [3] meta deactivating.
a. –Br
b. –CH₂CH₂CH₂Br
c. –COOH
d. –NHCOCH₂CH₃
e. –N(CH₂)₆COCH₃
f. –CCl₃

5. What reagents are needed to convert toluene (C₆H₅CH₃) to each compound?
a. C₆H₅COOH
b. C₆H₅CH₂Br
c. *p*-bromotoluene
d. *o*-nitrotoluene
e. *p*-ethyltoluene
f.

Chapter 18–6

Answers to Practice Test

1. a. 3
   b. 4
   c. 4

2. a. [p-butyl toluene + o-butyl toluene structures]
   b. m-aminobenzonitrile (H₂N, CN)
   c. phthalic acid (1,2-benzenedicarboxylic acid, CO₂H, CO₂H)
   d. CH₃O— with H₂N substituent
   e. 2-bromo-4'-methoxy-3'-sulfonic acid biphenyl (Br, OCH₃, SO₃H)
   f. N-(2-bromophenyl)benzamide + N-(4-bromophenyl)benzamide
   g. O₂N—C₆H₄—OCH₂CH₃ (para)
   h. 4-isopropylphenol + 3-isopropylphenol

3. a. **A**
   b. **C**

4. a. 2
   b. 1
   c. 3
   d. 1
   e. 1
   f. 3

5. a. KMnO₄
   b. Br₂, $h\nu$
   c. Br₂, FeBr₃
   d. HNO₃, H₂SO₄
   e. CH₃CH₂Cl, AlCl₃
   f. CH₃CH₂COCl, AlCl₃

Answers to Problems

18.1 The π electrons of benzene are delocalized over the six atoms of the ring, increasing benzene's stability and making them less available for electron donation. With an alkene, the two π electrons are localized between the two C's, making them more nucleophilic and thus more reactive with an electrophile than the delocalized electrons in benzene.

18.2

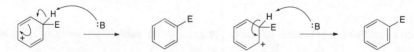

Reactions of Aromatic Compounds 18–7

18.3 Reaction with Cl_2 and $FeCl_3$ as the catalyst occurs in two parts. First is the formation of an electrophile, followed by a two-step substitution reaction.

[1]   :Cl—Cl: + FeCl$_3$ ⟶ :Cl—Cl—FeCl$_3$
       Lewis base   Lewis acid   electrophile

[2]   (benzene ring) + :Cl—Cl—FeCl$_3$ ⟶ (arenium ion intermediates) + FeCl$_4^-$

resonance-stabilized carbocation

[3]   (arenium ion) + :Cl—FeCl$_3$ ⟶ (chlorobenzene) + HCl + FeCl$_3$

18.4 There are two parts in the mechanism. The first part is formation of an electrophile. The second part is a two-step substitution reaction.

A = R—(benzene)—H

[1]   (SO$_3$) + H—OSO$_3$H ⟶ (H—O—SO$_3$) = $^+SO_3H$ + HSO$_4^-$
                                                       electrophile

[2]   (R—benzene) + $^+SO_3H$ ⟶ (resonance-stabilized carbocation intermediates)

resonance-stabilized carbocation

[3]   (arenium ion) + HSO$_4^-$ ⟶ (R—benzene—SO$_3$H) + H$_2$SO$_4$

B

18.5 Friedel–Crafts alkylation results in the transfer of an alkyl group from a halogen to a benzene ring. In Friedel–Crafts acylation an acyl group is transferred from a halogen to a benzene ring.

a. (benzene) + (isopropyl chloride) —AlCl$_3$→ (isopropylbenzene)

c. (benzene) + (propanoyl chloride) —AlCl$_3$→ (phenyl ethyl ketone)

b. (benzene) + (cyclohexyl chloride) —AlCl$_3$→ (cyclohexylbenzene)

Chapter 18–8

18.6 Remember that an acyl group is transferred from a Cl atom to a benzene ring. To draw the acid chloride, substitute a Cl for the benzene ring.

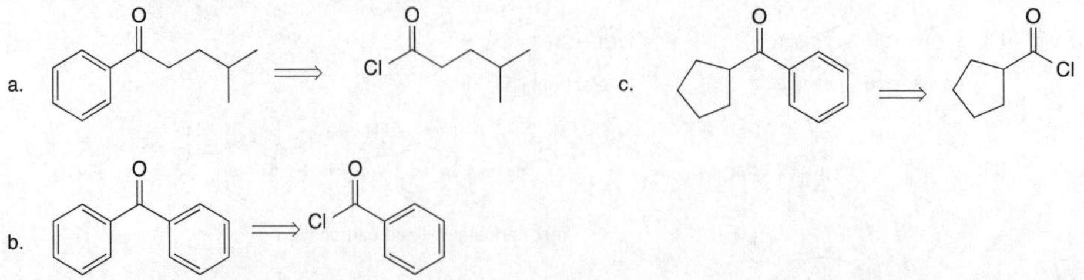

a.

c.

b.

18.7 To be reactive in a Friedel–Crafts alkylation reaction, the X must be bonded to an sp^3 hybridized carbon atom.

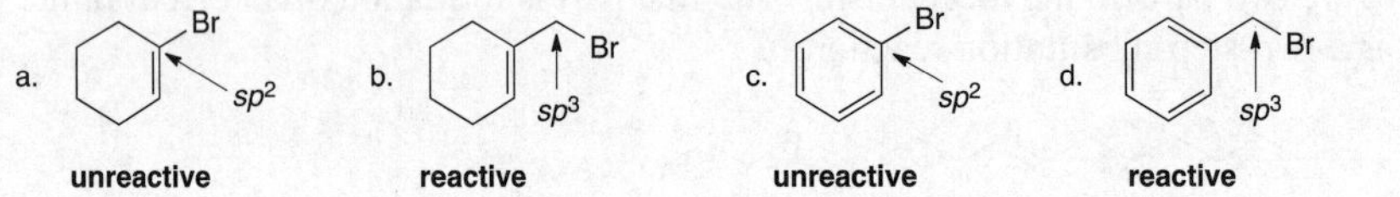

a.

Br

sp^2

unreactive

b.

Br

sp^3

reactive

c.

Br

sp^2

unreactive

d.

Br

sp^3

reactive

18.8 The product has an "unexpected" carbon skeleton, so rearrangement must have occurred.

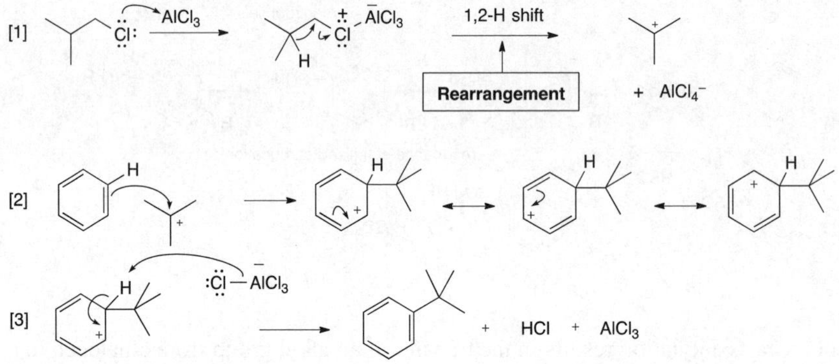

[1]

$AlCl_3$

$\overset{-}{A}lCl_3$

H

1,2-H shift

Rearrangement

+ $AlCl_4^-$

[2]

H

H

H

H

:Cl — $AlCl_3$

[3]

H

+ HCl + $AlCl_3$

18.9 Both alkenes and alcohols can form carbocations for Friedel–Crafts alkylation reactions.

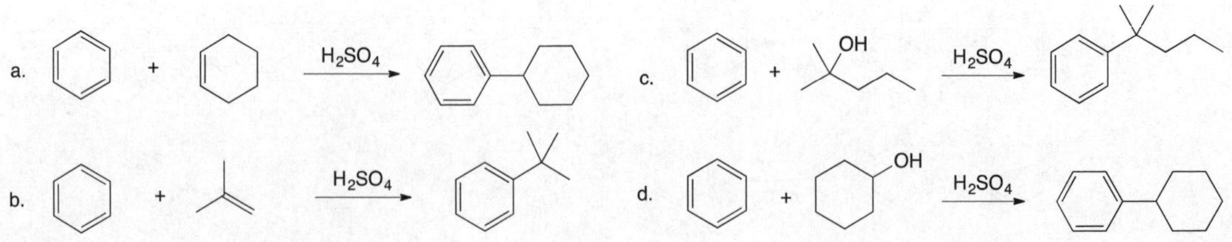

a.

+

H_2SO_4

c.

+

OH

H_2SO_4

b.

+

H_2SO_4

d.

+

OH

H_2SO_4

Reactions of Aromatic Compounds 18–9

18.10

[1]

[2]

[3]

B

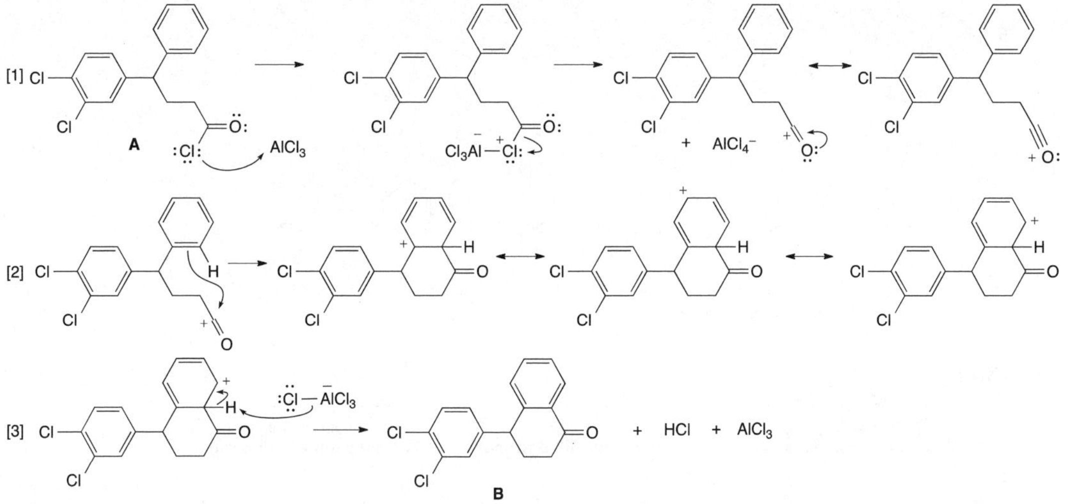

18.11 In parts (b) and (c), a 1,2-shift occurs to afford a rearrangement product.

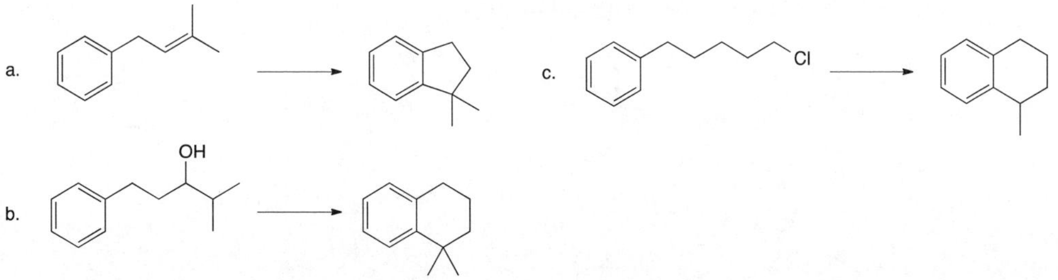

a.

c.

b.

18.12

a.

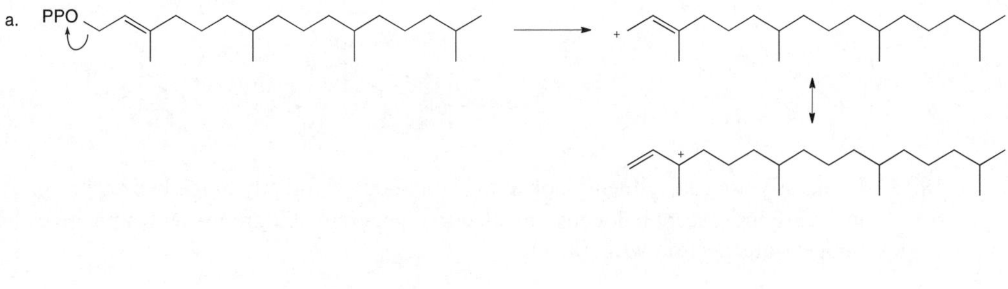

Chapter 18–10

b.

[structures]

[+ 7 resonance structures]   + HB⁺   **X**

18.13

a. $-CH_2CH_2CH_2CH_3$
    alkyl group
    electron donating

b. $-Br$
    halide
    electron withdrawing

c. $-OCH_2CH_3$
    electronegative O
    electron withdrawing

18.14 Electron-donating groups place a negative charge in the benzene ring. Draw the resonance structures to show how $-OCH_3$ puts a negative charge in the ring. Electron-withdrawing groups place a positive charge in the benzene ring. Draw the resonance structures to show how $-COCH_3$ puts a positive charge in the ring.

a.

b.

18.15 To classify each substituent, look at the atom bonded directly to the benzene ring. All R groups and Z groups (except halogens) are electron donating. All groups with a positive charge, $\delta+$, or halogens are electron withdrawing.

a. OCH_3
    lone pair on O
    electron donating

b. I
    halogen
    electron withdrawing

c.
    R group
    electron donating

Reactions of Aromatic Compounds 18–11

18.16 **Electron-donating groups** make the compound ***react faster*** than benzene in electrophilic aromatic substitution. **Electron-withdrawing groups** make the compound ***react more slowly*** than benzene in electrophilic aromatic substitution.

a.

electron withdrawing
reacts slower

b.

electron withdrawing
reacts slower

c.

lone pairs on O
electron donating
reacts faster

d.

halogen
electron withdrawing
reacts slower

e.

R group
electron donating
reacts faster

18.17 **Electron-donating groups** make the compound ***more reactive*** than benzene in electrophilic aromatic substitution. **Electron-withdrawing groups** make the compound ***less reactive*** than benzene in electrophilic aromatic substitution.

a.

R group
electron donating
more reactive

b.

two OH's
electron donating
more reactive

c.

C with 2 electronegative O's
electron withdrawing
less reactive

d.

electron withdrawing
less reactive

18.18 Chlorine inductively withdraws electron density and decreases the rate of electrophilic aromatic substitution. The closer the Cl is to the ring, the larger the effect it has. The larger the number of Cl's, the larger the effect.

least reactive

**intermediate
reactivity**

most reactive

18.19 Especially stable resonance structures have all atoms with an octet. Carbocations with additional electron donor R groups are also more stable structures. Especially unstable resonance structures have adjacent like charges.

Chapter 18–12

a.

especially stable with additional R group
stabilized carbocation

b.

especially stable
All atoms have an octet.

c.

especially unstable
2 adjacent (+) charges

18.20 Polyhalogenation occurs with highly activated benzene rings containing OH, NH$_2$, and related groups with a catalyst.

a. $\xrightarrow[\text{FeCl}_3]{\text{Cl}_2}$

b. $\xrightarrow{\text{Cl}_2}$ +

c. $\xrightarrow[\text{FeCl}_3]{\text{Cl}_2}$ +

18.21 Friedel–Crafts reactions do not occur with strongly deactivating substituents including NO$_2$, or with NH$_2$, NR$_2$, or NHR groups.

a. $-\text{SO}_3\text{H}$ $\xrightarrow[\text{AlCl}_3]{\text{CH}_3\text{Cl}}$ **no reaction**

strongly deactivating

b. $-\text{Cl}$ $\xrightarrow[\text{AlCl}_3]{\text{CH}_3\text{Cl}}$ $-\text{Cl}$ + $-\text{Cl}$

Cl is an o,p director.

c. $\xrightarrow[\text{AlCl}_3]{\text{CH}_3\text{Cl}}$ **no Friedel–Crafts reaction**

d. $\xrightarrow[\text{AlCl}_3]{\text{CH}_3\text{Cl}}$ +

18.22 To draw the product of a reaction with these disubstituted benzene derivatives and HNO$_3$, H$_2$SO$_4$, remember the following:
- If the two directing effects reinforce each other, the new substituent will be on the position reinforced by both.
- If the directing effects oppose each other, the stronger activator wins.
- No substitution occurs between two meta substituents.

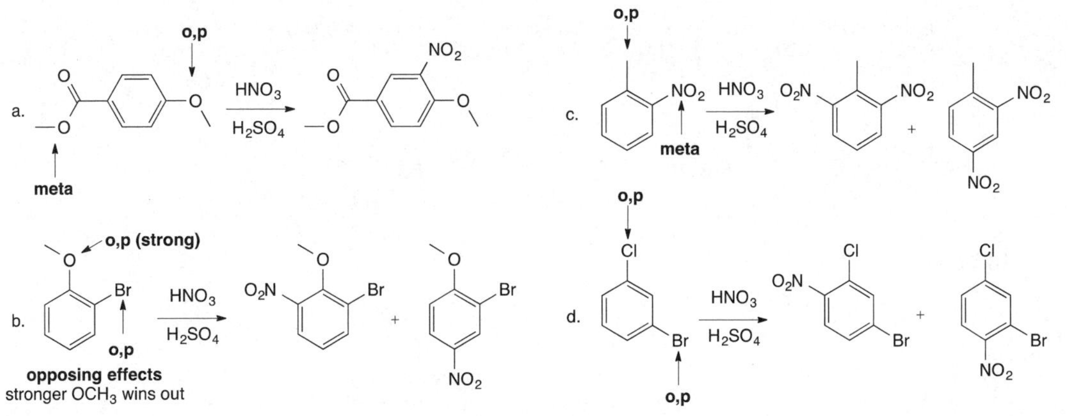
a.
o,p
meta
HNO_3
H_2SO_4
NO_2
c.
o,p
NO_2
meta
HNO_3
H_2SO_4
O_2N
NO_2
+
NO_2
NO_2
b.
o,p (strong)
Br
o,p
opposing effects
stronger OCH_3 wins out
HNO_3
H_2SO_4
O_2N
Br
+
Br
NO_2
d.
o,p
Cl
Br
o,p
HNO_3
H_2SO_4
O_2N
Cl
Br
+
Cl
Br
NO_2

18.23

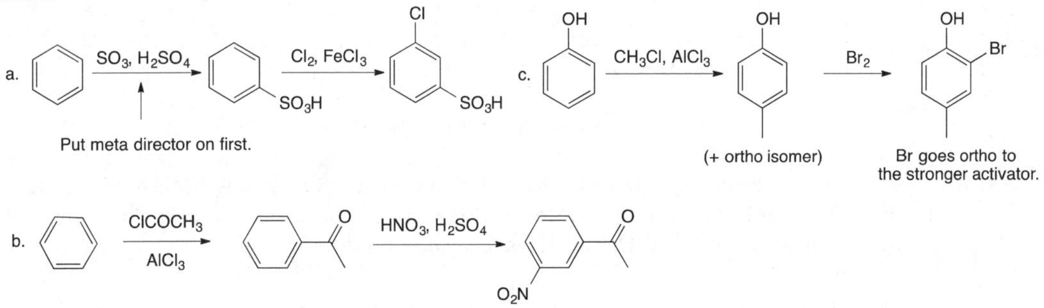
a.
SO_3, H_2SO_4
Put meta director on first.
SO_3H
Cl_2, $FeCl_3$
Cl
SO_3H
c.
OH
CH_3Cl, $AlCl_3$
OH
(+ ortho isomer)
Br_2
OH
Br
Br goes ortho to
the stronger activator.
b.
$ClCOCH_3$
$AlCl_3$
O
HNO_3, H_2SO_4
O
O_2N

18.24

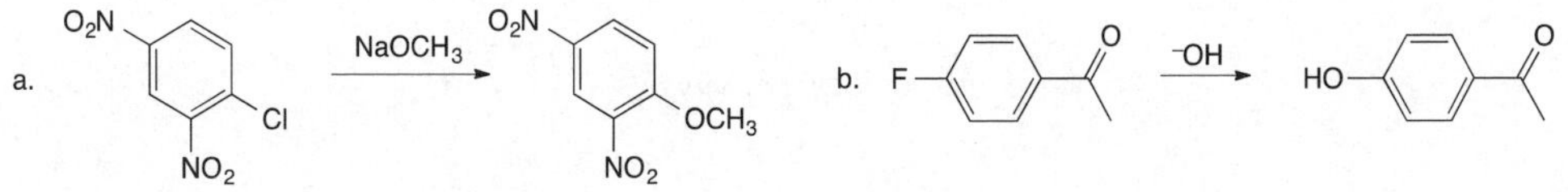
a.
O_2N
Cl
NO_2
$NaOCH_3$
O_2N
OCH_3
NO_2
b.
F
O
^{-}OH
HO
O

18.25

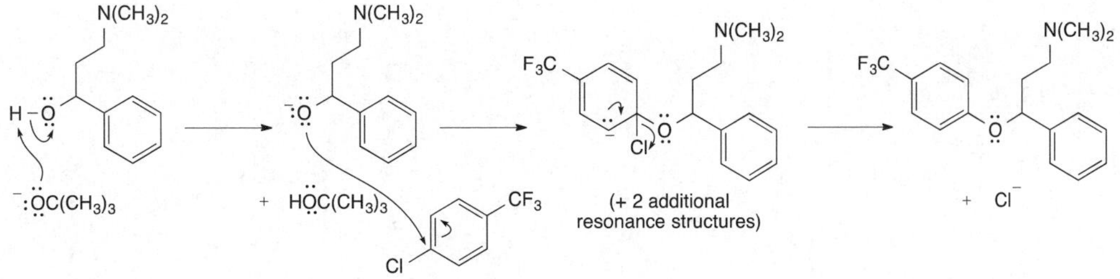
$N(CH_3)_2$
H–O
$^{-}:OC(CH_3)_3$
$N(CH_3)_2$
+ $HOC(CH_3)_3$
CF_3
Cl
F_3C
$N(CH_3)_2$
Cl
(+ 2 additional
resonance structures)
F_3C
$N(CH_3)_2$
+ Cl^-

18.26

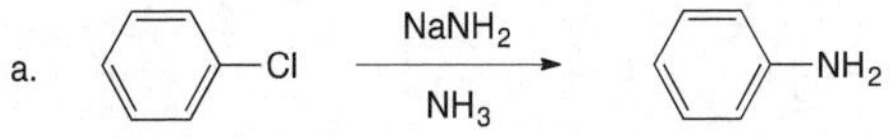
a.
Cl
$NaNH_2$
NH_3
NH_2

Chapter 18–14

b.

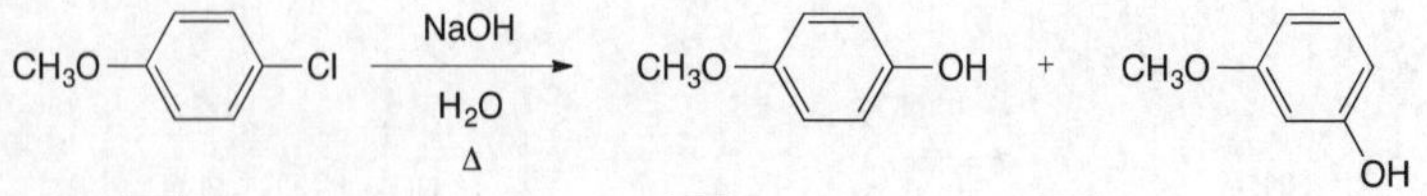

c.

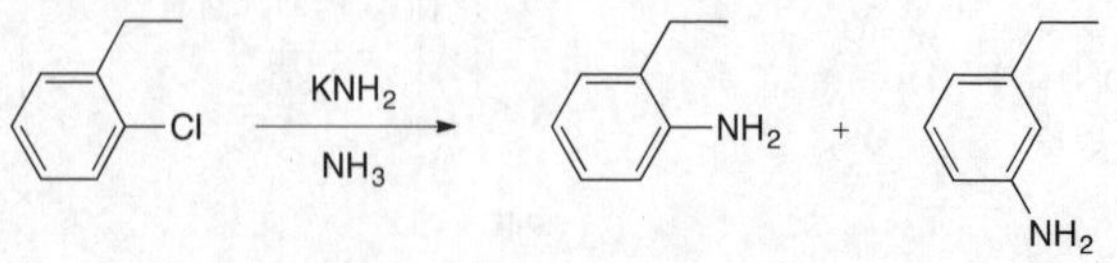

18.27

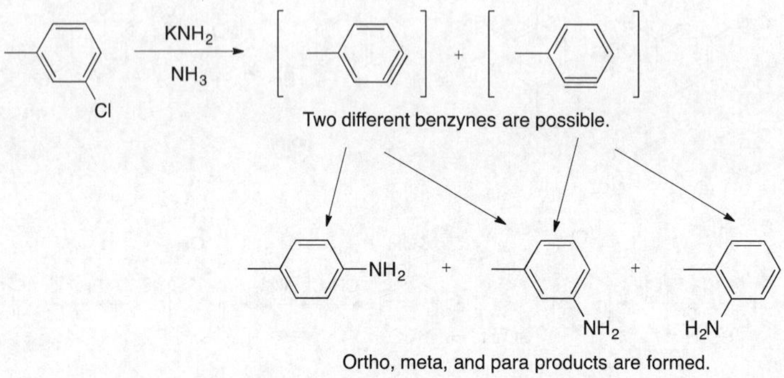

Two different benzynes are possible.

Ortho, meta, and para products are formed.

18.28 This reaction proceeds via a radical bromination mechanism and two radicals are possible: **A** (2° and benzylic) and **B** (1°). **B** (which leads to $C_6H_5CH_2CH_2Br$) is much less stable, so this radical is not formed and only $C_6H_5CH(Br)CH_3$ is formed as product.

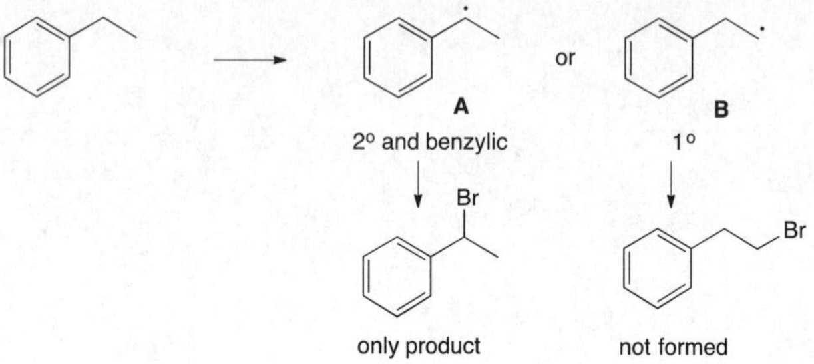

18.29

a.

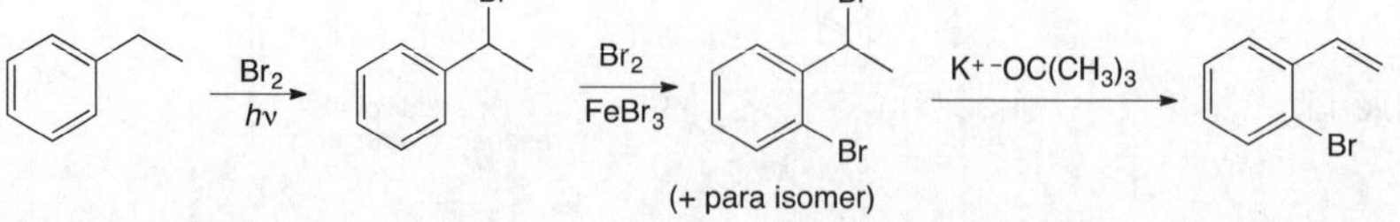

(+ para isomer)

b.

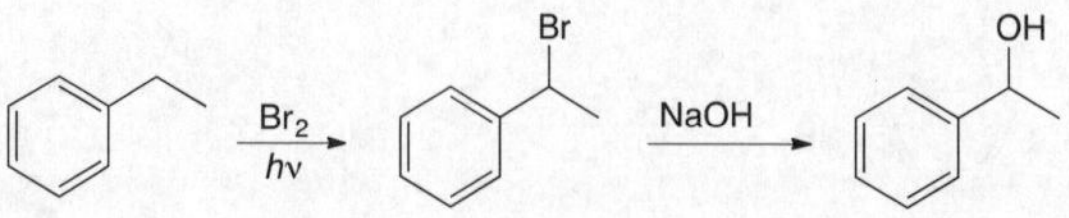

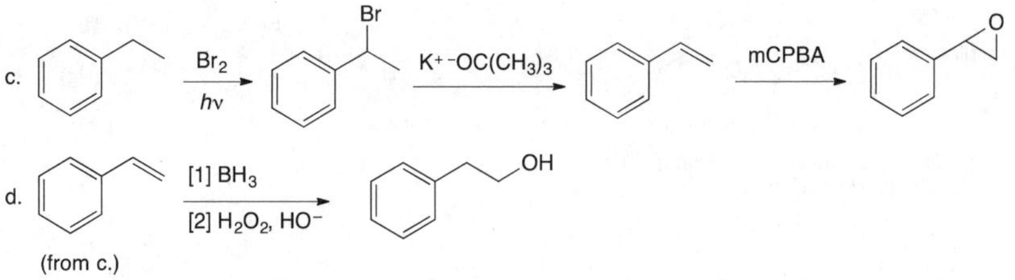

Reactions of Aromatic Compounds 18–15

c.

Br$_2$ / $h\nu$

K$^+$ $^-$OC(CH$_3$)$_3$

mCPBA

d.

[1] BH$_3$

[2] H$_2$O$_2$, HO$^-$

(from c.)

18.30 First use an acylation reaction, and then reduce the carbonyl group to form the alkyl benzenes.

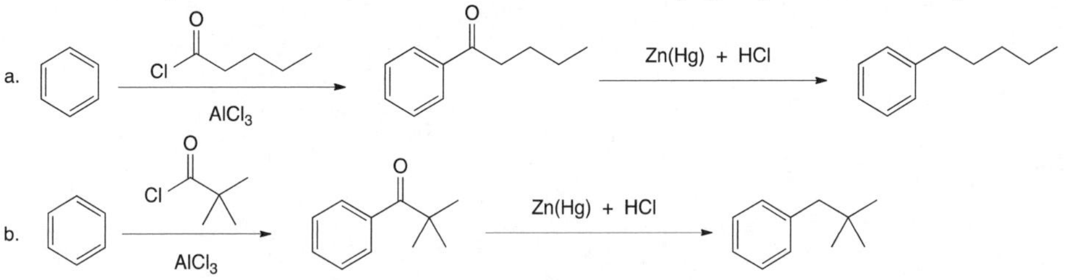

a.

AlCl$_3$

Zn(Hg) + HCl

b.

AlCl$_3$

Zn(Hg) + HCl

18.31

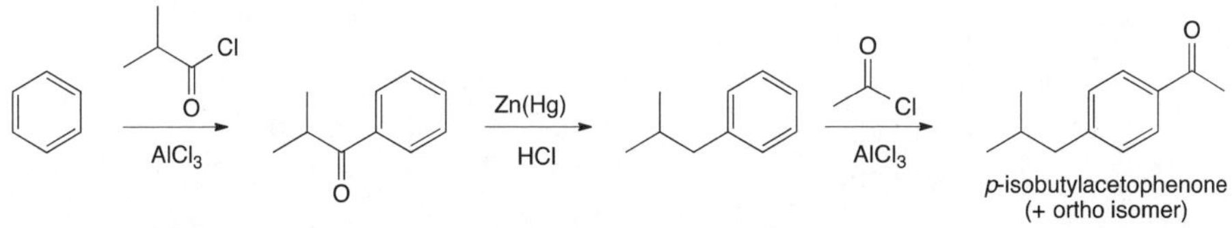

AlCl$_3$

Zn(Hg) / HCl

AlCl$_3$

p-isobutylacetophenone
(+ ortho isomer)

18.32

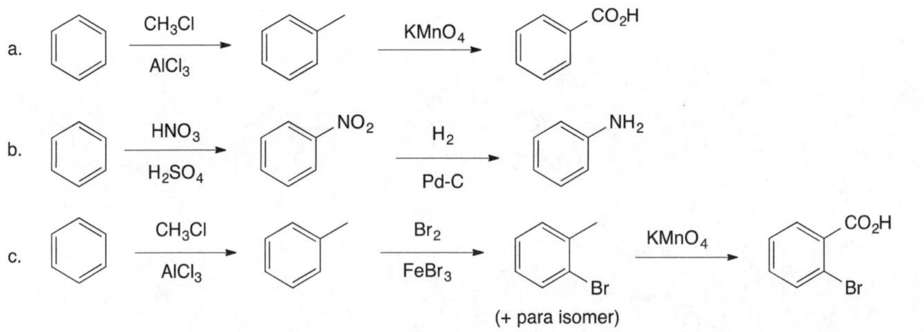

a.

CH$_3$Cl / AlCl$_3$

KMnO$_4$

b.

HNO$_3$ / H$_2$SO$_4$

H$_2$ / Pd-C

c.

CH$_3$Cl / AlCl$_3$

Br$_2$ / FeBr$_3$

(+ para isomer)

KMnO$_4$

18.33

a.

ClCH$_2$CH$_3$ / AlCl$_3$

AlCl$_3$

(+ ortho isomer)

SO$_3$ / H$_2$SO$_4$

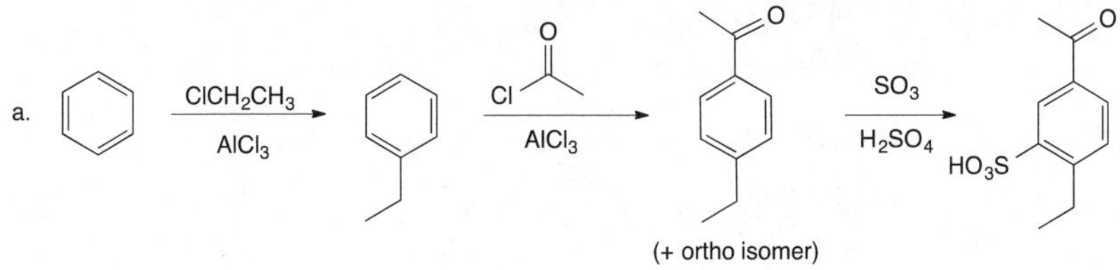

Chapter 18–16

b.

Br ← o,p director

o,p director

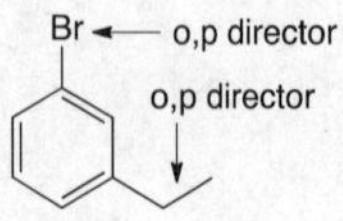

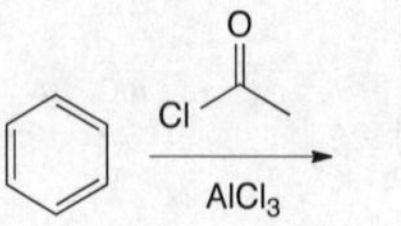

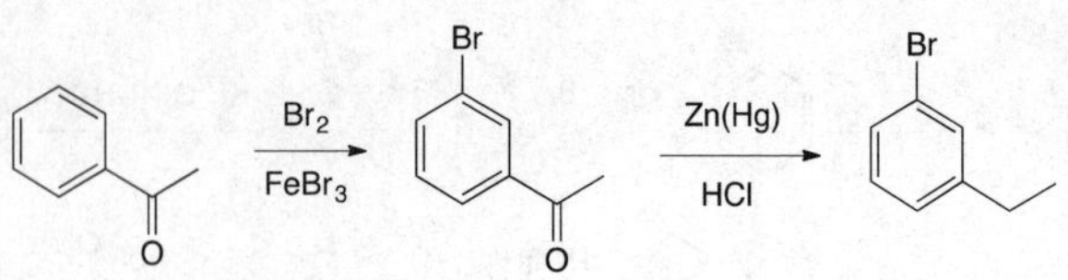

Both are o,p directors, but they are meta to each other. The alkyl group must be obtained by reduction of a carbonyl.

c.

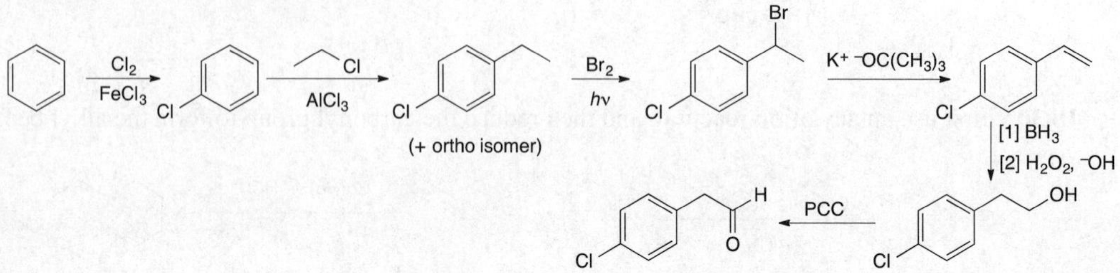

(+ ortho isomer)

18.34

A

B

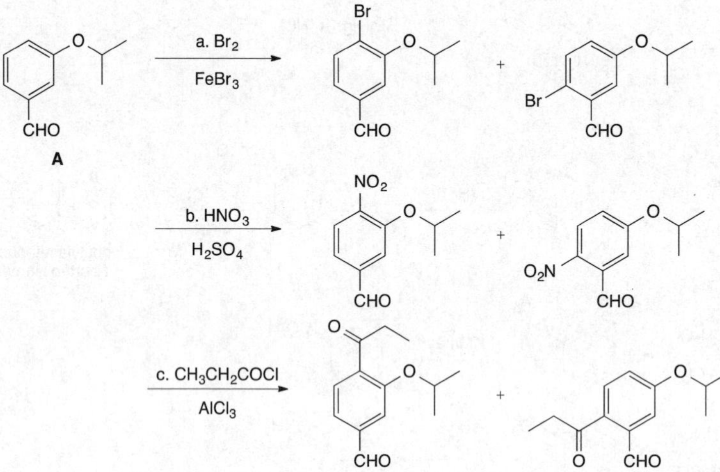

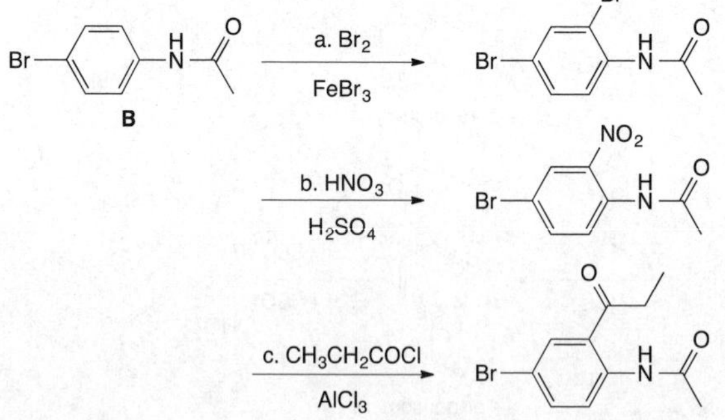

Reactions of Aromatic Compounds 18–17

18.35 Intramolecular Friedel–Crafts acylation occurs on the more activated aromatic ring.

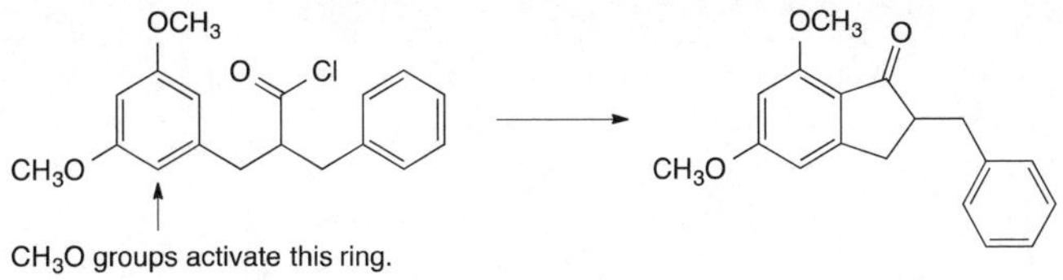

CH$_3$O groups activate this ring.

18.36 OH is an ortho, para director.

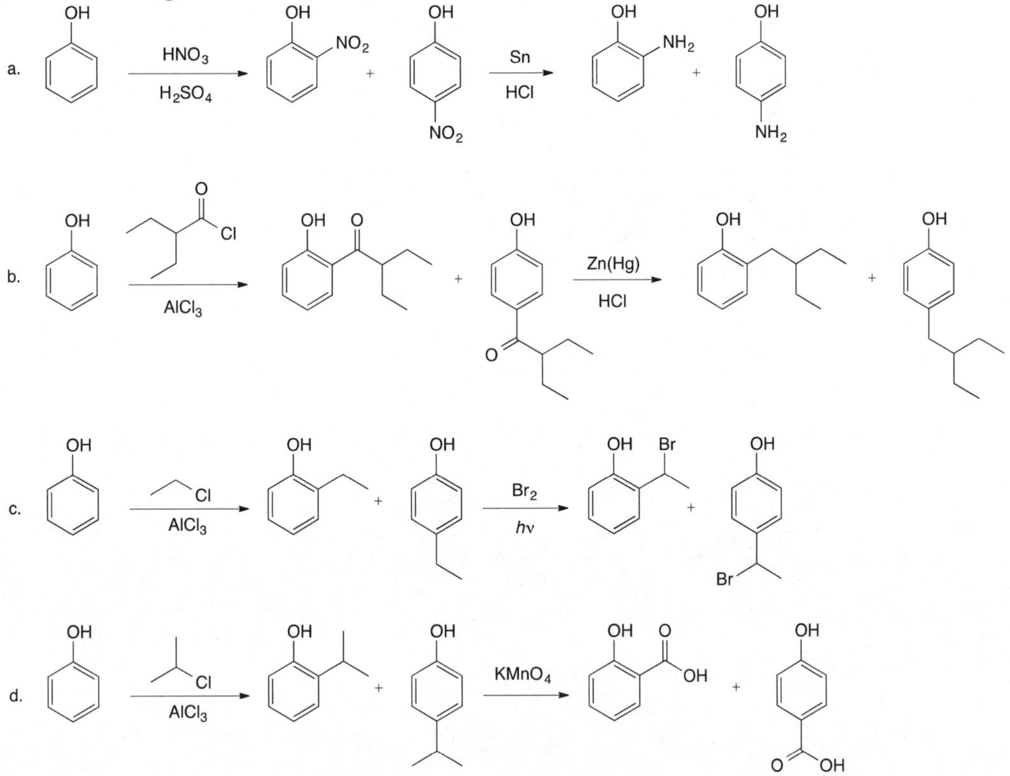

a.

b.

c.

d.

18.37

a. No reaction

b. No Friedel–Crafts reaction

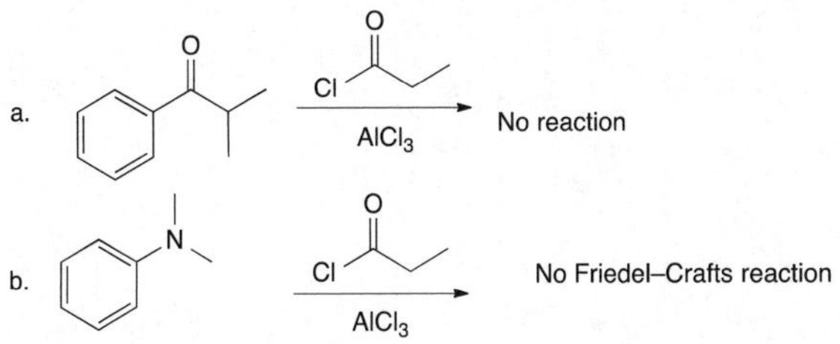

Chapter 18–18

c.

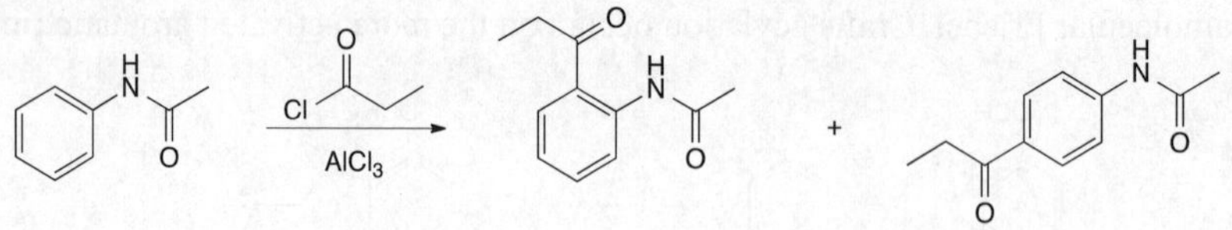

18.38

a.

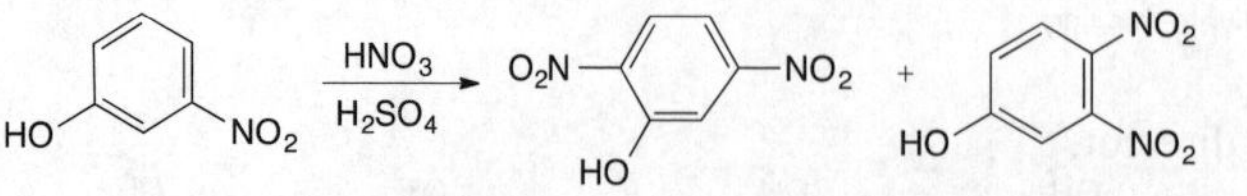

b.

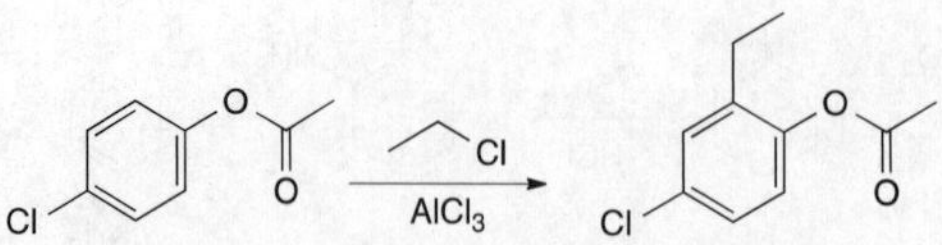

c.

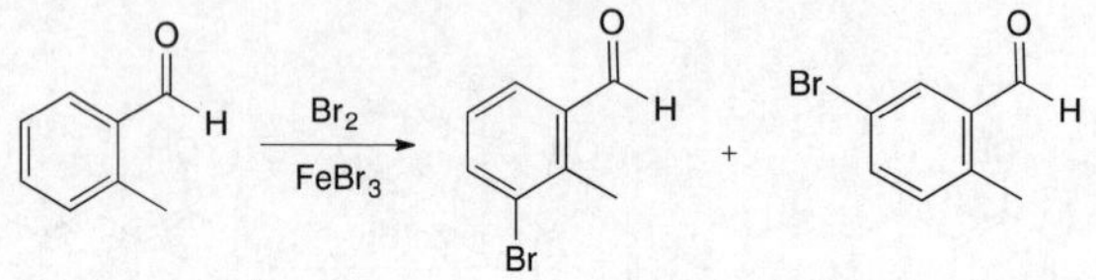

d.

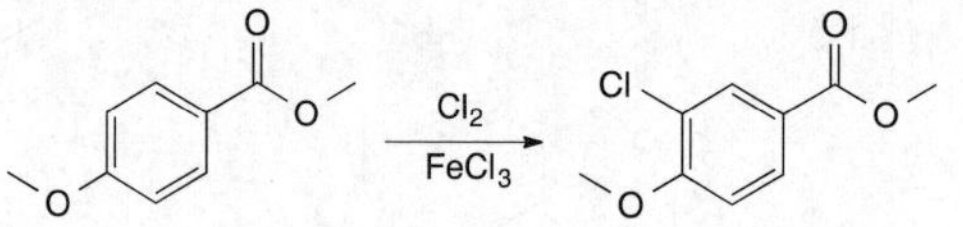

e.

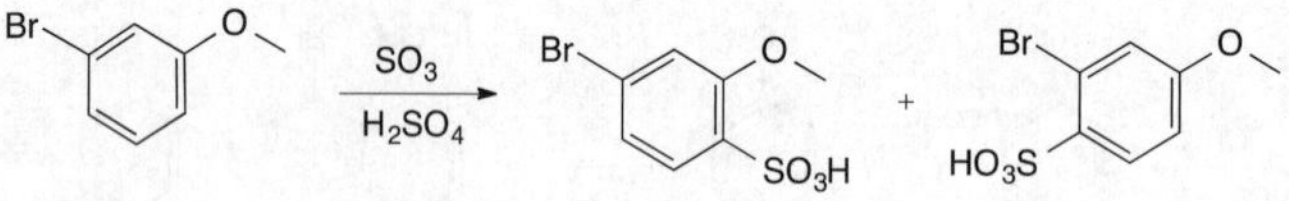

f.

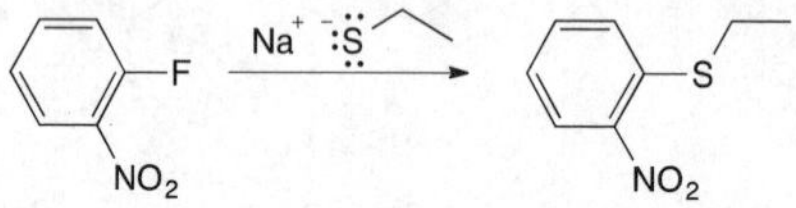

18.39 Watch out for rearrangements.

a.

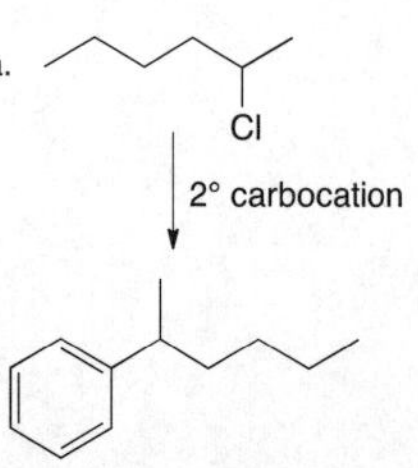

2° carbocation

b.

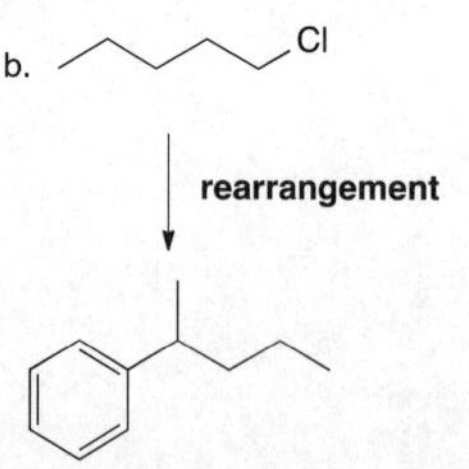

rearrangement

c.

rearrangement

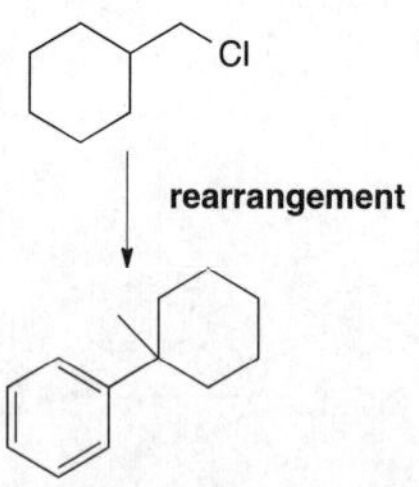

18.40

a.

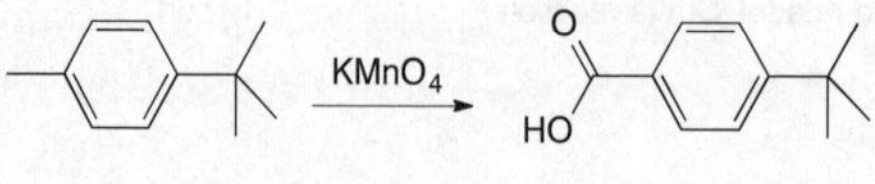

Reactions of Aromatic Compounds 18–19

b.

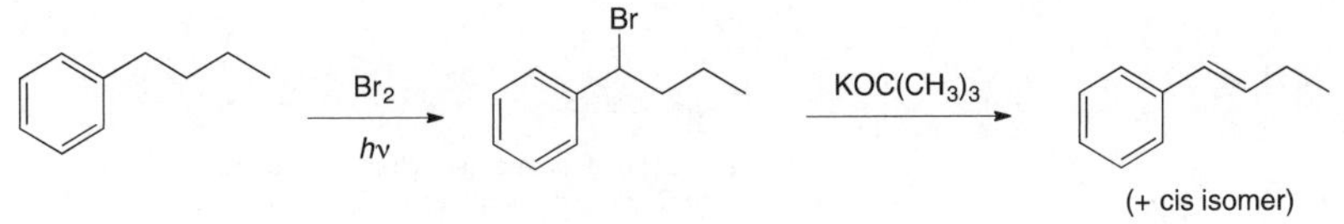

c.

[1] Cl$_2$, FeCl$_3$

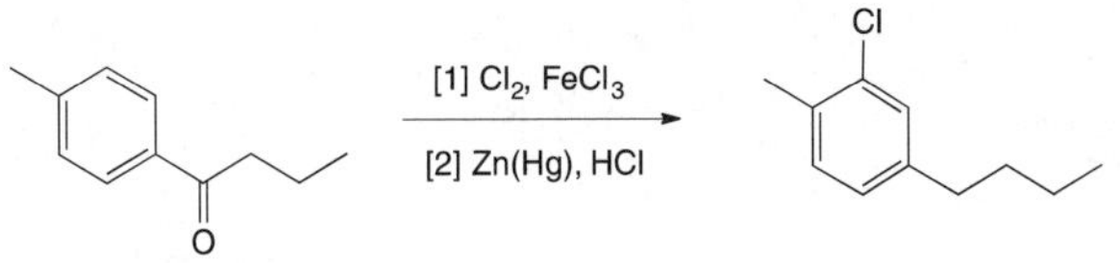

[2] Zn(Hg), HCl

d.

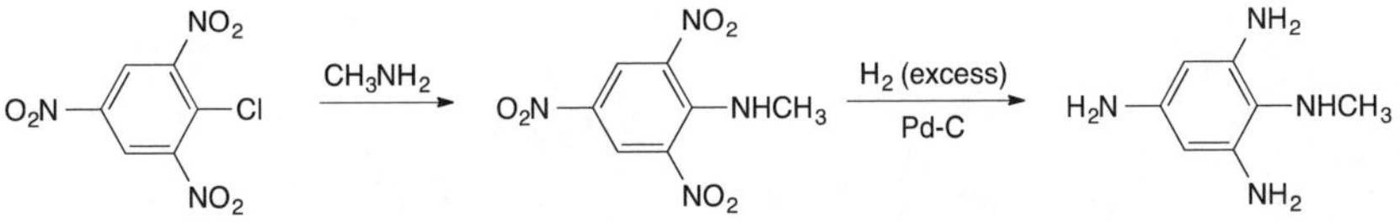

CH$_3$NH$_2$

H$_2$ (excess)

Pd-C

18.41

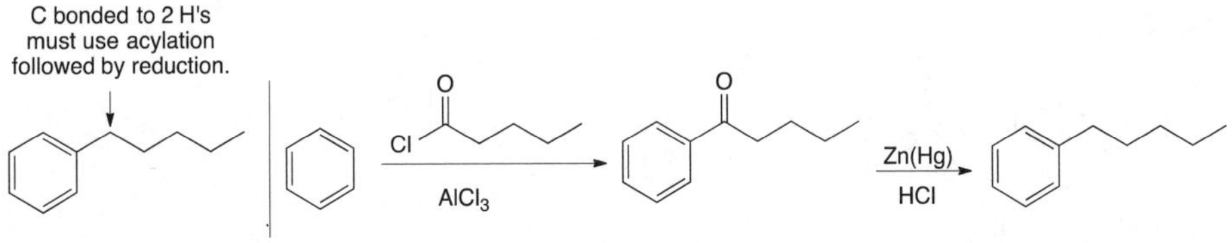

C bonded to 2 H's
must use acylation
followed by reduction.

a.

AlCl$_3$

Zn(Hg)
HCl

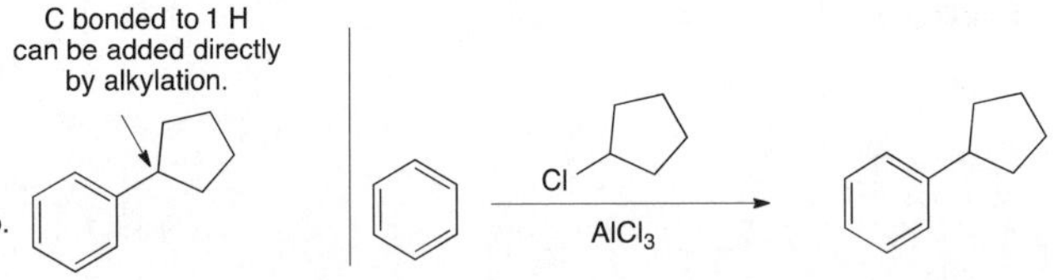

C bonded to 1 H
can be added directly
by alkylation.

b.

AlCl$_3$

c.

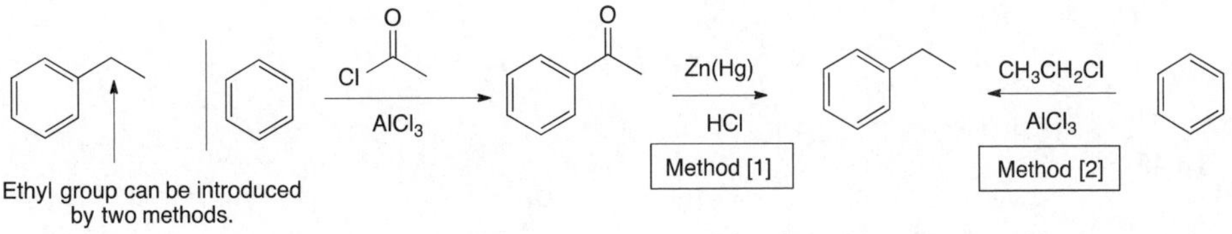

AlCl$_3$

Zn(Hg)
HCl

Method [1]

CH$_3$CH$_2$Cl

AlCl$_3$

Method [2]

Ethyl group can be introduced
by two methods.

18.42

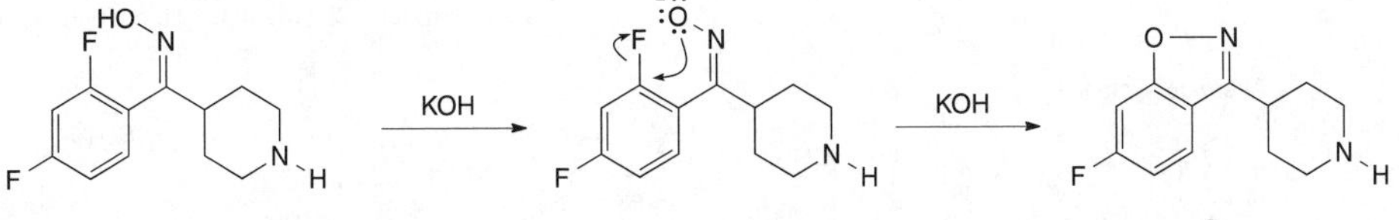

KOH

KOH

A

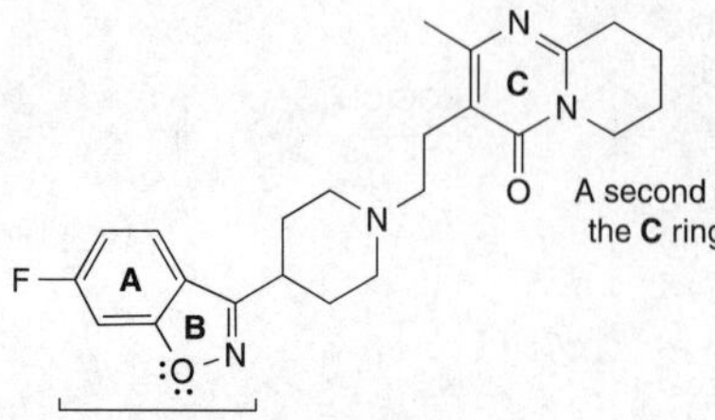
F
A
B
N
O
C
N
N
O

A second resonance structure can be drawn for the **C** ring, showing that it has six π electrons.

Rings **A** and **B** contain 10 π electrons, eight from the double bonds and two from O, making this ring system aromatic.

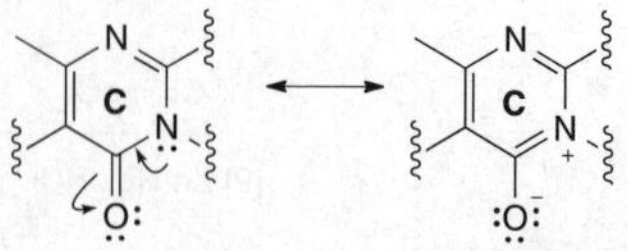
N
C
N
O
N
C
N
O

6 π electrons

18.43

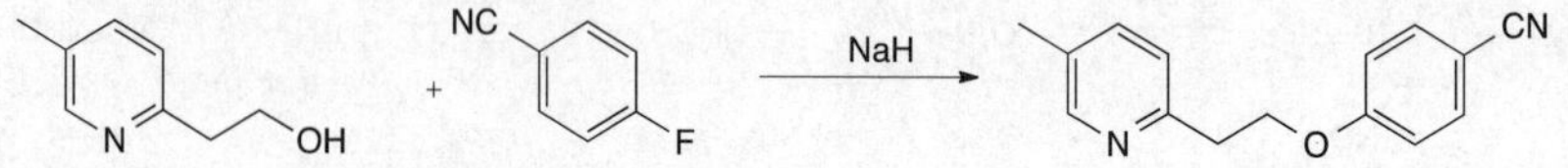
N
OH
NC
F
NaH
N
O
CN

A **B** **C**

18.44

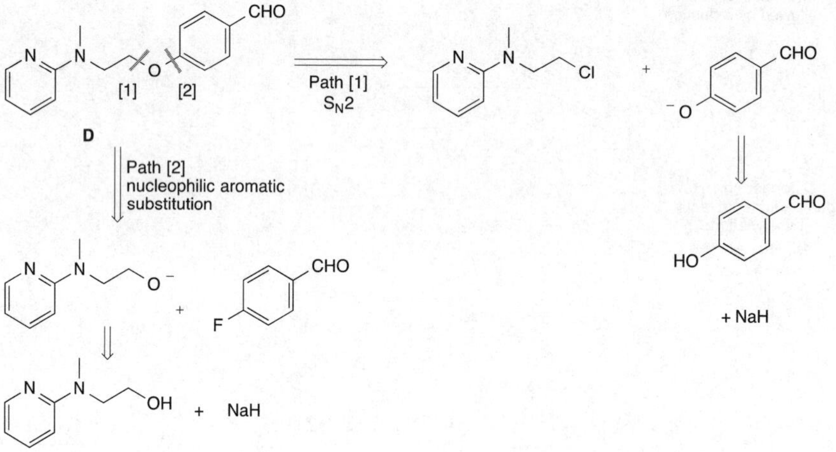
N
N
O
CHO
[1]
[2]
D
Path [1]
S_N2
N
N
Cl
+
−O
CHO
CHO
HO
+ NaH
Path [2]
nucleophilic aromatic
substitution
N
N
O−
+
F
CHO
N
N
OH
+
NaH

18.45

a.

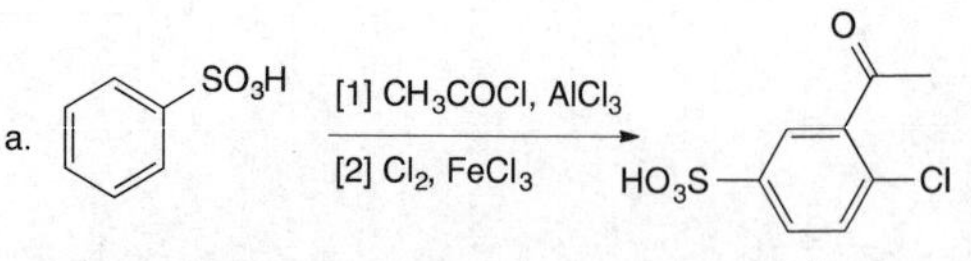
SO_3H
[1] CH_3COCl, $AlCl_3$
[2] Cl_2, $FeCl_3$
O
HO_3S
Cl

A

Step [1] won't work because a Friedel–Crafts reaction can't be done on a deactivated benzene ring, as is the case with the SO_3H substituent. Even if Step [1] did work, the second step would introduce Cl meta to SO_3H, not para as drawn.

Alternate synthesis:

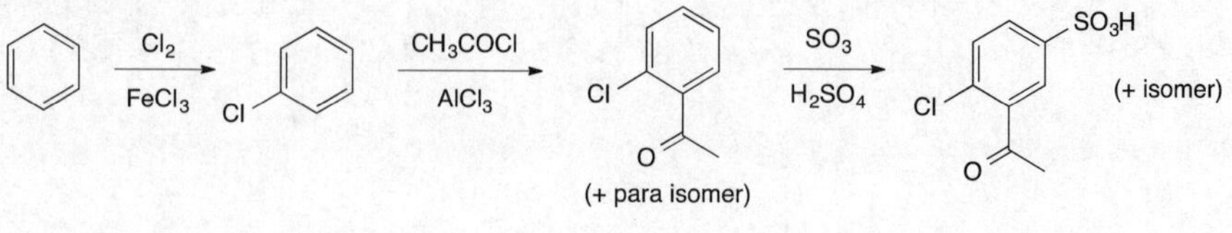
Cl_2
$FeCl_3$
Cl
CH_3COCl
$AlCl_3$
Cl
O
(+ para isomer)
SO_3
H_2SO_4
SO_3H
Cl
O
(+ isomer)

Reactions of Aromatic Compounds 18–21

b.

[1] CH$_3$CH$_2$CH$_2$CH$_2$Cl, AlCl$_3$

[2] HNO$_3$, H$_2$SO$_4$

= **B**

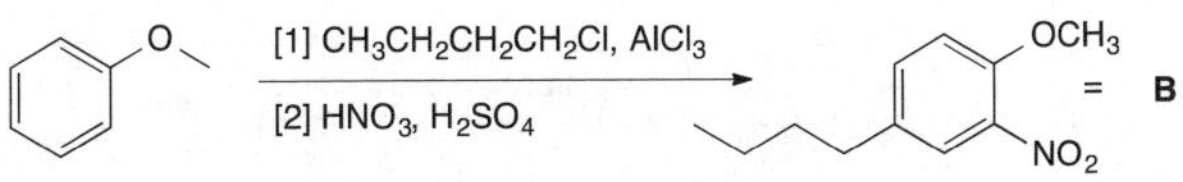

Step [1] involves a Friedel–Crafts alkylation using a 1° alkyl halide that will undergo rearrangement, so that a butyl group will not be introduced as a side chain.

Alternate synthesis:

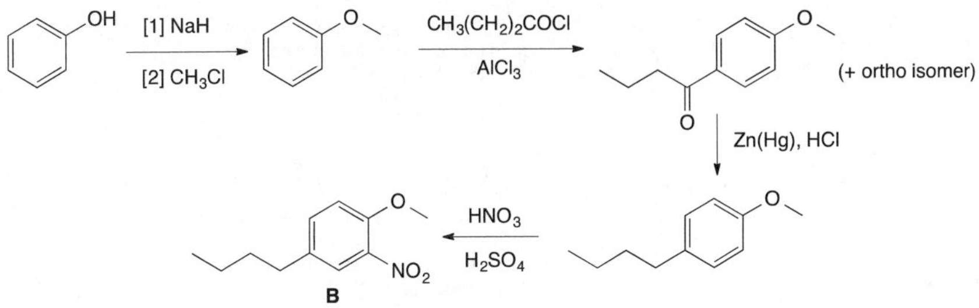

OH

[1] NaH

[2] CH$_3$Cl

CH$_3$(CH$_2$)$_2$COCl

AlCl$_3$

(+ ortho isomer)

Zn(Hg), HCl

HNO$_3$

H$_2$SO$_4$

B

18.46 Use the directions from Answer 18.17 to rank the compounds.

a.

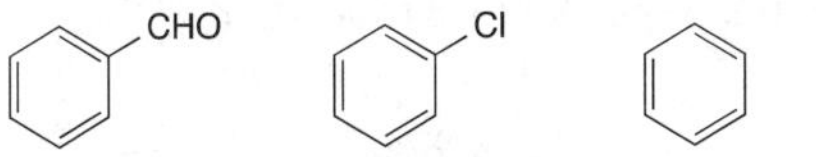

CHO

least reactive

Cl

intermediate reactivity

most reactive

b.

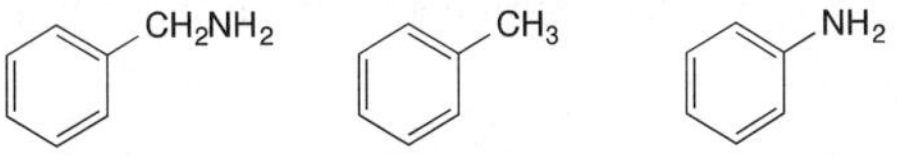

CH$_2$NH$_2$

least reactive

CH$_3$

intermediate reactivity

NH$_2$

most reactive

18.47

[1]
Br

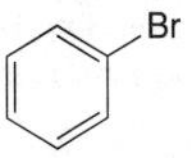

a. withdraw
b. donate
c. less
d. deactivate

[2]
C≡N

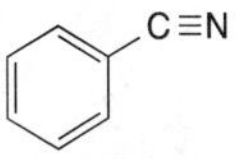

a. withdraw
b. withdraw
c. less
d. deactivate

[3]
O

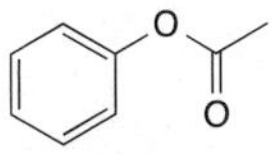

a. withdraw
b. donate
c. more
d. activate

18.48

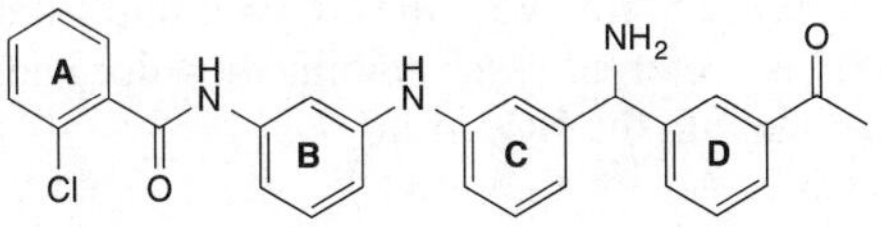

A

H
N

B

H
N

C

NH$_2$

D

O

Cl

O

a. **B**   b. **A**   c.

H
N

Cl   O
Br

H
N

NH$_2$

O

+

H
N

Cl   O

H
N

Br

NH$_2$

O

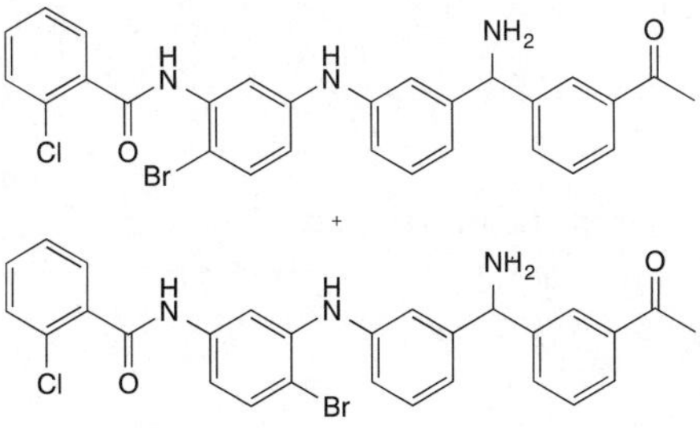

Chapter 18–22

18.49

a. [structure: phenyl–N-piperidine]
more electron rich
due to N atom
faster

→ [ortho E product with N-piperidine] + [para E product with N-piperidine]

b. [structure: phenyl–$\overset{+}{N}H(CH_3)_2$]
less electron rich
due to (+) charge on N
slower

→ [meta E product with $\overset{+}{N}H(CH_3)_2$]

c. electron withdrawing
$O_2N \leftarrow$ [structure with $\overset{+}{N}(CH_3)_3$]
less electron rich
due to (+) charge on N and electron-
withdrawing NO_2 group
slower

→ [product with O_2N, $\overset{+}{N}(CH_3)_3$ and E]

d. O_2N– [structure with N-piperidine]
electron   electron donating
withdrawing

→ O_2N– [product with E and N-piperidine]

Effects cancel out.
similar in reactivity to benzene

18.50 Electrophilic addition of HBr proceeds by the more stable carbocation.

[reaction scheme: nitro-stilbene with OCH₃ + H–Br → benzylic cation intermediate → resonance structure]

especially stable resonance structure
having all atoms with an octet
[+ 4 additional resonance structures]

[arrow down with Br⁻]

[major product structure]
major product

When the (+) charge is benzylic to the benzene ring with the OCH₃ group, additional resonance stabilization is present, so this pathway is preferred. Such stabilization does not result when the (+) charge is benzylic to the benzene ring with the NO_2 group.

18.51

[structure: $:\overset{..}{O}=\overset{..}{N}$–phenyl]
ortho, para
director

$\xrightarrow{E^+}$ [structure: $:\overset{..}{O}=\overset{..}{N}$– ortho-E product] + [structure: $:\overset{..}{O}=\overset{..}{N}$– para-E product]

Ortho and para products are isolated.

Reactions of Aromatic Compounds 18–23

With ortho and para attack there is additional resonance stabilization that delocalizes the positive charge onto the nitroso group. Such additional stabilization is not possible with meta attack. This makes –NO an ortho, para director. Since the N atom bears a partial (+) charge (because it is bonded to a more electronegative O atom), the –NO group inductively withdraws electron density, thus deactivating the benzene ring towards electrophilic attack. In this way, the –NO group resembles the halogens. Thus, the electron-donating resonance effect makes –NO an o,p director, but the electron-withdrawing inductive effect makes it a deactivator.

Ortho attack:

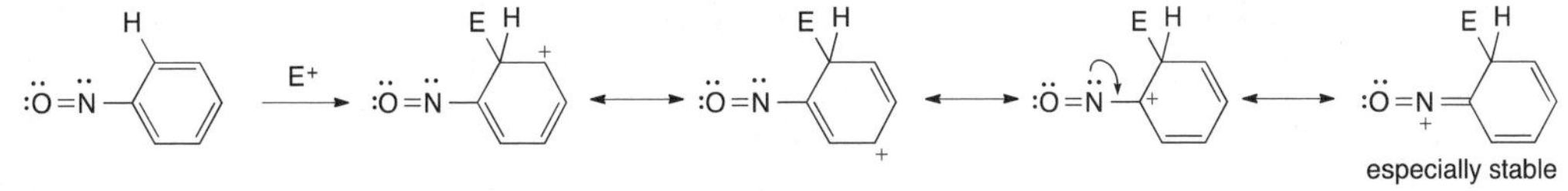

Meta attack:

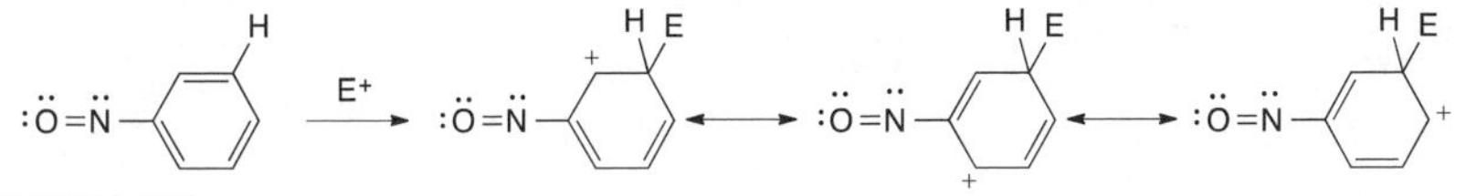

Para attack:

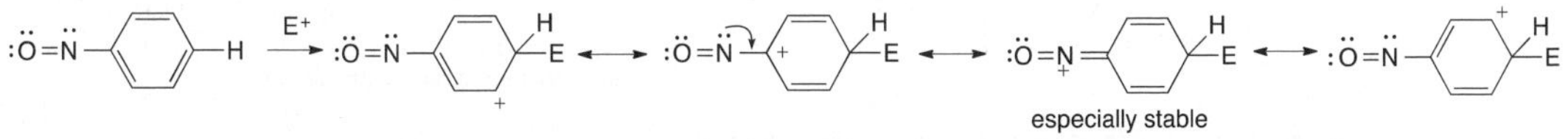

18.52

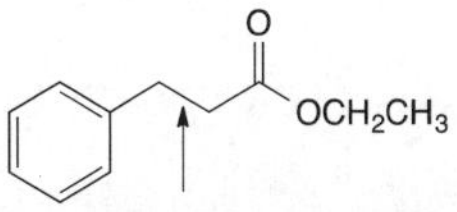

alkyl group on the benzene ring
R stabilizes (+) charges on the o,p positions by an electron-
donating inductive effect. This group behaves like any other R
group so that ortho and para products are formed in electrophilic
aromatic substitution.

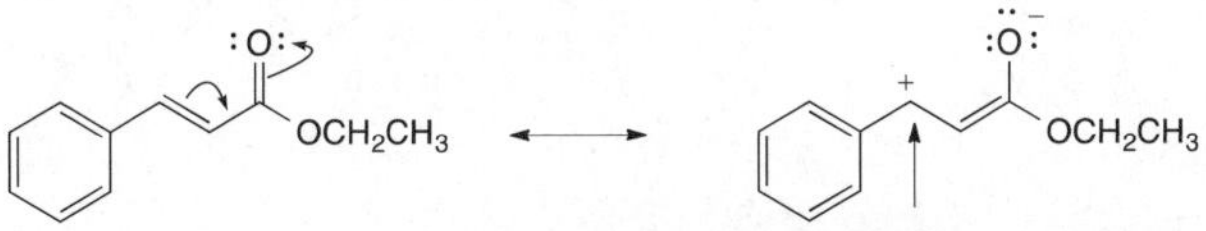

(+) charge on atom bonded to the benzene ring
Drawing resonance structures in electrophilic aromatic substitution results in
especially unstable structures for attack at the o,p positions—two (+) charges on
adjacent atoms. This doesn't happen with meta attack, so meta attack is preferred.
This is identical to the situation observed with all meta directors.

18.53 Increasing the number of electron-withdrawing groups (especially at the ortho and para positions to the leaving group) increases the rate of nucleophilic aromatic substitution. Increasing the electronegativity of the halogen increases the rate.

Chapter 18–24

a.

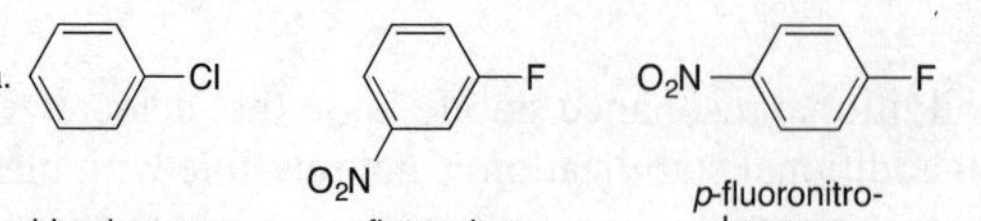

chlorobenzene
least reactive

m-fluoronitro-
benzene

p-fluoronitro-
benzene
most reactive

b.

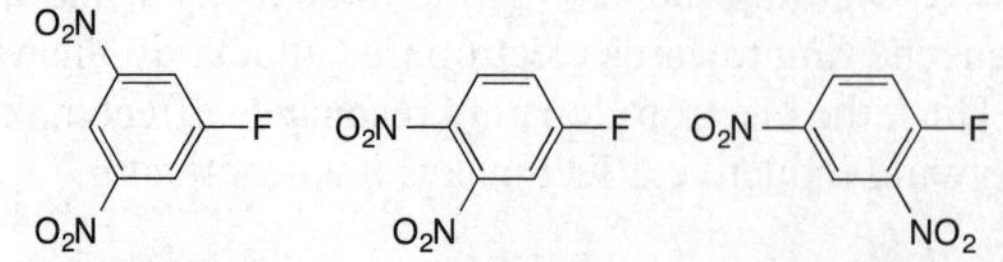

1-fluoro-3,5-dinitro-
benzene
least reactive

1-fluoro-3,4-
dinitrobenzene

1-fluoro-2,4-dinitro-
benzene
most reactive

c.

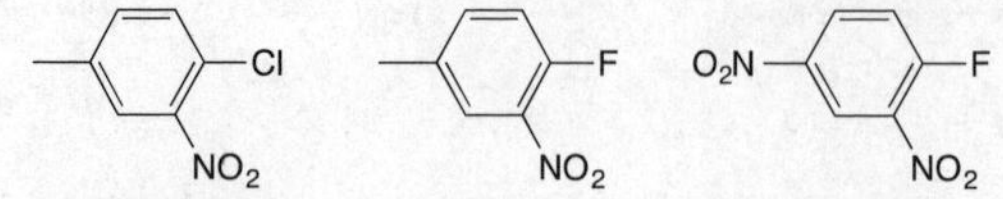

4-chloro-3-nitro-
toluene
least reactive

4-fluoro-3-
nitrotoluene

1-fluoro-2,4-dinitro-
benzene
most reactive

18.54

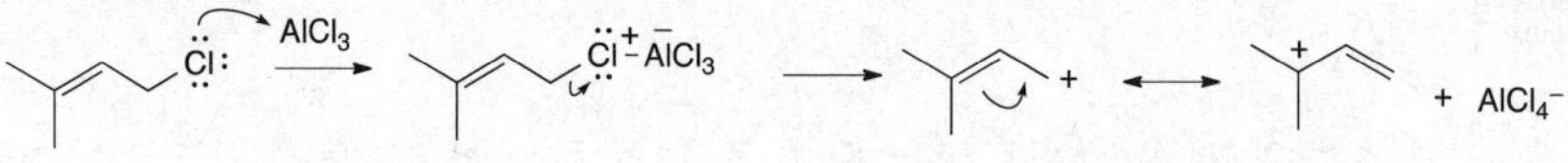

resonance-stabilized carbocation

Use both resonance forms to show how two products are formed.

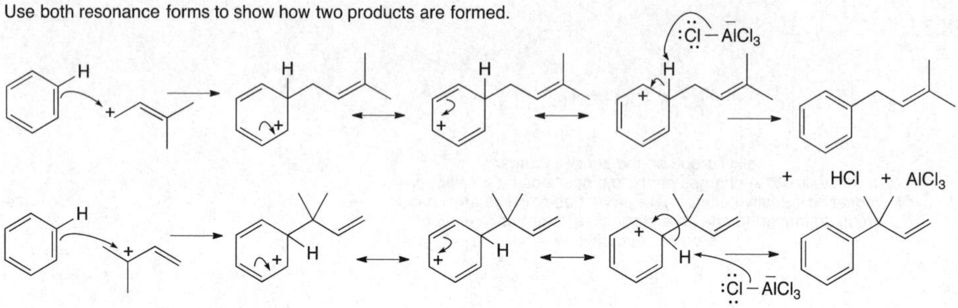

+ HCl + AlCl$_3$

18.55

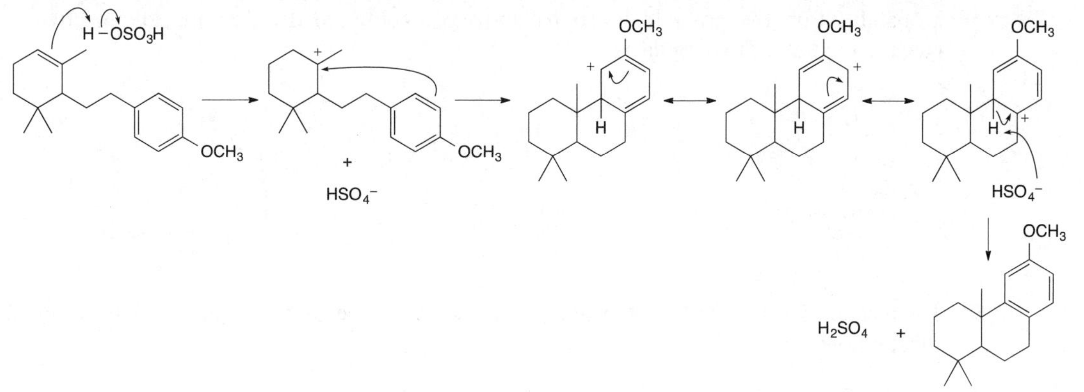

H–OSO₃H $H-OSO_3H$

OCH₃   OCH₃   OCH₃

+ HSO_4^-

HSO_4^-

OCH₃

H_2SO_4 +

18.56

:ÖH $H-OSO_3H$

$CH_3\ddot{O}$ $CH_3\ddot{O}$ $\overset{+}{:}ÖH_2$

+ HSO_4^-

$CH_3\ddot{O}$ + $H_2\ddot{O}$

1,2-CH₃ shift

$CH_3\overset{+}{O}$ $CH_3\ddot{O}$ $CH_3\ddot{O}$ $CH_3\ddot{O}$

$CH_3\ddot{O}$

HSO_4^-

$CH_3\ddot{O}$ + H_2SO_4

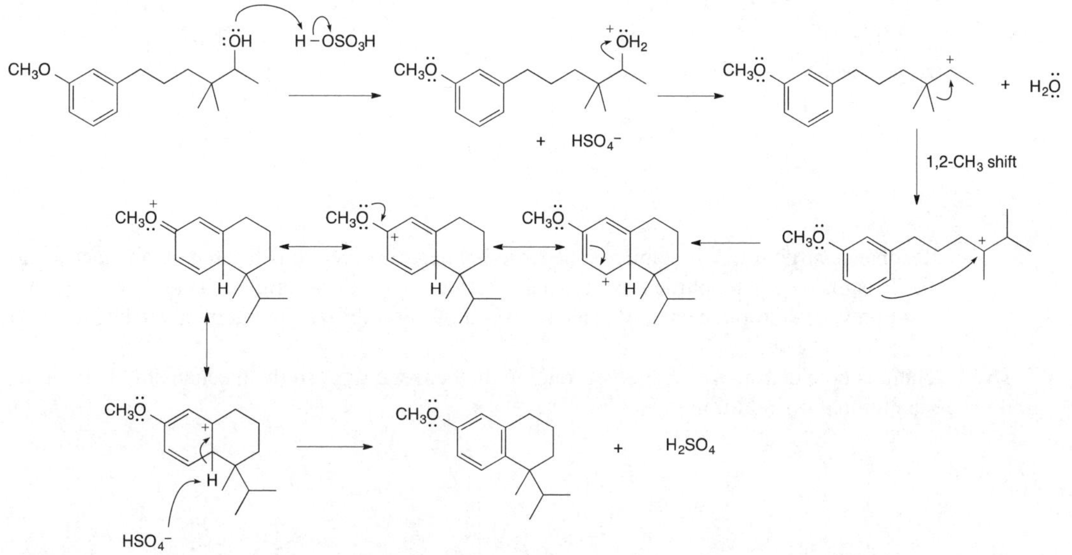

18.57

a. The product has one stereogenic center.

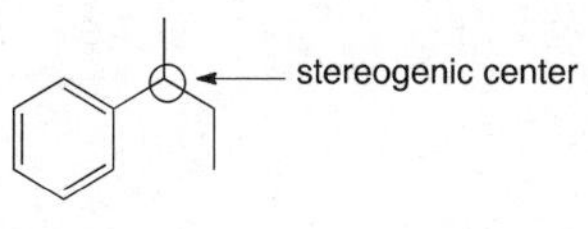

stereogenic center

Chapter 18–26

b. The mechanism for Friedel–Crafts alkylation with this 2° halide involves formation of a trigonal planar carbocation. Because the carbocation is achiral, it reacts with benzene with equal probability from two possible directions (above and below) to afford an optically inactive, racemic mixture of two products.

(R)-2-chlorobutane

trigonal planar
achiral carbocation

racemic mixture
optically inactive

18.58 The reaction follows the two-step addition–elimination mechanism for nucleophilic aromatic substitution.

2-chloropyridine

A

Resonance structure **A** is stabilized because the negative charge is located on an electronegative N. This makes nucleophilic aromatic substitution on 2-chloropyridine faster than a similar reaction with chlorobenzene, which has no N atom to stabilize the intermediate negative charge.

18.59 There is no electron-withdrawing group on the benzene ring, so the mechanism likely proceeds via elimination–addition.

+ HB$^+$

+ :B

+ Cl$^-$

Reactions of Aromatic Compounds 18–27

18.60

A This product is formed.

B This product is *not* formed.

Attack to form **A** proceeds via a carbocation for which **7** resonance structures can be drawn. Four resonance structures contain an intact benzene ring.

Attack to form **B** proceeds via a carbocation for which **6** resonance structures can be drawn. Only two resonance structures contain an intact benzene ring.

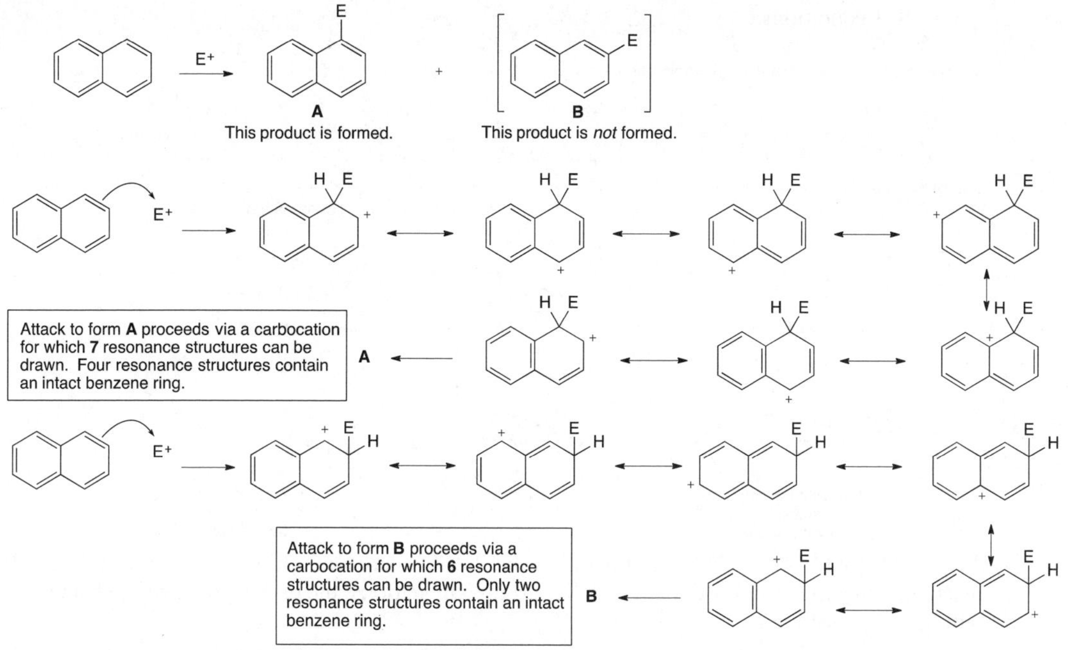

A reaction that occurs by way of the more stable carbocation is preferred, so product **A** is formed.

18.61

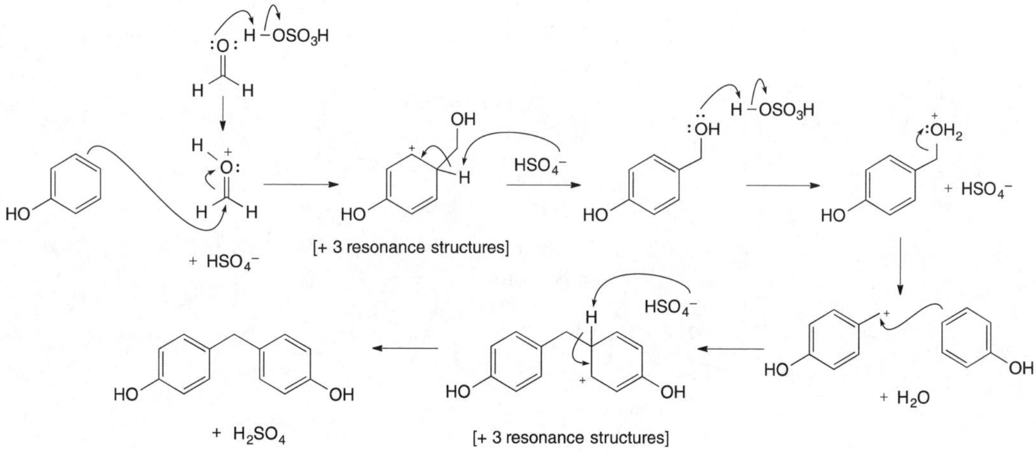

[+ 3 resonance structures]

[+ 3 resonance structures]

Chapter 18–28

18.62 Benzyl bromide forms a resonance-stabilized intermediate that allows it to react rapidly under S_N1 conditions.

Formation of a resonance-stabilized carbocation:

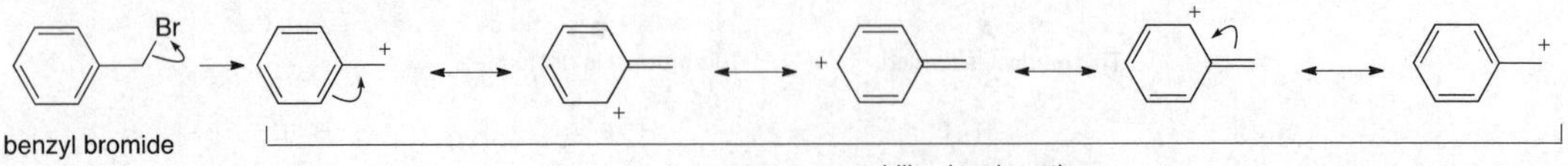

benzyl bromide

resonance-stabilized carbocation

benzyl methyl ether

$+ \; CH_3\ddot{O}H$

$+ \; :\ddot{Br}:^-$

$+ \; HBr$

electron-withdrawing group
destabilizing
slower reaction

electron-donor group
stabilizing
faster reaction

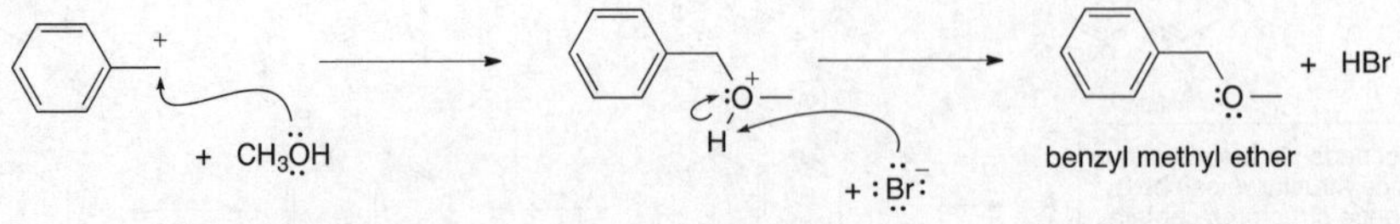

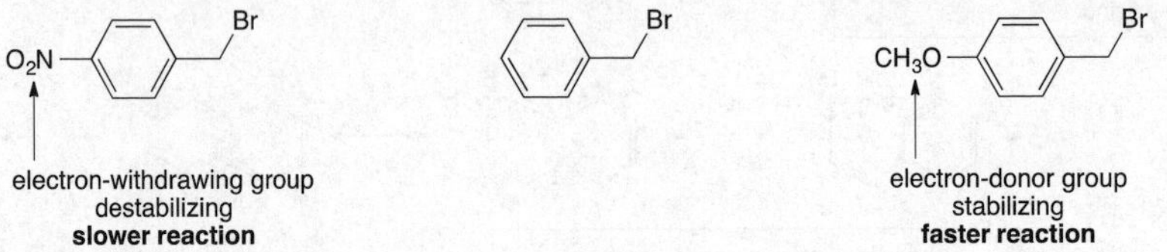

The electron-withdrawing NO_2 group will destabilize the carbocation, so the benzylic halide will be *less* reactive, while the electron-donating OCH_3 group will stabilize the carbocation, so the benzylic halide will be *more* reactive.

18.63

a.

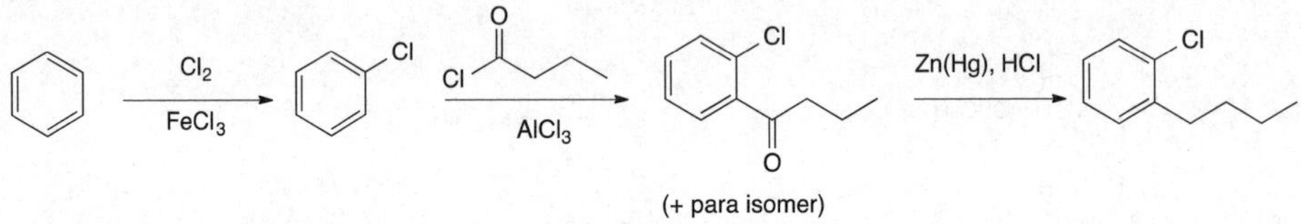

(+ para isomer)

b.

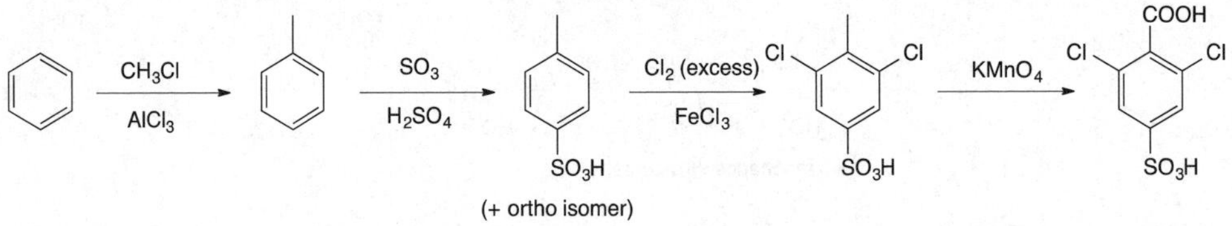

(+ ortho isomer)

c.

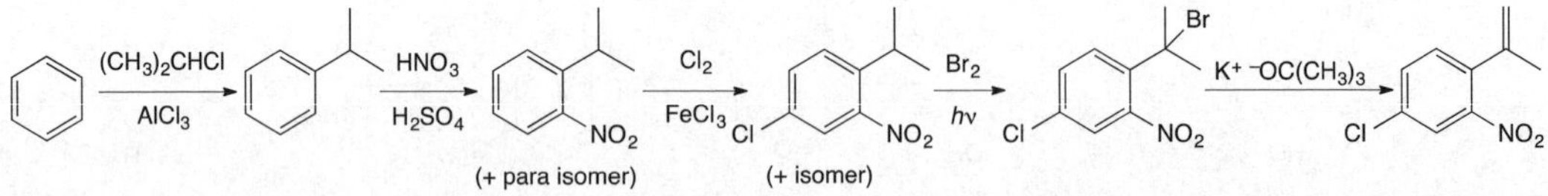

(+ para isomer)

(+ isomer)

d.

(+ ortho isomer)

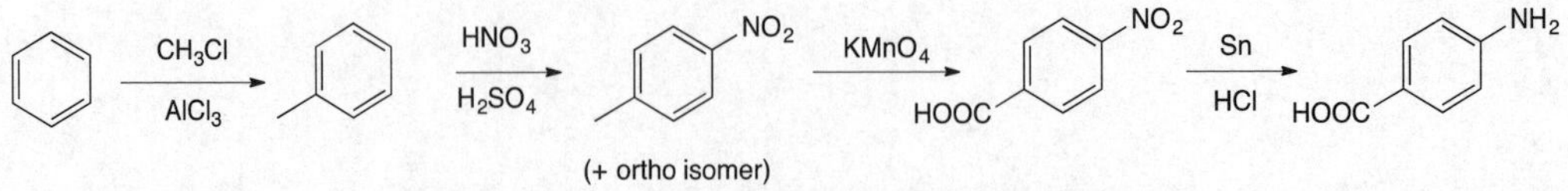

Reactions of Aromatic Compounds 18–29

e.

f.

g.

h.

18.64

a.

Chapter 18–30

b.

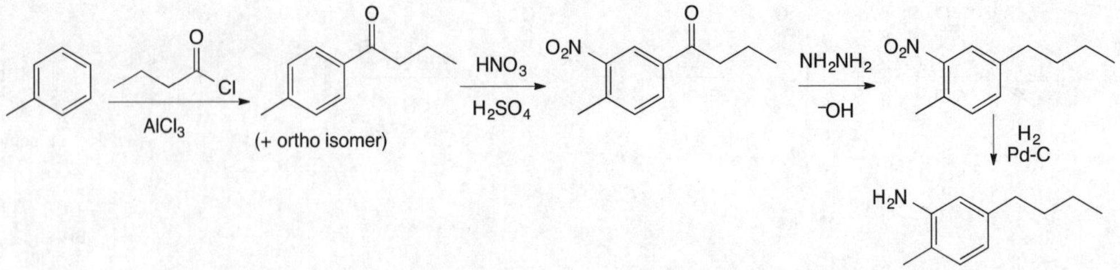

c.

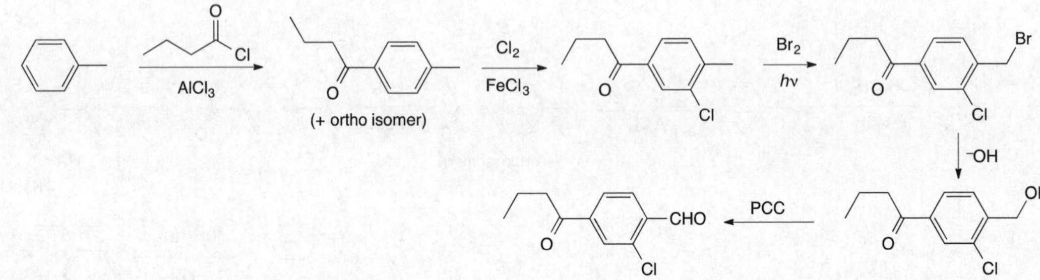

18.65

a.

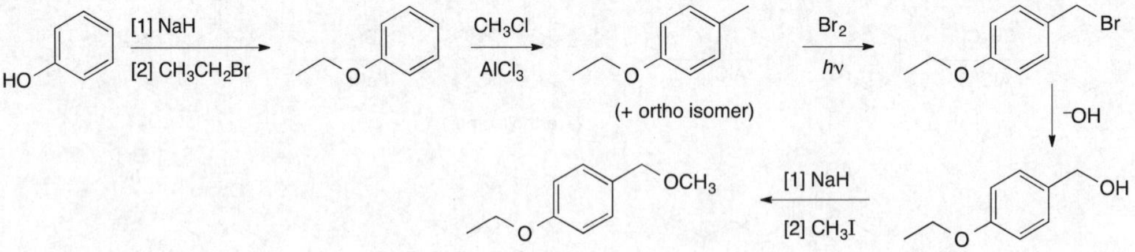

b.

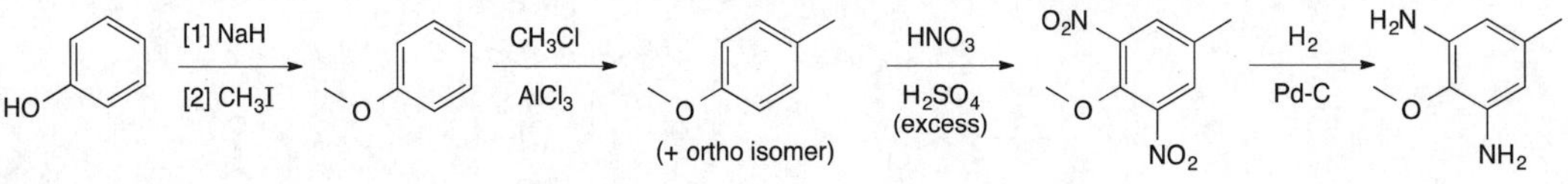

18.66

a.

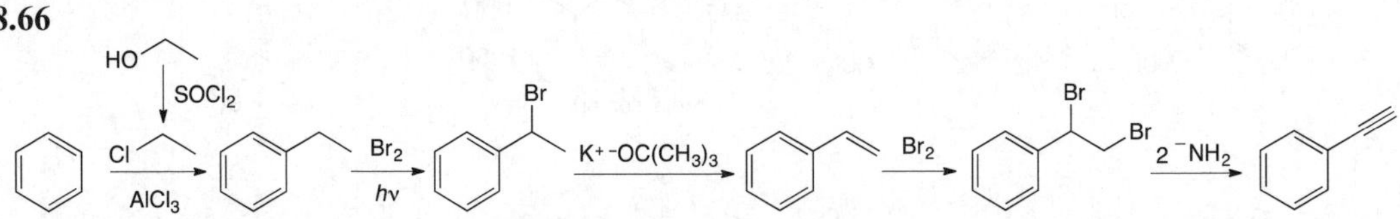

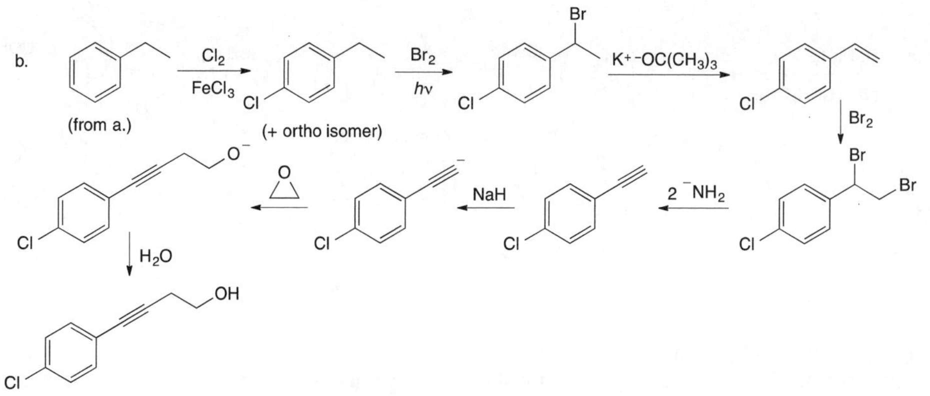

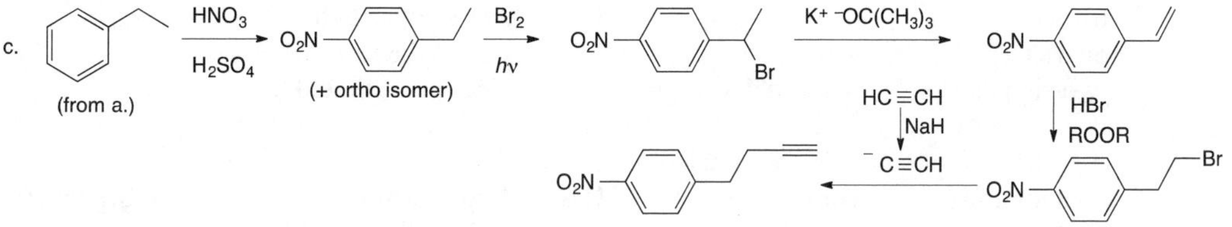

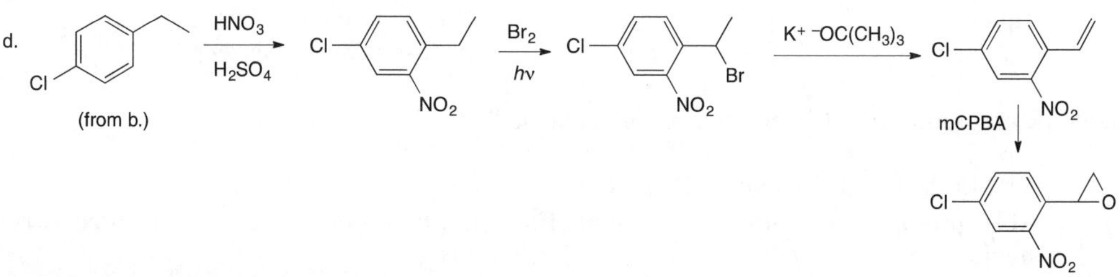

b.

(from a.)   (+ ortho isomer)

H_2O

c.

(from a.)   (+ ortho isomer)

d.

(from b.)

18.67 One possibility:

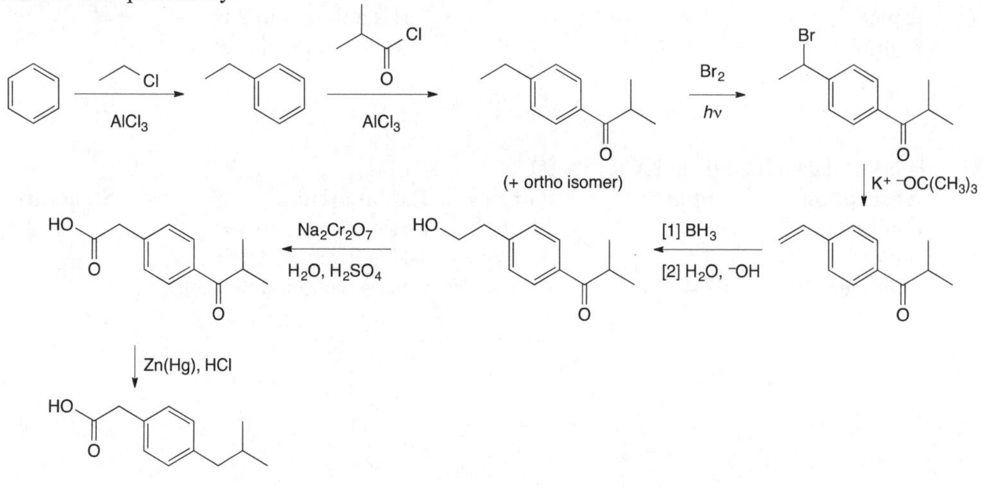

(+ ortho isomer)

ibufenac

Chapter 18–32

18.68

(+ ortho isomer)

X

18.69

^{1}H NMR data of compound **A** (C$_8$H$_9$Br):

Absorption	ppm	# of H's	Explanation	Structure:
triplet	1.2	3	3 H's adjacent to 2 H's	
quartet	2.6	2	2 H's adjacent to 3 H's	
two signals	7.1 and 7.4	4	para disubstituted benzene	

^{1}H NMR data of compound **B** (C$_8$H$_9$Br):

Absorption	ppm	# of H's	Explanation	Structure:
triplet	3.1	2	2 H's adjacent to 2 H's	
triplet	3.5	2	2 H's adjacent to 2 H's	
multiplet	7.1–7.4	5	monosubstituted benzene	

18.70 IR absorption at 1717 cm^{-1} means compound **C** has a C=O.

^{1}H NMR data of compound **C** (C$_{10}$H$_{12}$O):

Absorption	ppm	# of H's	Explanation	Structure:
singlet	2.1	3	3 H's	
triplet	2.8	2	2 H's adjacent to 2 H's	
triplet	2.9	2	2 H's adjacent to 2 H's	
multiplet	7.1–7.4	5	monosubstituted benzene	

18.71

^{1}H NMR data of compound **X** (C$_{10}$H$_{12}$O):

Absorption	ppm	# of H's	Explanation	Structure:
doublet	1.3	6	6 H's adjacent to 1 H	
septet	3.5	1	1 H adjacent to 6 H's	
multiplet	7.4–8.1	5	monosubstituted benzene	

Reactions of Aromatic Compounds 18–33

^{1}H NMR data of compound **Y** ($C_{10}H_{14}$):

Absorption	ppm	# of H's	Explanation	Structure:
doublet	0.9	6	6 H's adjacent to 1 H	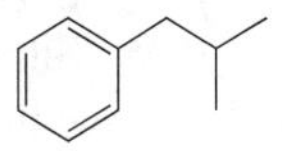
multiplet	1.8	1	1 H adjacent to many H's	
doublet	2.5	2	2 H's adjacent to 1 H	
multiplet	7.1–7.3	5	monosubstituted benzene	

18.72

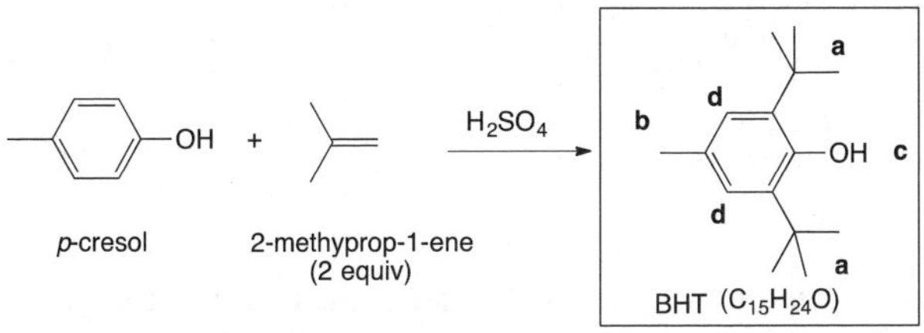

^{1}H NMR spectral data:

1.4 (singlet, 18 H) (a)
2.27 (singlet, 3 H) (b)
5.0 (singlet, 1 H) (c)
7.0 (singlet, 2 H) (d) ppm

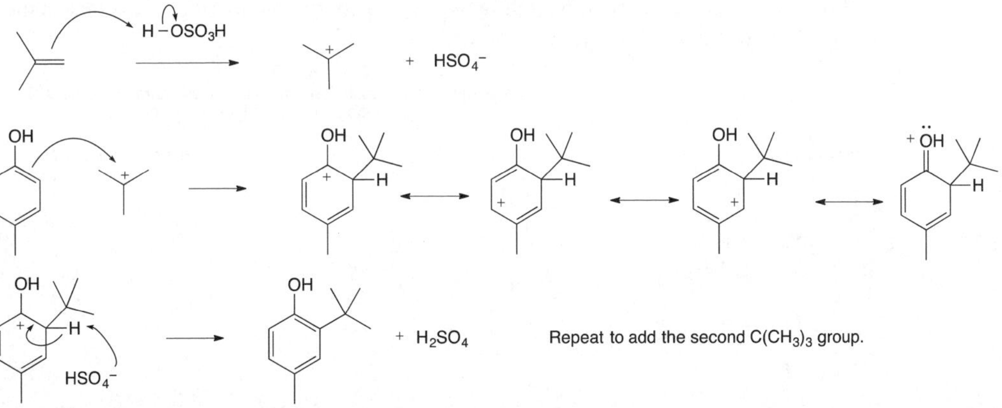

18.73

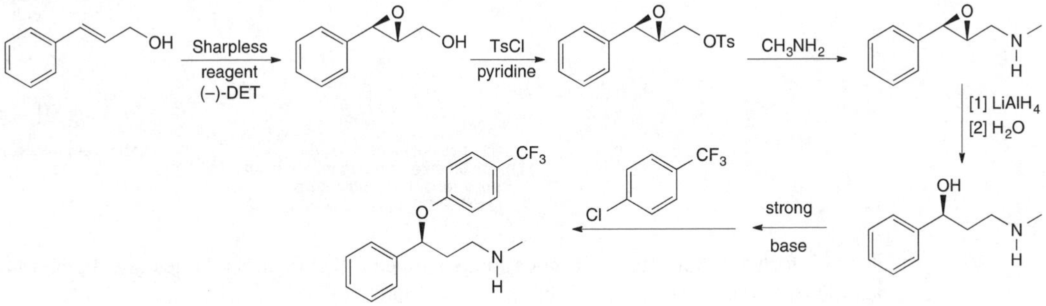

Chapter 18–34

18.74 Five resonance structures can be drawn for phenol, three of which place a negative charge on the ortho and para carbons. These illustrate that the electron density at these positions is increased, thus shielding the protons at these positions and shifting the absorptions to lower chemical shift. Similar resonance structures cannot be drawn with a negative charge at the meta position, so it is more deshielded and absorbs farther downfield, at higher chemical shift.

(–) charges on the ortho and para positions

18.75

a. Pyridine: The electron-withdrawing inductive effect of N makes the ring electron poor. Also, electrophiles E^+ can react with N, putting a positive charge on the ring. This makes the ring less reactive with another positively charged species.

To understand why substitution occurs at C3, compare the stability of the carbocation formed by attack at C2 and C3.

Electrophilic attack on N:

less reactive
than benzene

Electrophilic attack at C2:

N does not have an octet.
(+) charge on an electronegative N atom
poor resonance structure
Attack at C2 does not occur.

Electrophilic attack at C3:

better resonance structures

Attack at C3 forms a more stable carbocation, so attack at C3 occurs. Attack at C4 generates a carbocation of similar stability to attack at C2, so attack at C4 does *not* occur.

Reactions of Aromatic Compounds 18–35

b. Pyrrole is more reactive than benzene because the C's are more electron rich. The lone pair on N has an electron-donating resonance effect.

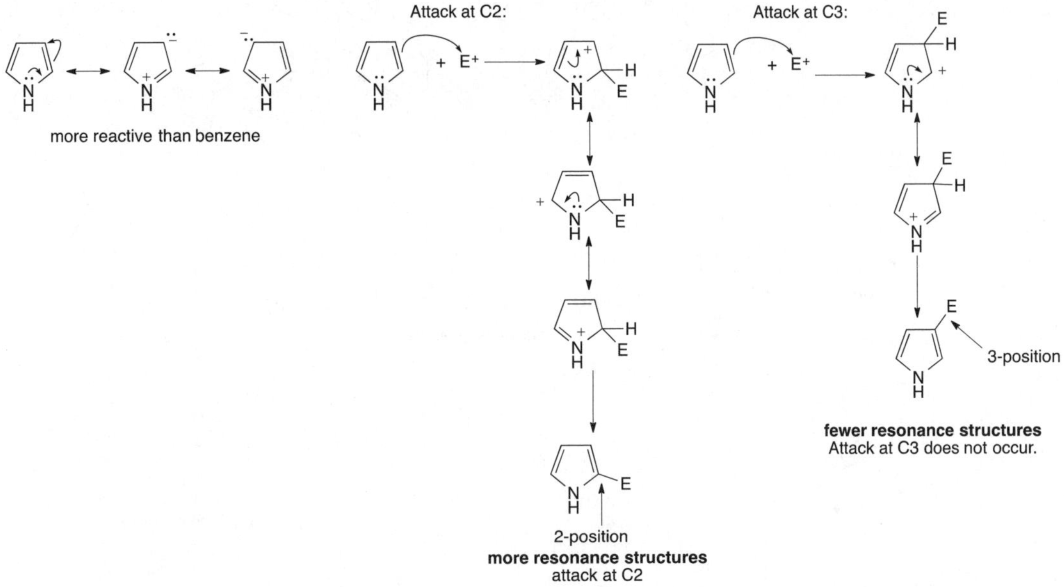

more reactive than benzene

Attack at C2 forms a more stable carbocation, so electrophilic substitution occurs at C2.

18.76

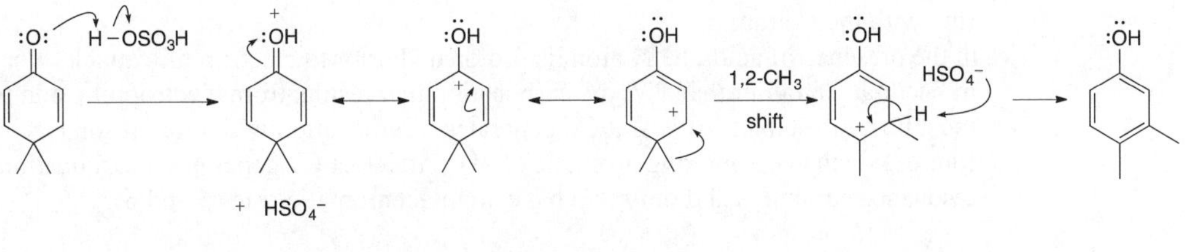

Chapter 18–36

18.77 Draw a stepwise mechanism for the following intramolecular reaction, which was used in the synthesis of the female sex hormone estrone.

overall reaction

Lewis acid

or HA

A

The steps:

+ :A⁻

+ H₂Ö:

(+ 1 resonance structure)

H−A +

A

A:⁻

H−A +

A

(+ 3 resonance structures)

18.78

a. In quinoline the lone pair on N occupies an sp^2 hybrid orbital, so it can never be donated to the ring by resonance. The N atom decreases the electron density of the ring in which it is located by an electron-withdrawing inductive effect, so substitution occurs on the other ring. In indole, the N atom donates its electron pair (which is contained in a p orbital) to the five-membered ring, increasing its electron density, so substitution occurs on the five-membered ring with the N atom.

b. In the presence of acid, the N atom is protonated prior to electrophilic attack. For substitution to occur at C8 rather than C7, the carbocation that results from electrophilic addition at C8 must be more stable. Attack at C8 generates a carbocation with more resonance structures, four of which keep one ring aromatic (**1–4**). Attack at C7 generates a carbocation with fewer resonance structures and only two have an intact aromatic ring (**5** and **6**).

c. With indole, attack at C3 forms the more highly resonance-stabilized carbocation.

Chapter 18–38

Attack at C3 forms resonance structures, all of which have an intact aromatic ring, and two of which have all atoms with octets. Attack at C2 forms a cation with more resonance structures, but only two have an intact aromatic ring, and only one has complete octets.

18.79

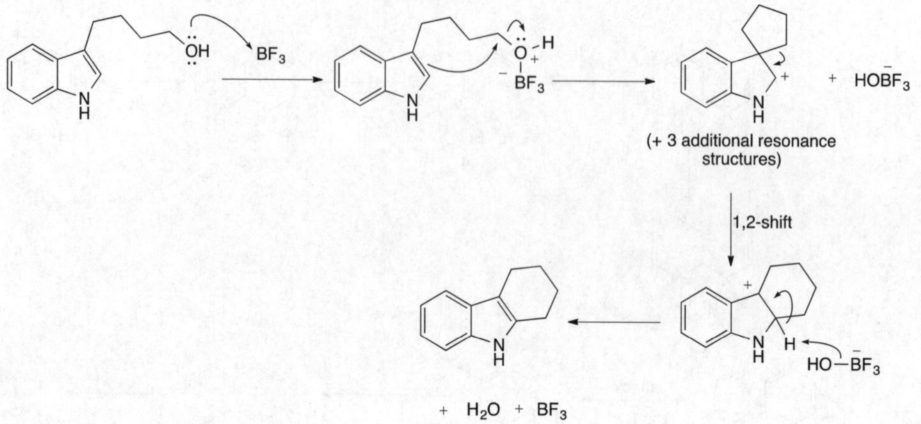

(+ 3 additional resonance structures)

1,2-shift

+ H₂O + BF₃

Chapter 19 Carboxylic Acids and the Acidity of the O–H Bond

Chapter Review

General facts

- Carboxylic acids contain a carboxy group (COOH). The central carbon is sp^2 hybridized and trigonal planar (19.1).
- Carboxylic acids are identified by the suffixes *-oic acid, carboxylic acid,* or *-ic acid* (19.2).
- Carboxylic acids are polar compounds that exhibit hydrogen bonding interactions (19.3).

Summary of spectroscopic absorptions (19.4)

IR absorptions	C=O	~1710 cm^{-1}
	O–H	3500–2500 cm^{-1} (very broad and strong)
^{1}H NMR absorptions	O–H	10–12 ppm (highly deshielded proton)
	C–H α to COOH	2–2.5 ppm (somewhat deshielded Csp^3–H)
^{13}C NMR absorption	C=O	170–210 ppm (highly deshielded carbon)

General acid–base reaction of carboxylic acids (19.9)

- Carboxylic acids are especially acidic because carboxylate anions are resonance stabilized.
- For equilibrium to favor the products, the base must have a conjugate acid with a pK_a > 5. Common bases are listed in Table 19.3.

Factors that affect acidity

Resonance effects. A carboxylic acid is more acidic than an alcohol or phenol because its conjugate base is more effectively stabilized by resonance (19.9).

ROH
pK_a = 16–18

pK_a = 10

R–COOH
pK_a ≈ 5

Increasing acidity

Inductive effects. Acidity increases with the presence of electron-withdrawing groups (like the electronegative halogens) and decreases with the presence of electron-donating groups (like polarizable alkyl groups) (19.10).

Chapter 19–2

Substituted benzoic acids.
- Electron-donor groups (D) make a substituted benzoic acid *less* acidic than benzoic acid.
- Electron-withdrawing groups (W) make a substituted benzoic acid *more* acidic than benzoic acid.

less acidic
higher pK_a

pK_a > 4.2

pK_a = 4.2

more acidic
lower pK_a

pK_a < 4.2

Increasing acidity

Other facts
- Extraction is a useful technique for separating compounds having different solubility properties. Carboxylic acids can be separated from other organic compounds by extraction, because aqueous base converts a carboxylic acid into a water-soluble carboxylate anion (19.12).
- A sulfonic acid (RSO_3H) is a strong acid because it forms a weak, resonance-stabilized conjugate base on deprotonation (19.13).
- Amino acids have an amino group on the α carbon to the carboxy group [$RCH(NH_2)COOH$]. Amino acids exist as zwitterions at pH ≈ 6. Adding acid forms a species with a net (+1) charge [$RCH(NH_3)COOH$]$^+$. Adding base forms a species with a net (−1) charge [$RCH(NH_2)COO$]$^−$ (19.14).

Practice Test on Chapter Review

1.a. Give the IUPAC name for the following compound.

b. Draw the structure corresponding to the following name: sodium *m*-bromobenzoate.

2.a. Which of the labeled atoms is least acidic?

1. H$_a$ 2. H$_b$ 3. H$_c$ 4. H$_d$ 5. H$_e$

b. Which of the following carboxylic acids has the lowest pK_a?

c. Which compound(s) can be converted to **A** by an oxidation reaction?

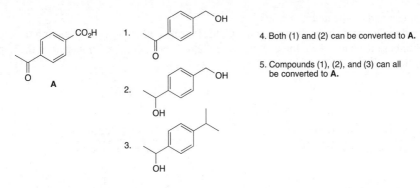

4. Both (1) and (2) can be converted to **A**.

5. Compounds (1), (2), and (3) can all be converted to **A**.

3. Rank the following compounds in order of increasing basicity. Label the *least* basic compound as **1**, the *most* basic compound as **3**, and the compound of *intermediate* basicity as **2**.

        A   B   C

4. Draw the organic products formed in each of the following reactions.

a. [structure: CH₃, H, C, NH₃⁺, COOH] + NaOH (1 equiv) →

b. HO—⟨⟩—CH₂CH₂—COOH + NaH (1 equiv) →

c. [structure: phenol with CH₂CH₂N(CH₃)H side chain] [1] NaH [2] CH₃I →

Answers to Practice Test

1.a. *cis*-2-methylcyclopentanecarboxylic acid

2.a. 1
  b. 5
  c. 5

3. A–2
  B–1
  C–3

4.a. [structure: CH₃, H, C, NH₃⁺, COO⁻]

b. HO—⟨⟩—CH₂CH₂—COO⁻

c. [structure: benzene with OCH₃ and CH₂CH₂N(CH₃)₂ groups]

1.b. [structure: 3-bromobenzoate, COO⁻Na⁺, Br]

Chapter 19–4

Answers to Problems

19.1   To name a carboxylic acid:
[1] Find the longest chain containing the COOH group and change the -*e* ending to -*oic acid*.
[2] Number the chain to put the COOH carbon at C1, but omit the number from the name.
[3] Follow all other rules of nomenclature.

a.

Number the chain to put COOH at C1.
6 carbon chain = **hexanoic acid**
3,3-dimethylhexanoic acid

c.

Number the chain to put COOH at C1.
6 carbon chain = **hexanoic acid**
2,4-diethylhexanoic acid

b.

Number the chain to put COOH at C1.
5 carbon chain = **pentanoic acid**
3-bromo-4-chloro-2-fluoropentanoic acid

d.

Number the chain to put COOH at C1.
9 carbon chain = **nonanoic acid**
4-isopropyl-6,8-dimethylnonanoic acid

19.2

a. 2-bromo**butanoic acid**

c. 3,3,4-trimethyl**heptanoic acid**

e. 3,4-diethyl**cyclohexanecarboxylic acid**

b. 2,3-dimethyl**pentanoic acid**

d. 2-*sec*-butyl-4,4-diethyl**nonanoic acid**

f. 1-isopropyl**cyclobutane-carboxylic acid**

19.3

a. α-methoxy**valeric acid**

c. α,β-dimethyl**caproic acid**

b. β-phenyl**propionic acid**

d. α-chloro-β-methyl**butyric acid**

19.4

a. **lithium benzoate**

b. **sodium formate**

c. **potassium 2-methylpropanoate**

d. **sodium 4-bromo-6-ethyl-octanoate**

Carboxylic Acids and the Acidity of the O–H Bond 19–5

19.5

2-propyl**pentanoic acid**   sodium 2-propyl**pentanoate**

19.6   More polar molecules have a higher boiling point and are more water soluble.

least polar
lowest boiling point
least H$_2$O soluble

intermediate polarity
intermediate boiling point

most polar
highest boiling point
most H$_2$O soluble

19.7   Look for functional group differences to distinguish the compounds by IR. Besides sp^3 hybridized C–H bonds at 3000–2850 cm^{-1} (which all three compounds have), the following functional group absorptions are seen:

carboxylic acid
2 strong absorptions
~1710 (C=O)
~2500–3500 (OH) cm^{-1}

ester
1 strong absorption
~1700 (C=O) cm^{-1}

alcohol
1 strong absorption
~3600–3200 (OH) cm^{-1}

19.8

19.9

PGF$_{2\alpha}$
a prostaglandin

enantiomer

There are five tetrahedral stereogenic centers labeled with (*). Both double bonds can exhibit cis–trans isomerism. Therefore, there are $2^7 = 128$ stereoisomers.

Chapter 19–6

19.10 1° Alcohols are converted to carboxylic acids by oxidation reactions.

a. [structure: hexanoic acid] $\Longrightarrow$ [structure: hexanol]   c. [structure: cyclohexanecarboxylic acid] $\Longrightarrow$ [structure: cyclohexylmethanol]

b. [structure: 2-methylpropanoic acid] $\Longrightarrow$ [structure: 2-methylpropanol]

19.11

a. [structure: benzyl alcohol] $\xrightarrow[\text{H}_2\text{SO}_4,\ \text{H}_2\text{O}]{\text{Na}_2\text{Cr}_2\text{O}_7}$ [structure: benzoic acid]

A

c. O_2N—[structure: p-nitrotoluene] $\xrightarrow{\text{KMnO}_4}$ O_2N—[structure: p-nitrobenzoic acid]

C
(Any R group with benzylic H's
can be present para to NO$_2$.)

b. [structure: alkyne] $\xrightarrow[\text{[2] H}_2\text{O}]{\text{[1] O}_3}$ [structure: carboxylic acid]
  B   (2 equiv)

d. [structure: diol with OH ← 2° OH and 1° OH labels] $\xrightarrow[\text{H}_2\text{SO}_4,\ \text{H}_2\text{O}]{\text{CrO}_3}$ [structure: keto acid]

D

19.12

[mechanism structure] :O: R–C(OH) + H–A $\longrightarrow$:O: R–C(OH$_2^+$) + A:⁻
                    not resonance stabilized

Protonation on the hydroxy O gives a
product that is not resonance stabilized,
so this pathway does not occur.

19.13

a. [cyclohexane]–COOH $\xrightarrow{\text{NaOH}}$ [cyclohexane]–COO⁻ Na⁺
   + H$_2$O

c. [structure]–OH $\xrightarrow{\text{NaH}}$ [structure]–O⁻ Na⁺ + H$_2$

b. [structure]–OH $\xrightarrow{\text{NaOCH}_3}$ [structure]–O⁻ Na⁺
   + HOCH$_3$

d. [benzene]–COOH $\xrightarrow{\text{NaHCO}_3}$ [benzene]–COO⁻ Na⁺
   + H$_2$CO$_3$

19.14 CH$_3$COOH has a pK_a of 4.8. Any base having a conjugate acid with a pK_a higher than 4.8 can deprotonate it.

a. F⁻ pK_a (HF) = 3.2 **not strong enough**
b. (CH$_3$)$_3$CO⁻ pK_a [(CH$_3$)$_3$COH] = 18 **strong enough**
c. CH$_3$⁻ pK_a (CH$_4$) = 50 **strong enough**
d. ⁻NH$_2$ pK_a (NH$_3$) = 38 **strong enough**
e. Cl⁻ pK_a (HCl) = –7.0 **not strong enough**

19.15

Increasing acidity: $H_a < H_b < H_c$

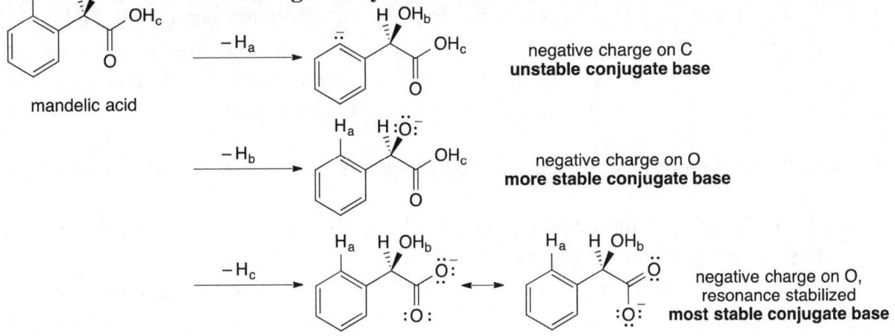

mandelic acid

$-H_a$ → negative charge on C **unstable conjugate base**

$-H_b$ → negative charge on O **more stable conjugate base**

$-H_c$ → negative charge on O, resonance stabilized **most stable conjugate base**

19.16 Electron-withdrawing groups make an acid more acidic, lowering its pK_a.

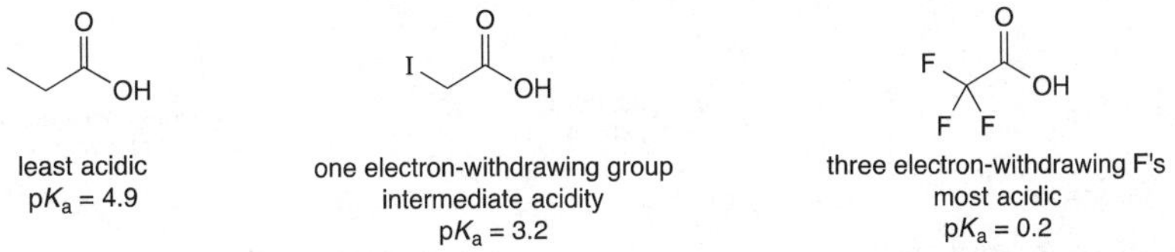

least acidic
$pK_a = 4.9$

one electron-withdrawing group
intermediate acidity
$pK_a = 3.2$

three electron-withdrawing F's
most acidic
$pK_a = 0.2$

19.17

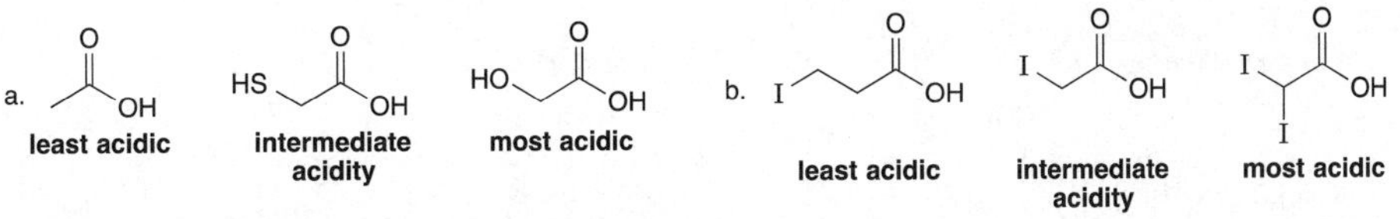

a. least acidic | intermediate acidity | most acidic

b. least acidic | intermediate acidity | most acidic

19.18

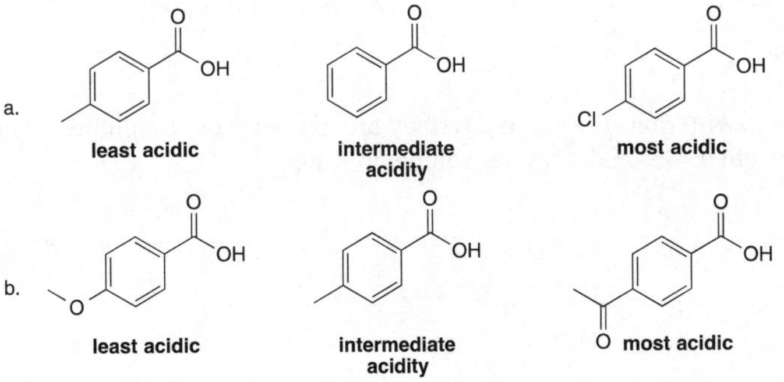

a. least acidic | intermediate acidity | most acidic

b. least acidic | intermediate acidity | most acidic

Chapter 19–8

19.19

Phenol **A** has a higher pK_a than phenol because of its substituents. Both OH and $C_{15}H_{31}$ are electron-donating groups, which make the conjugate base less stable. Therefore, the acid is **less acidic**.

19.20 To separate compounds by an extraction procedure, they must have different solubility properties.
  a. $CH_3(CH_2)_6COOH$ and $CH_3CH_2CH_2CH_2CH=CH_2$: **YES**. The acid can be extracted into aqueous base, while the alkene will remain in the organic layer.
  b. $CH_3CH_2CH_2CH_2CH=CH_2$ and $(CH_3CH_2CH_2)_2O$: **NO**. Both compounds are soluble in organic solvents and insoluble in water. Neither is acidic enough to be extracted into aqueous base.
  c. $CH_3(CH_2)_6COOH$ and NaCl: one carboxylic acid, one salt: **YES**. The carboxylic acid is soluble in an organic solvent while the salt is soluble in water.
  d. NaCl and KCl: two salts: **NO**.

19.21

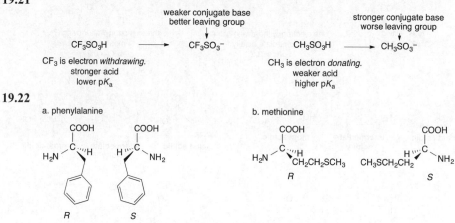

19.22

a. phenylalanine

b. methionine

19.23 Amino acids exist as zwitterions (i.e., salts), so they are too polar to be soluble in organic solvents like diethyl ether. Instead, they are soluble in water.

19.24

Carboxylic Acids and the Acidity of the O–H Bond 19–9

19.25

$$pI = \frac{pK_a(\text{COOH}) + pK_a(\text{NH}_3{}^+)}{2} = \frac{(2.58) + (9.24)}{2} = 5.91$$

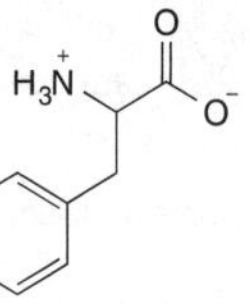

19.26

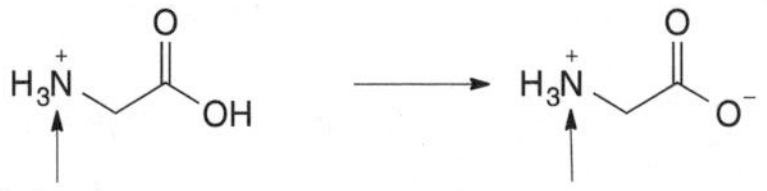

electron-withdrawing group

The nearby (+) stabilizes the conjugate base by an electron-withdrawing inductive effect, thus making the starting acid more acidic.

19.27

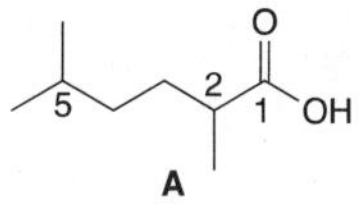

A

a. 2,5-dimethylhexanoic acid

b. NaOH

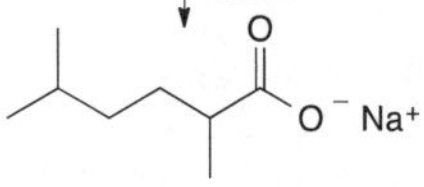

c. sodium 2,5-dimethylhexanoate

d. An alcohol or ether would have a much higher pK_a than a carboxylic acid.

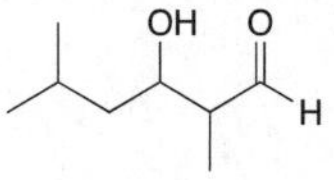

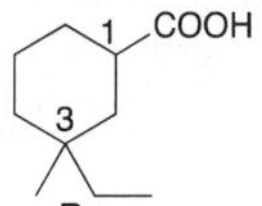

B

a. 3-ethyl-3-methylcyclohexanecarboxylic acid

b. NaOH

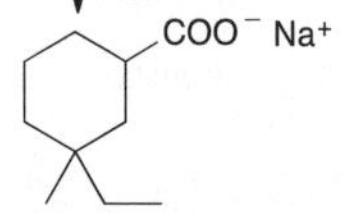

c. sodium 3-ethyl-3-methylcyclohexanecarboxylate

d.

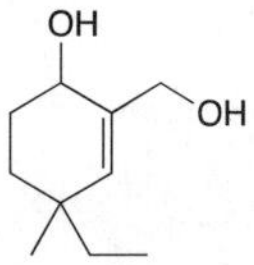

19.28

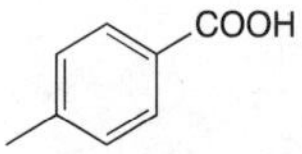

least acidic

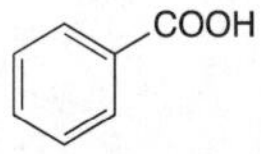

intermediate acidity

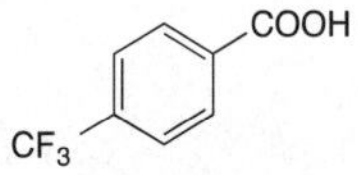

most acidic

Chapter 19–10

19.29 Use the directions from Answer 19.1 to name the compounds.

a. 4,4,5,5-tetramethyloctanoic acid

b. lithium 2-ethylpentanoate

c. 1-ethyl-3-isobutylcyclopentane-carboxylic acid

d. 5-bromo-4-ethyl-2-nitrobenzoic acid

e. sodium 2-methylhexanoate

f. 7-ethyl-5-isopropyl-3-methyldecanoic acid

19.30

a. 3,3-dimethylpentanoic acid

b. 4-chloro-3-phenylheptanoic acid

c. (R)-2-chloropropanoic acid

d. *m*-hydroxybenzoic acid

e. potassium acetate

f. sodium α-bromobutyrate

g. 2,2-dichloropentanedioic acid

h. 4-isopropyl-2-methyloctanedioic acid

19.31

a. OH — C2 or α carbon
CO₂H

IUPAC: 2-hydroxy**propanoic acid**
common: α-hydroxy**propionic acid**

b. C5 or δ carbon → OH
HO — CO₂H
↑ C3 or β carbon

IUPAC: 3,5-dihydroxy-3-methyl**pentanoic acid**
common: β,δ-dihydroxy-β-methyl**valeric acid**

19.32

lowest boiling point   intermediate boiling point   highest boiling point

Carboxylic Acids and the Acidity of the O–H Bond 19–11

19.33

a.

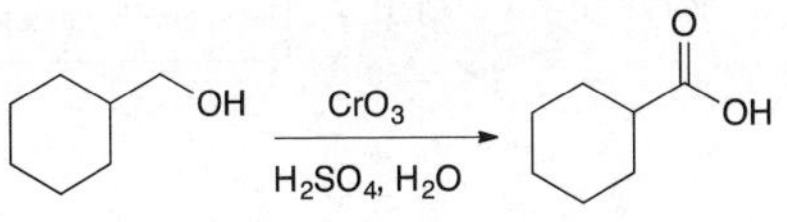

b.

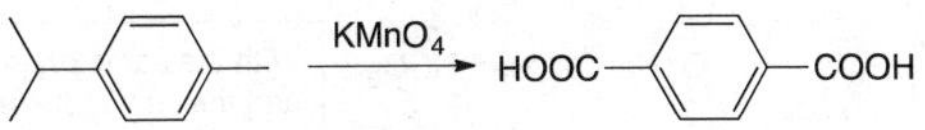

c.

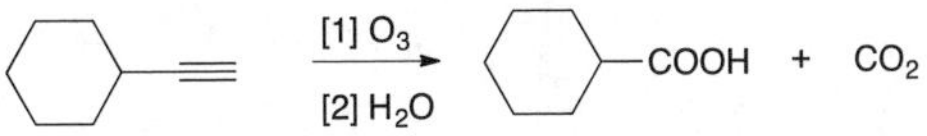

d.

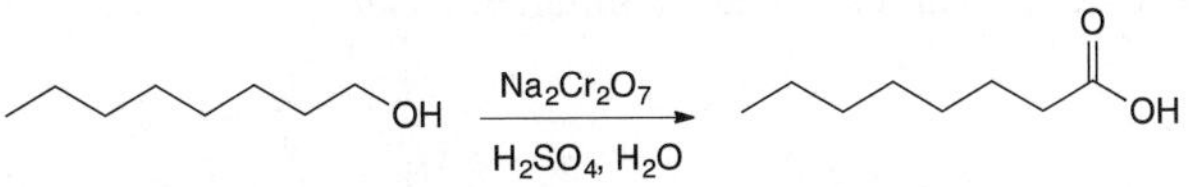

19.34

a.

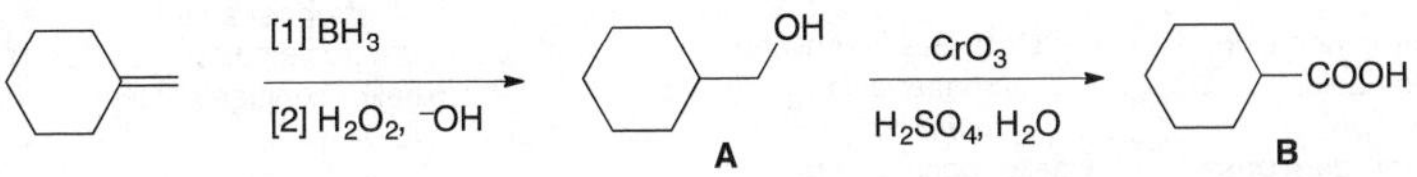

b.

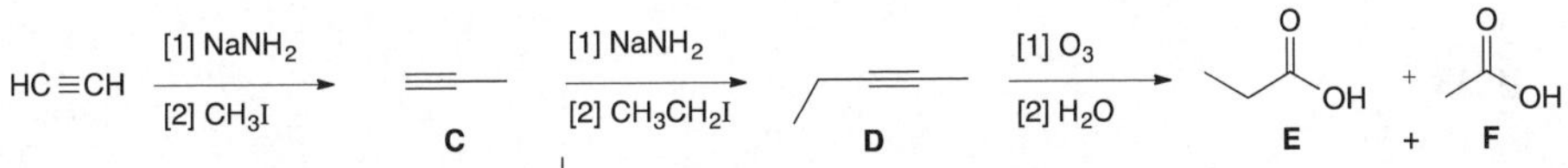

c.

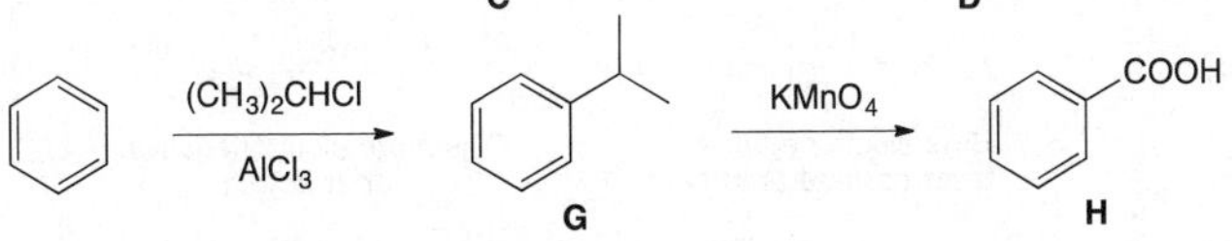

19.35

Bases: [1] $^-$OH pK_a (H$_2$O) = 15.7; [2] CH$_3$CH$_2^-$ pK_a (CH$_3$CH$_3$) = 50; [3] $^-$NH$_2$ pK_a (NH$_3$) = 38; [4] NH$_3$ pK_a (NH$_4^+$) = 9.4; [5] HC≡C$^-$ pK_a (HC≡CH) = 25.

a.

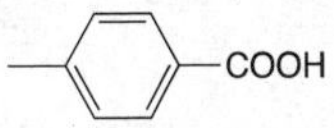

pK_a = 4.3
All of the bases
can deprotonate this RCO$_2$H.

b.

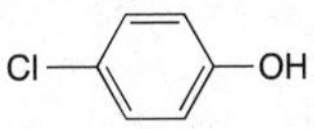

pK_a = 9.4
$^-$OH, CH$_3$CH$_2^-$, $^-$NH$_2$, and HC≡C$^-$
can deprotonate this phenol.

c. (CH$_3$)$_3$COH

pK_a = 18
CH$_3$CH$_2^-$, $^-$NH$_2$, and HC≡C$^-$
can deprotonate this ROH.

19.36

a.

pK_a ≈ 16

pK_a = 9.4

Reaction favors reactants.

b.

pK_a = 10

pK_a = 38

Reaction favors products.

Chapter 19–12

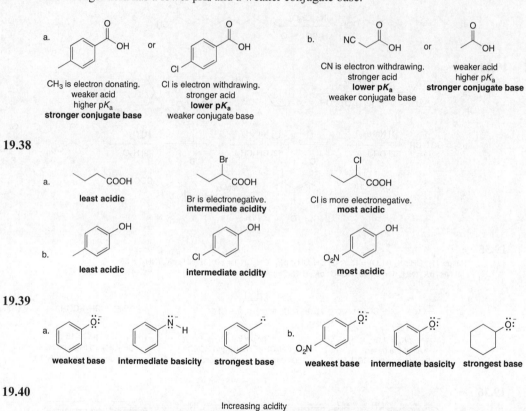

Carboxylic Acids and the Acidity of the O–H Bond 19–13

19.41 The OH of the phenol group in morphine is more acidic than the OH of the alcohol (p$K_a \approx 10$ versus p$K_a \approx 16$). KOH is basic enough to remove the phenolic OH, the most acidic proton.

most acidic proton
The OH is part of a phenol.
Methylation occurs here.

an alcohol

morphine

[1] KOH

Many resonance structures
stabilize the conjugate base.

codeine

[2] CH$_3$I

19.42 The closer the electron-withdrawing CH$_3$CO– group is to the carboxylic acid, the more it will stabilize the conjugate base, making the acid stronger.

pyruvic acid
stronger acid

farther away

acetoacetic acid
weaker acid

19.43

a. The negative charge on the conjugate base of *p*-nitrophenol is delocalized on the NO$_2$ group, stabilizing the conjugate base, and making *p*-nitrophenol more acidic than phenol (where the negative charge is delocalized only around the benzene ring).

p-nitrophenol
pK_a = 7.2

two of the possible resonance structures for the conjugate base
[See part (b) for all the possible resonance structures.]

phenol
pK_a = 10

b. In the para isomer, the negative charge of the conjugate base is delocalized over both the benzene ring and onto the NO$_2$ group, whereas in the meta isomer it cannot be delocalized onto the NO$_2$ group. This makes the conjugate base from the para isomer more highly resonance stabilized, and the para-substituted phenol more acidic than its meta isomer.

Chapter 19–14

pK$_a$ = 7.2
p-nitrophenol

negative charge on
two O atoms
very good resonance structure
more stable conjugate base
stronger acid

pK$_a$ = 8.3
m-nitrophenol

19.44 A CH$_3$O group has an electron-withdrawing inductive effect and an electron-donating resonance effect. In 2-methoxyacetic acid, the OCH$_3$ group is bonded to an sp^3 hybridized C, so there is no way to donate electron density by resonance. The CH$_3$O group withdraws electron density because of the electronegative O atom, stabilizing the conjugate base, and making CH$_3$OCH$_2$COOH a stronger acid than CH$_3$COOH.

more acidic acid   more stable conjugate base

In p-methoxybenzoic acid, the CH$_3$O group is bonded to an sp^2 hybridized C, so it can donate electron density by a resonance effect. This destabilizes the conjugate base, making the starting material less acidic than C$_6$H$_5$COOH.

less acidic acid   like charges nearby
                                           less stable conjugate base

19.45 Phenol has a pK$_a$ of 10, making p-methylthiophenol (pK$_a$ = 9.53) the stronger acid. A substituent that increases the acidity of a phenol must withdraw electron density to stabilize the negative charge of the conjugate base. An electron-withdrawing substituent deactivates a benzene ring towards electrophilic aromatic substitution, making p-methylthiophenol less reactive than phenol.

stronger acid
less reactive in electrophilic
substitution

Carboxylic Acids and the Acidity of the O–H Bond 19–15

19.46

The O in **A** is more electronegative than the N in **C**, so there is a stronger electron-withdrawing inductive effect. This stabilizes the conjugate base of **A**, making **A** more acidic than **C**.

A

$pK_a = 3.2$

B

$pK_a = 3.9$

C

$pK_a = 4.4$

Since the O in **A** is closer to the COOH group than the O atom in **B**, there is a stronger electron-withdrawing inductive effect. This makes **A** more acidic than **B**.

19.47

D

$-H^+$

Since the benzene ring is bonded to the α carbon (not the carbonyl carbon), this compound is not much different than any alkyl-substituted carboxylic acid.
least acidic

E

$-H^+$

The electron-withdrawing inductive effect of the NO$_2$ group helps stabilize the COO$^-$ group.
intermediate acidity

C

$-H^+$

Since the NO$_2$ group is bonded to a benzene ring that is bonded directly to the carbonyl group, inductive effects and resonance effects stabilize the conjugate base. For example, a resonance structure can be drawn that places a (+) charge close to the COO$^-$ group.
most acidic

Two of the resonance structures for the conjugate base of **C**:

unlike charges nearby
stabilizing

19.48 a. The pK_{a1} of phthalic acid is lower than the pK_{a1} of isophthalic acid because the electron-withdrawing CO$_2$H is closer to the negatively charged CO$_2^-$ of the conjugate base.

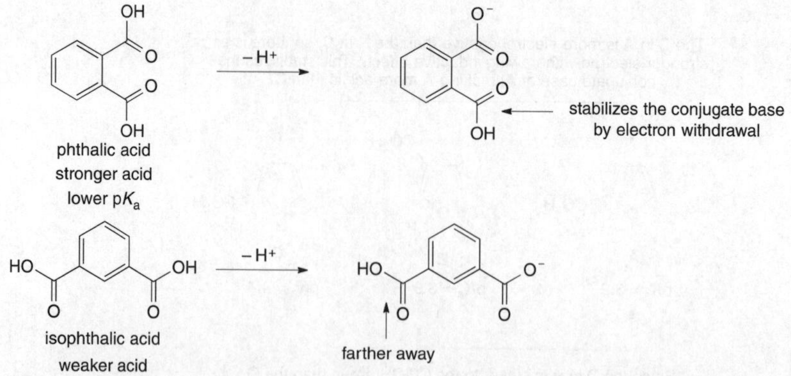

Chapter 19–16

OH

O

O

OH

phthalic acid
stronger acid
lower pK_a

— H$^+$ →

O$^-$

O

O

OH

← stabilizes the conjugate base
by electron withdrawal

HO

O

OH

O

isophthalic acid
weaker acid

— H$^+$ →

HO

O

O$^-$

O

farther away

b. After loss of the first proton, each compound now has one CO$_2$H group and one CO$_2^-$. The CO$_2^-$ is electron donating, so the closer it is located to the CO$_2$H, the more it destabilizes the resulting conjugate base.

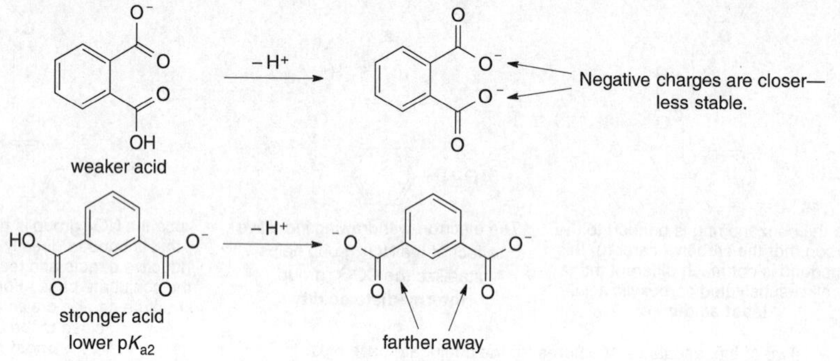

O$^-$

O

O

O

OH

weaker acid

— H$^+$ →

O

O$^-$

O$^-$

O

← Negative charges are closer—
less stable.

HO

O

O$^-$

O

stronger acid
lower pK_{a2}

— H$^+$ →

$^-$O

O

O$^-$

O

farther away

19.49

O

OH
*
labeled O atom

NaOH →

O

O$^-$
*

↕

O$^-$

O
*

The resonance-stabilized carboxylate anion can now be protonated on either O atom, the one with the label and the one without the label.

↓ H$_3$O$^+$

↓ H$_3$O$^+$

O

OH
*

OH

O
*

The label is now in two different locations.

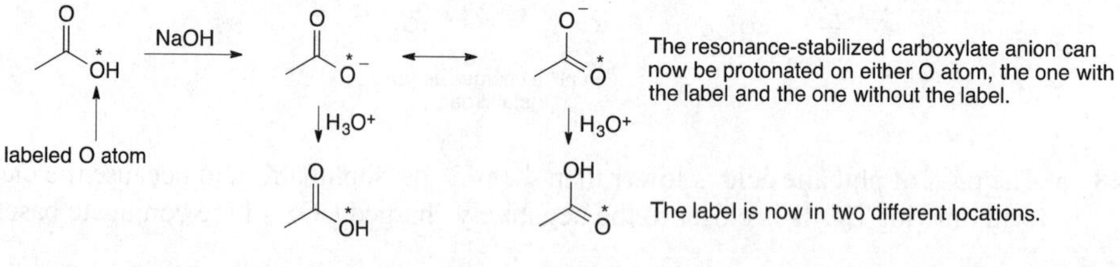

Carboxylic Acids and the Acidity of the O–H Bond 19–17

19.50

a.

cyclohexane-1,3-dione
increasing acidity: H$_b$ < H$_a$ < H$_c$

loss of H$_b$:

The most acidic proton forms the most stable conjugate base.

one Lewis structure
least stable conjugate base

loss of H$_a$:

2 resonance structures
intermediate stability

loss of H$_c$:

3 resonance structures
most stable conjugate base

b.

acetanilide
increasing acidity: H$_a$ < H$_c$ < H$_b$

loss of H$_b$:

7 resonance structures
most stable conjugate base

loss of H$_a$:

one Lewis structure
least stable conjugate base

loss of H$_c$:

2 resonance structures that delocalize the negative charge
intermediate stability

19.51

The conjugate base is resonance stabilized.
Two of the structures place a negative charge on an O atom.
weaker conjugate base
stronger acid

The conjugate base has only one Lewis structure.
stronger conjugate base
weaker acid

19.52

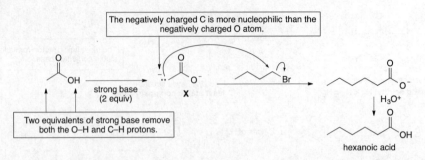

19.53

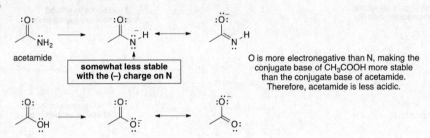

19.54

- Dissolve both compounds in CH_2Cl_2.
- Add 10% $NaHCO_3$ solution. This makes a carboxylate anion ($C_{10}H_7COO^-$) from **B**, which dissolves in the aqueous layer. The other compound (**A**) remains in the CH_2Cl_2.
- Separate the layers.

Carboxylic Acids and the Acidity of the O–H Bond 19–19

19.55

19.56 To separate two compounds in an aqueous extraction, one must be water soluble (or be able to be converted into a water-soluble ionic compound by an acid–base reaction), and the other insoluble. Octan-1-ol has greater than 5 C's, making it insoluble in water. Octane is an alkane, so it, too, is insoluble in water. Neither compound is acidic enough to be deprotonated by a base in aqueous solution. Because their solubility properties are similar, they cannot be separated by an extraction procedure.

19.57

a. Molecular formula: $C_3H_5ClO_2$ ⟶ one double bond or ring
   IR: 3500–2500 cm^{-1}, 1714 cm^{-1} ⟶ C=O and O–H
   NMR data: 2.87 (triplet, 2 H), 3.76 (triplet, 2 H), and 11.8 (singlet, 1 H) ppm

b. Molecular formula: $C_8H_8O_3$ ⟶ 5 double bonds or rings
   IR: 3500–2500 cm^{-1}, 1688 cm^{-1} ⟶ C=O and O–H
   NMR data: 3.8 (singlet, 3 H), 7.0 (doublet, 2 H), 7.9 (doublet, 2 H), and 12.7 (singlet, 1 H) ppm

   para disubstituted benzene ring

c. Molecular formula: $C_8H_8O_3$ ⟶ 5 double bonds or rings
   IR: 3500–2500 cm^{-1}, 1710 cm^{-1} ⟶ C=O and O–H
   NMR data: 4.7 (singlet, 2 H), 6.9–7.3 (multiplet, 5 H), and 11.3 (singlet, 1 H) ppm

   monosubstituted benzene ring

Chapter 19–20

19.58

Compound **A:** Molecular formula $C_4H_8O_2$ (one degree of unsaturation)
IR absorptions at 3600–3200 (O–H), 3000–2800 (C–H), and 1700 (C=O) cm^{-1}
^{1}H NMR data:

Absorption	ppm	# of H's	Explanation	Structure:
singlet	2.2	3	a CH$_3$ group	
singlet	2.55	1	1 H adjacent to none or OH	
triplet	2.7	2	2 H's adjacent to 2 H's	
triplet	3.8	2	2 H's adjacent to 2 H's	

Compound **B:** Molecular formula $C_4H_8O_2$ (one degree of unsaturation)
IR absorptions at 3500–2500 (O–H) and 1700 (C=O) cm^{-1}
^{1}H NMR data:

Absorption	ppm	# of H's	Explanation	Structure:
doublet	1.0	6	6 H's adjacent to 1 H	
septet	2.3	1	1 H adjacent to 6 H's	
singlet (very broad)	10.6	1	OH of RCOOH	

19.59

Compound **C:** Molecular formula $C_4H_8O_3$ (one degree of unsaturation)
IR absorptions at 3600–2500 (O–H) and 1734 (C=O) cm^{-1}
^{1}H NMR data:

Absorption	ppm	# of H's	Explanation	Structure:
triplet	1.3	3	a CH$_3$ group adjacent to 2 H's	
quartet	3.6	2	2 H's adjacent to 3 H's	
singlet	4.2	2	2 H's	
singlet	11.3	1	OH of COOH	

19.60

Compound **D:** Molecular formula $C_9H_9ClO_2$ (five degrees of unsaturation)
^{13}C NMR data: 30, 36, 128, 130, 133, 139, 179 = 7 different types of C's
^{1}H NMR data:

Absorption	ppm	# of H's	Explanation	Structure:
triplet	2.7	2	2 H's adjacent to 2 H's	
triplet	2.9	2	2 H's adjacent to 2 H's	
two signals	7.2	4	on benzene ring	
singlet	11.7	1	OH of COOH	

Carboxylic Acids and the Acidity of the O–H Bond 19–21

19.61

A   3 different C's
Spectrum [2]: peaks at 27, 39, 186 ppm

B   5 different C's
Spectrum [1]: peaks at 14, 22, 27, 34, 181 ppm

C   4 different C's
Spectrum [3]: peaks at 22, 26, 43, 180 ppm

19.62

GBL: Molecular formula $C_4H_6O_2$ (two degrees of unsaturation)
IR absorption at 1770 (C=O) cm^{-1}
^{1}H NMR data:

Absorption	ppm	# of H's	Explanation	Structure:
multiplet	2.28	2	2 H's adjacent to several H's	
triplet	2.48	2	2 H's adjacent to 2 H's	
triplet	4.35	2	2 H's adjacent to 2 H's	**GBL**

19.63

threonine

2R,3S   2S,3S   2R,3R   2S,3R
naturally
occurring

19.64

proline   enantiomer   zwitterion

19.65

a. methionine

pH = 1   pH = 6
form at isoelectric point   pH = 11

b. serine

pH = 1   pH = 6
form at isoelectric point   pH = 11

19.66

a. cysteine $pI = \dfrac{pK_a(COOH) + pK_a(NH_3^+)}{2}$ $= (2.05) + (10.25) / 2 = \mathbf{6.15}$

b. methionine $pI = \dfrac{pK_a(COOH) + pK_a(NH_3^+)}{2}$ $= (2.28) + (9.21) / 2 = \mathbf{5.75}$

Chapter 19–22

19.67

H₂N lysine
This lone pair is localized on the N atom, making it a base.

tryptophan
This lone pair is delocalized in the π system to give 10 π electrons, making it aromatic. This is similar to pyrrole (Chapter 17). Since these electrons are delocalized in the aromatic system, this N atom in tryptophan is not basic.

19.68

a. At pH = 1, the net charge is (+1).

b. increasing pH: As base is added, the most acidic proton is removed first, then the next most acidic proton, and so forth.

base (1 equiv)

base (2nd equiv)

base (3rd equiv)

c. monosodium glutamate

19.69 The first equivalent of NaH removes the most acidic proton—that is, the OH proton on the phenol. The resulting phenoxide can then act as a nucleophile to displace I to form a substitution product. With two equivalents, both OH protons are removed. In this case the more nucleophilic O atom is the stronger base—that is, the alkoxide derived from the alcohol (not the phenoxide), so this negatively charged O atom reacts first in a nucleophilic substitution reaction.

Carboxylic Acids and the Acidity of the O–H Bond 19–23

most acidic proton

HO—⟨benzene⟩—(CH₂)₃—OH

NaH
(1 equiv)

nucleophile

⁻O—⟨benzene⟩—(CH₂)₃—OH

[1] CH₃I
[2] H₂O

⟨methoxy⟩—⟨benzene⟩—(CH₂)₃—OH

NaH
(2 equiv)

⁻O—⟨benzene⟩—(CH₂)₃—O⁻

nucleophile

[1] CH₃I
[2] H₂O

HO—⟨benzene⟩—(CH₂)₃—O—CH₃

19.70

HO—⟨benzene⟩—COOH

p-hydroxybenzoic acid
less acidic than benzoic acid

The OH group donates electron density by its resonance effect and this destabilizes the conjugate base, making the acid less acidic than benzoic acid.

like charges on nearby atoms
destabilizing

⟨structure⟩ OH / COOH (*o*-hydroxybenzoic acid)

o-hydroxybenzoic acid
more acidic than benzoic acid

Intramolecular hydrogen bonding stabilizes the conjugate base, making the acid more acidic than benzoic acid.

19.71

2-hydroxybutanedioic acid
increasing acidity:
$H_d < H_c < H_b < H_e < H_a$

H_a and H_e must be the two most acidic protons because they are part of a carboxy group. Loss of a proton forms a resonance-stabilized carboxylate anion that has the negative charge delocalized on two O atoms. H_a is more acidic than H_e because the nearby OH group on the α carbon increases acidity by an electron-withdrawing inductive effect. H_b is the next most acidic proton because the conjugate base places a negative charge on the electronegative O atom, but it is not resonance stabilized.

The least acidic H's are H_c and H_d because these H's are bonded to C atoms. The electronegative O atom further acidifies H_c by an electron-withdrawing inductive effect.

Chapter 19–24

19.72

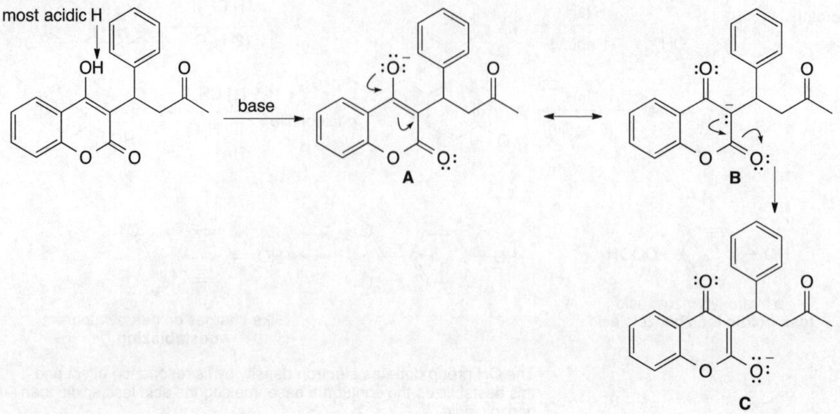

The conjugate base has three resonance structures, two of which place a negative charge on the oxygens. In this way the conjugate base resembles a carboxylate anion. In addition, the C=C's in **A** and **C** are conjugated.

Introduction to Carbonyl Chemistry 20–1

Chapter 20 Introduction to Carbonyl Chemistry; Organometallic Reagents; Oxidation and Reduction

Chapter Review

Reduction reactions

[1] Reduction of aldehydes and ketones to 1° and 2° alcohols (20.4)

R–C(=O)–R' (R' = H or alkyl) → [NaBH₄, CH₃OH or [1] LiAlH₄ [2] H₂O or H₂, Pd-C] → R–CH(OH)–R' (1° or 2° alcohol)

[2] Reduction of α,β-unsaturated aldehydes and ketones (20.4C)

R–C(=O)–CH=CH₂

- NaBH₄, CH₃OH → allylic alcohol • reduction of the C=O only
- H₂ (1 equiv), Pd-C → saturated ketone • reduction of the C=C only
- H₂ (excess), Pd-C → saturated alcohol • reduction of both π bonds

[3] Enantioselective ketone reduction (20.6)

PhC(=O)R → [1] (S)- or (R)- CBS reagent [2] H₂O → (R) 2° alcohol or (S) 2° alcohol
• A single enantiomer is formed.

[4] Reduction of acid chlorides (20.7A)

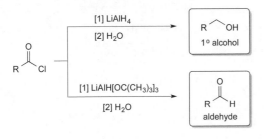

- LiAlH₄, a strong reducing agent, reduces an acid chloride to a 1° alcohol.
- With LiAlH[OC(CH₃)₃]₃, a milder reducing agent, reduction stops at the aldehyde stage.

Chapter 20–2

[5] Reduction of esters (20.7A)

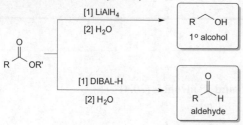

- LiAlH$_4$, a strong reducing agent, reduces an ester to a 1° alcohol.

- With DIBAL-H, a milder reducing agent, reduction stops at the aldehyde stage.

[6] Reduction of carboxylic acids to 1° alcohols (20.7B)

$$\text{RCOOH} \xrightarrow[\text{[2] H}_2\text{O}]{\text{[1] LiAlH}_4} \text{R-CH}_2\text{-OH} \quad (1° \text{ alcohol})$$

[7] Reduction of amides to amines (20.7B)

$$\text{R-C(O)-N(CH}_3)_2 \xrightarrow[\text{[2] H}_2\text{O}]{\text{[1] LiAlH}_4} \text{R-CH}_2\text{-N(CH}_3)_2 \quad \text{amine}$$

Oxidation reactions
Oxidation of aldehydes to carboxylic acids (20.8)

$$\text{RCHO} \xrightarrow[\text{or Ag}_2\text{O, NH}_4\text{OH}]{\text{CrO}_3, \text{Na}_2\text{Cr}_2\text{O}_7, \text{K}_2\text{Cr}_2\text{O}_7, \text{KMnO}_4} \text{RCOOH (carboxylic acid)}$$

- All Cr^{6+} reagents except PCC oxidize RCHO to RCOOH.
- Tollens reagent (Ag$_2$O + NH$_4$OH) oxidizes RCHO only. Primary (1°) and secondary (2°) alcohols do not react with Tollens reagent.

Preparation of organometallic reagents (20.9)

[1] Organolithium reagents: R–X + 2 Li ⟶ R–Li + LiX

[2] Grignard reagents: R–X + Mg $\xrightarrow{(\text{CH}_3\text{CH}_2)_2\text{O}}$ R–Mg–X

[3] Organocuprate reagents: R–X + 2 Li ⟶ R–Li + LiX

2 R–Li + CuI ⟶ R$_2$Cu$^-$ Li$^+$ + LiI

Introduction to Carbonyl Chemistry 20–3

[4] Lithium and sodium acetylides:

$$R-C\equiv C-H \xrightarrow{Na^+ \ ^-NH_2} R-C\equiv C^- \ Na^+ \text{ (a sodium acetylide)} + NH_3$$

$$R-C\equiv C-H \xrightarrow{R-Li} R-C\equiv C-Li \text{ (a lithium acetylide)} + R-H$$

Reactions with organometallic reagents

[1] Reaction as a base (20.9C)

$$R-M + H-\ddot{O}-R \longrightarrow R-H + M^+ + :\ddot{O}-R$$

- RM = RLi, RMgX, R$_2$CuLi
- This acid–base reaction occurs with H$_2$O, ROH, RNH$_2$, R$_2$NH, RSH, RCOOH, RCONH$_2$, and RCONHR.

[2] Reaction with aldehydes and ketones to form 1°, 2°, and 3° alcohols (20.10)

RCOR' $\xrightarrow{\text{[1] R"MgX or R"Li} \\ \text{[2] H}_2\text{O}}$ R(OH)(R')(R") (1°, 2°, or 3° alcohol)

R' = H or alkyl

[3] Reaction with esters to form 3° alcohols (20.13A)

RCOOR' $\xrightarrow{\text{[1] R"Li or R"MgX (2 equiv)} \\ \text{[2] H}_2\text{O}}$ R(OH)(R")(R") (3° alcohol)

[4] Reaction with acid chlorides (20.13)

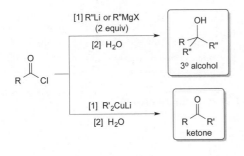

- More reactive organometallic reagents—R"Li and R"MgX—add two equivalents of R" to an acid chloride to form a 3° alcohol with two identical R" groups.

- Less reactive organometallic reagents—R'$_2$CuLi—add only one equivalent of R' to an acid chloride to form a ketone.

Chapter 20–4

[5] Reaction with carbon dioxide—Carboxylation (20.14A)

$$R-MgX \xrightarrow[\text{[2] } H_3O^+]{\text{[1] } CO_2} \underset{\text{carboxylic acid}}{R-\overset{\displaystyle O}{\underset{\displaystyle OH}{\|}}}$$

[6] Reaction with epoxides (20.14B)

$$\xrightarrow[\text{[2] } H_2O]{\text{[1] RLi, RMgX, or } R_2CuLi} \underset{\text{alcohol}}{}$$

[7] Reaction with α,β-unsaturated aldehydes and ketones (20.15B)

[1] R'Li or R'MgX
[2] H₂O
→ allylic alcohol

[1] R'₂CuLi
[2] H₂O
→ ketone

- More reactive organometallic reagents—R'Li and R'MgX—react with α,β-unsaturated carbonyls by 1,2-addition.

- Less reactive organometallic reagents—R'₂CuLi— react with α,β-unsaturated carbonyls by 1,4-addition.

Protecting groups (20.12)

[1] Protecting an alcohol as a *tert*-butyldimethylsilyl ether

$$R-O-H \ + \ Cl-Si \xrightarrow{\quad} R-O-Si$$

[Cl—TBDMS]

[R—O—TBDMS]
tert-butyldimethylsilyl ether

[2] Deprotecting a *tert*-butyldimethylsilyl ether to re-form an alcohol

$$R-O-Si \xrightarrow{Bu_4N^+F^-} R-O-H \ + \ F-Si$$

[R—O—TBDMS]

[F—TBDMS]

Introduction to Carbonyl Chemistry 20–5

Practice Test on Chapter Review

1. Which compounds undergo nucleophilic addition and which undergo substitution?

   a.    b. 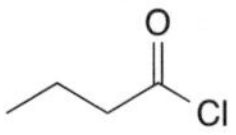   c. 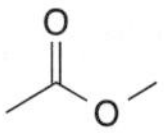   d.

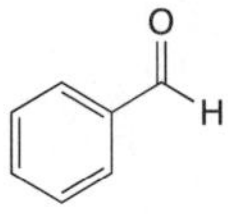

2. What product is formed when $CH_3CH_2CH_2Li$ reacts with each compound, followed by quenching with water and acid?

   a. CH_3CH_2CHO
   b. $(CH_3)_2CO$
   c. $CH_3CH_2CO_2CH_3$
   d. CH_3CH_2COCl

   e. CO_2
   f. $CH_2=CHCOCH_3$
   g. ethylene oxide
   h. CH_3COOH

3. What product is formed when $HO(CH_2)_4CHO$ is treated with each reagent?

   a. $NaBH_4$, CH_3OH
   b. PCC

   c. Ag_2O, NH_4OH
   d. $Na_2Cr_2O_7$, H_2SO_4, H_2O

4. What reagent is needed to convert $(CH_3CH_2)_2CHCOCl$ into each compound?

   a. $(CH_3CH_2)_2CHCOCH_2CH_3$
   b. $(CH_3CH_2)_2CHCHO$

   c. $(CH_3CH_2)_2CHC(OH)(CH_2CH_3)_2$
   d. $(CH_3CH_2)_2CHCH_2OH$

5. Draw the organic products formed in the following reactions.

   a. 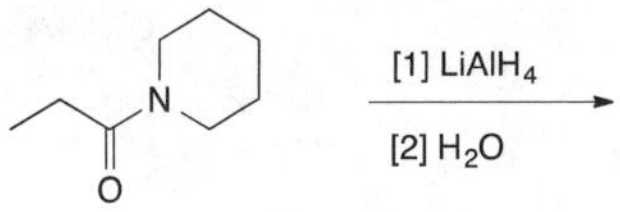 [1] LiAlH$_4$ [2] H$_2$O

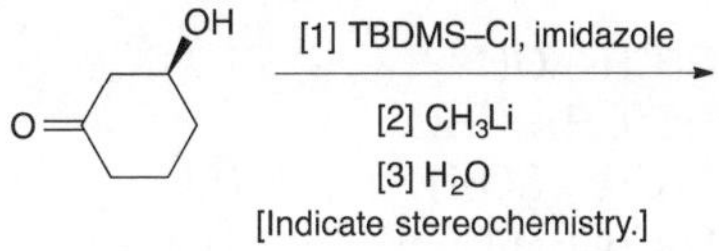

   b. 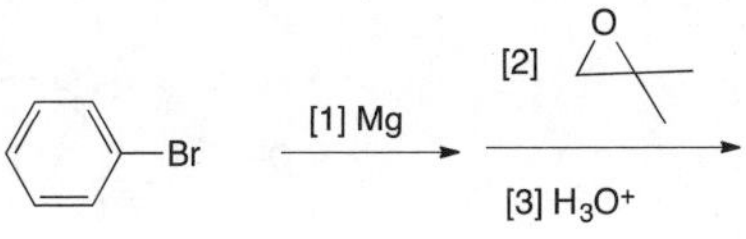 [1] Mg [2] (epoxide) [3] H$_3$O$^+$

   c. [1] TBDMS–Cl, imidazole [2] CH$_3$Li [3] H$_2$O [Indicate stereochemistry.]

   d. 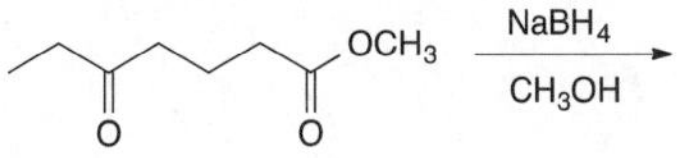 NaBH$_4$ CH$_3$OH

Chapter 20–6

6. What starting materials are needed to synthesize each compound using the indicated reagent or functional group?

a. Synthesize:

from an ester

c. Synthesize:

using a Grignard reagent

b. Synthesize:

using an organocuprate reagent

Answers to Practice Test

1.a. addition
   b. substitution
   c. substitution
   d. addition

2. a. $CH_3CH_2CH(OH)CH_2CH_2CH_3$
   b. $(CH_3)_2C(OH)CH_2CH_2CH_3$
   c. $CH_3CH_2C(OH)(CH_2CH_2CH_3)_2$
   d. $CH_3CH_2C(OH)(CH_2CH_2CH_3)_2$
   e. $CH_3CH_2CH_2CO_2H$
   f. $CH_2=CHC(OH)(CH_3)CH_2CH_2CH_3$
   g. $CH_3(CH_2)_4OH$
   h. $CH_3CH_2CH_3 + CH_3CO_2H$

3.a. $HO(CH_2)_5OH$
   b. $OHC(CH_2)_3CHO$
   c. $HO(CH_2)_4CO_2H$
   d. $HO_2C(CH_2)_3CO_2H$

4.a. $(CH_3CH_2)_2CuLi$
   b. $LiAlH[OC(CH_3)_3]_3$
   c. CH_3CH_2MgBr
   d. $LiAlH_4$

5.

a.

b.

c.

d.

6.

a.

+ CH_3MgBr

b.

c.

Answers to Problems

20.1

a.

[1] $C_{sp^2}-C_{sp^2}$

[3]

$C_{sp^2}-C_{sp^2}$

α-sinensal

[2] σ: $C_{sp^2}-O_{sp^2}$
    π: C_p-O_p

b. The O is sp^2 hybridized.
   Both lone pairs occupy sp^2 hybrid orbitals.

Introduction to Carbonyl Chemistry 20–7

20.2 A carbonyl compound with a leaving group (NR_2 or OR bonded to the C=O) undergoes substitution reactions. Those without leaving groups undergo addition.

no good leaving group
addition reactions

All other C=O's have leaving groups.
substitution reactions

20.3 Aldehydes are more reactive than ketones. In carbonyl compounds with leaving groups, the better the leaving group, the more reactive the carbonyl compound.

a. or

less hindered carbonyl
more reactive

c. or

better leaving group
more reactive

b. or

less hindered carbonyl
more reactive

d. or

better leaving group
more reactive

20.4 $NaBH_4$ reduces aldehydes to 1° alcohols and ketones to 2° alcohols.

a. $\xrightarrow[\text{CH}_3\text{OH}]{\text{NaBH}_4}$

c. $\xrightarrow[\text{CH}_3\text{OH}]{\text{NaBH}_4}$

b. $\xrightarrow[\text{CH}_3\text{OH}]{\text{NaBH}_4}$

20.5 1° Alcohols are prepared from aldehydes and 2° alcohols are from ketones.

a. $\Longrightarrow$ b. $\Longrightarrow$ c. $\Longrightarrow$

20.6

3° Alcohols cannot be made by reduction of a carbonyl group, because they do not contain a H on the C with the OH.

1-methylcyclohexanol

Chapter 20–8

20.7

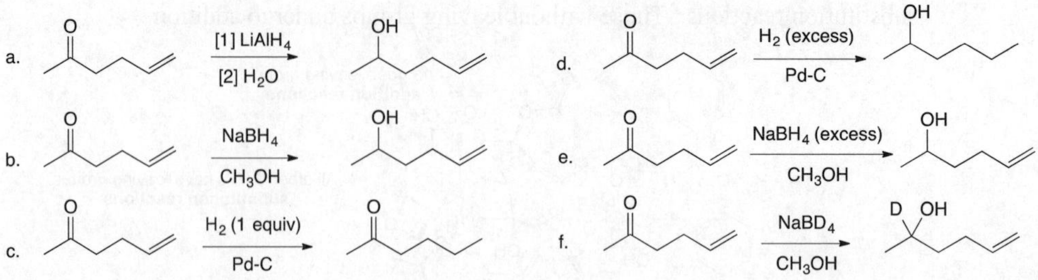

a. [1] LiAlH₄ / [2] H₂O

d. H₂ (excess) / Pd-C

b. NaBH₄ / CH₃OH

e. NaBH₄ (excess) / CH₃OH

c. H₂ (1 equiv) / Pd-C

f. NaBD₄ / CH₃OH

20.8

a. NaBH₄ / CH₃OH

b. NaBH₄ / CH₃OH

c. NaBH₄ / CH₃OH

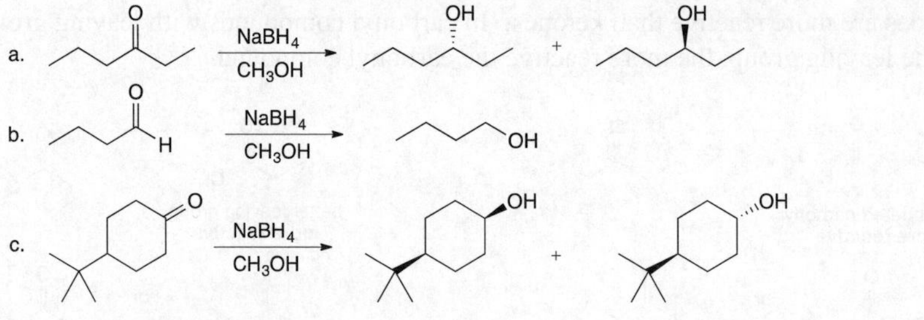

20.9 The 2° alcohol comes from a carbonyl group. Since hydride was delivered from the back side, the (*R*)-CBS reagent must be used.

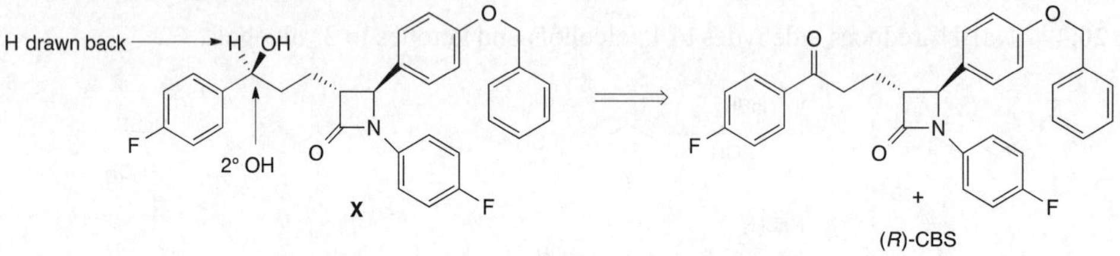

H drawn back ⟶ H OH

2° OH

X

(*R*)-CBS

20.10

Part [1]: Nucleophilic substitution of H for Cl

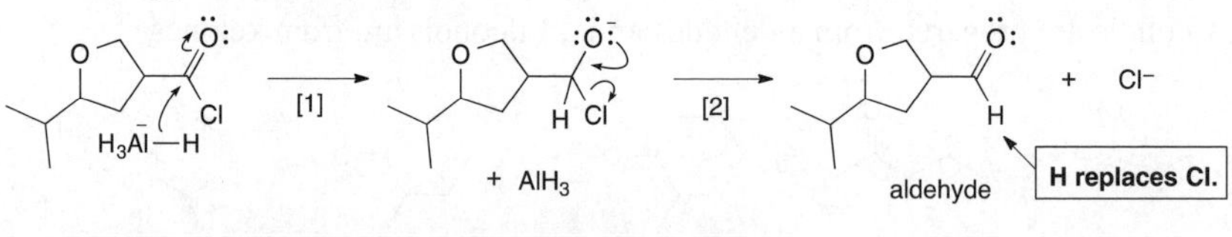

[1]

[2]

+ Cl⁻

+ AlH₃

aldehyde | H replaces Cl.

Part [2]: Nucleophilic addition of H⁻ to form an alcohol

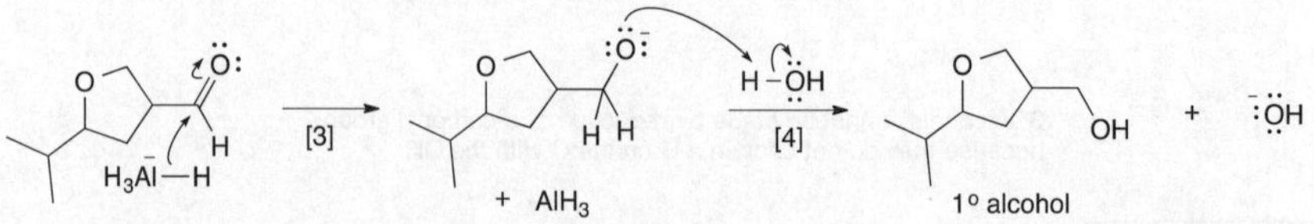

[3]

[4]

+ AlH₃

1° alcohol

+ ⁻OH

20.11 Acid chlorides and esters can be reduced to 1° alcohols. Keep the carbon skeleton the same in drawing an ester and acid chloride precursor.

Introduction to Carbonyl Chemistry 20–9

a. [structure: cyclopentylmethanol] ⟹ [structure: cyclopentanecarbonyl chloride] or [structure: methyl cyclopentanecarboxylate]

c. [structure: 4-methoxybenzyl alcohol] ⟹ [structure: 4-methoxybenzoyl chloride] or [structure: methyl 4-methoxybenzoate]

b. [structure: 2,3-dimethylbutanol] ⟹ [structure: 2,3-dimethylbutanoyl chloride] or [structure: methyl ester]

20.12

a. [structure with COOH] $\xrightarrow[\text{[2] H}_2\text{O}]{\text{[1] LiAlH}_4}$ [structure with OH]

c. [cyclohexyl amide] $\xrightarrow[\text{[2] H}_2\text{O}]{\text{[1] LiAlH}_4}$ [cyclohexylmethyl amine]

b. [amide with NH$_2$] $\xrightarrow[\text{[2] H}_2\text{O}]{\text{[1] LiAlH}_4}$ [amine with NH$_2$]

d. [structure: piperidinone] $\xrightarrow[\text{[2] H}_2\text{O}]{\text{[1] LiAlH}_4}$ [structure: piperidine]

20.13

a. [benzylamine] ⟹ [benzamide]

c. [structure] ⟹ [structure]

b. [cyclohexylmethyl diethylamine] ⟹ [amide] or [amide]

20.14

a. [keto ester with COOCH$_3$] $\xrightarrow[\text{[2] H}_2\text{O}]{\text{[1] LiAlH}_4}$ [diol] + HOCH$_3$

c. [CH$_3$O cyclohexanone] $\xrightarrow[\text{[2] H}_2\text{O}]{\text{[1] LiAlH}_4}$ [CH$_3$O cyclohexanol]

$\xrightarrow[\text{CH}_3\text{OH}]{\text{NaBH}_4}$ [hydroxy ester with COOCH$_3$]

$\xrightarrow[\text{CH}_3\text{OH}]{\text{NaBH}_4}$ [CH$_3$O cyclohexanol]

b. CH$_3$O [diacid/ester structure] OH $\xrightarrow[\text{[2] H}_2\text{O}]{\text{[1] LiAlH}_4}$ HO [diol] OH + HOCH$_3$

$\xrightarrow[\text{CH}_3\text{OH}]{\text{NaBH}_4}$ **Neither functional group reduced**

20.15 Tollens reagent reacts only with aldehydes.

a. [benzyl alcohol]
Ag$_2$O, NH$_4$OH → **No reaction**
$\xrightarrow[\text{H}_2\text{SO}_4, \text{H}_2\text{O}]{\text{Na}_2\text{Cr}_2\text{O}_7}$ [benzoic acid]

b. [hydroxy aldehyde]
Ag$_2$O, NH$_4$OH → [hydroxy acid]
$\xrightarrow[\text{H}_2\text{SO}_4, \text{H}_2\text{O}]{\text{Na}_2\text{Cr}_2\text{O}_7}$ [keto acid]

Chapter 20–10

20.16

a. **B** $\xrightarrow[\text{CH}_3\text{OH}]{\text{NaBH}_4}$

b. **B** $\xrightarrow[\text{[2] H}_2\text{O}]{\text{[1] LiAlH}_4}$

c. **B** $\xrightarrow{\text{PCC}}$

d. **B** $\xrightarrow{\text{Ag}_2\text{O, NH}_4\text{OH}}$

e. **B** $\xrightarrow[\text{H}_2\text{SO}_4,\ \text{H}_2\text{O}]{\text{CrO}_3}$

20.17

a. $CH_3CH_2Br + 2\ Li \longrightarrow CH_3CH_2Li + LiBr$

b. $CH_3CH_2Br + Mg \longrightarrow CH_3CH_2MgBr$

c. $CH_3CH_2Br + 2\ Li \longrightarrow CH_3CH_2Li + LiBr$

$2\ CH_3CH_2Li + CuI \longrightarrow LiCu(CH_2CH_3)_2 + LiI$

20.18

+ NaH $\longrightarrow$ Na$^+$ + H$_2$ ⟵ hydrogen gas

+ CH$_3$MgBr $\longrightarrow$ BrMg + CH$_4$ ⟵ methane gas

20.19 Organometallic compounds have a carbon–metal bond.

a. $BrMgC\equiv CCH_2CH_3$

organometallic
compound

b. $NaOCH_2CH_3$

NOT an
organometallic
compound

c. $KOC(CH_3)_3$

NOT an
organometallic
compound

d. PhLi =

organometallic
compound

20.20

a. Li + H$_2$O $\longrightarrow$ + LiOH

b. $\overset{\displaystyle}{}$MgBr + H$_2$O $\longrightarrow$ + HOMgBr

c. MgBr + H$_2$O $\longrightarrow$ + HOMgBr

d. Li + H$_2$O $\longrightarrow$ + LiOH

Introduction to Carbonyl Chemistry 20–11

20.21 To draw the products, add the alkyl or phenyl group to the carbonyl carbon and protonate the oxygen.

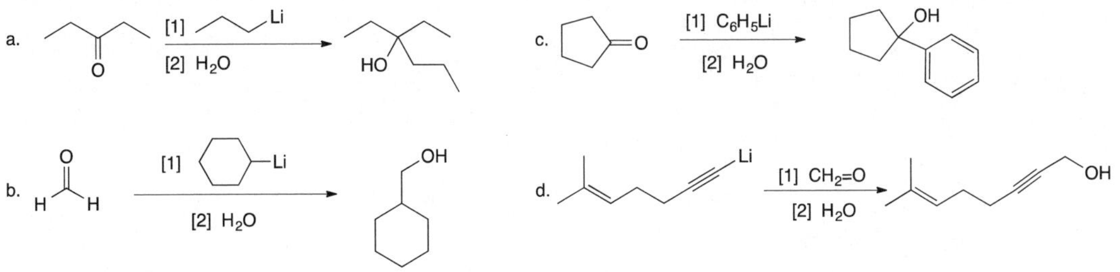

a. [1] (propyl)Li / [2] H_2O

b. [1] (cyclohexyl)Li / [2] H_2O

c. [1] C_6H_5Li / [2] H_2O

d. [1] $CH_2=O$ / [2] H_2O

20.22 Addition of RM always occurs from above and below the plane of the molecule.

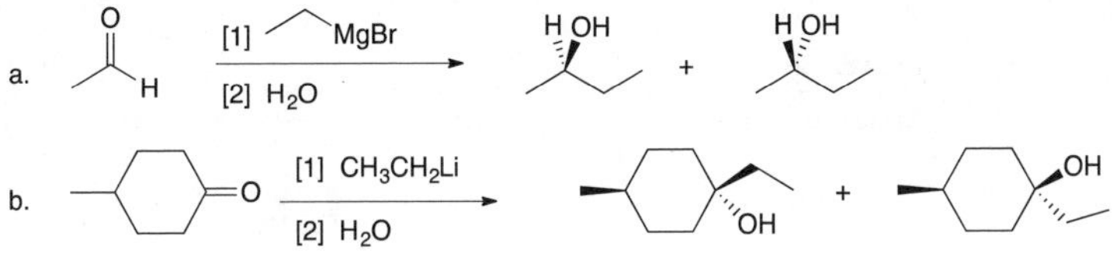

a. [1] (allyl)MgBr / [2] H_2O

b. [1] CH_3CH_2Li / [2] H_2O

20.23

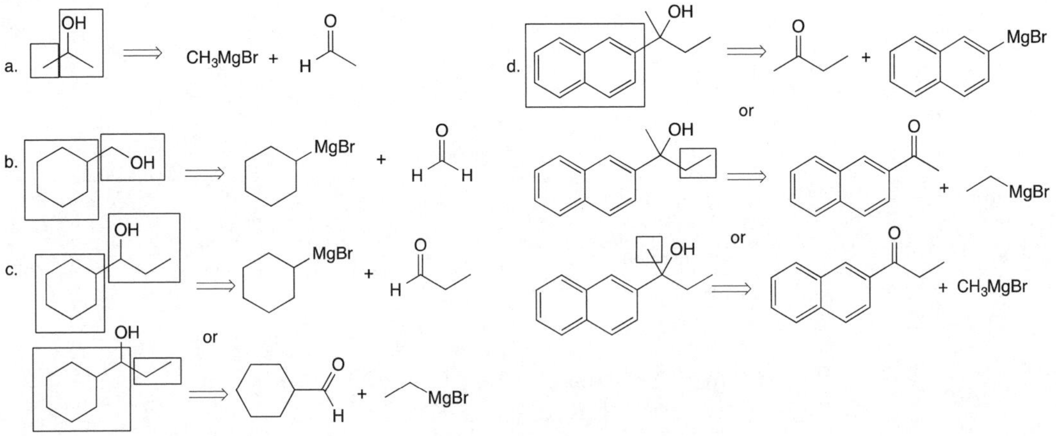

a. CH_3MgBr +

b.

c.

d.

20.24

a. linalool (three methods) ⟹ + CH_3Li

Li +

+

lavandulol ⟹

+ $CH_2=O$

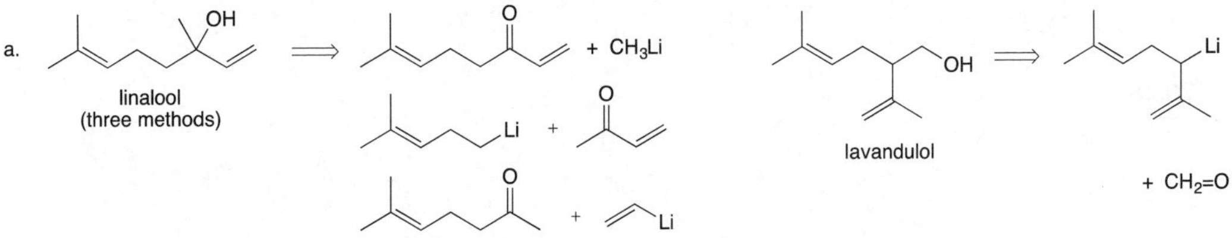

b. 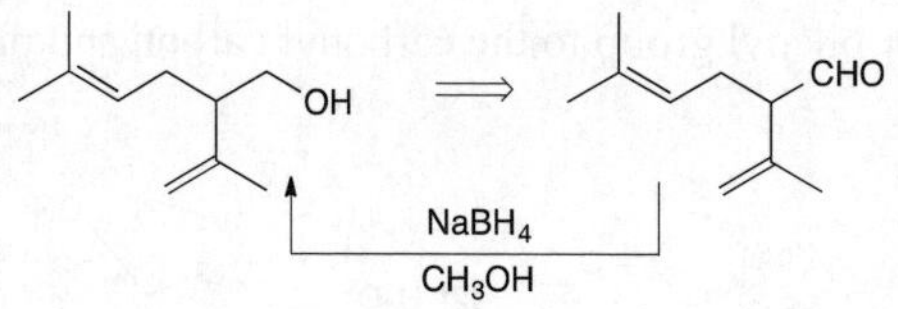
OH
CHO
NaBH4
CH3OH

c. Linalool is a 3° ROH. Therefore, it has no H on the carbon with the OH group, and cannot be prepared by reduction of a carbonyl compound.

20.25

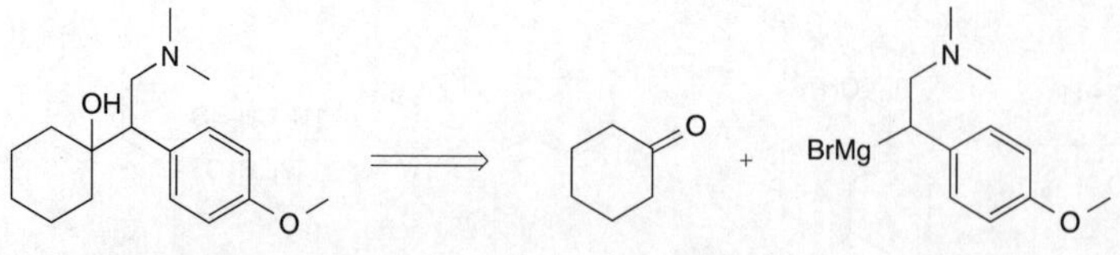
N
OH
O
+
BrMg
N
O

venlafaxine

20.26

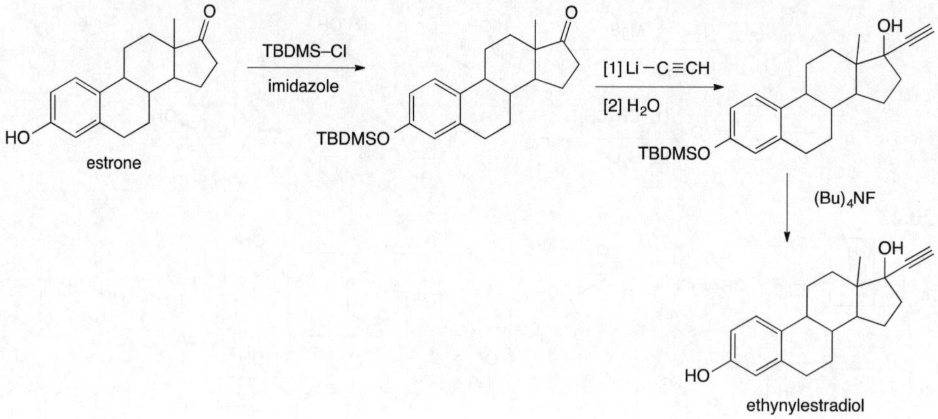
O
HO
estrone
TBDMS–Cl
imidazole
O
TBDMSO
[1] Li–C≡CH
[2] H2O
OH
TBDMSO
(Bu)4NF
OH
HO
ethynylestradiol

20.27

a. 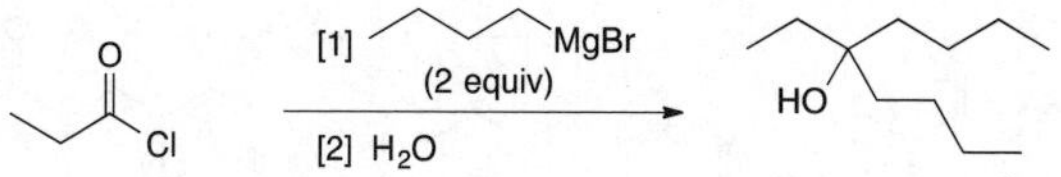
O
Cl
[1] MgBr
(2 equiv)
[2] H2O
HO

b. 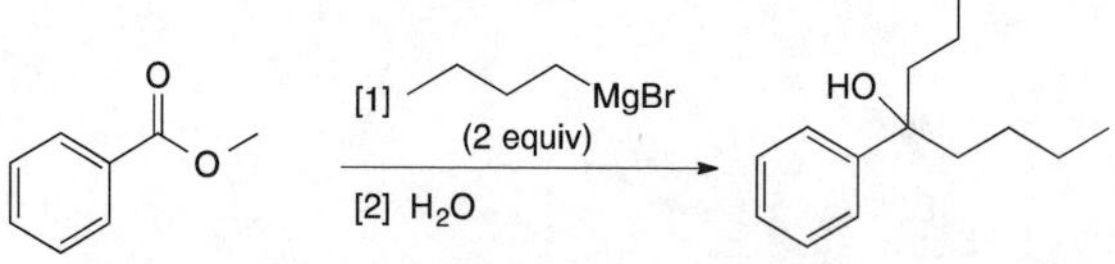
O
O
[1] MgBr
(2 equiv)
[2] H2O
HO

c. 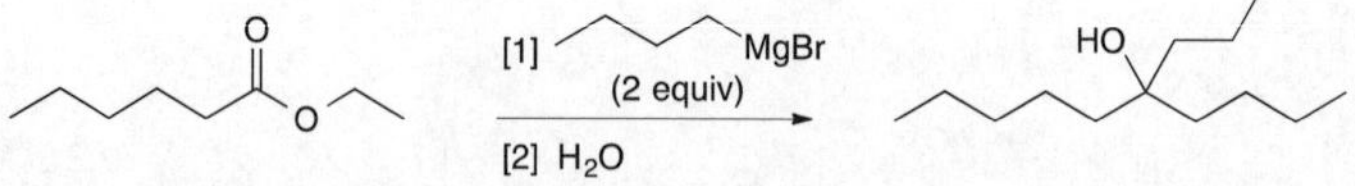
O
O
[1] MgBr
(2 equiv)
[2] H2O
HO

20.28

a. 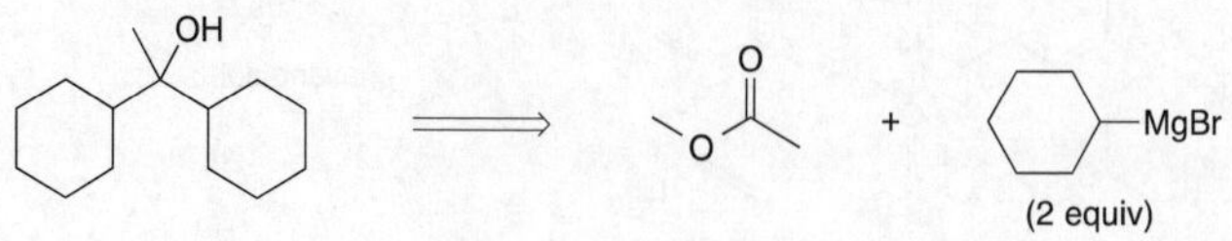
OH
O
O
+
MgBr
(2 equiv)

Introduction to Carbonyl Chemistry 20–13

b.

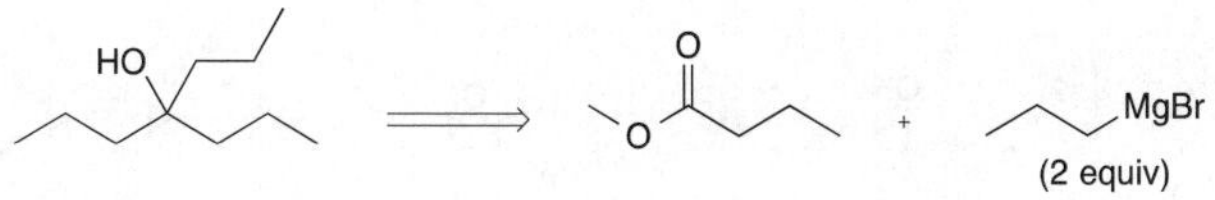

c.

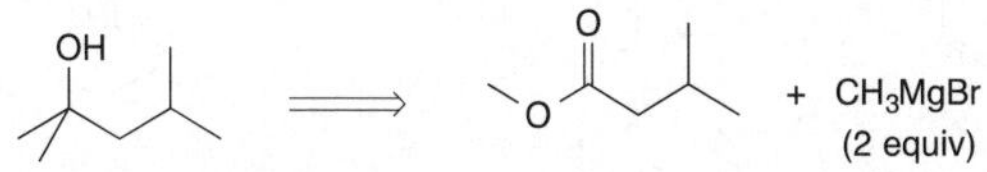

20.29 The R group of the organocuprate has replaced the Cl on the acid chloride.

a. 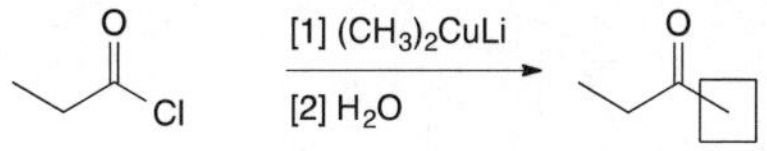 [1] (CH₃)₂CuLi [2] H₂O →

c. 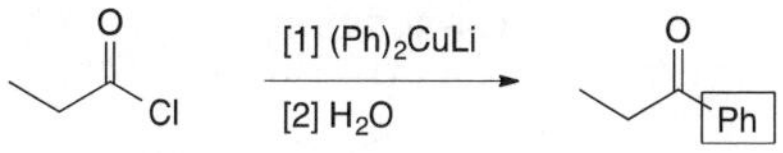 [1] (Ph)₂CuLi [2] H₂O →

b. 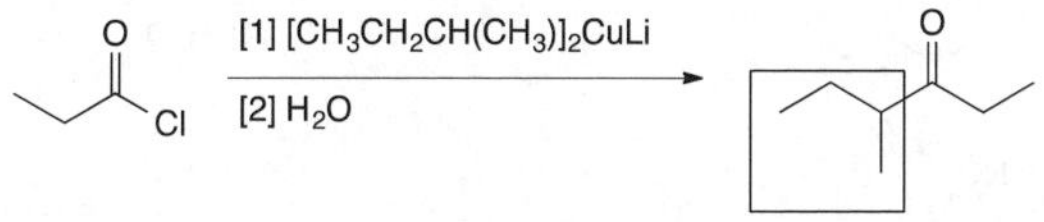 [1] [CH₃CH₂CH(CH₃)]₂CuLi [2] H₂O →

20.30

a. 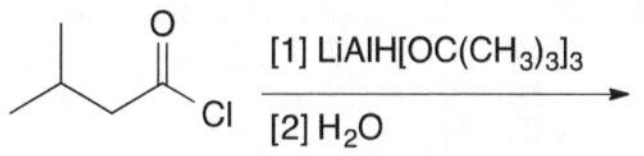 [1] LiAlH[OC(CH₃)₃]₃ [2] H₂O →

c. 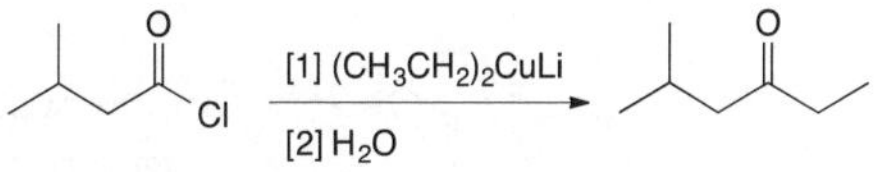 [1] (CH₃CH₂)₂CuLi [2] H₂O →

b. 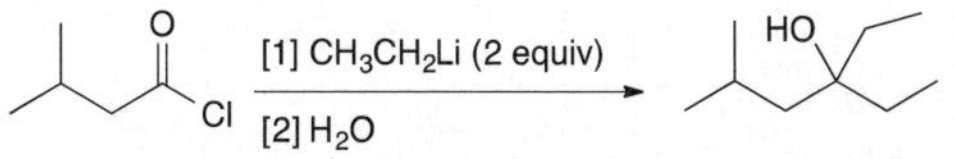 [1] CH₃CH₂Li (2 equiv) [2] H₂O →

d. 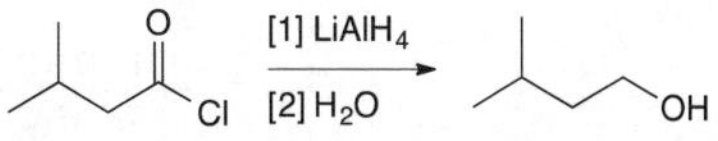 [1] LiAlH₄ [2] H₂O →

20.31

a. 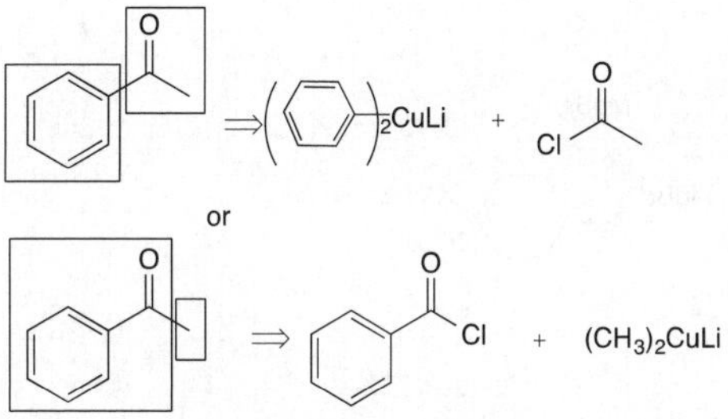 ⟹ ( )₂CuLi +

or

⟹ + (CH₃)₂CuLi

b. 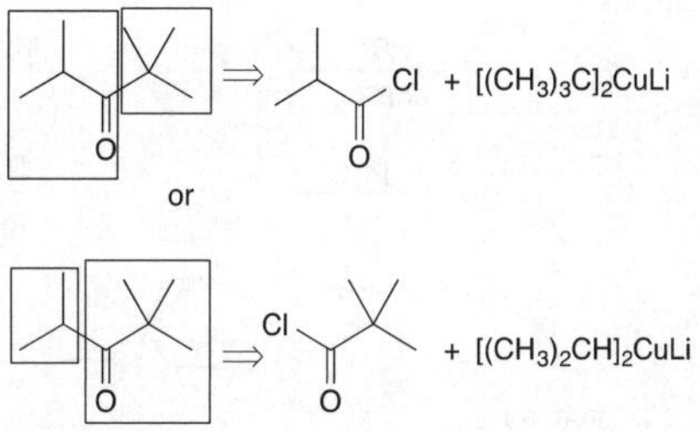 ⟹ + [(CH₃)₃C]₂CuLi

or

⟹ + [(CH₃)₂CH]₂CuLi

20.32

a. 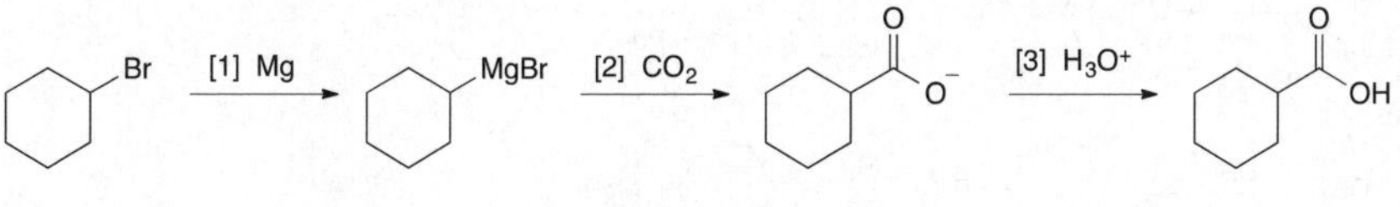 [1] Mg → MgBr [2] CO₂ → [3] H₃O⁺ →

b. 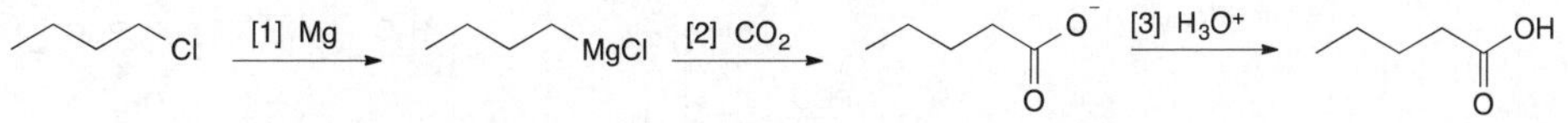 [1] Mg → MgCl [2] CO₂ → [3] H₃O⁺ →

c. 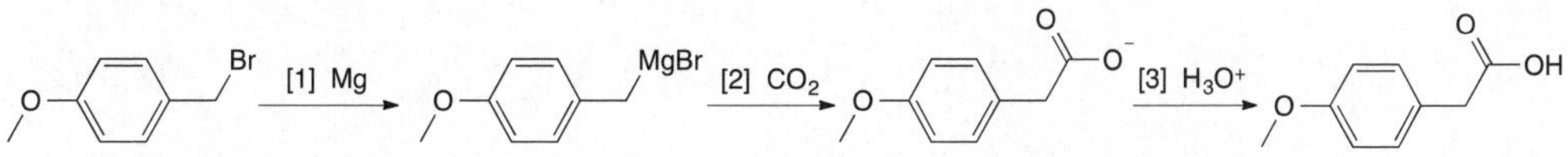 [1] Mg → MgBr [2] CO₂ → [3] H₃O⁺ →

Chapter 20–14

20.33

a. (+ enantiomer)

b.

c.

d.

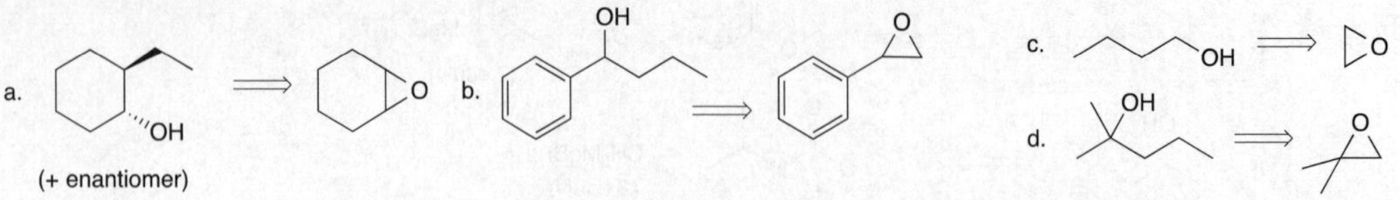

20.34 The characteristic reaction of α,β-unsaturated carbonyl compounds is nucleophilic addition. Grignard and organolithium reagents react by 1,2-addition and organocuprate reagents react by 1,4-addition.

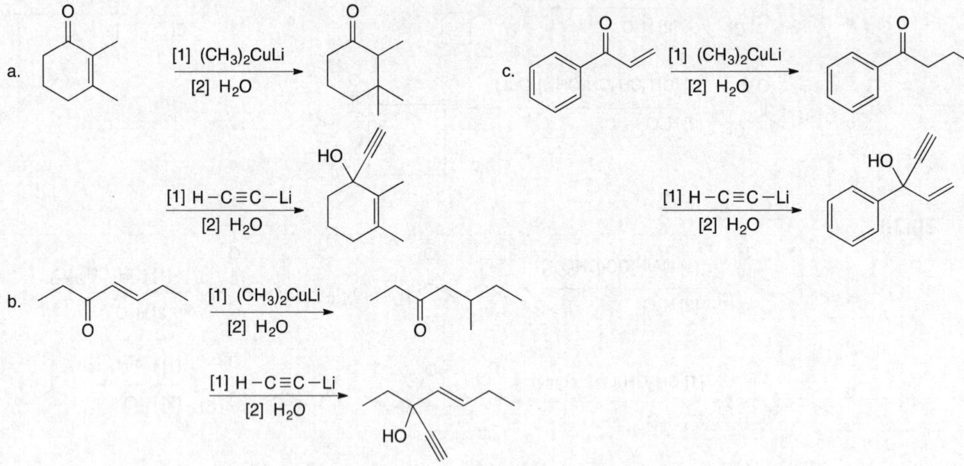

a.
[1] $(CH_3)_2CuLi$
[2] H_2O

[1] $H-C\equiv C-Li$
[2] H_2O

b.
[1] $(CH_3)_2CuLi$
[2] H_2O

[1] $H-C\equiv C-Li$
[2] H_2O

c.
[1] $(CH_3)_2CuLi$
[2] H_2O

[1] $H-C\equiv C-Li$
[2] H_2O

20.35

a.
OH
HBr or PBr₃
Br
Mg
MgBr

OH
PCC
O
[1] MgBr
[2] H_2O
OH

b. (from a.)
OH
HBr
Br

c. (from a.)
OH
H_2SO_4
+
H_2
Pd-C

d.
OH
HBr or PBr₃
Br
Mg
MgBr

OH
PCC
H

H_2O
OH
PCC
O

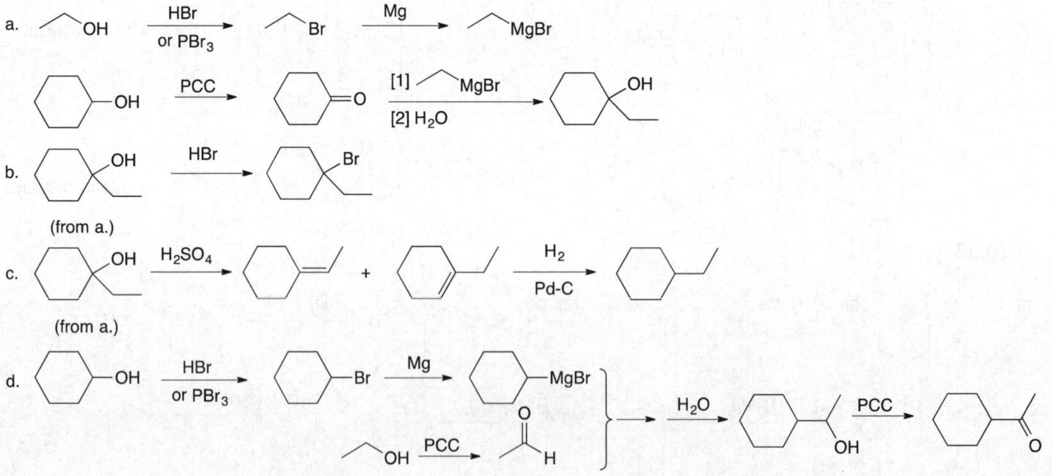

Introduction to Carbonyl Chemistry 20–15

e. (cyclohexyl)—MgBr $\xrightarrow{\triangle O}$ $\xrightarrow{H_2O}$ (cyclohexyl-CH₂CH₂—OH)

(from d.)

$\sim$OH $\xrightarrow[]{H_2SO_4}$ $CH_2{=}CH_2$ $\xrightarrow{mCPBA}$ $\triangle O$

20.36

A (2-pentanone)

a. NaBH₄ / CH₃OH → (2-pentanol, OH)

b. [1] LiAlH₄ / [2] H₂O → (2-pentanol, OH)

c. [1] CH₃MgBr (excess) / [2] H₂O → (OH tertiary alcohol)

d. [1] C₆H₅Li (excess) / [2] H₂O → (HO, C₆H₅ alcohol)

e. Na₂Cr₂O₇ / H₂SO₄ / H₂O → No reaction

B

a. NaBH₄ / CH₃OH → No reaction

b. [1] LiAlH₄ / [2] H₂O → (CH₃O— cyclohexane —OH)

c. [1] CH₃MgBr (excess) / [2] H₂O → (CH₃O— cyclohexane —C(CH₃)₂OH)

d. [1] C₆H₅Li (excess) / [2] H₂O → (CH₃O— cyclohexane —C(C₆H₅)(C₆H₅)OH)

e. Na₂Cr₂O₇ / H₂SO₄ / H₂O → No reaction

20.37

a. (2-butanol) —OH $\xrightarrow{PBr_3}$ ($\sim$Br) $\xrightarrow{Mg}$

(2-butanol) —OH $\xrightarrow{PCC}$ ($\underset{H}{\overset{O}{\|}}$ acetaldehyde) $\xrightarrow[{[2]\ H_2O}]{[1]\ BrMg{-}CH_2CH_3}$ (2-pentanol, OH)

Chapter 20–16

b.

CH$_3$OH
↓ PBr$_3$
CH$_3$Br
↓ Mg

[1] CH$_3$MgBr / [2] H$_2$O → (OH) —PCC→ (O) —[1] CH$_3$MgBr / [2] H$_2$O→ (OH)

(from a.)

c.

(OH) (from b.) —PBr$_3$→ (Br) —[1] Mg / [2] H$_2$C=O / [3] H$_2$O→ (OH)

CH$_3$OH ——PCC——

d.

(OH) —H$_2$SO$_4$→ CH$_2$=CH$_2$ —mCPBA→ (epoxide) —[1] BrMg—CH$_2$CH$_3$ (from a.) / [2] H$_2$O→ (OH)

20.38

a. (pentanal) —NaBH$_4$ / CH$_3$OH→ (pentanol)

b. (pentanal) —[1] LiAlH$_4$ / [2] H$_2$O→ (pentanol)

c. (pentanal) —H$_2$ / Pd-C→ (pentanol)

d. (pentanal) —PCC→ **No reaction**

e. (pentanal) —Na$_2$Cr$_2$O$_7$ / H$_2$SO$_4$, H$_2$O→ (pentanoic acid)

f. (pentanal) —Ag$_2$O / NH$_4$OH→ (pentanoic acid)

g. (pentanal) —[1] CH$_3$MgBr / [2] H$_2$O→ (2-hexanol)

h. (pentanal) —[1] C$_6$H$_5$Li / [2] H$_2$O→ (1-phenyl-1-pentanol)

i. (pentanal) —[1] (CH$_3$)$_2$CuLi / [2] H$_2$O→ **No reaction**

j. (pentanal) —[1] HC≡CNa / [2] H$_2$O→ (alkynol)

k. (pentanal) —[1] CH$_3$C≡CLi / [2] H$_2$O→ (alkynol)

l. (pentanol) —TBDMSCl / imidazole→ (O–TBDMS)

20.39

a. (benzoyl chloride) —(CH$_3$CH$_2$CH$_2$CH$_2$)$_2$CuLi→ (phenyl pentyl ketone)

b. (methyl benzoate) —(CH$_3$CH$_2$CH$_2$CH$_2$)$_2$CuLi→ **No reaction**

c. 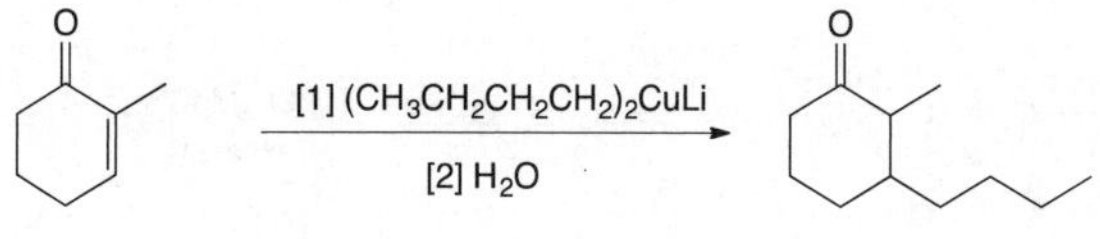
O
[1] $(CH_3CH_2CH_2CH_2)_2CuLi$
[2] H_2O
O

d. 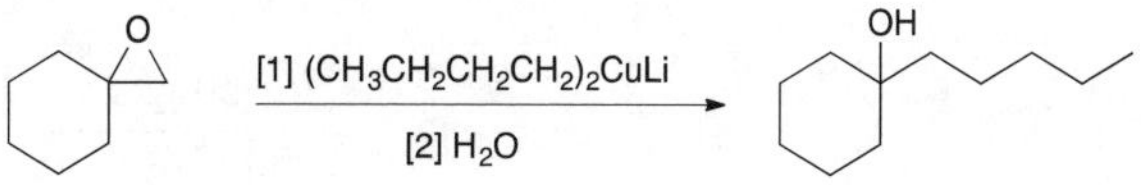
O
[1] $(CH_3CH_2CH_2CH_2)_2CuLi$
[2] H_2O
OH

20.40 Arrange the larger group [$(CH_3)_3C$–] on the left side of the carbonyl.

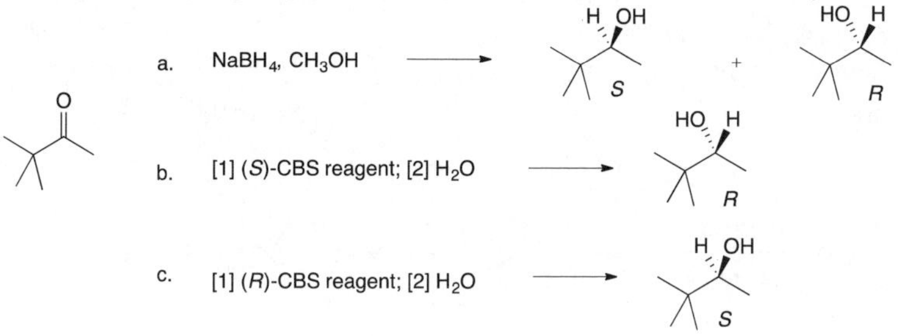
a. $NaBH_4$, CH_3OH
H OH
S
+
HO H
R
O
b. [1] (*S*)-CBS reagent; [2] H_2O
HO H
R
c. [1] (*R*)-CBS reagent; [2] H_2O
H OH
S

20.41

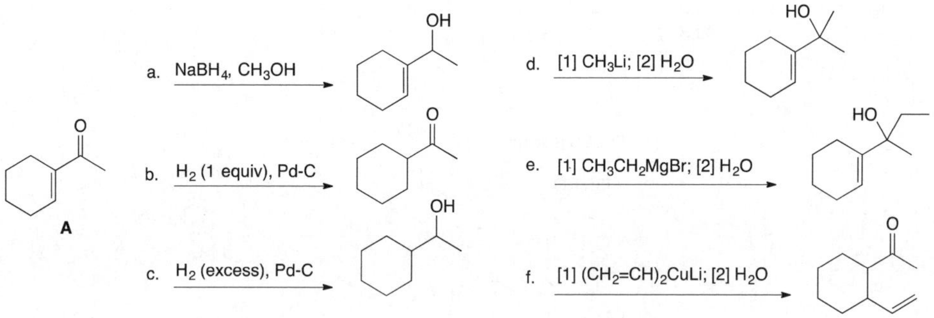
O
A
a. $NaBH_4$, CH_3OH
OH
b. H_2 (1 equiv), Pd-C
O
c. H_2 (excess), Pd-C
OH
d. [1] CH_3Li; [2] H_2O
HO
e. [1] CH_3CH_2MgBr; [2] H_2O
HO
f. [1] $(CH_2{=}CH)_2CuLi$; [2] H_2O
O

20.42

a. 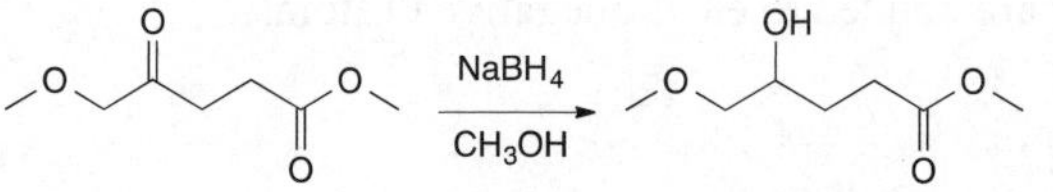
O
O
O
O
$NaBH_4$
CH_3OH
OH
O
O
O

c. 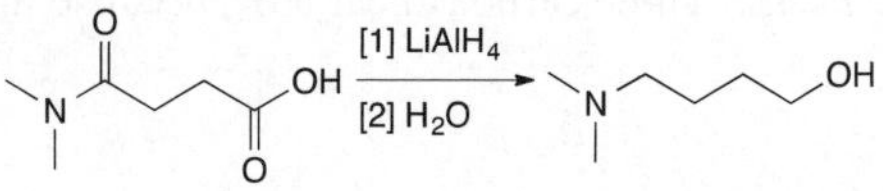
O
N
OH
O
[1] $LiAlH_4$
[2] H_2O
N
OH

b. 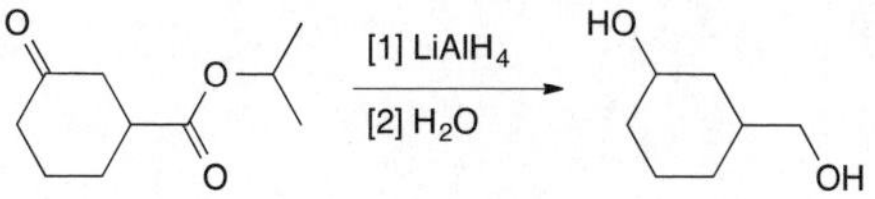
O
O
O
[1] $LiAlH_4$
[2] H_2O
HO
OH

d. 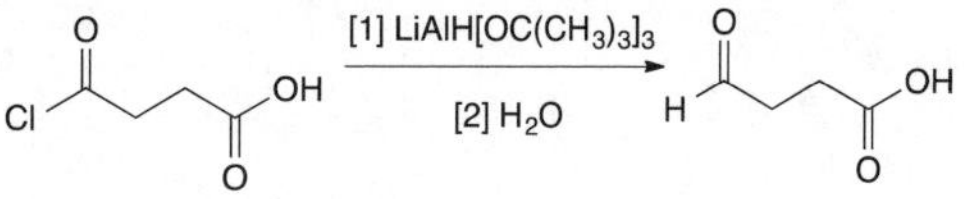
O
Cl
OH
O
[1] $LiAlH[OC(CH_3)_3]_3$
[2] H_2O
O
H
OH
O

Chapter 20–18

20.43

a.

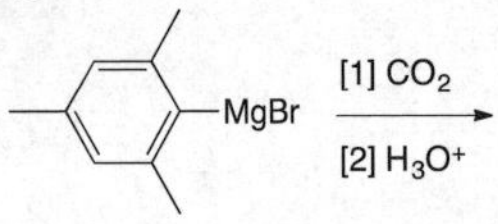

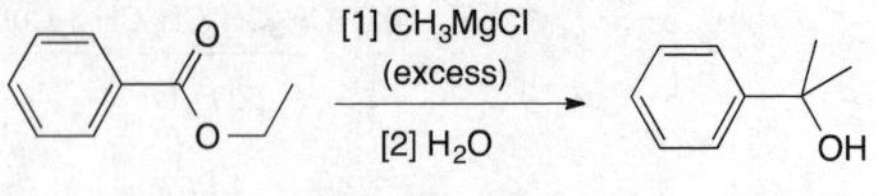

b.

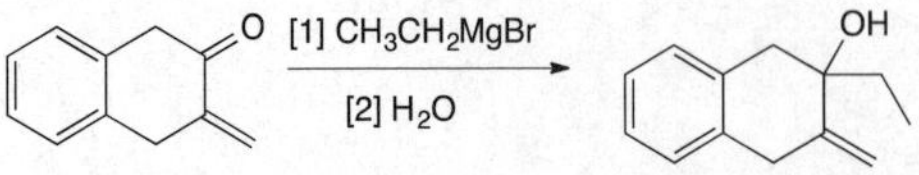

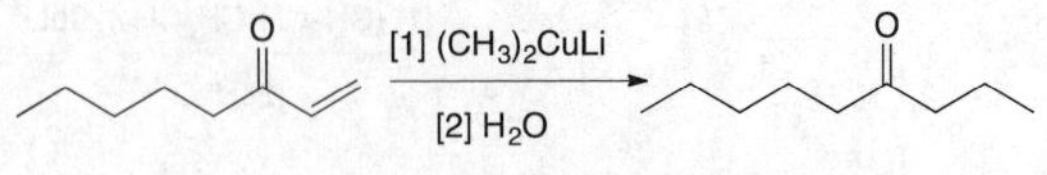

c.

d.

e.

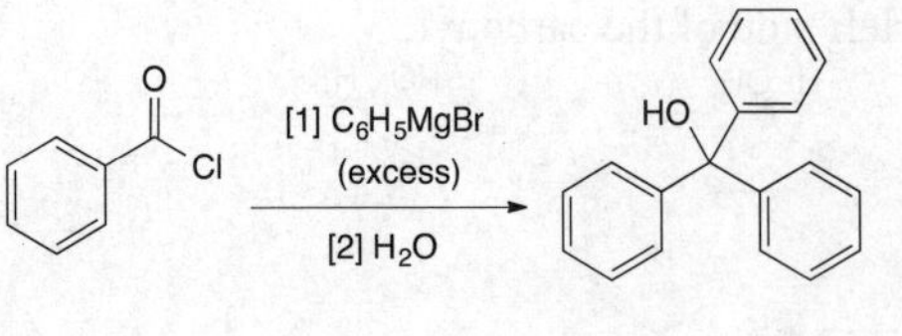

f.

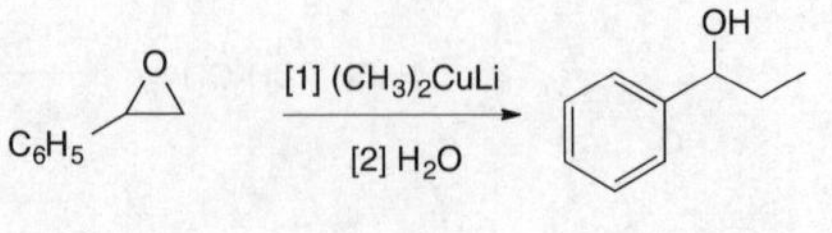

20.44

a.

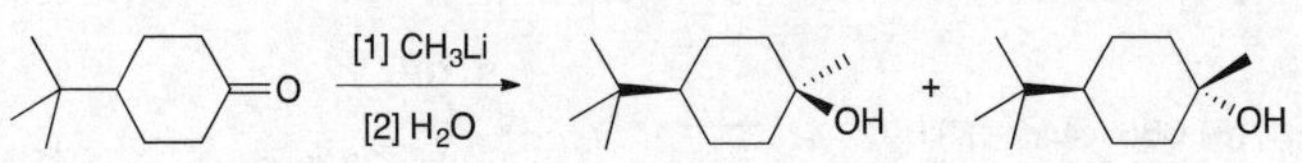

b.

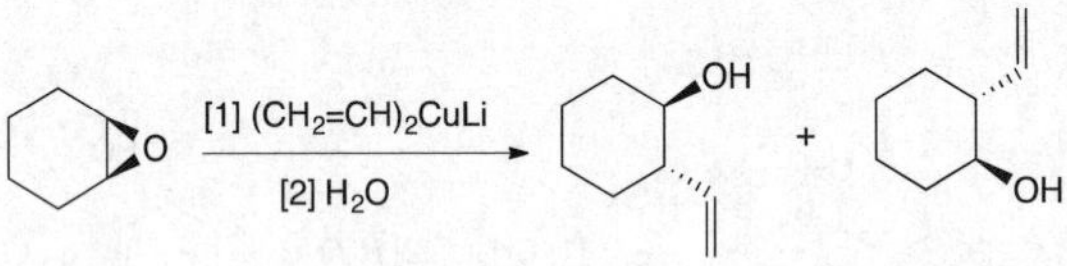

c.

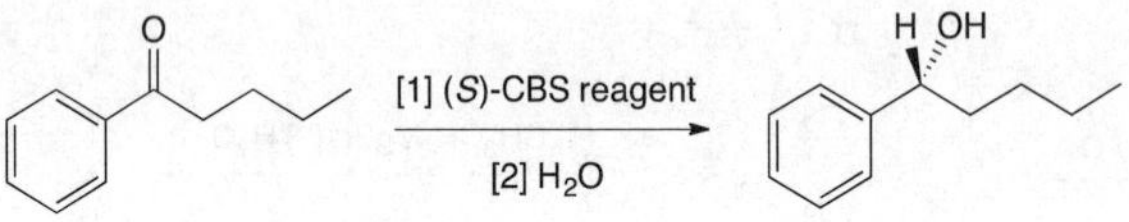

d.

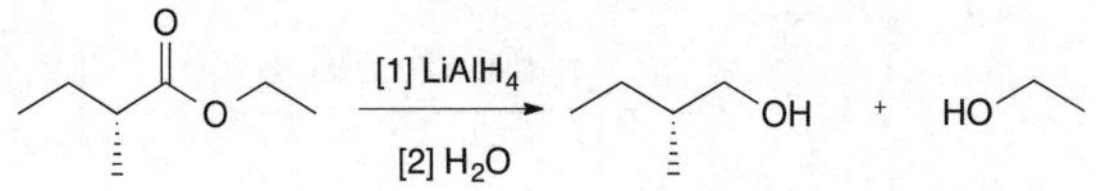

20.45 Three carbons bear a $\delta+$ because they are bonded to electronegative O atoms.

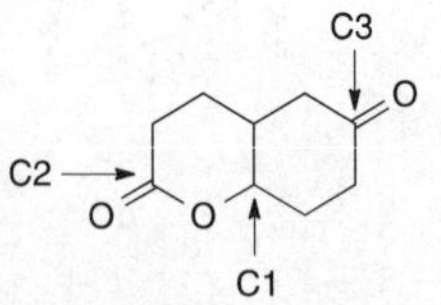

C1 is an sp^3 hybridized C bonded to an O so it bears a $\delta+$. There are no additional resonance structures that affect C1. C2 is part of a carbonyl that has three resonance structures, only one of which places a (+) charge on C. The O atom of the ester donates its electron pair, making the carbonyl C less electrophilic than C3. C3 is most electrophilic. Its carbonyl is stabilized by two resonance structures, one of which places a (+) charge on carbon.

Introduction to Carbonyl Chemistry 20–19

O donates electron density.

full (+) charge
most electrophilic site

20.46 The organolithium reagent is a nucleophile and a base. As a base it can remove the most acidic proton (between the benzene ring and C=O) to form an enolate that is protonated by D_3O^+.

A

B
$C_{12}H_{13}DO_3$

20.47 Both ketones are chiral molecules with carbonyl groups that have one side more sterically hindered than the other. In both reductions, hydride approaches from the less hindered side.

2 H's
less hindered
Attack comes from above.

The CH_3 groups on the bridgehead carbon make the top more hindered. H^- attacks from below to afford an exo OH group.
Attack comes from below.

[1] LiAlH₄
[2] H₂O

from above

endo OH group

The concave shape of the six-membered ring makes the bottom face of the C=O more sterically hindered. Addition of the H⁻ occurs from above to place the new C–H bond exo, making the OH endo.

[1] LiAlH₄
[2] H₂O

from below

exo OH group

20.48 A Grignard reagent contains a carbon atom with a partial negative charge, so it acts as a base and reacts with the OH of the starting halide, $BrCH_2CH_2CH_2CH_2OH$. This acid–base reaction destroys the Grignard reagent, so that addition cannot occur. To get around this problem, the OH group can be protected as a *tert*-butyldimethylsilyl ether, from which a Grignard reagent can be made.

Chapter 20–20

basic site   acidic site

Br～～～OH —Mg→ BrMg～～～OH —proton transfer→ ～～～O⁻ + HOMgBr
δ–
These will react.

INSTEAD: Use a protecting group.

Br～～～OH —TBDMS–Cl / imidazole→ Br～～～OTBDMS —Mg / ether→ BrMg～～～OTBDMS

protected OH

[1] cyclohexanone (C=O)
[2] H_2O

(Bu)₄NF

A

20.49 Compounds **F, G,** and **K** are all alcohols with aromatic rings so there will be many similarities in their proton NMR spectra. These compounds, however, will show differences in absorptions due to the CH protons on the carbon bearing the OH group. **F** has a CH_2OH group, which will give a singlet in the 3–4 ppm region of the spectrum. **G** is a 3° alcohol that has no protons on the C bonded to the OH group so it will have no peak in the 3–4 ppm region of the spectrum. **K** is a 2° alcohol that will give a doublet in the 3–4 ppm region of the spectrum for the CH proton on the carbon with the OH group.

[1] C_6H_5MgBr
[2] H_2O
A

H_2SO_4
B

[1] O_3
[2] CH_3SCH_3
C

HCl
D

[1] Mg
[2] $CH_2=O$
[3] H_2O
F

mCPBA
E

[1] $(CH_3)_2CuLi$
[2] H_2O
G

[1] LiAlH₄
[2] H_2O

OH
H

PBr₃

Br
I

Mg

MgBr
J

[1] C_6H_5CHO
[2] H_2O

K

20.50

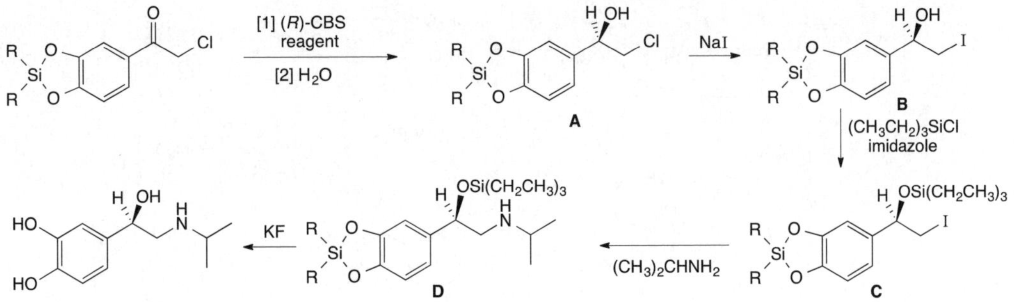

[1] (*R*)-CBS reagent
[2] H₂O

... A ...

NaI

... B ...

(CH₃CH₂)₃SiCl
imidazole

... C ...

(CH₃)₂CHNH₂

... D ...

KF

20.51

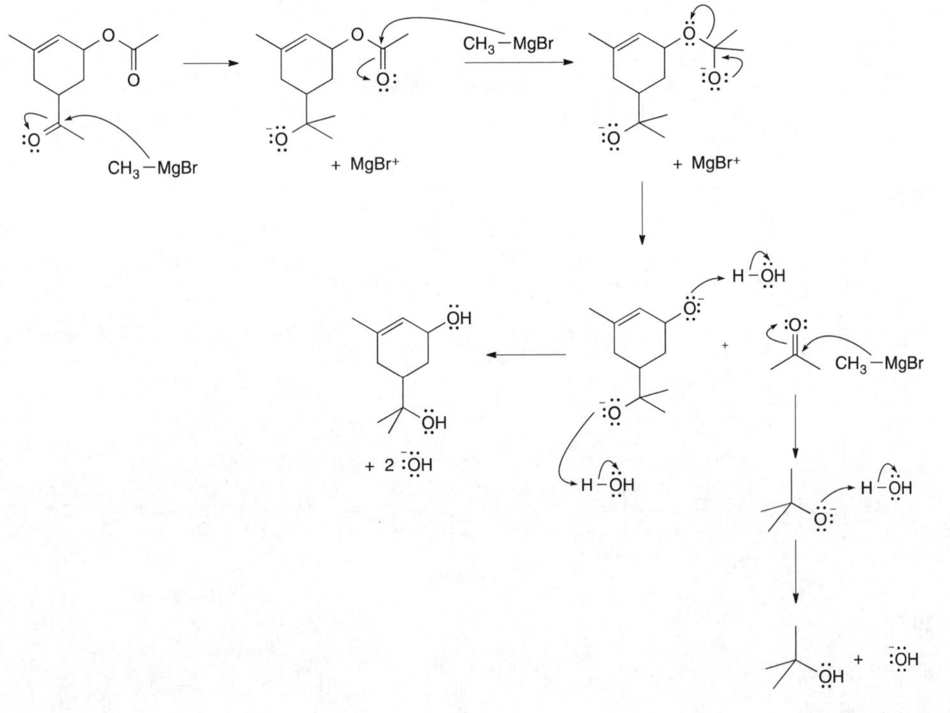

CH₃—MgBr

CH₃—MgBr

+ MgBr⁺

+ MgBr⁺

H—OH

CH₃—MgBr

H—OH

H—OH

+ 2 :ŌH

H—OH

20.52

BrMg

MgBr

+ MgBr

BrMg

+ MgBr

H—ŌH   H—ŌH

HŌMgBr +

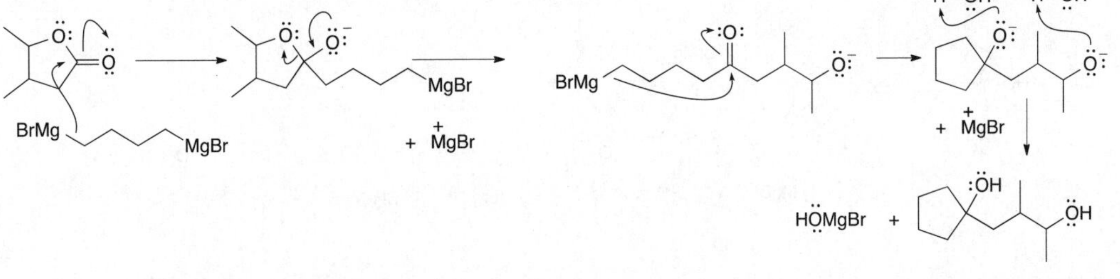

Chapter 20–22

20.53

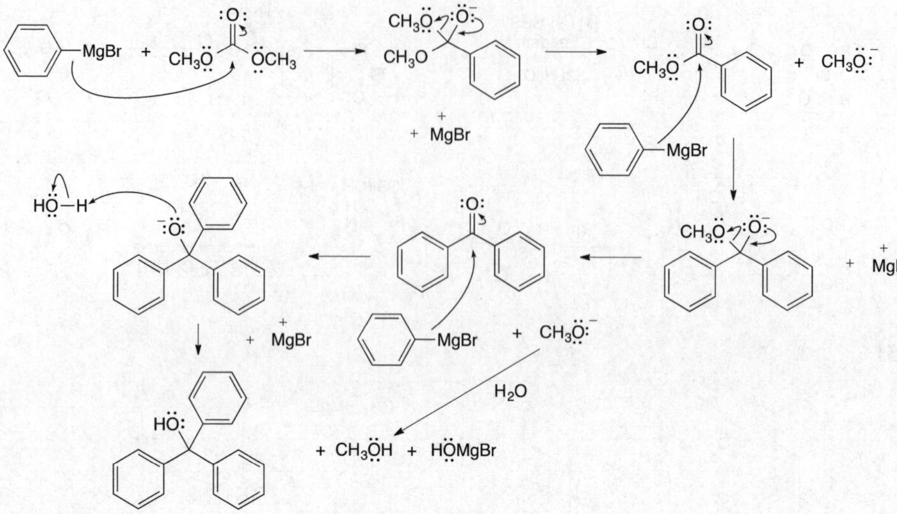

20.54

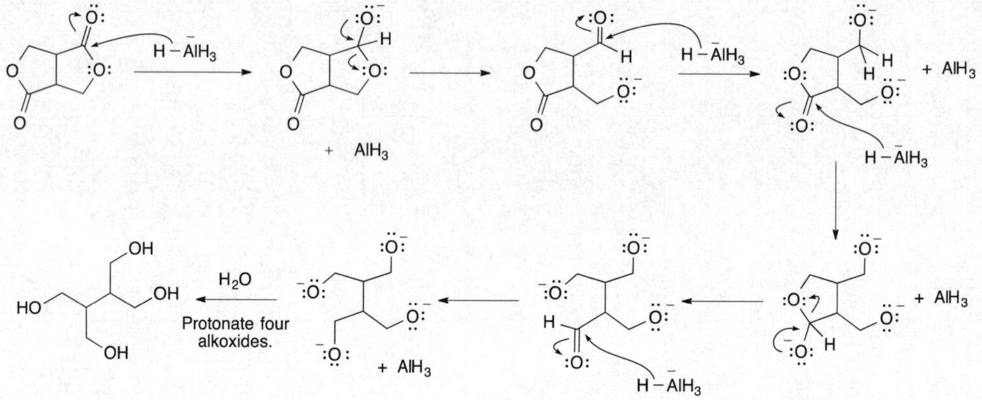

Protonate four alkoxides.

20.55

a.

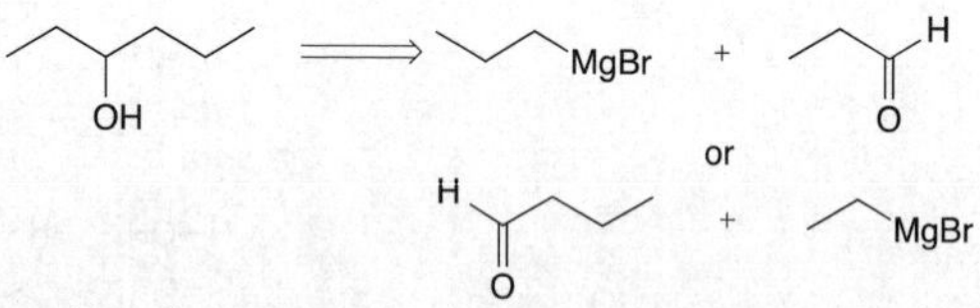

c.

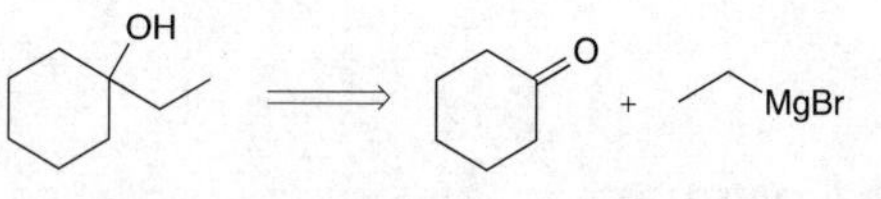

or

b.

or

or

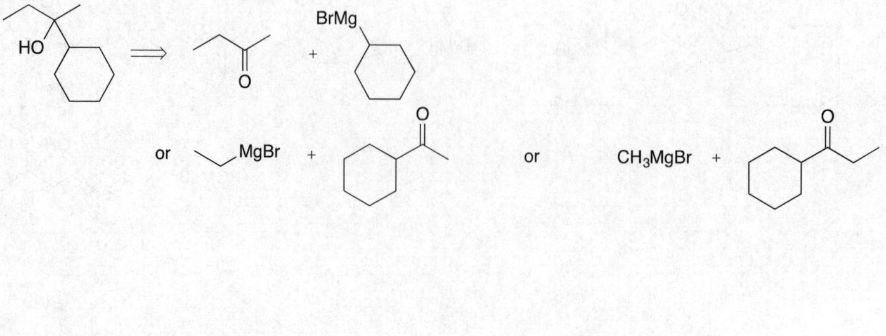

Introduction to Carbonyl Chemistry 20–23

20.56

(a)

(b)

(c)

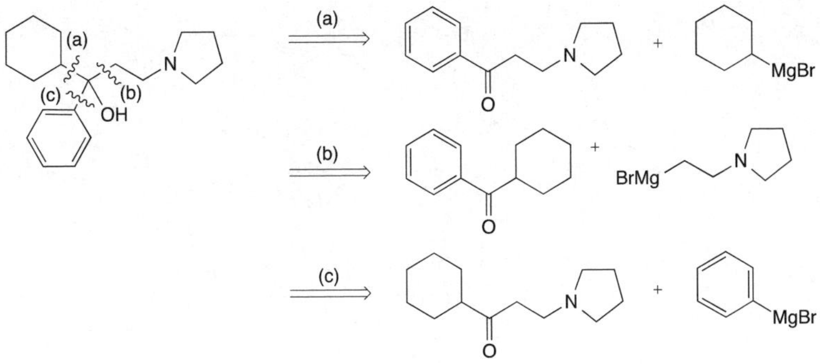

20.57

a.

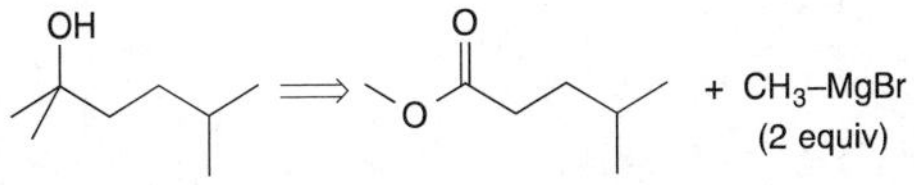

b.

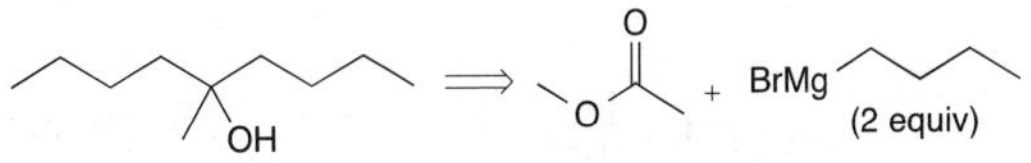

20.58

a. (two ways)

or

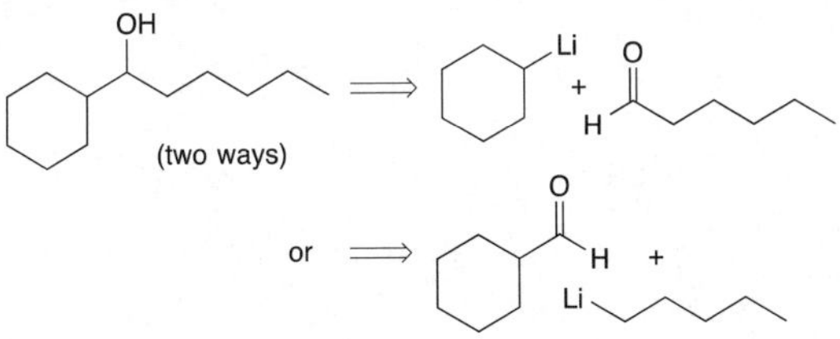

b. (three ways)

or (2 equiv)

or

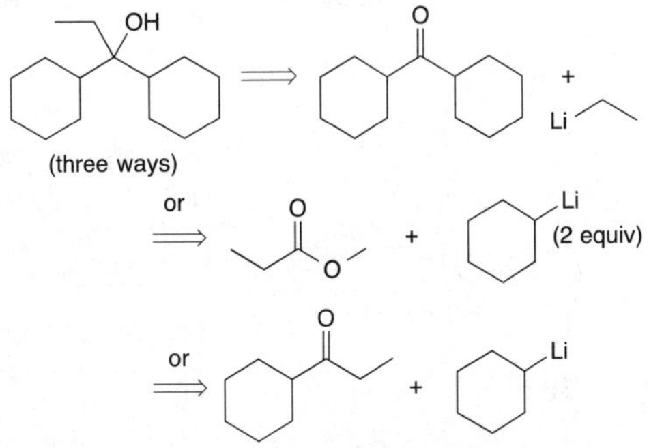

20.59

a.

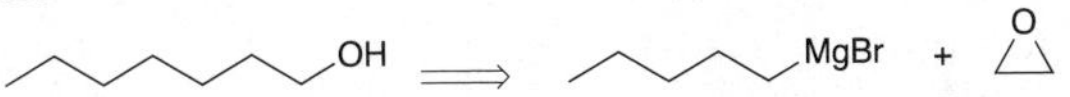

b.

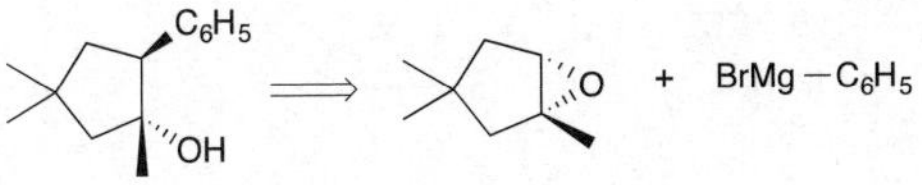

c.

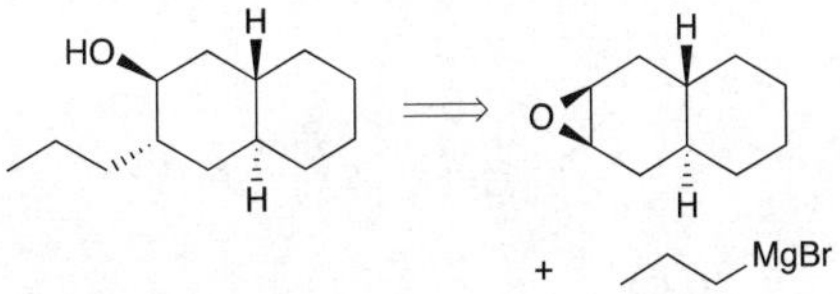

20.60

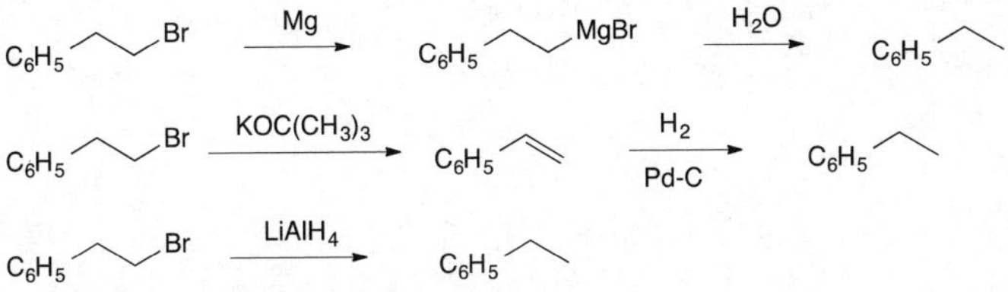

C_6H_5 — Br $\xrightarrow{Mg}$ C_6H_5 — MgBr $\xrightarrow{H_2O}$ C_6H_5

C_6H_5 — Br $\xrightarrow{KOC(CH_3)_3}$ C_6H_5 $\xrightarrow[Pd-C]{H_2}$ C_6H_5

C_6H_5 — Br $\xrightarrow{LiAlH_4}$ C_6H_5

Chapter 20–24

20.61

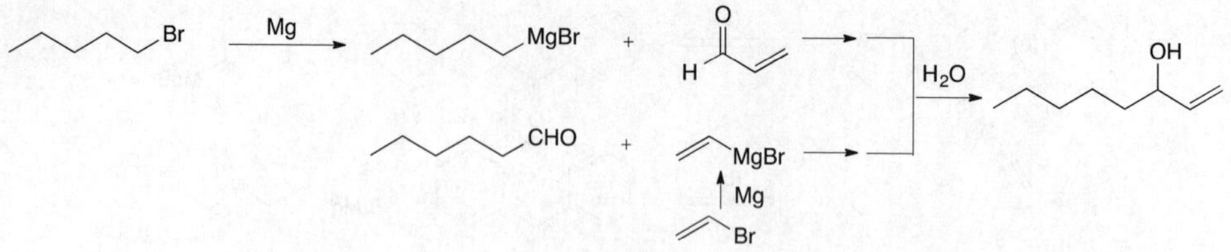

20.62

a.

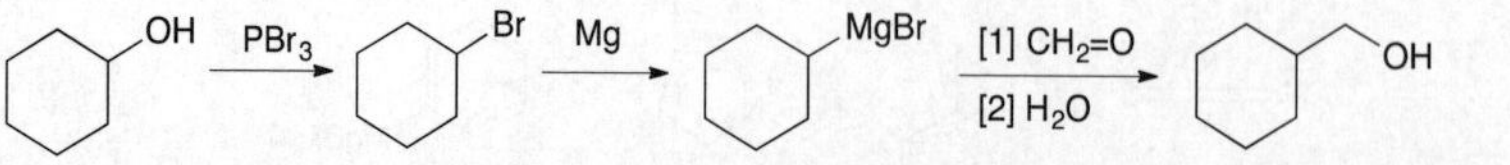

b.

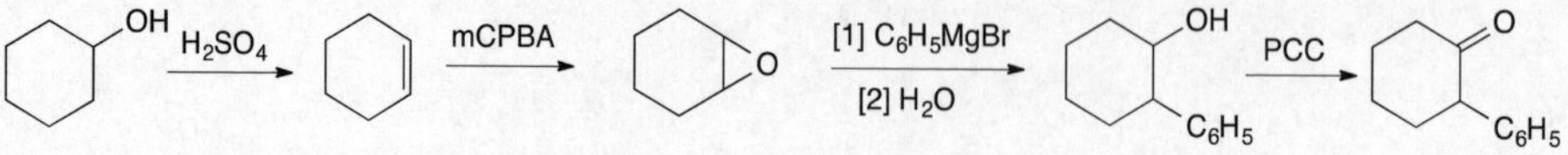

c.

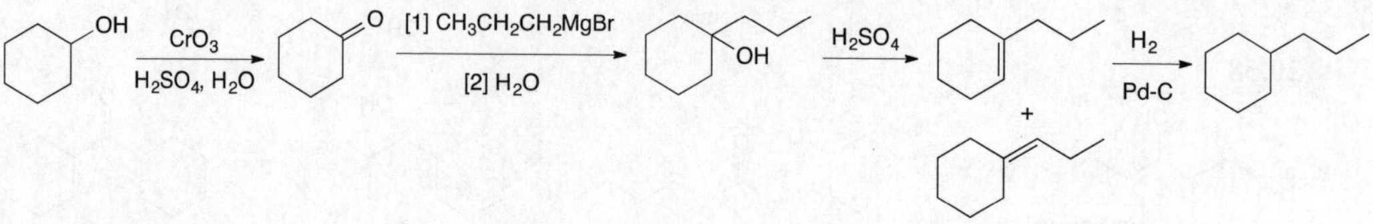

d.

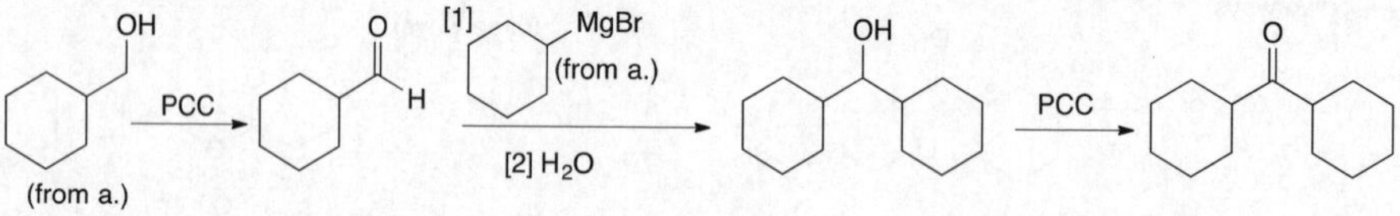

20.63

a.

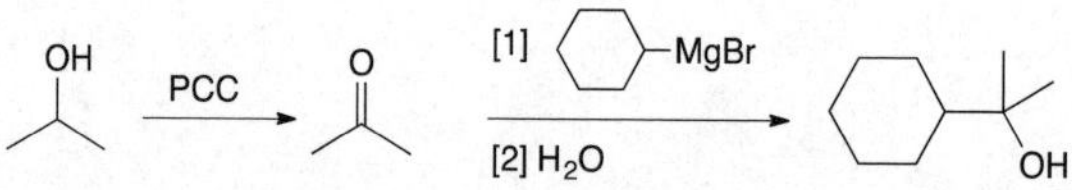

b.

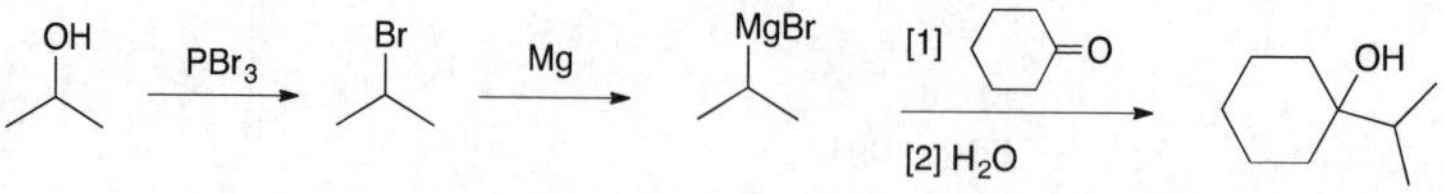

20.64

a.

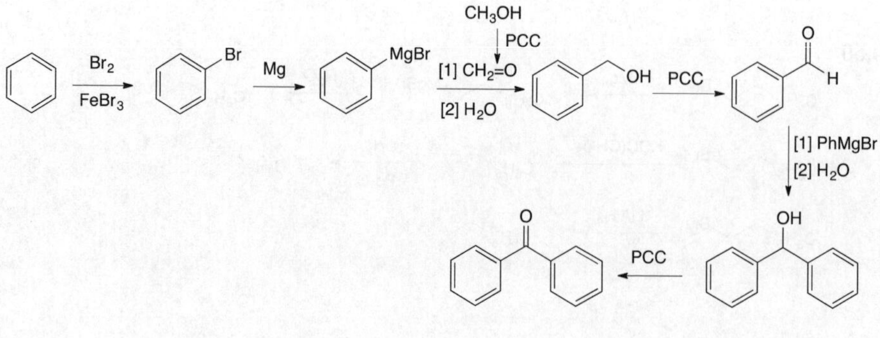

20.65

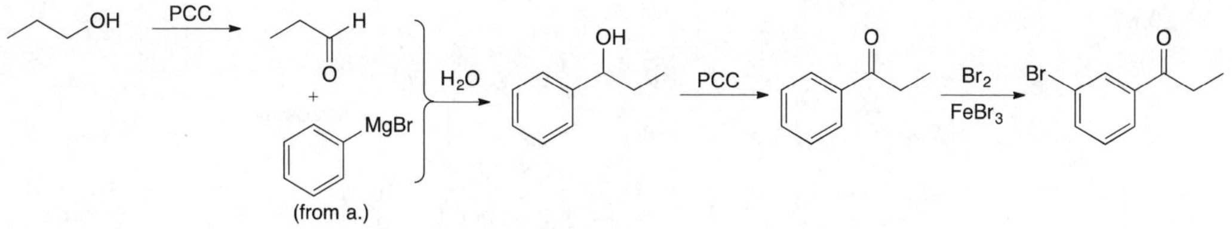

20.66

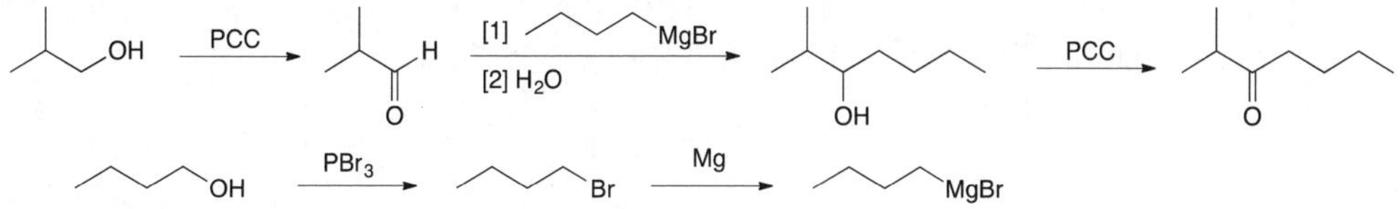

(from a.)

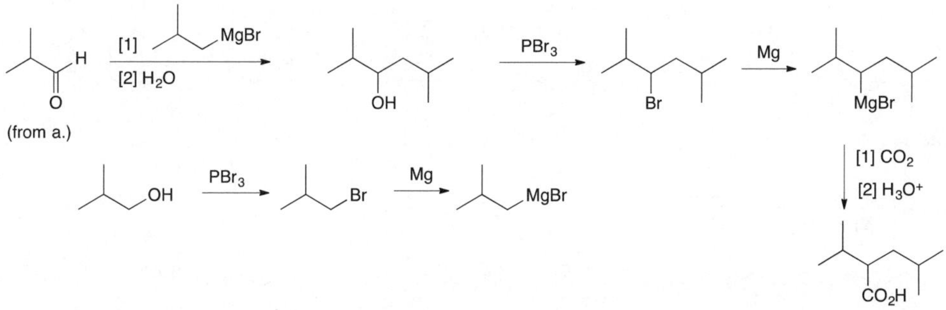

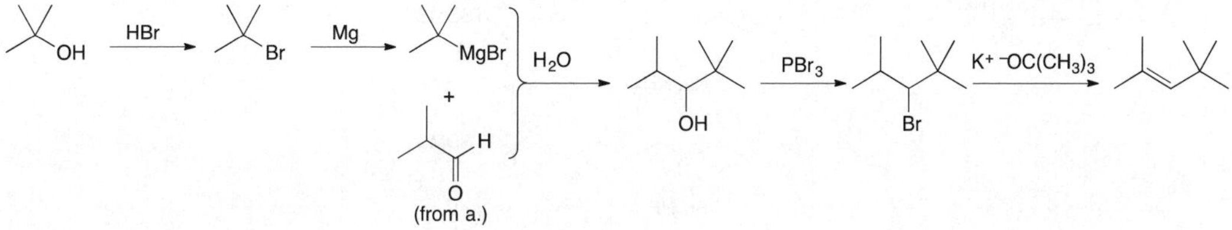

(Bu)₄NF

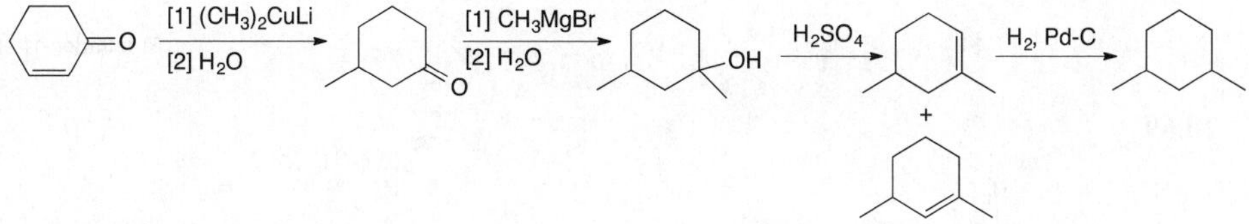

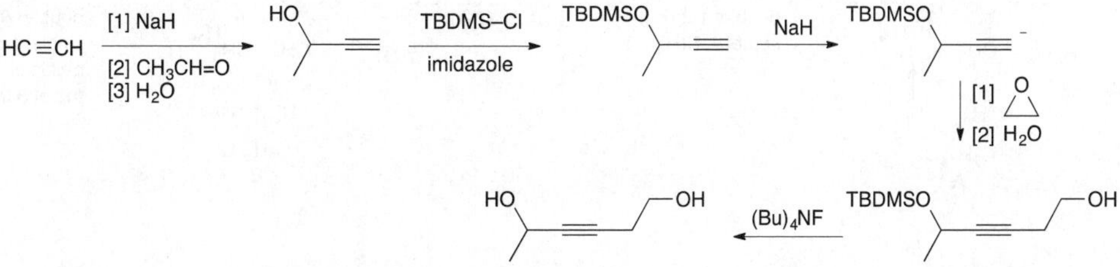

Chapter 20–26

20.67

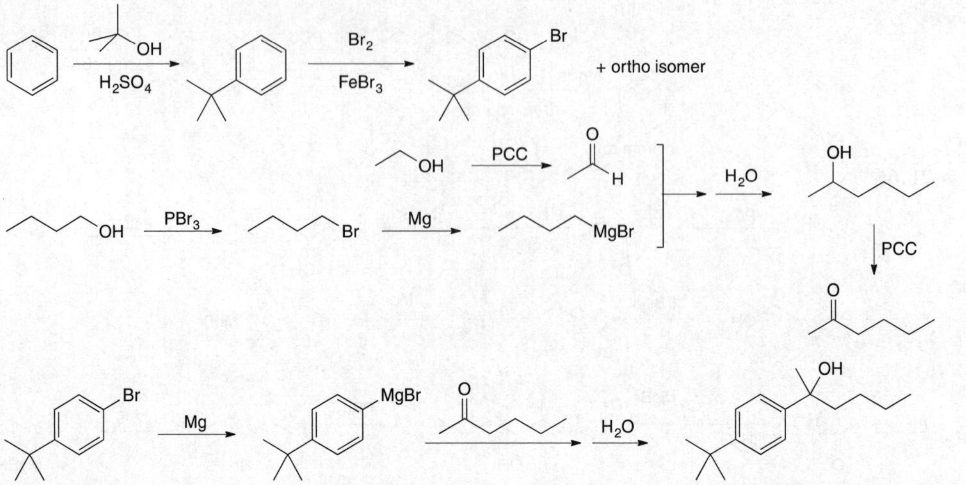

20.68

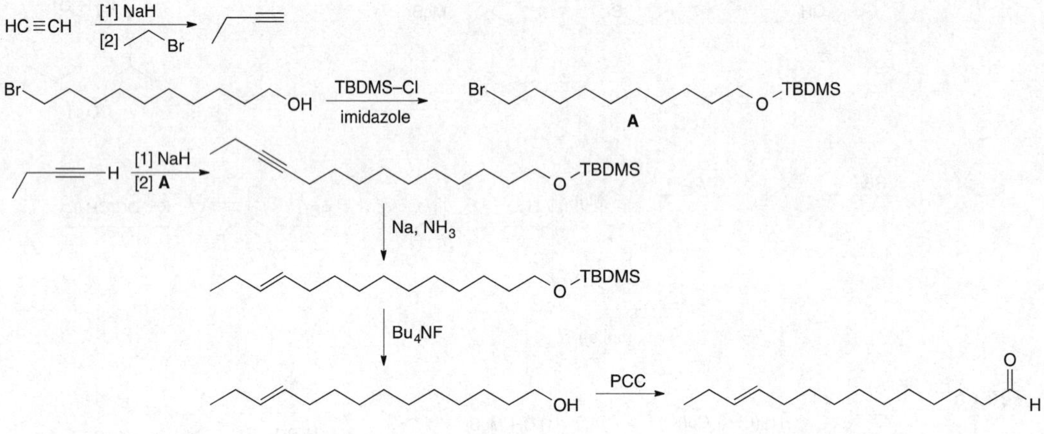

(E)-tetradec-11-enal

20.69

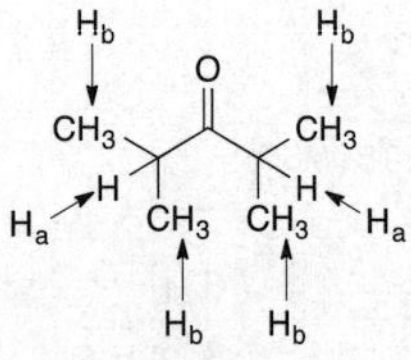

IR peak: 1716 cm^{-1} (C=O)
^{1}H NMR: 2 signals (ppm)
  doublet 1.2 (H$_b$)
  septet 2.7 (H$_a$)

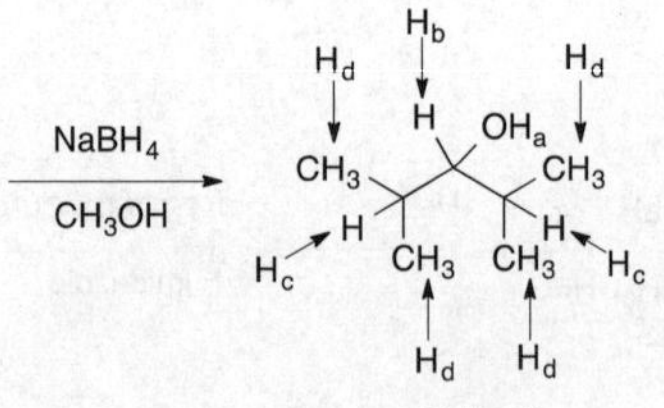

$\xrightarrow[\text{CH}_3\text{OH}]{\text{NaBH}_4}$

IR peak: 3600–3200 cm^{-1} (OH)
^{1}H NMR: 4 signals (ppm)
  doublet 0.9 (H$_d$)
  singlet 1.5 (H$_a$)
  multiplet 1.7 (H$_c$)
  triplet 3.0 (H$_b$)

C$_7$H$_{14}$O
A

C$_7$H$_{16}$O
B

20.70

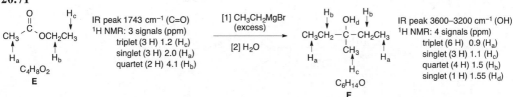

20.71

20.72 Molecular ion at $m/z = 86$: $C_5H_{10}O$ (possible molecular formula).

20.73 Molecular ion at $m/z = 86$: $C_5H_{10}O$ (possible molecular formula).

Chapter 20–28

20.74

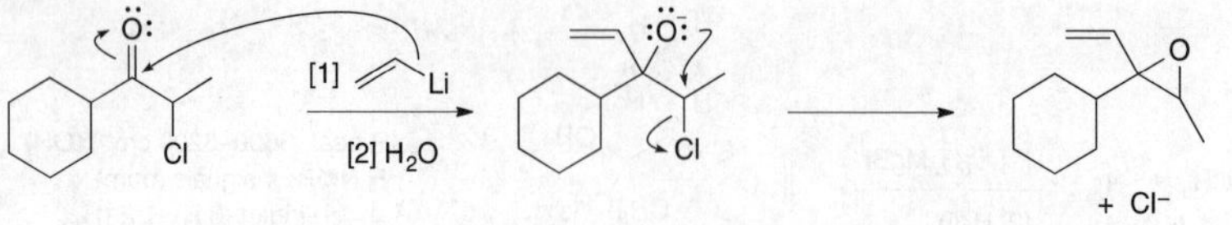

20.75

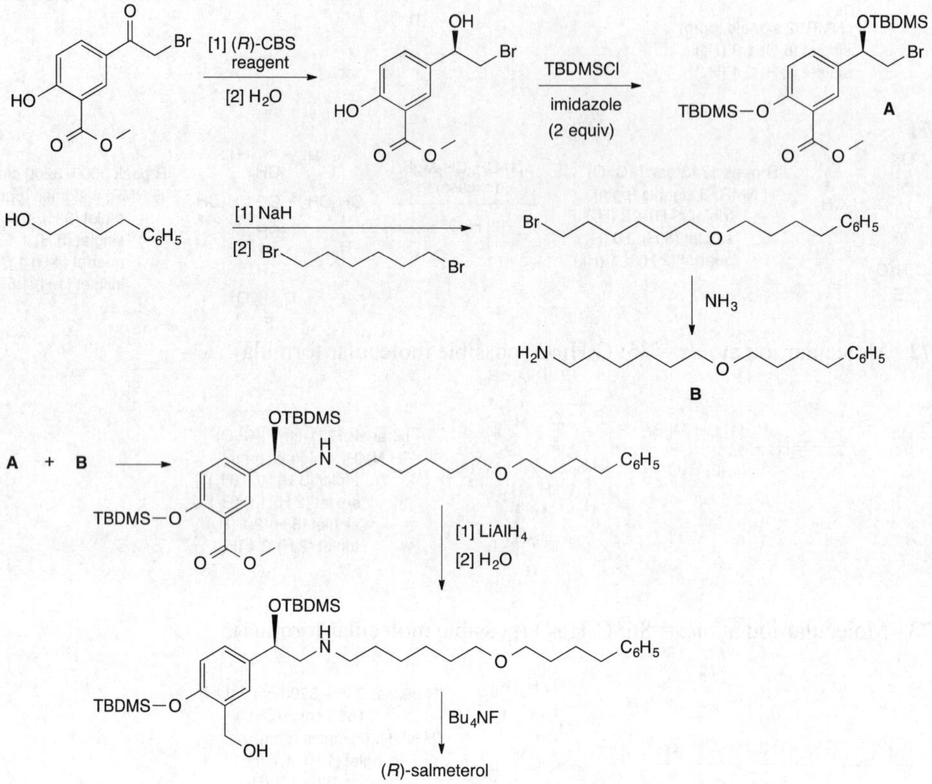

20.76

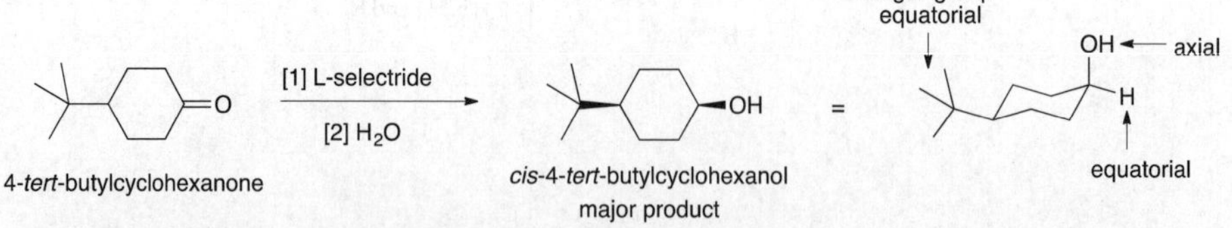

4-*tert*-butylcyclohexanone

cis-4-*tert*-butylcyclohexanol
major product

L-Selectride adds H⁻ to a C=O group. There are two possible reduction products—cis and trans isomers—but the cis isomer is favored. The key element is that the three *sec*-butyl groups make L-selectride a large, bulky reducing agent that attacks the carbonyl group from the less hindered direction.

Introduction to Carbonyl Chemistry 20–29

When H⁻ adds from the equatorial direction, the product has an axial OH and a new equatorial H. Since the equatorial direction is less hindered, this mode of attack is favored with large bulky reducing agents like L-selectride. In this case, the product is cis.

The axial H's hinder H⁻ attack from the axial direction. As a result, this mode of attack is more difficult with larger reducing agents. In this case the product is trans. This product is not formed to any appreciable extent.

20.77 The β carbon of an α,β-unsaturated carbonyl compound absorbs farther downfield in the ^{13}C NMR spectrum than the α carbon, because the β carbon is deshielded and bears a partial positive charge as a result of resonance. Because three resonance structures can be drawn for an α,β-unsaturated carbonyl compound, one of which places a positive charge on the β carbon, the decrease of electron density at this carbon deshields it, shifting the ^{13}C absorption downfield. This is not the case for the α carbon.

20.78

20.79

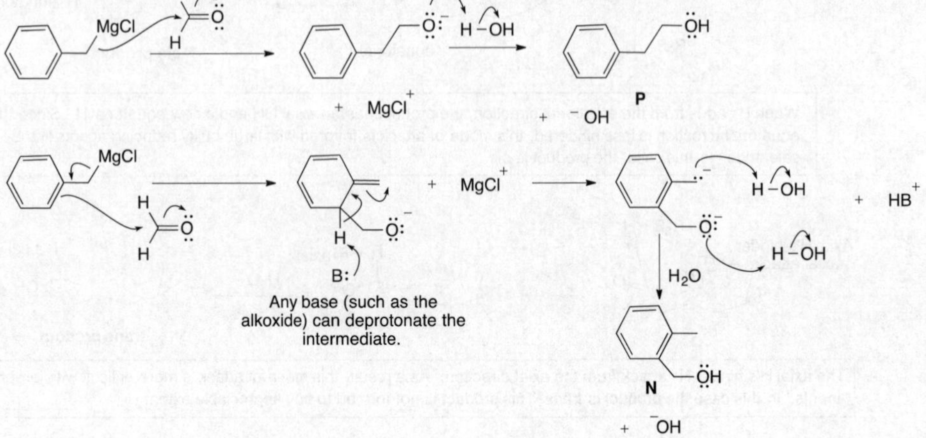
MgCl
P
MgCl
HB
B:
Any base (such as the alkoxide) can deprotonate the intermediate.
N

20.80

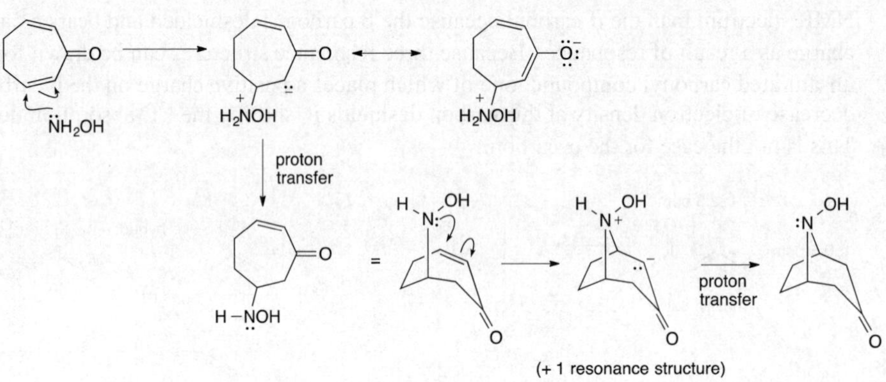
proton transfer
proton transfer
(+ 1 resonance structure)

20.81

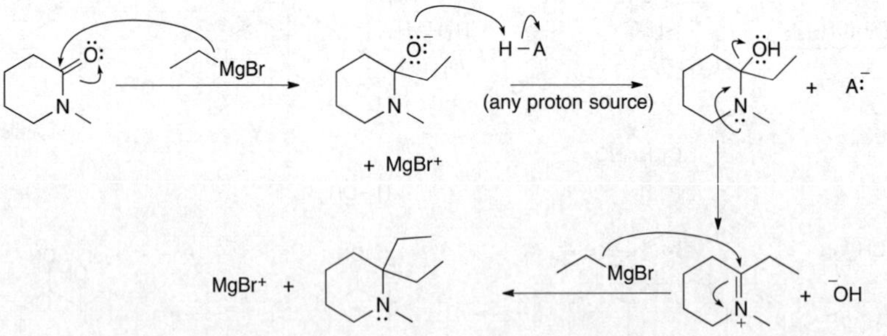
MgBr
(any proton source)
+ MgBr⁺
MgBr
MgBr⁺ +

Aldehydes and Ketones 21–1

Chapter 21 Aldehydes and Ketones—Nucleophilic Addition

Chapter Review

General facts

- Aldehydes and ketones contain a carbonyl group bonded to only H atoms or R groups. The carbonyl carbon is sp^2 hybridized and trigonal planar (21.1).
- Aldehydes are identified by the suffix -al, whereas ketones are identified by the suffix -one (21.2).
- Aldehydes and ketones are polar compounds that exhibit dipole–dipole interactions (21.3).

Summary of spectroscopic absorptions of RCHO and R₂CO (21.4)

IR absorptions	C=O	~1715 cm^{-1} for ketones
		• increasing frequency with decreasing ring size
		~1730 cm^{-1} for aldehydes
		• For both RCHO and R₂CO, the frequency decreases with conjugation.
	C_{sp^2}–H of CHO	~2700–2830 cm^{-1} (one or two peaks)
^{1}H NMR absorptions	CHO	9–10 ppm (highly deshielded proton)
	C–H α to C=O	2–2.5 ppm (somewhat deshielded C_{sp^3}–H)
^{13}C NMR absorption	C=O	190–215 ppm

Nucleophilic addition reactions

[1] Addition of hydride (H⁻) (21.8)

- The mechanism has two steps.
- H:⁻ adds to the planar C=O from both sides.

[2] Addition of organometallic reagents (R⁻) (21.8)

- The mechanism has two steps.
- R:⁻ adds to the planar C=O from both sides.

[3] Addition of cyanide (⁻CN) (21.9)

- The mechanism has two steps.
- ⁻CN adds to the planar C=O from both sides.

Chapter 21–2

[4] Wittig reaction (21.10)

- The reaction forms a new C–C σ bond and a new C–C π bond.
- $Ph_3P=O$ is formed as by-product.

[5] Addition of 1° amines (21.11)

- The reaction is fastest at pH 4–5.
- The intermediate carbinolamine is unstable, and loses H_2O to form the C=N.

[6] Addition of 2° amines (21.12)

- The reaction is fastest at pH 4–5.
- The intermediate carbinolamine is unstable, and loses H_2O to form the C=C.

[7] Addition of H₂O—Hydration (21.13)

- The reaction is reversible. Equilibrium favors the product only with less stable carbonyl compounds (e.g., H_2CO and Cl_3CCHO).
- The reaction is catalyzed with either H^+ or ^-OH.

[8] Addition of alcohols (21.14)

- The reaction is reversible.
- The reaction is catalyzed with acid.
- Removal of H_2O drives the equilibrium to favor the products.

Other reactions
[1] Synthesis of Wittig reagents (21.10A)

- Step [1] is best with CH_3X and RCH_2X because the reaction follows an S_N2 mechanism.
- A strong base is needed for proton removal in Step [2].

Aldehydes and Ketones 21–3

[2] Conversion of cyanohydrins to aldehydes and ketones (21.9)

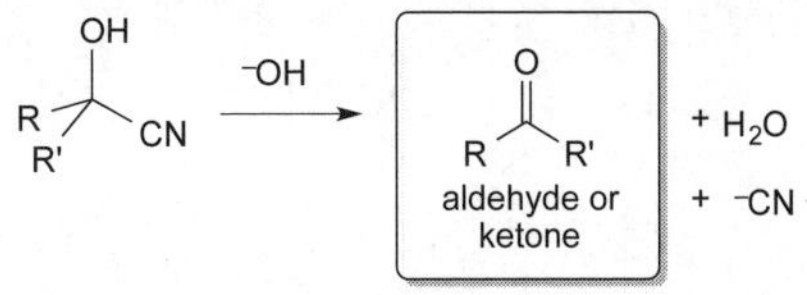

- This reaction is the reverse of cyanohydrin formation.

[3] Hydrolysis of nitriles (21.9)

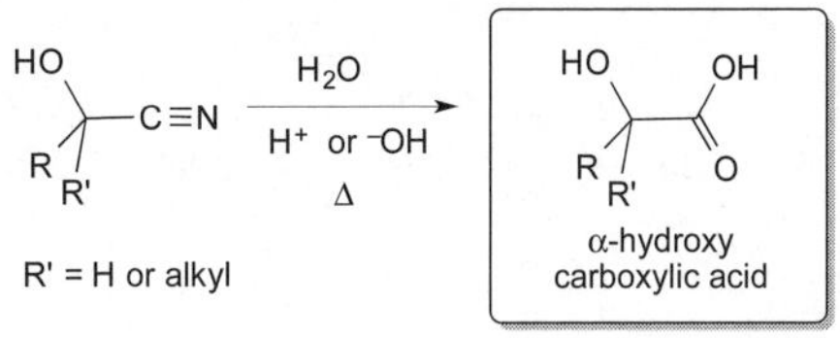

R' = H or alkyl

α-hydroxy
carboxylic acid

[4] Hydrolysis of imines and enamines (21.12)

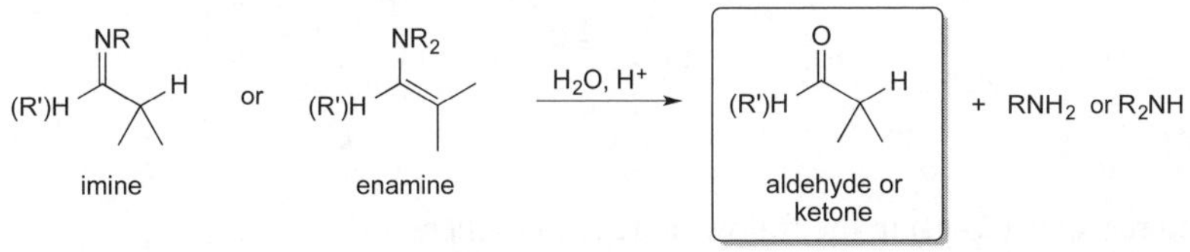

imine

enamine

aldehyde or
ketone

+ RNH$_2$ or R$_2$NH

[5] Hydrolysis of acetals (21.14)

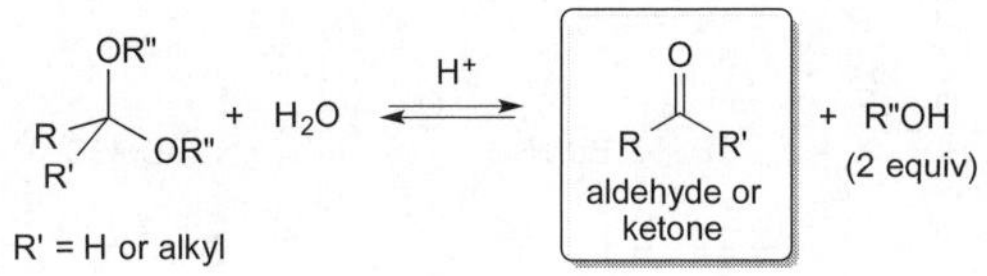

R' = H or alkyl

aldehyde or
ketone

+ R"OH
(2 equiv)

- The reaction is acid catalyzed and is the reverse of acetal synthesis.
- A large excess of H$_2$O drives the equilibrium to favor the products.

Chapter 21–4

Practice Test on Chapter Review

1. Give the IUPAC name for the following compounds.

   a. 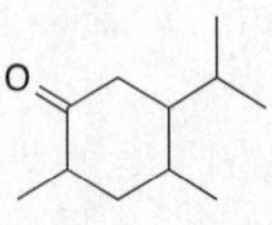   b. 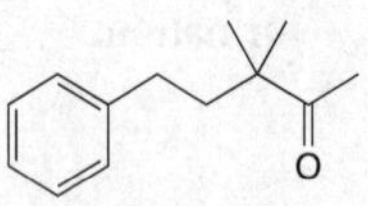   c.

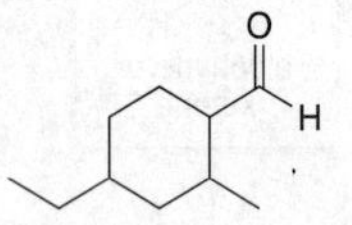

2. (a) Considering compounds **A–D,** which compound forms the smallest amount of hydrate? (b) Which compound forms the largest amount of hydrate?

          A   **B**   **C**   **D**

3. (a) Considering compounds **A–D,** which compound absorbs at the *lowest* wavenumber in its IR spectrum? (b) Which compound absorbs at the *highest* wavenumber in its IR spectrum?

          A   **B**   **C**   **D**

4. Fill in the lettered reagents **(A–G)** in the following reaction scheme.

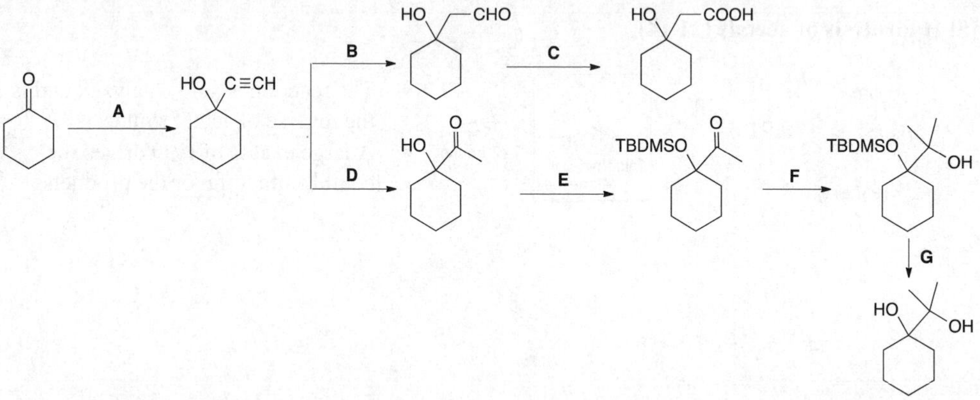

Aldehydes and Ketones 21–5

5. Draw the organic products formed in the following reactions.

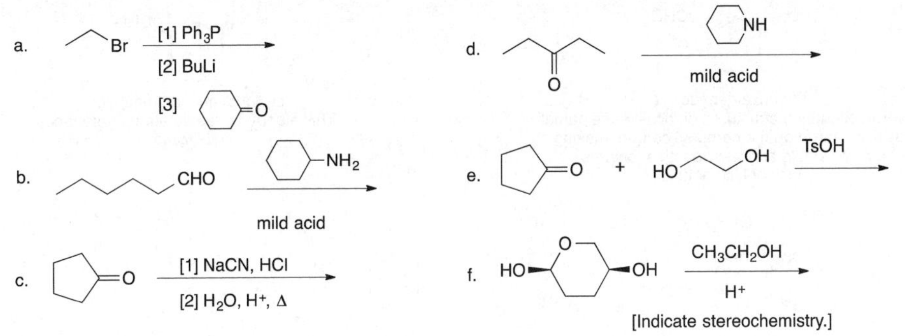

a. [structure: CH₂Br with chain] [1] Ph₃P / [2] BuLi / [3] cyclohexanone (=O)

b. [structure: pentanal chain CHO] + cyclohexyl-NH₂ → mild acid

c. [cyclopentanone =O] [1] NaCN, HCl / [2] H₂O, H⁺, Δ

d. [pentan-2-one structure with O] + piperidine (NH) → mild acid

e. [cyclopentanone =O] + HO—CH₂CH₂—OH → TsOH

f. HO—[tetrahydropyran ring]—OH + CH₃CH₂OH → H⁺
[Indicate stereochemistry.]

Answers to Practice Test

1.a. 5-isopropyl-2,4-
dimethyl-
cyclohexanone
 b. 3,3-dimethyl-5-
phenylpentan-2-one
 c. 4-ethyl-2-methyl-
cyclohexane-
carbaldehyde

4.
A = [1] LiC≡CH; [2] H₂O
B = [1] R₂BH; [2] H₂O₂, ⁻OH
C = Ag₂O, NH₄OH
D = H₂O, H₂SO₄, HgSO₄
E = TBDMS–Cl, imidazole
F = [1] CH₃Li; [2] H₂O
G = Bu₄N⁺F⁻

5.
a. [methylenecyclohexane with ethyl]
b. [chain with =N–cyclohexyl]
c. [cyclopentane with OH and CO₂H]
d. [structure with N-piperidine] (E + Z)
e. [spirocyclic dioxolane]
f. [tetrahydropyran with OEt and OH] + [tetrahydropyran with OEt and OH]

2.a. **D**
 b. **C**
3.a. **B**
 b. **C**

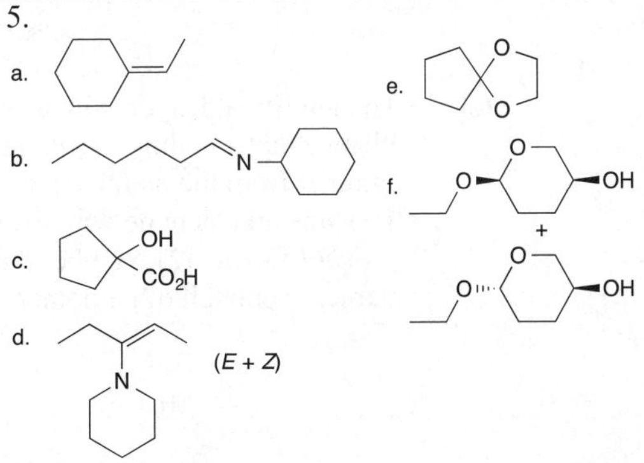

Answers to Problems

21.1 As the number of R groups bonded to the carbonyl C increases, reactivity towards nucleophilic attack decreases. Steric hindrance decreases reactivity as well.

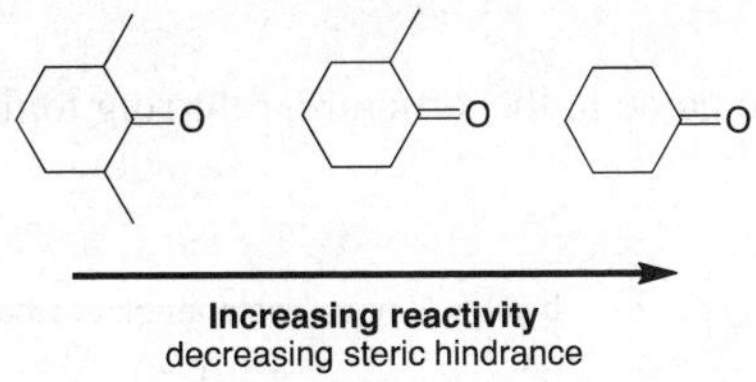

[structures: 2,6-dimethylcyclohexanone, 2-methylcyclohexanone, cyclohexanone]

→ **Increasing reactivity**
decreasing steric hindrance

Chapter 21–6

21.2 More stable aldehydes are less reactive towards nucleophilic attack.

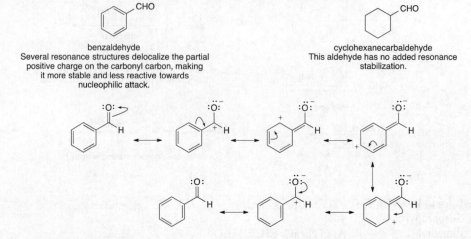

21.3
- To name an aldehyde with a chain of atoms: [1] Find the longest chain with the CHO group and change the *-e* ending to *-al*. [2] Number the carbon chain to put the CHO at C1, but omit this number from the name. Apply all other nomenclature rules.
- To name an aldehyde with the CHO bonded to a ring: [1] Name the ring and add the suffix *-carbaldehyde*. [2] Number the ring to put the CHO group at C1, but omit this number from the name. Apply all other nomenclature rules.

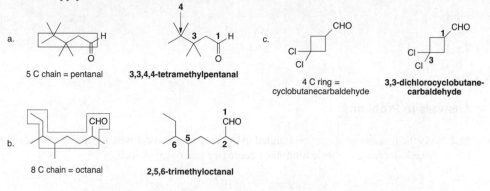

21.4 Work backwards from the name to the structure, referring to the nomenclature rules in Answer 21.3.

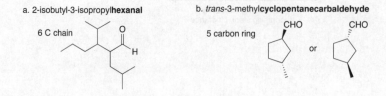

Aldehydes and Ketones 21–7

c. 1-methyl**cyclopropanecarbaldehyde**

3 carbon ring

d. 3,6-diethyl**nonanal**

9 C chain

21.5 • To name an acyclic ketone: [1] Find the longest chain with the carbonyl group and change the *-e* ending to *-one*. [2] Number the carbon chain to give the carbonyl C the lower number. Apply all other nomenclature rules.
• To name a cyclic ketone: [1] Name the ring and change the *-e* ending to *-one*. [2] Number the C's to put the carbonyl C at C1 and give the next substituent the lower number. Apply all other nomenclature rules.

a.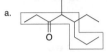
8 C chain = octanone

5-ethyl-4-methyloctan-3-one

c.
5 C chain = pentanone

2,2,4,4-tetramethylpentan-3-one

b.
5 C ring = cyclopentanone

3-*tert*-butyl-2-methylcyclopentanone

21.6 Most common names are formed by naming both alkyl groups on the carbonyl C, arranging them alphabetically, and adding the word ketone.

a. *sec*-butyl ethyl ketone

b. methyl vinyl ketone

c. *p*-ethylacetophenone

d. 3-benzoyl-2-benzylcyclopentanone

benzyl group: benzoyl group: 5 C ketone

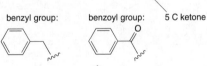

e. 6,6-dimethylcyclohex-2-enone

6 C ketone

f. 3-ethylhex-5-enal

21.7 Compounds with both a C–C double bond and an aldehyde are named as enals.

a. (*Z*)-3,7-dimethylocta-2,6-dienal

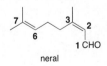
neral

b. (2*E*,6*Z*)-nona-2,6-dienal

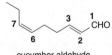
cucumber aldehyde

Chapter 21–8

21.8 Even though both compounds have polar C–O bonds, the electron pairs around the sp^3 hybridized O atom of diethyl ether are more crowded and less able to interact with electron-deficient sites in other diethyl ether molecules. The O atom of the carbonyl group of butan-2-one extends out from the carbon chain, making it less crowded. The lone pairs of electrons on the O atom can more readily interact with the electron-deficient sites in the other molecules, resulting in stronger forces and a higher boiling point.

butan-2-one   diethyl ether

21.9 For cyclic ketones, the carbonyl absorption shifts to higher wavenumber as the size of the ring decreases and the ring strain increases. Conjugation of the carbonyl group with a C=C or a benzene ring shifts the absorption to lower wavenumber.

a.   CHO   or   CHO

conjugated C=O   **higher wavenumber**
lower wavenumber

b.   =O   or   =O

smaller ring
higher wavenumber

21.10 The number of lines in their ^{13}C NMR spectra can distinguish the constitutional isomers.

pentan-2-one   pentan-3-one   3-methylbutan-2-one
5 lines   3 lines   4 lines

21.11

a.   [1] DIBAL-H   [2] H₂O

c.   [1] R₂BH   [2] H₂O₂, HO⁻

b.   PCC

d.   [1] O₃   [2] Zn, H₂O

21.12

a.   Cl / AlCl₃

c.   H₂O / H₂SO₄ HgSO₄

b.   Cl / [1] (CH₃)₂CuLi / [2] H₂O

Aldehydes and Ketones 21–9

21.13

Cleave this C=C with O_3.

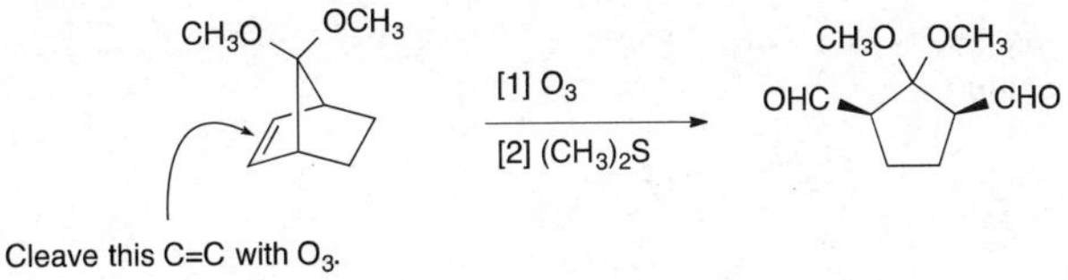

21.14 Addition of hydride or R–M occurs at a planar carbonyl C, so two different configurations at a new stereogenic center are possible.

a.

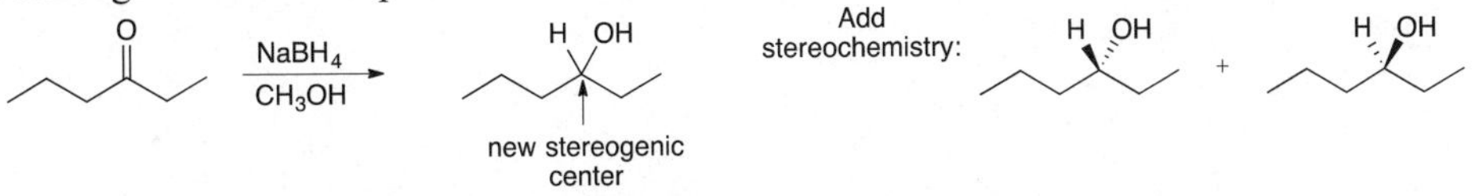

new stereogenic center

Add stereochemistry:

b.

Add stereochemistry:

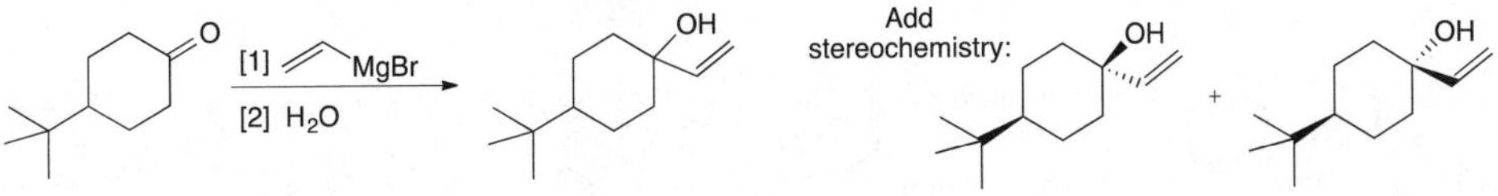

21.15 Treatment of an aldehyde or ketone with NaCN, HCl adds HCN across the double bond. Cyano groups are hydrolyzed by H_3O^+ to replace the three C–N bonds with three C–O bonds.

a.

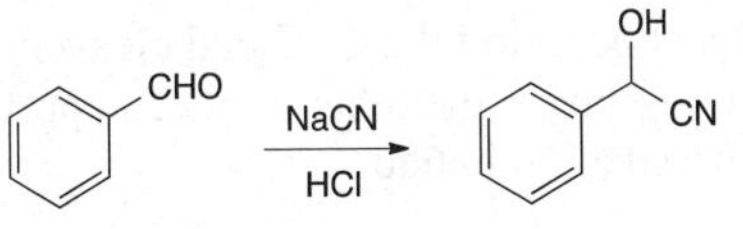

b.

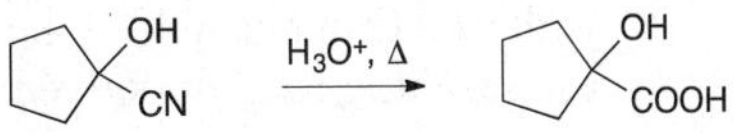

21.16

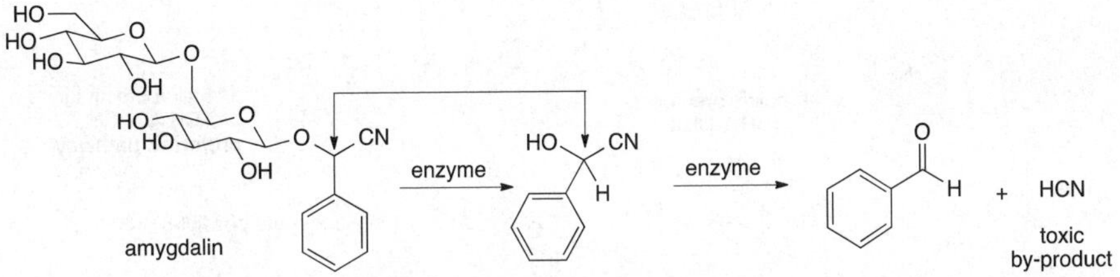

amygdalin

enzyme

enzyme

HCN

toxic by-product

21.17

a.

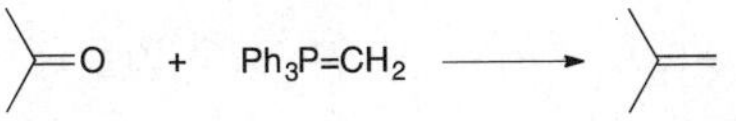

b.

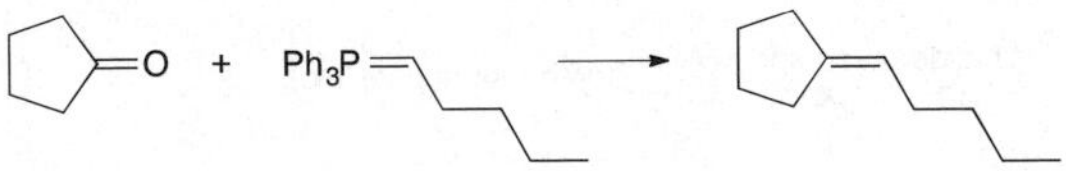

Chapter 21–10

21.18

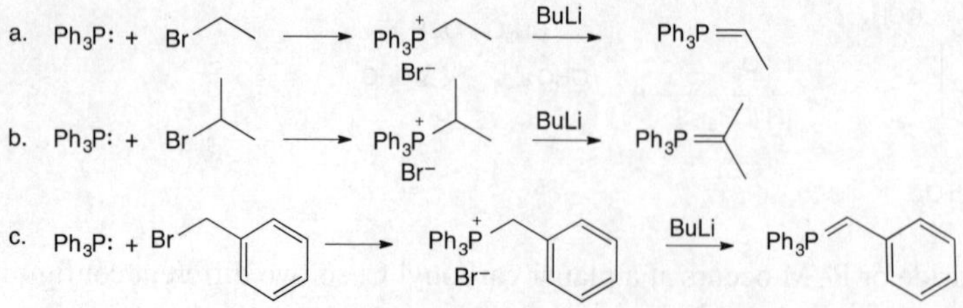

a. Ph₃P: + Br⁀ ⟶ Ph₃P⁺⁀ Br⁻ ⟶ (BuLi) Ph₃P=⁀

b. Ph₃P: + Br⁀ ⟶ Ph₃P⁺⁀ Br⁻ ⟶ (BuLi) Ph₃P=⁀

c. Ph₃P: + Br⁀⟶ Ph₃P⁺⁀ Br⁻ ⟶ (BuLi) Ph₃P=⁀

21.19

a.

b.

c.

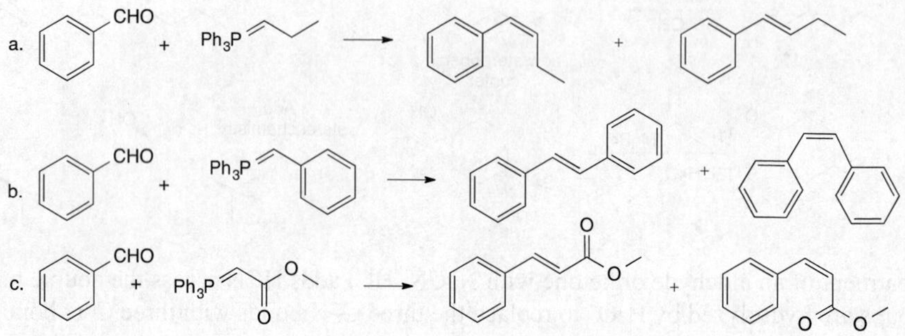

21.20 To draw the starting materials of the Wittig reactions, find the C=C and cleave it. Replace it with a C=O in one half of the molecule and a C=PPh₃ in the other half. The preferred pathway uses a Wittig reagent derived from a less hindered alkyl halide.

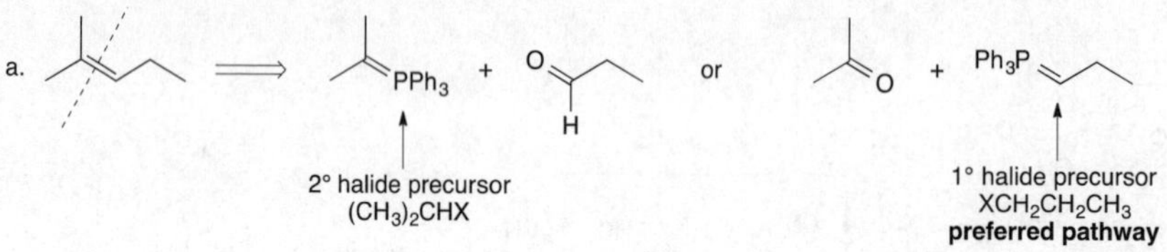

a.

2° halide precursor
(CH₃)₂CHX

1° halide precursor
XCH₂CH₂CH₃
preferred pathway

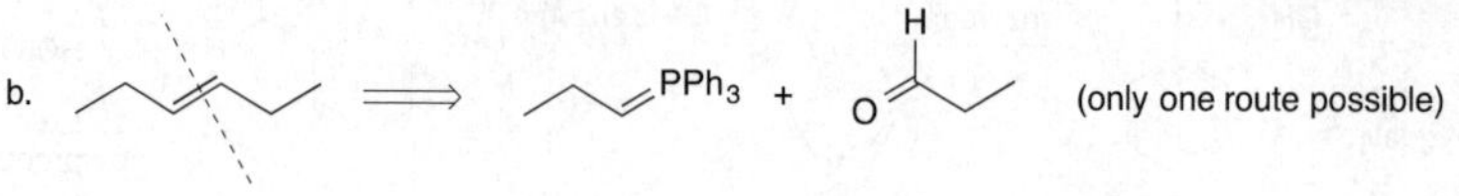

b. (only one route possible)

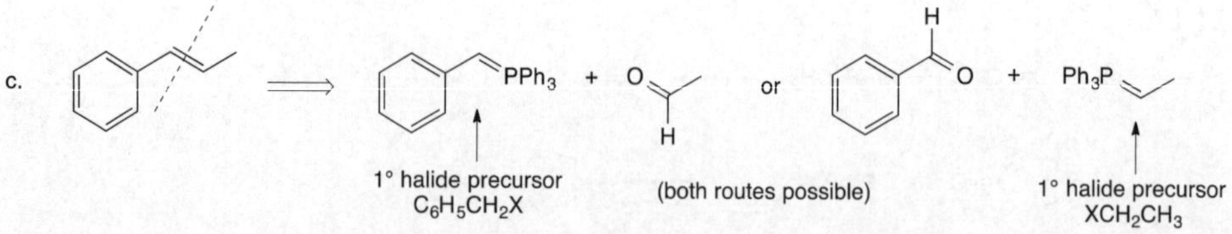

c.

1° halide precursor
C₆H₅CH₂X

(both routes possible)

1° halide precursor
XCH₂CH₃

Aldehydes and Ketones 21–11

21.21

a. Two-step sequence:

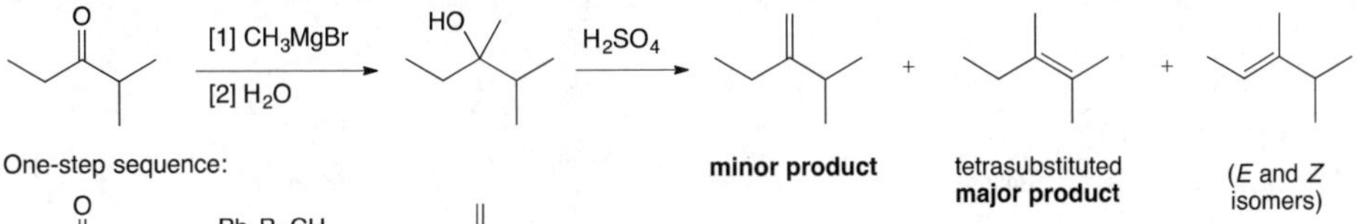

[1] CH$_3$MgBr
[2] H$_2$O

HO

H$_2$SO$_4$

minor product   tetrasubstituted   (*E* and *Z*
major product   isomers)

One-step sequence:

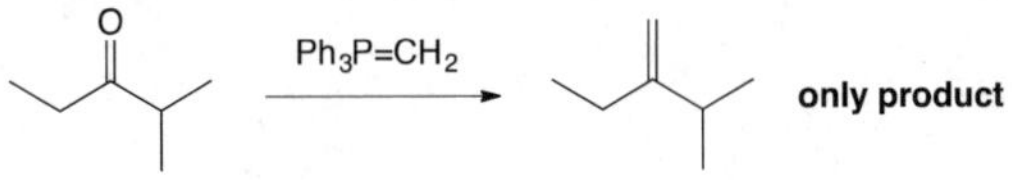

Ph$_3$P=CH$_2$

only product

b. Two-step sequence:

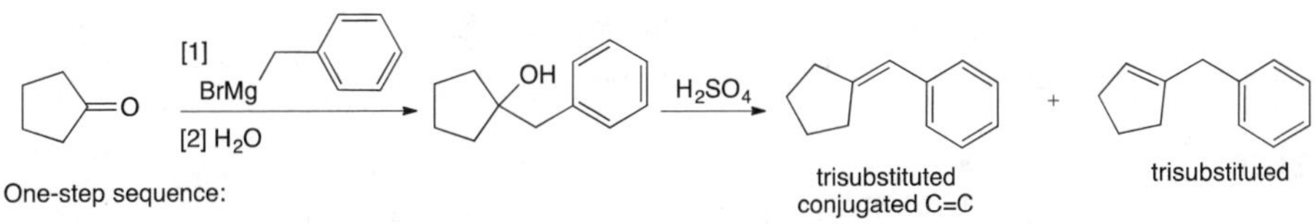

[1]

BrMg

[2] H$_2$O

OH

H$_2$SO$_4$

trisubstituted
conjugated C=C

trisubstituted

One-step sequence:

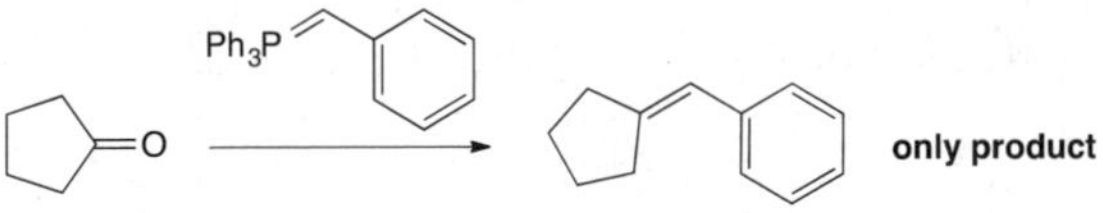

Ph$_3$P

only product

21.22 When a 1° amine reacts with an aldehyde or ketone, the C=O is replaced by C=NR.

a. —CHO

NH$_2$

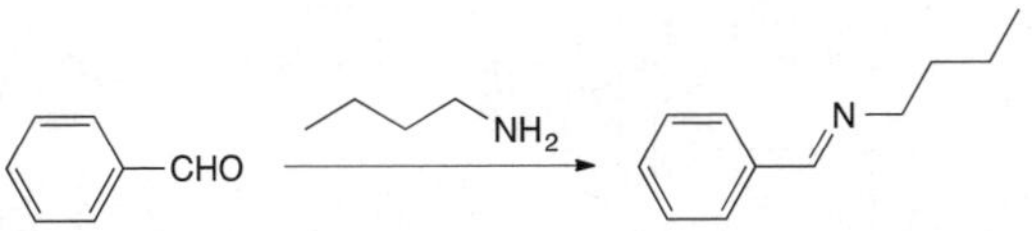

b.

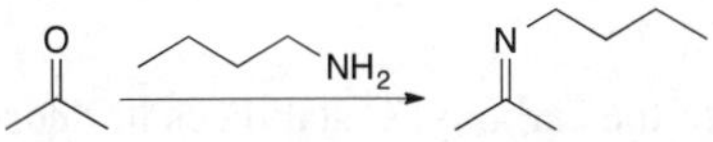

NH$_2$

c.

NH$_2$

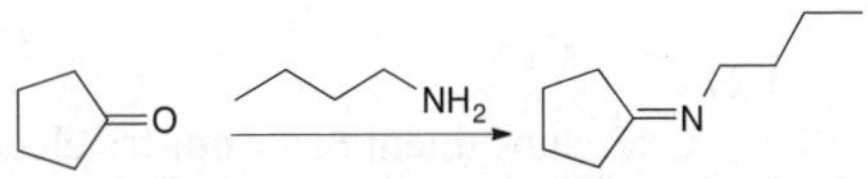

21.23 Remember that the C=NR is formed from a C=O and an NH$_2$ group of a 1° amine.

a.

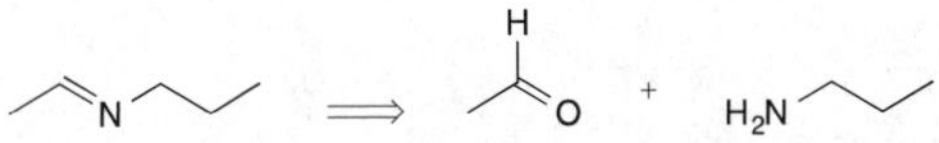

⟹

+   H$_2$N

b.

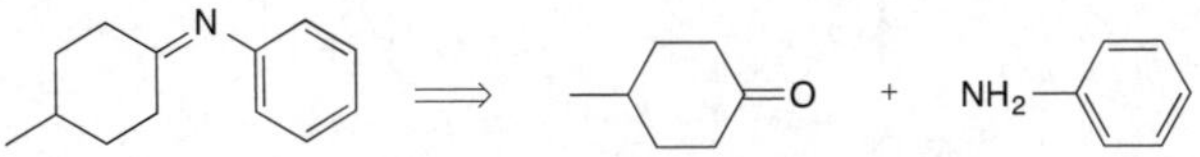

⟹

+   NH$_2$—

21.24

+

CH$_3$
N—H
CH$_3$

N(CH$_3$)$_2$

+

N(CH$_3$)$_2$

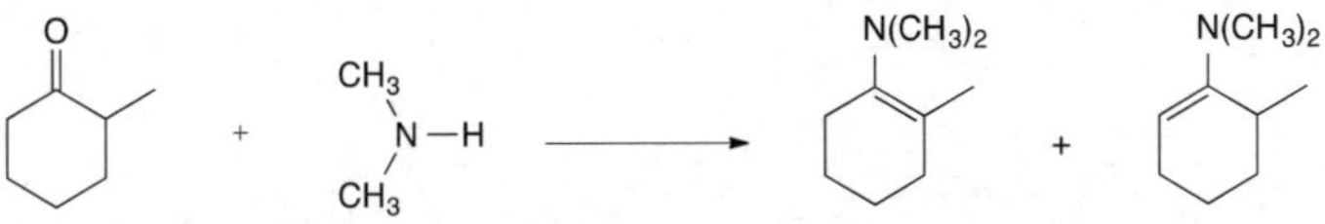

Chapter 21–12

21.25 • Imines are hydrolyzed to 1° amines and a carbonyl compound.
• Enamines are hydrolyzed to 2° amines and a carbonyl compound.

a.
imine $\xrightarrow[\text{H}^+]{\text{H}_2\text{O}}$ + H_2N— (1° amine)

b.
enamine $\xrightarrow[\text{H}^+]{\text{H}_2\text{O}}$ 2° amine   +

c.
enamine $\xrightarrow[\text{H}^+]{\text{H}_2\text{O}}$ $(\text{CH}_3)_2\text{NH}$ +
   2° amine

21.26

proton transfer

$\text{H}_2\ddot{\text{O}}:$ +

21.27

• A substituent that **donates** electron density to the carbonyl C stabilizes it, **decreasing** the percentage of hydrate at equilibrium.
• A substituent that **withdraws** electron density from the carbonyl C destabilizes it, **increasing** the percentage of hydrate at equilibrium.

a.   or

one R group on C=O   2 R groups
higher percentage of hydrate   on C=O

b.   or

F atoms are electron withdrawing.
higher percentage of hydrate

21.28

(+ 1 resonance structure)   + $\text{H}_3\ddot{\text{O}}^+$

Aldehydes and Ketones 21–13

21.29 Treatment of an aldehyde or ketone with two equivalents of alcohol results in the formation of an acetal (a C bonded to two OR groups).

a. [structure: cyclopentanone] + 2 CH₃OH →(TsOH)→ [structure: cyclopentane with OCH₃, OCH₃]

b. [structure: ketone] + HO–CH₂CH₂–OH →(TsOH)→ [structure: cyclic acetal]

21.30

a. [structure: cyclohexane with OCH₃ and OCH₃ on different carbons]

2 OR groups
on different C's
2 ethers

b. [structure: cyclohexane with OCH₃, OCH₃ on same C]

2 OR groups
on same C
acetal

c. [structure: bicyclic dioxolane]

2 OR groups
on same C
acetal

d. [structure: chain with OCH₃ and OH on same C]

1 OR group and
1 OH group
on same C
hemiacetal

21.31 The mechanism has two parts: [1] nucleophilic addition of ROH to form a hemiacetal; [2] conversion of the hemiacetal to an acetal.

[mechanism scheme: overall reaction]

overall reaction

+ H₂O

+ TsO⁻

hemiacetal

+ TsO–H

+ TsO⁻

+ H₂O:

carbocation
re-drawn

+ TsO–H

acetal

21.32

a. CH₃O OCH₃ [structure] + H₂O →(H₂SO₄)→ [ketone] + 2 CH₃OH

b. [bicyclic acetal structure] + H₂O →(H₂SO₄)→ [ketone] + [cyclohexane with OH, OH]

21.33

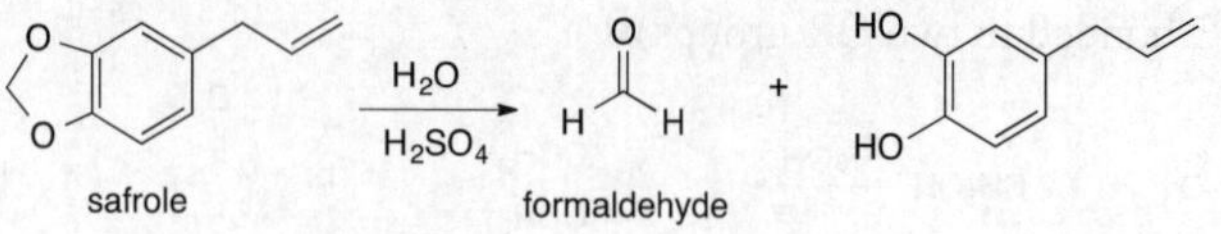
safrole
H_2O
H_2SO_4
formaldehyde
+

21.34

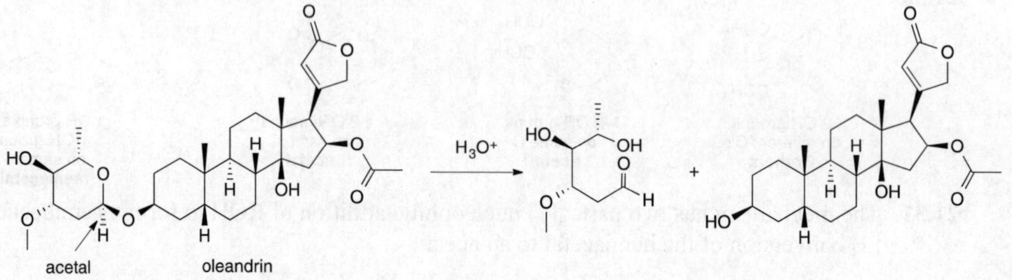
acetal
oleandrin
H_3O^+
+

21.35 Use an acetal protecting group to carry out the reaction.

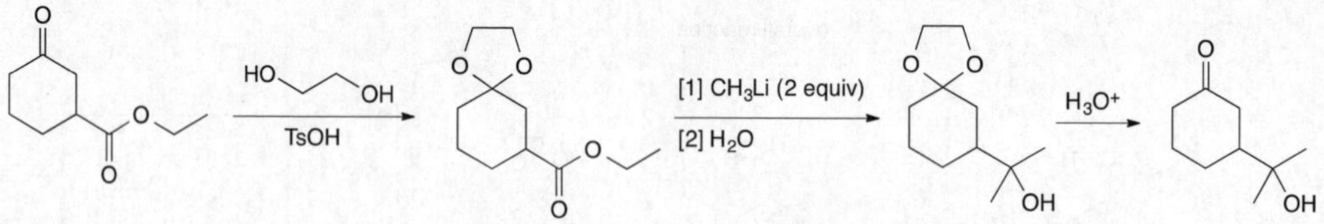
TsOH
[1] CH_3Li (2 equiv)
[2] H_2O
H_3O^+

21.36

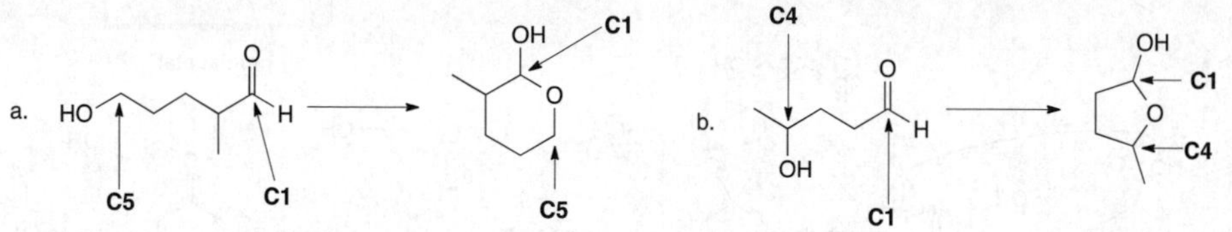
a.
C5
C1
b.
C4
C1

21.37

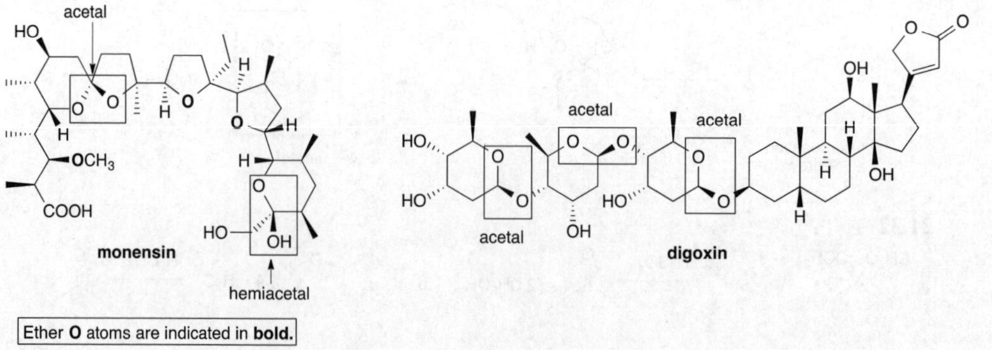
acetal
monensin
hemiacetal
acetal
acetal
acetal
digoxin
Ether O atoms are indicated in bold.

21.38 The hemiacetal OH is replaced by an OR group to form an acetal.

a. 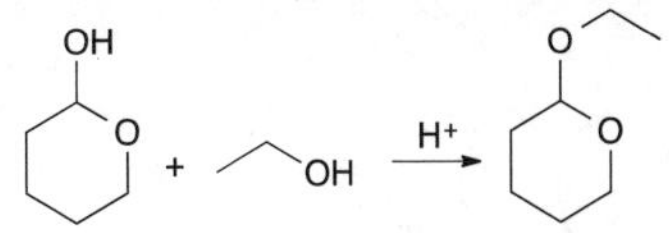 (tetrahydropyran-2-ol) + CH₃CH₂OH $\xrightarrow{H^+}$ (2-ethoxytetrahydropyran)

b. 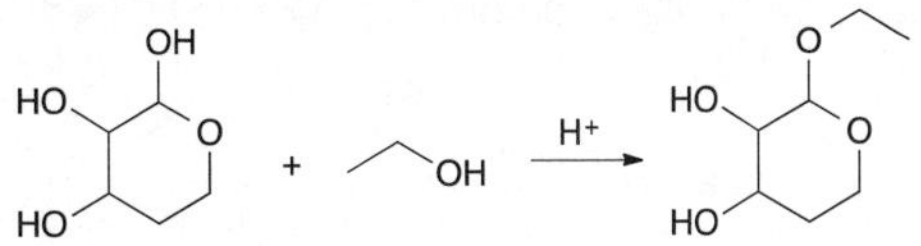 (trihydroxy pyranose) + CH₃CH₂OH $\xrightarrow{H^+}$ (2-ethoxy trihydroxy pyran)

21.39

a. 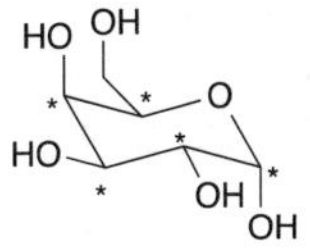 5 stereogenic centers (labeled with *)

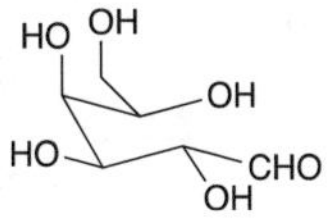

b. 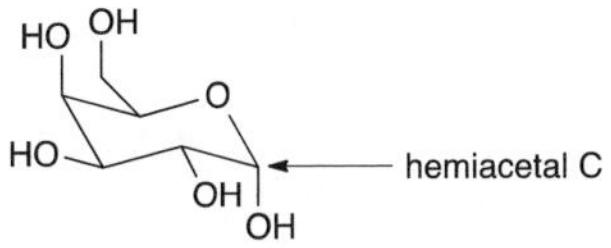 α-D-galactose — hemiacetal C

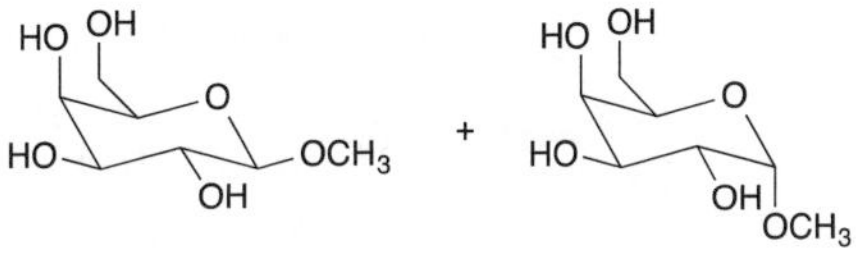

c. 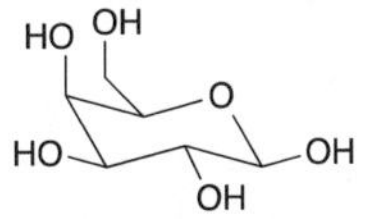 β-D-galactose

d.

e. (OCH₃ anomer) + (OCH₃ anomer)

21.40

a. 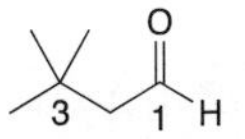 3,3-dimethylbutanal **A**

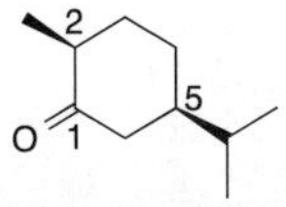

cis-5-isopropyl-2-methylcyclohexanone **B**

b. [1] **A** $\xrightarrow[\text{CH}_3\text{OH}]{\text{NaBH}_4}$ (alcohol)

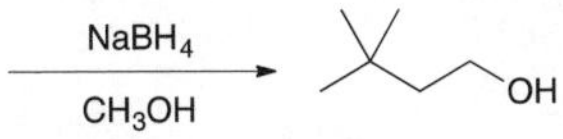

[2] **A** $\xrightarrow{\text{CH}_3\text{MgBr}}$ $\xrightarrow{\text{H}_2\text{O}}$ (alcohol) + (alcohol)

[3] **A** $\xrightarrow{\text{Ph}_3\text{P=CHOCH}_3}$ (enol ether OCH₃) (+ cis isomer)

[4] **A** $\xrightarrow[\text{mild H}^+]{\text{CH}_3\text{CH}_2\text{CH}_2\text{NH}_2}$ (imine)

[5] **A** $\xrightarrow[\text{H}^+]{\text{HOCH}_2\text{CH}_2\text{CH}_2\text{OH}}$ (cyclic acetal)

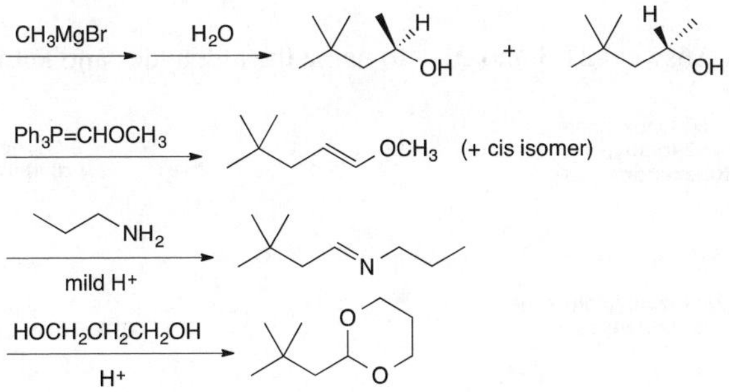

Chapter 21–16

[1] **B** $\xrightarrow[\text{CH}_3\text{OH}]{\text{NaBH}_4}$ [structures] +

[2] **B** $\xrightarrow{\text{CH}_3\text{MgBr}}$ $\xrightarrow{\text{H}_2\text{O}}$ [structures] +

[3] **B** $\xrightarrow{\text{Ph}_3\text{P=CHOCH}_3}$ [structures] + CH$_3$O [structure]

[4] **B** $\xrightarrow[\text{mild H}^+]{\text{NH}_2}$ [structure]

[5] **B** $\xrightarrow[\text{H}^+]{\text{HOCH}_2\text{CH}_2\text{CH}_2\text{OH}}$ [structure]

21.41 The least hindered carbonyl group is the most reactive.

least reactive   intermediate reactivity   most reactive

21.42

a. [structure] $\Longrightarrow$ [structure] CHO + HO—OH

b. [structure] $\Longrightarrow$ [structure] + [structure]

21.43 Use the rules from Answers 21.3 and 21.5 to name the aldehydes and ketones.

a. [structure]   6 C ring = cyclohexanone
5-ethyl-2-methyl-cyclohexanone

d. [structure] CHO   6 C = hexanal
3,4-diethylhexanal

b. [structure] CHO   **trans-2-benzylcyclohexane-carbaldehyde**

e. [structure] *E*   8 C = octenone
(E)-2,5-dimethyloct-5-en-4-one

c. [structure] NO$_2$ **o-nitroacetophenone**

f. [structure] CHO   6 C = hexenal
3,4-diethyl-2-methylhex-3-enal

Aldehydes and Ketones 21–17

21.44

a. 2-methyl-3-phenylbutanal

b. 3,3-dimethylcyclohexanecarbaldehyde

c. 3-benzoylcyclopentanone

d. 2-formylcyclopentanone

e. (R)-3-methylheptan-2-one

f. m-acetylbenzaldehyde

g. 2-sec-butylcyclopent-3-enone

h. 5,6-dimethylcyclohex-1-enecarbaldehyde

21.45

a.

$\begin{array}{c}\text{[1] Ph}_3\text{P} \\ \hline \text{[2] BuLi} \\ \text{[3] C}_6\text{H}_5\text{CH}_2\text{CH}_2\text{CHO}\end{array}$

(+ Z isomer)

b.

$\begin{array}{c}\text{[1] Ph}_3\text{P} \\ \hline \text{[2] BuLi} \\ \text{[3] CH}_3\text{CH}_2\text{CH}_2\text{CHO}\end{array}$

(+ Z isomer)

21.46

a.

$+ \text{ H}_2\text{N}-$

$\xrightarrow{\text{mild} \atop \text{acid}}$

b.

$\xrightarrow[\text{H}^+]{\text{HO} \quad \text{OH}}$

c.

$\xrightarrow{\text{H}_3\text{O}^+}$

$+ \text{ H}_2\text{N}-$

d. C$_6$H$_5$

$+$

$\xrightarrow{\text{mild} \atop \text{acid}}$

(E and Z isomers)

C$_6$H$_5$

e.

$\xrightarrow{\text{H}_3\text{O}^+, \Delta}$

Chapter 21–18

f.

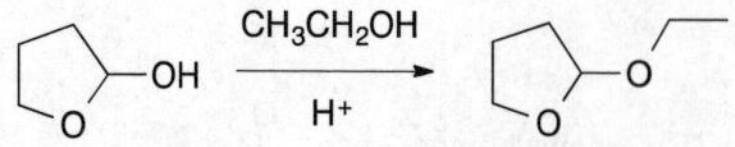

g.

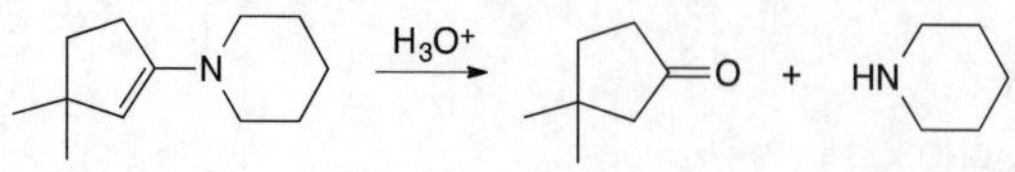

h.

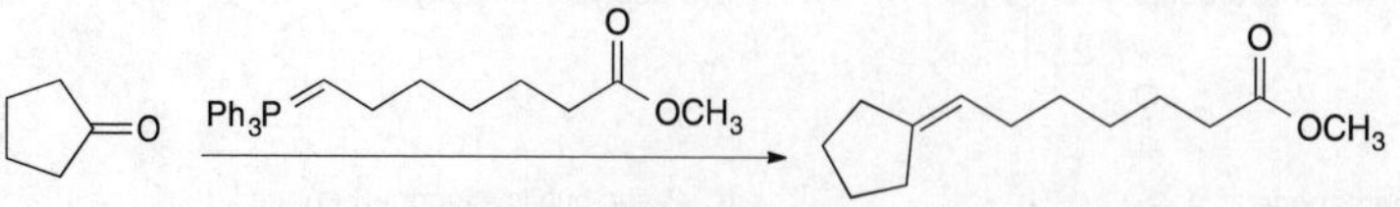

21.47

a.

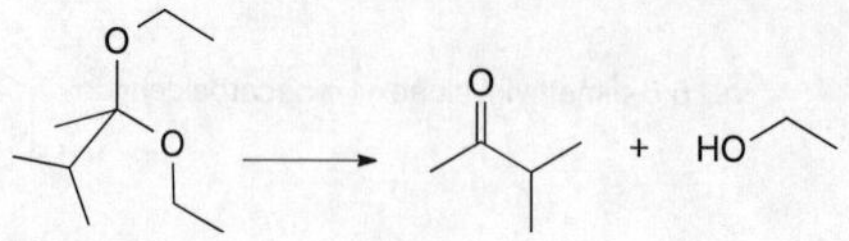

c.

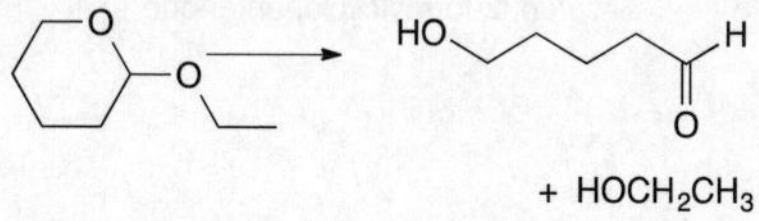

b.

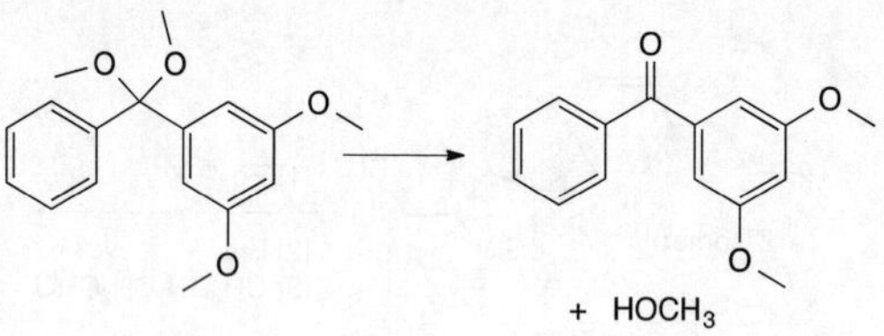

21.48

a.

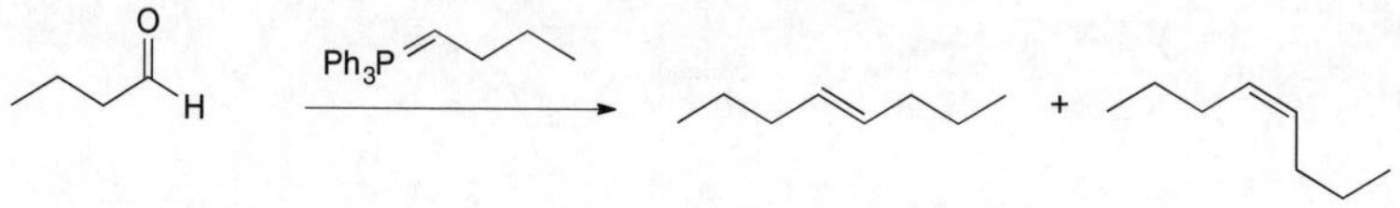

b.

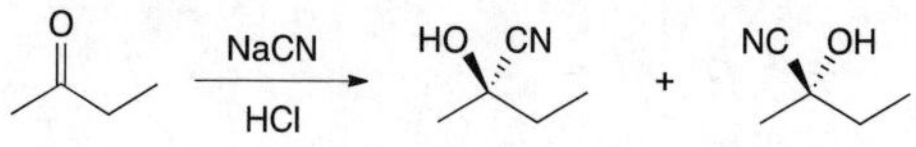

c.

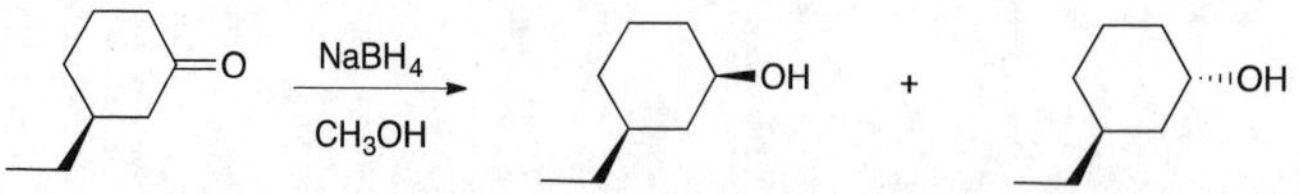

d.

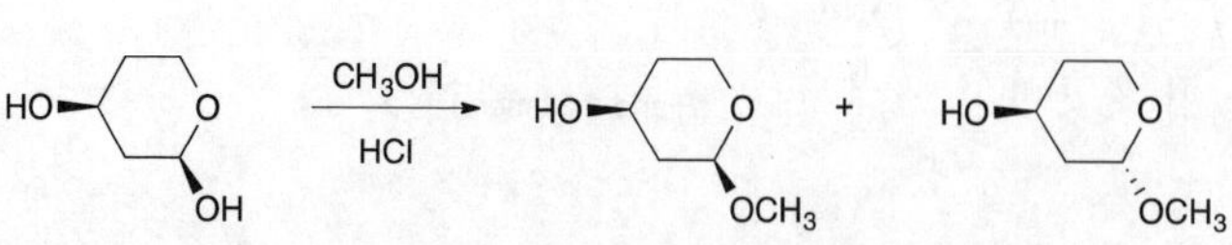

Aldehydes and Ketones 21–19

21.49

new stereogenic center

HO—CHO

A

achiral

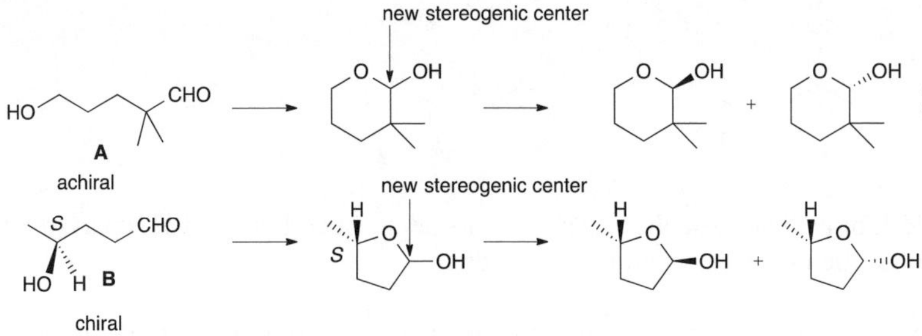

An equal mixture of enantiomers results, so the product is optically inactive.

new stereogenic center

S CHO

HO H **B**

chiral

A mixture of diastereomers results. Both compounds are chiral and they are not enantiomers, so the mixture is optically active.

21.50

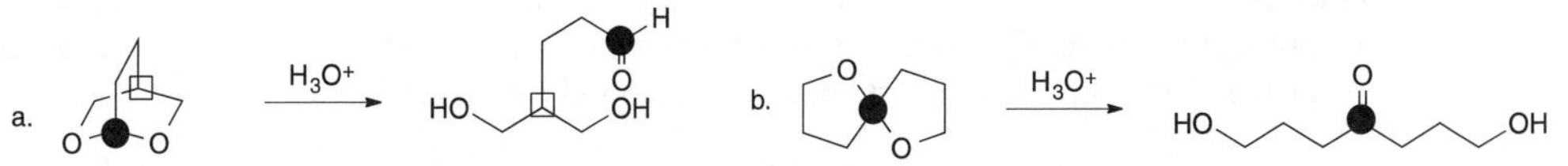

a. H_3O^+

b. H_3O^+

21.51

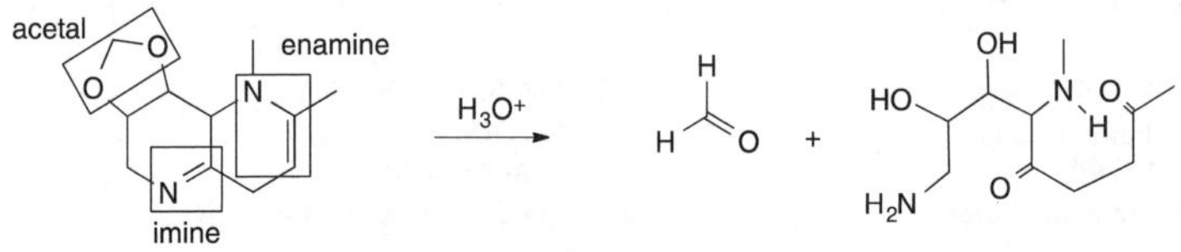

acetal   enamine

H_3O^+

imine

21.52

a.

acetal

acetal

acetal

etoposide

CH_3O OCH_3

OH

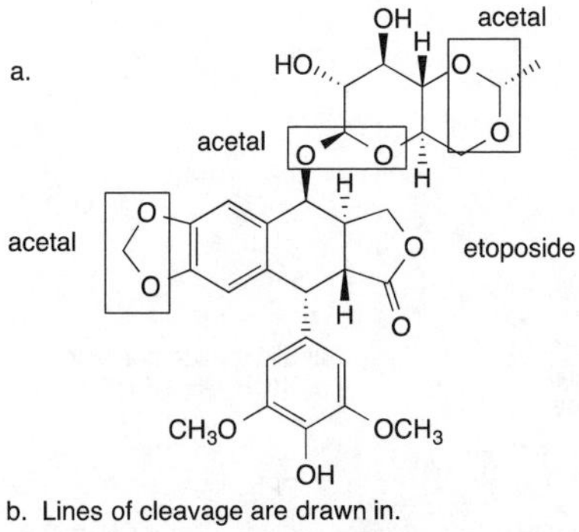

b. Lines of cleavage are drawn in.

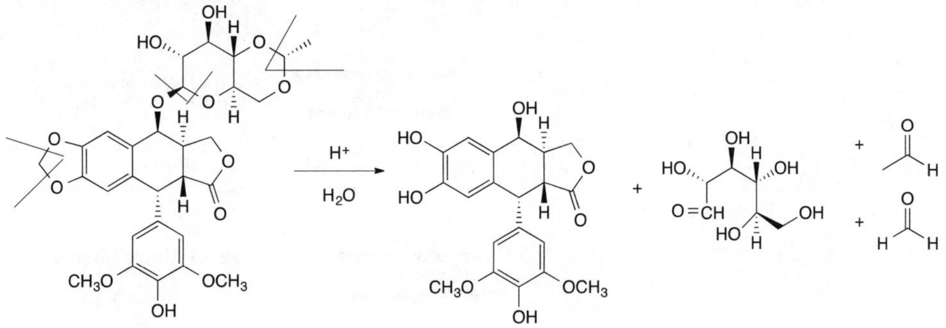

H^+
H_2O

CH_3O OCH_3

OH

CH_3O OCH_3

OH

Chapter 21–20

21.53 The less stable carbonyl compound forms the higher percentage of hydrate.

The aldehyde is destabilized because there is a δ+ on its adjacent carbon, which is part of a C=O. Thus, PhCOCHO has the higher concentration of hydrate.

21.54 Electron-donating groups decrease the amount of hydrate at equilibrium by stabilizing the carbonyl starting material. Electron-withdrawing groups increase the amount of hydrate at equilibrium by destabilizing the carbonyl starting material. Electron-donating groups make the IR absorption of the C=O shift to lower wavenumber because they stabilize the charge-separated resonance form, giving the C=O more single bond character.

	O_2N—⬡—$COCH_3$	CH_3O—⬡—$COCH_3$
	p-nitroacetophenone	*p*-methoxyacetophenone
a.	NO_2 withdrawing group **less stable**	CH_3O donating group **more stable**
b.	**higher percentage of hydrate**	**lower percentage of hydrate**
c.	**higher wavenumber**	**lower wavenumber**

21.55 Use the principles from Answer 21.20.

a.

    1° alkyl halide precursor
    ($XCH_2CH_2CH_2CH_3$)
    preferred pathway

    2° alkyl halide precursor
    [$(CH_3CH_2)_2CHX$]

b.

    methyl halide precursor
    (CH_3X)
    preferred pathway

    2° alkyl halide precursor
    ($CH_3CH_2CH_2CHXCH_3$)

c.

    1° alkyl halide precursor
    ($C_6H_5CH_2X$)
    preferred pathway

    2° alkyl halide precursor

Aldehydes and Ketones 21–21

21.56

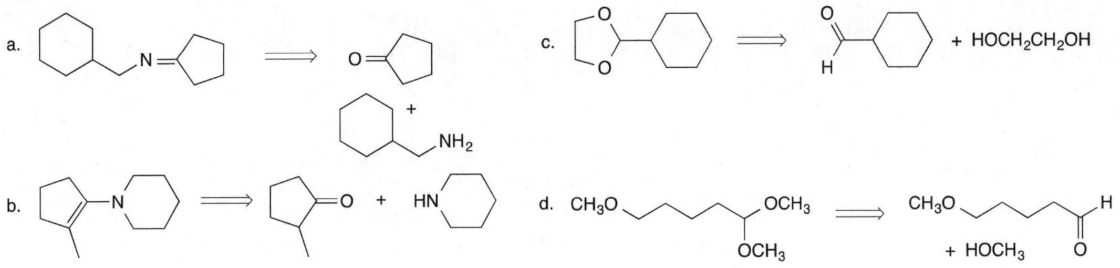

a.

c.

b.

d.

21.57

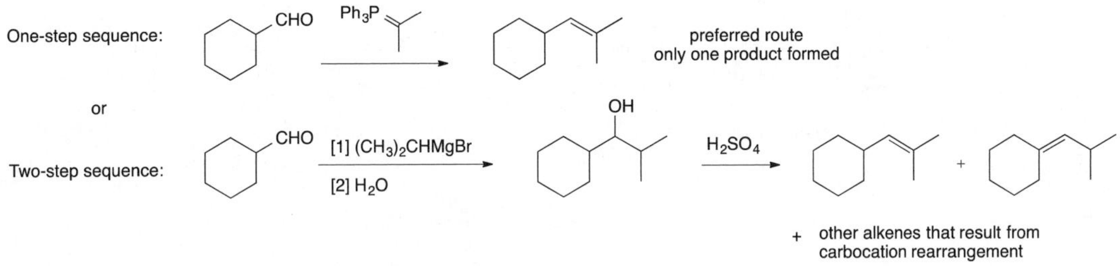

One-step sequence:

preferred route
only one product formed

or

Two-step sequence:

+ other alkenes that result from
carbocation rearrangement

21.58

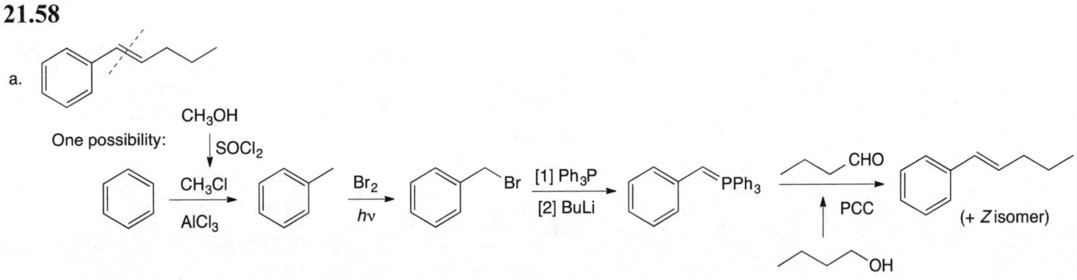

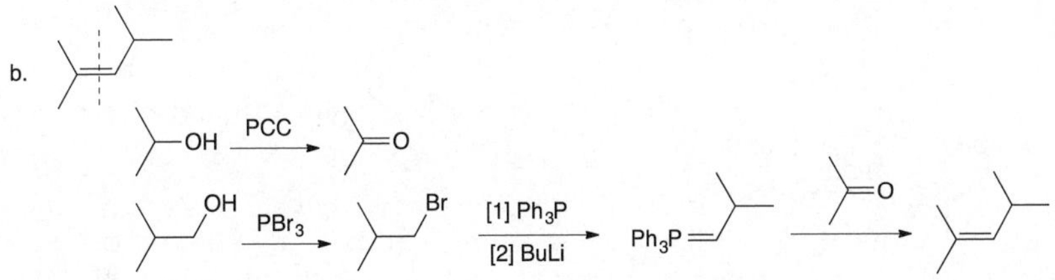

a.

One possibility:

(+ Z isomer)

b.

21.59

a.

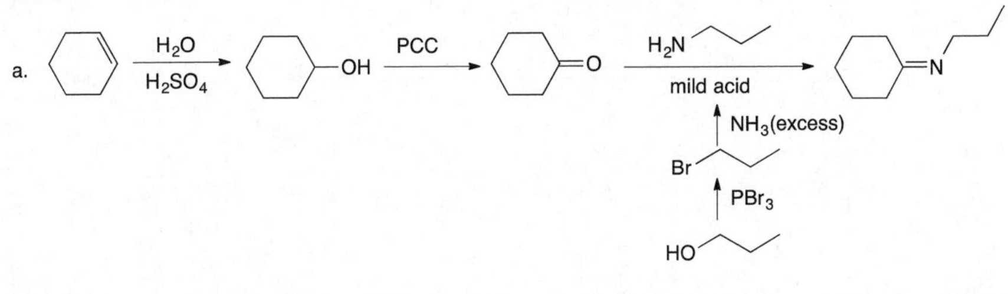

Chapter 21–22

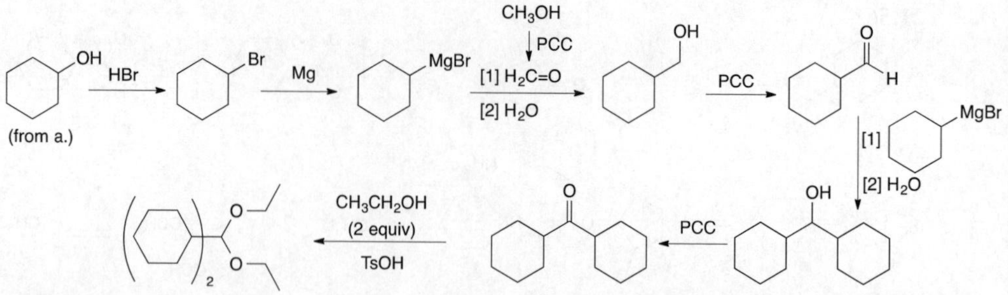

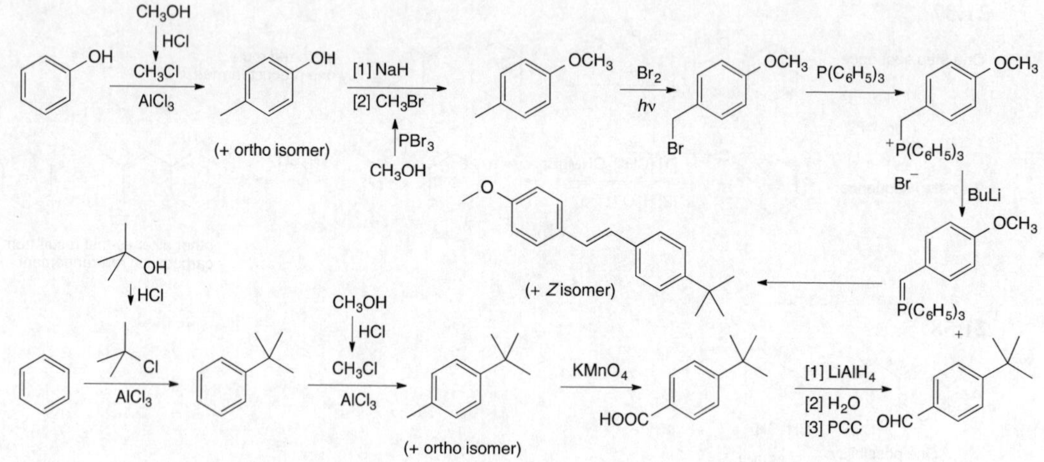

b.

21.60

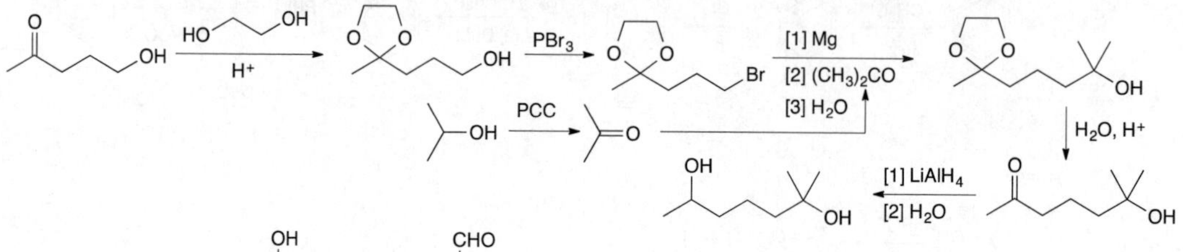

a.

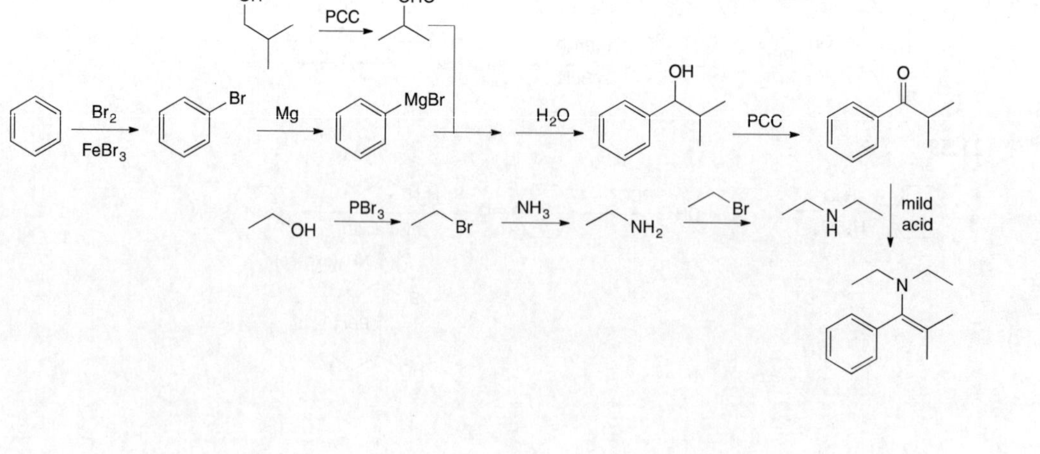

b.

c.

Aldehydes and Ketones 21–23

21.61

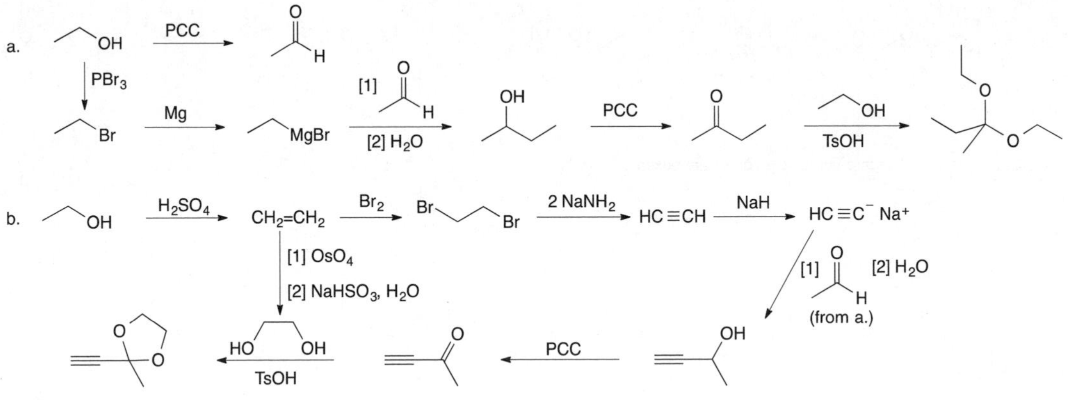

a.

PCC

PBr₃

Mg

[1]

[2] H₂O

PCC

TsOH

b.

H₂SO₄   CH₂=CH₂   Br₂   2 NaNH₂   HC≡CH   NaH   HC≡C⁻ Na⁺

[1] OsO₄

[2] NaHSO₃, H₂O

[1]   [2] H₂O

(from a.)

PCC

TsOH

21.62

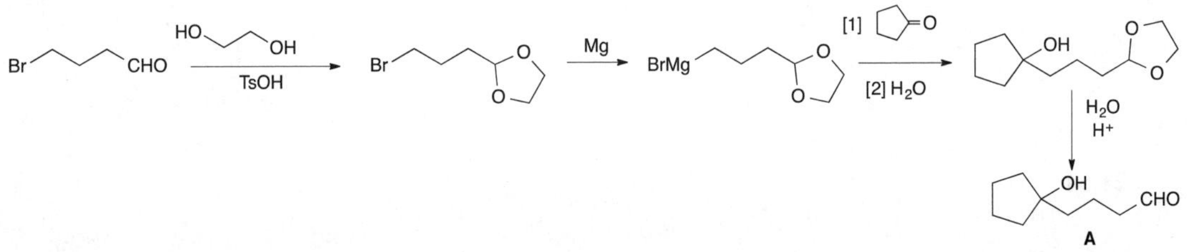

HO   OH

TsOH

Mg

[1]

[2] H₂O

H₂O
H⁺

A

21.63

a.   + Na⁺H⁻ ⟶   + H₂ + Na⁺   ⟶   + Cl⁻

methoxy methyl ether

b.   acetal

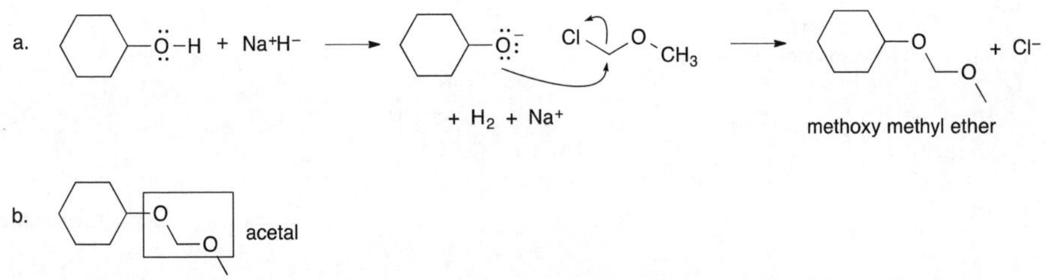

Chapter 21–24

c.

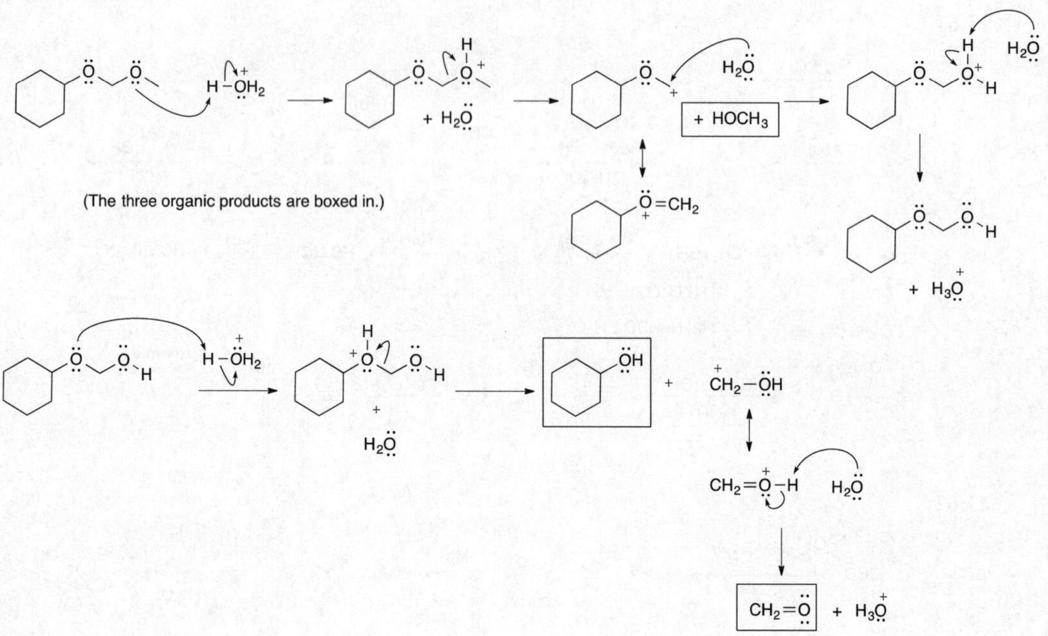

(The three organic products are boxed in.)

21.64

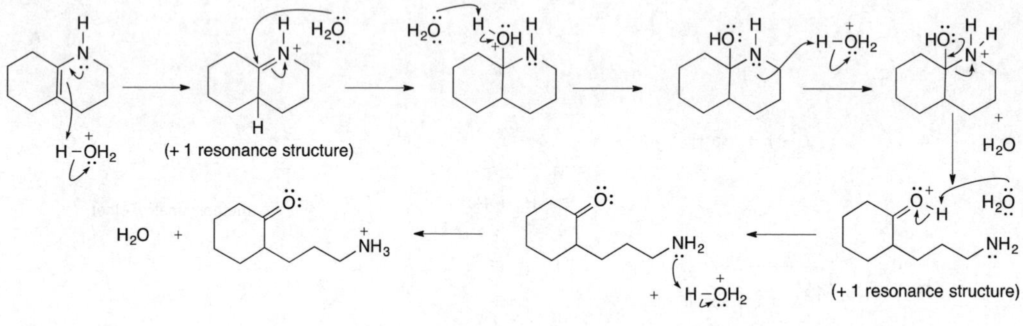

(+ 1 resonance structure)

(+ 1 resonance structure)

Aldehydes and Ketones 21–25

21.65

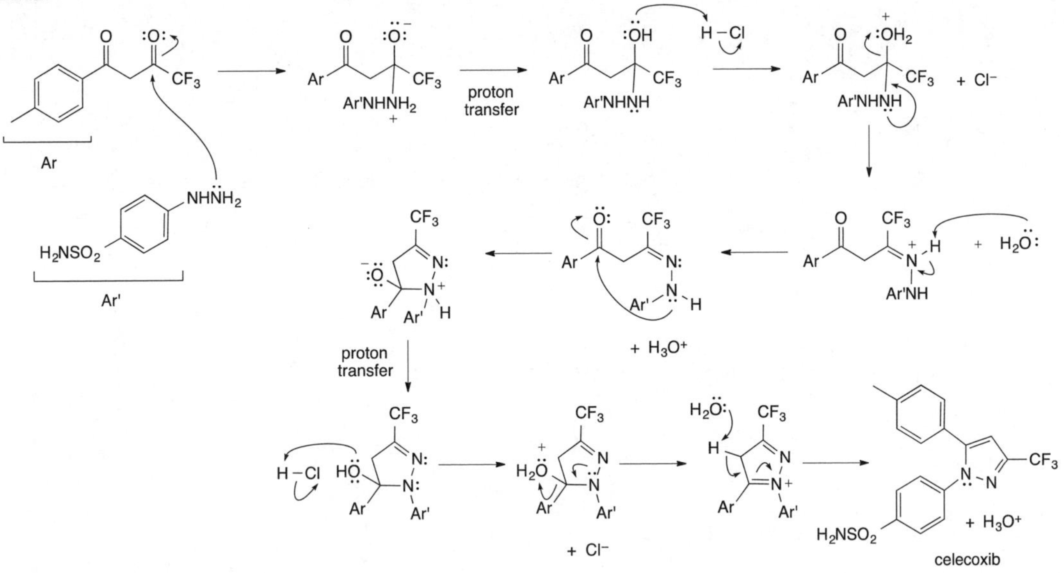

21.66 The OH groups react with the C=O in an intramolecular reaction, first to form a hemiacetal, and then to form an acetal.

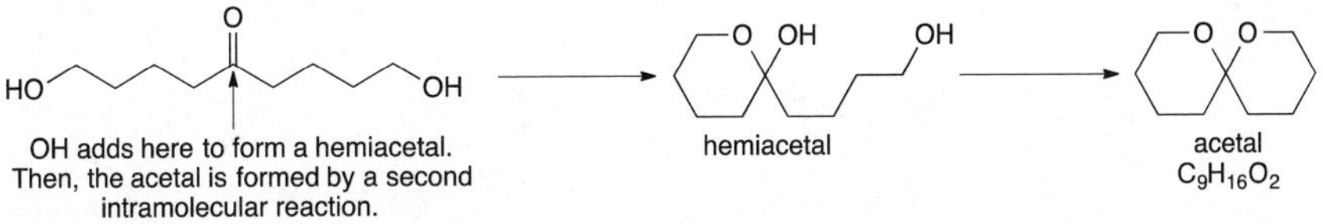

OH adds here to form a hemiacetal.
Then, the acetal is formed by a second
intramolecular reaction.

hemiacetal

acetal
$C_9H_{16}O_2$

21.67

a.

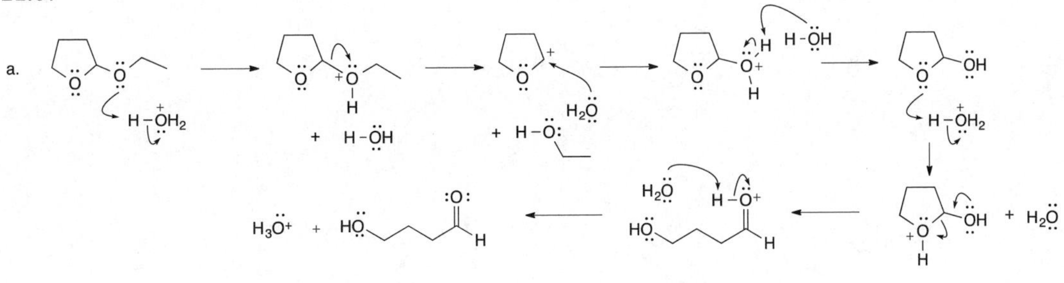

Chapter 21–26

b.

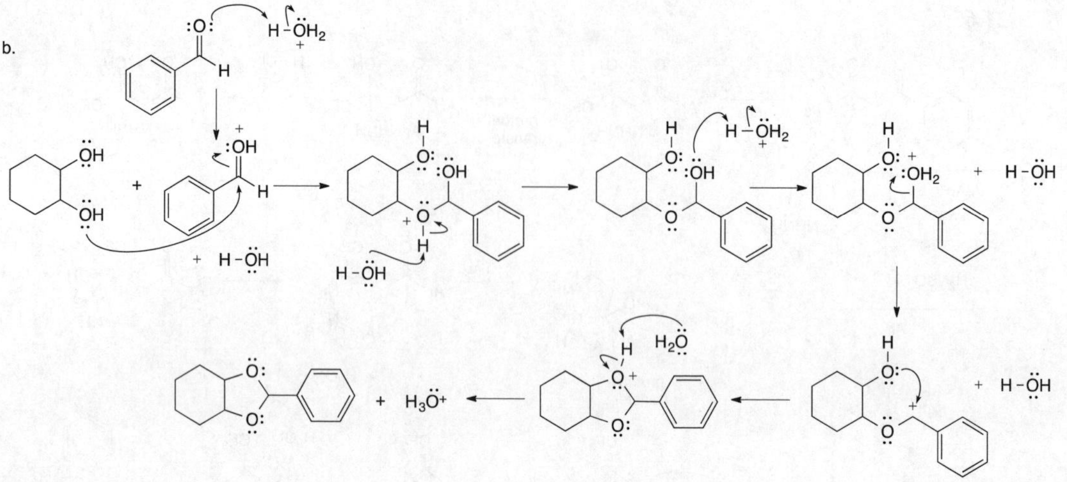

21.68

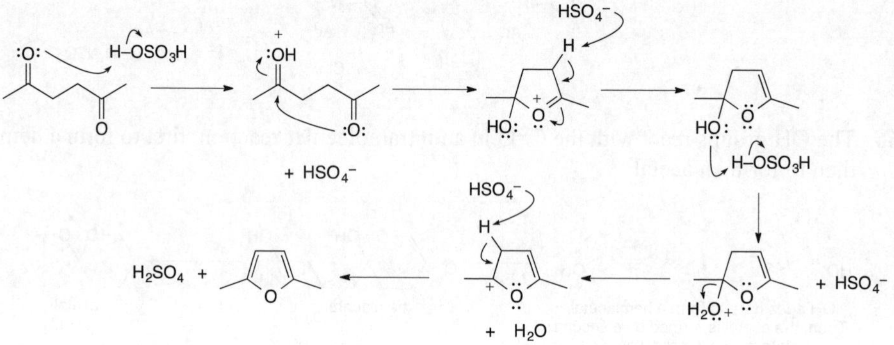

21.69

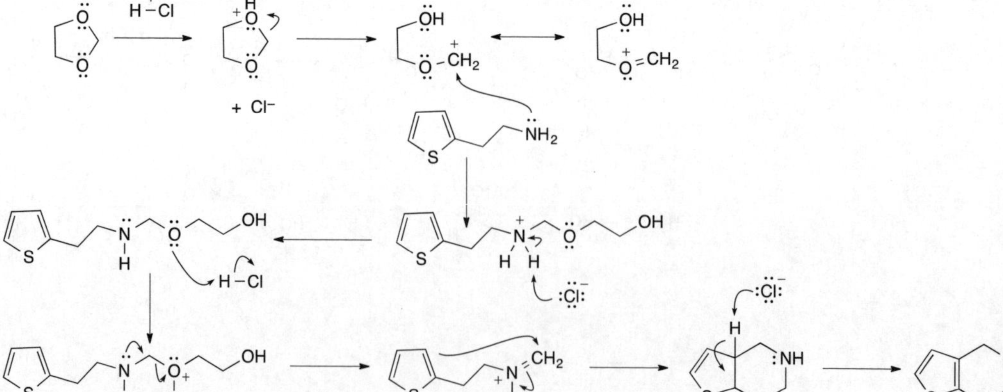

Aldehydes and Ketones 21–27

21.70

dopamine

proton
transfer

salsolinol

(+ 3 more resonance
structures)

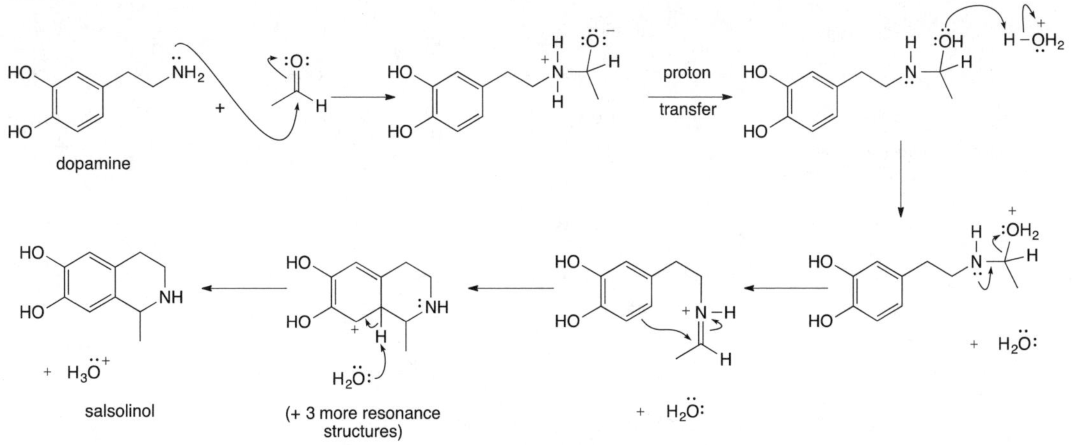

21.71 Hemiacetal **A** is in equilibrium with its acyclic hydroxy aldehyde. The aldehyde can undergo hydride reduction to form butane-1,4-diol and a Wittig reaction to form an alkene.

a.

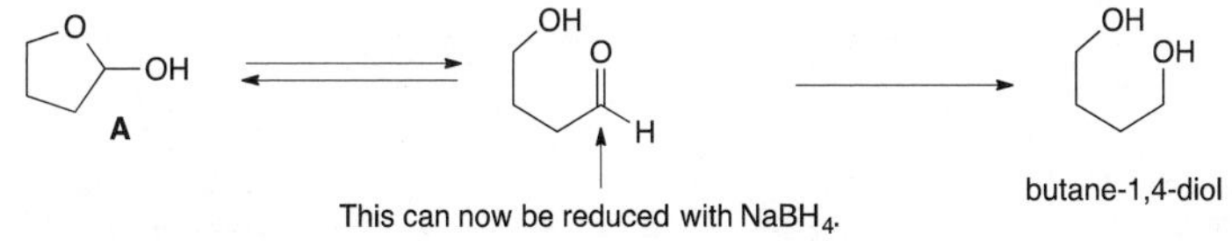

A

This can now be reduced with NaBH$_4$.

butane-1,4-diol

b.

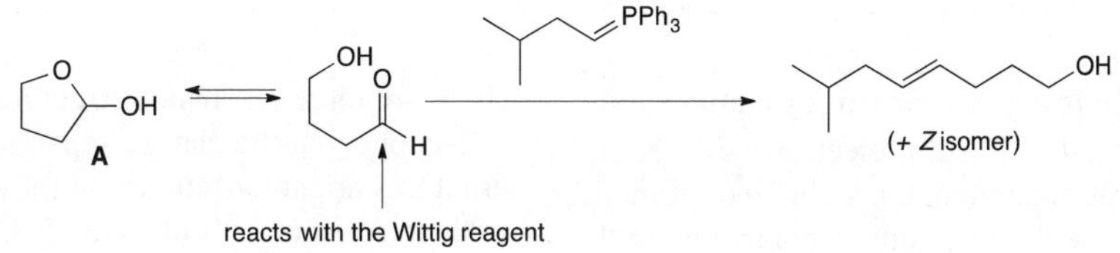

A

reacts with the Wittig reagent

(+ *Z* isomer)

c.

Y

NaOCH$_2$CH$_3$

X

aldehyde for Wittig
reaction

New C=C could be cis or
trans. The more stable trans
C=C is drawn.

isotretinoin

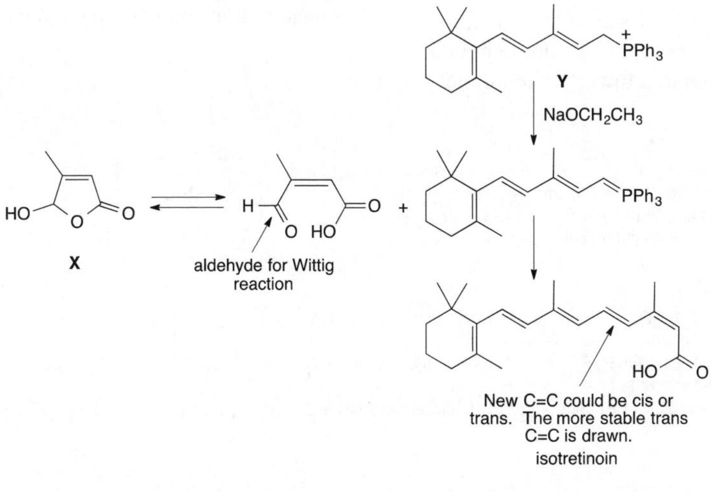

Chapter 21–28

21.72

[Mechanism scheme showing reaction of 5,5-dimethoxypentan-2-one with CH₃–MgI, followed by protonation and loss of methanol steps, leading through intermediates with (+1 resonance structure) notations and H₂O:, H–Ö–CH₃ byproducts, culminating in cyclic product Y with H₃O⁺.]

21.73

cyclopropenone
(1640 cm⁻¹)

[Four resonance structures of cyclopropenone showing cyclopropenyl cation with O⁻]

These three resonance structures include an aromatic ring; $4n + 2 = 2\,\pi$ electrons. Although they are charge separated, the stabilized aromatic ring makes these three structures contribute to the hybrid more than usual. Because these three resonance contributors have a C–O single bond, the absorption is shifted to a lower wavenumber.

cyclohex-2-enone
(1685 cm⁻¹)

[Three resonance structures of cyclohex-2-enone]

There are three resonance structures for cyclohex-2-enone, but the charge-separated resonance structures are not aromatic, so they contribute less to the resonance hybrid. The C=O absorbs in the usual region for a conjugated carbonyl.

21.74

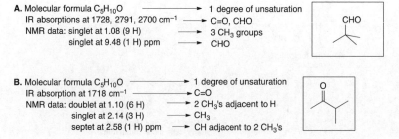

A. Molecular formula C₅H₁₀O ⟶ 1 degree of unsaturation
  IR absorptions at 1728, 2791, 2700 cm⁻¹ ⟶ C=O, CHO
  NMR data: singlet at 1.08 (9 H) ⟶ 3 CH₃ groups
         singlet at 9.48 (1 H) ppm ⟶ CHO

B. Molecular formula C₅H₁₀O ⟶ 1 degree of unsaturation
  IR absorption at 1718 cm⁻¹ ⟶ C=O
  NMR data: doublet at 1.10 (6 H) ⟶ 2 CH₃'s adjacent to H
         singlet at 2.14 (3 H) ⟶ CH₃
         septet at 2.58 (1 H) ppm ⟶ CH adjacent to 2 CH₃'s

C. Molecular formula $C_{10}H_{12}O$ ⟶ 5 degrees of unsaturation (4 due to a benzene ring)
IR absorption at 1686 cm^{-1} ⟶ C=O
NMR data:

    triplet at 1.21 (3 H) ⟶ CH_3 adjacent to 2 H's
    singlet at 2.39 (3 H) ⟶ CH_3
    quartet at 2.95 (2 H) ⟶ CH_2 adjacent to 3 H's
    doublet at 7.24 (2 H) ⟶ 2 H's on benzene ring
    doublet at 7.85 (2 H) ppm ⟶ 2 H's on benzene ring

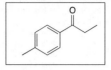

D. Molecular formula $C_{10}H_{12}O$ ⟶ 5 degrees of unsaturation (4 due to a benzene ring)
IR absorption at 1719 cm^{-1} ⟶ C=O
NMR data:

    triplet at 1.02 (3 H) ⟶ CH_3 adjacent to 2 H's
    quartet at 2.45 (2 H) ⟶ 2 H's adjacent to 3 H's
    singlet at 3.67 (2 H) ⟶ CH_2
    multiplet at 7.06–7.48 (5 H) ppm ⟶ a monosubstituted benzene ring

21.75

$C_7H_{16}O_2$: 0 degrees of unsaturation
IR: 3000 cm^{-1}: **C–H bonds**
NMR data (ppm):
    H_a: quartet at 3.8 (**4 H**), split by 3 H's
    H_b: singlet at 1.5 (**6 H**)
    H_c: triplet at 1.2 (**6 H**), split by 2 H's

$$CH_3CH_2-O-\underset{CH_3}{\overset{CH_3}{C}}-O-CH_2CH_3$$

(with labels $H_c\ H_a$ on left ethyl, H_b on the two CH_3 groups, $H_a\ H_c$ on right ethyl)

21.76

A. Molecular formula $C_9H_{10}O$
5 degrees of unsaturation
IR absorption at 1700 cm^{-1} → C=O
IR absorption at ~2700 cm^{-1} → CH of RCHO
NMR data (ppm):
    triplet at 1.2 (2 H's adjacent)
    quartet at 2.7 (3 H's adjacent)
    doublet at 7.3 (2 H's on benzene)
    doublet at 7.7 (2 H's on benzene)
    singlet at 9.9 (CHO)

B. Molecular formula $C_9H_{10}O$
5 degrees of unsaturation
IR absorption at 1720 cm^{-1} → C=O
IR absorption at ~2700 cm^{-1} → CH of RCHO
NMR data (ppm):
    2 triplets at 2.85 and 2.95 (suggests –CH_2CH_2–)
    multiplet at 7.2 (benzene H's)
    signal at 9.8 (CHO)

21.77

C. Molecular formula $C_6H_{12}O_3$
1 degree of unsaturation
IR absorption at 1718 cm^{-1} → C=O
NMR data (ppm):
    singlet at 2.1 (3 H's)
    doublet at 2.7 (2 H's)
    singlet at 3.3 (6 H's – 2 OCH_3 groups)
    triplet at 4.8 (1 H)

Chapter 21–30

21.78

D. Molecular ion at $m/z = 150$: $C_9H_{10}O_2$ (possible molecular formula)
  5 degrees of unsaturation
  IR absorption at 1692 cm$^{-1} \rightarrow$ C=O
  NMR data (ppm):
    triplet at 1.5 (3 H's – CH$_3$CH$_2$)
    quartet at 4.1 (2 H's – CH$_3$CH$_2$)
    doublet at 7.0 (2 H's – on benzene ring)
    doublet at 7.8 (2 H's – on benzene ring)
    singlet at 9.9 (1 H – on aldehyde)

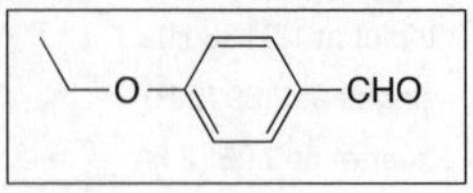

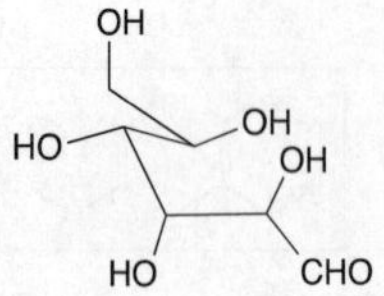

21.79

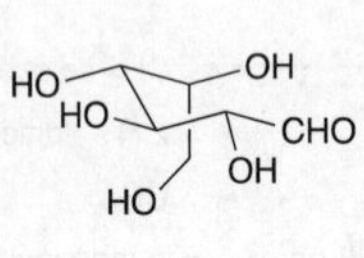

21.80

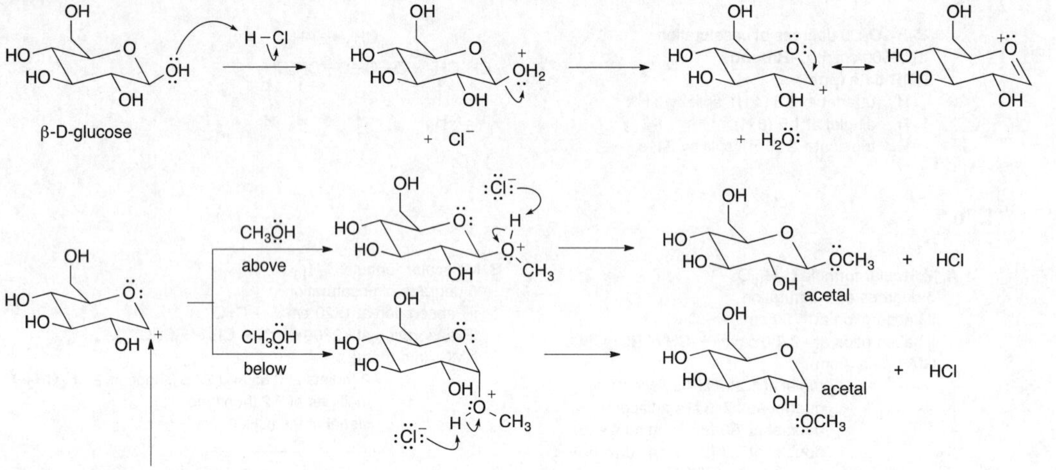

The carbocation is trigonal planar, so CH$_3$OH attacks from two different
directions, and two different acetals are formed.

21.81

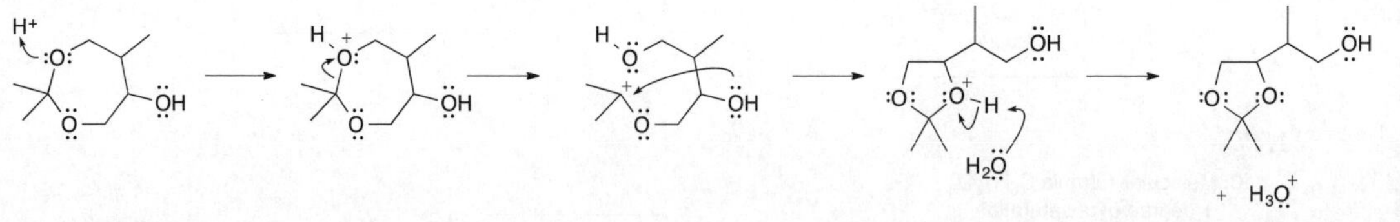

21.82

a.

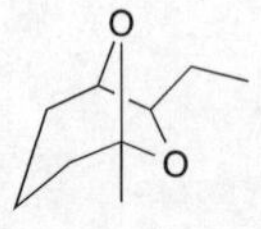

brevicomin

b.

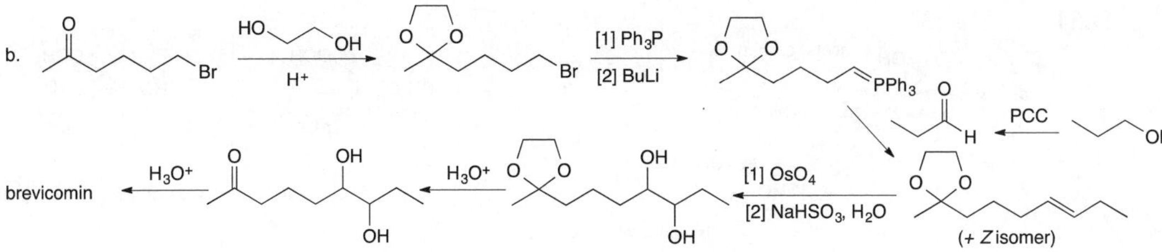

[1] Ph$_3$P
[2] BuLi

PCC

H$_3$O$^+$

[1] OsO$_4$
[2] NaHSO$_3$, H$_2$O

H$_3$O$^+$

brevicomin

(+ Z isomer)

21.83

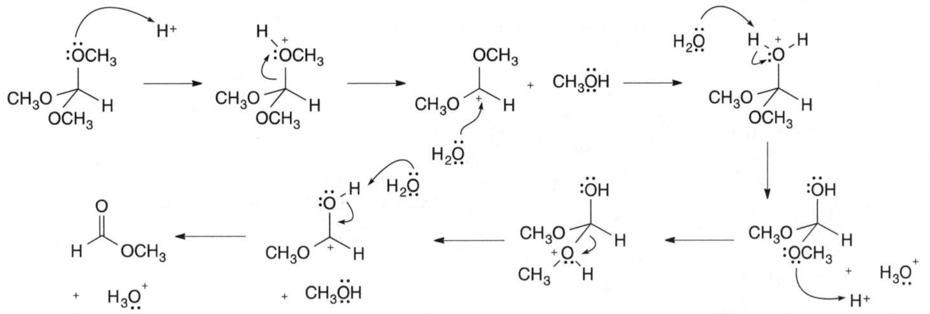

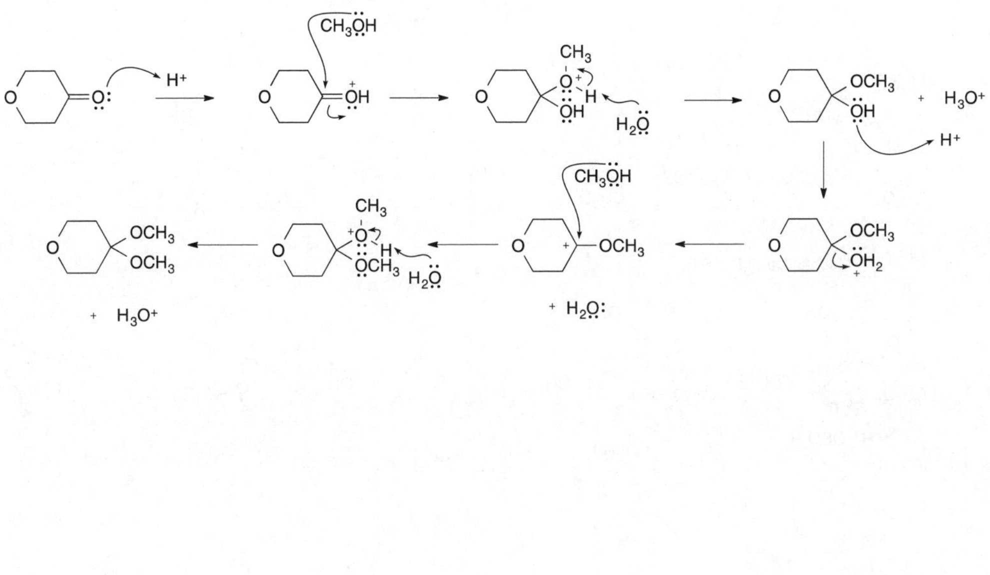

Chapter 21–32

21.84

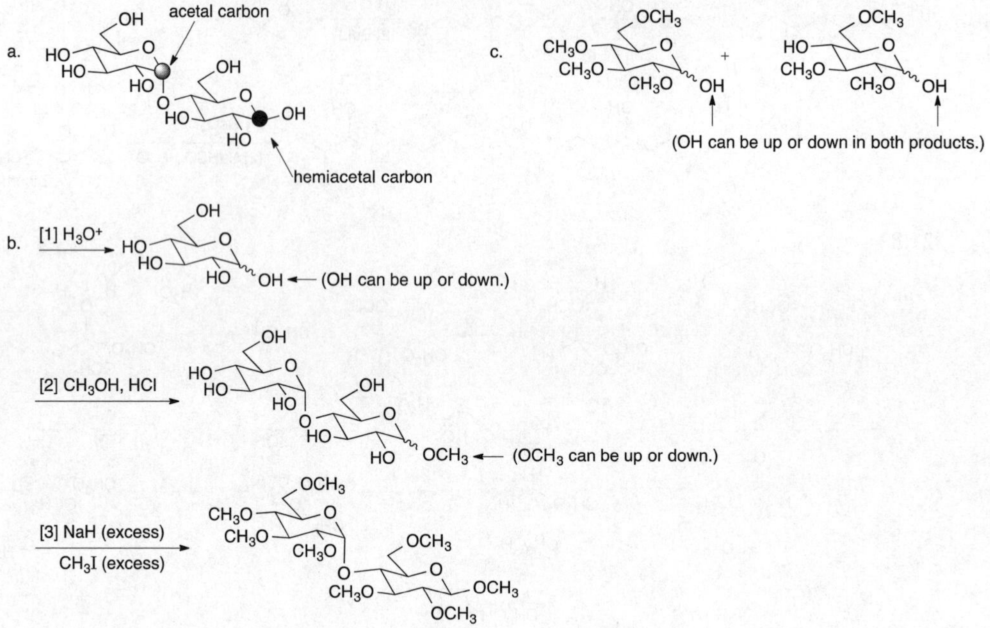

a. acetal carbon / hemiacetal carbon

c. (OH can be up or down in both products.)

b. [1] H_3O^+ → (OH can be up or down.)

[2] CH_3OH, HCl → (OCH_3 can be up or down.)

[3] NaH (excess) / CH_3I (excess) →

21.85

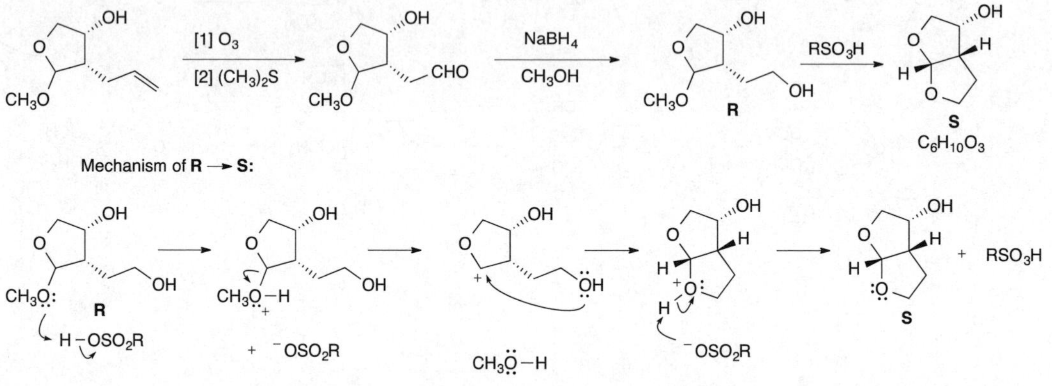

[1] O_3
[2] $(CH_3)_2S$

$NaBH_4$
CH_3OH

RSO_3H

R

S
$C_6H_{10}O_3$

Mechanism of **R** → **S**:

R

$CH_3\ddot{O}:$

$H-OSO_2R$

$CH_3\ddot{O}-H$

$+ \quad ^-OSO_2R$

$CH_3\ddot{O}-H$

$^-OSO_2R$

S

$+ \quad RSO_3H$

Aldehydes and Ketones 21–33

21.86 The mechanism involves S_N2 displacement of Br, followed by intramolecular enamine formation.

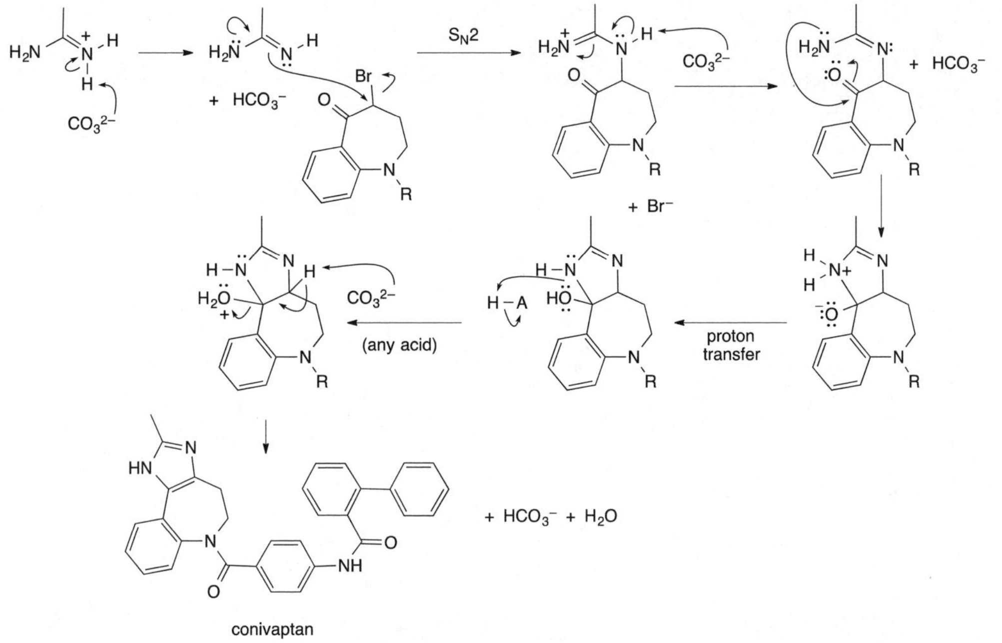

conivaptan

Carboxylic Acids and Their Derivatives 22–1

Chapter 22 Carboxylic Acids and Their Derivatives—Nucleophilic Acyl Substitution

Chapter Review

Summary of spectroscopic absorptions of RCOZ (22.5)

IR absorptions
- All RCOZ compounds have a C=O absorption in the region $1600-1850$ cm^{-1}.
  - RCOCl: 1800 cm^{-1}
  - $(RCO)_2O$: 1820 and 1760 cm^{-1} (two peaks)
  - RCOOR': $1735-1745$ cm^{-1}
  - RCONR'$_2$: $1630-1680$ cm^{-1}
- Additional amide absorptions occur at $3200-3400$ cm^{-1} (N–H stretch) and 1640 cm^{-1} (N–H bending).
- Decreasing the ring size of a cyclic lactone, lactam, or anhydride increases the frequency of the C=O absorption.
- Conjugation shifts the C=O to lower wavenumber.

^{1}H NMR absorptions
- C–H α to the C=O absorbs at $2-2.5$ ppm.
- N–H of an amide absorbs at $7.5-8.5$ ppm.

^{13}C NMR absorption
- C=O absorbs at $160-180$ ppm.

Summary of spectroscopic absorptions of RCN (22.5)

IR absorption
- C≡N absorption at 2250 cm^{-1}

^{13}C NMR absorption
- C≡N absorbs at $115-120$ ppm.

Summary: The relationship between the basicity of Z⁻ and the properties of RCOZ

- **Increasing basicity of the leaving group** (22.2)
- **Increasing resonance stabilization** (22.2)

acid chloride anhydride carboxylic acid ester amide

- **Increasing leaving group ability** (22.7B)
- **Increasing reactivity** (22.7B)
- **Increasing frequency of the C=O absorption in the IR** (22.5)

General features of nucleophilic acyl substitution

- The characteristic reaction of compounds having the general structure RCOZ is nucleophilic acyl substitution (22.1).
- The mechanism consists of two steps (22.7A):
  [1] Addition of a nucleophile to form a tetrahedral intermediate
  [2] Elimination of a leaving group
- More reactive acyl compounds can be used to prepare less reactive acyl compounds. The reverse is not necessarily true (22.7B).

Chapter 22–2

Nucleophilic acyl substitution reactions

[1] Reaction that synthesizes acid chlorides (RCOCl)

From RCOOH (22.10A):

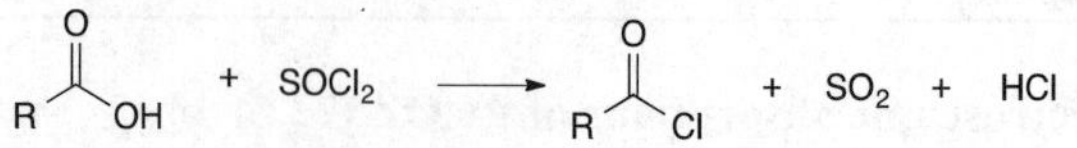

$$R\text{COOH} + SOCl_2 \longrightarrow R\text{COCl} + SO_2 + HCl$$

[2] Reactions that synthesize anhydrides [(RCO)₂O]

a. From RCOCl (22.8):

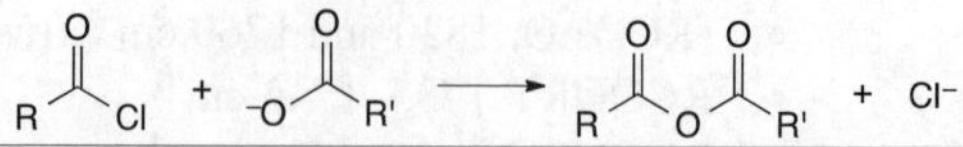

b. From dicarboxylic acids (22.10B):

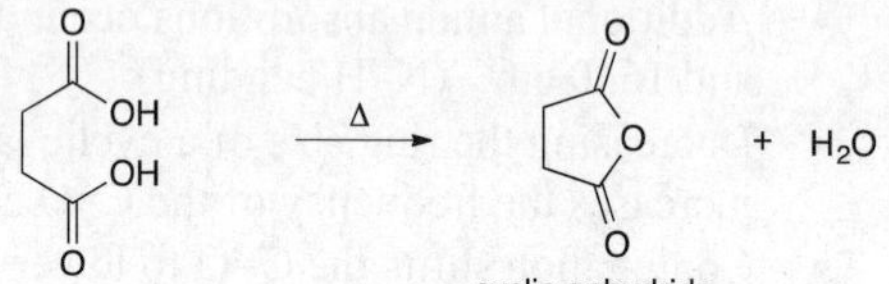

$$\xrightarrow{\Delta}$$ cyclic anhydride $+ H_2O$

[3] Reactions that synthesize carboxylic acids (RCOOH)

a. From RCOCl (22.8):

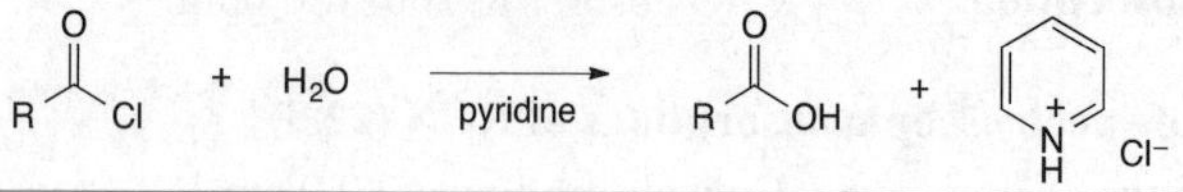

$$R\text{COCl} + H_2O \xrightarrow{\text{pyridine}} R\text{COOH} + \quad Cl^-$$

b. From (RCO)₂O (22.9):

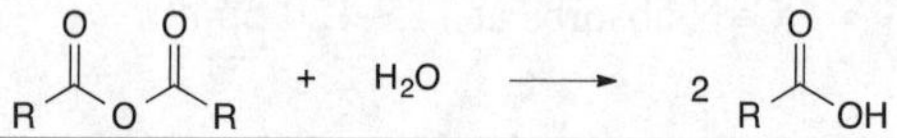

$$ + H_2O \longrightarrow 2 \; R\text{COOH}$$

c. From RCOOR' (22.11):

$$R\text{COOR'} + H_2O \xrightarrow{(H^+ \text{ or } ^-OH)} R\text{COOH} \; \text{or} \; R\text{COO}^- + R'\text{OH}$$

(with acid)   (with base)

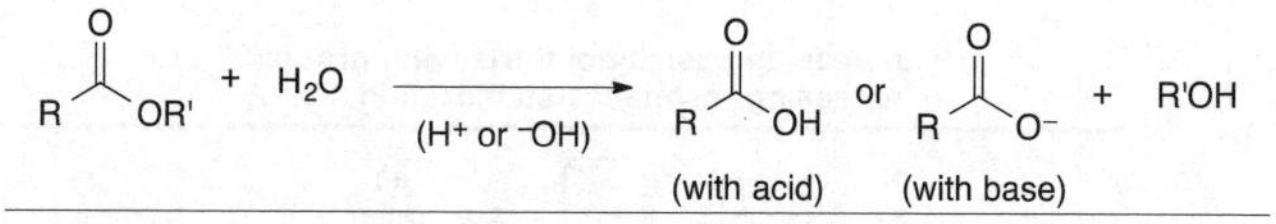

d. From RCONR'₂ (R' = H or alkyl, 22.13):

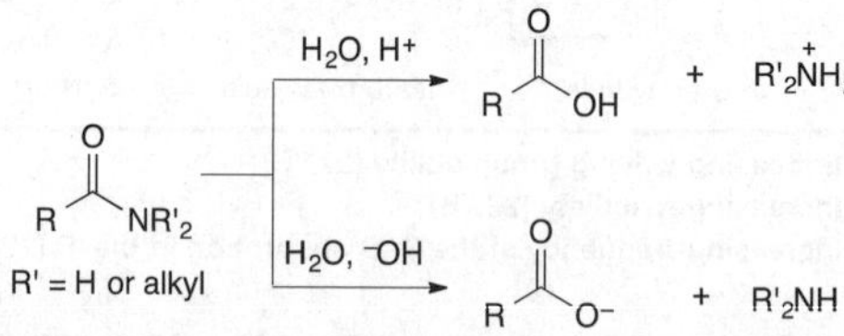

R' = H or alkyl

$\xrightarrow{H_2O, \; H^+}$ $R\text{COOH} + R'_2\overset{+}{N}H_2$

$\xrightarrow{H_2O, \; ^-OH}$ $R\text{COO}^- + R'_2NH$

[4] Reactions that synthesize esters (RCOOR')

a. From RCOCl (22.8):

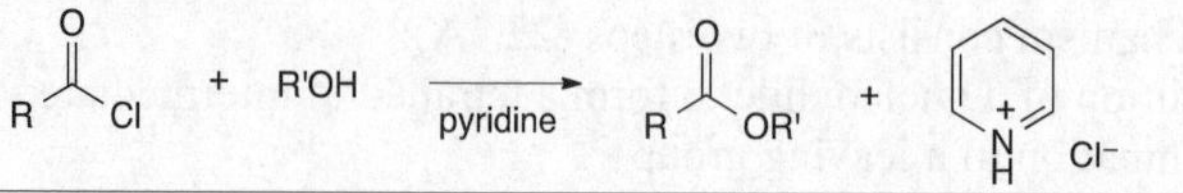

$$R\text{COCl} + R'\text{OH} \xrightarrow{\text{pyridine}} R\text{COOR'} + \quad Cl^-$$

Carboxylic Acids and Their Derivatives 22–3

b. From (RCO)₂O
(22.9):

$$\underset{R}{\overset{O}{\|}}\!\!-\!\!O\!\!-\!\!\underset{R}{\overset{O}{\|}} \ + \ R'OH \ \longrightarrow \ \underset{R}{\overset{O}{\|}}\!\!-\!\!OR' \ + \ RCOOH$$

c. From RCOOH
(22.10C):

$$\underset{R}{\overset{O}{\|}}\!\!-\!\!OH \ + \ R'OH \ \xrightarrow{H_2SO_4} \ \underset{R}{\overset{O}{\|}}\!\!-\!\!OR' \ + \ H_2O$$

---

[5] Reactions that synthesize amides (RCONH₂) [The reactions are written with NH₃ as the nucleophile to form RCONH₂. Similar reactions occur with R'NH₂ to form RCONHR', and with R'₂NH to form RCONR'₂.]

a. From RCOCl
(22.8):

$$\underset{R}{\overset{O}{\|}}\!\!-\!\!Cl \ + \ \underset{(2\ equiv)}{NH_3} \ \longrightarrow \ \underset{R}{\overset{O}{\|}}\!\!-\!\!NH_2 \ + \ NH_4^+Cl^-$$

b. From (RCO)₂O
(22.9):

$$\underset{R}{\overset{O}{\|}}\!\!-\!\!O\!\!-\!\!\underset{R}{\overset{O}{\|}} \ + \ \underset{(2\ equiv)}{NH_3} \ \longrightarrow \ \underset{R}{\overset{O}{\|}}\!\!-\!\!NH_2 \ + \ RCOO^-NH_4^+$$

c. From RCOOH
(22.10D):

$$\underset{R}{\overset{O}{\|}}\!\!-\!\!OH \ \xrightarrow[\text{[2] }\Delta]{\text{[1] }NH_3} \ \underset{R}{\overset{O}{\|}}\!\!-\!\!NH_2 \ + \ H_2O$$

$$\underset{R}{\overset{O}{\|}}\!\!-\!\!OH \ + \ R'NH_2 \ \xrightarrow{DCC} \ \underset{R}{\overset{O}{\|}}\!\!-\!\!NHR' \ + \ H_2O$$

d. From RCOOR'
(22.11):

$$\underset{R}{\overset{O}{\|}}\!\!-\!\!OR' \ + \ NH_3 \ \longrightarrow \ \underset{R}{\overset{O}{\|}}\!\!-\!\!NH_2 \ + \ R'OH$$

Nitrile synthesis (22.18)

Nitriles are prepared by S_N2 substitution using unhindered alkyl halides as starting materials.

$$R-X \ + \ {}^-CN \ \xrightarrow{S_N2} \ R-C\equiv N \ + \ X^-$$
$$R = CH_3,\ 1°$$

Reactions of nitriles
[1] Hydrolysis (22.18A)

$$R-C\equiv N \ \xrightarrow[(H^+\ or\ {}^-OH)]{H_2O} \ \underset{R}{\overset{O}{\|}}\!\!-\!\!OH \ \text{ or } \ \underset{R}{\overset{O}{\|}}\!\!-\!\!O^-$$
$$\text{(with acid)} \qquad \text{(with base)}$$

Chapter 22–4

[2] Reduction (22.18B)

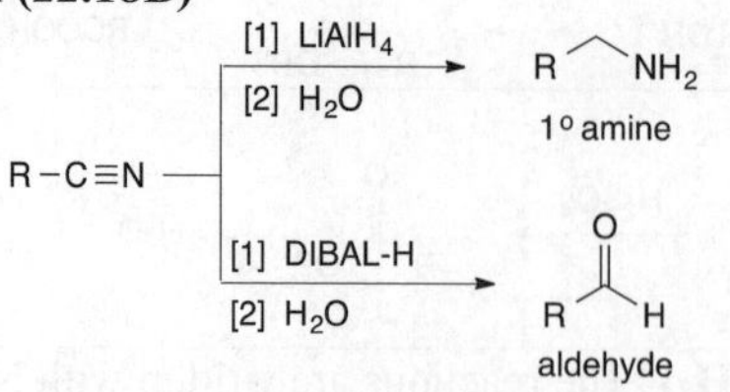

$$R-C\equiv N \quad \begin{array}{l} \xrightarrow{\text{[1] LiAlH}_4,\ \text{[2] H}_2\text{O}} \quad R\diagup\text{NH}_2 \quad 1°\ \text{amine} \\[2em] \xrightarrow{\text{[1] DIBAL-H},\ \text{[2] H}_2\text{O}} \quad \underset{\text{aldehyde}}{R\diagup\text{(C=O)}\diagdown H} \end{array}$$

[3] Reaction with organometallic reagents (22.18C)

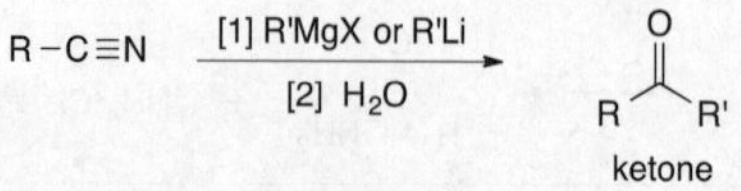

$$R-C\equiv N \xrightarrow[\text{[2] H}_2\text{O}]{\text{[1] R'MgX or R'Li}} \underset{\text{ketone}}{R\diagup\text{(C=O)}\diagdown R'}$$

Practice Test on Chapter Review

1. Give the IUPAC name for each of the following compounds.

a.

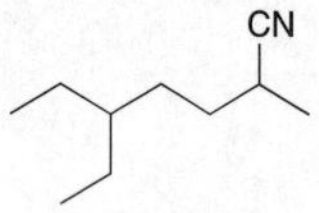

b.

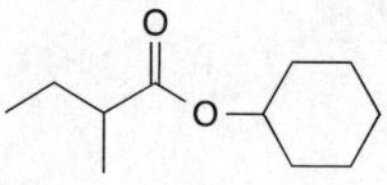

c.

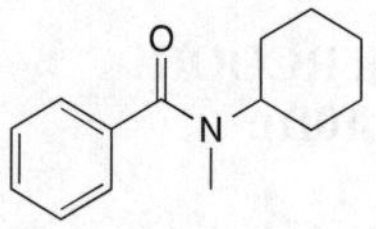

2. (a) Which compound absorbs at the *lowest* wavenumber in the IR? (b) Which compound absorbs at the *highest* wavenumber?

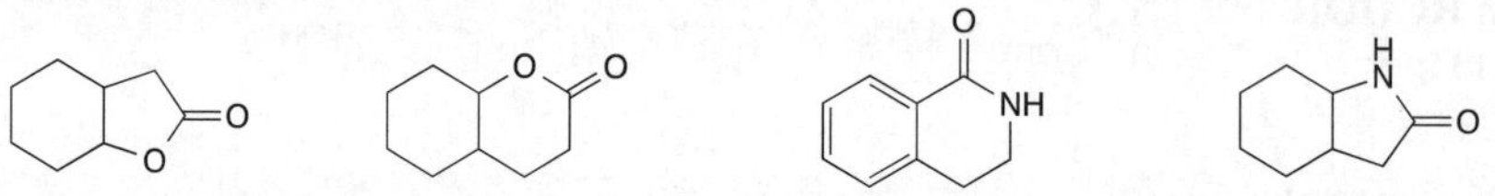

A   B   C   D

3.a. Which of the following reaction conditions can be used to synthesize an ester?

    1. RCOCl + R'OH + pyridine
    2. RCOOH + R'OH + H_2SO_4
    3. RCOOH + R'OH + NaOH
    4. Both methods (1) and (2) can be used to synthesize an ester.
    5. Methods (1), (2), and (3) can all be used to synthesize an ester.

b. Which of the following compounds is most reactive in nucleophilic acyl substitution?

    1. CH_3COCl 3. $CH_3CON(CH_3)_2$ 5. CH_3COOH
    2. CH_3COOCH_3 4. $(CH_3CO)_2O$

Carboxylic Acids and Their Derivatives 22–5

4. What reagent is needed to convert $(CH_3CH_2)_2CHCOOH$ into each compound?
   a. $(CH_3CH_2)_2CHCOO^-Na^+$
   b. $(CH_3CH_2)_2CHCOCl$
   c. $(CH_3CH_2)_2CHCON(CH_3)_2$
   d. $(CH_3CH_2)_2CHCO_2CH_2CH_3$
   e. $[(CH_3CH_2)_2CHCO]_2O$

5. What reagent is needed to convert $CH_3CH_2CH_2CN$ to each compound?
   a. $CH_3CH_2CH_2COOH$
   b. $CH_3CH_2CH_2CH_2NH_2$
   c. $CH_3CH_2CH_2COCH_2CH_3$
   d. $CH_3CH_2CH_2CHO$

6. Draw the organic products formed in the following reactions.

a. [1] NaCN [2] LiAlH₄ [3] H₂O [Indicate stereochemistry.]

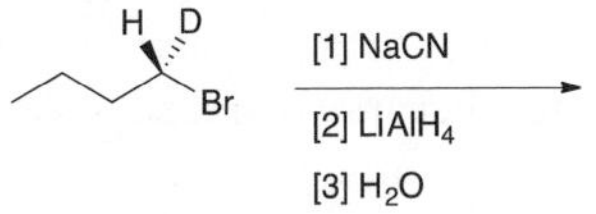

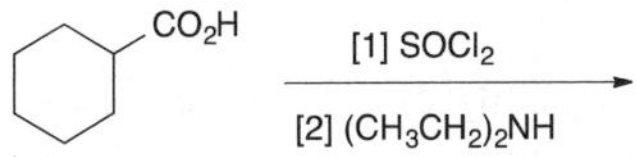

b. NaOH, H₂O

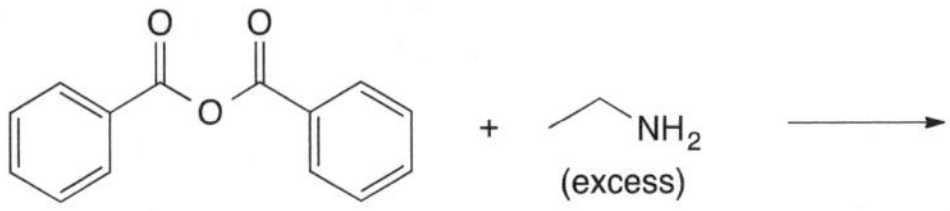

c.

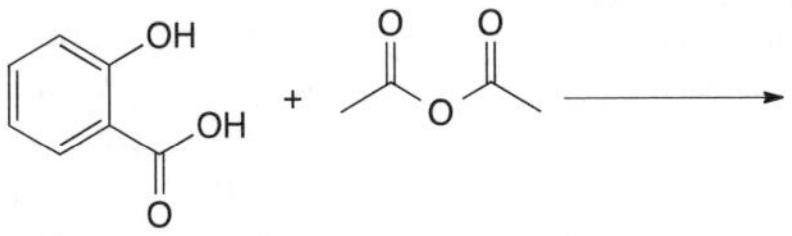

d. [1] SOCl₂ [2] (CH₃CH₂)₂NH

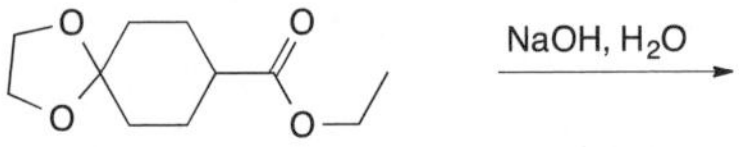

e. + NH₂ (excess) ⟶

f. [1] NaCN [2] CH₃CH₂MgBr [3] H₂O

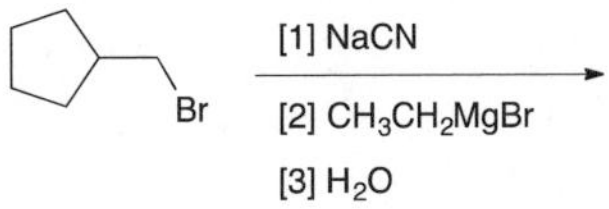

Answers to Practice Test

1.a. 5-ethyl-2-methyl-heptanenitrile
   b. cyclohexyl 2-methylbutanoate
   c. N-cyclohexyl-N-methylbenzamide
2.a. **C**
   b. **A**
3.a. 4
   b. 1
4.a. NaOH
   b. SOCl₂
   c. HN(CH₃)₂, DCC
   d. CH₃CH₂OH, H₂SO₄
   e. heat

5.a. H_3O^+
   b. [1] LiAlH₄; [2] H₂O
   c. [1] CH₃CH₂Li; [2] H₂O
   d. [1] DIBAL-H; [2] H₂O
6.
a.

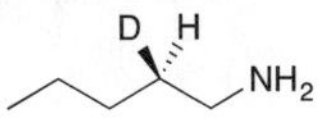

b.

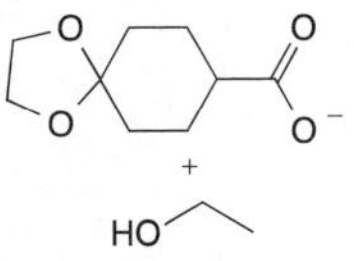

c.

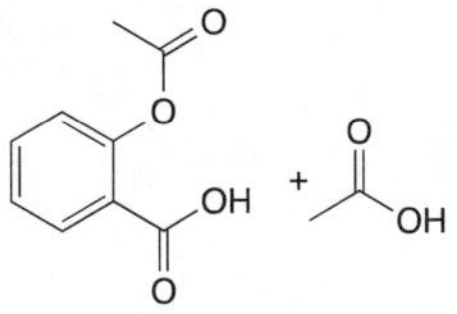

d.

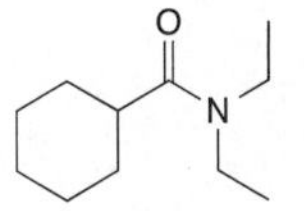

e.

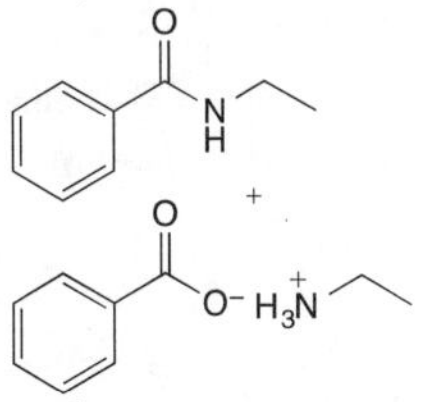

f.

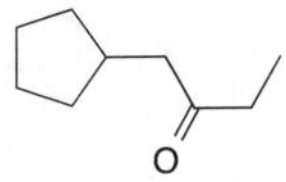

Chapter 22–6

Answers to Problems

22.1 The number of C–N bonds determines the classification as a 1°, 2°, or 3° amide.

1° amide

1° amide

NH$_2$ — 1° amide

oxytocin
All seven others are 2° amides.

1° amide

3° amide

HO

22.2 As the basicity of Z increases, the stability of RCOZ increases because of added resonance stabilization.

The **basicity of Z** determines how much this structure contributes to the hybrid. Br$^-$ is less basic than $^-$OH, so RCOBr is less stable than RCOOH.

22.3

CH$_3$—Cl

This resonance structure contributes little to the hybrid because Cl$^-$ is a weak base. Thus, the C–Cl bond has little double bond character, making it similar in length to the C–Cl bond in CH$_3$Cl.

CH$_3$—NH$_2$

This resonance structure contributes more to the hybrid because $^-$NH$_2$ is more basic. Thus, the C–N bond in HCONH$_2$ has more double bond character, making it shorter than the C–N bond in CH$_3$NH$_2$.

22.4

a.

Cl **2-ethylbutanoyl chloride**

2-ethyl

b.

OCH$_3$ **methyl benzoate**

alkyl group = methyl

acyl group = benzoate

Carboxylic Acids and Their Derivatives 22–7

c. acyl group = propanamide   **N-ethyl-N-methylpropanamide**

e. 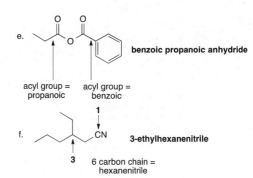 **benzoic propanoic anhydride**
acyl group = propanoic   acyl group = benzoic

d. acyl group = formate   alkyl group = ethyl   **ethyl formate**

f. 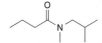 **3-ethylhexanenitrile**
6 carbon chain = hexanenitrile

22.5

a. 5-methylheptanoyl chloride

b. isopropyl propanoate

c. acetic formic anhydride

d. N-isobutyl-N-methylbutanamide

e. 3-methylpentanenitrile

f. o-cyanobenzoic acid

g. sec-butyl 2-methylhexanoate

h. N-ethylhexanamide

22.6 CH₃CONH₂ has a higher boiling point than CH₃CO₂H because it has more opportunities for hydrogen bonding. Each CH₃CONH₂ can hydrogen bond to three other molecules (at its O atom and two N–H bonds), whereas CH₃CO₂H has only two intermolecular hydrogen bonds possible (at its carbonyl O atom and OH bond).

22.7

a. ... and ... amide: C=O at lower wavenumber

b. ... and ... smaller ring: C=O at a higher wavenumber

c. ... and ... 2° amide: 1 N–H absorption at 3200–3400 cm⁻¹   1° amide: 2 N–H absorptions

d. ... and ... anhydride: 2 C=O peaks

Chapter 22–8

22.8

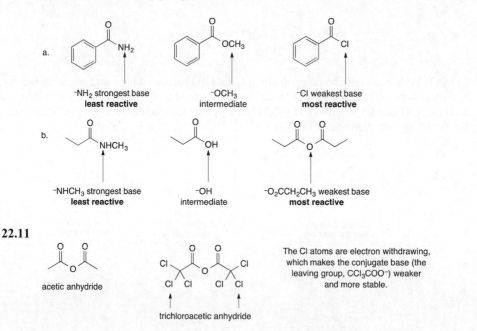

22.9 More reactive acyl compounds can be converted to less reactive acyl compounds.

a. CH$_3$COCl (more reactive) → CH$_3$COOH (less reactive) **YES**

b. CH$_3$CONHCH$_3$ (less reactive) → CH$_3$COOCH$_3$ (more reactive) **NO**

c. CH$_3$COOCH$_3$ (less reactive) → CH$_3$COCl (more reactive) **NO**

d. (CH$_3$CO)$_2$O (more reactive) → CH$_3$CONH$_2$ (less reactive) **YES**

22.10 The better the leaving group is, the more reactive the carboxylic acid derivative. The weakest base is the best leaving group.

a.
- PhCONH$_2$: –NH$_2$ strongest base, **least reactive**
- PhCOOCH$_3$: –OCH$_3$ intermediate
- PhCOCl: –Cl weakest base, **most reactive**

b.
- CH$_3$CH$_2$CONHCH$_3$: –NHCH$_3$ strongest base, **least reactive**
- CH$_3$CH$_2$COOH: –OH intermediate
- (CH$_3$CH$_2$CO)$_2$O: –O$_2$CCH$_2$CH$_3$ weakest base, **most reactive**

22.11

acetic anhydride ; trichloroacetic anhydride

The Cl atoms are electron withdrawing, which makes the conjugate base (the leaving group, CCl$_3$COO$^-$) weaker and more stable.

Carboxylic Acids and Their Derivatives 22–9

22.12

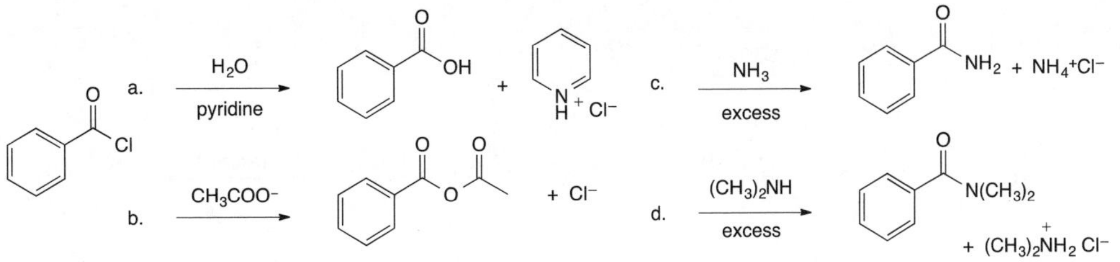

22.13 The mechanism has three steps: [1] nucleophilic attack by O; [2] proton transfer; and [3] elimination of the Cl⁻ leaving group to form the product.

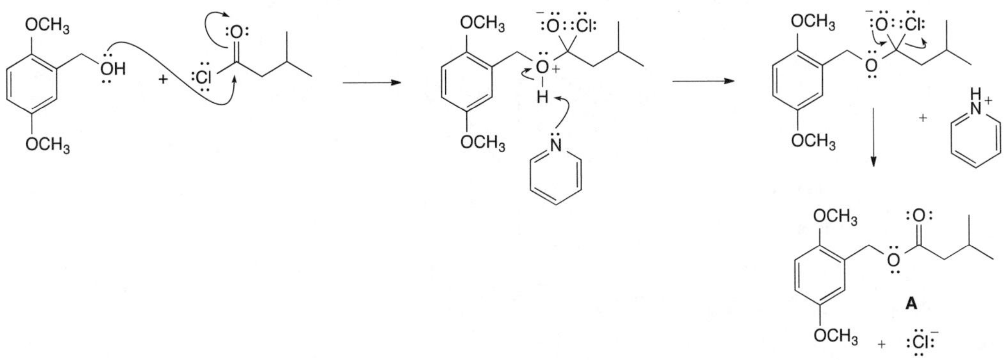

22.14

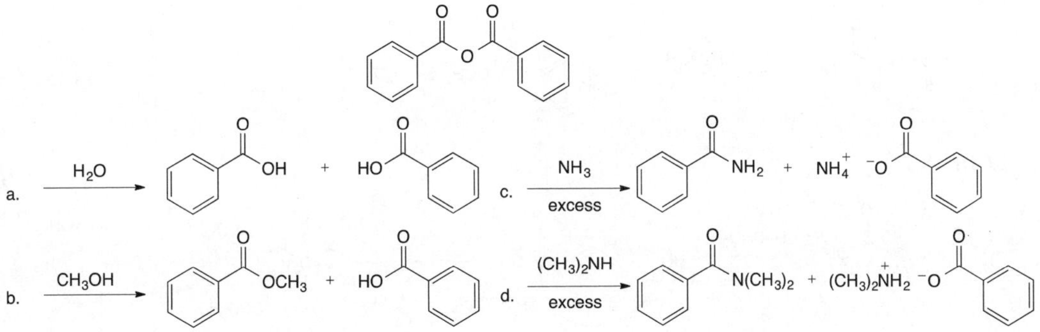

Chapter 22–10

22.15 Reaction of a carboxylic acid with thionyl chloride converts it to an acid chloride.

a.

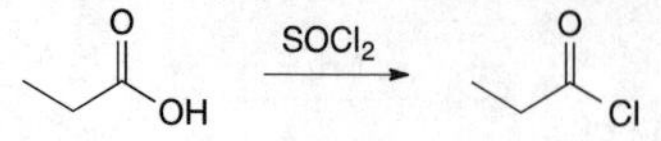

b.

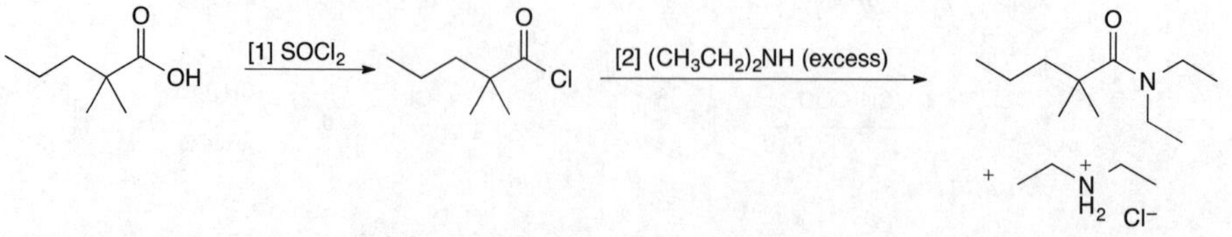

22.16

a.

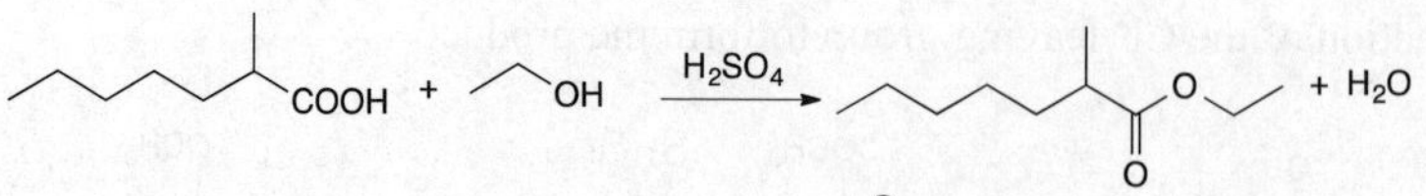

b.

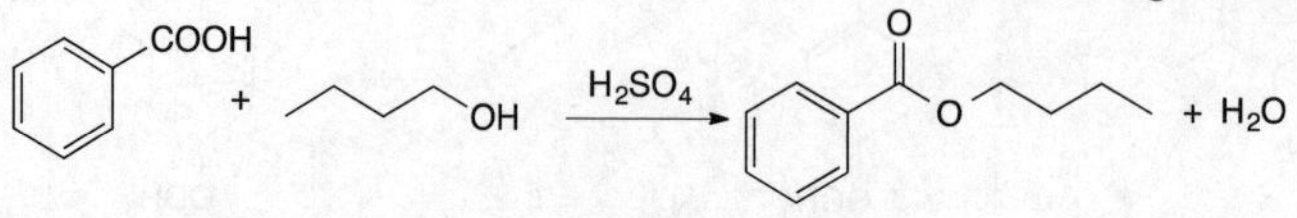

c.

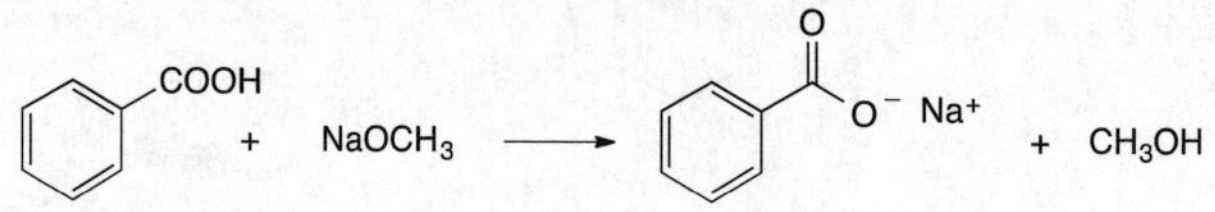

d.

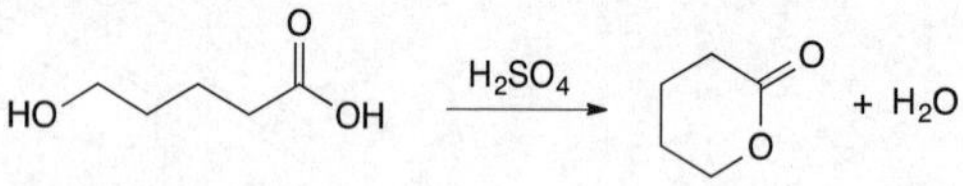

22.17

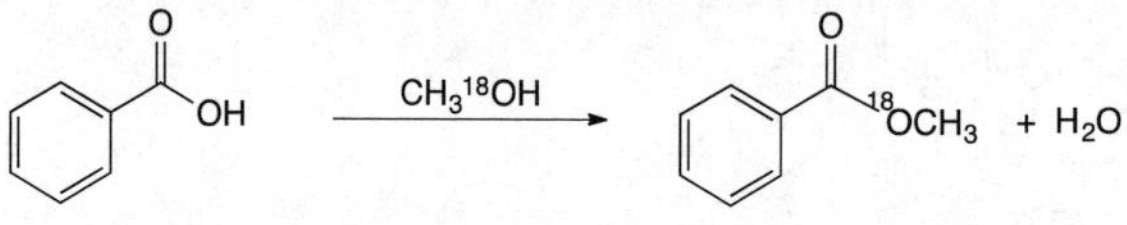

22.18

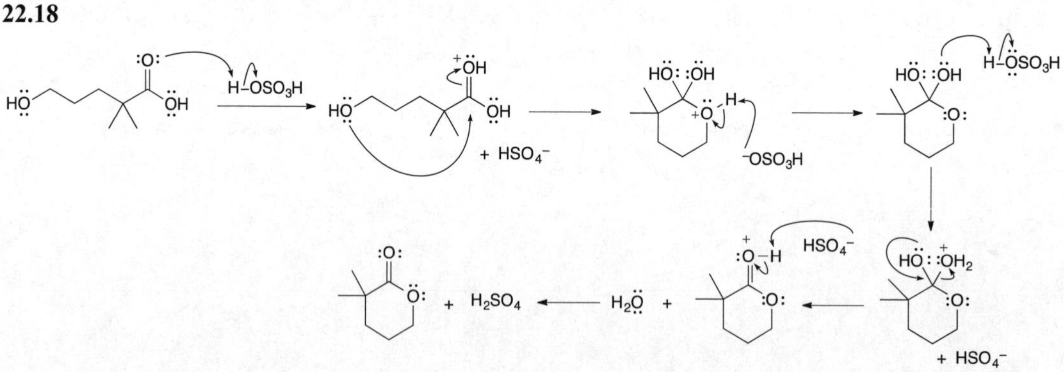

Carboxylic Acids and Their Derivatives 22–11

22.19

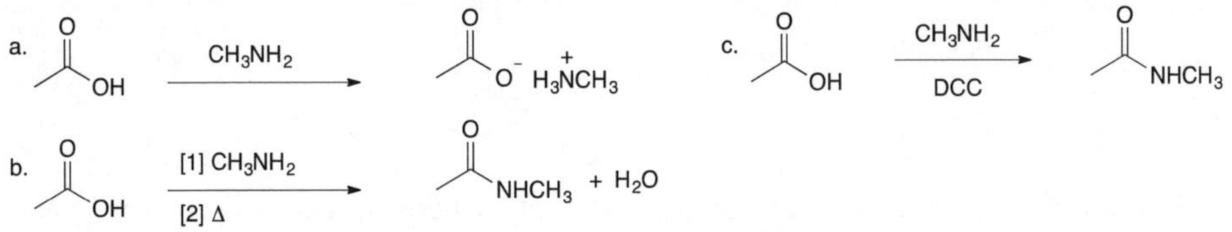

22.20

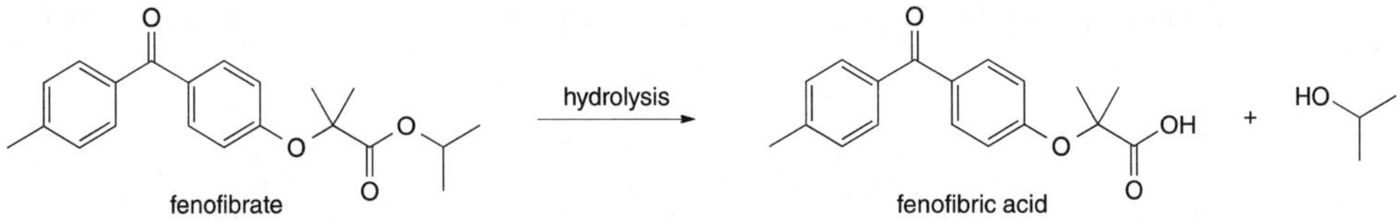

22.21 The three bonds broken during hydrolysis are indicated.

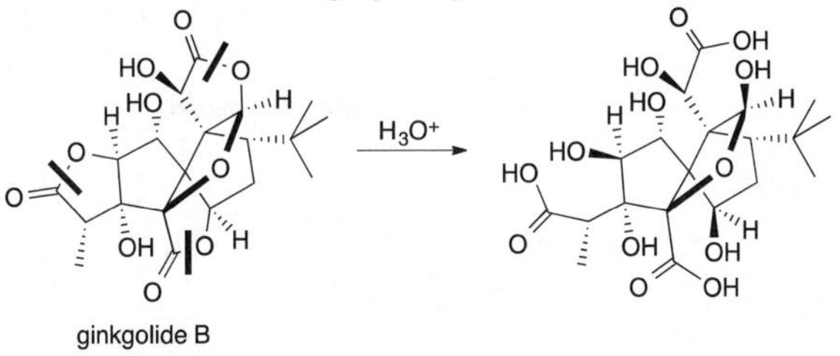

22.22

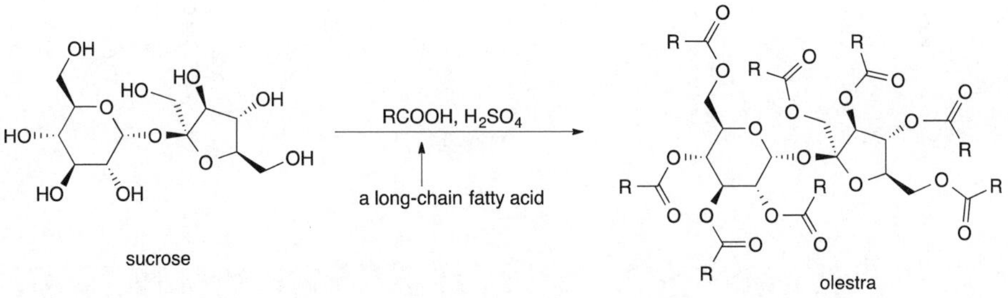

22.23

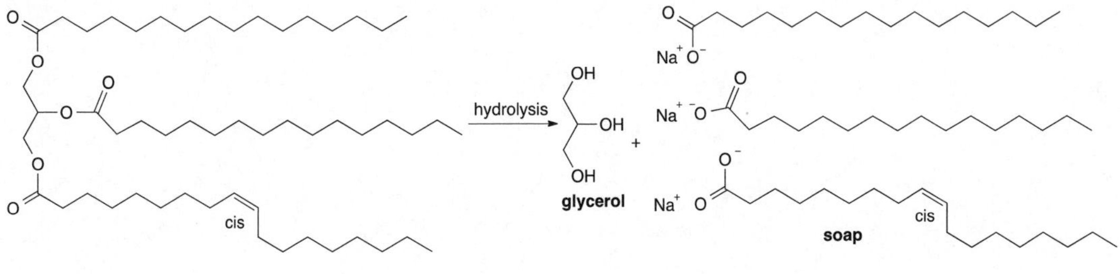

Chapter 22–12

22.24

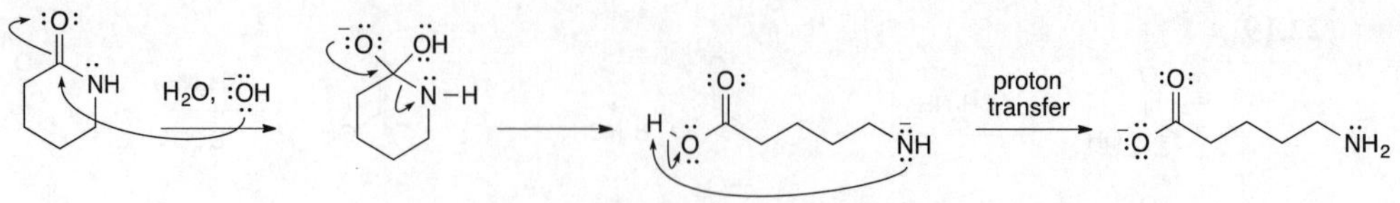

22.25 Aspirin has an ester, a more reactive acyl group, whereas acetaminophen has an amide, a less reactive acyl group. The ester makes aspirin more easily hydrolyzed with water from the air than acetaminophen. Therefore, Tylenol can be kept for many years, whereas aspirin decomposes.

ester

acetylsalicylic acid

amide

acetaminophen

22.26

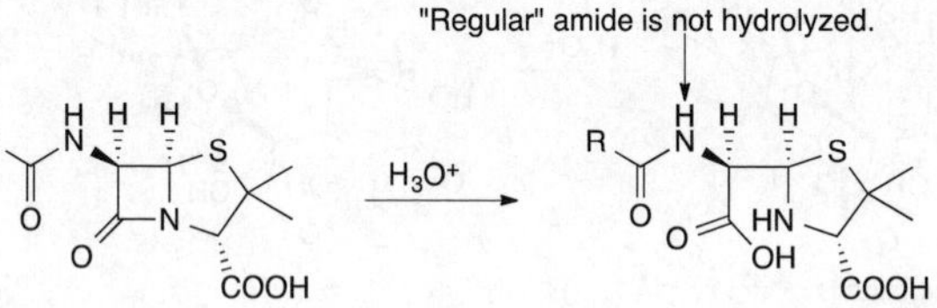

"Regular" amide is not hydrolyzed.

H_3O^+

22.27

nylon 6,10

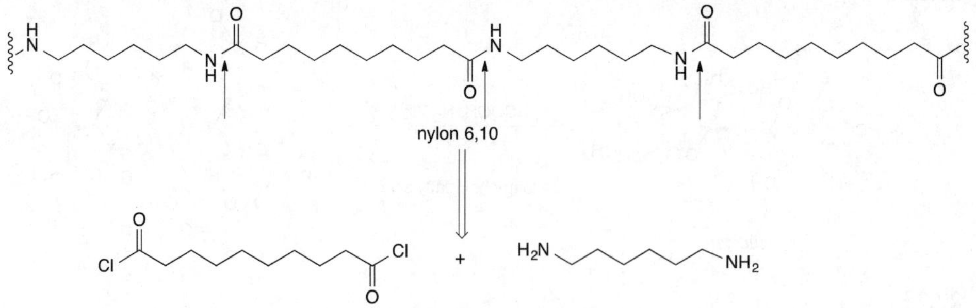

Carboxylic Acids and Their Derivatives 22–13

22.28

1,4-dihydroxymethylcyclohexane

terephthalic acid

Kodel

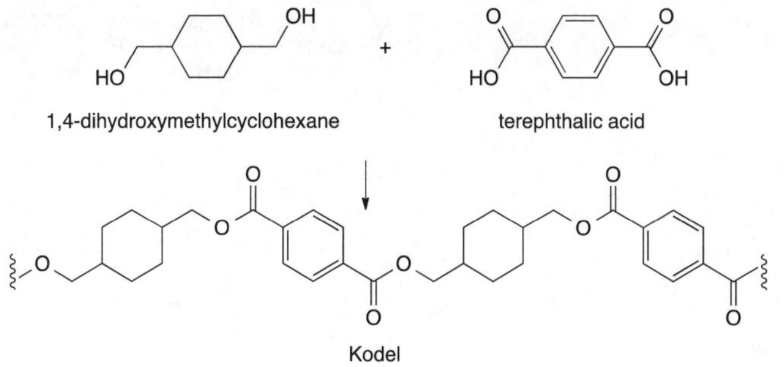

In the polyester Kodel, most of the bonds in the polymer backbone are part of a ring, so there are fewer degrees of freedom. Fabrics made from Kodel are stiff and crease resistant, due to these less flexible polyester fibers.

22.29

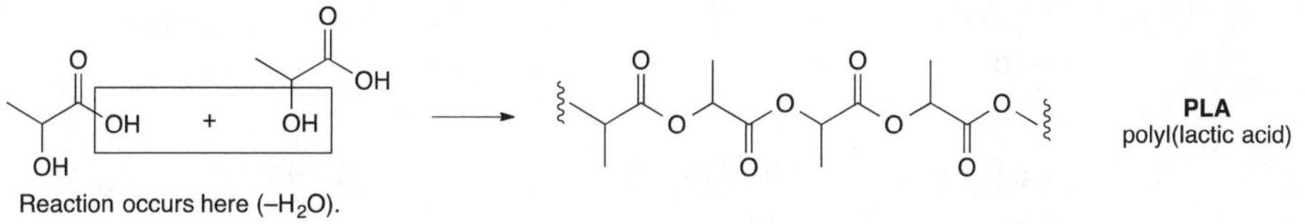

Reaction occurs here ($-H_2O$).

PLA
polyl(lactic acid)

22.30 Acetyl CoA acetylates the NH_2 group of glucosamine, because the NH_2 group is the most nucleophilic site.

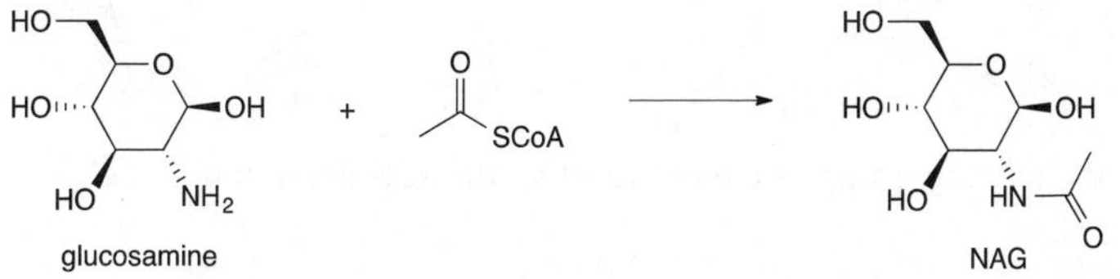

glucosamine

NAG

22.31

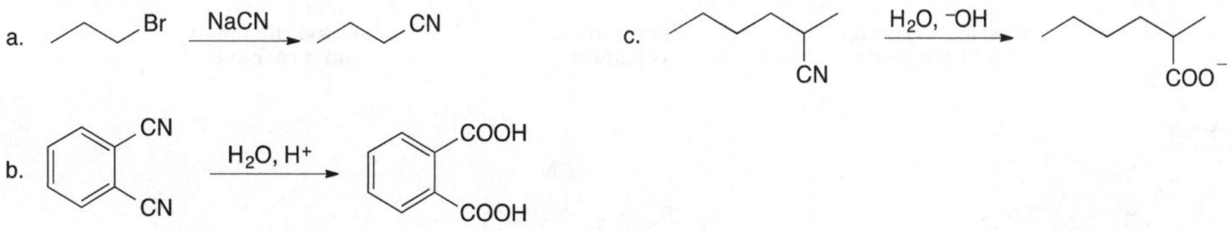

a.

c.

b.

22.32

a.

c.

b.

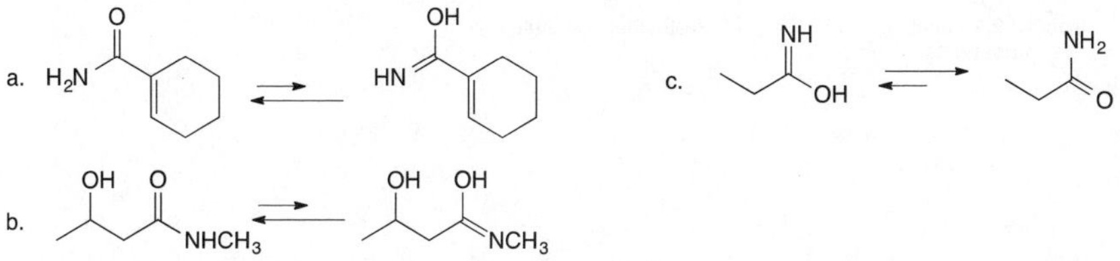

Chapter 22–14

22.33

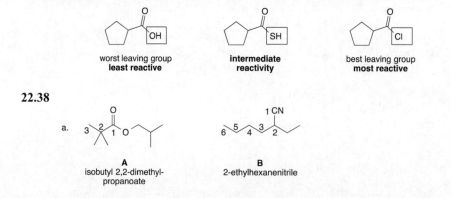

22.37 The better the leaving group, the more reactive the acyl compound.

Carboxylic Acids and Their Derivatives 22–15

b. A $\xrightarrow{[1]\ H_3O^+}$ COOH + HO⟍⟋

A $\xrightarrow[H_2O]{[2]\ ^-OH}$ COO⁻ + HO⟍⟋

A $\xrightarrow[\ ]{[3]\ \diagup\diagdown\diagup MgBr}\ \xrightarrow{H_2O}$ (product with OH) + HO⟍⟋

A $\xrightarrow[\ ]{[4]\ LiAlH_4}\ \xrightarrow{H_2O}$ ⟍⟋OH + HO⟍⟋

B $\xrightarrow{[1]\ H_3O^+}$ COOH

B $\xrightarrow[H_2O]{[2]\ ^-OH}$ COO⁻

B $\xrightarrow[\ ]{[3]\ \diagup\diagdown\diagup MgBr}\ \xrightarrow{H_2O}$ (ketone product)

B $\xrightarrow[\ ]{[4]\ LiAlH_4}\ \xrightarrow{H_2O}$ (amine product, NH₂)

22.39 Better leaving groups make acyl compounds more reactive. **C** has an electron-withdrawing NO_2 group, which stabilizes the negative charge of the leaving group, whereas **D** has an electron-donating OCH_3 group, which destabilizes the leaving group.

C

an electron-withdrawing substituent

D

an electron-donating substituent

leaving group from **C**   one possible resonance structure

leaving group from **D**   one possible resonance structure

Delocalizing the negative charge on the NO_2 stabilizes the leaving group, making **C** more reactive than **D**.

Adjacent negative charges destabilize the leaving group.

Chapter 22–16

22.40

a.

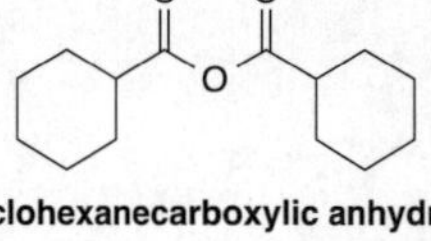

cyclohexanecarboxylic anhydride

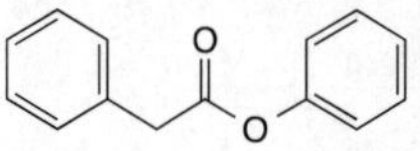

d. 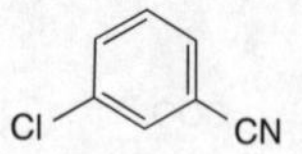 **m-chlorobenzonitrile**

b.

phenyl phenylacetate

e. 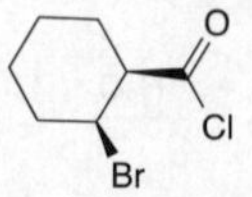 **cis-2-bromocyclohexane-carbonyl chloride**

c.

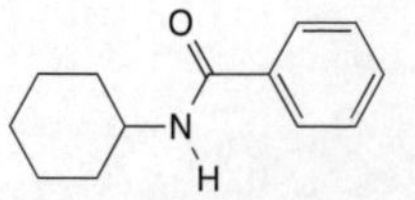

N-cyclohexylbenzamide

f. 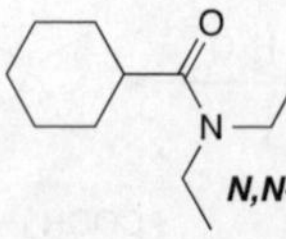 **N,N-diethylcyclohexanecarboxamide**

22.41

a. cyclohexyl propanoate

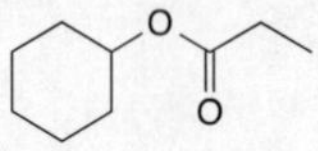

d. vinyl acetate

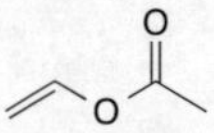

g. octyl butanoate

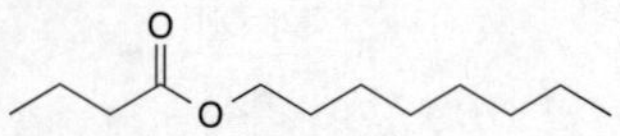

b. cyclohexanecarboxamide

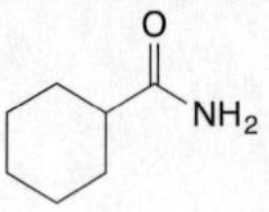

e. benzoic propanoic anhydride

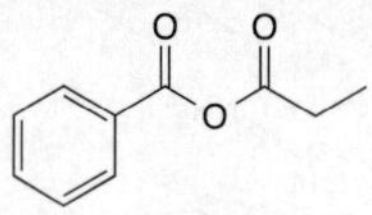

h. **N,N-dibenzylformamide**

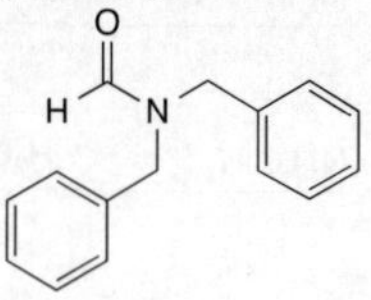

c. 4-methylheptanenitrile

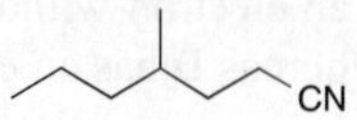

f. 3-methylhexanoyl chloride

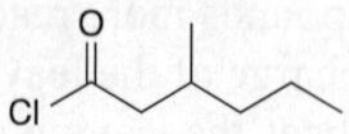

22.42

resonance structures for the leaving group

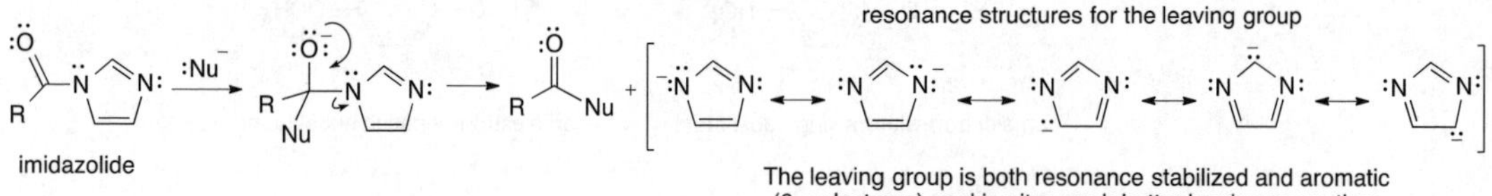

imidazolide

The leaving group is both resonance stabilized and aromatic (6 π electrons), making it a much better leaving group than exists in a regular amide.

Carboxylic Acids and Their Derivatives 22–17

22.43

Reaction as an acid:

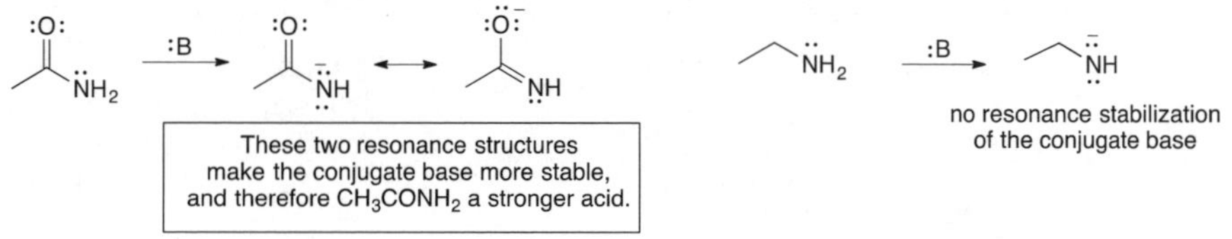

These two resonance structures make the conjugate base more stable, and therefore CH₃CONH₂ a stronger acid.

no resonance stabilization of the conjugate base

Reaction as a base:

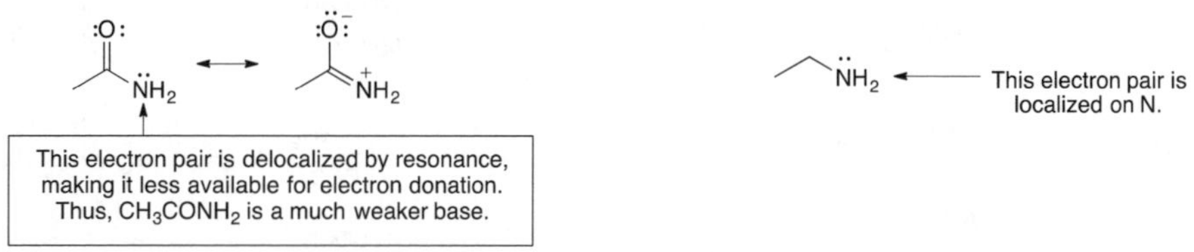

This electron pair is delocalized by resonance, making it less available for electron donation. Thus, CH₃CONH₂ is a much weaker base.

This electron pair is localized on N.

22.44 a. The electron pair on the O atom in phenyl acetate is delocalized on both the carbonyl group and the benzene ring. Thus, the C=O has less single bond character than the C=O of cyclohexyl acetate, so the absorption occurs at higher wavenumbers.

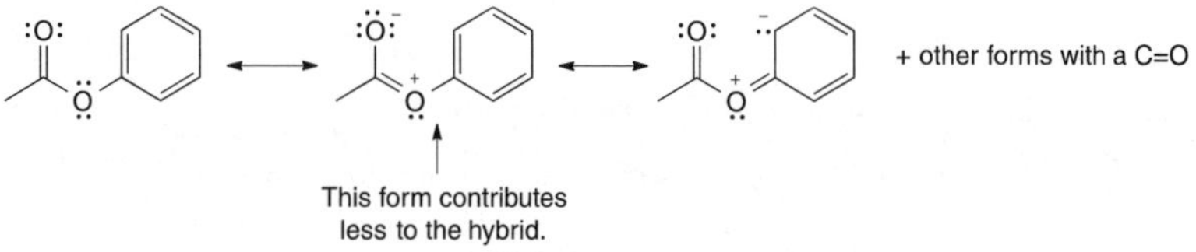

+ other forms with a C=O

This form contributes less to the hybrid.

b. Because the lone pair on the O atom in cyclohexyl acetate can only be delocalized on the C=O, the C=O of cyclohexyl acetate is more effectively stabilized by resonance.

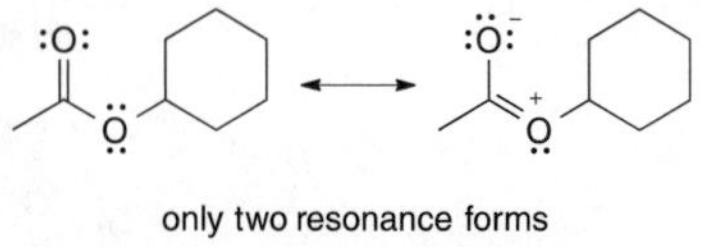

only two resonance forms

c. Phenyl acetate has a better leaving group and is more reactive.

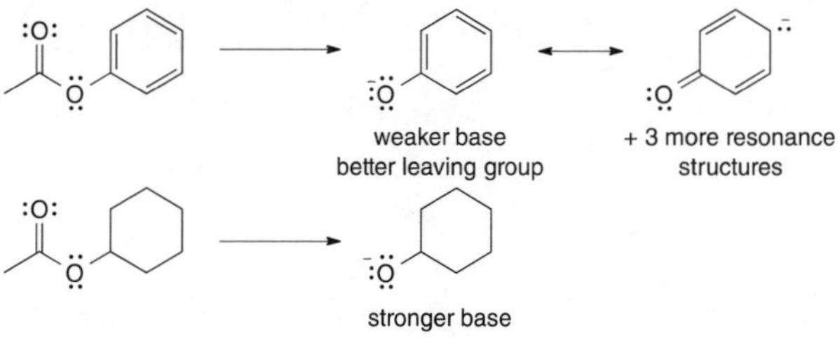

weaker base
better leaving group

+ 3 more resonance structures

stronger base

Chapter 22–18

22.45

$C_6H_5CH_2COOH$

a. $\xrightarrow{\text{NaHCO}_3}$ [structure: $C_6H_5CH_2COO^- Na^+$] + H_2CO_3

g. $\xrightarrow[\text{H}_2\text{SO}_4]{\text{CH}_3\text{OH}}$ [structure: $C_6H_5CH_2COOCH_3$]

b. $\xrightarrow{\text{NaOH}}$ [structure: $C_6H_5CH_2COO^- Na^+$] + H_2O

h. $\xrightarrow[^-\text{OH}]{\text{CH}_3\text{OH}}$ [structure: $C_6H_5CH_2COO^-$]

c. $\xrightarrow{\text{SOCl}_2}$ [structure: $C_6H_5CH_2COCl$]

i. $\xrightarrow[\text{[2] CH}_3\text{COCl}]{\text{[1] NaOH}}$ [structure: mixed anhydride $C_6H_5CH_2CO-O-COCH_3$]

d. $\xrightarrow{\text{NaCl}}$ no reaction

j. $\xrightarrow[\text{DCC}]{\text{CH}_3\text{NH}_2}$ [structure: $C_6H_5CH_2CONHCH_3$]

e. $\xrightarrow[\text{(1 equiv)}]{\text{NH}_3}$ [structure: $C_6H_5CH_2COO^- NH_4^+$]

k. $\xrightarrow[\text{[2] CH}_3\text{CH}_2\text{CH}_2\text{NH}_2]{\text{[1] SOCl}_2}$ [structure: $C_6H_5CH_2CONH-CH_2CH_2CH_3$]

f. $\xrightarrow[\text{[2] }\Delta]{\text{[1] NH}_3}$ [structure: $C_6H_5CH_2CONH_2$]

l. $\xrightarrow[\text{[2] (CH}_3\text{)}_2\text{CHOH}]{\text{[1] SOCl}_2}$ [structure: $C_6H_5CH_2COO-CH(CH_3)_2$]

22.46

[structure: $C_6H_5CH_2CN$]

a. $\xrightarrow{\text{H}_3\text{O}^+}$ [structure: $C_6H_5CH_2COOH$]

d. $\xrightarrow[\text{[2] H}_2\text{O}]{\text{[1] CH}_3\text{CH}_2\text{Li}}$ [structure: $C_6H_5CH_2COCH_2CH_3$]

b. $\xrightarrow[^-\text{OH}]{\text{H}_2\text{O,}}$ [structure: $C_6H_5CH_2COO^-$]

e. $\xrightarrow[\text{[2] H}_2\text{O}]{\text{[1] DIBAL-H}}$ [structure: $C_6H_5CH_2CHO$]

c. $\xrightarrow[\text{[2] H}_2\text{O}]{\text{[1] CH}_3\text{MgBr}}$ [structure: $C_6H_5CH_2COCH_3$]

f. $\xrightarrow[\text{[2] H}_2\text{O}]{\text{[1] LiAlH}_4}$ [structure: $C_6H_5CH_2CH_2NH_2$]

Carboxylic Acids and Their Derivatives 22–19

22.47

a.

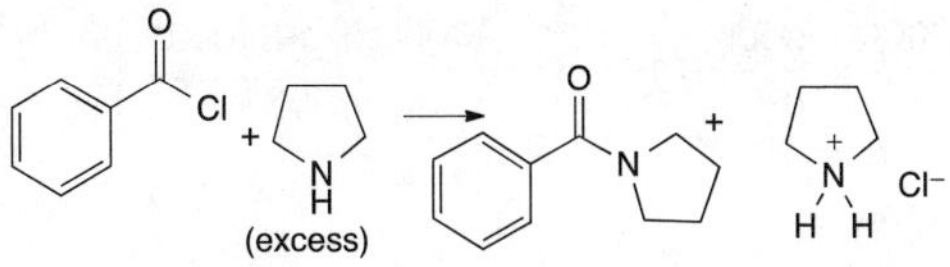

b.

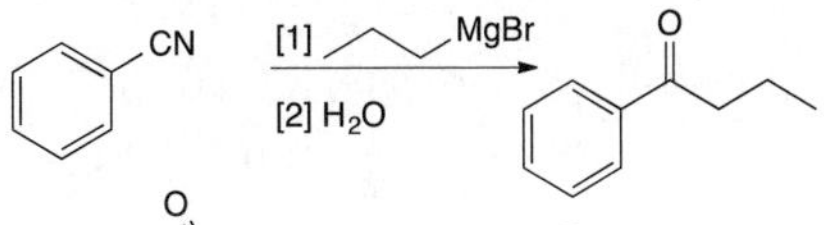

c.

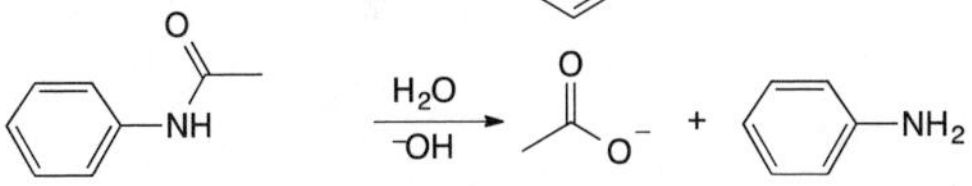

d.

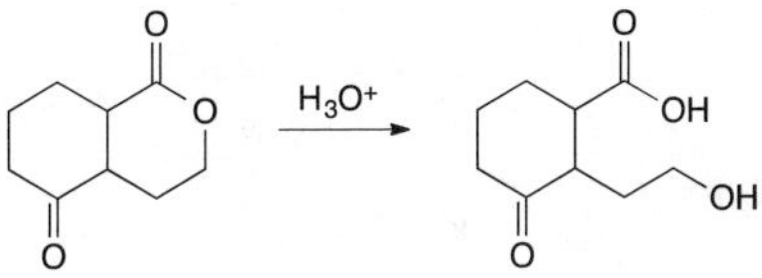

e.

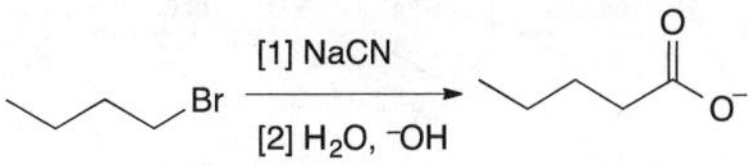

f.

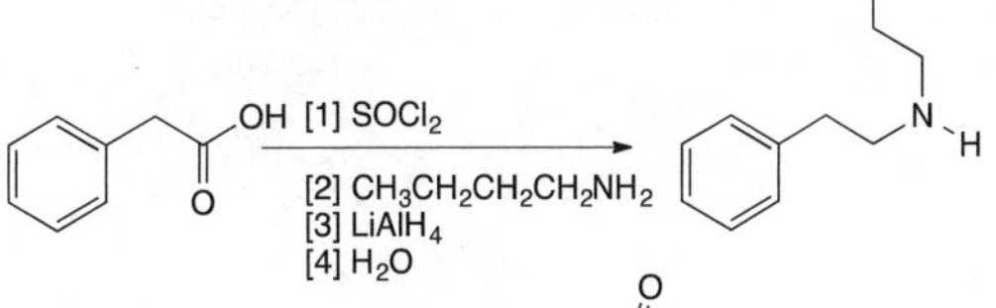

g.

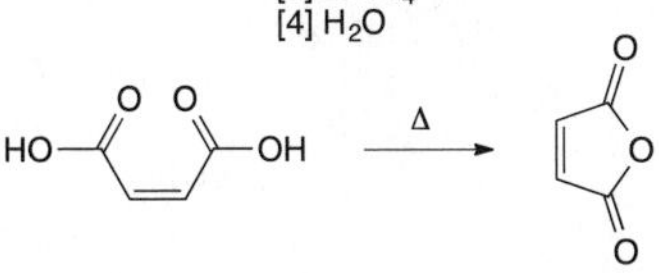

h.

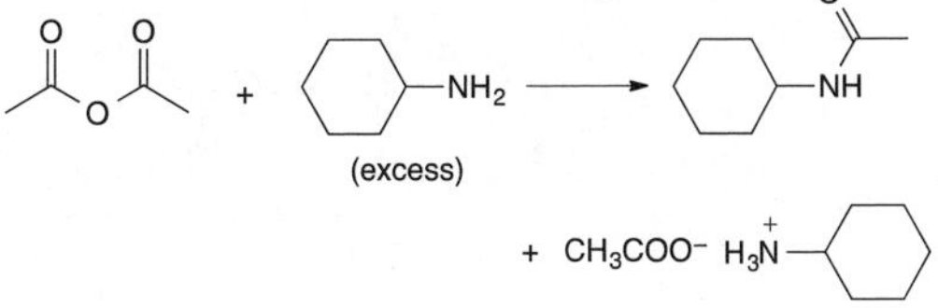

22.48

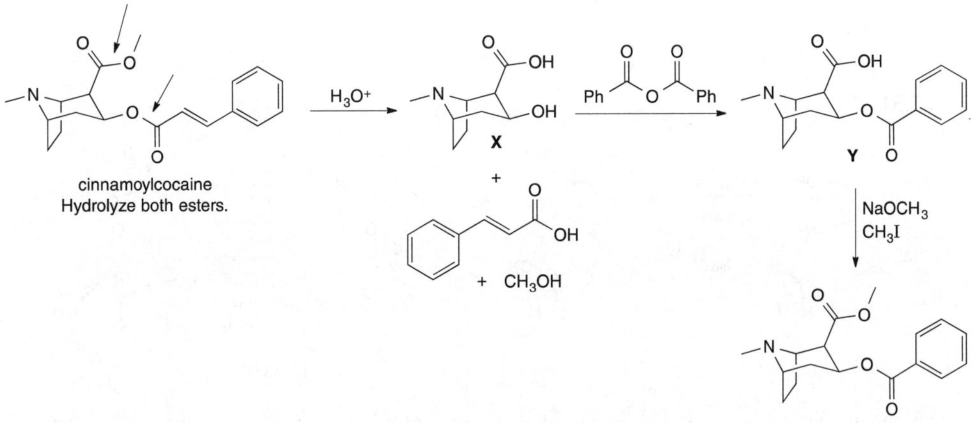

cinnamoylcocaine
Hydrolyze both esters.

cocaine

22.49

a.

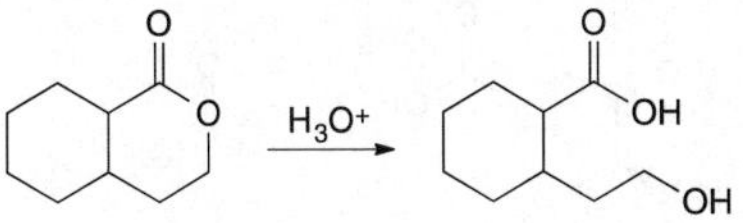

b.

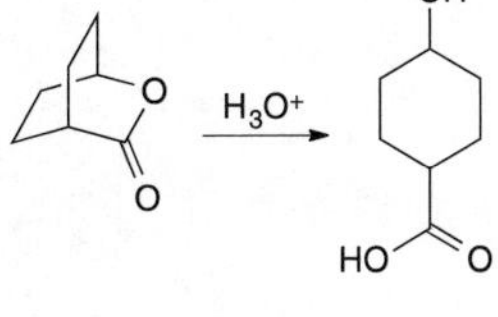

c.

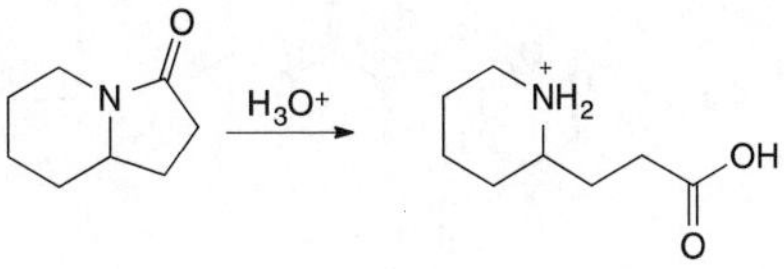

d.

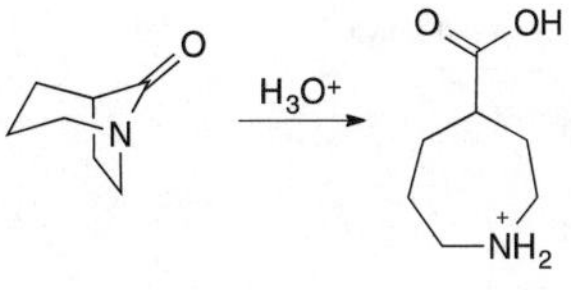

Chapter 22–20

22.50

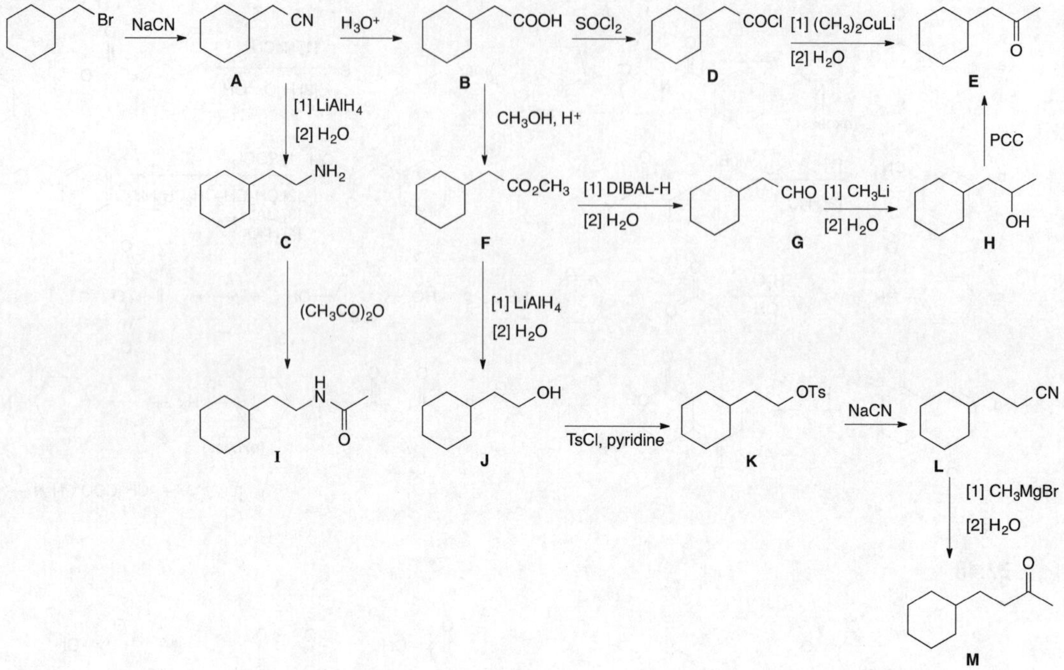

22.51

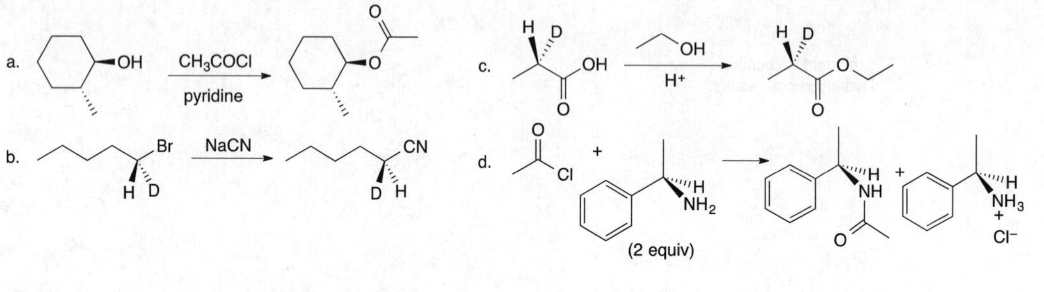

22.52 Hydrolyze the amide and ester bonds in both starting materials to draw the products.

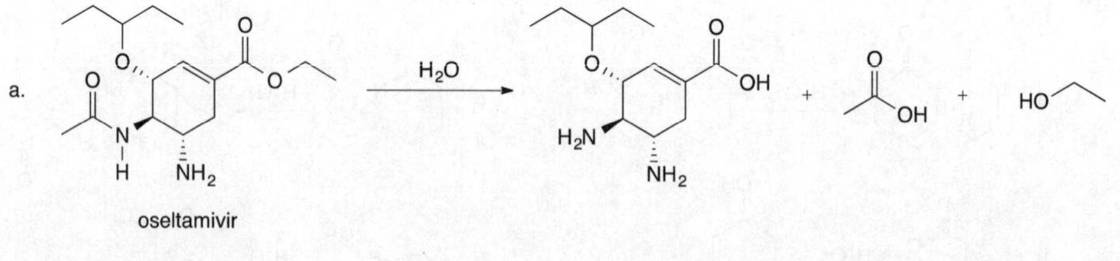

oseltamivir

b.

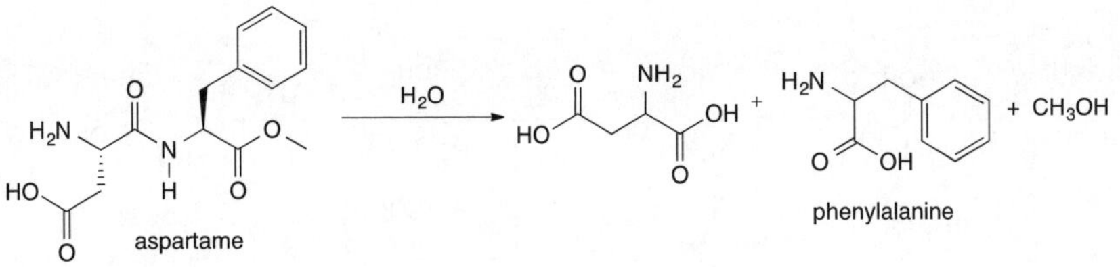

H2N
O
N
H
O
HO
O
aspartame
H2O
O
NH2
HO
OH
O
+
H2N
O
OH
phenylalanine
+ CH3OH

22.53

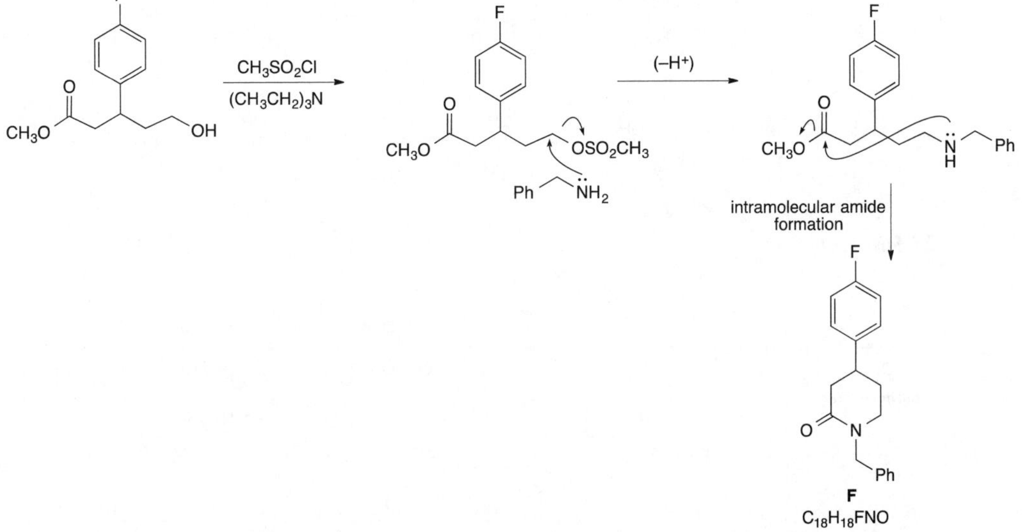

F
O
CH3O
OH
CH3SO2Cl
(CH3CH2)3N
OSO2CH3
Ph
NH2
(–H+)
N
H
Ph
intramolecular amide formation
O
N
Ph
F
C18H18FNO

22.54

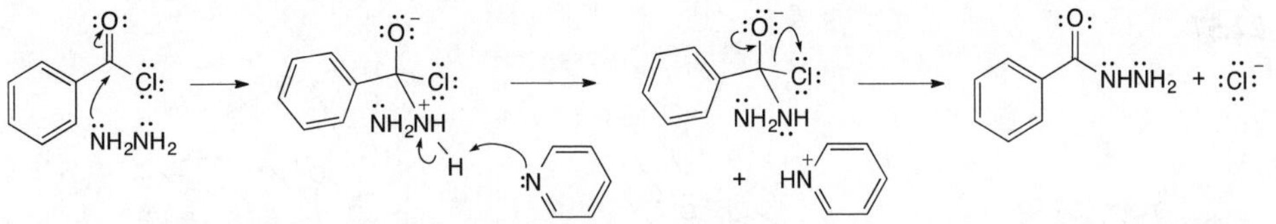

:O:
Cl:
NH2NH2
NH2NH
H
:N
+
HN
NHNH2
+ :Cl:

Chapter 22–22

22.55

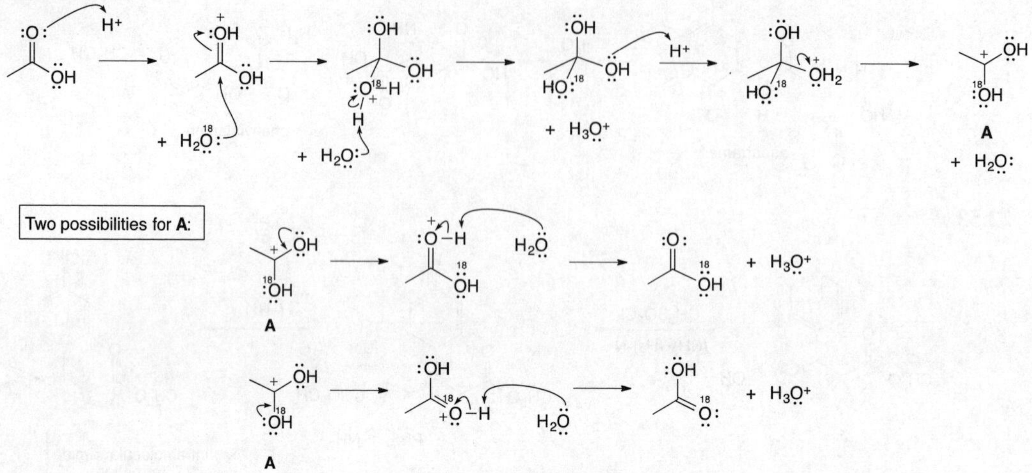

Two possibilities for **A**:

22.56

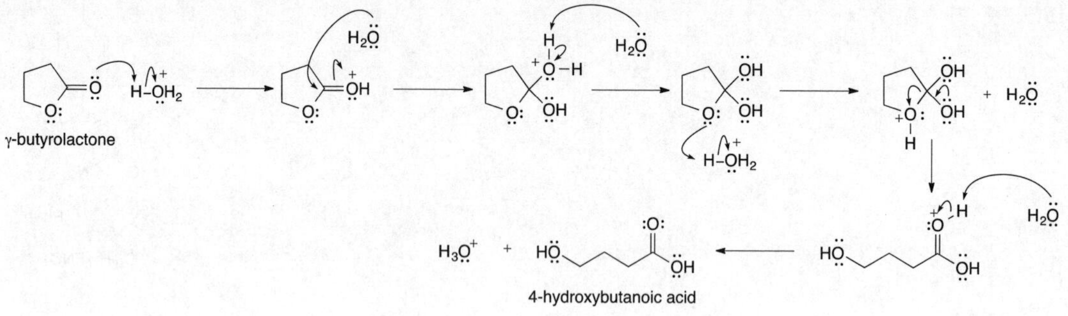

γ-butyrolactone

4-hydroxybutanoic acid
GHB

22.57

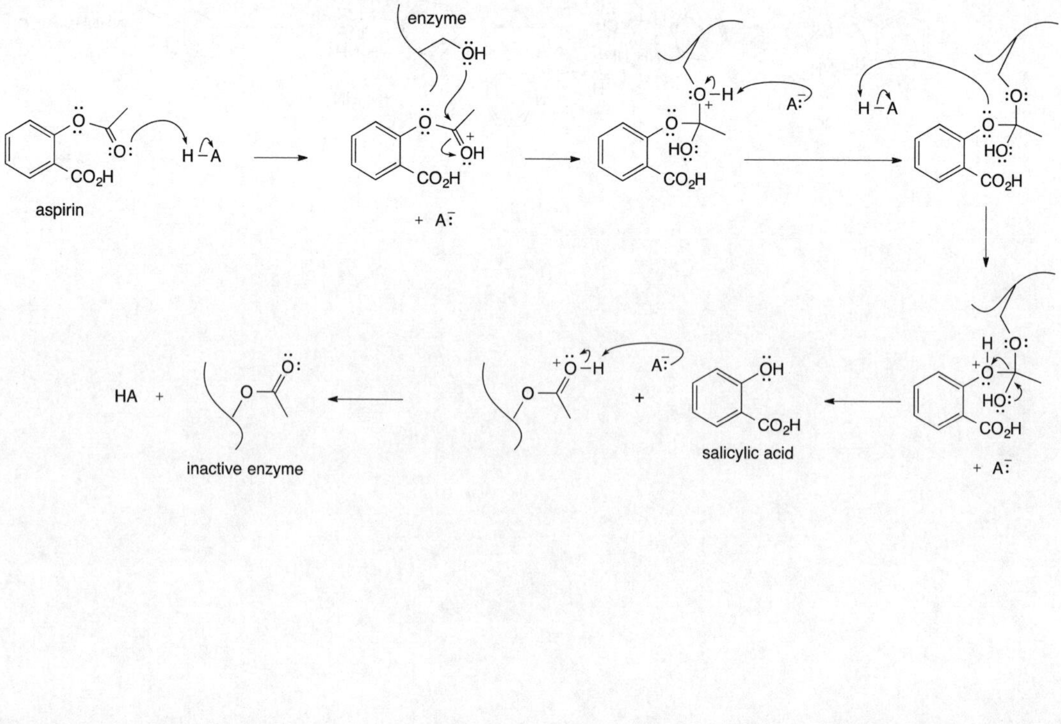

aspirin

enzyme

salicylic acid

inactive enzyme

Carboxylic Acids and Their Derivatives 22–23

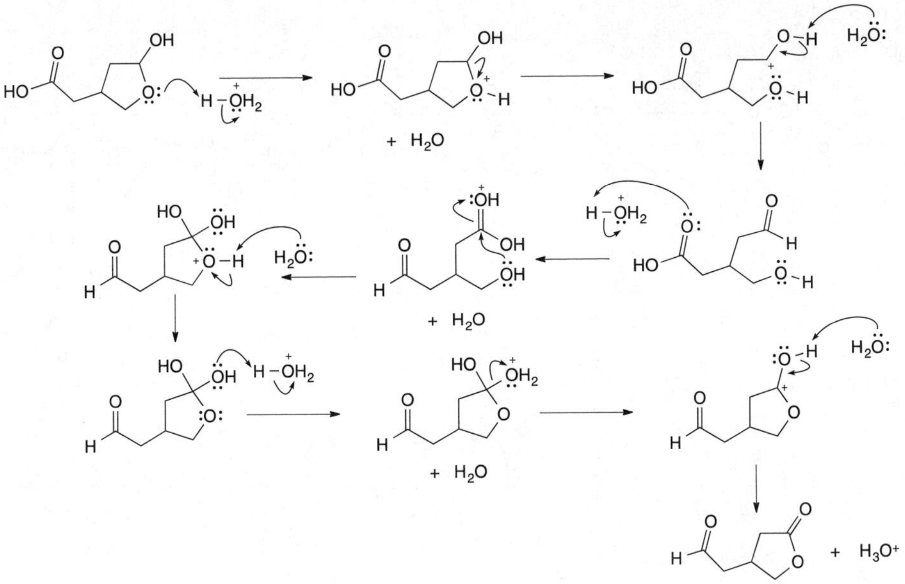

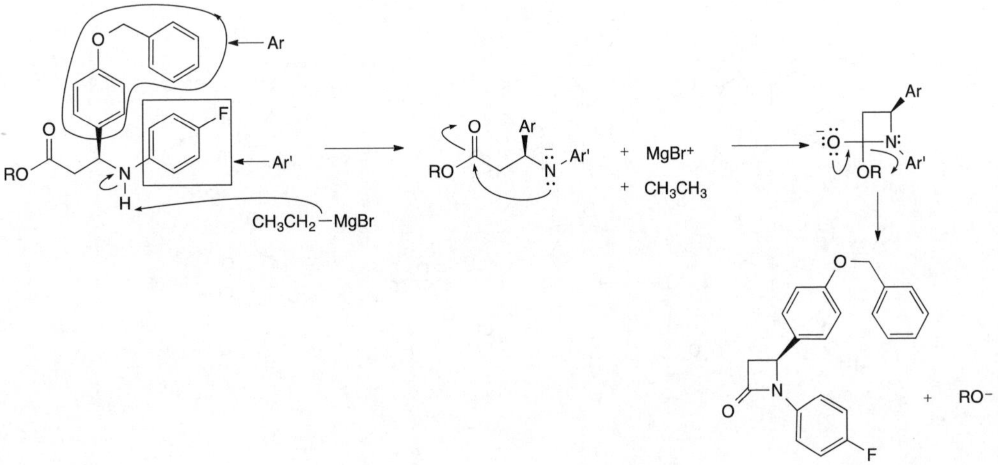

22.58

22.59

Chapter 22–24

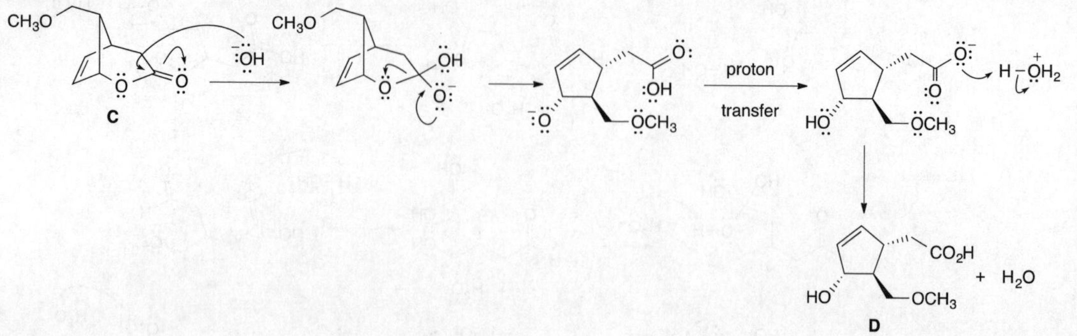

22.60

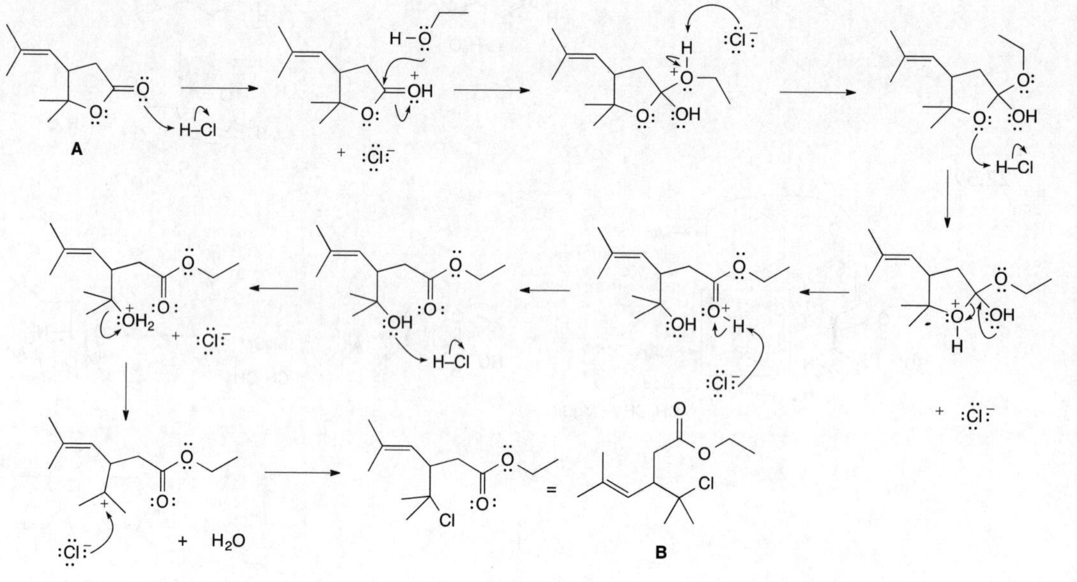

22.61

22.62 The mechanism is composed of two parts: hydrolysis of the acetal and intramolecular Fischer esterification of the hydroxy carboxylic acid.

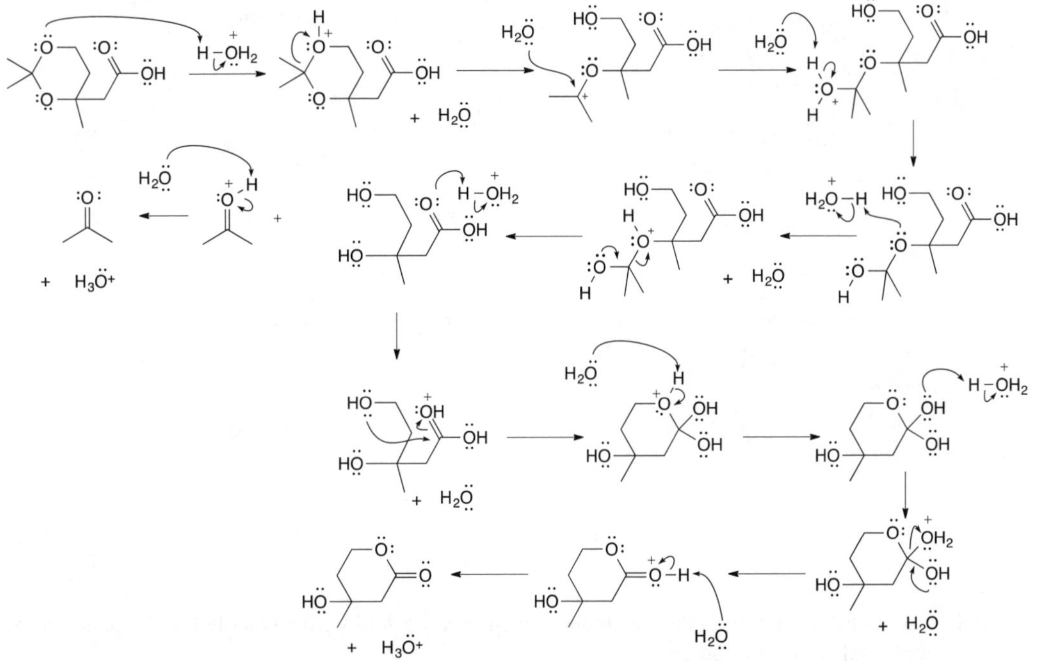

22.63

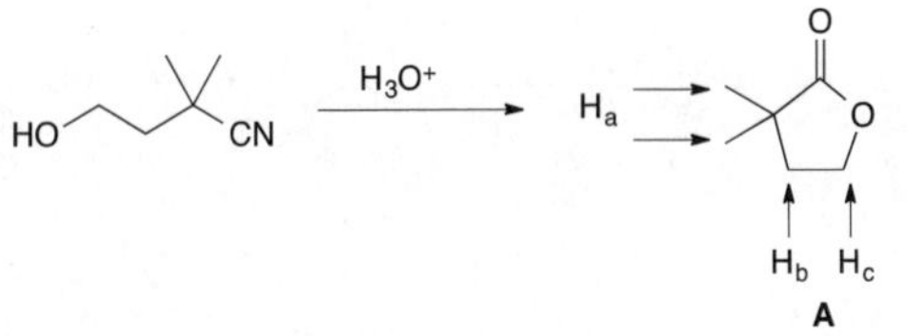

A
$C_6H_{10}O_2$: 2 degrees of unsaturation

IR: 1770 cm^{-1} from ester C=O in a five-membered ring
^{1}H NMR:

H_a = 1.27 ppm (singlet, 6 H) – 2 CH$_3$ groups
H_b = 2.12 ppm (triplet, 2 H) – CH$_2$ bonded to CH$_2$
H_c = 4.26 ppm (triplet, 2 H) – CH$_2$ bonded to CH$_2$

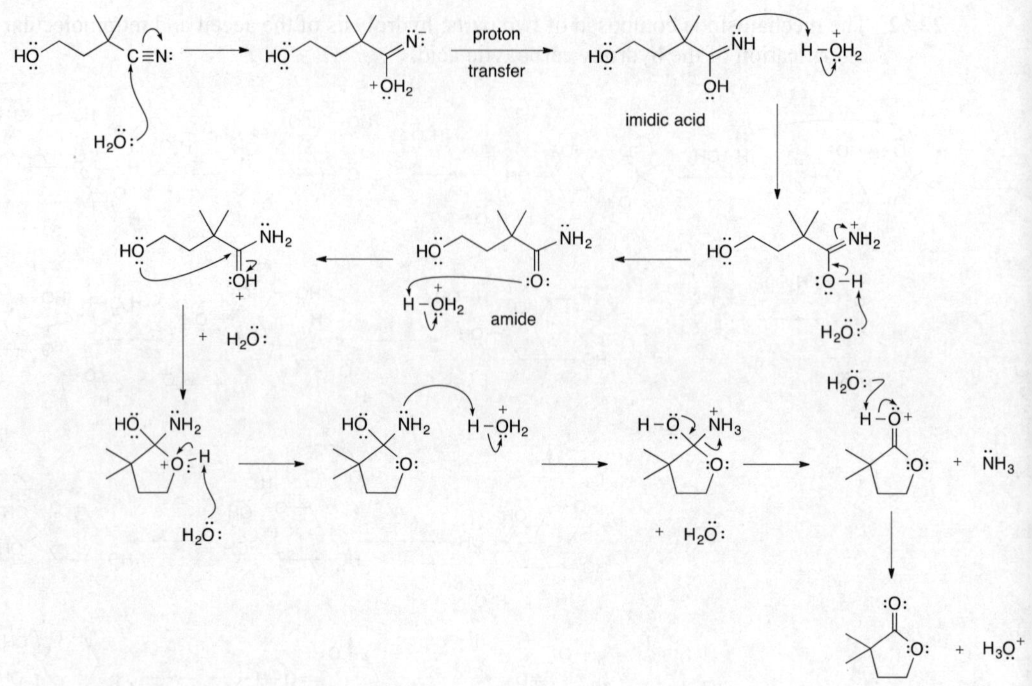

proton transfer

imidic acid

amide

22.64 Fischer esterification is the treatment of a carboxylic acid with an alcohol in the presence of an acid catalyst to form an ester.

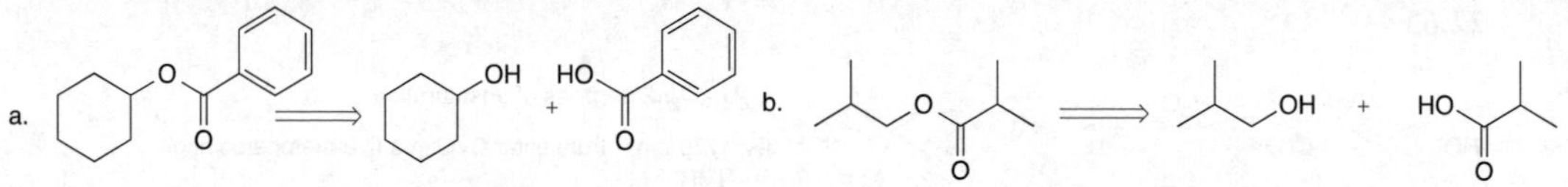

a.

b.

22.65

a.

b.

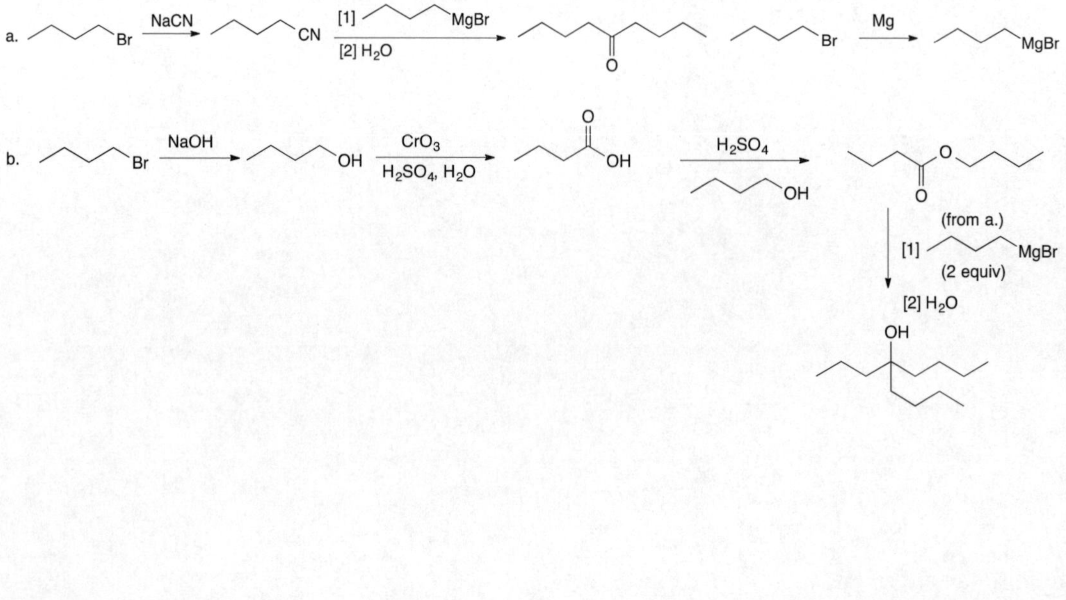

22.66

a. CH_3Cl + NaCN $\longrightarrow$ CH_3-CN $\xrightarrow{H_3O^+}$ CH_3-COOH

CH_3-Cl + Mg $\longrightarrow$ CH_3-MgCl $\xrightarrow[\text{[2] } H_3O^+]{\text{[1] } CO_2}$ CH_3-COOH

b. (bromobenzene, sp^2) + NaCN $\longrightarrow$ This method can't be used because an S_N2 reaction can't be done on an sp^2 hybridized C.

(bromobenzene) + Mg $\longrightarrow$ (phenylMgBr) $\xrightarrow[\text{[2] } H_3O^+]{\text{[1] } CO_2}$ (benzoic acid, COOH)

c. HO–(chain)–Br + NaCN $\longrightarrow$ HO–(chain)–CN $\xrightarrow{H_3O^+}$ HO–(chain)–COOH

HO–(chain)–Br + Mg $\longrightarrow$ This method can't be used because you can't make a Grignard reagent with an acidic OH group.

22.67

(benzene) $\xrightarrow[\text{AlCl}_3]{\overset{CH_3OH \,|\, SOCl_2}{CH_3Cl}}$ (toluene) $\xrightarrow[H_2SO_4]{HNO_3}$ (nitrotoluene, O_2N) (+ ortho isomer) $\xrightarrow{KMnO_4}$ (O_2N–C6H4–COOH) $\xrightarrow[H_2SO_4]{CH_3CH_2OH}$ (ethyl ester, O_2N) $\xrightarrow[\text{Pd-C}]{H_2}$ (ethyl ester, H_2N)

22.68

a. (phenol, OH) $\xrightarrow[\text{AlCl}_3]{\text{Cl (ethyl)}}$ (ethylphenol, OH) (+ para isomer) $\xrightarrow{KMnO_4}$ (salicylic acid, COOH, OH) $\xrightarrow{SOCl_2}$ (acid chloride, Cl, OH) $\xrightarrow[\text{(2 equiv)}]{NH_3}$ (salicylamide, NH_2, OH)

salicylamide

b. (phenol, OH) $\xrightarrow[H_2SO_4]{HNO_3}$ (HO–nitrophenol, NO_2) (+ ortho isomer) $\xrightarrow{H_2, \text{Pd-C}}$ (HO–aminophenol, NH_2) (More nucleophilic NH_2 reacts first.) + (acetyl chloride, Cl) $\longrightarrow$ (acetaminophen, HO) acetaminophen

c. (acetaminophen, HO) (from b.) $\xrightarrow{NaH}$ (^-O–phenyl–NHAc) $\xrightarrow{\text{Br (ethyl)}}$ (p-acetophenetidin, ethoxy)

acetaminophen (from b.)

p-acetophenetidin

Chapter 22–28

22.69

a.

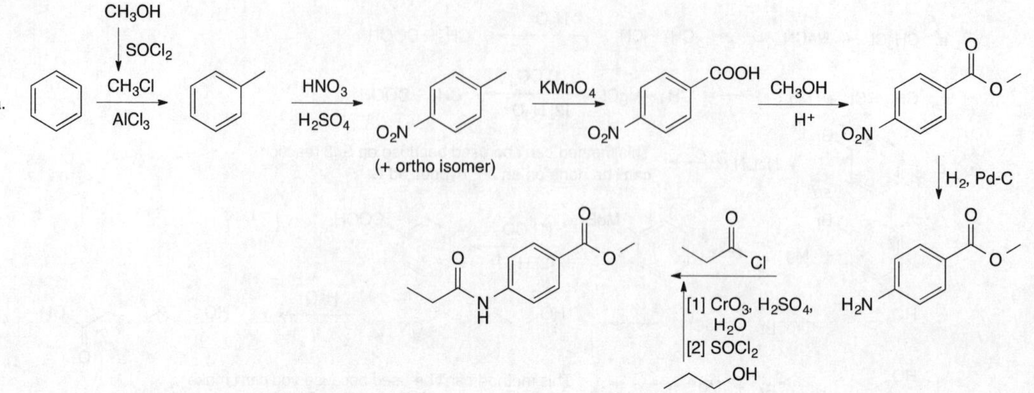

b.

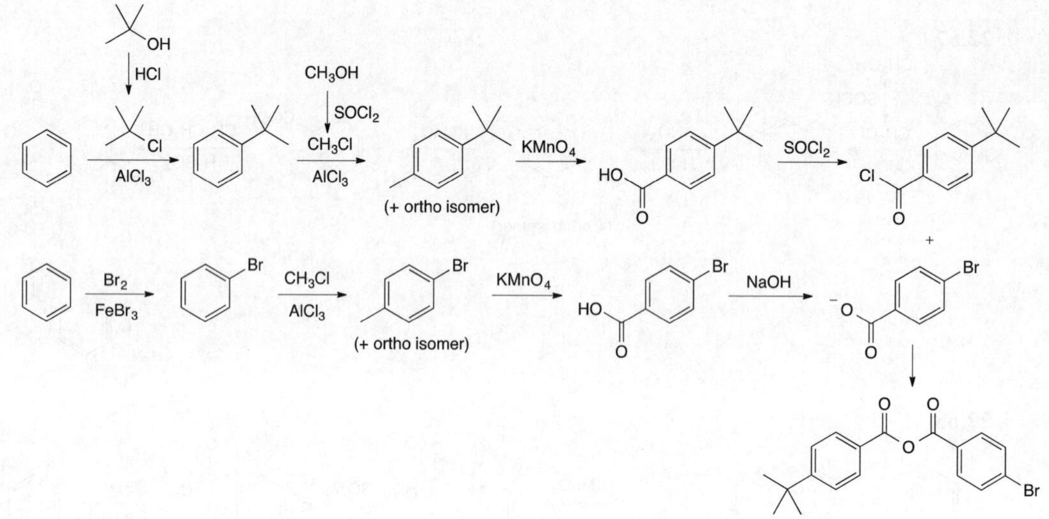

22.70

a.

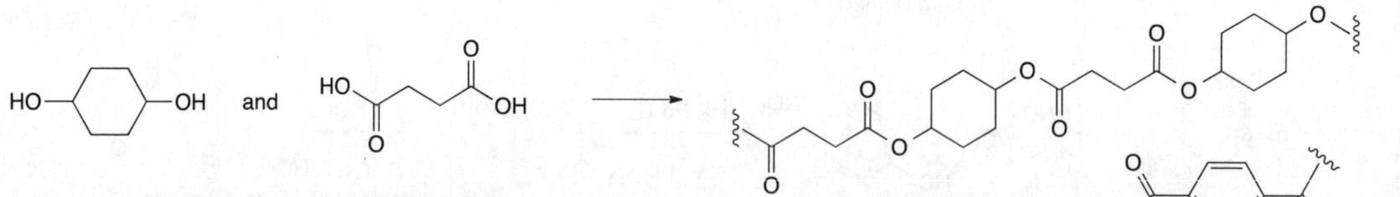

b.

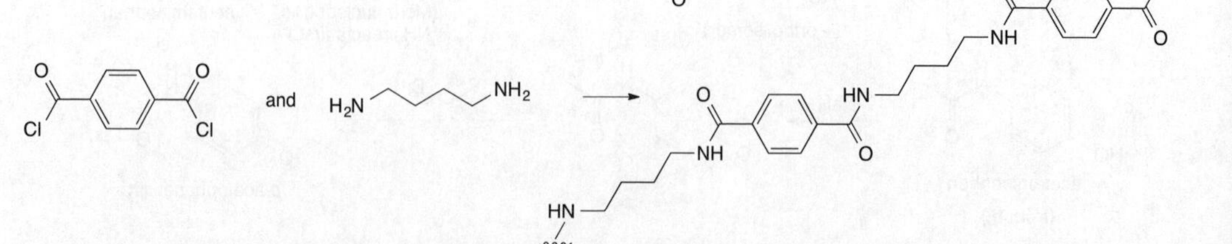

Carboxylic Acids and Their Derivatives 22–29

22.71

a.

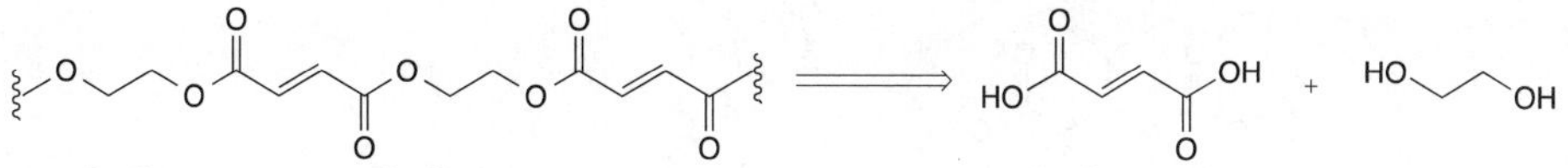

b.

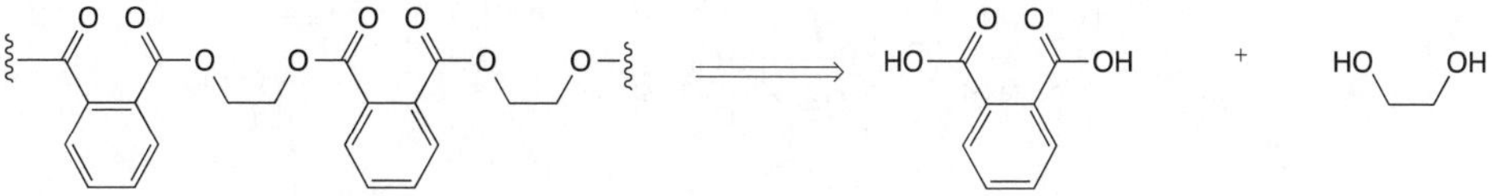

22.72

a. Docetaxel has fewer C's and one more OH group than paclitaxel. This makes docetaxel more water soluble than paclitaxel.

b.

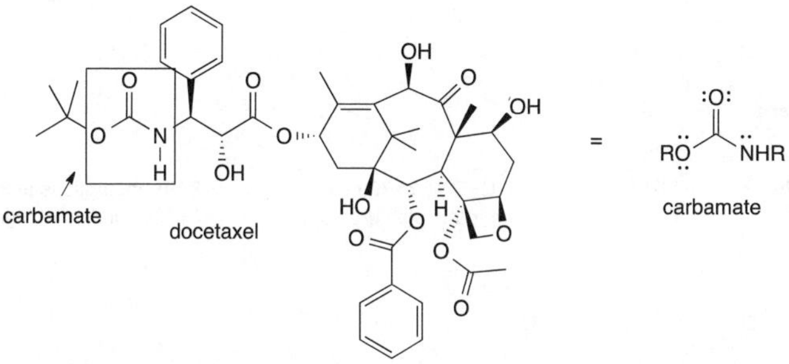

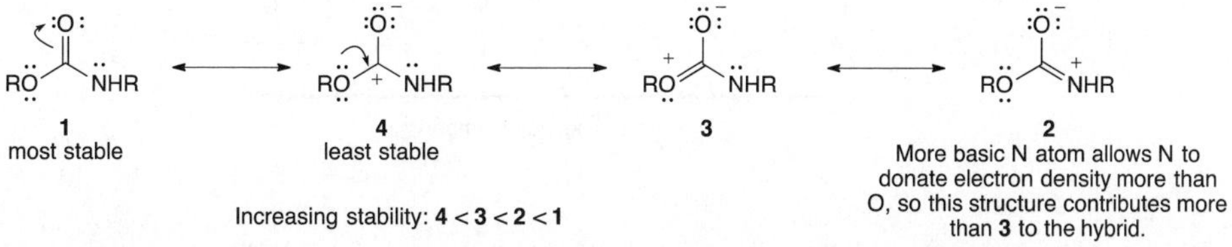

Increasing stability: **4 < 3 < 2 < 1**

c.

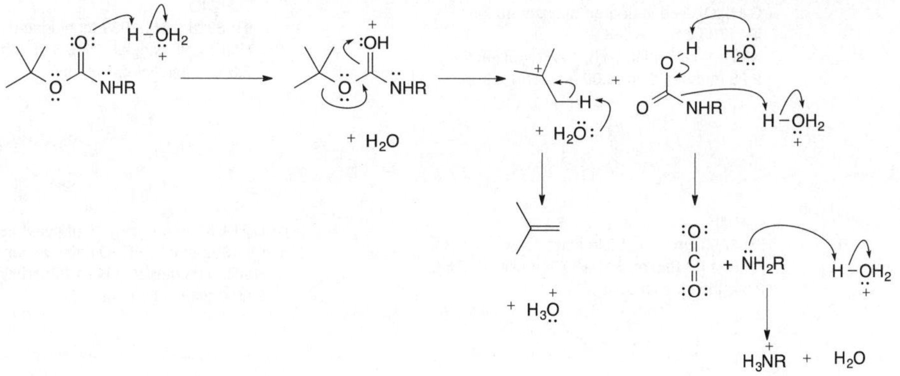

Chapter 22–30

d.

docetaxel

$$H_3O^+ \longrightarrow$$

+ CO_2 +

COOH

+ CH_3CO_2H

22.73

a.

and

Acid chloride CO absorbs at
much higher wavenumber.

ketone

b.

and

$C=O$ at <1700 cm^{-1}
due to the stabilized amide

2 NH absorptions at 3200–3400 cm^{-1}
$C=O$ absorption at higher wavenumber

22.74

Increasing wavenumber

22.75

a. $C_6H_{12}O_2 \rightarrow$ one degree of unsaturation
IR: 1738 cm$^{-1} \rightarrow$ C=O
NMR: 1.12 (triplet, 3 H), 1.23 (doublet, 6 H),
2.28 (quartet, 2 H), 5.00 (septet, 1 H) ppm

c. C_8H_9NO
IR: 3328 (NH), 1639 (conjugated amide C=O) cm^{-1}
NMR: 2.95 (singlet, 3 H), 6.95 (singlet, 1 H),
7.3–7.7 (multiplet, 5 H) ppm

b. C_4H_7N
IR: 2250 cm$^{-1} \rightarrow$ triple bond
NMR: 1.08 (triplet, 3 H), 1.70 (multiplet, 2 H),
2.34 (triplet, 2 H) ppm

d. $C_4H_7ClO \rightarrow$ one degree of unsaturation
IR: 1802 cm$^{-1} \rightarrow$ C=O (high wavenumber, RCOCl)
NMR: 0.95 (triplet, 3 H), 1.07 (multiplet, 2 H),
2.90 (triplet, 2 H) ppm

Carboxylic Acids and Their Derivatives 22–31

e. $C_{10}H_{12}O_2 \rightarrow$ five degrees of unsaturation
IR: 1740 cm^{-1} $\rightarrow$ C=O
NMR: 1.2 (triplet, 3 H), 2.4 (quartet, 2 H),
5.1 (singlet, 2 H), 7.1–7.5 (multiplet, 5 H) ppm

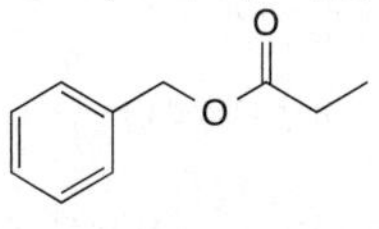

22.76

A. Molecular formula $C_{10}H_{12}O_2 \rightarrow$ five degrees of unsaturation
IR absorption at 1718 cm^{-1} $\rightarrow$ C=O
NMR data (ppm):

    triplet at 1.4 (CH$_3$ adjacent to 2 H's)
    singlet at 2.4 (CH$_3$)
    quartet at 4.4 (CH$_2$ adjacent to CH$_3$)
    doublet at 7.2 (2 H's on benzene ring)
    doublet at 7.9 (2 H's on benzene ring)

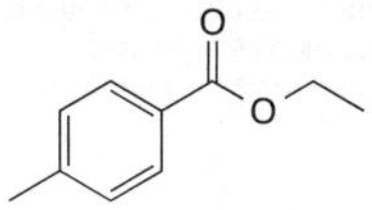

B. IR absorption at 1740 cm^{-1} $\rightarrow$ C=O
NMR data (ppm):

    singlet at 2.1 (CH$_3$)
    triplet at 2.9 (CH$_2$ adjacent to CH$_2$)
    triplet at 4.3 (CH$_2$ adjacent to CH$_2$)
    multiplet at 7.3 (5 H's, monosubstituted benzene)

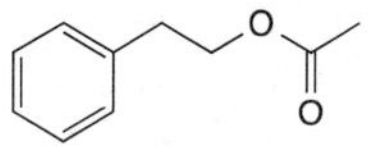

22.77

Molecular formula $C_{10}H_{13}NO_2 \rightarrow$ five degrees of unsaturation
IR absorptions at 3300 (NH) and 1680 (C=O, amide or conjugated) cm^{-1}
NMR data (ppm):

    triplet at 1.4 (CH$_3$ adjacent to CH$_2$)
    singlet at 2.2 (CH$_3$C=O)
    quartet at 3.9 (CH$_2$ adjacent to CH$_3$)
    doublet at 6.8 (2 H's on benzene ring)
    singlet at 7.1 (NH)
    doublet at 7.3 (2 H's on benzene ring)

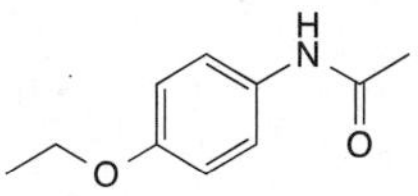

phenacetin

22.78

Molecular formula $C_{11}H_{15}NO_2 \rightarrow$ five degrees of unsaturation
IR absorption 1699 (C=O, amide or conjugated) cm^{-1}
NMR data (ppm):

    triplet at 1.3 (3 H) (CH$_3$ adjacent to CH$_2$)
    singlet at 3.0 (6 H) (2 CH$_3$ groups on N)
    quartet at 4.3 (2 H) (CH$_2$ adjacent to CH$_3$)
    doublet at 6.6 (2 H) (2 H's on benzene ring)
    doublet at 7.9 (2 H) (2 H's on benzene ring)

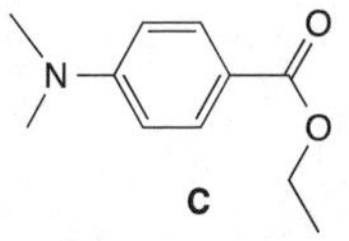

C

22.79

a. Molecular formula $C_6H_{12}O_2 \rightarrow$ one degree of unsaturation
IR absorption at 1743 cm^{-1} $\rightarrow$ C=O
^{1}H NMR data (ppm):

    triplet at 0.9 (3 H) – CH$_3$ adjacent to CH$_2$
    multiplet at 1.35 (2 H) – CH$_2$
    multiplet at 1.60 (2 H) – CH$_2$
    singlet at 2.1 (3 H – from CH$_3$ bonded to C=O)
    triplet at 4.1 (2 H) – CH$_2$ adjacent to the electronegative O atom and another CH$_2$

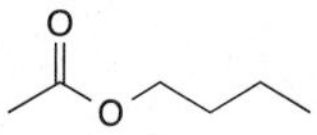

D

Chapter 22–32

    b. Molecular formula $C_6H_{12}O_2$ → one degree of unsaturation
        IR absorption at 1746 cm–1 → C=O
        ^{1}H NMR data (ppm):
            doublet at 0.9 (6 H) – 2 CH$_3$'s adjacent to CH
            multiplet at 1.9 (1 H)
            singlet at 2.1 (3 H) – CH$_3$ bonded to C=O
            doublet at 3.85 (2 H) – CH$_2$ bonded to electronegative O and CH

22.80 The extent of resonance stabilization affects the position of the C=O absorption in the IR of an amide.

A: This resonance structure does not contribute significantly to the hybrid because it places a double bond at the bridgehead N, an impossible geometry in small rings. The C=O has more double bond character, so the absorption is at a higher wavenumber.

B: This resonance structure contributes significantly to the hybrid, so the C=O has more single bond character. The absorption shifts to a lower wavenumber.

22.81

4.02 ppm

7.35 and 7.60 ppm

This resonance structure makes a significant contribution to the resonance hybrid.

There is restricted rotation around the amide C–N bond. The 2 H's are in different environments (one is cis to an O atom, and one is cis to CH$_2$Cl), so they give different NMR signals.

different environments

22.82

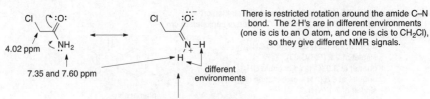

ethyl benzoate

Two OH groups are now equivalent and either can lose H$_2$O to form labeled or unlabeled ethyl benzoate.

Unlabeled starting material was recovered.

Carboxylic Acids and Their Derivatives 22–33

22.83

22.84 Both acetals are hydrolyzed by the usual mechanism for acetal hydrolysis (Steps [1 –[6]), forming four new OH groups. Intramolecular esterification forms a lactone (Steps [10]–[15]), followed by conversion of a carbonyl tautomer to an enol (Steps [16]–[17]). Both acetals are hydrolyzed at once in the given mechanism. For ease in drawing this long mechanism, the stereochemistry of intermediates is omitted.

Chapter 22–34

22.85

Substitution Reactions of Carbonyl Compounds 23–1

Chapter 23 Substitution Reactions of Carbonyl Compounds at the α Carbon

Chapter Review

Kinetic versus thermodynamic enolates (23.4)

kinetic enolate

Kinetic enolate
- The less substituted enolate
- Favored by strong base, polar aprotic solvent, low temperature: LDA, THF, –78 °C

thermodynamic enolate

Thermodynamic enolate
- The more substituted enolate
- Favored by strong base, protic solvent, higher temperature: NaOCH$_2$CH$_3$, CH$_3$CH$_2$OH, room temperature

Halogenation at the α carbon
[1] Halogenation in acid (23.7A)

$$R\text{-CO-CHRH} \xrightarrow[\text{CH}_3\text{COOH}]{X_2} R\text{-CO-CRX}$$

$X_2 = Cl_2, Br_2, \text{or } I_2$ → α-halo aldehyde or ketone

- The reaction occurs via enol intermediates.
- Monosubstitution of X for H occurs on the α carbon.

[2] Halogenation in base (23.7B)

$$R\text{-CO-CH}_2\text{R} \xrightarrow[\text{-OH}]{X_2 \text{ (excess)}} R\text{-CO-CX}_2\text{R}$$

$X_2 = Cl_2, Br_2, \text{or } I_2$

- The reaction occurs via enolate intermediates.
- Polysubstitution of X for H occurs on the α carbon.

[3] Halogenation of *methyl* ketones in base—The haloform reaction (23.7B)

$$R\text{-CO-CH}_3 \xrightarrow[\text{-OH}]{X_2 \text{ (excess)}} R\text{-COO}^- + HCX_3 \text{ (haloform)}$$

$X_2 = Cl_2, Br_2, \text{or } I_2$

- The reaction occurs with methyl ketones and results in cleavage of a carbon–carbon σ bond.

Chapter 23–2

Reactions of α-halo carbonyl compounds (23.7C)
[1] Elimination to form α,β-unsaturated carbonyl compounds

- Elimination of the elements of Br and H forms a new π bond, giving an α,β-unsaturated carbonyl compound.

[2] Nucleophilic substitution

- The reaction follows an S_N2 mechanism, generating an α-substituted carbonyl compound.

Alkylation reactions at the α carbon
[1] Direct alkylation at the α carbon (23.8)

- The reaction forms a new C–C bond to the α carbon.
- LDA is a common base used to form an intermediate enolate.
- The alkylation in Step [2] follows an S_N2 mechanism.

[2] Malonic ester synthesis (23.9)

- The reaction is used to prepare carboxylic acids with one or two alkyl groups on the α carbon.
- The alkylation in Step [2] follows an S_N2 mechanism.

[3] Acetoacetic ester synthesis (23.10)

- The reaction is used to prepare ketones with one or two alkyl groups on the α carbon.
- The alkylation in Step [2] follows an S_N2 mechanism.

Substitution Reactions of Carbonyl Compounds 23-3

Practice Test on Chapter Review

1.a. Which of the following compounds can be prepared using either the acetoacetic ester synthesis or the malonic ester synthesis?

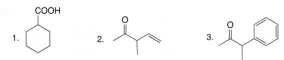

4. Both (1) and (2) can be prepared by one of these routes.
5. Compounds (1), (2), and (3) can all be prepared by these routes.

b. Which of the following compounds is *not* an enol form of dicarbonyl compound **A**?

5. Compounds (1)–(4) are all enols of **A**.

2.a. Which proton in compound **A** has the *lowest* pK_a?
  b. Which proton in compound **A** has the *highest* pK_a?

  c. Which proton in compound **B** is the least acidic?
  d. Which proton in compound **B** is the most acidic?

3. What reagents are needed to convert heptan-4-one to each compound?
   a. 3-bromoheptan-4-one
   b. 3-methylheptan-4-one
   c. 3,5-dimethylheptan-4-one
   d. hept-2-en-4-one
   e. 2-methylheptan-4-one

Chapter 23–4

4. Draw the organic products formed in each of the following reactions.

a. 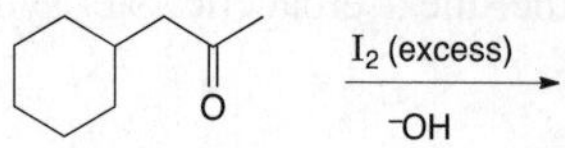 [cyclohexyl-CH₂-C(=O)-CH₃] $\xrightarrow[^-OH]{I_2 \text{ (excess)}}$

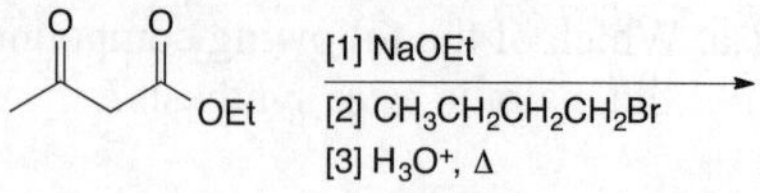

b. 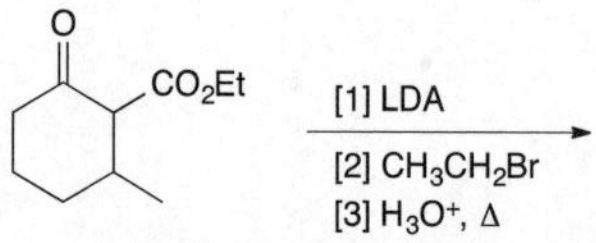 [structure with CO₂Et] $\xrightarrow{\substack{[1]\ LDA \\ [2]\ CH_3CH_2Br \\ [3]\ H_3O^+,\ \Delta}}$

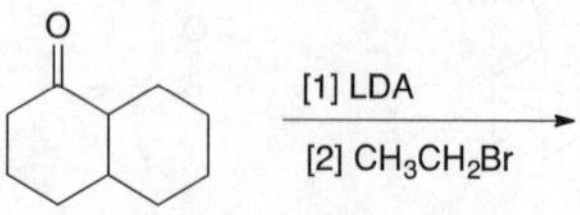

c. $CH_2(CO_2Et)_2$ $\xrightarrow{\substack{[1]\ NaOEt \\ [2]\ CH_3CH_2Br}}$ $\xrightarrow{\substack{[1]\ NaOEt \\ [2]\ C_6H_5CH_2Cl \\ [3]\ H_3O^+,\ \Delta}}$

d. [structure with OEt] $\xrightarrow{\substack{[1]\ NaOEt \\ [2]\ CH_3CH_2CH_2CH_2Br \\ [3]\ H_3O^+,\ \Delta}}$

e. [decalone structure] $\xrightarrow{\substack{[1]\ LDA \\ [2]\ CH_3CH_2Br}}$

5.a. What starting materials are needed to synthesize carboxylic acid **A** by a malonic ester synthesis?

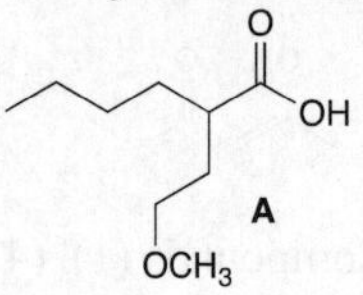

A [structure with OH, OCH₃]

b. What starting materials are needed to synthesize ketone **B** by the acetoacetic ester synthesis?

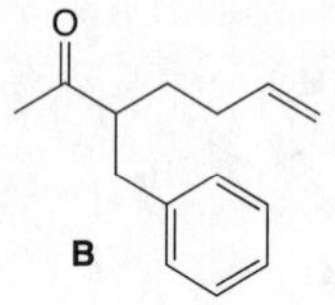

B [structure]

Answers to Practice Test

1.a. 1
  b. 4
2.a. H_b
  b. H_d
  c. H_b
  d. H_c
3.a. Br_2, CH_3CO_2H
  b. [1] LDA; [2] CH_3I
  c. []1 LDA; [2] CH_3I; [3] LDA; [4] CH_3I
  d. [1] Br_2, CH_3CO_2H; [2] Li_2CO_3, LiBr, DMF
  e. [1] Br_2, CH_3CO_2H; [2] Li_2CO_3, LiBr, DMF; [3] $(CH_3)_2CuLi$; [4] H_2O

4.
a. 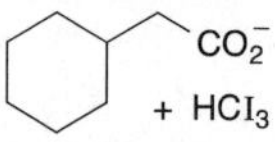 [cyclohexyl-CH₂-CO_2^-] + HCI_3
b. 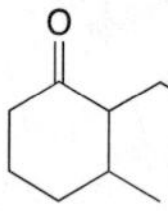 [structure]
c. 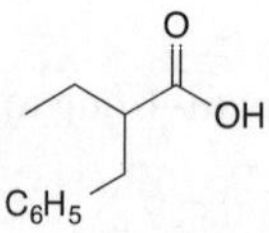 [structure] C_6H_5 ...OH

d. 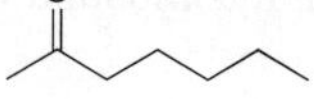 [structure]
e. 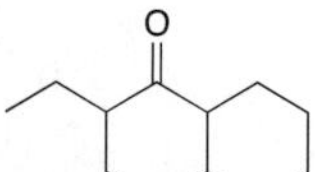 [decalone structure]

5.
a.  [structure with Br] ; CH_3O...Br ; $CH_2(CO_2Et)_2$
b. 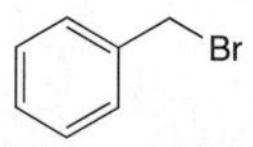 [benzyl Br structure] ; [structure with Br] ; [structure with CO_2Et]

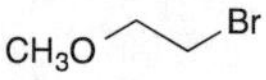

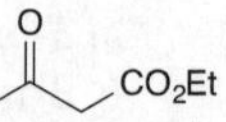

Substitution Reactions of Carbonyl Compounds 23–5

Answers to Problems

23.1 • To convert a ketone to its enol tautomer, change the C=O to C–OH, make a new double bond to an α carbon, and remove a proton at the other end of the C=C.
  • To convert an enol to its keto form, find the C=C bonded to the OH. Change the C–OH to a C=O, add a proton to the other end of the C=C, and delete the double bond.

[In cases where *E* and *Z* isomers are possible, only one stereoisomer is drawn.]

a.

b.

c.

d.

e.

f. Draw mono enol tautomers only.

(Conjugated enols are preferred.)

23.2 The mechanism has two steps for each part: protonation followed by deprotonation.

Chapter 23–6

23.3

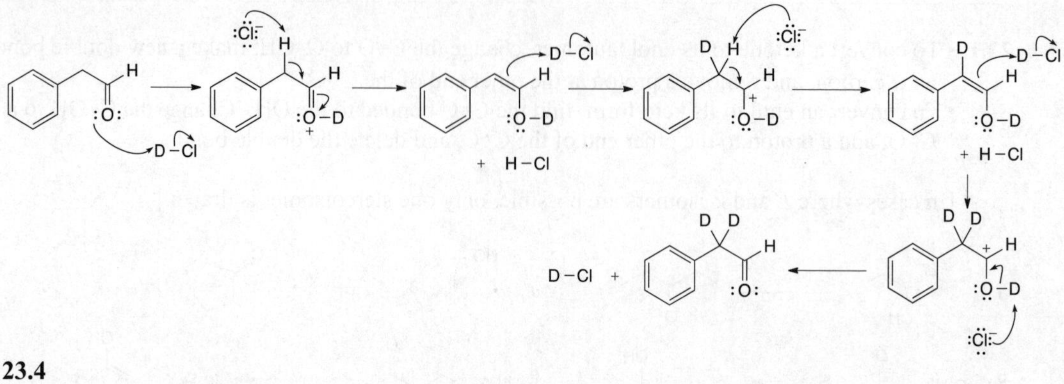

23.4

a.

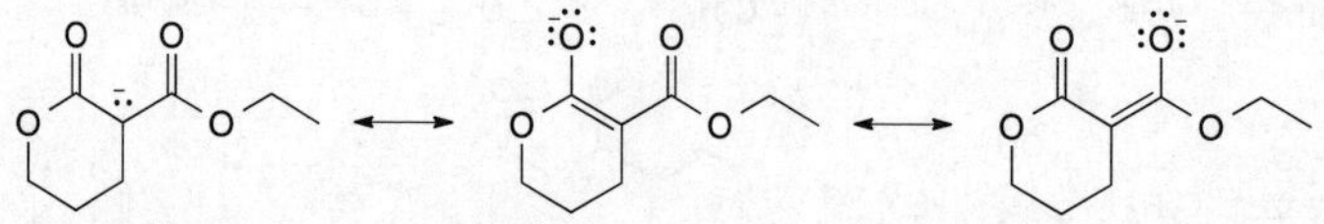

b.

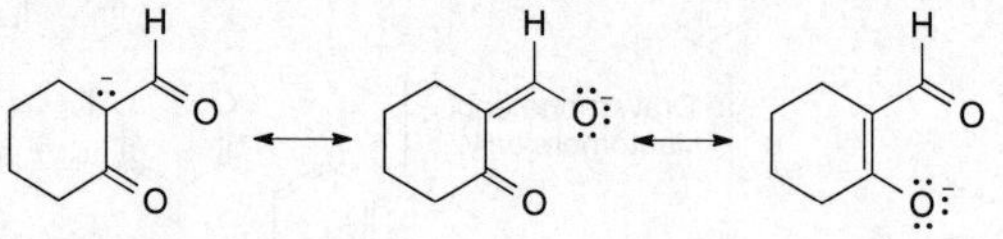

c.

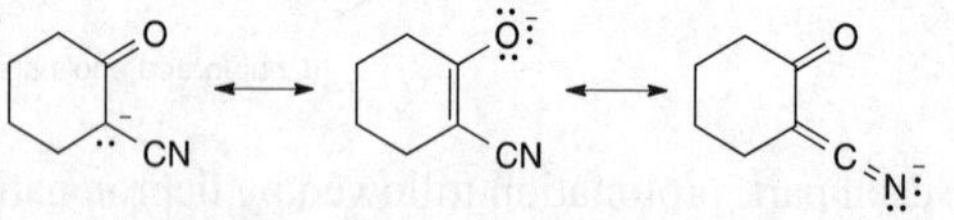

23.5   The indicated H's are α to a C=O or C≡N group, making them more acidic because their removal forms conjugate bases that are resonance stabilized.

a.

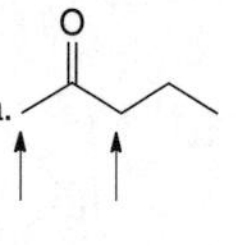

b.

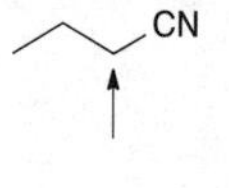

c.

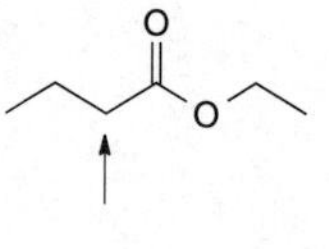

d.

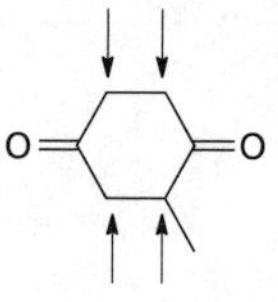

23.6

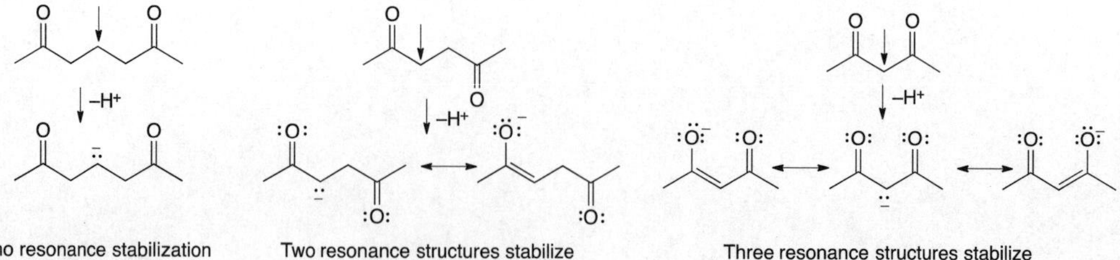

no resonance stabilization
least acidic

Two resonance structures stabilize
the conjugate base.
intermediate acidity

Three resonance structures stabilize
the conjugate base.
most acidic

Substitution Reactions of Carbonyl Compounds 23–7

23.7 In each of the reactions, the LDA pulls off the most acidic proton.

a. [structure] $\xrightarrow[\text{THF}]{\text{LDA}}$ [structure]

c. [structure] $\xrightarrow[\text{THF}]{\text{LDA}}$ [structure]

b. [structure] $\xrightarrow[\text{THF}]{\text{LDA}}$ [structure]

d. [structure] $\xrightarrow[\text{THF}]{\text{LDA}}$ [structure]

23.8 In addition to being strong bases, organolithiums are good nucleophiles that can add to a carbonyl group instead of pulling off a proton to generate an enolate.

23.9 • LDA, THF forms the kinetic enolate by removing a proton from the less substituted C.
• Treatment with NaOCH₃, CH₃OH forms the thermodynamic enolate by removing a proton from the more substituted C.

a. [structure] $\xrightarrow{\text{LDA, THF}}$ [structure]
$\xrightarrow[\text{CH}_3\text{OH}]{\text{NaOCH}_3}$ [structure]

c. [structure] $\xrightarrow{\text{LDA, THF}}$ [structure]
$\xrightarrow[\text{CH}_3\text{OH}]{\text{NaOCH}_3}$ [structure]

b. [structure] $\xrightarrow{\text{LDA, THF}}$ [structure]
$\xrightarrow[\text{CH}_3\text{OH}]{\text{NaOCH}_3}$ [structure]

23.10

a. This acidic H is removed with base to form an achiral enolate.

[structure] (R)-2-methylcyclohexanone $\xrightarrow{\text{NaOH}}$ [structure] achiral $\xrightarrow{\text{H}_2\text{O}}$ [structure] + [structure]

Protonation of the planar achiral enolate occurs with equal probability from two sides, so a racemic mixture is formed. The racemic mixture is optically inactive.

b. [structure] (R)-3-methylcyclohexanone $\xrightarrow{\text{NaOH}}$ [structure] or [structure] $\xrightarrow{\text{H}_2\text{O}}$ [structure]

This stereogenic center is not located at the α carbon, so it is not deprotonated with base. Its configuration is retained in the product, and the product remains optically active.

Chapter 23–8

23.11

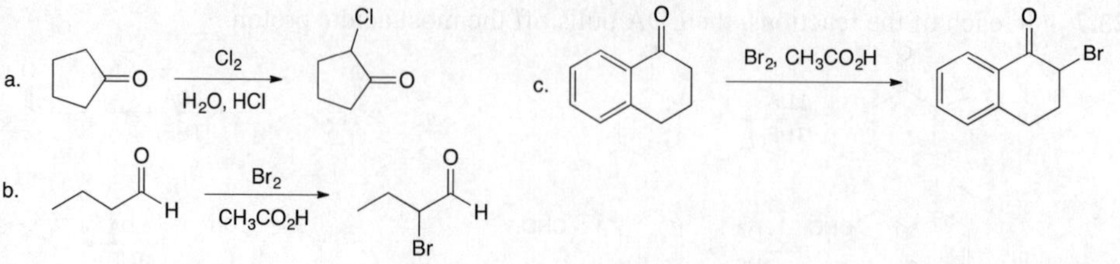

a.
$$Cl_2$$
$$H_2O, HCl$$

b.
$$Br_2$$
$$CH_3CO_2H$$
Br

c.
$$Br_2, CH_3CO_2H$$
Br

23.12

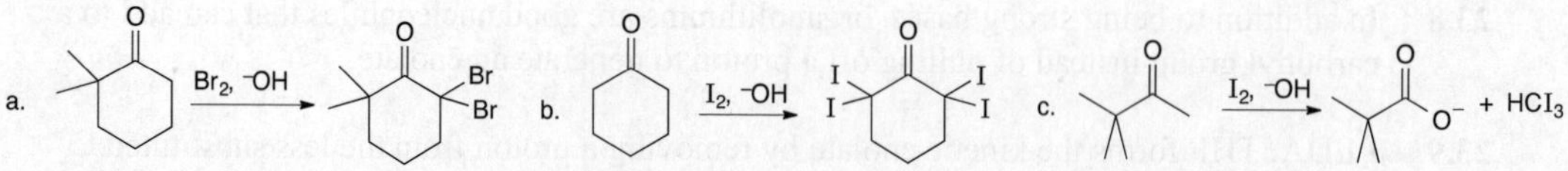

a.
$$Br_2, \ ^-OH$$
Br
Br

b.
$$I_2, \ ^-OH$$
I
I
I

c.
$$I_2, \ ^-OH$$
$$O^- + HCI_3$$

23.13

a.
Br
$$\dfrac{Li_2CO_3 \quad LiBr}{DMF}$$

c.
Br
$$CH_3SH$$
SCH_3

b.
Br
$$CH_3CH_2NH_2$$
NHCH_2CH_3

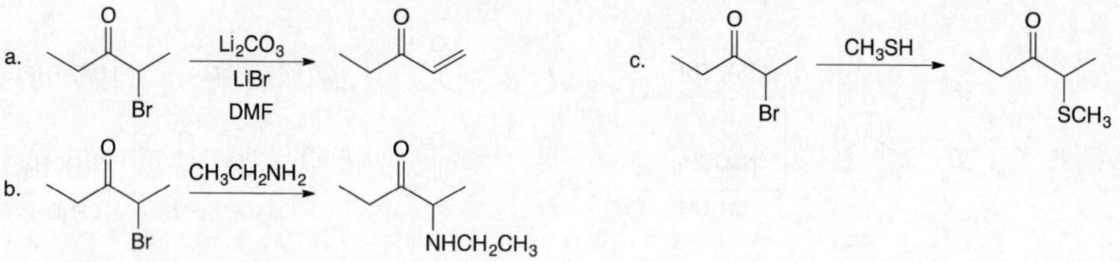

23.14 Bromination takes place on the α carbon to the carbonyl, followed by an S$_N$2 reaction with the nitrogen nucleophile.

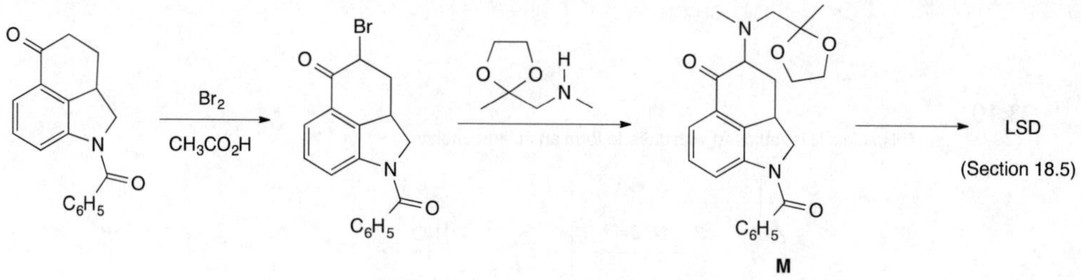

$$Br_2$$
$$CH_3CO_2H$$
C_6H_5

Br
C_6H_5

C_6H_5
M

LSD
(Section 18.5)

23.15

a.
$$\dfrac{[1] \ LDA, \ THF}{[2] \ CH_3CH_2I}$$

c.
$$\dfrac{[1] \ LDA, \ THF}{[2] \ CH_3CH_2I}$$

b.
$$\dfrac{[1] \ LDA, \ THF}{[2] \ CH_3CH_2I}$$

d.
CN
$$\dfrac{[1] \ LDA, \ THF}{[2] \ CH_3CH_2I}$$
CN

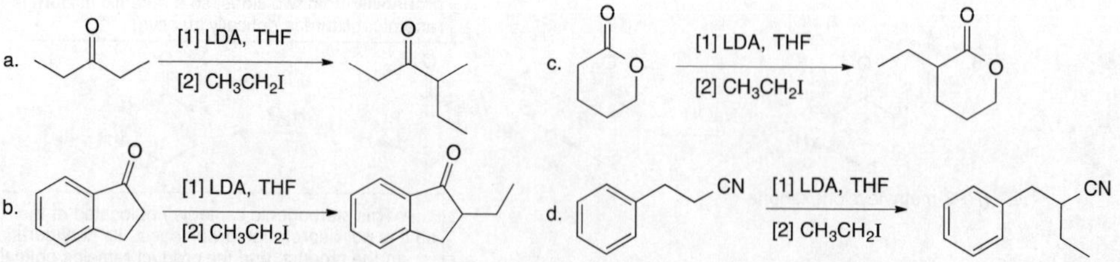

Substitution Reactions of Carbonyl Compounds 23–9

23.16

a.

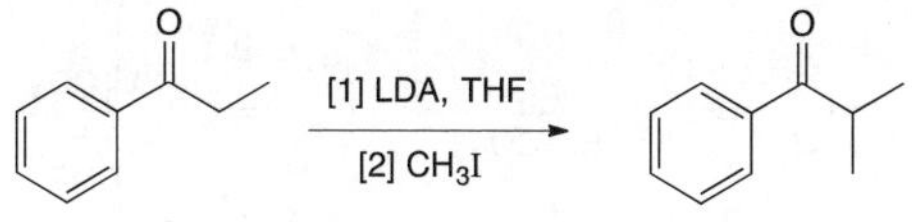

b.

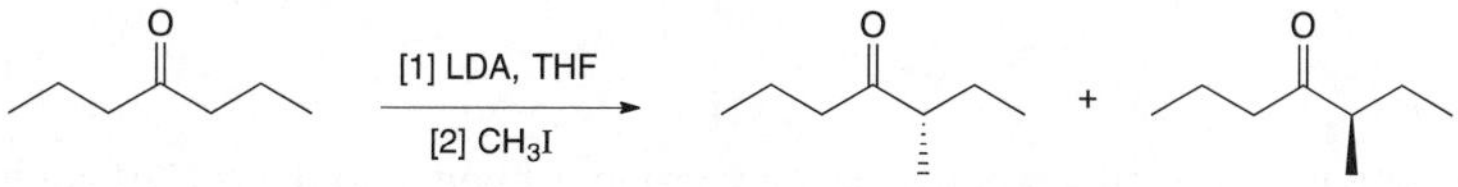

c.

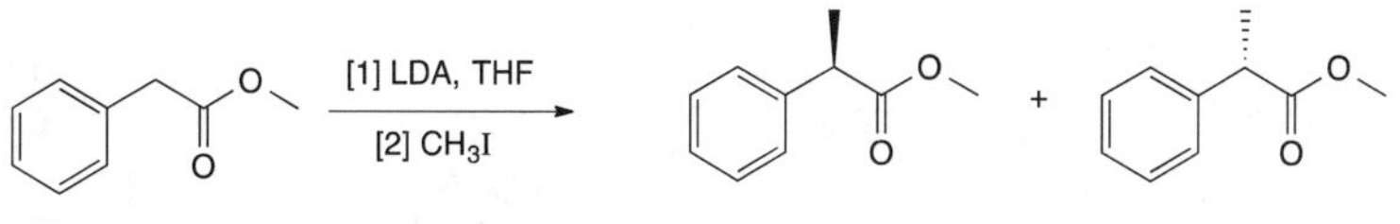

23.17 Three steps are needed: [1] formation of an enolate; [2] alkylation; [3] hydrolysis of the ester.

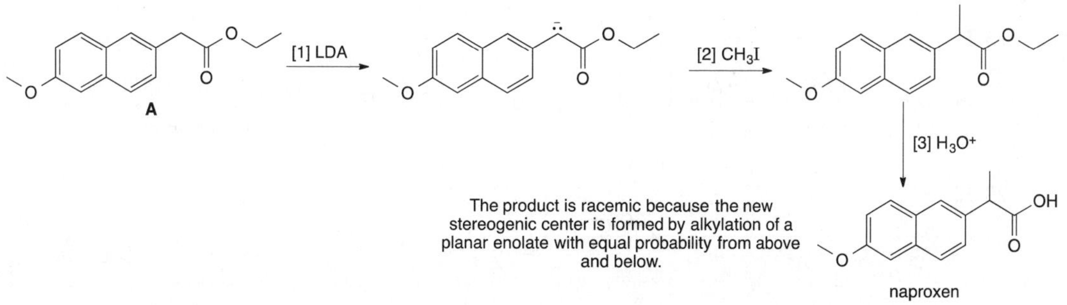

The product is racemic because the new stereogenic center is formed by alkylation of a planar enolate with equal probability from above and below.

naproxen

23.18

a.

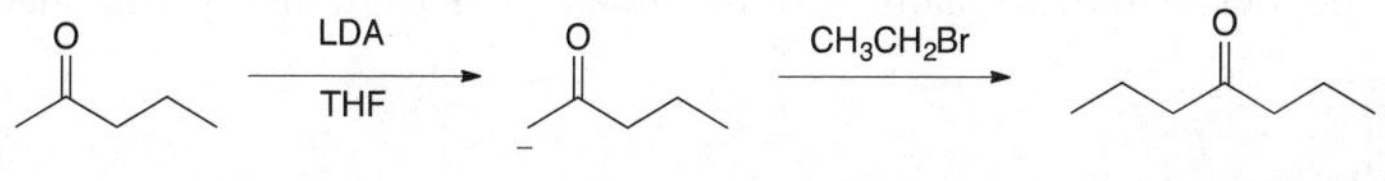

b.

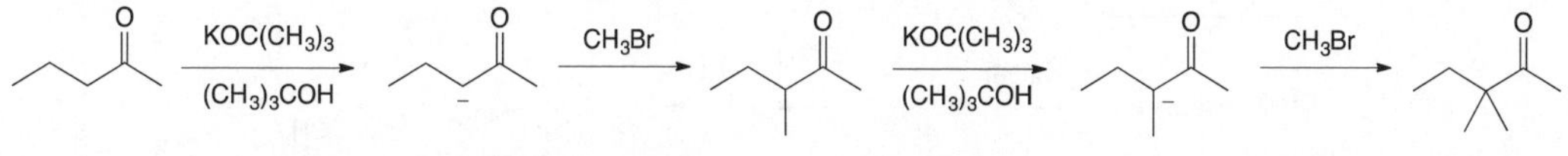

c.

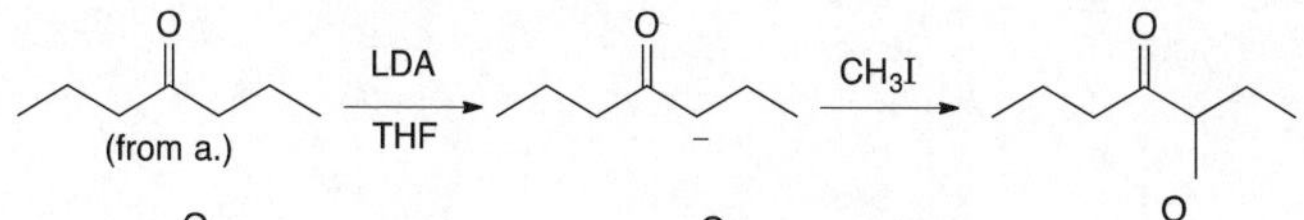

(from a.)

d.

(from b.)

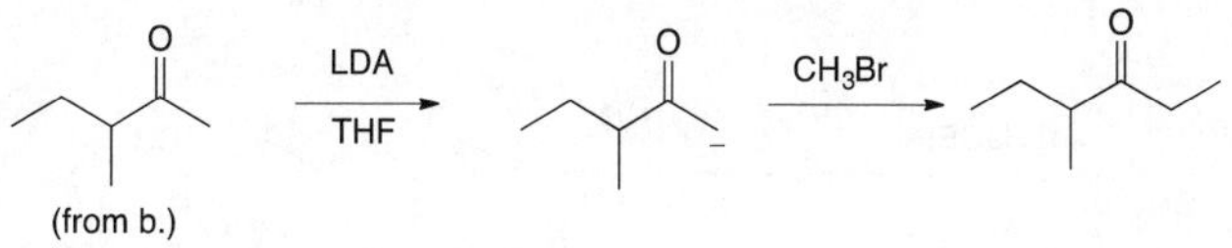

Chapter 23–10

23.19

23.20 Decarboxylation occurs only when a carboxy group is bonded to the α C of another carbonyl group.

a. YES b. NO c. YES d. NO

23.21

a. $CH_2(CO_2Et)_2$ $\xrightarrow[\text{[2]}]{\text{[1] NaOEt}}$ ... $\xrightarrow[\Delta]{H_3O^+}$ ...

b. $CH_2(CO_2Et)_2$ $\xrightarrow[\text{[2] CH}_3\text{Br}]{\text{[1] NaOEt}}$ $\xrightarrow[\text{[2] CH}_3\text{Br}]{\text{[1] NaOEt}}$ $\xrightarrow[\Delta]{H_3O^+}$ ...

23.22

a. Cl⌒⌒Cl ⟶ (cyclobutane COOH)

b. Br⌒O⌒Br ⟶ (tetrahydropyran COOH)

23.23 Locate the α C to the COOH group, and identify all of the alkyl groups bonded to it. These groups are from alkyl halides, and the remainder of the molecule is from diethyl malonate.

a.

$CH_2(CO_2Et)_2$ $\xrightarrow[\text{[2]}]{\text{[1] NaOEt}}$ (alkyl bromide) $\xrightarrow[\Delta]{H_3O^+}$ ...

b.

$CH_2(CO_2Et)_2$ $\xrightarrow[\text{[2]}]{\text{[1] NaOEt}}$ (Br) $\xrightarrow[\text{[2]}]{\text{[1] NaOEt}}$ (Br) $\xrightarrow[\Delta]{H_3O^+}$ ...

c.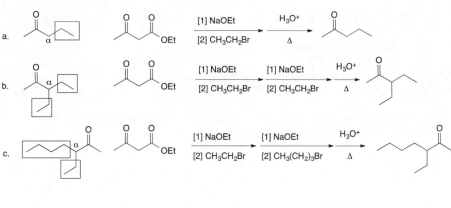

23.24 The reaction works best when the alkyl halide is 1° or CH₃X, because this is an S_N2 reaction.

a. (CH₃)₃CX
3° alkyl halide
(too crowded)

b. aryl halide
(leaving group on an sp² hybridized C)
Aryl halides are unreactive in S_N2 reactions.

c. This compound has 3 CH₃ groups on the α carbon to the COOH. The malonic ester synthesis can be used to prepare mono- and disubstituted carboxylic acids only: RCH₂COOH and R₂CHCOOH, but not R₃CCOOH.

23.25

23.26 Locate the α C. All alkyl groups on the α C come from alkyl halides, and the remainder of the molecule comes from ethyl acetoacetate.

23.27

Chapter 23–12

23.28

a. [1] NaOEt / [2] (with CH₃O-naphthalene-CH₂Br) → H₃O⁺, Δ → nabumetone

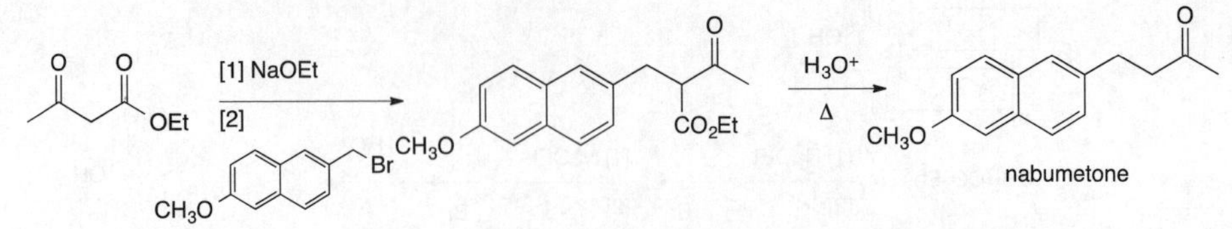

b. [1] LDA, THF / [2] (CH₃O-naphthalene-CH₂Br) → nabumetone

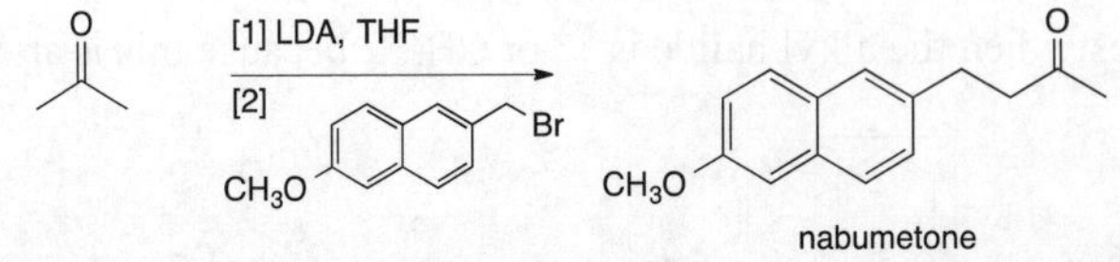

23.29 Use the directions from Answer 23.1 to draw the enol tautomer(s). In cases where E and Z isomers can form, only one isomer is drawn.

a.

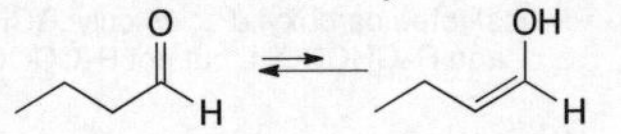

b. unconjugated enol (less stable) +

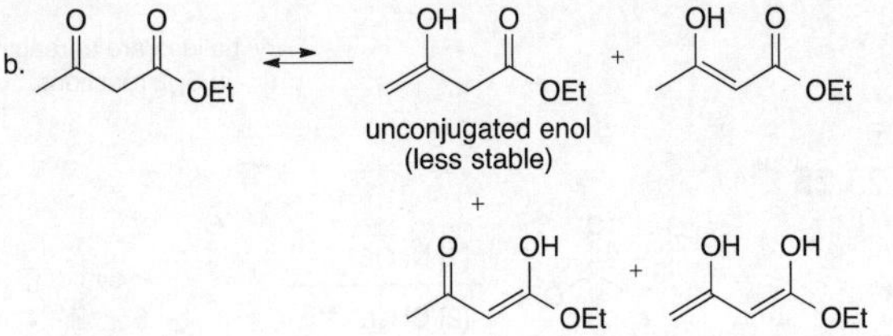

23.30

a. A →(NaOH, H₂O)→ B   b. C

A: one axial and one equatorial group, **less stable**

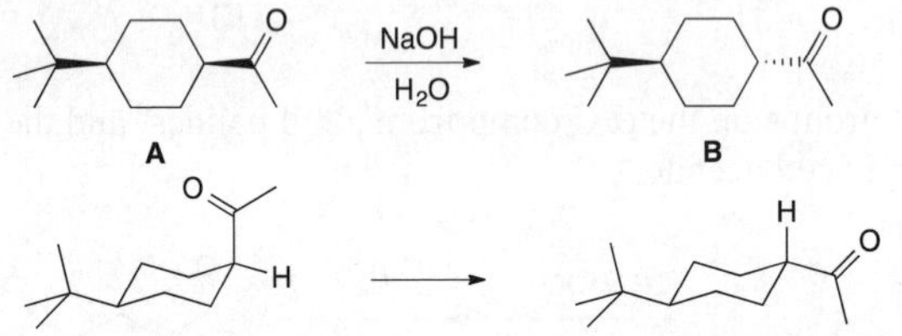

B: Both groups are equatorial. **more stable**

This isomerization will occur because it makes a more stable compound.

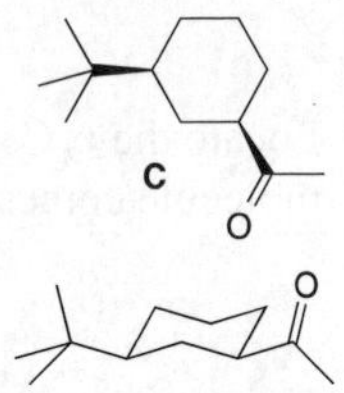

C: Both groups are equatorial = cis.

Compound **C** will not isomerize because it already has the more stable arrangement of substituents.

23.31 Use the directions from Answer 23.1 to draw the enol tautomer(s). In cases where E and Z isomers can form, only one isomer is drawn.

a. conjugated enol (more stable) +

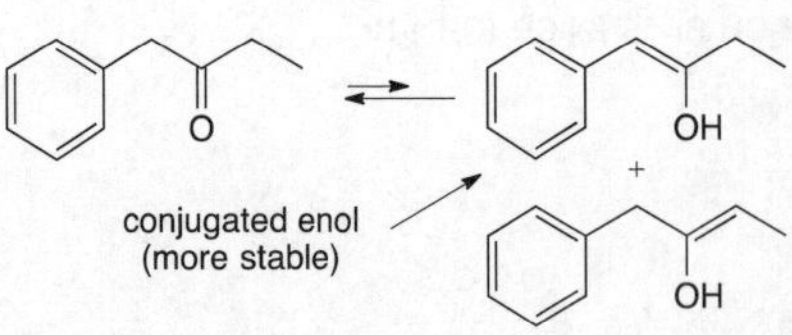

c. (mono enol form)   conjugated enol (more stable)

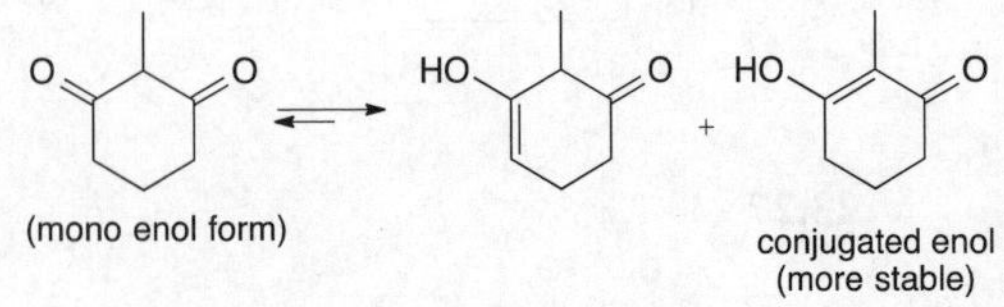

b.

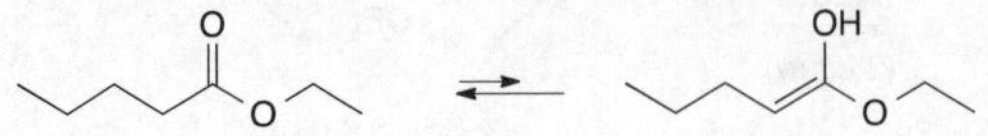

Substitution Reactions of Carbonyl Compounds 23–13

23.32

ethyl acetoacetate

The ester C=O is resonance stabilized, and is therefore less available for tautomerization. Since the carbonyl form of the ester group is stabilized by electron delocalization, less enol is present at equilibrium.

23.33

a.

Conjugation stabilizes this enol.
higher percentage of enol

not conjugated

b.

not conjugated

conjugated C=C
higher percentage of enol

23.34

a.

b.

c.

23.35

a.

H_c is bonded to an sp^2 hybridized C = **least acidic.**
H_a is bonded to an α C = **intermediate acidity.**
H_b is bonded to an α C, and is adjacent
to a benzene ring = **most acidic.**

b.

H_b is bonded to an sp^3 hybridized
C = **least acidic.**
H_c is bonded to an α C =
intermediate acidity.
H_a is bonded to O = **most acidic.**

23.36

a.

$\xrightarrow[\text{THF}]{\text{LDA}}$

b.

$\xrightarrow[\text{THF}]{\text{LDA}}$

c.

$\xrightarrow[\text{THF}]{\text{LDA}}$

d.

$\xrightarrow[\text{THF}]{\text{LDA}}$

Chapter 23–14

23.37

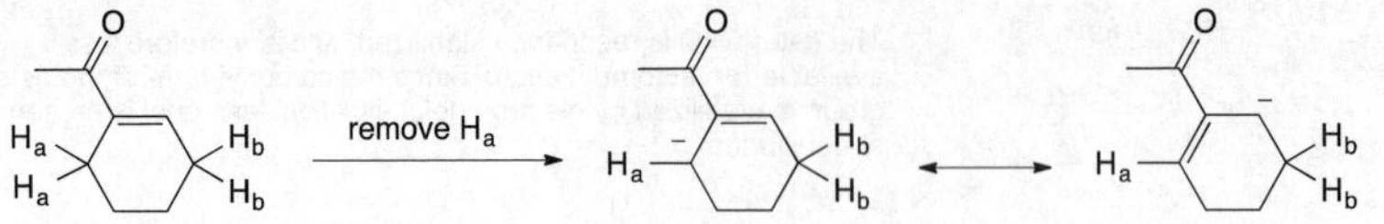

Removal of H_a gives two resonance structures. The negative charge is never on O.

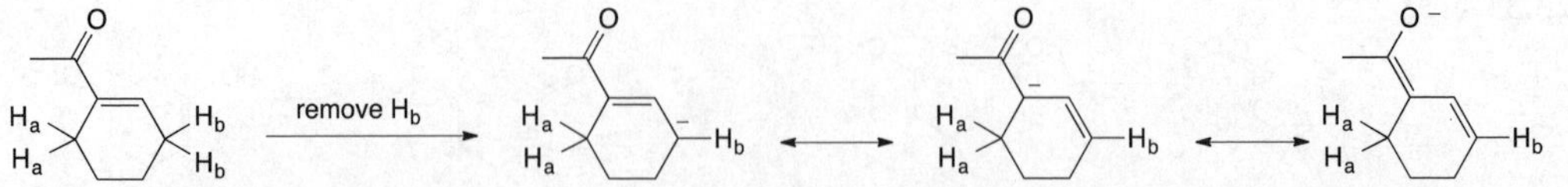

remove H_a

Removal of H_b gives three resonance structures. The negative charge is on O in one resonance structure, making the conjugate base more stable and H_b more acidic (lower pK_a).

23.38 a.

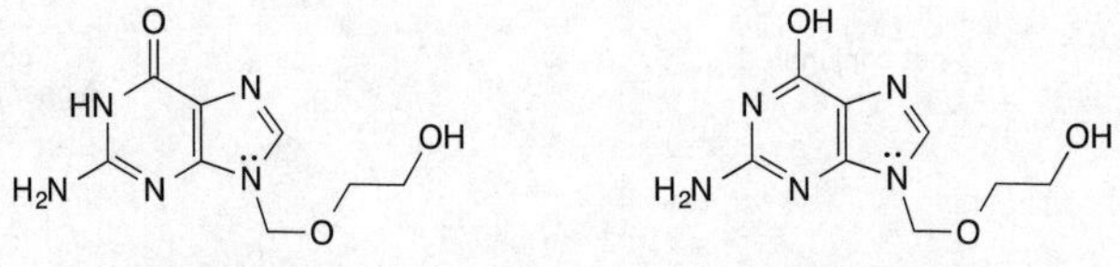

keto form
acyclovir

enol form

In the enol form, the bicyclic ring system has four π bonds (eight π electrons) and a lone pair on N, for a total of 10 π electrons. This makes it aromatic by Hückel's rule (Section 17.7).

b. The keto form of acyclovir can also be drawn in a resonance form that gives it 10 π electrons, making it aromatic as well.

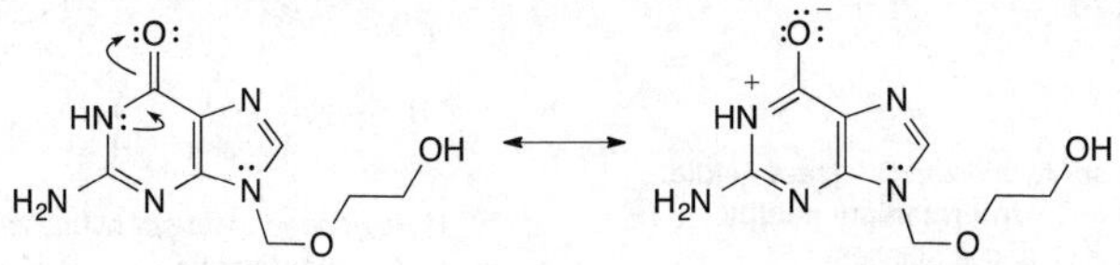

aromatic
completely conjugated
10 π electrons

Substitution Reactions of Carbonyl Compounds 23–15

23.39

pentane-2,4-dione

base (1 equiv)

A

[1] CH$_3$I
[2] H$_2$O

base
(2nd equiv)

more nucleophilic site

[1] CH$_3$I
[2] H$_2$O

B

One equivalent of base removes the most
acidic proton between the two C=O's, to
form **A** on alkylation with CH$_3$I.

With a second equivalent of base a dianion is formed.
Since the second enolate is less resonance stabilized,
it is more nucleophilic and reacts first in an alkylation
with CH$_3$I, forming **B** after protonation with H$_2$O.

23.40

enediol

keto tautomer **A**

keto tautomer **B**

The enediol is more stable than either of the two keto tautomers because it is more highly
conjugated. More resonance structures can be drawn, which delocalize the lone pairs of the two
OH groups bonded to the C=C. Such delocalization is not possible with either keto tautomer.

23.41 The mechanism of acid-catalyzed halogenation consists of two parts: **tautomerization** of the
carbonyl compound to the enol form and **reaction of the enol with halogen.**

A higher percentage of the more stable enol is present.

pentan-2-one

CH$_3$COOH

C=C has
1 bond to C

C=C has
2 bonds to C
more stable
(*E* and *Z* isomers)

Br$_2$

A

B

major product formed
from the more stable enol

23.42 In the haloform reaction, the three H's of the CH$_3$ group are successively replaced by X, to form
an intermediate that is oxidatively cleaved with base.

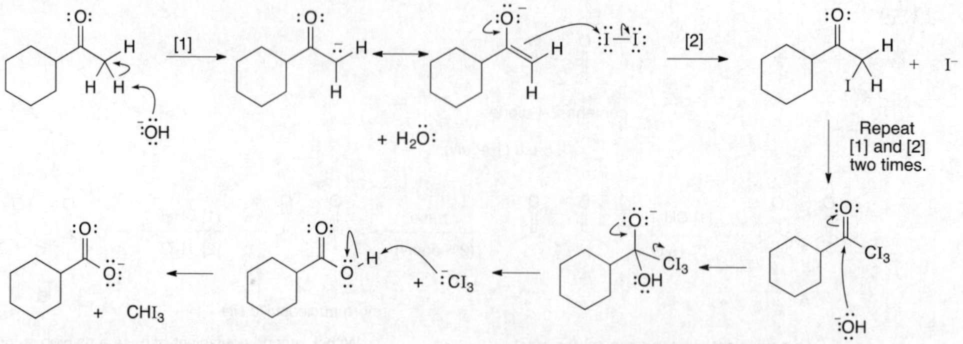

Chapter 23–16

23.43 Use the directions from Answer 23.23.

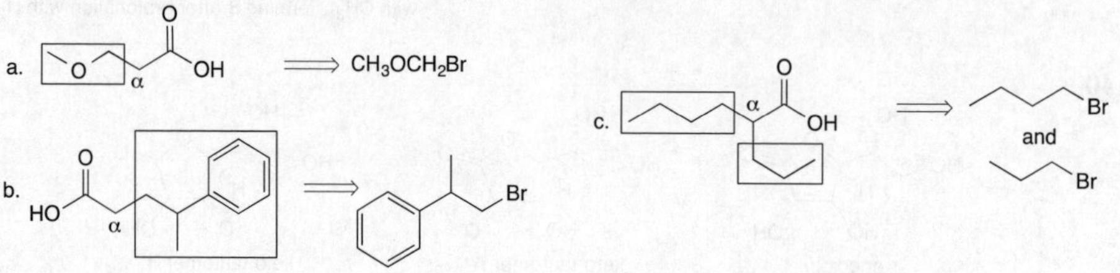

a. $\Longrightarrow$ CH₃OCH₂Br

b.

c.

23.44

a.

b.

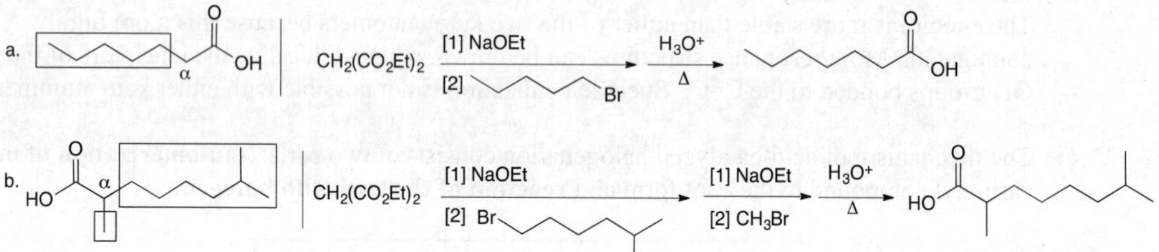

23.45

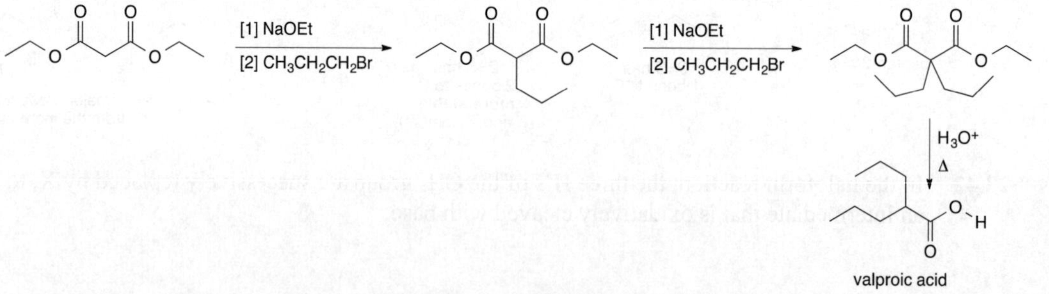

valproic acid

Substitution Reactions of Carbonyl Compounds 23–17

23.46

a. $CH_2(CO_2Et)_2$ [1] NaOEt / [2] $BrCH_2CH_2CH_2CH_2CH_2Br$ / [3] NaOEt → H_3O^+, Δ → (cyclohexyl)–COOH → [1] $LiAlH_4$ / [2] H_2O → (cyclohexyl)–CH$_2$OH

b. (cyclohexyl)–COOH (from a.) → CH_3OH / H_2SO_4 → (cyclohexyl)–CO_2CH_3 → [1] CH_3MgBr (2 equiv) / [2] H_2O → (cyclohexyl)–C(OH)(CH$_3$)$_2$

23.47

a. (methyloxirane) nucleophilic attack here → [1] Na^+ $^-CH(COOEt)_2$ / [2] H_2O → EtO–CO–CH(–CO–OEt)–CH$_2$–CH(OH)CH$_3$

b. (formaldehyde, H–CHO) → [1] Na^+ $^-CH(COOEt)_2$ / [2] H_2O → EtO–CO–CH(–CO–OEt)–CH$_2$OH

c. CH_3–CO–Cl → [1] Na^+ $^-CH(COOEt)_2$ / [2] H_2O → EtO–CO–CH(–CO–OEt)–CO–CH$_3$

d. (acetic anhydride) → [1] Na^+ $^-CH(COOEt)_2$ / [2] H_2O → EtO–CO–CH(–CO–OEt)–CO–CH$_3$ + CH_3COOH

23.48 Use the directions from Answer 23.26.

a. (ketone with α) | (ethyl acetoacetate, OEt) [1] NaOEt / [2] Br–(isopentyl) → H_3O^+, Δ → (product ketone)

b. (ketone with α) | (ethyl acetoacetate, OEt) [1] NaOEt / [2] CH_3CH_2Br → [1] NaOEt / [2] Br–CH$_2$–(cyclohexyl) → H_3O^+, Δ → (product ketone)

c. (methylcyclopentyl ketone with α) | (ethyl acetoacetate, OEt) [1] NaOEt / [2] Br–CH$_2$–CH(CH$_3$)–CH$_2$–Br / [3] NaOEt → H_3O^+, Δ → (product)

23.49

a. (ethyl acetoacetate, OEt) [1] NaOEt / [2] CH_3Br → [1] NaOEt / [2] CH_3Br → H_3O^+, Δ → (methyl isopropyl ketone)

b. (methyl isopropyl ketone, from a.) → LDA / THF → (enolate) → CH_3I → (ethyl isopropyl ketone)

Chapter 23–18

23.50

a.

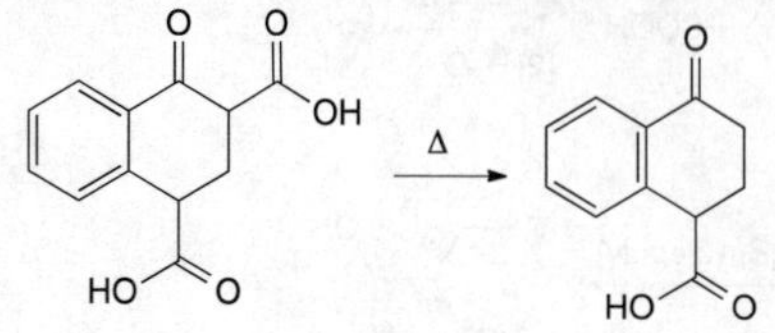

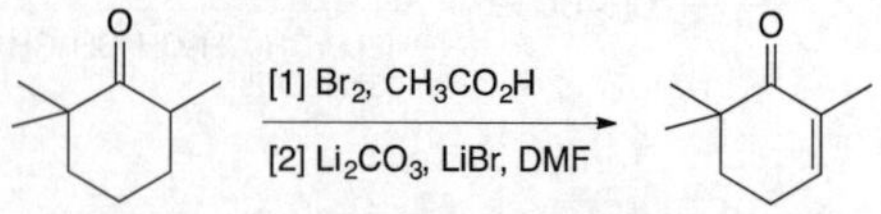

b.

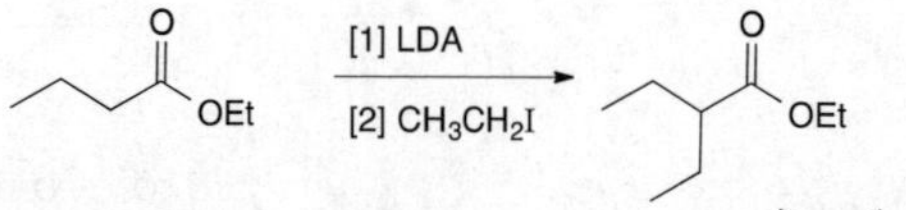

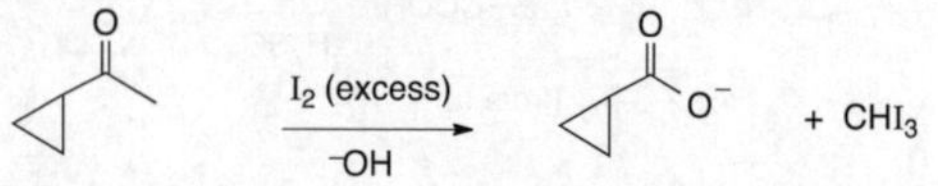

c.

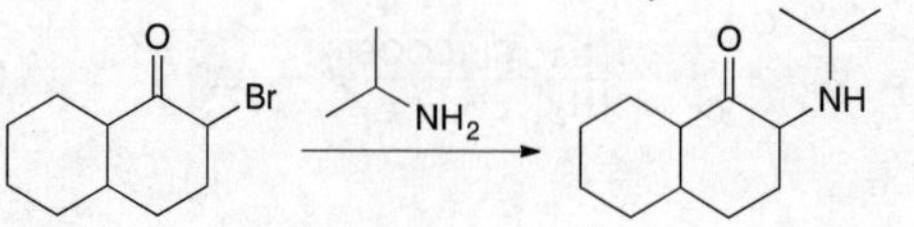

d.

e.

f.

g.

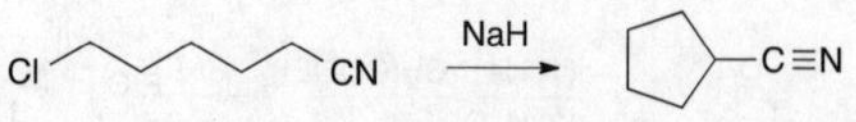

h.

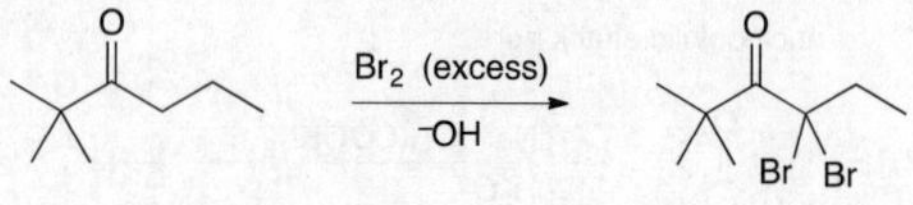

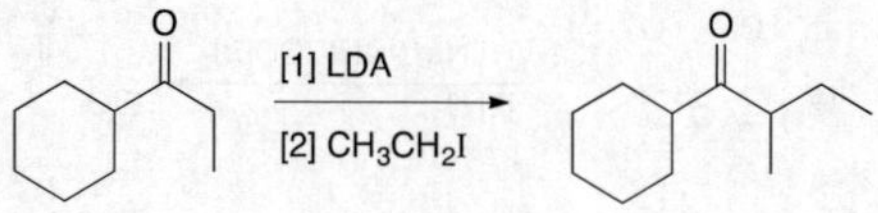

23.51

a.

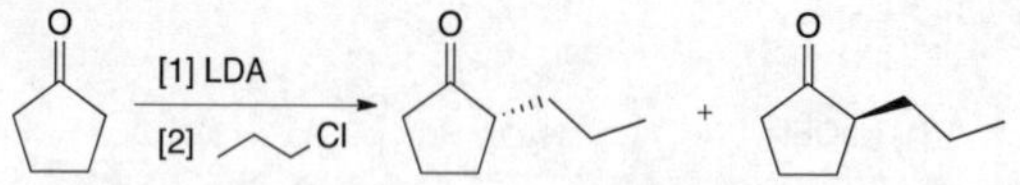

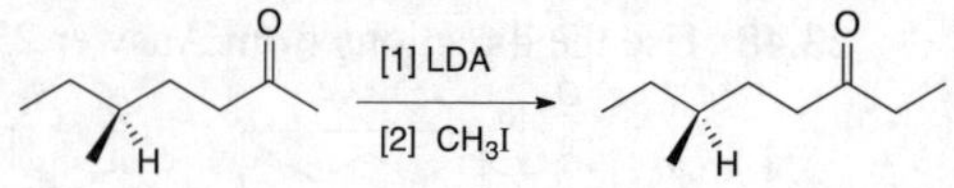

b.

c.

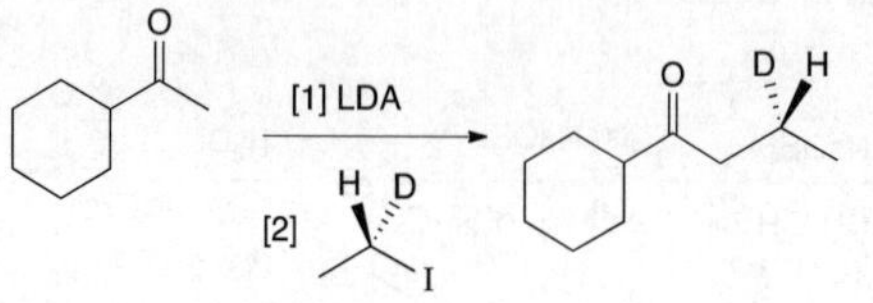

23.52

a.

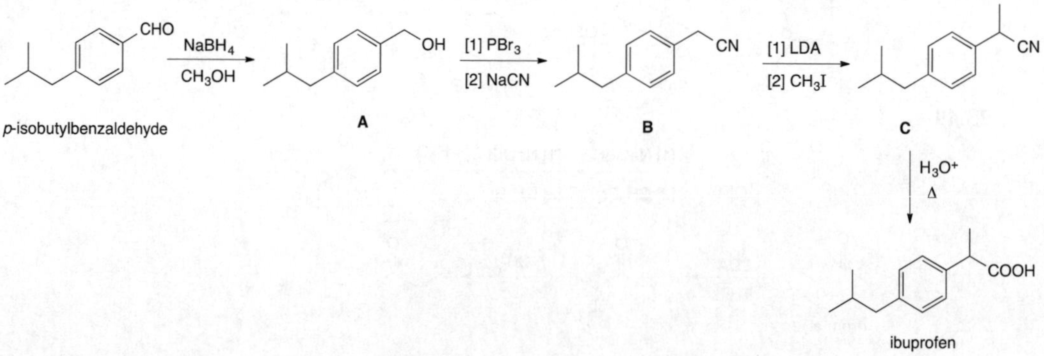

Substitution Reactions of Carbonyl Compounds 23–19

b.

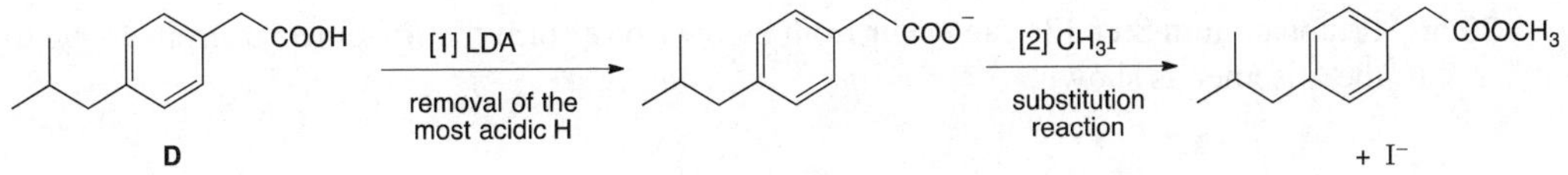

[1] LDA
removal of the
most acidic H

[2] CH₃I → [2] CH_3I
substitution
reaction

+ I⁻

D

Removal of the most acidic proton with LDA forms a carboxylate anion that reacts as a nucleophile with CH₃I to form an ester as substitution product.

23.53

a.

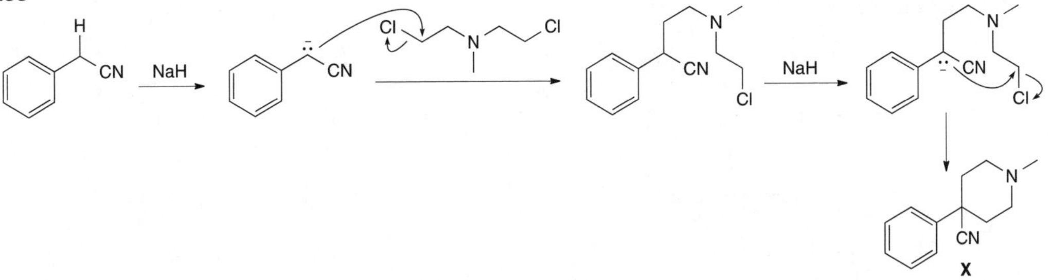

NaH

NaH

X

b. **X** can be converted to meperidine by hydrolysis of the nitrile and esterification.

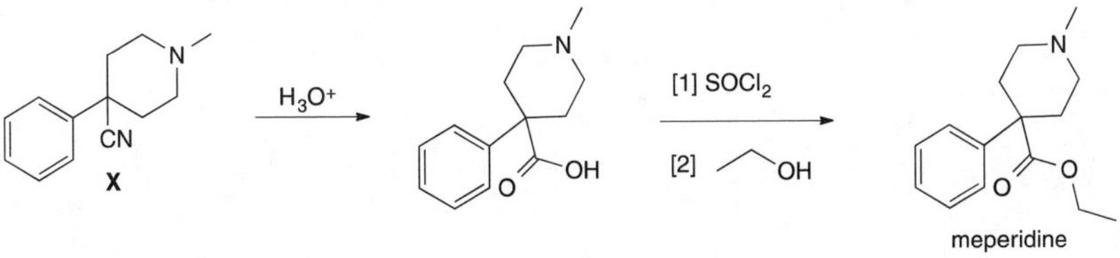

X

H_3O^+

[1] $SOCl_2$

[2] ⟍OH

meperidine

23.54

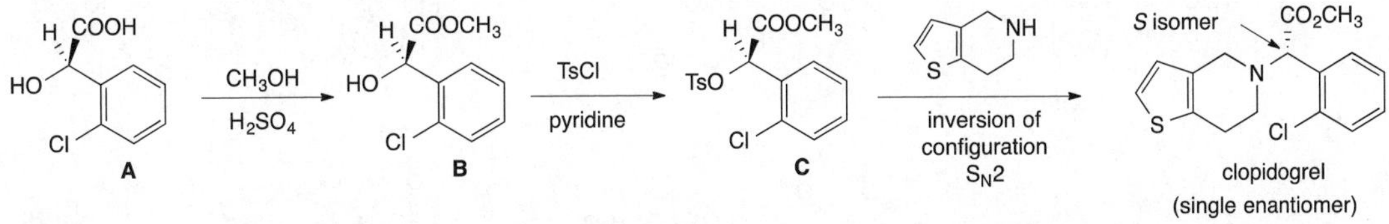

A

CH_3OH
H_2SO_4

B

TsCl
pyridine

C

inversion of
configuration
S_N2

S isomer

clopidogrel
(single enantiomer)

23.55

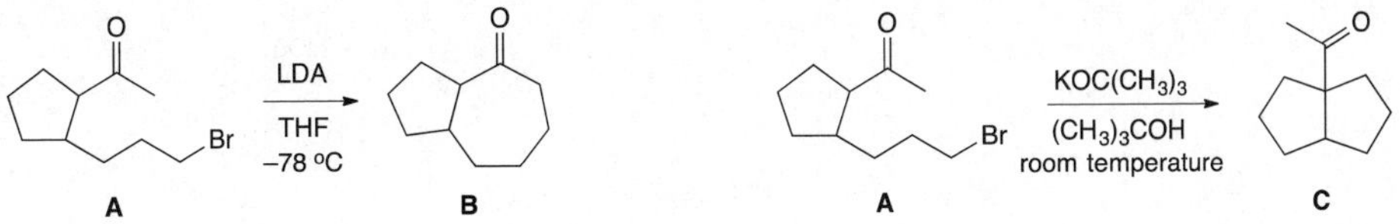

A

LDA
THF
−78 °C

B

A

$KOC(CH_3)_3$
$(CH_3)_3COH$
room temperature

C

Chapter 23–20

23.56 Protonation in Step [3] can occur from below (to re-form the *R* isomer) or from above to form the *S* isomer as shown.

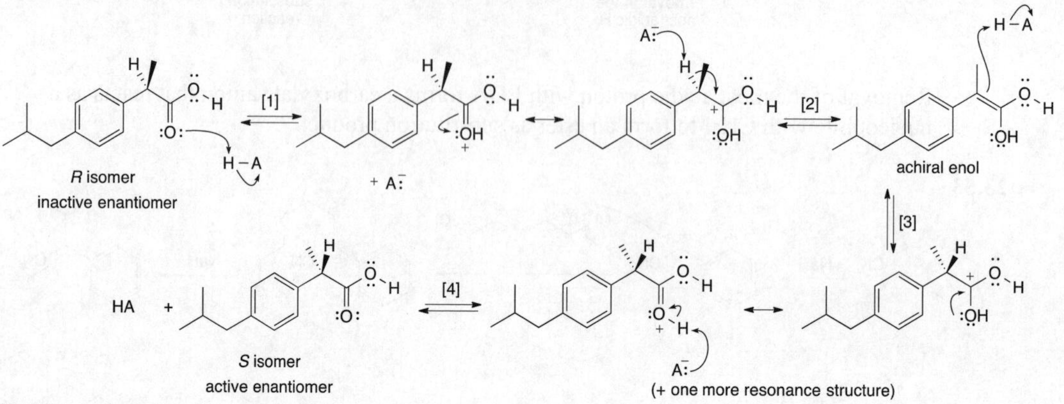

23.57

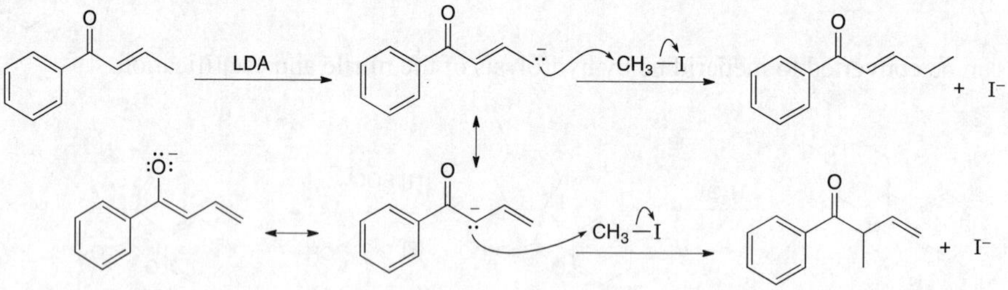

23.58

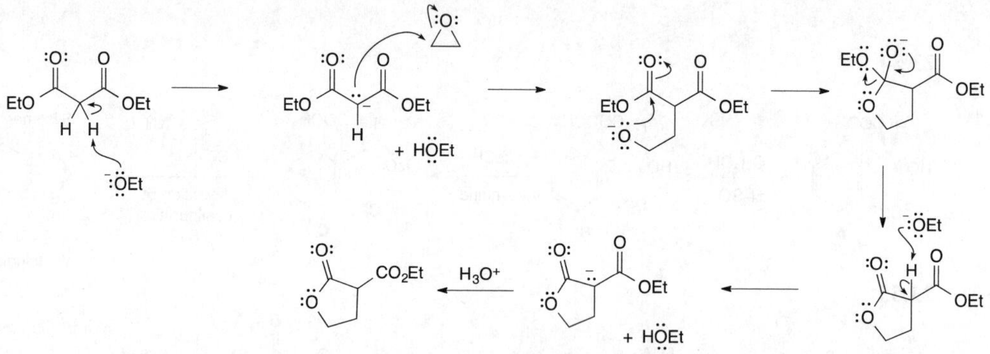

Substitution Reactions of Carbonyl Compounds 23–21

23.59 Protons on the γ carbon of an α,β-unsaturated carbonyl compound are acidic because of resonance.

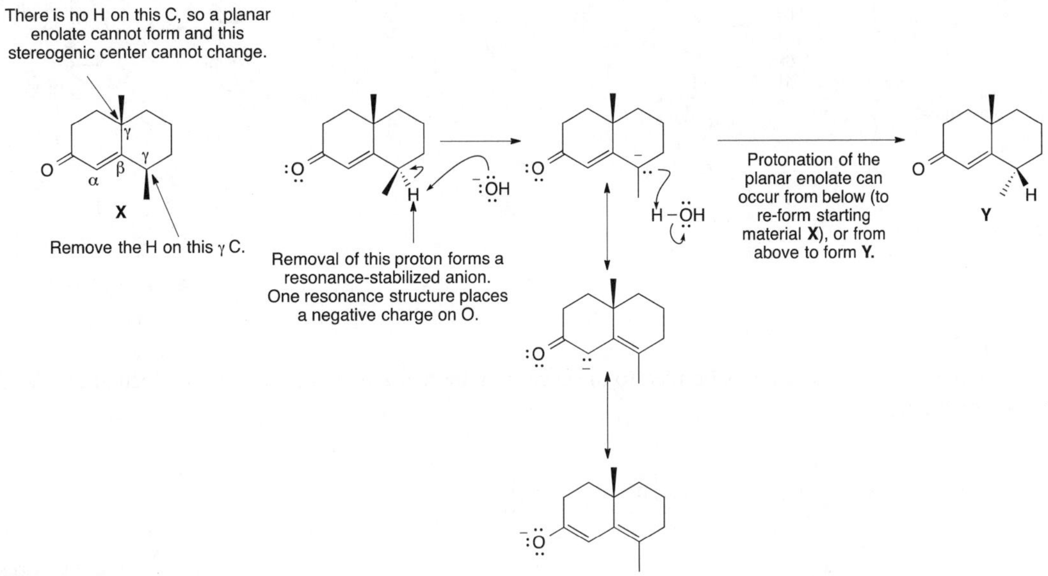

There is no H on this C, so a planar enolate cannot form and this stereogenic center cannot change.

X

Remove the H on this γ C.

Removal of this proton forms a resonance-stabilized anion. One resonance structure places a negative charge on O.

Protonation of the planar enolate can occur from below (to re-form starting material **X**), or from above to form **Y**.

Y

23.60

LDA = B:

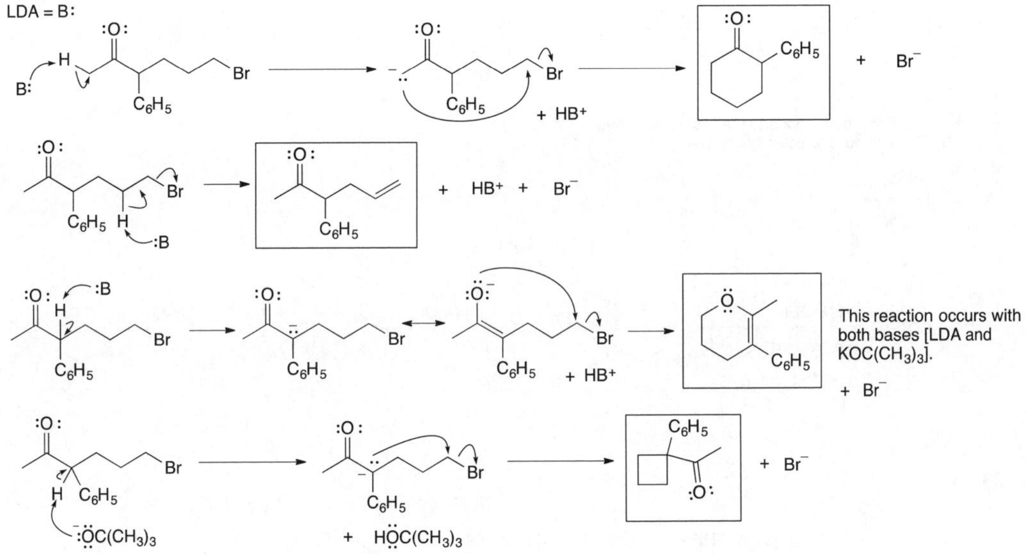

+ HB⁺

+ HB⁺ + Br⁻

This reaction occurs with both bases [LDA and KOC(CH₃)₃].

+ Br⁻

+ HÖC(CH₃)₃

+ Br⁻

Chapter 23–22

23.61

1st new C–C bond

LDA
THF
–78 °C
[1]

Remove
proton here.

A

B

[2]

+ Cl⁻

LDA
THF
–78 °C
[3]

[4]

C

2nd new C–C bond

+ Cl⁻

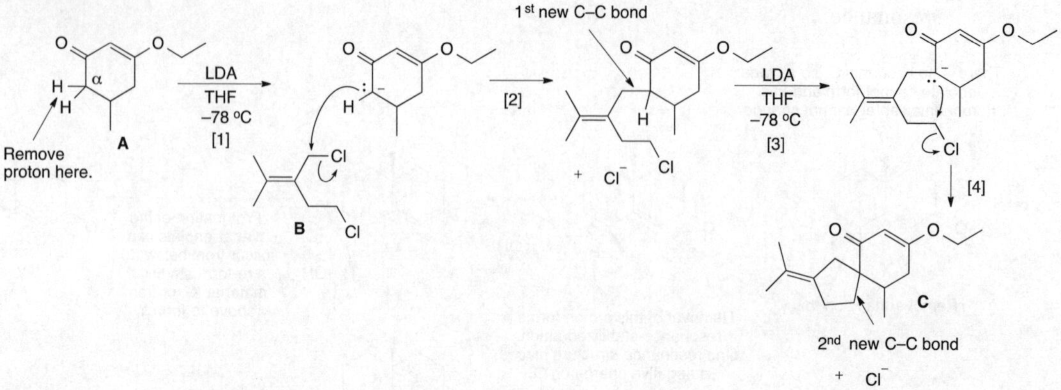

23.62

a. Because there are two C's bonded to the α carbon, there are two possible intramolecular alkylation
  reactions.

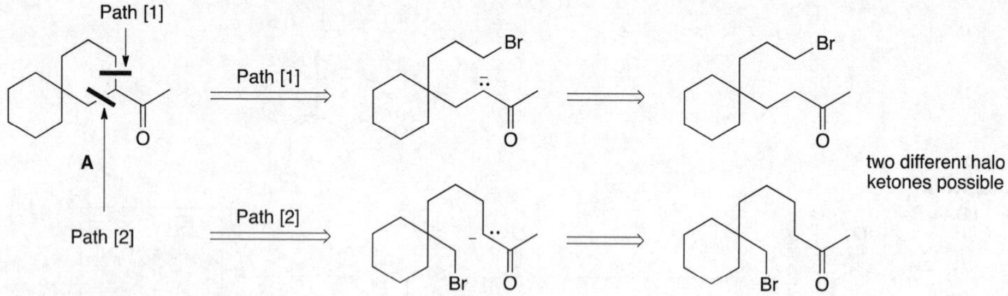

Path [1]

Path [1]

A

Path [2]

Path [2]

two different halo
ketones possible

b.

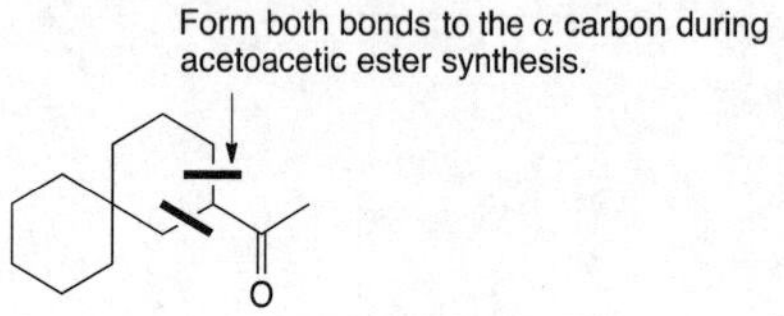

Form both bonds to the α carbon during
acetoacetic ester synthesis.

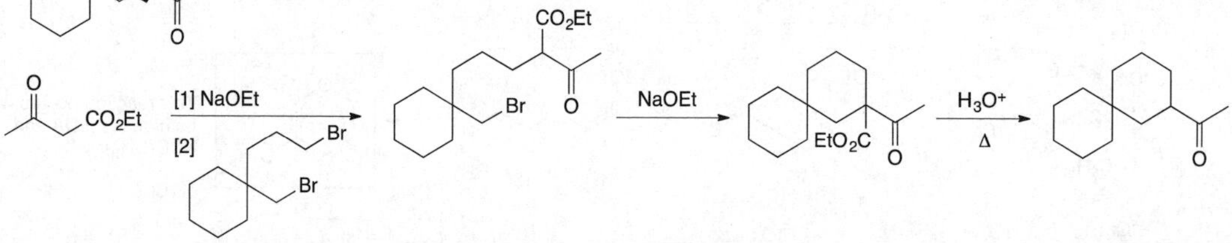

CO₂Et

[1] NaOEt

[2]

Br

Br

CO₂Et

Br

NaOEt

EtO₂C

H₃O⁺
Δ

23.63

a.

[1] LDA, THF

[2] CH₃Br

[1] LDA, THF

[2] CH₃CH₂Br

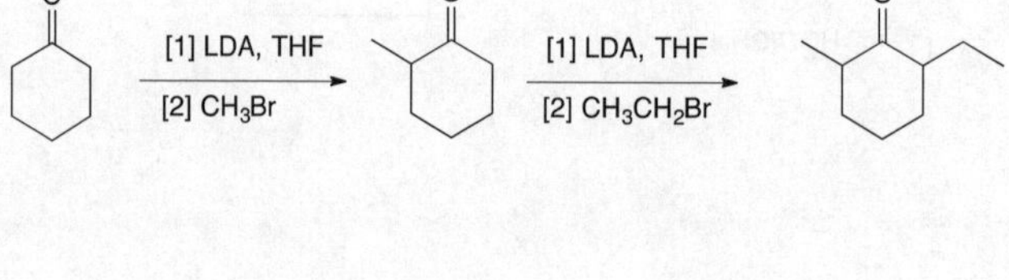

Substitution Reactions of Carbonyl Compounds 23–23

b.

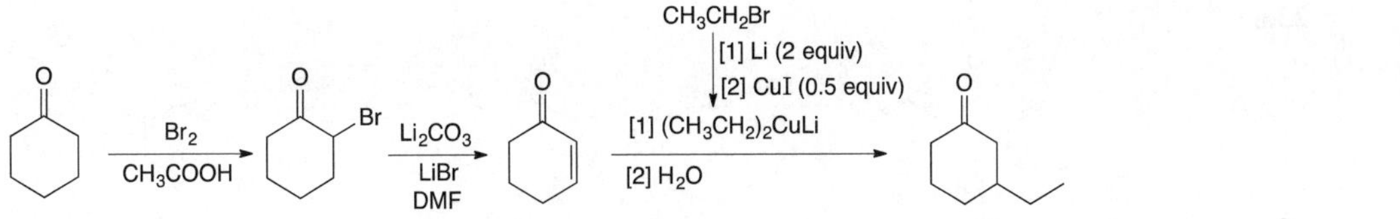

c.

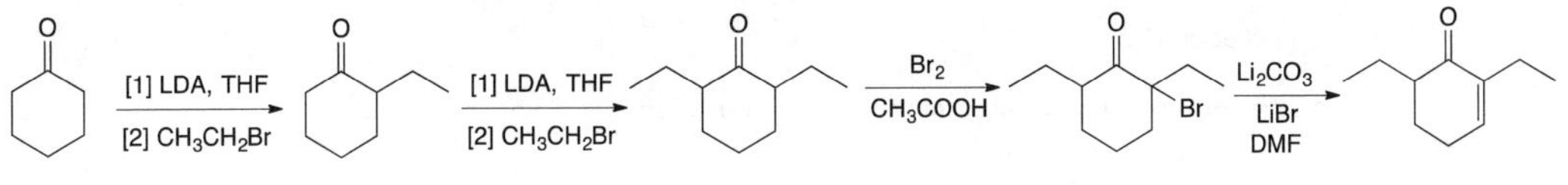

23.64

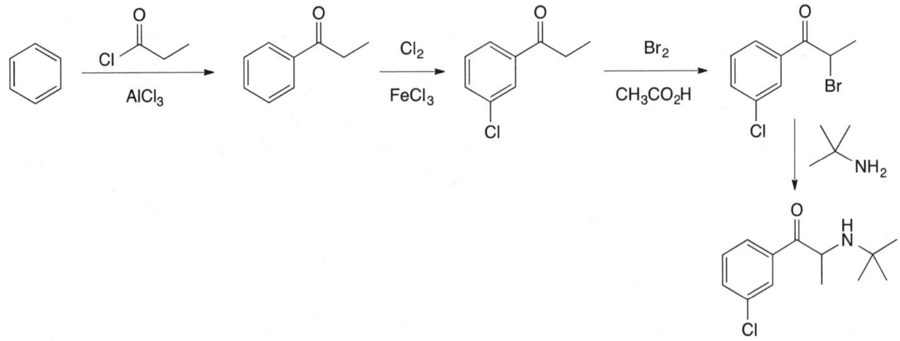

bupropion

23.65

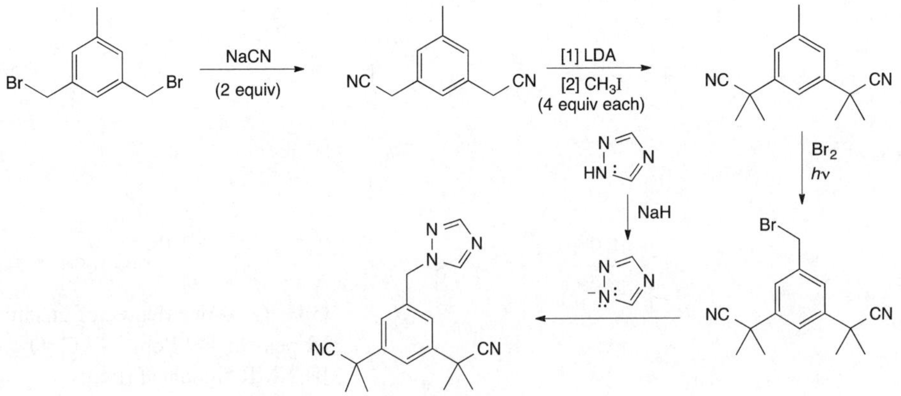

anastrozole

23.66

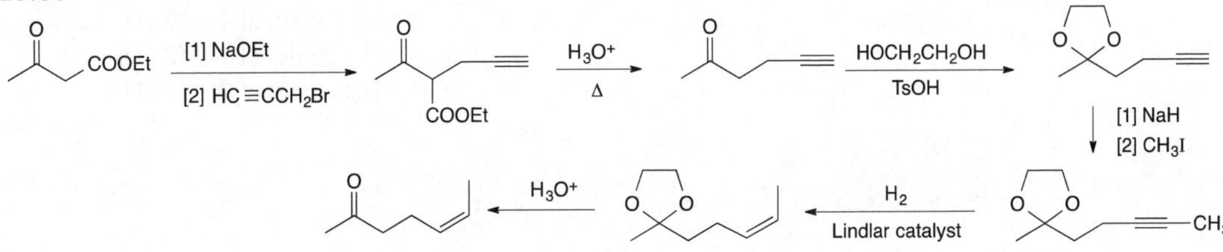

Chapter 23–24

23.67

most acidic H

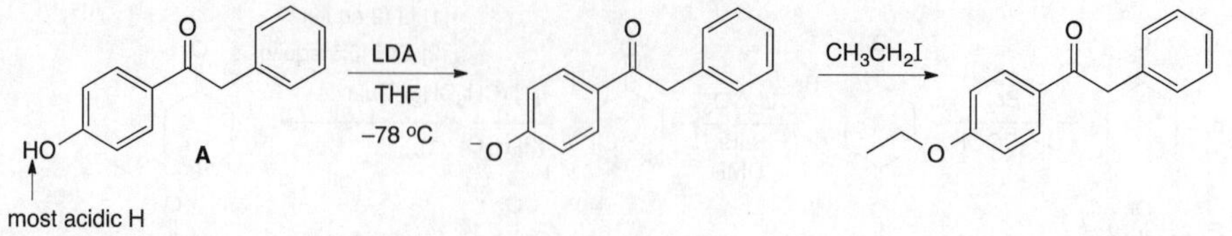

To synthesize the desired product, a protecting group is needed:

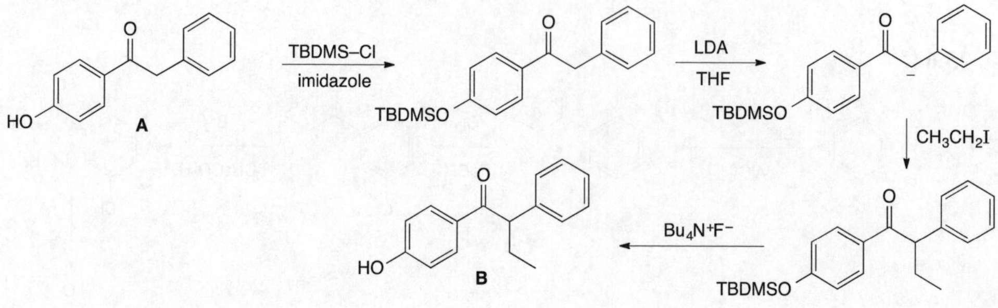

23.68

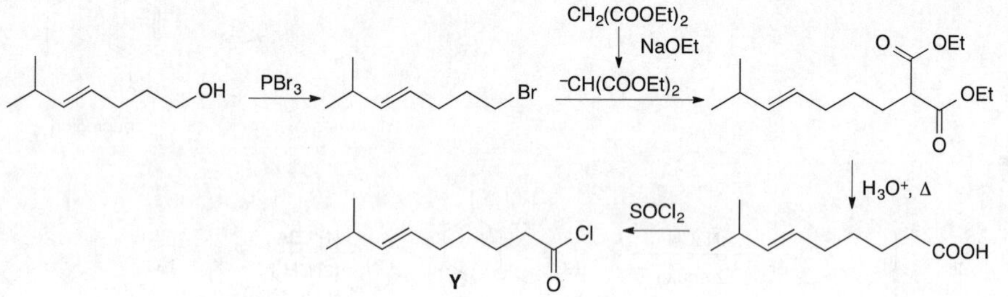

23.69

a.

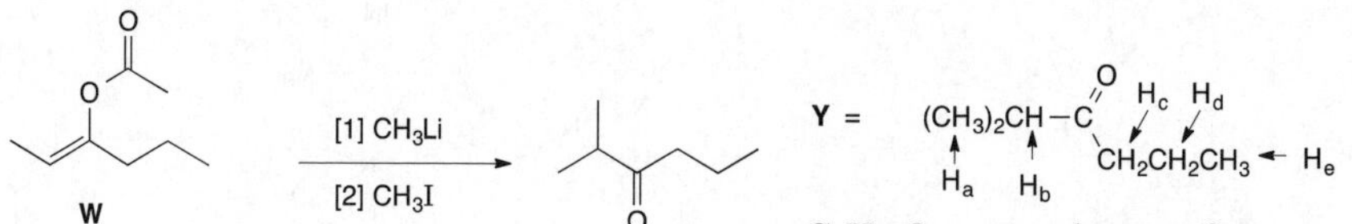

$Y = (CH_3)_2CH-\overset{\displaystyle O}{\overset{\|}{C}}-CH_2CH_2CH_3$

$C_7H_{14}O \rightarrow$ one degree of unsaturation
IR peak at 1713 cm$^{-1} \rightarrow$ C=O
^{1}H NMR signals at (ppm)
        H_e: triplet at 0.7 (3 H)
        H_a: doublet at 0.9 (6 H)
        H_d: sextet at 1.3 (2 H)
        H_c: triplet at 1.9 (2 H)
        H_b: septet at 2.1 (1 H)

b.

23.70 Removal of H_a with base does not generate an anion that can delocalize onto the carbonyl O atom, whereas removal of H_b generates an enolate that is delocalized on O.

Delocalization of this sort can't occur by removal of H_a, making H_a less acidic.

Removal of H_b gives an anion that is resonance stabilized, so H_b is more acidic.

Mechanism:

Chapter 23–26

23.71

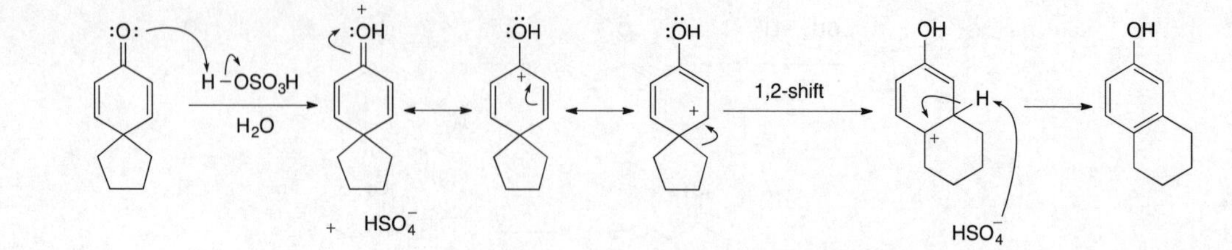

23.72

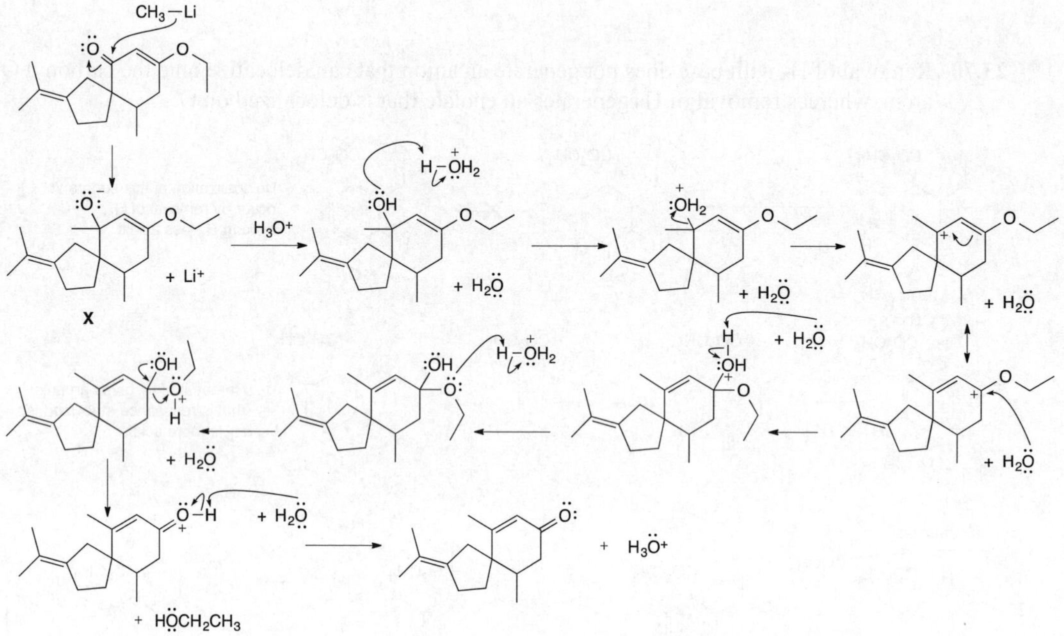

23.73

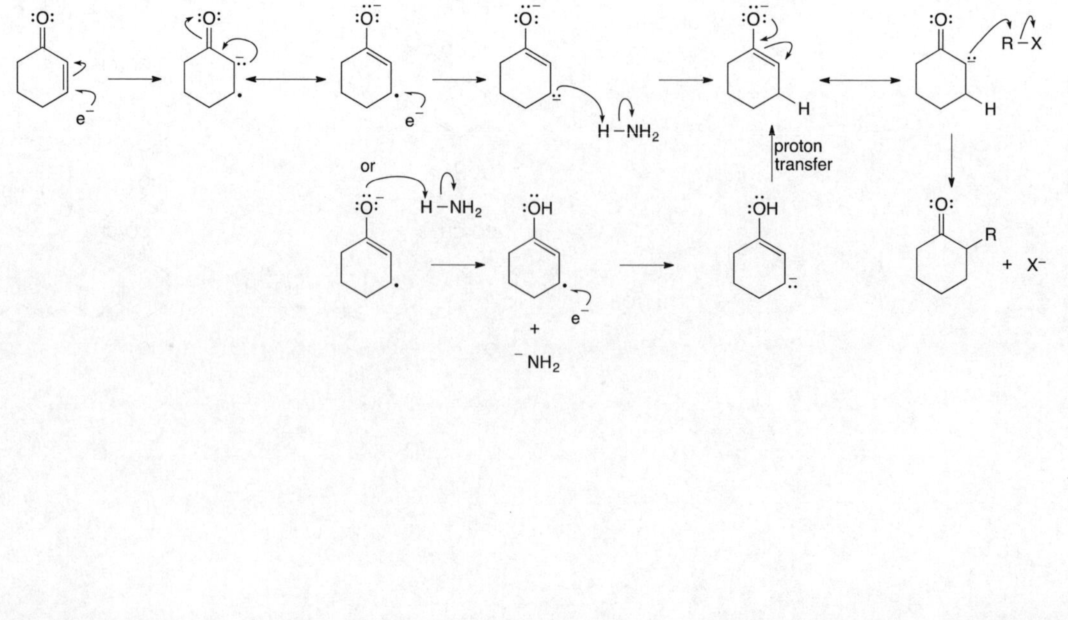

Substitution Reactions of Carbonyl Compounds 23–27

23.74 a. In the presence of base an achiral enolate that can be protonated from both sides is formed.

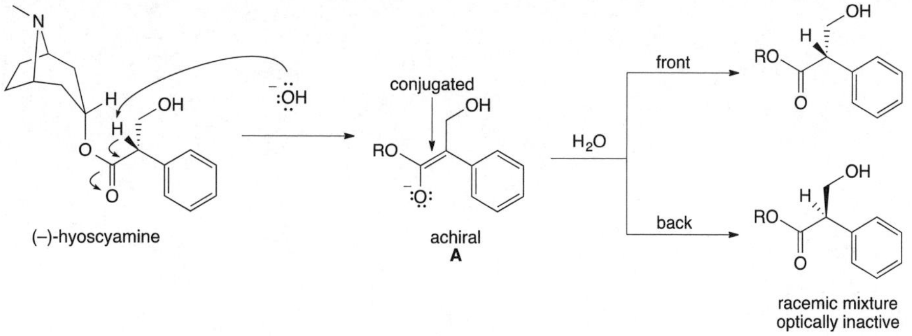

(–)-hyoscyamine

conjugated

achiral
A

H_2O

front

back

racemic mixture
optically inactive

b. The enolate **A** formed from (–)-hyoscyamine is conjugated with the benzene ring, making it easier to form. The enolate **B** formed from (–)-littorine is not conjugated, so it is less readily formed.

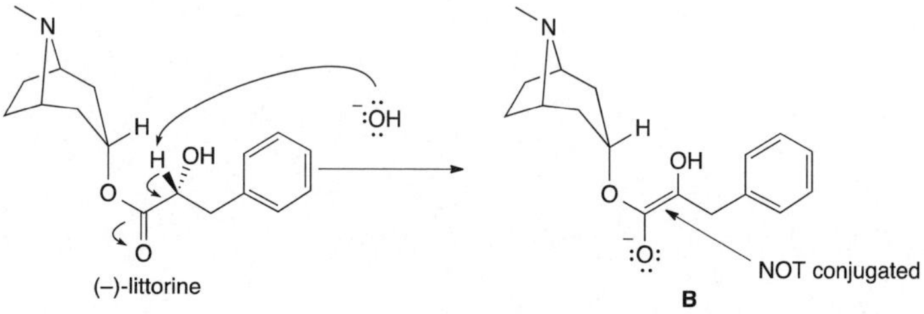

(–)-littorine

NOT conjugated

B

Carbonyl Condensation Reactions 24–1

Chapter 24 Carbonyl Condensation Reactions

Chapter Review

The four major carbonyl condensation reactions

Reaction type	Reaction
[1] Aldol reaction (24.1)	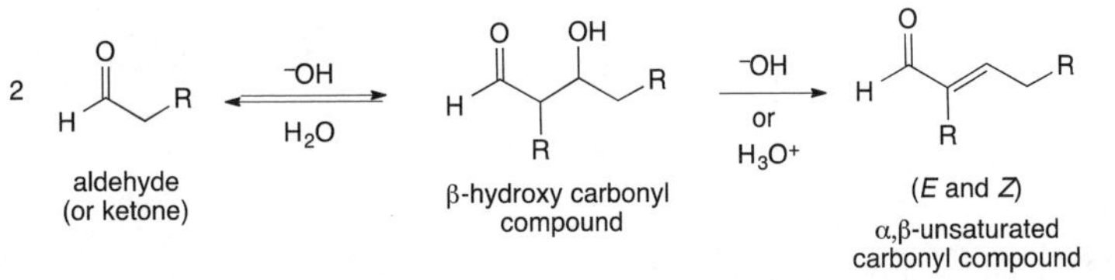 aldehyde (or ketone) β-hydroxy carbonyl compound (E and Z) α,β-unsaturated carbonyl compound
[2] Claisen reaction (24.5)	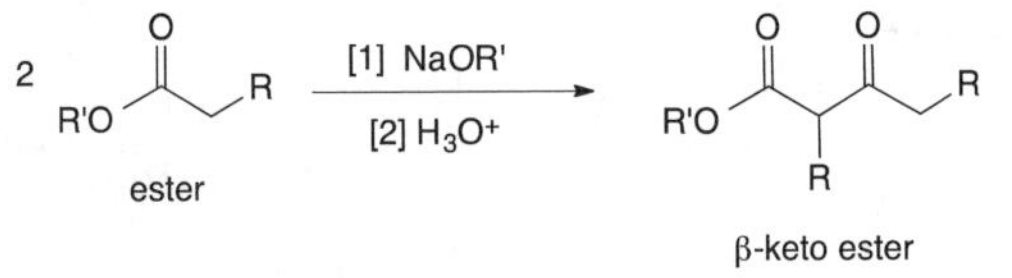 ester β-keto ester
[3] Michael reaction (24.8)	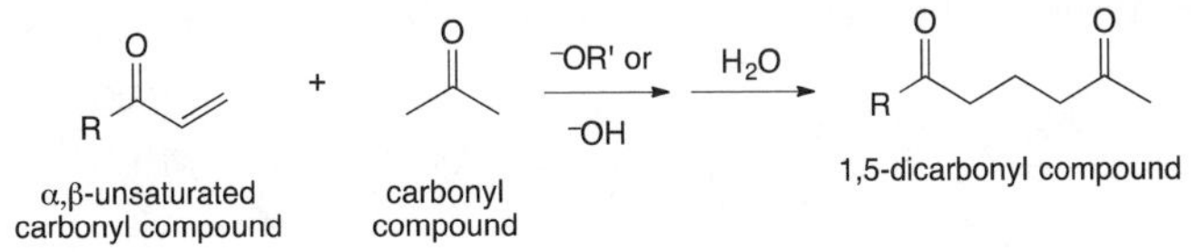 α,β-unsaturated carbonyl compound + carbonyl compound 1,5-dicarbonyl compound
[4] Robinson annulation (24.9)	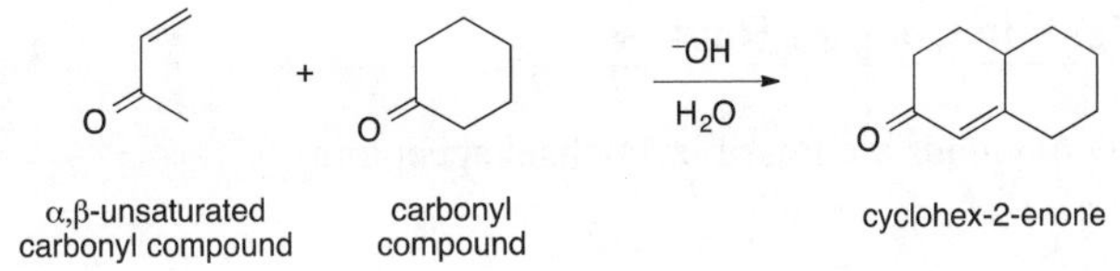 α,β-unsaturated carbonyl compound + carbonyl compound cyclohex-2-enone

Useful variations

[1] Directed aldol reaction (24.3)

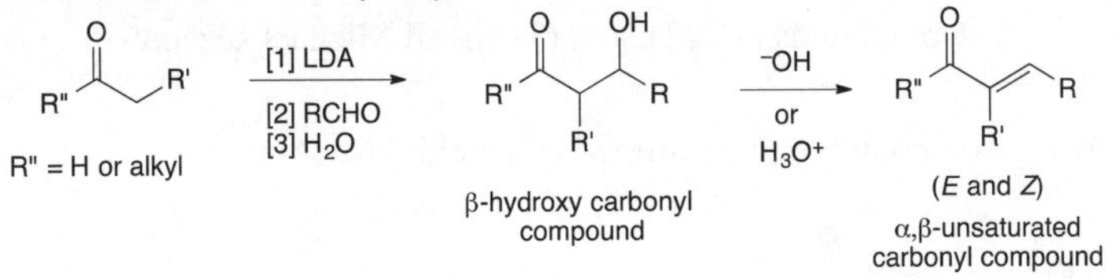

R" = H or alkyl

[1] LDA
[2] RCHO
[3] H_2O

β-hydroxy carbonyl compound

$^-$OH or H_3O^+

(E and Z)
α,β-unsaturated carbonyl compound

Chapter 24–2

[2] Intramolecular aldol reaction (24.4)

a. With 1,4-dicarbonyl compounds:

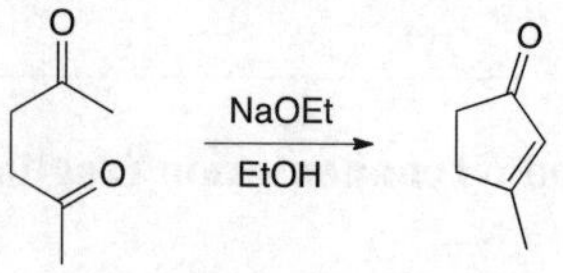

NaOEt / EtOH

b. With 1,5-dicarbonyl compounds:

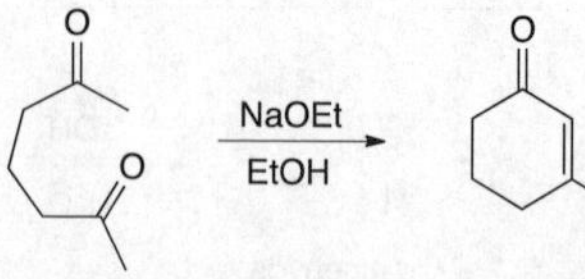

NaOEt / EtOH

[3] Dieckmann reaction (24.7)

a. With 1,6-diesters:

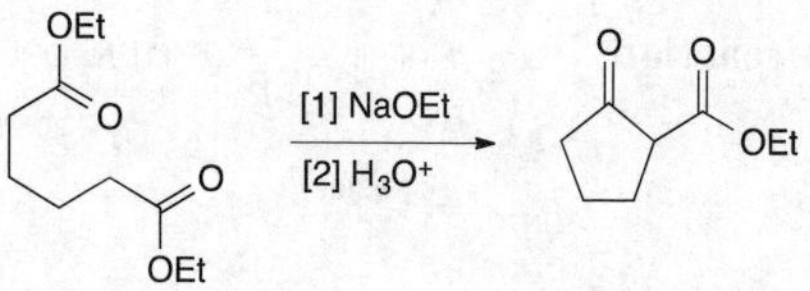

[1] NaOEt
[2] H_3O^+

b. With 1,7-diesters:

[1] NaOEt
[2] H_3O^+

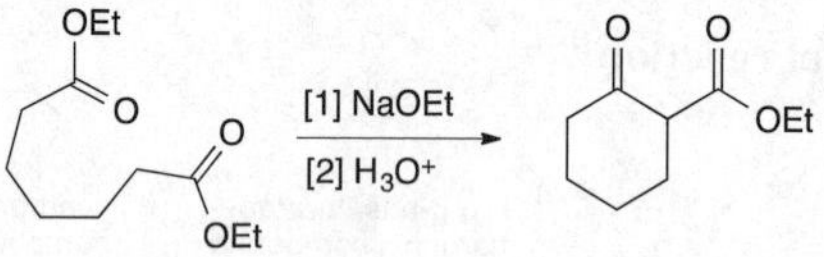

Practice Test on Chapter Review

1.a. Which compounds are possible Michael acceptors?

1.

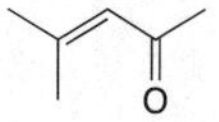

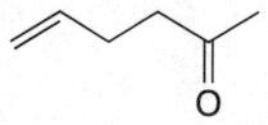

2.

3.

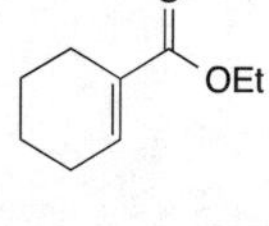

4. Both compounds (1) and (2) are Michael acceptors.

5. Compounds (1), (2), and (3) are all Michael acceptors.

b. Which of the following compounds can be formed by an aldol reaction?

1. 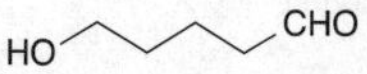 HO ... CHO

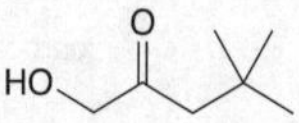

2.

3. HO ...

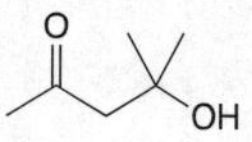

4. Both (1) and (2) can be formed by aldol reactions.

5. Compounds (1), (2), and (3) can all be formed by aldol reactions.

Carbonyl Condensation Reactions 24–3

c. Which compounds can be formed in a Robinson annulation?

1.

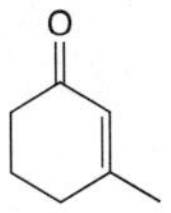

3.

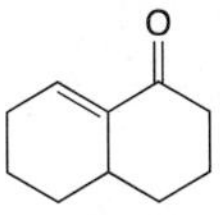

2. 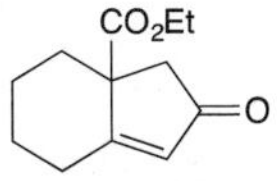   CO₂Et

4. Compounds (1) and (2) can be formed by Robinson annulations.

5. Compounds (1), (2), and (3) can all be formed by Robinson annulations.

d. What compounds can be used to form **A** by a condensation reaction?

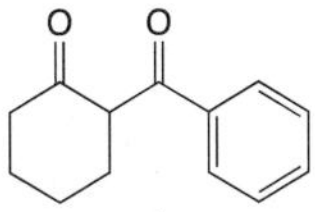

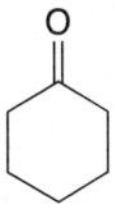

A

1. 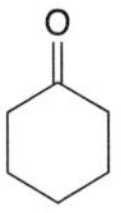   and   .CO₂Et

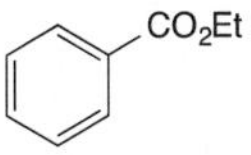

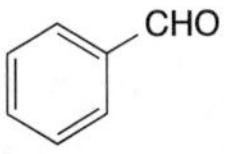

2.   .OEt

3.   and   .CHO

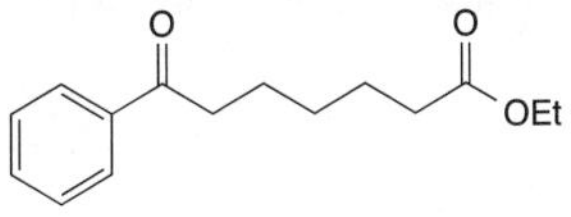

4. Compounds (1) and (2) can be used to form **A**.

5. Compounds (1), (2), and (3) can all be used to form **A**.

2. Give the reagents required for each step.

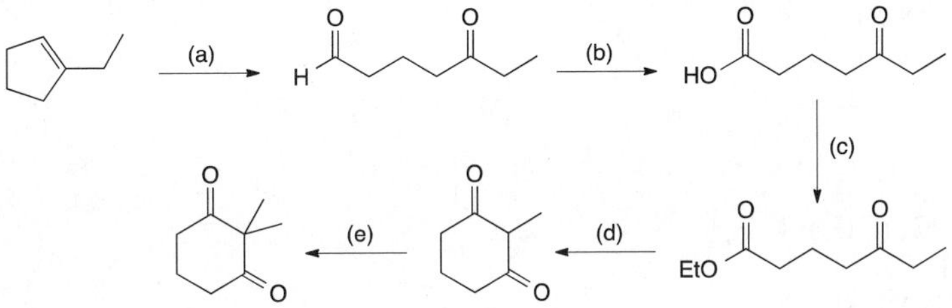

(a)

(b)

(c)

(e)

(d)

EtO

3. Draw the organic products formed in the following reactions.

a. 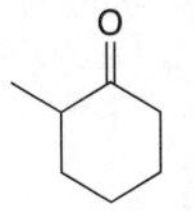   [1] LDA
        [2] CH₃CH₂CHO
        [3] H₂O

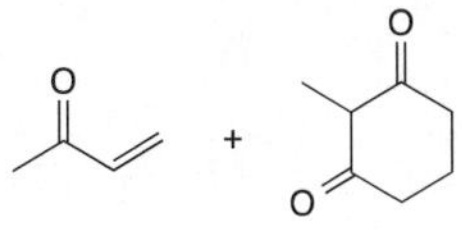

b. 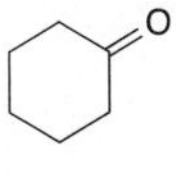   +   HCO₂Et   [1] NaOEt, EtOH
                            [2] H₃O⁺

c.   +   NaOEt
               EtOH

d. 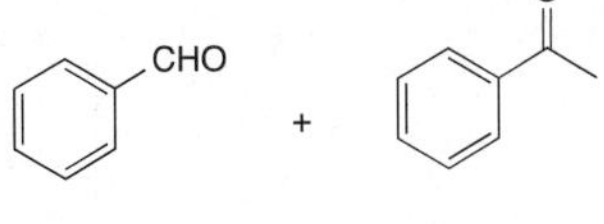   .CHO   +   ⁻OH, H₂O

Chapter 24–4

4.a. What organic starting materials are needed to synthesize **D** by a Robinson annulation reaction?

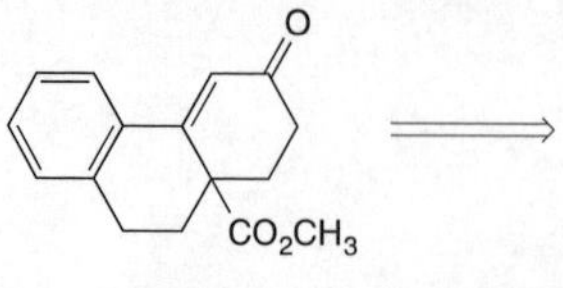

D

b. What organic starting materials are needed to synthesize β-keto ester **B** by a Dieckmann reaction?

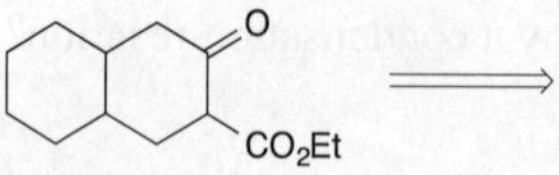

B

c. What starting materials are needed to synthesize **A** by an aldol reaction?

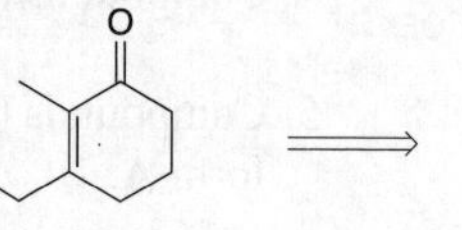

A

Answers to Practice Test

1.a. 4
  b. 2
  c. 1
  d. 4
2.a. [1] O_3; [2] $(CH_3)_2S$
  b. CrO_3, H_2SO_4, H_2O
  c. CH_3CH_2OH, H_2SO_4
  d. [1] NaOEt, EtOH; [2] H_3O^+
  e. [1] LDA; [2] CH_3I

3.

a.

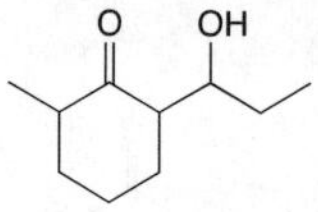

b.

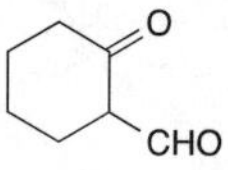

c.

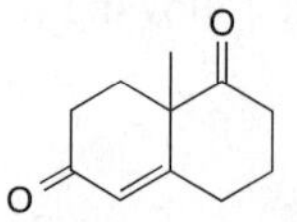

d.

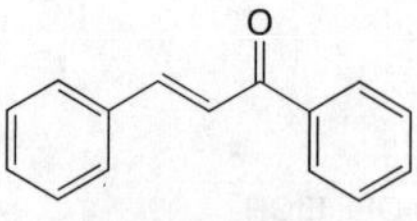

(*E* and *Z* isomers)

4.

a.

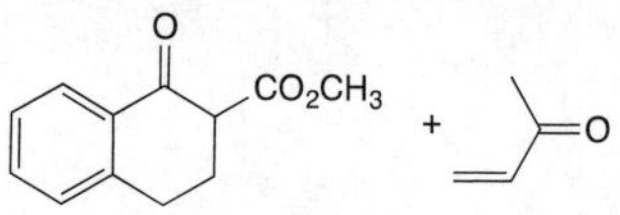

b.

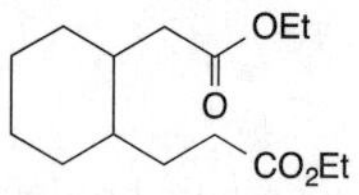

c.

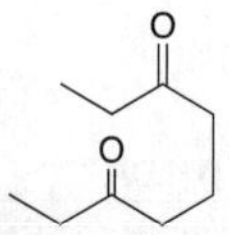

Carbonyl Condensation Reactions 24–5

Chapter 24: Answers to Problems

24.1

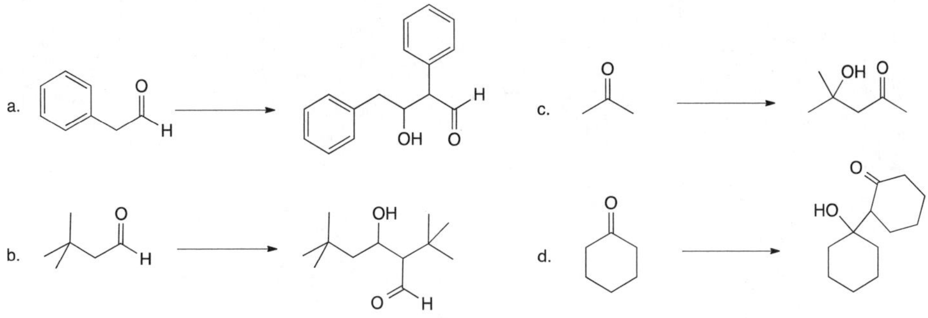

a.

b.

c.

d.

24.2

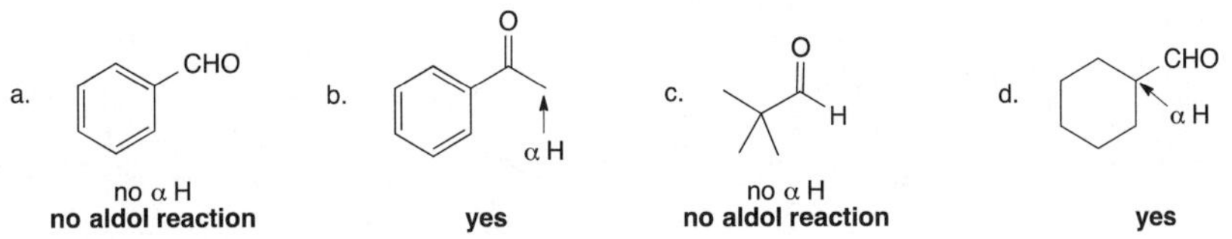

a.
no α H
no aldol reaction

b.
α H
yes

c.
no α H
no aldol reaction

d.
α H
yes

24.3

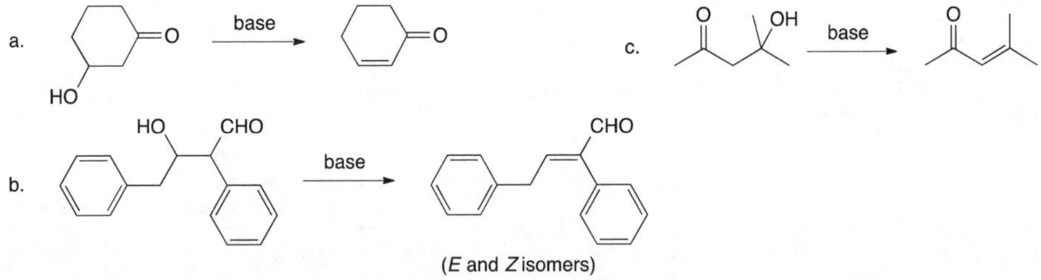

a.
base

c.
base

b.
base

(*E* and *Z* isomers)

24.4

+ $H_2\ddot{O}$:

HSO_4^-

+ H_2SO_4

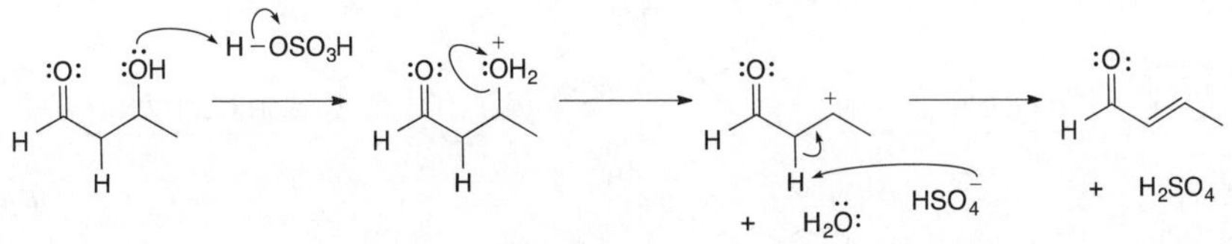

24.5 Locate the α and β C's to the carbonyl group, and break the molecule into two halves at this bond. The α C and all of the atoms bonded to it belong to one carbonyl component. The β C and all of the atoms bonded to it belong to the other carbonyl component.

Chapter 24–6

a.

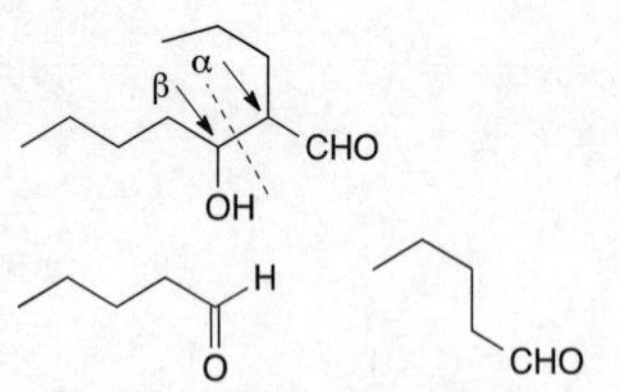

b.

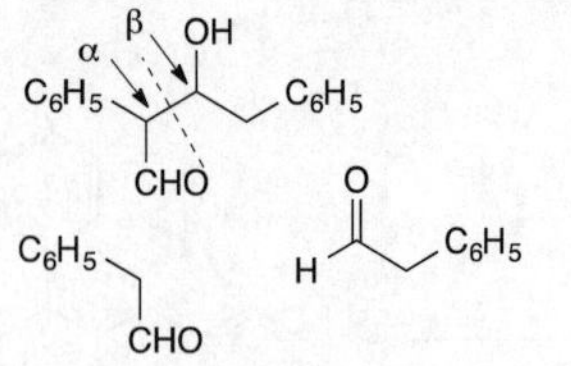

c.

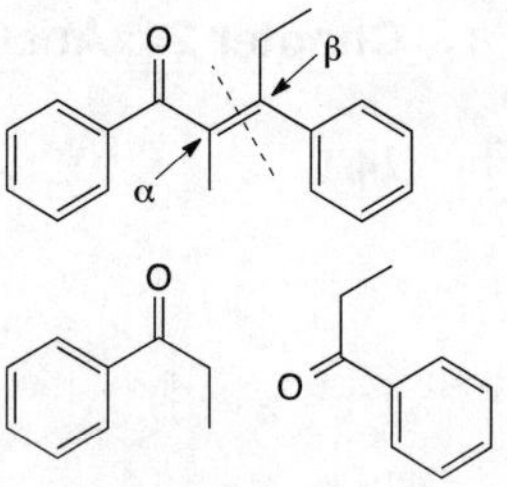

24.6

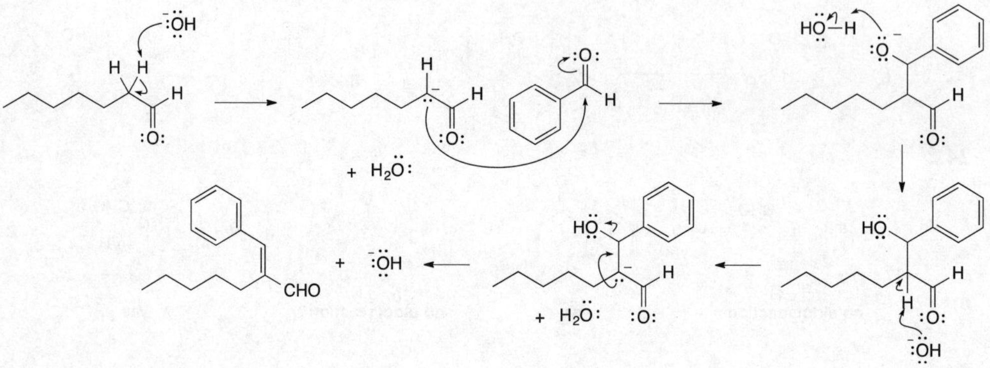

$+ H_2O$

24.7

a.

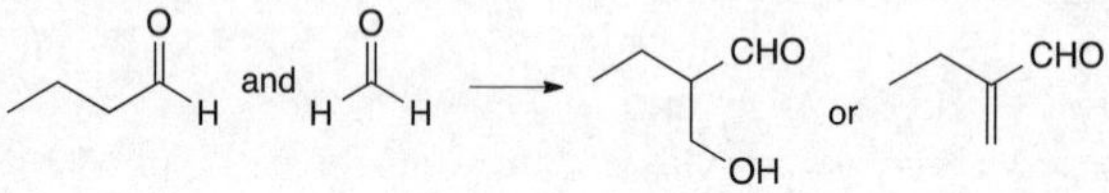

b.

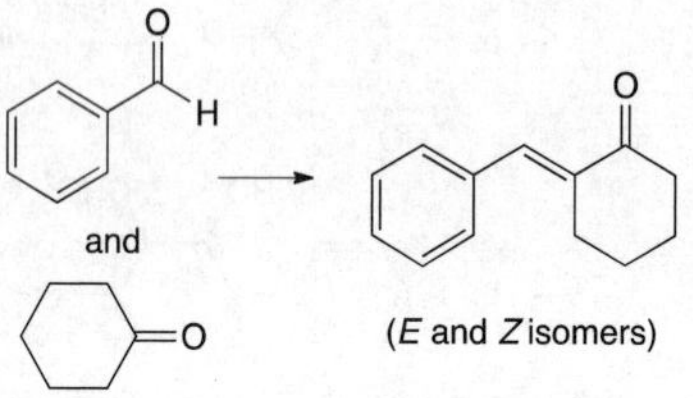

(E and Z isomers)

24.8

a.

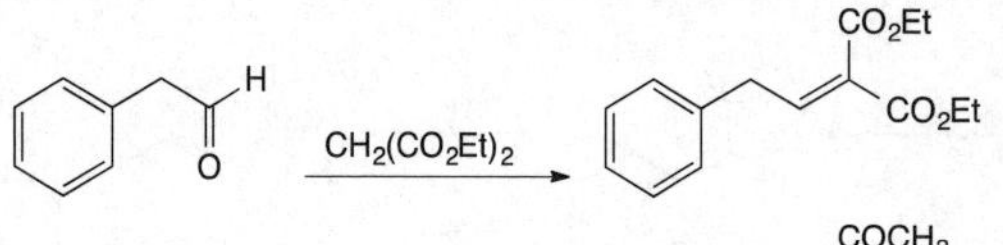

$CH_2(CO_2Et)_2$

b.

$CH_2(COCH_3)_2$

c.

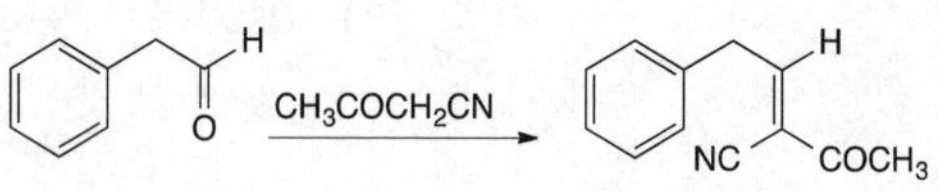

CH_3COCH_2CN

(E and Z isomers)

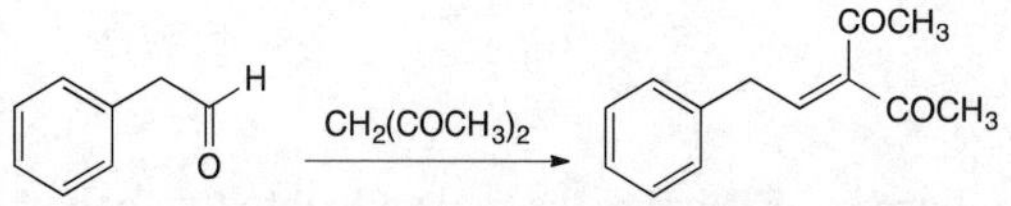

Carbonyl Condensation Reactions 24–7

24.9

a. (scheme)

b. (scheme) (E and Z mixture)

c. (from b.) (E and Z)

d. (from b.) (E and Z)

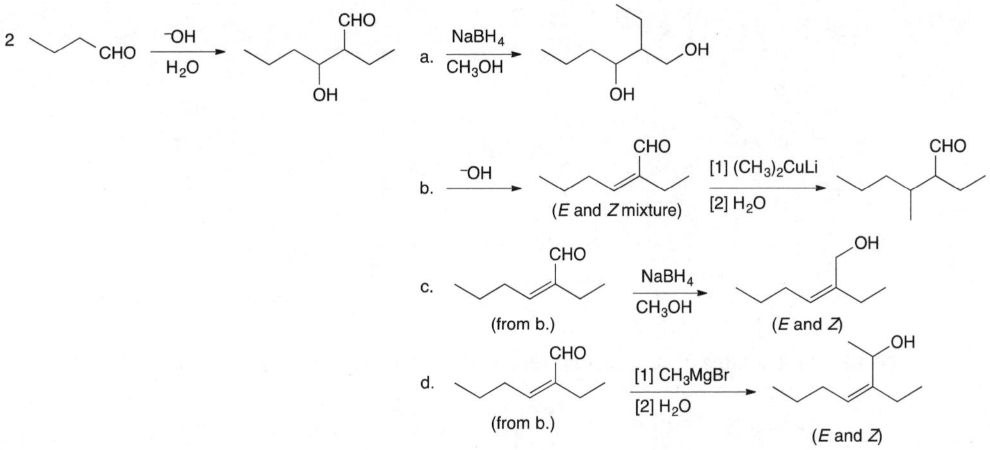

24.10 Find the α and β C's to the carbonyl group and break the bond between them.

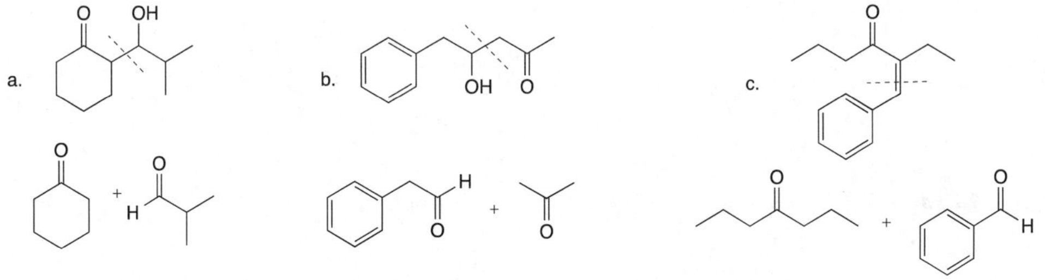

a.

b.

c.

24.11

gingerol

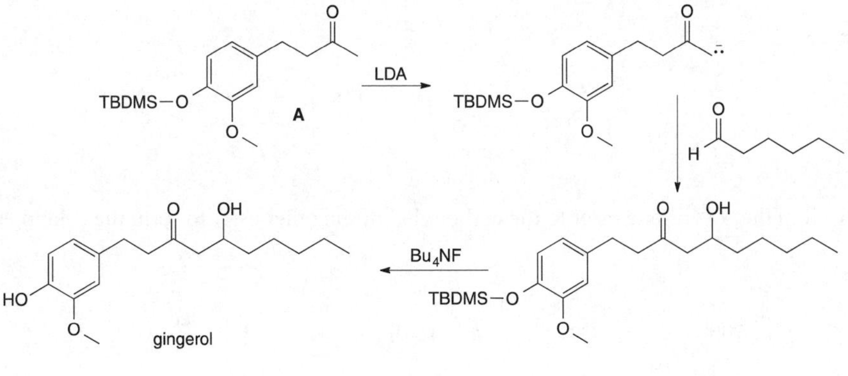

Chapter 24–8

24.12

H_2
Pd-C

CH$_3$O—

X

CH$_3$O—

donepezil

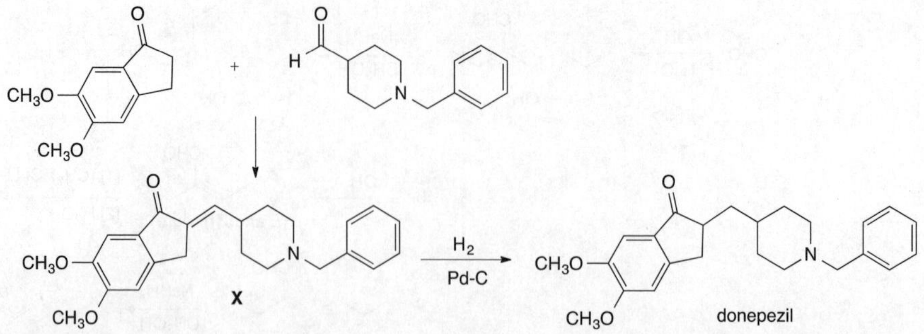

24.13 All enolates have a second resonance structure with a negative charge on O.

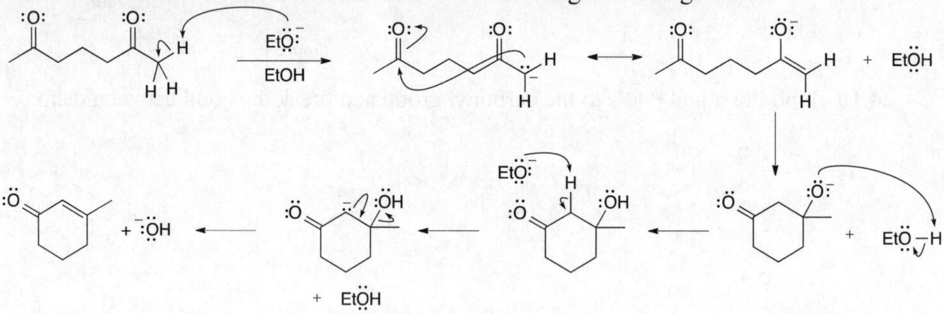

EtO$^-$
EtOH

+ EtÖH

EtÖ$^-$

EtÖ—H

+ EtÖH

24.14

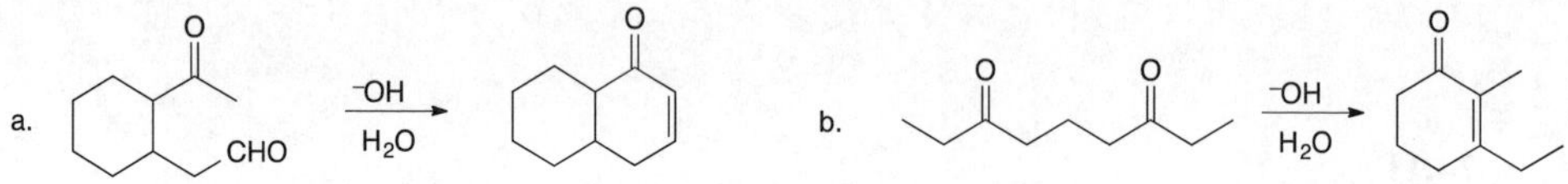

a.

$^-$OH
H$_2$O

b.

$^-$OH
H$_2$O

24.15

A

[1] O$_3$
[2] (CH$_3$)$_2$S

CHO

$^-$OH
H$_2$O

B

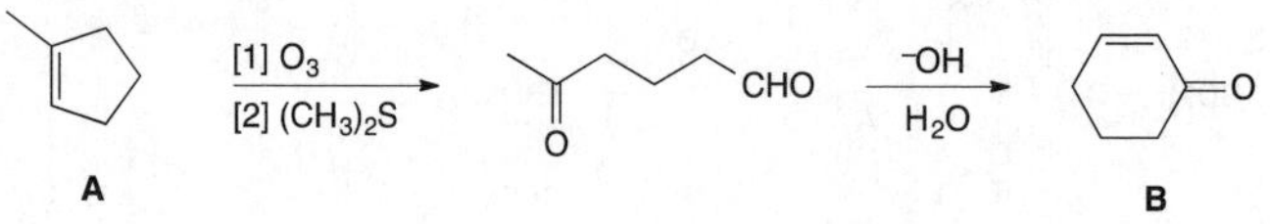

24.16 Join the α C of one ester to the carbonyl C of the other ester to form the β-keto ester.

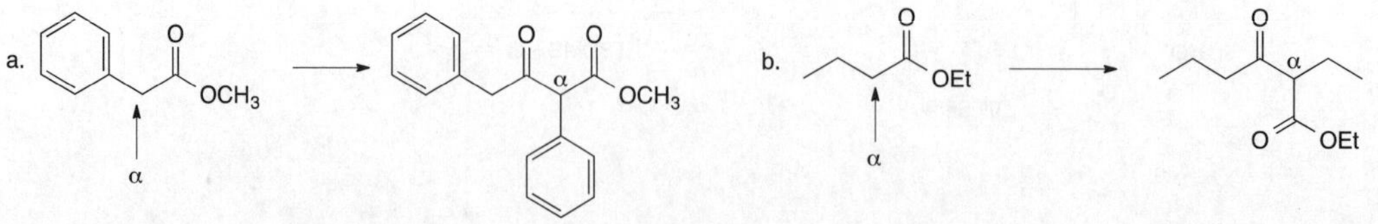

a.

OCH$_3$

α

OCH$_3$

α

b.

OEt

α

OEt

α

24.17 In a crossed Claisen reaction between an ester and a ketone, the enolate is formed from the ketone, and the product is a β-dicarbonyl compound.

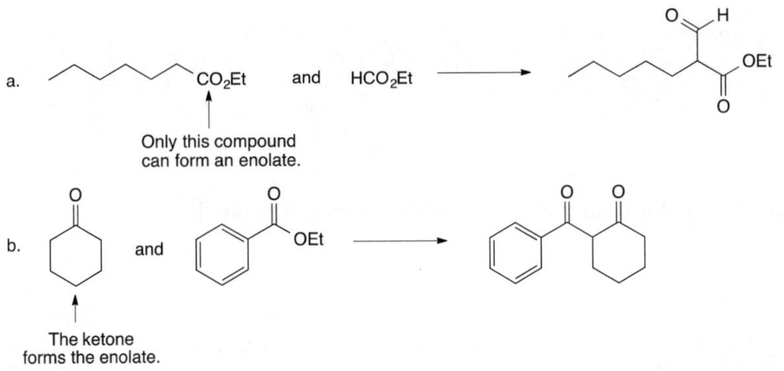

Carbonyl Condensation Reactions 24–9

a. [structure] and HCO_2Et ⟶ [structure]

Only this compound
can form an enolate.

b. [structure] and [structure] ⟶ [structure]

The ketone
forms the enolate.

24.18 A β-dicarbonyl compound like avobenzone is prepared by a crossed Claisen reaction between a ketone and an ester.

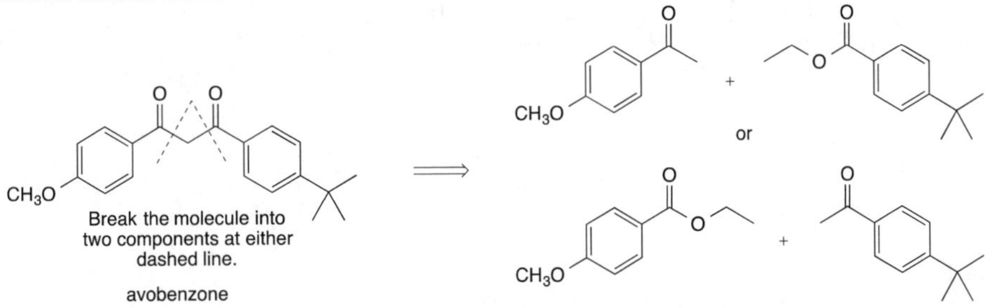

[structure]

Break the molecule into
two components at either
dashed line.

avobenzone

or

24.19

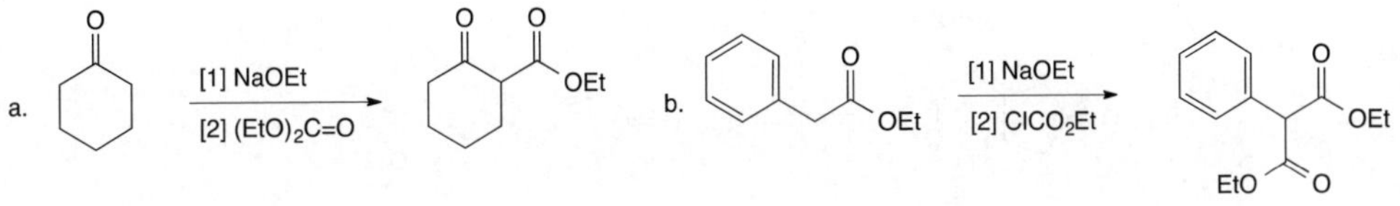

a. [structure] $\xrightarrow{\text{[1] NaOEt} \\ \text{[2] (EtO)}_2\text{C=O}}$ [structure]

b. [structure] $\xrightarrow{\text{[1] NaOEt} \\ \text{[2] ClCO}_2\text{Et}}$ [structure]

24.20

[structure] $\xrightarrow{\text{[1] NaOEt} \\ \text{[2] (EtO)}_2\text{C=O}}$ **A** $\xrightarrow{\text{[1] NaOEt} \\ \text{[2] CH}_3\text{I}}$ **B**

$\xrightarrow{\text{H}_3\text{O}^+ \\ \Delta}$ [structure]

ibuprofen

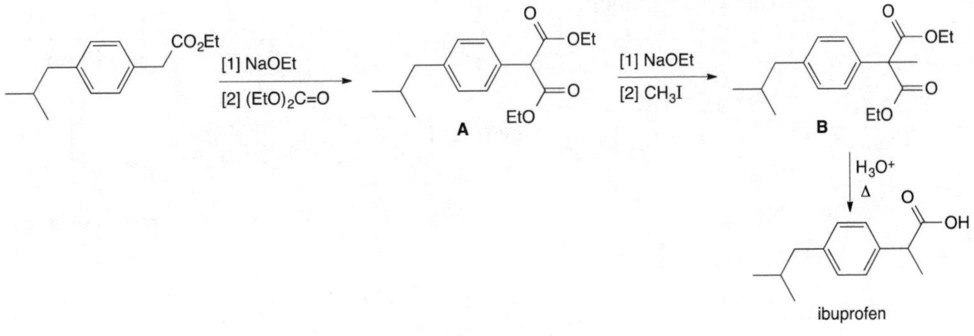

Chapter 24–10

24.21

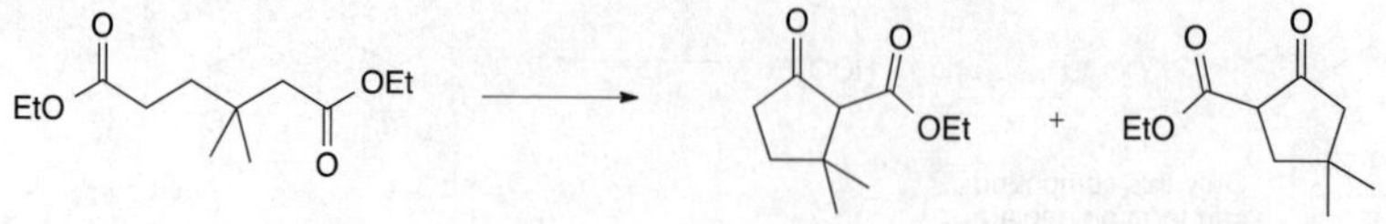

24.22 A Michael acceptor is an α,β-unsaturated carbonyl compound.

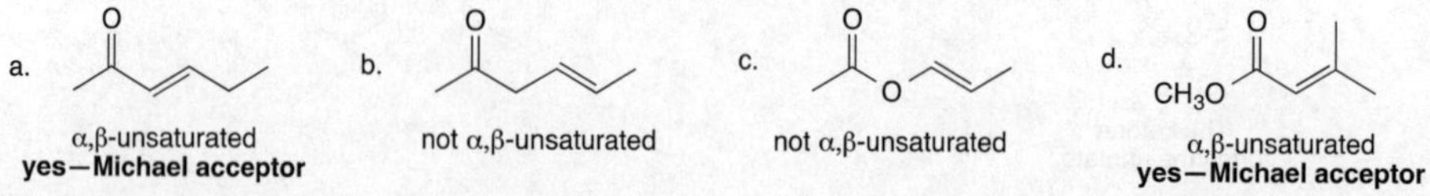

a.

α,β-unsaturated
yes—Michael acceptor

b.

not α,β-unsaturated

c.

not α,β-unsaturated

d.

α,β-unsaturated
yes—Michael acceptor

24.23

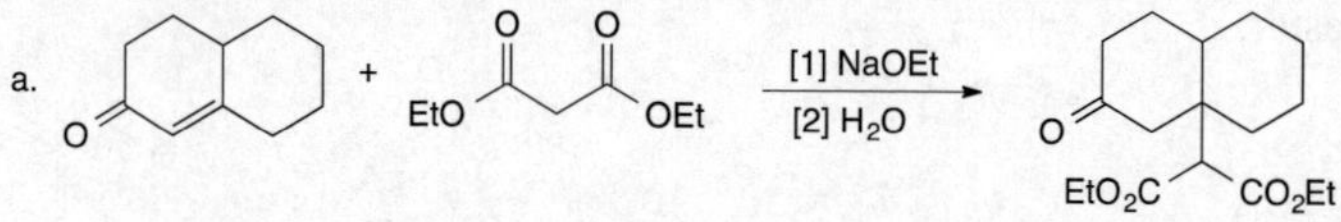

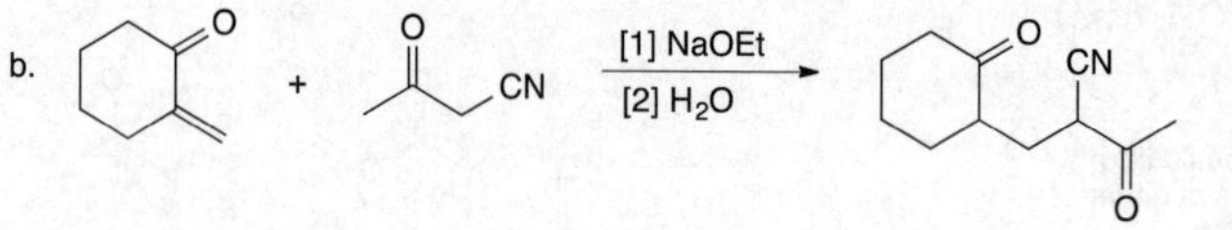

a.

b.

24.24

a.

b.

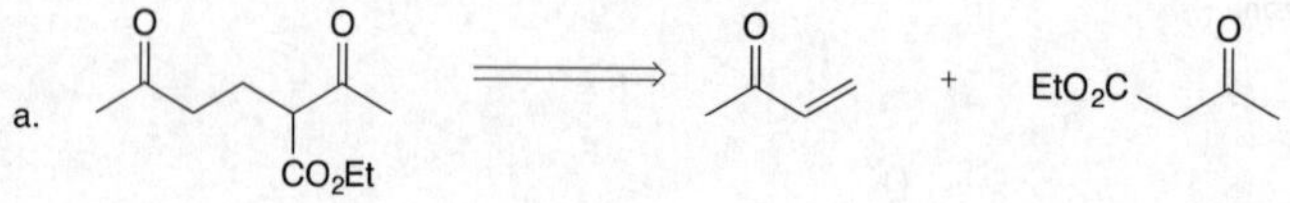

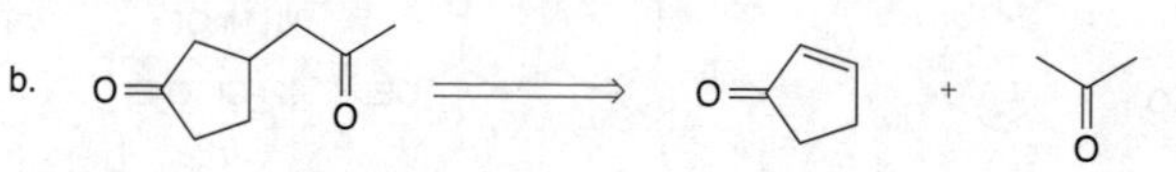

24.25 The Robinson annulation forms a six-membered ring and three new carbon–carbon bonds: two σ bonds and one π bond.

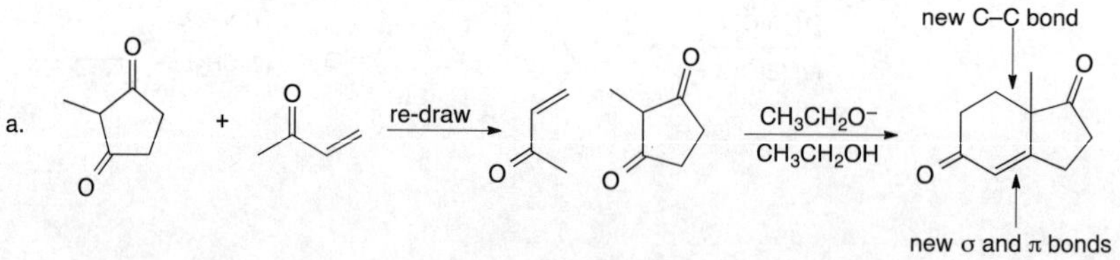

a.

Carbonyl Condensation Reactions 24–11

b. 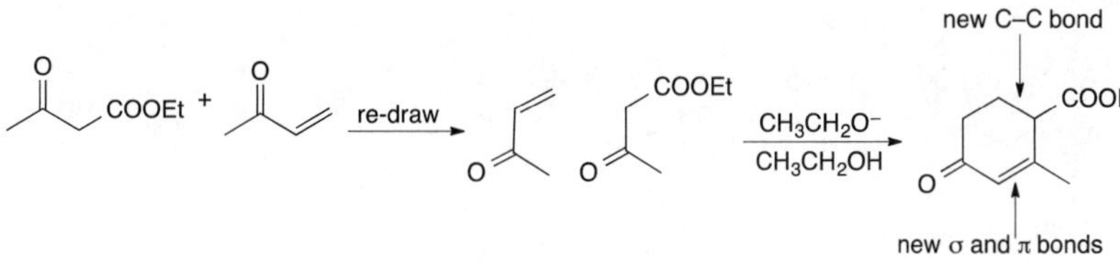 [chemical structures with "re-draw" and reaction conditions $CH_3CH_2O^-$ / CH_3CH_2OH, labels: new C–C bond, new σ and π bonds, COOEt]

c. [chemical structures with "re-draw" and reaction conditions $CH_3CH_2O^-$ / CH_3CH_2OH, labels: new C–C bond, new σ and π bonds]

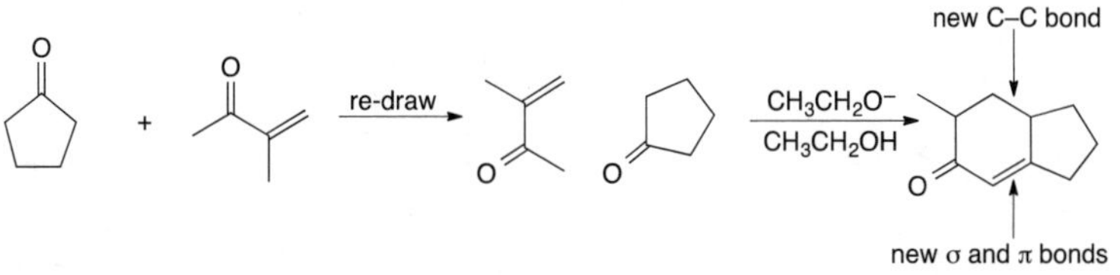

d. 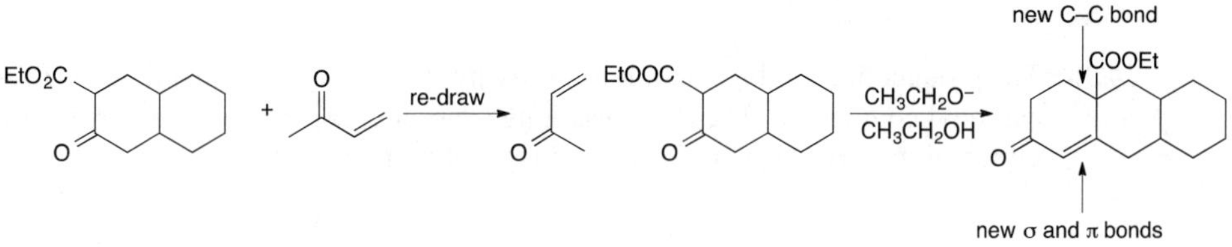 [chemical structures with "re-draw" and reaction conditions $CH_3CH_2O^-$ / CH_3CH_2OH, labels: new C–C bond, new σ and π bonds, COOEt, EtO$_2$C, EtOOC]

24.26 A Robinson annulation forms a conjugated cyclohex-2-enone. In a bicyclic product, one carbon of the C=C must be shared by both rings.

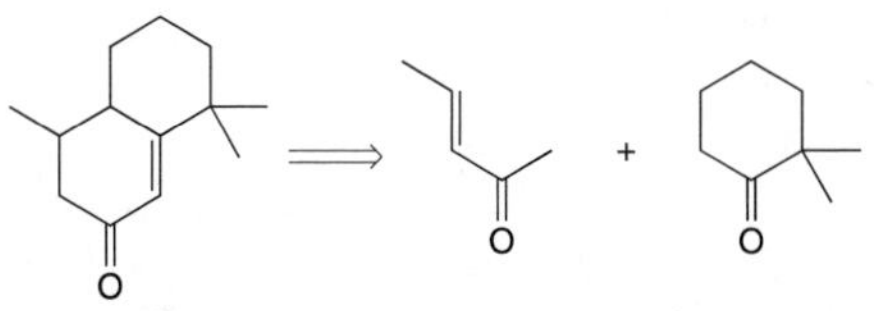

A
yes

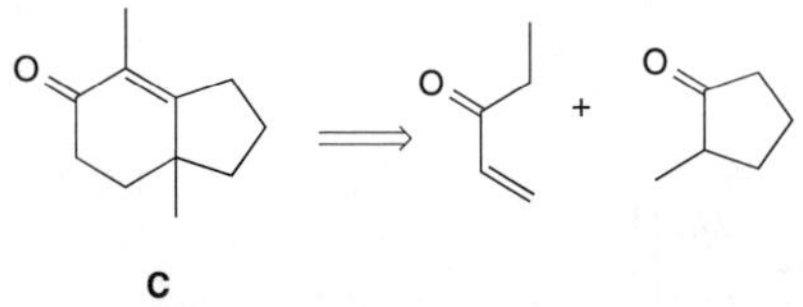

C
yes

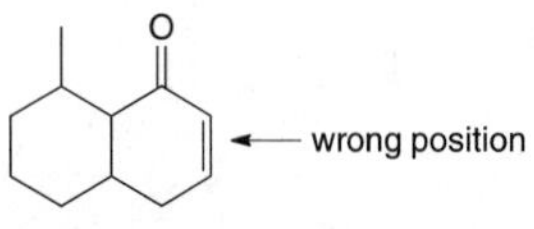

B
no
← wrong position

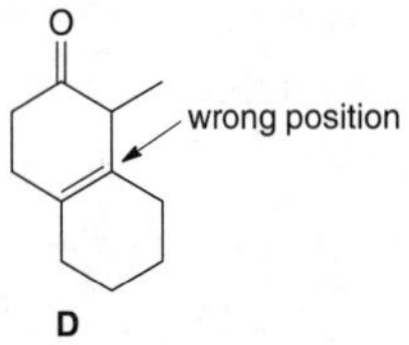

D
no
wrong position

24.27

a. 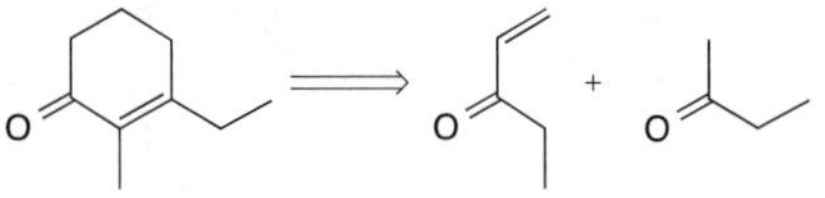 [chemical structures]

c. 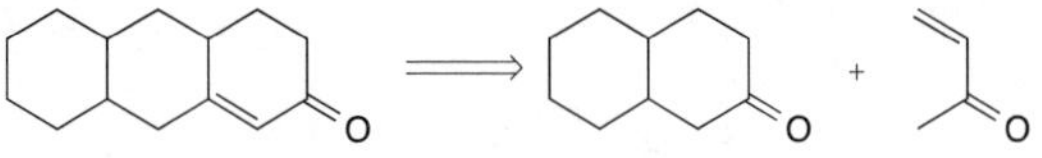 [chemical structures]

b. 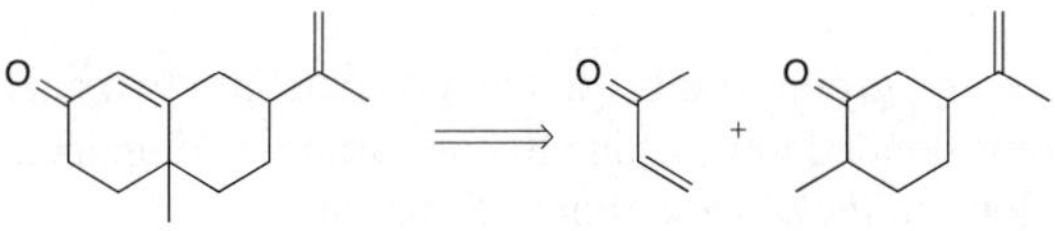 [chemical structures]

Chapter 24–12

24.28

a.   (reaction) $\xrightarrow[H_2O]{^-OH}$ or

b.   + C_6H_5CHO $\xrightarrow[H_2O]{^-OH}$ (E and Z isomers)

24.29

A $\xrightarrow[\text{[2] (CH}_3)_2S]{\text{[1] O}_3}$ CHO CHO $\xrightarrow[\text{EtOH}]{\text{NaOEt}}$ CHO   B

24.30 The product of an aldol reaction is a β-hydroxy carbonyl compound or an α,β-unsaturated carbonyl compound. The α,β-unsaturated carbonyl compound is drawn as product unless elimination of H_2O cannot form a conjugated system.

a. $\xrightarrow[H_2O]{^-OH}$

b.   + $\xrightarrow[H_2O]{^-OH}$

c.   + $\xrightarrow[H_2O]{^-OH}$ CHO   (E and Z isomers)

24.31

a. $\xrightarrow[\text{[3] H}_2O]{\substack{\text{[1] LDA} \\ \text{[2] CH}_3\text{CH}_2\text{CH}_2\text{CHO}}}$

b.   OEt $\xrightarrow[\substack{\text{[2]} \\ \text{[3] H}_2O}]{\text{[1] LDA}}$ OEt

24.32

a.   CHO $\longrightarrow$

b.   OHC   CHO $\longrightarrow$ CHO

c. $\longrightarrow$

24.33 Locate the α and β C's to the carbonyl group, and break the molecule into two halves at this bond. The α C and all of the atoms bonded to it belong to one carbonyl component. The β C and all the atoms bonded to it belong to the other carbonyl component.

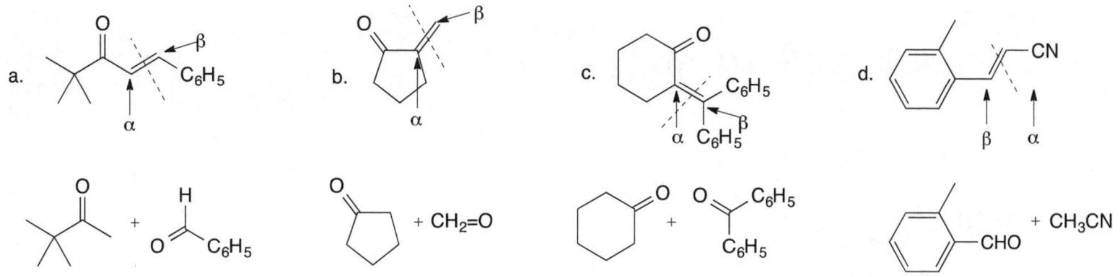

Carbonyl Condensation Reactions 24–13

a.   b.   c.   d.

24.34 Base removes the most acidic proton between the two C=O's in **B**. This enolate reacts with the aldehyde in **A** to form a product that loses H_2O.

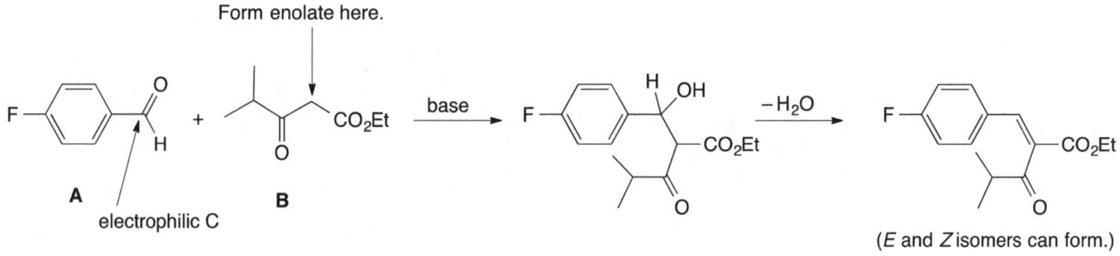

Form enolate here.

base

$-H_2O$

A
electrophilic C

B

(*E* and *Z* isomers can form.)

24.35

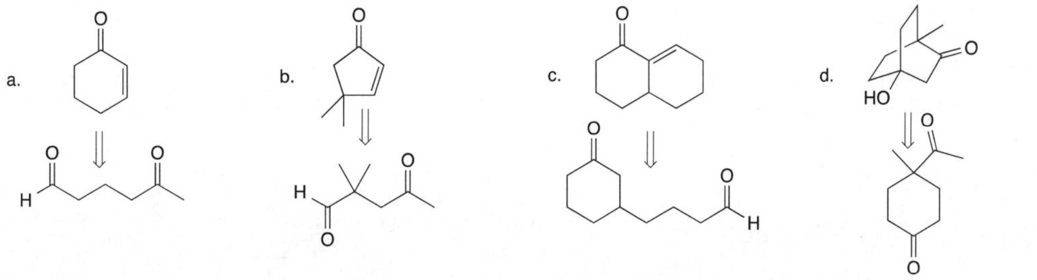

a.   b.   c.   d.

24.36 Ozonolysis cleaves the C=C, and base catalyzes an intramolecular aldol reaction.

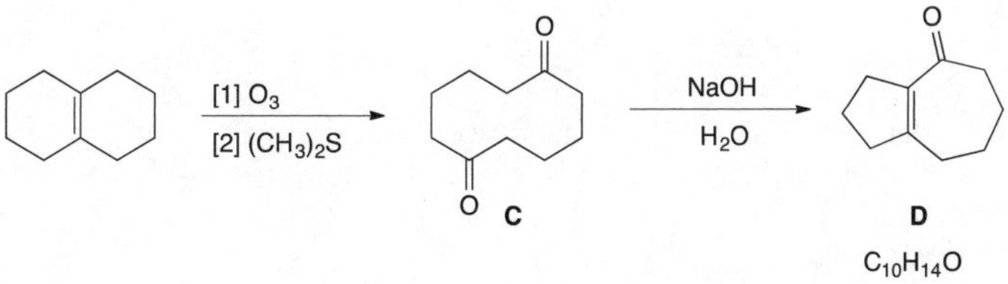

[1] O_3
[2] $(CH_3)_2S$

C

NaOH
H_2O

D
$C_{10}H_{14}O$

24.37 Aldol reactions proceed via resonance-stabilized enolates. **K** can form an enolate that allows for delocalization of the negative charge on O. Delocalization is not possible in **J,** because a double bond would be placed at a bridgehead carbon, which is geometrically impossible.

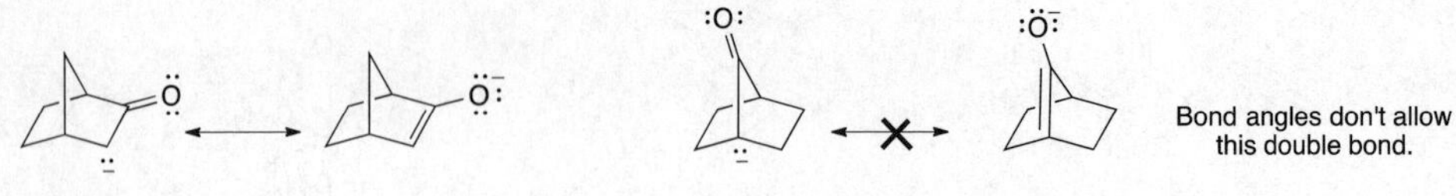

Chapter 24–14

Bond angles don't allow this double bond.

from **K**   from **J**

24.38

a.

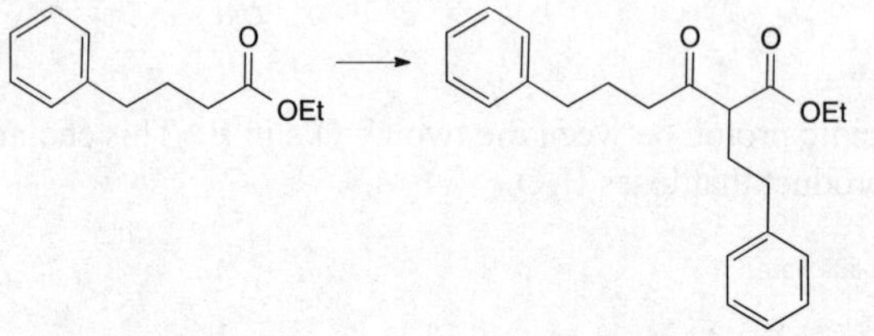

b.

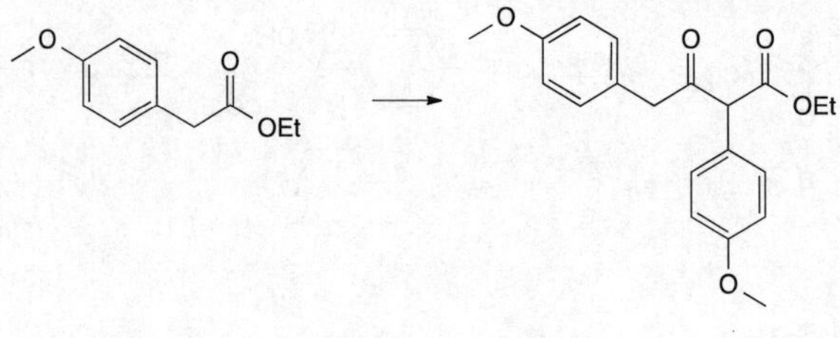

24.39

a.

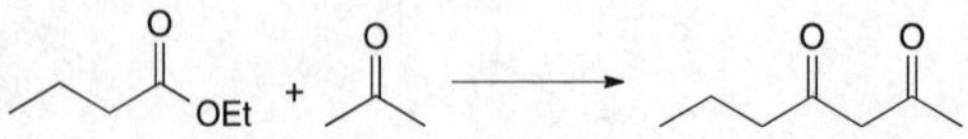

c.

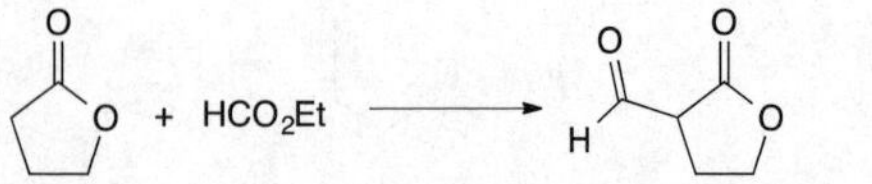

b.

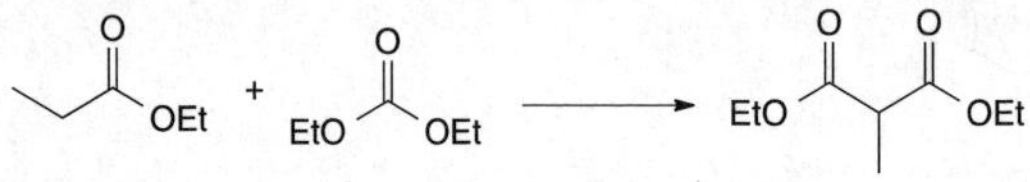

d.

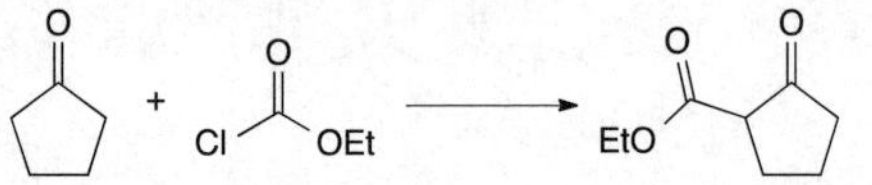

24.40

a.

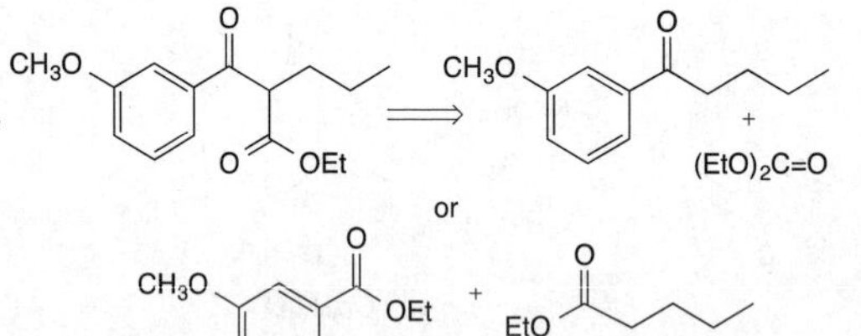

or

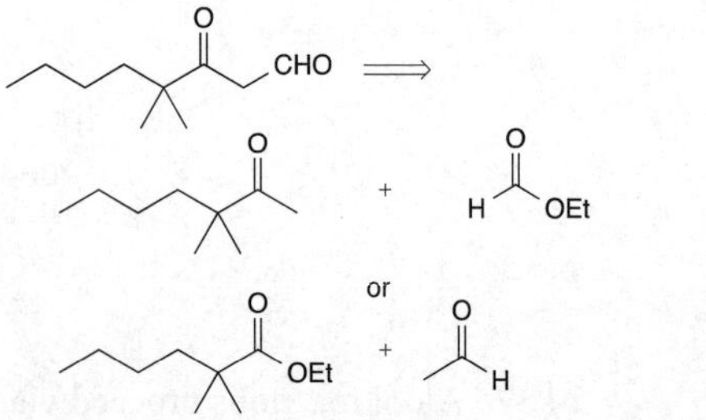

b.

c.

or

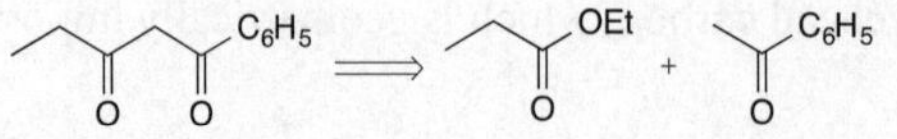

d.

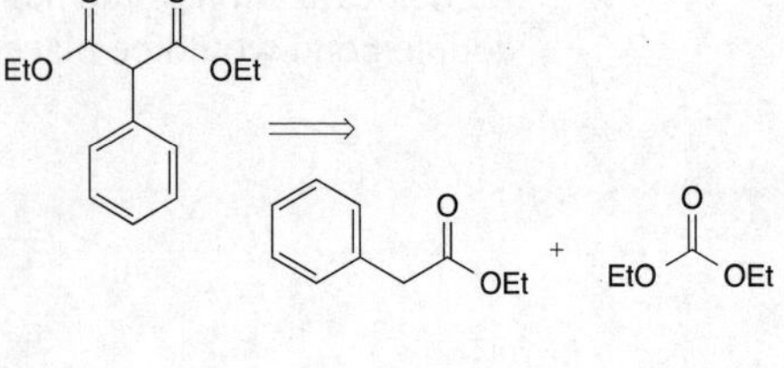

Carbonyl Condensation Reactions 24–15

24.41 Only esters with two H's or three H's on the α carbon form enolates that undergo Claisen reaction to form resonance-stabilized enolates of the product β-keto ester. Thus, the enolate forms on the CH₂ α to one ester carbonyl, and cyclization yields a five-membered ring.

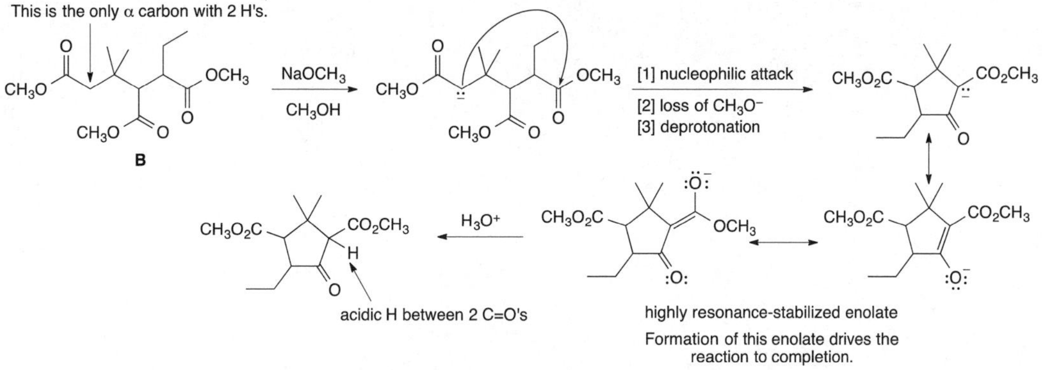

This is the only α carbon with 2 H's.

CH₃O OCH₃ → NaOCH₃ / CH₃OH → CH₃O OCH₃ → [1] nucleophilic attack [2] loss of CH₃O⁻ [3] deprotonation → CH₃O₂C CO₂CH₃

B

CH₃O₂C CO₂CH₃ ← H₃O⁺ ← CH₃O₂C OCH₃ ↔ CH₃O₂C CO₂CH₃

acidic H between 2 C=O's

highly resonance-stabilized enolate
Formation of this enolate drives the reaction to completion.

24.42

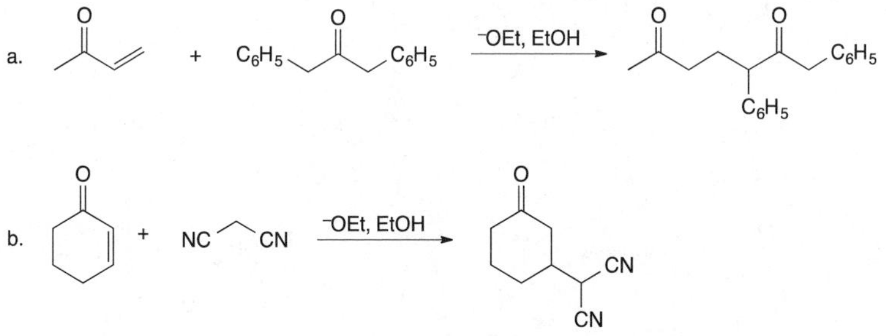

a.   + C₆H₅ C₆H₅   —OEt, EtOH→   C₆H₅

b.   + NC CN   —OEt, EtOH→   CN CN

24.43

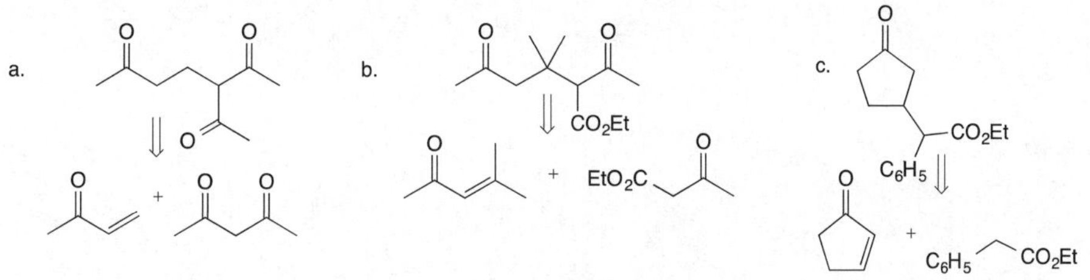

a.   b.   CO₂Et   EtO₂C   c.   CO₂Et   C₆H₅

C₆H₅   CO₂Et

24.44

a.   **A**   +   [H ... O]   —Michael reaction→

(*E* or *Z* isomer can be used.)

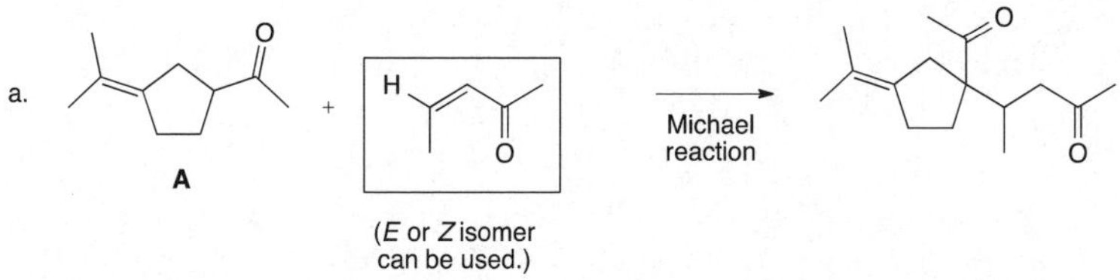

Chapter 24–16

b.

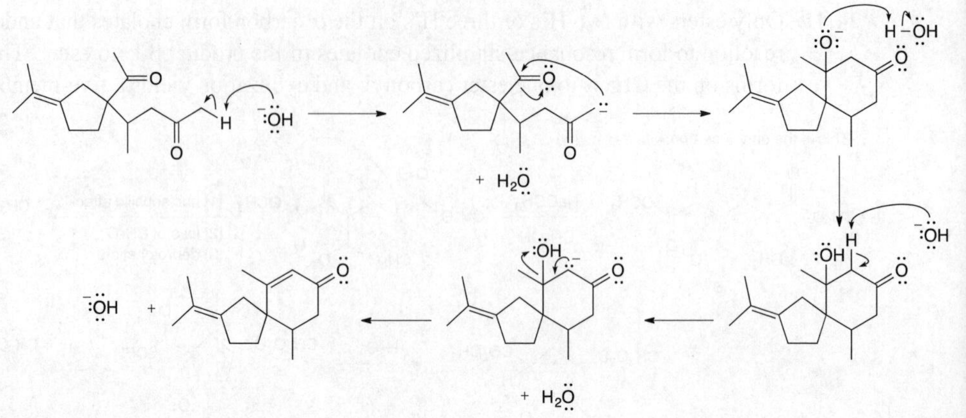

24.45

a.

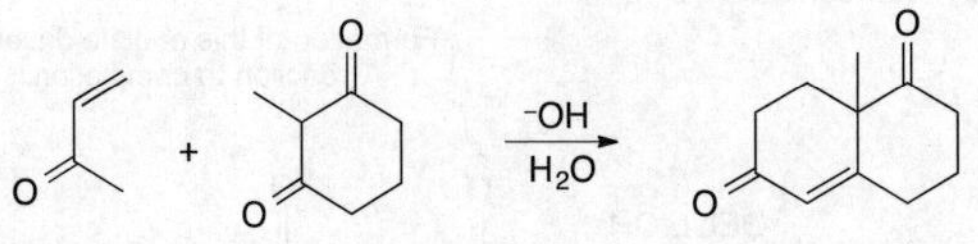

b.

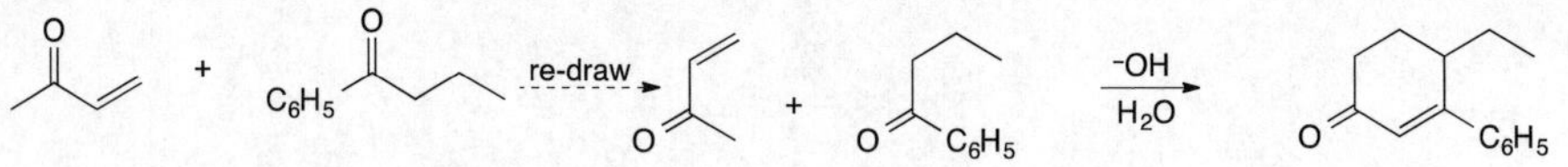

c.

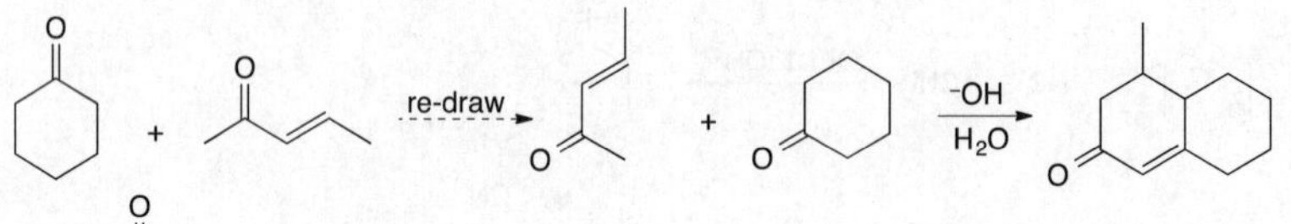

d.

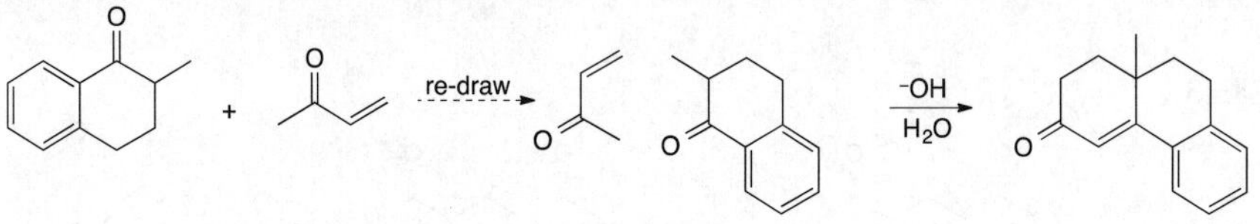

24.46

a.

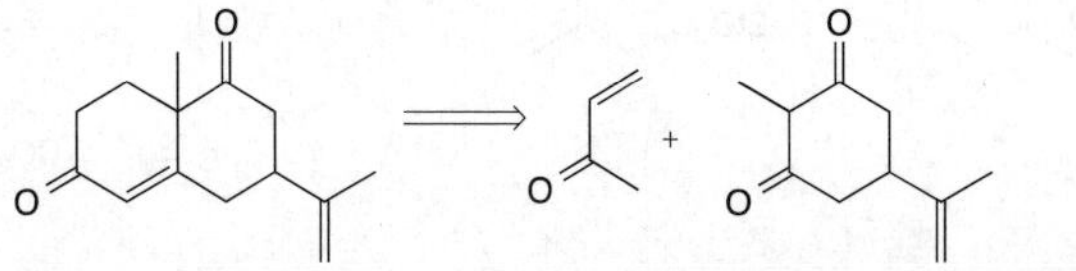

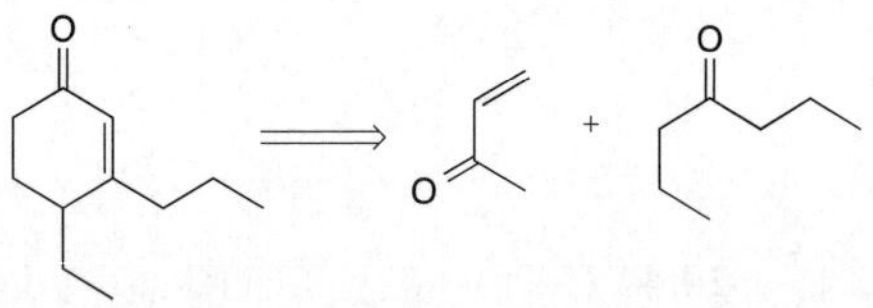

b.

c.

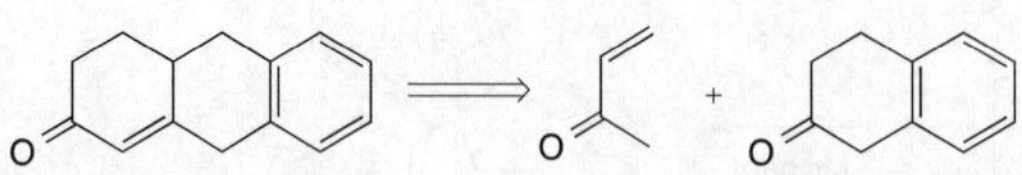

Carbonyl Condensation Reactions 24–17

24.47

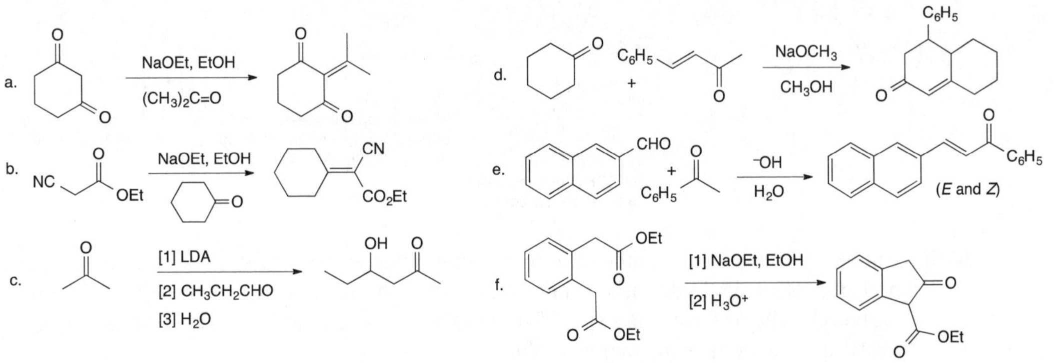

24.48

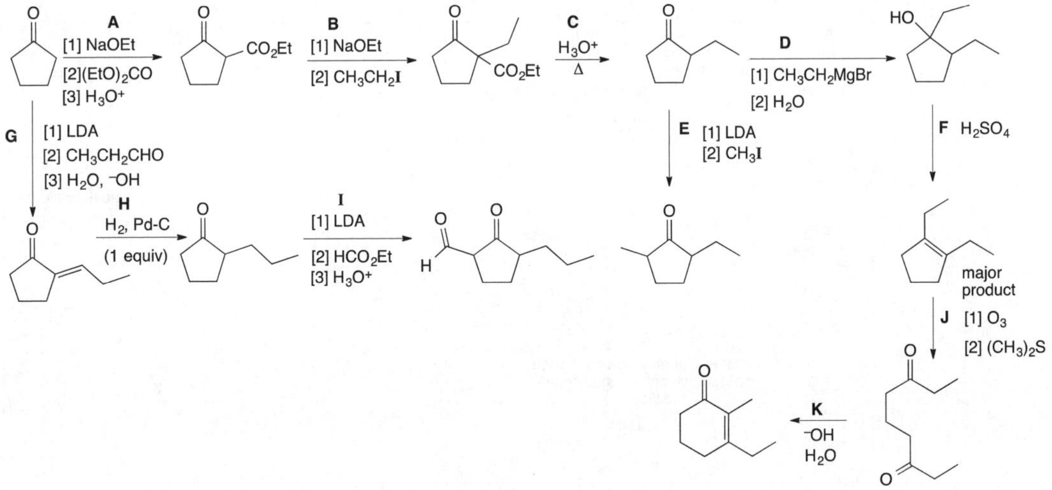

24.49

a. This reaction is a crossed Claisen between an ester and ketone. Cyclization forms the more stable six-membered ring.

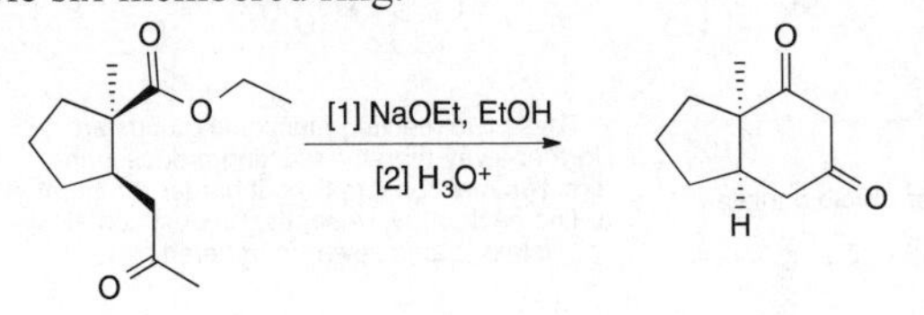

b.

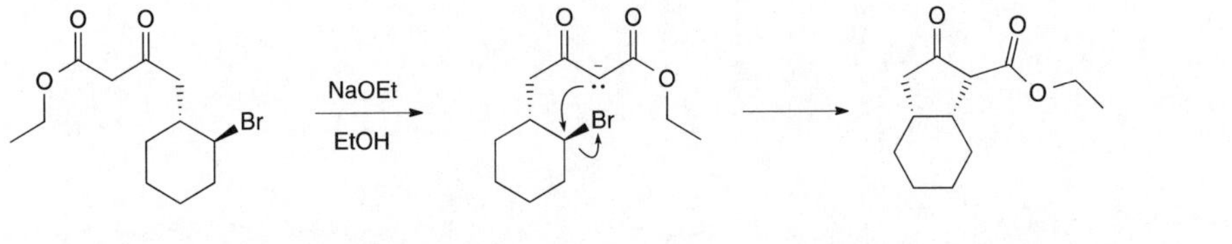

Chapter 24–18

24.50

[1] O$_3$
[2] (CH$_3$)$_2$S

NaOH
EtOH

A

B

Form the enolate here to
generate a five-membered ring
in the product.

new C–C bond

The RCHO has the more
accessible carbonyl.

24.51 Enolate **A** is more substituted (and more stable) than either of the other two possible enolates
and attacks an aldehyde carbonyl group, which is sterically less hindered than a ketone
carbonyl. The resulting ring size (five-membered) is also quite stable. That is why 1-
acetylcyclopentene is the major product.

most stable enolate

A

less hindered carbonyl

+ H$_2$O

+ H$_2$O

1-acetylcyclopentene
major product

B

The more hindered ketone carbonyl
makes nucleophilic attack more
difficult.

+ H$_2$O

+ H$_2$O

CHO

C

less stable enolate

+ H$_2$O

These two reacting functional groups are
farther away than the reacting groups in the
first two reactions, making it harder for them
to find each other. Also, the product contains
a less stable seven-membered ring.

+ H$_2$O

Carbonyl Condensation Reactions 24–19

24.52 Removal of a proton from CH_3NO_2 forms an anion for which three resonance structures can be drawn.

24.53 All enolates have a second resonance structure with a negative charge on O.

Chapter 24–20

24.54

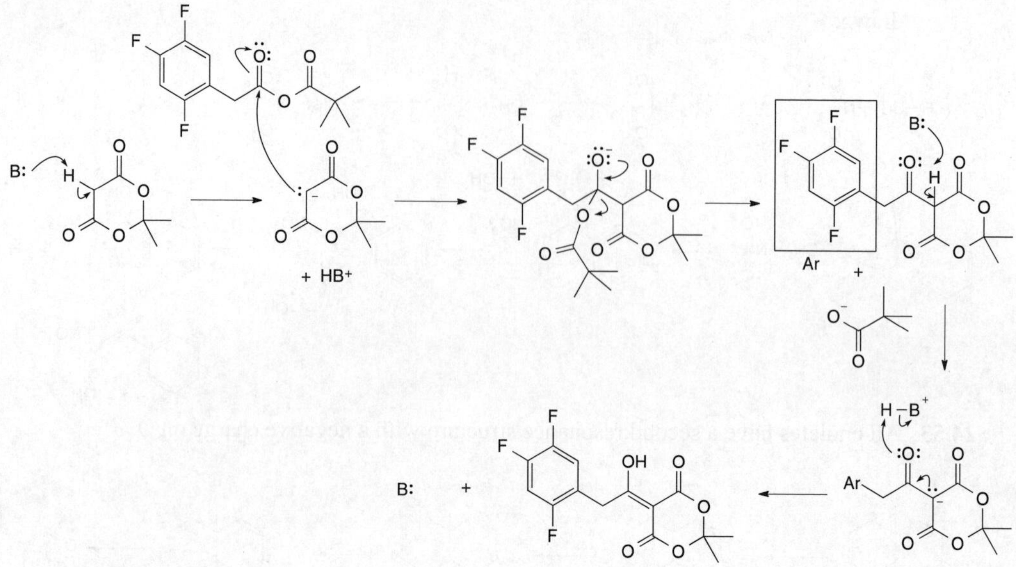

24.55 Et$_3$N reacts with phenylacetic acid to form a carboxylate anion that acts as a nucleophile to displace Br, forming **Y**. Then an intramolecular crossed Claisen reaction yields rofecoxib.

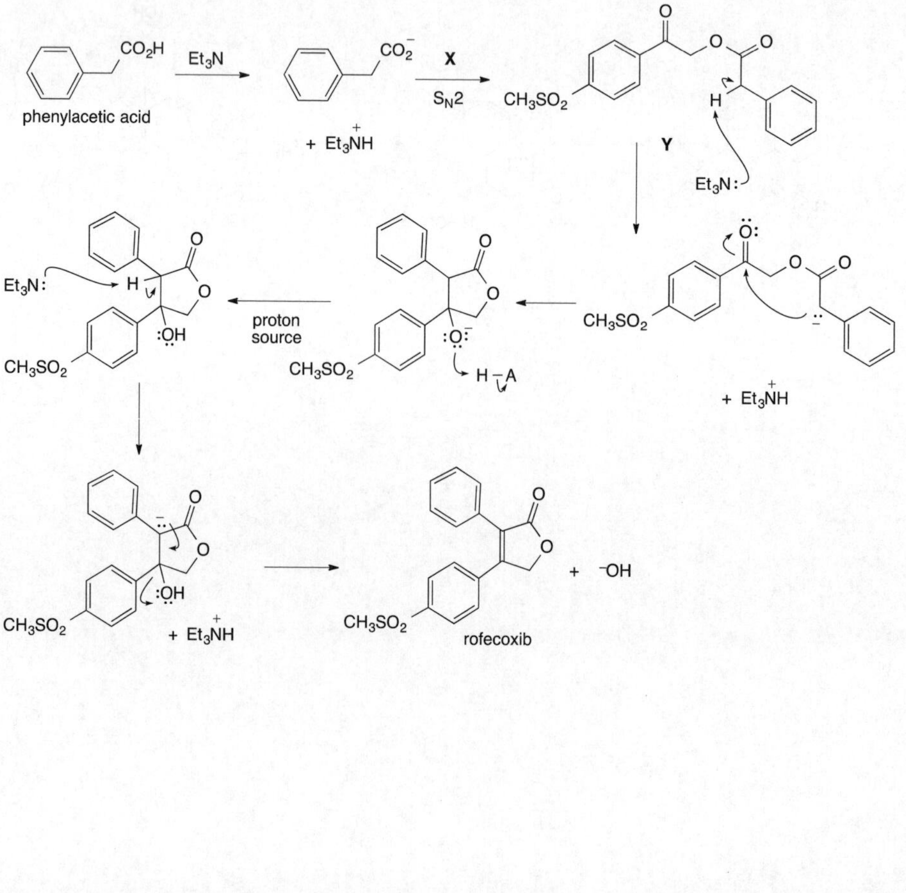

24.56

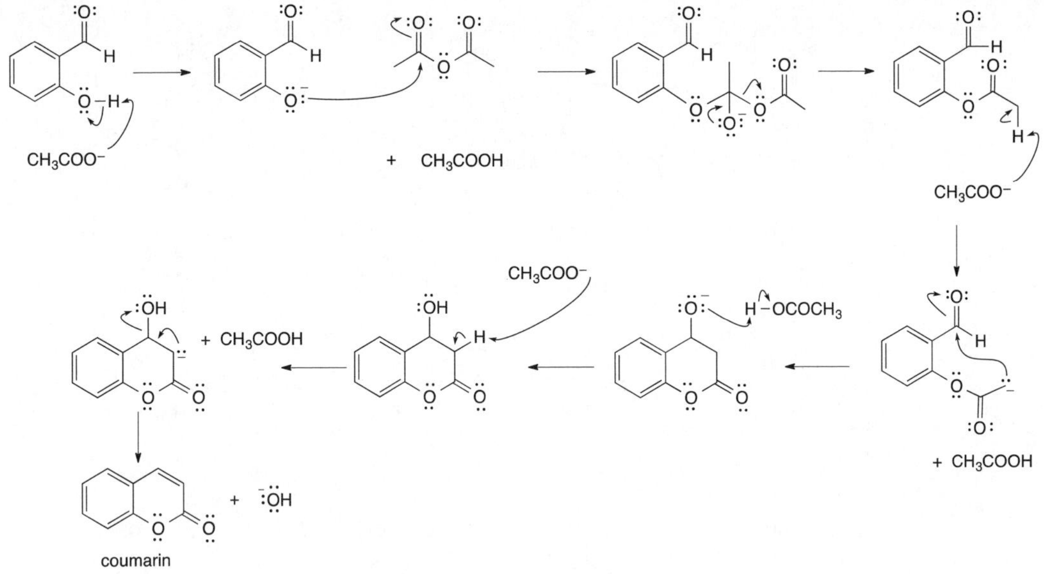

+ CH$_3$COOH

CH$_3$COO$^-$

+ CH$_3$COOH

:O: H–OCOCH$_3$

CH$_3$COO$^-$

:OH + CH$_3$COOH

:OH

coumarin + :OH

24.57 The mechanism consists of an intramolecular aldol reaction using an enolate formed by removal of a γ proton of an α,β-unsaturated carbonyl. The β-hydroxy carbonyl compound loses H$_2$O by the two-step E1cB mechanism to generate the final product.

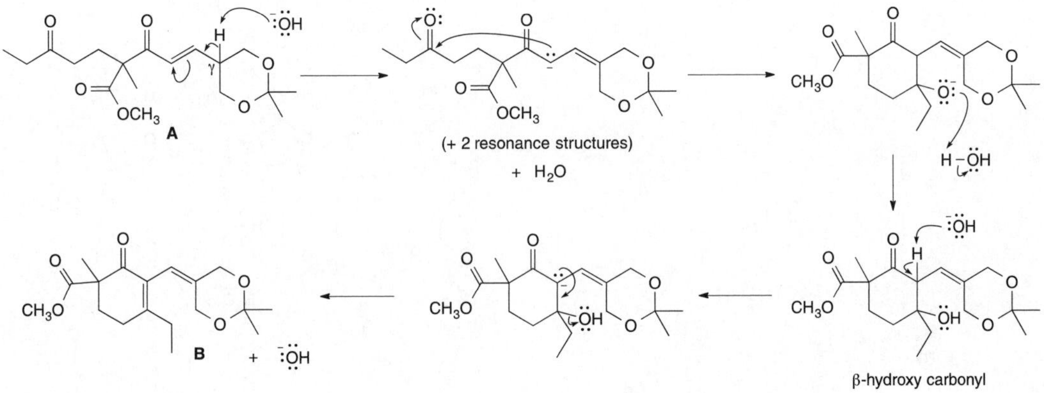

A

(+ 2 resonance structures)

+ H$_2$O

H–ÖH

:ÖH

β-hydroxy carbonyl

B + :ÖH

24.58 Because the reaction is carried out in acid, enols (not enolates) must be intermediates.

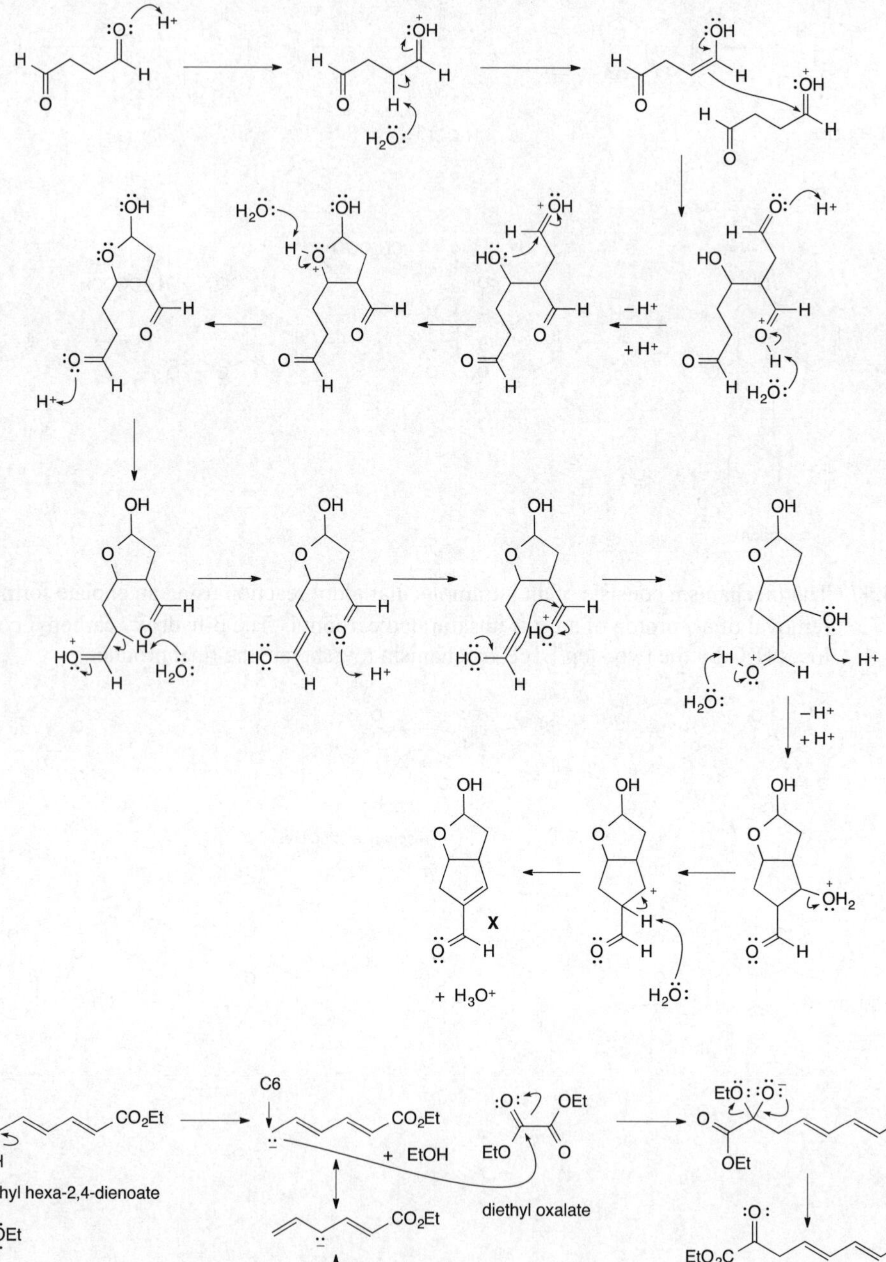

24.59

a.

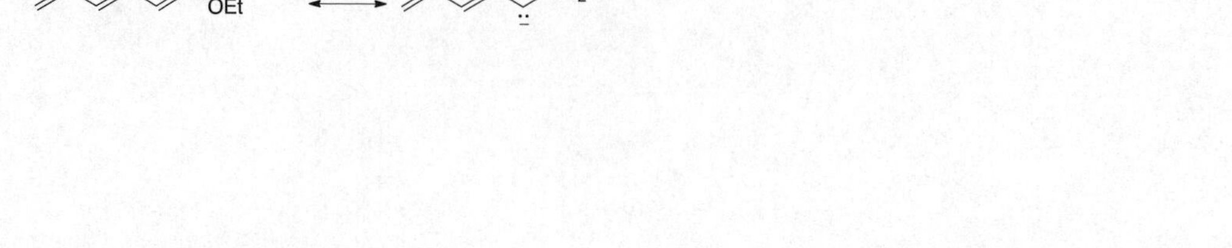
C6
ethyl hexa-2,4-dienoate
diethyl oxalate
+ EtOH

Carbonyl Condensation Reactions 24–23

b. The protons on C6 are more acidic than other sp^3 hybridized C–H bonds because a highly resonance-stabilized carbanion is formed when a proton is removed. One resonance structure places a negative charge on the carbonyl O atom. This makes the protons on C6 similar in acidity to the α H's to a carbonyl.

c. This is a crossed Claisen because it involves the enolate of a conjugated ester reacting with the carbonyl group of a second ester.

24.60

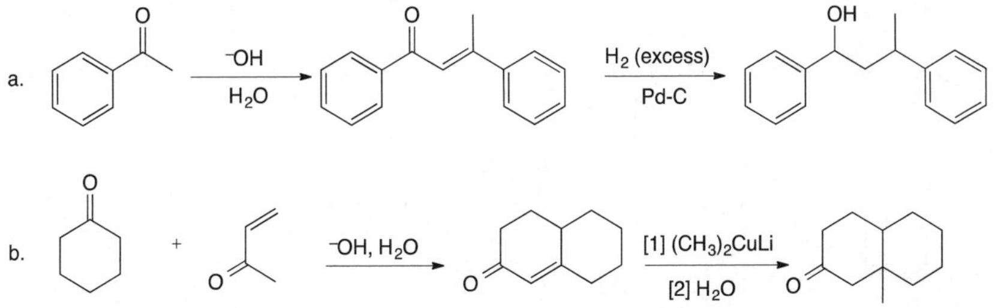

24.61

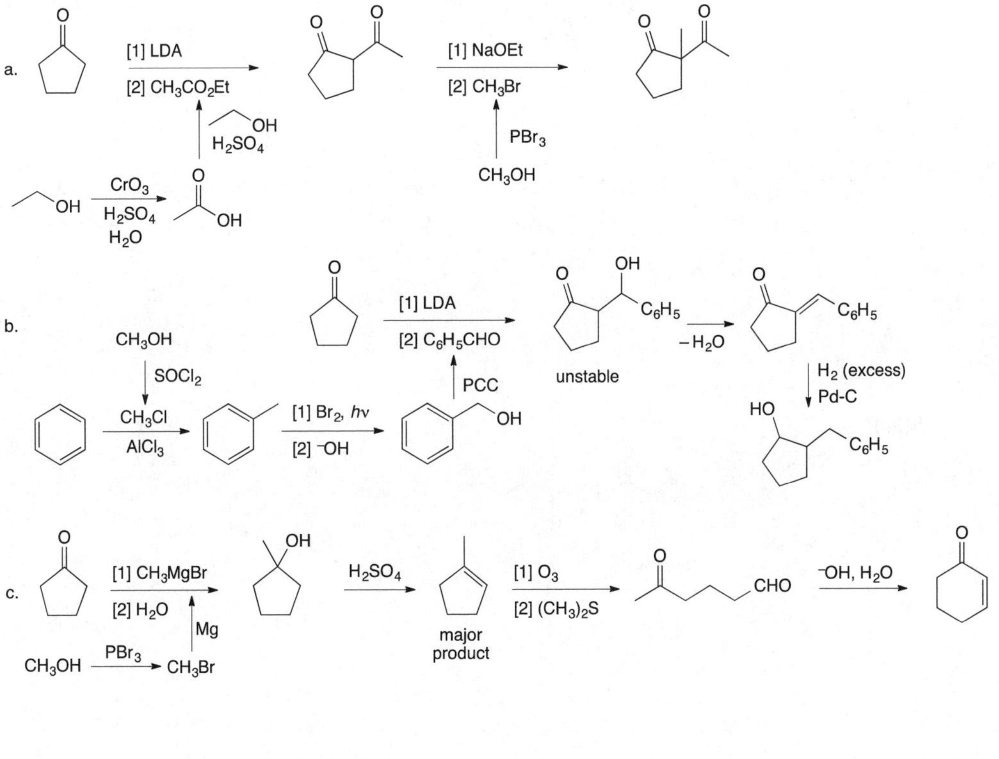

Chapter 24–24

24.62

a.

b.

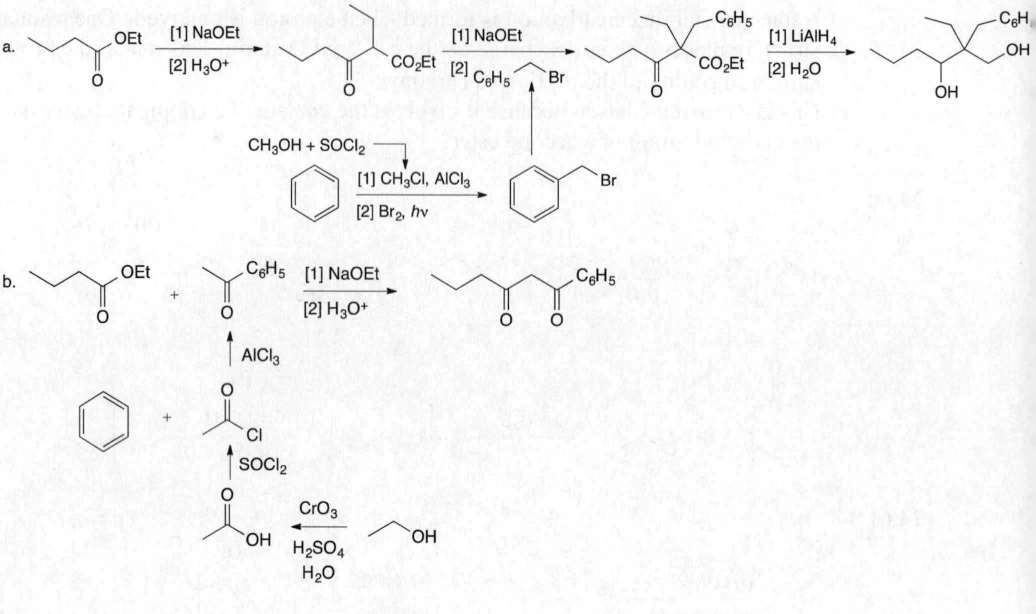

24.63

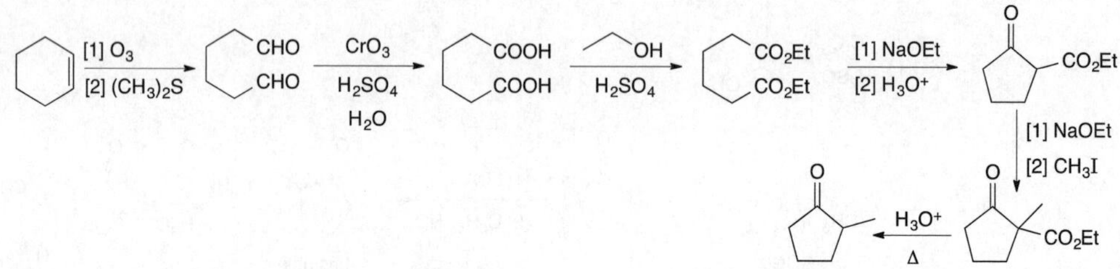

24.64

a.

octinoxate

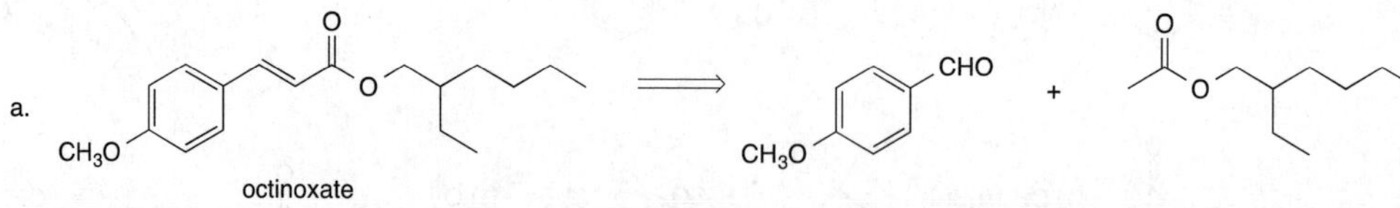

b.

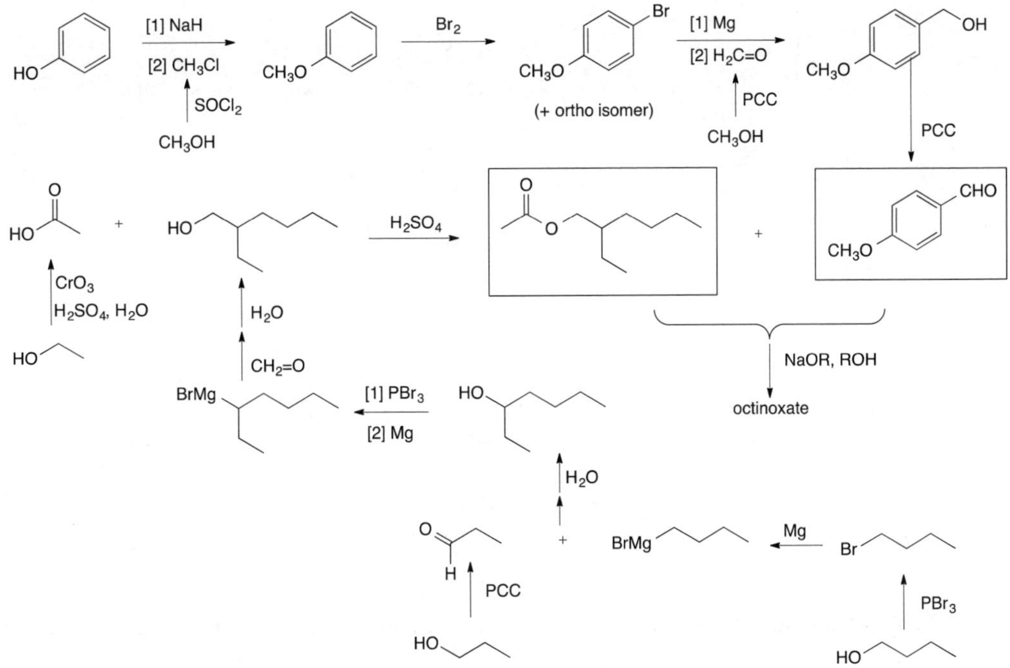

[1] NaH
[2] CH$_3$Cl

SOCl$_2$

CH$_3$OH

Br$_2$

(+ ortho isomer)

[1] Mg
[2] H$_2$C=O

PCC

CH$_3$OH

PCC

NaOR, ROH

octinoxate

CrO$_3$
H$_2$SO$_4$, H$_2$O

H$_2$SO$_4$

H$_2$O

CH$_2$=O

[1] PBr$_3$
[2] Mg

H$_2$O

H$_2$O

PCC

Mg

PBr$_3$

24.65

a. EtO

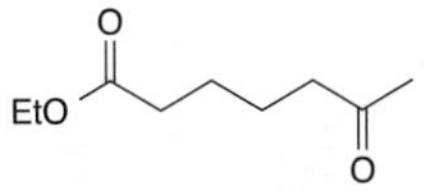

b. + EtO CH$_3$

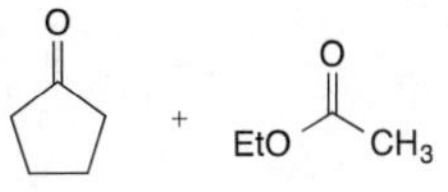

c. [1] NaOEt
   [2] CH$_3$I

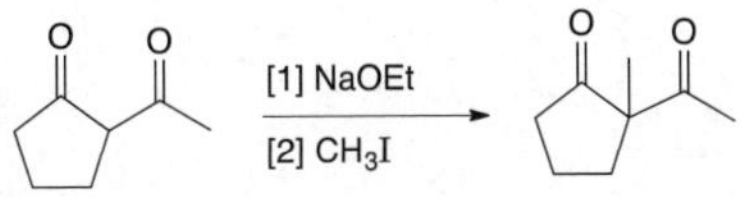

d. + base

   base

Two products
are possible.

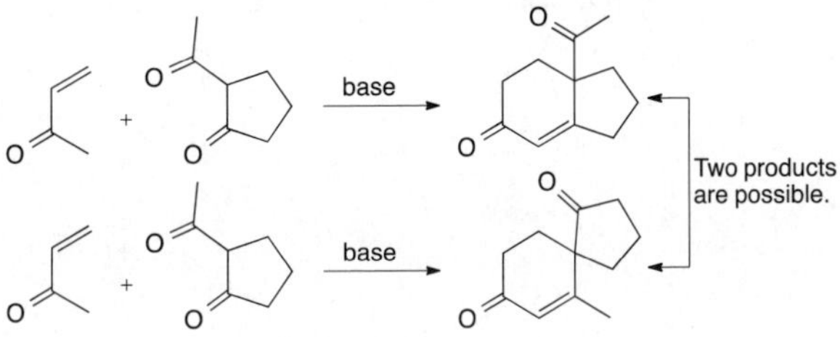

e.

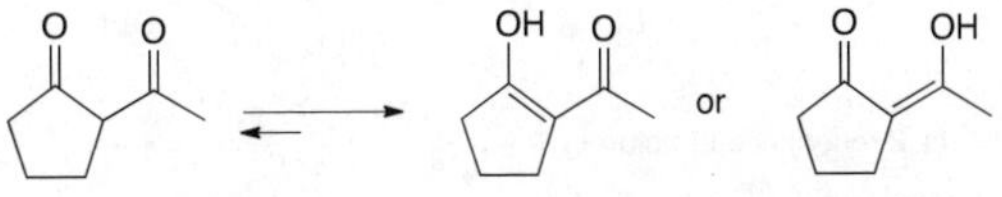

   or

conjugated tautomers favored

Chapter 24–26

24.66

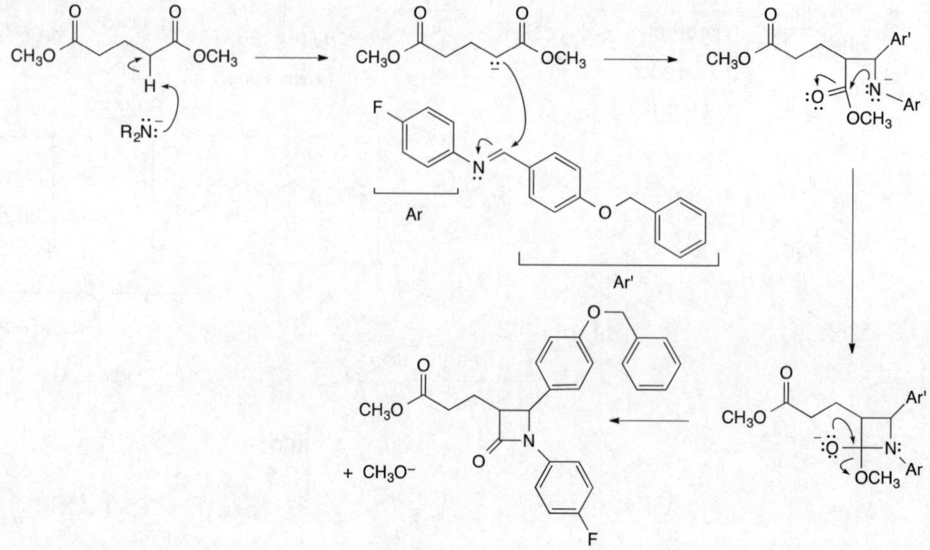

24.67 Rearrangement generates a highly resonance-stabilized enolate between two carbonyl groups.

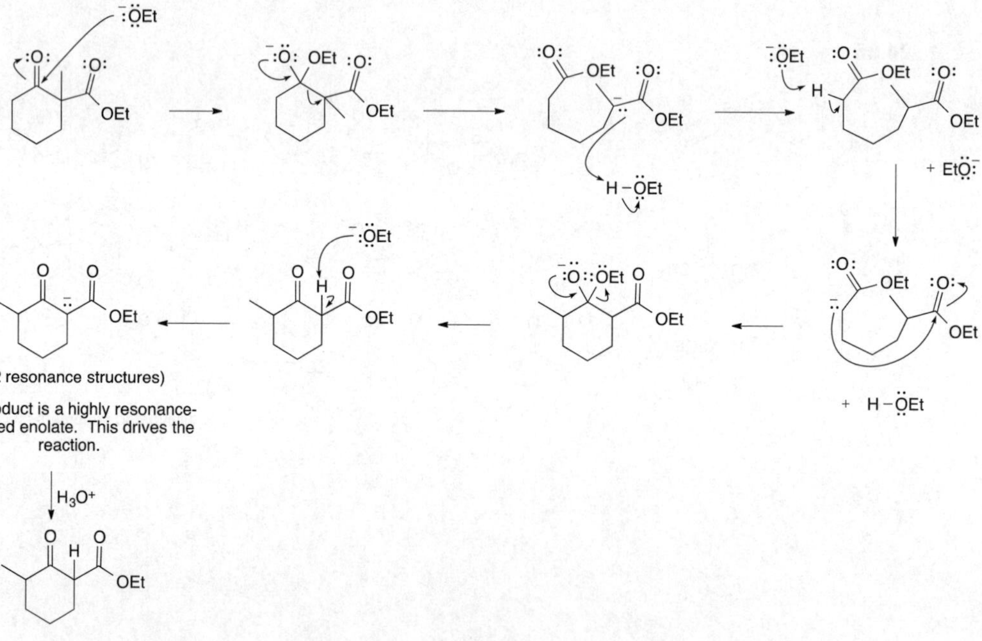

(+ 2 resonance structures)

This product is a highly resonance-stabilized enolate. This drives the reaction.

Carbonyl Condensation Reactions 24–27

24.68 All enolates have a second resonance structure with a negative charge on O.

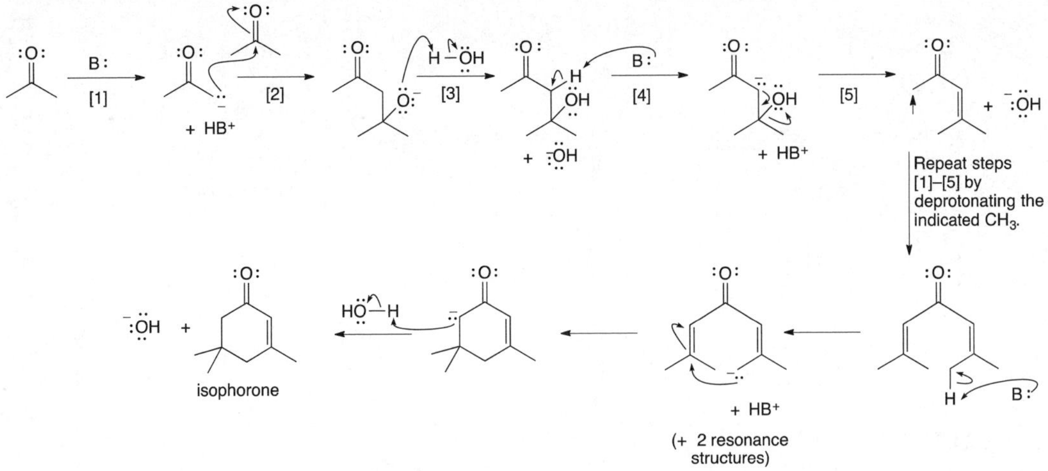

(reaction mechanism scheme)

Repeat steps [1]–[5] by deprotonating the indicated CH$_3$.

isophorone

(+ 2 resonance structures)

24.69

(reaction mechanism scheme)

1,4-addition

proton transfer

β-alkoxy carbonyl

1,4-dicarbonyl compound

intramolecular aldol

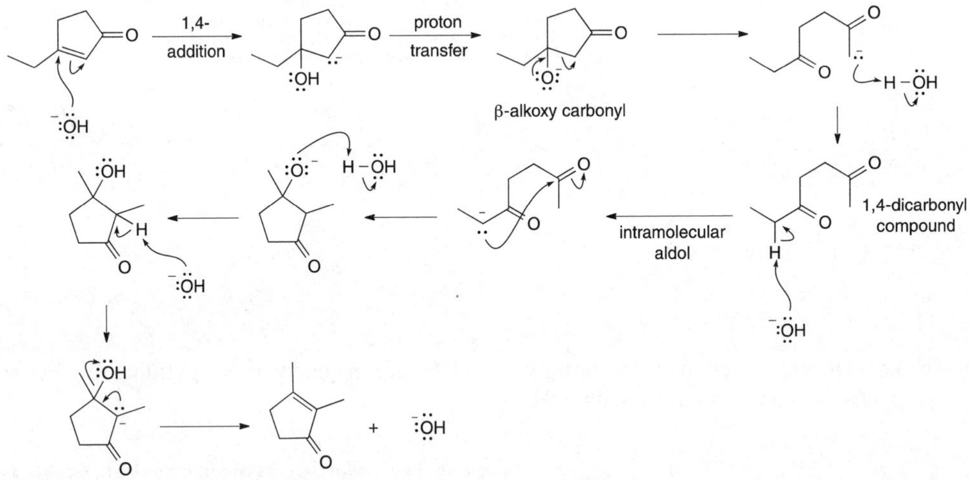

Chapter 24–28

24.70 All enolates have a second resonance structure with a negative charge on O.

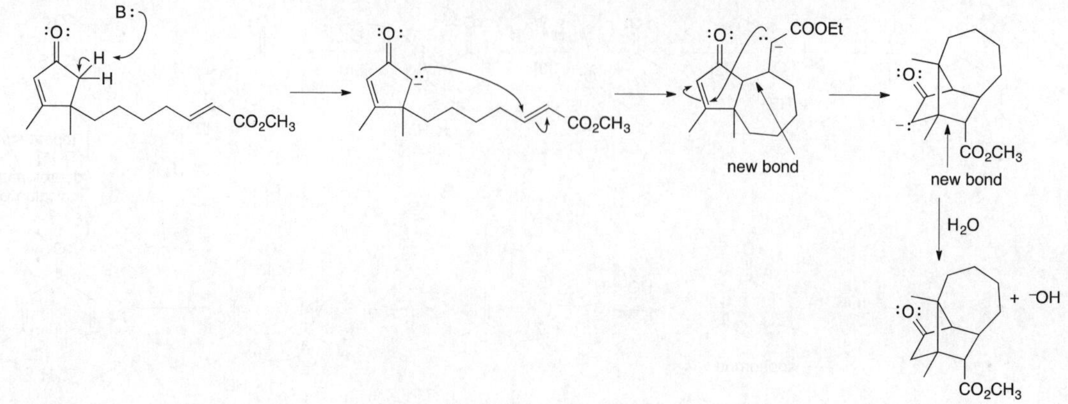

new bond

new bond

H₂O

+ ⁻OH

24.71

a.

(+ other resonance structures)

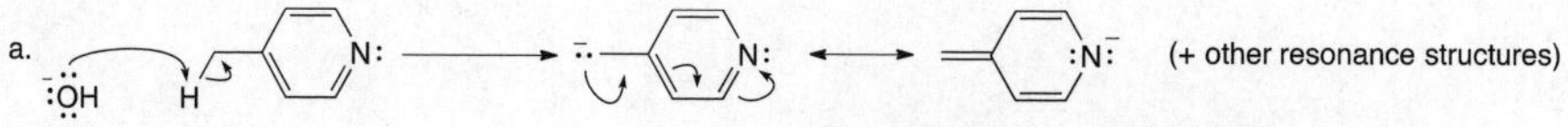

one possible resonance structure
The negative charge is delocalized on the electronegative N
atom. This factor is what makes the CH₃ group bonded to the
pyridine ring more acidic, and allows the condensation to occur.

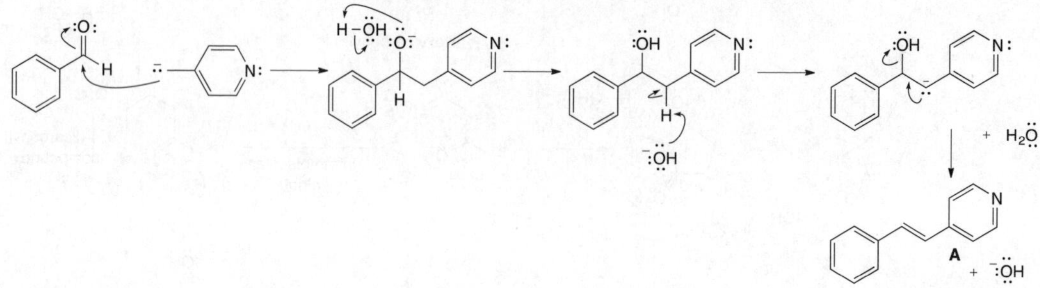

+ H₂O

A

+ ⁻OH

b. The condensation reaction can occur only if the CH₃ group bonded to the pyridine ring has acidic
hydrogens that can be removed with ⁻OH.

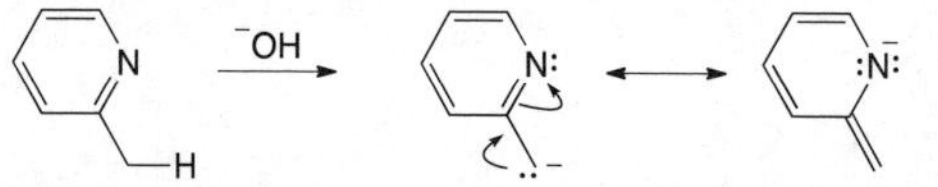

2-methylpyridine

(+ other resonance structures)

Since the negative charge is delocalized on the N, the CH₃ contains
acidic H's and reaction will occur.

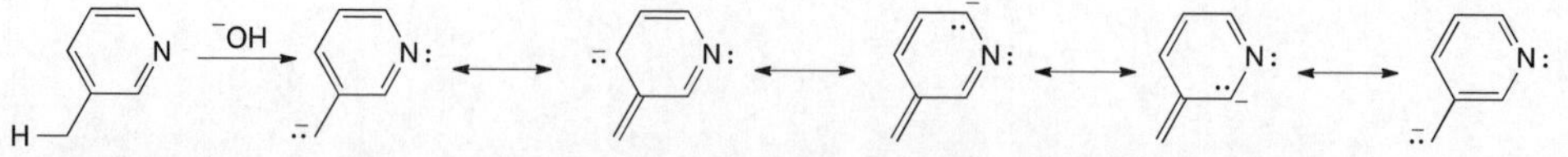

3-methylpyridine

No resonance structure places the negative charge on the N, so the CH₃
is not acidic and condensation does not occur.

24.72 The reaction takes place in acid, so enols are involved. After the initial condensation reaction, the NH₂ and C=O groups form an imine by an intramolecular reaction.

24.73

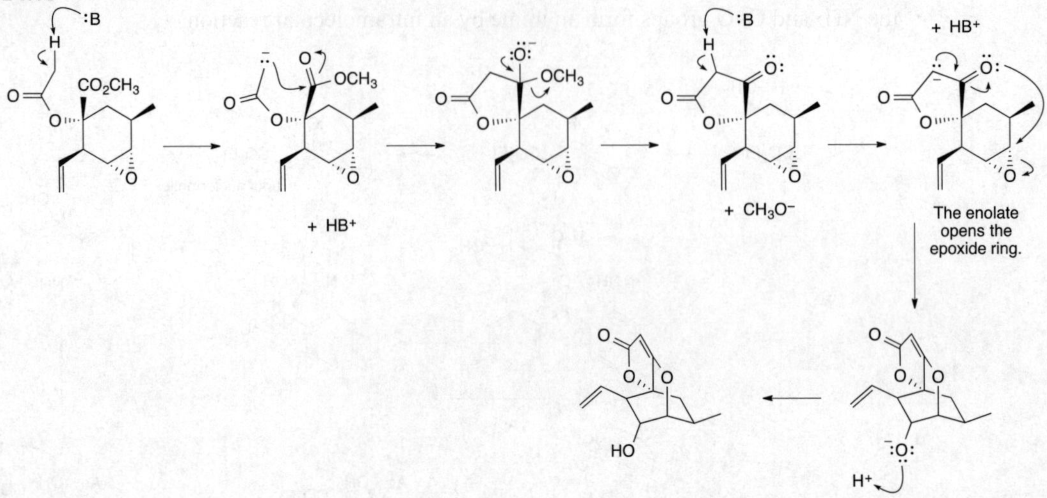
:B
H
CO_2CH_3
O
+ HB^+
OCH_3
:B
+ CH_3O^-
+ HB^+
The enolate opens the epoxide ring.
HO
H+

Chapter 25 Amines

Chapter Review

General facts
- Amines are organic nitrogen compounds having the general structure RNH$_2$, R$_2$NH, or R$_3$N, with a lone pair of electrons on N (25.1).
- Amines are named using the suffix -*amine* (25.3).
- All amines have polar C–N bonds. Primary (1°) and 2° amines have polar N–H bonds and are capable of intermolecular hydrogen bonding (25.4).
- The lone pair on N makes amines strong organic bases and nucleophiles (25.8).

Summary of spectroscopic absorptions (25.5)

Mass spectra	Molecular ion	Amines with an odd number of N atoms give an odd molecular ion.
IR absorptions	N–H	3300–3500 cm^{-1} (two peaks for RNH$_2$, one peak for R$_2$NH)
^{1}H NMR absorptions	NH	0.5–5 ppm (no splitting with adjacent protons)
	CH–N	2.3–3.0 ppm (deshielded Csp^3–H)
^{13}C NMR absorption	C–N	30–50 ppm

Comparing the basicity of amines and other compounds (25.10)
- Alkylamines (RNH$_2$, R$_2$NH, and R$_3$N) are more basic than NH$_3$ because of the electron-donating R groups (25.10A).
- Alkylamines (RNH$_2$) are more basic than arylamines (C$_6$H$_5$NH$_2$), which have a delocalized lone pair from the N atom (25.10B).
- Arylamines with electron-donor groups are more basic than arylamines with electron-withdrawing groups (25.10B).
- Alkylamines (RNH$_2$) are more basic than amides (RCONH$_2$), which have a delocalized lone pair from the N atom (25.10C).
- Aromatic heterocycles with a localized electron pair on N are more basic than those with a delocalized lone pair from the N atom (25.10D).
- Alkylamines with a lone pair in an sp^3 hybrid orbital are more basic than those with a lone pair in an sp^2 hybrid orbital (25.10E).

Preparation of amines (25.7)
[1] Direct nucleophilic substitution with NH$_3$ and amines (25.7A)

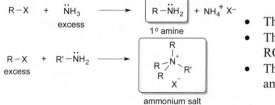

- The mechanism is S$_N$2.
- The reaction works best for CH$_3$X or RCH$_2$X.
- The reaction works best to prepare 1° amines and ammonium salts.

Chapter 25–2

[2] Gabriel synthesis (25.7A)

- The mechanism is S_N2.
- The reaction works best for CH_3X or RCH_2X.
- Only 1° amines can be prepared.

[3] Reduction methods (25.7B)

a. From nitro compounds

$R-NO_2 \xrightarrow{H_2, Pd-C \text{ or } Fe, HCl \text{ or } Sn, HCl} R-NH_2$ (1° amine)

b. From nitriles

$R-C\equiv N \xrightarrow{[1] LiAlH_4 \quad [2] H_2O} R-CH_2-NH_2$ (1° amine)

c. From amides

amide $\xrightarrow{[1] LiAlH_4 \quad [2] H_2O}$ 1°, 2°, and 3° amines

R' = H or alkyl

[4] Reductive amination (25.7C)

ketone/aldehyde + $R_2''NH \xrightarrow{NaBH_3CN}$ 1°, 2°, and 3° amines

R', R" = H or alkyl

- Reductive amination adds one alkyl group (from an aldehyde or ketone) to a nitrogen nucleophile.
- Primary (1°), 2°, and 3° amines can be prepared.

Reactions of amines

[1] Reaction as a base (25.9)

$R-\ddot{N}H_2 + H-A \rightleftharpoons R-\overset{+}{N}H_3 + :A^-$

[2] Nucleophilic addition to aldehydes and ketones (25.11)

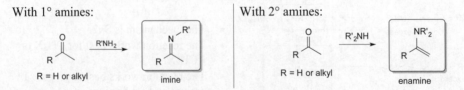

With 1° amines: ketone + R'NH_2 → imine

With 2° amines: ketone + R'_2NH → enamine

R = H or alkyl

[3] Nucleophilic substitution with acid chlorides and anhydrides (25.11)

$$R-C(=O)-Z + R'_2NH \text{ (2 equiv)} \longrightarrow R-C(=O)-NR'_2$$

Z = Cl or OCOR
R' = H or alkyl

1°, 2°, and 3° amides

[4] Hofmann elimination (25.12)

$$\underset{H}{\overset{}{\text{R}_2\text{CH}-\text{CR}_2-\text{NH}_2}} \xrightarrow[\text{[2] Ag}_2\text{O}]{\text{[1] CH}_3\text{I (excess)}} \text{alkene}$$

[3] Δ

- The less substituted alkene is the major product.

[5] Reaction with nitrous acid (25.13)

With 1° amines:

$$R-NH_2 \xrightarrow[HCl]{NaNO_2} R-\overset{+}{N}\equiv N: \; Cl^-$$

alkyl diazonium salt

With 2° amines:

$$R_2NH \xrightarrow[HCl]{NaNO_2} R_2N-N=O$$

N-nitrosamine

Reactions of diazonium salts
[1] Substitution reactions (25.14)

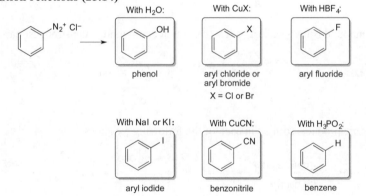

With H₂O: phenol
With CuX: aryl chloride or aryl bromide (X = Cl or Br)
With HBF₄: aryl fluoride
With NaI or KI: aryl iodide
With CuCN: benzonitrile
With H₃PO₂: benzene

[2] Coupling to form azo compounds (25.15)

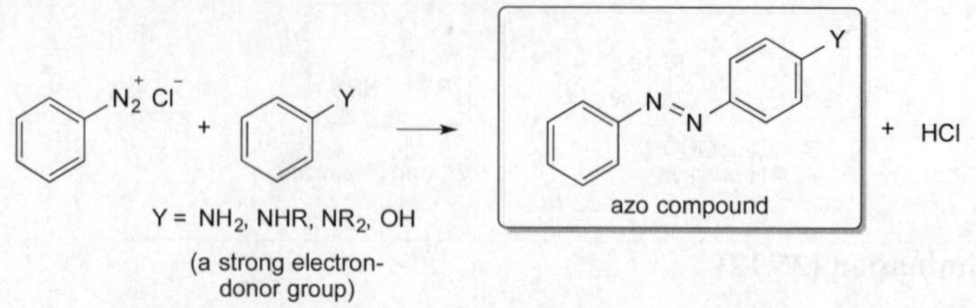

N_2^+ Cl^-
+
Y
N
N
Y
azo compound
+ HCl
Y = NH_2, NHR, NR_2, OH
(a strong electron-donor group)

Practice Test on Chapter Review

1. Give a systematic name for each of the following compounds.

a. 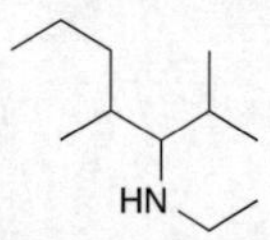

HN

b. 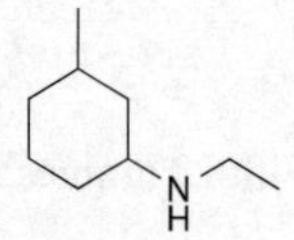

N
H

2. (a) Which compound is the weakest base? (b) Which compound is the strongest base?

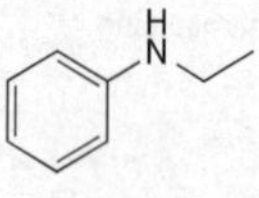

H
N

A

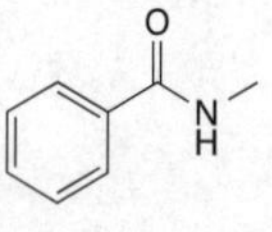

O
N
H

B

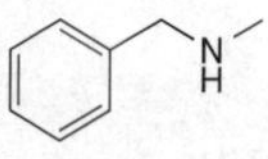

N
H

C

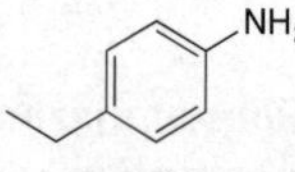

NH_2

D

3. (a) Which compound is the weakest base? (b) Which compound is the strongest base?

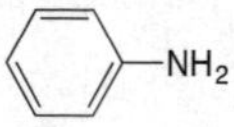

NH_2

A

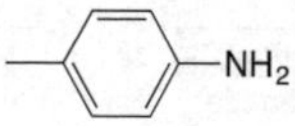

NH_2

B

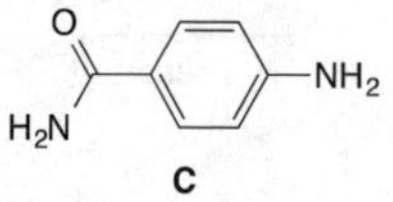

O
H_2N
NH_2

C

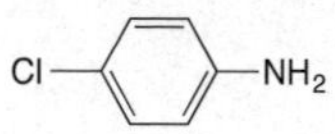

Cl
NH_2

D

4. Draw the organic products formed in each of the following reactions.

a. 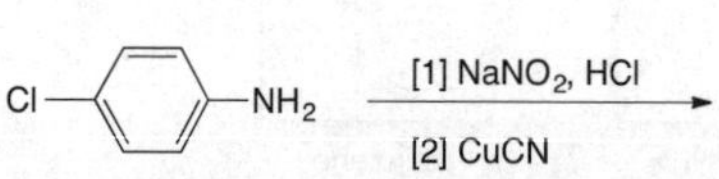

Cl
NH_2
[1] $NaNO_2$, HCl
[2] CuCN

d. 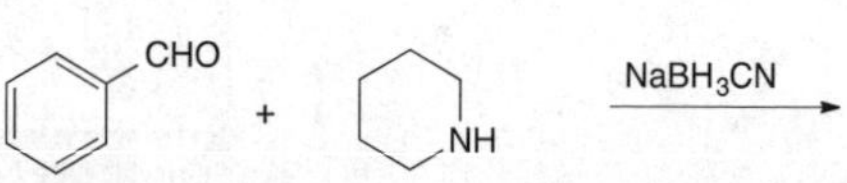

CHO
+
NH
$NaBH_3CN$

b. 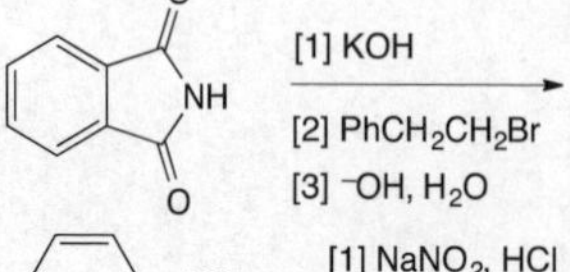

O
NH
O
[1] KOH
[2] $PhCH_2CH_2Br$
[3] ^-OH, H_2O

e. 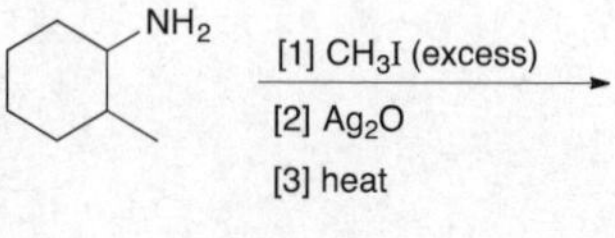

NH_2
[1] CH_3I (excess)
[2] Ag_2O
[3] heat

c. 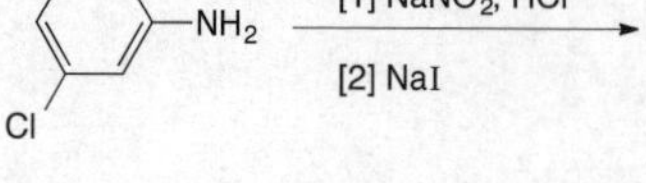

NH_2
Cl
[1] $NaNO_2$, HCl
[2] NaI

Amines 25–5

5. Draw the products formed when the given amine is treated with [1] CH₃I (excess); [2] Ag₂O; [3] Δ, and indicate the major product. You need not consider any stereoisomers formed in the reaction.

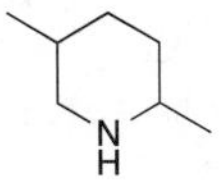

6. What organic starting materials are needed to synthesize **B** by reductive amination?

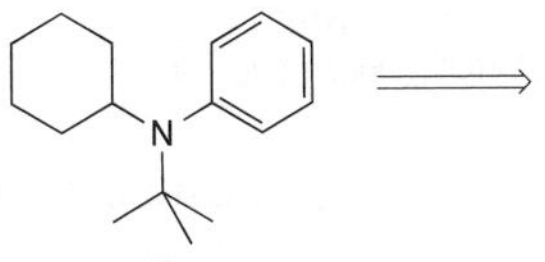

B

Answers to Practice Test

1.a. *N*-ethyl-2,4-dimethyl-heptan-3-amine
 b. *N*-ethyl-3-methylcyclohexanamine

2.a. **B**
 b. **C**

3.a. **C**
 b. **B**

4.
 a. 
 b. 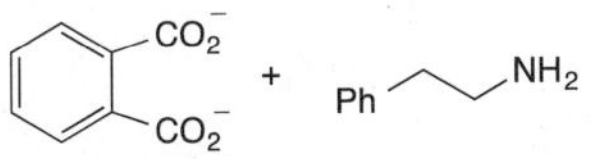
 c. 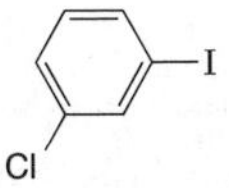
 d. 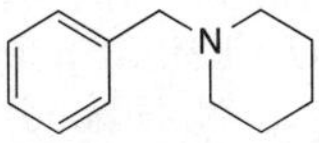
 e.

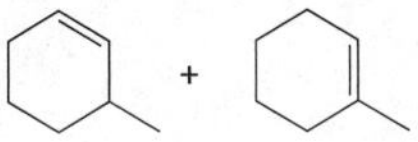

5.

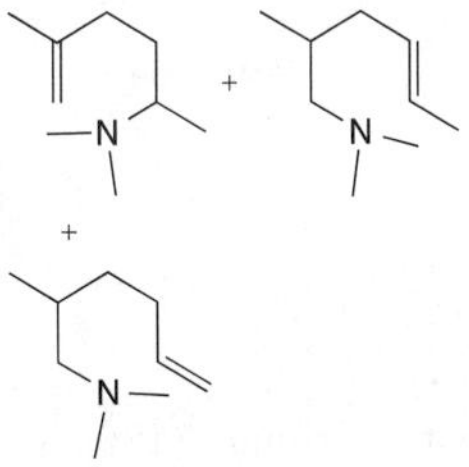

major

6.

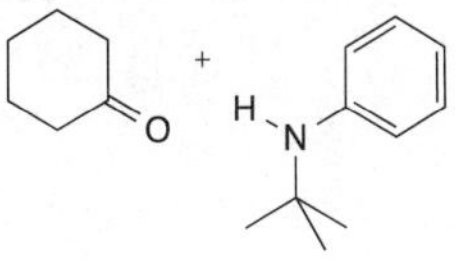

Answers to Problems

25.1 The N atom of an ammonium salt is a stereogenic center when the N is surrounded by four different groups. All stereogenic centers are circled.

a.

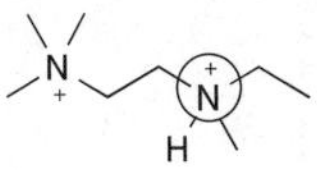

b.

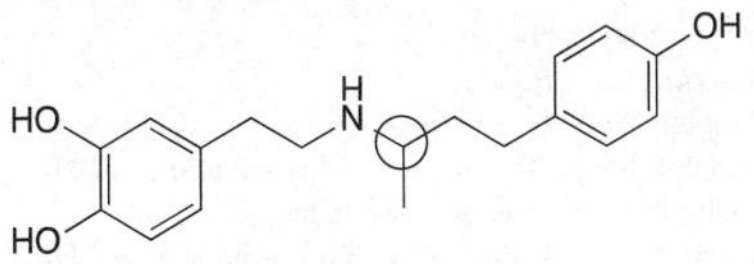

Chapter 25–6

25.2

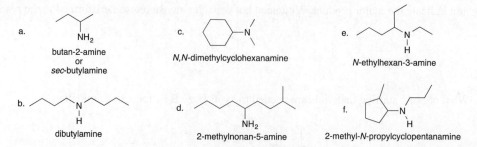

25.3 An **NH₂** group named as a substituent is called an **amino group**.

a. 2,4-dimethylhexan-3-amine   c. N-isopropyl-p-nitroaniline   e. N,N-dimethylethanamine   g. N-methylaniline

b. N-methylpentan-1-amine   d. N-methylpiperidine   f. 2-aminocyclohexanone   h. m-ethylaniline

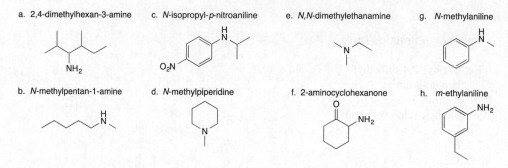

25.4 Primary (1°) and 2° amines have higher bp's than similar compounds (like ethers) incapable of hydrogen bonding, but lower bp's than alcohols, which have stronger intermolecular hydrogen bonds. Tertiary amines (3°) have lower boiling points than 1° and 2° amines of comparable molecular weight because they have no N–H bonds with which to form hydrogen bonds.

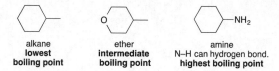

alkane
lowest boiling point

ether
intermediate boiling point

amine
N–H can hydrogen bond.
highest boiling point

25.5 **The NH signal occurs between 0.5 and 5.0 ppm.** The protons on the carbon bonded to the amine nitrogen are deshielded and typically absorb at 2.3–3.0 ppm. The NH protons are not split.

molecular formula C₆H₁₅N
¹H NMR absorptions (ppm):
  0.9 (singlet, 1 H) ⟶ NH
  1.10 (triplet, 3 H) ⟶ CH₃ adjacent to CH₂
  1.15 (singlet, 9 H) ⟶ (CH₃)₃C
  2.6 (quartet, 2 H) ⟶ CH₂ adjacent to CH₃

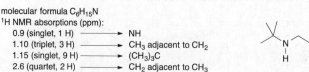

Amines 25–7

25.6 The atoms of 2-phenylethanamine are in bold.

a.

LSD
lysergic acid diethyl amide

b.

codeine

25.7 S_N2 reaction of an alkyl halide with NH₃ or an amine forms an amine or an ammonium salt.

a.

b.

25.8 The Gabriel synthesis converts an alkyl halide into a 1° amine by a two-step process: nucleophilic substitution followed by hydrolysis.

a.

b.

c.

25.9 The Gabriel synthesis prepares 1° amines from alkyl halides. Because the reaction proceeds by an S_N2 mechanism, the halide must be CH₃ or 1°, and X can't be bonded to an *sp²* hybridized C.

a.
aromatic
An S_N2 does not occur on an aryl halide. cannot be made by Gabriel synthesis

b.
can be made by Gabriel synthesis

c.
2° amine
cannot be made by Gabriel synthesis

d.
N on 3° C
An S_N2 does not occur on a 3° RX. cannot be made by Gabriel synthesis

25.10 **Nitriles are reduced to 1° amines with LiAlH₄. Nitro groups are reduced to 1° amines** using a variety of reducing agents. **Primary (1°), 2°, and 3° amides are reduced to 1°, 2°, and 3° amines** respectively, using LiAlH₄.

a.

b.

Chapter 25–8

c. [structure] ~~~NH₂ ⟹ ~~~NO₂   ~~~CN   ~~~C(=O)NH₂

25.11 Primary (1°), 2°, and 3° amides are reduced to 1°, 2°, and 3° amines respectively, using LiAlH₄.

a. [benzamide structure] →

b. [N,N-diethyl butanamide] →

c. [amide structure] →

25.12 Only amines with a CH₂ or CH₃ bonded to the N can be made by reduction of an amide.

a. [structure with NH₂]
N bonded to benzene cannot be made by reduction of an amide

b. [structure with NH₂]
N bonded to CH₂ can be made by reduction of an amide

c. [structure with NH₂]
N bonded to a 3° C cannot be made by reduction of an amide

d. [structure]
N with 2° C on both sides cannot be made by reduction of an amide

25.13 Reductive amination is a two-step method that converts aldehydes and ketones into 1°, 2°, and 3° amines. Reductive amination replaces a C=O by a C–H and C–N bond.

a. [benzaldehyde] $\xrightarrow[\text{NaBH}_3\text{CN}]{\text{CH}_3\text{NH}_2}$ [product]

b. [ketone] $\xrightarrow[\text{NaBH}_3\text{CN}]{\text{NH}_3}$ [product with NH₂]

c. [cyclohexanone] $\xrightarrow{\text{NaBH}_3\text{CN}}$ [product]

d. [acetone] + [amine with NH₂] $\xrightarrow{\text{NaBH}_3\text{CN}}$ [product HN]

25.14 Reductive amination occurs using the ketone in E and the amine in D.

E   +   D   →   enalapril

25.15

a. [adamantane structure with NH₂] ⟹ [adamantane ketone structure] + NH₃

Amines 25–9

b.

25.16

a.

phentermine

Only amines that have a C bonded to a H and N atom can be made by reductive amination; that is, an amine must have the following structural feature:

In phentermine, the C bonded to N is not bonded to a H, so it cannot be made by reductive amination.

b. systematic name: 2-methyl-1-phenylpropan-2-amine

25.17 The pK_a of many protonated amines is 10–11, so the pK_a of the starting acid must be **less than 10** for equilibrium to favor the products. Amines are thus readily protonated by strong inorganic acids (e.g., HCl and H_2SO_4) and by carboxylic acids.

a.

b.

c.

25.18 An amine can be separated from other organic compounds by converting it to a water-soluble ammonium salt by an acid–base reaction. In each case, the extraction procedure would employ the following steps:
- Dissolve the amine and either **X** or **Y** in CH_2Cl_2.
- Add a solution of 10% HCl. The amine will be protonated and dissolve in the aqueous layer, while **X** or **Y** will remain in the organic layer as a neutral compound.
- Separate the layers.

a.

- **soluble in H_2O**
- **insoluble in CH_2Cl_2**

- insoluble in H_2O
- soluble in CH_2Cl_2

Chapter 25–10

b. [structure] and [structure labeled Y] $\xrightarrow{\text{H}-\text{Cl}}$ [ammonium chloride structure] + [ether structure labeled Y]

- soluble in H_2O
- insoluble in CH_2Cl_2

- insoluble in H_2O
- soluble in CH_2Cl_2

25.19 Primary (1°), 2°, and 3° alkylamines are more basic than NH_3 because of the electron-donating inductive effect of the R groups.

a. [structure] and NH_3

2° alkylamine
CH_3 groups are electron donating.
stronger base

b. [structure]

1° alkylamine
stronger base

[structure with Cl]

1° alkylamine
Cl is electron withdrawing.
weaker base

25.20 Arylamines are less basic than alkylamines because the electron pair on N is delocalized. Electron-donor groups add electron density to the benzene ring making the arylamine more basic than aniline. Electron-withdrawing groups remove electron density from the benzene ring, making the arylamine less basic than aniline.

a. [structure with CH_3O and carbonyl, NH_2] [structure NH_2] [structure CH_3O, NH_2]

electron-
withdrawing group
least basic

arylamine
**intermediate
basicity**

electron-
donating group
most basic

b. [structure O_2N, NH_2] [structure NH_2] [cyclohexyl NH_2]

electron-
withdrawing group
least basic

arylamine
**intermediate
basicity**

alkylamine
most basic

25.21 Amides are much less basic than amines because the electron pair on N is highly delocalized.

[amide structure NH_2] [arylamine structure NH_2] [alkylamine structure NH_2]

amide
least basic

arylamine
intermediate basicity

alkylamine
most basic

25.22

a. Electron pair on N occupies an sp^2 hybrid orbital.

sp^2 hybridized
more basic

This N is also sp^2 hybridized but the electron pair occupies a p orpital, so it can delocalize onto the aromatic ring. Delocalization makes this N less basic.

[pyridine structure]

DMAP
4-(N,N-dimethylamino)pyridine

b. [nicotine structure]

sp^3 hybridized N
stronger base

nicotine

sp^2 hybridized N
higher percent s-character
weaker base

Amines 25–11

25.23 HCl protonates the more basic N atom.

a.

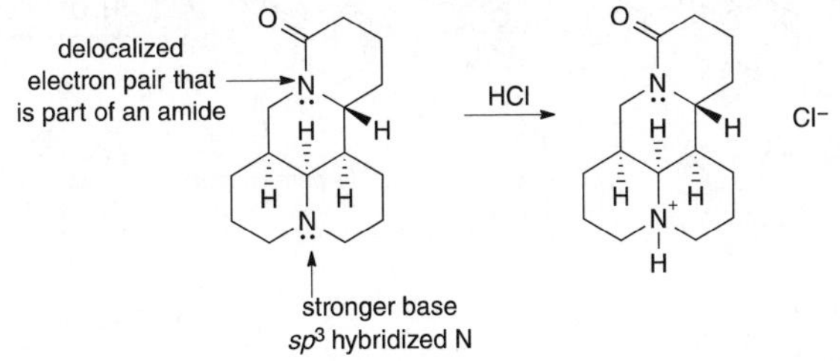

delocalized
electron pair that
is part of an amide

stronger base
sp^3 hybridized N

b.

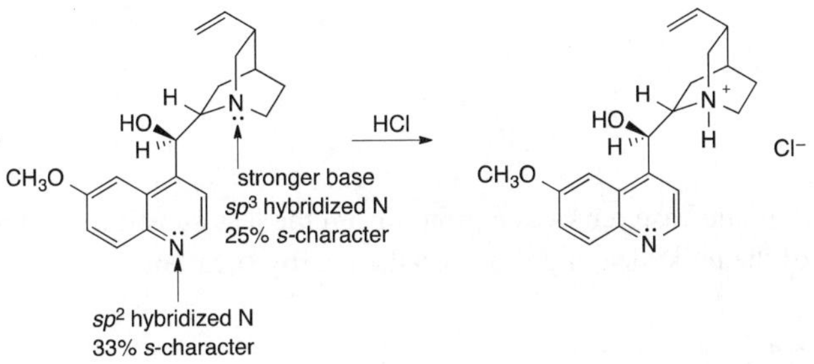

stronger base
sp^3 hybridized N
25% *s*-character

sp^2 hybridized N
33% *s*-character

25.24 Amines attack carbonyl groups to form products of nucleophilic addition or substitution.

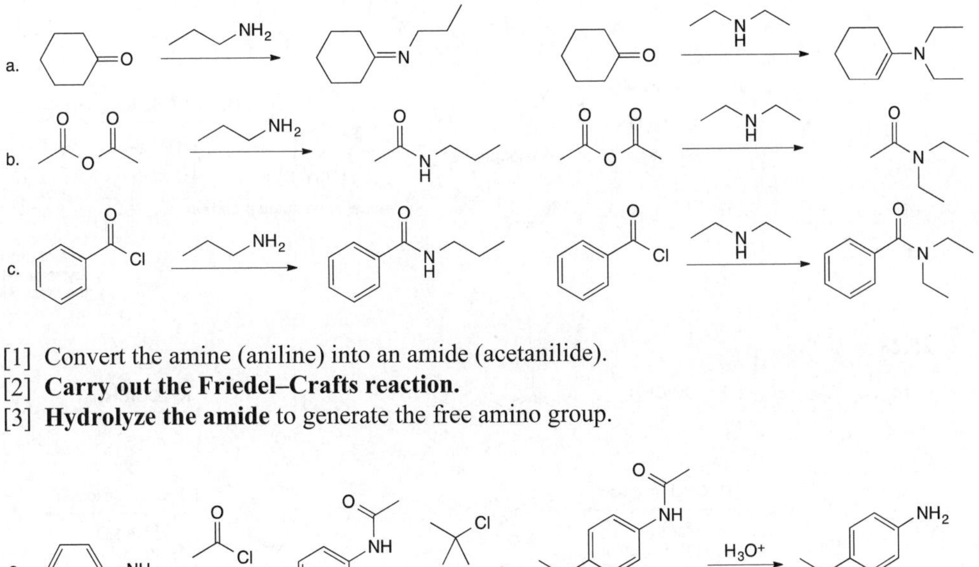

a.

b.

c.

25.25 [1] Convert the amine (aniline) into an amide (acetanilide).
[2] **Carry out the Friedel–Crafts reaction.**
[3] **Hydrolyze the amide** to generate the free amino group.

a.

AlCl$_3$

(+ ortho isomer)

H$_3$O$^+$

Chapter 25–12

b.

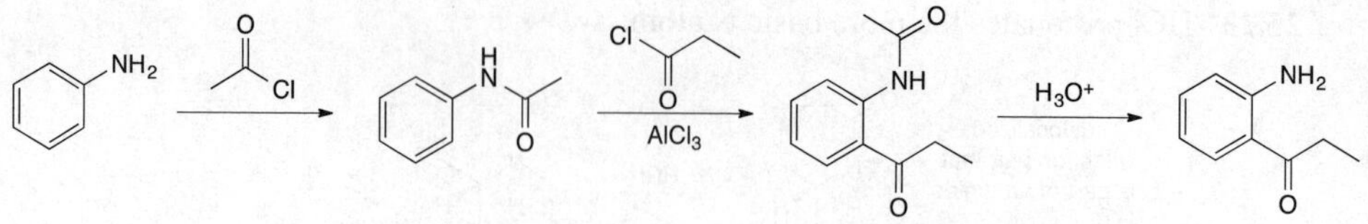

(+ para isomer)

25.26

a. 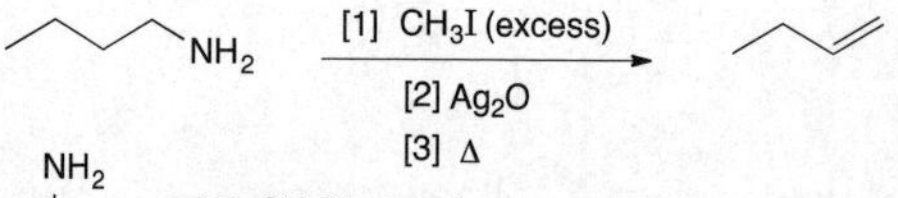 [1] CH₃I (excess) / [2] Ag₂O / [3] Δ

c. 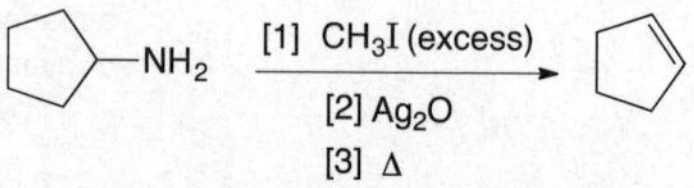 [1] CH₃I (excess) / [2] Ag₂O / [3] Δ

b. 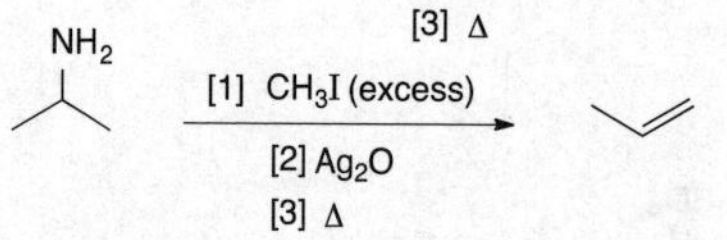 [1] CH₃I (excess) / [2] Ag₂O / [3] Δ

25.27 In a Hofmann elimination, the base removes a proton from the less substituted, more accessible β carbon atom, because of the bulky leaving group on the nearby α carbon.

a. 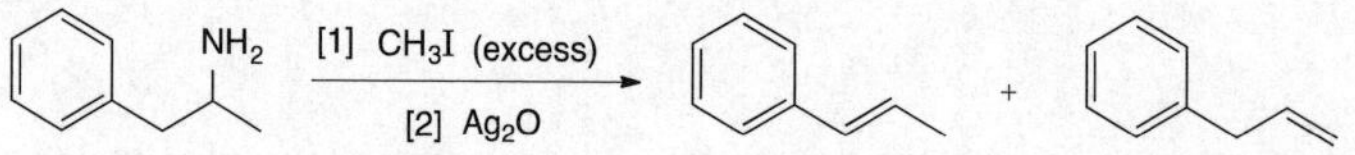 [1] CH₃I (excess) / [2] Ag₂O / [3] Δ

(+ Z isomer)   major product

b. 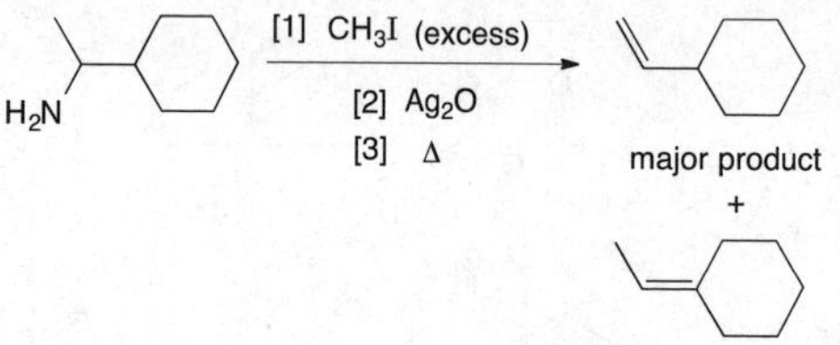 [1] CH₃I (excess) / [2] Ag₂O / [3] Δ

major product

c. 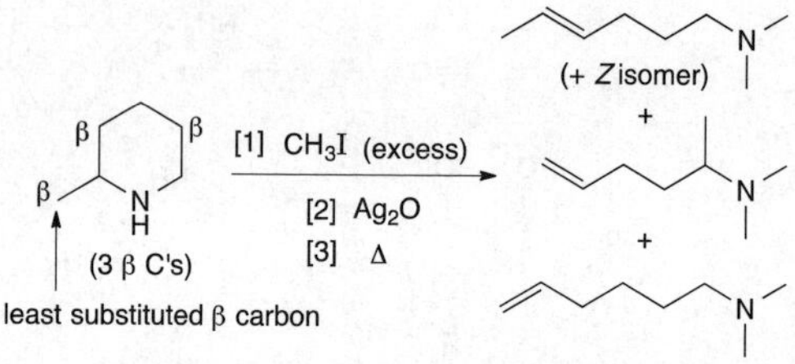 (3 β C's) / least substituted β carbon   [1] CH₃I (excess) / [2] Ag₂O / [3] Δ

(+ Z isomer)

major product, formed by removal of a H from the least substituted β C

25.28

a. 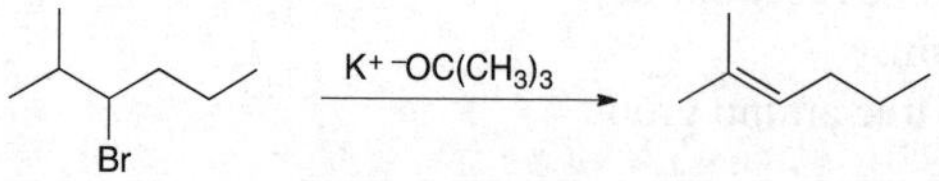 K⁺ ⁻OC(CH₃)₃

c. 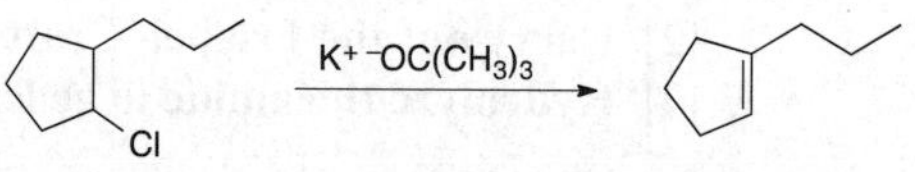 K⁺ ⁻OC(CH₃)₃

b. 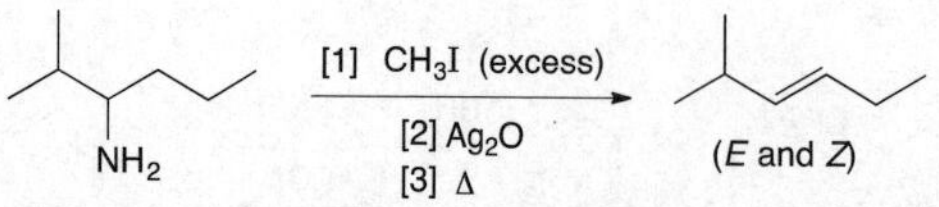 [1] CH₃I (excess) / [2] Ag₂O / [3] Δ

(E and Z)

d. 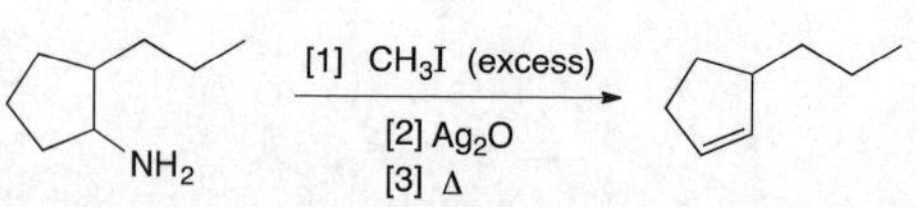 [1] CH₃I (excess) / [2] Ag₂O / [3] Δ

Amines 25–13

25.29

a.

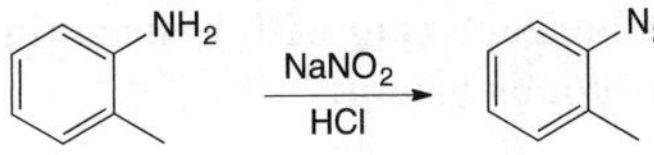

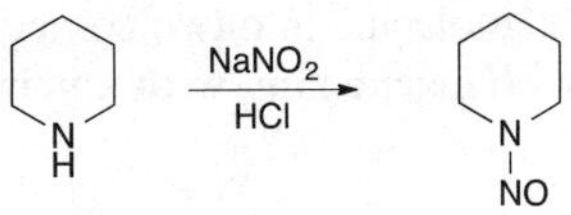

b.

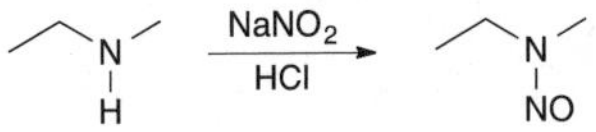

c.

d.

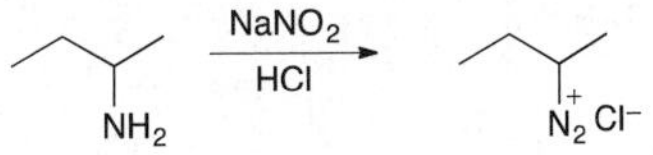

25.30

a.

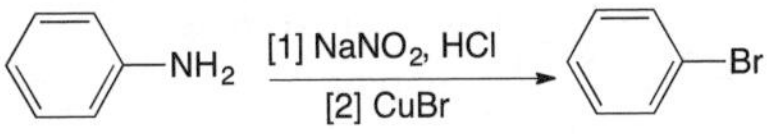

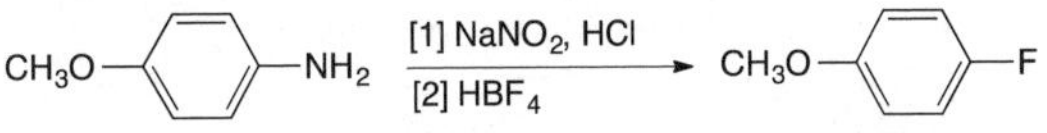

b.

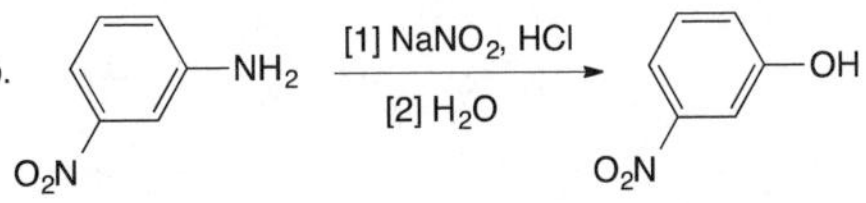

c.

d.

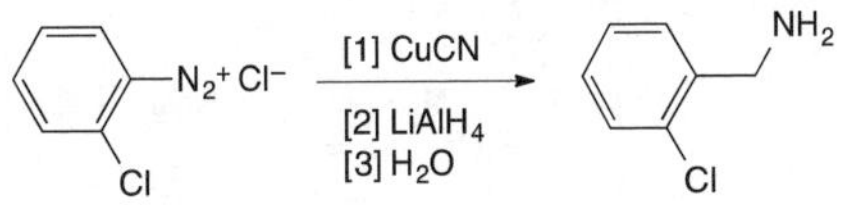

25.31

a.

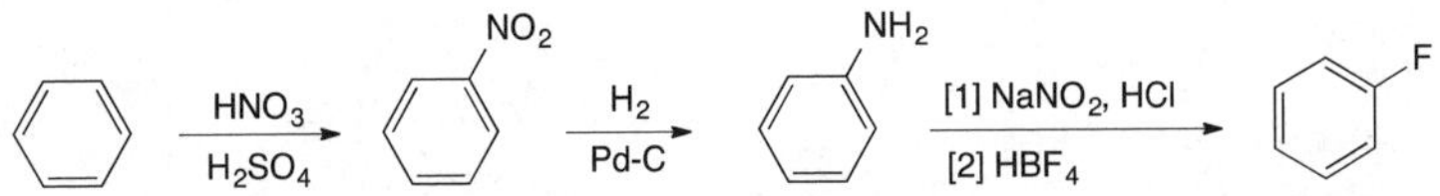

b.

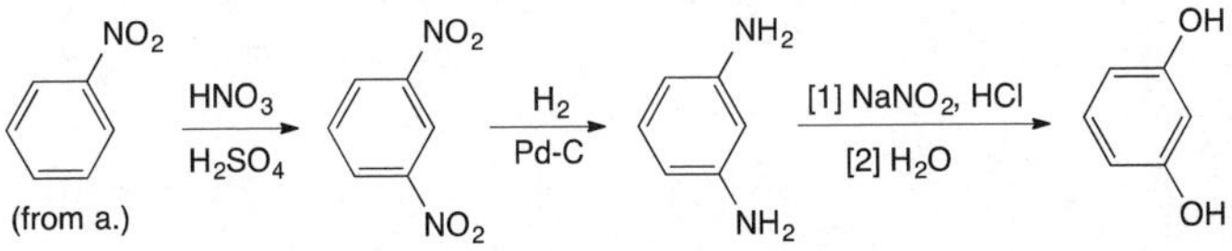

c.

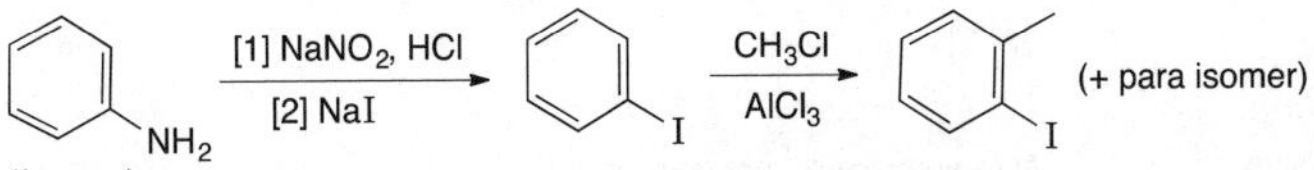

(from a.)

d.

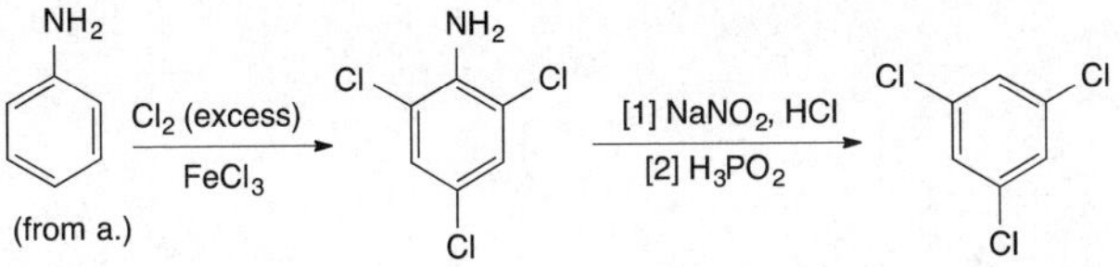

(from a.)

25.32

a.

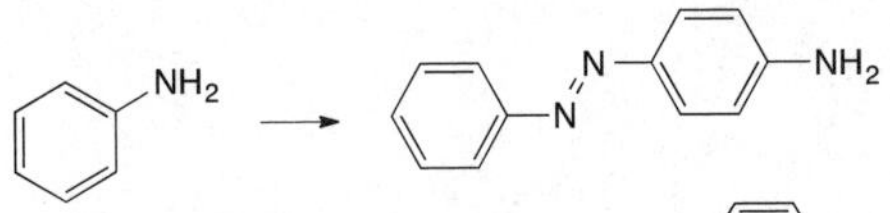

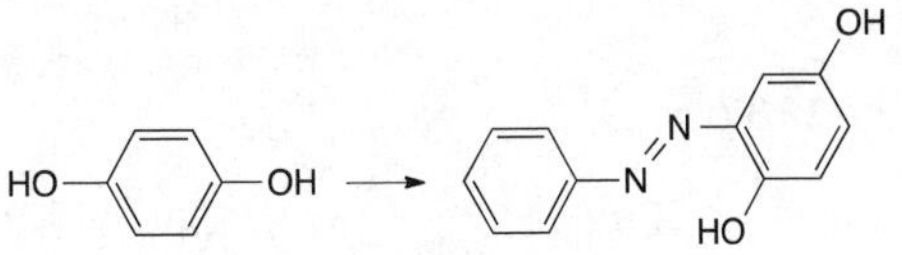

b.

c.

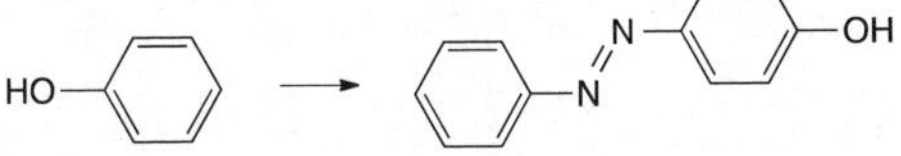

Chapter 25–14

25.33 To determine what starting materials are needed to synthesize a particular azo compound, always divide the molecule into two components: **one has a benzene ring with a diazonium ion, and one has a benzene ring with a very strong electron-donor group.**

a. H_2N- [structure]

b. $HO-$ [structure]

H_2N- [structure] $+$ Cl^- $^+N_2-$ [structure]

$HO-$ [structure] $+$ Cl^- $^+N_2-$ [structure]

25.34

a. [structure] **para red**

b. O_2N- [structure] **alizarine yellow R**

[structure] $+$ $:N\equiv N^+-$ [structure] $-NO_2$ Cl^-

O_2N- [structure] $\overset{+}{N}=N:$ Cl^- $+$ [structure]

25.35

a. [structure] NH_2
4,6-dimethylheptan-1-amine

b. [structure]
N,N-diethylcycloheptanamine

25.36

[structure] **or** [structure]

A
weaker base
(delocalized electron pair on N)

B
stronger base

25.37

a, b. [structure] NH $\xrightarrow{HCl}$ [structure] $^+NH_2$ $+$ Cl^-

varenicline

most basic
only *sp*^{3} hybridized N

25.38

a. 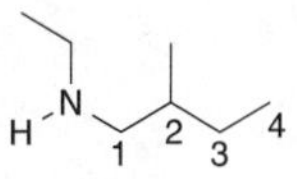 N-ethyl-2-methylbutan-1-amine

b. 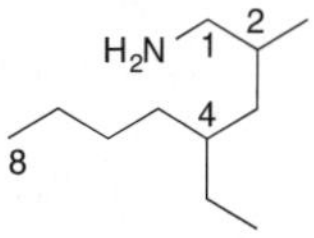 4-ethyl-2-methyloctan-1-amine

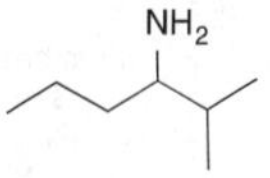

c. 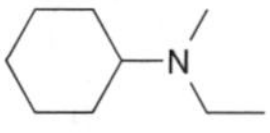 N-ethyl-N-methylcyclohexanamine

d. N-tert-butyl-N-ethylaniline

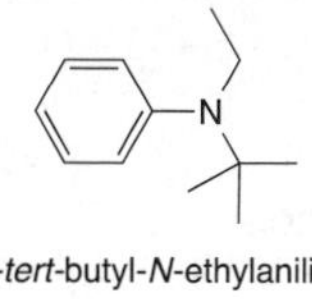

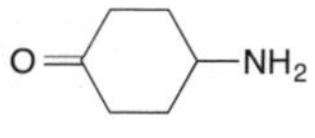

e. 4-aminocyclohexanone

f. 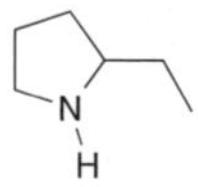 2-ethylpyrrolidine

g. 2-methylhexan-3-amine

h. 3-ethyl-2-methylcyclohexanamine

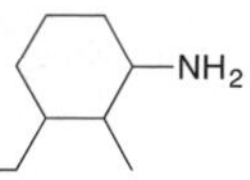

25.39

a. N-isobutylcyclopentanamine

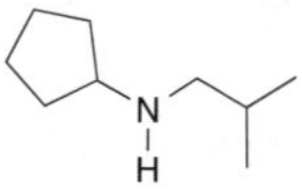

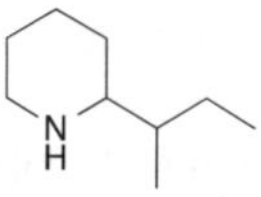

b. tri-tert-butylamine
N[C(CH₃)₃]₃

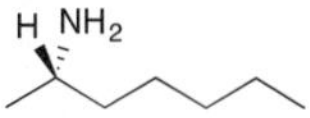

c. N,N-diisopropylaniline

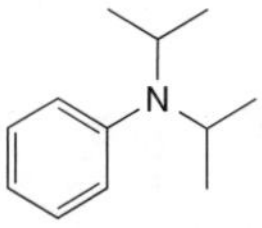

d. N-methylpyrrole

e. N-methylcyclopentanamine

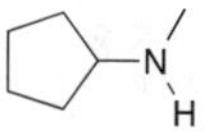

f. 3-methylhexan-2-amine

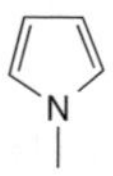

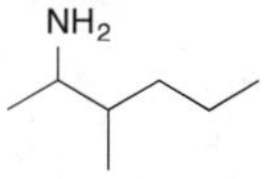

g. 2-sec-butylpiperidine

h. (S)-heptan-2-amine

25.40

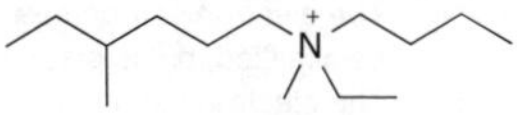

2 stereogenic centers
4 stereoisomers

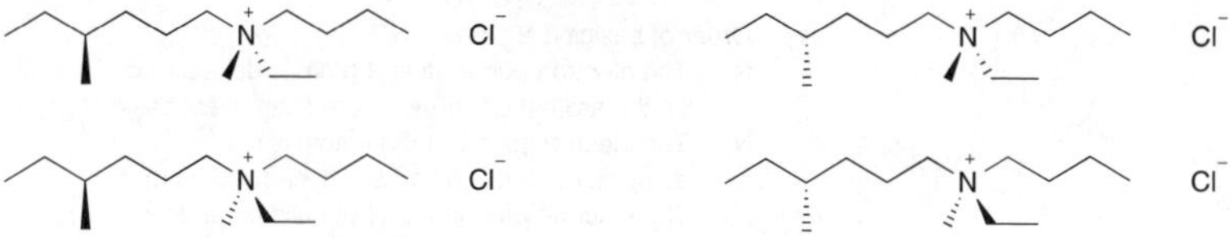

25.41

a. 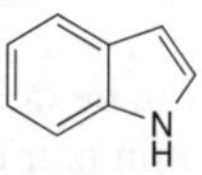
- delocalized electron pair on N — **least basic**
- sp² hybridized N — **intermediate basicity**
- sp³ hybridized N — **most basic**

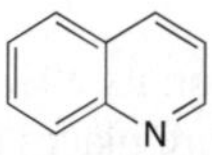

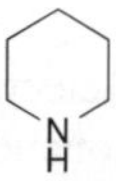

b. 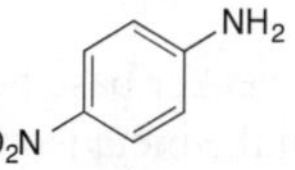
- electron-withdrawing group — **least basic** 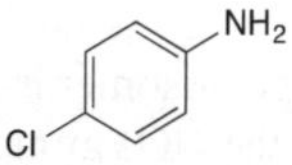
- **intermediate basicity** 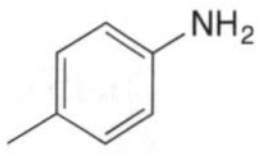
- electron-donating group — **most basic**

Chapter 25–16

25.42 The most basic N atom is protonated on treatment with acid.

a.

most basic

zolpidem

a 3° alkylamine with an sp^3
hybridized N
most basic

CH_3CO_2H →

$CH_3CO_2^-$

b.

aripiprazole

CH_3CO_2H →

$CH_3CO_2^-$

25.43 Pyrimidine has a second electronegative N atom that withdraws electron density from the other N, making the electron pair less available for electron donation.

This N pulls electron density
from the other N atom.

pyrimidine

25.44

a.

$N_b < N_a < N_c$

Order of basicity: $N_b < N_a < N_c$
N_b – The electron pair on this N atom is delocalized
on the O atom; least basic.
N_a – The electron pair on this N atom is not
delocalized, but is on an sp^2 hybridized atom.
N_c – The electron pair on this N atom is on an sp^3
hybridized N; most basic.

b.

$N_b < N_a < N_c$

Order of basicity: $N_b < N_a < N_c$
N_b – The electron pair on this N atom is delocalized
on the aromatic five-membered ring; least basic.
N_a – The electron pair on this N atom is not
delocalized, but is on an sp^2 hybridized atom.
N_c – The electron pair on this N atom is on an sp^3
hybridized N; most basic.

25.45 The para isomer is the weaker base because the electron pair on its NH_2 group can be delocalized onto the NO_2 group. In the meta isomer, no resonance structure places the electron pair on the NO_2 group, and fewer resonance structures can be drawn:

Amines 25–17

meta

para

25.46

A

pK_a of the conjugate acid = 5.2
stronger conjugate acid
weaker base
The electron pair of this arylamine
is delocalized on the benzene ring,
decreasing its basicity.

B

This two-carbon bridge makes it difficult
for the lone pair on N to delocalize on
the aromatic ring.

pK_a of the conjugate acid = 7.29
weaker conjugate acid
stronger base

Resonance structures that place a double bond between the N
atom and the benzene ring are destabilized. Since the electron
pair is more localized on N, compound **B** is more basic.

B

Geometry makes it difficult to have
a double bond here.

25.47

pyrrole
pK_a = 23
stronger acid

weaker conjugate base
The electron pair is delocalized, decreasing the
basicity. The N atom is sp^2 hybridized.

pyrrolidine
pK_a = 44
weaker acid

stronger conjugate base
The electron pair is not
delocalized on the ring. The N
atom is sp^3 hybridized.

25.48

a.

b.

or

c.

Chapter 25–18

25.49 In reductive amination, one alkyl group on N comes from the carbonyl compound. The remainder of the molecule comes from NH_3 or an amine.

a.

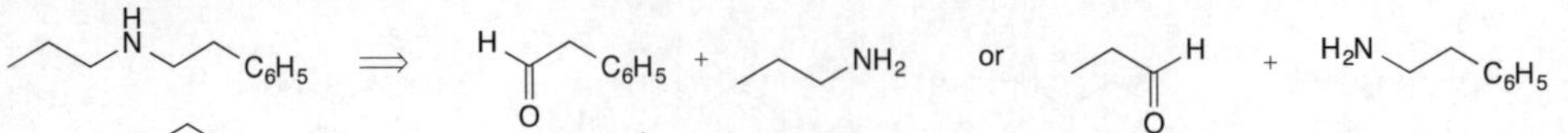

b.

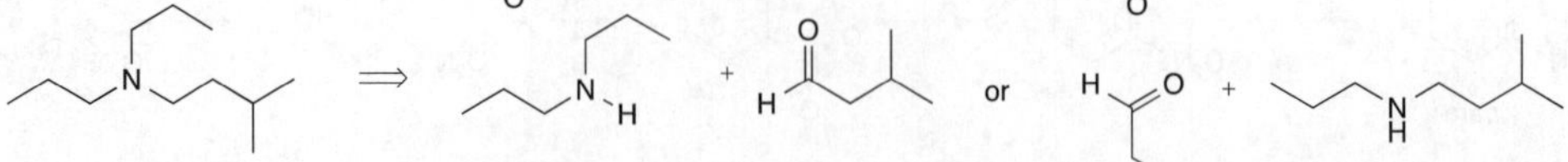

c.

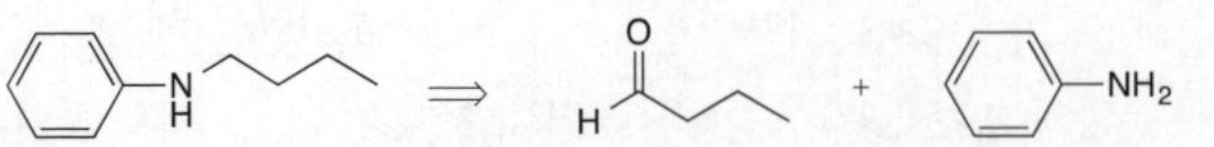

25.50

a.

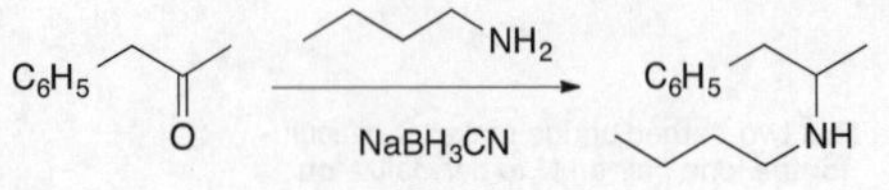

b.

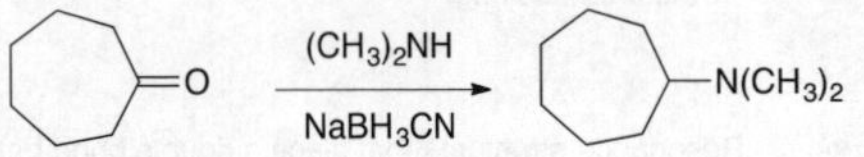

c.

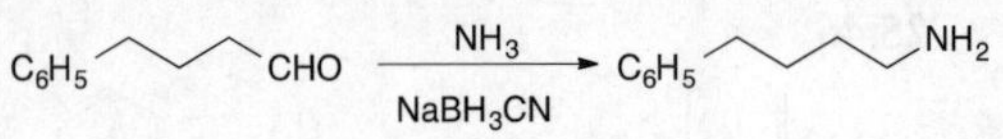

d.

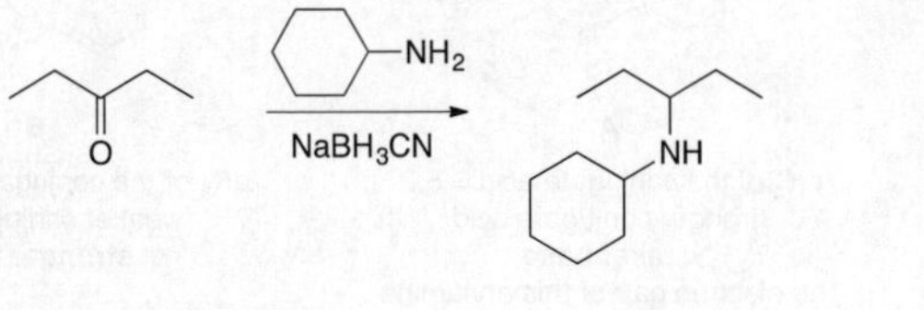

25.51 Use the directions from Answer 25.18. Separation can be achieved because benzoic acid reacts with aqueous base and aniline reacts with aqueous acid according to the following equations:

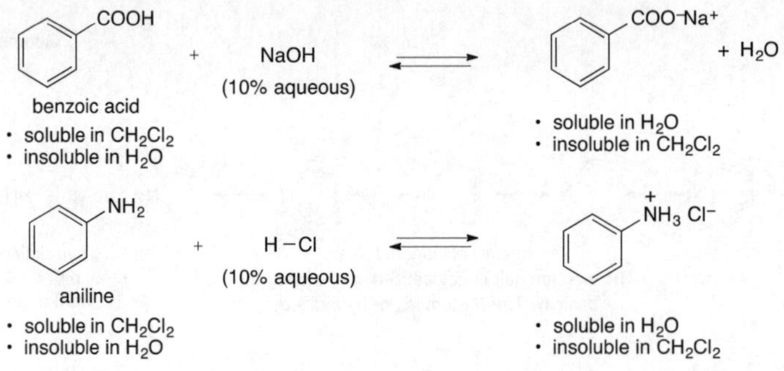

benzoic acid
• soluble in CH_2Cl_2
• insoluble in H_2O

+ NaOH
(10% aqueous)

$\rightleftharpoons$

COO^-Na^+ + H_2O
• soluble in H_2O
• insoluble in CH_2Cl_2

aniline
• soluble in CH_2Cl_2
• insoluble in H_2O

+ H−Cl
(10% aqueous)

$\rightleftharpoons$

$\overset{+}{N}H_3\ Cl^-$
• soluble in H_2O
• insoluble in CH_2Cl_2

Toluene ($C_6H_5CH_3$), on the other hand, is not protonated or deprotonated in aqueous solution, so it is always soluble in CH_2Cl_2 and insoluble in H_2O. The following flow chart illustrates the process.

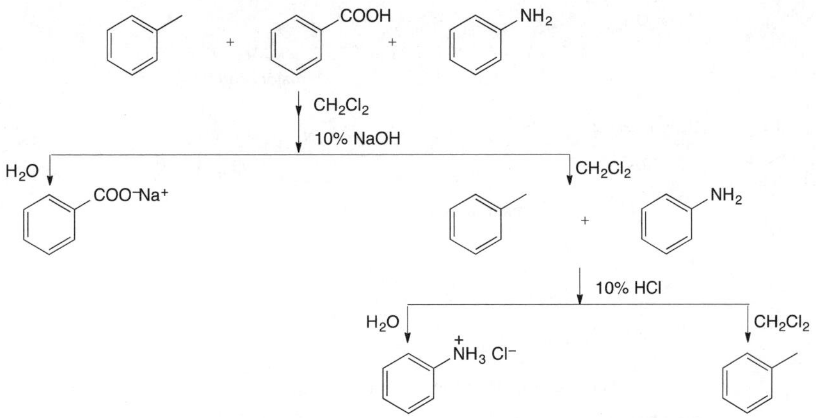
COOH
NH2
CH2Cl2
10% NaOH
H2O
COO⁻Na⁺
CH2Cl2
NH2
10% HCl
H2O
NH3 Cl⁻
CH2Cl2

25.52

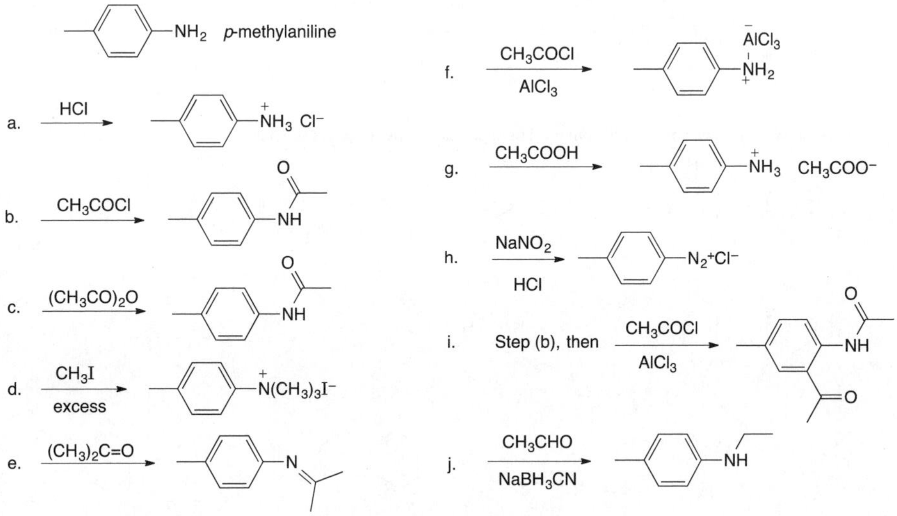
NH2 p-methylaniline
a. HCl
NH3 Cl⁻
b. CH3COCl
NH
c. (CH3CO)2O
NH
d. CH3I excess
N(CH3)3I⁻
e. (CH3)2C=O
N
f. CH3COCl AlCl3
AlCl3
NH2
g. CH3COOH
NH3 CH3COO⁻
h. NaNO2 HCl
N2⁺Cl⁻
i. Step (b), then CH3COCl AlCl3
NH
j. CH3CHO NaBH3CN
NH

25.53

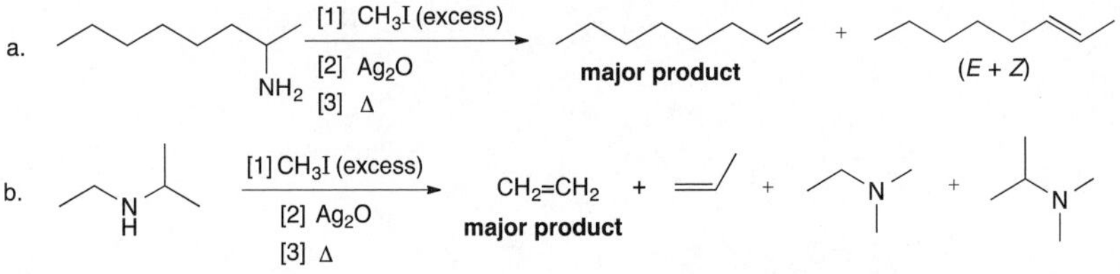
a.
NH2
[1] CH3I (excess)
[2] Ag2O
[3] Δ
major product
(E + Z)
b.
N
H
[1] CH3I (excess)
[2] Ag2O
[3] Δ
CH2=CH2
major product
N
N

Chapter 25–20

c.

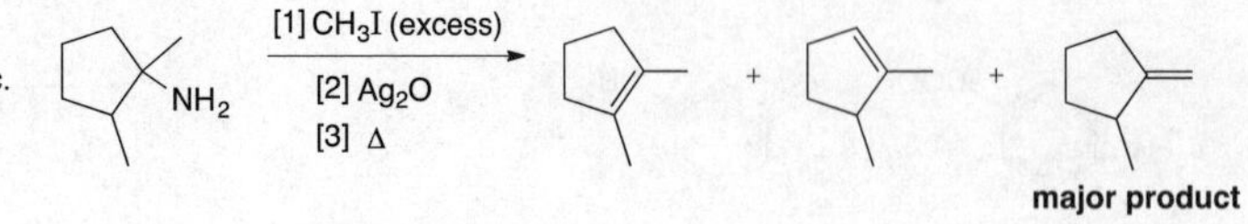

[1] CH$_3$I (excess)
[2] Ag$_2$O
[3] Δ

major product

d.

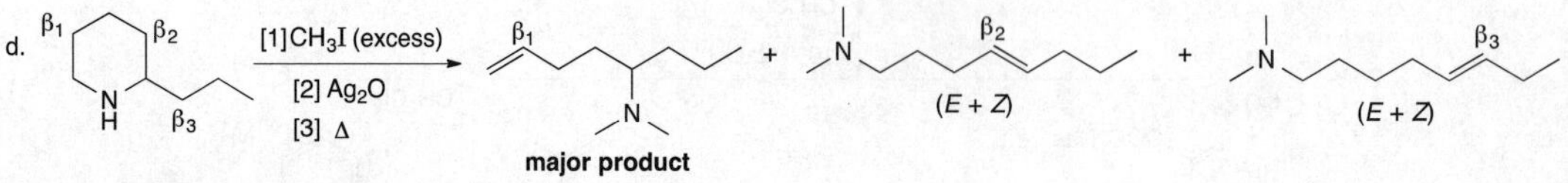

β$_1$ β$_2$

N   β$_3$
H

[1] CH$_3$I (excess)
[2] Ag$_2$O
[3] Δ

β$_1$

major product

+   β$_2$ (E + Z)   +   β$_3$ (E + Z)

25.54

a.

one stereogenic center

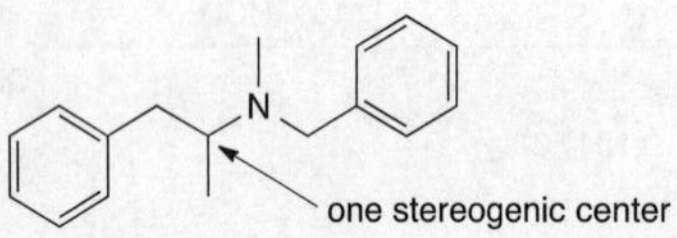

benzphetamine

b. Amides that can be reduced to benzphetamine:

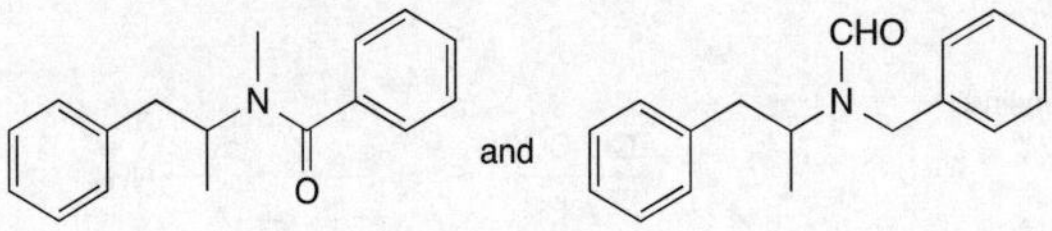

and

c. Amines + carbonyl compounds that form benzphetamine by reductive amination:

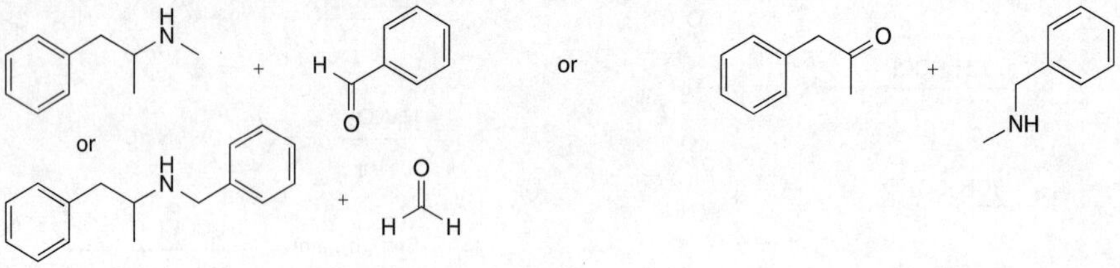

+   or   +

or

+

d.

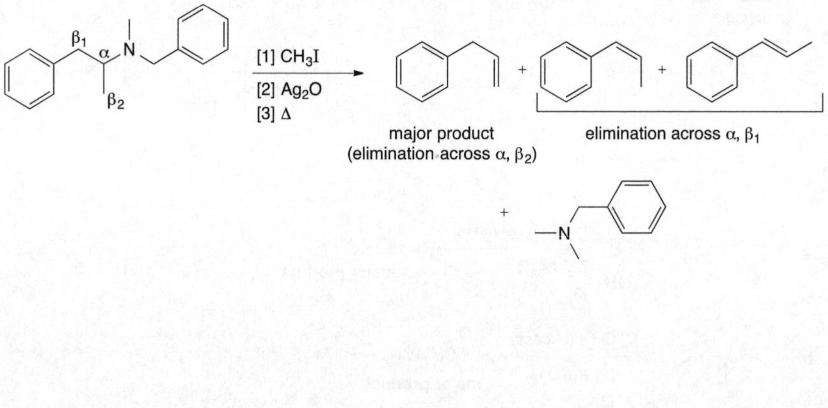

β$_1$ α   N

β$_2$

[1] CH$_3$I
[2] Ag$_2$O
[3] Δ

major product
(elimination across α, β$_2$)

elimination across α, β$_1$

+

Amines 25–21

25.55

a.

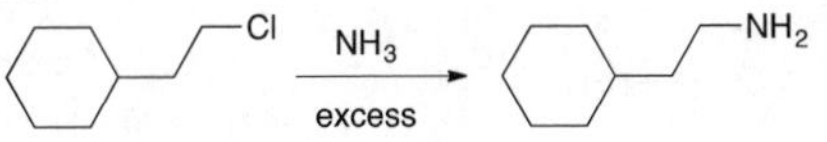

b.

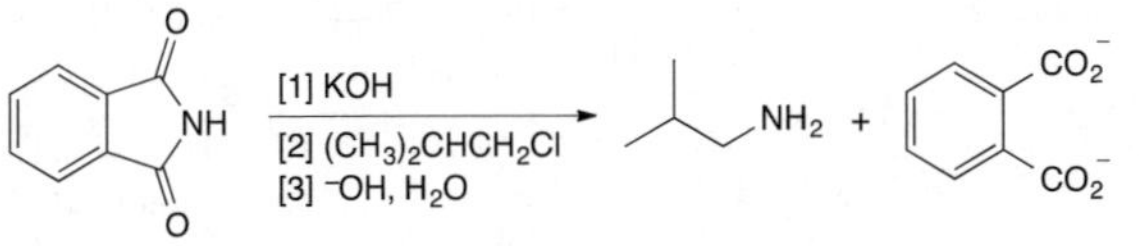

c.

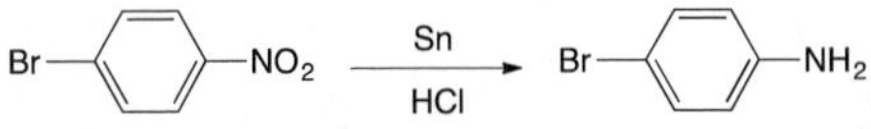

d.

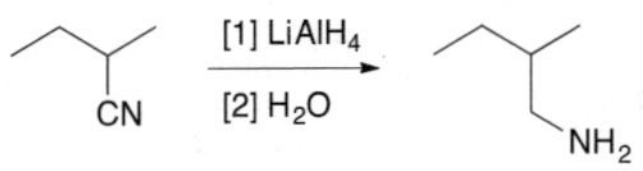

e.

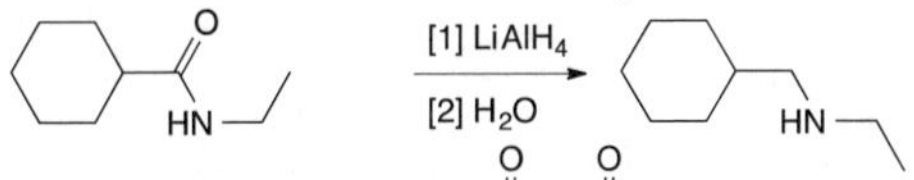

f.

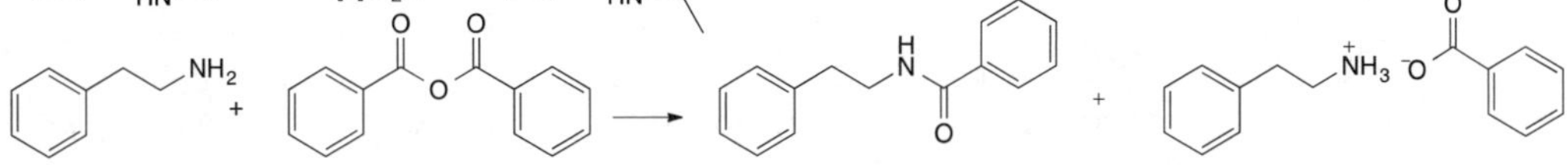

g.

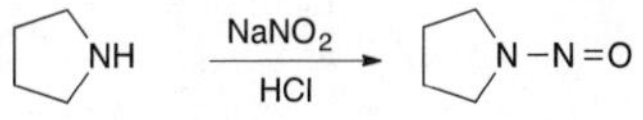

h.

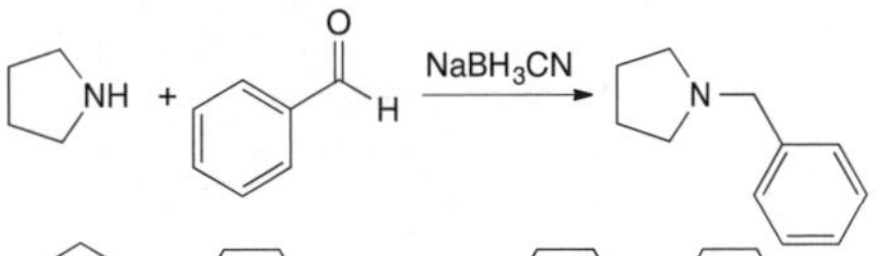

i.

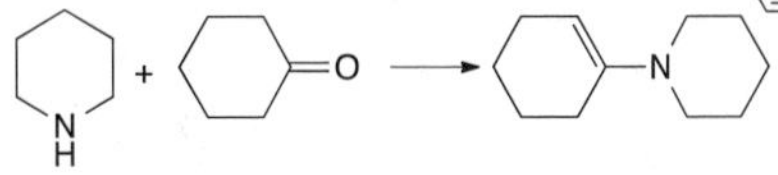

j.

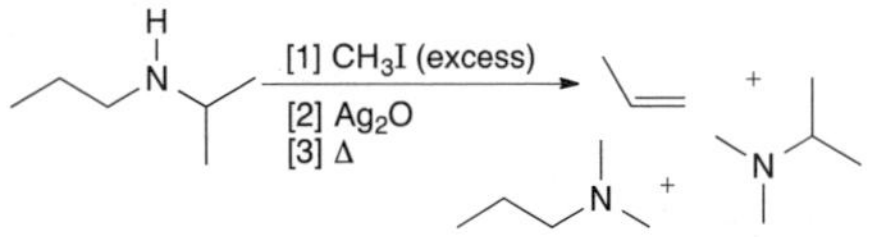

25.56 NH$_2$ and H must be anti for the Hofmann elimination. Rotate around the C–C bond so the NH$_2$ and H are anti.

a.

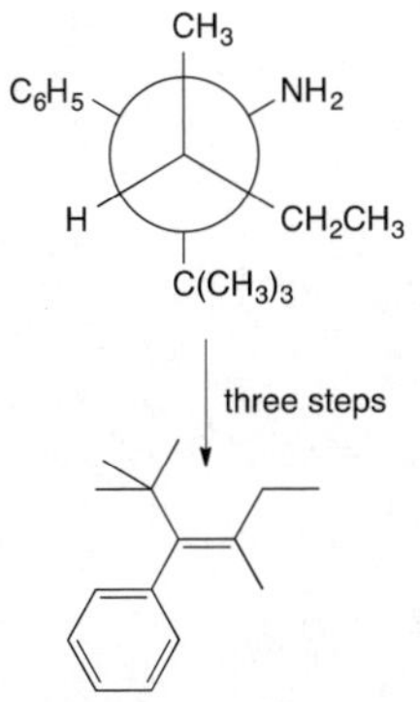

three steps

b.

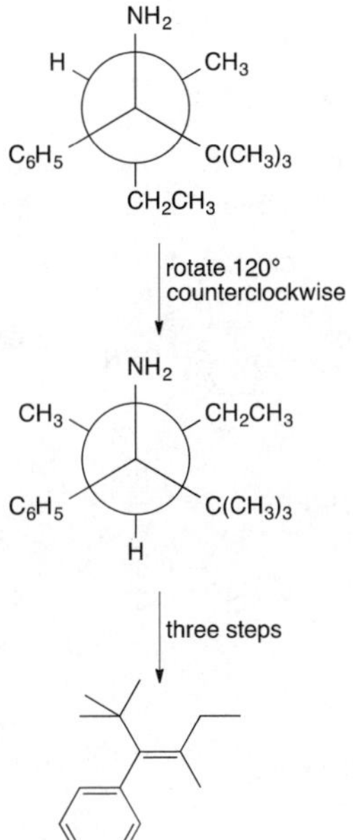

rotate 120° counterclockwise

three steps

c.

rotate 120° counterclockwise

three steps

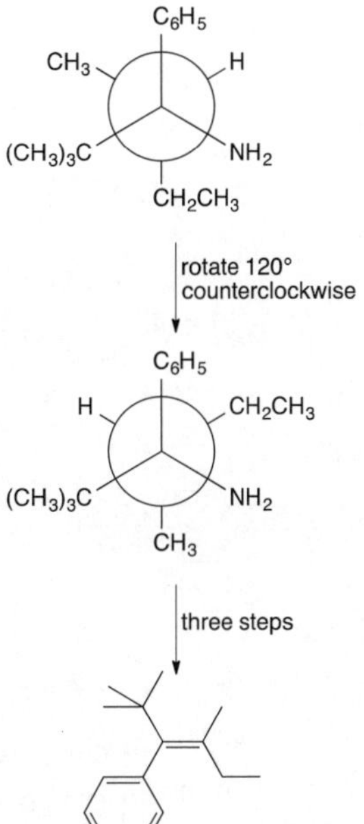

Chapter 25–22

25.57

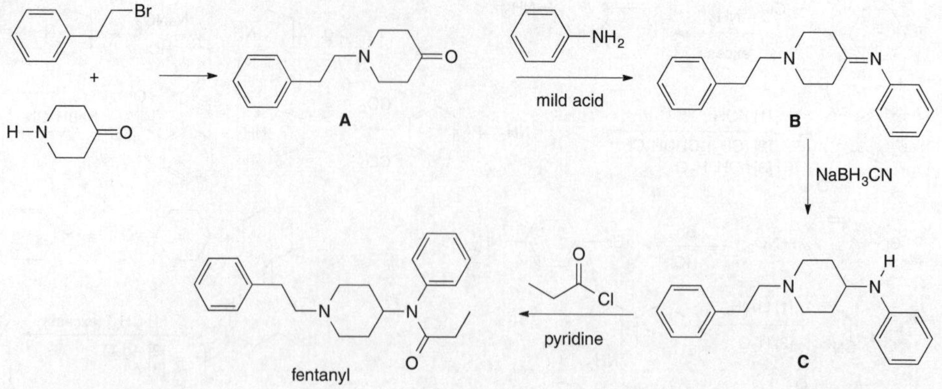

25.58

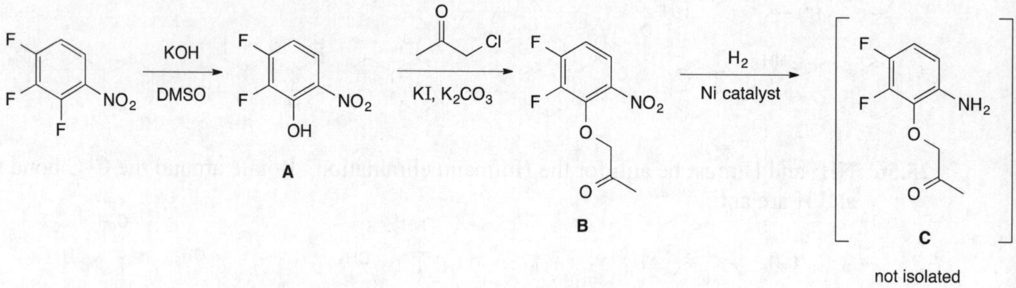

25.59

a.

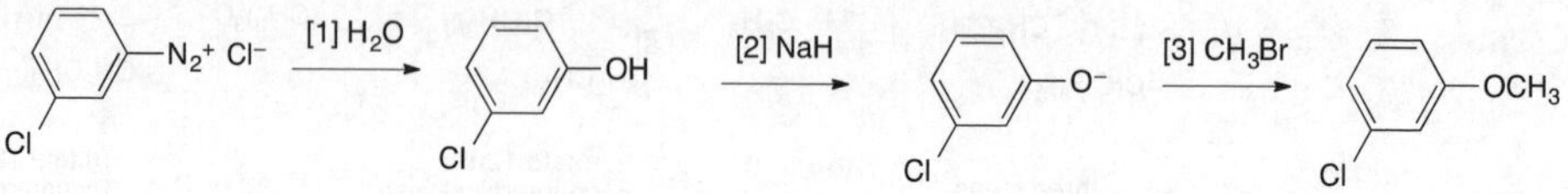

b.

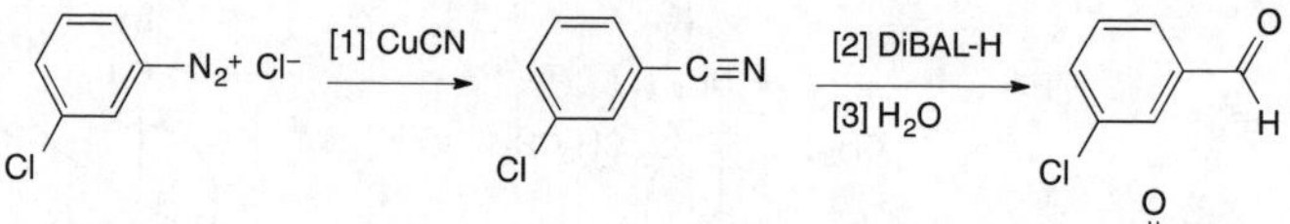

c.

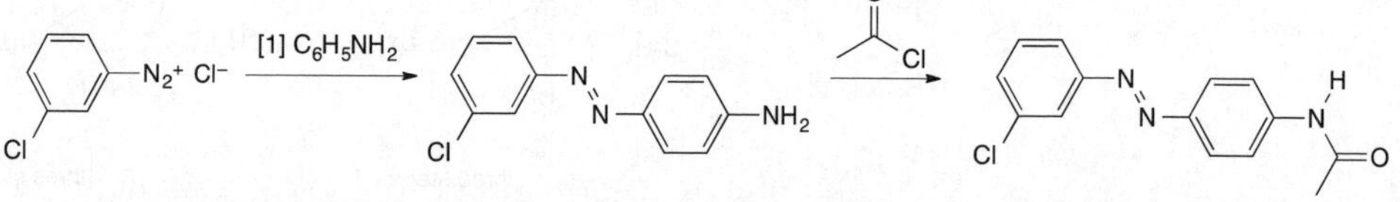

25.60

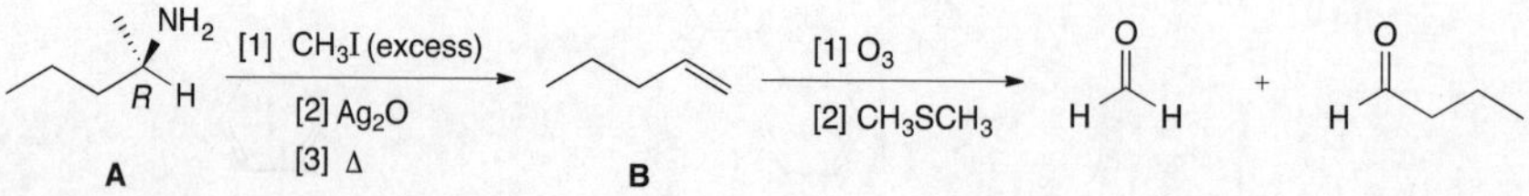

Amines 25–23

25.61

a.

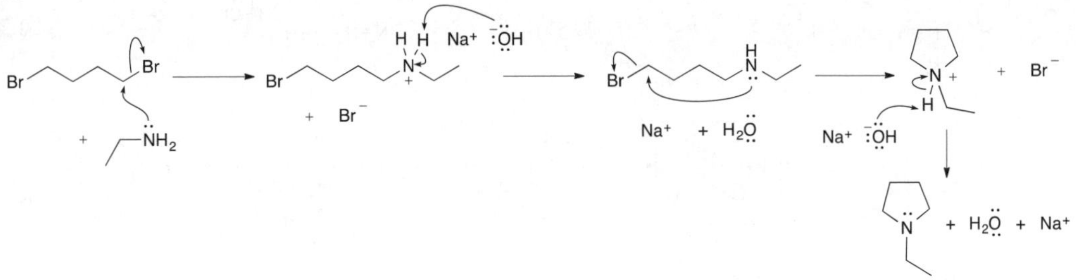

b.

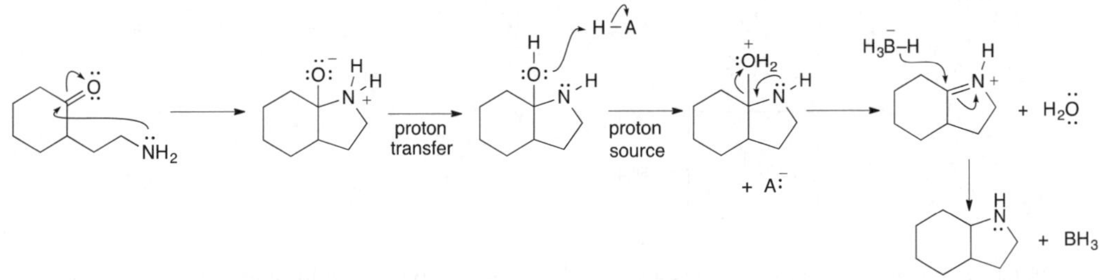

25.62

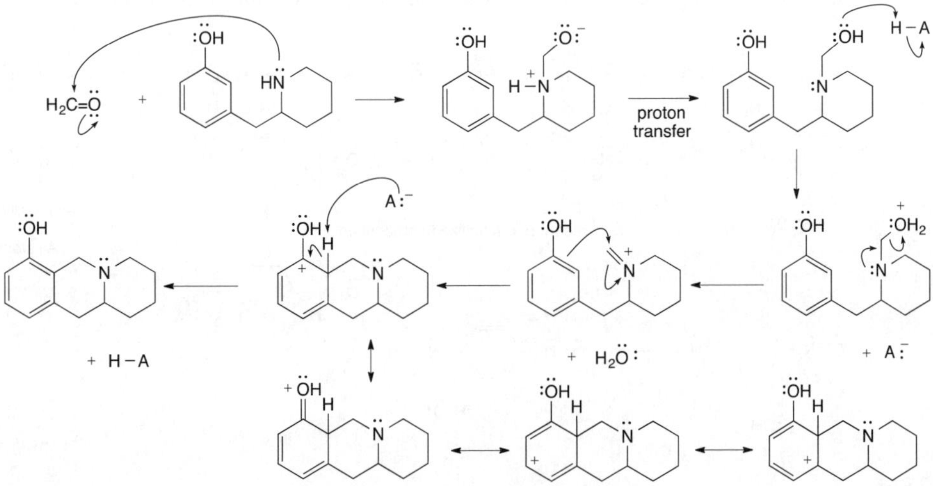

proton
transfer

proton
transfer

25.63

Overall reaction:

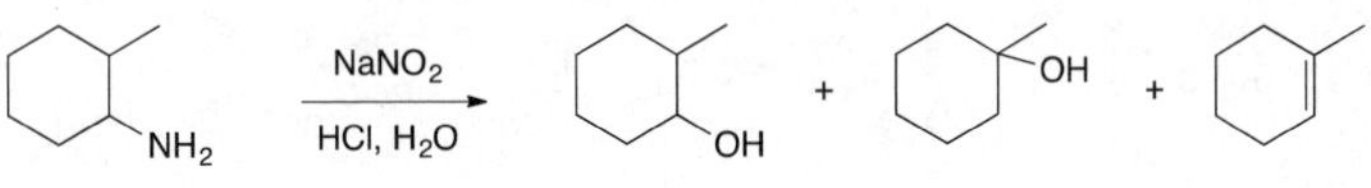

NaNO₂
$\xrightarrow{\text{HCl, H}_2\text{O}}$

The steps:

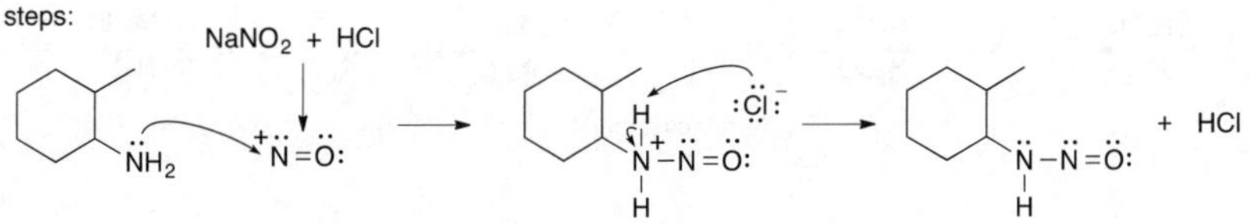

NaNO₂ + HCl

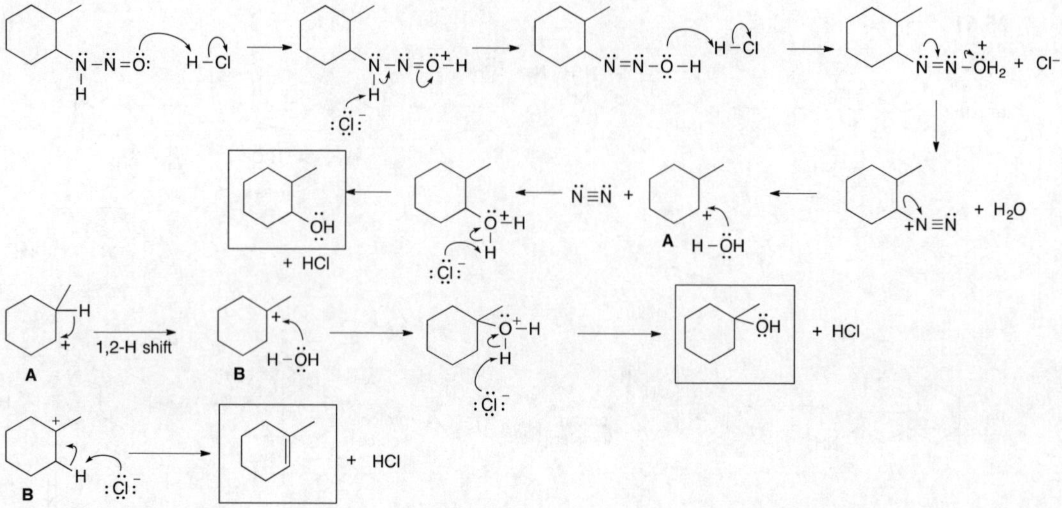

Chapter 25–24

25.64 A nitrosonium ion ($^+$NO) is a weak electrophile, so electrophilic aromatic substitution occurs only with a strong electron-donor group that stabilizes the intermediate carbocation.

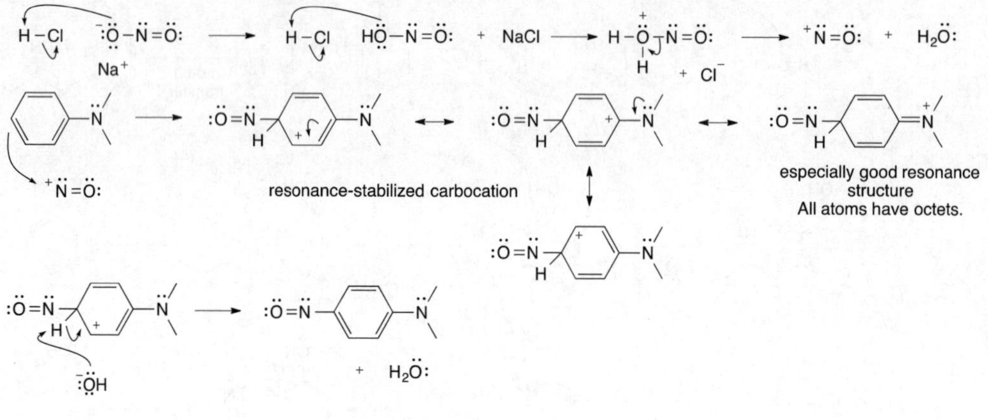

resonance-stabilized carbocation

especially good resonance structure
All atoms have octets.

25.65

a.

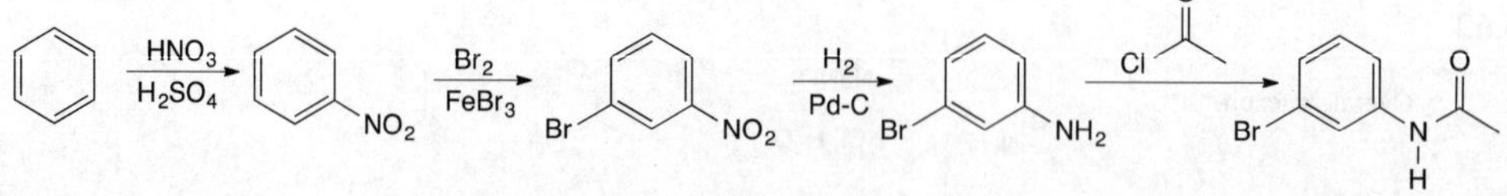

b.

(+ ortho isomer)

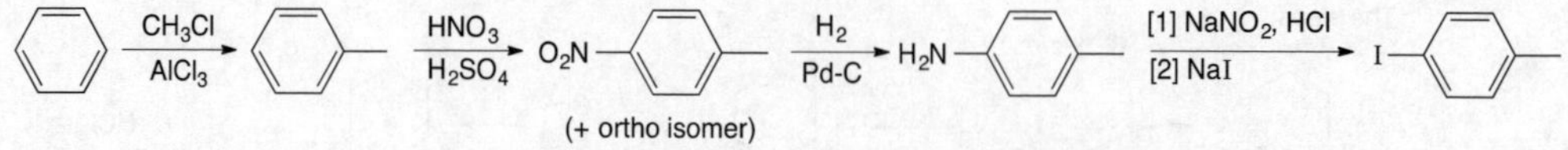

Amines 25–25

c.

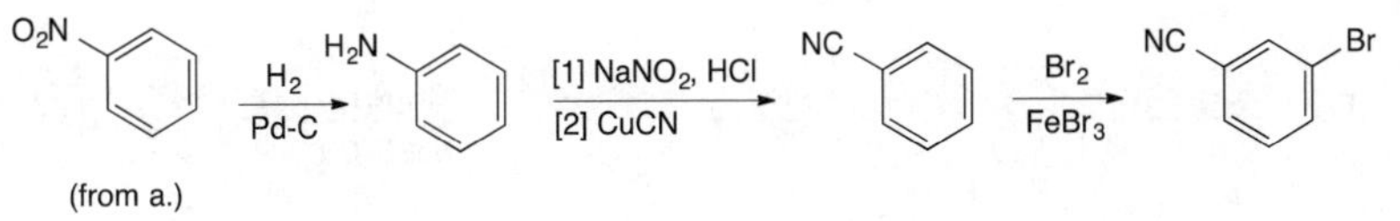

(from a.)

d.

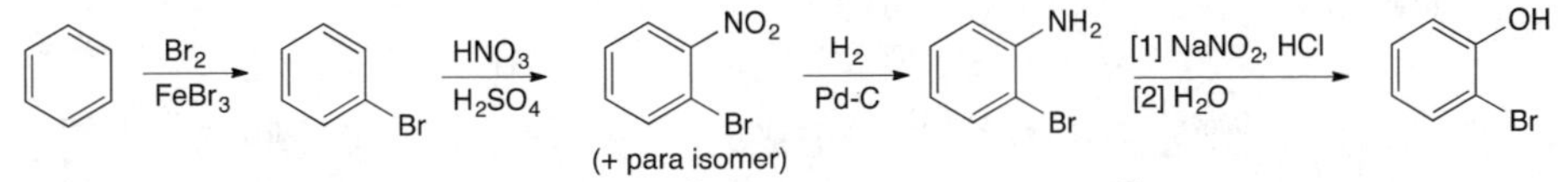

(+ para isomer)

e.

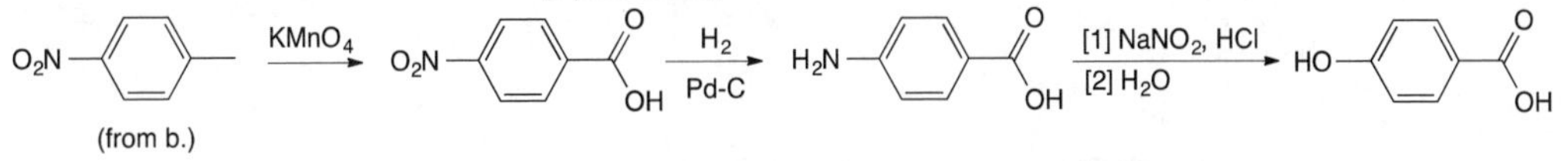

(from b.)

f.

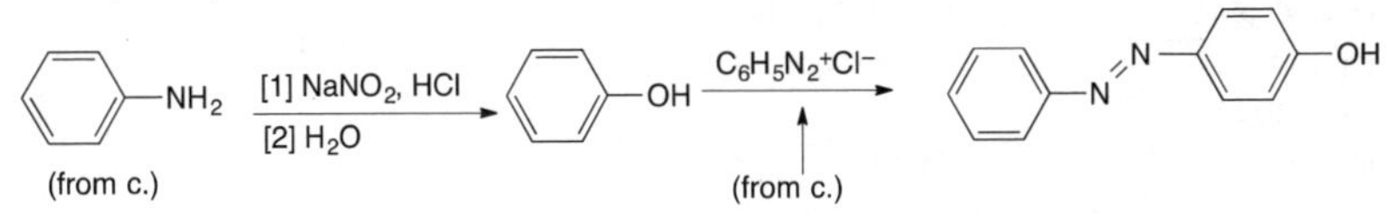

(from c.)   (from c.)

25.66

a.

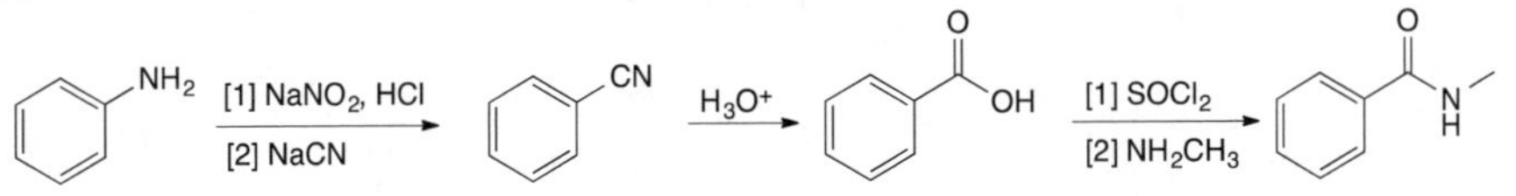

b.

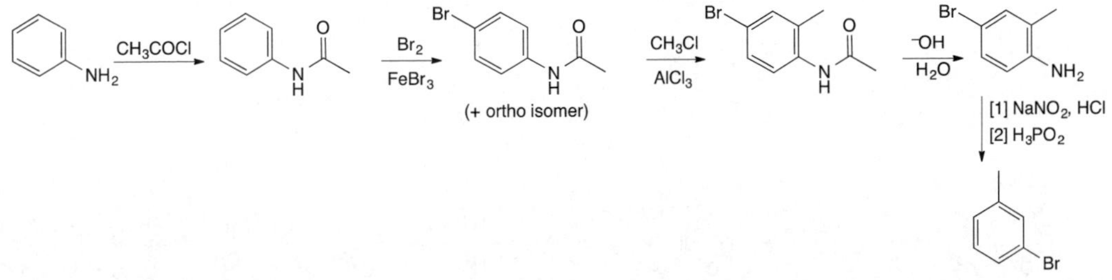

(+ ortho isomer)

c.

(from a., Step [1])

(+ ortho isomer)

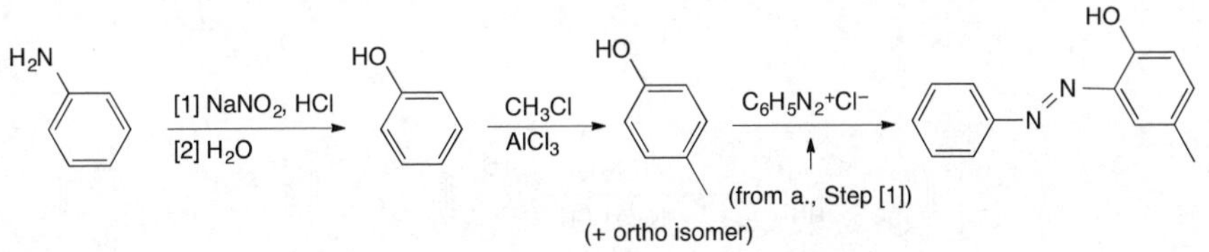

25.67

[1]

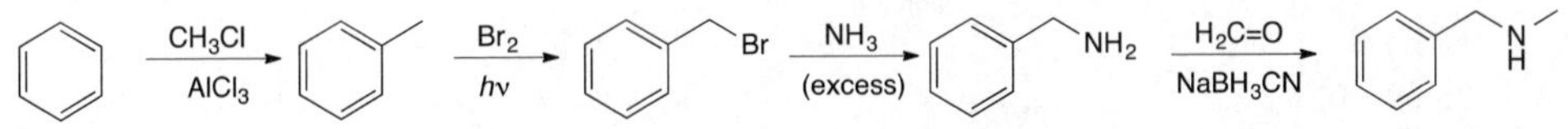

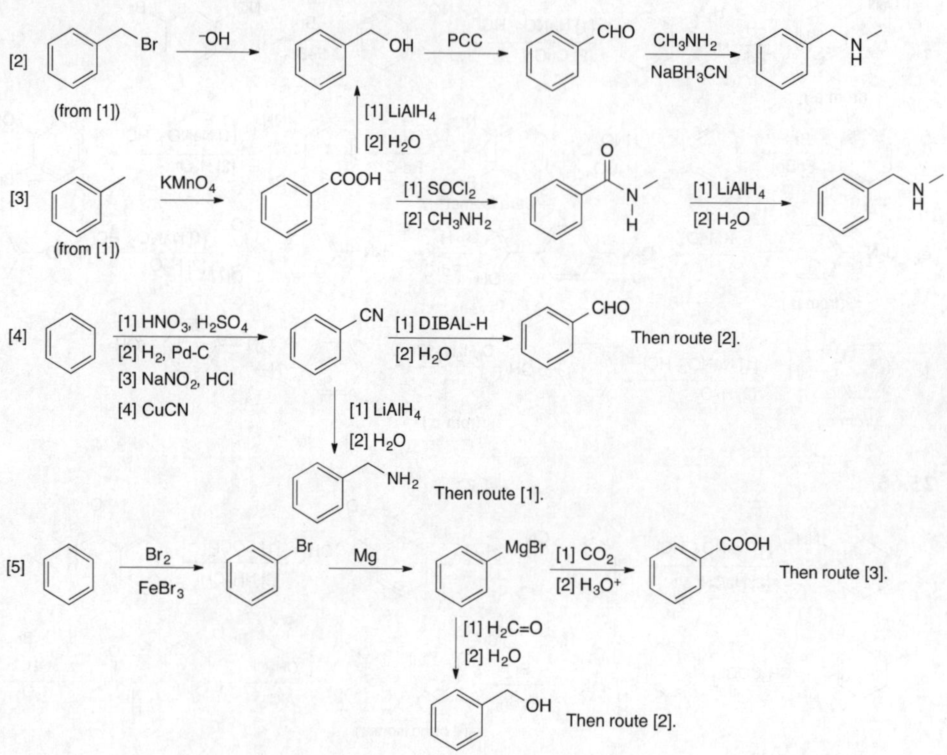

Chapter 25–26

[2]

(from [1])

[3]

(from [1])

[1] LiAlH$_4$
[2] H$_2$O

$^-$OH

PCC

CH$_3$NH$_2$
NaBH$_3$CN

KMnO$_4$

[1] SOCl$_2$
[2] CH$_3$NH$_2$

[1] LiAlH$_4$
[2] H$_2$O

[4]

[1] HNO$_3$, H$_2$SO$_4$
[2] H$_2$, Pd-C
[3] NaNO$_2$, HCl
[4] CuCN

[1] DIBAL-H
[2] H$_2$O

Then route [2].

[1] LiAlH$_4$
[2] H$_2$O

Then route [1].

[5]

Br$_2$
FeBr$_3$

Mg

[1] CO$_2$
[2] H$_3$O$^+$

Then route [3].

[1] H$_2$C=O
[2] H$_2$O

Then route [2].

25.68

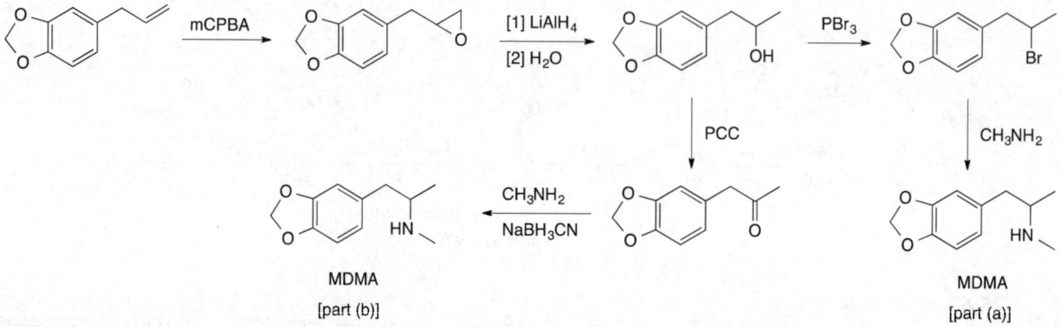

mCPBA

[1] LiAlH$_4$
[2] H$_2$O

PBr$_3$

PCC

CH$_3$NH$_2$

CH$_3$NH$_2$
NaBH$_3$CN

MDMA
[part (b)]

MDMA
[part (a)]

25.69

a.

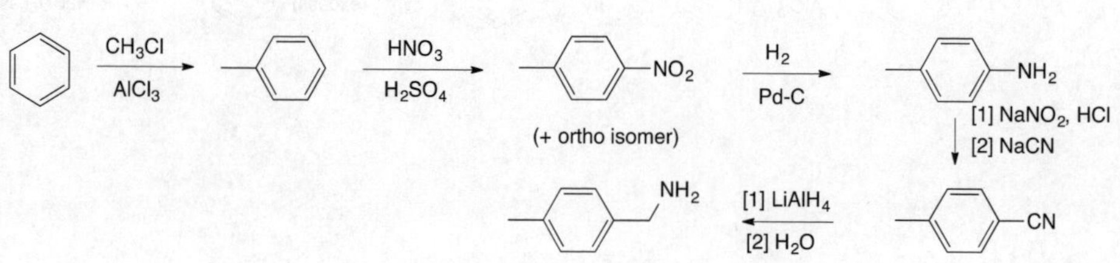

CH$_3$Cl
AlCl$_3$

HNO$_3$
H$_2$SO$_4$

(+ ortho isomer)

H$_2$
Pd-C

[1] NaNO$_2$, HCl
[2] NaCN

[1] LiAlH$_4$
[2] H$_2$O

b.

benzene $\xrightarrow[\text{AlCl}_3]{\text{CH}_3\text{CH}_2\text{Cl}}$ ethylbenzene $\xrightarrow[\text{H}_2\text{SO}_4]{\text{HNO}_3}$ ethyl-NO_2 (+ ortho isomer) $\xrightarrow[\text{Pd-C}]{\text{H}_2}$ ethyl-NH_2

$\xrightarrow[\text{[2] CuCN}]{\text{[1] NaNO}_2,\ \text{HCl}}$

ethyl-COOH $\xleftarrow{\text{H}_3\text{O}^+}$ ethyl-CN

c.

benzene $\xrightarrow[\text{H}_2\text{SO}_4]{\text{HNO}_3}$ NO_2 $\xrightarrow[\text{Pd-C}]{\text{H}_2}$ NH_2 $\xrightarrow{\text{CH}_3\text{COCl}}$ NHCOCH$_3$ $\xrightarrow[\text{H}_2\text{SO}_4]{\text{HNO}_3}$ NHCOCH$_3$, NO_2 $\xrightarrow[\text{FeCl}_3\ (2X)]{\text{Cl}_2}$ Cl NHCOCH$_3$ Cl, NO_2 $\xrightarrow[\text{H}_2\text{O}]{^-\text{OH}}$ Cl NH$_2$ Cl, NO$_2$

$\xrightarrow[\text{[2] H}_3\text{PO}_2]{\text{[1] NaNO}_2,\ \text{HCl}}$

Cl—Cl, NO$_2$ $\xrightarrow[\text{Pd-C}]{\text{H}_2}$ Cl—Cl, NH$_2$ $\xrightarrow[\text{[2]}\ \text{—NH}_2\ \text{(from above)}]{\text{[1] NaNO}_2,\ \text{HCl}}$ Cl—N=N—NH$_2$, Cl

25.70

a.

benzene $\xrightarrow[\text{H}_2\text{SO}_4]{\text{HNO}_3}$ O$_2$N— $\xrightarrow[\text{Pd-C}]{\text{H}_2}$ H$_2$N— $\xrightarrow[\text{[2] H}_2\text{O}]{\text{[1] NaNO}_2,\ \text{HCl}}$ HO— $\xrightarrow[\text{H}_2\text{SO}_4]{\text{HNO}_3}$ HO—NO$_2$ (+ ortho isomer) $\xrightarrow[\text{Pd-C}]{\text{H}_2}$ HO—NH$_2$

HO—N(H)COCH$_3$

b.

benzene $\xrightarrow[\text{AlCl}_3]{\text{ClCOCH}_2\text{CH}_3}$ → $\xrightarrow[\text{H}_2\text{O, HCl}]{\text{Cl}_2}$ →Cl $\xrightarrow{\text{NH}_2\text{CH}_3}$ →N(H)— $\xrightarrow[\text{CH}_3\text{OH}]{\text{NaBH}_4}$ OH, N(H)—

Chapter 25–28

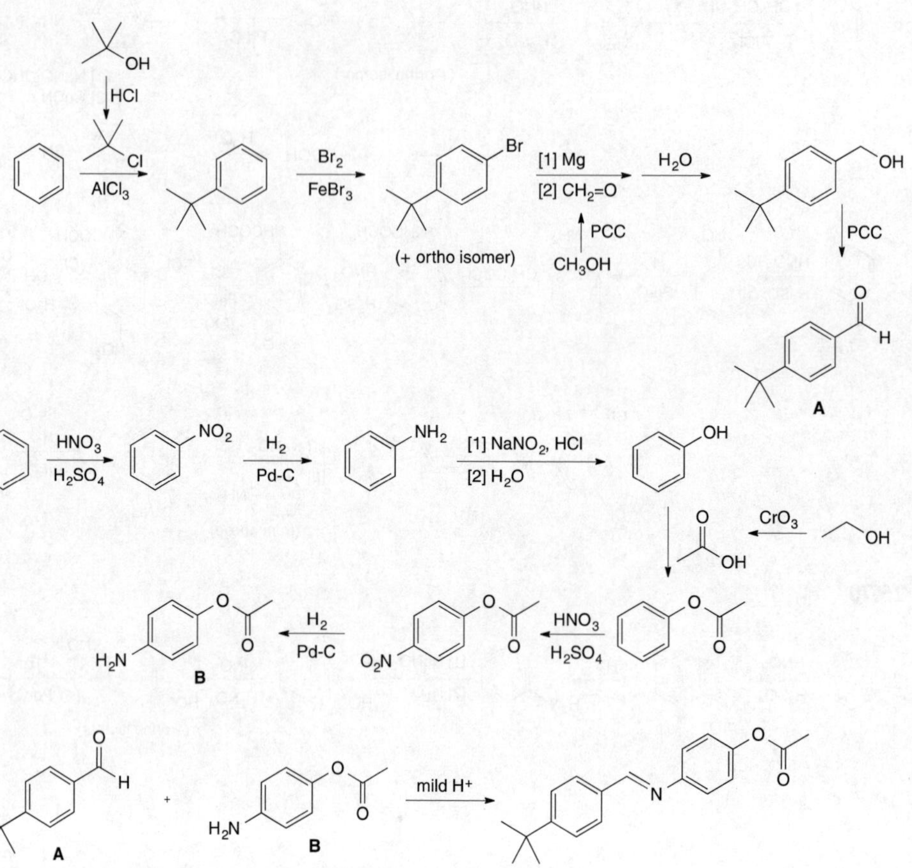

25.71

a.

(+ ortho isomer)

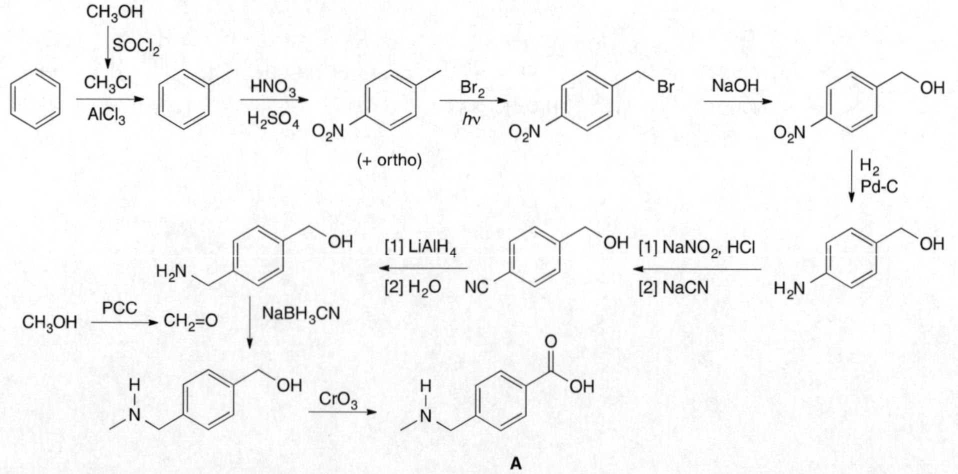

A

b.

(+ ortho)

A

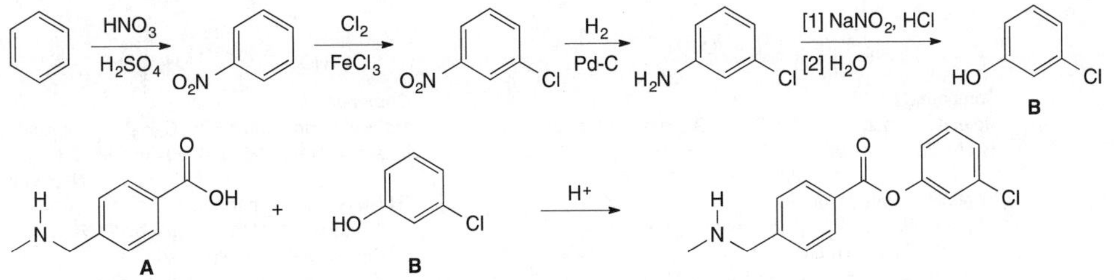

Amines 25–29

c. Probably a strong enough activator that the Friedel–Crafts reaction will still occur.

Make two parts:

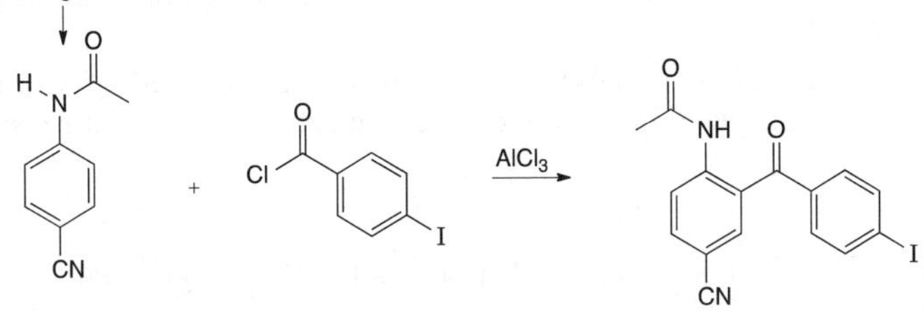

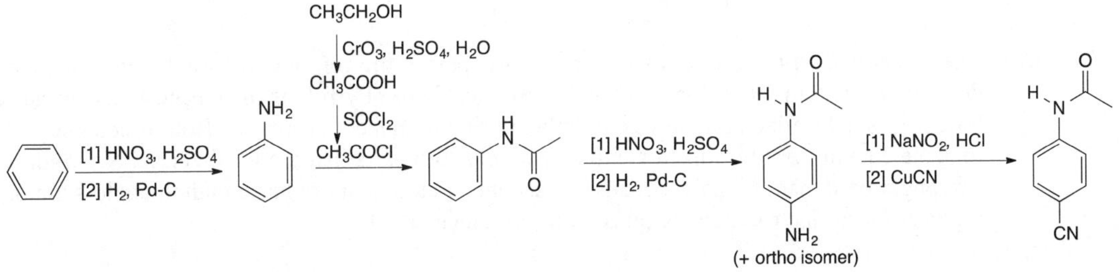

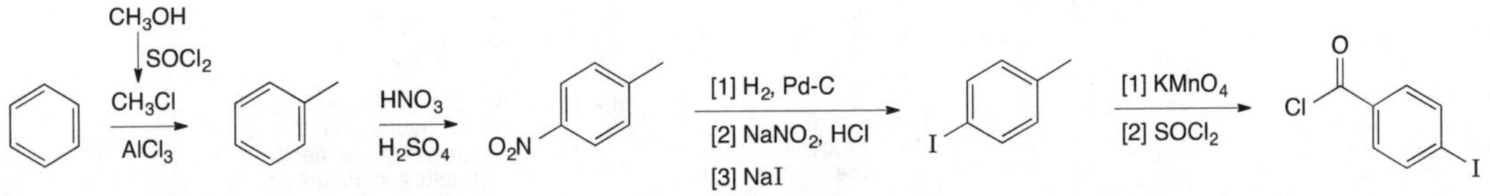

25.72

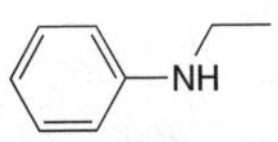

Compound A: $C_8H_{11}N$
IR absorption at 3400 cm^{-1} → 2° amine
^{1}H NMR signals at (ppm):
 1.3 (triplet, 3 H) CH$_3$ adjacent to 2 H's
 3.2 (quartet, 2 H) CH$_2$ adjacent to 3 H's
 3.6 (singlet, 1 H) amine H
 6.8–7.2 (multiplet, 5 H) benzene ring

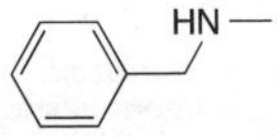

Compound B: $C_8H_{11}N$
IR absorption at 3310 cm^{-1} → 2° amine
^{1}H NMR signals at (ppm):
 1.4 (singlet, 1 H) amine H
 2.4 (singlet, 3 H) CH$_3$
 3.8 (singlet, 2 H) CH$_2$
 7.3 (multiplet, 5 H) benzene ring

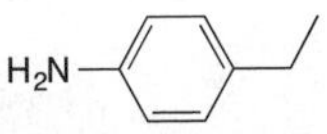

Compound C: $C_8H_{11}N$
IR absorption at 3430 and
 3350 cm^{-1} → 1° amine
^{1}H NMR signals at (ppm):
 1.3 (triplet, 3 H) CH$_3$ near CH$_2$
 2.5 (quartet, 2 H) CH$_2$ near CH$_3$
 3.6 (singlet, 2 H) amine H's
 6.7 (doublet, 2 H) ⎤ para disubstituted
 7.0 (doublet, 2 H) ⎦ benzene ring

Chapter 25–30

25.73

HO–CH₂–C≡N

Compound D:
Molecular ion at m/z = 71: C_3H_5NO (possible formula)
IR absorption at 3600–3200 cm⁻¹ → OH
     2263 cm⁻¹ → CN
¹H NMR signals at (ppm):
   2.5 (triplet, 2 H) CH₂ adjacent to 2 H's
   3.1 (singlet, 1 H) OH
   3.8 (triplet, 2 H) CH₂ adjacent to 2 H's

HO–CH₂CH₂CH₂–NH₂

Compound E:
Molecular ion at m/z = 75: C_3H_9NO (possible formula)
IR absorption at 3600–3200 cm⁻¹ → OH
     3636 cm⁻¹ → N–H of amine
¹H NMR signals at (ppm):
   1.6 (quintet, 2 H) CH₂ split by 2 CH₂'s
   2.5 (singlet, 3 H) NH₂ and OH
   2.8 (triplet, 2 H) CH₂ split by CH₂
   3.7 (triplet, 2 H) CH₂ split by CH₂

25.74 Guanidine is a strong base because its conjugate acid is stabilized by resonance. This resonance delocalization makes guanidine easily donate its electron pair; thus it's a strong base.

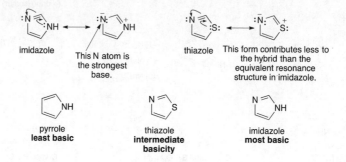

25.75 The compound with the most available electron pair or the compound with the highest electron density on an atom (N in this case) is the strongest base. Pyrrole is the weakest base because its lone pair is delocalized on the five-membered ring to make it aromatic. Both imidazole and thiazole contain sp^2 hybridized N atoms with electron pairs that are localized on N. Imidazole is a stronger base than thiazole, because its second N atom is more basic than thiazole's S atom, so it places more electron density on N by a resonance effect.

25.76

Amines 25–31

25.77

One possibility:

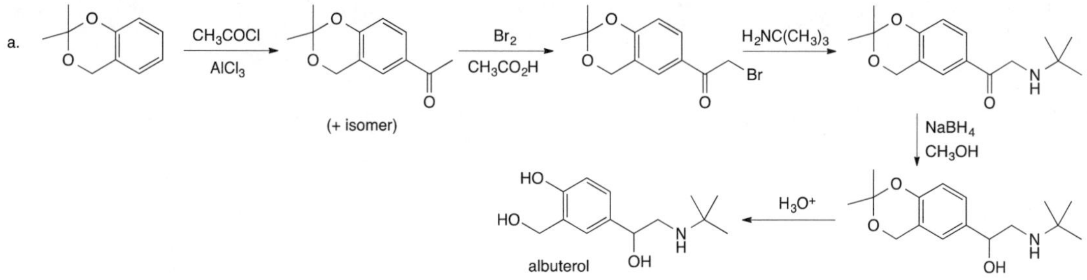

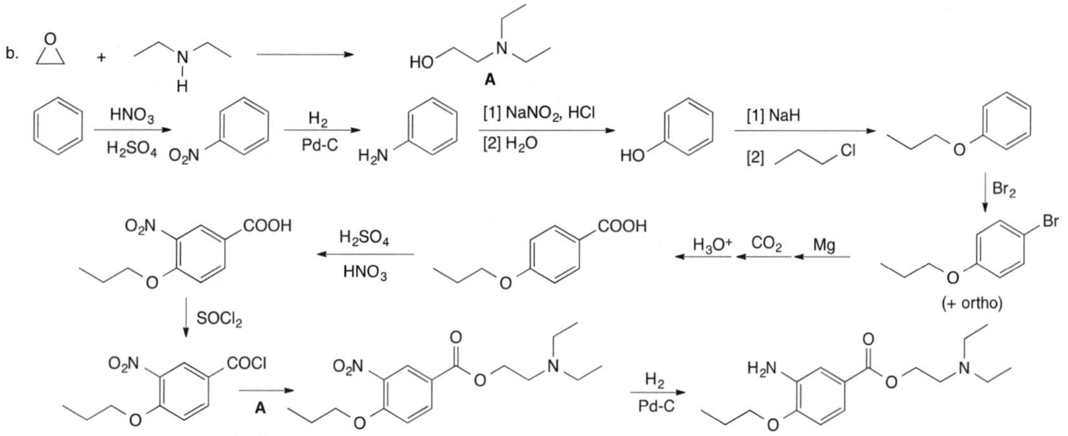

a.

(+ isomer)

albuterol

b.

A

A

(+ ortho)

A

25.78 $CH_2=O$ reacts with the amine to form an intermediate imine, which undergoes an intramolecular Diels–Alder reaction.

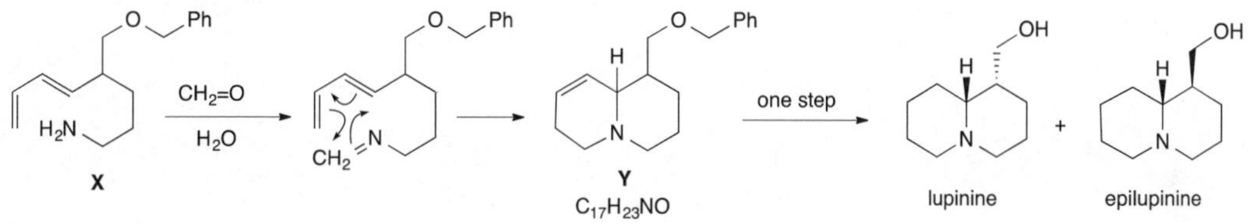

X

Y
$C_{17}H_{23}NO$

one step

lupinine   epilupinine

Chapter 26 Carbon–Carbon Bond-Forming Reactions in Organic Synthesis

Chapter Review

Coupling reactions

[1] Coupling reactions of organocuprate reagents (26.1)

R'—X + R₂CuLi ⟶ R'—R + RCu + LiX

X = Cl, Br, I

- R'X can be CH₃X, RCH₂X, 2° cyclic halides, vinyl halides, and aryl halides.
- X may be Cl, Br, or I.
- With vinyl halides, coupling is stereospecific.

[2] Suzuki reaction (26.2)

R'—X + R—BY₂ —[Pd(PPh₃)₄, NaOH]→ R'—R + HO—BY₂ + NaX

X = Br, I

- R'X is most often a vinyl halide or aryl halide.
- With vinyl halides, coupling is stereospecific.

[3] Heck reaction (26.3)

R'—X + CH₂=CHZ —[Pd(OAc)₂, P(o-tolyl)₃, Et₃N]→ R'CH=CHZ + Et₃NH⁺ X⁻

X = Br or I

- R'X is a vinyl halide or aryl halide.
- Z = H, Ph, COOR, or CN
- With vinyl halides, coupling is stereospecific.
- The reaction forms trans alkenes.

Cyclopropane synthesis

[1] Addition of dihalocarbenes to alkenes (26.4)

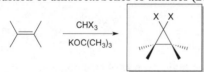

- The reaction occurs with syn addition.
- The position of substituents in the alkene is retained in the cyclopropane.

[2] Simmons–Smith reaction (26.5)

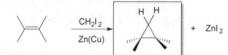

- The reaction occurs with syn addition.
- The position of substituents in the alkene is retained in the cyclopropane.

Metathesis (26.6)

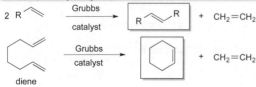

- Metathesis works best when CH₂=CH₂, a gas that escapes from the reaction mixture, is formed as one product.

Chapter 26–2

Practice Test on Chapter Review

1.a. Which functional groups react with lithium dialkyl cuprates?
   1. epoxides
   2. vinyl halides
   3. acid chlorides
   4. Compounds (1) and (2) both react with R_2CuLi.
   5. Compounds (1), (2), and (3) all react with R_2CuLi.

 b. Which of the following statements is (are) true for the Suzuki reaction?
   1. Arylboranes can serve as one reactant.
   2. The reaction is stereospecific.
   3. The reaction occurs between a vinyl or aryl halide and an alkene in the presence of a palladium catalyst.
   4. Statements (1) and (2) are both true.
   5. Statements (1), (2), and (3) are all true.

 c. Which of the following compounds can react with $CH_2=CHCN$ in a Heck reaction?

   1.　　　Br   4. Both (1) and (2) can react.

   2.　　　I   5. Compounds (1), (2), and (3) can all react.

   3. $CH_2=CH_2$

 d. Which of the following compounds yield(s) a pair of enantiomers on reaction with the Simmons–Smith reagent?

   1.   3.

                4. Compounds (1) and (2) both yield a pair of enantiomers.
   2.
                5. Compounds (1), (2), and (3) all yield a pair of enantiomers.

 e. Which of the following compounds can be made by a ring-closing metathesis reaction?

   1.   3.

                4. Compounds (1) and (2) can both be prepared.
   2.
                5. Compounds (1), (2), and (3) can all be prepared.

C–C Bond-Forming Reactions 26–3

f. Which of the following compounds can be prepared from $CH_3-C\equiv C-H$ by a Suzuki reaction? You may use other organic compounds or inorganic reagents.

1.

3.

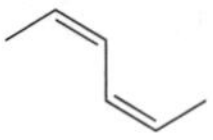

2.

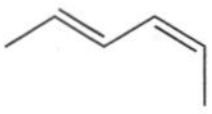

4. Compounds (1) and (2) can both be prepared.

5. Compounds (1), (2), and (3) can all be prepared.

2. Draw the product formed in each reaction. Indicate the stereochemistry around double bonds and stereogenic centers when necessary.

a. 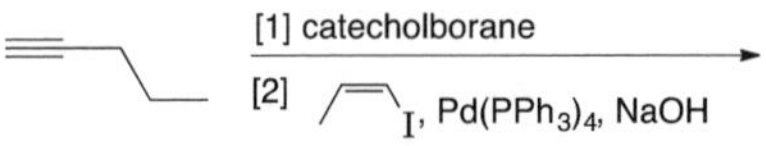
[1] catecholborane
[2] I, Pd(PPh$_3$)$_4$, NaOH

b. 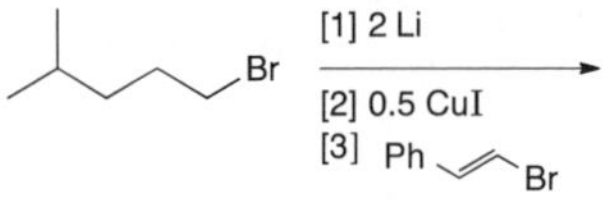
Br
[1] 2 Li
[2] 0.5 CuI
[3] Ph Br

c. 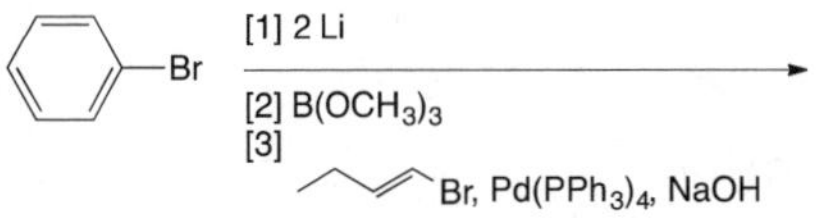
Br
[1] 2 Li
[2] B(OCH$_3$)$_3$
[3]
Br, Pd(PPh$_3$)$_4$, NaOH

d. 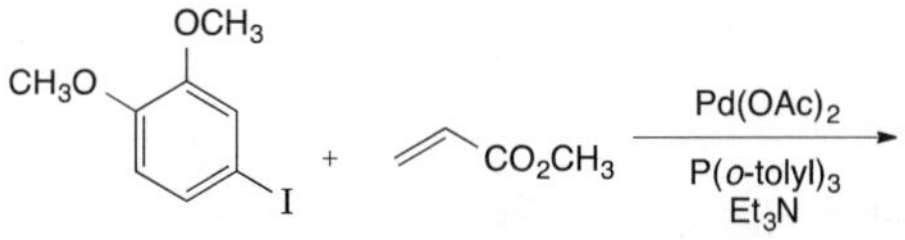
CH$_3$O
OCH$_3$
I
+ CO$_2$CH$_3$
Pd(OAc)$_2$
P(o-tolyl)$_3$
Et$_3$N

e. 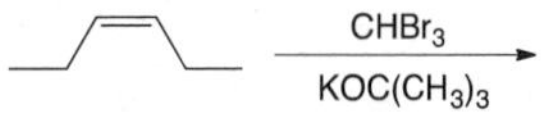
CHBr$_3$
KOC(CH$_3$)$_3$

f. 
Grubbs
cataylst

g. 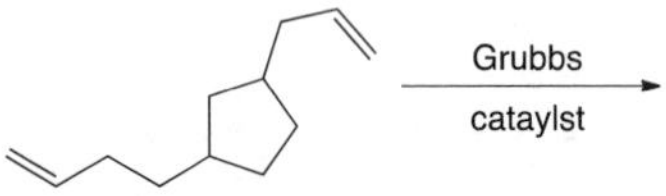
Grubbs
cataylst

3. What starting material is needed to synthesize each compound by ring-closure metathesis?

a.

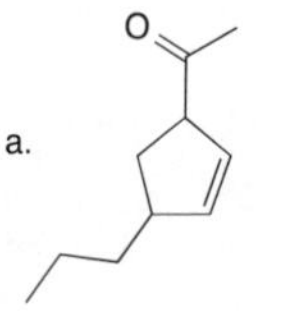

b. 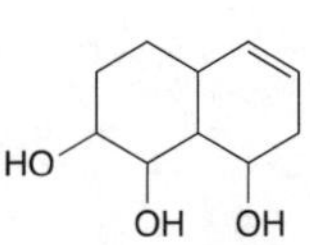
HO
OH   OH

c. 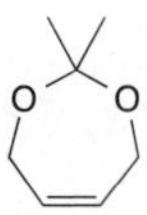
O   O

Chapter 26–4

Answers to Practice Test

1.a. 5   2.
 b. 4
 c. 4
 d. 2
 e. 3
 f. 4

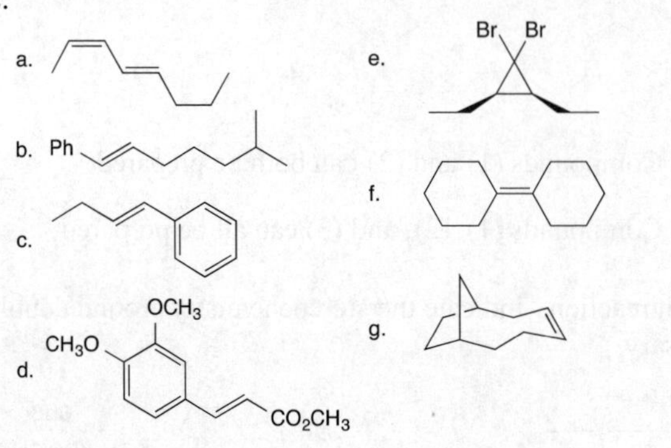

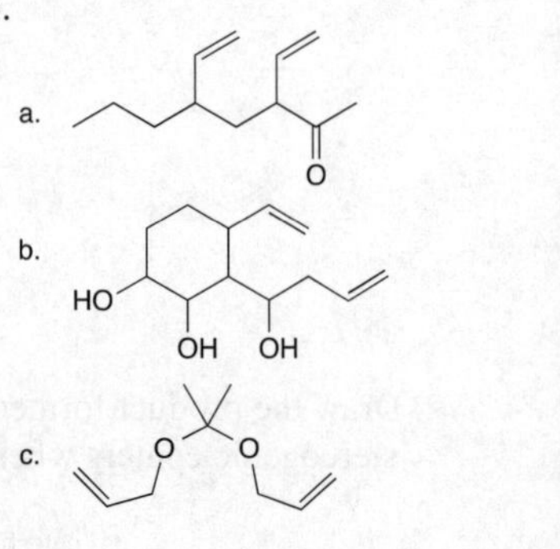

Answers to Problems

26.1 A new C–C bond is formed in each coupling reaction.

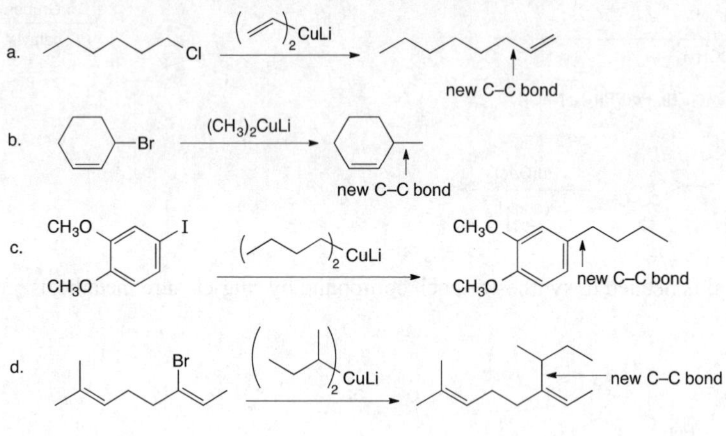

26.2

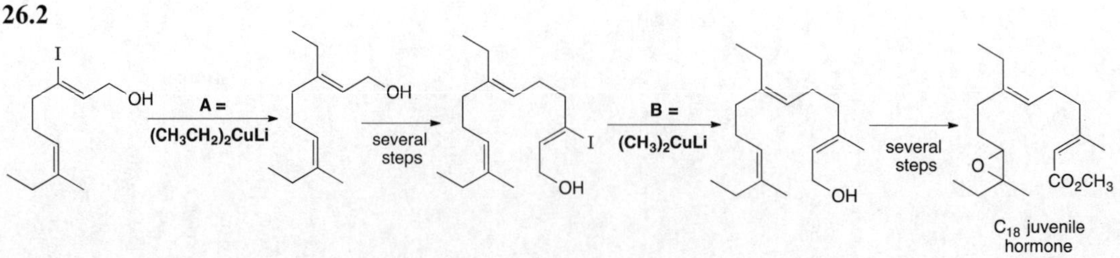

C–C Bond-Forming Reactions 26–5

26.3

a.

or

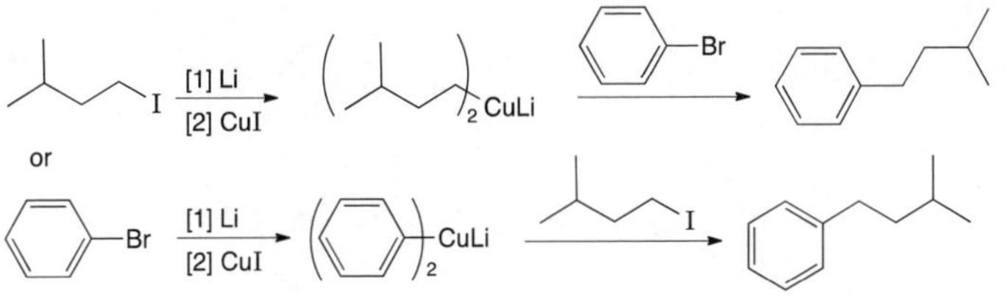

b.

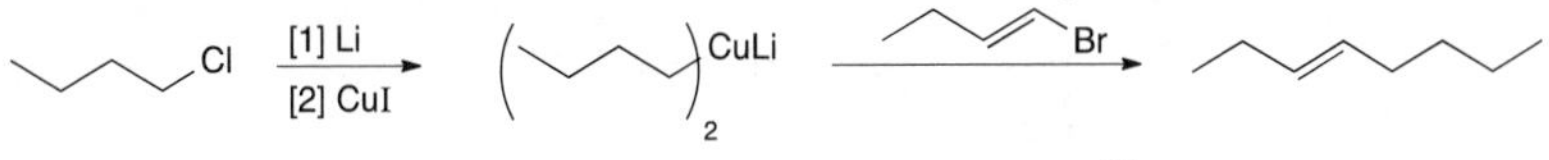

c.

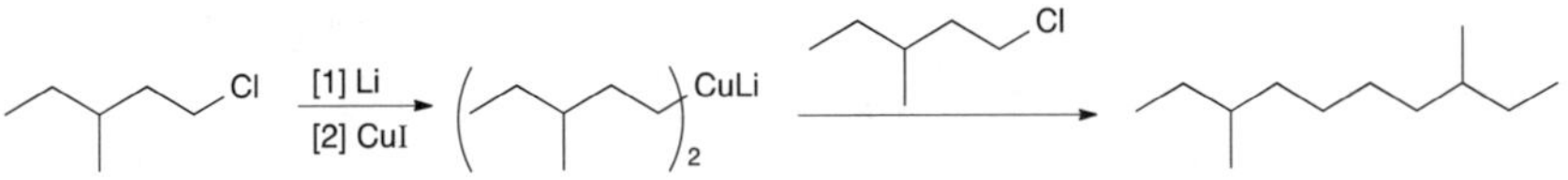

26.4

a.

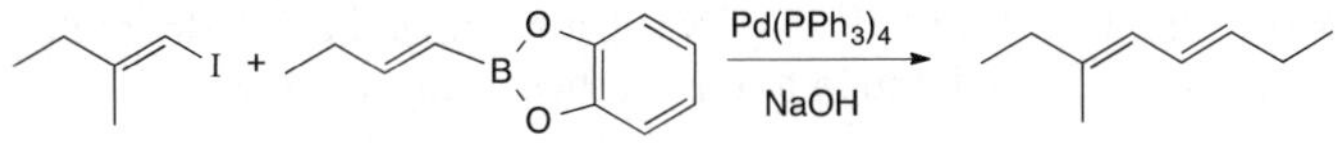

b.

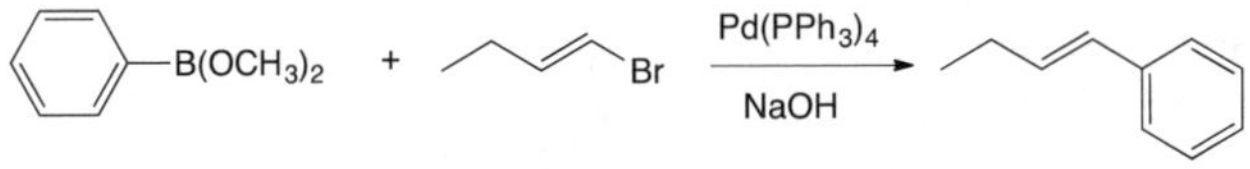

c.

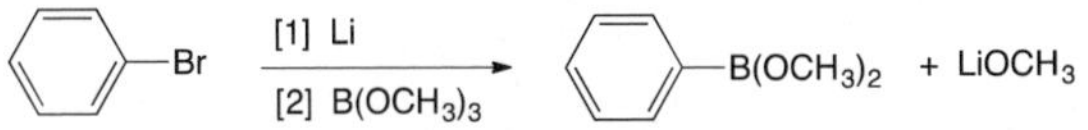

d.

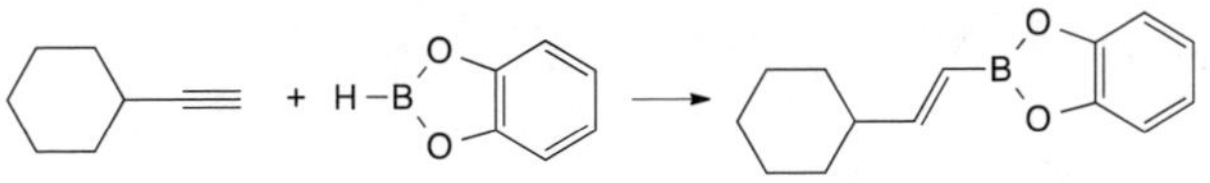

26.5 The Suzuki reaction forms a new carbon–carbon bond between a vinyl halide and an arylborane.

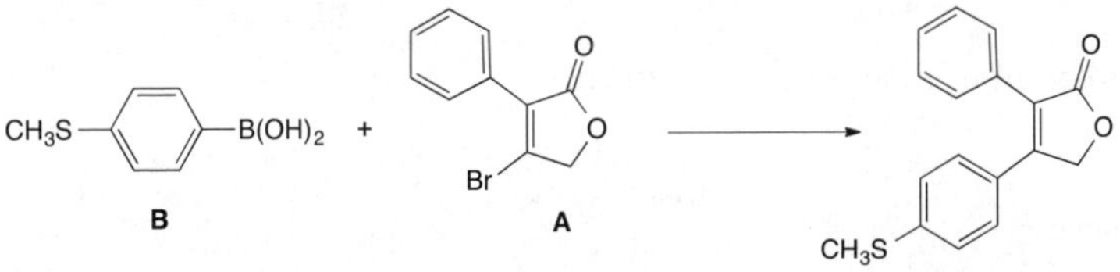

B   A

26.6

a.

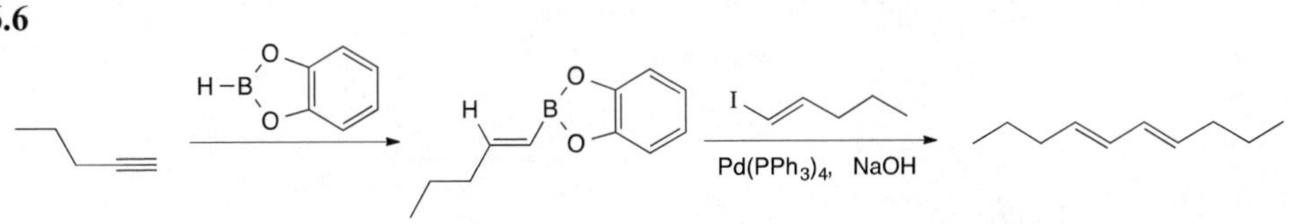

b.

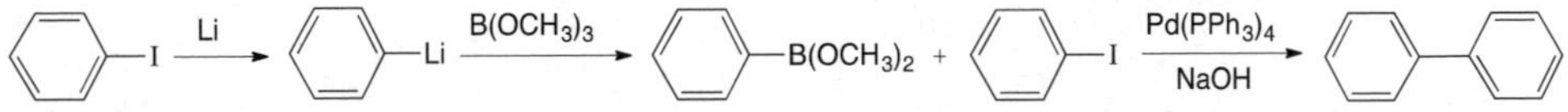

c.

(+ ortho isomer)

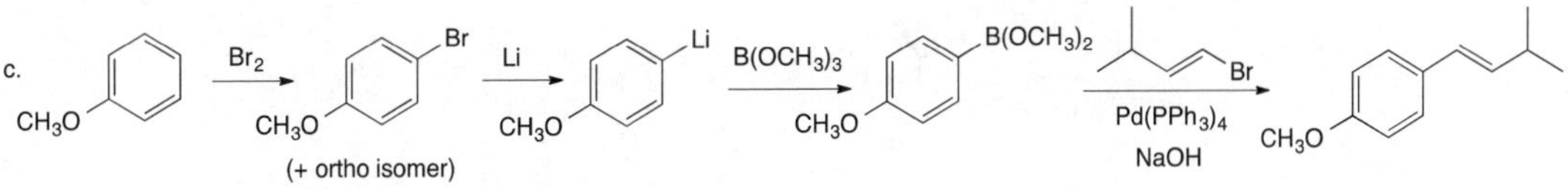

Chapter 26–6

26.7

a. [Br-substituted benzene] + [styrene] →(Heck reaction) [stilbene]

b. [2-methyl-1-propenyl iodide structure] I + [vinyl arene with OCH₃] →(Heck reaction) [product with OCH₃]

c. [acrylonitrile] CN + [allyl bromide structure] Br →(Heck reaction) [diene with CN]

d. [methyl acrylate] OCH₃ / O + [styryl bromide] Br →(Heck reaction) [product with OCH₃ and O]

26.8 Locate the double bond with the aryl, COOR, or CN substituent, and break the molecule into two components at the end of the C=C not bonded to one of these substituents.

a. [aryl cinnamate with OCH₃, O] ⟹ [aryl bromide] + [methyl acrylate OCH₃, O]

b. [stilbene] ⟹ [phenyl bromide] Br + [styrene]

c. [pyran-substituted chain with CO₂CH₃] ⟹ [pyran chain with Br] + [CO₂CH₃ acrylate]

26.9 Add the carbene carbon from either side of the alkene.

a. [propene] →(CHCl₃ / KOC(CH₃)₃) [dichlorocyclopropane, Cl Cl, H] + [dichlorocyclopropane enantiomer, Cl Cl, H]
enantiomers

b. [cis-alkene] →(CHCl₃ / KOC(CH₃)₃) [dichlorocyclopropane Cl Cl, H] + [dichlorocyclopropane Cl Cl, H]
identical

c. [methylcyclohexene] →(CHCl₃ / KOC(CH₃)₃) [bicyclic Cl Cl, H] + [bicyclic Cl Cl, H]
enantiomers

26.10

a. 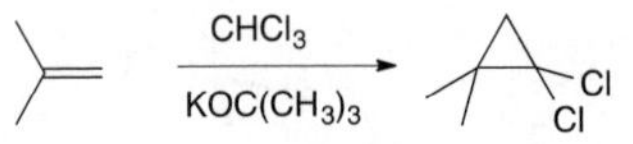
$CHCl_3$
$KOC(CH_3)_3$
Cl
Cl

c. 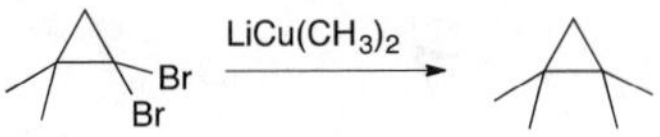
Br
Br
$LiCu(CH_3)_2$

(from b.)

b. 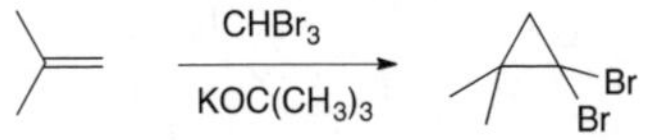
$CHBr_3$
$KOC(CH_3)_3$
Br
Br

26.11

a. 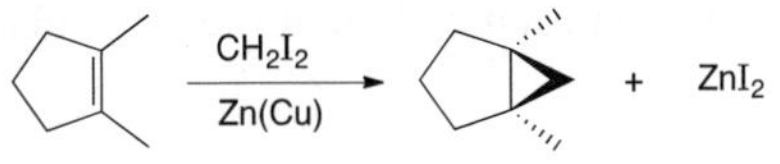
CH_2I_2
Zn(Cu)
+ ZnI_2

c. 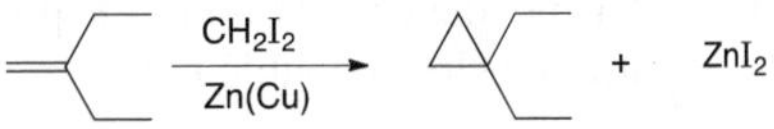
CH_2I_2
Zn(Cu)
+ ZnI_2

b. 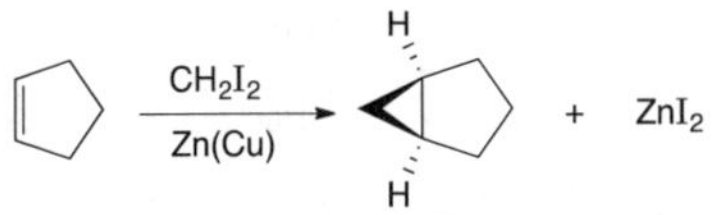
CH_2I_2
Zn(Cu)
H
H
+ ZnI_2

26.12 The relative position of substituents in the reactant is retained in the product.

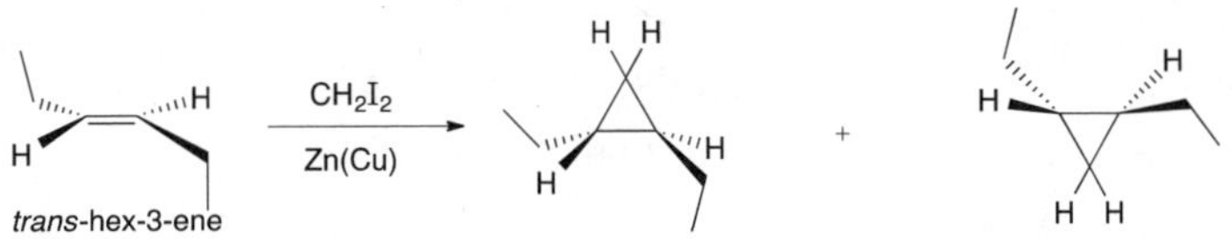
H
H
CH_2I_2
Zn(Cu)
H H
H
H
+
H
H
H H

trans-hex-3-ene

two enantiomers of *trans*-1,2-diethylcyclopropane

26.13

a. 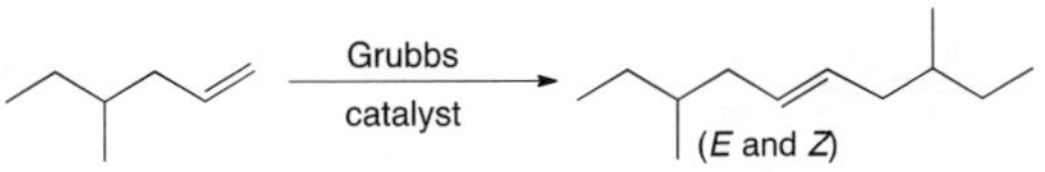
Grubbs
catalyst
(*E* and *Z*)

c. 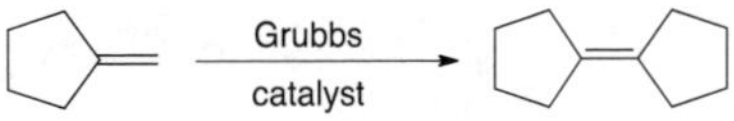
Grubbs
catalyst

b. 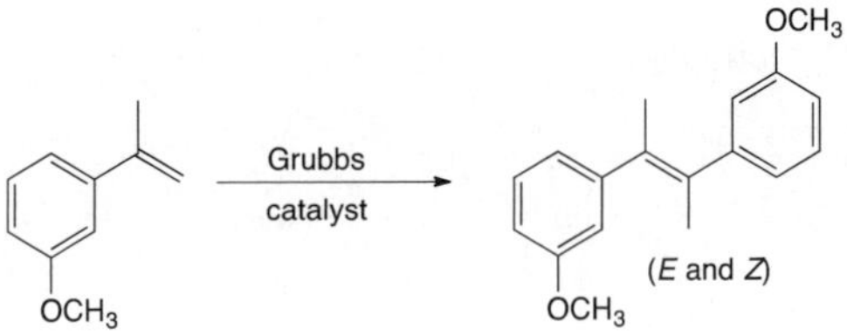
Grubbs
catalyst
OCH_3
OCH_3
OCH_3
(*E* and *Z*)

($CH_2{=}CH_2$ is also formed in each reaction.)

26.14

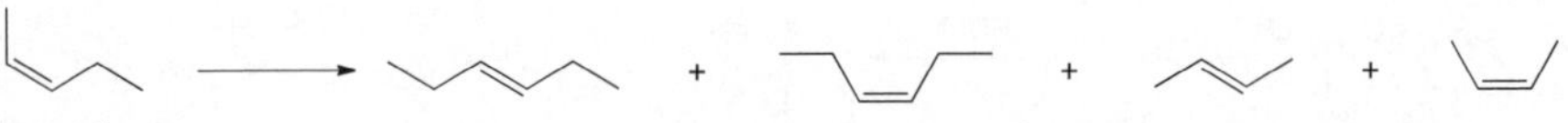
+
+
+

cis-pent-2-ene

There are four products formed in this reaction including stereoisomers, and therefore, it is not a practical method to synthesize 1,2-disubstituted alkenes.

26.15

a. 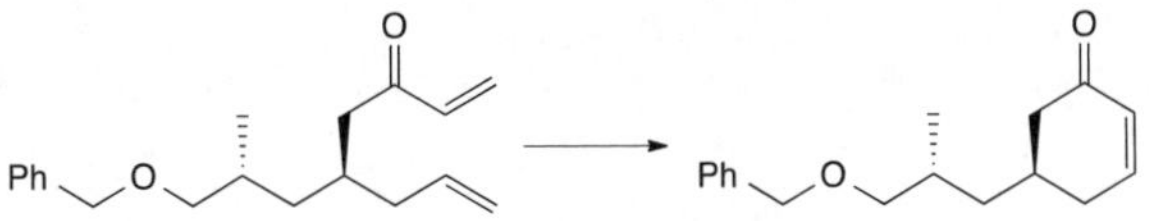
O
Ph
O
O
Ph
O

b. 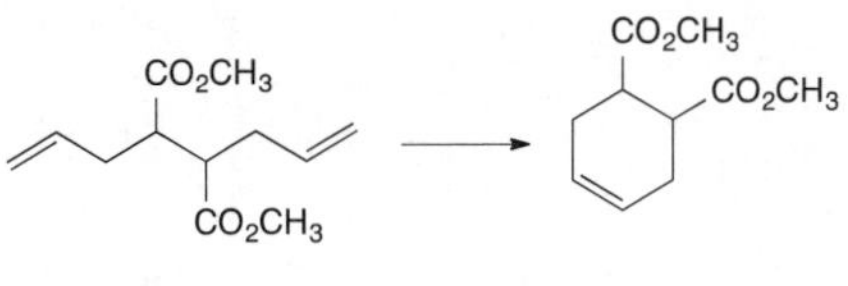
CO_2CH_3
CO_2CH_3
CO_2CH_3
CO_2CH_3

Chapter 26–8

26.16

RCM

new C–C bond here

V

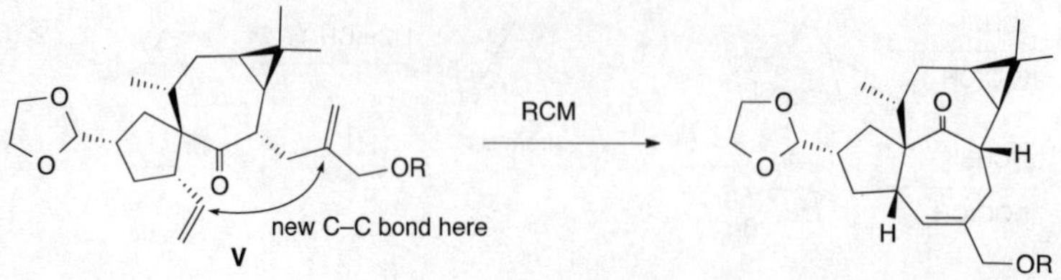

26.17 Cleave the C=C bond in the product, and then bond each carbon of the original alkene to a CH_2 group using a double bond.

a.

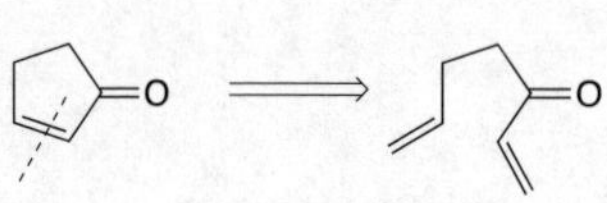

c.

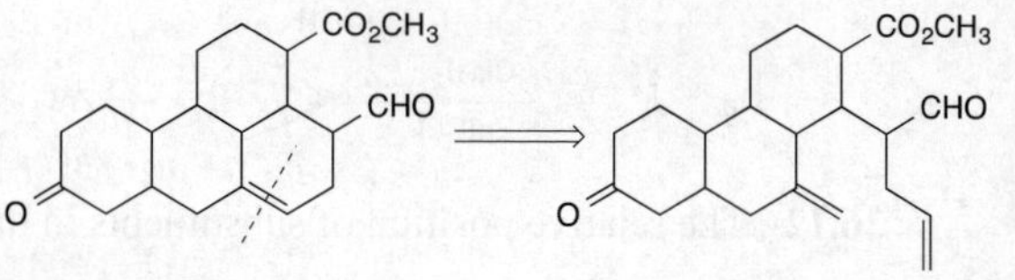

b.

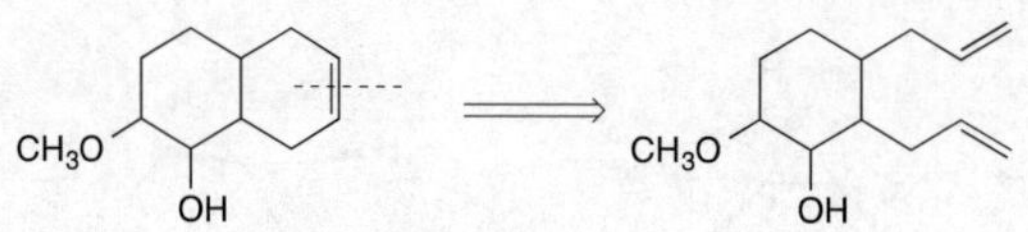

26.18 Inversion of configuration occurs with the substitution of the methyl group for the tosylate.

a.

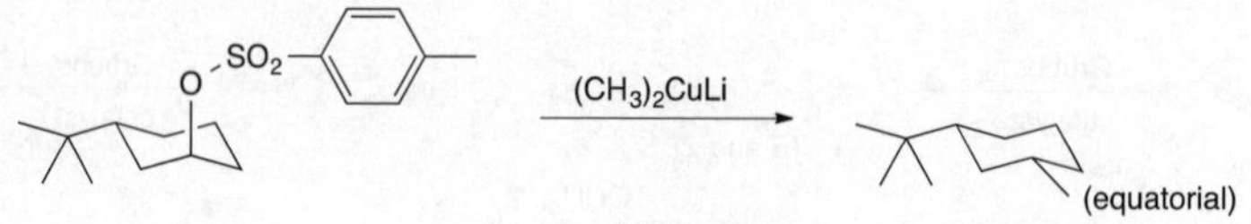

$(CH_3)_2CuLi$

(equatorial)

b.

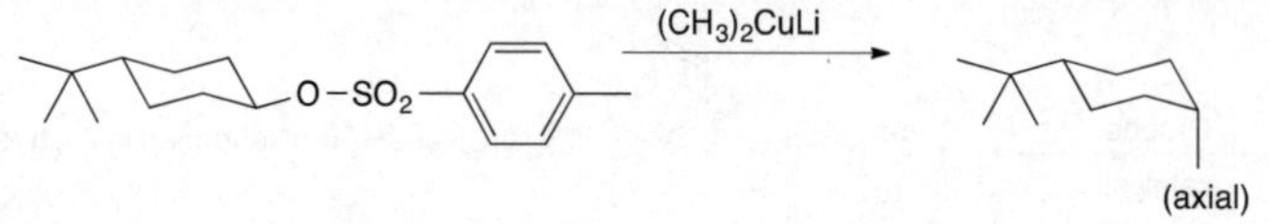

$(CH_3)_2CuLi$

(axial)

26.19

a.

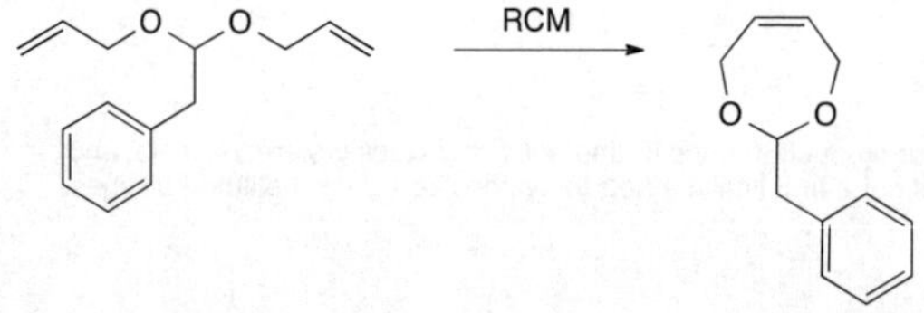

RCM

b.

RCM

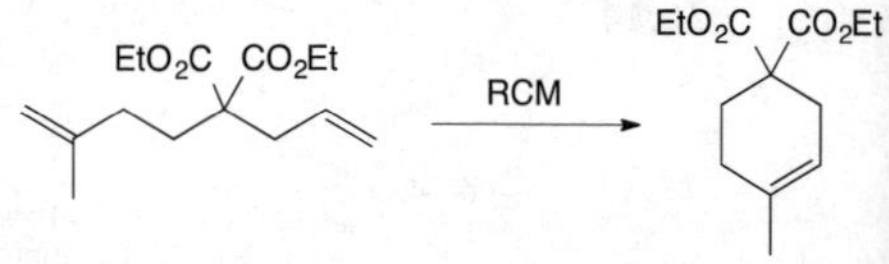

26.20

a. 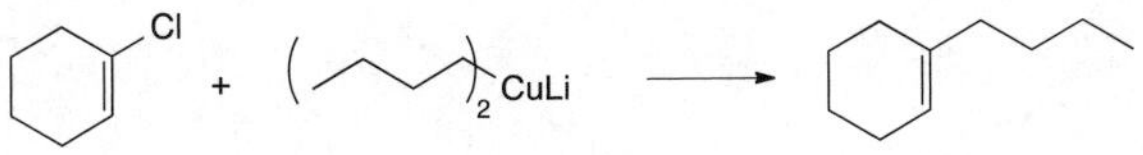
Cl
+
CuLi
2

b. 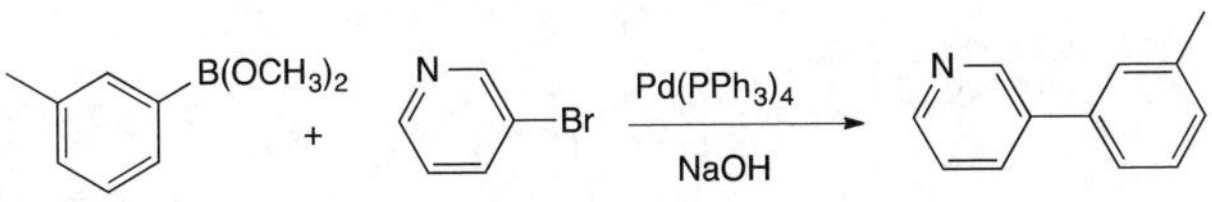
B(OCH3)2
+
N
Br
Pd(PPh3)4
NaOH
N

c. 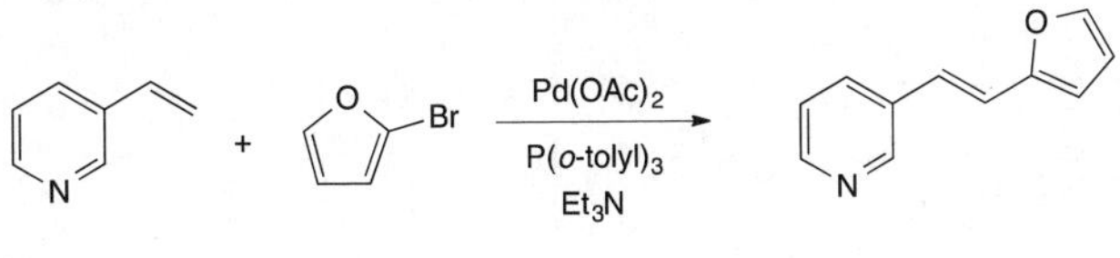
N
+
O
Br
Pd(OAc)2
P(o-tolyl)3
Et3N
O
N

d. 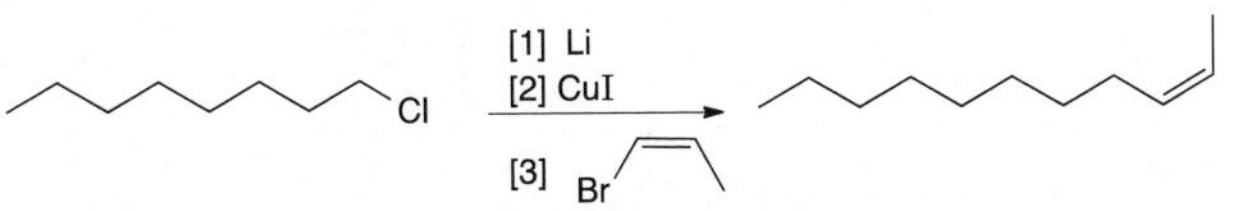
Cl
[1] Li
[2] CuI
[3] Br

e. 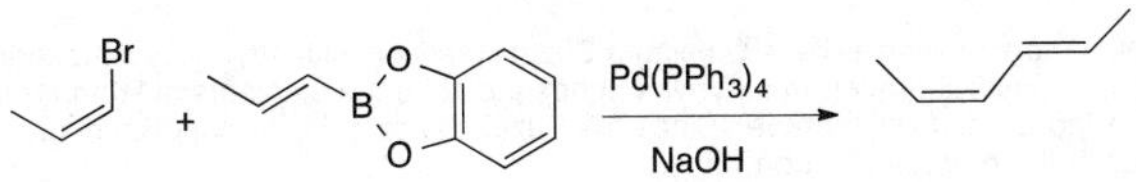
Br
+
O
B
O
Pd(PPh3)4
NaOH

f. 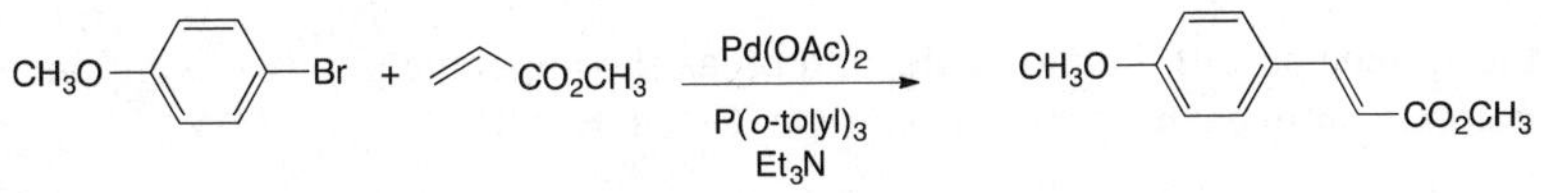
CH3O
Br
+
CO2CH3
Pd(OAc)2
P(o-tolyl)3
Et3N
CH3O
CO2CH3

g. 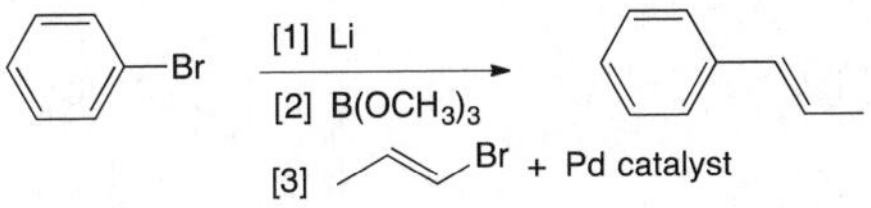
Br
[1] Li
[2] B(OCH3)3
[3] Br + Pd catalyst

h. 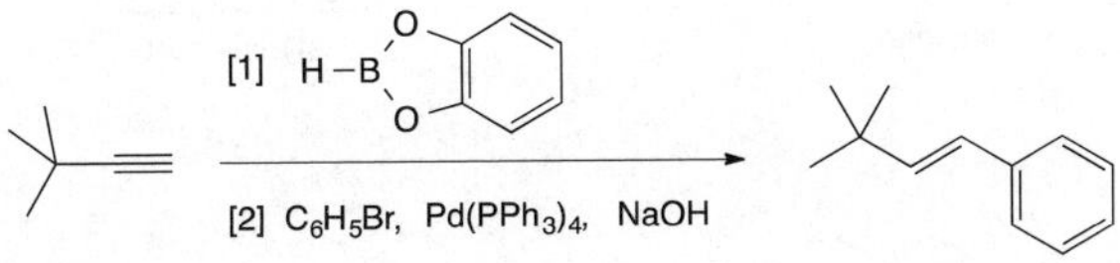
[1] H–B
O
O
[2] C6H5Br, Pd(PPh3)4, NaOH

26.21

a. 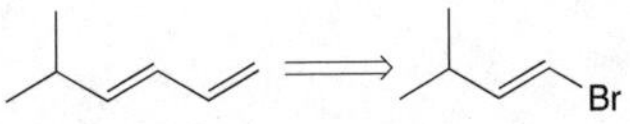
Br

c. 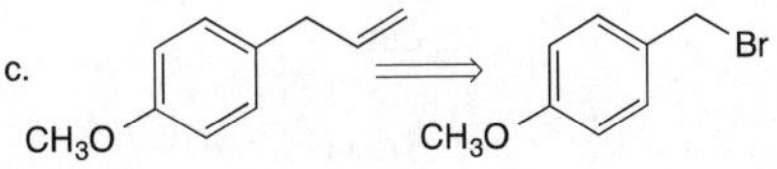
CH3O
Br
CH3O

b. 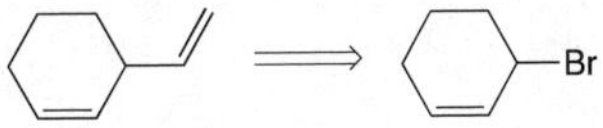
Br

Chapter 26–10

26.22

Each coupling reaction uses Pd(PPh$_3$)$_4$ and NaOH to form the conjugated diene.

ethynylcyclohexane

A

B

C

D

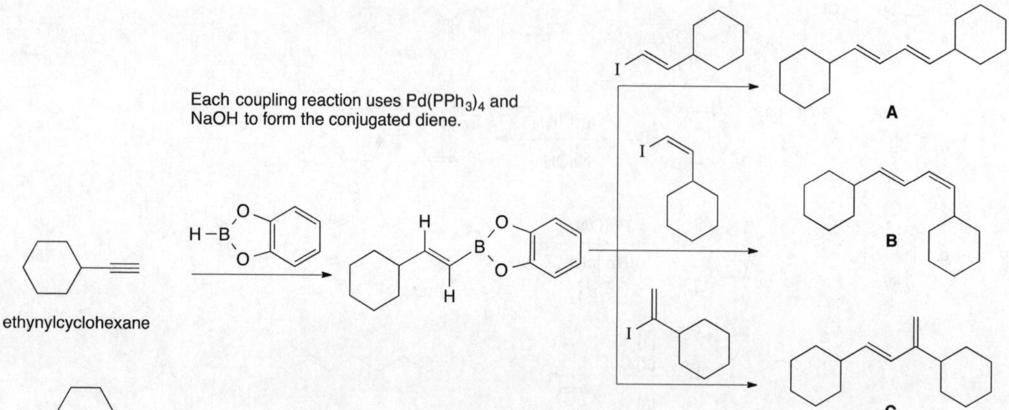

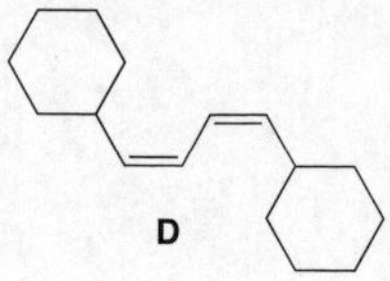

It is not possible to synthesize diene **D** using a Suzuki reaction with ethynylcyclohexane as starting material. Hydroboration of ethynylcyclohexane adds the elements of H and B in a syn fashion, affording a trans vinylborane. Since the Suzuki reaction is stereospecific, one of the double bonds in the product must be trans.

26.23 Locate the styrene part of the molecule, and break the molecule into two components. The second component in each reaction is styrene, C$_6$H$_5$CH=CH$_2$.

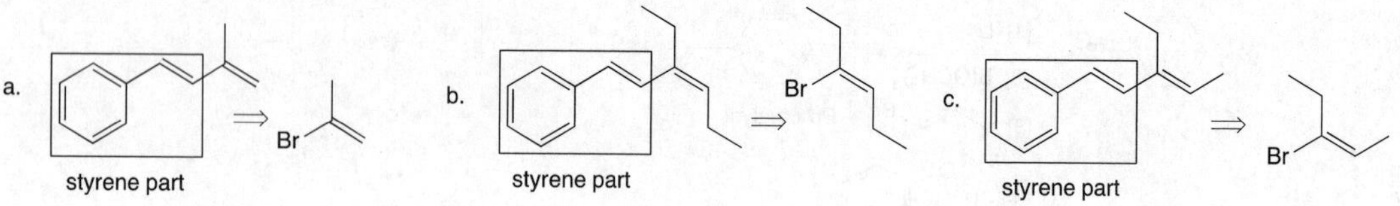

a.

styrene part

b.

styrene part

c.

styrene part

26.24

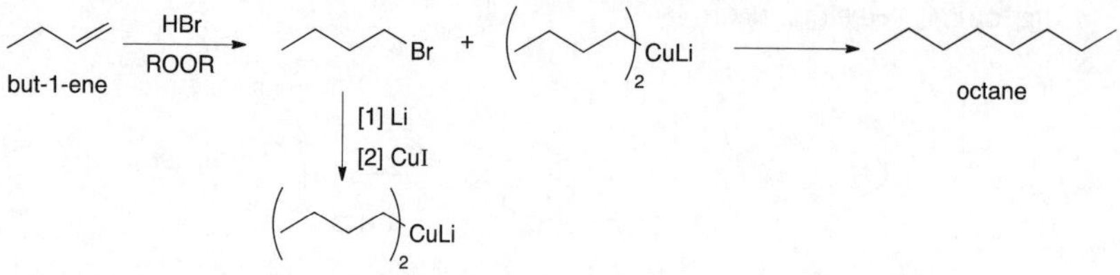

but-1-ene

$\xrightarrow[\text{ROOR}]{\text{HBr}}$

octane

[1] Li
[2] CuI

26.25

A

B

$\xrightarrow{\text{Suzuki reaction}}$

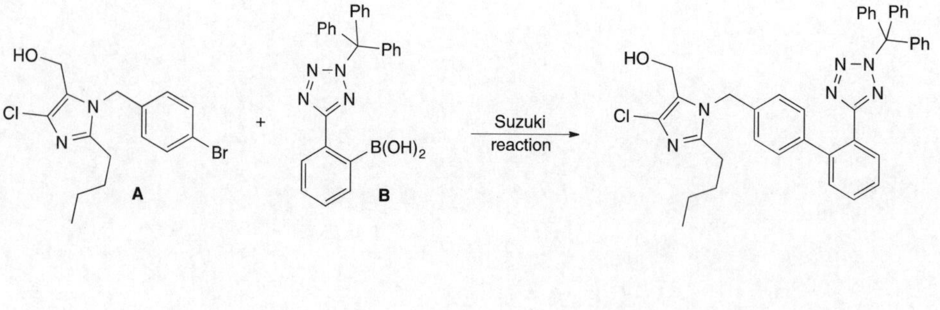

C–C Bond-Forming Reactions 26–11

26.26

26.27 Add the carbene carbon from either side of the alkene.

a.

b.

c.

d.

26.28 The new three-membered ring has a stereogenic center on the C bonded to the phenyl group, so the phenyl group can be oriented in two different ways to afford two stereoisomers. These products are diastereomers of each other.

26.29 High-dilution conditions favor intramolecular metathesis.

a.

b.

c.

Chapter 26–12

26.30 Retrosynthetically break the double bond in the cyclic compound and add a new =CH$_2$ at each end to find the starting material.

a.

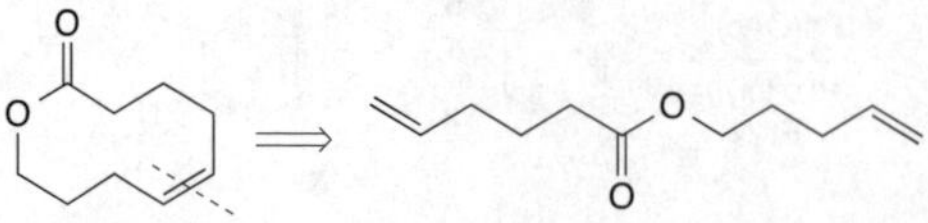

b.

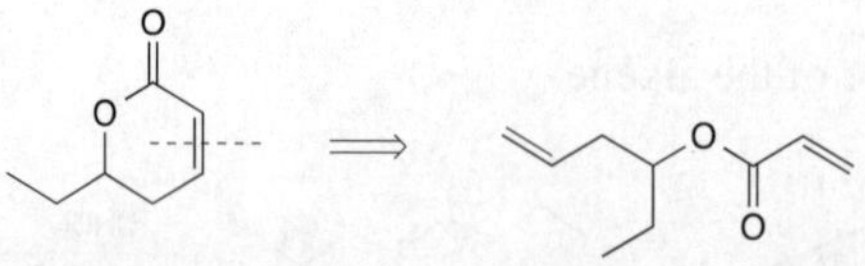

c.

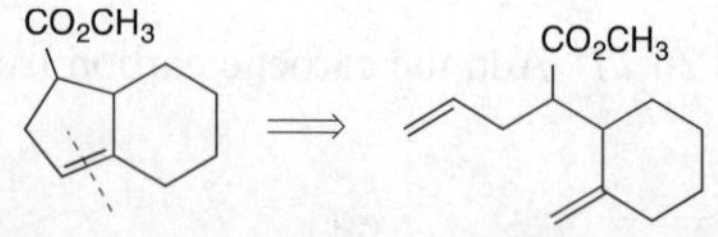

26.31 Alkene metathesis with two different alkenes is synthetically useful only when both alkenes are symmetrically substituted; that is, the two groups on each end of the double bond are identical to the two groups on the other end of the double bond.

a.

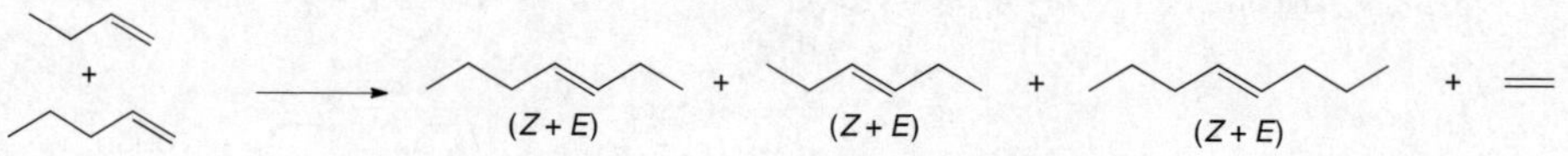

(Z + E)   (Z + E)   (Z + E)

b.

This reaction is synthetically useful because it yields only one product.

c.

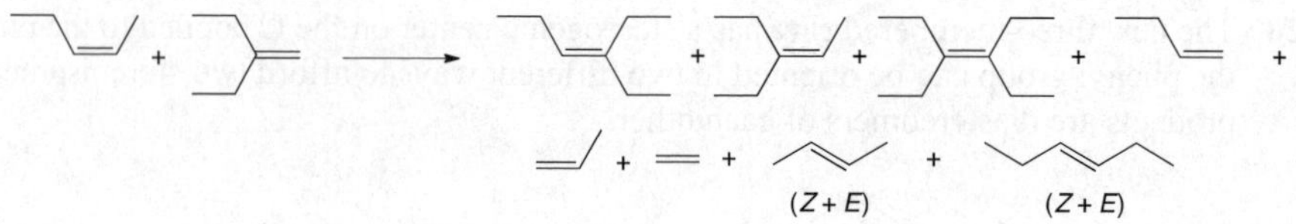

(Z + E)   (Z + E)

26.32

M

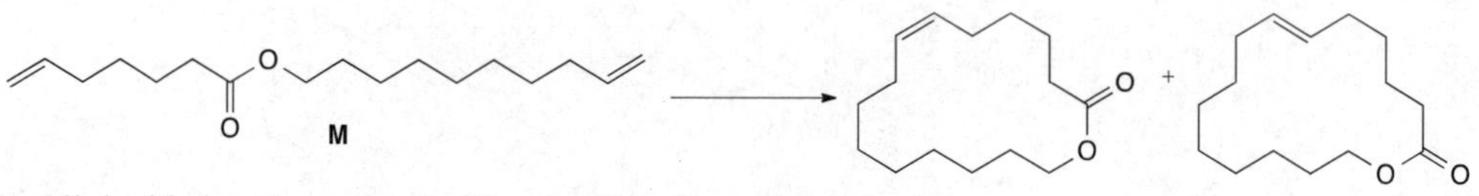

26.33 All double bonds can have either the E or Z configuration.

a.

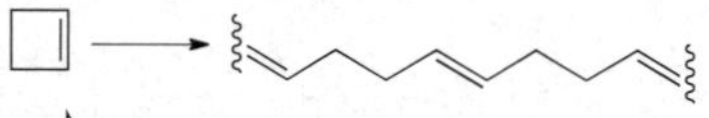

b.

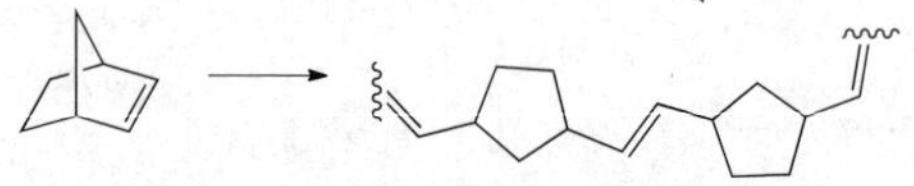

c.

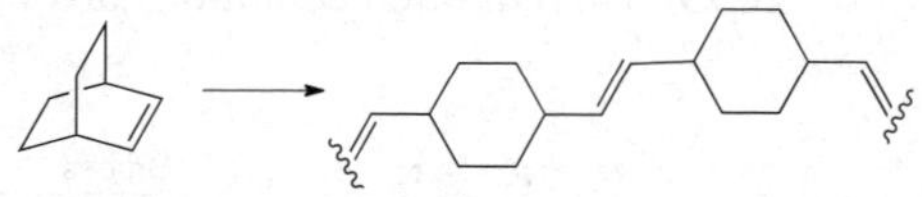

26.34

a.

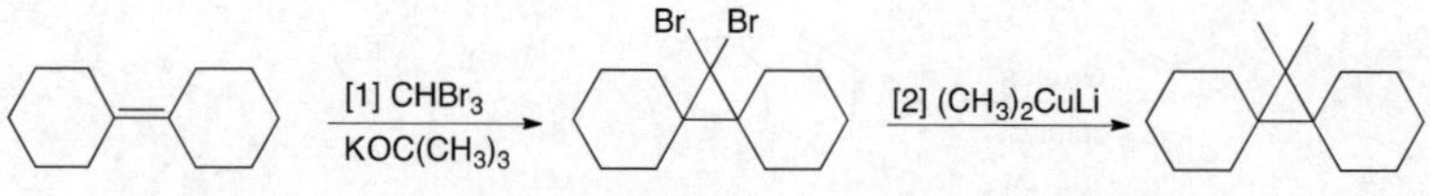

Br   Br

[1] CHBr$_3$
KOC(CH$_3$)$_3$

[2] (CH$_3$)$_2$CuLi

b.

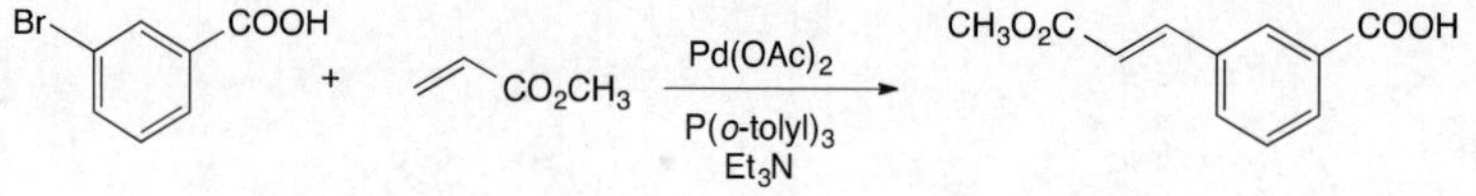

Br   COOH

+   CO$_2$CH$_3$

Pd(OAc)$_2$
P(o-tolyl)$_3$
Et$_3$N

CH$_3$O$_2$C   COOH

C–C Bond-Forming Reactions 26–13

c.

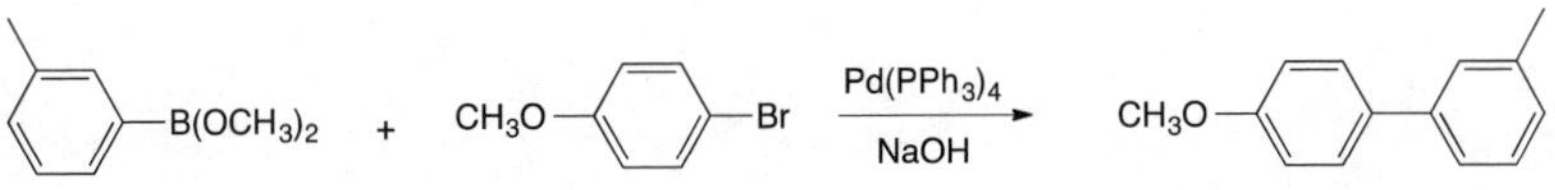

d.

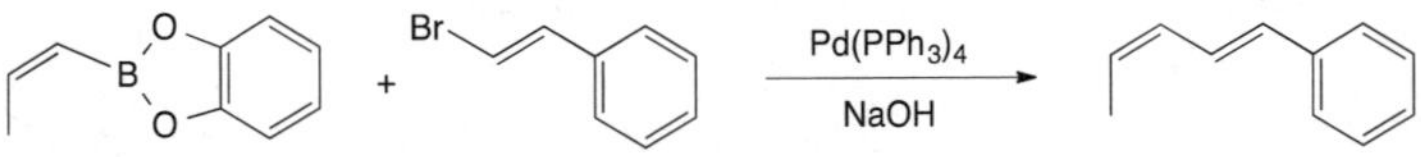

e.

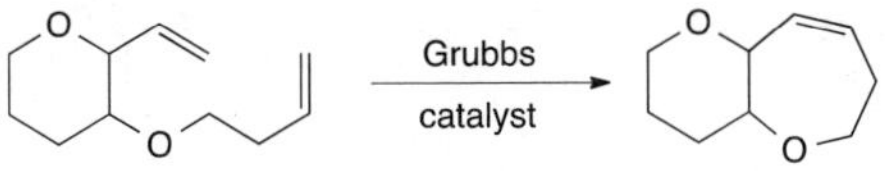

f.

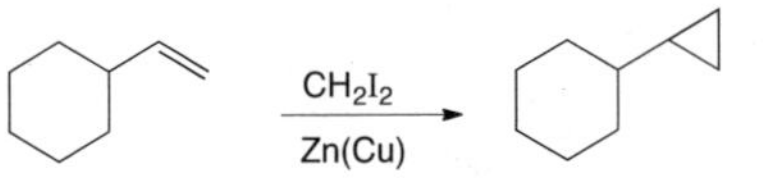

g.

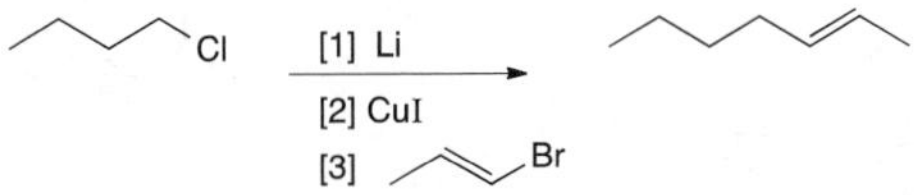

h.

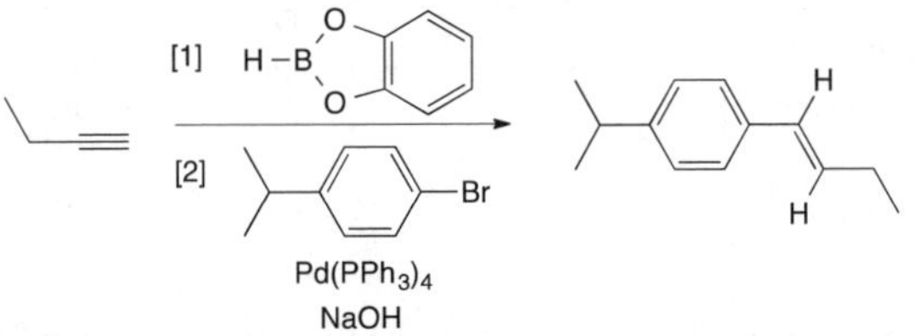

26.35

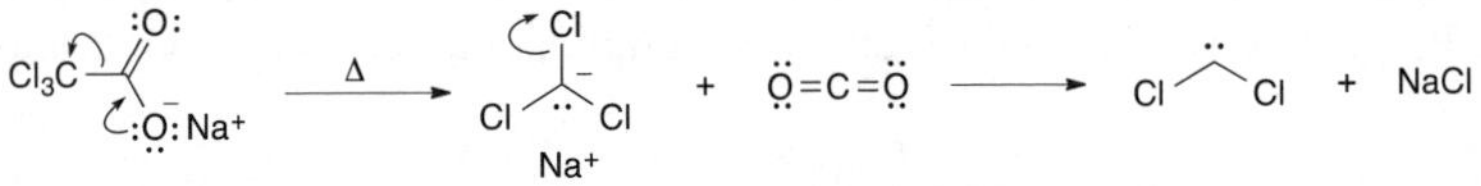

26.36 This reaction follows the Simmons–Smith reaction mechanism illustrated in Mechanism 26.5.

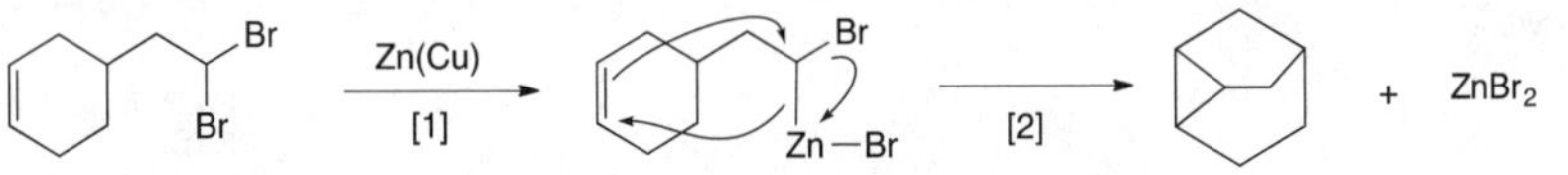

26.37

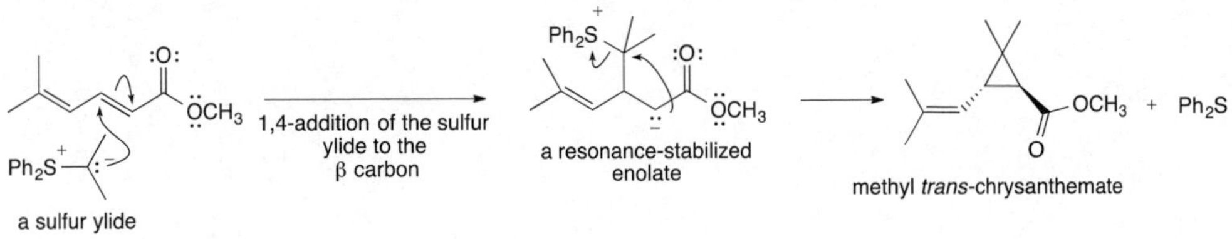

Chapter 26–14

26.38

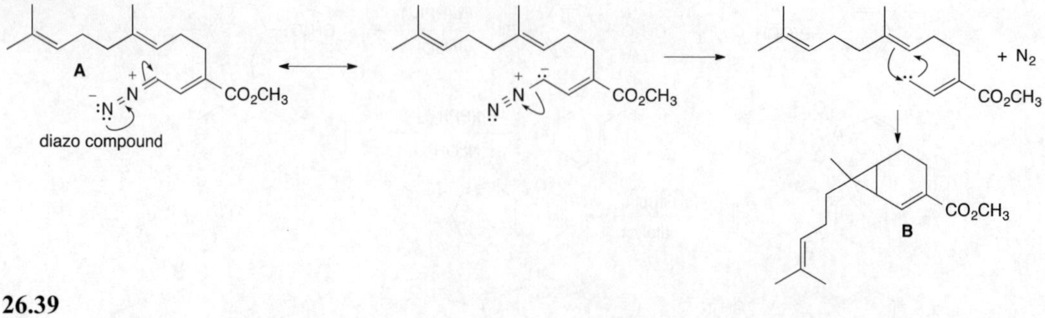

26.39

a.

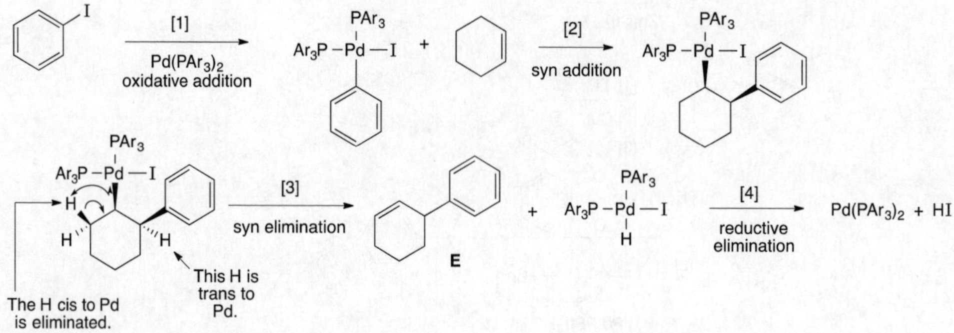

b. This suggests that the stereochemistry in Step [3] must occur with syn elimination of H and Pd to form **E**. Product **F** cannot form because the only H on the C bonded to the benzene ring is trans to the Pd species, so it cannot be removed if elimination occurs in a syn fashion.

26.40

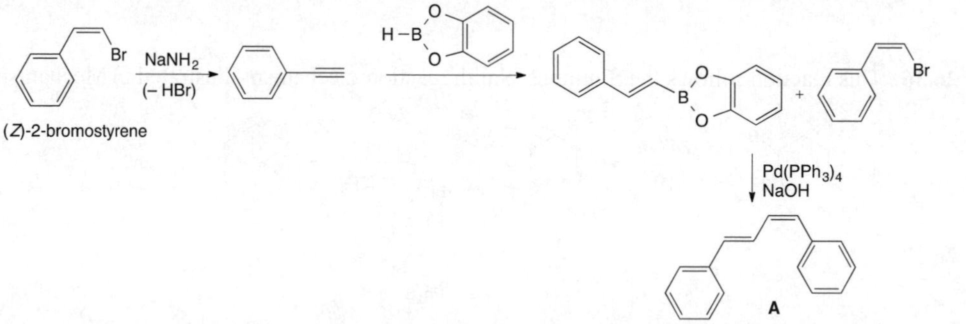

26.41

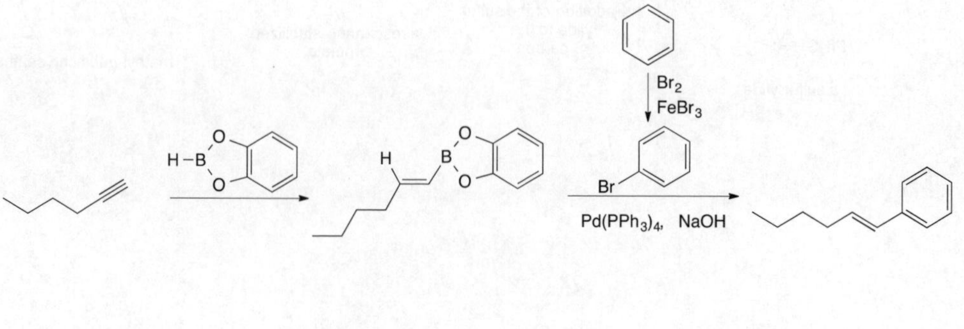

26.42

HO/\/\/\/\OH → (SOCl₂) → HO/\/\/\/\Cl

TBDMS–Cl
imidazole

HC≡CH → (NaH) → HC≡C⁻ + TBDMSO/\/\/\/\Cl

/\/\/\/\OTBDMS → (catecholborane BH) → (catecholboronate)–CH=CH–/\/\/\/\OTBDMS

Bu₄NF

(catecholboronate)–CH=CH–/\/\/\/\OH

26.43

a. CH_3O–C₆H₄–CH=CH–C₆H₅ ⟹ CH_3O–C₆H₄–Br + CH₂=CH–C₆H₅

Synthesize these two components, and then use a Heck reaction to synthesize the product.

C₆H₆ → (HNO₃, H₂SO₄) → C₆H₅–NO₂ → (H₂, Pd-C) → C₆H₅–NH₂ → (NaNO₂, HCl) → C₆H₅–N₂⁺Cl⁻ → (H₂O) → C₆H₅–OH

C₆H₅–OH → [1] NaH [2] CH₃Br → CH_3O–C₆H₅ → (Br₂) → CH_3O–C₆H₄–Br

PBr₃
CH₃OH

C₆H₆ → (CH₃CH₂OH; SOCl₂ → CH₃CHClCH₃; AlCl₃) → C₆H₅–CH₂CH₃ → (Br₂, hν) → C₆H₅–CHBr–CH₃ → (KOC(CH₃)₃) → C₆H₅–CH=CH₂

CH_3O–C₆H₄–Br + CH₂=CH–C₆H₅ → (Pd(OAc)₂, P(o-tolyl)₃, Et₃N) → CH_3O–C₆H₄–CH=CH–C₆H₅

b. CH_3O–C₆H₄–Br (from a.) → (CH₃CH₂OH; SOCl₂ → CH₃CHClCH₃; AlCl₃) → ethyl-CH_3O-C₆H₃-Br

ethyl-CH_3O-C₆H₃-Br + CH₂=CH–CO_2Et → (Pd(OAc)₂, P(o-tolyl)₃, Et₃N) → ethyl-CH_3O-C₆H₃-CH=CH–CO_2Et → (H₂, Pd-C) → ethyl-CH_3O-C₆H₃-CH₂CH₂–CO_2Et

C₆H₆ → (Br₂, FeBr₃) → C₆H₅–Br → (Mg) → C₆H₅–MgBr → [1] (2 equiv) [2] H₂O → (tertiary alcohol product)

Chapter 26–16

26.44

a.

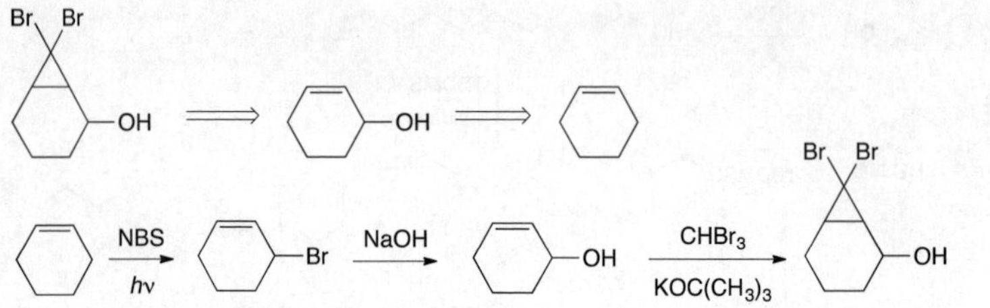

b.

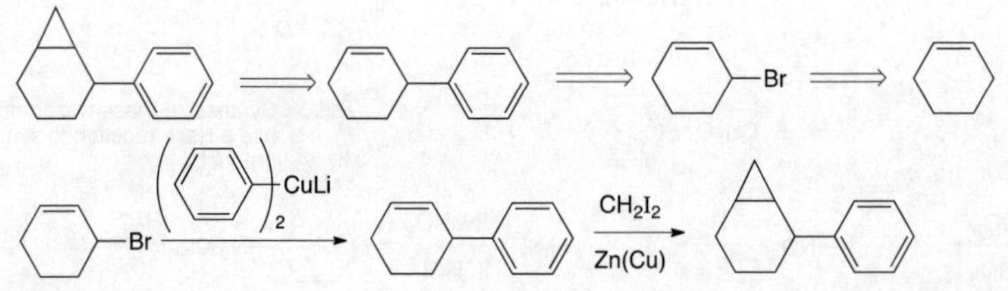

(from a.)

26.45

a.

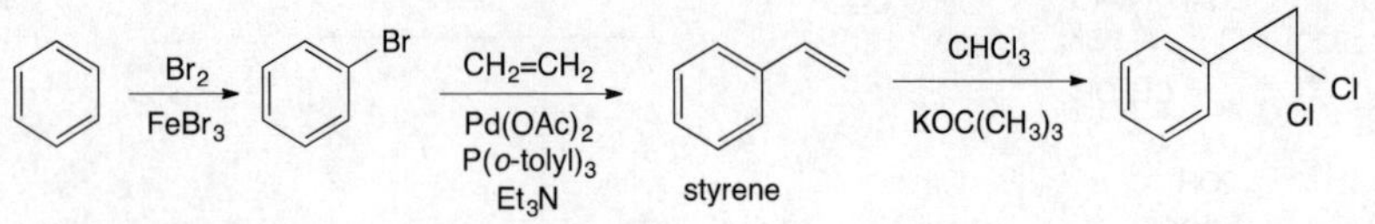

b.

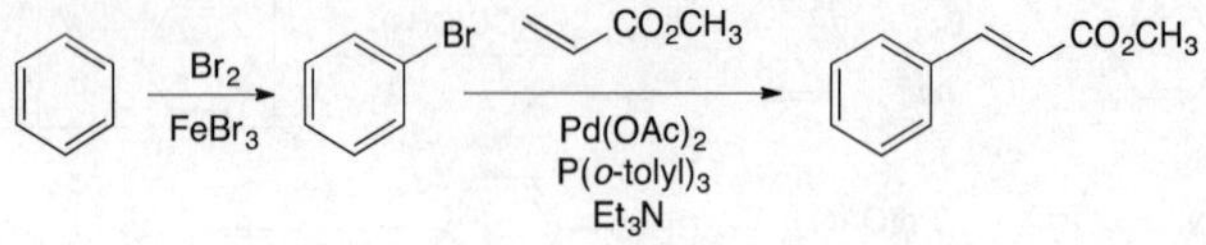

26.46

a.

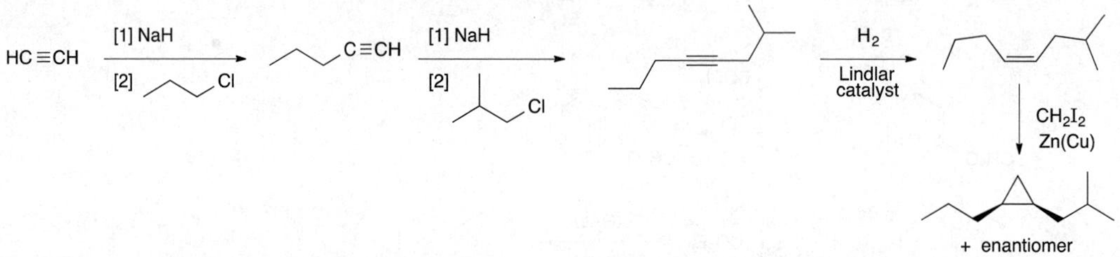

b.

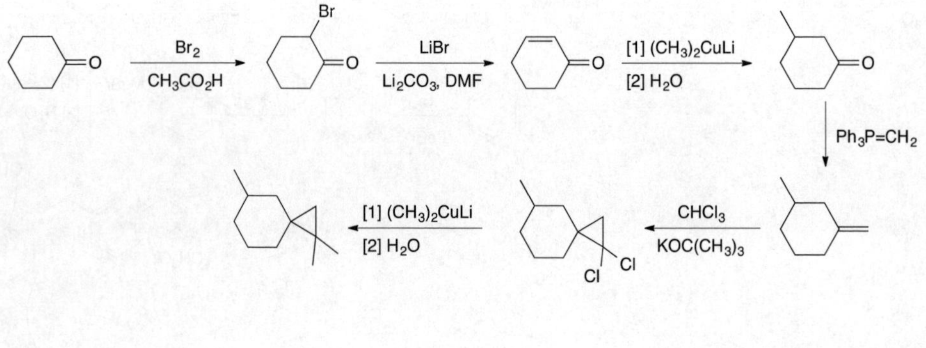

C–C Bond-Forming Reactions 26–17

26.47

a.

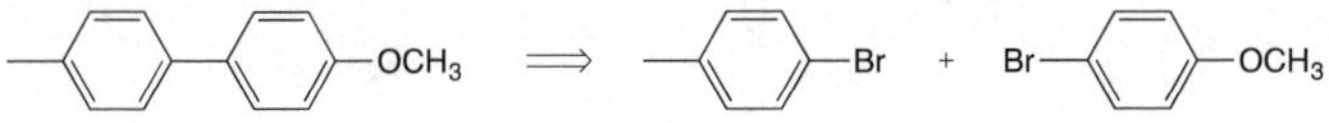

Either compound can be used to synthesize the organoborane, so two routes are possible.

Possibility [1]:

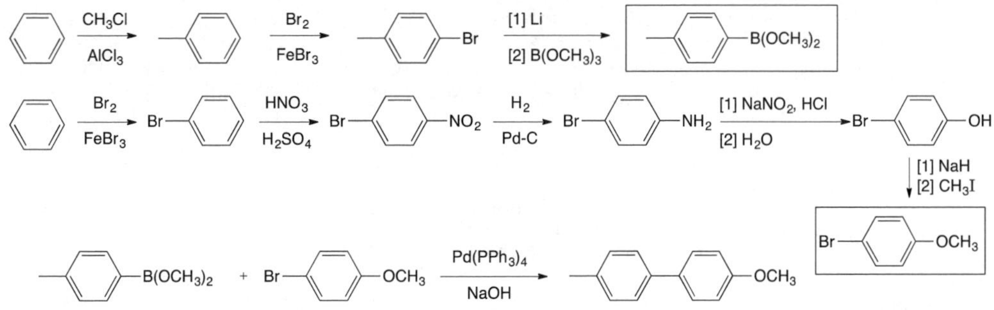

Possibility [2]:

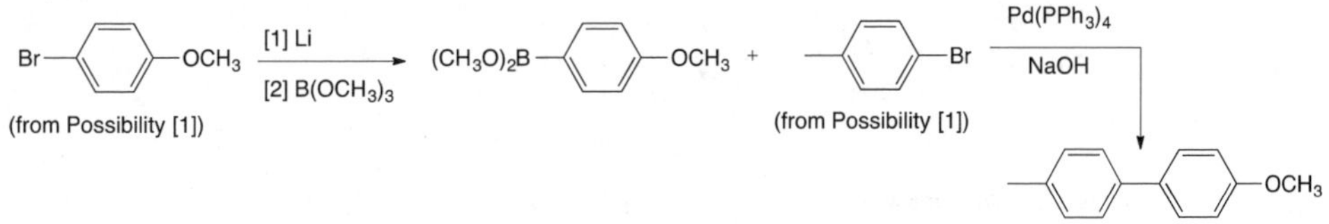

(from Possibility [1])

(from Possibility [1])

b.

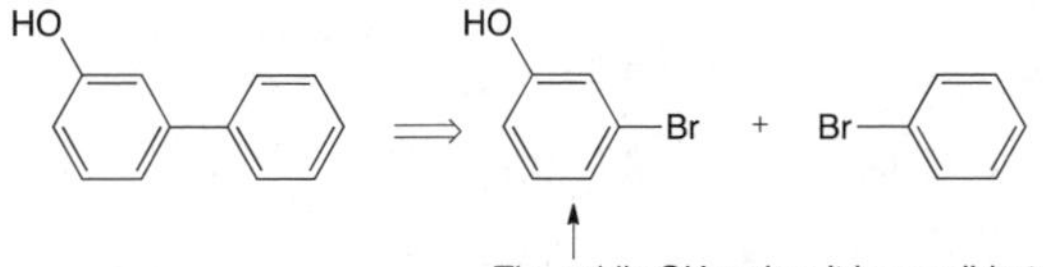

The acidic OH makes it impossible to prepare an organolithium reagent from this aryl halide, so this compound must be used as the aryl halide that couples with the organoborane from bromobenzene.

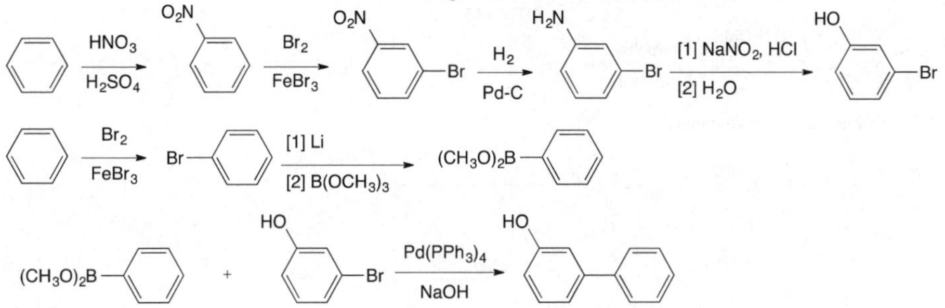

c.

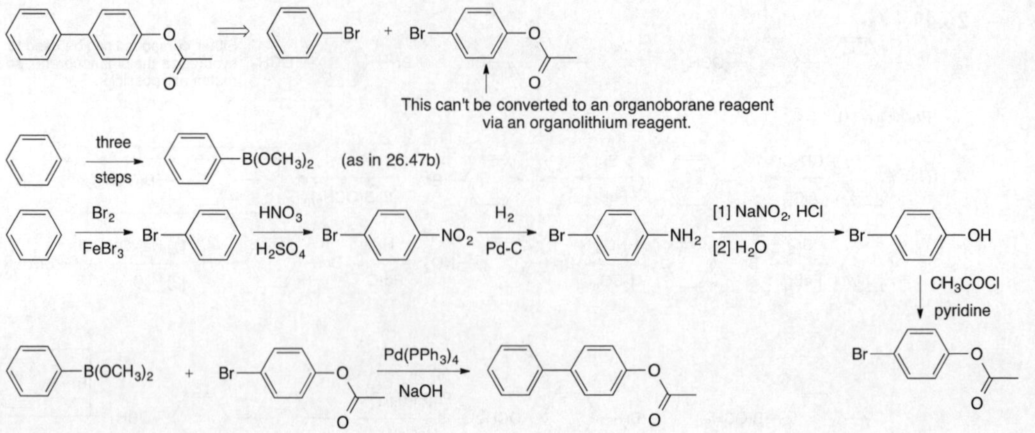

O
Br
Br
O
O
O
This can't be converted to an organoborane reagent via an organolithium reagent.
three steps
B(OCH3)2
(as in 26.47b)
Br2
FeBr3
HNO3
H2SO4
NO2
H2
Pd-C
NH2
[1] NaNO2, HCl
[2] H2O
OH
CH3COCl
pyridine
Pd(PPh3)4
NaOH

26.48

a.

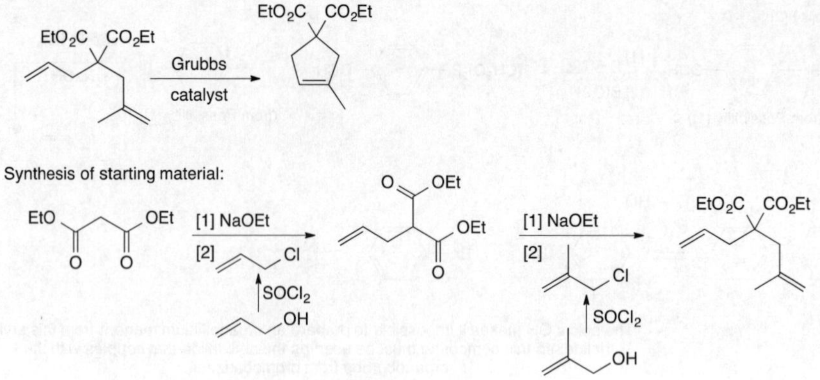

EtO2C CO2Et
Grubbs catalyst
Synthesis of starting material:
EtO
OEt
[1] NaOEt
[2]
Cl
SOCl2
OH

b.

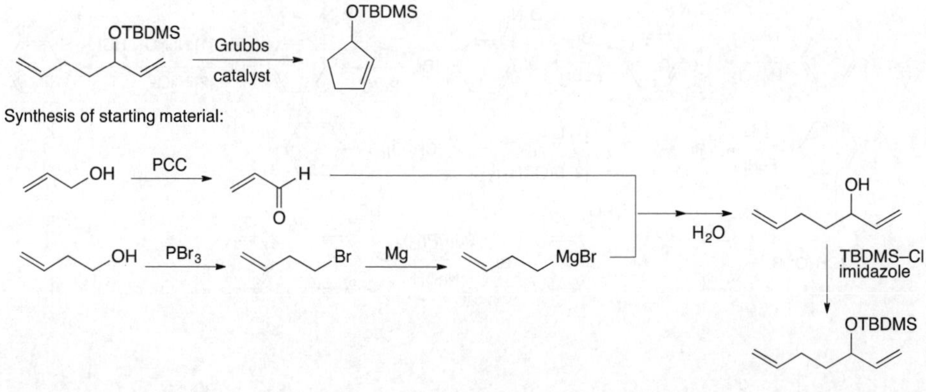

OTBDMS
Grubbs catalyst
Synthesis of starting material:
OH
PCC
H
O
PBr3
Br
Mg
MgBr
H2O
TBDMS–Cl imidazole

26.49

a.

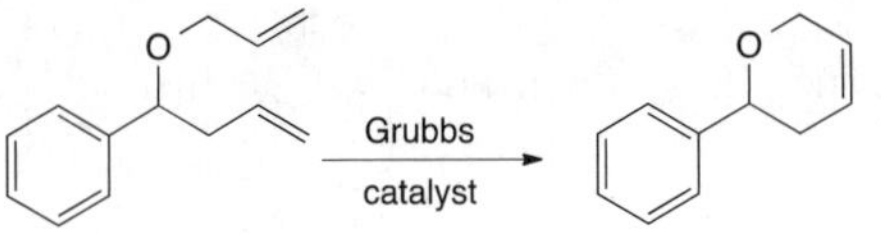

Grubbs catalyst

Synthesis of starting material:

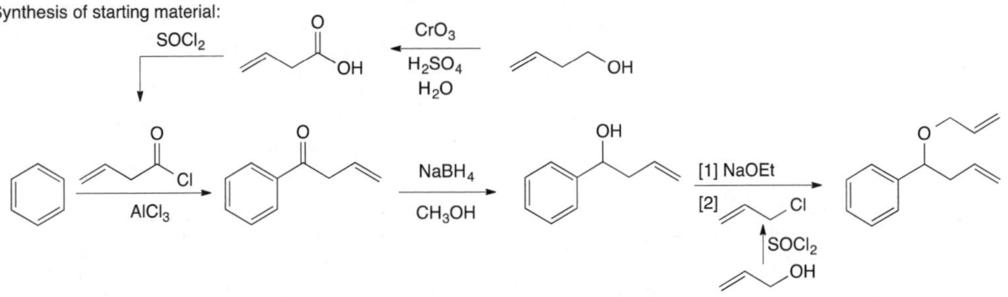

SOCl$_2$

CrO$_3$
H$_2$SO$_4$
H$_2$O

AlCl$_3$

NaBH$_4$
CH$_3$OH

[1] NaOEt
[2]

Cl

SOCl$_2$

OH

b.

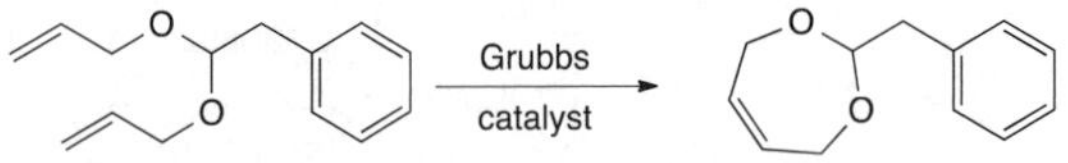

Grubbs catalyst

Synthesis of starting material:

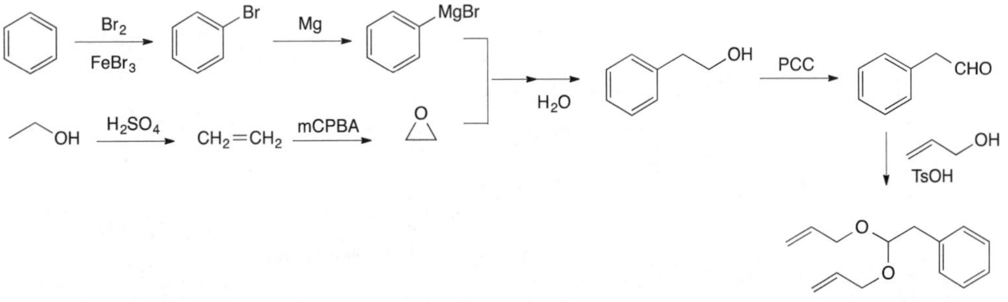

Br$_2$
FeBr$_3$

Mg

MgBr

H$_2$O

PCC

CHO

OH
H$_2$SO$_4$
CH$_2$=CH$_2$
mCPBA

OH
TsOH

26.50

CH$_3$O

Cl

N

O

LiCu

OTBDMS

)$_2$

CH$_3$O

Cl

N

O

OTBDMS

[1] Bu$_4$N$^+$ F$^-$
[2] PCC

CH$_3$O

Cl

N

O

CHO

directed aldol
[1] CH$_3$CHO + LDA → $^-$CH$_2$CHO
[2] H$_3$O$^+$

several steps

CH$_3$O

Cl

N

O

CHO

A

maytansine

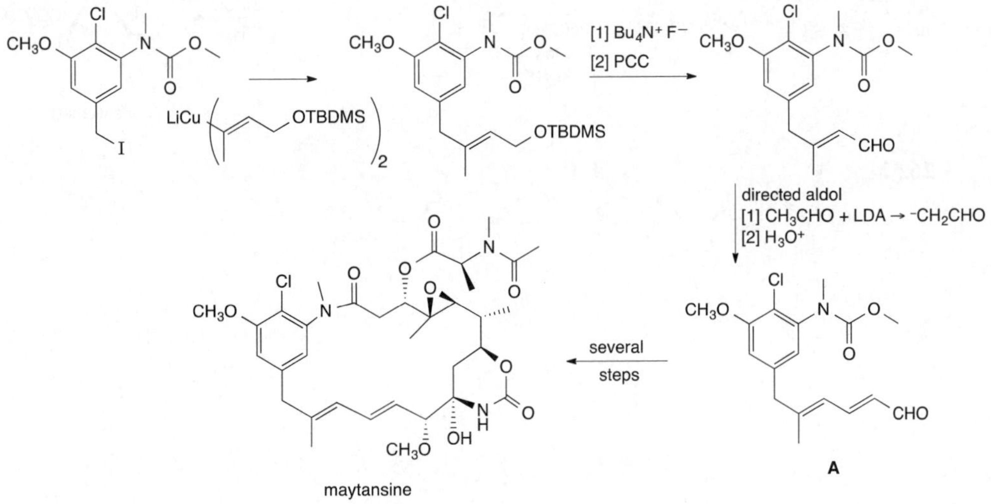

Chapter 26–20

26.51

a.

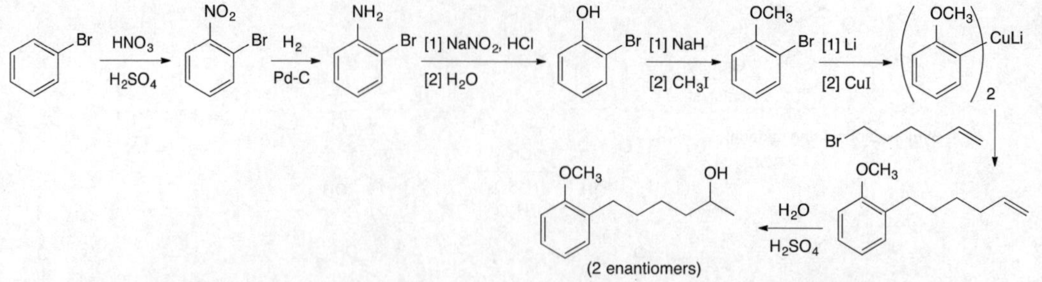

b.

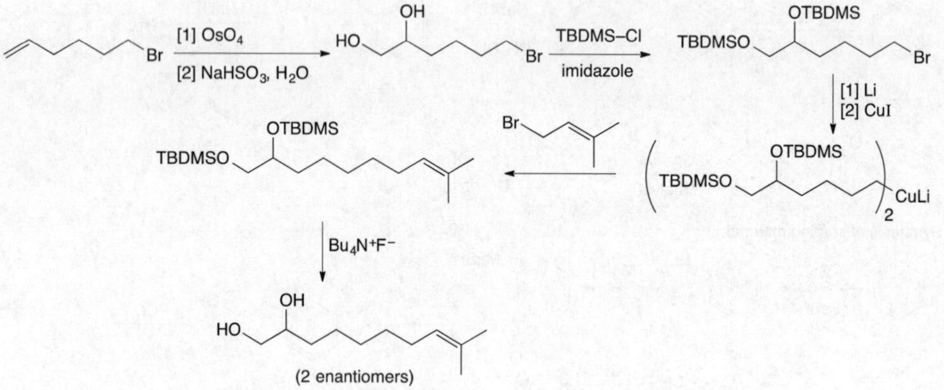

c.

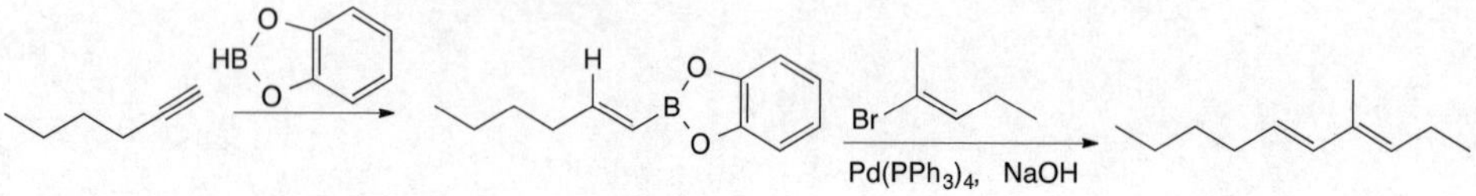

d.

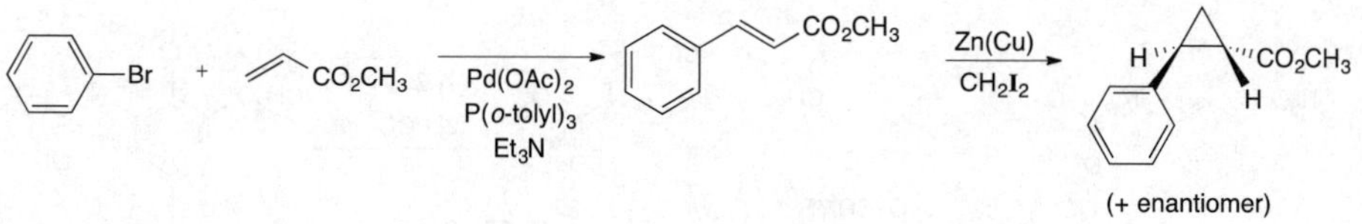

(+ enantiomer)

26.52

a.

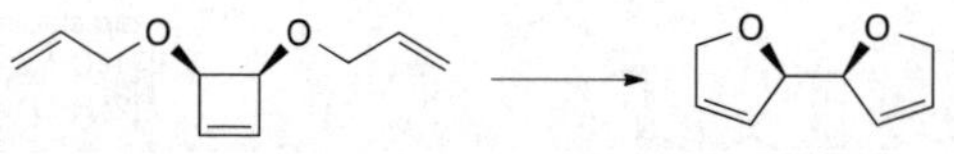

b.

26.53 There is more than one way to form **Z** by metathesis reactions. One possibility involves ring opening of the bicyclic alkene followed by successive ring closures to generate the five- and seven-membered rings.

Chapter 26–22

26.54

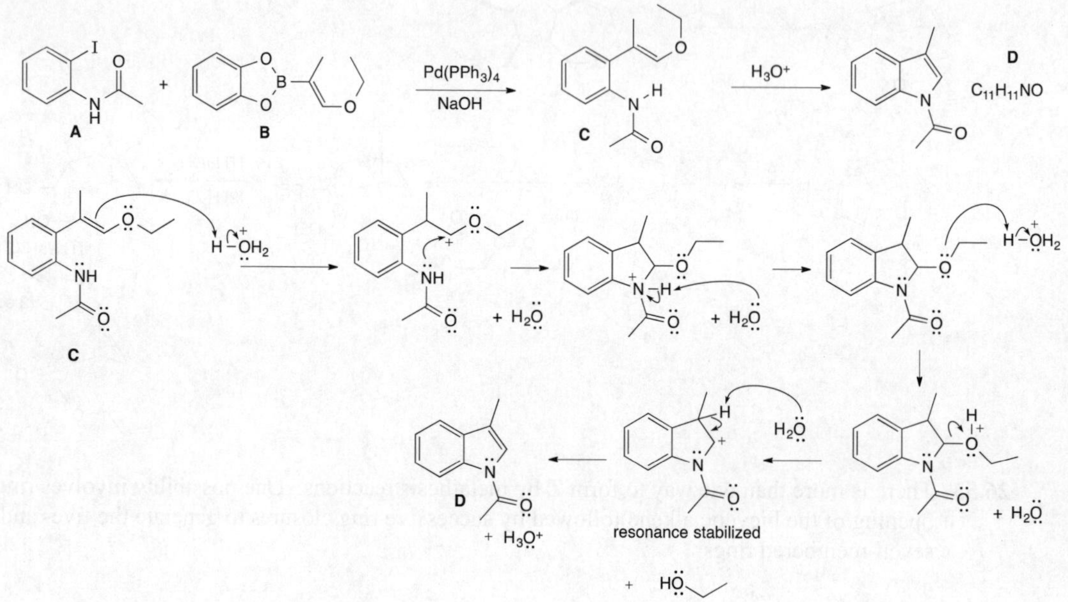

26.55

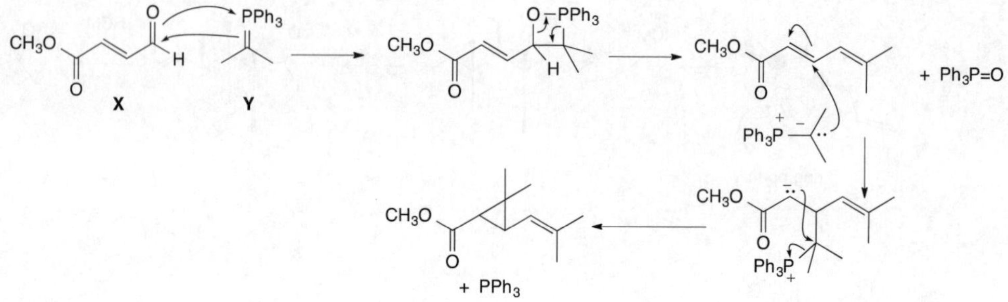

26.56

a. Reaction of a terminal alkene with the catalyst forms a metal–carbene that undergoes an intramolecular reaction with the triple bond, generating a new metal–carbene. A second intramolecular reaction forms the bicyclic product.

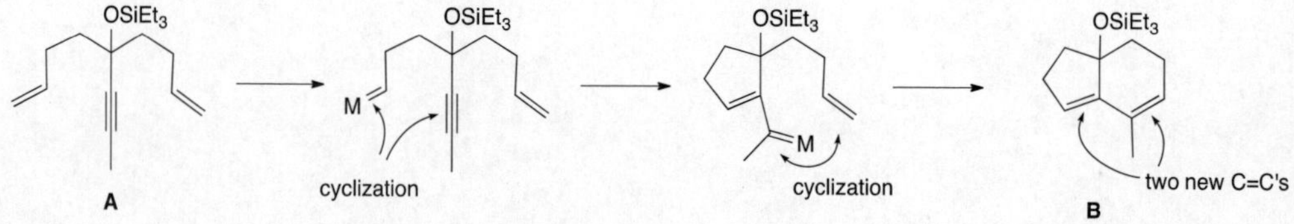

b. Two products are possible because the cascade of reactions can begin at two different double bonds.

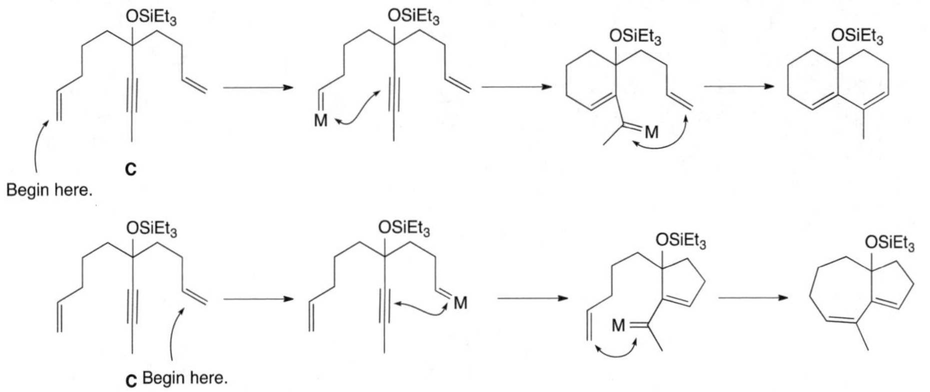

$OSiEt_3$ $OSiEt_3$ $OSiEt_3$ $OSiEt_3$

M

C

Begin here.

$OSiEt_3$ $OSiEt_3$ $OSiEt_3$ $OSiEt_3$

M

C Begin here.

Chapter 27 Pericyclic Reactions

Chapter Review

Electrocyclic reactions (27.3)

Woodward–Hoffmann rules for electrocyclic reactions

Number of π bonds	Thermal reaction	Photochemical reaction
Even	Conrotatory	Disrotatory
Odd	Disrotatory	Conrotatory

Examples
The stereochemistry of a thermal electrocyclic reaction is opposite to that of a photochemical electrocyclic reaction.

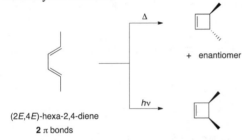

(2E,4E)-hexa-2,4-diene
2 π bonds

- A thermal electrocyclic reaction with an even number of π bonds occurs in a conrotatory fashion.

- A photochemical electrocyclic reaction with an even number of π bonds occurs in a disrotatory fashion.

Cycloaddition reactions (27.4)

Woodward–Hoffmann rules for cycloaddition reactions

Number of π bonds	Thermal reaction	Photochemical reaction
Even	Antarafacial	Suprafacial
Odd	Suprafacial	Antarafacial

Examples
[1] A thermal [4 + 2] cycloaddition takes place in a suprafacial fashion with an odd number of π bonds. An antarafacial photochemical [4 + 2] cycloaddition to form a six-membered ring cannot occur, because of the geometrical constraints of forming a six-membered ring.

$$\text{diene} + CH_2=CH_2 \xrightarrow[\text{suprafacial}]{\Delta} \text{cis product only}$$

Chapter 27–2

[2] A photochemical [2 + 2] cycloaddition takes place in a suprafacial fashion with an even number of π bonds. An antarafacial thermal [2 + 2] cycloaddition to form a four-membered ring cannot occur, because of the geometrical constraints of forming a four-membered ring.

Sigmatropic rearrangements (27.5)

Woodward–Hoffmann rules for sigmatropic rearrangements

Number of electron pairs	Thermal reaction	Photochemical reaction
Even	Antarafacial	Suprafacial
Odd	Suprafacial	Antarafacial

Examples

[1] A **Cope rearrangement** is a thermal [3,3] sigmatropic rearrangement that converts a 1,5-diene into an isomeric 1,5-diene.

[2] An **oxy-Cope rearrangement** is a thermal [3,3] sigmatropic rearrangement that converts a 1,5-dien-3-ol into a δ,ε-unsaturated carbonyl compound, after tautomerization of an intermediate enol.

[3] A **Claisen rearrangement** is a thermal [3,3] sigmatropic rearrangement that converts an unsaturated ether into a γ,δ-unsaturated carbonyl compound.

Practice Test on Chapter Review

1. a. Which of the following pericyclic reactions is symmetry allowed and readily occurs?
   1. a photochemical conrotatory electrocyclic ring closure of a conjugated triene
   2. a disrotatory thermal electrocyclic ring opening of a substituted cyclohexadiene
   3. a thermal [2 + 2] cycloaddition
   4. Reactions (1) and (2) will both occur.
   5. Reactions (1), (2), and (3) will all occur.

   b. Which of the following reactions requires suprafacial stereochemistry to be symmetry allowed?
   1. a photochemical [1,5] sigmatropic rearrangement
   2. a thermal [8 + 2] cycloaddition
   3. a photochemical [4 + 2] cycloaddition
   4. Reactions (1) and (2) are both suprafacial.
   5. Reactions (1), (2), and (3) are all suprafacial.

   c. What product(s) are formed from the photochemical [2 + 2] cycloaddition of (*E*)-hex-3-ene?

   A   B   C

   1. **A** only
   2. **B** only
   3. **C** only
   4. **A** and **B**
   5. **A, B,** and **C**

   d. What product(s) are formed from the photochemical electrocyclic ring opening of *cis*-3,4-dimethylcyclobutene?

   1. (2*E*,4*E*)-hexa-2,4-diene
   2. (2*E*,4*Z*)-hexa-2,4-diene
   3. (*Z*)-hexa-1,3,5-triene
   4. Compounds (1) and (2) are both formed.
   5. Compounds (1), (2), and (3) are all formed.

e. What product(s) are formed by the [3,3] sigmatropic rearrangement of cyclodeca-1,5-dien-3-ol?

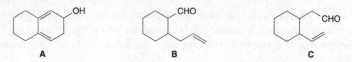

1. A only
2. B only
3. C only
4. A and B
5. A, B, and C

2. Consider the *p* orbitals of the terminal carbons of a conjugated polyene with like phases on the same side of the molecule (as in **A**) or opposite sides of the molecule (as in **B**), and answer each question.

a. Which drawing is consistent with the ground state HOMO of a conjugated triene?
b. Which drawing is consistent with the excited state LUMO of a conjugated diene?
c. Which drawing is consistent with the ground state LUMO for a conjugated tetraene?

3. What type of sigmatropic rearrangement is depicted in each reaction?

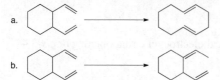

Answers to Practice Test

1. a. 4	2. a. **A**	3. a. [3,3]
b. 2	b. **B**	b. [1,3]
c. 4	c. **A**	
d. 1		
e. 3		

Pericyclic Reactions 27–5

Answers to Problems

27.1 Use the following definitions:
- An **electrocyclic ring closure** is an intramolecular reaction that forms a cyclic product containing one more σ bond and one fewer π bond than the reactant. An **electrocyclic ring opening** is a reaction in which a σ bond of a cyclic reactant is cleaved to form a conjugated product with one more π bond.
- A **cycloaddition** is a reaction between two compounds with π bonds that forms a cyclic product with two new σ bonds.
- A **sigmatropic rearrangement** is a reaction in which a σ bond is broken in the reactant, the π bonds rearrange, and a σ bond is formed in the product.

a. 2 π bonds   σ bond broken   3 π bonds   **electrocyclic reaction**

b.   + $H_2C = CH_2$ σ bond formed   **cycloaddition**
          σ bond formed

c. 4 π bonds   3 π bonds   σ bond formed   **electrocyclic reaction**

d. 1 π bond   σ bond broken   1 π bond   σ bond formed   **sigmatropic rearrangement**

27.2

a. For a bonding molecular orbital, the number of bonding interactions is greater than the number of nodes.

b. For an antibonding molecular orbital, the number of bonding interactions is less than the number of nodes.

For buta-1,3-diene:

	Bonding	Nodes	Type of MO
ψ_1	3	0	bonding MO
ψ_2	2	1	bonding MO
ψ_3^*	1	2	antibonding MO
ψ_4^*	0	3	antibonding MO

Chapter 27–6

27.3 The molecular orbitals of all conjugated dienes look similar.

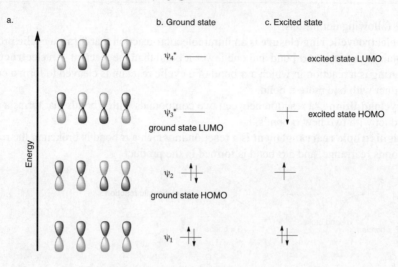

27.4 a. There are 10 molecular orbitals from the 10 p orbitals of the five π bonds.
b. Five molecular orbitals are bonding and five molecular orbitals are antibonding.
c. The lowest energy molecular orbital (ψ_1) has zero nodes.

d. The highest energy molecular orbital (ψ_{10}^*) has nine nodes.

27.5 To draw the product of an electrocyclic reaction, use curved arrows and begin at a π bond. Move the π electrons to an adjacent carbon–carbon bond, and continue in a cyclic fashion.

Pericyclic Reactions 27–7

27.6 Thermal electrocyclic reactions occur in a *disrotatory* fashion for a conjugated polyene with an *odd* number of π bonds, and in a *conrotatory* fashion for a conjugated polyene with an *even* number of π bonds.

a. [structure] C$_6$H$_5$ ... C$_6$H$_5$ — Δ, disrotatory → C$_6$H$_5$ ... C$_6$H$_5$
3 π bonds

b. [structure] hν — Δ, conrotatory → [structure]
2 π bonds

27.7 For an *even* number of π bonds, thermal electrocyclic reactions occur in a *conrotatory* fashion.

a. [structure] — re-draw → [structure] — Δ → [structure] + enantiomer

b. [structure] — re-draw → [structure] — Δ → [structure]

27.8 Photochemical electrocyclic reactions occur in a *conrotatory* fashion for a conjugated polyene with an *odd* number of π bonds, and in a *disrotatory* fashion for a conjugated polyene with an *even* number of π bonds.

[structure] C$_6$H$_5$ ... C$_6$H$_5$ — hν, conrotatory → C$_6$H$_5$ ... C$_6$H$_5$
+ enantiomer

b. [structure] hν — hν, disrotatory → [structure]
[The (*E,E*) diene is favored over the (*Z,Z*) diene.]

27.9 The photochemical electrocyclic reaction cleaves a six-membered ring to form a hexatriene.

[structure] HO — 7-dehydrocholesterol — hν → HO — provitamin D$_3$

27.10 Use the rules for electrocyclic reactions found in Answers 27.6 and 27.8.

a. [structure] — re-draw → [structure] 3 π bonds — Δ, disrotatory → [structure]
— hν, conrotatory → [structure] + enantiomer

Chapter 27–8

b. [structure] --re-draw--> [structure with 3 π bonds]
— Δ, disrotatory → [product] + enantiomer
— hν, conrotatory → [product]

27.11 Use the rules for electrocyclic reactions found in Answer 27.6. A reaction with three π bonds and a disrotatory cyclization is thermal.

[cyclic triene] → [bicyclic product with H's shown cis]

27.12 Count the number of π electrons in each reactant to classify the cycloaddition.

a. [tropone] (2 π electrons) + CH₂=CH₂ (2 π electrons) → [product] (one possibility)   [2 + 2] cycloaddition

b. [cyclopentadienone] (4 π electrons) + CH₂=CH₂ (2 π electrons) → [product]   [4 + 2] cycloaddition

c. [tropone] (6 π electrons) + CH₂=CH₂ (2 π electrons) → [product]   [6 + 2] cycloaddition

27.13 A thermal suprafacial addition is symmetry allowed in a [4 + 2] cycloaddition because like phases interact.

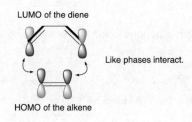

LUMO of the diene
HOMO of the alkene
Like phases interact.

Pericyclic Reactions 27–9

27.14 A thermal [4 + 2] cycloaddition is suprafacial.

a.

a. (structure) + CH₂=CH₂ —re-draw→ (structure) (E,E) diene —Δ→ (structure)

b. (structure) + (structure with CN, CN) —re-draw→ (structure) (E,Z) diene —Δ→ (structure with CN, CN) + enantiomer

27.15

a. (structure with CO₂CH₃, H) —Δ, endo addition→ **X** (structure with CO₂CH₃, H, O)

b. (diene HOMO structure) diene HOMO / dienophile LUMO → **X** (structure with CO₂CH₃)

The dienophile is under the diene, by the rule of endo addition (Section 16.13). The H's at the ring fusion are cis to each other, but trans to the CO₂CH₃ group.

27.16 A photochemical [2 + 2] cycloaddition is suprafacial.

a. (structure with C₆H₅, C₆H₅) + (structure) —hv→ (structure with C₆H₅, C₆H₅) + enantiomer

b. (structure with O, H, CN) + CH₂=CH₂ → (structure with O, H, CN) + enantiomer

27.17

a. The photochemical [6 + 4] cycloaddition involves five π bonds (the total number of π electrons divided by two) and is antarafacial.

b. A thermal [8 + 2] cycloaddition involves five π bonds and is suprafacial.

Chapter 27–10

27.18 A photochemical [4 + 2] cycloaddition like the Diels–Alder reaction must proceed by an antarafacial pathway. This would require either the 1,3-diene or the alkene component to twist 180° in order for the like phases of the *p* orbitals to overlap. Such a rotation is not possible in the formation of a six-membered ring.

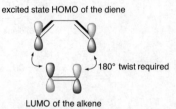

27.19 Locate the σ bonds broken and formed, and count the number of atoms that connect them.

a.

σ bond broken
[1,3] sigmatropic rearrangement
σ bond formed

b.

σ bond broken
[3,3] sigmatropic rearrangement
σ bond formed

27.20

a. re-draw [1,7]

b, c. The reaction involves four electron pairs (three π bonds and one σ bond), so it proceeds by an antarafacial pathway under thermal conditions, and by a suprafacial pathway under photochemical conditions.

27.21 Draw the products of each reaction.

a. re-draw [3,3]

Pericyclic Reactions 27–11

b. [3,3]

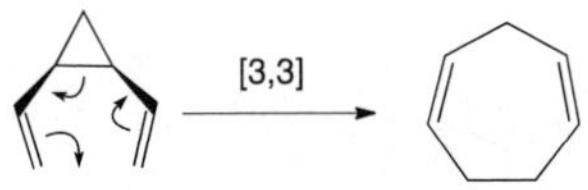

c. [3,3]   tautomerize

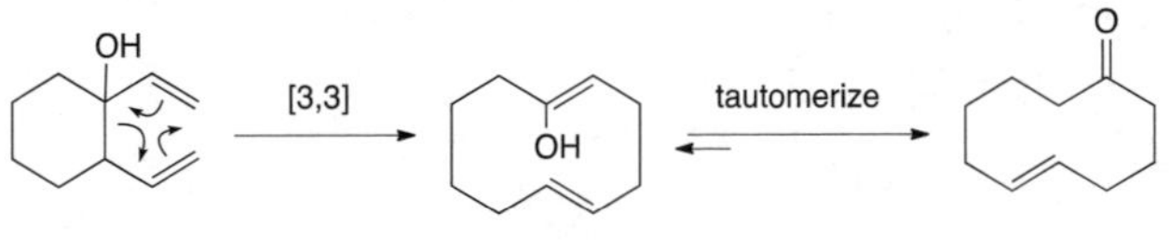

27.22 Draw the product after protonation.

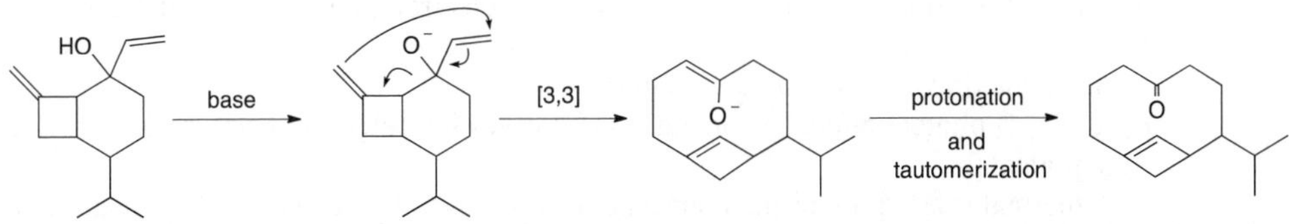

base   [3,3]   protonation

and
tautomerization

27.23 Re-draw geranial to put the ends of the 1,5-diene close together. Then draw three curved arrows, beginning at a π bond.

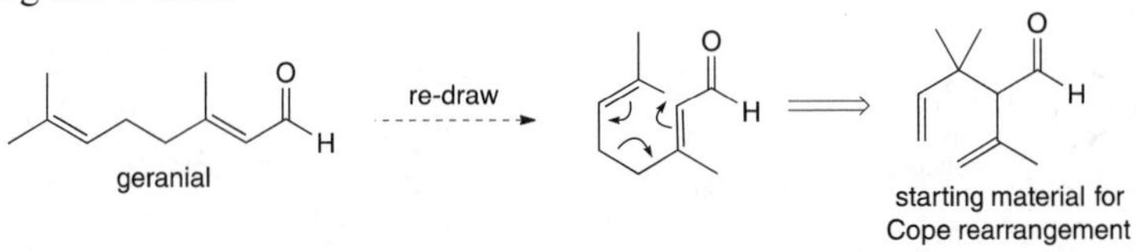

re-draw

geranial

starting material for
Cope rearrangement

27.24 Draw the product of Claisen rearrangement.

a. [3,3]

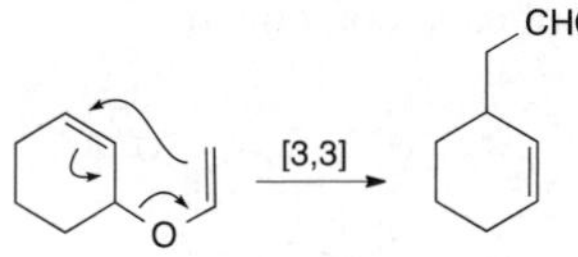

c. [3,3]

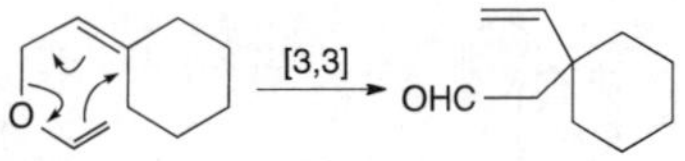

b. [3,3]

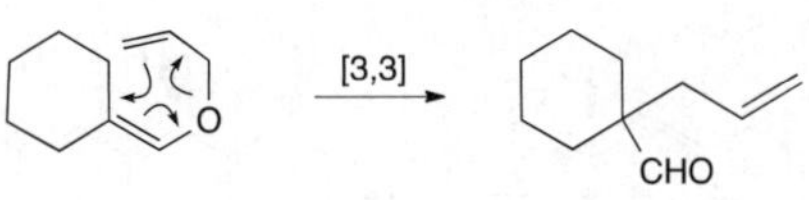

27.25

a.

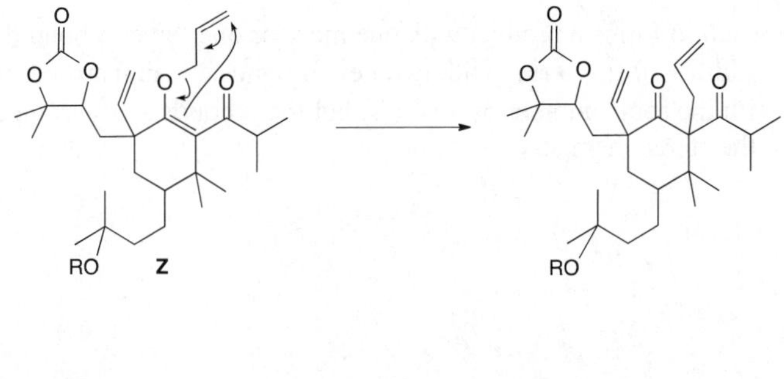

Chapter 27–12

b.

metathesis

+ $CH_2=CH_2$

27.26 Predict the stereochemistry of each reaction using Table 27.4.

a. A [6 + 4] thermal cycloaddition involves five electron pairs, making the reaction suprafacial.

b. A photochemical electrocyclic ring closure of deca-1,3,5,7,9-pentaene involves five electron pairs, making the reaction conrotatory.

c. A [4 + 4] photochemical cycloaddition involves four electron pairs, making the reaction suprafacial.

d. A thermal [5,5] sigmatropic rearrangement involves five electron pairs, making the reaction suprafacial.

27.27 Use the rules found in Answers 27.6 and 27.8.

27.28 Draw the product of [3,3] sigmatropic rearrangement of each compound.

a.

[3,3]

tautomerization

b.

[3,3]

tautomerization

27.29 An electrocyclic reaction forms a product with one more or one fewer π bond than the starting material. A cycloaddition forms a ring with two new σ bonds. A sigmatropic rearrangement forms a product with the same number of π bonds, but the π bonds are rearranged. Use Table 27.4 to determine the stereochemistry.

Pericyclic Reactions 27–13

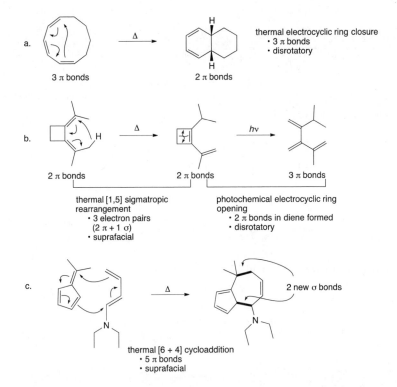

27.30 Use the rules for thermal electrocyclic reactions found in Answer 27.6.

27.31 Use the rules for photochemical electrocyclic reactions found in Answer 27.8.

Although conrotatory ring opening could also form, at least in theory, an all-(Z) triene, steric hindrance during ring opening would cause the terminal CH$_3$'s to crash into one another, making this process unlikely.

Chapter 27–14

b.

+
enantiomer

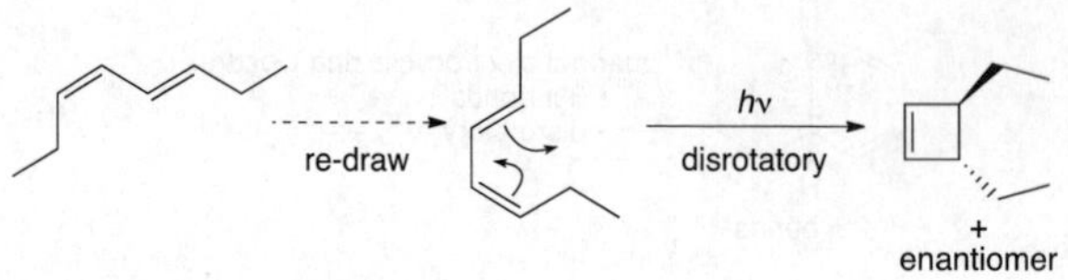

27.32 Use the rules found in Answer 27.8.

a.

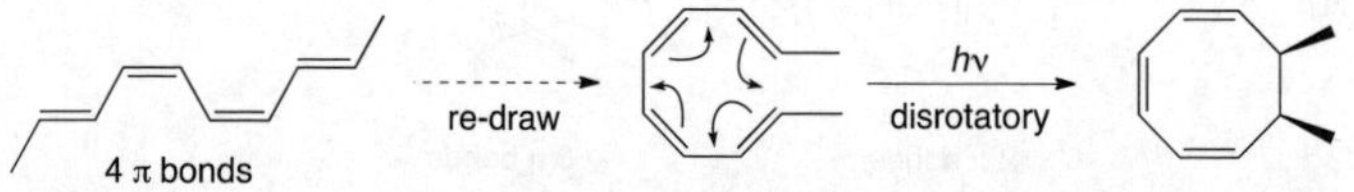

4 π bonds

b.

4 π bonds

+ enantiomer

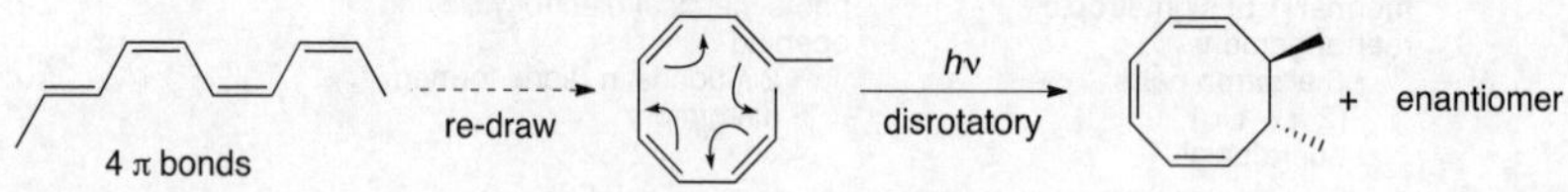

27.33 Use the rules found in Answers 27.6 and 27.8.

a, b.

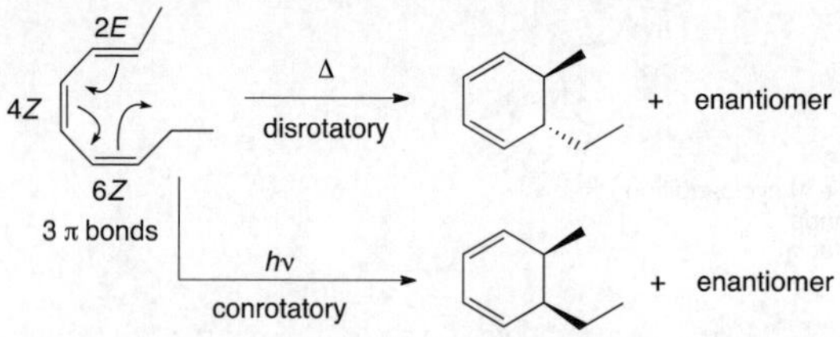

2E

4Z

6Z

3 π bonds

+ enantiomer

+ enantiomer

c.

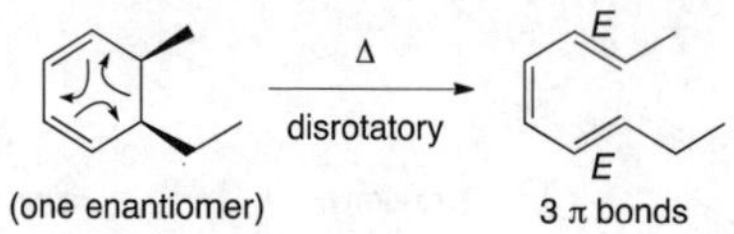

(one enantiomer)

3 π bonds

d.

(one enantiomer)

3 π bonds

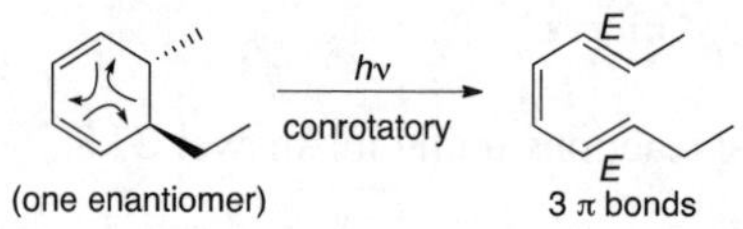

27.34 The trans product is indicative of a disrotatory ring closure from the cyclic triene with the given stereochemistry at the double bonds. A disrotatory ring closure with a polyene having three π bonds must occur under thermal conditions.

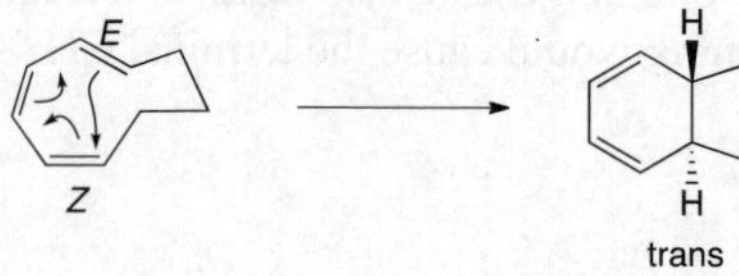

trans

Pericyclic Reactions 27–15

27.35 A disrotatory cyclization of a reactant with an even number of π bonds must occur under photochemical conditions.

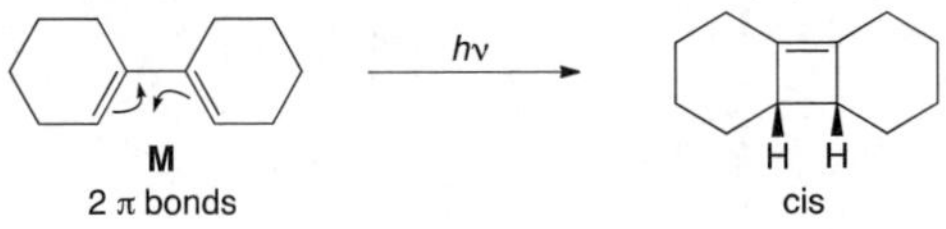

M
2 π bonds

hν →

cis

27.36 Use the rules found in Answers 27.6 and 27.8.

a, c.

N
3 π bonds

Δ
disrotatory →

cis

b, c.

N
3 π bonds

hν
conrotatory →

trans
+ enantiomer

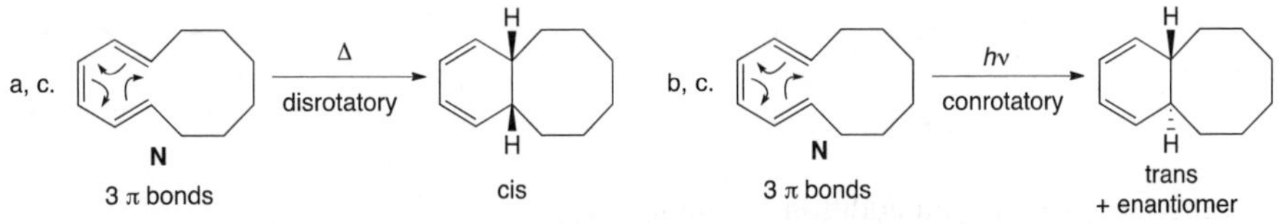

27.37 Use the rules found in Answers 27.6 and 27.8.

P

Δ
conrotatory ←

Z **Q**
2 π bonds

hν
disrotatory ↑

R

(A 10-membered ring cannot contain two *E* double bonds.)

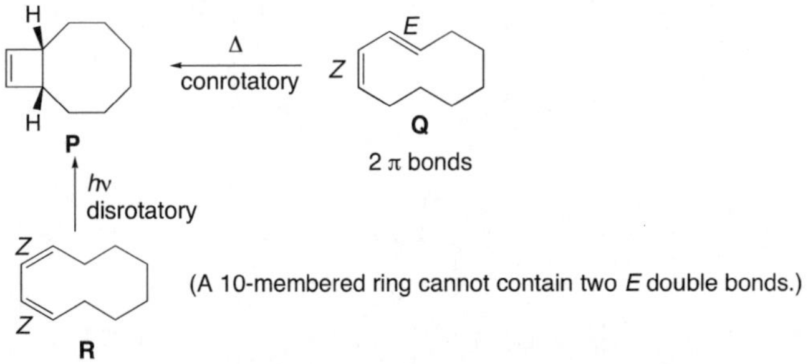

27.38 The reaction involves three π bonds in one reactant and two π bonds in the second reactant, so the reaction is a [6 + 4] cycloaddition. A suprafacial cycloaddition with five π bonds must proceed under thermal conditions.

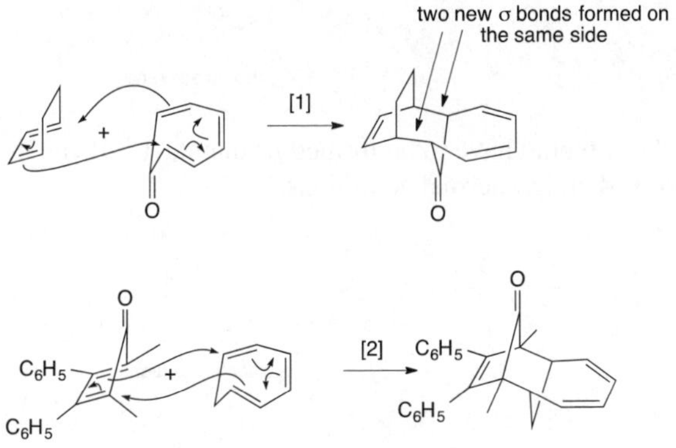

two new σ bonds formed on the same side

+ [1] →

C_6H_5
C_6H_5
+ [2] → C_6H_5
C_6H_5

Chapter 27–16

27.39 The Diels–Alder reaction is a thermal, suprafacial [4 + 2] cycloaddition.

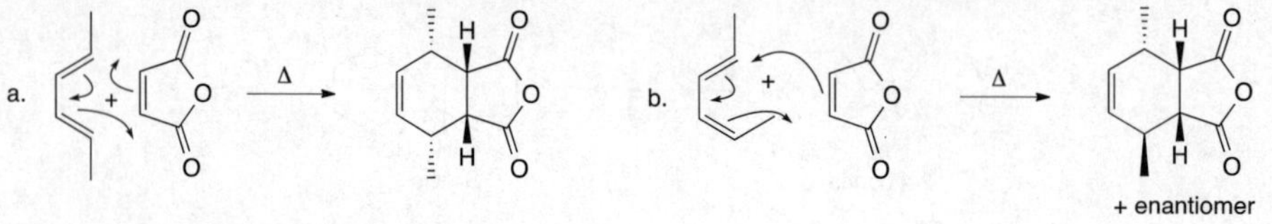

a. [reaction scheme]

b. [reaction scheme]

+ enantiomer

27.40 A photochemical [2 + 2] cycloaddition is suprafacial.

a. [reaction scheme]

b. [reaction scheme]

27.41 A thermal [4 + 2] cycloaddition is suprafacial.

a. [reaction scheme]

b. [reaction scheme]

c. [reaction scheme]

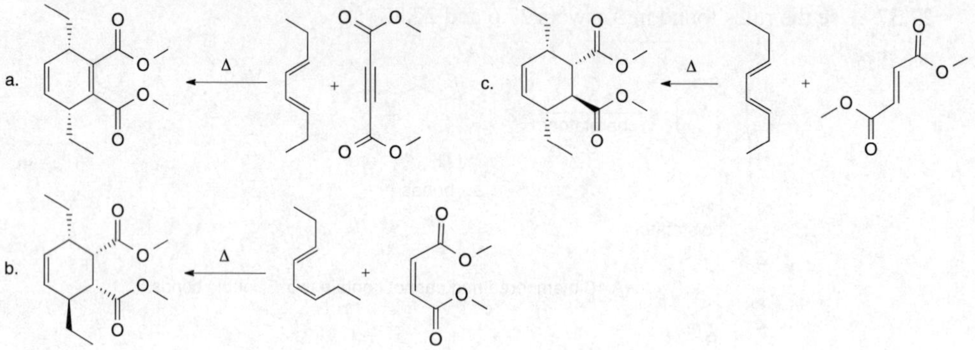

27.42 Buta-1,3-diene can react with itself in a symmetry-allowed thermal [4 + 2] cycloaddition to form 4-vinylcyclohexene.

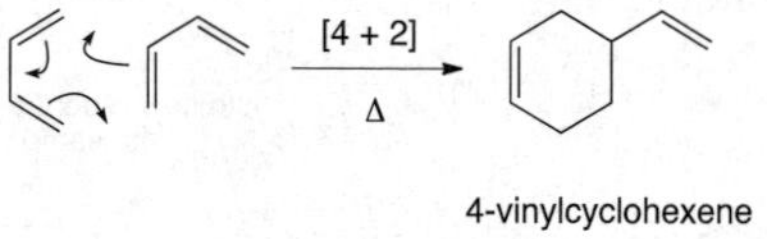

[reaction scheme]

4-vinylcyclohexene

Cycloocta-1,5-diene would have to be formed from buta-1,3-diene by a [4 + 4] cycloaddition, which is not allowed under thermal conditions.

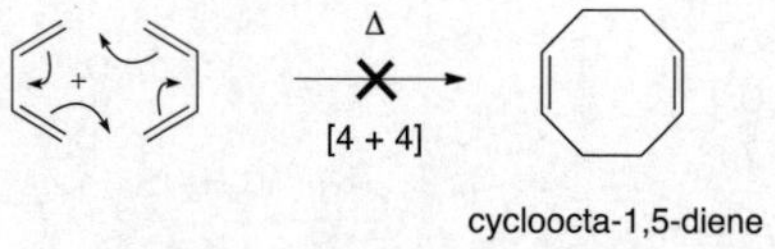

[reaction scheme]

cycloocta-1,5-diene

Pericyclic Reactions 27–17

27.43 A series of three [2 + 2] cycloadditions with *E* alkenes forms **X**.

27.44 Re-draw the reactant and product to more clearly show the relative location of the bonds broken and formed.

a.

b.

27.45 Draw the products of each reaction.

a.

c.

b.

27.46 a. Two [1,5] sigmatropic rearrangements occur.

5-methyl-
cyclopenta-1,3-diene

1-methyl-
cyclopenta-1,3-diene

2-methyl-
cyclopenta-1,3-diene

Chapter 27–18

b.

5-methyl isomer   [1,3]   2-methyl isomer

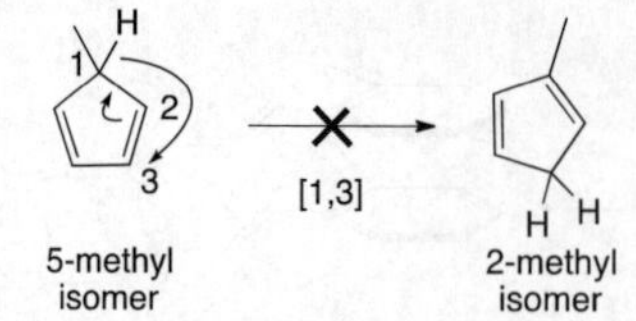

A [1,3] sigmatropic rearrangement requires photochemical conditions not thermal conditions, so 5-methylcyclopenta-1,3-diene cannot rearrange directly to its 2-methyl isomer by a [1,3] shift.

27.47

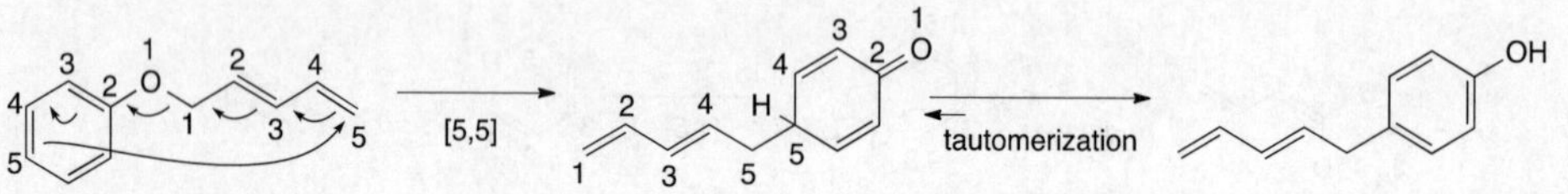

27.48 Re-draw **A** to put the ends of allyl vinyl ether close together, and use curved arrows to draw the Claisen product. Then re-draw the Claisen product to put the ends of the 1,5-diene close together to draw the product of the Cope rearrangement.

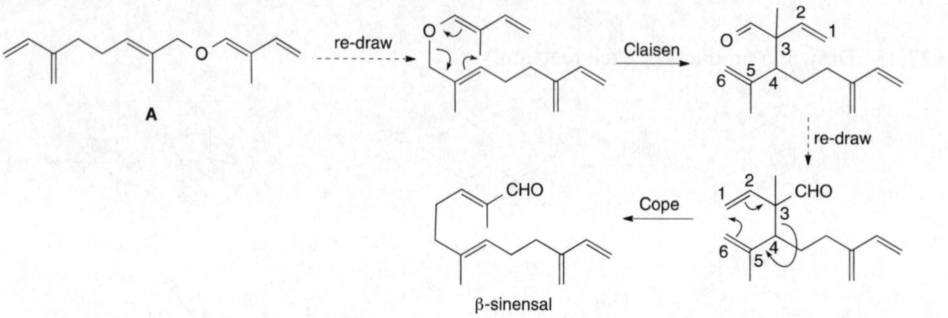

27.49

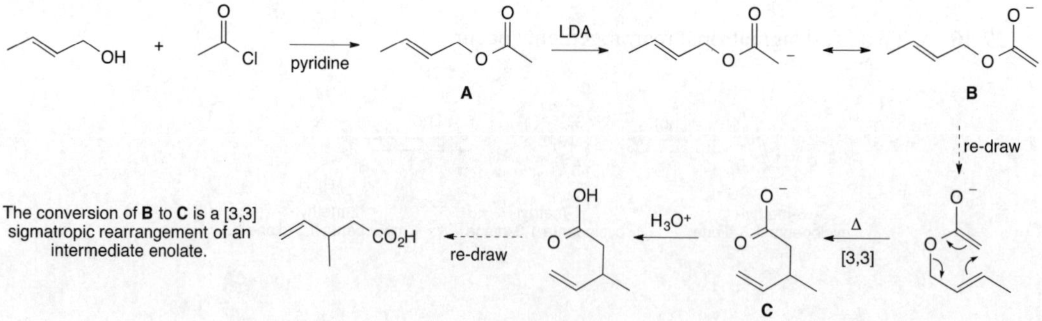

The conversion of **B** to **C** is a [3,3] sigmatropic rearrangement of an intermediate enolate.

Pericyclic Reactions 27–19

27.50 Use the definitions found in Answer 27.1.

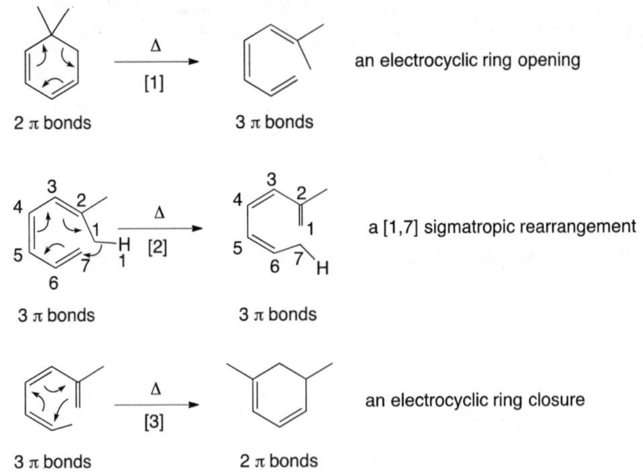

2 π bonds   Δ [1]   3 π bonds   an electrocyclic ring opening

3 π bonds   Δ [2]   3 π bonds   a [1,7] sigmatropic rearrangement

3 π bonds   Δ [3]   2 π bonds   an electrocyclic ring closure

27.51

a. 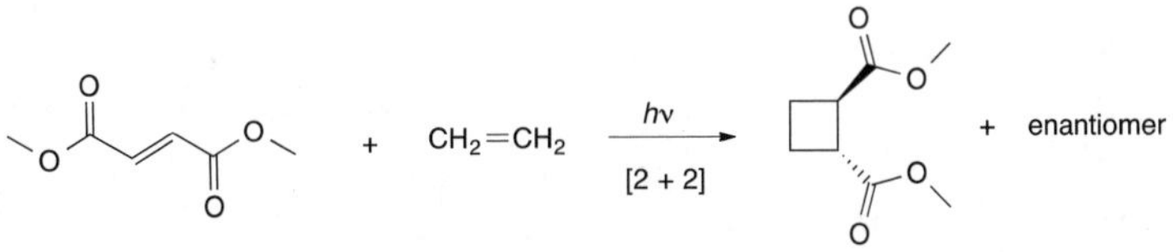   + $CH_2{=}CH_2$ hv [2 + 2]   + enantiomer

b. 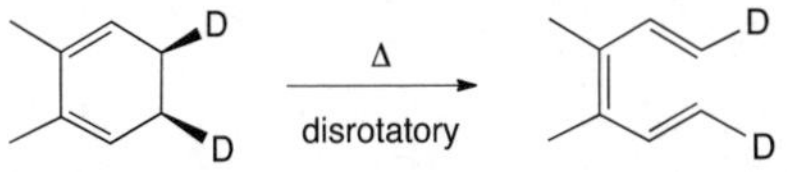   Δ disrotatory

c. 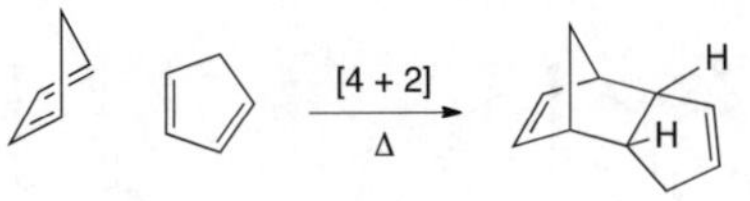   [4 + 2] Δ

d. 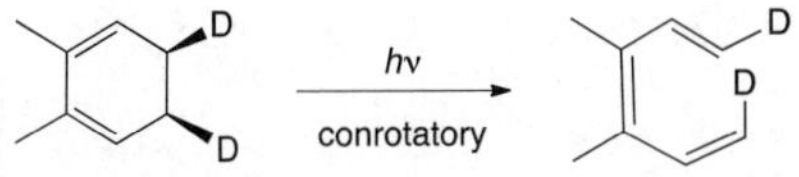   hv conrotatory

27.52

a. 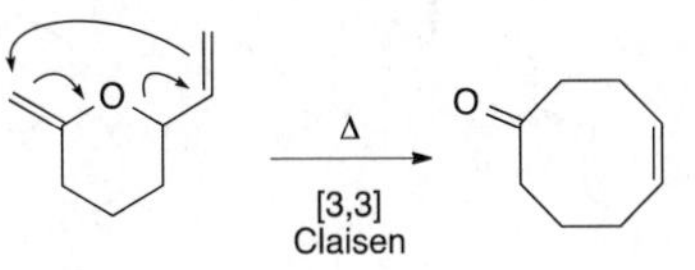   Δ [3,3] Claisen

b. 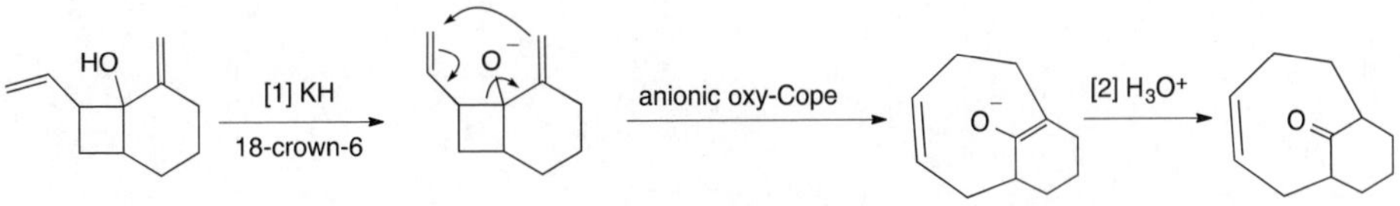   [1] KH 18-crown-6   anionic oxy-Cope   [2] H_3O^+

Chapter 27–20

c.

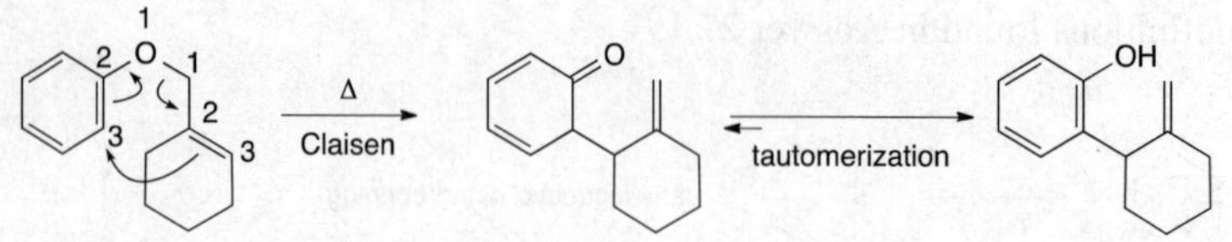

d.

C_7H_8

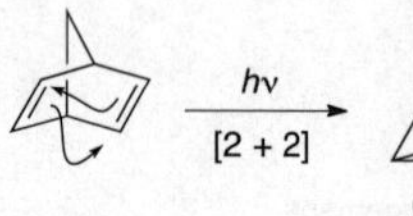

27.53 The mechanism consists of sequential [3,3] sigmatropic rearrangements, followed by tautomerization.

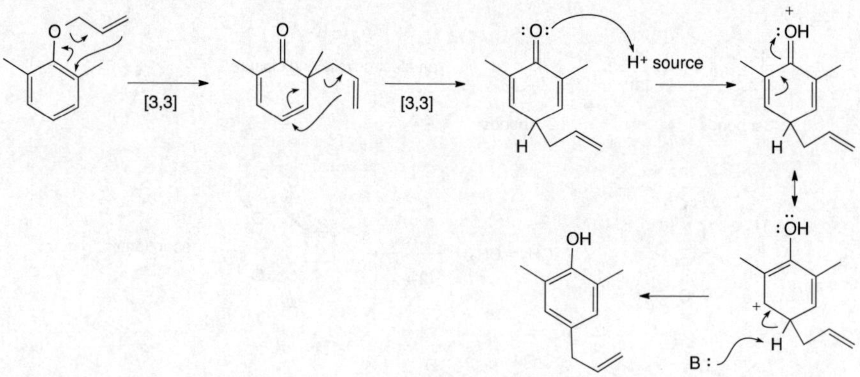

27.54

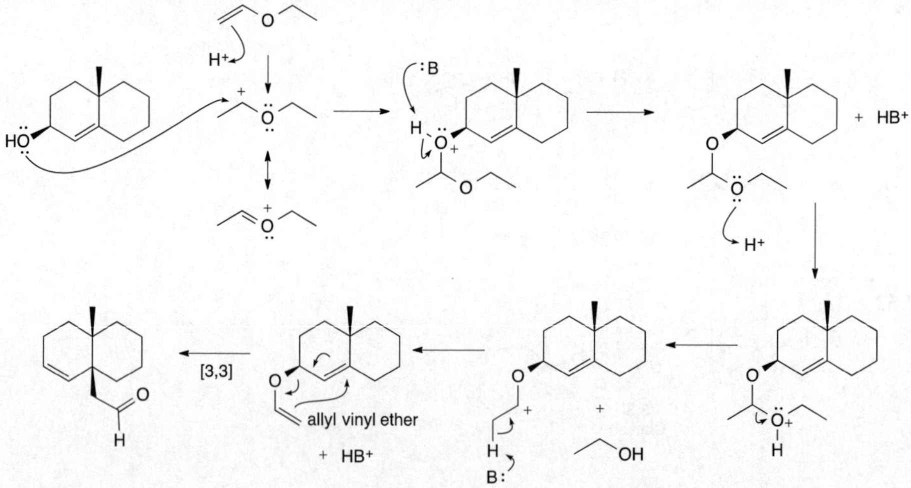

Pericyclic Reactions 27–21

27.55

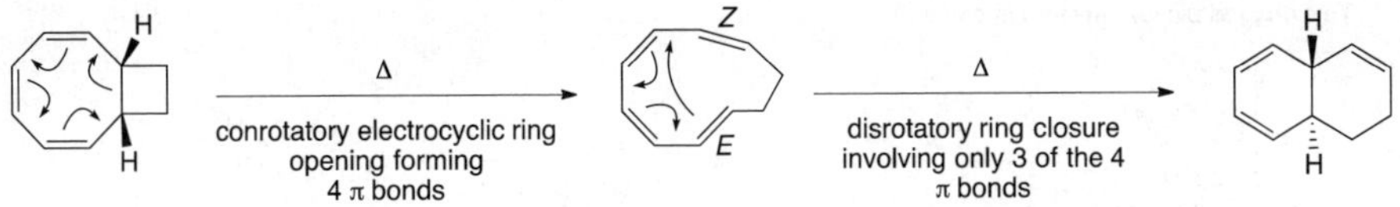

Δ

conrotatory electrocyclic ring
opening forming
4 π bonds

Δ

disrotatory ring closure
involving only 3 of the 4
π bonds

27.56

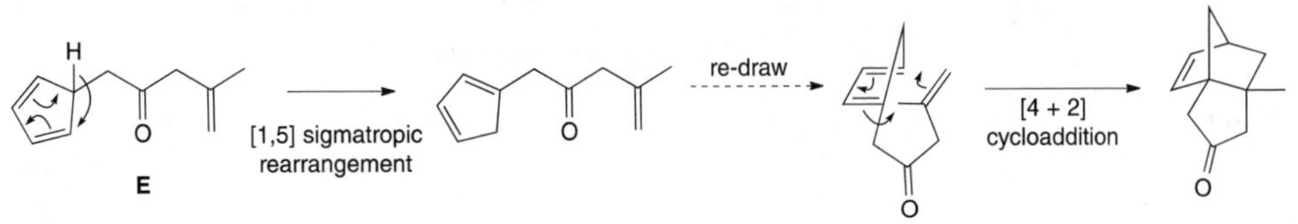

[1,5] sigmatropic
rearrangement

re-draw

[4 + 2]
cycloaddition

27.57

Δ

electrocyclic ring opening
forming two π bonds

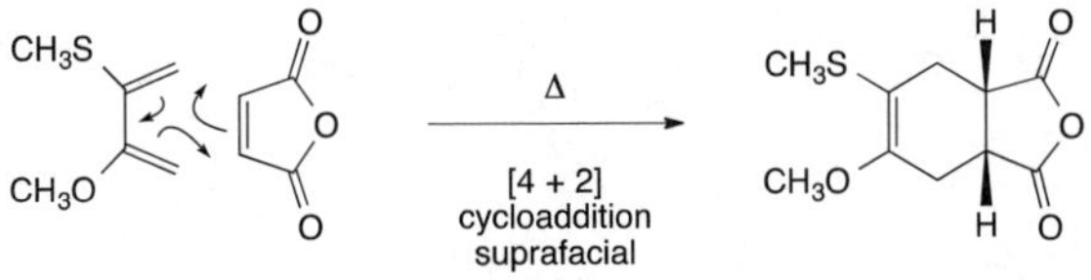

Δ

[4 + 2]
cycloaddition
suprafacial

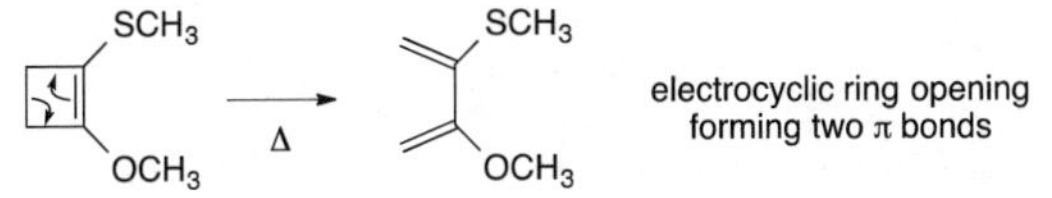

27.58 The mechanism consists of a [4 + 2] cycloaddition, followed by intramolecular imine formation.

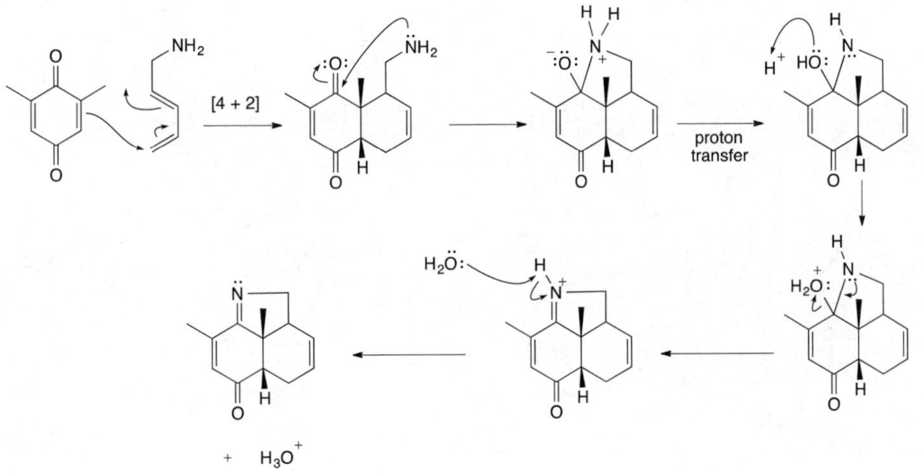

[4 + 2]

proton
transfer

+ H_3O^+

27.59

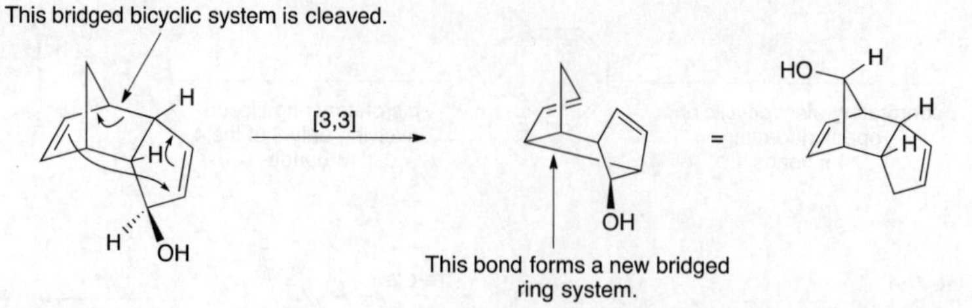
This bridged bicyclic system is cleaved.
[3,3]
This bond forms a new bridged ring system.
OH
HO
H
=

27.60

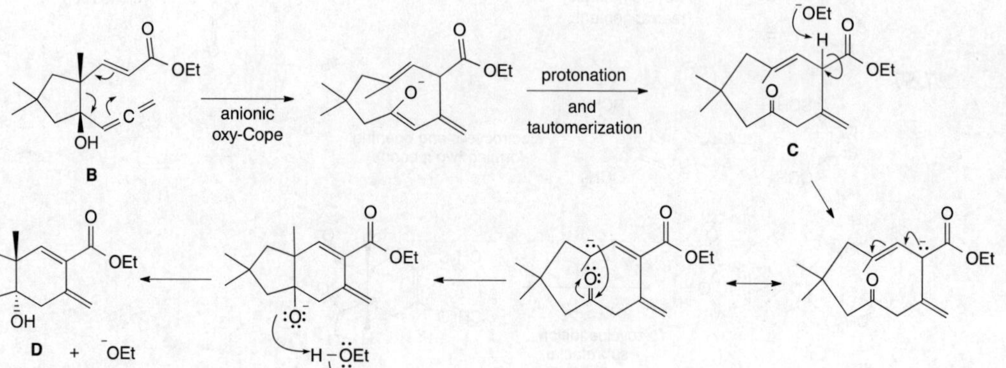
OEt
anionic
oxy-Cope
protonation
and
tautomerization
B
C
D
+ ⁻OEt
H–ÖEt

27.61

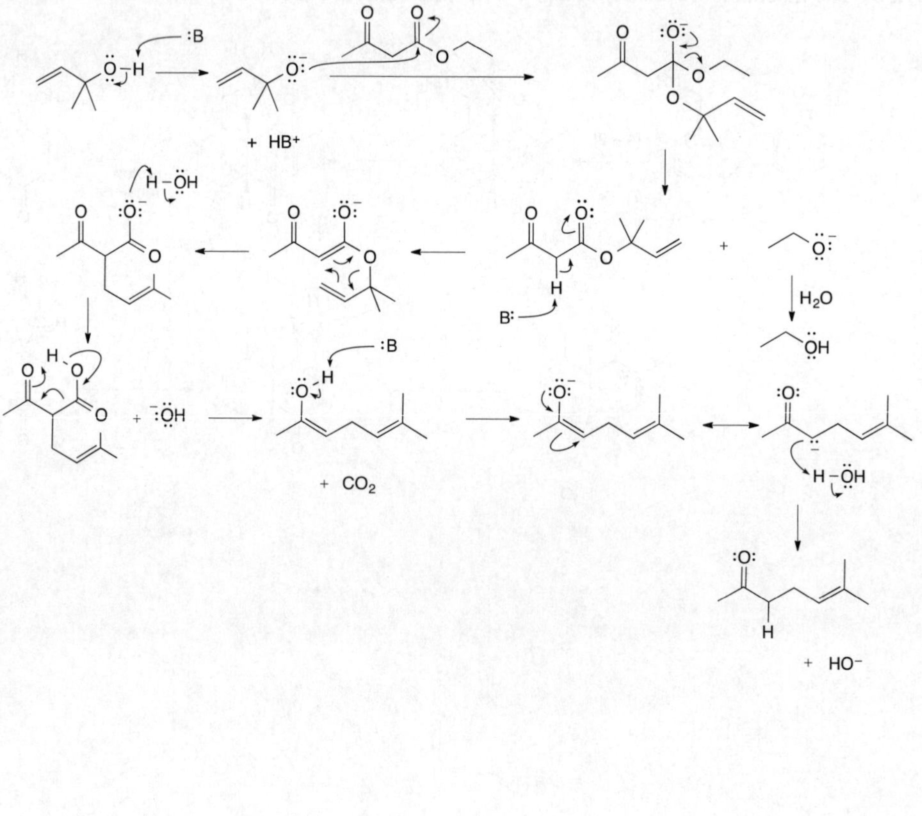
:B
+ HB+
H–ÖH
B:
H2O
+ CO2
+ HO−

Pericyclic Reactions 27–23

27.62 Conrotatory cyclization of **Y** using four π bonds forms **X**. Disrotatory ring closure of **X** can occur in two ways—on the top face or the bottom face of the eight-membered ring to form diastereomers.

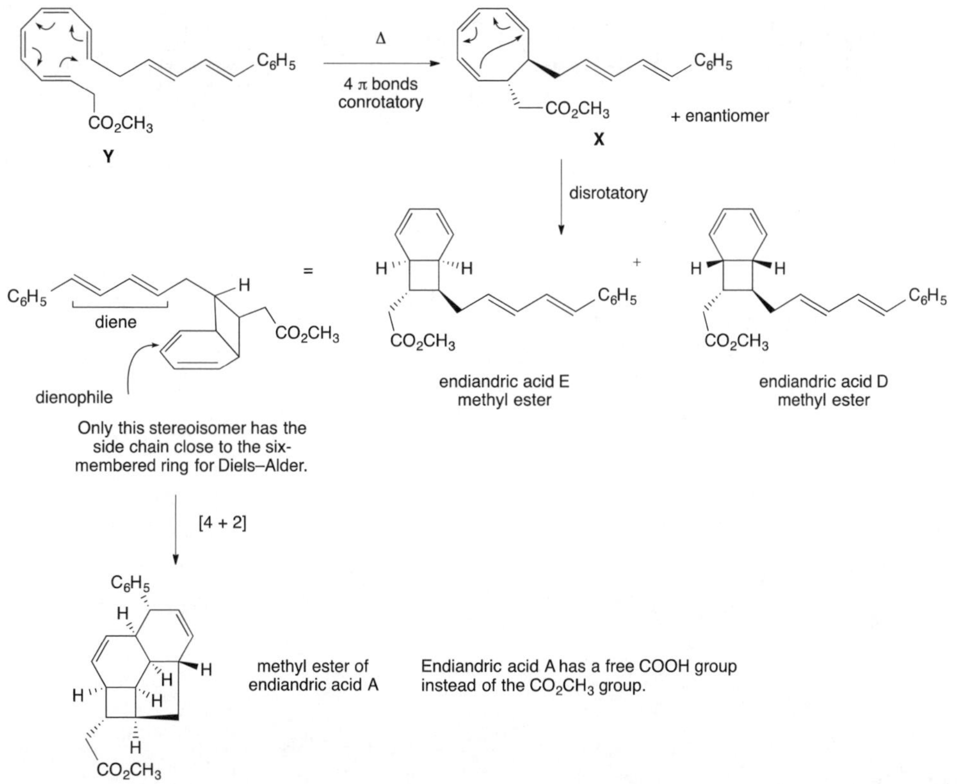

Δ
4 π bonds
conrotatory

C_6H_5

CO_2CH_3

Y

CO_2CH_3 + enantiomer

X

disrotatory

C_6H_5

diene

CO_2CH_3

dienophile

Only this stereoisomer has the side chain close to the six-membered ring for Diels–Alder.

endiandric acid E
methyl ester

endiandric acid D
methyl ester

[4 + 2]

C_6H_5

CO_2CH_3

methyl ester of
endiandric acid A

Endiandric acid A has a free COOH group
instead of the CO_2CH_3 group.

27.63

a.

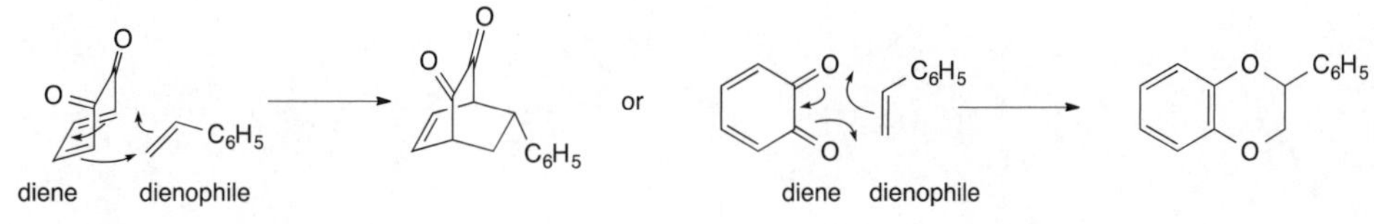

diene   dienophile

or

diene   dienophile

b.

diene   dienophile

or

diene   dienophile

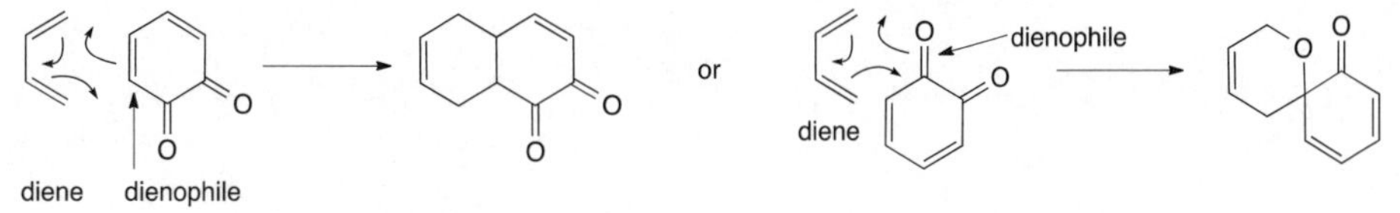

Chapter 27–24

c.

Chapter 28 Carbohydrates

Chapter Review

Important terms

- **Aldose** A monosaccharide containing an aldehyde (28.2)
- **Ketose** A monosaccharide containing a ketone (28.2)
- **D-Sugar** A monosaccharide with the O bonded to the stereogenic center farthest from the carbonyl group drawn on the right in the Fischer projection (28.2C)
- **Epimers** Two diastereomers that differ in configuration around one stereogenic center only (28.3)
- **Anomers** Monosaccharides that differ in configuration at only the hemiacetal OH group (28.6)
- **Glycoside** An acetal derived from a monosaccharide hemiacetal (28.7)

Acyclic, Haworth, and 3-D representations for D-glucose (28.6)

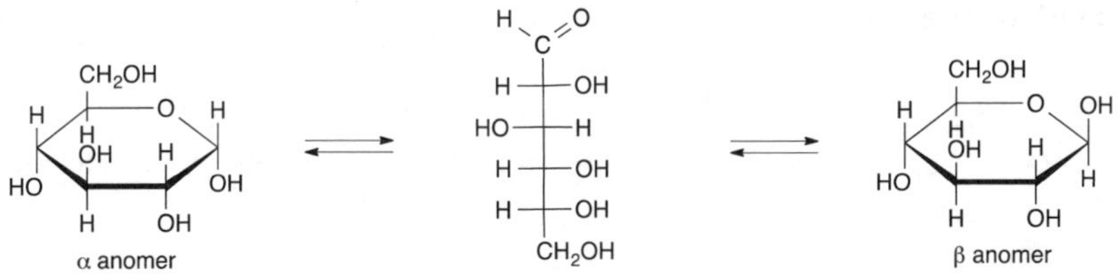

α anomer
α-D-glucose

acyclic aldehyde

β anomer
β-D-glucose

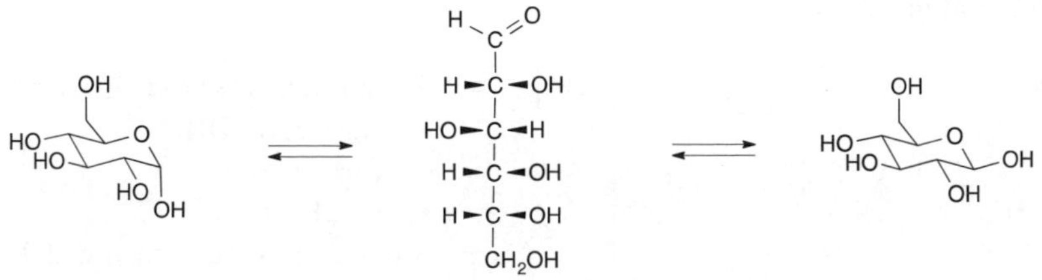

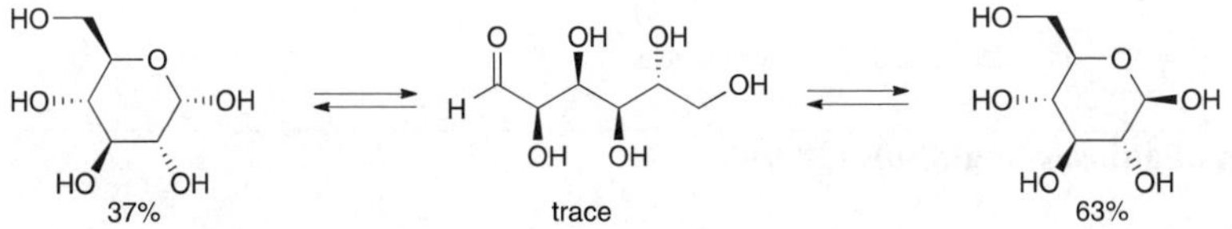

37% trace 63%

Reactions of monosaccharides involving the hemiacetal

[1] Glycoside formation (28.7A)

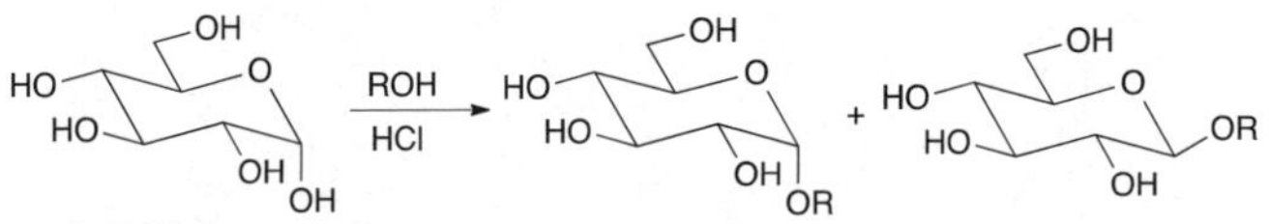

α-D-glucose α glycoside β glycoside

- Only the hemiacetal OH reacts.
- A mixture of α and β glycosides forms.

Chapter 28–2

[2] Glycoside hydrolysis (28.7B)

[structure with OR] →(H₃O⁺) α anomer + β anomer + ROH

- A mixture of α and β anomers forms.

Reactions of monosaccharides at the OH groups
[1] Ether formation (28.8)

[structure] →(Ag₂O, RX) [per-O-alkylated structure]

- All OH groups react.
- The stereochemistry at all stereogenic centers is retained.

[2] Ester formation (28.8)

[structure] →(Ac₂O or AcCl, pyridine) [per-acetylated structure]

- All OH groups react.
- The stereochemistry at all stereogenic centers is retained.

Reactions of monosaccharides at the carbonyl group
[1] Oxidation of aldoses (28.9B)

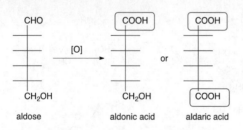

- Aldonic acids are formed using:
  - Ag₂O, NH₄OH
  - Cu²⁺
  - Br₂, H₂O
- Aldaric acids are formed with HNO₃, H₂O.

[2] Reduction of aldoses to alditols (28.9A)

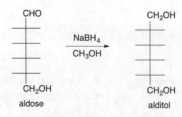

Carbohydrates 28–3

[3] Wohl degradation (28.10A)

- The C1–C2 bond is cleaved to shorten an aldose chain by one carbon.
- The stereochemistry at all other stereogenic centers is retained.
- Two epimers at C2 form the same product.

[4] Kiliani–Fischer synthesis (28.10B)

- One carbon is added to the aldehyde end of an aldose.
- Two epimers at C2 are formed.

Other reactions
[1] Hydrolysis of disaccharides (28.11)

A mixture of anomers is formed.

[2] Formation of *N*-glycosides (28.13B)

- Two anomers are formed.

Practice Test on Chapter Review

1.a. How are the following two representations related to each other?

1. **A** and **B** are anomers of each other.
2. **A** and **B** are epimers of each other.
3. **A** and **B** are diastereomers of each other.
4. Statements (1) and (2) are both true.
5. Statements (1), (2), and (3) are all true.

Chapter 28–4

b. Which of the following statements is (are) true about monosaccharide **C**?

1. **C** is a D-sugar.
2. The β anomer is drawn.
3. **C** is an aldohexose.
4. Statements (1) and (2) are both true.
5. Statements (1), (2), and (3) are all true.

c. Which of the following are different representations for monosaccharide **D**?

4. Both (1) and (2) are representations for **D**.
5. Compounds (1), (2), and (3) all represent **D**.

d. Which aldoses give an optically active compound upon reaction with NaBH₄ in CH₃OH?

4. Both (1) and (2) give an optically active product.
5. Compounds (1), (2), and (3) all give optically active products.

2. Answer each question about monosaccharide **D** as True (T) or False (F).

a. **D** is a D-sugar.
b. **D** is drawn as an α anomer.
c. **D** is an aldohexose.
d. Reduction of **D** with NaBH₄ in CH₃OH forms an optically inactive alditol.
e. Oxidation of **D** with Br₂, H₂O forms an optically active aldonic acid.

f. Oxidation of **D** with HNO₃ forms an optically active aldaric acid.
g. C2 has the *R* configuration.
h. Treatment of **D** with CH₃OH, HCl forms two products.
i. Treatment of **D** with Ag₂O, and CH₃I (excess) forms two products.
j. An epimer of **D** at C3 has an axial OH group.

3. Answer the following questions about the three monosaccharides (**A**–**C**) drawn below.

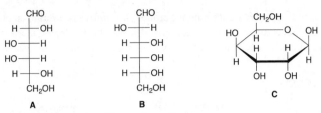

a. Draw the α anomer of **A** in a Haworth projection.
b. Draw the β anomer of **B** in a three-dimensional representation using a chair conformation.
c. Convert **C** into the acyclic form of the monosaccharide using a Fischer projection.
d. Which two aldoses yield **A** in a Wohl degradation?

4. Draw the product of each reaction with the starting material D-xylose.

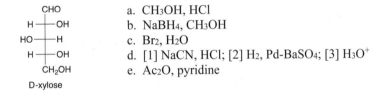

a. CH$_3$OH, HCl
b. NaBH$_4$, CH$_3$OH
c. Br$_2$, H$_2$O
d. [1] NaCN, HCl; [2] H$_2$, Pd-BaSO$_4$; [3] H$_3$O$^+$
e. Ac$_2$O, pyridine

Answers to Practice Test

1. a. 5
   b. 5
   c. 4
   d. 1

2. a. T
   b. F
   c. T
   d. F
   e. T
   f. T
   g. F
   h. T
   i. F
   j. T

3. [structures shown for a, b, c]

4. [structures shown for a, b, c, d, e]

Chapter 28–6

Answers to Problems

28.1 A *ketose* is a monosaccharide containing a ketone. An *aldose* is a monosaccharide containing an aldehyde. A monosaccharide is called: a *triose* if it has three C's, a *tetrose* if it has four C's, a *pentose* if it has five C's, a *hexose* if it has six C's, and so forth.

a. a ketotetrose

$$CH_2OH$$
$$C=O$$
$$H-C-OH$$
$$CH_2OH$$

b. an aldopentose

$$CHO$$
$$H-C-OH$$
$$H-C-OH$$
$$H-C-OH$$
$$CH_2OH$$

c. an aldotetrose

$$CHO$$
$$H-C-OH$$
$$H-C-OH$$
$$CH_2OH$$

28.2 Rotate and re-draw each molecule to place the horizontal bonds in front of the plane and the vertical bonds behind the plane. Then use a cross to represent the stereogenic center in a Fischer projection formula.

28.3 For each molecule:
[1] Convert the Fischer projection formula to a representation with wedges and dashes.
[2] Assign priorities (Section 5.6).
[3] Determine *R* or *S* in the usual manner. Reverse the answer if priority group [4] is oriented forward (on a wedge).

a.

$$CH_2NH_2$$
$$Cl——CH_2Br$$
$$H$$

[1] →

$$CH_2NH_2$$
$$Cl—C—CH_2Br$$
$$H$$

[2] →

3
$$CH_2NH_2$$
1 $$Cl—C—CH_2Br$$ 2
$$H$$
4

[3] →

3
$$CH_2NH_2$$
1 $$Cl—C—CH_2Br$$ 2
$$H$$

S configuration

b.

$$CHO$$
$$Cl——H$$
$$CH_2NH_2$$

[1] →

$$CHO$$
$$Cl—C—H$$
$$CH_2NH_2$$

[2] →

2
$$CHO$$
1 $$Cl—C—H$$ 4
$$CH_2NH_2$$
3

[3] →

2
$$CHO$$
1 $$Cl—C—H$$ — H forward
$$CH_2NH_2$$
3

S configuration

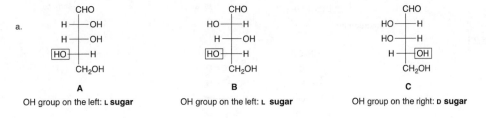

28.6 A D sugar has the OH group on the stereogenic center farthest from the carbonyl on the right. An L sugar has the OH group on the stereogenic center farthest from the carbonyl on the left.

b. **A** and **B** are diastereomers.
A and **C** are enantiomers.
B and **C** are diastereomers.

Chapter 28–8

28.7 There are 32 aldoheptoses; 16 are D sugars.

C2 R, CHO
C3 R H—OH
H—OH
H—OH
H—OH
H—OH
CH₂OH

CHO
H—OH
H—OH
HO—H
HO—H
H—OH
CH₂OH

CHO
H—OH
H—OH
HO—H
H—OH
H—OH
CH₂OH

CHO
H—OH
H—OH
H—OH
HO—H
H—OH
CH₂OH

28.8 *Epimers* are two diastereomers that differ in the configuration around only one stereogenic center.

epimers

CHO
H—OH
H—OH
CH₂OH

D-erythrose

CHO
HO—H
H—OH
CH₂OH

D-threose

and

CHO
H—OH
HO—H
CH₂OH

L-threose

28.9 a. D-allose and L-allose: **enantiomers**
b. D-altrose and D-gulose: **diastereomers** but not epimers
c. D-galactose and D-talose: **epimers**
d. D-mannose and D-fructose: **constitutional isomers**
e. D-fructose and D-sorbose: **diastereomers** but not epimers
f. L-sorbose and L-tagatose: **epimers**

28.10

a.
CH₂OH
C=O
HO—H
H—OH
H—OH
CH₂OH

D-fructose

CH₂OH
C=O
H—OH
HO—H
HO—H
CH₂OH

L-fructose

b.
CH₂OH
C=O
HO—H
HO—H
H—OH
CH₂OH

D-tagatose

c.
CH₂OH
C=O
HO—H
H—OH
HO—H
CH₂OH

L-sorbose

enantiomers

28.11

S
CH₂OH
C=O
HO—H
H—OH
H—OH
CH₂OH

D-fructose

S
CH₂OH
C=O
HO—H
HO—H
H—OH
CH₂OH

D-tagatose

28.12

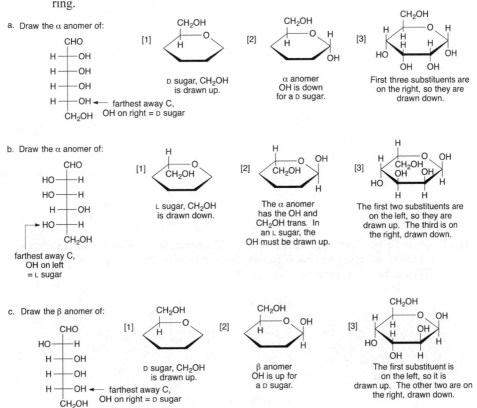

28.13 Step [1]: Place the O atom in the upper right corner of a hexagon, and add the CH₂OH group on the first carbon counterclockwise from the O atom.
Step [2]: Place the anomeric carbon on the first carbon clockwise from the O atom.
Step [3]: Add the substituents at the three remaining stereogenic centers, clockwise around the ring.

Chapter 28–10

28.14 To convert each Haworth projection into its acyclic form:
[1] Draw the C skeleton with the CHO on the top and the CH₂OH on the bottom.
[2] Draw in the OH group farthest from the C=O.
   A CH₂OH group drawn up means a D sugar; a CH₂OH group drawn down means an L sugar.
[3] Add the three other stereogenic centers.
   "Up" groups go on the left, and "down" groups go on the right.

28.15 To convert a Haworth projection into a 3-D representation with a chair cyclohexane:
[1] Draw the pyranose ring as a chair with the O as an "up" atom.
[2] Add the substituents around the ring.

28.16 Cyclization always forms a new stereogenic center at the anomeric carbon, so two different anomers are possible.

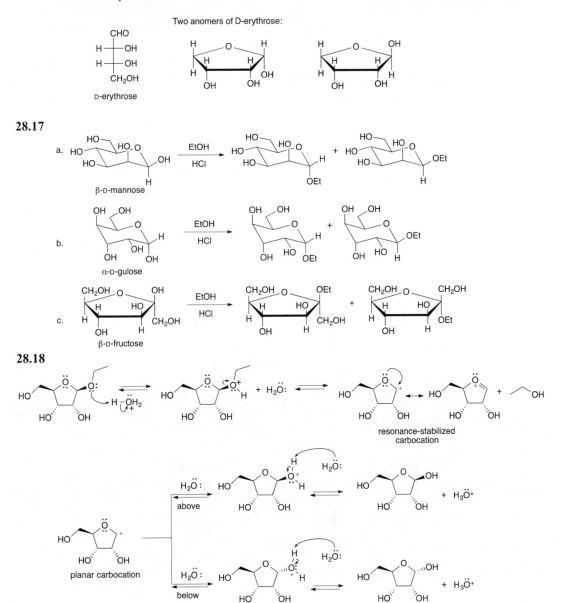

Chapter 28–12

28.19

a. All circled O atoms are part of a glycoside.

rebaudioside A

Trade name Truvia

b. Hydrolysis of rebaudioside A breaks each bond indicated with a dashed line and forms four molecules of glucose and the aglycon drawn.

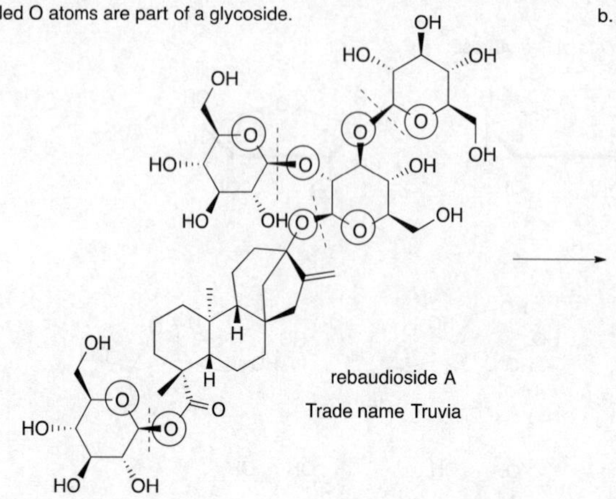

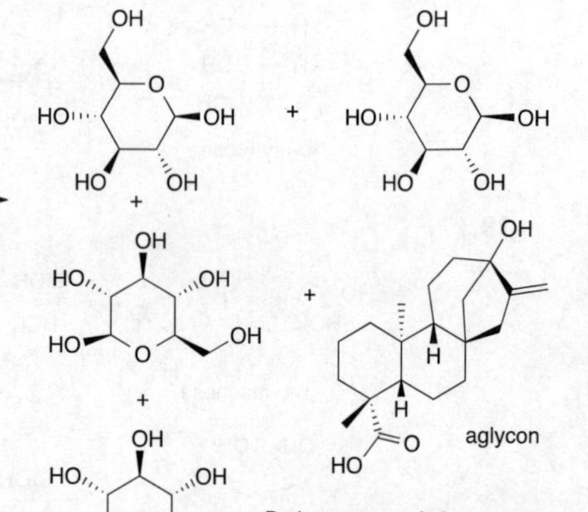

aglycon

Both anomers of glucose are formed, but only the β anomer is drawn.

28.20

a.

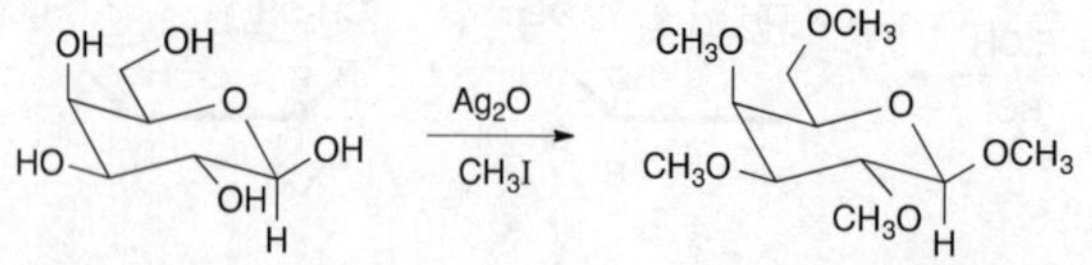

$$Ag_2O \quad CH_3I$$

b.

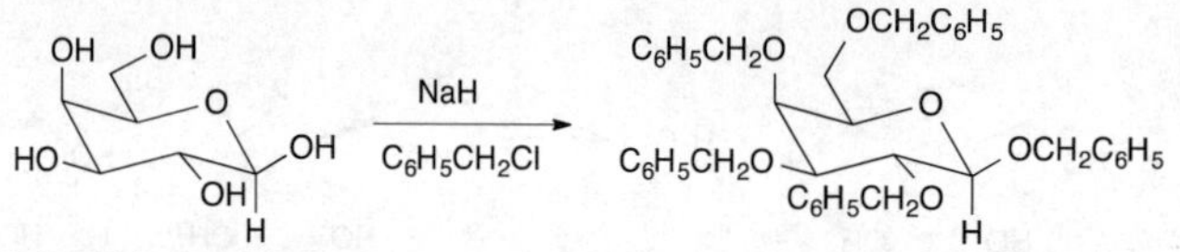

$$NaH \quad C_6H_5CH_2Cl$$

c.

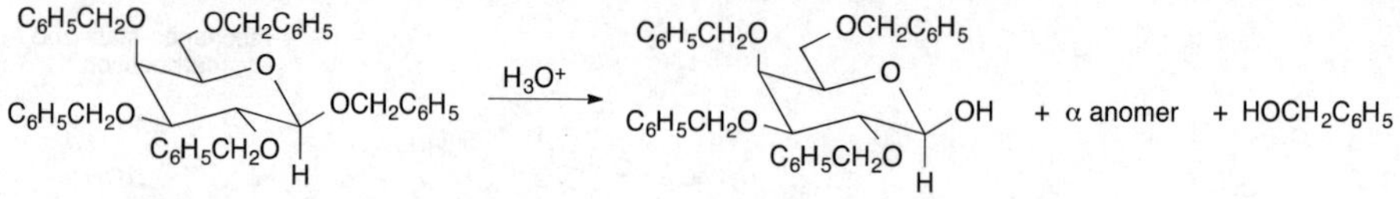

$$H_3O^+$$

+ α anomer + $HOCH_2C_6H_5$

d.

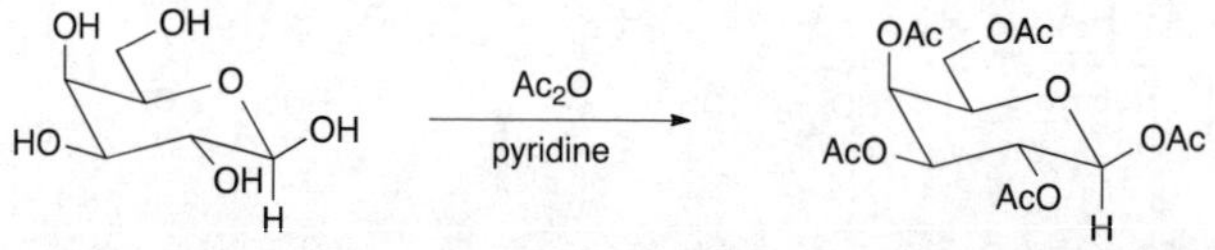

$$Ac_2O \quad pyridine$$

e.

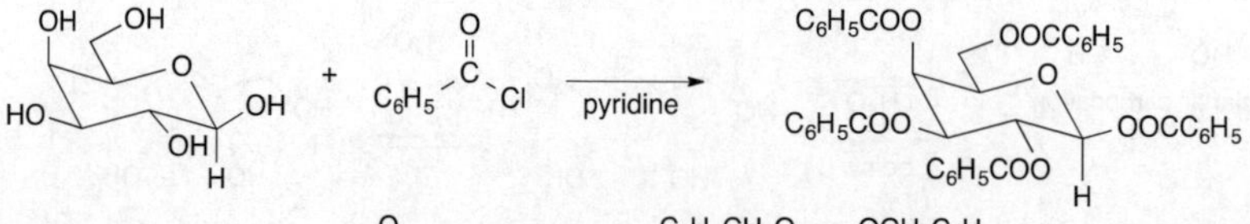

+ $C_6H_5\overset{O}{\underset{}{C}}Cl$ pyridine

f.   product in (c) + $C_6H_5\overset{O}{\underset{}{C}}Cl$ pyridine

+ α anomer

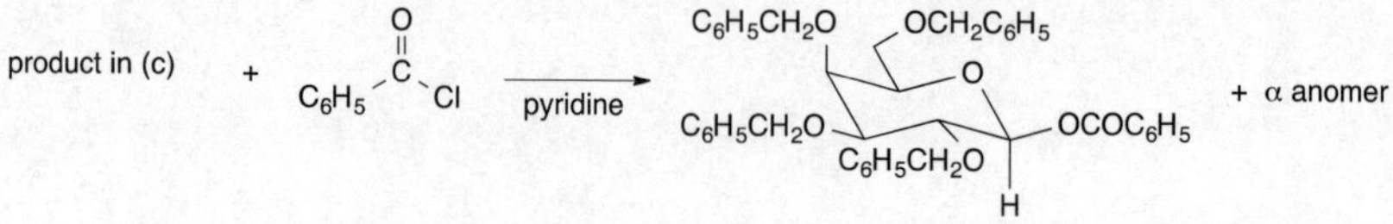

Carbohydrates 28–13

28.21

CH$_2$OH   ═O   HO──H   HO──H   H──OH   CH$_2$OH
D-tagatose

$\xrightarrow[\text{CH}_3\text{OH}]{\text{NaBH}_4}$

CH$_2$OH   H──OH   HO──H   HO──H   H──OH   CH$_2$OH
D-galactitol

+

CH$_2$OH   HO──H   HO──H   HO──H   H──OH   CH$_2$OH
D-talitol

28.22 Carbohydrates containing a hemiacetal are in equilibrium with an acyclic aldehyde, making them reducing sugars. Glycosides are acetals, so they are not in equilibrium with any acyclic aldehyde, making them nonreducing sugars.

a. hemiacetal
reducing sugar

b. acetal
nonreducing sugar

c. hemiacetal
lactose
reducing sugar

28.23

a.

CHO   HO──H   H──OH   H──OH   CH$_2$OH

$\xrightarrow[\text{NH}_4\text{OH}]{\text{Ag}_2\text{O}}$

COOH   HO──H   H──OH   H──OH   CH$_2$OH

c.

CHO   HO──H   H──OH   H──OH   CH$_2$OH

$\xrightarrow[\text{H}_2\text{O}]{\text{HNO}_3}$

COOH   HO──H   H──OH   H──OH   COOH

b.

CHO   HO──H   H──OH   H──OH   CH$_2$OH

$\xrightarrow[\text{H}_2\text{O}]{\text{Br}_2}$

COOH   HO──H   H──OH   H──OH   CH$_2$OH

28.24 Molecules with a plane of symmetry are optically inactive.

a.

CHO   H──OH   H──OH   CH$_2$OH
D-erythrose

$\longrightarrow$

COOH   H──OH   H──OH   COOH
optically inactive

c.

CHO   H──OH   HO──H   HO──H   H──OH   CH$_2$OH
D-galactose

$\longrightarrow$

COOH   H──OH   HO──H   HO──H   H──OH   COOH
optically inactive

b.

CHO   HO──H   HO──H   H──OH   CH$_2$OH
D-lyxose

$\longrightarrow$

COOH   HO──H   HO──H   H──OH   COOH
optically active

28.25

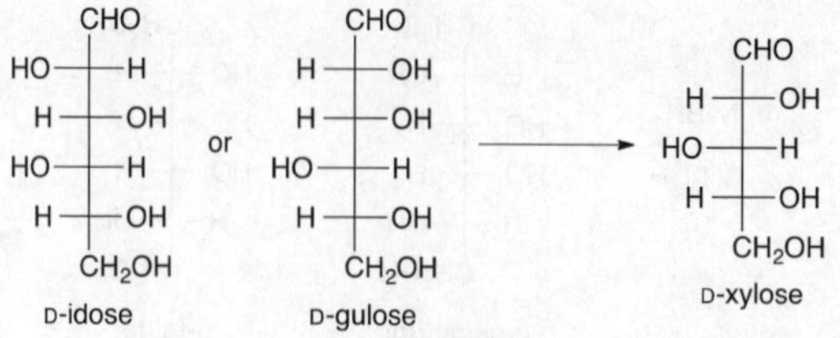

or
D-idose
D-gulose
D-xylose

28.26

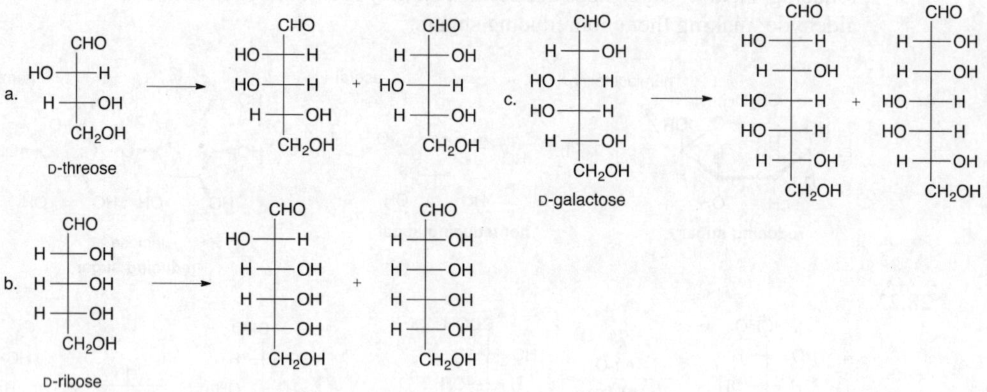

a.
D-threose
c.
D-galactose
b.
D-ribose

28.27

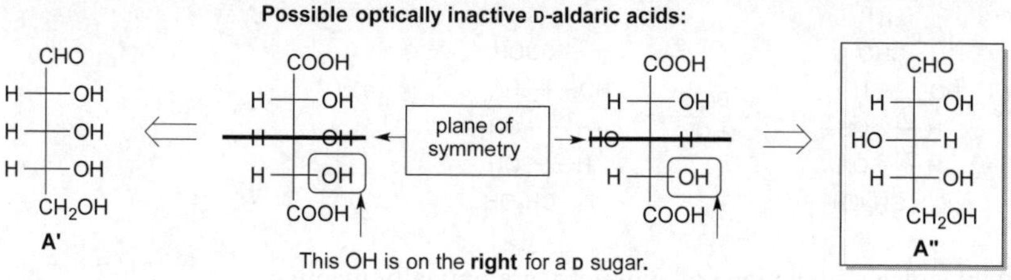

Possible optically inactive D-aldaric acids:
plane of symmetry
A'
A"
This OH is on the **right** for a D sugar.

There are two possible structures for the D-aldopentose (**A'** and **A"**), and the Wohl degradation determines which structure corresponds to **A**.

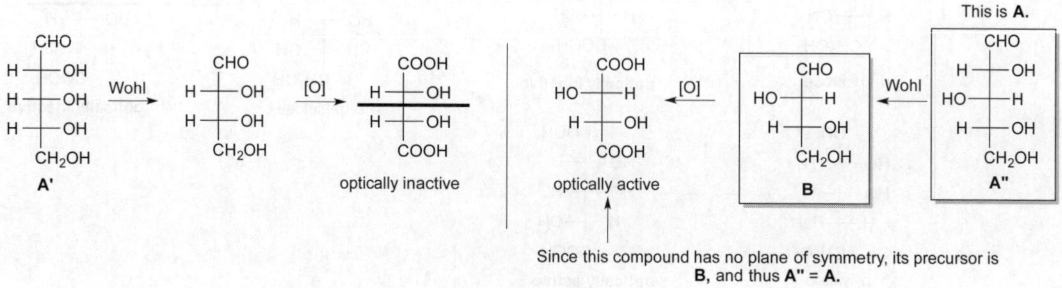

This is **A**.
Wohl
[O]
A'
optically inactive
optically active
B
A"
Since this compound has no plane of symmetry, its precursor is **B**, and thus **A" = A**.

28.28

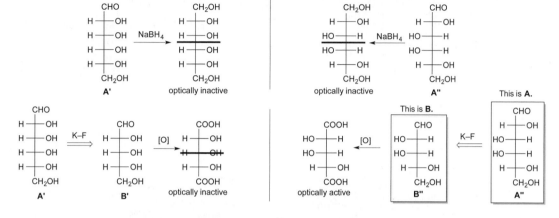

Two D-aldohexoses (**A'** and **A"**) give optically inactive alditols on reduction. **A"** is formed from **B"** by Kiliani–Fischer synthesis. Because **B"** affords an optically active aldaric acid on oxidation, **B"** is **B** and **A"** is **A**. The alternate possibility (**A'**) is formed from an aldopentose **B'** that gives an optically inactive aldaric acid on oxidation.

28.29

The same products are formed on hydrolysis of the α and β anomers of maltose.

28.30

cellobiose

Two possible anomers here. β OH is drawn.

28.31

a.

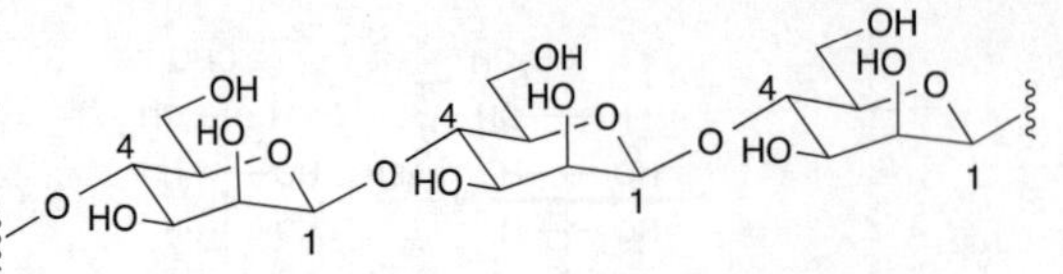

b.

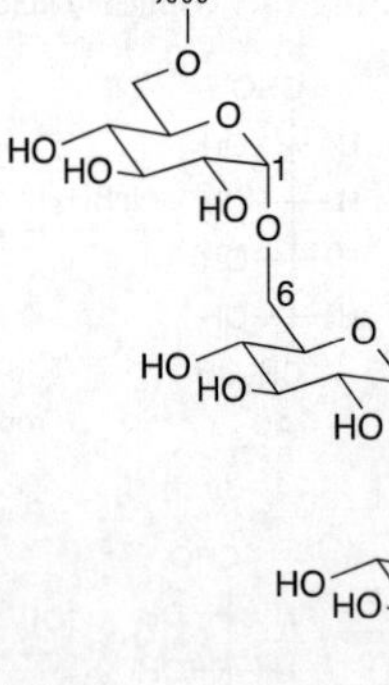

dextran

28.32

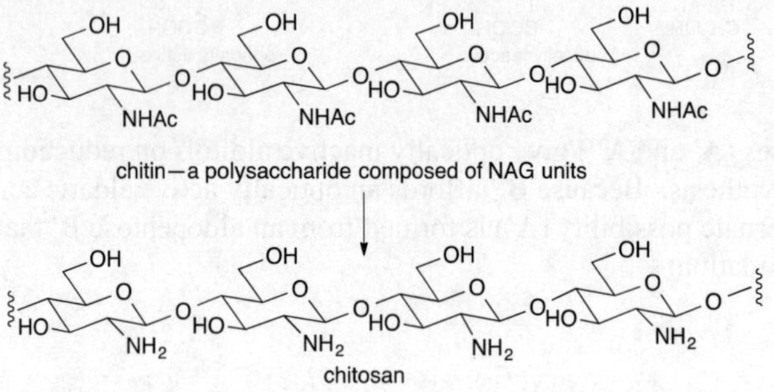
chitin—a polysaccharide composed of NAG units
chitosan

28.33

a.

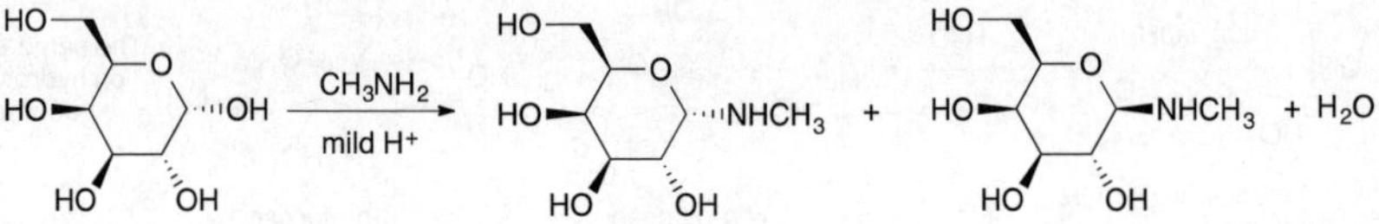
CH3NH2
mild H+
+ H2O

b.

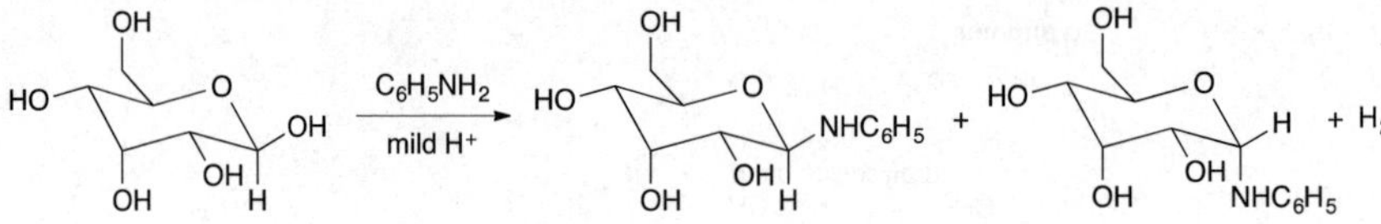
C6H5NH2
mild H+
+ H2O

Carbohydrates 28–17

28.34

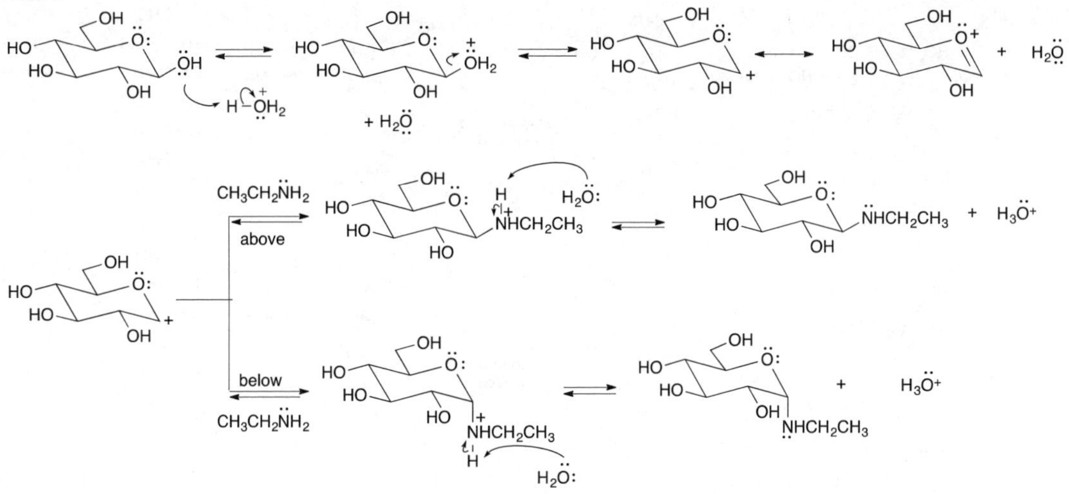

28.35

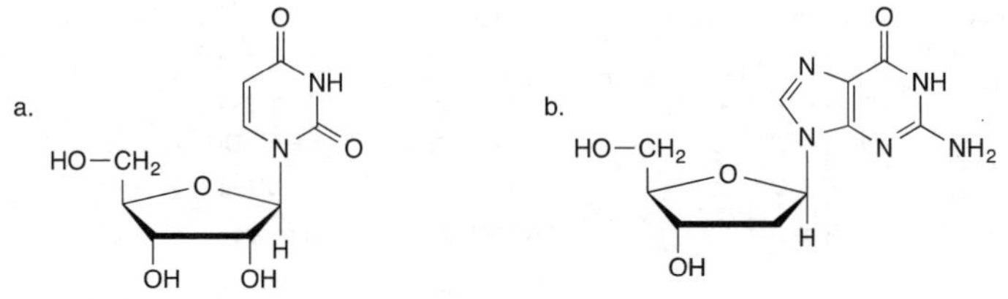

a.

b.

28.36

a. Two purine bases (A and G) are both bicyclic bases. Therefore they are too big to hydrogen bond to each other on the inside of the DNA double helix.

b. Hydrogen bonding between guanine and cytosine has three hydrogen bonds, whereas between guanine and thymine there are only two. This makes hydrogen bonding between guanine and cytosine more favorable.

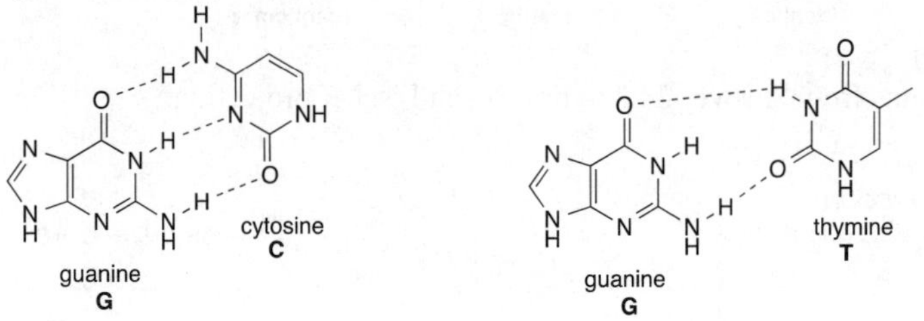

Chapter 28–18

28.37

a. 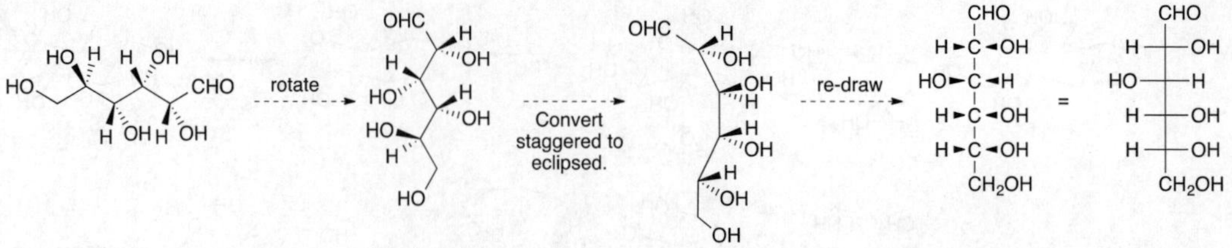 [structure] rotate → [structure] Convert staggered to eclipsed. → [structure] re-draw → [structure] = [structure]

b. 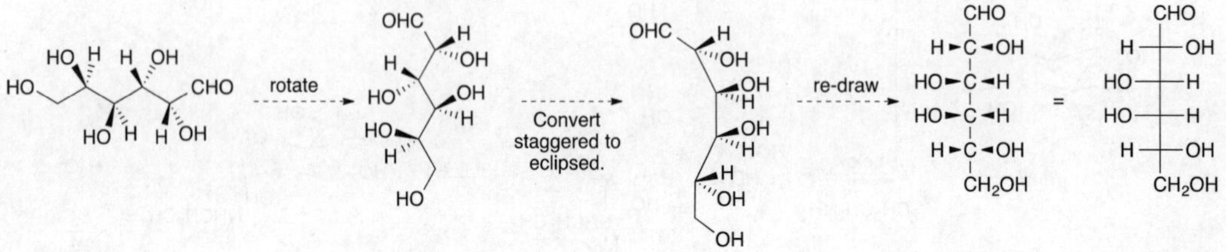 [structure] rotate → [structure] Convert staggered to eclipsed. → [structure] re-draw → [structure] = [structure]

28.38

[structure] = [structure]   [structure] = [structure]

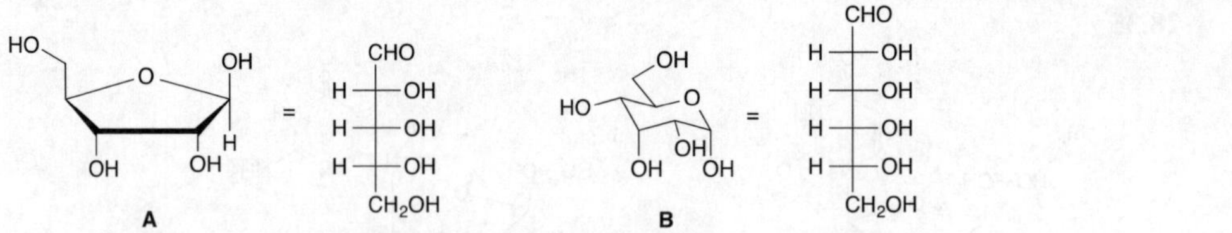

A
β anomer   D-ribose   **B**
α anomer   D-allose

28.39 Label the compounds with *R* or *S* and then classify.

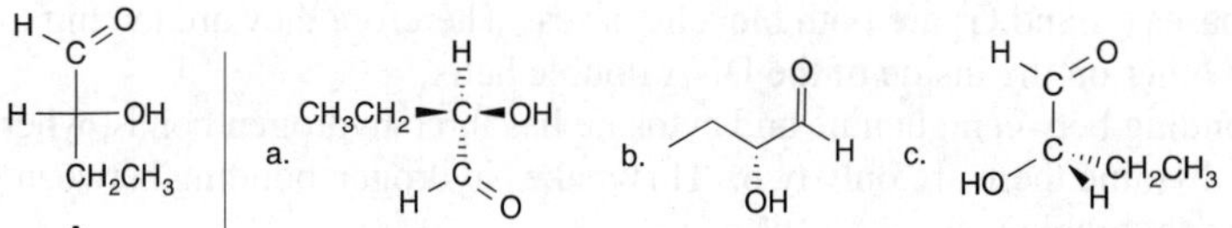

[structure] **A** *R* | a. [structure] **R** identical   b. [structure] **R** identical   c. [structure] **S** enantiomer

28.40 Use the directions from Answer 28.2 to draw each Fischer projection.

a. 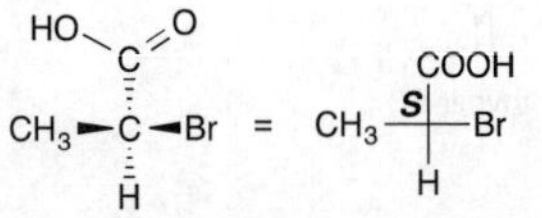 [structure] = [structure]

c. 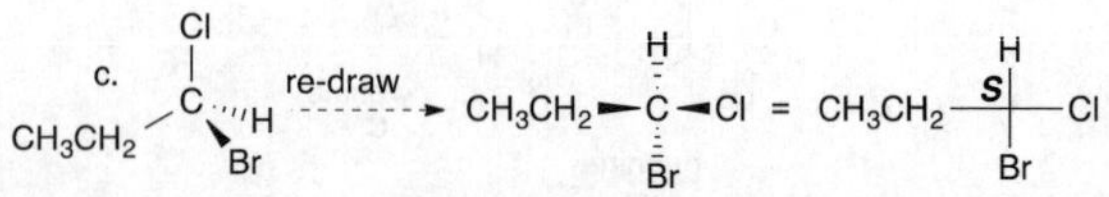 [structure] re-draw → [structure] = [structure]

b. 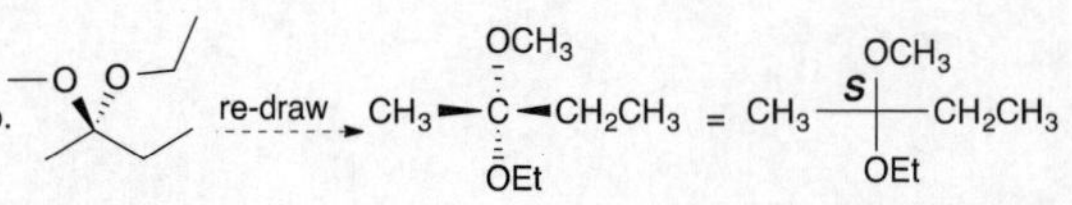 [structure] re-draw → [structure] = [structure]   d. 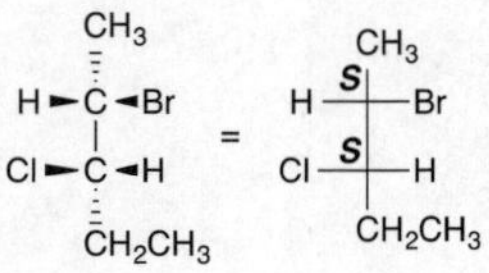 [structure] = [structure]

e. 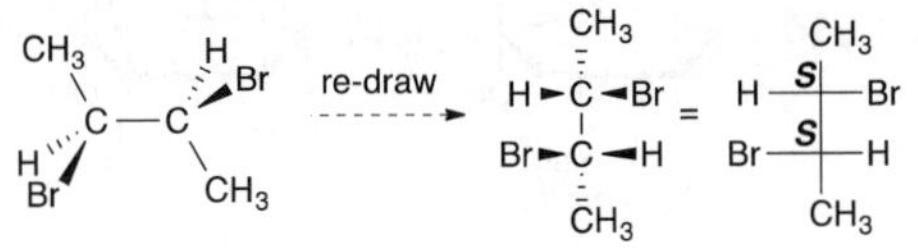 re-draw

f. 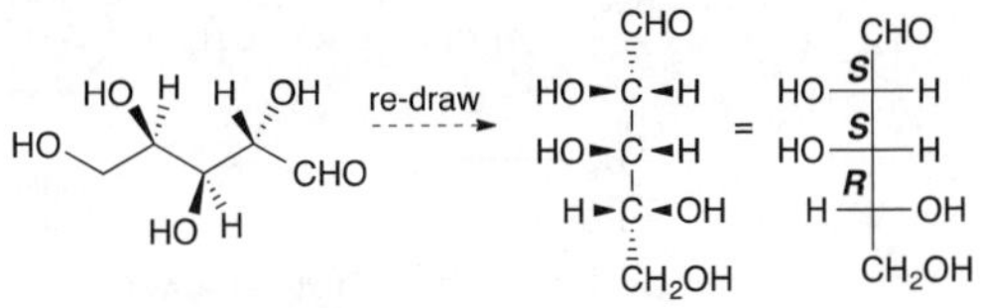 re-draw

28.41

a. 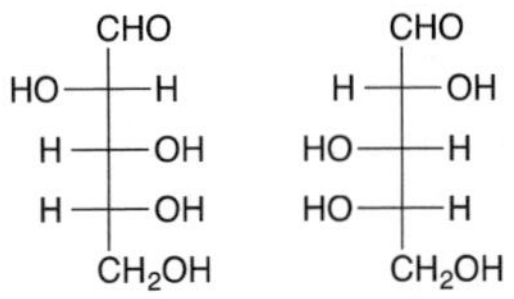 D-arabinose   enantiomer   b. 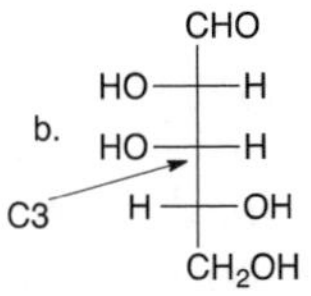 epimer (C3)   c. 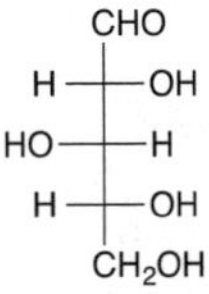 diastereomer (but not epimer)   d. 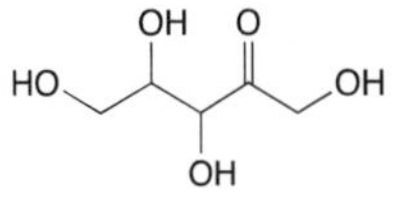 constitutional isomer

28.42

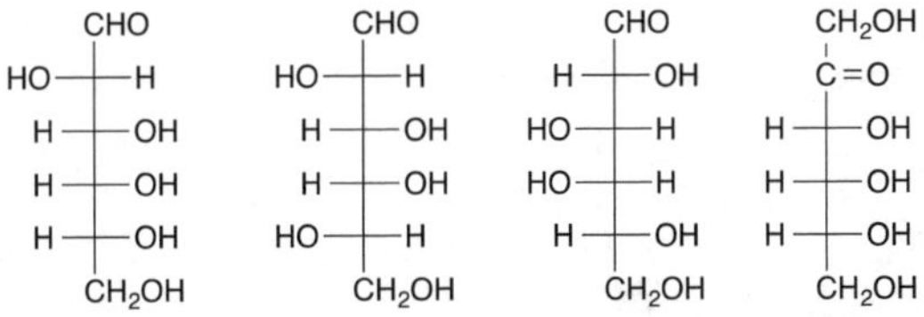

A   B   C   D 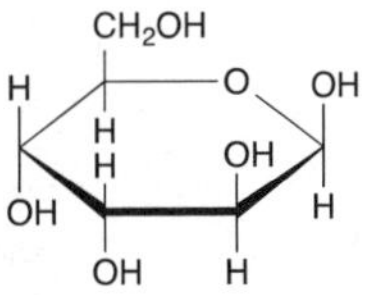   E 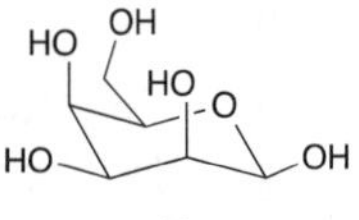   F

a. **A** and **B**   epimers
b. **A** and **C**   diastereomers

c. **B** and **C**   enantiomers
d. **A** and **D**   constitutional isomers

e. **E** and **F**   diastereomers

28.43   Use the directions from Answer 28.13.

a. β-D-talopyranose

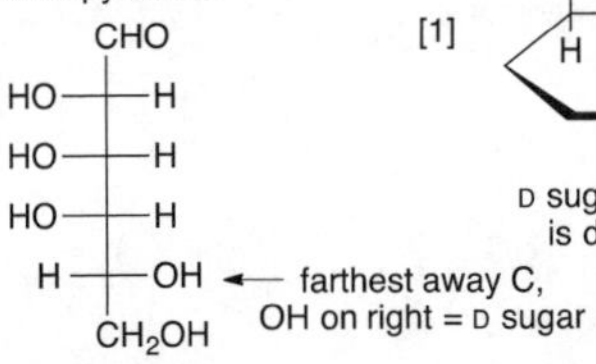

D-talose

[1] 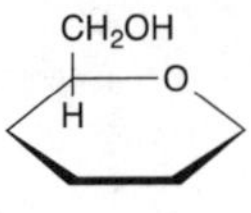   D sugar, CH$_2$OH is drawn up.   farthest away C, OH on right = D sugar

[2] 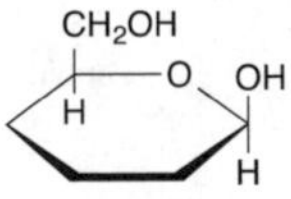   β anomer OH is up for a D sugar.

[3]

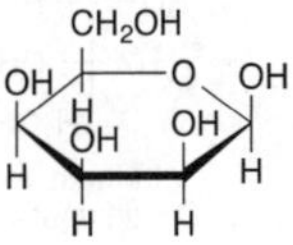

b. α-D-galactopyranose

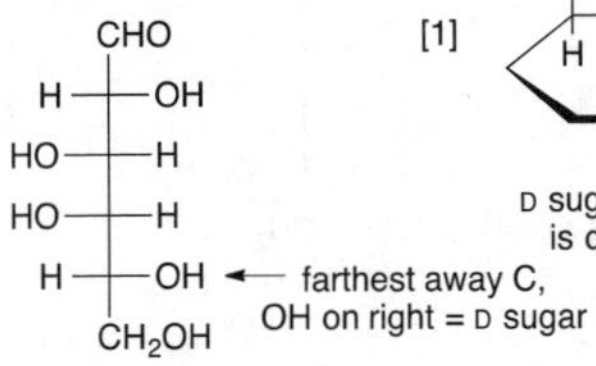

D-galactose

[1] 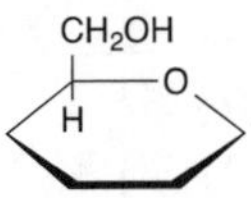   D sugar, CH$_2$OH is drawn up.   farthest away C, OH on right = D sugar

[2] 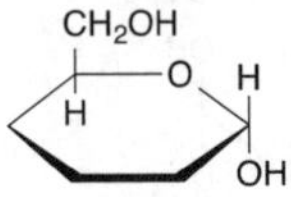   α anomer OH is down.

[3]

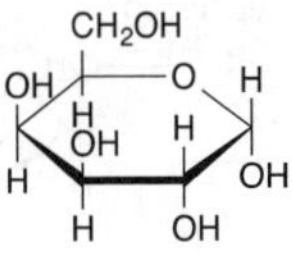

Chapter 28–20

c. α-D-tagatofuranose

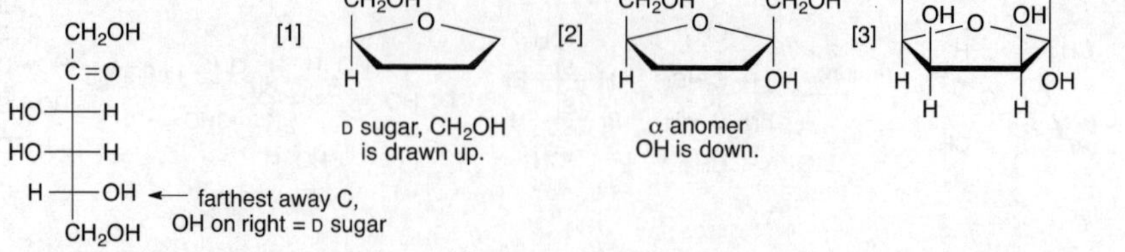

CH$_2$OH
C=O
HO——H
HO——H
H——OH ← farthest away C, OH on right = D sugar
CH$_2$OH

D-tagatose

[1] D sugar, CH$_2$OH is drawn up.

[2] α anomer OH is down.

[3]

28.44

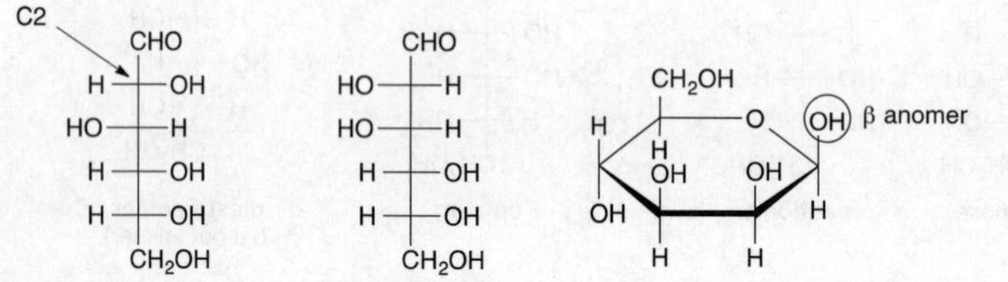

C2
CHO
H——OH
HO——H
H——OH
H——OH
CH$_2$OH

D-glucose

CHO
HO——H
HO——H
H——OH
H——OH
CH$_2$OH

epimer at C2

(OH) β anomer

28.45

a.
CHO 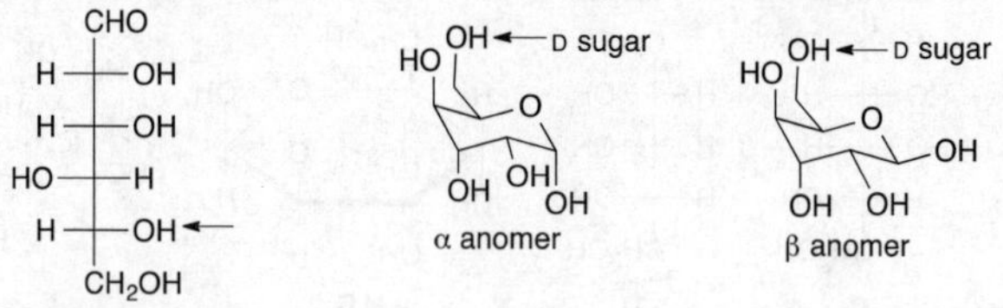
H——OH
H——OH
HO——H
H——OH ←
CH$_2$OH
farthest away C, OH on right = D sugar

HO OH ← D sugar
α anomer

HO OH ← D sugar
β anomer

b.
CHO 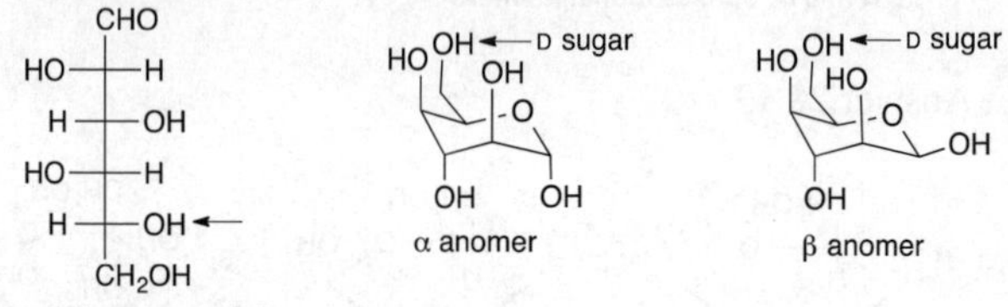
HO——H
H——OH
HO——H
H——OH ←
CH$_2$OH
farthest away C, OH on right = D sugar

HO OH ← D sugar
α anomer

HO OH ← D sugar
β anomer

28.46 Use the directions from Answer 28.14.

"up" group on left
a.

CH$_2$OH ← CH$_2$OH is up = D sugar
"up" group on left
"down" group on right
"up" group on left

[1]
CHO

CH$_2$OH

[2]
CHO

H——OH ←
CH$_2$OH
OH on right = D sugar

[3]
CHO
H——OH
HO——H
HO——H
H——OH
CH$_2$OH

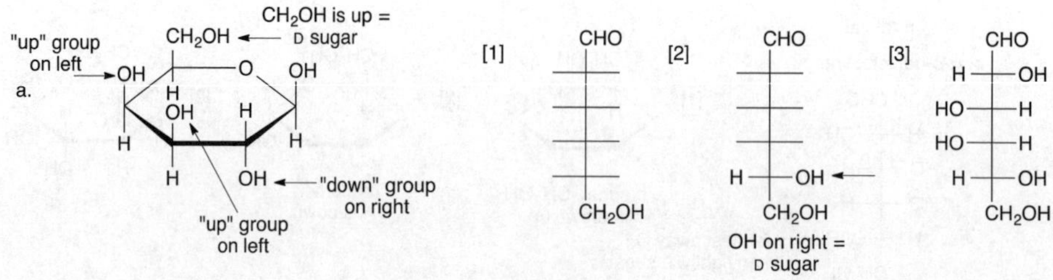

Carbohydrates 28–21

b.

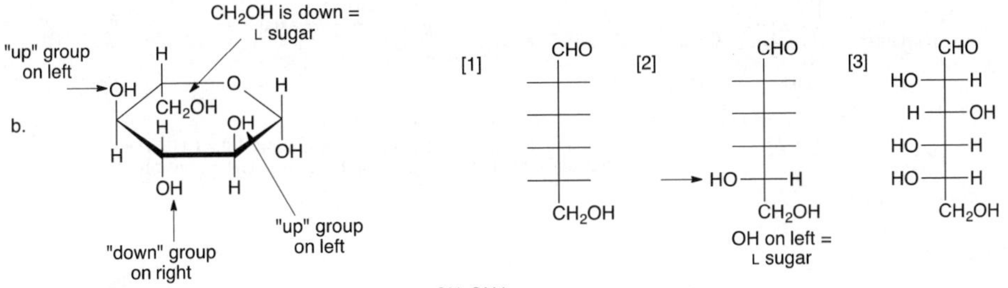

"up" group on left

CH$_2$OH is down = L sugar

"down" group on right

"up" group on left

[1]

CHO
|
|
|
CH$_2$OH

[2]

CHO
|
|
HO—H
|
CH$_2$OH

OH on left = L sugar

[3]

CHO
HO—H
H—OH
HO—H
HO—H
CH$_2$OH

c.

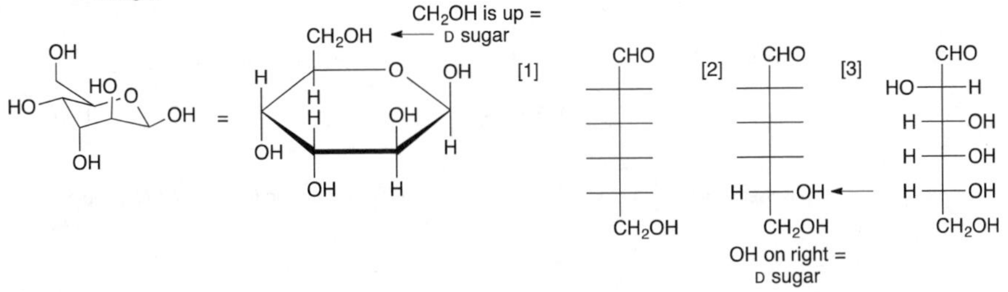

CH$_2$OH is up = D sugar

[1]

CHO
|
|
|
|
CH$_2$OH

[2]

CHO
|
|
H—OH
CH$_2$OH

OH on right = D sugar

[3]

CHO
HO—H
H—OH
H—OH
H—OH
CH$_2$OH

d. D sugar

HOCH$_2$

CH$_2$OH
|
C=O
H—OH
H—OH
H—OH
CH$_2$OH

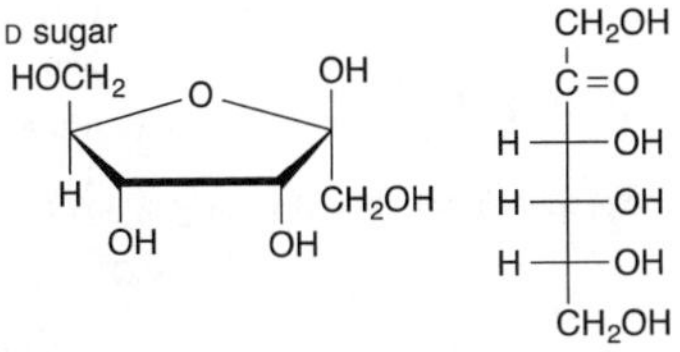

28.47

CHO
HO—H
H—OH
H—OH
CH$_2$OH

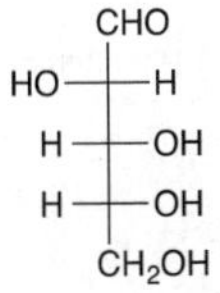

D-arabinose

a.

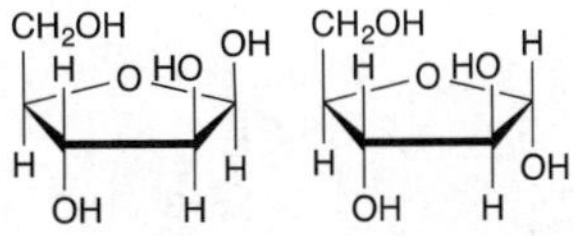

β anomer

α anomer

b.

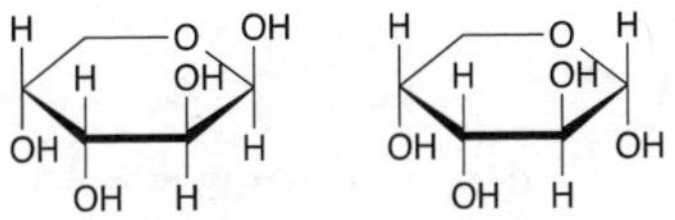

two anomers in the pyranose form

28.48

Two anomers of D-idose, as well as two conformations of each anomer:

α anomer

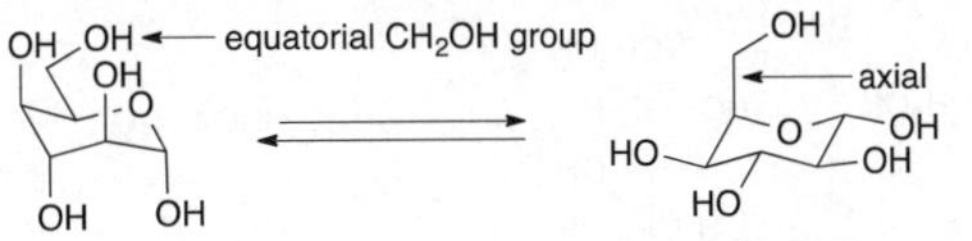

equatorial CH$_2$OH group

axial

4 axial substituents

4 equatorial OH groups

More stable conformation for the α anomer—the CH$_2$OH is axial, but all other groups are equatorial.

Chapter 28–22

β anomer

OH OH ← equatorial CH$_2$OH group

OH

O

OH

OH

3 axial substituents

OH — axial

OH

← axial

HO

HO

O

OH

3 equatorial OH groups

The more stable conformation for the β anomer—the CH$_2$OH is axial, as is the anomeric OH, but three other OH groups are equatorial.

28.49

a. CH$_3$I, Ag$_2$O

CH$_3$O

OCH$_3$

O

CH$_3$O

OCH$_3$

d. The product in (a), then H$_3$O$^+$

OCH$_3$ OCH$_3$

O

CH$_3$O

CH$_3$O OH

+ β anomer

b. CH$_3$OH, HCl

HO

OH

HO

OH

O

HO

HO

OH

HO

HO OCH$_3$

+ β anomer

D-gulose

e. The product in (b), then Ac$_2$O, pyridine

OAc OAc

O

OAc

OAc OCH$_3$

+ β anomer

c. Ac$_2$O, pyridine

OAc OAc

O

OAc OAc

f. The product in (d), then C$_6$H$_5$CH$_2$Cl, Ag$_2$O

CH$_3$O OCH$_3$

O

CH$_3$O

CH$_3$O OCH$_2$C$_6$H$_5$

+ β anomer

28.50

CHO

HO—H

H—OH

H—OH

H—OH

CH$_2$OH

D-altrose

a. (CH$_3$)$_2$CHOH, HCl

OH

OH

HO

O

OH OCH(CH$_3$)$_2$

+ β anomer

b. NaBH$_4$, CH$_3$OH

CH$_2$OH

HO—H

H—OH

H—OH

H—OH

CH$_2$OH

c. Br$_2$, H$_2$O

COOH

HO—H

H—OH

H—OH

H—OH

CH$_2$OH

d. HNO$_3$, H$_2$O

COOH

HO—H

H—OH

H—OH

H—OH

COOH

Carbohydrates 28–23

e. [1] NH₂OH
   [2] (CH₃CO)₂O, NaOCOCH₃
   [3] NaOCH₃

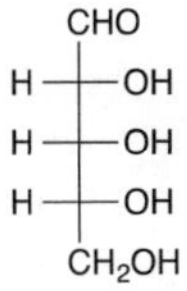

$$\begin{array}{c} \text{CHO} \\ \text{H} - \text{OH} \\ \text{H} - \text{OH} \\ \text{H} - \text{OH} \\ \text{CH}_2\text{OH} \end{array}$$

g. CH₃I, Ag₂O

+ β anomer

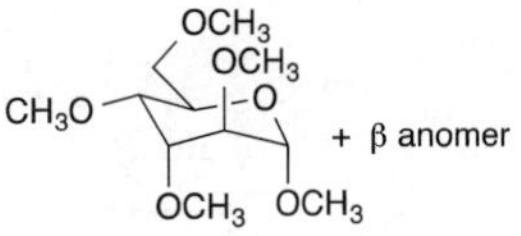

f. [1] NaCN, HCl
   [2] H₂, Pd-BaSO₄
   [3] H₃O⁺

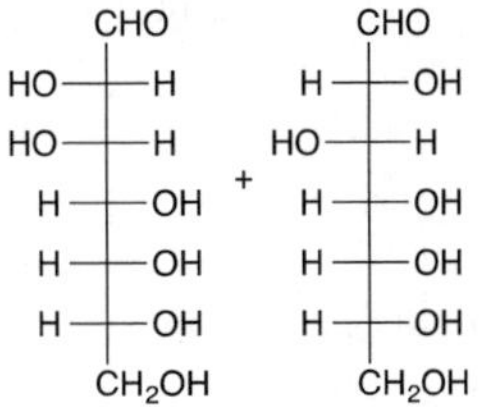

$$\begin{array}{c} \text{CHO} \\ \text{HO} - \text{H} \\ \text{HO} - \text{H} \\ \text{H} - \text{OH} \\ \text{H} - \text{OH} \\ \text{H} - \text{OH} \\ \text{CH}_2\text{OH} \end{array}$$
+
$$\begin{array}{c} \text{CHO} \\ \text{H} - \text{OH} \\ \text{HO} - \text{H} \\ \text{H} - \text{OH} \\ \text{H} - \text{OH} \\ \text{H} - \text{OH} \\ \text{CH}_2\text{OH} \end{array}$$

h. C₆H₅CH₂NH₂, mild H⁺

+ β anomer

28.51

salicin → (H₃O⁺) → monosaccharide (both anomers) + aglycon

solanine → (H₃O⁺) → monosaccharide (both anomers) + aglycon + monosaccharide (both anomers) + monosaccharide (both anomers)

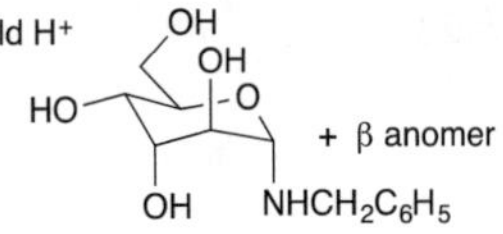

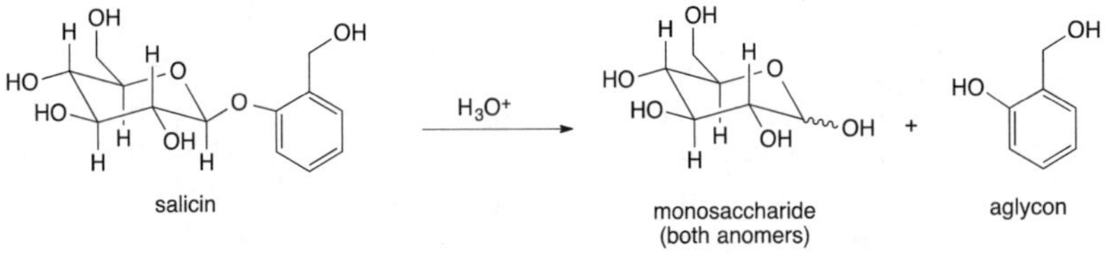

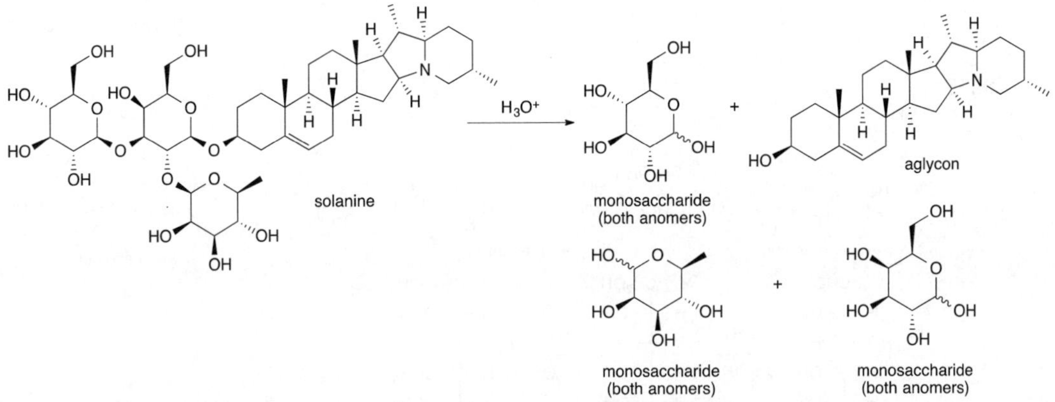

28.52

a.

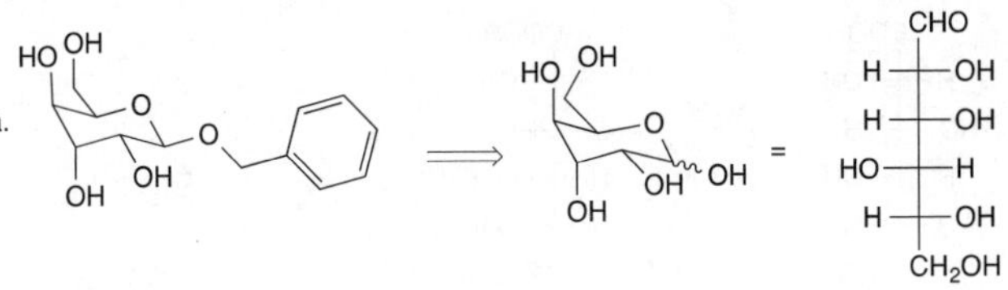

b. 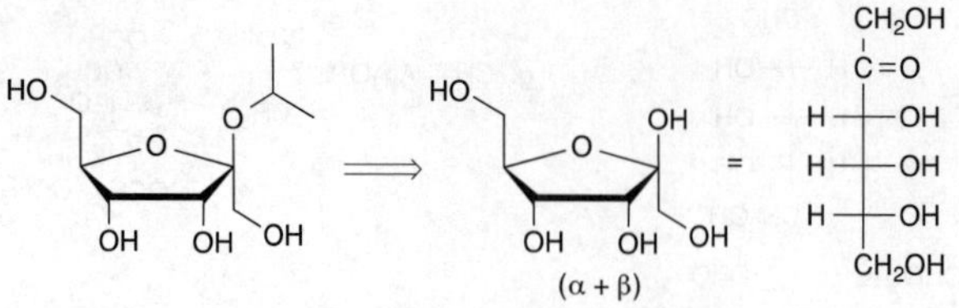
(α + β)

28.53

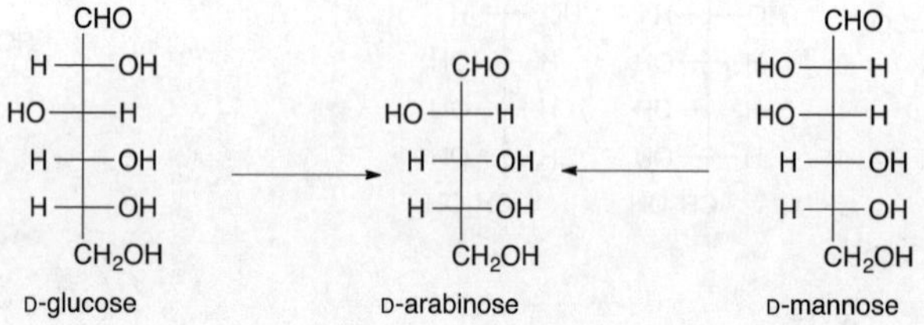
D-glucose
D-arabinose
D-mannose

28.54

a. 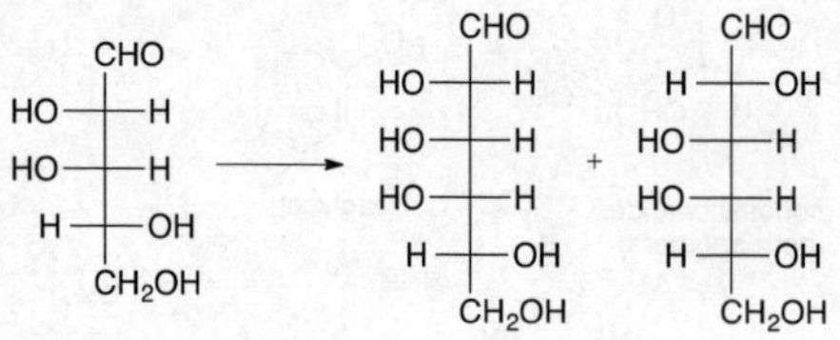

b. 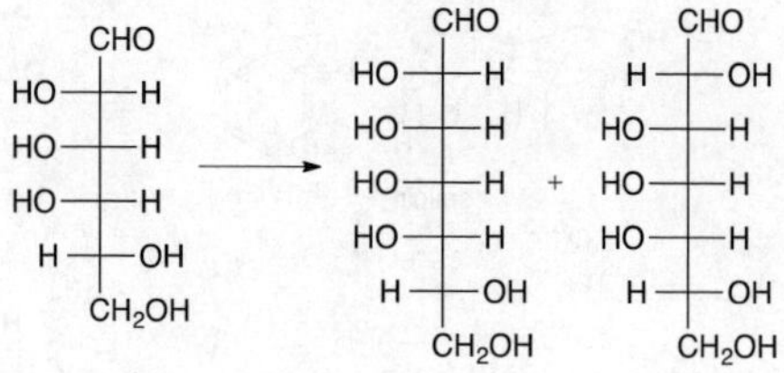

28.55

a. 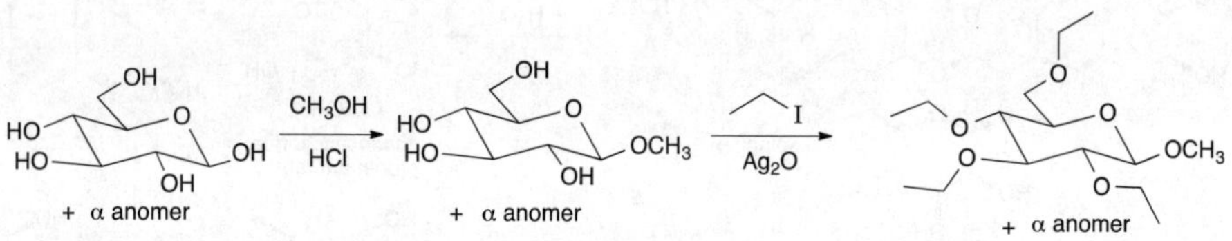
CH3OH
HCl
Ag2O
+ α anomer
+ α anomer
+ α anomer

b. 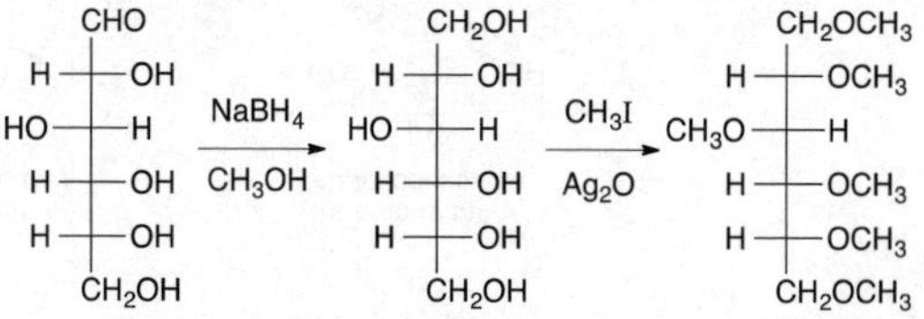
NaBH4
CH3OH
CH3I
Ag2O

c. 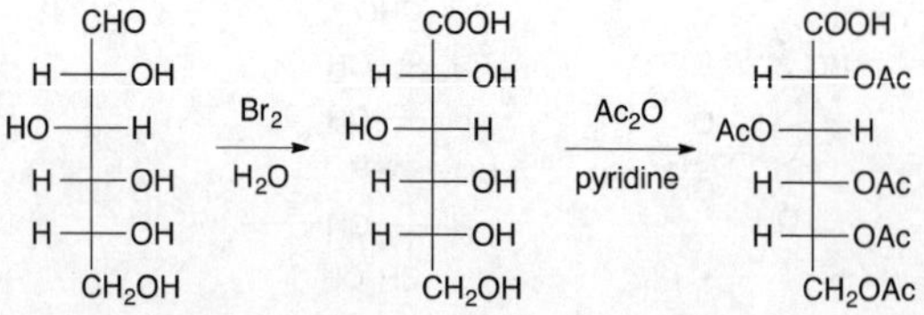
Br2
H2O
Ac2O
pyridine

Carbohydrates 28–25

28.56 Molecules with a plane of symmetry are optically inactive.

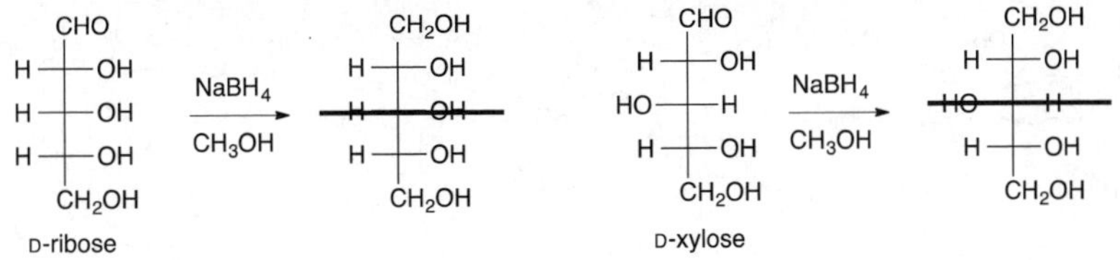

28.57

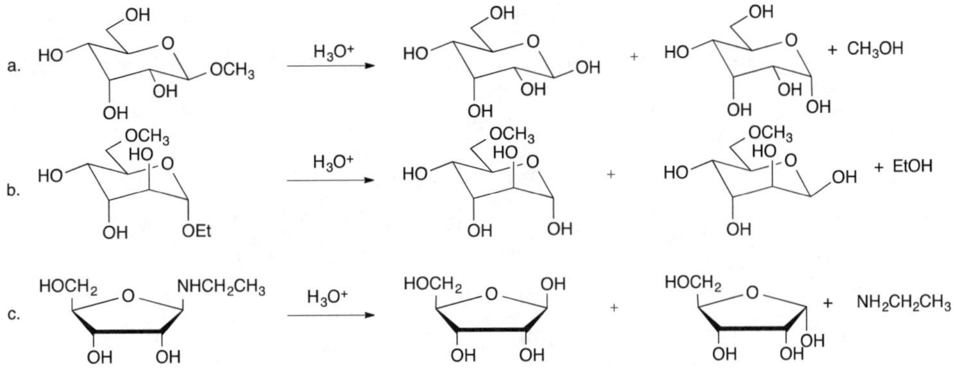

28.58

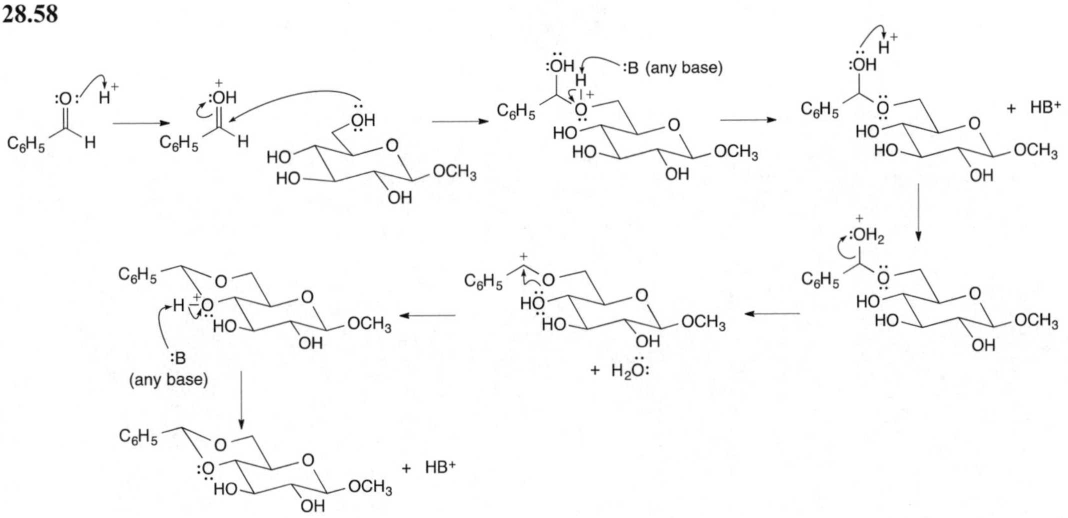

Chapter 28–26

28.59

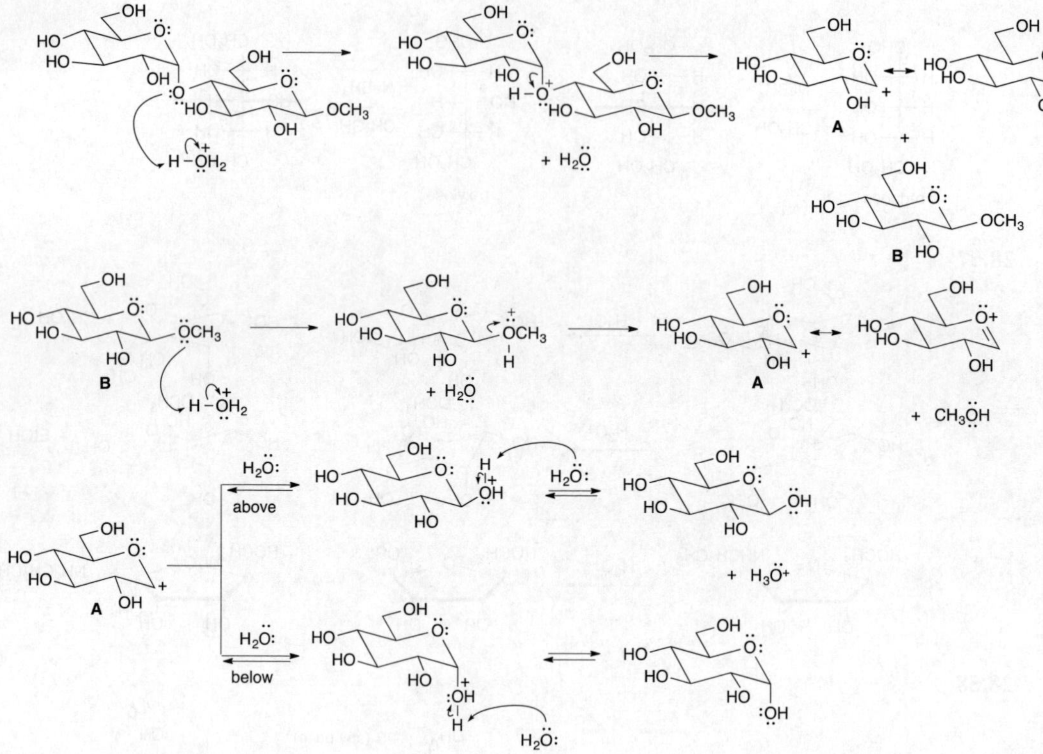

28.60

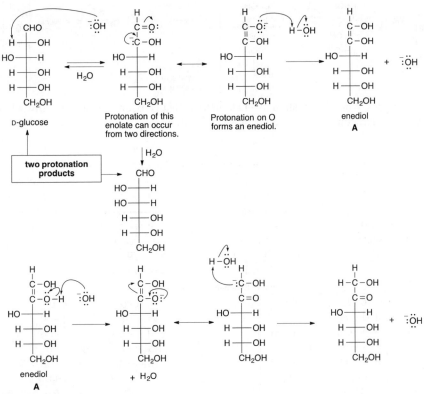

28.61

Two D-aldopentoses (**A'** and **A"**) yield optically active aldaric acids when oxidized.
Optically active D-aldaric acids:

Chapter 28–28

Only **A″** undergoes Wohl degradation to an aldotetrose that is oxidized to an optically active aldaric acid, so **A″** is the structure of the D-aldopentose in question.

28.62

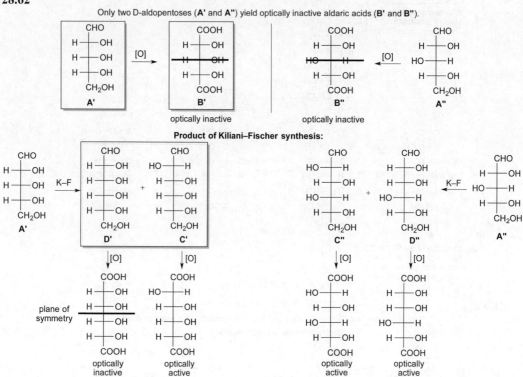

Only **A′** fits the criteria. Kiliani–Fischer synthesis of **A′** forms **C′** and **D′**, which are oxidized to one optically active and one optically inactive aldaric acid. A similar procedure with **A″** forms two optically active aldaric acids. Thus, the structures of **A–D** correspond to the structures of **A′–D′**.

28.63

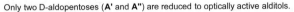
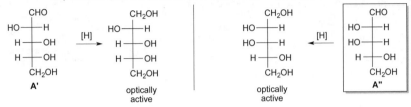

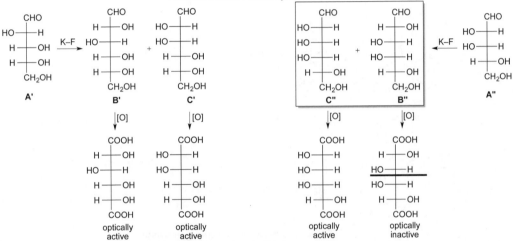

Only **A″** fits the criteria. Kiliani–Fischer synthesis of **A″** forms **B″** and **C″**, which are oxidized to one optically inactive and one optically active diacid. A similar procedure with **A′** forms two optically active diacids. Thus, the structures of **A–C** correspond to **A″–C″**.

28.64 A disaccharide formed from two mannose units in a 1→4-α-glycosidic linkage:

28.65

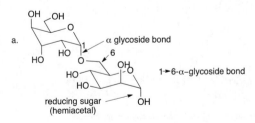

b.

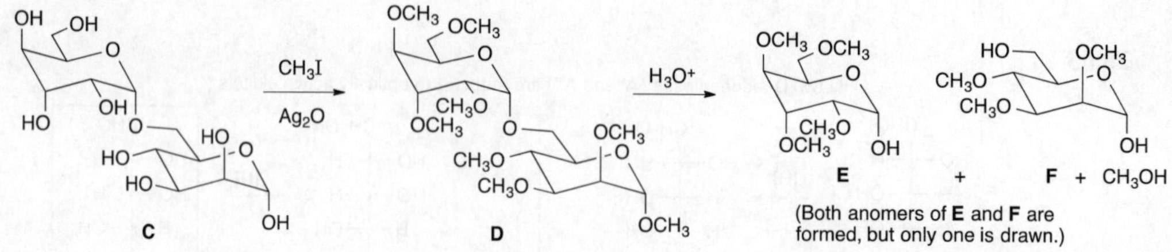
C
D
E
F
CH_3I
Ag_2O
H_3O^+
CH_3OH

(Both anomers of **E** and **F** are formed, but only one is drawn.)

28.66

a, b.

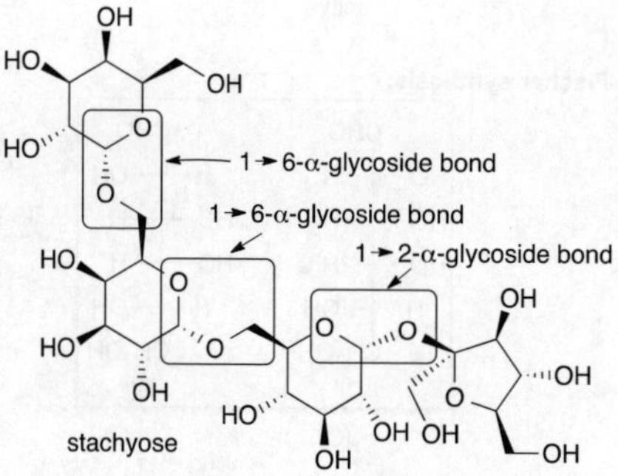
1→6-α-glycoside bond
1→6-α-glycoside bond
1→2-α-glycoside bond

stachyose

c.

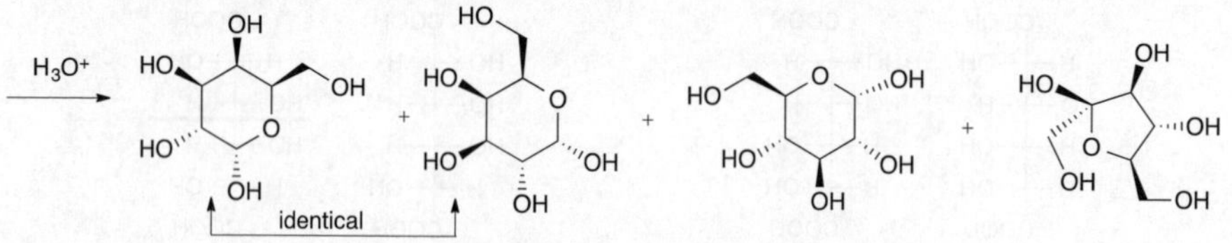
H_3O^+
identical

Two anomers of each monosaccharide are formed, but only one anomer is drawn.

d. Stachyose is not a reducing sugar because it contains no hemiacetal.

e.

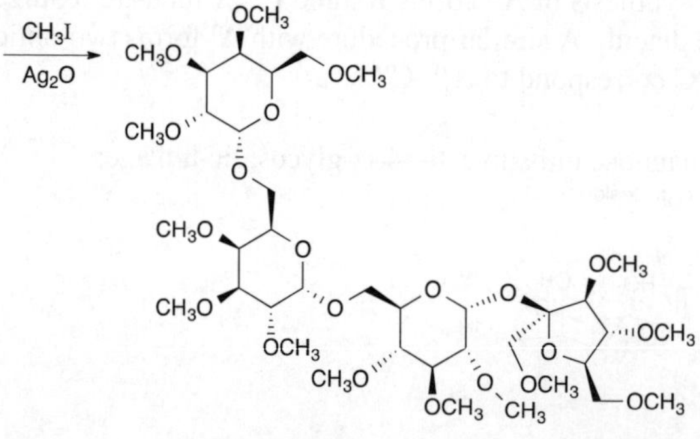
CH_3I
Ag_2O

f.

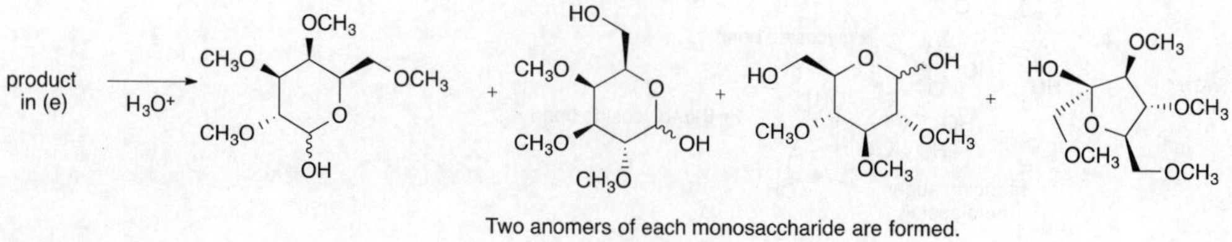
product in (e)
H_3O^+

Two anomers of each monosaccharide are formed.

Carbohydrates 28–31

28.67

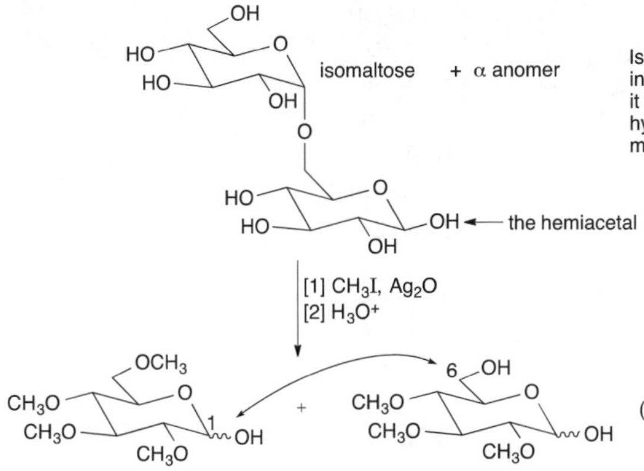

Isomaltose must be composed of two glucose units in an α-glycosidic linkage. Since it is a reducing sugar it contains a hemiacetal. The free OH groups in the hydrolysis products show where the two monosaccharides are joined.

[1] CH$_3$I, Ag$_2$O
[2] H$_3$O$^+$

(Both anomers are present.)

28.68

a.

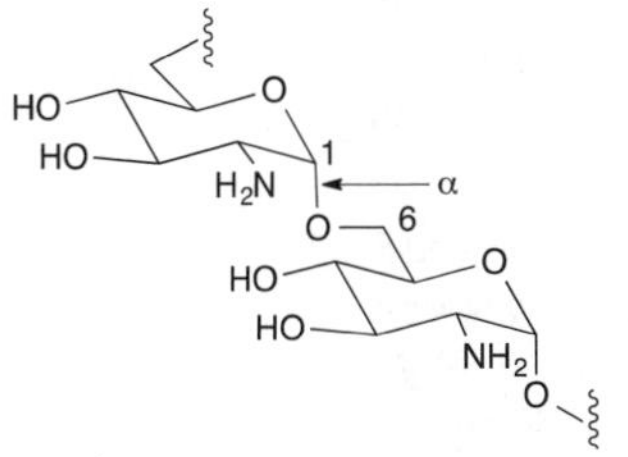

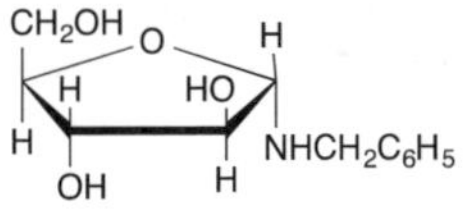

b.

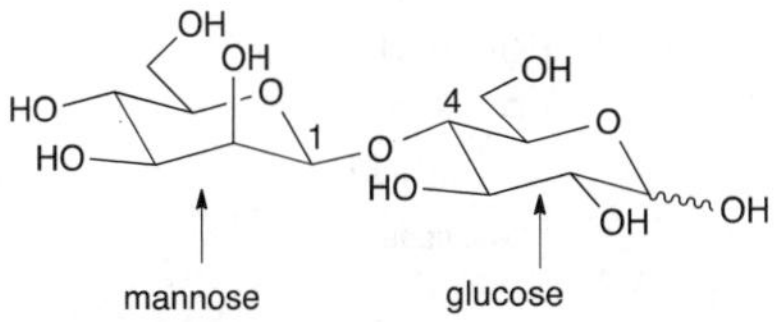

mannose   glucose

c.

d.

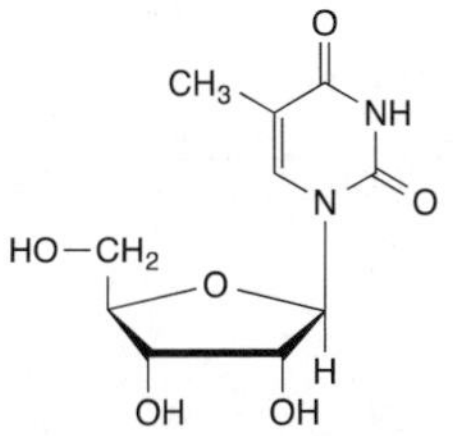

28.69

a.

rotate   re-draw

b. OH on left in Fischer
L-monosaccharide

more stable chair

Ring
flip.

The α anomer has the CH$_3$ on C5 and the anomeric OH trans.

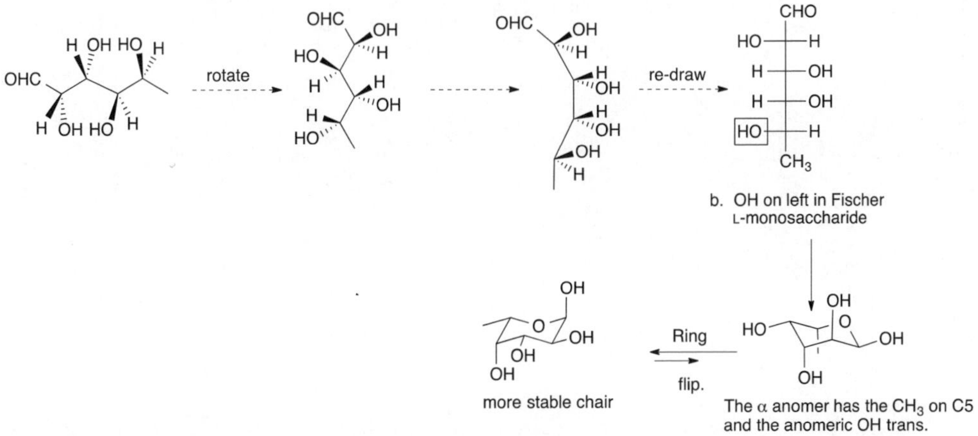

c. Fucose is unusual because it is an L-monosaccharide and it contains a CH$_3$ group rather than a CH$_2$OH group on its terminal carbon.

Chapter 28–32

28.70

a. 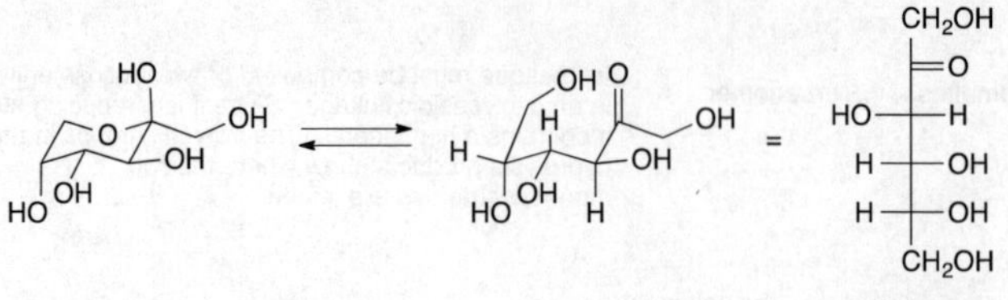 D-fructose

b. 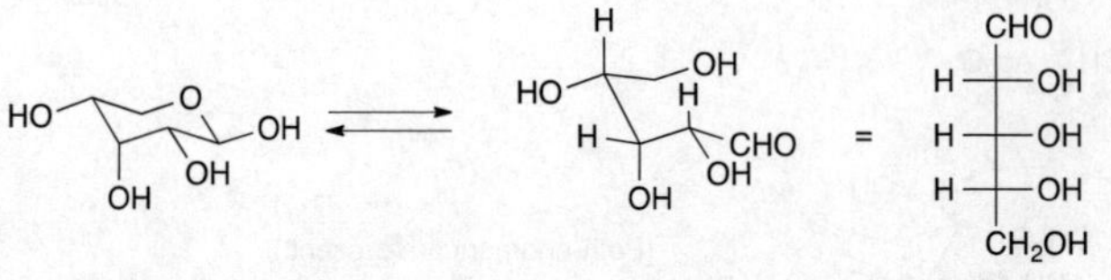 D-ribose

c. 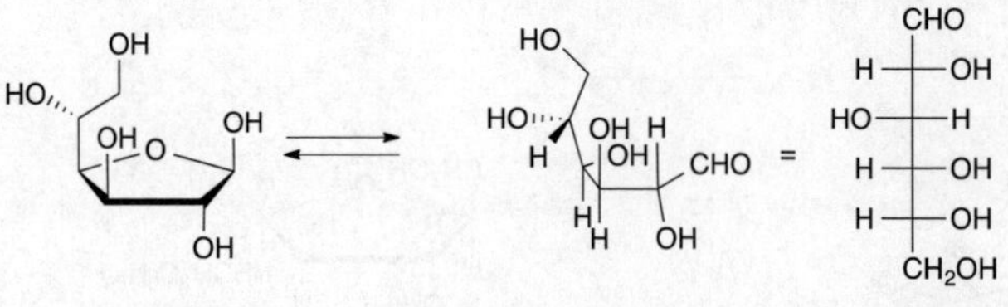 D-glucose

d. 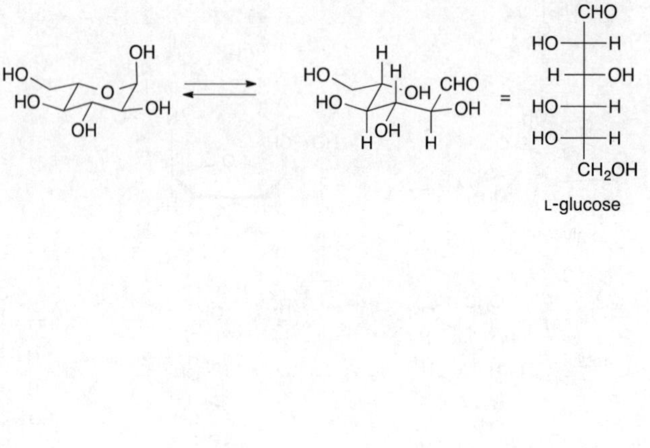 L-glucose

Carbohydrates 28–33

28.71

Ignoring stereochemistry along the way:

Amino Acids and Proteins 29–1

Chapter 29 Amino Acids and Proteins

Chapter Review

Synthesis of amino acids (29.2)

[1] From α-halo carboxylic acids by S_N2 reaction

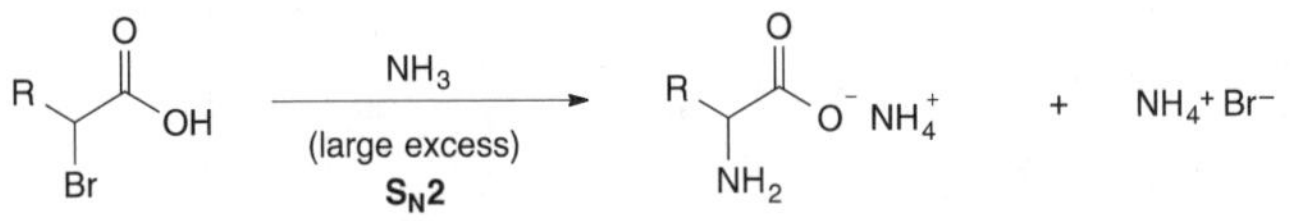

[2] By alkylation of diethyl acetamidomalonate

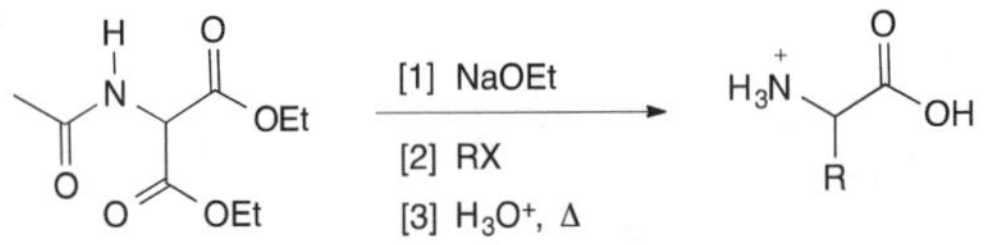

- Alkylation works best with unhindered alkyl halides—that is, with CH_3X and RCH_2X.

[3] Strecker synthesis

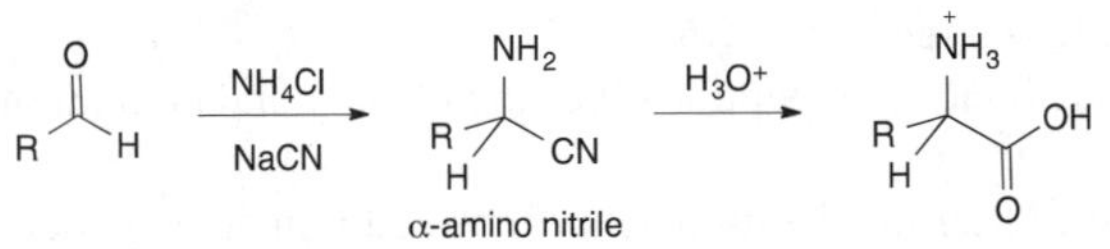

Preparation of optically active amino acids

[1] Resolution of enantiomers by forming diastereomers (29.3A)

- Convert a racemic mixture of amino acids into a racemic mixture of *N*-acetyl amino acids [(*S*)- and (*R*)-CH₃CONHCH(R)COOH].
- React the enantiomers with a chiral amine to form a mixture of diastereomers.
- Separate the diastereomers.
- Regenerate the amino acids by protonation of the carboxylate salt and hydrolysis of the *N*-acetyl group.

Chapter 29–2

[2] Kinetic resolution using enzymes (29.3B)

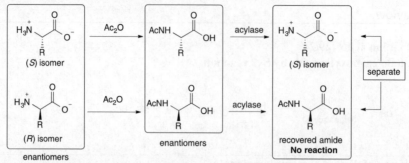

[3] By enantioselective hydrogenation (29.4)

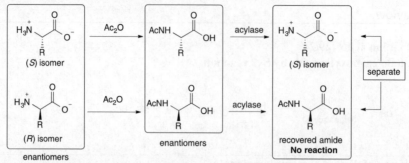

Summary of methods used for peptide sequencing (29.6)
- Complete hydrolysis of all amide bonds in a peptide gives the identity and amount of the individual amino acids.
- Edman degradation identifies the N-terminal amino acid. Repeated Edman degradations can be used to sequence a peptide from the N-terminal end.
- Cleavage with carboxypeptidase identifies the C-terminal amino acid.
- Partial hydrolysis of a peptide forms smaller fragments that can be sequenced. Amino acid sequences common to smaller fragments can be used to determine the sequence of the complete peptide.
- Selective cleavage of a peptide occurs with trypsin and chymotrypsin to identify the location of specific amino acids (Table 29.2).

Adding and removing protecting groups for amino acids (29.7)
[1] Protection of an amino group as a Boc derivative

Amino Acids and Proteins 29–3

[2] Deprotection of a Boc-protected amino acid

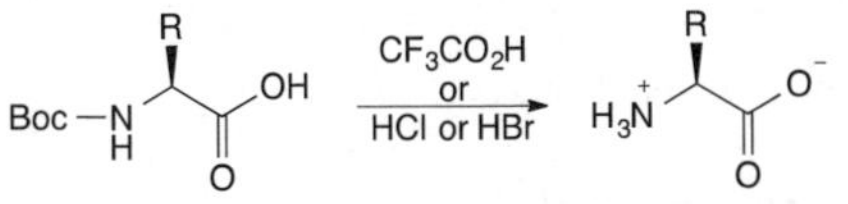

[3] Protection of an amino group as an Fmoc derivative

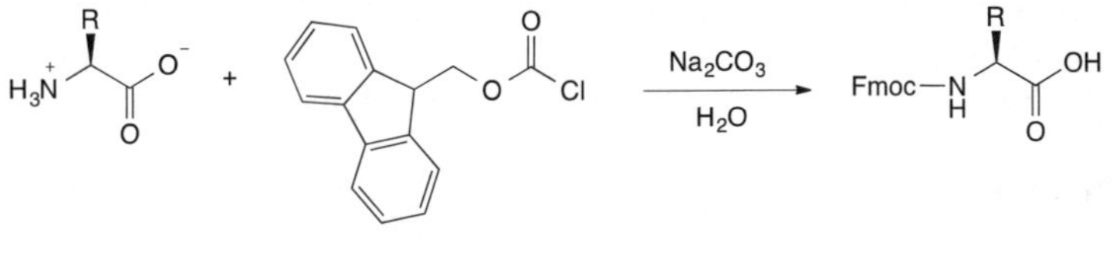

[4] Deprotection of an Fmoc-protected amino acid

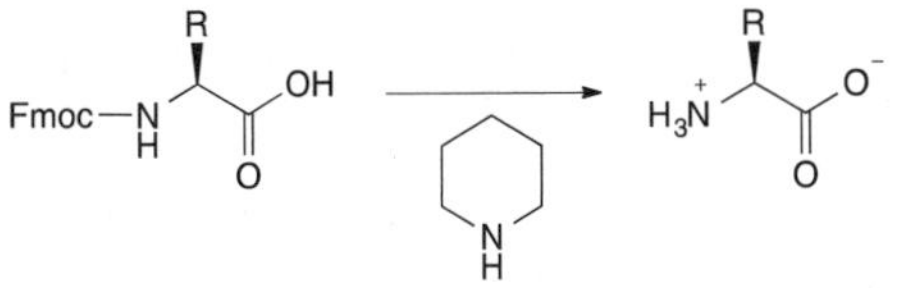

[5] Protection of a carboxy group as an ester

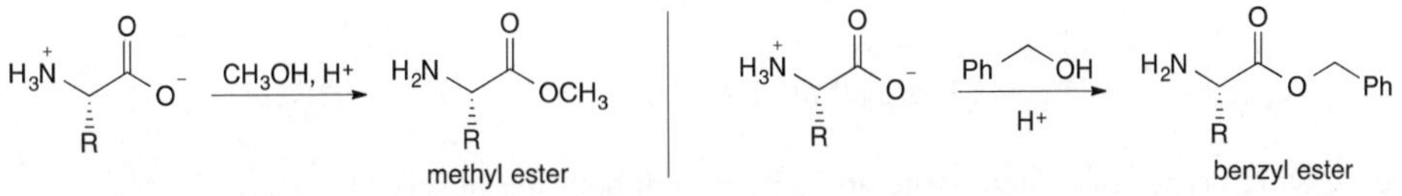

methyl ester

benzyl ester

[6] Deprotection of an ester group

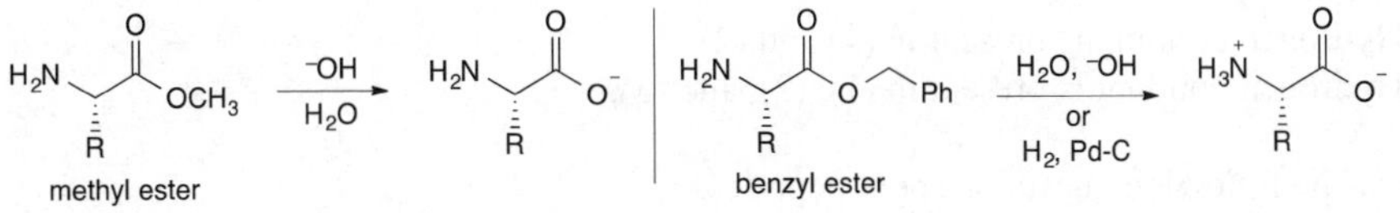

methyl ester

benzyl ester

Synthesis of dipeptides (29.7)
[1] Amide formation with DCC

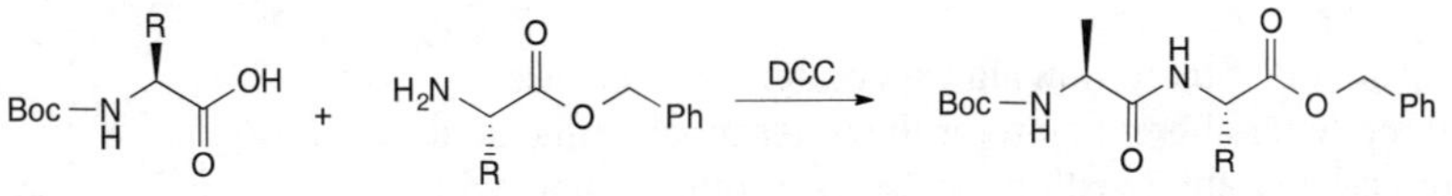

Chapter 29–4

[2] Four steps are needed to synthesize a dipeptide:
    a. **Protect** the amino group of one amino acid using a Boc or Fmoc group.
    b. **Protect** the carboxy group of the second amino acid using an ester.
    c. Form the amide bond with **DCC.**
    d. **Remove both protecting groups** in one or two reactions.

Summary of the Merrifield method of peptide synthesis (29.8)
[1] Attach an Fmoc-protected amino acid to a polymer derived from polystyrene.
[2] Remove the Fmoc protecting group.
[3] Form the amide bond with a second Fmoc-protected amino acid using DCC.
[4] Repeat steps [2] and [3].
[5] Remove the protecting group and detach the peptide from the polymer.

Practice Test on Chapter Review

1.a. Which statement is true about the peptide Ala–Gly–Tyr–Phe?
    1. The N-terminal amino acid is Ala.
    2. The N-terminal amino acid is Phe.
    3. The peptide contains four peptide bonds.
    4. Statements (1) and (3) are true.
    5. Statements (2) and (3) are true.

  b. Which of the following peptides is hydrolyzed by trypsin?
    1. Glu–Ser–Gly–Arg
    2. Arg–Gln–Trp–Asp
    3. Glu–Val–Leu–Lys
    4. Peptides (1) and (2) are hydrolyzed.
    5. Peptides (1), (2), and (3) are all hydrolyzed.

  c. In which types of protein structure is hydrogen bonding observed?
    1. α-helix
    2. β-pleated sheet
    3. 3° structure
    4. Hydrogen bonding is present in (1) and (2).
    5. Hydrogen bonding is present in (1), (2), and (3).

2. Answer the following questions about peptides.

Ala   Val   Ser

  a. Draw the structure of the following tripeptide: Val–Ser–Ala.
  b. Give the three-letter abbreviation for the N-terminal amino acid.
  c. Give the three-letter abbreviation for the C-terminal amino acid.

Amino Acids and Proteins 29–5

3. Answer the following questions about the amino acid leucine (2-amino-4-methylpentanoic acid), which has pK_a's of 2.33 and 9.74 for its ionizable functional groups.
   a. Draw a Fischer projection for L-leucine and label the stereogenic center as R or S.
   b. What is the pI of leucine?
   c. Draw the structure of the predominant form of leucine at its isoelectric point.
   d. Draw the structure of the predominant form of leucine at pH 10.
   e. Is leucine an acidic, basic, or neutral amino acid?

4. What product is formed when the amino acid phenylalanine is treated with each reagent?
   a. PhCH$_2$OH, H$^+$
   b. Ac$_2$O, pyridine
   c. PhCOCl, pyridine
   d. (Boc)$_2$O
   e. C$_6$H$_5$N=C=S

5. Draw the amino acids and peptide fragments formed when the octapeptide Tyr–Gly–Ala–Lys–Val–Ser–Phe–Met is treated with each reagent or enzyme:
   a. chymotrypsin
   b. trypsin
   c. carboxypeptidase
   d. C$_6$H$_5$N=C=S

Answers to Practice Test

1.a. 1
   b. 2
   c. 5
2.a

Val   Ser   Ala

   b. Val
   c. Ala
3.

   a.
   b. 6.04
   c.
   d.
   e. neutral

4.
   a.
   b.
   c.

   d.
   e.

5.a. Tyr, Gly–Ala–Lys–Val–Ser–Phe, Met
   b. Tyr–Gly–Ala–Lys, Val–Ser–Phe–Met
   c. Tyr–Gly–Ala–Lys–Val–Ser–Phe, Met
   d. Tyr, Gly–Ala–Lys–Val–Ser–Phe–Met

Chapter 29–6

Answers to Problems

29.1

L-isoleucine

29.2

a.   b.   c.   d. HO

29.3   In an amino acid, the electron-withdrawing carboxy group destabilizes the ammonium ion ($-NH_3^+$), making it more readily donate a proton; that is, it makes it a stronger acid. Also, the electron-withdrawing carboxy group removes electron density from the amino group ($-NH_2$) of the conjugate base, making it a weaker base than a 1° amine, which has no electron-withdrawing group.

29.4

zwitterionic form

29.5   The most direct way to synthesize an α-amino acid is by **S$_N$2 reaction of an α-halo carboxylic acid with a large excess of NH$_3$**.

a. $\xrightarrow[\text{large excess}]{NH_3}$ glycine

b. $\xrightarrow[\text{large excess}]{NH_3}$ isoleucine

c. $\xrightarrow[\text{large excess}]{NH_3}$ phenylalanine

29.6

a. $\xrightarrow{CH_3I}$ alanine

b. $\xrightarrow{}$ leucine

c. $\xrightarrow{}$ isoleucine

Amino Acids and Proteins 29–7

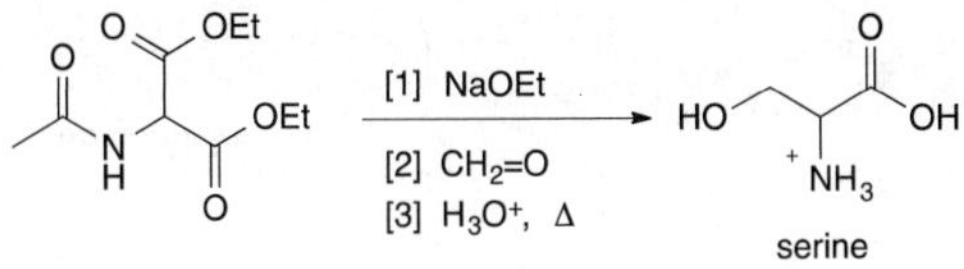

29.7

[1] NaOEt
[2] CH₂=O → [2] $CH_2=O$
[3] H₃O⁺, Δ → [3] H_3O^+, Δ

serine

29.8

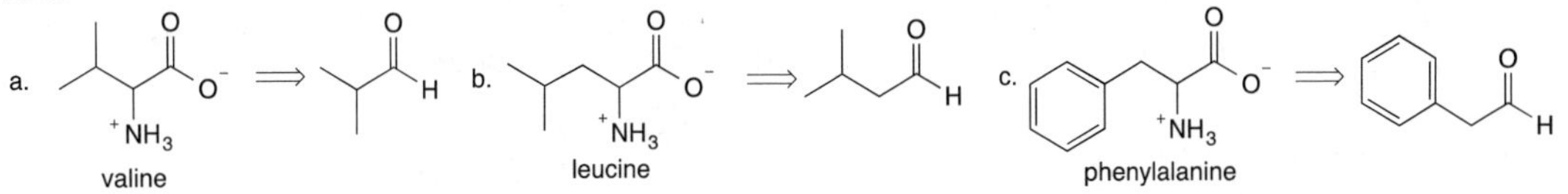

a. valine

b. leucine

c. phenylalanine

29.9

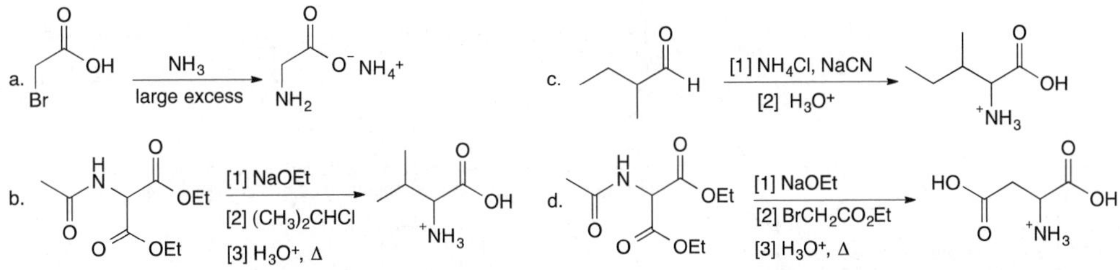

a. NH_3 large excess

c. [1] NH_4Cl, NaCN [2] H_3O^+

b. [1] NaOEt [2] $(CH_3)_2CHCl$ [3] H_3O^+, Δ

d. [1] NaOEt [2] $BrCH_2CO_2Et$ [3] H_3O^+, Δ

29.10 A chiral amine must be used to resolve a racemic mixture of amino acids.

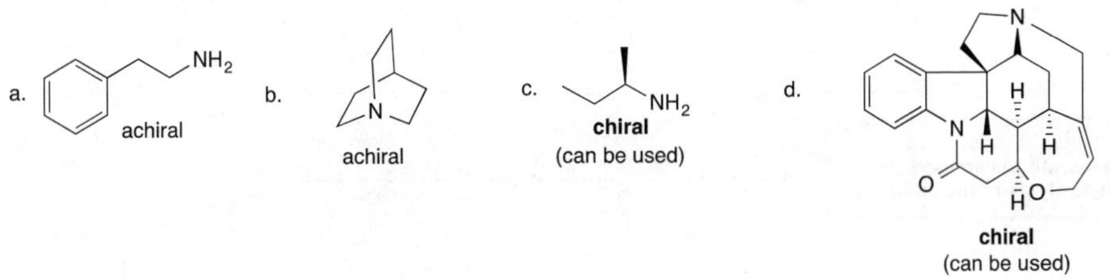

a. achiral

b. achiral

c. chiral (can be used)

d. chiral (can be used)

Chapter 29–8

29.11

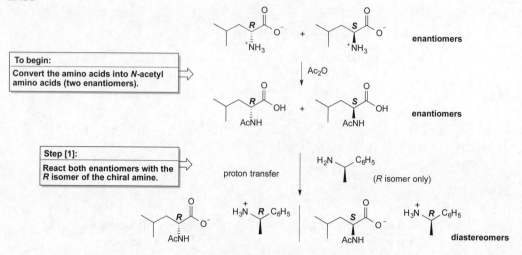

29.12

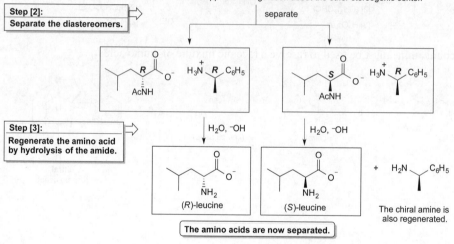

Amino Acids and Proteins 29–9

29.13

a.

c.

b.

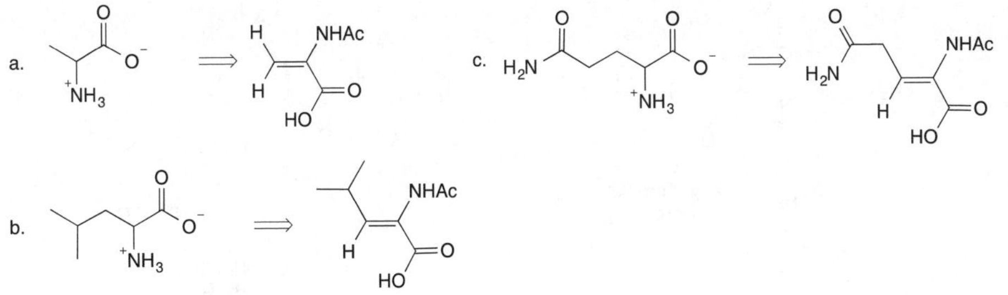

29.14 Draw the peptide by joining adjacent COOH and NH₂ groups in amide bonds.

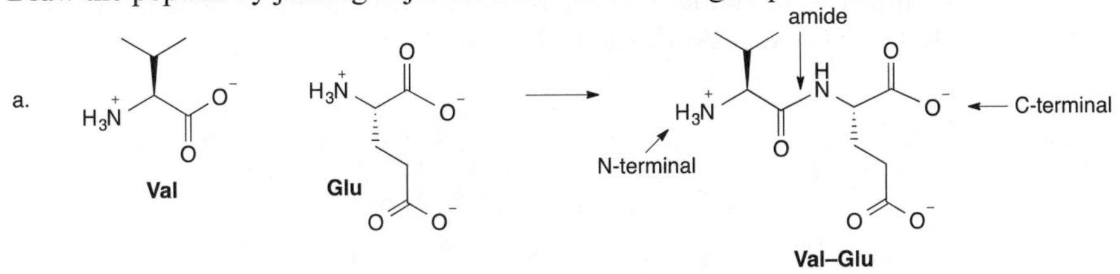

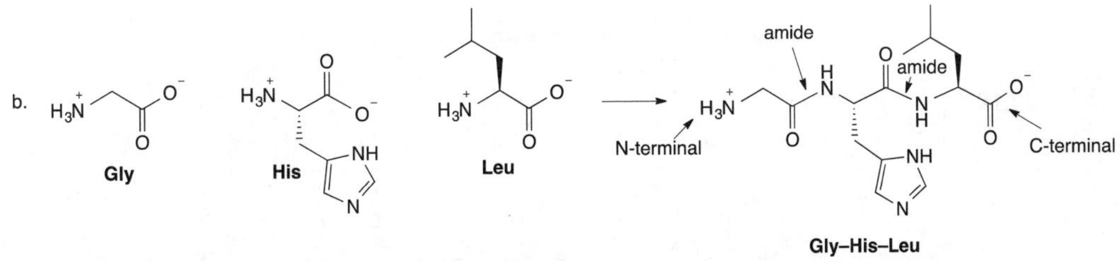

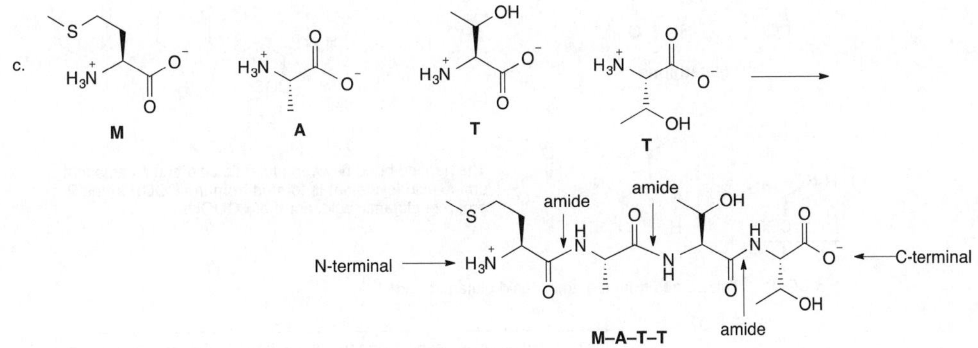

a.

Val   **Glu**   **Val–Glu**

b.

Gly   **His**   **Leu**   **Gly–His–Leu**

c.

M   **A**   **T**   **T**   **M–A–T–T**

Chapter 29–10

29.15

a.

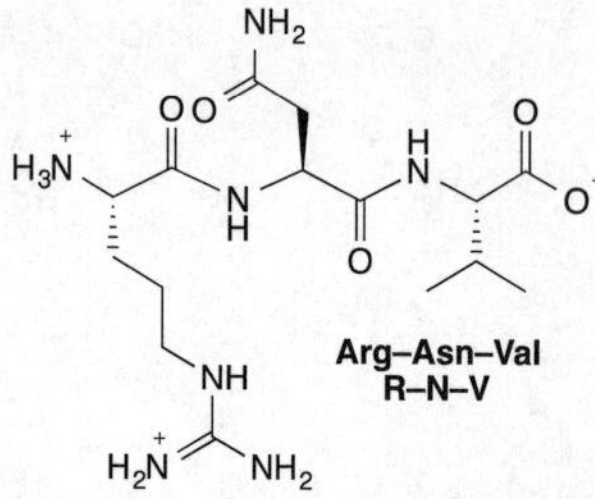

Arg–Asn–Val
R–N–V

b.

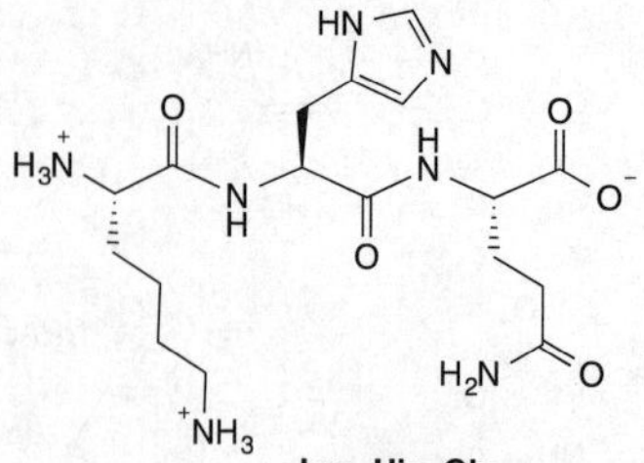

Lys–His–Gln
K–H–Q

29.16 There are six different tripeptides that can be formed from three amino acids (A, B, C): A–B–C, A–C–B, B–A–C, B–C–A, C–A–B, and C–B–A.

29.17

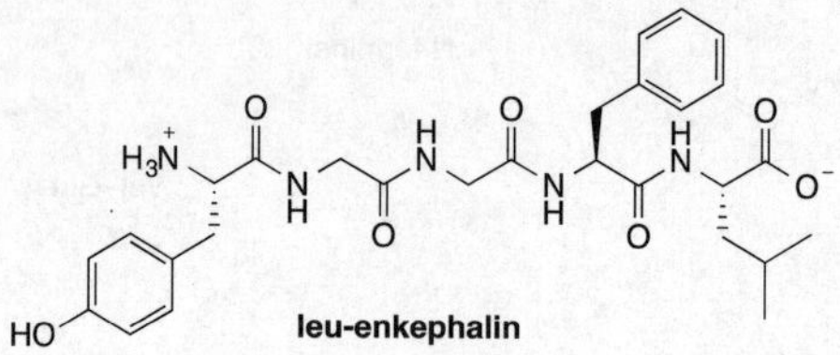

leu-enkephalin

29.18

a.

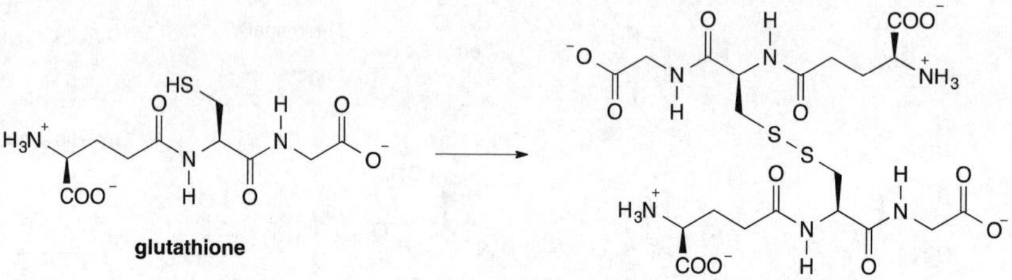

glutathione

b.

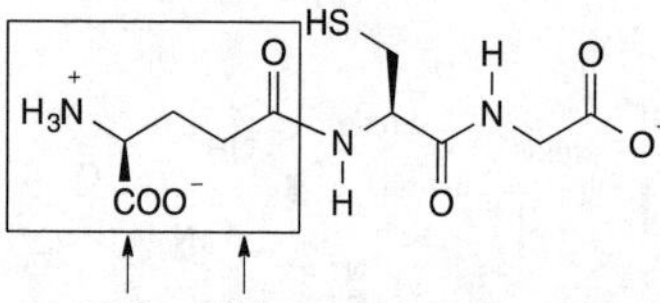

The peptide bond beween glutamic acid and its adjacent amino acid (cysteine) is formed from the COOH in the R group of glutamic acid, not the α COOH.

α COO⁻   This comes from the amino acid glutamic acid.

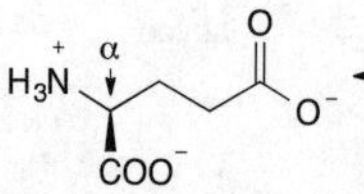

This carboxy group is used to form the amide bond in the peptide, not the α COOH, as is usual. That's what makes glutathione's structure unusual.

glutamic acid

Amino Acids and Proteins 29–11

29.19

a. [structure: C₆H₅-N, C=O, methyl group, C=S, NH — from Ala]

b. [structure: C₆H₅-N, C=O, isopropyl group, C=S, NH — from Val]

29.20 Determine the sequence of the octapeptide as in Sample Problem 29.2. Look for overlapping sequences in the fragments.

common amino acids

Ala–Leu–Tyr
Tyr–Leu–Val–Cys
Val–Cys–Gly–Glu

Answer:
Ala–Leu–Tyr–Leu–Val–Cys–Gly–Glu

29.21 Trypsin cleaves peptides at amide bonds with a carbonyl group from Arg and Lys. Chymotrypsin cleaves at amide bonds with a carbonyl group from Phe, Tyr, and Trp.

a. [1] Gly–Ala–Phe–Leu–Lys + Ala
   [2] Phe–Tyr–Gly–Cys–Arg + Ser
   [3] Thr–Pro–Lys + Glu–His–Gly–Phe–Cys–Trp–Val–Val–Phe
b. [1] Gly–Ala–Phe + Leu–Lys–Ala
   [2] Phe + Tyr + Gly–Cys–Arg–Ser
   [3] Thr–Pro–Lys–Glu–His–Gly–Phe + Cys–Trp + Val–Val–Phe

29.22

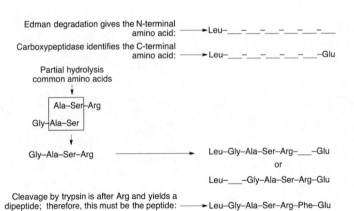

Chapter 29–12

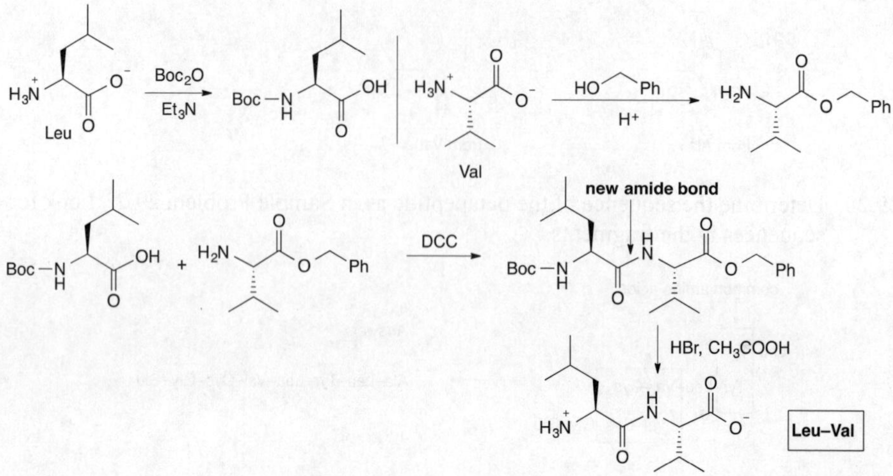

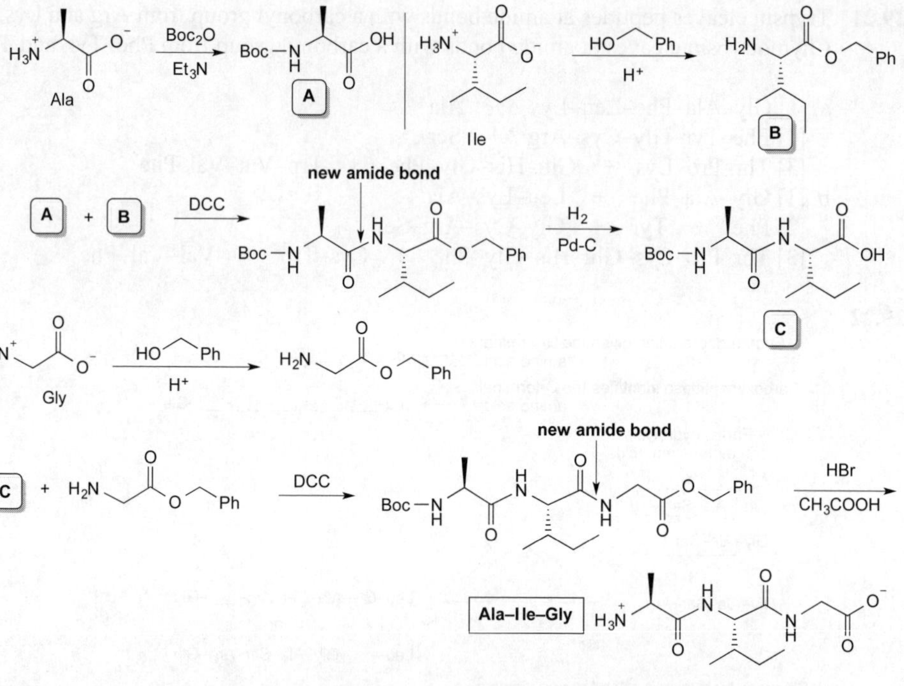

29.23

Amino Acids and Proteins 29–13

29.24 The dipeptide depicted in the 3-D model has alanine as the N-terminal amino acid and cysteine as the C-terminal amino acid.

29.25

All Fmoc-protected amino acids are made by the following general reaction:

Chapter 29–14

29.26 In a *parallel* β-pleated sheet, the strands run in the *same* direction from the N- to C-terminal amino acid. In an *antiparallel* β-pleated sheet, the strands run in the *opposite* direction.

Amino Acids and Proteins 29–15

29.27

a. Ser and Tyr

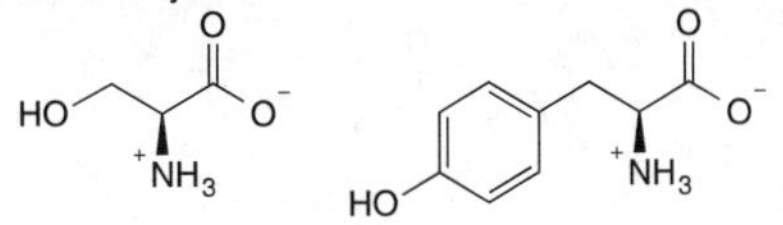

side chains with
OH groups
hydrogen bonding

c. 2 Phe residues

van der Waals forces

b. Val and Leu

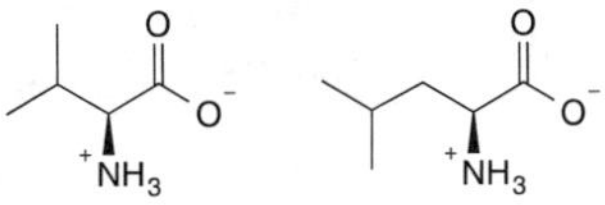

side chains with only
C–C and C–H bonds
van der Waals forces

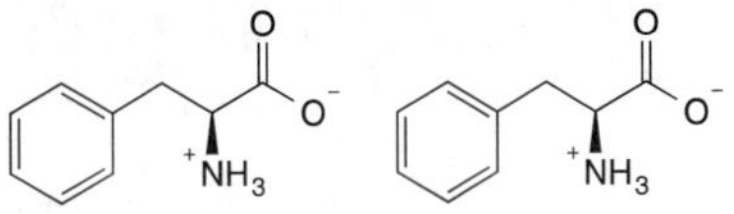

29.28 a. The R group for glycine is a hydrogen. The R groups must be small to allow the β-pleated sheets to stack on top of each other. With large R groups, steric hindrance prevents stacking.

b. Silk fibers are water insoluble because most of the polar functional groups are in the interior of the stacked sheets. The β-pleated sheets are stacked one on top of another, so few polar functional groups are available for hydrogen bonding to water.

29.29

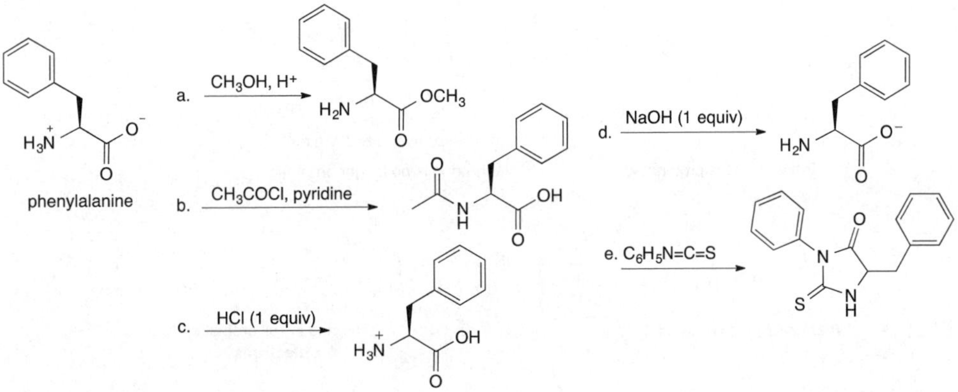

29.30

a. N-terminal amino acid: alanine
C-terminal amino acid: serine

b. A–Q–C–S

c. Amide bonds are **bold** (not wedges).

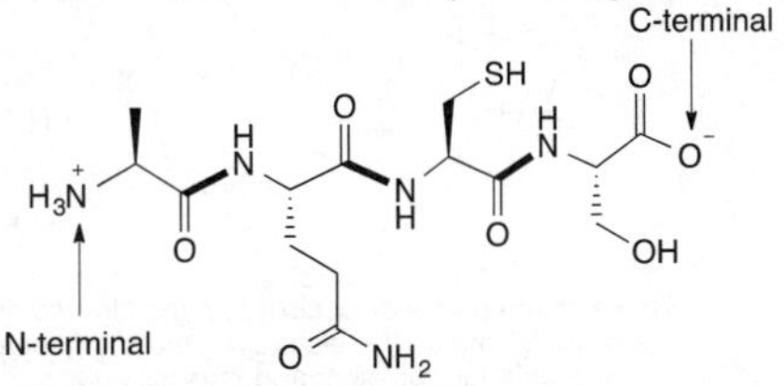

Ala–Gln–Cys–Ser
A–Q–C–S

29.31 The dipeptide is composed of phenylalanine and leucine.

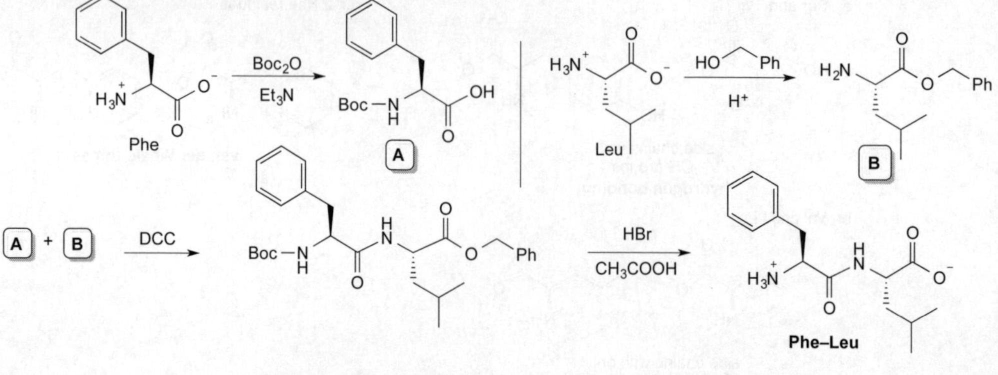
Boc$_2$O
Et$_3$N
Phe
A
Leu
B
DCC
HBr
CH$_3$COOH
Phe–Leu

29.32

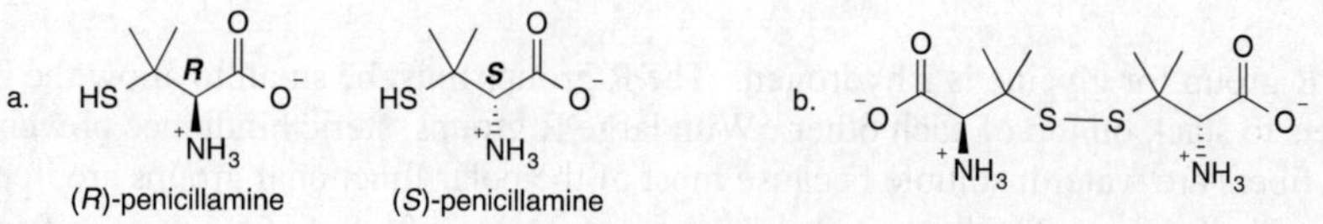
a.
(R)-penicillamine
(S)-penicillamine
b.

29.33 The electron pair on the N atom not part of a double bond is delocalized on the five-membered ring, making it less basic.

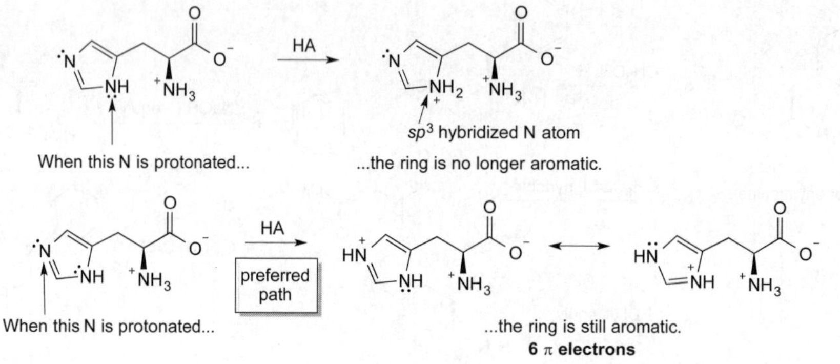
HA
When this N is protonated...
sp^3 hybridized N atom
...the ring is no longer aromatic.
preferred path
When this N is protonated...
...the ring is still aromatic.
6 π electrons

29.34

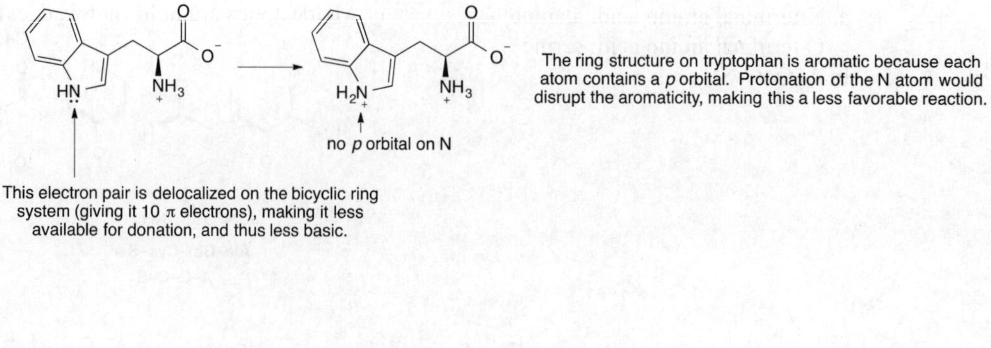
no p orbital on N
The ring structure on tryptophan is aromatic because each atom contains a p orbital. Protonation of the N atom would disrupt the aromaticity, making this a less favorable reaction.
This electron pair is delocalized on the bicyclic ring system (giving it 10 π electrons), making it less available for donation, and thus less basic.

Amino Acids and Proteins 29–17

29.35 At its isoelectric point, each amino acid is neutral.

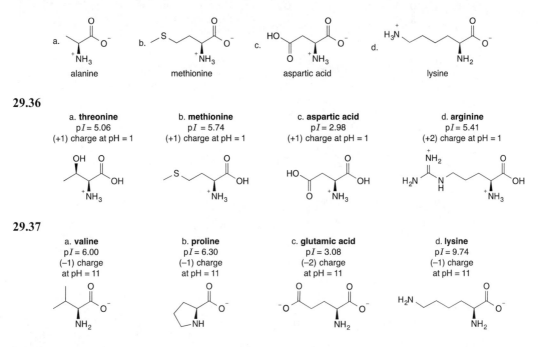

29.38 The terminal NH₂ and COOH groups are ionizable functional groups, so they can gain or lose protons in aqueous solution.

a. [structure of Ala zwitterion] → [tripeptide A–A–A drawn with all uncharged atoms]

b. At pH = 1 [tripeptide structure with H₃N⁺ and COOH termini]

c. The pK_a of the COOH of the tripeptide is higher than the pK_a of the COOH group of alanine, making it less acidic. This occurs because the COOH group in the tripeptide is farther away from the –NH₃⁺ group. The positively charged –NH₃⁺ group stabilizes the negatively charged carboxylate anion of alanine more than the carboxylate anion of the tripeptide because it is so much closer in alanine. The opposite effect is observed with the ionization of the –NH₃⁺ group. In alanine, the –NH₃⁺ is closer to the COO⁻ group, so it is more difficult to lose a proton, resulting in a higher pK_a. In the tripeptide, the –NH₃⁺ is farther away from the COO⁻, so it is less affected by its presence.

Chapter 29–18

29.39

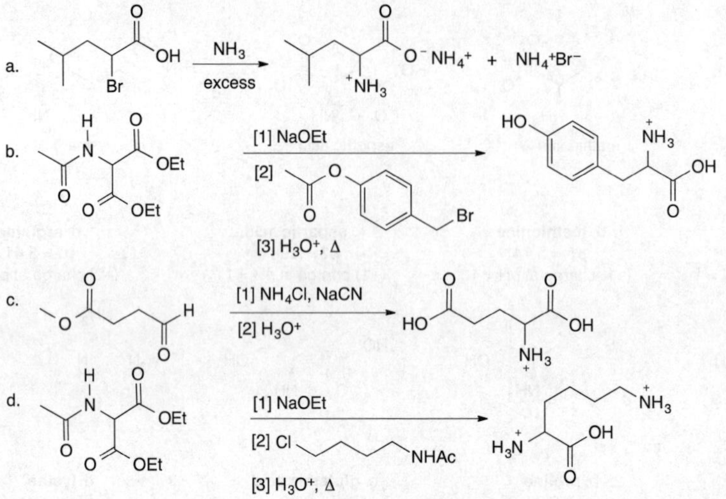

a.

b.

[1] NaOEt
[2]
[3] H₃O⁺, Δ

c.

[1] NH₄Cl, NaCN
[2] H₃O⁺

d.

[1] NaOEt
[2] Cl
[3] H₃O⁺, Δ

29.40

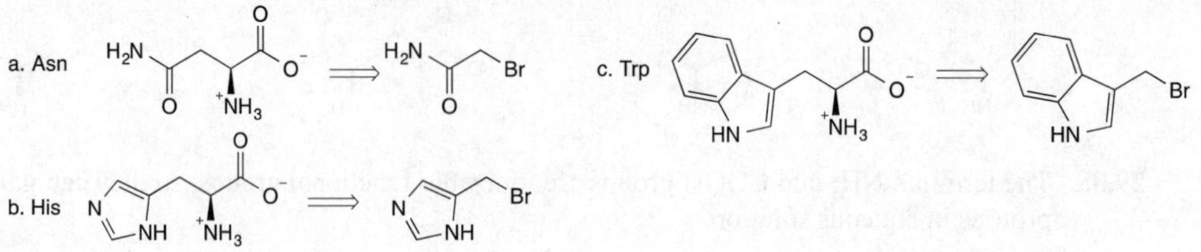

a. Asn

b. His

c. Trp

29.41

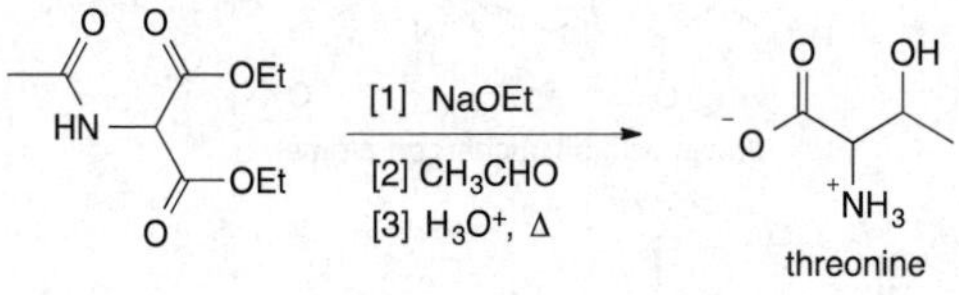

[1] NaOEt
[2] CH₃CHO
[3] H₃O⁺, Δ

threonine

29.42

a.

Br₂
CH₃CO₂H

CrO₃
H₂SO₄, H₂O

NH₃
excess

glycine

b.

[1] NH₄Cl, NaCN
[2] H₃O⁺

alanine

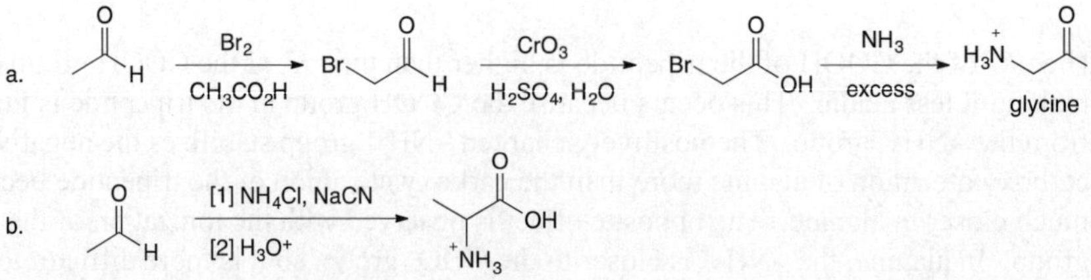

29.43

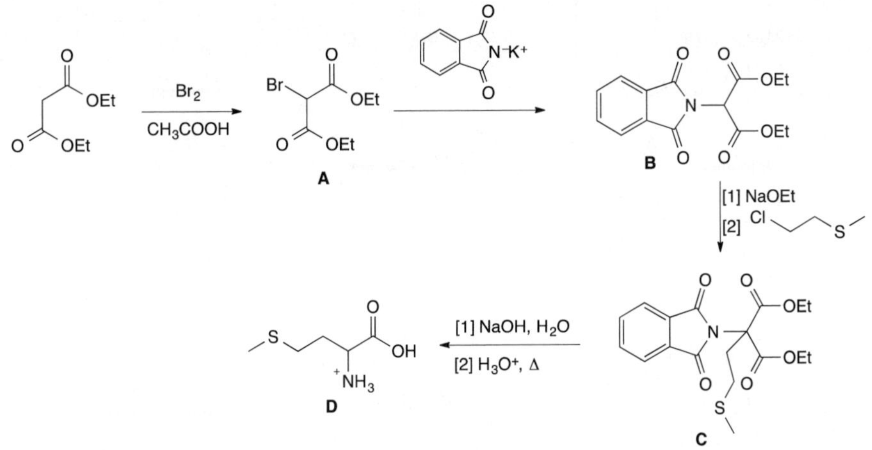

29.44

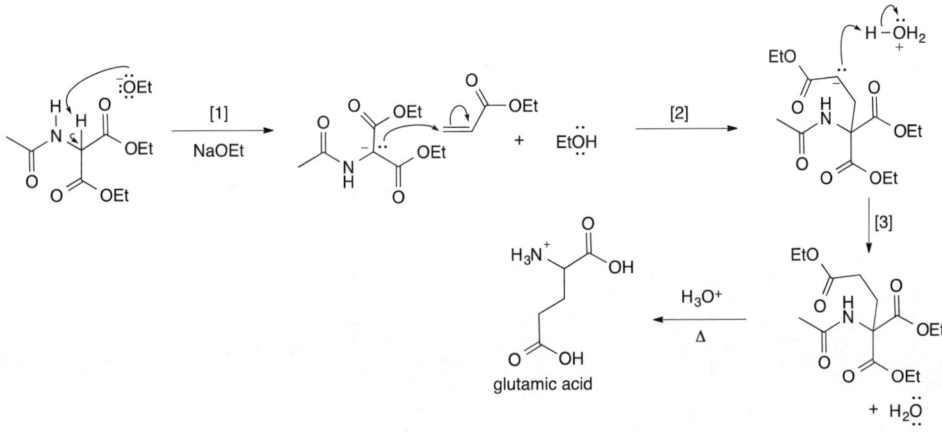

glutamic acid

29.45

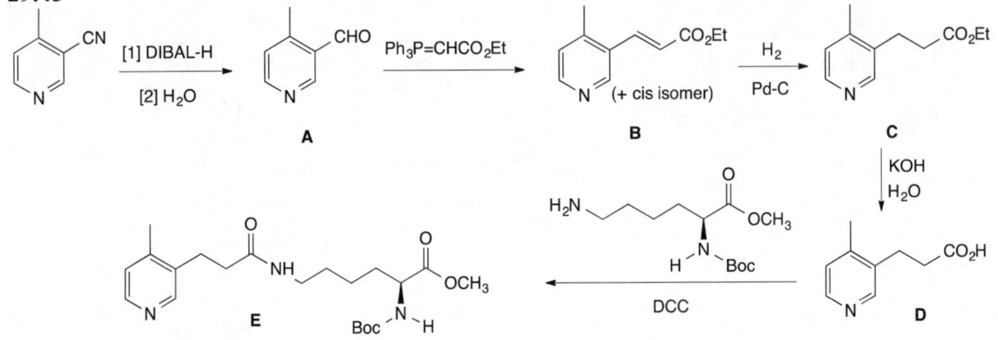

Chapter 29–20

29.46

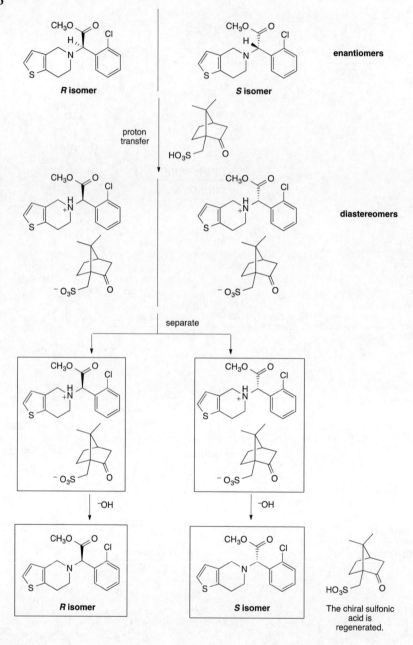

29.47

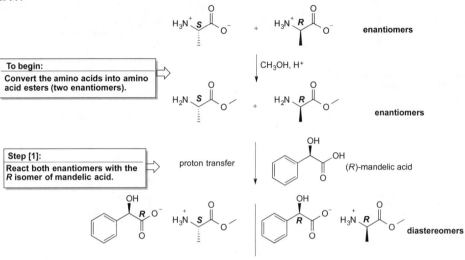

To begin: Convert the amino acids into amino acid esters (two enantiomers).

Step [1]: React both enantiomers with the R isomer of mandelic acid.

These salts have the *same* configuration around one stereogenic center, but the *opposite* configuration about the other stereogenic center.

Step [2]: Separate the diastereomers.

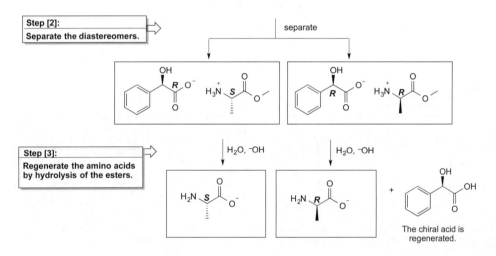

Step [3]: Regenerate the amino acids by hydrolysis of the esters.

The chiral acid is regenerated.

Chapter 29–22

29.48

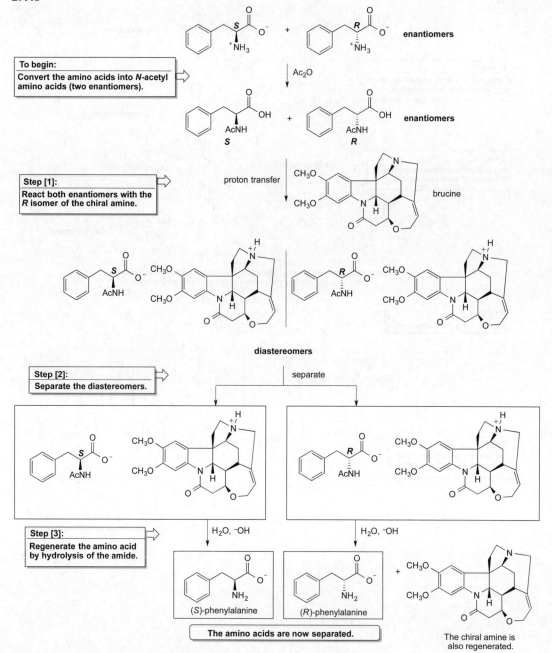

Amino Acids and Proteins 29–23

29.49

a.

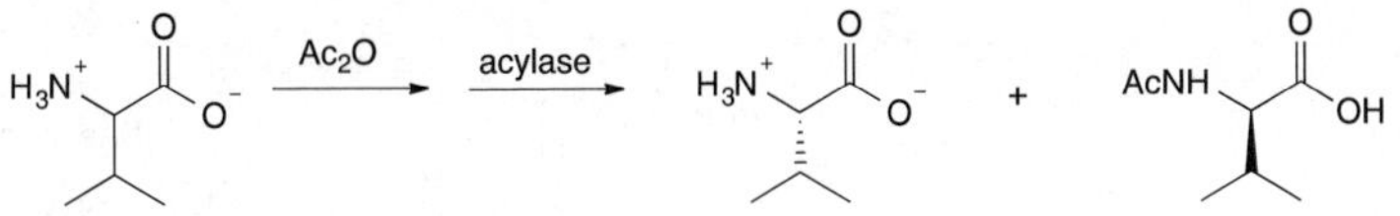

racemic mixture

b.

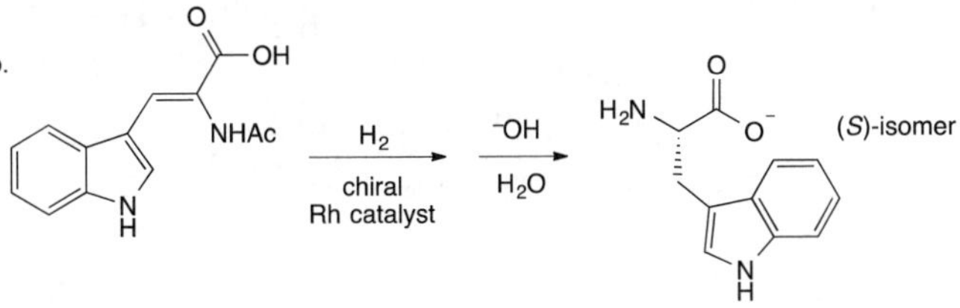

(S)-isomer

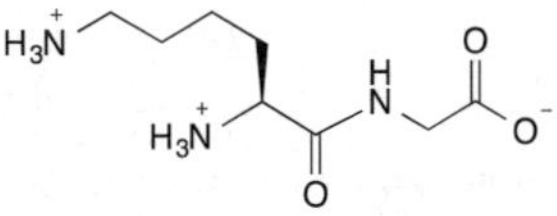

29.50

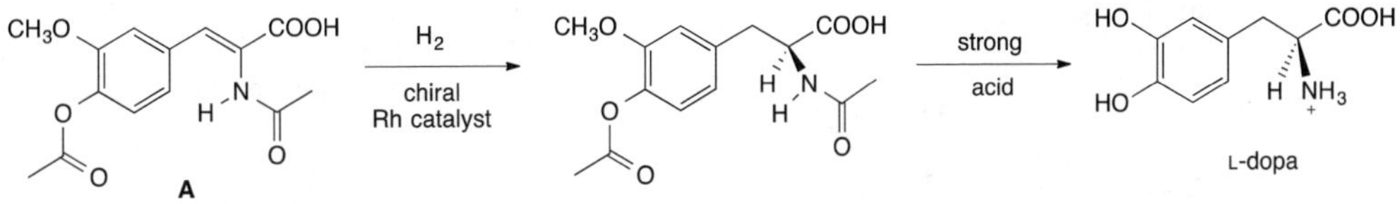

A

L-dopa

29.51

a.

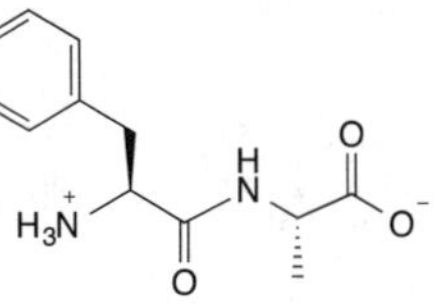

Phe–Ala

b.

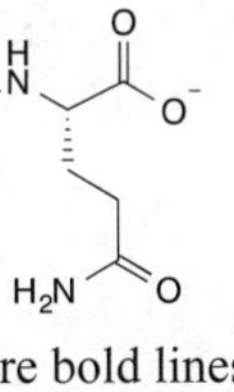

Gly–Gln

c.

Lys–Gly

d.

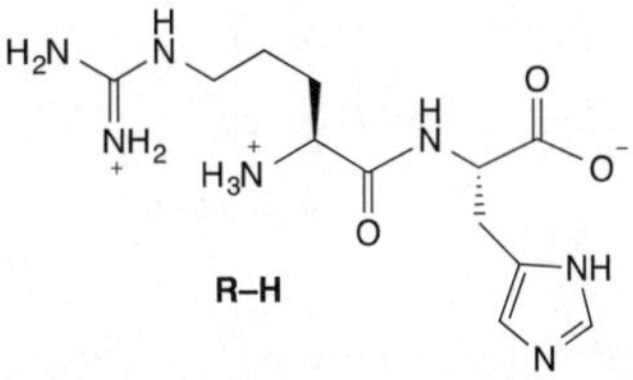

R–H

29.52 Amide bonds are bold lines (not wedges). The structure is drawn with all uncharged atoms.

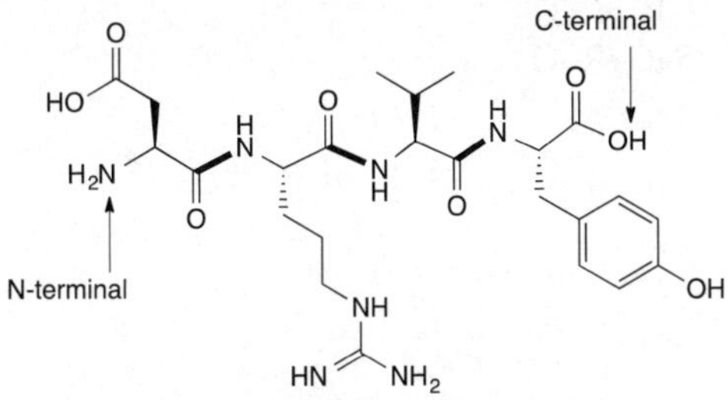

Asp–Arg–Val–Tyr
D–R–V–Y

Chapter 29–24

29.53 Name a peptide from the N-terminal to the C-terminal end.

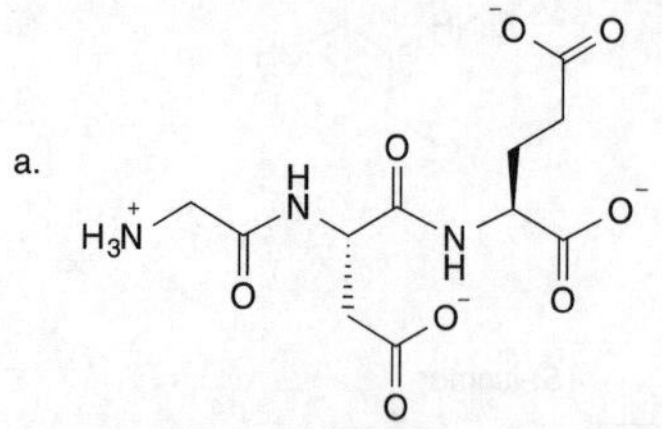

a.

Gly–Asp–Glu
G–D–E

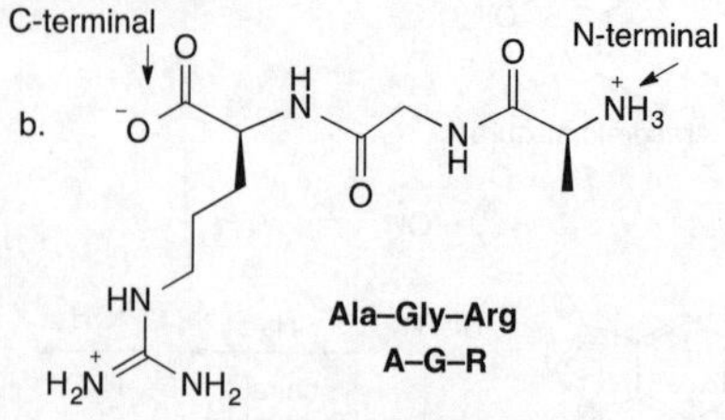

b.

Ala–Gly–Arg
A–G–R

29.54 The unusual features of gramicidin S are the presence of the uncommon enantiomer of phenylalanine (D-Phe) in the molecule, as well as the presence of an uncommon amino acid, labeled **X**.

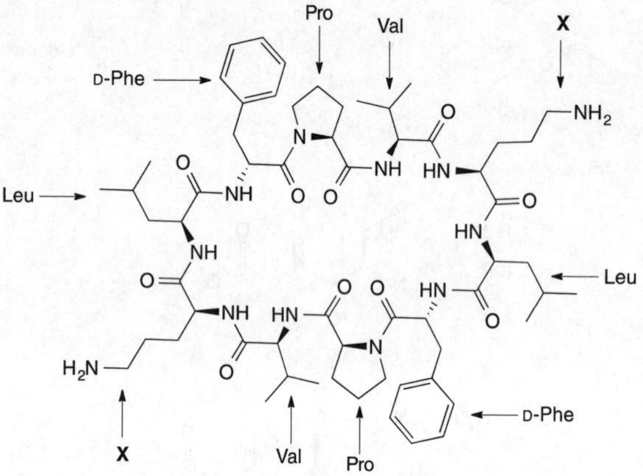

gramicidin S

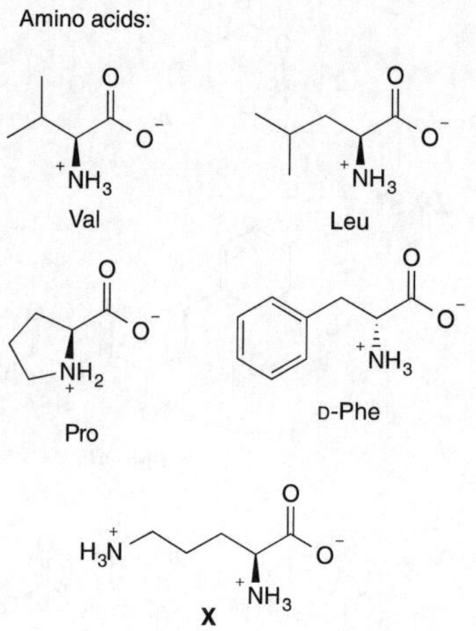

Amino acids:

Val

Leu

Pro

D-Phe

X

29.55

a. A–P–F + L–K–W + S–G–R–G
b. A–P–F–L–K + W–S–G–R + G
c. A–P–F–L–K–W–S–G–R + G
d. A + P–F–L–K–W–S–G–R–G

29.56

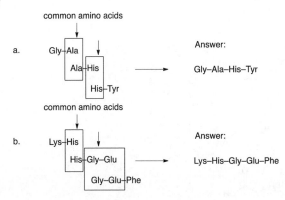

Three-letter abbreviation: Asp–Arg–Val–Tyr–Ile–His–Pro–Phe–His–Leu

a. Trypsin cleavage products: Asp–Arg + Val–Tyr–Ile–His–Pro–Phe–His–Leu

b. Chymotrypsin cleavage products: Asp–Arg–Val–Tyr + Ile–His–Pro–Phe + His–Leu

c. ACE cleavage products: Asp–Arg–Val–Tyr–Ile–His–Pro–Phe (angiotensin II) + His–Leu

29.57

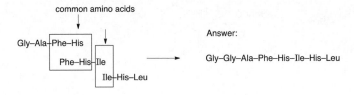

29.58 Gly is the N-terminal amino acid (from Edman degradation), and Leu is the C-terminal amino acid (from treatment with carboxypeptidase). Partial hydrolysis gives the rest of the sequence.

common amino acids

Gly–Ala–Phe–His
　　　Phe–His–Ile
　　　　　Ile–His–Leu

Answer:

Gly–Gly–Ala–Phe–His–Ile–His–Leu

Chapter 29–26

29.59 Edman degradation data give the N-terminal amino acid for the octapeptide and all smaller peptides.

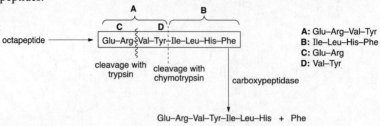

29.60

a.–f.

29.61

Chapter 29–28

29.62 Make all the Fmoc derivatives as described in Problem 29.25.

Amino Acids and Proteins 29–29

b.

(Fmoc-Ile)

[1] base
[2] Cl‒‒‒POLYMER

[1]

[2] DCC

(Fmoc-Ala)

[1]

[2] DCC +

(Fmoc-Gly)

[1]

[2] DCC +
(Fmoc-Phe)

[1]

[2] HF

Phe–Gly–Ala–Ile

+ F‒‒‒POLYMER

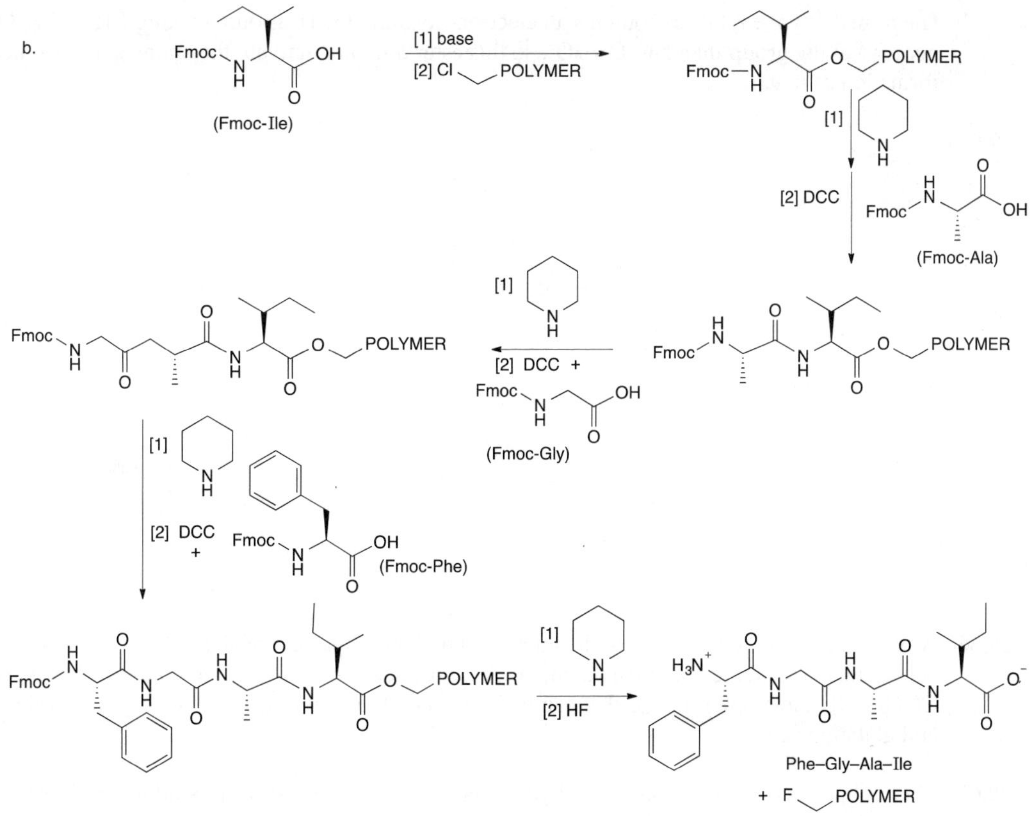

29.63

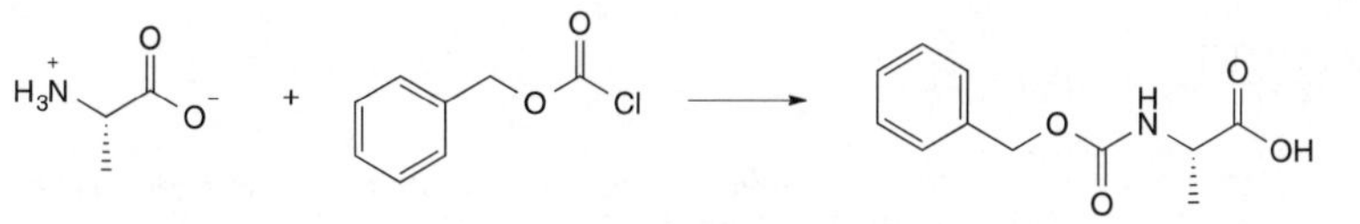

29.64

a. A *p*-nitrophenyl ester activates the carboxy group of the first amino acid to amide formation by converting the OH group into a good leaving group, the *p*-nitrophenoxide group, which is highly resonance stabilized. In this case the electron-withdrawing NO_2 group further stabilizes the leaving group.

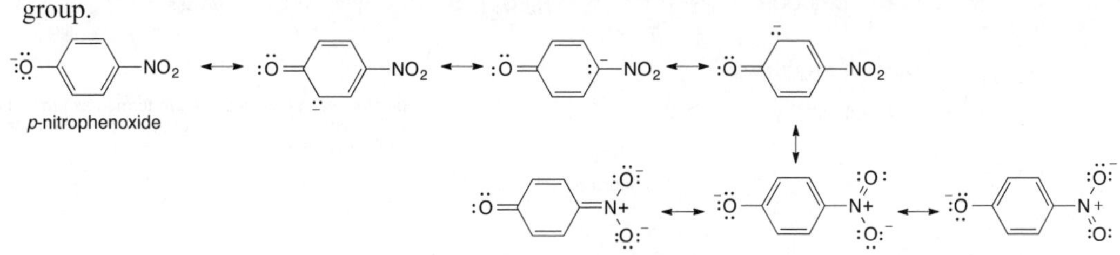

p-nitrophenoxide

The negative charge is delocalized on the O atom of the NO_2 group.

Chapter 29–30

b. The *p*-methoxyphenyl ester contains an electron-donating OCH_3 group, making $CH_3OC_6H_4O^-$ a poorer leaving group than $NO_2C_6H_4O^-$, so this ester does *not* activate the amino acid to amide formation as much.

29.65

Fmoc-protected amino acid

DMF

proton transfer

29.66 Amino acids commonly found in the interior of a globular protein have nonpolar or weakly polar side chains: isoleucine and phenylalanine. Amino acids commonly found on the surface have COOH, NH_2, and other groups that can hydrogen bond to water: aspartic acid, lysine, arginine, and glutamic acid.

29.67 The proline residues on collagen are hydroxylated to increase hydrogen bonding interactions.

[O]

The new OH group allows more hydrogen bonding interactions between the chains of the triple helix, thus stabilizing it.

29.68

Br_2
CH_3COOH

CrO_3
H_2SO_4, H_2O

NH_3
excess

valine

[1] NH_4Cl, NaCN
[2] H_3O^+, Δ

leucine

(Racemic valine and leucine are formed as products, but the synthesis of the tripeptide is drawn with one enantiomer only.)

Amino Acids and Proteins 29–31

valine → Boc₂O, Et₃N → **A** (Boc-Val-OH)

valine → C₆H₅CH₂OH, H⁺ → **C**

leucine → HO-CH₂Ph, H⁺ → **B**

A + **B** → DCC → Boc-Val-Leu-OBn → H₂, Pd-C → Boc-NH-Val-Leu-OH

Boc-Val-Leu-OH → DCC, **C** → Boc-Val-Leu-Val-OBn

→ HBr, CH₃COOH → H₃N⁺-Val-Leu-Val-O⁻   **Val–Leu–Val**

29.69 Perhaps using a chiral amine R*NH₂ (or related chiral nitrogen-containing compound) to make a chiral imine, will now favor formation of one of the amino nitriles in the Strecker synthesis. Hydrolysis of the CN group and removal of R* would then form the amino acid.

R-CHO → NH₂R* (chiral amine) → chiral imine (R-CH=NR*) → ⁻CN → amino nitrile (R-CH(NHR*)-CN) → Hydrolyze nitrile → R-CH(NHR*)-COOH → Remove R* → amino acid (R-CH(NH₃⁺)-COO⁻) enantiomerically enriched (?)

Perhaps a large excess of one stereoisomer will be formed.

29.70 This reaction is similar to the reaction of penicillin with the glycopeptide transpeptidase enzyme discussed in Section 22.14. Serine has a nucleophilic OH, which can open the strained β-lactone to form a covalently bound, inactive enzyme.

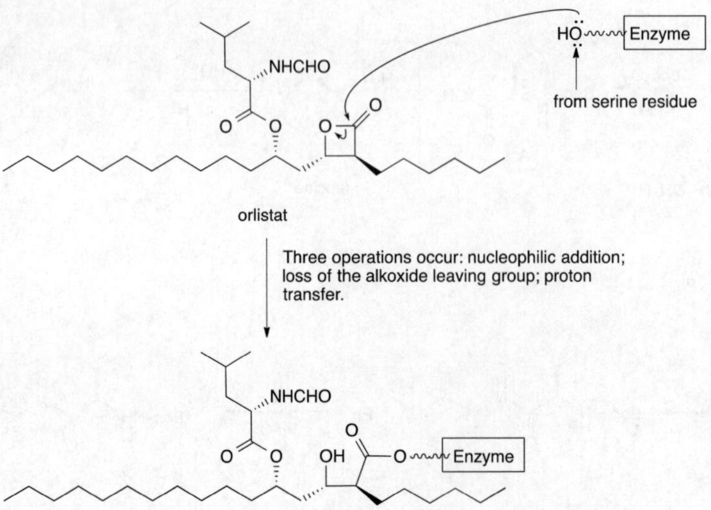
HO~~~~Enzyme
from serine residue
NHCHO
orlistat
Three operations occur: nucleophilic addition; loss of the alkoxide leaving group; proton transfer.
NHCHO
OH
O~~~~Enzyme

29.71

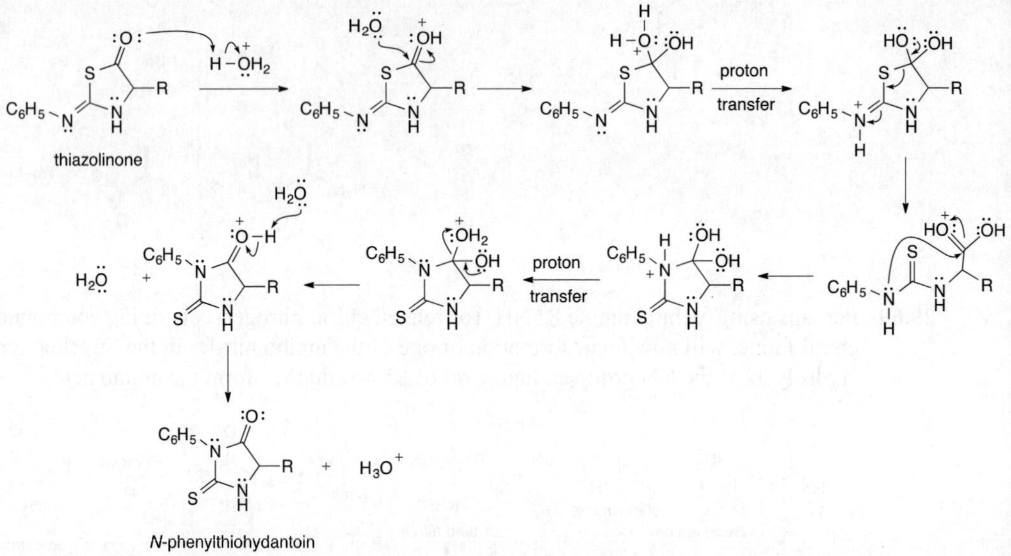
thiazolinone
proton
transfer
proton
transfer
N-phenylthiohydantoin

Synthetic Polymers 30–1

Chapter 30 Synthetic Polymers

Chapter Review

Chain-growth polymers—Addition polymers

[1] Chain-growth polymers with alkene starting materials (30.2)

- General reaction:

- Mechanism—three possibilities, depending on the identity of Z:

Type	Identity of Z	Initiator	Comments
[1] radical polymerization	Z stabilizes a radical. Z = R, Ph, Cl, etc.	A source of radicals (ROOR)	Termination occurs by radical coupling or disproportionation. Chain branching occurs.
[2] cationic polymerization	Z stabilizes a carbocation. Z = R, Ph, OR, etc.	H–A or a Lewis acid (BF$_3$ + H$_2$O)	Termination occurs by loss of a proton.
[3] anionic polymerization	Z stabilizes a carbanion. Z = Ph, COOR, COR, CN, etc.	An organolithium reagent (R–Li)	Termination occurs only when an acid or other electrophile is added.

[2] Chain-growth polymers with epoxide starting materials (30.3)

- The mechanism is S$_N$2.
- Ring opening occurs at the less substituted carbon of the epoxide.

Chapter 30–2

Examples of step-growth polymers—Condensation polymers (30.6)

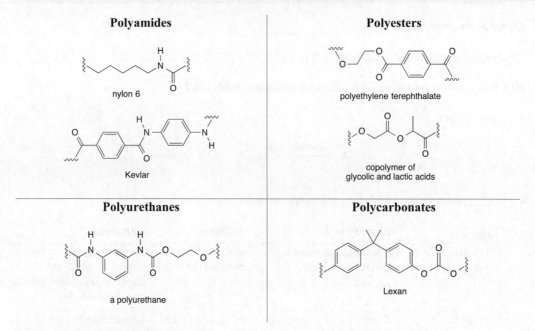

Structure and properties

- Polymers prepared from monomers having the general structure $CH_2=CHZ$ can be **isotactic, syndiotactic,** or **atactic,** depending on the identity of Z and the method of preparation (30.4).
- **Ziegler–Natta catalysts** form polymers without significant branching. Polymers can be isotactic, syndiotactic, or atactic, depending on the catalyst. Polymers prepared from 1,3-dienes have the *E* or *Z* configuration, depending on the monomer (30.4, 30.5).
- Most polymers contain ordered crystalline regions and less ordered amorphous regions (30.7). The greater the crystallinity, the harder the polymer.
- **Elastomers** are polymers that stretch and can return to their original shape (30.5).
- **Thermoplastics** are polymers that can be molded, shaped, and cooled such that the new form is preserved (30.7).
- **Thermosetting polymers** are composed of complex networks of covalent bonds, so they cannot be melted to form a liquid phase (30.7).

Practice Test on Chapter Review

1.a. Which of the following statements is (are) true about chain-growth polymers?
    1. The reaction mechanism involves initiation, propagation, and termination.
    2. The reaction may occur with anionic, cationic, or radical intermediates.
    3. Epoxides can serve as monomers.
    4. Statements (1) and (2) are both true.
    5. Statements (1), (2), and (3) are all true.

Synthetic Polymers 30–3

b. Which of the following alkenes is likely to undergo anionic polymerization?
  1. $CH_2=CHCO_2CH_3$
  2. $CH_2=CHOCH_3$
  3. $CH_2=CHCH_2CO_2CH_3$
  4. Both (1) and (2) will react.
  5. Compounds (1), (2), and (3) will all react.

c. Which of the following compounds can serve as an initiator in cationic polymerization?
  1. butyllithium
  2. $(CH_3)_3COOC(CH_3)_3$
  3. BF_3
  4. Both (1) and (2) can serve as initiators.
  5. Compounds (1), (2), and (3) can all serve as initiators.

d. Which of the following statements is (are) true about step-growth polymers?
  1. A small molecule such as H_2O or HCl is extruded during synthesis.
  2. Polycarbonates are an example of a step-growth polymer.
  3. Step-growth polymers are also called addition polymers.
  4. Statements (1) and (2) are both true.
  5. Statements (1), (2), and (3) are all true.

2. Label each statement as True (T) or False (F).
  a. A polyester is the most easily recycled polymer.
  b. Natural rubber is a polymer of repeating isoprene units in which all double bonds have the E configuration.
  c. A syndiotactic polymer has all Z groups bonded to the polymer chain on the same side.
  d. A polyether can be formed by anionic polymerization of an epoxide.
  e. An epoxy resin is a chain-growth polymer.
  f. A branched polymer is more amorphous, giving it a higher T_m.
  g. Polystyrene is a thermoplastic that can be melted and molded in shapes that are retained when the polymer is cooled.
  h. A polyurethane is a condensation polymer.
  i. Ziegler–Natta catalysts are used to form highly branched chain-growth polymers.
  j. Using a feedstock from a renewable source is one method of green polymer synthesis.

3. What monomer(s) are needed to synthesize each polymer?

Chapter 30–4

Answers to Practice Test

1.a. 5   2.a. T   f. F   3.
  b. 1   b. F   g. T
  c. 3   c. F   h. T   a.
  d. 4   d. T   i. F
             e. F   j. T   b.

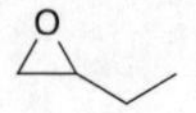

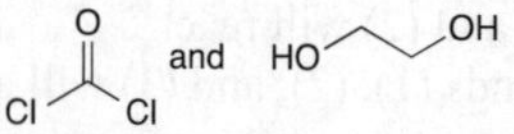

c.

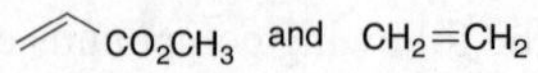

d.

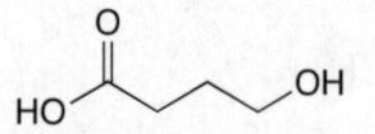

e.

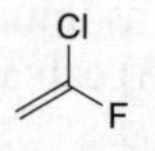

Answers to Problems

30.1   Place brackets around the repeating unit that creates the polymer.

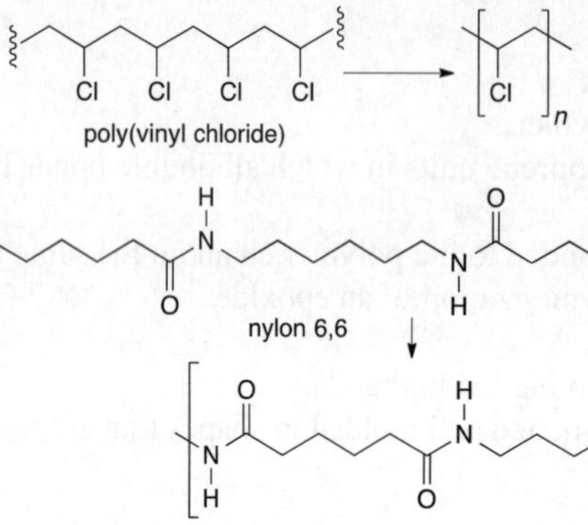

poly(vinyl chloride)

nylon 6,6

30.2   Draw each polymer formed by chain-growth polymerization.

a.

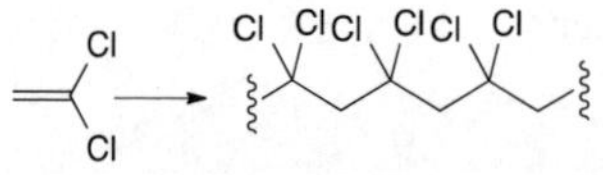

b.

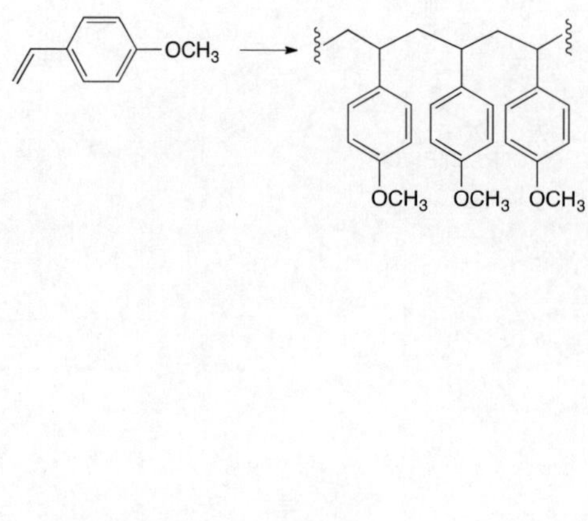

c.

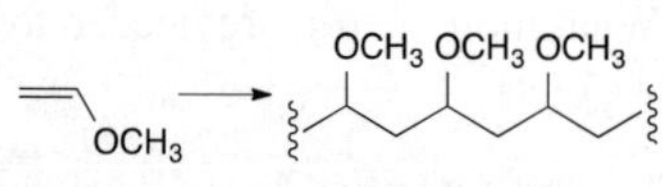

d.

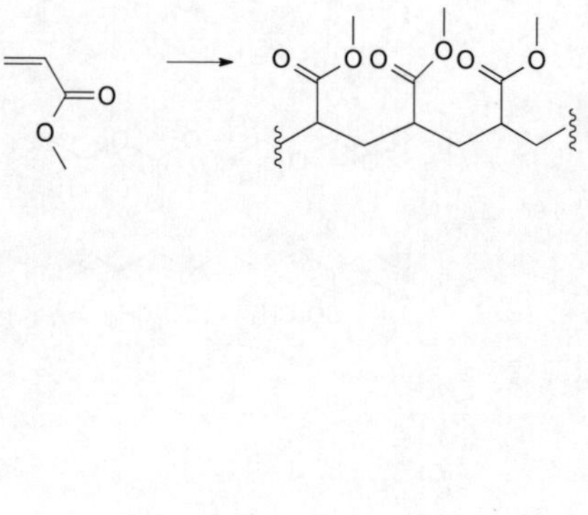

Synthetic Polymers 30–5

30.3 Draw each polymer formed by radical polymerization.

a. [structure] ⟶ [structure]

b. [structure] ⟶ [structure]

30.4 Use Mechanism 30.1 as a model of radical polymerization.

Initiation:

$(CH_3)_3CO\!-\!OC(CH_3)_3$ $\xrightarrow{[1]}$ 2 $(CH_3)_3CO\cdot$ [structure] $\xrightarrow{[2]}$ [structure]

Propagation:

[structure] + [structure] $\xrightarrow{[3]}$ [structure] Repeat Step [3] over and over.

Termination:

[structure] $\xrightarrow{[4]}$ [structure]

(Ac = CH₃CO–)

30.5 Radical polymerization forms a long chain of polystyrene with phenyl groups bonded to every other carbon. To form branches on this polystyrene chain, a radical on a second polymer chain abstracts a H atom. Abstraction of H_a forms a resonance-stabilized radical **A'**. The 2° radical **B'** (without added resonance stabilization) is formed by abstraction of H_b. Abstraction of H_a is favored, therefore, and this radical goes on to form products with 4° C's (**A**).

[reaction scheme with structures]

abstraction of H_b → 2° radical, no resonance stabilization → **B'** → **B** (3° C)

abstraction of H_a → resonance-stabilized benzylic radical → **A'** → **A** (4° C, **favored**)

[resonance structures]

Chapter 30–6

30.6 Cationic polymerization proceeds via a carbocation intermediate. Substrates that form more stable 3° carbocations react more readily in these polymerization reactions than substrates that form less stable 1° carbocations. $CH_2=C(CH_3)_2$ will form a more substituted carbocation than $CH_2=CH_2$.

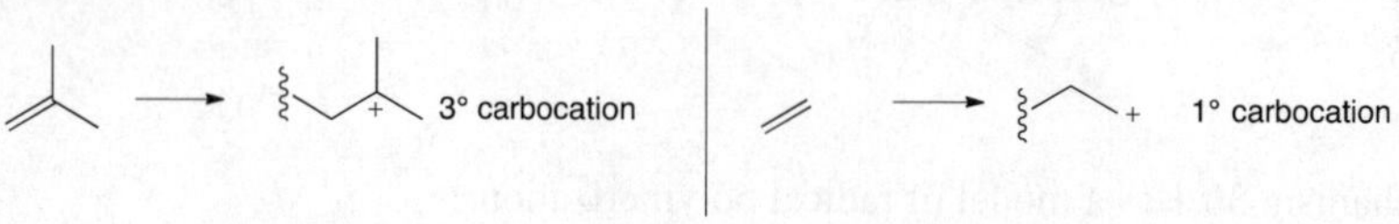

30.7 Cationic polymerization occurs with alkene monomers having substituents that can stabilize carbocations, such as alkyl groups and other electron-donor groups. Anionic polymerization occurs with alkene monomers having substituents that can stabilize a negative charge, such as COR, COOR, or CN.

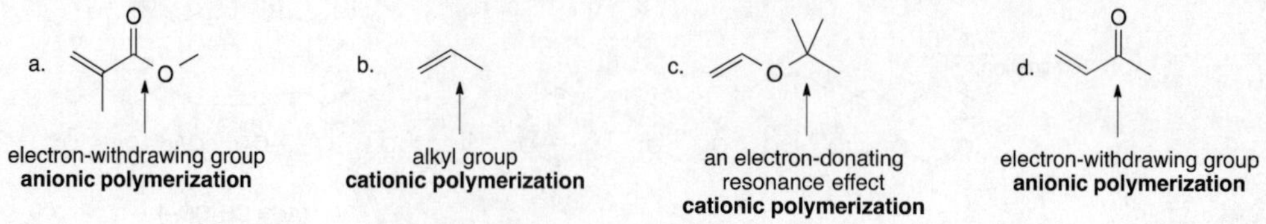

a.
electron-withdrawing group
anionic polymerization

b.
alkyl group
cationic polymerization

c.
an electron-donating
resonance effect
cationic polymerization

d.
electron-withdrawing group
anionic polymerization

30.8 Use Mechanism 30.4 as a model of anionic polymerization.

Initiation:

Propagation:

Termination:

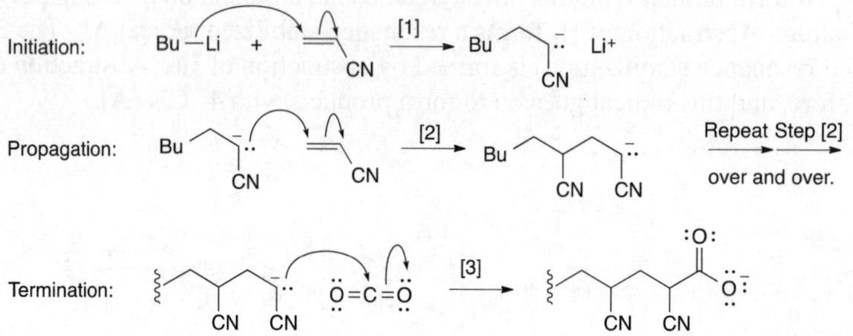

30.9 Styrene ($CH_2=CHPh$) can by polymerized by all three methods of chain-growth polymerization because a benzene ring can stabilize a radical, a carbocation, or a carbanion by resonance delocalization.

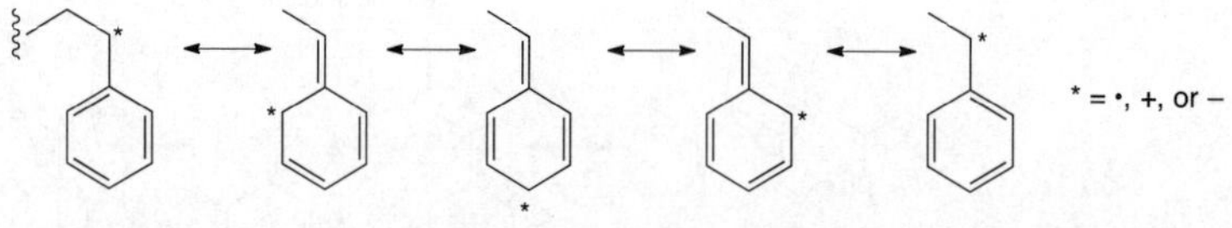

* = •, +, or −

Synthetic Polymers 30–7

30.10 Draw the copolymers formed in each reaction.

a.

b.

30.11

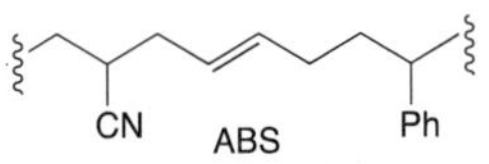

ABS

30.12

a.

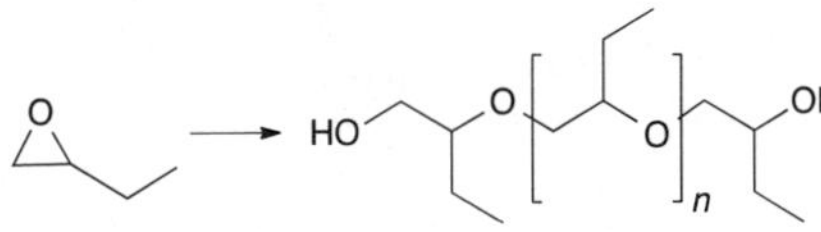

b.

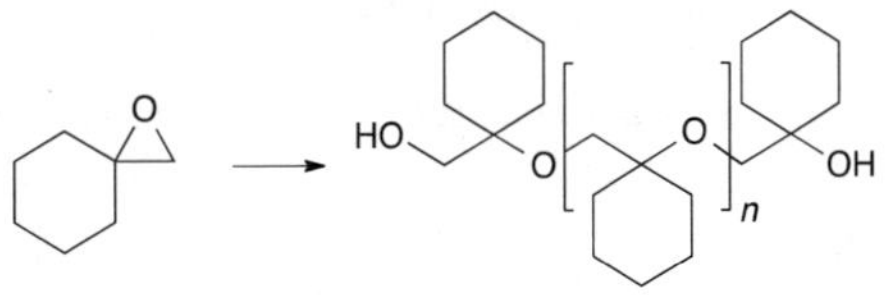

30.13

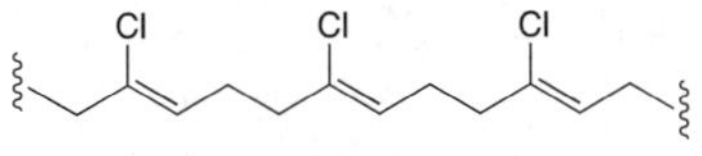

neoprene

All double bonds have the *Z* configuration.

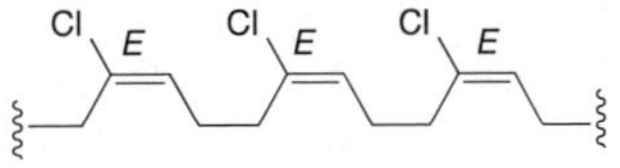

E configuration of each double bond

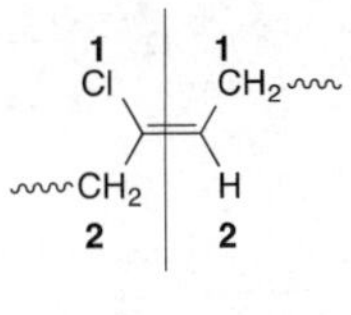

Two higher priority groups (1's)
are on the same side of the
double bond - - - > *Z* configuration.

30.14

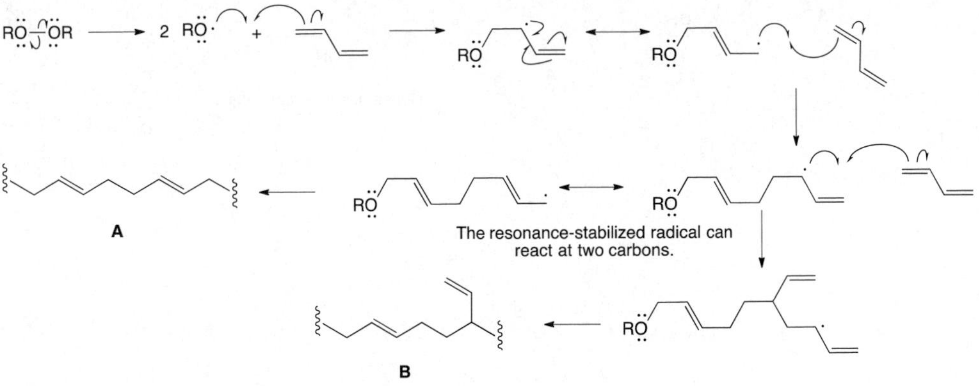

A

The resonance-stabilized radical can
react at two carbons.

B

30.15

a. 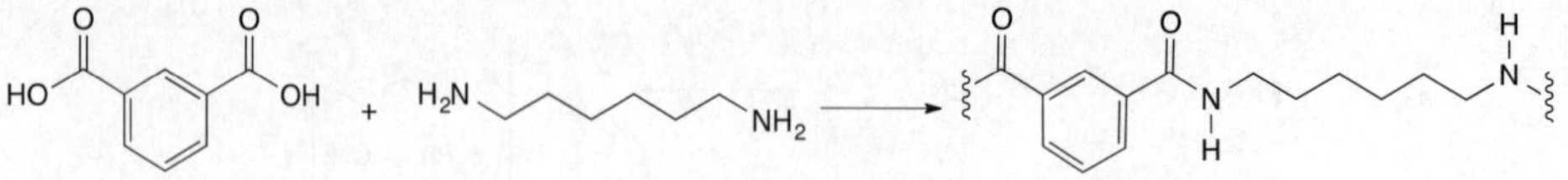

b. 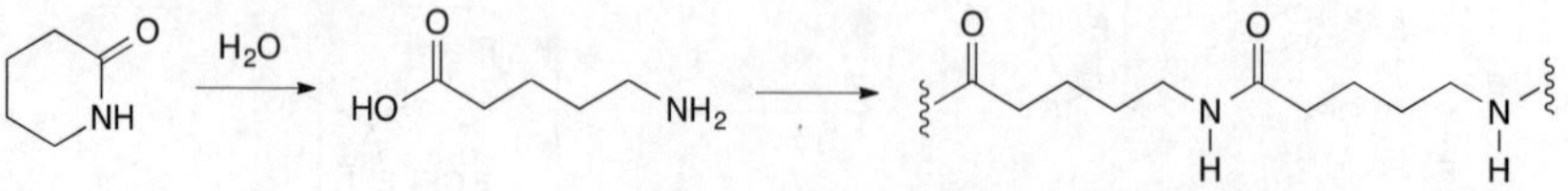
H2O

c. 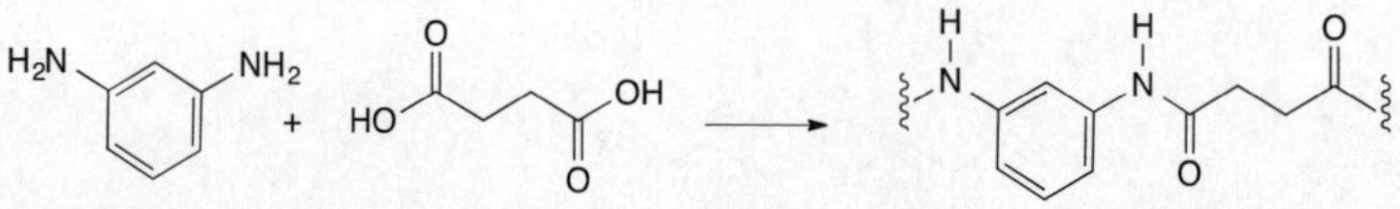

30.16

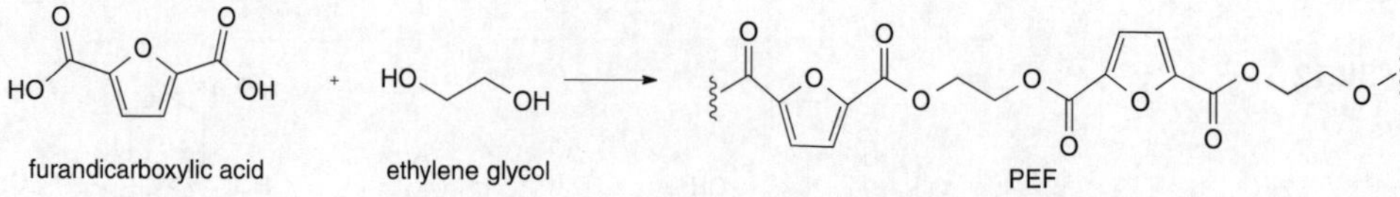
furandicarboxylic acid
ethylene glycol
PEF

30.17

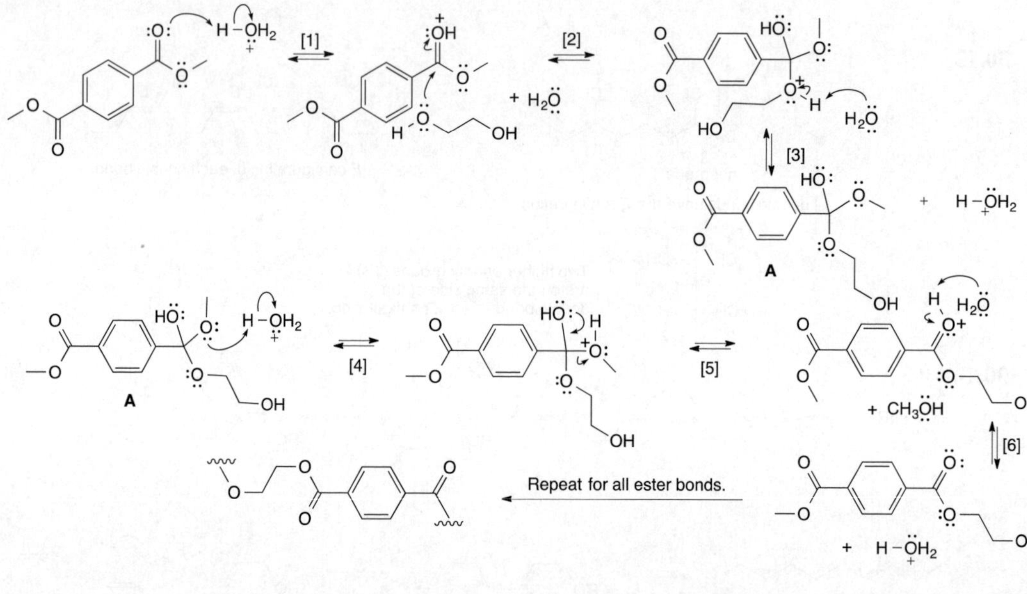
[1]
[2]
[3]
[4]
[5]
[6]
A
Repeat for all ester bonds.

30.18

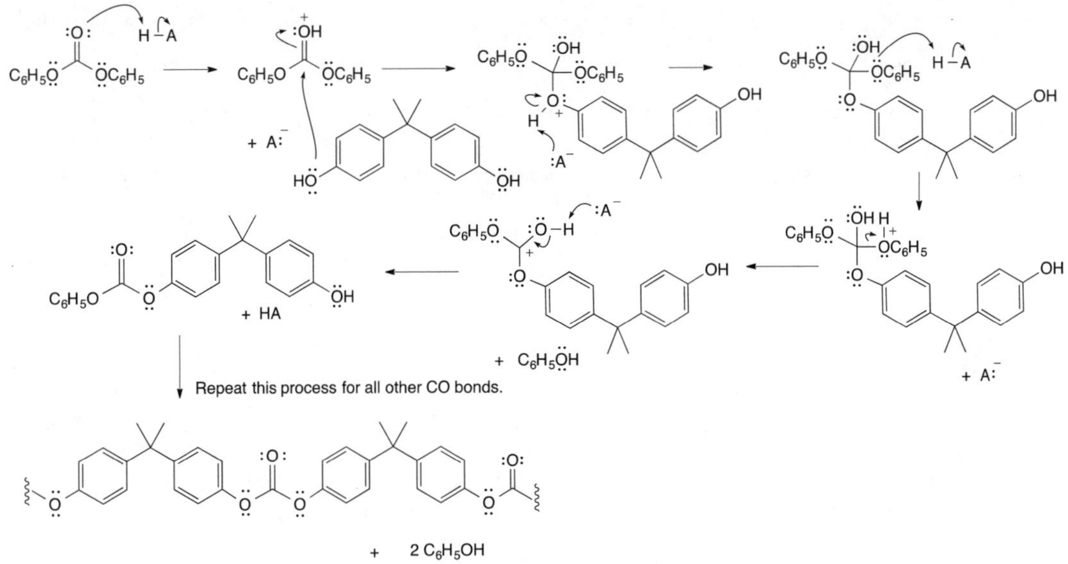

$C_6H_5\ddot{O}$ $\ddot{O}C_6H_5$ H—A C_6H_5O $\overset{+}{\ddot{O}H}$ $\ddot{O}C_6H_5$ $C_6H_5\ddot{O}$ $\ddot{O}H$ $\ddot{O}C_6H_5$ $C_6H_5\ddot{O}$ $\ddot{O}H$ $\ddot{O}C_6H_5$ H—A

+ A:⁻

HO OH   OH

:A⁻

C_6H_5O $\ddot{O}$ $\overset{+}{\ddot{O}}$—H   :A⁻ $C_6H_5\ddot{O}$ $\overset{\ddot{O}H\ H}{\underset{O C_6H_5}{|}}$

:O:

C_6H_5O $\ddot{O}$ OH   OH   OH

+ HA   + $C_6H_5\ddot{O}H$ + A:⁻

Repeat this process for all other CO bonds.

:O: :O:

O $\ddot{O}$ O $\ddot{O}$ O

+   2 C_6H_5OH

30.19

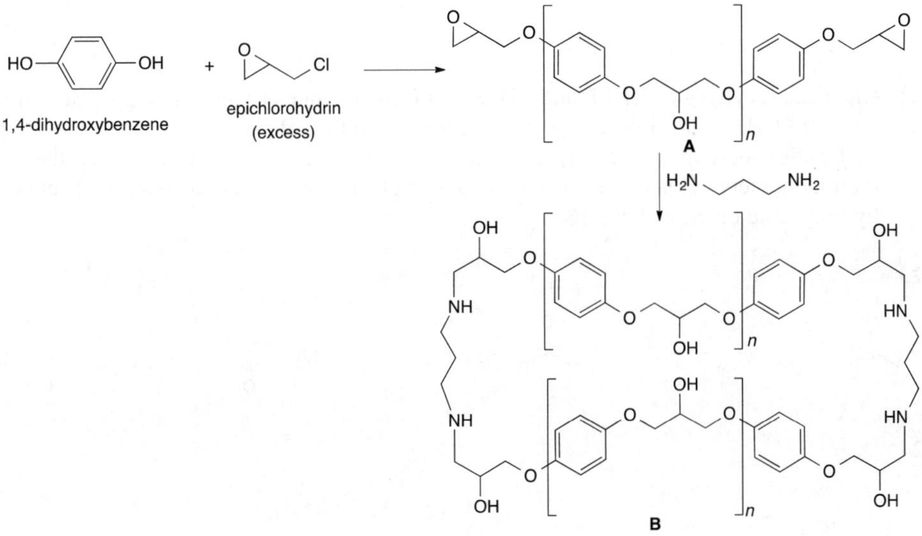

HO—◯—OH   +   epichlorohydrin (excess)   →   A

1,4-dihydroxybenzene

H_2N NH_2

→   B

Chapter 30–10

30.20

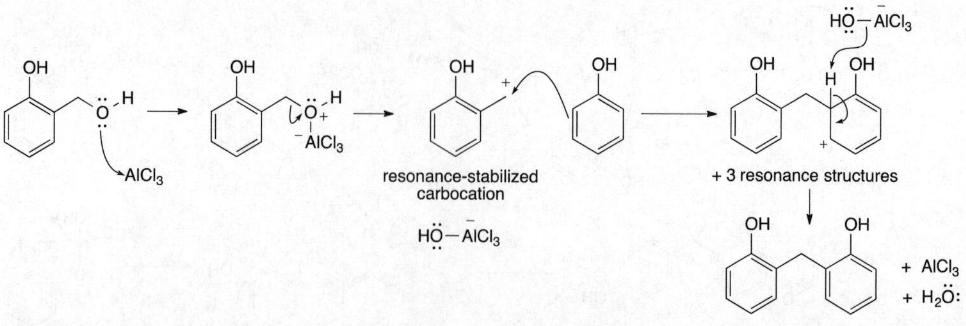

resonance-stabilized
carbocation

$H\ddot{O}-\overline{A}lCl_3$

$H\ddot{O}-\overline{A}lCl_3$

+ 3 resonance structures

+ $AlCl_3$
+ $H_2\ddot{O}$:

30.21

cardinol

$H_2C=O$

H^+

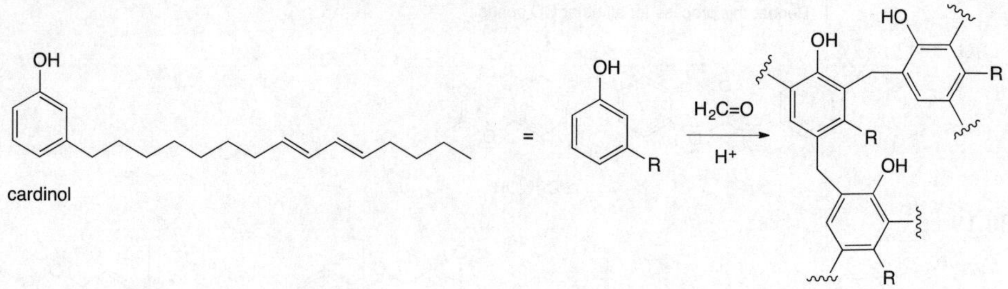

30.22 Chemical recycling of HDPE and LDPE is not easily done because these polymers are both long chains of CH_2 groups joined together in a linear fashion. Because there are only C–C bonds and no functional groups in the polymer chain, there are no easy methods to convert the polymers to their monomers. This process is readily accomplished only when the polymer backbone contains hydrolyzable functional groups.

30.23

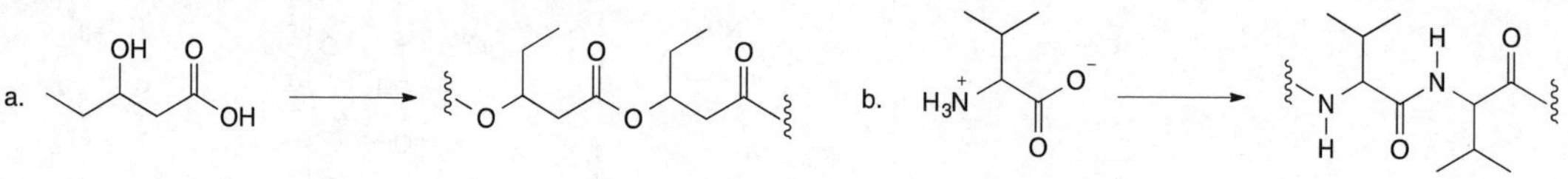

a.

b.

30.24

a.

b.

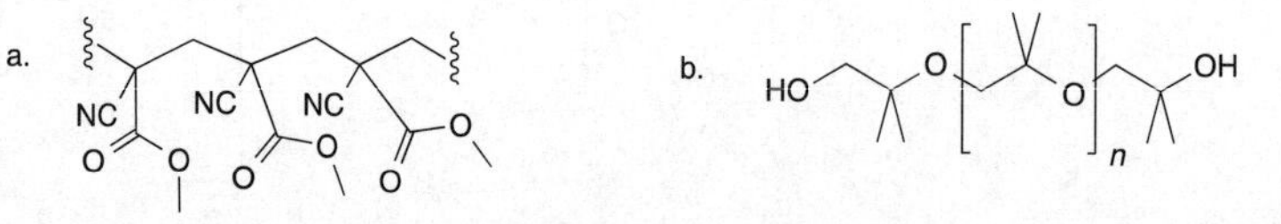

Synthetic Polymers 30–11

30.25

a.

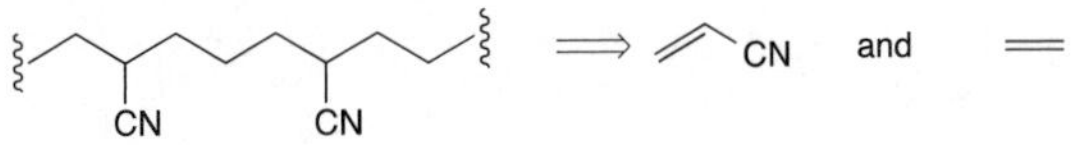

b.

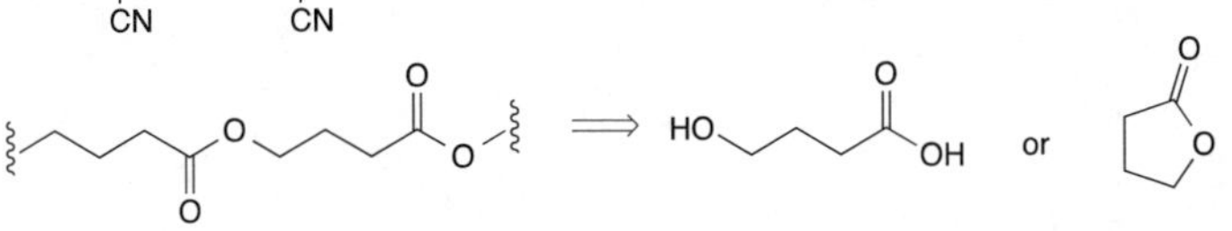

30.26

a.

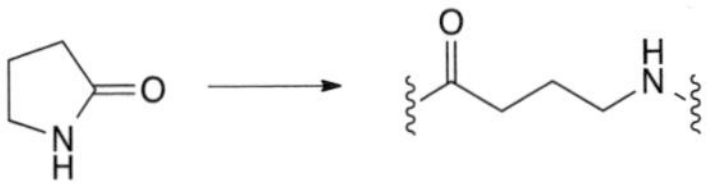

b. 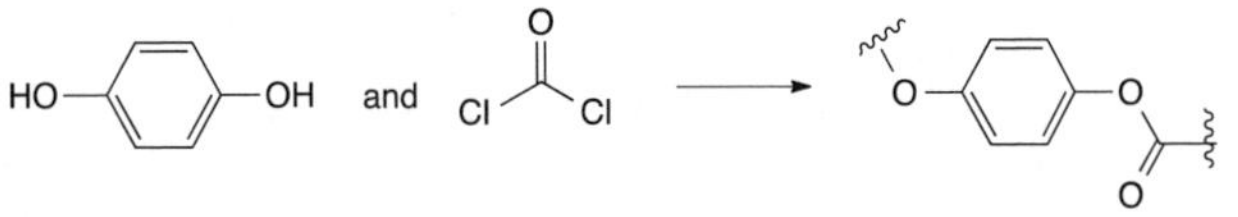 HO—⬡—OH and ClC(O)Cl →

30.27 Draw the polymer formed by chain-growth polymerization as in Answer 30.2.

a.

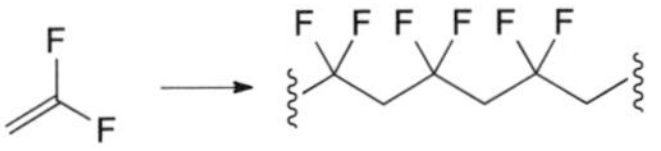

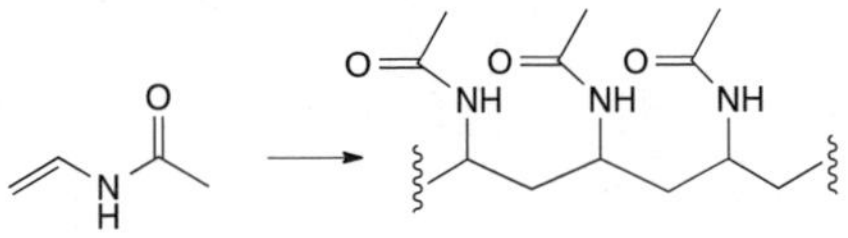

b.

c.

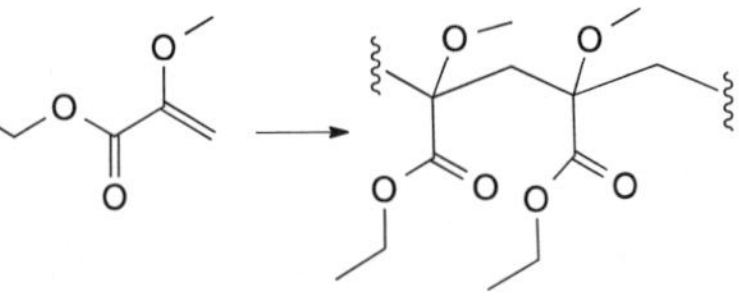

d.

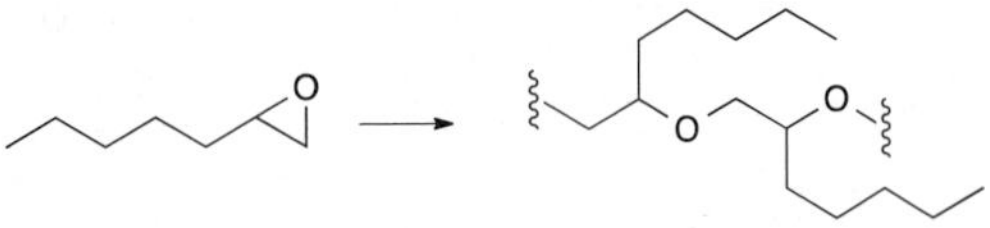

30.28 Draw the copolymers.

a. 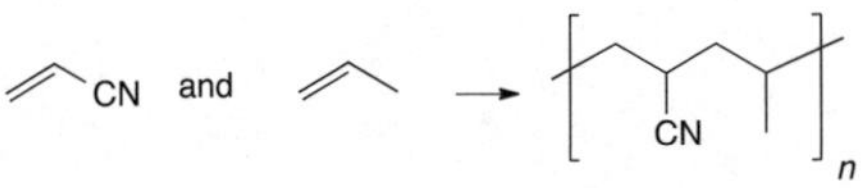 CN and →

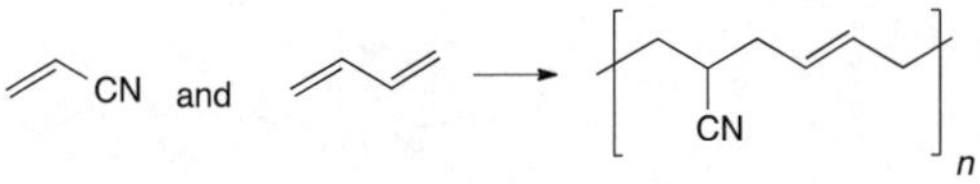

b. 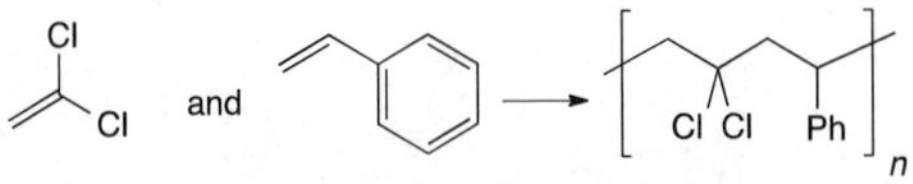 and →

c. CN and →

d.  and →

Chapter 30–12

30.29

a.

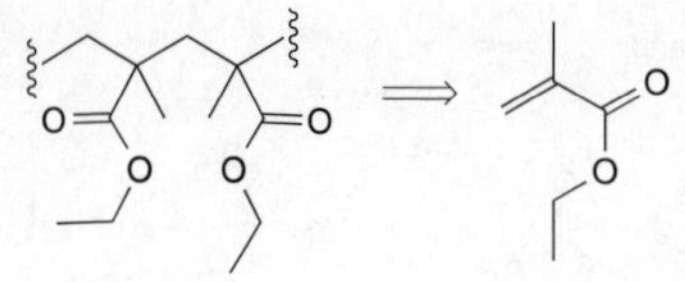

b.

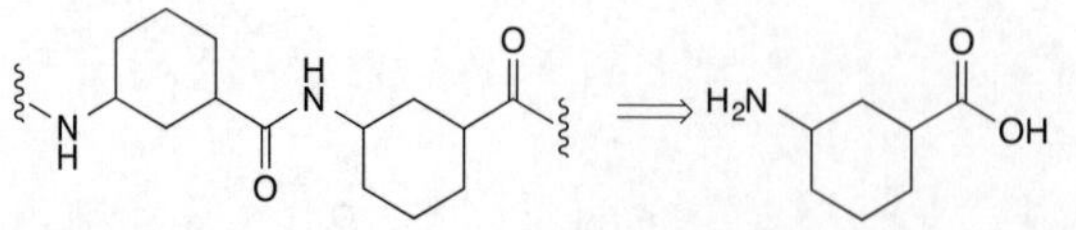

c.

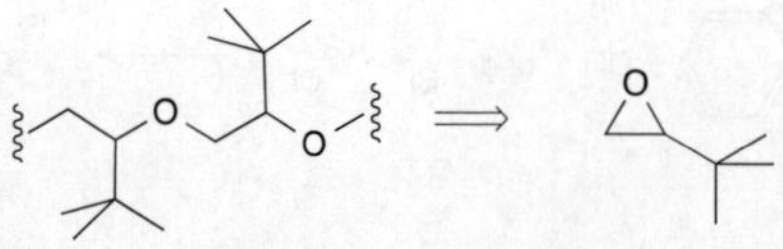

d.

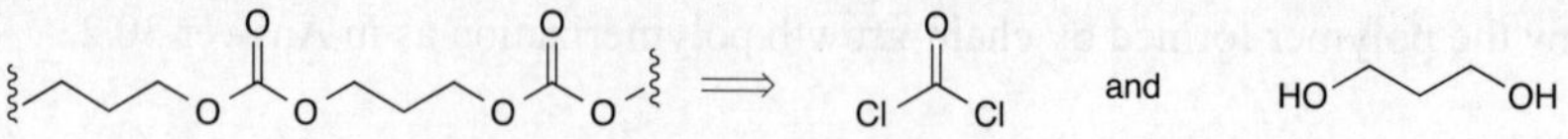

30.30

a.

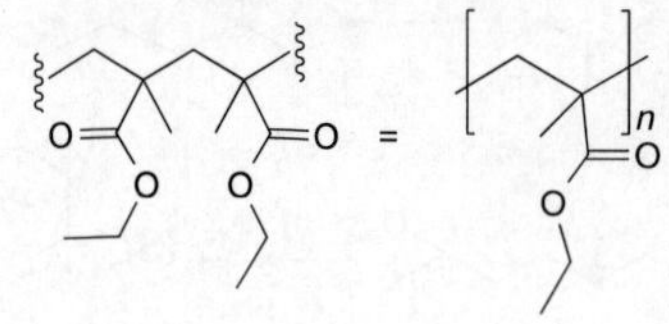

b.

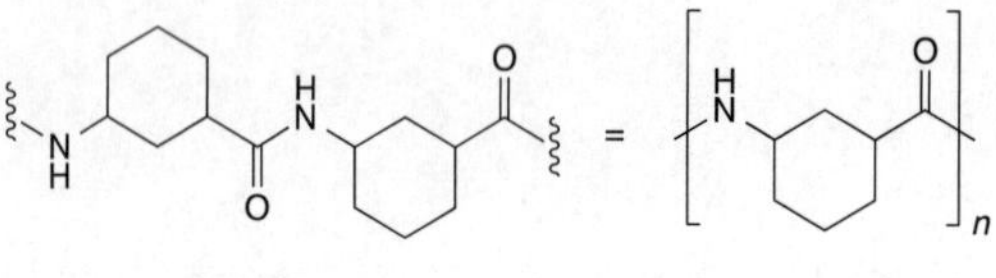

c.

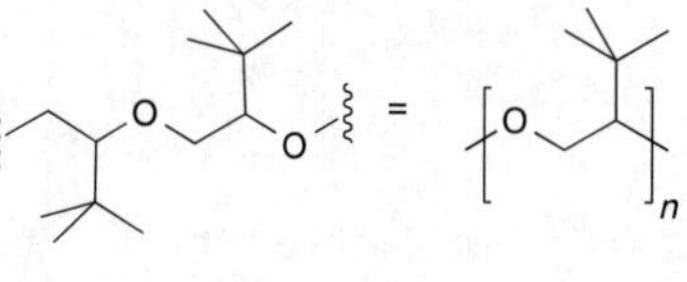

d.

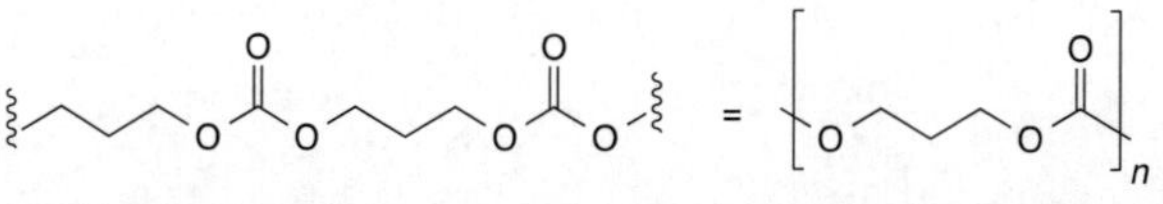

30.31 An **isotactic polymer** has all Z groups on the same side of the carbon backbone. A **syndiotactic polymer** has the Z groups alternating from one side of the carbon chain to the other. An **atactic polymer** has the Z groups oriented randomly along the polymer chain.

a.

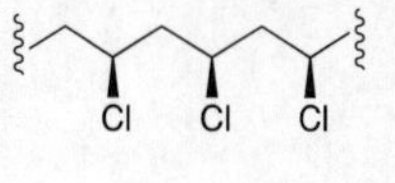

b.

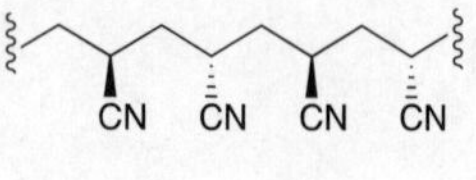

c.

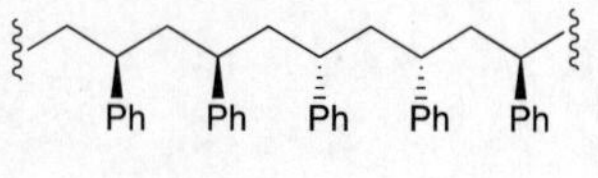

Synthetic Polymers 30–13

30.32

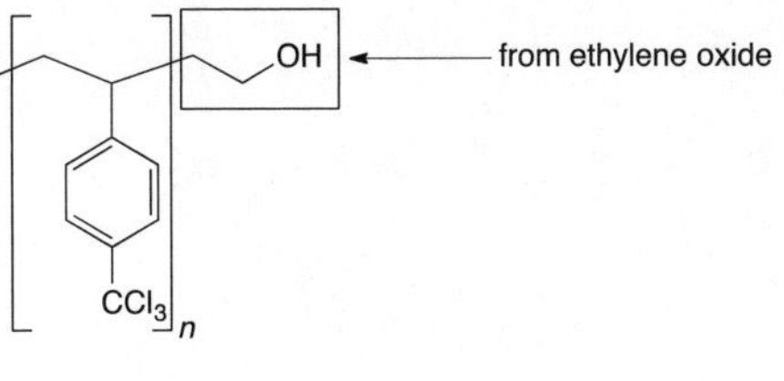

from ethylene oxide

30.33

a.

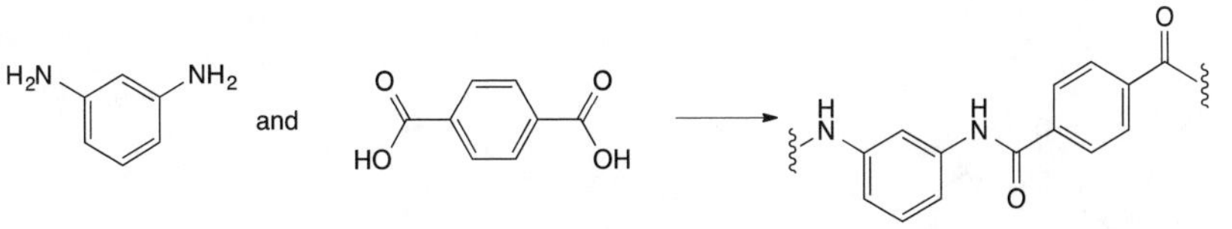

b.

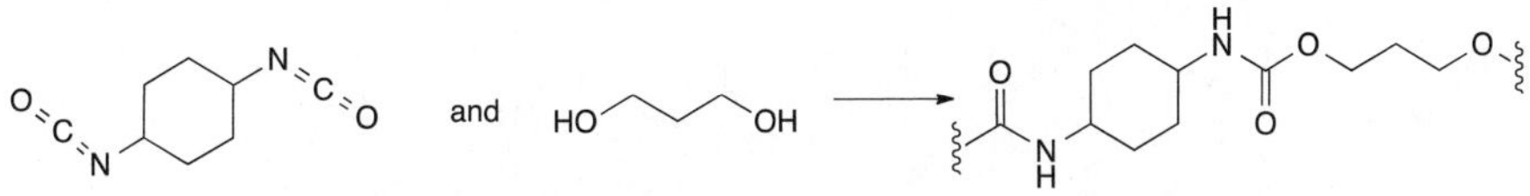

c.

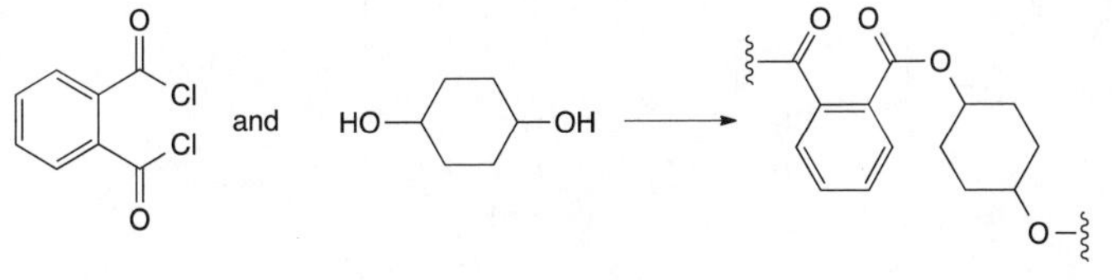

d.

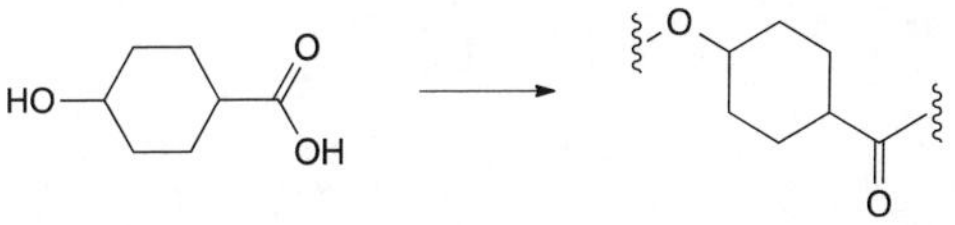

30.34

a.

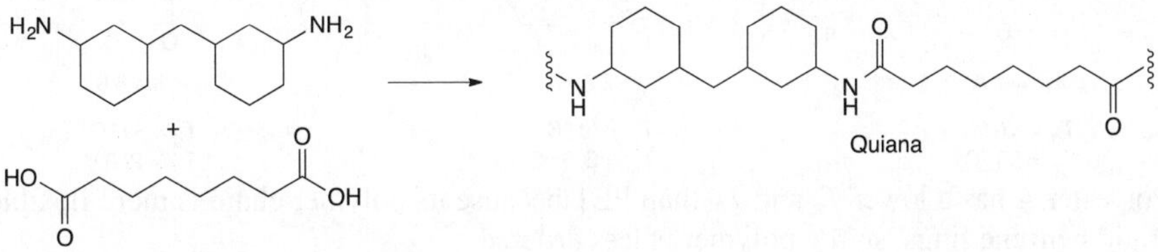

Quiana

b.

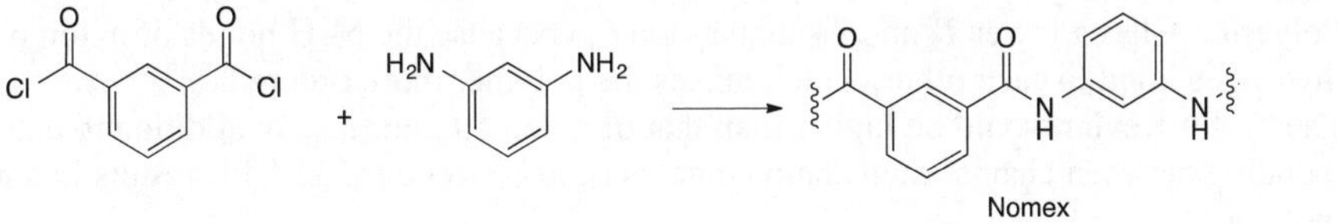

Nomex

Chapter 30–14

30.35

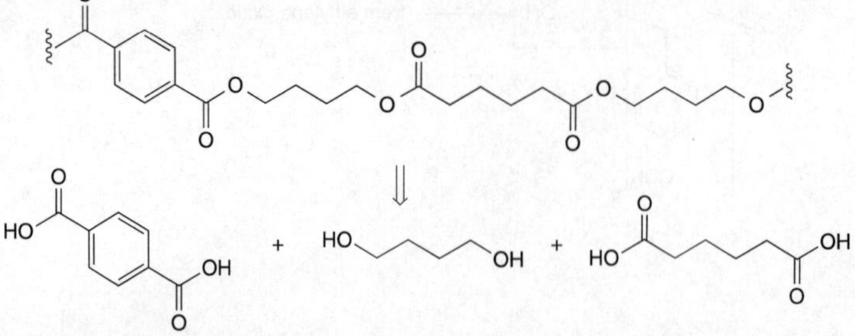

30.36

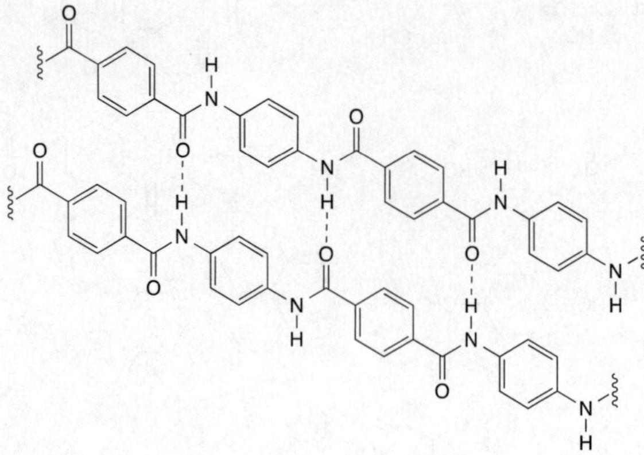

Kevlar

30.37

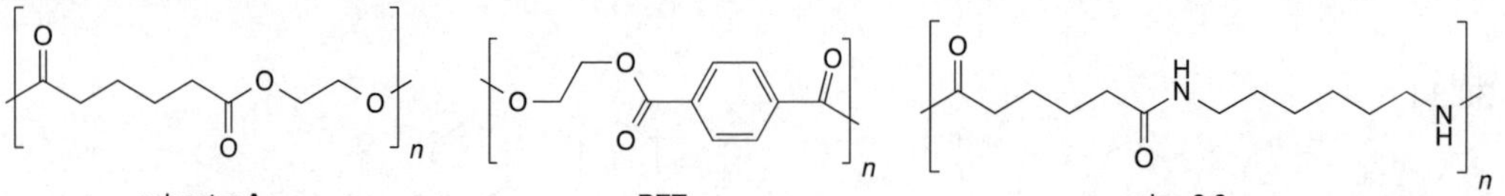

polyester **A**	PET	nylon 6,6
$T_g = <0\ ^oC$	$T_g = 70\ ^oC$	$T_g = 53\ ^oC$
$T_m = 50\ ^oC$	$T_m = 265\ ^oC$	$T_m = 265\ ^oC$

a. Polyester **A** has a lower T_g and T_m than PET because its polymer chain is more flexible. There are no rigid benzene rings, so the polymer is less ordered.

b. Polyester **A** has a lower T_g and T_m than nylon 6,6 because the N–H bonds of nylon 6,6 allow chains to hydrogen bond to each other, which makes the polymer more ordered.

c. The T_m for Kevlar would be higher than that of nylon 6,6, because in addition to extensive hydrogen bonding between chains, each chain contains rigid benzene rings. This results in a more ordered polymer.

Synthetic Polymers 30–15

30.38

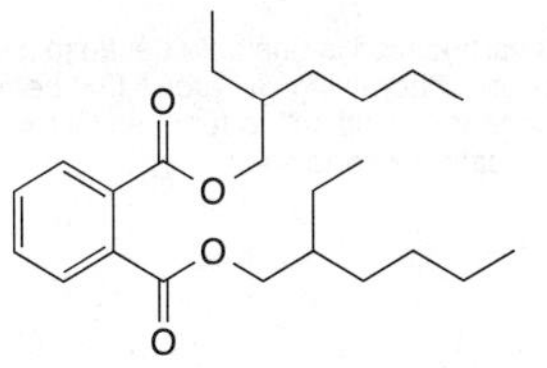

A

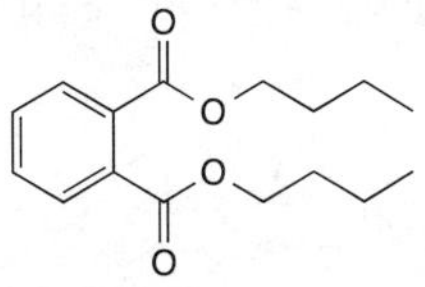

dibutyl phthalate

Diester **A** is often used as a plasticizer in place of dibutyl phthalate because it has a higher molecular weight, giving it a higher boiling point. **A** should therefore be less volatile than dibutyl phthalate, so it should evaporate from a polymer less readily.

30.39

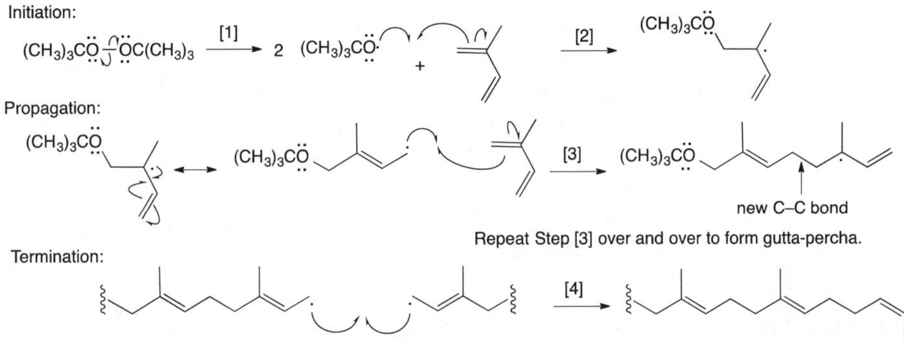

Initiation:

Propagation:

Repeat Step [3] over and over to form gutta-percha.

Termination:

30.40

a highly resonance-stabilized carbocation

Repeat Steps [3] and [4]. ⟶ **A**

new C–C bond

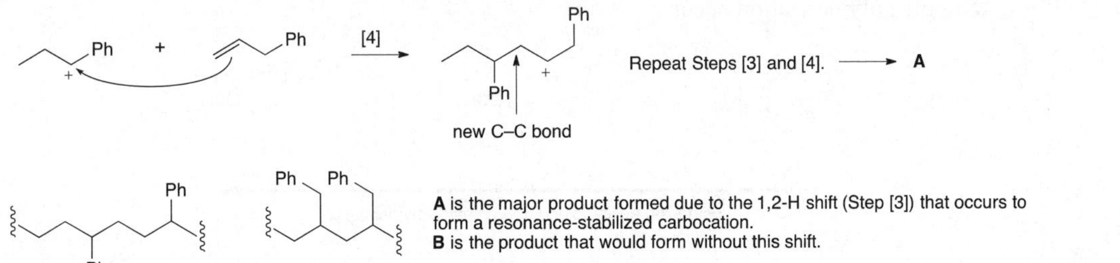

A

major product

B

A is the major product formed due to the 1,2-H shift (Step [3]) that occurs to form a resonance-stabilized carbocation.
B is the product that would form without this shift.

30.41

$F_3\bar{B}-O^+H_2$ ... CN → [cation CH₃-CH⁺-C≡N with δ+] — This carbocation is unstable because it is located next to an electron-withdrawing CN group that bears a δ+ on its C atom. This carbocation is difficult to form, so CH₂=CHCN is only slowly polymerized under cationic conditions.

$F_3\bar{B}-O^+H_2$ ... CH₂-CN → [2° carbocation CH₃-CH⁺-CH₂-C≡N]

This 2° carbocation is more stable because it is not directly bonded to the electron-withdrawing CN group. As a result, it is more readily formed. Thus, cationic polymerization can occur more readily.

30.42

Initiation: Bu—Li + CH₂=CHPh →[1] Bu-CH₂-CH(Ph)⁻ Li⁺

Propagation: Bu-CH₂-CH(Ph)⁻ + CH₂=CHPh →[2] Bu-CH₂-CH(Ph)-CH₂-CH(Ph)⁻ →(Repeat Step [2] over and over)

Termination: ~CH(Ph)-CH₂-CH(Ph)⁻ + O=C=O →[3] ~CH(Ph)-CH₂-CH(Ph)-C(=O)-O⁻

30.43

The substituent on styrene determines whether cationic or anionic polymerization is preferred. When the substituent stabilizes a carbocation, cationic polymerization will occur. When the substituent stabilizes a carbanion, anionic polymerization will occur.

a. CH₂=CH-C₆H₄-OCH₃ — **cationic polymerization**

b. CH₂=CH-C₆H₄-NO₂ — **anionic polymerization**

c. CH₂=CH-C₆H₄-CF₃ — **anionic polymerization**

d. CH₂=CH-C₆H₄-CH₂CH₃ — **cationic polymerization**

30.44

The rate of anionic polymerization depends on the ability of the substituents on the alkene to stabilize an intermediate carbanion: the better a substituent stabilizes a carbanion, the faster anionic polymerization occurs.

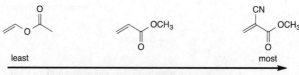

least → most

increasing ability to undergo anionic polymerization

Synthetic Polymers 30–17

30.45 The reason for this selectivity is explained in Figure 9.9. In the ring opening of an unsymmetrical epoxide under acidic conditions, nucleophilic attack occurs at the carbon atom that is more able to accept a δ+ in the transition state; that is, nucleophilic attack occurs at the more substituted carbon. The transition state having a δ+ on a C with an electron-donating CH_3 group is more stabilized (lower in energy), permitting a faster reaction.

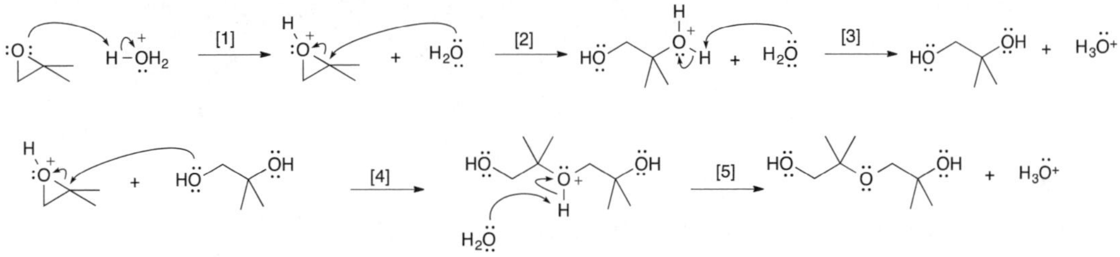

Repeat Steps [4] and [5] over and over.

30.46

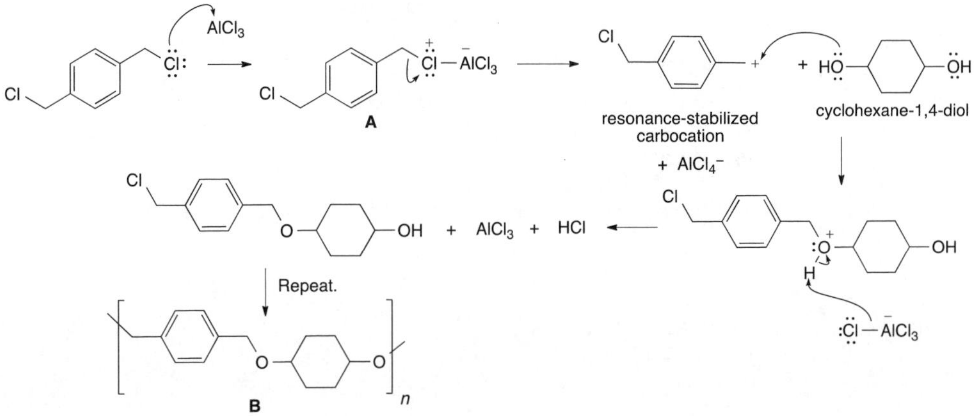

30.47

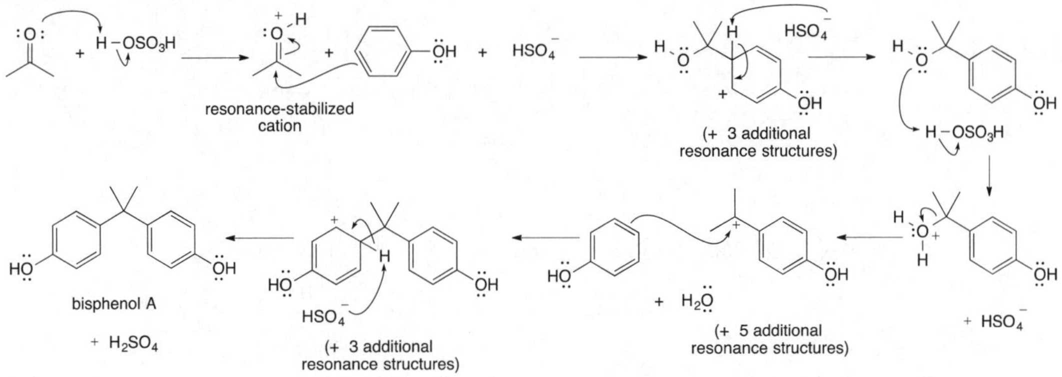

Chapter 30–18

30.48

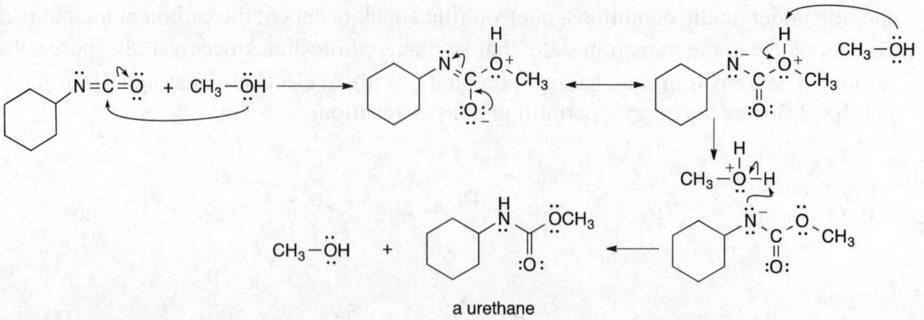

a urethane

30.49

a.

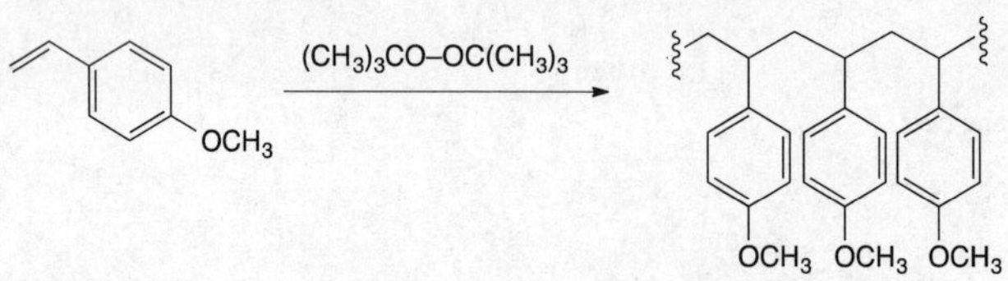

$(CH_3)_3CO-OC(CH_3)_3$

b.

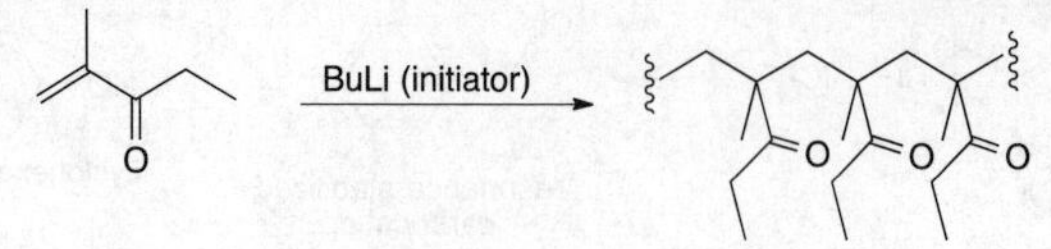

BuLi (initiator)

c.

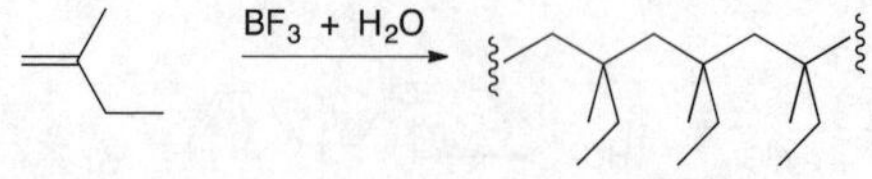

$BF_3 + H_2O$

d.

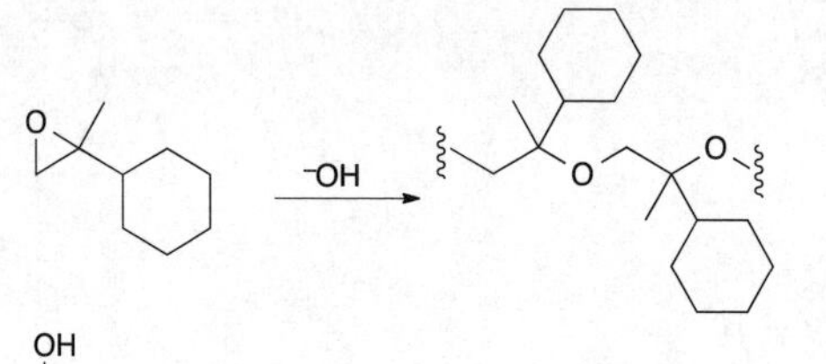

^-OH

e.

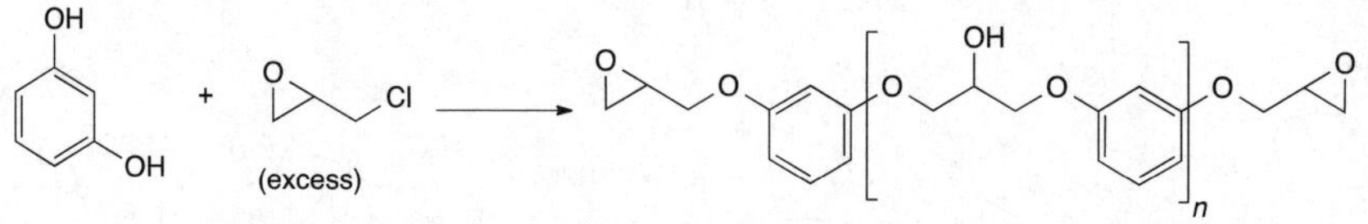

(excess)

f.

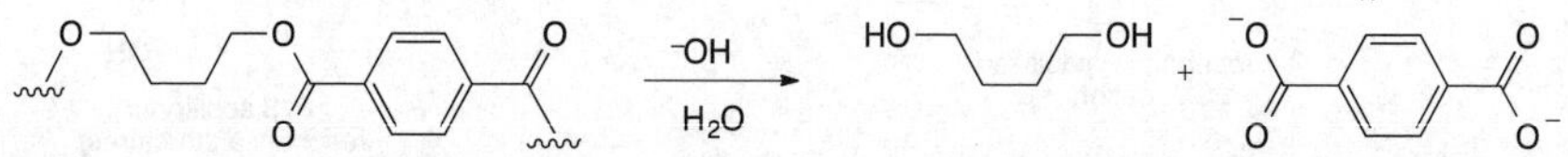

^-OH

H_2O

g.

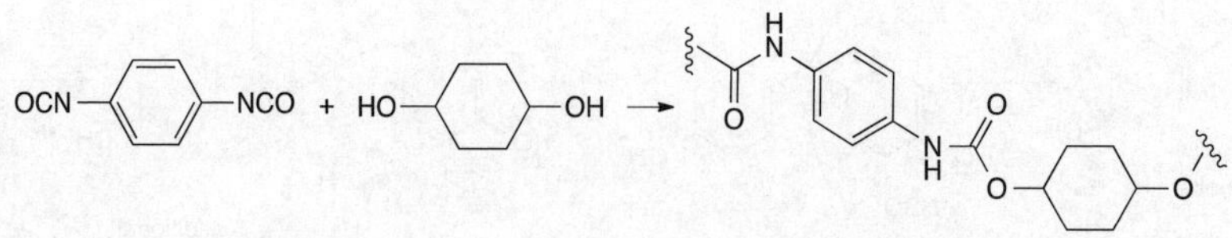

$OCN-\!\!\!-\!\!\!-NCO + HO-\!\!\!-\!\!\!-OH \rightarrow$

h.

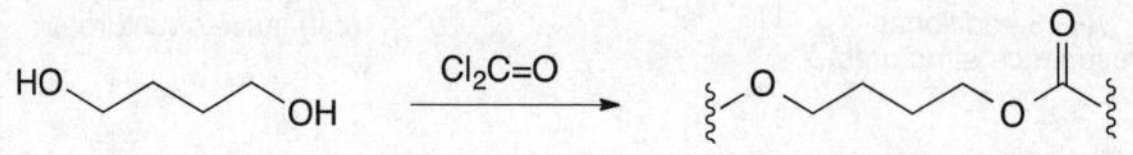

$Cl_2C=O$

Synthetic Polymers 30–19

30.50 Polyethylene bottles are resistant to NaOH because they are hydrocarbons with no reactive sites. Polyester shirts and nylon stockings both contain functional groups. Nylon contains amides and polyester contains esters, two functional groups that are susceptible to hydrolysis with aqueous NaOH. Thus, the polymers are converted to their monomer starting materials, creating a hole in the garment.

30.51

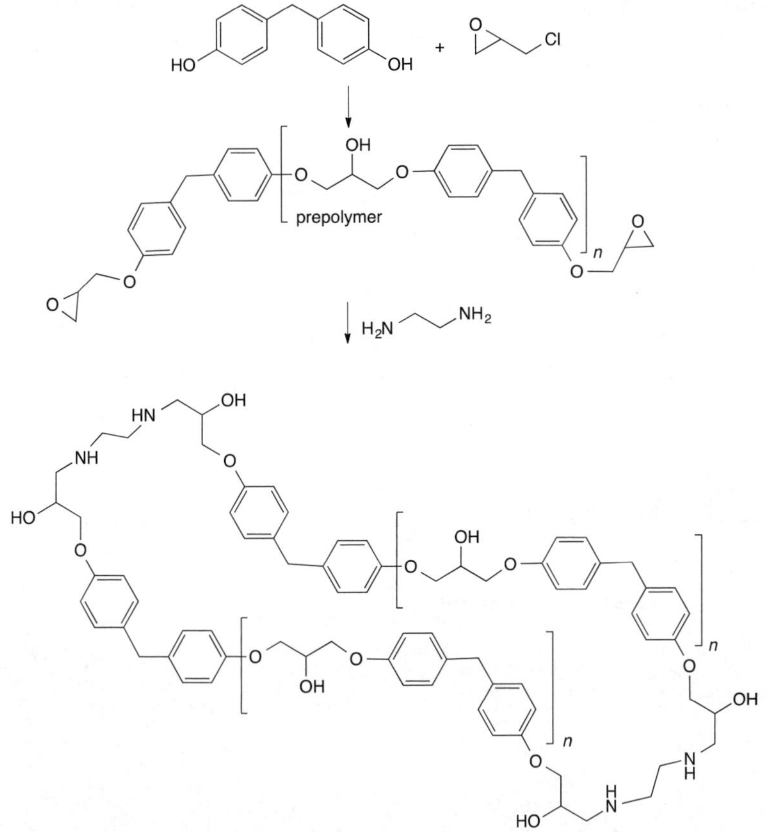

Chapter 30–20

30.52

a.

vinyl alcohol → poly(vinyl alcohol)

Poly(vinyl alcohol) cannot be prepared from vinyl alcohol because vinyl alcohol is not a stable monomer. It is the enol of acetaldehyde (CH_3CHO), and thus it can't be converted to poly(vinyl alcohol).

b.

vinyl acetate $\xrightarrow[\text{polymerization}]{\substack{\text{ROOR} \\ \text{radical}}}$ poly(vinyl acetate) $\xrightarrow[\substack{H_2O \\ \text{hydrolysis}}]{^-OH}$ poly(vinyl alcohol) $+ CH_3CO_2^-$

c.

poly(vinyl alcohol) $\xrightarrow{H^+}$ poly(vinyl butyral)

an acetal

30.53

$\xrightarrow[\text{AlCl}_3]{\text{CH}_3\text{Cl}}$ $\xrightarrow[\text{AlCl}_3]{\text{CH}_3\text{Cl}}$ $\xrightarrow{\text{KMnO}_4}$ terephthalic acid

$CH_2=CH_2$ $\xrightarrow[\text{[2] NaHSO}_3,\ H_2O]{\text{[1] OsO}_4}$ ethylene glycol

or

$\xrightarrow{\text{mCPBA}}$ $\xrightarrow{H_2O,\ ^-OH}$

30.54

phenol $+ CH_2=O \longrightarrow$

Since phenol has no substituents at any ortho or para position, an extensive network of covalent bonds can join the benzene rings together at all ortho and para positions to the OH groups.

Bakelite

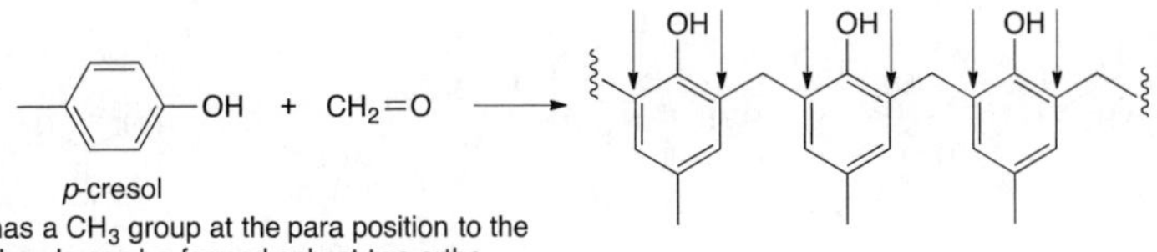

Synthetic Polymers 30–21

p-cresol

Since *p*-cresol has a CH$_3$ group at the para position to the OH group, new bonds can be formed only at two ortho positions, so that a less extensive three-dimensional network can form.

30.55

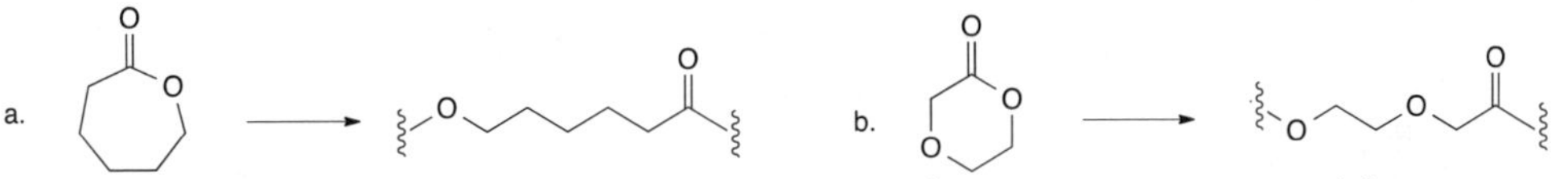

a.

ε-caprolactone → polycaprolactone

b.

p-dioxanone → polydioxanone

30.56

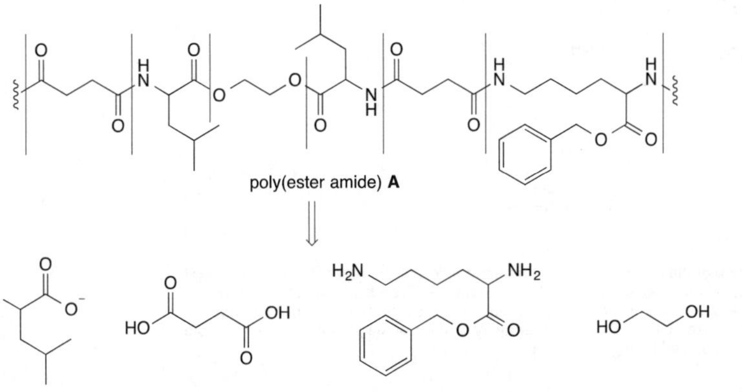

poly(ester amide) **A**

leucine

the benzyl ester of lysine

30.57

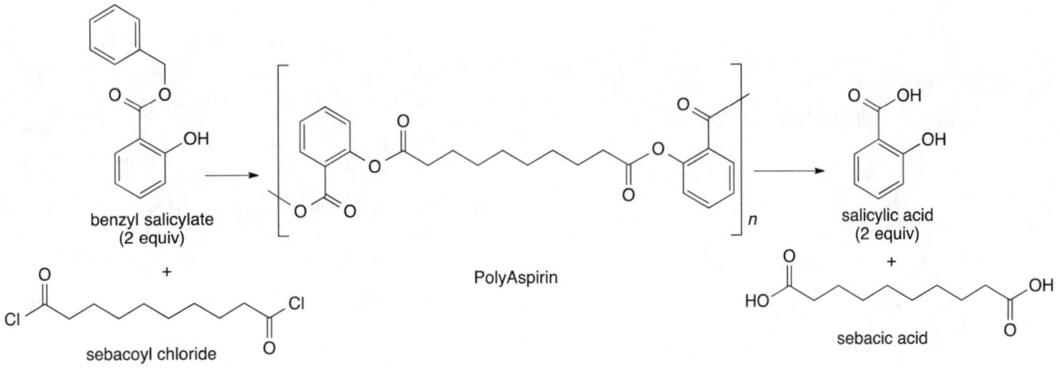

benzyl salicylate
(2 equiv)

+

sebacoyl chloride

PolyAspirin

salicylic acid
(2 equiv)

+

sebacic acid

Chapter 30–22

30.58

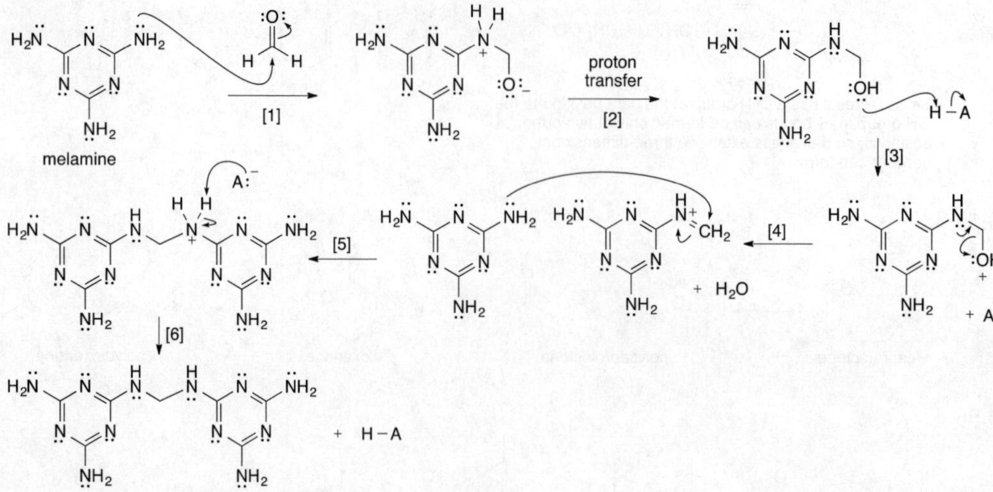

melamine

[1]

proton
transfer
[2]

[3]

[5]

+ H$_2$O

[4]

+ A:$^-$

[6]

+ H — A

30.59

a.

[1]
intramolecular
H abstraction

re-draw

[2]

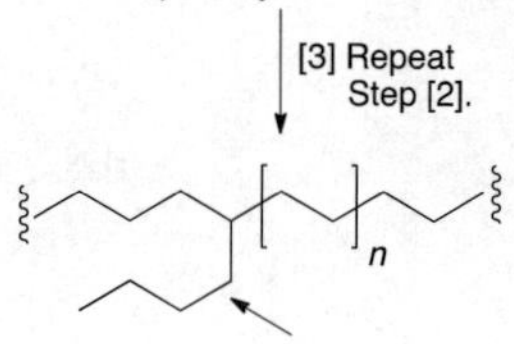

[3] Repeat
Step [2].

butyl substituent

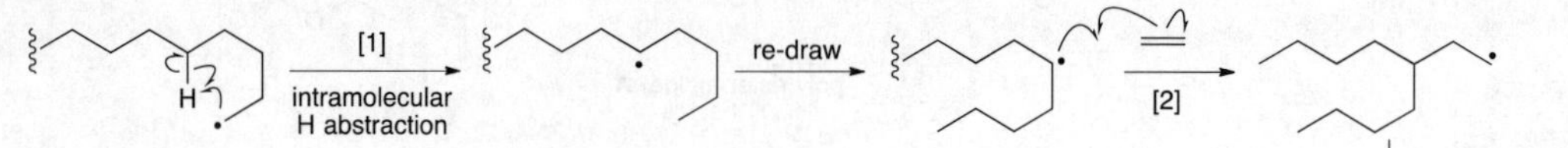

b. Abstraction of the H is more facile than abstraction of the other H's because the H
atom that is removed is six atoms from the radical. The transition state for this
intramolecular reaction is cyclic, and resembles a six-membered ring, the most
stable ring size. Other H's are too far away or the transition state would resemble
a smaller, less stable ring.

30.60

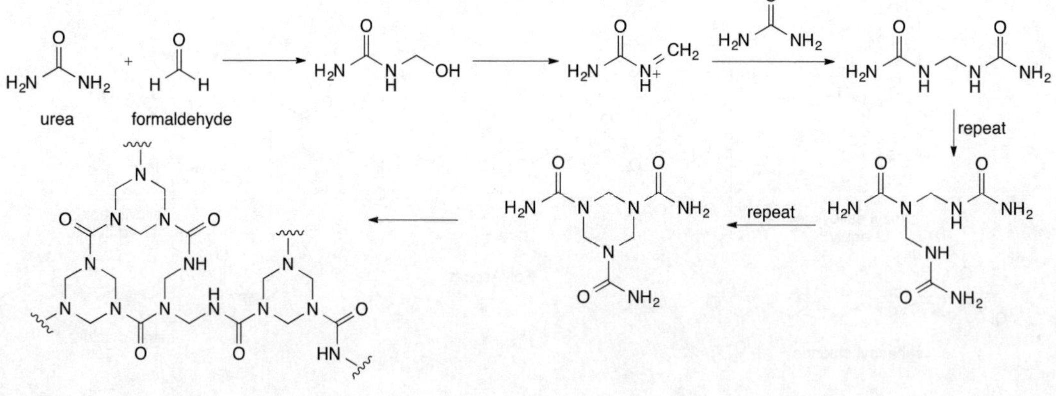

urea   formaldehyde

repeat

repeat

Chapter 31 Lipids

Chapter Review

Hydrolyzable lipids

[1] Waxes (31.2)—Esters formed from a long-chain alcohol and a long-chain carboxylic acid

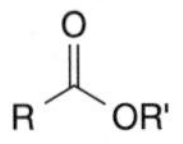

R, R' = long chains of C's

[2] Triacylglycerols (31.3)—Triesters of glycerol with three fatty acids

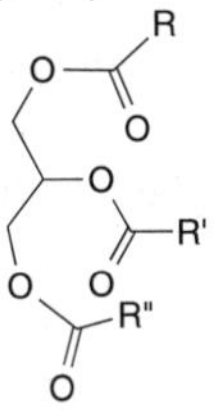

R, R', R" = alkyl groups with 11–19 C's

[3] Phospholipids (31.4)

a. Phosphatidylethanolamine (cephalin)

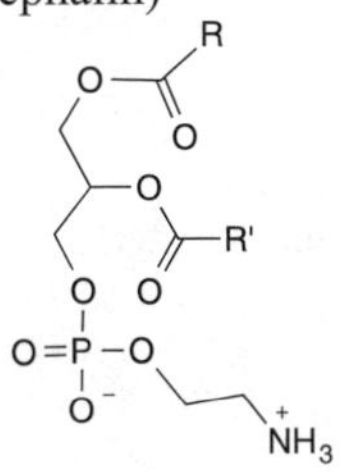

R, R' = long carbon chain

b. Phosphatidylcholine (lecithin)

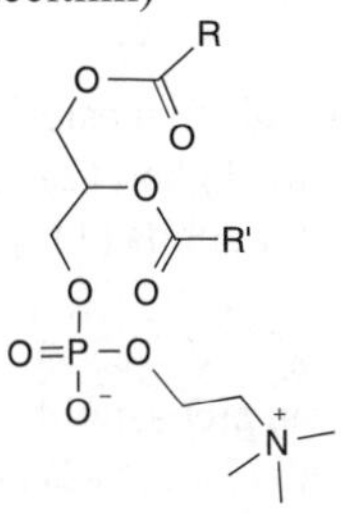

R, R' = long carbon chain

c. Sphingomyelin

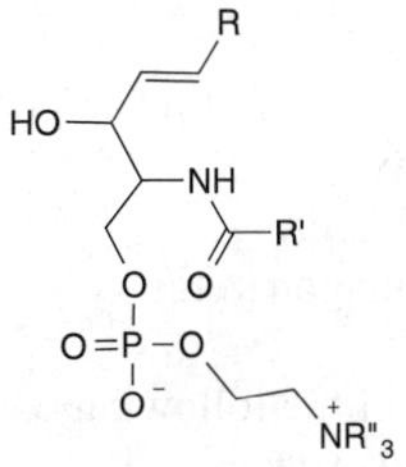

R = $CH_3(CH_2)_{12}$
R' = long carbon chain
R" = H or CH_3

Chapter 31–2

Nonhydrolyzable lipids

[1] Fat-soluble vitamins (31.5)—Vitamins A, D, E, and K

[2] Eicosanoids (31.6)—Compounds containing 20 carbons derived from arachidonic acid. There are four types: prostaglandins, thromboxanes, prostacyclins, and leukotrienes.

[3] Terpenes (31.7)—Lipids composed of repeating five-carbon units called isoprene units

Isoprene unit	Types of terpenes			
	[1] monoterpene	10 C's	[4] sesterterpene	25 C's
	[2] sesquiterpene	15 C's	[5] triterpene	30 C's
	[3] diterpene	20 C's	[6] tetraterpene	40 C's

[4] Steroids (31.8)—Tetracyclic lipids composed of three six-membered and one five-membered ring

Practice Test on Chapter Review

1.a. Which of the following compounds contains an ester?
    1. a wax   4. Both (1) and (2) contain esters.
    2. a cephalin   5. Compounds (1), (2), and (3) all contain esters.
    3. a sphingomyelin

  b. Which of the following compounds is an eicosanoid?
    1. a leukotriene   4. Both (1) and (2) are eicosanoids.
    2. a thromboxane   5. Compounds (1), (2), and (3) are all eicosanoids.
    3. a prostaglandin

  c. Which of the following statements is (are) true about **A?**

1. **A** is a lecithin.
2. **A** is a phosphoacylglycerol.
3. **A** has two tetrahedral stereogenic centers.
4. Both (1) and (2) are true.
5. Statements (1), (2), and (3) are all true.

2. Label each compound as a (1) hydrolyzable or (2) nonhydrolyzable lipid.
   a. a triacylglycerol
   b. vitamin A
   c. cholesterol
   d. a lecithin
   e. oleic acid
   f. PGE₁
   g. a wax
   h. a phosphoacylglycerol

3. For each compound: How many isoprene units does the compound contain? Classify the compound as a monoterpenoid, sesquiterpenoid, etc.

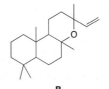
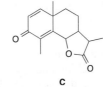

A   B   C

4. Answer True (T) or False (F) for each statement.
   a. Eicosanoids are biosynthesized from dimethylallyl diphosphate and isopentenyl diphosphate.
   b. Prostaglandins are biosynthesized from arachidonic acid.
   c. Fats have lower melting points than oils, and are generally solids at room temperature.
   d. A cephalin is one type of phosphoacylglycerol.
   e. Sphingomyelins possess a polar head and two nonpolar tails.
   f. A sesterterpene contains six isoprene units.
   g. The typical steroid skeleton has four six-membered rings joined with trans ring fusions.
   h. Waxes are high molecular weight lipids formed from a fatty acid and a long-chain amine.

5. Which terms describe each compound: (a) hydrolyzable lipid; (b) triacylglycerol; (c) phosphoacylglycerol; (d) phospholipid; (e) nonhydrolyzable lipid; (f) prostaglandin; (g) terpenoid? More than one term may apply to a compound.

Chapter 31–4

Answers to Practice Test

1.a. 4	2.a. 1	3. **A:** 2, monoterpenoid	4. a. F	5. **A:** e, f
b. 5	b. 2	**B:** 4, diterpenoid	b. T	**B:** e, g
c. 3	c. 2	**C:** 3, sesquiterpenoid	c. F	**C:** a, c, d
	d. 1		d. T	
	e. 2		e. T	
	f. 2		f. F	
	g. 1		g. F	
	h. 1		h. F	

Answers to Problems

31.1 Join the carboxylic acid and alcohol together to form an ester.

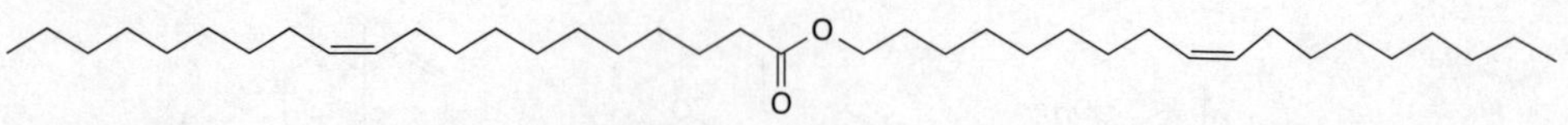

31.2 Eicosapentaenoic acid has 20 C's and 5 C=C's. An increasing number of double bonds decreases the melting point, so eicosapentaenoic acid should have a melting point lower than arachidonic acid; that is, $< -49 \,°C$.

31.3

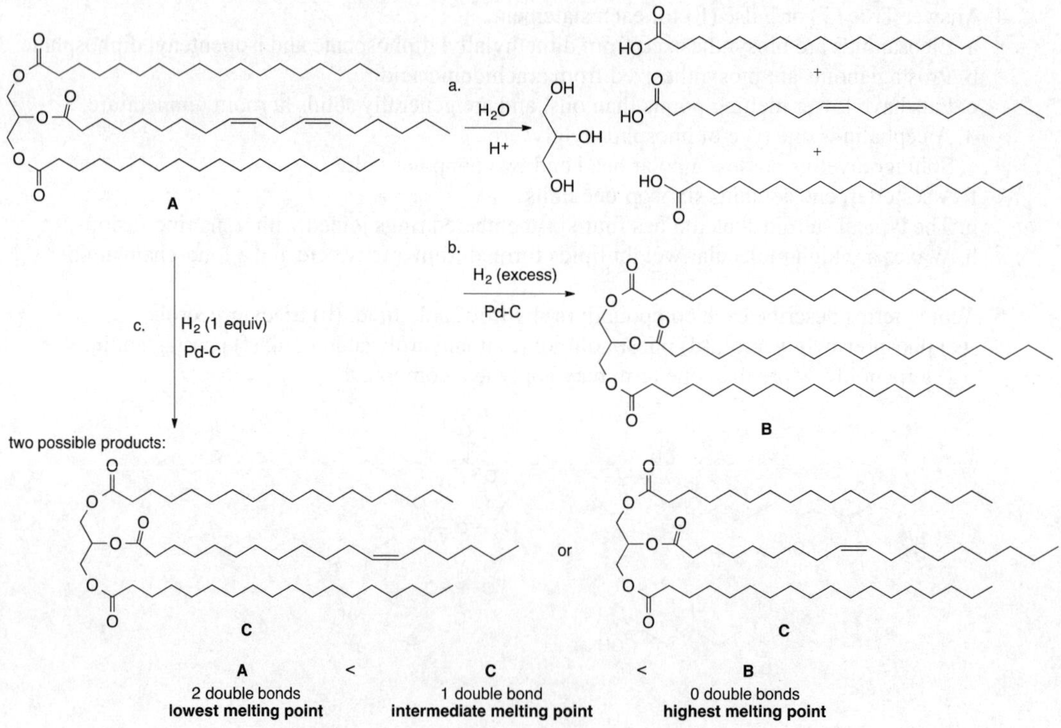

Lipids 31–5

31.4   Lauric acid is a saturated fatty acid but has only 12 C's. The carbon chain is much shorter than palmitic acid (16 C's) and stearic acid (18 C's), making coconut oil a liquid at room temperature.

31.5

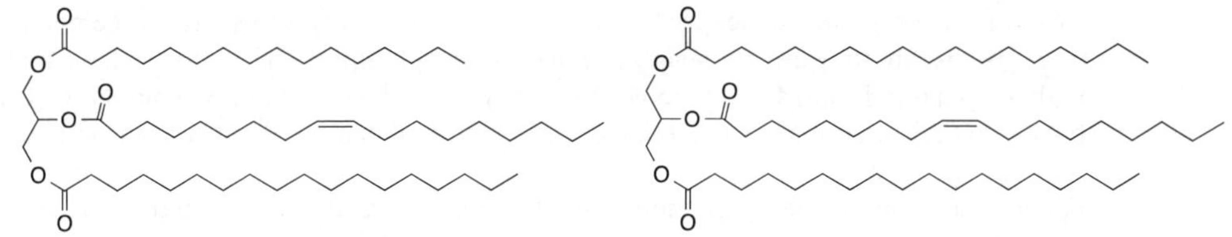

31.6   A lecithin is a type of phosphoacylglycerol. Two of the hydroxy groups of glycerol are esterified with fatty acids. The third OH group is part of a phosphodiester, which is also bonded to another low molecular weight alcohol.

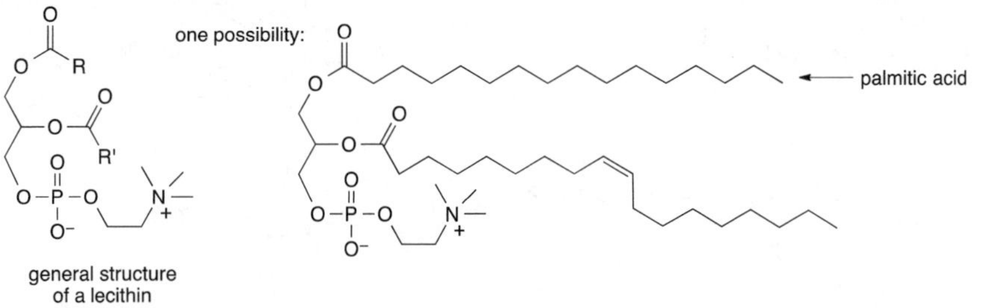

31.7   Soaps and phosphoacylglycerols have hydrophilic and hydrophobic components. Both compounds have an ionic "head" that is attracted to polar solvents like H$_2$O. This head is small in size compared to the hydrophobic region, which consists of one or two long hydrocarbon chains. These nonpolar chains consist of only C–C and C–H bonds and exhibit only van der Waals forces.

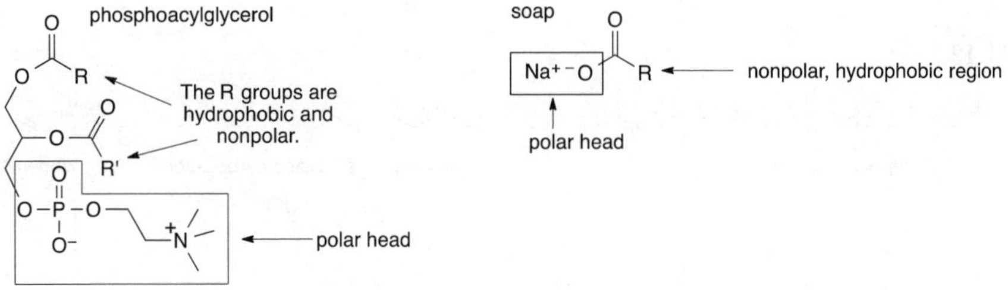

Chapter 31–6

31.8 Phospholipids have a polar (ionic) head and two nonpolar tails. These two regions, which exhibit very different forces of attraction, allow the phospholipids to form a bilayer with a central hydrophobic region that serves as a barrier to agents crossing a cell membrane, while still possessing an ionic head to interact with the aqueous environment inside and outside the cell. Two different regions are needed in the molecule. Triacylglycerols have three polar, uncharged ester groups, but they are not nearly as polar as phospholipids. They do not have an ionic head with nonpolar tails and so they do not form bilayers. They are largely nonpolar C–C and C–H bonds, so they are not attracted to an aqueous medium, making them H_2O insoluble.

31.9 Fat-soluble vitamins are hydrophobic and therefore are readily stored in the fatty tissues of the body. Water-soluble vitamins, on the other hand, are readily excreted in the urine, so large concentrations cannot build up in the body.

31.10 Isoprene units are shown in bold.

31.11

four isoprene units = diterpene

31.12

Lipids 31–7

31.13

resonance-stabilized carbocation

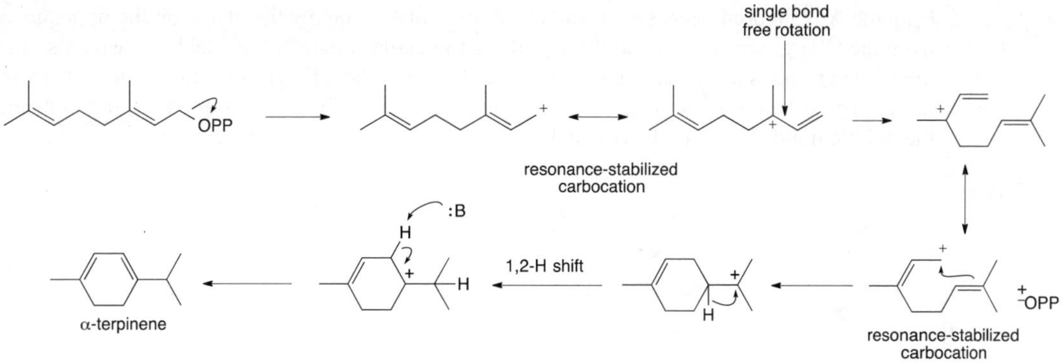

single bond free rotation

resonance-stabilized carbocation

:B

1,2-H shift

α-terpinene

31.14

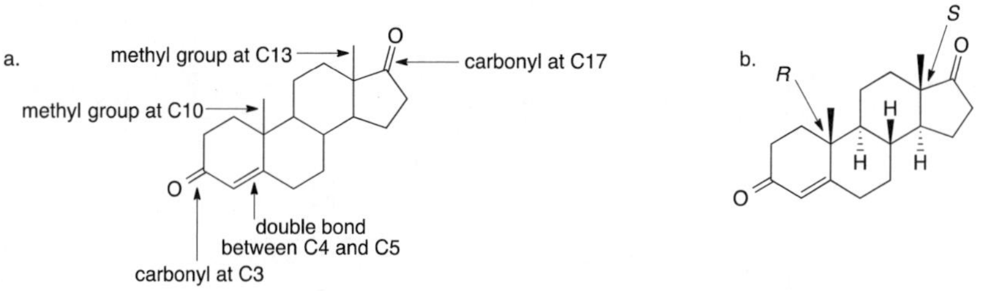

a.

methyl group at C13 → O ← carbonyl at C17

methyl group at C10

carbonyl at C3

double bond between C4 and C5

b.

R

S

31.15

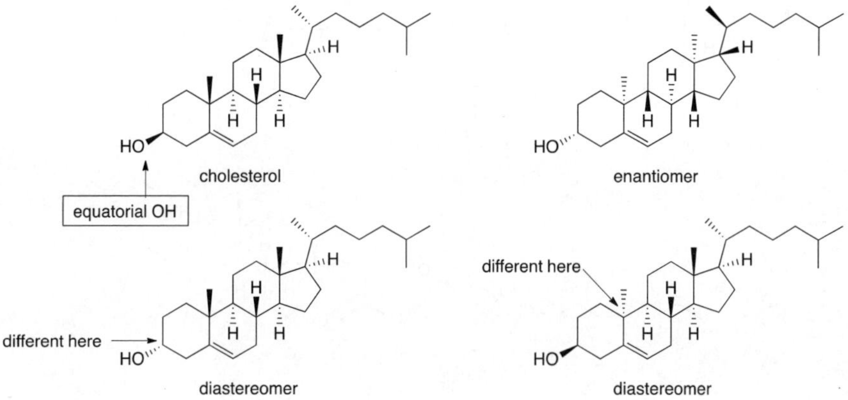

cholesterol

equatorial OH

enantiomer

different here → diastereomer

different here → diastereomer

31.16

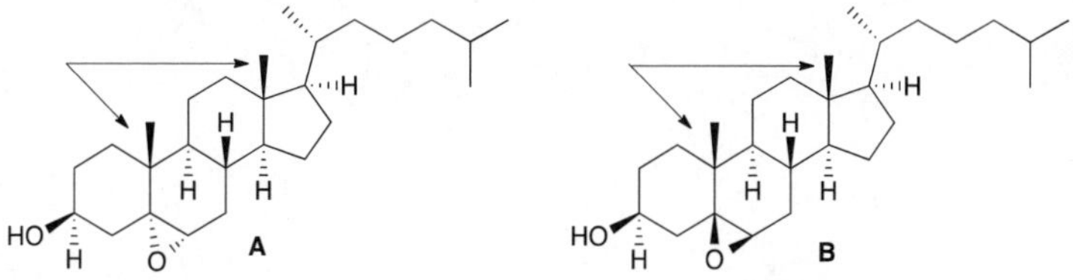

A

B

Chapter 31–8

All four rings are in the same plane. The bulky CH₃ groups (arrows) are located above the plane. Epoxide **A** is favored because it results from epoxidation below the plane, on the opposite side from the CH₃ groups that shield the top of the molecule somewhat to attack by reagents. In **B,** the epoxide ring is above the plane on the same side as the CH₃ groups. Formation of **B** would require epoxidation of the planar C=C from the less accessible, more sterically hindered side of the double bond. This path is thus disfavored.

31.17

a. 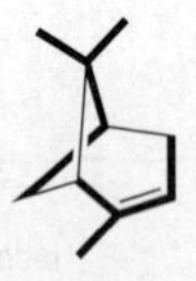 α-pinene   b. 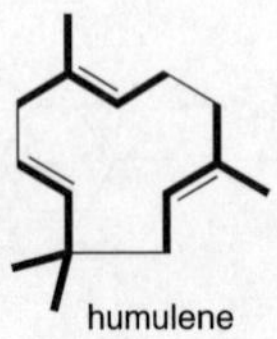 humulene

31.18

a. 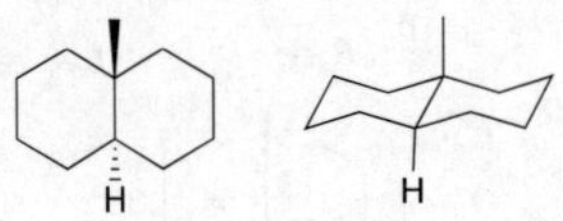 trans   b. 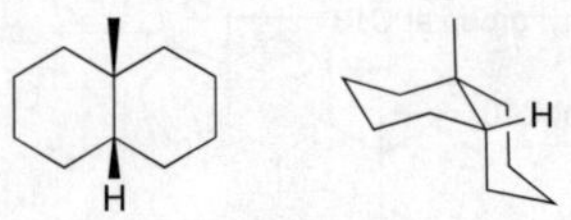 cis

31.19

a. 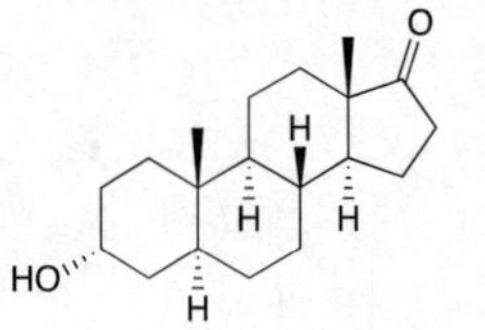   b.

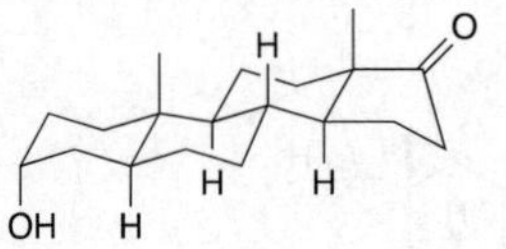

31.20

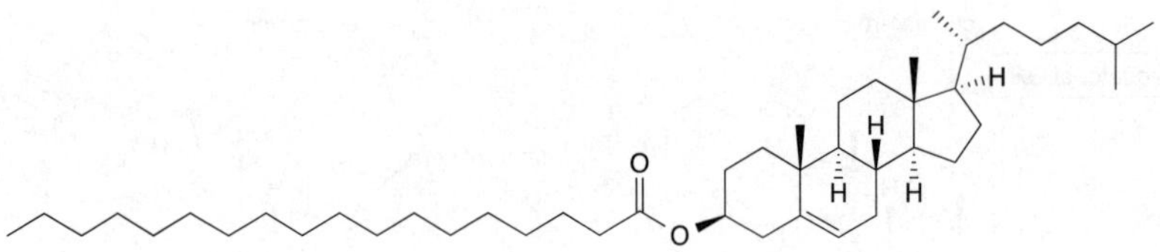

Lipids 31–9

31.21 Each compound has one tetrahedral stereogenic center (circled), so there are two stereoisomers (two enantiomers) possible. All C=C's have the *Z* configuration.

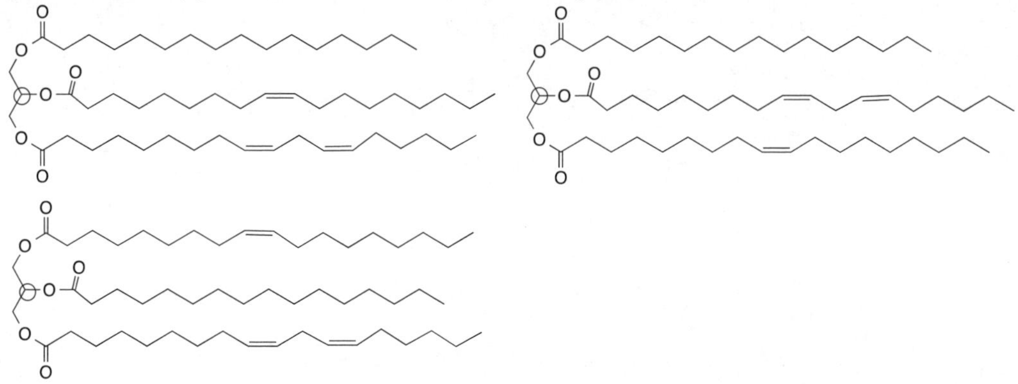

31.22

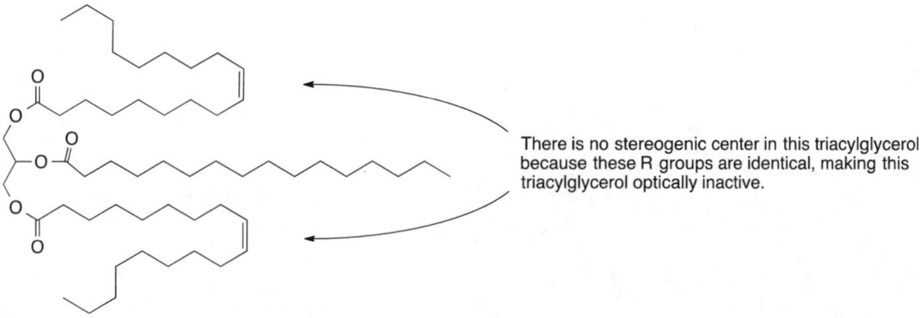

There is no stereogenic center in this triacylglycerol because these R groups are identical, making this triacylglycerol optically inactive.

31.23

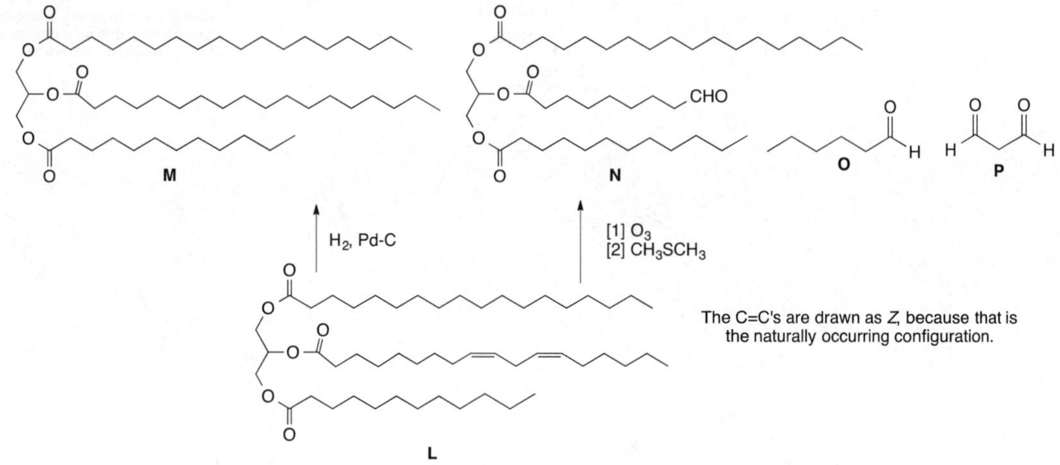

M

N

O

P

H$_2$, Pd-C

[1] O$_3$
[2] CH$_3$SCH$_3$

The C=C's are drawn as *Z*, because that is the naturally occurring configuration.

L

Chapter 31–10

31.24 When R" = CH2CH2NH3+, the compound is called a **phosphatidylethanolamine** or **cephalin.**

a.

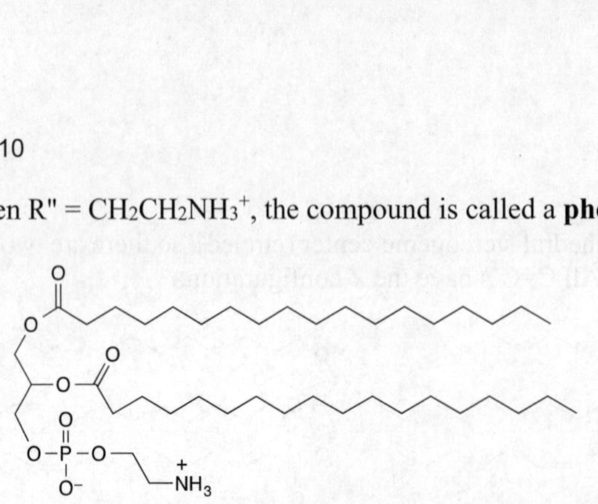

cephalin

b.

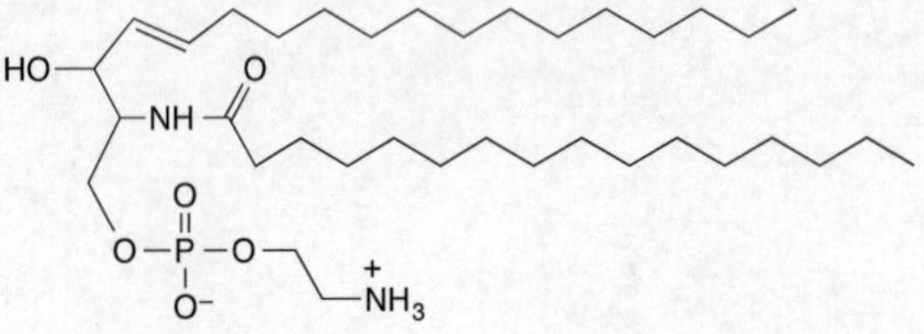

sphingomyelin

31.25

a.

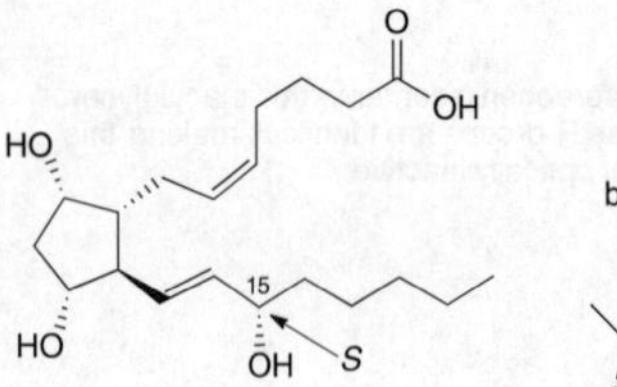

PGF$_{2\alpha}$

b.

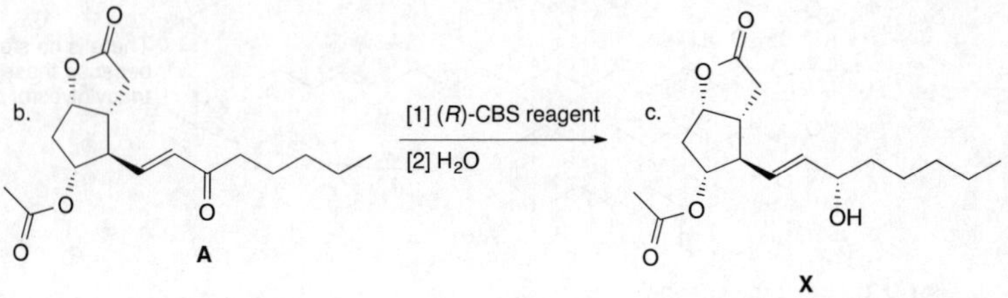

A

[1] (R)-CBS reagent

[2] H$_2$O

c.

X

Use a chiral reducing agent to add hydride from one side only to form a single diastereomer.

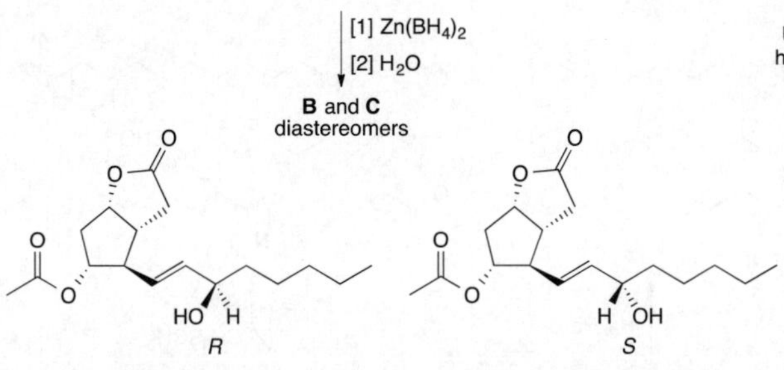

[1] Zn(BH$_4$)$_2$

[2] H$_2$O

B and **C**
diastereomers

R

S

Lipids 31–11

31.26

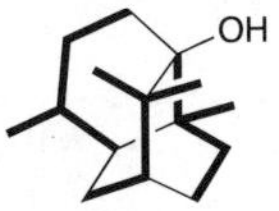

a. 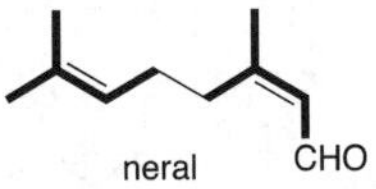 neral CHO

b. 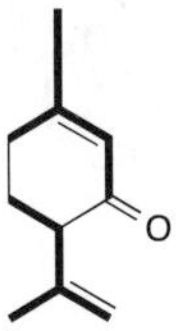 carvone

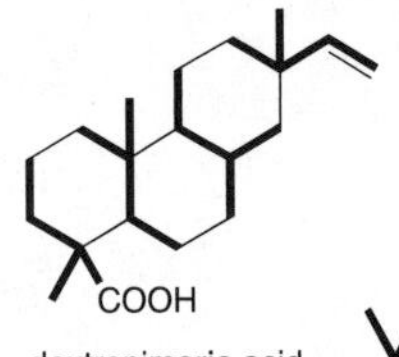

c. 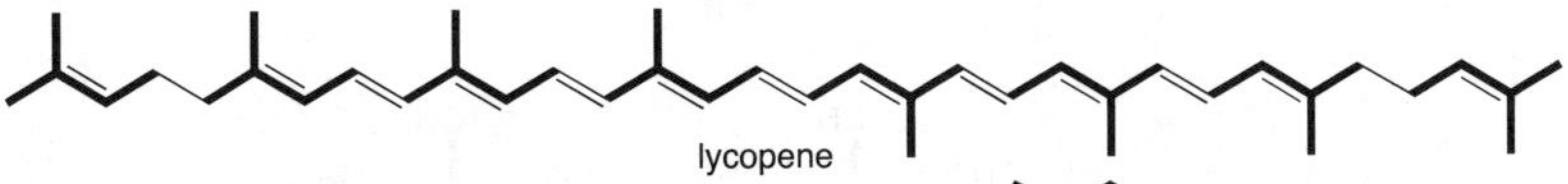 lycopene

d. 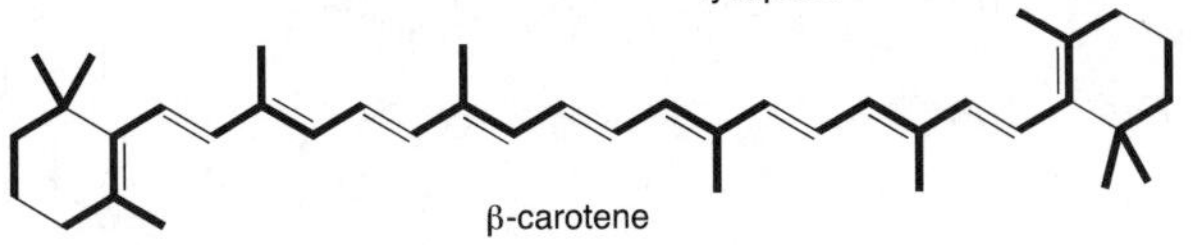 β-carotene

e. patchouli alcohol

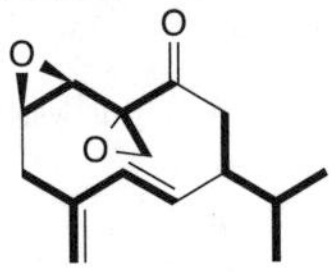

f. periplanone B

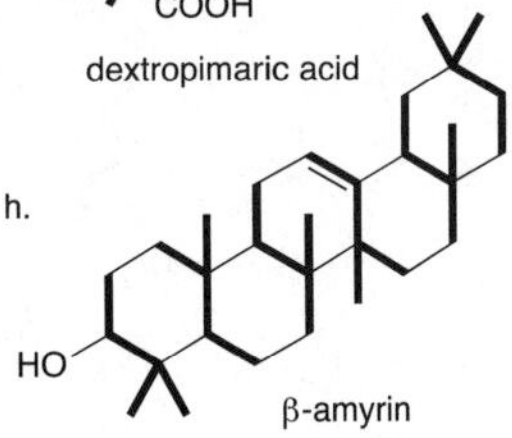

g. dextropimaric acid COOH

h. β-amyrin HO

31.27 A *monoterpene* **contains 10 carbons** and two isoprene units; a *sesquiterpene* **contains 15 carbons** and three isoprene units, etc. See Table 30.5.

a. 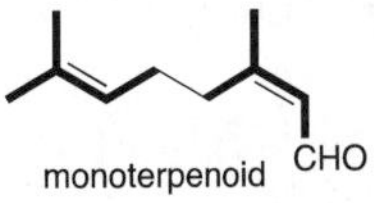 monoterpenoid CHO

b. 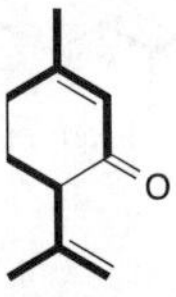 monoterpenoid

c. 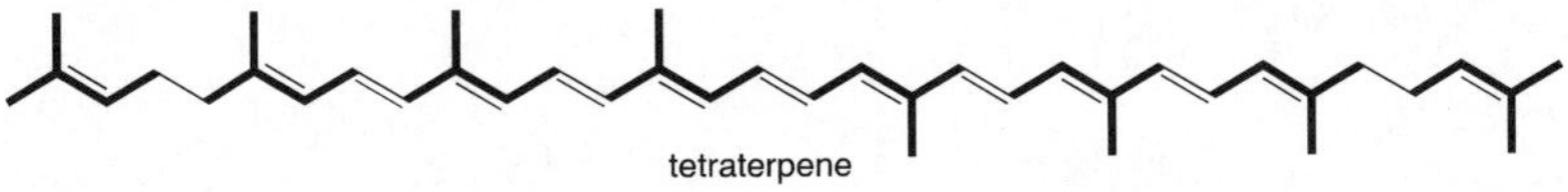 tetraterpene

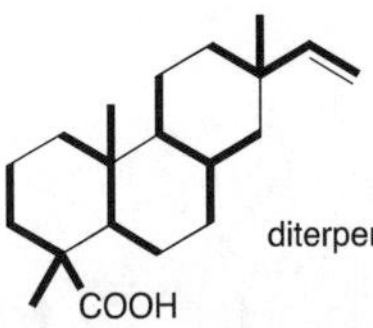

d. 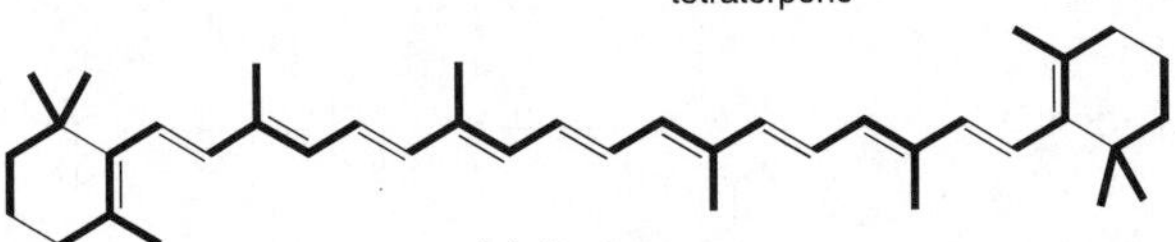 tetraterpene

e. sesquiterpenoid

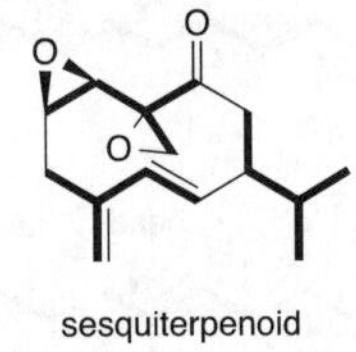

f. sesquiterpenoid

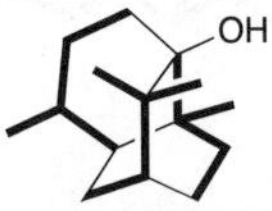

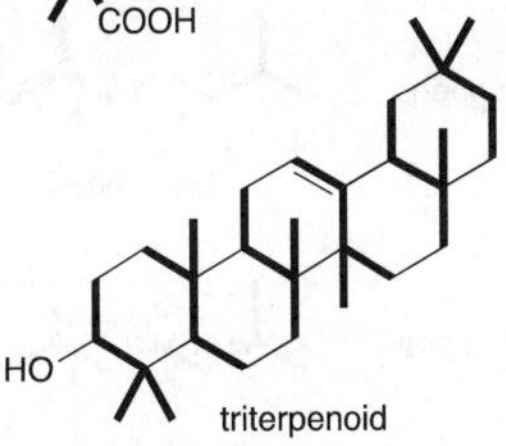

g. diterpenoid COOH

h. triterpenoid HO

Chapter 31–12

31.28 a. Cleave the two esters and the acetals.

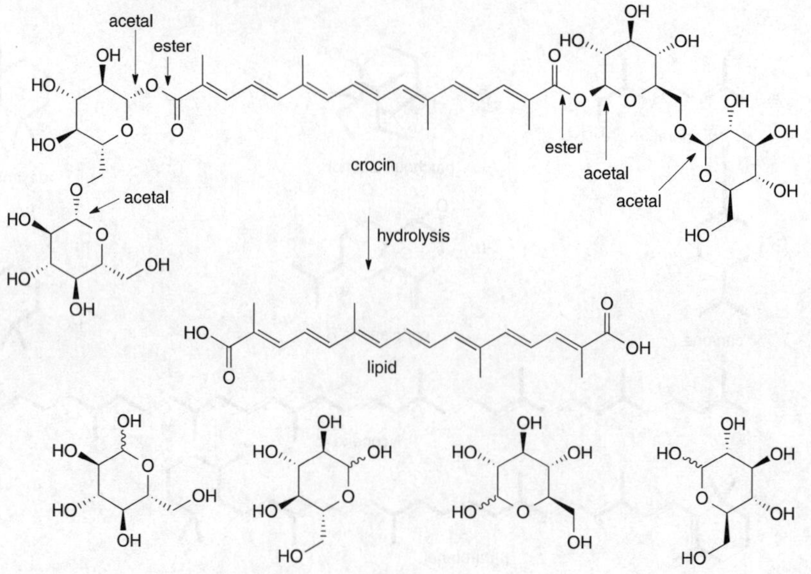

crocin

hydrolysis

lipid

b. The lipid portion has 20 C's, so it is a diterpenoid. The four isoprene units are labeled in bold.

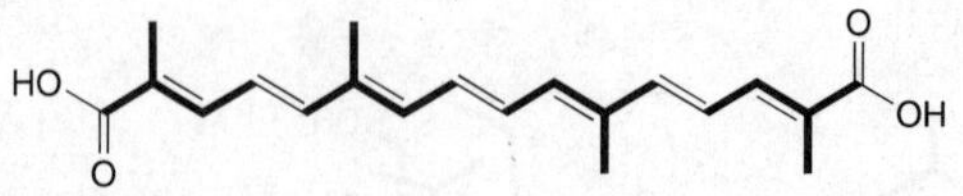

31.29

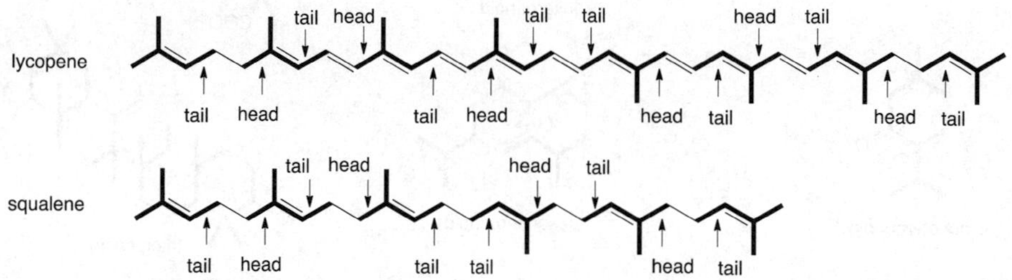

lycopene

squalene

31.30

resonance-stabilized
carbocation

α-pinene

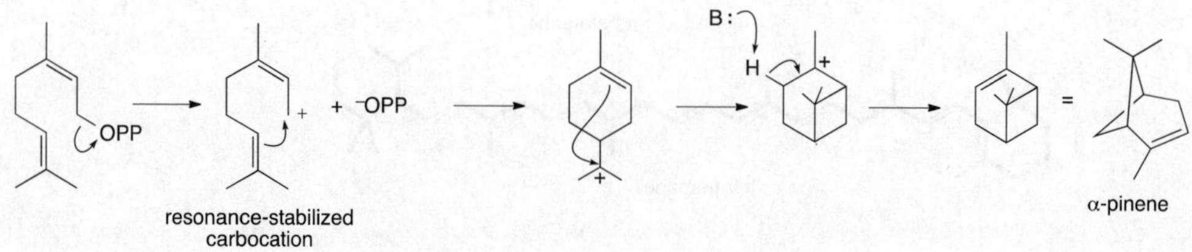

Lipids 31–13

31.31 The unusual feature in the cyclization that forms flexibilene is that a 2° carbocation rather than a 3° carbocation is generated. Cyclization at the other end of the C=C would have given a 3° carbocation and formed a 14-membered ring. In addition, the 2° carbocation does not rearrange to form a 3° carbocation.

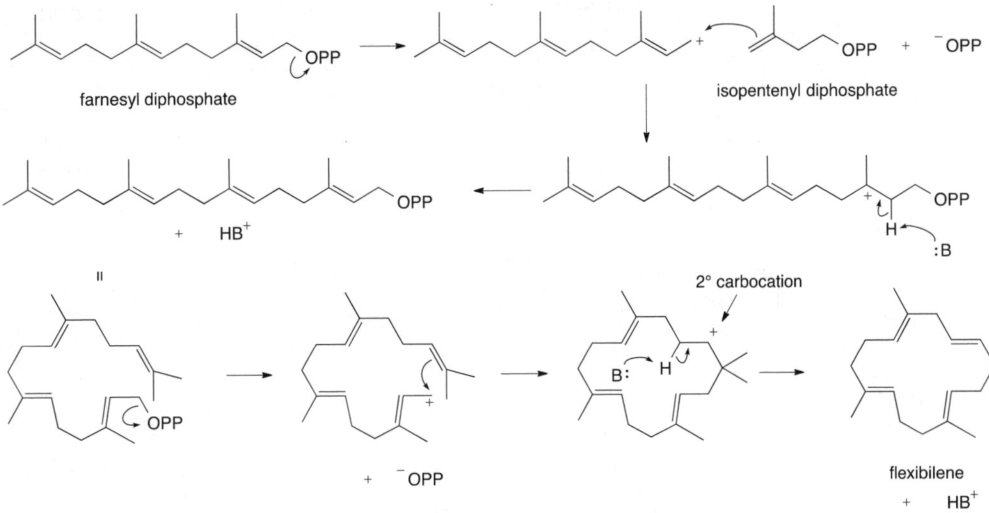

farnesyl diphosphate

isopentenyl diphosphate

2° carbocation

flexibilene

31.32

a.

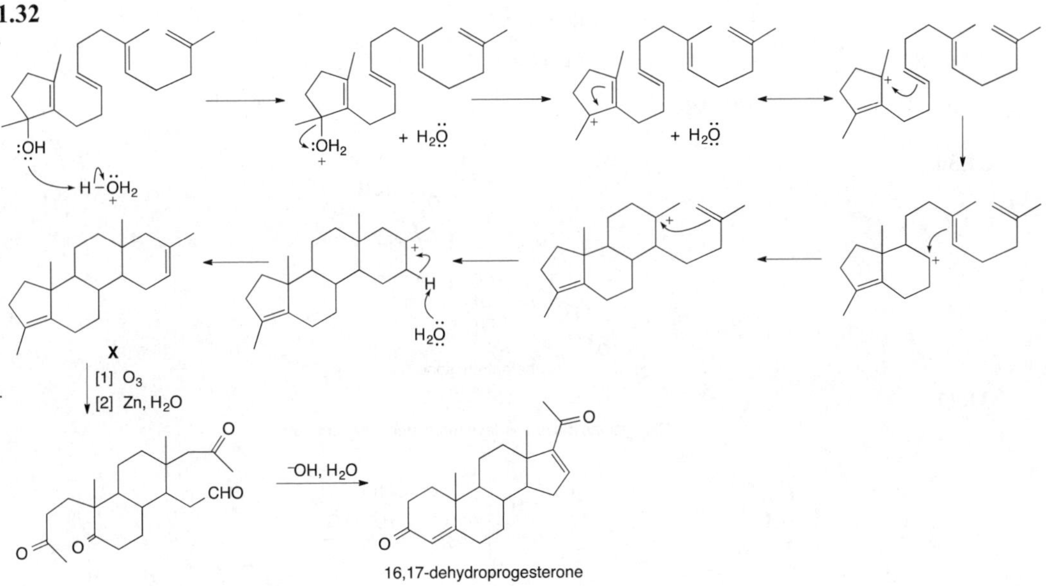

X

b.

[1] O₃
[2] Zn, H₂O

16,17-dehydroprogesterone

Chapter 31–14

31.33

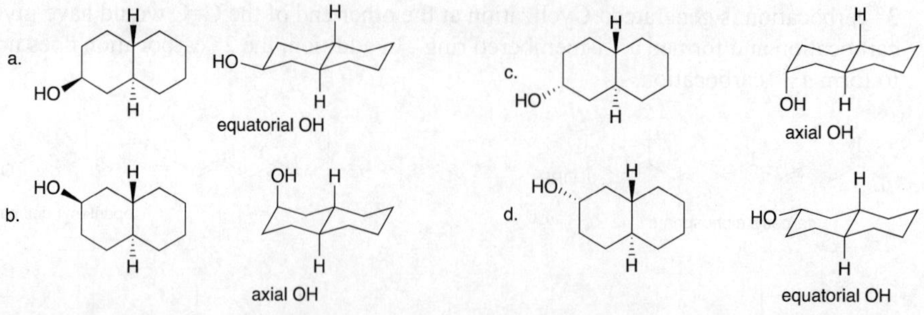

a.

HO⟋⟍ ... equatorial OH

c.

HO⟋⟍ ... axial OH

b.

HO⟋⟍ ... axial OH

d.

HO⟋⟍ ... equatorial OH

31.34

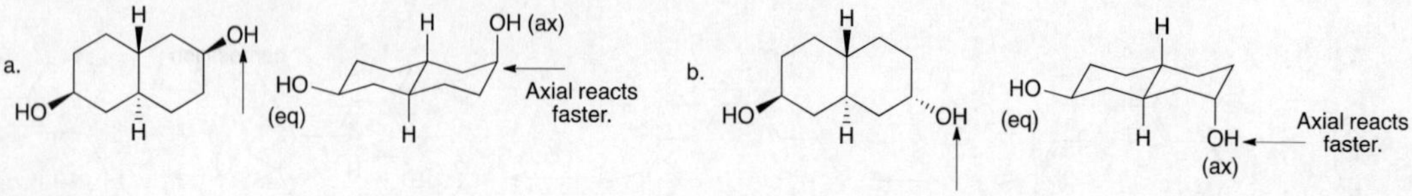

a.

OH

HO⟋⟍ (eq) ... OH (ax)

Axial reacts faster.

b.

HO⟋⟍ (eq) ... OH (ax)

Axial reacts faster.

31.35

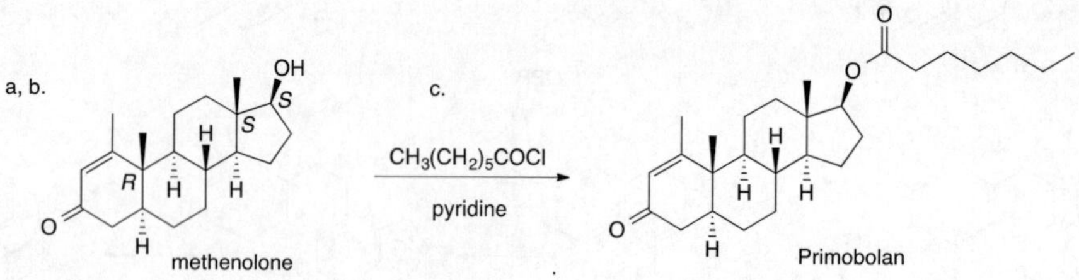

a, b.

OH
S
R

O

methenolone

c.

CH₃(CH₂)₅COCl
pyridine

O

Primobolan

31.36

HO

O OH

OH

1
2
9 H
16
F H

O

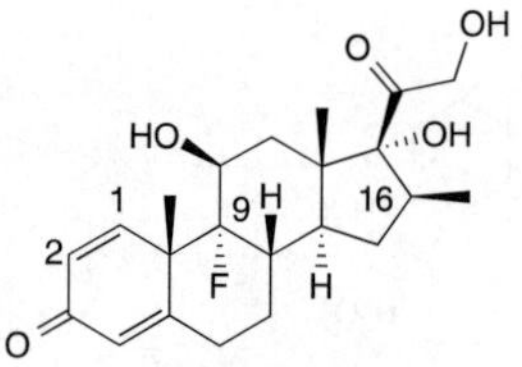

betamethasone

31.37

CH₃ groups make this face more sterically hindered.

a.

O

=

O

b.

H₂, Pd-C

O

H₂ added from below

The bottom face is more accessible, so the H₂ is added from this face to form an equatorial OH.

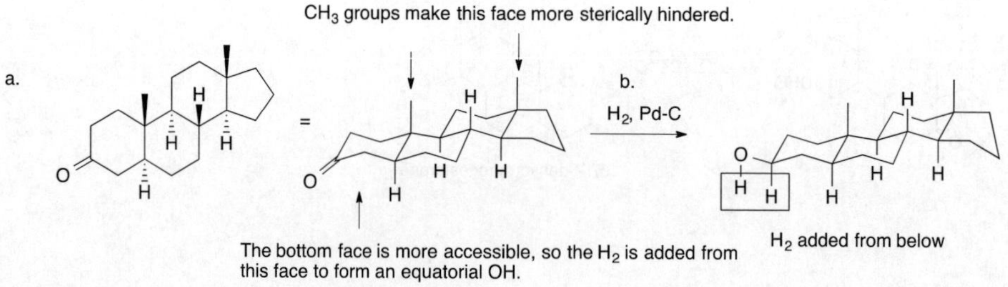

Lipids 31–15

31.38

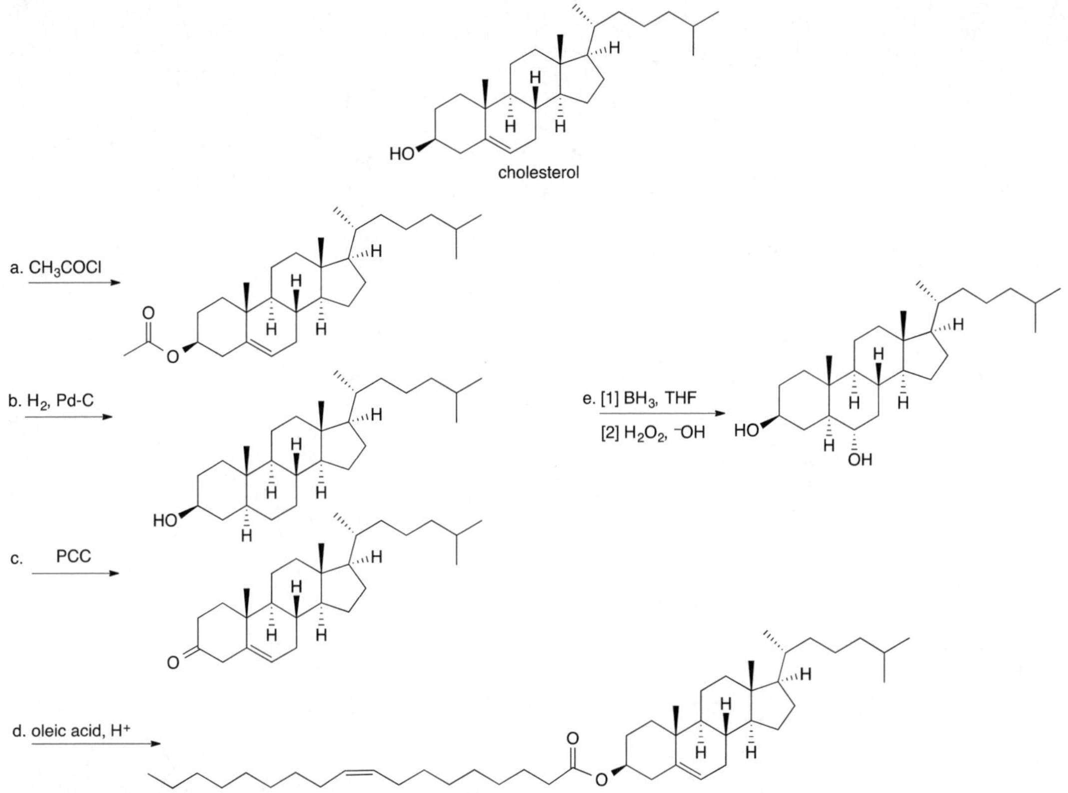

cholesterol

a. CH₃COCl →

b. H₂, Pd-C →

c. PCC →

d. oleic acid, H⁺ →

e. [1] BH₃, THF
   [2] H₂O₂, ⁻OH →

31.39

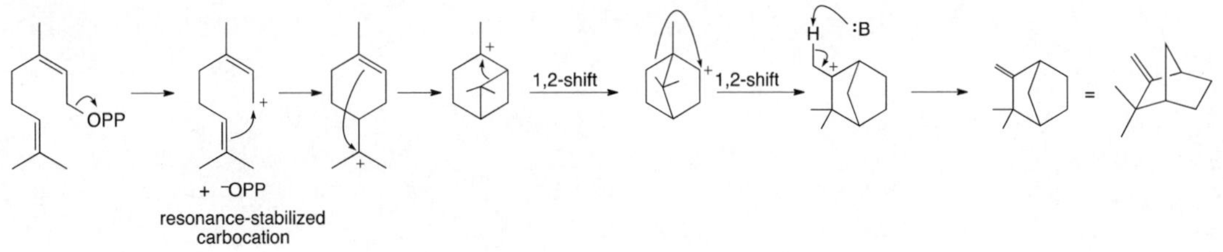

+ ⁻OPP
resonance-stabilized
carbocation

1,2-shift

1,2-shift

:B

Chapter 31–16

31.40 Re-draw the starting material in a conformation that suggests the structure of the product.

31.41

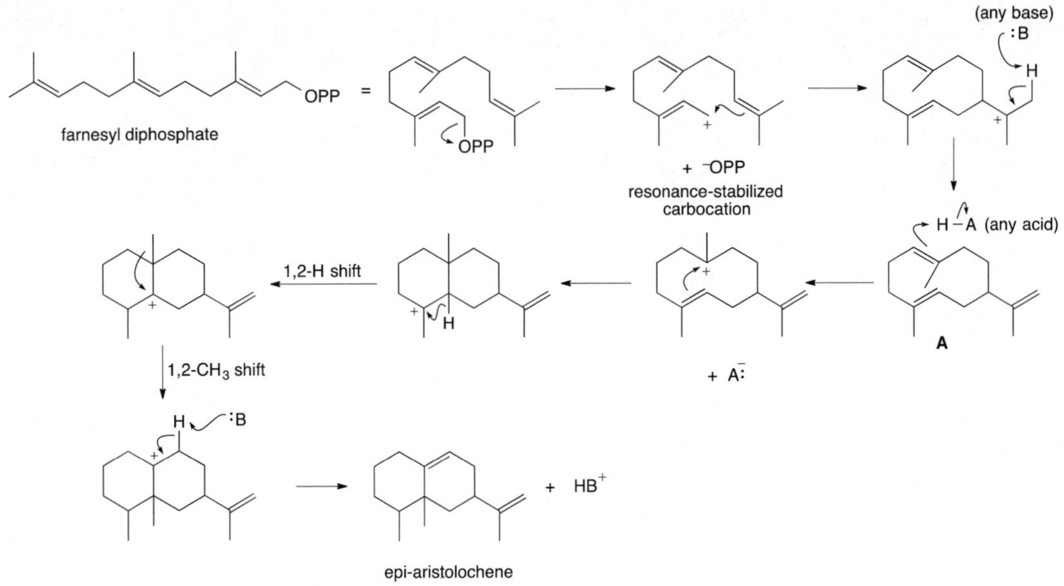

farnesyl diphosphate

+ ⁻OPP
resonance-stabilized
carbocation

(any base)
:B

H—A (any acid)

A

+ A:⁻

1,2-H shift

1,2-CH₃ shift

:B

+ HB⁺

epi-aristolochene

Student Study Guide/Solutions Manual to accompany
ORGANIC CHEMISTRY, FIFTH EDITION JANICE
GORZYNSKI SMITH, ERIN R. SMITH

Original edition Copyright ©2017
by The McGraw-Hill Education. All rights reserved.

English reprint rights arranged with The McGraw-Hill Global Education Holdings, LLC.
through Japan UNI Agency, Inc., Tokyo.

スミス有機化学
問題の解き方(第5版)[英語版]

著　　　者	Janice Gorzynski Smith
	Erin R. Smith

検印廃止

第 3 版　第 1 刷　2014 年 12 月 20 日
第 5 版　第 1 刷　2018 年 4 月 1 日
　　　　第12刷　2024 年 9 月 10 日

発 行 者　曽 根 良 介
発 行 所　㈱化 学 同 人
〒600-8074 京都市下京区仏光寺通柳馬場西入ル
編集部　TEL 075-352-3711　FAX 075-352-0371
企画販売　TEL 075-352-3373　FAX 075-351-8301
　　　　　　　　　　振 替　01010-7-5702
e-mail webmaster@kagakudojin.co.jp
URL https://www.kagakudojin.co.jp

JCOPY 〈出版者著作権管理機構委託出版物〉
本書の無断複写は著作権法上での例外を除き禁じられて
います．複写される場合は，そのつど事前に，出版者著作
権管理機構（電話 03-5244-5088，FAX 03-5244-5089，
e-mail: info@jcopy.or.jp）の許諾を得てください．

本書のコピー，スキャン，デジタル化などの無断複製は著
作権法上での例外を除き禁じられています．本書を代行
業者などの第三者に依頼してスキャンやデジタル化するこ
とは，たとえ個人や家庭内の利用でも著作権法違反です．

印刷・製本　モリモト印刷株式会社

Printed in Japan
乱丁・落丁本は送料小社負担にてお取りかえします．

ISBN978-4-7598-1947-2